THE OXFORD HANDBOOK OF

EVOLUTIONARY MEDICINE

THE OXFORD HANDBOOK OF

EVOLUTIONARY MEDICINE

Edited by

MARTIN BRÜNE

and

WULF SCHIEFENHÖVEL

OXFORD
UNIVERSITY PRESS

OXFORD
UNIVERSITY PRESS

Great Clarendon Street, Oxford, OX2 6DP,
United Kingdom

Oxford University Press is a department of the University of Oxford.
It furthers the University's objective of excellence in research, scholarship,
and education by publishing worldwide. Oxford is a registered trade mark of
Oxford University Press in the UK and in certain other countries

First Edition published in 2019
Impression: 1

Published in the United States of America by Oxford University Press
198 Madison Avenue, New York, NY 10016, United States of America

British Library Cataloguing in Publication Data
Data available

Library of Congress Control Number: 2018944186

ISBN 978–0–19–878966–6

Printed and bound by
CPI Group (UK) Ltd, Croydon, CRO 4YY

Foreword

When Charles Robert Darwin published, at long last, his ground-breaking theory of adaptation and natural selection—*On the Origin of Species*—he was clearly afflicted by doubts and qualms. These are evident in the full title '*On the Origin of Species by Means of Natural Selection, or the Preservation of Favoured Races in the Struggle for Life*'. Thus an explicit part of the mechanisms he described, and what we now call evolution, referred to the process of selection, by which the best-adapted individuals survive and reproduce whereas others are deselected. Darwin actually disliked the term 'evolution' and was also unhappy with Herbert Spencer's coinage, 'survival of the fittest'. Still, Darwin's achievement is all the more remarkable for he did not know the actual logistics of selection—Mendel's genetical research still lay intellectually fallow and, indeed, would for some decades. Darwin was, however, greatly influenced by the theory of population growth propounded by Thomas Malthus: geometric increases in populations run off cliffs of resources which grow only arithmetically. With this, he was also distraught to realise that what happened to the less well adapted was enfeeblement, debility, and even death (not only of the self, but also of the lineage) by deselection.

Put another way, from the very beginning of evolutionary theory, links to the realm of medicine were manifest though long ignored and, when attended to, often misconstrued. Such led to the dangerous blind alleys of Social Darwinism as well as even more awful social policies. It is beyond the present scope and aims to summarise, much less analyse, the tremendous, incalculable pain brought about by atrocious perversions of Darwin's masterful work, but all this is otherwise well chronicled. Here it is important to underscore that none of these was ever endorsed by Darwin. One hundred and fifty years or so later, the medical profession now well knows these misapplications were wrong—both scientifically and morally. So for what now can the theory of evolution be 'good' in contemporary medicine and health science? Evolution is usually construed as a slow and gradual process whereas medicine has a different mode and tempo, that is, to expeditiously diagnose and treat health conditions. Yet, might there be things evolutionary to which medicine may look to leverage for therapeutic progress or, at least, palliation and greater heuristic understanding? Yes, there are.

For one, the microbial world surrounding us is rapidly evolving: *Escherichia coli* bacteria can divide every 20 minutes. Our skin and gut are populated by 'aliens' in numbers that come close to the number of cells that carry our own genome. Most microbiota are commensal, that is, they help us keep alive and healthy, but a few are not. Indeed, the rapidity of bacterial reproduction drives equally rapid evolutionary dynamics by which pathogenic traits can newly arise. Thus, some bacteria cause severe medical problems under special conditions we cannot understand without a deep knowledge about how our bodies and 'they' interact, as well as the mechanisms of such interactions, both physiological and pathological.

Second, there is of course more than a kernel of truth in the notion that evolution is slow, at least with regard to species *bauplans*. For instance, for some millions of years, our ancestors and all members of our own species have walked on two legs. Verbal communications have been around for at least 500,000 years; our forebears started to domesticate animals some 30,000 years ago; and selection has endowed us (or, at least, some of us) with the means to digest milk beyond infancy. However, our body is, in other ways, a living fossil. Our vertebral column, for example, is not suited to bipedalism, much less more modern sedentary postures such as sitting at a desk or to watch TV. Nor, for that matter, is our brain. The consequences of exposing our young for extended periods of time to computer games and other technical gadgets has only begun to be recognised as a matter of medical concern. Likewise, the vast overshoot of calories that our bodies ingest can wreak havoc with our metabolism. In other words, our biological heritage is, to some degree, mismatched with our modern way of living.

Third, humans continue to evolve, for reasons and in ways that have only recently begun to be explored. Changing environments to which humans may adapt include exposure to pathogens, old and new, social contingencies, and the impact of modern medicine, which some posit as a 'relaxation' of natural selection.

All this needs to be better understood than is currently the case. Thus, medical curricula need the yeasts of evolutionary science to leaven dough often turgid with science not risen with phylogenetic perspectives. Indeed, virtually every organ system is involved in an ongoing 'arms race' between our bodies and our environments—gestational, developmental, social, cultural, microbial, nutritional, et al. The sooner the better evolutionary science is put into practice: future generations of physicians in all specialties, indeed, all health practitioners and scientists must understand the pertinence of evolution for medical research and practice.

The Oxford Handbook of Evolutionary Medicine is a fine example in the medical literature of 'punctuated evolution', that is, unusually rapid progress in form and function. As the first textbook to tackle the mammoth task of integrating evolutionary knowledge comprehensively into a format familiar to health science students and professionals, it is arranged by organ systems akin to a traditional text of medical physiology or pathophysiology. An incredible number of cross-references link chapters to one another and serve to underscore how organic beings are more than the sum of parts. Likewise, the evolutionary background knowledge so essential to this work is accessible and straightforward. Moreover, this is a well-edited book written by experts in their medical fields, such that the volume offers a state-of-the-art compilation of evolutionary insights into health and disease.

Much of this success is attributable to the editors, Professors Martin Brüne and Wulf Schiefenhövel, who bring a commanding range of expertise to this gargantuan task. Dr. Brüne graduated in medicine from the Westphalian Wilhelms University in Münster in 1988 and completed training first in neurology, then psychiatry in 1995, as well as a Visiting Research Scientist fellowship at the Centre for the Mind of the Australian National University and University of Sydney. These timelines are important since his deep interest in evolutionary neuropsychiatry evolved even as the field itself was emerging. He is currently Professor of Psychiatry at the LWL University-Hospital, Ruhr University Bochum, Germany, and has authored more than 250 articles and book chapters, including a rare profound monograph *Textbook of Evolutionary Psychiatry and Psychosomatic Medicine: The Origins of Psychopathology* (2nd edition, Oxford University Press, 2016). For his part,

Professor Schiefenhövel is Head of the Human Ethology Group at the Max-Planck-Institute for Ornithology. He studied medicine at the universities of Munich and Erlangen-Nuremberg and has long been Professor for Medical Psychology and Ethnomedicine at Munich. He has carried out fieldwork in ethnomedicine, evolutionary medicine, social anthropology, human ethology, linguistics, and population genetics in Mainland and Island New Guinea since 1965—notably among the Eipo, a then (1974) Neolithic Highland Papuan group in the Province of Papua/Indonesia. Among his many, many contributions sustained over decades, he is a founding member of the Human Sciences Centre, University of Munich, and former Professor of Medical Psychology at that university. For several decades, Professor Schiefenhövel has been a regular guest professor in Human Ethology at both the University of Innsbruck and the Institute for Behavioral Biology of the Royal University Groningen.

Above and beyond their compendious erudition and expertise in the subject, Professors Brüne and Schiefenhövel have assembled an exceptional team of chapter authors who write with considerable authority and a good degree of convergent substance and style.

I strongly commend *The Oxford Handbook of Evolutionary Medicine* to students of medicine, anthropology, biology, and psychology, as well as to scholars and clinicians. Regardless of where you are in your journey through medicine—pupil, teacher, specialist, perhaps patient, or even an interested layperson—*The Oxford Textbook of Evolutionary Medicine* is a wonderful new 'ecosystem' to which medicine can better adapt even as you can explore it in marvellous detail.

Daniel R. Wilson, MD, PhD
President, Western University of Health Sciences, USA

Preface

Our idea to edit a volume about evolutionary medicine dates back several years, when both of us agreed that evolutionary theory has much to contribute to the understanding, diagnosis, and treatment of medical conditions, while at the same time evolutionary theory was almost absent or negligibly dealt with in medical textbooks, let alone the curricula of medical schools. There were a few textbooks in special areas of medicine, though, like Wenda Trevathan's *Human Birth: An Evolutionary Perspective* (1987), and another by one of us (WS) published in German about *Birth-Giving Behaviour and Reproductive Strategies* (original title: *Geburtsverhalten und Reproduktive Strategien*, 1988). Even though these volumes and the ingenious classic book entitled *Why We Get Sick. The New Science of Darwinian Medicine* by Randolph Nesse and George C. Williams (1994) were well received in some academic communities, their main lessons went unheard at both bench and bedside. However, for the understanding of health and disease, it is tremendously important to acknowledge that bodies are not like machines, but compositions of adaptations to a vast complexity of environmental challenges that coined our evolutionary past, and that selection operates on human phenotypes to the present day.

The relevance of these messages resides in the fact that nature has produced design compromises and trade-offs between sometimes diametrically opposite biological requirements germane to survival and reproduction. Together with the observation that many ancient adaptations barely fulfil their biological purposes in modern environments, these insights have advanced the intellectual endeavour to come to grips with the question why body functions can go wrong, in spite of the long history of biological adaptation that should have selected against any kind of 'weakness' or 'design failure'. In fact, we now get to understand that human bodies, like other living things, are 'ecosystems' which coevolved with other organisms like viruses, bacteria, fungi, and helminths, most frequently to the benefit of both host and commensal guest, but sometimes detrimental to the human organism. Environmental changes, often in association with cultural evolution, have brought about new challenges to the *conditio humana*, many of which occurred so rapidly that biological adaptation could not keep pace—in evolutionary medicine, referred to as 'mismatch'. For example, the agricultural transition has produced not only new ways of food acquisition, but also new problems germane to physical and mental health, including ones that are commonly well known such as lactose intolerance (as the ancestral condition), and new infectious agents, which crossed species borders when domesticated animals came to live in close proximity with humans. Along similar lines, evolutionarily ancient germs like *Plasmodium falciparum* became more problematic for cattle herders, because *Anopheles* evolution led to the emergence of strains that preferably fed on human, not animal, blood. Similar processes have occurred more recently, including bird flu and swine flu epidemics in East Asia, as well as Ebola endemics.

So, in our recent history, new infectious diseases, new cancers, and other medical conditions have appeared, which can only be fully understood by adopting an evolutionary perspective. This view, we strongly believe, needs also to be taken to the medical classroom and to the clinics. It is not just an academic endeavour to explore how we can better deal with bacteria that have developed antibiotic resistance. This is clinical reality in almost every intensive care unit, and we need to find new approaches to deal with life-threatening conditions that arise from resistant microbes. The list of practical issues could easily be expanded, but this is not the right place to do so. The reader is referred to the individual chapters of this volume.

Interestingly, the link between evolutionary theory and medicine is not new. With regard to Charles Darwin's life, it is known that he suffered, throughout his life, from waxing and waning, often debilitating clinical symptoms, which in retrospect could have been signs of Crohn's disease, lactose intolerance, MELAS syndrome (a mitochondrial gene defect causing encephalomyopathy, lactic acidosis, and stroke-like episodes, passed on in the female line), or Chagas disease caused by *Trypanosoma cruzi*, which he might have caught during the long and strenuous horse-back travels in South America during his voyage around the world on board the *Beagle*. Charles Darwin's father, Robert Darwin, was a well-established medical doctor in Shrewsbury, England, and wanted his two sons to become doctors themselves; his younger son Charles, however, was appalled by the bloodiness of that profession and stopped his medical career at Edinburgh University. The marriage with his cousin Emma Wedgewood caused great concern in Darwin: he very much worried that this union on the fringe of incest would produce disease in his offspring, especially as he himself had carried out experiments on the effects of inbreeding in plants.

Given the degree of suffering (one corner of his study at his home Down House was separated off as a toilet, as vomiting and flatulence were, among other signs of disease, a life-long burden), it is interesting that Darwin did not extend, with one exception, the evolutionary principles he discovered to the field of medicine. The exception is his important contribution to psychology and psychiatry in the book *The Expression of Emotions in Man and Animals*, published in 1872, in which he convincingly showed that mammals and humans are connected through common descent, which explains the manifold similarities in perception and communication of emotions through non-verbal language. In this book, he used photographs of psychiatric patients who displayed, as Darwin writes, facial expressions in a more extreme, less-controlled way than non-patients.

We are well aware of the fact that Darwin's ideas were taken up by researchers and policy makers who believed that the idea of natural selection could be applied to large-scale societal issues—'Social Darwinism' paved the way for the most dreadful medical maltreatments of minorities, including compulsory sterilisation; on a broader scale, it also served as a basis for the holocaust. Darwin as a person and scientist actually advocated the opposite of what Social Darwinism stands for.

Since his ground-breaking work, a number of new scientific discoveries, including the double-helix structure of DNA by Watson and Crick, Hamilton's rules of kin selection, and life history theory described by Stearns, have greatly advanced the field of evolutionary biology. Moreover, Nikolaas Tinbergen, Nobel Laureate in Physiology and Medicine, gave a precise account of which questions need to be addressed (namely causation, ontogeny, evolution, and survival value) to fully understand any anatomical feature or biological process. These concepts are now reflected in several excellent textbooks about evolutionary medicine. A selection includes *Diseases and Human Evolution* by

Ethne Barnes (2005), *Evolution in Health and Disease* edited by Stephen C. Stearns and Jacob C. Koella, *Evolutionary Medicine and Health* edited by Wenda Trevathan, E.O. Smith, and James J. McKenna (2008), *Principles of Evolutionary Medicine* by Peter Gluckman, Allan Beedle, and Mark Hanson (2009), *Evolution and Medicine* by Robert Perlman (2013), *Evolutionary Thinking in Medicine* edited by Alexandra Alvergne, Crispin Jenkinson, and Charlotte Faurie (2016), and *Evolutionary Medicine* by Stephen C. Stearns and Ruslan Medzhitov (2016), all of which are extraordinarily useful and scholarly written for students of medicine, biology, psychology, anthropology, and related sciences. So why edit another one?

The approach taken in this volume differs in some ways from how the existing textbooks are organised. We believe it is particularly helpful for medical students and clinicians, as well as students and scholars from other life sciences, if evolutionary medicine is dealt with according to the systematic logic of a textbook of medical physiology and pathophysiology; that is, categorising the textbook according to organic systems may make it easier to capture the relevance of evolutionary concepts for human medicine as a whole, including ideas of how different organic systems interact with one another. It is our hope that this structure will facilitate a way for evolutionary medicine to gain an inroad into classrooms and curricula.

Following this thread, we divided *The Oxford Handbook of Evolutionary Medicine* into two larger parts. Part I deals with the general principles of evolutionary science and their relevance to the understanding of medical problems. That is, in Chapter 1, Randolph Nesse outlines the Core Principles for Evolutionary Medicine, including the widely acknowledged concept (in biology) of analysing both proximate and evolutionary dimensions of biological traits, which was first put forward by Nobel laureate Nicolaas Tinbergen in a seminal article published (in 1963) on the occasion of the 60th birthday of his fellow ethologist, Konrad Lorenz. Chapter 2, written by Diana Le Duc and Torsten Schöneberg, deals with Cellular Signalling Systems. Evolutionary principles of Genetics and Epigenetics are outlined in Chapter 3 by Paul Ewald and Holly Swain Ewald, exemplifying gene–environment interactions in relation to common diseases. In Chapter 4 on Growth and Development, Robin Bernstein and Barry Bogin describe human ontogeny and growth patterns from an evolutionary perspective. One particular life-stage, characterised by the deterioration of body functions, is the topic of Chapter 5 on Senescence and Ageing written by Xiaqing Zhao and Daniel Promislow. Part I is then completed by Chapter 6 about Nutrition, Energy Expenditure, Physical Activity, and Body Composition by Ann Caldwell, Stanley Boyd Eaton, and Melvin Konner.

Part II entitled 'Specific Systems' focuses on clinical theory and practice; in contrast to classic textbooks of human physiology and pathophysiology, specific emphasis is placed on the conceptualisation of disease pathology in an evolutionary framework. Chapter 7 by Martin Häusler, Nicole Bender, Lafi Aldakak, Francesco Galassi, Patrick Eppenberger, Maciej Henneberg, and Frank Rühli deals with the Musculoskeletal System. In Chapter 8, Mark Hill describes the Skin and Integument, including its embryological development and relevance for the understanding of skin diseases. Chapter 9 about the Haematopoetic System, by Eric Pietras and James DeGregori, concerns the evolution of the blood system and vulnerabilities of this system to diseases like leukaemia. In Chapter 10, Graham Rook takes us through the complicated matter of the Immune System and explains how early developmental exposure to germs impacts on the maturation of immunological function, and why dysfunction may occur much more often in our contemporary environments than it probably did in the past. In Chapter 11, Kevin Shah, Kalyanam Shivkumar, Mehdi Nojoumi, and Barbara Natterson-Horowitz deal with some of the most frequent diseases of our modern world affecting the Cardiovascular System. Chapter 12, written by Olga Carvalho and John Maina, describes the evolution of the Respiratory System and how lung diseases may occur in relation to specific adaptations of the mammalian respiratory tract. In Chapter 13, John Furness, Eve Boyle, Josiane Fakhry, Joanna Gajewski, and Linda Fothergill portray the evolution of the human Digestive System, including the manifold and only recently researched ways the gut microbiome interacts with other organ systems. Chapter 14 about the Excretory System, by Paola Romagnani and Hans-Joachim Anders, teaches us that an evolutionary point of view, looking at the individual nephron rather than the kidney as a whole, opens new avenues to the understanding of diseases of the Excretory System. In Chapter 15, Richard Bribiescas provides us with new insights about Endocrinology, particularly differences between male and female bodies in regard to sex hormones, and how this helps our understanding of endocrinological diseases. Chapter 16: Sexuality, Reproduction, and Birth by Wulf Schiefenhövel and Wenda Trevathan delves into human sexual behaviour, parturition, and mother–infant interaction, how and why these spheres of life are shaped by strong evolutionary forces, and governed by biological principles and cultural traditions, and what evolutionary insights can contribute to modern obstetrics and paediatrics. In Chapter 17 about the Brain, Spinal Cord, and Sensory Systems, Martin Brüne describes why such an incredibly complex organ evolved, how it interacts with other organ systems, and how it deals with challenges from 'stressful' environments, sometimes in less than optimal ways which may give rise to neuropsychiatric disease.

Chapter 18, The Future of Medicine, stands for itself. Here, we explore how evolutionary medicine can contribute to the prevention, diagnosis, and treatment of human medical conditions. We also try to envisage or speculate how medicine may look in a hundred years from now.

We are proud that we were able to win over internationally highly regarded scholars and clinicians as authors for this *Handbook*. Without their outstanding dedication, commitment, and support, this book would have never been realised. A multi-author volume has many advantages, but also some disadvantages. Having experts from different disciplines on board is invaluable with regard to efforts to include the latest and most up-to-date views on evolutionary medicine. On the other hand, a few redundancies are unavoidable; even slightly different slants and opinions on specific topics may occur here and there—however, science is the evolution of concepts and ideas based on empirical evidence, and it essentially lives on controversies and debate. So we encourage our readers to read the book from its first to the last page. To facilitate this venture, we have put great emphasis on linking and cross referencing the chapters wherever possible.

One feature readers should not expect to appear in this *Handbook* concerns the inclusion of exhaustive graphical material and figures. That is, we sacrificed the goal of depicting anatomically detailed images in favour of illustrative line drawings showing conceptually relevant elements of organ systems, because it has not been our aim to copy the work of the great master anatomists who so skilfully drew images of anatomical sections; nor is it the purpose of this volume to feature anatomical details which can better be found in modern anatomy textbooks based on modern imaging technology.

Finally, we would like to acknowledge several persons and institutions for their generous support. The Thyssen Foundation provided us with a generous grant to organise a conference about evolutionary medicine, which took place at the Hanse-Wissenschaftskolleg

(HWK), Institute for Advanced Studies, in Delmenhorst in October 2016. This enabled us to invite our contributors, the majority of whom were able to follow our invitation to present and discuss their fabulous work in an atmosphere of reverence, encouragement, and scholarship, which is not always found in academic meetings. Our special thanks go to Dr. Dorothe Poggel and Professor Reto Weiler who additionally invited us to a fellowship at the HWK in October 2017, which gave us the opportunity to discuss and work on relevant editorial issues. We are also grateful to Martin Baum, Commissioning Editor at Oxford University Press, Charlotte Holloway, Senior Assistant Commissioning Editor, Kumar Anbazhagan, Project Manager, and to Julie Musk, Copyeditor, for their wonderful support in getting *The Oxford Handbook of Evolutionary Medicine* published.

Martin Brüne and Wulf Schiefenhövel

Contents

List of Contributors

Lafi Aldakak Evolutionary Morphology and Adaptation Group, Institute of Evolutionary Medicine, Faculty of Medicine, University of Zurich, Switzerland

Hans-Joachim Anders Professor of Internal Medicine, Division of Nephrology, University Hospital, Ludwig Maximilians University, Munich, Germany

Nicole Bender Evolutionary Morphology and Adaptation Group, Institute of Evolutionary Medicine, Faculty of Medicine, University of Zurich, Switzerland

Robin M. Bernstein Associate Professor of Anthropology, Health and Society Program, Institute of Behavioral Science, University of Colorado, Boulder, United States of America

Barry Bogin Professor of Biological Anthropology, School of Sport, Exercise and Health Sciences, Loughborough University, United Kingdom

Eve K. Boyle Graduate Student, Center for the Advanced Study of Human Paleobiology, George Washington University, Washington, DC, United States of America

Richard G. Bribiescas Professor of Anthropology and Ecology and Evolutionary Biology, Deputy Provost for Faculty Development and Diversity, Yale University, New Haven, CT, United States of America

Martin Brüne Professor of Psychiatry, LWL University Hospital Bochum, Department of Psychiatry, Psychotherapy and Preventive Medicine, Division of Cognitive Neuropsychiatry, Ruhr University Bochum, Germany

Ann E. Caldwell Instructor/Fellow, Division of Endocrinology, Metabolism, and Diabetes, Anschutz Health and Wellness Center, University of Colorado Denver, Anschutz Medical Campus, Aurora, CO, United States of America

Olga Carvalho Invited Assistant Professor, Institute of Histology and Embryology, Faculty of Medicine, University of Coimbra, Portugal

James DeGregori Professor, Department of Biochemistry and Molecular Genetics, University of Colorado Denver, Anschutz Medical Campus, Aurora, CO, United States of America

Stanley Boyd Eaton Associate Professor of Anthropology and Radiology (Emeritus), Emory University and Emory School of Medicine, Emory University, Atlanta, GA, United States of America

Patrick Eppenberger Paleopathology and Mummy Studies Group, Institute of Evolutionary Medicine, Faculty of Medicine, University of Zurich, Switzerland

Paul W. Ewald Professor and Director of the Program in Evolutionary Medicine, Department of Biology, University of Louisville, KN, United States of America

Josiane Fakhry Graduate Student, Department of Anatomy and Neuroscience, University of Melbourne, Australia

Linda J. Fothergill Graduate Student, Department of Anatomy and Neuroscience, University of Melbourne, Australia

John B. Furness Professor, Digestive Physiology and Nutrition Laboratories, Florey Institute of Neuroscience and Mental Health, University of Melbourne, Australia

Joanna Gajewski Digestive Physiology and Nutrition Laboratories, Florey Institute of Neuroscience and Mental Health, Parkville, Australia

Francesco M. Galassi Paleopathology and Mummy Studies Group, Institute of Evolutionary Medicine, Faculty of Medicine, University of Zurich, Switzerland

Martin Häusler Evolutionary Morphology and Adaptation Group, Institute of Evolutionary Medicine, Faculty of Medicine, University of Zurich, Switzerland

Maciej Henneberg Wood Jones Professor of Anthropological and Comparative Anatomy, University of Adelaide, Adelaide Medical School, Australia

Mark A. Hill Department of Anatomy, School of Medical Sciences, University of New South Wales, Sydney, Australia

Melvin Konner Samuel Candler Dobbs Professor, Department of Anthropology, Program in Neuroscience and Behavioral Biology, Emory University, Atlanta, GA, United States of America

Diana Le Duc Human Genetics Fellow, Institute of Human Genetics, University Medical Center Leipzig; Max Planck Institute for Evolutionary Anthropology, Department of Evolutionary Genetics, Leipzig, Germany

John N. Maina Research Professor, Department of Zoology, Auckland Park Campus, University of Johannesburg, Johannesburg 2006, South Africa

Barbara Natterson-Horowitz Visiting Professor, Department of Human Evolutionary Biology, Harvard University, Cambridge, MA; David Geffen School of Medicine at UCLA; Adjunct Professor, UCLA Department of Ecology and Evolutionary Biology, Los Angeles, CA, United States of America

Randolph M. Nesse Foundation Professor of Life Sciences and Founding Director, Center for Evolution and Medicine, Arizona State University, Tempe, AZ, United States of America

Mehdi Nojoumi Medical Student, UC San Diego School of Medicine, San Diego, CA, United States of America

Eric M. Pietras Assistant Professor, Division of Hematology, Department of Medicine, University of Colorado Denver, Anschutz Medical Campus, Aurora, CO, United States of America

Daniel E. L. Promislow Professor, Department of Pathology and Department of Biology, University of Washington, Seattle, WA, United States of America

Paola Romagnani Professor of Nephrology, Department of Biomedical and Experimental Sciences 'Mario Serio', Excellence Centre for Research, Transfer and High Education for the Development of DE NOVO Therapies (DENOTHE), Paediatric Nephrology Unit, Meyer Children's Hospital, University of Florence, Italy

Graham A. W. Rook Professor of Medical Microbiology, Centre for Clinical Microbiology, Department of Infection, University College London, United Kingdom

Frank Rühli Founding Chair and Director, Institute of Evolutionary Medicine, Faculty of Medicine, University of Zurich, Switzerland

Wulf Schiefenhövel Professor of Medical Psychology and Ethnomedicine (Emeritus), Max-Planck-Institute for Ornithology, Human Ethology Group, Seewiesen, Germany

Torsten Schöneberg Professor of Molecular Biochemistry, Institute of Biochemistry, Faculty of Medicine, University of Leipzig, Germany

Kevin S. Shah Cardiology Fellow, University of California, Los Angeles, CA, United States of America

Kalyanam Shivkumar Professor of Medicine and Radiology, UCLA Cardiac Arrhythmia Center and EP Programs, Adult Cardiac Catheterization Laboratories, RR UCLA Medical Center, UCLA Health System, Los Angeles, CA, United States of America

Holly A. Swain Ewald Department of Biology, University of Louisville, Kentucky, United States of America

Wenda Trevathan Regents Professor of Anthropology (Emerita), New Mexico State University, United States of America

Xiaqing Zhao Post-doctoral Fellow, Department of Pathology, University of Washington, Seattle, WA, United States of America

PART I

GENERAL PRINCIPLES

CHAPTER 1

CORE PRINCIPLES FOR EVOLUTIONARY MEDICINE

RANDOLPH M. NESSE

Abstract

New interest in evolution and medicine arose late in the twentieth century from the recognition that there are several possible kinds of evolutionary explanation for aspects of the body that leave it vulnerable to disease, in addition to the inevitability of mutations. Investigations of related hypotheses have led to rapid growth of evolutionary medicine, and its expansion to integrate demographic, phylogenetic, and population genetic methods. Evolutionary approaches to understanding disease are part of a major transition in biology, from viewing the body as a designed machine to a fully biological view of the body's organic complexity as fundamentally different from that of designed machines.

Keywords

evolutionary medicine, Darwinian medicine, proximate, natural selection, vulnerability, organic complexity, mechanism, phylogeny, population genetics

1.1 Introduction

Fast-growing new interest in the intersection between evolutionary biology and medicine has been spurred by three ideas that developed in the late twentieth century. The first is the recognition that all traits need an evolutionary explanation in addition to an explanation of mechanisms. This idea was promoted by Ernst Mayr (Mayr 1982), but was given its fullest expression in what are now widely recognised as Tinbergen's Four Questions (Tinbergen 1963;

Bateson and Laland 2013; Nesse 2013; Medicus 2015). The second development was the recognition that natural selection does not mainly shape traits to benefit groups and species, but rather it increases the frequency of alleles that are transmitted to future generations more rapidly than other alleles, by whatever method (Williams 1966; Dawkins 1976). The third development, related to the second, was the recognition that alleles that harm an individual's health and reproductive success can nonetheless be selected for if they give sufficient advantages to kin (Hamilton 1964; Crespi et al. 2014). The intersection of these three ideas suggested that aspects of bodies that leave them vulnerable to disease have evolutionary explanations in addition to the widely recognised inevitability of mutations (Williams and Nesse 1991; Nesse and Williams 1994; Stearns 1999).

Seeking explanations for suboptimal traits is by no means new. It was a major focus for William Paley's 1802 book, *Natural Theology: or, Evidences of the Existence and Attributes of the Deity; Collected from the Appearances of Nature* (Paley 1802). Paley explained the body's suboptimal 'contrivances' as puzzles posed by a deity to impress and occupy scientists. The book inspired Charles Darwin, whose discovery of natural selection eventually provided a scientific explanation for why bodies are the way they are (Darwin 1859; Ayala 2007); Darwin understandably focused, however, on traits well-suited to their functions. After natural selection was integrated with genetics in the mid-twentieth-century 'modern synthesis', suboptimal traits were routinely attributed to mutations and genetic drift. 'Natural selection just can't do any better' was the most common explanation, and talk about other possible reasons for vulnerability to disease was often dismissed as speculation.

The new development, arising late in the twentieth century, was the recognition that natural selection can help explain apparent maladaptations as well as adaptations. Several possible kinds of explanation for traits that leave bodies vulnerable to disease are recognised, in addition to the limits of natural selection. They include mismatch between the environment in which a trait evolved and the current environment to which organisms are exposed, coevolution with other organisms, trade-offs, and reproductive success at the cost of health as well as from the limits of natural selection, and defences with major costs (Crespi 2000; Nesse 2005a).

As is the case for bodies, the field of evolutionary medicine has vulnerabilities that are closely associated with its strengths. The early focus on the adaptive significance of traits that seem maladaptive created great interest; resolving the paradox of the persistence of apparently harmful traits is inherently fascinating. This fascination led to enthusiastic attempts to find evolutionary explanations for things that were not shaped by natural selection; misguided attempts to provide evolutionary explanations for diseases themselves remain all too common. The field as a whole has dealt with this problem relatively well; however, exposure to dramatic speculations aroused general scepticism among some scientists who focus exclusively on proximate mechanisms, especially those unfamiliar with methods for testing evolutionary hypotheses. The challenge of finding the best ways to frame and test evolutionary hypotheses about disease continues (Nesse 2011a), along with the challenge of encouraging interest in such questions without also encouraging wild speculation.

Just as the body has suboptimal traits because of canalised developmental pathways laid down early in the course of evolution, such as the eye's blind spot, the field of evolutionary medicine is somewhat constrained by its origins in ways that make it suboptimal. It has emphasised only one of Darwin's two discoveries—natural selection as the process that accounts for why traits are the way they are. Darwin also showed the unity of all life from a common phylogenetic origin. This second discovery has been neglected, and phylogenetic, and population genetic methods more generally, remain to be fully integrated.

1.2 Core Principles for Evolutionary Medicine

1.2.1 What Core Principles Are

Describing core principles of evolutionary medicine must begin by defining core principles in general, as well as the field of evolutionary medicine. Education researchers have encouraged the formulation of core principles for fields as a way to focus on the big ideas that endure and organise thinking. Niemi and Phelan (2008) define core concepts as being 'organized around central concepts or principles, or "big ideas". The nature of these concepts differs from domain to domain, but in general they are abstract principles that can be used to organize broad areas of knowledge and make inferences in the domain, as well as determining strategies for solving a wide range of problems.'

Evolutionary medicine is the field that uses principles of evolutionary biology to better understand, prevent, and treat disease, and that uses studies of disease to advance basic evolutionary biology. It includes all work at the intersection of the basic science of evolutionary biology with the professions of medicine and public health. The phrase 'evolutionary medicine' gives the mistaken impression that it is a special kind of medical practice. This is an unfortunate result of the early history of the field, and cannot be readily corrected. 'Darwinian medicine' is a more accurate synonym, with the same disadvantage; it is infrequently used now because 'Darwinian' has negative connotations for so many members of the general public. 'Evolution and medicine' is a useful phrase to describe the overlap between the fields that define evolutionary medicine, but this leaves out public health, nursing, psychotherapy, and veterinary medicine. 'Evolution and the health professions' is accurate—but unlikely to catch on. 'Evolutionary medicine' is the keyword that will likely endure, despite its limitations.

Several textbooks and many review papers describe principles of evolutionary medicine (Nesse and Williams 1994; Stearns 1999; Trevathan et al. 2007; Nesse and Stearns 2008; Stearns and Koella 2008; Gluckman et al. 2009a; Nesse et al. 2010; Stearns 2012; Perlman 2013; Stearns and Medzhitov 2016). The challenge of synthesising them into a single list is formidable. To meet this challenge, a recent study used the Delphi method to organise the recommendations of thirty-seven evolutionary medicine experts into fourteen core principles for the field (Grunspan et al. 2018). This study posed the question: What are the core principles for evolutionary medicine? After four waves of voting and revisions, fourteen principles were endorsed by at least 80% of the respondents. The survey respondents also suggested, but did not reach agreement on, fourteen additional possible core principles, some of which were overlapping, superordinate, or subcategories.

This chapter relies heavily on the principles formulated by the Delphi study. In the course of that study, it became clear that the task of organising the core principles for evolutionary medicine poses special challenges. Some are nested within others, some overlap, and some fit within several other categories. This chapter provides and reviews an expanded list of core principles, showing, where possible, how they are related to each other, to evolutionary biology, and to evolutionary medicine. It considers principles of evolutionary biology that are especially useful in evolutionary medicine, rather than principles that are specific to evolutionary medicine. The result is the list of core principles in Box 1.1. Most come straight from evolutionary biology, but many take on a new slant when used to understand pathology instead of normal function. A few are more specialised principles that are particularly closely associated with the evolutionary medicine, some of which have emerged from efforts to achieve a better understanding of disease. Boundaries between categories of core principles are fuzzy, so, instead of attempting to classify each one explicitly, this chapter will instead make note of relevant issues where appropriate.

Box 1.1 Core Principles Useful for Evolutionary Medicine

** Indicates one of the final core principles from the Delphi study
* Indicates principles suggested by Delphi study respondents that did not reach 80% agreement

1. **All traits need both proximate and evolutionary explanations
2. A full explanation for any trait requires answers to all four of Tinbergen's Questions
3. Traits that leave bodies vulnerable to disease have several possible kinds of evolutionary explanations
   - 3.1. **The limitations of natural selection can help to explain disease vulnerabilities
     - 3.3.1. Selection minimises mutations but cannot completely prevent them
     - 3.3.2. Path dependence is responsible for many suboptimal traits
   - 3.2. **Mismatch between bodies and changing environments accounts for much disease.
   - 3.3. Coevolution with pathogens explains several kinds of host vulnerability
     - 3.3.1. **Coevolution explains patterns of virulence
     - 3.3.2. Antibiotic resistance is a product of natural selection
     - 3.3.3. *Microbiomes are useful and disruptions cause disease
     - 3.3.4. **Coevolution causes arms races that shape dangerous defences
   - 3.4. **Trade-offs characterise all aspects of bodies and they explain many traits that leave bodies vulnerable to disease
   - 3.5. **Natural selection maximises allele transmission at the expense of health
     - 3.5.1. **Sexual selection increases reproduction at the expense of health
     - 3.5.2. Alleles that bias transmisssion may account for some diseases
   - 3.6. **Defences provide protection in the face of threats and damage, but at considerable costs
     - 3.6.1. Defences are aversive for good reasons
     - 3.6.2. Negative emotions are useful defensive responses
     - 3.6.3. *The smoke detector principle explains unnecessary expression of defence responses

4. **Selection shapes mechanisms that mediate plasticity in various time frames
   4.1. *Developmental Origins of Health and Disease (DOHaD) is an important cause of disease vulnerability
   4.2. **Selection has shaped fast and slow life histories with implications for health
5. **Natural selection works mainly at the level of the gene
   5.1. Group selection is a viable explanation only under constrained circumstances
   5.2. A multigenerational perspective is important
6. *Kin selection can explain some traits that reduce individual reproductive success
   6.1. Natural selection continues to act after menopause
   6.2. Weaning conflicts are inevitable
   6.3. Conflicts between maternal and paternal genomes can cause disease
7. Control of cell replication is crucial for metazoan life
8. *Intragenomic conflicts can influence health
9. *Somatic selection changes cell genotypes during the lifetime of an individual
10. Natural selection shapes life history traits
11. Genes with deleterious effects can be selected for if they offer compensating benefits
12. Cliff-edged fitness landscapes can account for the persistence of some genetic diseases
13. *Attention to ethics is important
14. Races are not biological categories
15. Genetic differences between human subgroups influence health
16. It is a mistake to assume that what is is what ought to be
17. Natural selection is not over for humans
18. **Genetic methods for tracing relationships and phylogenies have many applications in evolutionary medicine
    18.1. Tracing human ancestry is medically relevant
    18.2. Phylogenetic methods can trace the origins and spread of pathogens
19. Methods for framing and testing evolutionary hypotheses remain under development
20. Organic complexity is different in kind from the complexity in machines

1.3 Specific Core Principles

Box 1.1 lists 20 core principles and additional subprinciples relevant to evolution, health, and disease. Each receives a brief description below, along with thoughts about its relevance and common misunderstandings.

1.3.1 All Traits Need Both Proximate and Evolutionary Explanations

This principle is increasingly recognised, but still widely misunderstood. Proximate explanations describe traits and how they work. Evolutionary explanations explain how traits came to be the way they are. A brief conversation about the distinction is rarely sufficient to get the idea clear for many scientists. Comprehension by students, in my experience, requires discussion of many examples over several hours.

The distinction was described by Ernst Mayr, in several articles (Mayr 1961, 1974) and his magisterial book *The Growth of Biological Thought* (Mayr 1982), which portrayed biology as two intersecting enterprises, one describing mechanisms, the other the evolution of those mechanisms. He called these explanations 'proximate' and 'ultimate', but associations of the word 'ultimate' with philosophical and religious traditions have led most authors to instead simply call them 'evolutionary explanations'. On occasion, they have been referred to as different 'levels of explanation' (Reeve and Sherman 1993), but this risks confusion with the more usual use of levels to refer to levels of organisation nested within each other. Some have questioned the utility of the distinction, noting that many evolutionary processes involve reciprocal causation in which proximate mechanisms themselves influence future selection forces, sexual selected traits and preferences for those traits being an example (Laland et al. 2011); however, few would argue that either a proximate explanation of mechanisms or an evolutionary explanation of a sequence of traits is sufficient alone. Both are necessary.

Many clinicians and scientists in the health sciences remain unaware of the need for both proximate and evolutionary explanations. Those that are aware of the distinction sometimes view them as alternatives, although they are synergistic complementary explanations. A further challenge is posed because methods for testing evolutionary hypotheses differ substantially from those used to test hypotheses about proximate mechanisms (Nesse 2011a). In short, the transition from relying exclusively on proximate explanations in medicine, to routine recognition of the need to also pose and test evolutionary hypotheses, is still in progress. Evolutionary medicine is helping to advance this transition.

1.3.2 A Full Explanation for Any Trait Requires Answers to All Four of Tinbergen's Questions

Expanding the focus to Tinbergen's Four Questions transcends some of the difficulty. In an article to honour his friend and colleague Konrad Lorenz, Nico Tinbergen (1963) suggested four different questions that must all be addressed to fully explain any behaviour. They became the widely accepted foundation for the field of animal behaviour, and they have inspired much work in evolutionary medicine (Nesse 2013), including psychiatry (Brüne 2014b). I spent several months trying to understand their full import before finally grasping that two are about proximate mechanisms, and two are about evolution. Furthermore, proximate questions are of two kinds: description of a current mechanism and description of the how mechanisms develop in an individual, from a DNA code to an adult organism. Evolutionary questions are also of two kinds: description of the adaptive significance of a trait and description of the trait's phylogeny. Questions about mechanism and adaptive significance are answered by descriptions of an organism at a cross-section in time. Questions about development and phylogeny require consideration of an historical sequence of events. These distinctions organise the four questions as shown in Table 1.1 (Nesse 2013).

Table 1.1 Tinbergen's Four Questions Organised

Four areas of biology: four questions		Two objects of explanation	
		Developmental/historical A sequence that results in the trait	**Single form** The trait at one slice in time
Two kinds of explanation	**Proximate** Explains how organisms work by describing mechanisms and their ontogeny	**Ontogeny** Q: **How does the trait develop in individuals?** A: Description of the trait's forms at sequential life stages, and the mechanisms that control development	**Mechanism** Q: **What is the structure of the trait? How does it work?** A: Description of the trait's anatomy, physiology, regulation, and how the trait works to accomplish a function
	Evolutionary Explains how a trait came to its current form by describing a sequence of forms, and how variations were influenced by selection and other evolutionary factors	**Phylogeny** Q: **What is the phylogenetic history of the trait?** A: Description of the history of the trait as reconstructed from its phenotype and genotype precursors	**Adaptive significance** Q: **How have variations in the trait interacted with environments to influence fitness in ways that help to explain the trait's form?** A: Description of how variations in the trait have influenced fitness

Source: Nesse (2013); see also Medicus (2015).

Adapted from *Trends in Ecology & Evolution*, 28 (12), Randolph M. Nesse, Tinbergen's four questions, organized: a response to Bateson and Laland, pp. 681–2, doi.org/10.1016/j.tree.2013.10.008

1.3.3 Traits that Leave Bodies Vulnerable to Disease Have Several Possible Kinds of Evolutionary Explanation

Mutations, genetic drift, and the general limits of natural selection were long the accepted general explanations for aspects of the body that seem suboptimal. This reflected a tacitly creationist view of organisms, as if they were designed and perfectible, as if there were one normal blueprint that creates optimal organisms. A more fully evolutionary view recognises that alleles shape somas that maximise genetic transmission, and that variation is intrinsic to the genome and phenotypes.

Much of what is new in evolutionary medicine emerges from considering five other kinds of explanation for why genetically 'normal' individuals are nonetheless vulnerable to diseases (Williams and Nesse 1991; Nesse and Williams 1994; Nesse 2005a; Gluckman et al. 2009a; Stearns and Medzhitov 2016). The most commonly cited ones are mismatch with environments, coevolution with pathogens, trade-offs that limit perfection, reproductive success at the expense of health, and defences that seem like diseases. This list of possible kinds of explanation for disease vulnerability has proved serviceable, but it is by no means the only alternative. Some, especially those studying the Developmental Origins of Health and Disease (DOHaD) (Kuzawa et al. 2008; Godfrey et al. 2010; Bateson and Gluckman 2011; Hanson 2015), have suggested adding development as a separate category. This can be useful; however, some diseases with developmental origins are mainly due to constraints, some mainly from trade-offs, and many from mismatch with environments, so more specific categories can draw closer attention to causal factors.

The list can be collapsed or expanded into fewer or more categories. Mismatch and coevolution both result in disease vulnerability because natural selection is too slow to adapt a species to a fast-changing environment. Constraints arise for several reasons: the limitations imposed by the laws of physics, the impossibility of maintaining a completely accurate code, and path dependence, the impossibility of starting fresh with a fundamentally new design. Trade-offs limit the optimality of all traits. Defensive responses are not a reason for vulnerability to disease; however, they are costly, they often cause problems, and responses such as pain and fever can seem like diseases, especially since they give rise to many normal false alarms that lead to requests for medical treatments. The last and conceptually most important principle is that selection does not shape organisms for health, longevity, or happiness, but only for maximum transmission of genes.

More than one such explanation often is relevant. For instance, vulnerability to atherosclerosis is explained by mismatch with modern environments, the benefits of inflammatory cells in the lining of arteries, and coevolution with pathogens (Nesse and Weder 2007). A tendency to emphasise one kind of explanation at the expense of others, or to treat two different kinds of explanation as alternatives, results in much confusion.

Most of these categories apply equally to machines and bodies. Errors in blueprints or the DNA code cannot be completely prevented, and trade-offs are inevitable for all traits and machines or bodies. Several factors are, however, distinctive to bodies. In particular, while engineers can start from scratch to redesign a component of a machine, natural selection is limited to tinkering, resulting in extreme jury-rigged designs such as the path of the recurrent laryngeal nerve, from the brainstem down into the thorax, then ascending again behind the oesophagus.

The other major difference is that machines and their parts are designed to serve specific functions, while bodies are shaped to maximise the transmission of their genes, even at a cost to health and longevity, and their parts have overlapping functions. This makes bodies profoundly different from machines. Smaller but still important differences include the nature of redundancy. Machines have backup systems that kick in when the primary system is not working. Bodies have some similar redundant systems, but they are better protected by the tight integration of multiple systems so that the failure of a single component often has little effect on function.

1.3.3.1 *The Limitations of Natural Selection Can Help to Explain Disease Vulnerabilities*

Two main kinds of constraints are especially important for evolutionary medicine: limited ability to preserve the information code, and path dependence. Bodies and machines both are vulnerable to failure because of code errors; however, bodies differ from machines in that bodies have no one definitive perfect plan, of the sort that a blueprint provides for machine. There is no one normal genome, and genomes are not essentialised kinds. Instead, genomes are collections of diverse alleles competing for representation in future generations. Other more general constraints apply to any body or machine. Space and time are limited. Energy is conserved. Entropy is, in the long run, unyielding.

1.3.3.1.1 SELECTION MINIMISES MUTATIONS BUT CANNOT ELIMINATE THEM

Mutations cannot be completely prevented or repaired, and it takes time to purge mutations, so disease is inevitable. Usually, however, genetic variations are proposed as an explanation for why some individuals get sick, not for why all members of a species have a trait, such as a windpipe that intersects with the food passageway, that leaves them all vulnerable to a disease. A prevalent general model assumes there is some optimum for each trait that provides robustness and efficient function, and that mutations constantly spread the distribution, while stabilising selection narrows it. While correct and relevant, this principle is of limited utility for explaining why all individuals in a species share traits that make them vulnerable to disease.

Natural selection reduces mutation rates for higher organisms to the minimum possible, given the costs of maintaining genomic integrity, and physical constraints. This principle requires mention because the idea persists that natural selection maintains a higher mutation rate to benefit the evolution of species. This mistake arises from the misconception that selection shapes traits that benefit the species at the expense of an individual's alleles. An allele that increases mutation rates will be selected against because offspring with the allele will tend to have defects, and the higher mutation rate will further degrade the code with every subsequent generation. Selection shapes mechanisms that reduce mutation rates to the level where their costs equal the costs of DNA replication and repair mechanisms, or to the level where fidelity is limited by physical constraints, such as inevitable damage from chemical factors and environmental radiation (Sung et al. 2016). Possible exceptions to the general principle are found in facultative mechanisms in bacteria that increase mutation rates in response to certain kinds of severe stress (Rosenberg et al. 2012). In such cases, alleles that increase the mutation rate temporarily can give an advantage because the number of descendants is huge, and success depends on a winner-takes-all-lottery in which the chances of winning increase if the tickets have different numbers.

1.3.3.1.2 PATH DEPENDENCE IS RESPONSIBLE FOR MANY SUBOPTIMAL TRAITS

Bodies are more constrained by path dependence than machines. Natural selection can make only small changes because large ones are likely to be fatal or physically impossible. Even in automobiles, however, major changes, such as relocating the gasoline tank, have large costs, so engineers do not undertake them lightly. For bodies, such radical redesign is rarely possible. Natural selection works by tinkering. This results in suboptimal traits such as the opening of the windpipe into a space shared by the food passage, and the long winding paths of the vas deferens and the recurrent laryngeal nerve. Such traits inspired William Paley to extraordinary flights of creative argument to try to reconcile such suboptimal traits with divine design (Paley 1802).

1.3.3.2 *Mismatch between Bodies and Changing Environments Accounts for Much Disease*

Vulnerability to disease resulting from bodies ill-adapted to their current environments is a major theme in evolutionary medicine (Eaton et al. 1988; Gluckman and Hanson 2006). On occasion, it has been viewed incorrectly as the only focus of evolutionary medicine.

This principle sometimes causes misunderstanding for those who assume it implies that people were healthier in ancestral environments. They did not suffer from modern diseases, but they nonetheless suffered a huge burden of disease. Overall, health is vastly better for people living in modern settings; however, it is also true that most chronic disease today results from exposure to aspects of modern societies that were absent for our ancestors. Evidence is now coming in to confirm that atherosclerotic disease is far less common in horticulturalists (Kaplan et al. 2017). Longevity is also dramatically greater now; however, this is not mainly because of slower ageing or dramatically lower mortality rates in adulthood—the big difference is lower mortality rates for children. In the past, the average lifespan may have been 30 years because of high infant mortality, but many who survived to age 30 lived on for decades more (Hill and Hurtado 1996).

The exact burden of disease caused by mismatch is not certain, but is definitely large. Rates of cardiovascular disease and breast cancer are at least an order of magnitude higher now than they were for our foraging ancestors (Kaplan et al. 2017). Obesity and diabetes are epidemic (Flegal et al. 2012). Allergies have increased exponentially in the past 50 years for reasons that are in urgent need of more study (Armelagos and Barnes 1999; Brüne and Hochberg 2013). Disorders related to gluten sensitivity suggest that we are still adapting to agriculture and a grain-based diet (Lindeberg 2009; Brüne and Hochberg 2013). Drug abuse was uncommon until pure drugs and novel means of administration became readily available (Nesse and Berridge 1997). Eating disorders have increased dramatically in recent decades (Rosenvinge and Pettersen 2015). Disorders related to preoccupation with electronic devices are growing fast. Preferences to limit reproduction by using birth control are heritable (Mealey and Segal 1993), but slow to change. Given enough time, selection could presumably adapt bodies and minds to cope better, but environments change too fast. (For further discussion, see Chapter 6: Nutrition, Energy Expenditure, Physical Activity, and Body Composition.)

Autoimmune disorders pose a particularly dramatic example and challenge. Crohn's disease, type 1 diabetes, multiple sclerosis, and other autoimmune diseases have been increasing rapidly in just the past few decades (Bach 2002). What accounts for this sudden change remains uncertain; however, antibiotics that disrupt microbiomes are a likely suspect (Blaser 2014). (For further discussion, see Chapter 10: Immune System.)

Mismatch results from migration as well as from changing environments. This explains phenomena such as vulnerability to rickets in people with dark skin living in the north (Jablonski 2004), and vulnerability to skin cancer in people with light skin living closer to the equator (Greaves 2014). Selection shapes differences in HLA alleles and other genes depending on the exposure of a population to pathogens (Karlsson et al. 2014). The loss of the protein that malaria parasites use to enter blood cells is common in some human subpopulations that have evolved in conjunction with malaria (Miller et al. 1975; Lentsch 2002), and a strong selection coefficient of 0.043 has led to near fixation of this loss in some areas since its origins about 40,000 years before present (Karlsson et al. 2014; McManus et al. 2017). Migration between malarial and non-malarial environments can make such genotypes adaptive or maladaptive.

1.3.3.3 *Coevolution with Pathogens Explains Several Kinds of Host Vulnerability*

Recognition of the special dynamics that arise from host–pathogen coevolution was one of the early major advances in evolutionary medicine (May and Anderson 1983; Ewald 1994;

Ebert and Hamilton 1996). In its simplest form, coevolution results in vulnerability because a pathogen may have tens of thousands of generations during a single host lifetime. From this perspective, it is amazing that large multicellular organisms with long lifespans exist. They are possible only because of sophisticated immune defences, some of which also use somatic selection among immune cells to adapt antigen-recognising cells to the challenges of fast-changing pathogens.

1.3.3.3.1 COEVOLUTION EXPLAINS PATTERNS OF VIRULENCE

As recently as a few decades ago, the generalisation 'pathogens don't want to kill their hosts' was widely accepted. Recognition that natural selection acts to benefit species only to the extent that this benefits individuals has transformed microbiology, with increasingly sophisticated evolutionary models of how natural selection shapes virulence (Read 1994; Ewald 1995; Frank 1996; Schmid-Hempel and Frank 2007). It was recognised early that modes of transmission have major effects. Pathogens that can only be transmitted in person get advantages by ensuring their hosts are up and about and not killed off soon. Pathogens that can be transmitted by needles, clinician's hands, mosquitos, or impure water tend to gain advantages from fast mass replication with less regard for the host's longevity (Frank 1992; Levin and Bull 1994; Read 1994; Ewald 1995; Levin 1996).

Studies of cooperation among microbes offers good examples of how clinically relevant investigations can advance by the application of basic evolutionary biology (Velicer 2003; West, Griffin and Gardner 2007; Schmid-Hempel 2011; Foster and Bell 2012). Biofilms continue to provide both a clinically difficult problem and major challenges for evolutionary explanation of how traits that harm the reproductive success of an individual cell can persist by their benefits to nearby cells (Queller 1994; Hansen et al. 2007; Oliveria et al. 2015). The basic principle remains the same: alleles can increase in frequency only if bearers with the allele have more surviving kin, on average, than others (West et al. 2007). This is an area where group selection and kin selection models have both proved useful (Queller 1994; Redfield 2002; Kreft 2004; Dugatkin et al. 2005; Crespi et al. 2014).

1.3.3.3.2 ANTIBIOTIC RESISTANCE IS A PRODUCT OF NATURAL SELECTION

Antibiotic resistance is often described as the most practical application of evolution to medicine. Lack of appreciation for the power of natural selection led, in the middle of the twentieth century, to the tragically overly optimistic prediction that antibiotics would lead to elimination of infectious disease (Neu 1992; Salmond and Welch 2008). Pathogens turned out to be capable of evolving resistance to every possible molecule aimed at them. This is less surprising when it is recognised that antibiotics themselves are mostly products of bacteria and fungi shaped over billions of years to compete effectively with other organisms (D'Costa et al. 2006). Antibiotic resistance is not exactly coevolution, but it is similar, because every time an organism develops resistance, researchers attempt to find a new agent to get around that resistance, and the organisms soon adapt.

Despite these insights, a fully sophisticated evolutionary approach to antibiotic resistance is still developing; the word 'evolution' is avoided in many medical articles (Antonovics et al. 2007). If new strategies to combat infection are to be found, they will need to rely on more sophisticated evolutionary models (Pepper 2008; Vale et al. 2016; Huijben and Paaijmans 2017). Even such basic questions as whether it is wise to continue antibiotics for a full 10-day course remain controversial (Read et al. 2011; Day and Read 2016; Bouglé et al. 2017) but increasingly studied, with longer courses now recognised as usually unnecessary

(Uranga et al. 2016). The *BMJ* recently published an article entitled 'The antibiotic course has had its day' (Llewelyn et al. 2017). It argued that taking the full course of antibiotics was unnecessary in many cases and did not prevent resistance; however, it did not mention evolution, natural selection, or the work of evolutionary biologists who study antibiotic resistance.

1.3.3.3.3 MICROBIOMES ARE USEFUL AND DISRUPTIONS CAUSE DISEASE

Just a few years ago, most doctors tended to think that bacteria were bad and best avoided or killed. In the 1980s, I was called to consult on many patients with intractable diarrhoea who were said to have a psychogenic condition because they had received multiple courses of antibiotics and their stool was clear of identifiable pathogens. In retrospect, many of these patients had infections with *Clostridium difficile* or some other overgrowth organism, freed from the constraints imposed by a normal microbiome. Now they would be treated effectively with a microbiome transplant (Agrawal et al. 2016). The magnitude of the error is hard to comprehend.

Research on the 'old friends' hypothesis revealed that we are dependent on many microbes, and disrupting them causes disease (Rook et al. 2017). The role of antibiotics in disrupting microbiomes is becoming clear (Blaser 2014). Recent methods use global sequencing to measure microbiomes reveals the myriads of microbes interacting in our microbiomes, some of which influence diseases, especially obesity (Turnbaugh et al. 2006). They also reveal changes induced in just a few days by diet changes (David et al. 2014), and how seasonal dietary changes influence the microbiome of Hadza (Smits et al. 2017).

The opposite schema is that we are a holobiont consisting of many well-coordinated cooperating organisms (Bordenstein and Theis 2015). Some would even say that it is a mistake to think about individuals without their microbiomes. Certainly, coexistence with microbes is our natural state, and disruptions are responsible for much disease, perhaps even most of the current epidemic of autoimmune disease. However, those microbes inevitably evolve to whatever phenotypes best preserve and spread them. Much cooperation results, but also much competition. (For further discussion, see Chapter 13: Digestive System.)

1.3.3.3.4 COEVOLUTION CAUSES ARMS RACES THAT SHAPE DANGEROUS DEFENCES

Selection that shapes a new host defence initiates new selection pressure on the pathogen for ways to get around the defence, which initiates new selection for an improved defence; the result is both host and the pathogen 'running as fast as they can' to keep up with changes in the other, like the Red Queen in *Alice in Wonderland* (Ridley 1994; Morran et al. 2011). This is a likely explanation for the maintenance of sex, and its very substantial costs (Ridley 1994; Auld et al. 2016; Metzger et al. 2016; Neiman et al. 2017). The implications for human disease are substantial.

The protections shaped have their own dangers. Thus, it is not surprising that diseases caused by the immune system are becoming more prevalent as those caused by infections decline (Bach 2002). Selection can shape defences that often harm hosts because the selection pressure from pathogens will push a defensive capacity to the point where the reliable marginal benefits are less than the occasional marginal costs. This has practical implications for decisions about whether to use medications that block inflammation. In influenza, fatal outcomes may result either from the direct action of the virus or from the inflammatory response, or a combination. Multiple trials of steroids in influenza patients have yielded inconsistent outcomes (Salomon et al. 2007). An evolutionary perspective suggests that it should be possible to predict which patients will benefit from down-regulating the inflammatory response, and which ones will be harmed.

1.3.3.4 *Trade-offs Characterise All Aspects of Bodies*

No trait can be perfect, because changes improve one trait will compromise other traits or other adaptive aspects of the same trait. For instance, higher levels of stomach acid will provide increased protection against infection at the cost of increased risk of ulcers. Immune surveillance for a wider range of antigens provides better protection against infection but more risk of autoimmune responses (Bergstrom and Antia 2006). Easier initiation of apoptosis for potentially malignant cells will protect against cancer, at the cost of decreased capacity for tissue repair (Abegglen et al. 2015). Maintaining a high body weight offers protection during periods of food shortage, at the cost of slower locomotion and needing more calories to sustain the body.

Trade-offs are central to life history theory, especially the benefits and costs of early reproduction versus a longer lifespan, and the relative benefits and costs to males versus females of soma maintenance versus competitive ability (Stearns 1989; Hill 1993; Brüne 2014a). Understanding the regulation of defences requires understanding their costs and benefits versus the alternative of less or no response. The central role of trade-offs in every aspect of life is the single most important principle of evolutionary medicine. It encourages clinicians and researchers to think not about if a trait is perfect or flawed, but instead about how natural selection shapes trade-offs to maximise inclusive fitness with much resulting vulnerability to disease.

1.3.3.5 *Natural Selection Maximises Allele Transmission at the Expense of Health*

This may be the deepest and most surprising principle, and the one most specific to evolutionary medicine. Natural selection shapes health and longevity only to the extent that they increase reproductive success. Alleles that harm health or shorten lifespan are selected for if they increase inclusive fitness. Examples include sexual selection, reviewed here, and ageing, fast versus slow life histories, and intragenomic competition, discussed subsequently.

1.3.3.5.1 SEXUAL SELECTION INCREASES REPRODUCTION AT THE EXPENSE OF HEALTH

Sex differences in lifespan illustrate the costs of ability to compete for mates. In species where ability to compete for mates gives males major payoffs, selection shapes males for high investment in competitive ability despite the costs of risk-taking and limited ability to repair tissue damage (Trivers 1972; Liker and Székely 2005). Females are subject to the same trade-offs in general; however, they benefit relatively more from investments in tissue repair. This results in mortality rates for human males at sexual maturity in developed countries about three times higher than those for females (Kruger and Nesse 2006). While some of the excess is due to risk-taking and behaviour, males also have increased risks for infection, cancer, and metabolic disease (Kruger and Nesse 2004).

Sexual selection explains other medically relevant sex differences. For instance, the two-fold increased risk of anxiety disorders for women versus men is fairly consistent worldwide

(Ruscio et al. 2017). This has prompted many to ask why women have too much anxiety. An evolutionarily informed view instead considers the possibility that anxiety regulation mechanisms for women are set to thresholds close to the optimum to benefit individual women, while higher anxiety thresholds for men increase competitive ability at the expense of increased rates of harm (Stein and Nesse 2015). Sex differences in the effects of an allele on fitness are illustrated by haemochromatosis: the resulting liver damage is less severe for women because of regular blood loss with menstruation (Moirand 1997).

1.3.3.5.2 ALLELES THAT BIAS TRANSMISSION MAY ACCOUNT FOR SOME DISEASES

Alleles can advance their own replication at the expense of the individual. Such phenomena are usually well controlled by mechanisms shaped by natural selection that advance the interests of the 'parliament of genes', as described further in Section 1.3.8.

1.3.3.6 *Defences Provide Protection in the Face of Threats and Damage, but at Considerable Costs*

Many problems people bring to medical attention are not direct products of disease, they are protective defences shaped by natural selection in conjunction with mechanisms that monitor for situations in which they can be useful (Nesse and Williams 1994). Defences are especially obvious in the face of infection (Ewald 1980). Expulsion of pathogens by means of rhinorrhoea, cough, vomiting, and diarrhoea offers powerful protection. Inflammation provides graded specific responses to specific infectious challenges. As noted already, coevolution with pathogens explains why defences are maintained despite their extreme costs and risks.

Selection has shaped systems to defend against many other risks (Harvell 1990). Some defences are fixed. Skin pigmentation, for instance, protects against cutaneous damage and skin cancer, with the trade-off of increasing vulnerability to vitamin D deficiency and rickets (Jablonski and Chaplin 2010; Greaves 2014; Jablonski and Chaplin 2017) (see Chapter 8). Other defences are responses expressed when they are needed. Reflexive withdrawal from heat and tissue damage is useful. Blinking protects against foreign matter in the eye, sneezing against foreign matter in nasal passages, and itching against skin parasites. Shivering protects against hypothermia, sweating against hyperthermia. These systems are usually called 'facultative responses' because they are associated with systems that monitor for situations in which the response is needed. Or, more exactly, when the benefits are greater than the costs, explaining why false alarms are normal and common in such systems.

1.3.3.6.1 DEFENCES ARE AVERSIVE FOR GOOD REASONS

Activation of most defensive responses is associated with subjective pain or other aversive experiences that make them seem like disorders themselves. The aversiveness motivates escape and avoidance. Recognising that aversive defences are useful is a major contribution of an evolutionary approach in clinical medicine. This can guide clinical decision-making about when it is, and is not, appropriate to use medications to block such responses.

1.3.3.6.2 NEGATIVE EMOTIONS ARE USEFUL DEFENSIVE RESPONSES

The cognitive tendency to attribute specific functions to specific things is on display in the history of emotions research. For instance, fear is said to serve the function of promoting escape, anger to defend against attack. A more explicitly evolutionary approach recognises that emotions have multiple functions, and that different emotions are distinguished from each other, not by their functions, but by the situations in which they are useful (Plutchik 1980; Wierzbicka 1986; Nesse 1990; Nesse and Ellsworth 2009). This approach transcends unresolvable debates about how many basic emotions there are by recognising that emotional responses evolve from prior responses, so we should expect them to be overlapping, not separate. It also helps to explain why emotional states are almost all associated with subjective experiences of pleasure or pain. No response would be shaped for situations that did not involve opportunities or threats.

The burden of disease posed by anxiety and depressive disorders is huge (Kessler et al. 2009). Approaches to those disorders have mostly looked only at events and mechanisms that account for differences in vulnerability. Evolutionary medicine instead encourages analysis of how the expression of emotions is regulated, and the reasons why negative emotions so often seem to be expressed excessively, or in situations where they are not essential (Nesse 2011b). This knowledge is useful for clinical evaluations looking for the origins of such emotions, and making clinical decisions about when it is safe to use medications to block them.

1.3.3.6.3 THE SMOKE DETECTOR PRINCIPLE EXPLAINS UNNECESSARY EXPRESSION OF DEFENCE RESPONSES

Many defences are relatively inexpensive compared with the enormous costs of not expressing a defence when it is needed. For instance, vomiting might cost only a few hundred calories; however, failure to vomit can be fatal when a toxin or a pathogen is in the gut. This is observed regularly and tragically in the deaths of college students who fail to vomit after drinking an entire bottle of liquor, and patients who take an overdose of pills that include agents that inhibit vomiting.

The 'smoke detector principle' is especially relevant to emotional disorders. For instance, the cost of a panic attack may be only 100 calories, while the absence of a panic response in the face of a predator may be death, about 100,000 calories. Applying a standard signal detection analysis to the situation allows calculation of how intense the signal should be before it is optimal to flee. The ratio of the costs is 1000:1. So, if a noise is loud enough to indicate that the likelihood of the presence of a lion is greater than 1 in 1000, then flight is optimal, even though it will turn out to be unnecessary 999 times out of 1000. False alarms are normal and expected in such systems.

The 'smoke detector principle' designation is appropriate because false alarms from smoke detectors are recognised as necessary and normal to ensure full protection against any actual real fire (Nesse 2001, 2005b). The idea is not new. In the seventeenth century, Blaise Pascal argued that if the existence of God was unlikely but possible, then belief was still worthwhile, because the costs are low, while eternal damnation is painful for an infinity (Hacking 1972). The smoke detector principle has been adapted by evolutionary psychologists to the cognitive domain as 'error management theory', to analyse the adaptive significance of apparently erroneous decisions and beliefs (Haselton and Buss 2000). The larger framework is signal detection theory, first described by Green and Swets (1966), and now used by psychologists to analyse experiments and by engineers to design circuits and machines. Its full range of applications in evolutionary medicine and public health remains to be explored.

1.3.4 Selection Shapes Mechanisms that Mediate Plasticity in Various Time Frames

Defences are only a few of many plastic responses shaped by natural selection (West-Eberhard 2003). The misconception that an evolutionary approach emphasises fixed responses or 'genetic determinism' remains prevalent. This is surprising, because one of the main differences between machines and bodies is that bodies have myriad systems that monitor internal and external states and adjust physiology and behaviour to cope with varying circumstances. These range from the instantaneous blink response in a fraction of a second, cardiovascular responses in seconds, metabolic adjustments over minutes, skin tanning over days, life history characteristics over years, and even adjusting levels of fat storage and stress responses across generations.

1.3.4.1 *Developmental Origins of Health and Disease (DOHaD) is an Important Cause of Disease Vulnerability*

David Barker and colleagues discovered that low birth weight predicts later obesity and vulnerability to atherosclerosis and other inflammatory diseases (Barker et al. 1993). They described the idea as the 'thrifty phenotype hypothesis' (Hales and Barker 2001), making a connection to the 'thrifty genotype' hypothesis proposed by James Neel (1962). Peter Gluckman and colleagues extended this line of thinking with the proposal that the thrifty phenotype might represent a 'predictive adaptive response' that adjusts metabolism for a lifetime based on cues about future environments that mothers transmit to their fetuses (Gluckman et al. 2005). This idea has developed into Developmental Origins of Health and Disease (DOHaD), a vibrant area of research that is particularly important as economic transitions in developing countries create ever-growing epidemics of obesity and inflammatory diseases (Hanson 2015). Specific epigenetic mechanisms mediate the effects (Gluckman et al. 2009b), and some epigenetic marks can be transmitted across generations, explaining non-genomic familial transmission of obesity (Gluckman et al. 2007). Whether or not the phenomenon is an adaptation or an epiphenomenon remains controversial (Wells 2012). In one particularly interesting test, baboons that were subject to caloric deprivation *in utero* turned out to be inferior at surviving a subsequent famine when they were adults (Tung et al. 2016). (For further discussion, see Chapter 4: Growth and Development.)

1.3.4.2 *Selection Has Shaped Fast and Slow Life Histories with Implications for Health*

DOHaD emphasises the plasticity of metabolic responses to early environments. A related principle studies how early experiences influence life history characteristics, especially so-called fast and slow life history strategies (Dobson and Oli 2007). In harsh environments, where life is likely to be short, investments in tissue repair and maintenance tend to give lower payoffs than investing in early frequent production of offspring, despite the costs to health. Some evidence suggests that mechanisms monitor levels of stress early in life, perhaps as indicated by cortisol levels, and adjust behaviour and metabolic systems accordingly to cope with the prevailing environment (Del Giudice et al. 2011). Systems set to a fast life history mode have reduced defences against infectious diseases and ageing, and more of a tendency to take risks and reproduce early. This has been proposed as an overarching framework for understanding mental disorders (Del Giudice 2014) and other disorders more generally. It is supported by evidence that specific epigenetic mechanisms initiated by cortisol exposure can transmit stress sensitivity from mother to fetus and between generations (Meaney 2010).

The role of epigenetic effects is a fast-developing area with important medical applications (Feinberg 2007; Esteller 2008; Keverne 2014). Exposure to stress *in utero* can increase stress reactivity not only in the offspring, but also in the grand-offspring (Skinner 2014). Similarly, prenatal exposure to famine influences body size and risks of diabetes and schizophrenia (Lumey et al. 2011). Early exposure to licking and grooming changes methylation of a cortisol receptor in the rat hippocampus (Weaver et al. 2004), but can be reversed by administration of methionine (Weaver et al. 2005). The effect has been confirmed for humans exposed to abuse in childhood (McGowan et al. 2009), with its multiple documented pathogenic effects mediated by neural mechanisms (Nemeroff 2016). (For further discussion, see Chapter 17: Brain, Spinal Cord, and Sensory Systems.)

Learning is invoked remarkably often as an alternative to evolutionary explanations, but it is just one more plasticity mechanism that adapts individuals to their specific environments. The tendency to frame debates as nature versus nurture has been decried for decades, but it persists (Weatherall 1995; Ridley 2003). The situation is complicated by tendencies to emphasise possible political implications of the role of genetic or environmental variations that account for differences between individuals and groups (Gould 1981). This adds to innate cognitive biases to suggest that these debates will persist despite all efforts to emphasise that genes interacting with environments shape all traits.

Like learning, human capacities for culture were shaped by natural selection in a process that resembles domestication (Nesse 2010; Wilkins et al. 2014; Henrich 2015). Imitation, sensitivity to mores, moral emotions, conformity, and other psychological characteristics are well suited to living in a complex social group with stable cultures. Natural selection shapes minds that make culture possible, and cultures create selection forces with many effects (Richerson and Boyd 2005). For instance, the ability to digest lactose as an adult gives big benefits to individuals in dairying cultures that, in turn, make dairying more beneficial (Tishkoff et al. 2006). The spread of lactose-tolerance alleles reflects migration as well as mutation and selection; the fast growth of herding populations may have spurred migrations (Itan et al. 2009).

1.3.5 Natural Selection Works Mainly at the Level of the Gene

Recognition that selection does not mainly shape traits to benefit groups or species was a milestone, not only for understanding behaviour, but also for understanding the body and disease. George C. Williams brought it to wide attention in 1966 with the publication of *Adaptation and Natural Selection*. The insight synergised beautifully with William Hamilton's recognition of kin selection at about the same time: alleles that decrease individual reproductive success can nonetheless be selected for if they give sufficient benefits to relatives who have the same alleles (Hamilton 1964a, b). These ideas came to wider recognition with publication of Dawkins' *The Selfish Gene* in 1982. Resulting further controversies and elaborations continue without let-up (West et al. 2007; Queller and Strassmann 2009; Leigh 2010; Nowak et al. 2017). It is useless to try to resolve the debate here, but essential to describe the relevant core principles that have been so essential to the development of evolutionary medicine, and the misunderstandings that result from confusion about these principles.

1.3.5.1 *Group Selection is a Viable Explanation Only under Constrained Circumstances*

An allele that reduces an individual's reproductive success relative to others in a group can nonetheless persist, or even become more common, if it provides sufficient benefits to the group. If other members of the group share alleles in common by descent, the process is well described by kin selection. There is general agreement that kin selection and group selection models can be different ways of viewing the same process (Lehmann et al. 2007; Frank 2012), but much disagreement remains about whether one framework is routinely superior (West et al. 2008; Wilson 2008). Even the utility of kin selection has been challenged recently (Nowak et al. 2017). Despite all the controversy, the core principle remains important: alleles whose bearers have fewer offspring than average tend to become less frequent in a population over time unless the alleles provide benefits to kin, or if their costs to individuals are small relative to effects that substantially influence the growth rates of groups with a constrained set of characteristics; these latter conditions are uncommon.

The exemplar for group selection was evidence reported by Wynne-Edwards that groups of animals experiencing inadequate food supplies tend to reduce reproduction in ways likely to ensure the continuation of the group (Wynne-Edwards 1962). This work inspired the critique by Williams (1966), who pointed out that individuals who reduced their reproduction for the good of the group would pass on fewer alleles to the next generation, so the trait would be selected against. This has led to enduring debates about the circumstances in which benefits to a group would be sufficient to account for the persistence or increase in frequency of alleles that were associated with tendencies to such 'altruism' (Wilson and Sober 1994; Foster et al. 2006; West et al. 2007; Leigh 2010; Nowak et al. 2017).

Some of the controversy results from using the term 'group selection' for phenomena quite different from the original concept (West et al. 2007). In particular, the effect of selective association of individuals with certain traits has been called group selection, with much resulting confusion. Any process that results in association of altruists with other altruists give mutual benefits that generate selection forces. The mechanism can be as simple as viscosity that results in offspring staying in proximity, or as complex as humans making choices about whom they want to partner with in joint enterprises. As Stuart West and colleagues said succinctly about calling such models group selection, 'An alternative is to state as simply as possible what they are—models of non-random assortment of altruistic genes' (West et al. 2007, p. 11).

1.3.5.2 *A Multigenerational Perspective is Important*

A trait that harms individual reproductive success can persist if it protects the group from complete collapse at some future time. This may well be an explanation for the maintenance of sex despite its substantial costs compared with asexual reproduction (Stearns 1987; Hamilton et al. 1990; Lehtonen et al. 2012). A population of lizards that gains the ability to reproduce asexually will grow twice as fast as other groups that retain sexual reproduction; however, the resulting lack of genetic diversity increases the risk of infection (Moritz et al. 1991). Whether this is best described as group selection or clade selection is a point of discussion (Williams 1996).

Studies of sex ratios offer another useful perspective. That they are close to 1:1 for many species provides evidence against consistent strong effects of group selection. Sex ratios are subject to natural selection, and groups with mostly females expand nearly twice as fast than groups with equal numbers of males and females. If group selection were pervasive and powerful, female-biased sex ratios would be common. The equal sex ratio persists because, as Fisher observed, individual parents gain a fitness advantage by having offspring of whatever sex is in relatively short supply, because those offspring will, on average, themselves have more offspring (Fisher 1930). The principle is nicely and commonly illustrated by deciding what pub to go to. Choosing whatever venue has a higher proportion of the opposite sex maximises reproductive success, or at least mating success. The equal sex ratio is a classic example of a trait stabilised by negative frequency-dependent selection (Parker and Smith 1990).

Extraordinary sex ratios provide deeper insight, based on Hamilton's recognition that fitness is maximised by putting equal effort into producing males and females (Hamilton 1967) and extensive subsequent work on specific cases (Trivers and Willard 1973). Controversy continues about whether such cases are best viewed as products of group selection, kin selection, social selection, or multilevel selection, or if all frameworks can contribute (Kramer and Meunier 2016).

Artificial breeding provides convincing demonstrations of group selection. For instance, aggressive pecking among multiple chickens in a typical small cage reduces egg production. Breeding chickens from the cages with the least pecking and the most eggs selects, within a few generations, for further reduced pecking and increased egg production (Ortman and Craig 1968). This fine example of group selection demonstrates how breeding can contravene the state of nature for chickens, in which selection shapes behaviours that maximise individual reproductive success.

Attempts to document cases of group selection among unrelated individuals in the wild continue. For instance, a case in social spiders appeared promising (Pruitt and Goodnight 2014), but soon received serious criticism (Grinsted et al. 2015). These controversies are fascinating and relevant to medicine, but they should not distract from the core principle that explanations based on group selection are problematic, and should be invoked only with caution and full considerations of alternative explanations.

1.3.6 Kin Selection Can Explain Some Traits that Reduce Individual Reproductive Success

As noted already, the mystery of altruistic behaviour in honeybees led to William Hamilton's recognition that kin have genes identical by descent, so an allele that reduces the fitness of an individual could nonetheless be selected for if it sufficiently increases the fitness of kin (Hamilton 1964a, b). This discovery revolutionised the study of animal behaviour (Alcock 2001), and spurred hundreds of studies showing that organisms, including humans, help kin more than others (West et al. 2002; Lehmann and Keller 2006). It has many important implications for medicine, as outlined in an important review of 'Hamiltonian medicine' (Crespi et al. 2014).

1.3.6.1 *Natural Selection Continues to Act after Menopause*

Many medical researchers and clinicians think that selection cannot influence anything after menopause. This misunderstanding reflects two mistakes: failure to recognise that menopause itself is a trait that needs explanation, and failure to recognise the role of kin selection.

Menopause is not a universal trait for mammals. It is observed in only a few species, mostly ones where social relationships strongly influence fitness (Peccei 2001). Attempts to explain menopause began with the suggestion by Williams (1957) that continuing to reproduce beyond a certain age could risk the success of existing offspring, so maximising fitness might be accomplished by stopping individual reproduction and instead investing in existing offspring and other kin. The idea has been described as 'the grandmother hypothesis' to reflect the benefits grandmothers provide to grandchildren (Hawkes et al. 1998). Studies showing increased survival for children with grandparents support the idea (Hawkes 2003), but it remains uncertain if the benefits are large enough to compensate for the loss of direct reproductive success (Hill and Hurtado 1991; Rogers 1993; Austad 1994; Shanley and Kirkwood 2001; Kachel et al. 2011). An alternative explanation views menopause as an epiphenomenon, perhaps resulting from rapidly expanding lifespans for humans compared with other primates, perhaps from competition between oocytes, only some of which will ever be released (Reiber 2010). (For further discussion, see Chapters 15 and 16.)

1.3.6.2 *Weaning Conflicts are Inevitable*

A mother is related to her offspring by $r = 0.5$, but related to herself by $r = 1.0$. A new-born infant is completely dependent on her care, so maximising her reproductive success requires fulfilling all the infant's needs in its first months. There comes a point, however, when she would gain greater genetic representation in future generations by stopping nursing and becoming pregnant again. At that point, however, the baby's fitness is still maximised by continuing to nurse; the resulting weaning conflicts are universal for mammals. At a slightly older age, the genetic interests of the toddler are best served by giving up on trying to get milk from the mother, freeing her to have younger siblings who will have alleles in common with the toddler. This simple principle is but one aspects of the more general theory proposed by Robert Trivers to explain parent offspring conflicts (Trivers 1974). Other conflicts between paternal and maternal genomes that increase vulnerability to diseases and are covered below.

1.3.6.3 *Conflicts between Maternal and Paternal Genomes Can Cause Disease*

In species where a female may mate with different males for different reproductive episodes, the interests of the paternal genome are advanced by inducing additional investments from the female in this offspring, while the maternal genome's interests are advanced by retaining resources to maintain the soma and reserve resources for future reproduction. David Haig has suggested that such mechanisms may be important factors in diseases of pregnancy, and that patterns of genomic imprinting match the predictions (Haig 1993). Eggs have epigenetic marks that reduce the expression of the insulin-like growth factor 2 gene (*IGF2*), somewhat reducing fetal growth, to the benefit of the maternal genome. Sperm have imprinted marks that reduce the expression of *IGF2r*, a gene with effects antagonistic to IGF2. If imprinting works normally, the effects balance each other, but if one is missing, an offspring will be larger or smaller than average (Haig 2004).

The role of these competing systems in human diseases is illustrated by Prader–Willi syndrome, caused by failure of expression of genes on chromosome 15q11–q13 that are normally expressed only when paternally derived. As predicted, infants with Prader–Willi syndrome have behavioural phenotypes characterised by weak early suckling and other behaviours that would benefit maternal genomes relative to paternal genomes (Haig and Wharton 2003). Williams syndrome has a related interpretation (Crespi and Procyshyn 2017). Beckwith–Wiedemann syndrome and Silver–Russell syndrome also appear to result from sexually antagonistic imprinting errors (Eggermann et al. 2008). The clinical implications of such systems for imprinting are especially important for *in vitro* fertilisation (IVF) methods that bypass developmental steps that deposit imprinting marks. IVF-produced babies are 14 times more likely than others to have Beckwith–Wiedemann syndrome (Halliday et al. 2004). Possible effects of drugs that influence DNA methylation, such as high-dose folic acid administered during pregnancy, deserve close attention (Smith et al. 2008). (For further discussion, see Chapter 16: Sexuality, Reproduction, and Birth.)

Bernard Crespi and colleagues have developed related ideas into a hypothesis about schizophrenia and autism (Crespi et al. 2007; Crespi and Badcock 2008; Crespi 2010). They view autism as associated with a hypermale brain (Baron-Cohen 2002) resulting from of excessive influence from the paternal genome, and schizophrenia as resulting from excessive influence of the maternal genome (Dinsdale et al. 2016). This seems to be consistent with patterns of methylation on genes that influence the disorders. It also predicts that birth weight will be a bit higher for children who go on to develop autism, compared with schizophrenia; this prediction was confirmed in a study using a database of 700,000 Danish subjects (Byars et al. 2014). This by no means confirms that theory definitively, but the line of investigation illustrates how and evolutionary perspective can open new avenues of investigation.

1.3.7 Control of Cell Replication is Crucial for Metazoan Life

The challenge of controlling cell replication postponed the origins of multicellular life for billions of years (Valentine 1978; Queller and Strassmann 2013). Selection has met that challenge remarkably well, as evidenced by remarkably low rates of cancer, especially in young organisms (DeGregori 2011); however, the limitations of such systems mean that cancer will always be a problem (Frank 2007). Cells initially aggregated to defend themselves and to control their environments better. Because all cells in such groups are not genetically identical, alleles that induce cells to replicate faster than other cells are selected for, even if this harms the larger group. The conflict is usually described using the human terms 'cooperation' and 'defection'. It took several billion years before ways to enforce cooperation evolved. The key to a solution was to ensure that all cells in an individual were genetically identical at the start of development, and to ensure that most cells in the body, the somatic line, cannot create new individuals, and therefore can advance their own genetic interests only by contributing to the welfare of the individual. This required sequestering a special line of cells, the germline, devoted only to reproduction under carefully controlled circumstances.

1.3.8 Intragenomic Conflicts Can Influence Health

Conflicts at the genome level (Austin et al. 2009) have been a major focus of evolutionary medicine investigations. If reproduction were initiated from a group of non-identical cells, the age-old conflict would rise again, at a large risk to the individual. Cells with alleles that made them more likely to become a future germline would be selected for, even if that harmed the individual. The solution is meiosis. The process of slimming the genome down to a single strand of DNA ensures that all cells in the next soma are genetically identical (Hurst and Nurse 1991). Identical cells can advance their genetic interests only by doing what is good for the individual's inclusive fitness.

Even this is not protection enough, however. An allele can become more common in future generations by disrupting the development of adjacent cells that do not contain the same allele. Such systems based on meiotic drive usually require both a toxin and an antidote at separate sites (Sandler and Novitski 1957; Lyttle 1993). It has been suggested that crossing over during meiosis separates such pairs of alleles, protecting the system from intragenomic conflict (Haig and Grafen 1991; Hurst 1998), but at the risk of creating genetic errors.

The inheritance of multiple copies of mitochondrial genomes poses a special case. Inheritance of mitochondria is only from the mother, and only a few copies are transmitted in the oocyte. They multiply in the individual to 30 billion in number by mid-gestation (Haig 2016). In the process, mitochondria with alleles that speed replication become more common, even if they contribute less to the welfare of the individual. Across generations, those with alleles that result in preferential access to the germline become more common. A variety of mechanisms evolved to control such inefficiency. The bottleneck at conception increases the variation among oocytes so that the more cooperative mitochondria can be selected. The fusion of mitochondria at cell replication shares soluble products in a way that 'levels the playing field' and allows a future fission to select for mitochondria that benefit the individual (Haig 2016). Studies of how competing selection forces shape mitochondria, and how mitochondrial traits that benefit mitochondrial genomes can increase disease vulnerability, offer an important opportunity for evolutionary medicine (Wallace 2005; Ma and O'Farrell 2016).

1.3.9 Somatic Selection Changes Cell Genotypes during the Lifetime of an Individual

Somatic selection is illustrated dramatically by competition within a malignant tumour between cells with different genotypes: those that reproduce the fastest and persist the longest soon take over (Greaves and Maley 2012). Recognition of the role of somatic selection is encouraging application to cancer chemotherapy of analytic methods developed to study antibiotic resistance. It is also encouraging bringing in ecological methods to investigate the tissues that surround a malignancy and how variations in microenvironments speed or slow tumour growth (DeGregori 2017). Together, these new perspectives are suggesting ways to make chemotherapy more effective, often by reducing doses in 'adaptive chemotherapy' (Gillies et al. 2012; Enriquez-Navas et al. 2016). (For further discussion, see Chapter 9: Haematopoetic System.)

Somatic selection also occurs in normal adaptive immunity when lymphocytes that recognise an epitope divide faster, become more prevalent, and develop memory, ready for any subsequent episode of possible reinfection. Trade-offs in such systems balance the benefits of having cell populations that recognise many antigens versus the risks of autoimmune disease (Schmid-Hempel 2011).

1.3.10 Natural Selection Shapes Life History Traits

Traits such as number of offspring, size of offspring, timing of reproduction, age of maturity, and rates of aging are shaped by selection to maximise inclusive fitness (Stearns 1989; Chisholm 1993; Hill and Kaplan 1999). All involve trade-offs. Having larger offspring necessarily means having fewer. Starting reproduction earlier necessarily means offspring will be smaller and of lower quality. A short interbirth interval sacrifices maternal health and ability to invest in existing offspring. Differing patterns of investment in male or female offspring can be analysed using life history theory (Hinde 2009). Selection shapes these and other life history traits in ways that maximise average inclusive fitness (Kaplan et al. 2000).

Life history theory (LHT) was developed to account for differences between species, but has been adapted to consider variations in life history traits among individuals. Some such differences, such as the duration of gestation, or the age at onset of reproduction, are heritable traits acted on by selection to shape the average and distribution of such traits for the species. As with all other traits, life history traits can be poorly suited to modern conditions. For instance, mechanisms that initiate menarche interact with modern environments to initiate cycling four or more years earlier than for our ancestors, creating millions of very young mothers whose bodies and brains have not developed as fast as their reproductive capacities (Allsworth et al. 2005). For further details, see Chapter 4.

More recently, interest has focused on variations in life history traits induced by environmental influences. In particular, early exposure to stress induces a 'fast life history strategy' characterised by risk-taking and decreased investment in tissue maintenance and health (Kaplan et al. 2000; Dobson and Oli 2007; Austad and Finch 2016; Shalev and Belsky 2016; Wells et al. 2017). Such mechanisms have been proposed to help account for increased health risks in people subject to adversity or discrimination (Promislow and Harvey 1990; Bielby et al. 2007; Dobson and Oli 2007). They have also been said to provide an organising principle for understanding mental disorders (Del Giudice and Ellis 2016).

Whether these variations are facultative adaptations shaped by selection or epiphenomena of other processes is an important question that needs to be addressed for each proposed mechanism. Separate approaches will likely be needed for proposals about influences within a single lifespan, and influences that transcend generations.

1.3.11 Genes with Deleterious Effects Can Be Selected for if They Offer Compensating Benefits

Antagonistic pleiotropy refers to single genes with some effects that are beneficial in one situation or time in the life and other effects that are harmful in others. It has been especially relevant for understanding the evolution of senescence (Williams 1957; Hamilton 1966; Kirkwood and Austad 2000), where multiple studies now document the trade-offs between early reproduction and faster ageing (Kirkwood and Rose 1991). Strong effects of senescence on fitness in wild populations provide evidence that accumulation of mutations is an insufficient explanation (Nesse 1988; Nussey et al. 2013).

Trade-offs at other times can also be relevant. For instance, alleles that increase the likelihood of fertilisation or uterine implantation will be strongly selected for, even if they impose substantial fitness costs later. Sexually antagonistic selection can be viewed as a special kind of antagonistic pleiotropy (Rice 1992). For instance, alleles that increase iron absorption tend to cause haemochromatosis in men, but few problems in women, who lose some blood with monthly menstrual cycling (Adams et al. 1997). (For further discussion, see Chapter 5: Senescence and Ageing.)

Balancing selection, as described by Dobzhansky (1963) describes the process that maintains alternative alleles at a locus because of frequency- or situation-dependent selection. The maintenance of the sickle cell haemoglobin allele in balance with the regular haemoglobin allele is the classic example (Allison 1954; Livingstone 1960). Homozygous individuals with regular haemoglobin are vulnerable to malaria; those homozygous for sickle cell haemoglobin get severe sickle cell disease with vastly decreased reproductive success. Heterozygote individuals are somewhat protected against death from malaria but do not have severe symptoms of sickle cell disease, so, in areas where malaria is prevalent, heterozygote individuals have greater fitness and the sickle cell allele increases in frequency until it becomes so common that homozygotes become common. In environments free from malaria, sickle cell alleles are selected against. Sickle cell disease is sometimes held up as an exemplar for evolutionary medicine. However, on an evolutionary timescale, the sickle cell allele is relatively new, and examples of diseases explained by balancing selection are few and most are related to red blood cell variations that protect against malaria. The likely explanation is that the costs of heterozygote advantage impose strong selection for alternative solutions. (For further discussion, see Chapter 9: Haematopoetic System.)

Disease vulnerabilities are sometimes attributed to balancing selection when authors actually mean antagonistic pleiotropic effects of single alleles, or effects of trade-offs at the level of a trait. However, balancing selection is occasionally a viable explanation for disease vulnerability. Genome-wide scans have found loci where balancing selection seems to be occurring (Asthana et al. 2005), and new methods are identifying more sites that suggest balancing selection (Charlesworth 2006; de Filippo et al. 2016; Gloss and Whiteman 2016), making this an area of opportunity for exploration.

Balancing selection is often incorrectly invoked to explain the persistence of rare alleles that combine to cause disease, such as the scores of alleles with tiny effects that increase vulnerability to autism, epilepsy, and schizophrenia. In general, there is little evidence that such alleles are maintained at some intermediate frequency by selective advantages that depend on their frequency or on effects in different situations, although this remains possible. Mutation–selection balance is often a more plausible explanation (Keller and Miller 2006), along with complications arising from epistasis or antagonistic pleiotropy.

1.3.12 Cliff-Edged Fitness Landscapes May Account for the Persistence of Some Genetic Diseases

Sequencing the human genome brought hope that we would soon find the alleles that cause highly heritable disorders such as autism, schizophrenia, and epilepsy. Genome-wide association studies have found, however, that no common alleles account for more than a tiny fraction of vulnerability to most of these diseases (Woo et al. 2017). One possible explanation arises when fitness for a value of a trait increases rapidly close to a fitness cliff, beyond which catastrophic failure is likely. For instance, selecting horses for speed shapes a long thin leg bone that is vulnerable to breakage. The bone morphology that maximises speed (and breeding potential) for the average individual results in catastrophic failure for a few individuals. This is a dramatic example of selection maximising fitness of alleles at a cost to the health of individuals. It may help to explain the persistence of alleles that increase vulnerability to schizophrenia (Nesse 2004), and it has been applied to the trade-off between fetal head size and the pelvic opening (Mitteroecker et al. 2016). (For further discussion, see Chapter 16: Sexuality, Reproduction, and Birth.)

If cliff-edged fitness functions are common, they may help to explain the persistence of other highly heritable disorders for which no common alleles with major effects can be found. The benefits of pushing a trait close to a cliff edge combine with the costs of catastrophic failure for a few individuals to create stabilising selection that should narrow the range of variation around the point of maximum fitness. If this stabilisation involves the effects of many alleles with small effects, the result would help to account for what has been called 'missing heritability'. Note that this model is consistent with the absence of major advantages for individuals whose relatives have the disease, and that the responsible alleles might not be abnormal. Note also that the alleles are not necessarily mutations, nor are they necessarily abnormal; together, multiple alleles shape a trait to a stable phenotype that maximises average fitness but makes catastrophic disease or injury inevitable for some individuals (Nesse, in preparation).

1.3.13 Attention to Ethics is Important

A short history of evolutionary approaches in medicine shows how misunderstandings of race and genetics lead to a 'medical Darwinism' that was associated with serious mistakes, and, more peripherally, with the moral catastrophe of eugenics and the Holocaust (Zampieri 2009). Modern evolutionary medicine is a fundamentally different enterprise, one that uses evolutionary principles to improve the health of individuals; it has so far avoided major ethical compromises. Continued wariness is indicated, however, especially as growing genomic data reveals additional details of differences between human subgroups, and as techniques for editing genomes come into wide use. New genetic technologies now pose greater ethical challenges than evolutionary applications in medicine, but vigilance is called for to ensure that evolutionary ideas about disease are not used to derogate human subgroup or individuals.

1.3.14 Races are not Biological Categories

Tendencies run deep to view people from different geographical origins as members of different circumscribed groups with definable distinctive characteristics. Anthropological studies in the early twentieth century describing the characteristics of races as essentialised biologically separate groups amplified these tendencies (Smedley and Smedley 2005). The use of words designating different races further encourages the persistence of incorrectly thinking about human subgroups as subspecies. New investigations documenting genetic differences between human subpopulations risk reawakening these mistaken kinds of categorisation, along with the animus that flourishes when subgroups view each other as fundamentally different (Graves 2001). The reality for humans is that genetic differences between subgroups are relatively small as compared with subgroups of other primates, and skin colour is by no means a reliable trait to identify individuals from different geographical origins. Furthermore, recent data show that the genes that code for skin pigmentation variations are mostly far older than the emigration of humans from Africa, and selection on this standing variation has resulted in a panoply of skin tone variations that transcend racial categories (Crawford et al. 2017).

1.3.15 Genetic Differences between Human Subgroups Influence Health

Despite the relatively small genetic differences between humans with different geographical ancestry, genetic markers allow reliable identification of an individual's continent of origin (Elhaik et al. 2014), and a principal components analysis confirms the separation of human genome groups into those that correspond to different continents (Jorde and Wooding 2004). Genetic differences even allow pinpointing the village of origin for many people in the UK (Leslie et al. 2015). However, these differences are substantially smaller than those between groups of closely related primates (Long and Kittles 2009). The social nature of the idea of race should not impede consideration of genetic differences between human subgroups that influence vulnerability to disease.

Humans with ancestors from different geographical locations have genetic differences that influence health. The obvious example is variations in skin pigmentation that protect against skin cancer and destruction of folic acid, at the cost of decreased ability to synthesise vitamin D and the associated risk of rickets in northern climates (Greaves 2014; Jablonski and Chaplin 2017). The relatively high levels of bone mineral density in people of African descent despite the high prevalence of vitamin D deficiency has prompted investigations that discovered selection for alleles that increase vitamin D binding (Powe et al. 2013). Other gross observable differences include shorter limbs in people from cold climates, where loss of heat was a significant selective factor. However, mutation, migration, and genetic drift are also potent explanations for genetic differences between subpopulations.

Of special significance is the loss of about half of human genetic variation in the process of the migration out of Africa. This global bottleneck has been augmented by additional bottlenecks that result in genetic differences without adaptive significance.

Genetic differences related to pathogen exposure are also significant. A tendency to make antibodies to schistosomiasis in areas where schistosomiasis is a selection force is associated with tendencies to protective immune responses that also cause asthma in response to exposure to cockroach antigens (Barnes et al. 1999). As noted already, haemoglobin S causes red blood cells exposed to the stress of malaria infection to shift into a shape that is associated with increased clearance in the spleen. This does not provide protection against contracting malaria, but it does reduce mortality rates by speeding clearance of infected red blood cells (Luzzatto 2012). Protection is also provided by selection in malarial areas for the absence of the Duffy antigen/chemokine receptor (DARC), used by plasmodia to enter cells. The majority of people from areas where malaria is prevalent lack the normal DARC protein (Lentsch 2002; McManus et al. 2017). New inexpensive sequencing methods are speeding-up investigations into genetic variations likely selected by other pathogen exposures (Penman and Gupta 2017). (For further discussion, see Chapter 9: Haematopoetic System.)

Decreased activity of alcohol dehydrogenase and aldehyde dehydrogenase is especially common in people of East Asian descent, causing symptoms including flushing (Lin and Cheng 2002). The prevalence of these alleles in certain areas could be attributed to their benefit of preventing alcohol dependence in a culture with long exposure to distilled spirits, and people with these alleles do tend to drink less than others. Also, the location of the gene in the middle of the longest haplotype in Asians seems to provide supporting evidence for selection providing protection against alcoholism. However, new genetic evidence suggests that differences in alcohol dehydrogenase result from the many influences of these variations on traits other than exposure to alcohol (Polimanti and Gelernter 2018).

1.3.16 It is a Mistake to Assume that What is, is What Ought to Be

Learning that a trait is 'natural' influences many people to believe that the trait in question is good, or at least acceptable (Elqayam and Evans 2011). This tendency is amplified by arguments for the adaptive functions that shaped a trait. The most salient example is about mating patterns. Males get greater fitness gains from additional matings than females. This insight seems to many people to help justify infidelity by males in mating relationships. The general tendency to withhold moral judgement from traits that seem natural has long been noted and criticised by philosophers, but such admonitions have little influence in the wider world, so vigilance is needed as evolutionary ideas are applied more widely in medicine. The 'is versus ought' fallacy is often described as 'the naturalistic fallacy', although this term has a slightly different meaning in philosophical circles (Greene 2003; Curry 2006).

1.3.17 Natural Selection is Not Over for Humans

Public health advances make death in childhood unlikely, so most children born now grow up to be reproductively capable adults. This had led some to conclude that natural selection is no longer influencing human evolution (Rose 2001). This mistake results from a simple misunderstanding. Natural selection does not require differences in mortality rates; it only requires genetic differences that influence the number of surviving offspring, and those differences persist. However, the dramatic decrease in childhood mortality rates, and in mortality rates more generally, certainly have greatly weakened the force of selection acting on modern populations, leading some to have concern about the implications for the genome (Kondrashov 2017).

1.3.18 Genetic Methods for Tracing Relationships and Phylogenies Have Many Applications in Evolutionary Medicine

As noted in the introduction, the origins of evolutionary medicine in attempts to explain disease vulnerability have led to the relative neglect of phylogenetic and population genetic methods. My own limitations in this area make it impossible for this chapter to do more than emphasise that extensive work using such methods exists (Kumar et al. 2012), and that it is important for evolutionary medicine. They are useful for identifying taxonomic relationships, estimating divergence times, and understanding why traits are gained or lost over time. Continued efforts to integrate such work with other aspects of evolutionary medicine will benefit all parties.

1.3.18.1 *Tracing Human Ancestry is Medically Relevant*

No one imagined just a few decades ago that we would be able to sequence DNA from Neanderthals, much less identify specific loci in modern humans (Green et al. 2010). The identification of a separate Denisovan lineage was a more unanticipated surprise. Some have considered the medical significance of genes recently incorporated into *Homo sapiens* matings with Denisovan (Tishkoff and Verrelli 2003; Simonti et al. 2016), but so far they are mostly inspiring closer looks. Additional insights are coming from comparisons of human and chimpanzee genomes (Olson and Varki 2003).

Specialised techniques of looking at patterns of X- and Y-chromosome variation have found strong support for anthropological theories that propose dramatic shifts in social organisation after agriculture made food storage and social hierarchies possible. A recent analysis found that reproductive skew of a few males contributing disproportionately to the average human genome increased dramatically just about the time agriculture was spreading rapidly (Webster and Wilson Sayres 2016).

1.3.18.2 *Phylogenetic Methods Can Trace the Origins and Spread of Pathogens*

A few decades ago it was a major accomplishment to use sequences to trace the spread of a specific food-borne pathogen to its source. Now, new methods are revealing findings with often-urgent public health implications (Zhao et al. 2014). Research on the SARS virus quickly traced its source to bats (Li et al. 2005). New studies on Ebola document not only its spread, but how many times it crossed over to humans and when (Gire et al. 2014).

Speculation about the origins of HIV in humans have been resolved thanks to phylogenetic methods that identify its spread from virus circulating in humans much longer than was previously suspected (Heeney et al. 2006; Wertheim and Worobey 2009).

Spread of a pathogen within human populations can be traced using genetic methods. Tuberculosis preceded European invasions into the New World via seals, followed by a later incursion of human-adapted lines (Bos et al. 2014; Honap et al. 2017). Methods for tracing pathogen phylogenies continue to develop rapidly (Hartfield et al. 2014).

1.3.19 Methods for Framing and Testing Evolutionary Hypotheses Remain under Development

As already pointed out, many scientists and most clinicians do not fully grasp the difference between proximate and evolutionary explanations. Those who do are often unaware of the breadth of methods for framing and testing evolutionary hypotheses and the benefits of systematically considering both (Hinde and Milligan 2011). The result is widespread misunderstanding and scepticism. Much scepticism is justified by the prevalence of elementary mistakes such as proposing adaptive functions for diseases or specific alleles associated with a disease. Patience is warranted, as is attention to the challenges of testing such hypotheses and the prevalence of elementary mistakes (Mace et al. 2003; Ellison and Jasienska 2007; Nesse 2011a). The core concepts in evolutionary medicine are many, and some are subtle. No chapter can convey them; a whole book is necessary—one with many examples, like this book.

Several common human cognitive glitches make this all the harder (Kahneman 2011; Nisbett 2015). The desire for monocausal explanations is strong, making it difficult for many people to see how a condition like atherosclerosis can require an explanation that includes mismatch, trade-offs, constraints, and defences. The human tendency to attribute specific functions to specific things makes it hard to help people see that it rarely makes sense to say, 'this is a gene for' or 'the function of this emotion is X'. These difficulties are not specific to evolutionary medicine, but they are particularly prominent, because the systems under study are organically complex.

1.3.20 Organic Complexity is Different in Kind from the Complexity in Machines

We are in the midst of a transition from viewing the body as a designed mechanism to viewing it as a product of natural selection characterised by organic complexity (Nesse et al. 2012; Hauser et al. 2017). Much of biology and medicine takes a tacitly creationist view of the body as if it had nicely separated modules with specific functions, connected in simple ways to other modules. Charts depicting biochemical pathways tend to show idealised systems that neglect myriad messy interactions of molecules with each other and receptors. We teach endocrine systems as if they have simple molecules acting on one target. Neuroscientists describe loci and tracts with functions that belie the underlying organic complexity.

Seeking simplification is understandable. Much of science's power and beauty derives from is ability to simplify, and we must describe principles and findings to other people in ways that human brains can comprehend. However, genes, molecules, hormones, and organs interact in ways very different from the components of a machine. Instead of one or few backup mechanisms, they have intertwined functions that maintain stability despite missing parts (Nijhout 2002; Hammerstein et al. 2006; Bateson and Gluckman 2011). Instead of being designed to serve some specific purpose, they were shaped to maximise reproduction, sometimes at the expense of health. This evolutionary view of the body as organically complex will spread slowly because it reveals the inadequacy of our beautiful simple models. However, as we increasingly view the body and disease through an evolutionary lens, disease will make more and more sense, and we will gain increasing abilities to prevent and treat it.

1.4 Conclusion

New interest in evolution and medicine was initiated by asking new questions about why natural selection left so many traits in the body vulnerable to disease. The resulting new field remains limited by these origins, but it is expanding so fast that no list of core principles can be complete. Readers should add their own, and teachers should not limit the content of their classes to the core principles listed here.

Acknowledgements

Thanks to Anne Stone, Benjamin Trumble, Daniel Grunspan, Jon Laman, and Martin Brüne for comments and suggestions that helped to improve this chapter.

References

Abegglen, L. M., Caulin, A. F., Chan, A., et al. (2015). Potential mechanisms for cancer resistance in elephants and comparative cellular response to DNA damage in humans. *JAMA* 314, 1850–60.

Adams, P. C., Deugnier, Y., Moirand, R., et al. (1997). The relationship between iron overload, clinical symptoms, and age in 410 patients with genetic hemochromatosis. *Hepatology* 25, 162–6.

Agrawal, M., Aroniadis, O. C., Brandt, L. J., et al. (2016). The long-term efficacy and safety of fecal microbiota transplant for recurrent, severe, and complicated *Clostridium difficile* infection in 146 elderly individuals. *J Clin Gastroenterol* 50, 403–7.

Alcock, J. (2001). *The Triumph of Sociobiology*. New York: Oxford University Press.

Allison, A. C. (1954). The distribution of the sickle cell train in East Africa and elsewhere, and its apparent relationship to the incidence of subtertian malaria. *Trans R Soc Trop Med Hyg* 48, 312–18.

Allsworth, J. E., Weitzen, S., Boardman, L. A. (2005). Early age at menarche and allostatic load: data from the Third National Health and Nutrition Examination Survey. *Ann Epidemiol* 15, 438–44.

Antonovics, J., Abbate, J. L., Baker, C. H., et al. (2007). Evolution by any other name: antibiotic resistance and avoidance of the E-word. *PLoS Biol* 5, e30.

Armelagos, G. J. and Barnes, K. (1999). The evolution of human disease and the rise of allergy: epidemiological transitions. *Med Anthropol* 18, 187–213.

Asthana, S., Schmidt, S., and Sunyaev, S. (2005). A limited role for balancing selection. *Trends Genet* 21, 30–2.

Auld, S. K. J. R., Tinkler, S. K., and Tinsley, M. C. (2016). Sex as a strategy against rapidly evolving parasites. *Proc Biol Sci* 283, 20162226.

Austad, S. N. (1994). Menopause: an evolutionary perspective. *Exp Gerontol* 29, 255–63.

Austad, S. N. and Finch, C. E. (2016). Human life history evolution: new perspectives on body and brain growth. In: Tibayrenc, M. and Ayala, F. J. (eds) *On Human Nature: Evolution, Diversity, Psychology, Ethics, Politics and Religion*. Boston: Elsevier.

Austin, B., Trivers, R., and Burt, A. (2009). *Genes in Conflict: The Biology of Selfish Genetic Elements*. Cambridge, MA: Harvard University Press.

Ayala, F. J. (2007). Darwin's greatest discovery: design without designer. *Proc Natl Acad Sci U S A* 104, 8567–73.

Bach, J. F. (2002). The effect of infections on susceptibility to autoimmune and allergic diseases. *N Engl J Med* 347, 911–20.

Barker, D. J., Gluckman, P. D., Godfrey, K. M., et al. (1993). Fetal nutrition and cardiovascular disease in adult life. *Lancet* 341, 938–41.

Barnes, K. C., Armelagos, G. J., and Morreale, S. C. (1999). Darwinian medicine and the emergence of allergy. In: Trevathan, W., McKenna, J., and Smith, E. O. (eds) *Darwinian Medicine*. New York: Oxford University Press, pp. 209–43.

Baron-Cohen, S. (2002). The extreme male brain theory of autism. *Trends Cogn Sci* 6, 248–54.

Bateson, P. and Gluckman, P. (2011). *Plasticity, Robustness, Development and Evolution*. Cambridge: Cambridge University Press.

Bateson, P. and Laland, K. N. (2013). Tinbergen's four questions: an appreciation and an update. *Trends Ecol Evol* 28, 712–18.

Bergstrom, C. T. and Antia, R. (2006). How do adaptive immune systems control pathogens while avoiding autoimmunity? *Trends Ecol Evol* 21, 22–8.

Bielby, J., Mace, G. M., Bininda-Emonds, O. R., et al. (2007). The fast–slow continuum in mammalian life history: an empirical reevaluation. *Am Nat* 169, 748–57.

Blaser, M. J. (2014). *Missing Microbes: How the Overuse of Antibiotics Is Fueling Our Modern Plagues*. New York: Macmillan.

Bordenstein, S. R. and Theis, K. R. (2015). Host biology in light of the microbiome: ten principles of holobionts and hologenomes. *PLoS Biol* 13, e1002226.

Bos, K. I., Harkins, K. M., Herbig, A., et al. (2014). Pre-Columbian mycobacterial genomes reveal seals as a source of New World human tuberculosis. *Nature* 514, 494–7.

Bouglé, A., Foucrier, A., Dupont, H., et al. (2017). Impact of the duration of antibiotics on clinical events in patients with *Pseudomonas aeruginosa* ventilator-associated pneumonia: study protocol for a randomized controlled study. *Trials* 18, 37.

Brüne, M. (2014a). Life history theory as organizing principle of psychiatric disorders: implications and prospects exemplified by borderline personality disorder. *Psychol Inq* 25, 311–21.

Brüne, M. (2014b). On aims and methods of psychiatry—a reminiscence of 50 years of Tinbergen's famous questions about the biology of behavior. *BMC Psychiatry* 14, 364.

Brüne, M. and Hochberg, Z. (2013). Secular trends in new childhood epidemics: insights from evolutionary medicine. *BMC Med* 11, 226.

Byars, S. G., Stearns, S. C., Boomsma, J. J. (2014). Opposite risk patterns for autism and schizophrenia are associated with normal variation in birth size: phenotypic support for hypothesized diametric gene-dosage effects. *Proc Biol Sci* 281, 20140604.

Charlesworth, D. (2006). Balancing selection and its effects on sequences in nearby genome regions. *PLoS Genet* 2, e64.

Chisholm, J. S. (1993). Death, hope, and sex: life-history theory and the development of reproductive strategies. *Curr Anthropol* 34, 1–24.

Crawford, N. G., Kelly, D. E., Hansen, M. E. B., et al. (2017). Loci associated with skin pigmentation identified in African populations. *Science* 358, eaan8433.

Crespi, B. J. (2000). The evolution of maladaptation. *Hered Edinb* 84, 623–9.

Crespi, B. J. (2010). Revisiting Bleuler: relationship between autism and schizophrenia. *Br J Psychiatry* 196, 495; author reply 495–6.

Crespi, B. J. and Badcock, C. R. (2008). Psychosis and autism as diametrical disorders of the social brain. *Behav Brain Sci* 31, 241–61; discussion 261–320.

Crespi, B. J. and Procyshyn, T. L. (2017). Williams syndrome deletions and duplications: genetic windows to understanding anxiety, sociality, autism, and schizophrenia. *Neurosci Biobehav Rev* 79, 14–26.

Crespi, B. J., Summers, K., and Dorus, S. (2007). Adaptive evolution of genes underlying schizophrenia. *Proc Biol Sci* 274, 2801–10.

Crespi, B. J., Foster, K., and Úbeda, F. (2014). First principles of Hamiltonian medicine. *Philos Trans R Soc B Biol Sci* 369, 20130366.

Curry, O. (2006). Who's Afraid of the Naturalistic Fallacy? *Evol Psychol* 4, 234–47.

Darwin, C. (1859). *On the Origin of Species*. London: John Murray.

David, L. A., Maurice, C. F., Carmody, R. N., et al. (2014). Diet rapidly and reproducibly alters the human gut microbiome. *Nature* 505, 559–63.

Dawkins, R. (1976). *The Selfish Gene*. Oxford: Oxford University Press.

Day, T. and Read, A. F. (2016). Does high-dose antimicrobial chemotherapy prevent the evolution of resistance? *PLoS Comput Biol* 12, e1004689.

D'Costa, V. M., McGrann, K. M., Hughes, D. W., et al. (2006). Sampling the antibiotic resistome. *Science* 311, 374–7.

de Filippo, C., Key, F. M., Ghirotto, S., et al. (2016). Recent selection changes in human genes under long-term balancing selection. *Mol Biol Evol* 33, 1435–47.

DeGregori, J. (2011). Evolved tumor suppression: Why are we so good at not getting cancer? *Cancer Res* 71, 3739.

DeGregori, J. (2017). Connecting cancer to its causes requires incorporation of effects on tissue microenvironments. *Cancer Res* 77, 6065–8.

Del Giudice, M. (2014). An evolutionary life history framework for psychopathology. *Psychol Inq* 25, 261–300.

Del Giudice, M. and Ellis, B. J. (2016). Evolutionary foundations of developmental psychopathology. In: Cicchetti, D (ed.) *Developmental Psychopathology, Vol. 2: Developmental Neuroscience*, 3rd ed. Hoboken, NJ: Wiley, pp. 1–58.

Del Giudice, M., Ellis, B. J., and Shirtcliff, E. A. (2011). The adaptive calibration model of stress responsivity. *Neurosci Biobehav Rev* 35, 1562–92.

Dinsdale, N., Mokkonen, M., and Crespi, B. (2016). The 'extreme female brain': increased cognitive empathy as a dimension of psychopathology. *Evol Hum Behav* 37, 323–36.

Dobson, F. S. and Oli, M. K. (2007). Fast and slow life histories of mammals. *Ecoscience* 14, 292–7.

Dobzhansky, T. (1963). Evolutionary and population genetics. *Science* 142, 1131–5.

Dugatkin, L. A., Perlin, M., Lucas, J. S., et al. (2005). Group-beneficial traits, frequency-dependent selection and genotypic diversity: an antibiotic resistance paradigm. *Proc Biol Sci* 272, 79–83.

Eaton, S. B., Konner, M. J., and Shostak, M. (1988). Stone agers in the fast lane: chronic degenerative diseases in evolutionary perspective. *Am J Med* 84, 739–49.

Ebert, D. and Hamilton, W. D. (1996). Sex against virulence: the coevolution of parasitic diseases. *Trends Ecol Evol* 11, 79–82.

Eggermann, T., Eggermann, K., and Schönherr, N. (2008). Growth retardation versus overgrowth: Silver–Russell syndrome is genetically opposite to Beckwith–Wiedemann syndrome. *Trends Genet* 24, 195–204.

Elhaik, E., Tatarinova, T., Chebotarev, D., et al. (2014). Geographic population structure analysis of worldwide human populations infers their biogeographical origins. *Nat Commun* 5, 3513.

Ellison, P. T. and Jasienska, G. (2007). Constraint, pathology, and adaptation: How can we tell them apart? *Am J Hum Biol* 19, 622–30.

Elqayam, S. and Evans, J. S. B. T. (2011). Subtracting 'ought' from 'is': descriptivism versus normativism in the study of human thinking. *Behav Brain Sci* 34, 233–48.

Enriquez-Navas, P. M., Kam, Y., Das, T., et al. (2016). Exploiting evolutionary principles to prolong tumor control in preclinical models of breast cancer. *Sci Transl Med* 8, 327ra24.
Esteller, M. (2008). Epigenetics in evolution and disease. *Lancet* 372, S90–6.
Ewald, P. W. (1980). Evolutionary biology and the treatment of signs and symptoms of infectious disease. *J Theor Biol* 86, 169–76.
Ewald, P. W. (1994). *Evolution of Infectious Disease*. New York: Oxford University Press.
Ewald, P. W. (1995). The evolution of virulence: a unifying link between parasitology and ecology. *J Parasitol* 81, 659–69.
Feinberg, A. (2007). Phenotypic plasticity and the epigenetics of human disease. *Nature* 447, 433–40.
Fisher, R. A. (1930). *The Genetical Theory of Natural Selection*. Oxford: Clarendon Press.
Flegal, K. M., Carroll, M. D., Kit, B. K., et al. (2012). Prevalence of obesity and trends in the distribution of body mass index among US adults, 1999–2010. *JAMA* 307, 491–7.
Foster, K. R. and Bell, T. (2012). Competition, not cooperation, dominates interactions among culturable microbial species. *Curr Biol* 22, 1845–50.
Foster, K. R., Wenseleers, T., Ratnieks, F. L. W., et al. (2006). There is nothing wrong with inclusive fitness. *Trends Ecol Evol* 21, 599–600.
Frank, S. A. (1992). A kin selection model for the evolution of virulence. *Proc Biol Sci* 250, 195–7.
Frank, S. A. (1996). Models of parasite virulence. *Q Rev Biol* 71, 37–78.
Frank, S. A. (2007). *Dynamics of Cancer: Incidence, Inheritance, and Evolution*. Princeton, NJ: Princeton University Press.
Frank, S. A. (2012). Natural selection. III. Selection versus transmission and the levels of selection. *J Evol Biol* 25, 227–43.
Gillies, R. J., Verduzco, D., and Gatenby, R. A. (2012). Evolutionary dynamics of carcinogenesis and why targeted therapy does not work. *Nat Rev Cancer* 12, 487–93.
Gire, S. K., Goba, A., Andersen, K. G., et al. (2014). Genomic surveillance elucidates Ebola virus origin and transmission during the 2014 outbreak. *Science* 345, 1369–72.
Gloss, A. D. and Whiteman, N. K. (2016). Balancing selection: walking a tightrope. *Curr Biol* 26, R73–6.
Gluckman, P. D. and Hanson, M. (2006). *Mismatch: Why Our World No Longer Fits Our Bodies*. New York: Oxford University Press.
Gluckman, P. D., Hanson, M. A., and Spencer, H. G. (2005). Predictive adaptive responses and human evolution. *Trends Ecol Evol* 20, 527–33.
Gluckman, P. D., Hanson, M. A., and Beedle, A. S. (2007). Non-genomic transgenerational inheritance of disease risk. *Bioessays* 29, 145–54.
Gluckman, P. D., Beedle, A., and Hanson, M. (2009a). *Principles of Evolutionary Medicine*. Oxford: Oxford University Press.
Gluckman, P. D., Hanson, M. A., Buklijas, T., et al. (2009b). Epigenetic mechanisms that underpin metabolic and cardiovascular diseases. *Nat Rev Endocrinol* 5, 401–8.
Godfrey, K. M., Gluckman, P. D., and Hanson, M. A. (2010). Developmental origins of metabolic disease: life course and intergenerational perspectives. *Trends Endocrinol Metab* 21, 199–205.
Gould, S. J. (1981). *The Mismeasure of Man*. New York: Norton.
Graves, J. L. (2001). *The Emperor's New Clothes: Biological Theories of Race at the Millennium*. New Brunswick, NJ: Rutgers University Press.
Greaves, M. (2014). Was skin cancer a selective force for black pigmentation in early hominin evolution? *Proc Biol Sci* 281, 20132955.
Greaves, M. and Maley, C. C. (2012). Clonal evolution in cancer. *Nature* 481, 306–13.
Green, D. M, and Swets, J. A. (1966). *Signal Detection Theory and Psycho-Physics*. New York: Wiley.
Green, R. E., Krause, J., Briggs, A. W., et al. (2010). A draft sequence of the Neanderthal genome. *Science* 328, 710–22.
Greene, J. (2003). From neural 'is' to moral 'ought': What are the moral implications of neuroscientific moral psychology? *Nat Rev Neurosci* 4, 846–50.
Grinsted, L., Bilde, T., Gilbert, J. D. J., et al. (2015). Questioning evidence of group selection in spiders. *Nature* 524, E1–5.

Grunspan, D., Nesse, R., Barnes, M. E., et al. (2018). Core principles of evolutionary medicine: A Delphi study. *Evol Med Public Health* 2018, 13–23.
Hacking, I. (1972). The logic of Pascal's wager. *Am Philos Q* 9, 186–92.
Haig, D. (1993). Genetic conflicts in human pregnancy. *Q Rev Biol* 68, 495–532.
Haig, D. (2004). Genomic imprinting and kinship: How good is the evidence? *Annu Rev Genet* 38, 553–85.
Haig, D. (2016). Intracellular evolution of mitochondrial DNA (mtDNA) and the tragedy of the cytoplasmic commons. *BioEssays* 38, 549–55.
Haig, D. and Grafen, A. (1991). Genetic scrambling as a defense against meiotic drive. *J Theor Biol* 153, 531–58.
Haig, D, and Wharton, R. (2003). Prader–Willi syndrome and the evolution of human childhood. *Am J Hum Biol* 15, 320–9.
Hales, C. N. and Barker, D. J. (2001). The thrifty phenotype hypothesis. *Br Med Bull* 60, 5–20.
Halliday, J., Oke, K., Breheny, S., et al. (2004). Beckwith–Wiedemann syndrome and IVF: a case-control study. *Am J Hum Genet* 75, 526–8.
Hamilton, W. D. (1964a). The genetical evolution of social behaviour. I and II. *J Theor Biol* 7, 1–16.
Hamilton, W. D. (1964b). The genetical evolution of social behaviour. II. *J Theor Biol* 7, 17–52.
Hamilton, W. D. (1966). The moulding of senescence by natural selection. *J Theor Biol* 12, 12–45.
Hamilton, W. D. (1967). Extraordinary sex ratios. *Science* 156, 477–88.
Hamilton, W. D., Axelrod, R., and Tanese, R. (1990). Sexual reproduction as an adaptation to resist parasites (a review). *Proc Natl Acad Sci U S A* 87, 3566–73.
Hammerstein, P., Hagen, E. H., Herz, A. V., et al. (2006). Robustness: a key to evolutionary design. *Biol Theory* 1, 90–3.
Hansen, S. K., Rainey, P. B., Haagensen, J. A., et al. (2007). Evolution of species interactions in a biofilm community. *Nature* 445, 533–6.
Hanson, M. (2015). The birth and future health of DOHaD. *J Dev Orig Health Dis* 6, 434–7.
Hartfield, M., Murall, C. L., and Alizon, S. (2014). Clinical applications of pathogen phylogenies. *Trends Mol Med* 20, 394–404.
Harvell, C. D. (1990). The ecology and evolution of inducible defenses. *Q Rev Biol* 65, 323–40.
Haselton, M. G. and Buss, D. M. (2000). Error management theory: a new perspective on biases in cross-sex mind reading. *J Pers Soc Psychol* 78, 81–91.
Hauser, D. J., Nesse, R. M., and Schwarz, N. (2017). Lay theories and metaphors of health and illness. In: Zedelius, C., Müller, B., and Schooler, J. (eds) *The Science of Lay Theories*. Cham: Springer, pp. 341–54.
Hawkes, K. (2003). Grandmothers and the evolution of human longevity. *Am J Hum Biol* 15, 380–400.
Hawkes, K., O'Connell, J. F., Blurton Jones, N. G., et al. (1998). Grandmothering, menopause, and the evolution of human life histories. *Proc Natl Acad Sci U S A* 95, 1336–9.
Heeney, J. L., Dalgleish, A. G., Weiss, R. A. (2006). Origins of HIV and the evolution of resistance to AIDS. *Science* 313, 462–6.
Henrich, J. (2015). *The Secret of Our Success: How Culture Is Driving Human Evolution, Domesticating Our Species, and Making Us Smarter*. Princeton, NJ: Princeton University Press.
Hill, K. (1993). Life history theory and evolutionary anthropology. *Evol Anthropol Issues News Rev* 2, 78–88.
Hill, K. and Hurtado, A. (1991). The evolution of premature reproductive senescence and menopause in human females. *Hum Nat* 2, 313–50.
Hill, K. and Hurtado, A. M. (1996). *Ache Life History: The Ecology and Demography of a Foraging People*. Hawthorne, NY: Aldine.
Hill, K. and Kaplan, H. (1999). Life history traits in humans: theory and empirical studies. *Annu Rev Anthropol* 28, 397–430.
Hinde, K. (2009). Richer milk for sons but more milk for daughters: sex-biased investment during lactation varies with maternal life history in rhesus macaques. *Am J Hum Biol* 21, 512–19.

Hinde, K. and Milligan, L. A. (2011). Primate milk: proximate mechanisms and ultimate perspectives. *Evol Anthropol Issues News Rev* 20, 9–23.

Honap, T. P., Vågene, A. J., Herbig, A., et al. (2017). Genomic analyses of ancient *Mycobacterium tuberculosis* complex strains from the Americas. Poster presentation.

Huijben, S. and Paaijmans, K. P. (2017). Putting evolution in elimination: winning our ongoing battle with evolving malaria mosquitoes and parasites. *Evol Appl* 11, 415–30.

Hurst, L. D. (1998). Selfish genes and meiotic drive. *Nature* 391, 223.

Hurst, L. D. and Nurse, P. (1991). A note on the evolution of meiosis. *J Theor Biol* 150, 561–3.

Itan, Y., Powell, A., Beaumont, M. A., et al. (2009). The origins of lactase persistence in Europe. *PLoS Comput Biol* 5, e1000491.

Jablonski, N. G. (2004). The evolution of human skin and skin color. *Annu Rev Anthropol* 33, 585–623.

Jablonski, N. G. and Chaplin, G. (2010). Human skin pigmentation as an adaptation to UV radiation. *Proc Natl Acad Sci U S A* 107, 8962–8.

Jablonski, N. G. and Chaplin, G. (2017). The colours of humanity: the evolution of pigmentation in the human lineage. *Philos Trans R Soc Lond B Biol Sci* 372, 20160349.

Jorde, L. B. and Wooding, S. P. (2004). Genetic variation, classification and 'race'. *Nat Genet* 36(11 Suppl), S28–33.

Kachel, A. F., Premo, L. S., and Hublin, J.-J. (2011). Grandmothering and natural selection. *Proc Biol Sci* 278, 384–91.

Kahneman, D. (2011). *Thinking, Fast and Slow*. London: Macmillan.

Kaplan, H., Hill, K., Lancaster, J., et al. (2000). A theory of human life history evolution: diet, intelligence, and longevity. *Evol Anthropol Issues News Rev* 9, 156–85.

Kaplan, H., Thompson, R. C., Trumble, B. C., et al. (2017). Coronary atherosclerosis in indigenous South American Tsimane: a cross-sectional cohort study. *Lancet* 389, 1730–9.

Karlsson, E. K., Kwiatkowski, D. P., and Sabeti, P. C. (2014). Natural selection and infectious disease in human populations. *Nat Rev Genet* 15, 379–93.

Keller, M. C. and Miller, G. (2006). Resolving the paradox of common, harmful, heritable mental disorders: Which evolutionary genetic models work best? *Behav Brain Sci* 29, 385–404.

Kessler, R. C., Aguilar-Gaxiola, S., Alonso, J., et al. (2009). The global burden of mental disorders: an update from the WHO World Mental Health (WMH) surveys. *Epidemiol Psychiatr Soc* 18, 23–33.

Keverne, E. B. (2014). Significance of epigenetics for understanding brain development, brain evolution and behaviour. *Neuroscience* 264, 207–17.

Kirkwood, T. B. and Austad, S. N. (2000). Why do we age? *Nature* 408, 233–8.

Kirkwood, T. B. L, and Rose, M. R. (1991). Evolution of senescence: late survival sacrificed for reproduction. *Philos Trans R Soc B Biol Sci* 332, 15–24.

Kondrashov, A. S. (2017). *Crumbling Genome: The Impact of Deleterious Mutations on Humans*. Hoboken, NJ: Wiley.

Kramer, J. and Meunier, J. (2016). Kin and multilevel selection in social evolution: a never-ending controversy? *F1000Res* 5(F1000 Faculty Rev), 776.

Kreft, J.-U. (2004). Conflicts of interest in biofilms. *Biofilms* 1, 265–76.

Kruger, D. J. and Nesse, R. M. (2004). Sexual selection and the male:female mortality ratio. *Evol Psychol* 2, 66–85.

Kruger, D. J. and Nesse, R. M. (2006). An evolutionary life-history framework for understanding sex differences in human mortality rates. *Hum Nat* 17, 74–97.

Kumar, S., Sanderford, M., Gray, V. E., et al. (2012). Evolutionary diagnosis method for variants in personal exomes. *Nat Methods* 9, 855–6.

Kuzawa, C. W., Gluckman, P. D., Hanson, M. A., et al. (2008). Evolution, developmental plasticity, and metabolic disease. In: Stearns, S. C. and Koella, J. C. (eds) *Evolution in Health and Disease*. Oxford: Oxford University Press, pp. 253–64.

Laland, K. N., Sterelny, K., Odling-Smee, J., et al. (2011). Cause and effect in biology revisited: Is Mayr's proximate-ultimate dichotomy still useful? *Science* 334, 1512–16.

Lehmann, L. and Keller, L. (2006). The evolution of cooperation and altruism—a general framework and a classification of models. *J Evol Biol* 19, 1365–76.

Lehmann, L., Keller, L., West, S., et al. (2007). Group selection and kin selection: two concepts but one process. *Proc Natl Acad Sci U S A* 104, 6736–9.

Lehtonen, J., Jennions, M. D., and Kokko, H. (2012). The many costs of sex. *Trends Ecol Evol* 27, 172–8.

Leigh, E. G. (2010). The group selection controversy. *J Evol Biol* 23, 6–19.

Lentsch, A. B. (2002). The Duffy antigen/receptor for chemokines (DARC) and prostate cancer. A role as clear as black and white? *FASEB J* 16, 1093–5.

Leslie, S., Winney, B., Hellenthal, G., et al. (2015). The fine scale genetic structure of the British population. *Nature* 519, 309–14.

Levin, B. R. (1996). The evolution and maintenance of virulence in microparasites. *Emerg Infect Dis* 2, 93–102.

Levin, B. R. and Bull, J. J. (1994). Short-sighted evolution and the virulence of pathogenic microorganisms. *Trends Microbiol* 2, 76–81.

Li, W., Shi, Z., Yu, M., et al. (2005). Bats are natural reservoirs of SARS-like coronaviruses. *Science* 310, 676–9.

Liker, A. and Székely, T. (2005). Mortality costs of sexual selection and parental care in natural populations of birds. *Evolution* 59, 890–7.

Lin, Y. P. and Cheng, T. J. (2002). Why can't Chinese Han drink alcohol? Hepatitis B virus infection and the evolution of acetaldehyde dehydrogenase deficiency. *Med Hypotheses* 59, 204–7.

Lindeberg, S. (2009). Modern human physiology with respect to evolutionary adaptations that relate to diet in the past. In: Hublin, J. J. and Richards, M. P. (eds) *The Evolution of Hominin Diets*. Dordrecht: Springer, pp. 43–57.

Livingstone, F. B. (1960). The wave of advance of an advantageous gene: the sickle cell gene in Liberia. *Hum Biol* 32, 197–202.

Llewelyn, M. J., Fitzpatrick, J. M., Darwin, E., et al. (2017). The antibiotic course has had its day. *BMJ* 358, j3418.

Long, J. C. and Kittles, R. A. (2009). Human genetic diversity and the nonexistence of biological races. *Hum Biol* 81, 777–98.

Lumey, L. H., Stein, A. D., and Susser, E. (2011). Prenatal famine and adult health. *Annu Rev Public Health* 32, 237–62.

Luzzatto, L. (2012). Sickle cell anaemia and malaria. *Mediterr J Hematol Infect Dis* 4, e2012065.

Lyttle, T. W. (1993). Cheaters sometimes prosper: distortion of Mendelian segregation by meiotic drive. *Trends Genet* 9, 205–10.

Ma, H. and O'Farrell, P. H. (2016). Selfish drive can trump function when animal mitochondrial genomes compete. *Nat Genet* 48, 798–802.

Mace, R., Jordan, F., and Holden, C. (2003). Testing evolutionary hypotheses about human biological adaptation using cross-cultural comparison. *Comp Biochem Physiol Mol Integr Physiol* 136, 85–94.

May, R. M. and Anderson, R. M. (1993). Epidemiology and genetics in the coevolution of parasites and hosts. *Proc Biol Sci* 219, 281–313.

Mayr, E. (1961). Cause and effect in biology. *Science* 134, 1501–6.

Mayr, E. (1974). Teleological and teleonomic, a new analysis. *Boston Stud Philos Sci* 14, 91–117.

Mayr, E. (1982). *The Growth of Biological Thought: Diversity, Evolution, and Inheritance*. Cambridge, MA: Belknap/Harvard.

McGowan, P. O., Sasaki, A., D'Alessio, A. C., et al. (2009). Epigenetic regulation of the glucocorticoid receptor in human brain associates with childhood abuse. *Nat Neurosci* 12, 342–8.

McManus, K. F., Taravella, A. M., Henn, B. M., et al. (2017). Population genetic analysis of the DARC locus (Duffy) reveals adaptation from standing variation associated with malaria resistance in humans. *PLoS Genet* 13, e1006560.

Mealey, L. and Segal, N. L. (1993). Heritable and environmental variables affect reproduction-related behaviors, but not ultimate reproductive success. *Personal Individ Differ* 14, 783–94.

Meaney, M. J. (2010). Epigenetics and the biological definition of gene × environment interactions. *Child Dev* 81, 41–79.
Medicus, G. (2015). *Being Human. Bridging the Gap between the Sciences of Body and Mind*. Berlin: Verlag für Wissenschaft und Bildung.
Metzger, C. M., Luijckx, P., Bento, G., et al. (2016) The Red Queen lives: epistasis between linked resistance loci. *Evolution* 70, 480–7.
Miller, L. H., Mason, S. J., Dvorak, J. A., et al. (1975). Erythrocyte receptors for (*Plasmodium knowlesi*) malaria: Duffy blood group determinants. *Science* 189, 561–3.
Mitteroecker, P., Huttegger, S. M., Fischer, B., et al. (2016). Cliff-edge model of obstetric selection in humans. *Proc Natl Acad Sci U S A* 113, 14680–5.
Moirand, R. (1997). Clinical features of genetic hemochromatosis in women compared with men. *Ann Intern Med* 127, 105.
Moritz, C., McCallum, H., Donnellan, S., et al. (1991). Parasite loads in parthenogenetic and sexual lizards (*Heteronotia binoei*): support for the Red Queen hypothesis. *Proc Biol Sci* 244, 145–9.
Morran, L. T., Schmidt, O. G., Gelarden, I. A., et al. (2011). Running with the Red Queen: host-parasite coevolution selects for biparental sex. *Science* 333, 216–18.
Neel, J. V. (1962). Diabetes mellitus: a 'thrifty' genotype rendered detrimental by 'progress'? *Am J Hum Genet* 14, 352–62.
Neiman, M., Lively, C. M., Meirmans, S. (2017). Why sex? A pluralist approach revisited. *Trends Ecol Evol* 32, 589–600.
Nemeroff, C. B. (2016). Paradise lost: the neurobiological and clinical consequences of child abuse and neglect. *Neuron* 89, 892–909.
Nesse, R. M. (1988). Life table tests of evolutionary theories of senescence. *Exp Gerontol* 23, 445–53.
Nesse, R. M. (1990). Evolutionary explanations of emotions. *Hum Nat* 1, 261–89.
Nesse, R. M. (2001). The smoke detector principle. Natural selection and the regulation of defensive responses. *Ann N Y Acad Sci* 935, 75–85.
Nesse, R. M. (2004). Cliff-edged fitness functions and the persistence of schizophrenia (commentary). *Behav Brain Sci* 27, 862–3.
Nesse, R. M. (2005a). Maladaptation and natural selection. *Q Rev Biol* 80, 62–70.
Nesse, R. M. (2005b). Natural selection and the regulation of defenses: a signal detection analysis of the smoke detector principle. *Evol Hum Behav* 26, 88–105.
Nesse, R. M. (2010). Social selection and the origins of culture. In: Schaller, M., Heine, S. J., Norenzayan, A., et al. (eds) *Evolution, Culture, and the Human Mind*. Philadelphia: Psychology Press, pp. 137–50.
Nesse, R. M. (2011a). Ten questions for evolutionary studies of disease vulnerability. *Evol Appl* 4, 264–77.
Nesse, R. M. (2011b). Why has natural selection left us so vulnerable to anxiety and mood disorders? *Can J Psychiatry* 56, 705–6.
Nesse, R. M. (2013). Tinbergen's four questions, organized: a response to Bateson and Laland. *Trends Ecol Evol*, 28, 681–2.
Nesse, R. M. (in preparation). Cliff-edged fitness functions inevitably cause genetic diseases.
Nesse, R. M. and Berridge, K. C. (1997). Psychoactive drug use in evolutionary perspective. *Science* 278, 63–6.
Nesse, R. M. and Ellsworth, P. C. (2009). Evolution, emotions, and emotional disorders. *Am Psychol* 64, 129–39.
Nesse, R. M. and Stearns, S. C. (2008). The great opportunity: evolutionary applications to medicine and public health. *Evol Appl* 1, 28–48.
Nesse, R. M. and Weder, A. (2007). Darwinian medicine: what evolutionary medicine offers to endothelium researchers. In: Aird, W. (ed.) *Endothelial Biomedicine*. Cambridge: Cambridge University Press, pp. 122–8.
Nesse, R. M. and Williams, G. C. (1994). *Why We Get Sick: The New Science of Darwinian Medicine*. New York: Vintage Books.

Nesse, R. M., Bergstrom, C. T., Ellison, P. T., et al. (2010). Making evolutionary biology a basic science for medicine. *Proc Natl Acad Sci U S A* 107(Suppl 1), 1800–7.

Nesse, R. M., Ganten, D., Gregory, T., et al. (2012). Evolutionary molecular medicine. *J Mol Med (Berl)* 90, 509–22.

Neu, H. C. (1992). The crisis in antibiotic resistance. *Science* 257, 1064–73.

Niemi, D. and Phelan, J. (2008). Eliciting big ideas in biology. Paper presented at the Conceptual Assessment in Biology II Conference, Asilomar, CA.

Nijhout, H. F. (2002). The nature of robustness in development. *Bioessays* 24, 553–63.

Nisbett, R. E. (2015). *Mindware: Tools for Smart Thinking*. New York: Farrar, Straus and Giroux.

Nowak, M. A., McAvoy, A., Allen, B., et al. (2017). The general form of Hamilton's rule makes no predictions and cannot be tested empirically. *Proc Natl Acad Sci U S A* 114, 5665–70.

Nussey, D. H., Froy, H., Lemaitre, J.-F., et al. (2013). Senescence in natural populations of animals: widespread evidence and its implications for bio-gerontology. *Ageing Res Rev* 12, 214–25.

Oliveria, N. M., Martinez-Garcia, E., Xavier, J., et al. (2015). Biofilm formation as a response to ecological competition. *PLoS Biol* 13, e1002191.

Olson, M. V. and Varki, A. (2003). Sequencing the chimpanzee genome: insights into human evolution and disease. *Nat Rev Genet* 4, 20–8.

Ortman, L. L. and Craig, J. V. (1968). Social dominance in chickens modified by genetic selection—physiological mechanisms. *Anim Behav* 16, 33–7.

Paley, W. (1802). *Natural Theology, or, Evidences of the Existence and Attributes of the Deity*. London: Printed for J. Faulder.

Parker, G. A. and Smith, J. M. (1990). Optimality theory in evolutionary biology. *Nature* 348, 27–33.

Peccei, J. S. (2001). Menopause: adaptation or epiphenomenon? *Evol Anthropol Issues News Rev* 10, 43–57.

Penman, B. S. and Gupta, S. (2017). Detecting signatures of past pathogen selection on human HLA loci: are there needles in the haystack? *Parasitology* 15 Aug, 1–12. doi: 10.1017/S0031182017001159 [Epub ahead of print].

Pepper, J. W. (2008). Defeating pathogen drug resistance: guidance from evolutionary theory. *Evolution* 62, 3185–91.

Perlman, R. (2013). *Evolution and Medicine*. Oxford: Oxford University Press.

Plutchik, R. (1980). *Emotion: A Psychoevolutionary Synthesis*. New York: Harper & Row.

Polimanti, R. and Gelernter, J. (2018). *ADH1B*: from alcoholism, natural selection, and cancer to the human phenome. *Am J Med Genet B Neuropsychiatr Genet* 177, 113–25.

Powe, C. E., Evans, M. K., Wenger J., et al. (2013). Vitamin D-binding protein and vitamin D Status of black Americans and white Americans. *N Engl J Med* 369, 1991–2000.

Promislow, D. E. and Harvey, P. H. (1990). Living fast and dying young: a comparative analysis of life-history variation among mammals. *J Zool* 220, 417–37.

Pruitt, J. N. and Goodnight, C. J. (2014). Site-specific group selection drives locally adapted group compositions. *Nature* 514, 359.

Queller, D. C. (1994). Genetic relatedness in viscous populations. *Evol Ecol* 8, 70–73.

Queller, D. C. and Strassmann, J. E. (2009). Beyond society: the evolution of organismality. *Philos Trans R Soc Lond B Biol Sci* 364, 3143–55.

Queller, D. C. and Strassmann, J. E. (2013). Experimental evolution of multicellularity using microbial pseudo-organisms. *Biol Lett* 9, 20120636.

Read, A. F. (1994). The evolution of virulence. *Trends Microbiol* 2, 73–6.

Read, A. F., Day, T., and Huijben, S. (2011). The evolution of drug resistance and the curious orthodoxy of aggressive chemotherapy. *Proc Natl Acad Sci U S A* 108, 10871–7.

Redfield, R. J. (2002). Is quorum sensing a side effect of diffusion sensing? *Trends Microbiol* 10, 365–70.

Reeve, H. K. and Sherman, P. W. (1993). Adaptation and the goals of evolutionary research. *Q Rev Biol* 68, 1–32.

Reiber, C. (2010). Female gamete competition: a new evolutionary perspective on menopause. *J Soc Evol Cult Psychol* 4, 215–40.

Rice, W. R. (1992). Sexually antagonistic genes: experimental evidence. *Science* 256, 1436–9.
Richerson, P. J. and Boyd, R. (2005). *Not by Genes Alone: How Culture Transformed Human Evolution*. Chicago: University of Chicago Press.
Ridley, M. (1994). *The Red Queen: Sex and the Evolution of Human Nature*. Harmondsworth, UK: Penguin.
Ridley, M. (2003). *Nature via Nurture: Genes, Experience, and What Makes Us Human*. New York: HarperCollins.
Rogers, A. R. (1993). Why Menopause? *Evol Ecol* 7, 406–20.
Rook, G., Bäckhed, F., Levin, B. R., et al. (2017). Evolution, human–microbe interactions, and life history plasticity. *Lancet* 390, 521–30.
Rose, M. R. (2001). *Darwin's Spectre: Evolutionary Biology in the Modern World*. Princeton, NJ: Princeton University Press.
Rosenberg, S. M., Shee, C., Frisch, R. L., et al. (2012). Stress-induced mutation via DNA breaks in *Escherichia coli*: a molecular mechanism with implications for evolution and medicine. *Bioessays* 34, 885–92.
Rosenvinge, J. H. and Pettersen, G. (2015). Epidemiology of eating disorders, part I: introduction to the series and a historical panorama. *Adv Eat Disord* 3, 76–90.
Ruscio, A. M., Hallion, L. S., Lim, C. C. W., et al. (2017). Cross-sectional comparison of the epidemiology of DSM-5 generalized anxiety disorder across the globe. *JAMA Psychiatry* 74, 465–75.
Salmond, G. P. C. and Welch, M. (2008). Antibiotic resistance: adaptive evolution. *Lancet* 372, S97–103.
Salomon, R., Hoffmann, E., and Webster, R. G. (2007). Inhibition of the cytokine response does not protect against lethal H5N1 influenza infection. *Proc Natl Acad Sci U S A* 104, 12479–81.
Sandler, L. and Novitski, E. (1957). Meiotic drive as an evolutionary force. *Am Nat* 91, 105–10.
Schmid-Hempel, P. (2011). *Evolutionary Parasitology: The Integrated Study of Infections, Immunology, Ecology, and Genetics*. New York: Oxford University Press.
Schmid-Hempel, P. and Frank, S. A. (2007). Pathogenesis, virulence, and infective dose. *PLoS Pathog* 3, e147.
Shalev, I. and Belsky, J. (2016). Early-life stress and reproductive cost: a two-hit developmental model of accelerated aging? *Med Hypotheses* 90, 41–7.
Shanley, D. P. and Kirkwood, T. B. L. (2001). Evolution of the human menopause. *Bioessays* 23, 282–7.
Simonti, C. N., Vernot, B., Bastarache, L., et al. (2016). The phenotypic legacy of admixture between modern humans and Neandertals. *Science* 351, 737–41.
Skinner, M. K. (2014). Environmental stress and epigenetic transgenerational inheritance. *BMC Med* 12, 153.
Smedley, A. and Smedley, B. D. (2005). Race as biology is fiction, racism as a social problem is real: anthropological and historical perspectives on the social construction of race. *Am Psychol* 60, 16–26.
Smith, A. D., Kim, Y.-I., and Refsum, H. (2008). Is folic acid good for everyone? *Am J Clin Nutr* 87, 517–33.
Smits, S. A., Leach, J., Sonnenburg, E. D., et al. (2017). Seasonal cycling in the gut microbiome of the Hadza hunter-gatherers of Tanzania. *Science* 357, 802–6.
Stearns, S. C. (1987). *The Evolution of Sex and Its Consequences*. Basel: Birkhäuser.
Stearns, S. C. (1989). Trade-offs in life-history evolution. *Funct Ecol* 3, 259–68.
Stearns, S. C. (ed.) (1999). *Evolution in Health and Disease*. Oxford: Oxford University Press.
Stearns, S. C. (2012). Evolutionary medicine: its scope, interest and potential. *Proc Biol Sci* 279, 4305–21.
Stearns, S. C. and Koella, J. C. (eds) (2008). *Evolution in Health and Disease*, 2nd ed. New York: Oxford University Press.
Stearns, S. C. and Medzhitov, R. (2016). *Evolutionary Medicine*. Sunderland, MA: Sinauer Associates.
Stein, D. J. and Nesse, R. M. (2015). Normal and abnormal anxiety in the age of DSM-5 and ICD-11. *Emot Rev* 7, 223–9.

Sung, W., Ackerman, M. S., Dillon, M. M., et al. (2016). Evolution of the insertion–deletion mutation rate across the Tree of Life. *G3 (Bethesda)* 6, 2583–91.

Tinbergen, N. (1963). On the aims and methods of ethology. *Z Tierpsychol* 20, 410–63.

Tishkoff, S. A. and Verrelli, B. C. (2003). Patterns of human genetic diversity: implications for human evolutionary history and disease. *Annu Rev Genomics Hum Genet* 4, 293–340.

Tishkoff, S., Reed, F., Ranciaro, A., et al. (2006). Convergent adaptation of human lactase persistence in Africa and Europe. *Nat Genet* 39, 31–40.

Trevathan, W. R., McKenna, J. J., Smith, E. O. (eds) (2007). *Evolutionary Medicine*, 2nd ed. New York: Oxford University Press.

Trivers, R. L. (1972). Parental investment and sexual selection. In: Campbell, B. (ed.) *Sexual Selection and the Descent of Man*. New York: Aldine de Gruyter, pp. 136–79.

Trivers, R. L. (1974). Parent–offspring conflict. *Integr Comp Biol* 14, 249–64.

Trivers, R. L. and Willard, D. E. (1973). Natural selection of parental ability to vary the sex ratio of offspring. *Science* 179, 90–2.

Tung, J., Archie, E. A., Altmann, J., et al. (2016). Cumulative early life adversity predicts longevity in wild baboons. *Nat Commun* 7, 11181.

Turnbaugh, P. J., Ley, R. E., Mahowald, M. A., et al. (2006). An obesity-associated gut microbiome with increased capacity for energy harvest. *Nature* 444, 1027–131.

Uranga, A., España, P. P., Bilbao, A., et al. (2016). Duration of antibiotic treatment in community-acquired pneumonia: a multicenter randomized clinical trial. *JAMA Intern Med* 176, 1257–65.

Vale, P. F., McNally, L., Doeschl-Wilson, A., et al. (2016). Beyond killing: Can we find new ways to manage infection? *Evol Med Public Health* 2016, 148–57.

Valentine, J. W. (1978). The evolution of multicellular plants and animals. *Sci Am* 239(3), 141–60.

Velicer, G. J. (2003). Social strife in the microbial world. *Trends Microbiol* 11, 330–7.

Wallace, D. (2005). A mitochondrial paradigm of metabolic and degenerative diseases, aging, and cancer: a dawn for evolutionary medicine. *Annu Rev Genet* 39, 359–407.

Weatherall, D. J. (1995). The role of nature and nurture in common diseases: Garrod's legacy. *J Nutr Environ Med* 5, 63–76.

Weaver, I. C. G., Cervoni, N., Champagne, F. A., et al. (2004). Epigenetic programming by maternal behavior. *Nat Neurosci* 7, 847–54.

Weaver, I. C. G., Champagne, F. A., Brown, S. E., et al. (2005). Reversal of maternal programming of stress responses in adult offspring through methyl supplementation, altering epigenetic marking later in life. *J Neurosci* 25, 11045–54.

Webster, T. H. and Wilson Sayres, M. A. (2016). Genomic signatures of sex-biased demography: progress and prospects. *Curr Opin Genet Dev* 41, 62–71.

Wells, J. C. (2012). A critical appraisal of the predictive adaptive response hypothesis. *Int J Epidemiol* 41, 229–35.

Wells, J. C. K., Nesse, R. M., Sear, R., et al. (2017). Evolutionary public health: introducing the concept. *Lancet* 390, 500–9.

Wertheim, J. O. and Worobey, M. (2009). Dating the age of the SIV lineages that gave rise to HIV-1 and HIV-2. *PLoS Comput Biol* 5, e1000377.

West, S. A., Griffin, A. S., and Gardner, A. (2007). Social semantics: altruism, cooperation, mutualism, strong reciprocity and group selection. *J Evol Biol* 20, 415–32.

West, S. A., Griffin, A. S., and Gardner, A. (2008). Social semantics: how useful has group selection been? *J Evol Biol* 21, 374.

West, S. A., Pen, I., and Griffin, A. S. (2002). Cooperation and competition between relatives. *Science* 296, 72–5.

West-Eberhard, M. J. (2003). *Developmental Plasticity and Evolution*. Oxford: Oxford University Press.

Wierzbicka, A. (1986). Human emotions: universal or culture-specific? *Am Anthropol* 88, 584–94.

Wilkins, A. S., Wrangham, R. W., and Fitch, W. T. (2014). The 'domestication syndrome' in mammals: a unified explanation based on neural crest cell behavior and genetics. *Genetics* 197, 795–808.

Williams, G. C. (1957). Pleiotropy, natural selection, and the evolution of senescence. *Evolution* 11, 398–411.
Williams, G. C. (1966). *Adaptation and Natural Selection: A Critique of Some Current Evolutionary Thought*. Princeton, NJ: Princeton University Press.
Williams, G. C. (1996). *Plan and Purpose in Nature*. London: Weidenfeld & Nicolson.
Williams, G. C. and Nesse, R. M. (1991). The dawn of Darwinian medicine. *Q Rev Biol* 66, 1–22.
Wilson, D. S. (2008). Social semantics: toward a genuine pluralism in the study of social behaviour. *J Evol Biol* 21, 368–73.
Wilson, D. S. and Sober, E. (1994). Reintroducing group selection to the human behavioral sciences. *Behav Brain Sci* 17, 585–607.
Woo, H. J., Yu, C., Kumar, K., et al. (2017). Large-scale interaction effects reveal missing heritability in schizophrenia, bipolar disorder and posttraumatic stress disorder. *Transl Psychiatry* 7, e1089.
Wynne-Edwards, V. C. (1962). *Animal Dispersion in Relation to Social Behaviour*. Edinburgh: Oliver & Boyd.
Zampieri, F. (2009). Medicine, evolution and natural selection: an historical overview. *Q Rev Biol* 84, 1–23.
Zhao, X., Lin, C.-W., Wang, J., et al. (2014). Advances in rapid detection methods for foodborne pathogens. *J Microbiol Biotechnol* 24, 297–312.

CHAPTER 2

CELLULAR SIGNALLING SYSTEMS

DIANA LE DUC AND TORSTEN SCHÖNEBERG

Abstract

Effective reception, delivery, and processing of information is fundamental to all life forms. Physical and chemical signals are perceived from both outside and inside an organism. The nature, duration, and intensity of signals are processed into information, mainly encoded as concentration differences of ions and molecules that ultimately lead to a reaction of the organism. Although the advent of the first and most primitive signalling system will remain unknown, it probably existed already in the first hours of life. Disturbances of well-orchestrated signalling systems are often the basis of diseases. Understanding the complexity of signalling networks is required for rational intervention in different disease stages. Understanding the evolutionary history of signalling systems can help us unveil the requirements for proper functioning of a given signalling network. This chapter provides an overview of how cellular communication evolved, works, and contributes to our understanding of human diseases in the light of evolution.

Keywords

information, signalling, disease, signalling network, evolutionary history, cellular communication, evolutionary medicine

2.1 Evolution of Communication

About 4.5 billion years ago (bya) our Solar System, including the Earth, was at the dawn of its existence. The young Earth, in the throes of its origin, with erupting volcanoes and impacting asteroids, did not welcome life for about 750 million years. Fossil data suggest that life on Earth emerged at least 3.5 bya (Schopf 2006). A crucial step in evolution was the formation of macromolecules with the essential characteristic that they became able to direct the synthesis of new copies of themselves. Only such a molecule would be capable of further reproduction and evolution. According to the 'RNA-world hypothesis', RNA was the informational polymer that supported the emergence of life (Joyce 2002). The first cell is presumed to have arisen after RNA was enclosed in a vesicle, thus avoiding dilution within the primordial sea (Koch and Silver 2005). The replication of RNA represented in fact the communication of genetic information. Thus, the history of life's evolution is incorporated within the genetic code of present-day living cells.

While, from a narrow perspective, cell communication is just a means of supporting complex biological processes like homoeostasis or development, in a broader view, cell communication is the basis of life's evolution, which started with the communication of genetic information. Cells are supposed to have originated in an ocean of organic molecules that enabled them to obtain energy directly from the environment. This might have provided the evolutionary pressure for developing communication mechanisms, since receiving information from the environment, about, for example, nutrient availability, would have been advantageous. Given the need for communication in the first hours after life appeared, this must have been an early invention. Indeed, we see communication in present-day prokaryotes, bacteria, and archaea, which are the simplest living organisms on Earth and representatives of the primitive life forms. Many species of bacteria use quorum sensing. Quorum sensing is defined as a response in gene expression in correlation with population density and environmental confinement. This communication system was first described in *Vibrio fischeri*, a luminescent marine bacterium that lives in symbiosis with the Hawaiian bobtail squid (Nealson et al. 1970). *V. fischeri* lives in the photophore (light-producing) organ of the squid, using nutrients supplied by the squid host; when a threshold cell density is met, of the order of 10^{11} cells/ml, the bacteria start transcribing luciferase, which leads to bioluminescence. The light thus produced helps the squid to camouflage by counterillumination, which prevents a shadow being cast on the seabed and makes it more difficult for predators located underneath to detect the squid. The association between the bacteria and the squid is initiated in each generation as newly hatched squid are colonised by bacterial symbionts from the surrounding seawater. Quorum-sensing coordinates the colonisation process by secretion of small signal molecules that accumulate in the surrounding environment. When the bacterial population reaches a threshold cell density, the concentration of the quorum-sensing signal is sufficient to induce gene expression. Colonisation factors are thus expressed only when they are beneficial to the bacterial cell (e.g. when they are associated with the host), such that costly processes happen only when the environment is appropriate (Lupp and Ruby 2005).

From a medical perspective, quorum sensing becomes important when bacteria like *Pseudomonas aeruginosa*, an opportunistic pathogen, coordinate the formation of biofilms and swarming motility, which enables colonisation and eventually overcomes the host immune system, leading to disease (Sauer et al 2002). The issue may be even more relevant in the present medical era given the increased prevalence of multidrug-resistant pathogens, a great concern for the medical and scientific community (Carvalhaes et al. 2013). To this end, quorum quenching, the process of interrupting quorum sensing by disrupting the signalling (Chan et al. 2010), has good potential in antimicrobial strategies (John et al. 2016). Additionally, quorum-sensing molecules have lately been related to multiple medical fields from oncology (Wynendaele et al. 2012; De Spiegeleer et al. 2015; John et al. 2016) to neurology (Wynendaele et al. 2015) and immunology (Khambati et al. 2017). The quorum-sensing system seems to regulate important ecological traits of bacteria, with wide implications. This wide involvement may be due to the ancient nature of the system, which has been continually present during evolution, starting with *Proteobacteria* and the *Firmicutes* (Lerat and Moran 2004).

While quorum sensing-driven colonisation may be seen as a response of the cell to the immediate surrounding environment, the same system has been implicated in sporulation, when cells anticipate the environmental stress and form spores, which are more resistant to adverse conditions (Van Gestel et al. 2012). To anticipate the changes in the environment, cell-to-cell communicating systems had to evolve, since time was another variable to consider. A cell would thus have a selective advantage when it can accurately time the developmental transition between two states (e.g. normal state in nutrient availability conditions and spore in nutrient scarcity). Such a model supports the evolution of individuals that specialise in different ecological conditions (Van Gestel et al. 2012). Cell communication is hence the basis of ecological diversification.

Given their communication abilities, it is reasonable to assume that some bacteria may have specialised in surviving in aerobic conditions or in light-dominated environments. About 2.7 bya, after 1–1.5 billion years of prokaryote evolution, the eukaryotes emerged. Unlike the prokaryotes, which comprise a cell wall with genetic material inside, the eukaryotes have major internal organisation, which makes them more complex, structured, and efficient. In 1910, the endosymbiotic theory was proposed by the Russian botanist Konstantin Mereschkowski (Mereschkowsky 1910). According to this theory, eukaryotic cells must have evolved from a symbiotic association of prokaryotes, such that an anaerobic cell allowed an aerobic one to live within itself, thus benefiting from oxidative metabolism of the latter. Indeed, in 1967, Lynn Margulis supported the theory with microbiological evidence and now it is widely accepted that eukaryotic cell organelles such as mitochondria originate from aerobic bacteria and chloroplasts are in fact photosynthetic bacteria, such as the cyanobacteria. In fact, these organelles contain their own genetic material, separated from the nucleus. The organelles' functions (e.g. ribosomal translation) are still close to those found in prokaryotes, a fact one should always remember when therapeutically eradicating prokaryotes. That is, the 'ancestral remnants' inside of us, the mitochondria, may limit specificity of antibiotic compounds and cause side-effects in the treated human (Pacheu-Grau et al. 2010; Singh et al. 2014).

According to Margulis and Sagan, 'Life did not take over the globe by combat, but by networking' (Margulis and Sagan 2001), which implies an important role of cell-to-cell communication. The metabolic cooperation of unicellular organisms finally paved the road towards multicellularity. Whether cells communicate through direct contact or by signalling molecules, there are always three major stages for communication (Figure 2.1):

- reception—the target cell detects a signal;
- transduction—the conversion of the signal to a form that leads to a specific cellular response;
- response—the specific cellular effect.

Given the long-term evolution of cellular communication, the basic stages of the system are conserved throughout all types of organisms, with very high similarities in the components among different species.

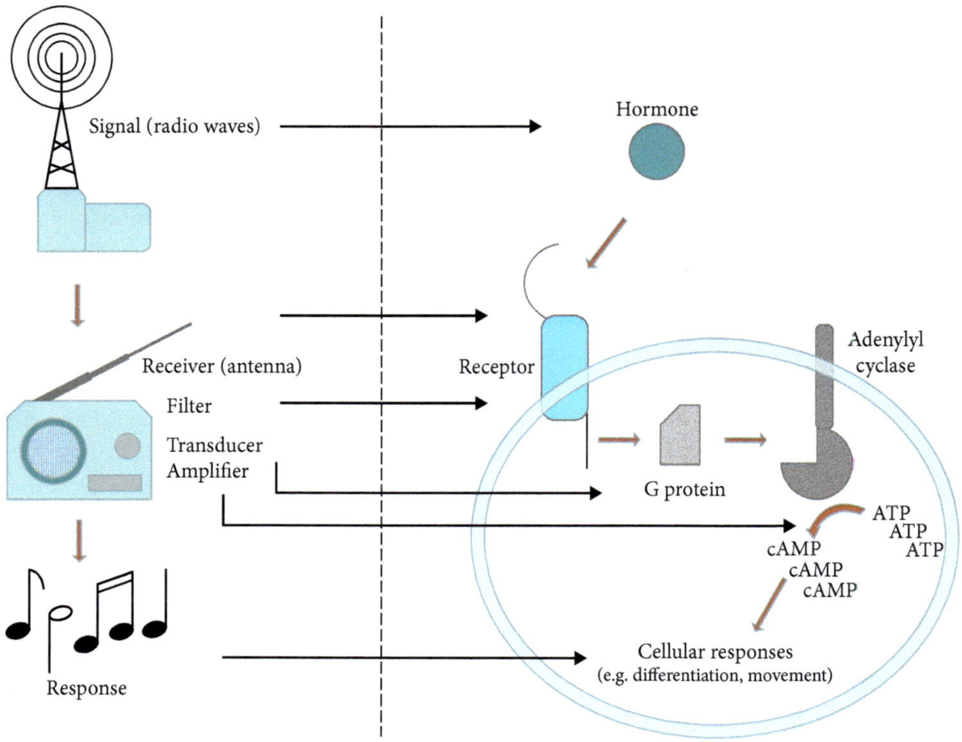

FIGURE 2.1 Different stages of cellular signalling. An extracellular signal is transmitted and specifically (through filtering) received by a receptor. The receptor transforms the extracellular signal via structural changes that are recognised by an adaptor protein (e.g. G protein). The adaptor protein transforms and transmits signal to an amplifier (e.g. an adenylyl cyclase), producing a metabotropic signalling molecule (e.g. cAMP) on a logarithmic scale. Differences in intracellular AMP levels induce multiple cellular responses by modulating the function of effector molecules such as protein kinases and transcription factors.

2.2 Evolution of Multicellularity

Multicellular organisms evolved from the unicellular eukaryotes about 1.7 bya. The eukaryotes comprise at least twenty-five examples of organisms that independently supported the evolution from unicellularity to multicellularity (Parfrey and Lahr 2013). Multicellular organisms start from a unicellular stage, and this single cell divides repeatedly, resulting in the final multicellular form. After generating the multicellular form, the cells can either remain attached to each other, or disperse to maximise their access to food. The transition from unicellular to multicellular organisms, such as animals, plants, and fungi, evolved through a stage termed alternatively aggregative or sorocarpic multicellularity. Here, the dispersed cells of the multicellular organism come together again under stressful conditions (like scarce nutrients) and form a multicellular body or sorocarp with resilient cysts or spores (Du et al. 2015). Intuitively, the communication between these cells plays a very important role in forming the multicellular body, which eventually helps survival in adverse conditions.

The cellular slime mould *Dictyostelium discoideum* is such an example; normally the amoeba spends most of its lifetime foraging alone in soils. However, in response to starvation, hundreds of thousands of amoebae aggregate and form a motile multicellular slug (Kessin 2001). The slug migrates to find a more suitable place, like the top layer of soil, and then a multicellular fruiting body composed of resistant spores develops. The entire process is organised by stalk cells located at the tip of the body, which show cell–cell connections akin to epithelia in animals (Dickinson et al. 2012). Moreover, many key features of animal multicellularity have been claimed to occur in the dictyostelid amoeba, including cell adhesion, communication and signalling, differentiation, and development (Parfrey and Lahr 2013). This is rather surprising given (1) the fundamentally different strategy of multicellularity in amoebae, which happens by aggregation of already differentiated, individual cells during one life cycle, compared with multicellularity in animals, which happens by the repeated division of the zygote, followed by differentiation and development; (2) the last common ancestor between dictyostelids and animals was about 1 bya, and many unicellular and multicellular lineages separate them (Parfrey and Lahr 2013), which renders the similarities as very unexpected. The possible scenarios for the observed similarities are that either they shared a common ancestor that possessed communication strategies found in both lineages, or the acquisition of the skills happened by convergent evolution.

To test the two scenarios, one would need to study deeper a similar trait in both lineages. One such important feature is the cell–cell junctions in the animal epithelium and *D. discoideum* tip cells. Epithelium forms the lining of all free body surfaces in animals. One type of junctions between the epithelial cells are adherens junctions, which are formed by protein complexes that consist of transmembrane cadherin proteins that bind cadherins of other cells extracellularly. Intracellularly, cadherins are connected to the actin cytoskeleton via a protein complex containing α- and β-catenin. Essentially, adherens junctions mechanically link the actin cytoskeleton of adjacent epithelial cells, which provides structural integrity (Grimson et al. 2000). Hence, both cadherins and catenins are essential to animal multicellularity. Thus, the discovery of catenin homologs in *D. discoideum* was a major breakthrough in understanding the origins of multicellularity (Grimson et al. 2000). Moreover, β-catenin also plays a role in cell signalling pathways, especially the Wnt pathway, suggesting an important role in cell communication (Wikramanayake et al. 2003). Yet, the presence of the catenin genes on both lineages does not necessarily imply the presence in their common ancestor. In fact, phylogenetic analyses proved that α-catenin in *D. discoideum* is not homologous to the metazoan α-catenin, but is a duplication of another gene, vinculin (Parfrey and Lahr 2013). This actually supports a functional similarity between the animal and dictyostelid catenins that was achieved through molecular convergence (Parfrey and Lahr 2013). Having proven a case of convergent evolution to multicellularity and given the widespread multicellularity across the eukaryotic tree, one may argue that this is a state easy to evolve towards (Grosberg and Strathmann 2007). It was suggested that, given strong selective pressures, multicellularity can be rapidly induced. Indeed, experimental data on algae show that predation pressure can induce multicellularity in just a few generations (Grosberg and Strathmann 2007).

Recent genomic sequencing projects have revealed that many genes involved in processes essential to multicellularity, such as cell adhesion and communication, are already present in unicellular organisms. Although the genetic toolbox for multicellularity is ancestral, the microbial communicating machinery was assigned to new and additional functions in multicellular organisms (Parfrey and Lahr 2013).

The presence of multiple differentiated cell types in an organism was shown to allow higher complexity with respect to division of labour and also an increase in size (Bonner 2003). The greatest impact on division of labour would be with regard to reproduction. While only a small fraction of the cells may reproduce, the rest have to selflessly support the reproductive cells by accomplishing somatic functions. This is very similar to what happens in an ant colony, where workers are forced to give up reproduction in favour of the reproductive queen. However, the reproductive division of labour leads to a potential conflict among ants, and cheaters may arise (Chapuisat 2009). By analogy, cheating cell variants that may become a reproductive cell could be selected, although such variants decrease individual fitness. In this regard, *cancer* has been understood as a *loss of multicellularity* (Davies and Lineweaver 2011). In this case, a so-called cheater cell acquires reproductive capacities and divides uncontrollably to the disadvantage of the entire organism. In an ant society, cheaters are recognised by smell, such that a simple hydrocarbon triggers an aggressive attitude of the workers towards egg-laying cheaters (Smith et al. 2009). In a multicellular organism, discovering a marker for cheater cells could unveil coercing mechanisms that may re-establish health. However, the best road we can follow towards this goal is the one shown by evolution. Evolution may reveal cell communication mechanisms that are essential for human health. The new genomic era has just begun uncovering the mere tip of the iceberg.

2.3 Genome Evolution—Prerequisite for Cellular Communication

A pertinent question that often arises is related to the 'uniqueness of life'. What we see in present-day organisms is a single genetic code that seems to be derived from one origin. The unique origin is indicated by the basic molecular biological and biochemical mechanisms present in all bacterial, archaeal, and eukaryotic cells. However, computer simulations suggest that the initial stages in biochemical evolution could have occurred many times in parallel in the oceans or atmosphere of the primordial Earth (Brown 2006). It is therefore quite possible that 'life' arose multiple times. Also, life could be based on informational molecules other than DNA and RNA, as Orgel has suggested (Orgel 2000). A primordial form of the RNA could have had threose rather than ribose in their sugar-phosphate backbones, retaining the properties of RNA, but with more stability (Orgel 2000). Nevertheless, the genetic code seen in modern life suggests that the present system first acquired the means to synthesise protein enzymes, and also to adopt a DNA genome. Initially, the RNA had catalytic activity through the ribozymes, but gradually the protein enzymes with greater catalytic potential took over this function. As a coding molecule, RNA exhibits relative instability of the RNA phosphodiester bond, because of the indirect effect of the 2′-OH group. Transferring the coding function to DNA was easy to achieve by a reduction of ribonucleotides to deoxyribonucleotides; adopting the double-stranded DNA as the coding molecule had yet another advantage of repairing DNA damage by copying the sister strand. The DNA–RNA–protein cells seemed to have had an advantage over the other forms of

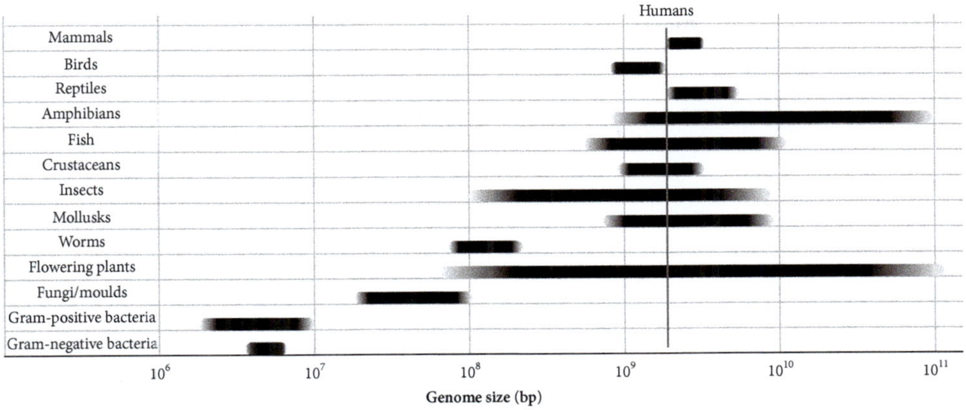

FIGURE 2.2 Range of haploid genome sizes in various life forms.

information communication; thus, they multiplied more rapidly and out-competed the RNA cells (and possibly other systems' cells) for nutrients (Brown 2006).

As adaptive evolution can be best considered from the perspective of genes (Dawkins 2016), the high diversity of life goes hand in hand with the evolution of genomes and consequently with the diversification of means of communication. Genome evolution accompanies the development of new functions; hence at the very base of communication mechanisms are the blueprints of genetic evolution and natural selection.

2.3.1 Gene Gain

The gene number increased dramatically in two evolutionary periods: the origin of eukaryotes and the origin of vertebrates (Bird 1995) (Figure 2.2). This suggests that gene expansion may be an important mechanism for organismic evolution. Susumu Ohno, an important figure in the history of genome evolution, argued more than 45 years ago that big leaps in evolution must have been accompanied by the acquisition of new gene loci with new functions (Ohno 1970).

New genes can be gained in a genome either (1) by duplication of already existing genes or (2) by acquisition from other genomes (Figure 2.3). Genome size varies considerably by a factor of a thousand in plants (Bennett and Leitch 2005), and even higher in animals (Gregory 2005) (Figure 2.2), such that an increased genome size does not correlate with the complexity of the organism.

2.3.1.1 *Gene Duplication*

According to how big the duplicated region of the genome is, duplication can refer to:

- the whole genome;
- a part of chromosome or an entire chromosome;
- a single gene or groups of genes.

For *whole-genome duplication*, an error must arise during meiosis and lead to the production of diploid gametes. Further, if two diploid gametes fuse, this will result in a tetraploid

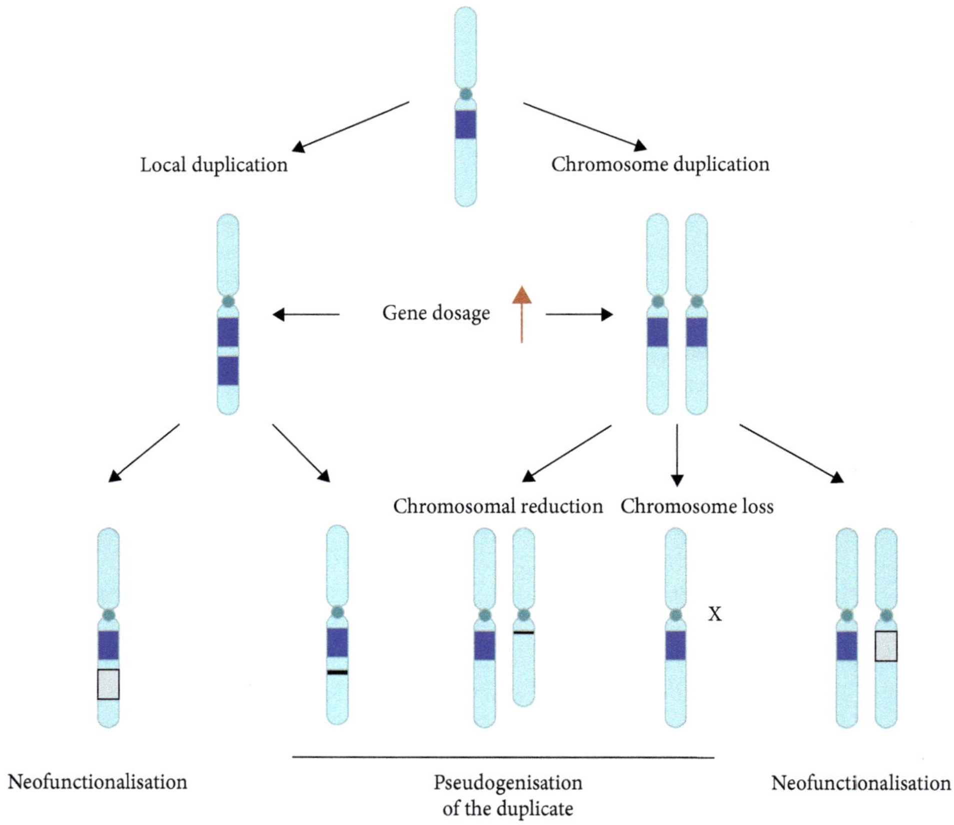

FIGURE 2.3 Possible fates of gene number evolution. Duplication can happen by duplicating either the gene region or the entire chromosome. There are three distinct outcomes if the two copies are retained. Different functions of genes are indicated here by blue and grey. In gene dosage, the two copies can keep performing the same function as the ancestral copy before the duplication event. In this way, redundancy is introduced and activity of the gene may be increased. In neofunctionalisation, one copy still performs the functions of the ancestral gene; the other copy acquires mutations that could generate a novel function (grey) or no function at all (pseudogenisation).

cell and the process is called autopolyploidy. The process is mostly present in plants, since their cells are often viable because each chromosome still has a homologous partner. However, among animals, autopolyploidy is less common, most probably because of the presence of more than a pair of chromosomes (Brown 2006). Also, autopolyploidy prevents interbreeding with an organism similar to the one it is derived from. Specifically, a cross between a tetraploid and diploid leads to a triploid offspring unable to reproduce because one full set of its chromosomes lack homologous partners. This means that autoploidy is a strategy of speciation, where the hallmark of two species is that they are unable to interbreed. *Saccharomyces cerevisiae*, the yeast, may be such an example of a whole-genome duplication that took place approximately 100 million years ago (mya) at an evolutionary split from *Kluyveromyces*, a genus of ascomycetous yeasts (Wolfe and Shields 1997). Also, some type of fish have increased numbers of the *Hox* gene cluster, which are also supposed to have arisen from a genome duplication event (Taylor et al. 2001). This is more relevant for metazoan biology, since *Hox* genes are a subset of very conserved homeotic genes, which are involved in cell-to-cell communication to form the craniocaudal (head–tail) axis of the embryo in most modern species, including humans.

Nevertheless, the whole-genome duplication strategy does not lead to gain of new genes, but rather gene expansion through extra copies of every gene. Still, this provides the potential for forming new genes, since the extra genes are backed up by the other copies and so, at least in theory, they could undergo mutational change without affecting the viability of the organism. Mostly the mutations will be deleterious, which will inactivate the genes—pseudogenisation—but occasionally may also lead to genes with new functions. On a long timescale, the inactivated genes will be excluded from the genomes and the duplicated genes with useful properties will be retained. However, it would be impossible to decide whether a gene arose by genome duplication or simply by duplication of individual genes. To this end, one would need to find regions that are duplicated and where the gene order is preserved. This may also be influenced by the recombination rate; so, overall, genome duplication is rather hard to prove in modern organisms (Piskur 2001).

In the case of *chromosomal duplication* (Figure 2.3), the result is a cell that contains three copies of a chromosome, which is called a trisomy in humans. The effects of trisomy in modern organisms suggest that this is not a major mechanism for gene gain. Specifically, in humans, the most common trisomy, trisomy 16 (Hassold et al. 1995), leads to an abortion in the first pregnancy trimester. Other trisomies may lead to genetic diseases like the Down syndrome (trisomy 21) or Edwards syndrome (trisomy 18), and similar deleterious effects have been observed in trisomic mutants of other species, for example *Drosophila melanogaster* (Devlin et al. 1988).

A major mechanism to acquire new genetic material is *gene duplication*. This is supported by the presence of multigene families, which are common components of all genomes. Using evolutionary analysis techniques that consider the mutation rate and selective pressure, one can identify the gene duplication events and the ancestral gene (Yang 2007). There are three main mechanisms for gene duplication:

- *Ectopic recombination or unequal crossing-over.* If similar nucleotide sequences are not present at identical places in a pair of homologous chromosomes, unequal crossing-over may take place. The event is mediated by sequence similarity at the duplicate breakpoints and the result is a duplication on one chromosome and a reciprocal deletion on the other.
- *Replication slippage.* This is usually involved in the duplication of short genetic sequences. Replication of DNA is mediated by DNA polymerase, which, under normal circumstances after replication pauses, dissociates from the DNA strand. However, if dissociation does not occur, the polymerase aligns the replicating strand to an incorrect position and can incidentally copy the same sequence more than once. The process is also mediated by the presence of repetitive sequences.
- *Retrotransposition.* This occurs when a replicating retro-element or retrovirus enters the cell and viral proteins attach to the cellular mRNA by mistake (instead of the viral genome that is supposed to be reverse-transcribed). The result is a reverse transcription of the retrogene, which may be integrated in the genome.

Gene duplications are considered the most important source of genetic novelty (Ohno 1970). The genetic redundancy created by gene duplication releases one of the copies of a gene from selective pressure. Essentially, one copy of the gene can undergo mutations without the risk of affecting the cellular function that is maintained by the non-mutated copy. This makes the acquisition of novel functions and characteristics possible. In general, two genes with the same function are very unlikely to be kept in the genome, unless an extra amount of gene product may be advantageous (Nowak et al. 1997). However, after a while, the two daughter genes may evolve to differentiate in some aspects of their function—this is called *subfunctionalisation*. One such example may be the glutamate dehydrogenase gene (*GLUD2*) present in humans, which arose through duplication in primates during a period of brain-size increase (Shashidharan and Plaitakis 2014). Glutamate dehydrogenase (GDH) is central to the metabolism of glutamate, the main excitatory neurotransmitter. After neuron firing, the neurotransmitter must be inactivated, and this happens in the astrocytes that metabolise it using different enzymes, among which is GDH. While *GLUD1* is a housekeeping gene found in multiple tissues (Sonnewald et al. 1997), *GLUD2* is specifically expressed in nerve tissue and testes (Shashidharan et al. 1994). Despite the high levels of guanosine 5′-triphosphate (GTP) in the brain, which inhibits the activity of GDH, the mutation of Gly456 to Ala made GLUD2 resistant to GTP. Moreover, GLUD2 has a 25 times higher activity than the basal level, it can function under low pH conditions, and it is instantly activated in the brain at high-frequency firing of the neurons (Plaitakis et al. 2000, 2003; Burki and Kaessmann 2004). Alterations in this gene have been implicated in human neurodegenerative disorders (Plaitakis. et al. 2000).

The escape from selective pressure can sometimes lead to an even more dramatic change in the gene sequence with the acquisition of new functions. To this end, the ability to discriminate a broader range of wavelengths is most often mediated by increasing the number of opsins through gene duplication (Porath-Krause et al. 2016). A photopigment can absorb only a portion of the light spectrum, such that colour discrimination can be achieved, if the photopigments have different light sensitivities. Changes in the amino acid residues at positions that interact with the chromophore (i.e. 'spectral tuning sites') shift the wavelength at which absorbance is greatest (λ_{max}) and the duplicated gene of the visual pigment acquires a new function of phototransduction of a different part of the light spectrum (Cronin and Porter 2014; Porath-Krause et al. 2016). This model of acquiring new functions is termed *neofunctionalisation* (Figure 2.3).

However, most of the duplicated genes will be pseudogenised, since it is generally not advantageous to carry two identical genes (Nowak et al. 1997) (Figure 2.3). Hence, if the duplicated gene is not under any selective pressure, pseudogenisation will occur in the first million years after the duplication event (Lynch and Conery 2000). Pseudogenisation usually occurs by nonsense or missense mutation in the coding region, promoter mutation, or loss of exon splicing junctions (Magadum et al. 2013). Gradually, while mutations accumulate, the gene becomes a pseudogene, which is either unexpressed or functionless (Zhang 2003). One classical example of pseudogenisation in humans concerns the olfactory receptor family. Of the about 600 genes encoding for olfactory receptors in humans, almost half are pseudogenes. This has been suggested to be a result of the limited relevance of smell for humans (Rouquier et al. 2000). Yet, some pseudogenes may also serve different functions. Some of them seem to have a role in regulating the expression of the ancestral gene. Such an example is *PTEN*, a tumour suppressor gene. Maintaining a normal level of PTEN protein is thus essential for preventing oncogenesis. The gene *PTEN* and the pseudogene *PTENP1* are both regulated by microRNAs (miRNAs). When *PTENP1* binds miRNA, the cellular concentration of miRNA decreases, allowing *PTEN* to escape the miRNA repression regulation, and thus be expressed (Pink et al. 2011). Thus, although pseudogenes per se do not contribute to cellular communication, they may promote and sustain the process.

2.3.1.2 *Gene Transfer*

Another way in which a genome can acquire new genes is by horizontal transfer from other species. While horizontal gene transfer is a very well-known mechanism of genetic material acquisition in the evolution of prokaryote genomes (Brown 2006), it has only recently been described as a way to shape metazoan genomes (Crisp et al. 2015). In bacteria, lateral gene transfer by plasmids is responsible for some cases of acquired antibiotic resistance. In animals, including humans, most of the transferred genes are from bacteria and the majority of the newly acquired genes are involved in metabolic pathways, suggesting that this mechanism may have contributed to biochemical diversification throughout evolution (Crisp et al. 2015).

2.3.2 Gene Loss

While Susumu Ohno pioneered the idea that gene duplication is essential to evolution, he also promoted the idea that a substantial fraction of the still-functioning mammalian genes are dispensable (Ohno 1985). He consequently argued that species with too accurate DNA polymerases and editing mechanisms cannot easily discard dispensable genes, and that the half-life of fish enzyme loci that are redundant and could be discarded is about 50 million years. Indeed, using CRISPR genome editing systems and a gene trap method in cancer cell lines, new studies have showed that approximately 90% of the genes are dispensable for cell proliferation and survival (Blomen et al. 2015; Wang et al. 2015).

Gene loss has classically been viewed as the loss of redundant gene duplicates without functional consequences, and hence without influence on the evolutionary process or cell communication. However, this may not be the whole picture. Olson formulated the 'less-is-more' hypothesis, in which he proposed that loss of function can be an adaptive response, usually under drastic shifts in environmental conditions (Olson 1999).

In the case of the human immunodeficiency virus type 1 (HIV–1), a loss-of-function mutation by deleting 32 base pairs of the C–C chemokine receptor type 5 gene (*CCR5*) confers resistance to AIDS (Dean et al. 1996). The gene loss of *CCR5* is positively selected, but this may be a result of other viruses' pressure, rather than HIV (Saxena, 2009). It thus seems reasonable to assume that, under certain conditions, loss of function may be advantageous and thus a useful tool for evolution. This implies that interrupting cellular communication may sometimes be beneficial, as in the case of CCR5 protein, which serves as a secondary receptor on $CD4^+$ T lymphocytes for certain strains of HIV–1.

Another loss-of-function mutation that has been suggested to be important for human evolution affected the myosin heavy chain 16 (*MYH16*). The mutation is estimated to have occurred approximately 2.4 mya, probably after a change in diet that did not require powerful masticatory jaw muscles. Thus, the proteomic distinction between humans and chimpanzees may be correlated with a traceable anatomic imprint of the jaw muscles in the fossil record (Stedman et al. 2004). The loss may also have allowed an increase in cranial capacity and brain size that occurred in the hominid lineage (Albalat and Canestro 2016). Other gene losses that proved beneficial in human evolution are the non-functionalisation of CMP-*N*-acetylneuraminic acid hydroxylase (*CMAH*) and *CASPASE12*, which confer resistance to some pathogens (Chou et al. 2002) and severe sepsis (Wang et al. 2006), respectively.

Together, these findings suggest that the loss of communication means may sometimes be beneficial for the organism.

2.4 Modules of Cellular Communication

Cellular communication enables a cell to receive and send information, a process that is advantageous to the cell or multicellular organism. The input can come from the cell's environment (endocrine, paracrine) or from itself (autocrine). Cellular communication is energy-consuming, which is why evolution selected a subset of inputs and outputs suitable for communication among the myriads of possible chemical and physical signals. Such signals are characterised by their energy (or amplitude or concentration), duration, and speed. The cell needs to filter (selectivity) and to decode (transduction) the input, and has to decide (intracellular signalling networks) whether a response is required. This pipeline seems to be universal and is the blueprint of all organismic communication systems. Physical inputs are electromagnetic energy and mechanical forces; chemical inputs are as diverse as the repertoire of different molecules. Communication systems are built in a way that signal inputs are computed convergently—different signals lead to the same response—or divergently—one signal is split into several responses. To realise communication, several conserved modules were established during evolution and were just adapted to the specific signal to which they respond. For example, ion channels are considered to mediate fast responses by changing ion concentrations. In humans, the neurotransmitter acetylcholine activates skeletal muscle contraction via a specific ion channel. In flies, this fast neuromuscular transmission is realised also via an ion channel, but the transmitter is glutamate (Collins and Diantonio 2007). In this example, the module 'ion channel' was selected from a bunch of neurotransmitter receptors to guarantee the required speed of communication. Signal specificity is mainly achieved at the receptor level; transducers decode and integrate signals of many receptors into metabotropic and ionotropic information and feed them into a complex network of cellular responses. There is a repertoire of such receptor, transducer, and signalling network modules that can be used in a block-wise manner to equip a cell. Although there are species-specific differences, most modules are universal and found in all eukaryotes. This communication inventory can be multiplied or reduced, making evolution faster. In principle, it seems that the number of components of cellular communication in pro- and eukaryotes increases relatively to genome size (Tamames et al. 1996; Konstantinidis and Tiedje 2004). However, this universality has disadvantages as well. Several bacterial, plant, and animal toxins (e.g. pertussis toxin, nicotine, and mastoparan) target such ubiquitous modules, equipping the toxin producer with species-independent weapons. It also teaches us that therapeutic specificity can mainly be achieved at the level of signal input.

It becomes more and more clear that the combinatorics of signalling components contributes to species differences. The quantitative and qualitative equipment with mainly receptor modules varies significantly among even closely related species and fine-tunes the fitness in their specific ecological niches. Some signalling components are absolutely essential and loss of function is not compatible with life. Such components are usually central signal transducers or involved in fundamental functions. The signal-transducing G proteins G_s and G_{13} are examples where a loss cannot be compensated by other components, leading to embryonal death (Offermanns 2001). In contrast, more than 45% of all G-protein-coupled receptor (GPCR) gene-deficient mice have no obvious phenotypic expression (Schoneberg et al. 2004), indicating that the receptor level contributes less to the organisms' fitness. The following subsections will give a short overview of the main receptor, transducer, and response modules necessary to understand how rapid changes in the communication system influence the fitness of modern humans.

2.4.1 Signalling Molecules

Cells in multicellular organisms communicate by means of many different kinds of molecules. These molecules serve as signal transmitters and exhibit considerable variation in their structure and function. Structure-wise, the signalling molecules range from amino acids, small peptides, or proteins, to steroids and fatty acid derivatives, and even nucleotides or dissolved gases such as nitric oxide and carbon monoxide. Most signalling molecules are secreted by the signalling cell into the extracellular space. Based on the distance over which the information is conveyed, there are generally three ways in which a signalling molecule can act:

- *endocrine signalling*, when stimulation happens over long distances;
- *paracrine signalling*, when neighbouring cells are stimulated;
- *autocrine signalling*, when the molecule stimulates the cell that produced it.

Signalling molecules also differ in the way they act on the target cells. In most cases, the signalling molecule acts extracellularly on receptors that are transmembrane proteins. However, sometimes the receptors are located inside the target cell, and the signal molecule has to enter the cell to activate them. This implies that the signalling molecule is hydrophobic in order to diffuse through the plasma membrane.

2.4.1.1 *Communication over Short or Long Distances*

Signal molecules may stay bound to the surface of the signalling cell and interfere only with cells that they come into direct contact with. This is usually the case with immune responses and antigen-presenting cells. However, signal molecules are usually secreted. A secreted molecule may act either on distant targets or as a local mediator in the immediate environment of the signalling cell (Figure 2.4). The latter represents paracrine signalling.

Paracrine signalling is a highly conserved way of communication among cells and it is usually implicated in developmental processes. At least four growth factor families, which are highly conserved among species, act in a paracrine fashion: the fibroblast growth factor (FGF) family, the hedgehog family, the Wnt family, and the TGF-β superfamily.

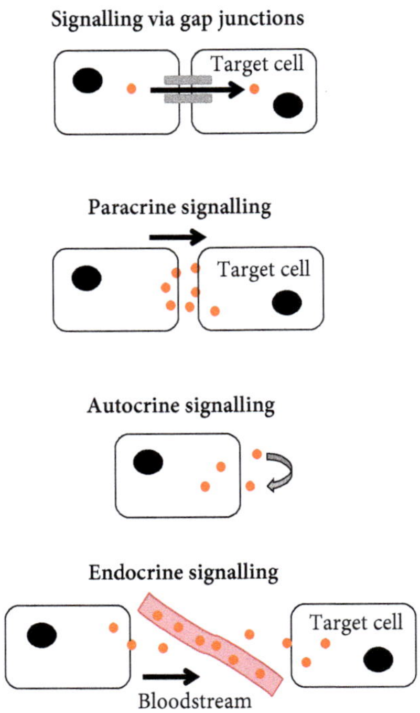

FIGURE 2.4 Ways of cellular communication. A cell can target another cell directly by gap junctions for, for example, ion exchange, a nearby cell (paracrine signalling), itself (autocrine signalling), or a distantly located cell using the circulatory system to transport signalling molecules (endocrine signalling).

FGFs have been implicated in limb patterning (Logan 2003), but they may also act on tumour cells, which aberrantly express FGF tyrosine kinase receptors that are paracrinely stimulated (Lappi 1995). Similarly, the hedgehog family is involved in digit patterning and the establishment of the body plan in organisms with bilateral symmetry (Lopez-Rios 2016). Paracrine signalling by Wnt is essential to multicellularity, since it allows a patterned spatial arrangement of cells and it also promotes cellular diversity or differentiation (Loh et al. 2016). The TGF-β superfamily of signalling molecules is also normally implicated in axial symmetry of the body and organ morphogenesis, while signalling disruption has been described in various human diseases ranging from cancer to chondrodysplasias and pulmonary hypertension (Attisano and Wrana 2002).

Nevertheless, for a multicellular organism, cells located in different parts of the body need to communicate in order to coordinate their actions. To this end, specialised cells have evolved to enable communication over long distances. One such type of cells are neurons, whose long processes called axons enable contact with target cells located far away. An active neuron sends electrical impulses (action potentials) at a speed of 100 m per second along the axon. When the impulse reaches the end of the axon, a chemical signal called a neurotransmitter is released. The location where the neurotransmitter is released is called a chemical synapse, and it is designed so to ensure that the signal is delivered specifically to the postsynaptic target cell. Nerve signalling has to be very precise in both time and space (Alberts et al. 2002).

Unlike neurons, endocrine cells perform long-distance signalling at relatively low speed. These cells secrete their signal molecules, called hormones, into the blood flow (Figure 2.4), so their signalling is also spatially less precise than neurotransmission. However, owing to dilution of hormone concentration in the circulating blood, hormones act at very low concentrations of $< 10^{-8}$ M, unlike neurotransmitters, which act at much higher concentrations: for example, acetylcholine in the neuromuscular junction acts at 5×10^{-4} M. This implies that hormone receptors must have a high affinity for their ligands (Alberts et al. 2002).

Autocrine signalling involves only one cell, which secretes signal molecules that bind back to its own receptors (Figure 2.4). This type of signalling is mainly important during development, helping cells to take on their functions and reinforce their identity. To this end, autocrine stimulation may be essential to the 'community effect' observed in early development. At that stage, a group of identical cells can respond to a differentiation factor, while a single isolated cell cannot. This implies that this mechanism enables cells of a kind to proliferate and make the same developmental decisions (Alberts et al. 2002). From a medical standpoint, autocrine signalling is important in cancer development, where tumour cells up-regulate growth and survival pathways through growth factors that bind back to the same cell's receptors (Grivennikov and Karin 2008). In this way, cancer cells can survive in places where normal cells of that type may not survive.

2.4.1.2 *Target Cell Stimulation on the Cell Surface or Inside*

The widest variety of signalling molecules in animals are peptides. Being mainly hydrophilic, they cannot cross the plasma membrane and thus have to stimulate the target cell on the surface. This group of signalling molecules includes peptide hormones, growth factors, neuropeptides, and neurotransmitters.

Neurotransmitters carry signals between neurons or from neurons to peripheral effector cells. This group of hydrophilic molecules includes acetylcholine, dopamine, epinephrine (adrenaline), serotonin, histamine, glutamate, glycine, and γ-aminobutyric acid (GABA). Neurotransmitters are released in the synaptic cleft and they target the postsynaptic cell. Although the system has been perfected in the nervous tissue of multicellular organisms, synaptic protein families are present in unicellular eukaryotes such as the yeast *Saccharomyces cerevisiae* and the amoeba *Dictyostelium discoideum* (Ryan and Grant 2009). The neurotransmitters diffuse across the synaptic cleft and bind to receptors on the surface of the cell, in many cases ligand-gated ion channels or GPCRs. Since neurotransmitters can usually reach high concentrations in the synaptic cleft, their receptors have a relatively low affinity, which means that the neurotransmitter can dissociate rapidly from the receptor to terminate and reinitiate the signalling. Moreover, neurotransmitters may also be actively terminated by enzyme degradation or their reuptake in the secreting neuron. The balance of the system has been fine-tuned over hundreds of millions of years of evolution (Ryan and Grant 2009). (For further discussion, see Chapter 17: Brain, Spinal Cord, and Sensory Systems.) The importance of the equilibrium is reflected in the pathology associated with the neurotransmitter imbalance. To this end, a disruption in the dopamine system results in Parkinson's disease, which is characterised by stiffness, tremors or shaking, and reduced motor control. Dopamine–glutamate interplay dysfunctions have been incriminated in many psychiatric disorders, including bipolar affective disorder and schizophrenia (Tomasetti et al. 2017). Similarly, drugs that inhibit serotonin reuptake from the synaptic cleft have been successfully used in the treatment of major depression (Jakobsen et al. 2017).

Unlike neurotransmitters, *neuropeptides* are larger molecules and they may also stimulate cells located further away, where they act as neurohormones. Examples are the enkephalins and endorphins. These peptides act on the same receptors as morphine, thus being analgesic by decreasing the pain responses in the central nervous system. The complete vertebrate opioid receptor system was established after some genome duplication events about 450 mya, already in the first jawed vertebrates (Dreborg et al. 2008). Enkephalins and endorphins play a plethora of roles, which include immune modulation (Wybran 1985), control of pituitary hormone release, paracrine stimulation in a variety of organs like pancreas and gonads, pubertal maturation, and reproduction (Petraglia et al. 1993). The long-standing presence of the opioid system during evolution, as well as its involvement in essential physiological functions like modulation of pain, response to stress, sexual behaviour, and food intake, suggests that imbalances in the system may result in pathological conditions. Indeed, endorphin withdrawal has been incriminated in premenstrual syndrome, postpartum depression (Halbreich and Endicott 1981), and major depression, in which endogenous opioid peptides may have some therapeutic benefit (Rouquier et al. 2000).

Nerve growth factor (NGF) was the first *polypeptide growth factor* to be discovered, by Rita Levi-Montalcini in the 1950s. It is a member of the neutrophin family that regulate not only nerve growth and survival, but also axial rotation during embryonic development (Manca et al. 2012). Following its discovery, other growth factors were characterised, including epidermal growth factor (EGF) and platelet-derived growth factor (PDGF), which play critical roles in controlling cell proliferation, both during embryonic development and in the adult organism, where they may promote wound healing. Other polypeptide growth factors, such as cytokines, regulate the differentiation of blood cells and modulate the activity of immune cells. These signalling molecules stimulate surface receptors, but some may act as membrane-anchored growth factors, rather than being secreted in the blood flow, and act as signalling molecules during direct cell–cell interactions. Imbalances in the signalling pathways of growth factors usually result in uncontrolled cell proliferation and tumours. EGF has been well characterised in the development of breast cancer and consequently new therapies are targeting its receptor (EGFR) (Manca et al. 2012). Autoimmune diseases such as inflammatory arthritis may arise from disruption of cytokine stimulation (Miossec 2017).

Another class of molecules that regulate inflammatory responses are the *eicosanoids*, which, unlike the polypeptide growth factors, have a lipid structure. Eicosanoids are synthesised from arachidonic acid and they include prostaglandins, prostacyclin, thromboxanes, and leukotrienes. Eicosanoids have been called 'the molecules of evolution' since they orchestrate a multitude of mechanisms, such as replication and reproduction, embryonic development, aging, and variation, and they also constitute a response system linking the environment to the internal machinery of cells (Lieb 2001). The synthesis of eicosanoids starts with the conversion of arachidonic acid to prostaglandin H_2, a reaction catalysed by an enzyme called cyclooxygenase. This enzyme is inhibited by aspirin and other non-steroidal anti-inflammatory drugs. Eicosanoids play very diverse roles in cell signalling, which may explain the very diverse effects of aspirin from reduction of inflammation and pain, inhibition of platelet aggregation, to cancer prevention (Mitrugno et al. 2017).

Signalling on the cell surface is also employed by *peptide hormones*, for example insulin, glucagon, and the hormones produced by the pituitary gland (which include, among others, growth hormone, follicle-stimulating hormone, and prolactin). These hormones are part of the endocrine systems present in animals. Apparently, the biochemical roots of the endocrine systems lie beyond the kingdom Metazoa, since communication between organs with different functions already necessitated some means of interaction to establish and maintain homeostasis (Kleine and Rossmanith 2016). As is usual in evolutionary terms, the older a system is, the better perfected and more sophisticated it gets, but also the worse are the disease states that arise from its malfunction. That is why modern medicine deals with many diseases that involve hormone signalling, from diabetes and hypertension to reproductive challenges.

The other type of hormones present in humans are *steroid hormones*. These are lipids synthesised from cholesterol and hence are lipophilic. Their ability to cross the plasma membrane allows binding to intracellular receptors. Steroid hormones are produced by the adrenal cortex (corticosteroids) or by the gonads and placenta (sex steroids). According to the receptors that they bind to, there are five classes of steroid hormones: glucocorticoids and mineralocorticoids (corticosteroids) and androgens, oestrogens, and progestogens (sex steroids). The steroid receptor family arose by a process of gene duplication and divergence. Later on, the androgen and progesterone receptors recruited as their ligands steroids that were intermediates in the biochemical reaction pathway of oestrogen synthesis (Eick and Thornton 2011). The pathologies associated with imbalances in steroid hormone production are medical textbook examples of diseases (Cushing's disease and Addison's disease). These usually dramatically affect the organism, from metabolic disorders to salt and water imbalance and disrupted sexual characteristics.

Even *gases* may act as signalling molecules. Gases are able to diffuse directly across the plasma membrane, but, unlike steroid hormones, they do not bind to receptors to regulate transcription, but rather to intracellular target enzymes. The simple gas nitric oxide (NO) is, for example, an important paracrine signalling molecule in the circulatory, nervous, and immune systems. NO can act only locally, because it is very unstable, with a half-life of only a few seconds. A well-known effect mediated by NO is vasodilation. To this end, NO enters smooth muscle cells, where it reacts with iron bound to the active site of guanylyl cyclase, which increases its activity and produces more cGMP (see Section 2.4.3.2). The final response is smooth muscle relaxation and vessel dilation. This is the mechanism that is the basis of penile erection. In the immune system, NO is produced by macrophages to kill bacteria. However, under certain circumstances, when the amount of NO produced is high following massive bacterial invasion (conditions that trigger sepsis, for example), systemic vasodilation can lead to a dramatic drop in blood pressure. In the brain, NO is involved in long-term potentiation that may contribute to learning and memory building (Hopper and Garthwaite 2006). Evolutionarily, NO is considered an ancient messenger molecule, being present in both prokaryotes and eukaryotes. Given the old presence of NO signalling, it may have acquired very diverse functions in different tissues. However, NO usually acts as a signalling molecule when the concentration of oxygen falls, such that in general NO 'ties the cell's different commitments to its metabolic budget' (Nathan 2004). NO promotes the formation of mitochondria and enhances the cell's capacity for oxidative metabolism (Nisoli et al. 2003), so disturbances in signalling can lead to mitochondrial damage and may result in neurodegenerative diseases (Bolanos and Heales 2010).

2.4.2 Receptors and Transducers

Although didactically the discrimination among receptors, transduces, and effectors helps to keep track of the signalling component, in many cases a single molecule can take over many functions in the signalling cascade (Figure 2.1). Since this chapter is dedicated to the impact of communication systems on human diseases, we focus only on the relevant molecular groups to understand the examples given below.

2.4.2.1 *G-Protein-Coupled Receptor Signalling*

The most successful signalling system is surely that of the GPCRs. Once introduced into the early eukaryotic genome, GPCRs evolved in various structural families, eventually resulting in hundreds of members in invertebrate and vertebrate genomes, including the human genome. The conserved molecular architecture formed of seven transmembrane helices and the highly preserved set of intracellular signal transducers have been retained over 1 billion years of eukaryotic evolution (Krishnan et al. 2012). Manifold combinations of amino acid residues within the transmembrane core and the loop regions have produced a versatile binding pocket for almost every natural compound that may serve as a signal. The balanced composition of conserved and variable structures is the key to the evolutionary success of GPCRs. Signals recognised by GPCRs include cations, amino acids, nucleotides, sugars, all kinds of metabolites, peptides, and proteins, but also physical signals such as photons and mechanical forces (Schoneberg et al. 2016). GPCRs convergently signal to transducers, called G proteins, which, when activated, initiate a variety of intracellular signalling cascades. The repertoire of GPCRs is adapted to the communicational needs of a given organism and can vary significantly among species. Humans have about 800 GPCR genes, more than 350 of which code for non-olfactory GPCRs (Gloriam et al. 2007). Their key roles in cellular communication have made them the target for more than 30% of all human drugs on the market (Lagerstrom and Schioth 2008). Mutations in GPCRs can cause more than thirty acquired and inherited diseases, such as retinitis pigmentosa, hypo- and hyperthyroidism, nephrogenic diabetes insipidus, several fertility disorders, and carcinomata (Schoneberg et al. 2004). Further, many toxins and addictive drugs, such as cannabinoids, opiates, LSD, psilocybin, and amphetamines, target GPCRs and their signalling pathways.

Termination of GPCR signalling occurs at different levels. Following activation, GPCRs are phosphorylated by specific kinases (GRKs) and uncoupled from the G protein by arrestins. Finally, the activated receptors become internalised, recycled, or degraded. This all terminates a signal and protects the cell from overstimulation. Continuous stimulation of a GPCR results in disappearance of the receptor protein, which we call in medicine 'tolerance development'. Permanent stimulation is not a situation a receptor-signalling system is made for, and the selective pressure on it is reduced when signal discrimination is no longer possible. As a virtual example, if all food were to taste bitter without being a risk to fitness, we would not taste it after a while and would lose this sense over generations.

Given their importance in communication, it is not surprising that variations in GPCRs have contributed to human evolution. Human migration into new environments essentially required the recognition of natural plant toxins through taste and smell, and more sensitive variants may have conferred an important selective advantage. Many bitter and specifically smelling glucopyranosides have a highly toxic cyanogenic activity. Signals of recent positive selection have been found for the bitter-taste receptor gene *TAS2R16*, harbouring a non-synonymous site K172N that makes this bitter-taste receptor more sensitive for cyanogenic glycosides (Soranzo et al. 2005). In contrast, many genes of the human *TAS2R* repertoire show relaxed constraints and even loss-of-function mutations suggesting, together with previous findings of large reduction in the human odorant receptor repertoire, reduced sensory capabilities of humans in comparison with many other mammals (Wang et al. 2004). Well-sorted and preselected food in our supermarkets labelled with expiry dates will further reduce the selective pressure to rely on taste and olfaction, further reducing our repertoire of sensory GPCRs. Indeed, 10% of the human population cannot smell hydrogen cyanide, which might have exerted a selective disadvantage in ancestral environments (Gross 2007).

Global variation in skin and hair pigmentation is one of the most striking examples of variation in humans. Over 200 genes and genomic loci have been identified as being involved in pigmentation, among them the melanocortin type 1 receptor—a GPCR that controls melanogenesis (Hofreiter and Schoneberg 2010). Unlike many other genomic loci exhibiting higher diversity in African populations, *MC1R* gene diversity is highest in Eurasian populations (Rana. et al. 1999; Harding et al. 2000), suggesting strong purifying selection in Africa. The higher frequency of loss-of-function mutations in European populations has been discussed as a loss of constraint because of lower UV-B light exposition. Selection of low-function MC1R variants is also discussed as giving advantage for 25-hydroxyvitamin D synthesis and reduced energy waste owing to lower pigment synthesis. However, red-haired or pale Europeans migrating to regions with higher UV index like Australia have a higher risk of skin cancer, including melanoma (Macgregor et al. 2011). This example provides an impressive demonstration of the way in which the rapid migration and exposure to other environments that come with civilisation lead to overwriting of long adaptation processes and a consequent increase in otherwise rare diseases.

2.4.2.2 *Receptor Tyrosine Kinases*

Growth factors, cytokines, and several hormones realise their signalling via receptor tyrosine kinases (RTKs). Most RTKs are homodimeric receptors, although some (e.g. the insulin receptor) exist as multimeric complexes. Each monomer consists of an extracellular ligand-binding domain, a transmembrane domain, and an intracellular tyrosine kinase domain. Upon extracellular ligand binding, receptor dimerisation is stabilised and triggers reciprocal *trans*-phosphorylation of a tyrosine residue in the cytoplasmic portion of each receptor monomer. This recruits adaptor proteins, finally leading to activation of mitogenic signalling cascades and phospholipase C (Lemmon and Schlessinger 2010). RTKs are mainly involved in paracrine communication tuning cell growth, development, and differentiation. Classic RTKs appear to be present only in animals; however, they share some structural relation to other plasma membrane receptor kinases also found in plants, suggesting a common evolutionary origin before the plant–animal split (Shiu and Bleecker 2001). As for GPCRs, RTK signalling systems are fine-tuned for an organism's needs and, for example, permanent stimulation of the insulin receptor by continuous nutritional-triggered release of insulin leads to first adaptive and later to manifest insulin resistance.

2.4.2.3 *Ligand-Gated Ion Channels*

One of the first evolutionary steps in establishing life included the exchange of ions between the environment and the cellular interior via ion channels and transporters. Transmembrane-spanning peptide helices and their self-assembly into a helical bundle, which can form a pore within the membrane, was an important evolutionary invention. At one point in early prokaryotic life, this transmembrane ion exchange became regulated by molecules that bind to the ion channel and modulate its open/close probability by dynamic rearrangement of the helix bundle. This simple principle is most probably the prototype of the first receptor in life's history, propagating in multiple forms of ligand-gated ion channels and ion pumps (Chen et al. 1999; Tasneem. et al. 2005; Jaiteh et al. 2016). One of the first signals regulating ion transfer through membranes were photons. Light-driven proton pumps are found in the halophilic archaea and marine bacteria that formed the molecular basis of phototrophy in early life on Earth (Beja et al. 2000). Most fast synaptic transmissions are mediated by ligand-gated ion channels. For example, both neuro–neuro and neuromuscular signal transduction use nicotinic acetylcholine receptors, which are ligand-gated sodium channels. Since ligand-gated ion channels are universal and important for instant communication, they are targeted by many toxins. For example, a number of plant (e.g. curare), frog (e.g. epibatidine), snake (e.g. viper peptides), and snail (e.g. α-conotoxins) toxins act on nicotinic acetylcholine receptors (Kudryavtsev et al. 2015), immediately paralysing the victims.

Reflecting their physiological importance, genetic variants in ion channels and transporters have been assigned to human evolution and diseases. Numerous studies have extracted signatures of selection in those genes related to recent human evolution. For example, variants in the calcium ion channel TRPV6 show signatures of selections in the European population. The date of selection in Europeans was estimated to be around 10,000 years before present, suggesting that TRPV6 may have coevolved with lactose tolerance, since fresh milk is a major source of calcium in European populations (Hughes et al. 2008).

Mutations in the epithelial chloride ion ATP-binding cassette (ABC) transporter, better known as the cystic fibrosis transmembrane conductance regulator (CFTR), are responsible for cystic fibrosis. The variant ΔF508 is very frequent in the European population (3–4%) and disease-causing when homozygous. Several studies provide evidence of recent selection of ΔF508 (Wiuf 2001) and it has been hypothesised that heterozygote individuals may have some advantage compared with the wild-type variant, for example with regard to resistance to asthma (Schroeder et al. 1995), dehydrating pathogens like *Vibrio cholerae* (Bertranpetit and Calafell 1996), pathogens from domesticated animals (Alfonso-Sanchez et al. 2010), or climate change (Borzan et al. 2014).

2.4.2.4 *Nuclear Hormone Receptors*

A number of lipophilic molecules can cross cell membranes and interact with intracellular hormone receptors, including steroids, retinoic acid, vitamins, and thyroid hormone. Steroid receptors are usually found in the cytoplasm in association with so-called heat shock proteins. Upon steroid binding, these receptor proteins dimerise, translocate into the nucleus, and, as transcription factors, initiate complex transcription programs. In contrast, receptors for vitamin D_3 and the thyroid hormone T_3 usually reside already in the nucleus and initiate their transcription programs upon ligand binding without translocation. Being directly or indirectly involved in the expression regulation of hundreds of genes, they orchestrate complex physiological pathways such as metabolic homeostasis, development, and detoxification. Nuclear hormone receptors are evolutionarily highly conserved, and there is little evidence for positive selection (Krasowski et al. 2005). Given the complexity of their actions, it is not surprising that ligands binding to nuclear receptors are among the most frequently prescribed drugs (cortisol, thyroxin, and oestrogen).

Besides the classical nuclear hormone receptors, there are nuclear receptors such as retinoid X receptor (RXR), peroxisome proliferator-activated receptors (PPARs), the pregnane and xenobiotic receptor (PXR), the constitutive androstane receptor (CAR), the liver X receptor (LXR), the aryl hydrocarbon receptor (AHR), and the farnesoid X receptor (FXR), which are involved in individual adaptation to nutrients and xenobiotics. For example, upon binding of fatty acids to PPARα, the receptor can heterodimerise with the activated RXR, with the complex then acting as a transcription factor regulating adaptation of liver metabolism depending on food intake and composition. Therefore, these nuclear receptors are prone to contribute to environmental species adaptation. Indeed, several haplotypes of PPARα are associated with high-altitude adaptation in Tibet and Ethiopia (Simonson et al. 2010; Scheinfeldt et al. 2012). It has been speculated that the metabolic demands imposed by high-altitude-dependent hypoxia shift fuel preference to glucose oxidation and glycolysis at the expense of fatty acid oxidation, which would improve adaptation to decreased oxygen availability (Ge et al. 2015). It is therefore very likely that humans developed specific subsets of nuclear receptor variants adapted to their specific xenobiotic environments.

2.4.3 Second Messengers and Effectors

Ions and metabolites are the 'currency' a cell can work with. Therefore, cells use intracellular concentrations of selected ions and metabolites as measures for extracellular signals. For ions, concentration gradients between extra- and intracellular are the driving forces, and ATP-consuming pumps regenerate these gradients after signal-induced changes (so-called ionotropic signal transduction). In the case of metabolites, the extracellular signal needs to be transduced to an enzyme generating or degrading the metabolite (so-called metabotropic signal transduction). However, in both, ionotropic and metabotropic signalling, the signal itself is encoded as a concentration difference between ions and/or metabolites.

2.4.3.1 *Calcium Ions—the Universal Signalling Molecule*

Calcium ions are ubiquitous and abundant in nature. Normally, the calcium ion concentration in the cell (100 nM) is significantly lower than that in the environment. The different intra- and extracellular calcium ion concentrations are tightly regulated by intracellular ion pumps (sarco/endoplasmic reticulum Ca^{2+}-ATPase, SERCA) and stores (endoplasmic reticulum and mitochondria) and extracellular hormones (parathyroid hormone (PTH), calcitonin, and calcitriol) in their respective target organs (kidney, bowel, and bone). This calcium gradient is used to amplify outside and inside signals and to change the functions of intracellular effector molecules. Thus, at certain intracellular concentrations, calcium ions can bind to proteins, subsequently causing massive changes in their three-dimensional protein structure. Binding of calcium cations is mediated mainly by the negatively charged carboxyl side chains of the amino acids glutamate and aspartate. There are also specific structures, the so-called EF-hand motif, coordinating calcium ions in a protein. The EF-hand motif is a helix–loop–helix structure and is found in a large family of calcium-binding proteins from bacteria to vertebrates. The most prominent example is troponin C, where calcium binding to its EF hands initiates muscle contraction through actin–myosin interaction. As a universal signalling molecule, changes in calcium ion balance ultimately lead to severe consequences and even death. Numerous biogenic toxins directly or indirectly act on the cellular calcium ion homeostasis—examples are thapsigargin and the digitalis glycosides.

2.4.3.2 *Cyclic Nucleotides—Success of Simplicity*

Nucleotides and their phosphates (e.g. ATP and GTP) are among the oldest biomolecules in living nature, forming the structural basis for RNA and DNA. They are useful not only to store information and energy but also as signals and metabotropic molecules in signalling pathways. Cyclic nucleotides, such as cAMP and cGMP, are simple intramolecular phosphodiesters generated by nucleotide triphosphate cyclases (Figure 2.5). There are membrane-anchored and soluble cyclases, which can be further divided into guanylyl and adenylyl cyclases, generating cGMP and cAMP, respectively.

Over the evolution of life, a great number of cyclases evolved, being present in bacteria, fungi, plants, and animals (Gancedo 2013). It has even been speculated that soluble adenylyl cyclases were introduced into eukaryotes by endosymbiosis (see Section 2.1) (Blackstone 2014). Cyclases are regulated by a variety of molecules, such as calmodulin, G proteins, nitric oxide (NO), calcium, and bicarbonate (Steegborn 2014). The generated cAMP and cGMP act as intracellular signalling molecules by interacting with kinases and cyclic nucleotide monophosphate-regulated ion channels and triggering numerous cellular programs. cAMP and cGMP levels are also controlled by the activity of phosphodiesterases, enzymes

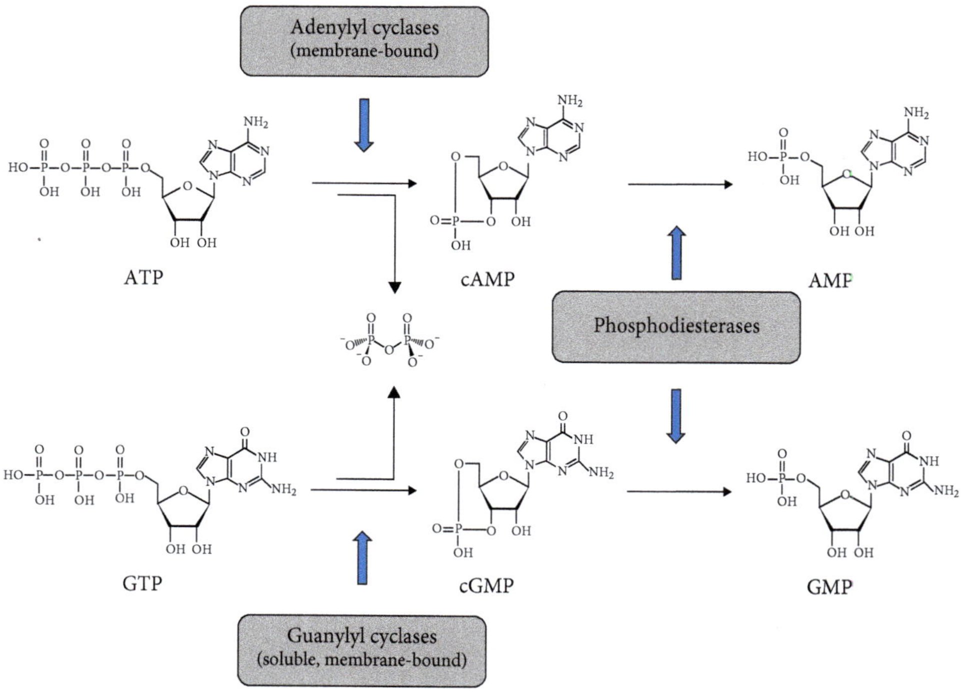

FIGURE 2.5 Cyclic nucleotides as universal metabotropic signalling molecules. Cyclic nucleotides (cAMP and cGMP) are produced by membrane-anchored and soluble cyclases from ATP and GTP. Phosphodiesterases degrade the cyclic nucleotides to AMP and GMP.

that hydrolyse cyclic nucleotide monophosphates. Being universal intracellular second messengers, a number of drugs and toxins target the homoeostasis of cAMP and cGMP. In almost every culture, humans use plant extracts, such as coffee, tea, mate, cola, cacao, and guaraná, containing methylxanthines (e.g. caffeine) to increase cAMP and cGMP levels via non-specific phosphodiesterase inhibition at higher concentrations. Inhibition of distinct cyclases is useful in treating erectile dysfunction (sildenafil) and asthma and chronic obstructive pulmonary disease (theophylline) (Boswell-Smith et al. 2006). Forskolin, a natural compound of *Plectranthus barbatus*, activates adenylyl cyclases and was used in native African and Asian cultures to treat cardiac and neurological diseases, although at present forskolin is used to reduce fat deposits (Godard et al. 2005). However, the cAMP signalling system is also a target for effective toxins. Cholera toxin activates the G_s protein, leading to activation of adenylyl cyclases and subsequently increasing cAMP with the well-known consequences for the gastrointestinal tract (Sack et al. 2004). *Bordetella pertussis* (a Gram-negative bacterium that causes whooping cough) and *Bacillus anthracis* (a Gram-positive bacterium that causes anthrax) even express adenylyl cyclases as components of their exotoxins, which cause disease pathology by production of high cAMP levels in the host organism (Baker and Kelly 2004).

As briefly reviewed here, cyclic nucleotide monophosphates are evolutionarily old signalling molecules, being relevant in almost all pro- and eukaryotes. This conserved nature helps to understand human pathologies but also makes it difficult to interfere directly in this signalling system for specific therapeutic purposes. Therefore, most of our current drugs targeting cyclic nucleotide monophosphate levels act indirectly via receptors and enzymes that modulate cAMP or cGMP levels—just to name a few: epinephrine, beta-adrenoceptor agonists and blockers, dopamine receptor agonists and antagonists, and nitroglycerine.

2.5 Does Noise Matter in Cellular Communication?

Noise is generally defined as random fluctuations of a signal that hinder perception of an expected informative signal. In cellular communication, noise occurs owing to stochastic changes between different structural stages—for example, in an enzyme between the active and inactive conformations or in an ion channel between the open and closed stages. As a result, levels of released neurotransmitters or second messengers are never zero. The probability of the stochastic appearance of functionally different stages is determined by many factors, such as temperature, solvents, and radiation, but also by the structure of the biomolecule. Natural selection can influence the level of noise produced by a signalling component at the level of its structural stability. If we consider 'noise' as the remnant of an energetic trade-off between the energy costs of allowing for an informative signal and the need to reduce the background signal due to random fluctuations, the answer is: 'noise matters'. Even if this sounds paradoxical, noise helps to save energy. In the 'real world', maximum power can be obtained faster from a warm car engine than from a cold one, because of the need to overcome the moment of inertia in the latter case. We all know that even a little bit of a warm-up aids sporting performance. Most predators have a body temperature higher than that of the environment. So, regulation in nature usually does not work in a simple binary 0 and 1 mode; rather, biological responses are determined by the relative frequency of the two stages 0 and 1. It is well established that the open/closed probabilities of ion channels contribute to the level of cellular membrane potentials. Ligands bound to ion channels just change the open/closed probability, and therefore the membrane potential sometimes reaches a certain threshold that depolarises the cell (the action potential). Although this follows an 'all or nothing principle', the effect on the organism is the result of the integrated sum of the number of such events over a given time.

An examination of thyroid function may illustrate how and why signalling noise matters at the physiological and pathophysiological level. Thyroid function is essential for life and is controlled by the pituitary hormone TSH acting on the thyroid TSH receptor. To maintain basal vital function, a certain amount of the thyroid hormones must be released from the thyroid. One regulatory option would be to continuously secrete TSH from the pituitary, but this would be energy-consuming. The second option is to increase the activity probability of the TSH receptor by structural adjustment (mutations) of the receptor protein. Indeed, a significant number of TSH receptor molecules display an active conformation even without TSH. This physiological threshold was selected during evolution and fixed in the amino acid structure of the receptor. Changes in the environment—for example changes of habitat temperature because of housing—conflict with this laboriously acquired threshold. Mutational activation and inactivation mirror impressively the effects on body temperature in hyper- and hypothyroidism, respectively (Gruters et al. 1999).

Thresholds and a certain basal activity can also be seen at more complex levels such as behaviour. The balance of escape (anxiety) and attraction (tameness) is a behavioural trade-off and varies individually (Albert et al. 2012). Recent genetic studies have revealed a genetic basis for the intraspecies variability that most probably detrains different thresholds (e.g. in the endocrine stress-response system) (Heyne et al. 2014). Changes in social structures can require changes in such thresholds. However, the threshold between noise and signal is genetically fixed by evolutionary selection, and this requires time and a number of generations. Obviously, conflicts between the environment and genetically fixed thresholds can cause diseases.

To the best of our knowledge, there is only one receptor/transducer system without significant noise, namely vision triggered by rhodopsin. With eyes closed—so that one sees just black—noise in the photosensory system will matter in the dark. Classically, activation of a signal transduction system leads to a gain of downstream signalling molecules (increases in cAMP, calcium ions, and phosphorylation). This almost completely noiseless photosystem is realised by an inverse signal transduction where activation of rhodopsin leads to a reduction in cGMP—a loss of a signalling molecule. The reduced cGMP levels lead to a decreased potassium current and subsequently to decreased concentrations of intracellular calcium ions and a reduction in glutamate concentration. Put simply, rhodopsin inactivity causes constant glutamate secretion, and rhodopsin activity turns it off. Inverse signalling seems to be advantageous to ensure a low signal-to-noise ratio, although constant release of glutamate from the inactive photoreceptor appears to be more energy-consuming than triggering the release upon activation. Even minor changes in this system eventually result in diseases such as retinitis pigmentosa. Interestingly, if one compares the amino acid conservation between non-opsin GPCRs and rhodopsin, rhodopsin is much more strongly conserved in vertebrate evolution than most other GPCRs (Schulz and

Schoneberg, 2003). This much stronger constraint at the genetic level (which keeps rhodopsin without noise) is highly energy-consuming. Loss of constraint—for example, by being trapped in a cave—leads to loss of vision (Niemiller et al. 2013; Cartwright et al. 2017).

In sum, noise matters in biological networks and represents a trade-off with respect to biophysical properties of biomolecules, to the energy needed to reduce noise in order to filter informative signals, and to environmentally adapted thresholds allowing the species to perform in the ecological niche. Rapid environmental changes can cause conflicts such that the threshold is insufficient for organismal fitness—which we call disease.

2.6 Medical Consequences of Conserved Signalling Systems

Cellular recognition and computing of signals is evolutionarily ancient and a prerequisite for multicellularity. Signals orchestrate the specific function of cells in a multicellular organism via conserved signalling pathways. But signals also suppress the 'selfishness' of the individual cell, for example, by inhibiting uncontrolled proliferation. Loss of signalling capacities usually reduces fitness and can cause disease, but it can also be advantageous in the case that specific informational input is no longer required or overloads the organism. Comparison of signalling systems throughout living species clearly reveals that humans are not special with regard to extra- and intra-organismal communication systems. It is rather the combination of signalling systems, minor differences in signal-to-noise ratios, and the supramolecular network and communication structure that defines and adapts a species to its ecological niche. In this game, evolution plays with gene numbers and rearrangement of conserved functional domains that can be fine-tuned by mutations. This reveals signalling components to be highly relevant for adaptation, but it also makes a species vulnerable to rapid environmental changes, which cannot be compensated by the relatively slow genetic processes. Therefore, many common 'diseases of civilisation' can be considered as resulting from a conflict between the evolutionary status quo of recent humans and the mostly human-made environmental changes of the past 150 years. In almost all of these 'modern' diseases, signalling networks have failed to adapt to environmental changes—examples include obesity (leptin–melanocortin-regulated appetite control) and hypertension (increased sympathetic tone, and salt load).

The universality of signalling systems is a threat, but also a gift for human health. Biogenic toxins often act on a multitude of species to protect the producer against a broad range of potential predators or to paralyse and kill prey of different size and origin. This is only possible because of conserved signalling molecules and pathways. A number of pathogens use conserved signalling machinery to enter, reprogram, and even kill their hosts. Moreover, bacteria and viruses incorporate signalling molecules of their hosts into their genomes to protect themselves or to control function of the host cells. However, the universality of many signalling systems has given rise to a wealth of natural compounds that influence signalling. Most of the lead structures of our current drugs are still based on natural compounds. During human history, we have mainly mined nature's chemical laboratory because the toolboxes, selection processes, and the time that nature has had available are far beyond our own current capabilities. The low number of exclusively synthetic or rationally designed drugs that have made it into clinical use supports this fact. Finally, conservation of signalling processes allows us to use non-human species to test the effectiveness of therapeutic approaches—a fact that made modern medicine possible.

References

Albalat, R., and Canestro, C. (2016). Evolution by gene loss. *Nat Rev Genet* 17, 379–91.

Albert, F. W., Somel, M., Carneiro, M., et al. (2012). A comparison of brain gene expression levels in domesticated and wild animals. *PLoS Genet* 8, e1002962.

Alberts, B., Johnson, A., Lewis, J., et al. (2002). *Molecular Biology of the Cell*, 4th ed. New York: Garland Science.

Alfonso-Sanchez, M. A., Perez-Miranda, A. M., Garcia-Obregon, S., et al. (2010). An evolutionary approach to the high frequency of the Delta F508 CFTR mutation in European populations. *Med Hypotheses* 74, 989–92.

Attisano, L. and Wrana, J. L. (2002). Signal transduction by the TGF-beta superfamily. *Science* 296, 1646–7.

Baker, D. A. and Kelly, J. M. (2004). Structure, function and evolution of microbial adenylyl and guanylyl cyclases. *Mol Microbiol* 52, 1229–42.

Beja, O., Aravind, L., Koonin, E. V., et al. (2000). Bacterial rhodopsin: evidence for a new type of phototrophy in the sea. *Science* 289, 1902–6.

Bennett, M. D. and Leitch, I. J. (2005). Genome size evolution in plants. In: Gregory, T. R. (ed.) *The Evolution of the Genome*. Amsterdam: Elsevier Academic Press, pp. 89–162.

Bertranpetit, J. and Calafell, F. (1996). Genetic and geographical variability in cystic fibrosis: evolutionary considerations. *Ciba Found Symp* 197, 97–114; discussion 114–18.

Bird, A. P. (1995). Gene number, noise reduction and biological complexity. *Trends Genet* 11, 94–100.

Blackstone, N. W. (2014). sAC as a model for understanding the impact of endosymbiosis on cell signaling. *Biochim Biophys Acta* 1842, 2548–54.

Blomen, V. A., Majek, P., Jae, L. T., et al. (2015). Gene essentiality and synthetic lethality in haploid human cells. *Science* 350, 1092–6.

Bolanos, J. P. and Heales, S. J. (2010). Persistent mitochondrial damage by nitric oxide and its derivatives: neuropathological implications. *Front Neuroenergetics* 2, 1.

Bonner, J. T. (2003). On the origin of differentiation. *J Biosci* 28, 523–8.

Borzan, V., Tomasevic, B., and Kurbel, S. (2014). Hypothesis: possible respiratory advantages for heterozygote carriers of cystic fibrosis linked mutations during dusty climate of last glaciation. *J Theor Biol* 363, 164–8.

Boswell-Smith, V., Spina, D., and Page, C. P. (2006). Phosphodiesterase inhibitors. *Br J Pharmacol*, 147(Suppl 1), S252–7.

Brown, T. A. (2006). *Genomes*. New York: Garland Science.

Burki, F. and Kaessmann, H. (2004). Birth and adaptive evolution of a hominoid gene that supports high neurotransmitter flux. *Nat Genet* 36, 1061–3.

Cartwright, R. A., Schwartz, R. S., Merry, A. L., et al. (2017). The importance of selection in the evolution of blindness in cavefish. *BMC Evol Biol* 17, 45.

Carvalhaes, C. G., Cayô, R., and Gales, A. C. (2013). *Klebsiella pneumoniae* carbapenemase-producing *Klebsiella pneumoniae* in the intensive care unit: a real challenge to physicians, scientific community, and society. *Shock* 39, 32–7.

Chan, K. G., Wong, C. S., Yin, W. F., et al. (2010). Rapid degradation of *N*-3-oxo-acylhomoserine lactones by a *Bacillus cereus* isolate from Malaysian rainforest soil. *Antonie Van Leeuwenhoek* 98, 299–305.

Chapuisat, M. (2009). Social evolution: the smell of cheating. *Curr Biol*, 19, R196–8.

Chen, G. Q., Cui, C., Mayer, M. L., et al. (1999). Functional characterization of a potassium-selective prokaryotic glutamate receptor. *Nature* 402, 817–21.

Chou, H. H., Hayakawa, T., Diaz, S., et al. (2002). Inactivation of CMP-*N*-acetylneuraminic acid hydroxylase occurred prior to brain expansion during human evolution. *Proc Natl Acad Sci U S A* 99, 11736–41.

Collins, C. A. and Diantonio, A. (2007). Synaptic development: insights from *Drosophila*. *Curr Opin Neurobiol* 17, 35–42.

Crisp, A., Boschetti, C., Perry, M., et al. (2015). Expression of multiple horizontally acquired genes is a hallmark of both vertebrate and invertebrate genomes. *Genome Biol* 16, 50.

Cronin, T. W. and Porter, M. L. (2014). The evolution of invertebrate photopigments and photoreceptors. In: Hunt D., Hankins M., Collin S., et al. (eds). *Evolution of Visual and Non-Visual Pigments*. Boston: Springer, pp. 105–35.

Davies, P. C. and Lineweaver, C. H. (2011). Cancer tumors as Metazoa 1.0: tapping genes of ancient ancestors. *Phys Biol* 8, 015001.

Dawkins, R. (2016). *The Selfish Gene*, 40th anniversary edition. Oxford: Oxford University Press.

Dean, M., Carrington, M., Winkler, C., et al. (1996). Genetic restriction of HIV-1 infection and progression to AIDS by a deletion allele of the CKR5 structural gene. Hemophilia Growth and Development Study, Multicenter AIDS Cohort Study, Multicenter Hemophilia Cohort Study, San Francisco City Cohort, ALIVE Study. *Science* 273, 1856–62.

De Spiegeleer, B., Verbeke, F., D'hondt, M., et al. (2015). The quorum sensing peptides PhrG, CSP and EDF promote angiogenesis and invasion of breast cancer cells in vitro. *PLoS One* 10, e0119471.

Devlin, R. H., Holm, D. G., and Grigliatti, T. A. (1988). The influence of whole-arm trisomy on gene expression in *Drosophila*. *Genetics* 118, 87–101.

Dickinson, D. J., Nelson, W. J., and Weis, W. I. (2012). An epithelial tissue in *Dictyostelium* challenges the traditional origin of metazoan multicellularity. *Bioessays* 34, 833–40.

Dreborg, S., Sundstrom, G., Larsson, T. A., et al. (2008). Evolution of vertebrate opioid receptors. *Proc Natl Acad Sci U S A* 105, 15487–92.

Du, Q., Kawabe, Y., Schilde, C., et al. (2015). The evolution of aggregative multicellularity and cell–cell communication in the Dictyostelia. *J Mol Biol* 427, 3722–33.

Eick, G. N. and Thornton, J. W. (2011). Evolution of steroid receptors from an estrogen-sensitive ancestral receptor. *Mol Cell Endocrinol* 334, 31–8.

Gancedo, J. M. (2013). Biological roles of cAMP: variations on a theme in the different kingdoms of life. *Biol Rev Camb Philos Soc* 88, 645–68.

Ge, R. L., Simonson, T. S., Gordeuk, V., et al. (2015). Metabolic aspects of high-altitude adaptation in Tibetans. *Exp Physiol* 100, 1247–55.

Gloriam, D. E., Fredriksson, R., and Schioth, H. B. (2007). The G protein-coupled receptor subset of the rat genome. *BMC Genomics* 8, 338.

Godard, M. P., Johnson, B. A., and Richmond, S. R. (2005). Body composition and hormonal adaptations associated with forskolin consumption in overweight and obese men. *Obes Res* 13, 1335–43.

Gregory, T. (2005). *Genome Size Evolution in Animals*, Vol. 1. London: Elsevier Academic Press.

Grimson, M. J., Coates, J. C., Reynolds, J. P., et al. (2000). Adherens junctions and β-catenin-mediated cell signalling in a non-metazoan organism. *Nature* 408, 727–31.

Grivennikov, S. and Karin, M. (2008). Autocrine IL-6 signaling: a key event in tumorigenesis? *Cancer Cell* 13, 7–9.

Grosberg, R. K. and Strathmann, R. R. (2007). The evolution of multicellularity: a minor major transition? *Annu Rev Ecol Evol Syst* 38, 621–54.

Gross, L. (2007). A genetic basis for hypersensitivity to 'sweaty' odors in humans. *PLoS Biol* 5, e298.

Gruters, A., Krude, H., Biebermann, H., et al. (1999). Alterations of neonatal thyroid function. *Acta Paediatr Suppl* 88, 17–22.

Halbreich, U. and Endicott, J. (1981). Possible involvement of endorphin withdrawal or imbalance in specific premenstrual syndromes and postpartum depression. *Med Hypotheses* 7, 1045–58.

Harding, R. M., Healy, E., Ray, A. J., et al. (2000). Evidence for variable selective pressures at MC1R. *Am J Hum Genet* 66, 1351–61.

Hassold, T., Merrill, M., Adkins, K., et al. (1995). Recombination and maternal age-dependent nondisjunction: molecular studies of trisomy 16. *Am J Hum Genet* 57, 867–74.

Heyne, H. O., Lautenschlager, S., Nelson, R., et al. (2014). Genetic influences on brain gene expression in rats selected for tameness and aggression. *Genetics* 198, 1277–90.

Hofreiter, M. and Schoneberg, T. (2010). The genetic and evolutionary basis of colour variation in vertebrates. *Cell Mol Life Sci* 67, 2591–603.

Hopper, R. A. and Garthwaite, J. (2006). Tonic and phasic nitric oxide signals in hippocampal long-term potentiation. *J Neurosci* 26, 11513–21.

Hughes, D. A., Tang, K., Strotmann, R., et al. (2008). Parallel selection on *TRPV6* in human populations. *PLoS One* 3, e1686.

Jaiteh, M., Taly, A., and Henin, J. (2016). Evolution of pentameric ligand-gated ion channels: pro-loop receptors. *PLoS One* 11, e0151934.

Jakobsen, J. C., Katakam, K. K., Schou, A., et al. (2017). Selective serotonin reuptake inhibitors versus placebo in patients with major depressive disorder. A systematic review with meta-analysis and Trial Sequential Analysis. *BMC Psychiatry* 17, 58.

John, J., Saranathan, R., Adigopula, L. N., et al. (2016). The quorum sensing molecule *N*-acyl homoserine lactone produced by *Acinetobacter baumannii* displays antibacterial and anticancer properties. *Biofouling* 32, 1029–47.

Joyce, G. F. (2002). The antiquity of RNA-based evolution. *Nature* 418, 214–21.

Kessin, R. H. (2001). *Dictyostelium: Evolution, Cell Biology, and the Development of Multicellularity.* Cambridge: Cambridge University Press.

Khambati, I., Han, S., Pijnenburg, D., Jang, H., et al. (2017). The bacterial quorum-sensing molecule, *N*-3-oxo-dodecanoyl-L-homoserine lactone, inhibits mediator release and chemotaxis of murine mast cells. *Inflamm Res* 66, 259–68.

Kleine, B. and Rossmanith, W. G. (2016). Evolution of the endocrine system. In: Kleine, B. and Rossmanith, W. G. (eds) *Hormones and the Endocrine System: Textbook of Endocrinology*. Cham: Springer, pp. 347–53.

Koch, A. L. and Silver, S. (2005). The first cell. *Adv Microb Physiol* 50, 227–59.

Konstantinidis, K. T. and Tiedje, J. M. (2004). Trends between gene content and genome size in prokaryotic species with larger genomes. *Proc Natl Acad Sci U S A* 101, 3160–5.

Krasowski, M. D., Yasuda, K., Hagey, L. R., et al. (2005). Evolutionary selection across the nuclear hormone receptor superfamily with a focus on the NR1I subfamily (vitamin D, pregnane X, and constitutive androstane receptors). *Nucl Recept* 3, 2.

Krishnan, A., Almen, M. S., Fredriksson, R., et al. (2012). The origin of GPCRs: identification of mammalian like *Rhodopsin, Adhesion, Glutamate* and *Frizzled* GPCRs in fungi. *PLoS One* 7, e29817.

Kudryavtsev, D., Shelukhina, I., Vulfius, C., et al. (2015). Natural compounds interacting with nicotinic acetylcholine receptors: from low-molecular weight ones to peptides and proteins. *Toxins (Basel)* 7, 1683–701.

Lagerstrom, M. C. and Schioth, H. B. (2008). Structural diversity of G protein-coupled receptors and significance for drug discovery. *Nat Rev Drug Discov* 7, 339–57.

Lappi, D. A. (1995). Tumor targeting through fibroblast growth factor receptors. *Semin Cancer Biol* 6, 279–88.

Lemmon, M. A. and Schlessinger, J. (2010). Cell signaling by receptor tyrosine kinases. *Cell* 141, 1117–34.

Lerat, E. and Moran, N. A. (2004). The evolutionary history of quorum-sensing systems in bacteria. *Mol Biol Evol* 21, 903–13.

Lieb, J. (2001). Eicosanoids: the molecules of evolution. *Med Hypotheses* 56, 686–93.

Logan, M. (2003). Finger or toe: the molecular basis of limb identity. *Development* 130, 6401–10.

Loh, K. M., Van Amerongen, R., and Nusse, R. (2016). Generating cellular diversity and spatial form: Wnt signaling and the evolution of multicellular animals. *Dev Cell* 38, 643–55.

Lopez-Rios, J. (2016). The many lives of SHH in limb development and evolution. *Semin Cell Dev Biol* 49, 116–24.

Lupp, C. and Ruby, E. G. (2005). *Vibrio fischeri* uses two quorum-sensing systems for the regulation of early and late colonization factors. *J Bacteriol* 187, 3620–9.

Lynch, M. and Conery, J. S. (2000). The evolutionary fate and consequences of duplicate genes. *Science* 290, 1151–5.

Macgregor, S., Montgomery, G. W., Liu, J. Z., et al. (2011). Genome-wide association study identifies a new melanoma susceptibility locus at 1q21.3. *Nat Genet* 43, 1114–18.

Magadum, S., Banerjee, U., Murugan, P., et al. (2013). Gene duplication as a major force in evolution. *J Genet* 92, 155–61.

Manca, A., Capsoni, S., Di Luzio, A., et al. (2012). Nerve growth factor regulates axial rotation during early stages of chick embryo development. *Proc Natl Acad Sci U S A* 109, 2009–14.

Margulis, L. and Sagan, D. (2001). Marvellous microbes. *Resurgence* 206, 10–12.

Mereschkowsky, C. (1910). Theorie der zwei Plasmaarten als Grundlage der Symbiogenesis, einer neuen Lehre von der Entstehung der Organismen. *Biol Centralbl* 30, 278–88.

Miossec, P. (2017). Update on interleukin-17: a role in the pathogenesis of inflammatory arthritis and implication for clinical practice. *RMD Open* 3, e000284.

Mitrugno, A., Sylman, J. L., Ngo, A. T., et al. (2017). Aspirin therapy reduces the ability of platelets to promote colon and pancreatic cancer cell proliferation: Implications for the oncoprotein c-MYC. *Am J Physiol Cell Physiol* 312, C176–89.

Nathan, C. (2004). The moving frontier in nitric oxide-dependent signaling. *Sci STKE* 2004(257), pe52.

Nealson, K. H., Platt, T., and Hastings, J. W. (1970). Cellular control of the synthesis and activity of the bacterial luminescent system. *J Bacteriol* 104, 313–22.

Niemiller, M. L., Fitzpatrick, B. M., Shah, P., et al. (2013). Evidence for repeated loss of selective constraint in rhodopsin of amblyopsid cavefishes (Teleostei: Amblyopsidae). *Evolution* 67, 732–48.

Nisoli, E., Clementi, E., Paolucci, C., et al. (2003). Mitochondrial biogenesis in mammals: the role of endogenous nitric oxide. *Science* 299, 896–9.

Nowak, M. A., Boerlijst, M. C., Cooke, J., et al. (1997). Evolution of genetic redundancy. *Nature* 388, 167–71.

Offermanns, S. (2001). *In vivo* functions of heterotrimeric G-proteins: studies in Gα-deficient mice. *Oncogene* 20, 1635–42.

Ohno, S. (1970). *Evolution by Gene Duplication*. New York: Springer.

Ohno, S. (1985). Dispensable genes. *Trends Genet* 1, 160–4.

Olson, M. V. (1999). When less is more: gene loss as an engine of evolutionary change. *Am J Hum Genet* 64, 18–23.

Orgel, L. (2000). Origin of life. A simpler nucleic acid. *Science* 290, 1306–7.

Pacheu-Grau, D., Gomez-Duran, A., Lopez-Perez, M. J., et al. (2010). Mitochondrial pharmacogenomics: barcode for antibiotic therapy. *Drug Discov Today* 15, 33–9.

Parfrey, L. W. and Lahr, D. J. (2013). Multicellularity arose several times in the evolution of eukaryotes (response to DOI 10.1002/bies.201100187). *Bioessays* 35, 339–47.

Petraglia, F., Comitini, G., and Genazzani, A. (1993). β-Endorphin in human reproduction. In: Herz, A., Akil, H., and Simon, E.J. (eds) *Opioids II*. Berlin: Springer, pp. 763–80.

Pink, R. C., Wicks, K., Caley, D. P., et al. (2011). Pseudogenes: pseudo-functional or key regulators in health and disease? *RNA* 17, 792–8.

Piskur, J. (2001). Origin of the duplicated regions in the yeast genomes. *Trends Genet* 17, 302–3.

Plaitakis, A., Metaxari, M., and Shashidharan, P. (2000). Nerve tissue-specific (GLUD2) and housekeeping (GLUD1) human glutamate dehydrogenases are regulated by distinct allosteric mechanisms: implications for biologic function. *J Neurochem* 75, 1862–9.

Plaitakis, A., Spanaki, C., Mastorodemos, V., et al. (2003). Study of structure–function relationships in human glutamate dehydrogenases reveals novel molecular mechanisms for the regulation of the nerve tissue-specific (GLUD2) isoenzyme. *Neurochem Int* 43, 401–10.

Porath-Krause, A. J., Pairett, A. N., Faggionato, D., et al. (2016). Structural differences and differential expression among rhabdomeric opsins reveal functional change after gene duplication in the bay scallop, *Argopecten irradians* (Pectinidae). *BMC Evol Biol* 16, 250.

Rana, B. K., Hewett-Emmett, D., Jin, L., et al. (1999). High polymorphism at the human melanocortin 1 receptor locus. *Genetics* 151, 1547–57.

Rouquier, S., Blancher, A., and Giorgi, D. (2000). The olfactory receptor gene repertoire in primates and mouse: evidence for reduction of the functional fraction in primates. *Proc Natl Acad Sci U S A* 97, 2870–4.

Ryan, T. J. and Grant, S. G. (2009). The origin and evolution of synapses. *Nat Rev Neurosci* 10, 701–12.

Sack, D. A., Sack, R. B., Nair, G. B., et al. (2004). Cholera. *Lancet* 363, 223–33.

Sauer, K., Camper, A. K., Ehrlich, G. D., Costerton, J. W., and Davies, D. G. (2002). *Pseudomonas aeruginosa* displays multiple phenotypes during development as a biofilm. *J Bacteriol* 184, 1140–54.

Saxena, S. K. (2009). Controversial role of smallpox on historical positive selection at the *CCR5* chemokine gene (*CCR5-Δ32*). *J Infect Dev Ctries* 3, 324–6.

Scheinfeldt, L. B., Soi, S., Thompson, S., et al. (2012). Genetic adaptation to high altitude in the Ethiopian highlands. *Genome Biol* 13, R1.

Schöneberg, T., Schulz, A., Biebermann, H., et al. (2004). Mutant G-protein-coupled receptors as a cause of human diseases. *Pharmacol Ther* 104, 173–206.

Schöneberg, T., Kleinau, G., and Bruser, A. (2016). What are they waiting for? Tethered agonism in G protein-coupled receptors. *Pharmacol Res* 108, 9–15.

Schopf, J. W. (2006). Fossil evidence of Archaean life. *Philos Trans R Soc Lond B Biol Sci* 361, 869–85.

Schroeder, S. A., Gaughan, D. M., and Swift, M. (1995). Protection against bronchial asthma by *CFTR* ΔF508 mutation: a heterozygote advantage in cystic fibrosis. *Nat Med* 1, 703–5.

Schulz, A. and Schöneberg, T. (2003). The structural evolution of a P2Y-like G-protein-coupled receptor. *J Biol Chem* 278, 35531–41.

Shashidharan, P. and Plaitakis, A. (2014). The discovery of human *GLUD2* glutamate dehydrogenase and its implications for cell function in health and disease. *Neurochem Res* 39, 460–70.

Shashidharan, P., Michaelidis, T. M., Robakis, N. K., et al. (1994). Novel human glutamate dehydrogenase expressed in neural and testicular tissues and encoded by an X-linked intronless gene. *J Biol Chem* 269, 16971–6.

Shiu, S. H. and Bleecker, A. B. (2001). Receptor-like kinases from *Arabidopsis* form a monophyletic gene family related to animal receptor kinases. *Proc Natl Acad Sci U S A* 98, 10763–8.

Simonson, T. S., Yang, Y., Huff, C. D., et al. (2010). Genetic evidence for high-altitude adaptation in Tibet. *Science* 329, 72–5.

Singh, R., Sripada, L. and Singh, R. (2014). Side effects of antibiotics during bacterial infection: mitochondria, the main target in host cell. *Mitochondrion* 16, 50–4.

Smith, A. A., Holldober, B., and Liebig, J. (2009). Cuticular hydrocarbons reliably identify cheaters and allow enforcement of altruism in a social insect. *Curr Biol* 19, 78–81.

Sonnewald, U., Westergaard, N., and Schousboe, A. (1997). Glutamate transport and metabolism in astrocytes. *Glia* 21, 56–63.

Soranzo, N., Bufe, B., Sabeti, P. C., et al. (2005). Positive selection on a high-sensitivity allele of the human bitter-taste receptor TAS2R16. *Curr Biol* 15, 1257–65.

Stedman, H. H., Kozyak, B. W., Nelson, A., et al. (2004). Myosin gene mutation correlates with anatomical changes in the human lineage. *Nature* 428, 415–18.

Steegborn, C. (2014). Structure, mechanism, and regulation of soluble adenylyl cyclases—similarities and differences to transmembrane adenylyl cyclases. *Biochim Biophys Acta* 1842, 2535–47.

Tamames, J., Ouzounis, C., Sander, C., et al. (1996). Genomes with distinct function composition. *FEBS Lett* 389, 96–101.

Tasneem, A., Iyer, L. M., Jakobsson, E., et al. (2005). Identification of the prokaryotic ligand-gated ion channels and their implications for the mechanisms and origins of animal Cys-loop ion channels. *Genome Biol* 6, R4.

Taylor, J. S., Van De Peer, Y., and Meyer, A. (2001). Genome duplication, divergent resolution and speciation. *Trends Genet* 17, 299–301.

Tomasetti, C., Iasevoli, F., Buonaguro, E. F., et al. (2017). Treating the synapse in major psychiatric disorders: the role of postsynaptic density network in dopamine-glutamate interplay and psychopharmacologic drugs molecular actions. *Int J Mol Sci* 18, 135.

Van Gestel, J., Nowak, M. A., and Tarnita, C. E. (2012). The evolution of cell-to-cell communication in a sporulating bacterium. *PLoS Comput Biol* 8, e1002818.

Wang, T., Birsoy, K., Hughes, N. W. et al. (2015). Identification and characterization of essential genes in the human genome. *Science* 350, 1096–101.

Wang, X., Thomas, S. D., and Zhang, J. (2004). Relaxation of selective constraint and loss of function in the evolution of human bitter taste receptor genes. *Hum Mol Genet* 13, 2671–8.

Wang, X., Grus, W. E., and Zhang, J. (2006). Gene losses during human origins. *PLoS Biol* 4, e52.

Wikramanayake, A. H., Hong, M., Lee, P. N., et al. (2003). An ancient role for nuclear β-catenin in the evolution of axial polarity and germ layer segregation. *Nature* 426, 446–50.

Wiuf, C. (2001). Do ΔF508 heterozygotes have a selective advantage? *Genet Res* 78, 41–7.

Wolfe, K. H. and Shields, D. C. (1997). Molecular evidence for an ancient duplication of the entire yeast genome. *Nature* 387, 708–13.

Wybran, J. (1985). Enkephalins and endorphins as modifiers of the immune system: present and future. *Fed Proc* 44, 92–4.

Wynendaele, E., Pauwels, E., Van De Wiele, C., et al. (2012). The potential role of quorum-sensing peptides in oncology. *Med Hypotheses* 78, 814–17.

Wynendaele, E., Verbeke, F., Stalmans, S. et al. (2015). Quorum sensing peptides selectively penetrate the blood–brain barrier. *PLoS One,* 10, e0142071.

Yang, Z. (2007). PAML 4: phylogenetic analysis by maximum likelihood. *Mol Biol Evol* 24, 1586–91.

Zhang, J. (2003). Evolution by gene duplication: an update. *Trends Ecol Evol* 18, 292–8.

CHAPTER 3

GENETICS AND EPIGENETICS

PAUL W. EWALD
AND HOLLY A. SWAIN EWALD

Abstract

Genetic influences on human disease can be understood in the context of essential and exacerbating causes. Alleles are essential causes when they are required for these diseases to occur. Severe diseases with essential genetic causes are rare unless they provide a compensating benefit in evolutionary fitness (accrued through increases in survival and/or reproduction) that offsets their fitness costs. Duchenne's muscular dystrophy illustrates that upper limit of incidence for diseases that kill before reproduction and have essential genetic causes that provide no compensating benefit: about one death per 10,000 births. Sickle cell anaemia exceeds this limit in many populations because the causal allele confers a compensating benefit: protection against *falciparum* malaria. Alleles that contribute to common, severe diseases such as atherosclerosis and Alzheimer's disease tend to be exacerbating causes, which act by increasing vulnerability to environmental causes. Inferring genetic causation from familial patterns is not straight-forward because genetic associations may be correlated with environmental exposures. Relatively high monozygotic twin cooncordances for schizophrenia, for example, are associated with *in utero* exposure to the parasite *Toxoplasma gondii*, which appears to cause a substantial portion of schizophrenia. Epigenetic contributions to disease involve relatively long-term but reversible modification of genetic influences that may occur in response to micro- or macroenvironmental exposures. Integration of these aspects of disease causation can be of practical use in the health sciences by clarifying interventions that can prevent, cure, or strongly ameliorate disease.

Keywords

Alzheimer's disease, ApoE, atherosclerosis, cystic fibrosis, evolutionary medicine, genetic, epigenetic, schizophrenia, sickle cell

3.1 Introduction

3.1.1 Genetic and Environmental Causation

The definition of genetic diseases has changed gradually over the past century (Magnus 2004). As currently defined, genetic disease results from 'an abnormality in an individual's genome' (http://www.medicinenet.com/genetic_disease/article.htm) or 'the absence of a gene or by products of a defective gene' (http://www.dictionary.com/browse/genetic-disease). The rapidly growing knowledge about effects of alleles on disease raises questions that make apparent the inadequacies of these standard definitions for the future of medical science. What is an abnormality? Is it any allele that deviates from the norm? What is defective? New alleles arise by mutations. Some are harmful in some environments and beneficial in others. If they are beneficial, 'defective' does not seem an appropriate term.

The trend in understanding genetic contributions to disease indicates that allelic variants contribute to almost every human disease. Considering the pervasiveness of genetic contributions to disease, it has been popular to conclude that 'all disease is genetic disease' (Magnus 2004), a mindset characterised as 'genetic imperialism' (Juengst 2004). Environmental causes are also pervasive, perhaps justifying the conclusion that all diseases are environmental diseases.

Revisions of the widespread definitions of genetic disease all seem to be problematic (Magnus 2004). Nevertheless, it has been conceptually and practically useful to separate diseases according to genetic and environmental causes, even though there has been concern that the concept of genetic causation has increasingly led to acceptance of the genetic basis as the main cause, while environmental contributions are overlooked or downplayed (Dekeuwer 2015). The broad range of deviation from a one-to-one correspondence between alleles and disease raises an important question: When is it justified to assign primary importance to genetic causation? (Dekeuwer 2015).

To account for the extent to which a particular causal allele is associated with a disease, geneticists use the term penetrance. Complete penetrance indicates a one-to-one association. Incomplete penetrance refers to the more common situation in which a causal allele does not always generate the disease. Incomplete penetrance can result from influences of other genes in the organism, epigenetic effects, which could cause a gene to be unexpressed, and environmental effects. Geneticists, perhaps with some degree of genetic imperialism, accommodate the variation in their ability to understand diseases with some genetic basis by referring to them as 'complex genetic diseases'.

This chapter attempts to provide an assessment of genetic causation of disease in a way that provides balanced consideration of all contributing causes. We specify categories of genetic causation and illustrate them with a small number of particular diseases (Table 3.1), with the expectation that a detailed understanding of a particular example provides a basis for recognising the value of each category for understanding the broad spectrum of genetic contributions to disease. We emphasise three organising principles: (1) the concepts of 'essential' and 'exacerbating' causes, (2) consideration of proximate and ultimate causes, and (3) the triad of disease causation, which highlights the need to understand the interplay between different causes of disease.

Table 3.1 Categories of Genetic Causation (see text for details)

Category of genetic causation	Negative effect on individual fitness	Prevalence	Example
Essential; without compensating benefits	High	Rare if severe	Duchenne's muscular dystrophy
	Low	Variable	Achromatopsia
Essential; with compensating benefit	High	Low to moderate	Resistance to falciparum malaria
	Low	Variable	Duffy antigen resistance to vivax malaria
Exacerbating; increased vulnerability to infectious or environmental cause	Variable	Variable	Epsilon 4 vulnerability to infectious agents of atherosclerosis and Alzheimer's disease

3.1.2 Essential and Exacerbating Causes

The concepts of 'essential' and 'exacerbating' causes of disease offer a framework for conceptually organising genetic contributions to disease. A cause is defined as essential if it is required for a disease to occur. A cause is defined as exacerbating if it is not required for the disease to occur.

The essential cause of sickle cell anaemia is a mutation in the allele for the beta chain of haemoglobin, which substitutes a valine for a glutamic acid at position six of the beta chain. It is the essential cause because sickle cell anaemia cannot occur in people who do not have this allele in their genome. The 'normal' allele for the beta chain makes a person vulnerable to *falciparum* malaria, and is therefore an exacerbating cause of this infectious disease. The essential cause of *falciparum* malaria is infection by the protozoan *Plasmodium falciparum*.

There is some circularity in categorising a cause as exacerbating or essential, because a disease can be defined in part on the basis of a cause. However, this problem of circularity already occurs more generally in the categorisation of disease causation. For instance, malaria was found to be caused by *Plasmodium* pathogens. Then it was concluded that any individuals who have manifestations that look like malaria but are not infected with plasmodia do not have malaria. Finally, once these two steps are taken, it can be definitively concluded that all malaria is caused by *Plasmodium*, and, hence, that *Plasmodium* is an essential cause of malaria. A positive depiction of this circularity is that the identification of disease categories and disease causation develops together. If the malarial symptoms that are actually caused by *Plasmodium* are distinct from similar symptoms in people without *Plasmodium* infection, the disease category of malaria and *Plasmodium* as the essential cause can be accepted together. Similarly, the sickle cell allele is not an essential cause of anaemia, but if the category of sickle cell anaemia is medically useful, then it is also useful to recognise that the sickle cell allele is the essential cause of sickle cell anaemia.

Broad consideration of essential and exacerbating causes is valuable because it prompts consideration of the full spectrum of causation. Without this prompting, essential or exacerbating causes could be overlooked. Identification of essential causes is important because prevention of an essential cause prevents the disease; however, much of the damage that occurs in diseases can be attributable to exacerbating causes. Low environmental oxygen concentration, for example, can contribute to the development of life-threatening sickling crises in individuals who are heterozygous for the sickle cell allele. Low oxygen concentration is an exacerbating cause because sickling crises can arise in individuals who are not in an environment with low oxygen concentration. Still the contribution of low oxygen concentration is important medically because heterozygous individuals can be counselled against intense physical activity at high altitudes, for example, where the risk of a sickling crisis is potentiated.

It is sometimes difficult to determine whether a genetic cause is essential or exacerbating. Determining the full scope of causal factors can help resolve this difficulty through clarification of causal networks for a given disease.

Recognising that a genetic cause is exacerbating can clarify other influences that need to be targeted to prevent, cure, or ameliorate disease. Berylliosis provides an illustration. It occurs among people who are exposed to beryllium in activities related to fluorescent lamp manufacture, aerospace, atomic energy, ceramics, metallurgy, metal machining, ceramics, automotives, and electronics (Lang 1994; Handa et al. 2009). Berylliosis is characterised by inflammation, formation of non-caseating granulomas, and, eventually, fibrosis of the lungs (Freiman and Hardy 1970). The essential cause is high levels of exposure to beryllium, but the disease occurs mostly among people who have glutamate (instead of lysine) at site 69 of the HLA-DPB1 protein (Saltini et al. 1998). Embedded in the membrane of macrophages, HLA-DPB1 presents antigens to helper T cells. HLA-DPB1 (glu) is not defective but rather contributes strongly to the pathology of chronic beryllium disease when beryllium is present. Prior to the technological developments that increased the exposure to beryllium, HLA-DPB1(glu) was probably comparable in effectiveness to HLA-DPB1(lys); otherwise natural selection would have kept it from increasing to its allele frequency of 30%. Descriptions of the pathogenesis of beryllium disease refer to this genetic variant in the context of beryllium-associated antigen presentation, autoimmunity, and the aggregation of helper T cells, macrophages, and plasma cells in granulomas (Dai et al. 2013; Falta et al. 2013). Although specific infections are not generally mentioned in this pathogenesis, the antigen-presenting function of HLA-DPB1 raises the possibility that the pathogenesis of chronic beryllium disease could involve an interaction between infectious agents, beryllium, and HLA-DPB1; particular pathogens, for example, could contribute to the granulomatous responses if their antigen presentation by HLA-DPB1(glu) is altered relative to HLA-DPB1(lys).

Although infections and inflammation are associated with berylliosis and its pathogenesis (Handa et al. 2009), increased vulnerability to infection conferred by HLA-DPB1(glu) has not been evaluated. It is presumed that infections result from treatment of berylliosis with anti-inflammatories, but infection may play a role earlier during pathogenesis, contributing to inflammation and granuloma development particularly in individuals with the HLA-DPB1(glu) allele. This uncertainty is a manifestation of the general recognition that autoimmunity may result from an interplay between HLA variants and infection, but scientific understanding of this interplay is complex and just beginning to emerge (Matzaraki et al. 2017).

Resolving the interplays between genetic, infectious, and environmental causation may have great impact on the care of individuals. With regard to berylliosis, for example, there is no cure, nor an ability to prevent the genetic cause: the HLA-DPB1(glu) allele. Health interventions therefore must focus on the other contributing causes. Individuals who carry HLA-DPB1(glu) could be counselled to reduce exposure to beryllium through job choice or emphasis on precautions that reduce exposure in the workplace. Similarly, if exacerbating pathogens are identified, particularly rigorous care could be directed to individuals with HLA-DPB1(glu) to prevent and control infections.

3.1.3 Ultimate and Proximate Causation

The distinction between ultimate and proximate causation is another dichotomy that is central to a balanced understanding of genetic contributions to disease. Proximate causes explain the mechanisms of pathogenesis—how molecular, biochemical, cellular, and physiological actions bring about disease. Ultimate causes explain the evolutionary origins of disease—why evolutionary processes have generated the proximate mechanisms of disease. Proximate and ultimate explanations are therefore often referred to as *how* and *why* explanations, respectively. Ultimate explanations are generally cast in terms of evolutionary selection (subsuming natural selection, sexual selection, kin selection, and artificial selection), with effects being formulated in the currency of evolutionary fitness (the net effect on the passing on of the corresponding alleles). When applied to genetic diseases, ultimate explanations evaluate effects of the disease on the differential contribution of the allelic basis for the disease into succeeding generations, through effects on survival and/or reproduction. Broadly educated biologists generally consider that a complete understanding of any life process requires both categories of explanation. (For further discussion, see Chapter 1: Core Principles for Evolutionary Medicine.)

Ultimate explanations are generally simplified to address why a characteristic is present as opposed to absent. With regard to genetic diseases, however, ultimate explanations must be cast in terms of the frequency of the causal alleles rather than the presence or absence of a characteristic. The reason is that a genetic cause of a disease is being continually regenerated by mutation. An ultimate explanation of a genetic disease therefore addresses why its causal allele may be sustained at a frequency that is very close to zero or at a relatively high frequency even though it may impose substantial fitness costs.

If an allele is an essential cause of a genetic disease, then natural selection will tend to reduce its prevalence in proportion to the disease's negative effects on fitness, unless there is some compensating fitness benefit to the allele. When a compensating benefit does not occur, the opposing effects of selection against an allele versus generation of an allele by mutations will generate an equilibrium. Alleles that always cause death before reproductive ages will be represented in the population solely as a result of mutations each generation. Mutations in a particular gene tend to occur at a very low frequency per generation. Lethal genetic diseases that are caused by such alleles will therefore tend to be rare.

Prior to modern interventions, sickle cell anaemia occurred in about one per 500 births among African-Americans, but prevalence can be as high as one in twenty births in countries with high prevalence of *falciparum* malaria, such as Nigeria. As these frequencies are far greater than the threshold that could be maintained by mutation alone, an ultimate explanation for sickle cell anaemia must invoke some compensating benefit. It is now generally accepted that the ultimate explanation for these high frequencies is increased resistance to *falciparum* malaria among heterozygotes.

3.1.4 The Triad of Disease Causation

The causes of all diseases can be subdivided into three categories: genetic, symbiotic, and environmental.

Genetic causes can be defined as the effects on disease of particular alleles. Symbiotic causes can be defined as effects on disease from replicative agents that live intimately with the host organism. If the symbiont is a parasite, its presence contributes to the disease. If the symbiont is a mutualist, its absence can contribute to disease. In an evolutionary context a parasite has a net negative effect on the evolutionary fitness of the host, whereas a mutualist has a net positive effect. We define parasites broadly in this chapter as multicellular, cellular, and subcellular replicative agents that live in or on a host organism and have a net negative effect on the host's evolutionary fitness. Pathogens are defined as parasites at or below the unicellular level of organisation. In biological and medical discussions, commensals are symbionts that are presumed to have neither a net positive nor a net negative effect on the host fitness; however, the labelling of an agent as a commensal reflects an inability to measure fitness effects on the host with sufficient accuracy to determine whether the overall net effect is positive or negative. In principle, therefore, commensalism is best considered as a conceptual dividing line between parasitism and mutualism. In practice it is often used as a category for organisms that have net effects on hosts that are too slight to be able to measure.

The increasingly detailed knowledge about contributors to health and disease leads to the conclusion that all diseases probably involve causes from at least two and almost always all three of these categories of causation: genetic, environmental, and symbiotic. Moreover, as indicated in Section 3.1.2, the interactions between different causes need to be considered before one can conclude that the causes of a particular disease are understood.

When correlates of disease are identified without certainty about causation, they are referred to as risk factors, which can be rigorously quantified and demonstrated. Reference to risk factors is therefore 'safe' for researchers because demonstration of causation is often difficult. Some investigators therefore focus on risk factors without delving into causation. This retreat from causation is, however, problematic. Not only has the principle of causation been centrally important to the advancement of science, but also assignment of causation provides guidance for medical practice—blocking a cause of disease will have a beneficial effect on the patient, but blocking a risk factor that is not causally involved will not.

The environmental category, narrowly defined, includes all environmental causes that are not replicative agents. This category includes effects of diet, lifestyle, chemicals, and radiation—too much of a bad thing (e.g. exposure to radioactivity) or too little of a good thing (e.g. lack of vitamins). To keep in mind that all risk factors can potentially interact to cause disease, it is useful to picture these causes as the vertices of a triangle, and to place a disease inside the triangle closest to the vertex that includes the essential cause(s) of the disease and furthest from the vertex that includes the least important exacerbating cause(s). In this context the traditional categorisation of diseases as genetic, infectious, or environmental is depicted by proximity to the corresponding vertex (Figure 3.1). In the case of sickle cell anaemia, for example, the disease would be placed close to but slightly below the genetic vertex because the sickle cell mutation is the essential cause of the disease and infectious agents and non-infectious environmental factors contribute importantly to the illness (see Sections 3.1.2 and 3.2.1).

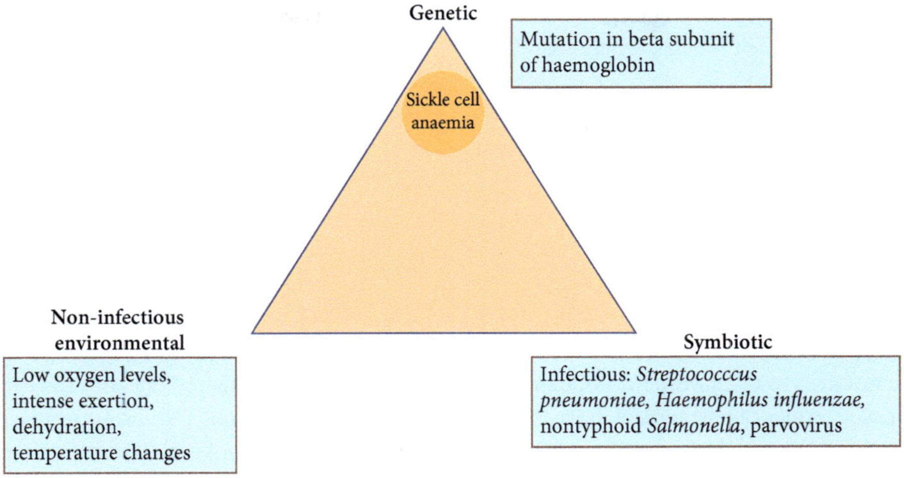

FIGURE 3.1 Representation of sickle cell anaemia within the triad of disease causation. Sickle cell anaemia is placed closest to the genetic causation vertex because the sickle cell allele is its essential cause. Its location below the vertex corresponds to the importance of infectious agents and both external and internal environments as exacerbating causes (see text for details).

3.1.5 Epigenetics

Epigenetics can be thought of as an aspect of causation that links together microenvironmental, macroenvironmental, symbiotic, and genetic causation. Epigenetics encompasses (potentially) heritable alterations to gene expression that exclude changes in the DNA sequence. The two major areas of epigenetic modification to gene-driven activity are methylation of DNA and modification of histones—chromatin proteins that structure and regulate access to genomic DNA. While seemingly limited in diversity of overall mechanisms, these epigenetic modifiers are essential to cell specificity/fate in multicellular organisms and to the ability to (epi)genetically respond to environmental change within the individual versus through adaptation across generations. Epigenetic alterations such as gene silencing or activation can be lasting in somatic cells and thus can be inherited across cell divisions, while germline epigenetic changes may be passed on to offspring. (For further discussion, see Section 3.6, Chapter 2: Cellular Signalling Systems.)

3.2 Alleles as Essential Causes

3.2.1 Genetic Diseases with a Compensating Benefit

Sickle cell anaemia is one of the best-studied and globally prevalent diseases for which the essential cause is an allele. It is characterised by low red blood cell density (anaemia) and sickle-shaped red blood cells. The essential cause of sickle cell anaemia is a point mutation in an allele encoding the beta chain of the haemoglobin molecule. If both copies of the beta haemoglobin gene in a person have the sickle cell mutation (i.e. the person is *homozygous* for the sickle cell allele) then the haemoglobin tends to polymerise, distorting the red blood cells into a sickle shape. The less pliable, sickle-shaped cells tend to cluster in small blood vessels, blocking blood flow, thereby depriving downstream tissues of oxygen and nutrients. Damage to the tissues can have life-threatening consequences if they are in vital organs, such as the heart and spleen. (For further discussion, see Chapter 9: Haematopoetic System.)

Although blocking the essential cause of a disease can prevent or cure it, this gold standard is generally not feasible for genetic diseases because the genetic make-up of an individual is difficult to change. In the case of sickle cell anaemia, elimination of the essential cause is possible in some cases through bone marrow transplantation. Donors with matching HLA (human leukocyte antigen) type but without the sickle cell mutation provide haematopoietic marrow cells to recolonise the bone marrow of the sickle cell anaemia patient after the haematopoietic cells of the patient have been eliminated (Walters 2015). Use of bone marrow transplant therapy, however, is limited by several constraints. Suitable HLA matches can be obtained for only about one-third of individuals with sickle cell anaemia (Walters 2015). Even when a donor is considered suitable, haematopoietic transplantation can cause infertility and life-threatening graft versus host disease (Walters 2015). Moreover, it is prohibitively expensive or not available for much of the world's population with sickle cell anaemia.

Sickle cell anaemia illustrates how understanding the interplay between an essential genetic cause and exacerbating causes can be useful because for most individuals with sickle cell anaemia interfering with exacerbating causes will be the best option. Infections are the immediate cause of death in most sickle cell patients worldwide (Booth et al. 2010). When blood vessels supplying the spleen are obstructed, the damage to the spleen increases vulnerability to life-threatening infections (Booth et al. 2010). The spleen removes pathogens from circulation through phagocytosis and fosters antibody production (Booth et al. 2010). Spleen function is important in controlling bacteria, especially those that are encapsulated (Booth et al. 2010). Being less effectively controlled by spleen-dependent immunological defences, encapsulated bacteria such as *Streptococcus pneumoniae* and *Haemophilus influenzae* are more likely to generate life-threatening sepsis (Booth et al. 2010).

A review of 244 autopsies conducted in the United States from 1929 to 1996 on individuals who were homozygous for the sickle cell allele and died from sickle cell disease identified infection as the immediate cause of about half of the deaths (Manci et al. 2003). Most of the deaths were from respiratory complications, which were mostly attributed to *S. pneumoniae* (30.1%), *H. influenzae* (6.1%), and 'viruses/atypical organisms' (38.6%). Several other pathogens are associated with sickle cell anaemia and probably contribute to morbidity and mortality: *Pseudomonas aeruginosa, Klebsiella pneumoniae, Escherichia coli, Mycoplasma pneumoniae, Chlamydia* (= *Chlamydophila*) *pneumoniae*, and non-typhoid *Salmonella* (Manci et al. 2003; Booth et al. 2010; McCavit et al. 2011).

Prophylactic treatment with penicillin and vaccination against *S. pneumoniae* and *H. influenzae* (Quinn et al. 2010) have been credited with significantly reduced mortality in children with sickle cell anaemia (Kavanagh et al. 2011), indicating that recognising and treating the exacerbating infectious causes of sickle cell anaemia can be important.

Consideration of genetic diseases in the context of contributions to infection may improve the rigour of interpretation of manifestations and treatment. Platt et al. (1994), for example, noted that high leukocyte densities were associated with increased probability of death in sickle cell anaemia patients. They suggested that reductions in leukocyte density might be useful therapeutically; however, if the leukocytes are elevated as a defence against a threatening infection, reduction of leukocyte density would be counterproductive.

Genetic diseases with a compensating benefit need not be severe. Mutations in the Duffy antigen receptor, for example, provide protection against *vivax* malaria because *Plasmodium vivax* attaches to the receptor to enter red blood cells. The mutation in the receptor prevents this entry. A non-functional Duffy antigen receptor provides this benefit with little negative effect. Its frequency therefore increased to very high frequencies in some populations. These high frequencies presumably drove *P. vivax* to extinction in West Africa, leaving the paradoxical outcome of high frequencies of a genetic defence against a pathogen that is no longer present. (For further discussion, see Chapter 9: Haematopoetic System.)

3.2.2 Genetic Diseases without a Compensating Benefit

Section 3.2.1 illustrates an important aspect of essential genetic causes of disease: alleles can be prevalent, essential causes of severe disease if they provide a large compensating benefit. When the harm occurs primarily among homozygotes, the compensating benefit allows the frequency of the damaging allele to rise in the population until the positive effects that derive from increased survival and reproduction of heterozygotes are offset by the negative effects in homozygotes, which increase disproportionately with increases in allele frequency. Without a compensating benefit, mutations that are directly responsible for the disease (i.e. essential genetic causes) will arise by mutation each generation and be selected against by the negative effects of the mutated gene on the person's survival and reproduction. If negative effects of an allele prohibit reproduction because, for example, it is genetically dominant and lethal during childhood, it will be cleared each generation by selection. The incidence of such a disease would be equal to the rate at which the allele is regenerated by mutation per generation.

If the associated gene is very long and the disease can be generated by mutations at many sites along the gene, then the equilibrium frequency can be greater than if the gene is short with few sites for disease-generating mutations. Duchenne's muscular dystrophy provides a sense of the maximum allele frequency for genetic diseases that reduce evolutionary fitness to zero because it can result from a variety of different mutations in the longest human gene. Individuals with Duchenne's muscular dystrophy experience a progressive degeneration of muscles and die without reproducing. The essential cause of Duchenne's muscular dystrophy is mutation in the gene that encodes dystrophin, which is a protein that anchors the cytoskeleton of muscle cells to the extracellular matrix. The mutations are recessive, but the gene is on the X chromosome, so males with one copy of the gene develop the disease and females with one copy do not. The disease will develop in about one in 5000 males and one in 10,000 people overall. The extraordinary length of the dystrophin gene and the concomitantly large number of mutations that can destroy the dystrophin protein indicate that the frequency of Duchenne's muscular dystrophy must rank among the highest of any human disease that extinguishes the evolutionary fitness of the afflicted without providing any compensating benefit. Generalising this idea, the equivalent of about one pre-reproductive death per 10,000 births should represent a maximum frequency permitted by evolutionary selection for a disease with an essential genetic cause that provides no compensating benefit. So, for example, if a genetic disease due to a dominant allele was 100% lethal prior to reproduction in half of the people carrying the allele, but caused a fitness reduction of about 50% in the other half, the maximum frequency of the disease that could be maintained in the population by mutation alone would be about 1.5 per 10,000 (i.e. one per 10,000 for half the population and two per 10,000 for the other half).

The threshold of about one pre-reproductive death equivalent is useful for understanding genetic causation because a fitness decrement above this threshold indicates that some other factor besides mutation must be in play. A compensating benefit (e.g. manifested as 'heterozygote advantage') is one such factor. Another factor involves historical accidents generally associated with small population size (e.g. genetic drift), which can allow a damaging allele to maintain itself temporarily at higher levels, until natural selection reduces its frequency to the equilibrium maintainable by mutation alone.

Sickle cell anaemia illustrates this idea. Prior to modern interventions, the fitness cost of sickle cell anaemia was about one per 500 births. A compensating benefit therefore must be presumed, unless some other process of population genetics can be invoked. Founder effects, for example, may sometimes explain an unusually high frequency of a damaging allele. A founder effect can be defined as the effect on allele frequency that derives from the particular genetic make-up of the small number of individuals from which the population multiplied. The small founding population could result from individuals that generated a population in a new geographic area or a temporary restriction in the population (i.e. a bottleneck).

Founder effects cannot explain the high prevalence of sickle cell anaemia, but they can be responsible for significant genetic problems, particularly in isolated populations. The complete achromatopsia (colour blindness due to complete absence of cones) that occurred in the Micronesian atoll of Pingelap provides an illustration. This autosomal recessive condition affects about 0.001–0.003% of the world's population, but about 5% of Pingelap inhabitants (Hussels and Morton 1972). The fitness costs are unknown but were probably substantial in hunter-gatherer settings because colour blindness and the associated myopia surely compromised a person's ability to distinguish food items and threats such as predators. The high frequency in Pingelap is attributed to typhoon Lengkieki, which occurred around 1775, and an ensuing famine, which reduced the population to about twenty survivors (Morton et al. 1972). The achromatopsia allele in the Pingelap population is caused by a mutation in the CNGB3 gene (Sundin et al. 2000) that was inherited from Nanmwarki Mwanenised, the ruler of the Pingalapese at the time of the typhoon. He was heterozygous for the mutation, survived the famine (Hussels and Morton 1972), and contributed disproportionately to the subsequent population growth. The high prevalence of achromatopsia in Pingelap is therefore best explained by the chance repopulation of Pingelap by people who were heterozygous for the allele. Although the allele must confer a fitness disadvantage when present homozygously, advantages associated with the ruling family's access to resources, the absence of natural selection acting against heterozygous individuals, and the limited time available for natural selection since the typhoon probably allowed the frequency of the allele to remain at an elevated level over the past two centuries.

If a genetic disease in a group results from a founder effect, we expect just one mutation in the founder and hence in the group. If the genetic disease results from a compensating benefit, it could result from a single mutation or multiple mutations, with each providing a selective advantage for the individuals that bear them heterozygotically. But even when multiple mutations are present, distinguishing founder effects from heterozygote advantage can be difficult.

A case in point is Tay-Sachs disease, which occurs in disproportionately high frequency in Ashkenazi Jews (about one in 3600 births). In this group the disease can result from several different mutations; moreover, two other severe disorders caused by mutated genes coding for enzymes in the same biochemical pathway also occur in the Ashkenazim: Gaucher's disease and Niemann–Pick disease. This occurrence of many mutations leading to similar biological effects is strong evidence implicating some compensating benefit rather than a founder effect (Knudsen 1973). Some epidemiological data indicate that Tay-Sachs heterozygotes are resistant to tuberculosis (Diamond 1988). Yet even with the strong circumstantial evidence implicating a compensating benefit, severe historical population bottlenecks of the Ashkenazim have diminished populations sufficiently that founder effect hypothesis cannot be excluded (Risch et al. 2003; Slatkin 2004). This ambiguity may eventually be resolved if additional evidence clarifies whether Tay-Sachs heterozygotes are protected against tuberculosis. Tay-Sachs heterozygotes produce unusually high levels of a protein, the beta subunit of hexosaminidase, which has antimicrobial effects on *Mycobacterium tuberculosis*, supporting the compensating benefit rather than the founder effect hypothesis (Fernandes Filho and Shapiro 2004; Koo et al. 2008).

3.2.3 Applying an Integrated Approach: Cystic Fibrosis

The principles presented above provide a framework for investigating and understanding diseases for which alleles are essential causes but knowledge is fragmentary. The proximate cause of cystic fibrosis is a mutation in the cystic fibrosis transmembrane conductance regulator (CFTR) protein. CFTR resides in cell membranes and transports chloride ions to the outside of the cell membrane. Cells of people who are homozygous for the defective allele do not export chloride ions sufficiently, causing the mucus lining the lungs to become more viscous, impairing clearance of the mucus and embedded pathogens, such as *S. pneumoniae* and *P. aeruginosa*. The eventual result is cumulative damage to the lung, which often is lethal. (For further discussion, see Chapter 12: Respiratory System.)

People who are heterozygous for the cystic fibrosis allele show little if any negative physiological effects. Among people with northern European ancestry, about one out of thirty is a carrier, and cystic fibrosis occurs in one out of every 2500 births. Why is cystic fibrosis more prevalent than is maintainable by mutation alone? The lack of a bottleneck in the affected populations and the presence of many causal mutations in the CFTR gene negate a founder effect. Several lines of evidence indicate that the CFTR allele, like the sickle cell allele, is maintained by heterozygote advantage, probably protection against infection (Poolman and Galvani 2007; Withrock et al. 2015). Attention has focused on three candidate pathogens: *Vibrio cholerae, Salmonella typhi* (also known as *Salmonella enterica typhi*), and *M. tuberculosis* (Poolman and Galvani 2007; Withrock et al. 2015).

V. cholerae has been suggested because its pathogenesis uses the CFTR protein. The B subunit of the cholera toxin causes the cell to activate the CFTR protein to pump into the gut lumen chloride ions, which draw out sodium ions by ionic attraction, and water by osmosis. The net result is diarrhoea, which allows *V. cholerae* to flush competing species out of the intestinal tract and to disperse itself in the external environment. The dehydration resulting from this diarrhoea is the main cause of cholera's lethality, through induction of hypovolaemic shock. If a mutation in the CFTR gene interferes with the binding of the B subunit then a person would be resistant to the often lethal effects of the diarrhoea. A major problem with this hypothesis for elevated prevalence of cystic fibrosis in northern Europeans is that the geographic patterns do not match up. CFTR mutations have not spread in South Asia, where cholera has been endemic for centuries and probably for millennia. Cholera first entered northern Europe in 1826, after improvements in transportation allowed for the toxigenic variants of *V. cholerae*, which causes non-persistent infections, to reach northern Europe from South Asia. Cholera therefore seems inadequate as an explanation for the presence of the cystic fibrosis allele in northern Europeans (Poolman and Galvani 2007).

The bacterium that causes typhoid fever, *S. typhi*, uses the normal CFTR protein as a receptor for entry into the cell but cannot use the mutated CFTR protein (Pier et al. 1998). It therefore may be more exposed to and controllable by immunological responses in people who have one mutated CFTR allele (Pier et al. 1998). *S. typhi* has apparently been present in northern Europe for centuries, but has also been widespread in other parts of the world where the CFTR allele is present at low frequencies. This geographic mismatch therefore also weakens protection against *S. typhi* as an explanation for the elevated frequency of CFTR mutations in northern Europeans.

Elevated prevalence of cystic fibrosis alleles corresponds more closely geographically with tuberculosis exposure than with typhoid fever or cholera (Poolman and Galvani 2007). Prevalence and mortality of tuberculosis are lower in cystic fibrosis heterozygotes. One possible mechanism of protection involves reduced availability of sulphur in cystic fibrosis heterozygotes, which inhibits virulence of *M. tuberculosis*. Estimates of selective pressure suggest that tuberculosis would have exerted a stronger selective pressure than cholera or typhoid fever on northern Europeans during relevant time periods (Poolman and Galvani 2007).

Evolutionary considerations of the genetics and distribution of cystic fibrosis therefore indicate that it is being maintained at an elevated frequency in northern Europeans because of some compensating fitness benefit, but unlike the situation with sickle cell anaemia, it is not yet clear what that benefit is. Resistance to each of the three suspected pathogens could have contributed to an elevated frequency of the CFTR allele, but at this point evidence favours resistance to tuberculosis as the most significant selective pressure.

3.3 Alleles as Exacerbating Causes

3.3.1 The Epsilon 4 Allele, Atherosclerosis, and Alzheimer's Disease

Heart attacks and strokes kill more people in prosperous western countries than any other general category of disease (for figures from the United States see https://www.cdc.gov/nchs/fastats/deaths.htm). Although Alzheimer's disease kills one person for every seven killed by heart attacks and strokes, it profoundly impairs the quality of life for the afflicted, friends, and families.

In 1993, the epsilon 4 allele of the apolipoprotein epsilon 4 gene was reported to be a risk factor for late-onset Alzheimer's disease (Saunders et al. 1993; Strittmatter et al. 1993; Hyman et al. 1996), an association that has been borne out by a meta-analysis of published studies of the association (Steel and Eslick 2015). Concurrently with the discovery of its association with Alzheimer's disease, the epsilon 4 allele was also becoming recognised as a risk factor for atherosclerosis, the main pathological process leading to heart attacks and strokes (Mahley 1988; Jarvik et al. 1995; Wilson et al. 1996; Ilveloski et al. 1999).

It would seem reasonable to portray the epsilon 4 allele as inherently damaging, like the alleles that cause cystic fibrosis or sickle cell anaemia. However, like the other epsilon proteins encoded by the apolipoprotein E gene, epsilon 4 helps transport fat and cholesterol to and from cells through the blood. It has been the principal allele for humans and other primates for many millions of years (Fullerton et al. 2000). During the evolution of *Homo sapiens*, however, its prevalence has declined to the point that it now occurs in a minority of people (Fullerton et al. 2000); it has persisted in some populations at frequencies of nearly 50%, but occurs with a frequency of about 5–10% in others (Corbo and Scacchi 1999). There is no generally accepted explanation for the loss of epsilon 4 human evolution nor the variation among populations. Epsilon 4 has declined the most in human populations with a long history of agriculture (Corbo and Scacchi 1999). This fact is consistent with epsilon 4 being a 'thrifty allele' that declined in the context of energy-rich agricultural diets (Corbo and Scacchi 1999). But this hypothesis does not account for the fact that most of the decline in the frequency of epsilon 4 occurred prior to the onset of agriculture—its frequency is less than 50% even in hunter-gatherers (Corbo and Scacchi 1999). Moreover, epsilon 4 has been associated with fitness benefits in rural subpopulations with increased exposure to gastrointestinal infections (Trumble et al. 2017; van Exel et al. 2017) and probably therefore with concomitant nutrient deficits. These findings together with the past predominance of epsilon 4 in *H. sapiens* and its persistence in all human populations argue against the possibility that it is an inherently bad allele.

3.3.2 Early Pathogenesis of Atherosclerosis

A focus on the earliest stage of atherosclerosis provides a baseline for understanding its causes and the role of epsilon 4. Early pathology of atherosclerosis is characterised by deposition of cholesterol clefts, minute spindle-shaped deposits of cholesterol within the walls of arteries. White blood cells are present, particularly neutrophils, which are specialists in attacking pathogens, and foam cells, which are macrophages that acquire a foamy appearance because they contain minute cholesterol 'bubbles'. When the foam cells die, the cholesterol that they release is thought to be the source of the cholesterol in the clefts. A complete explanation of the initiating causes of atherosclerosis must therefore account for the early presence of foam cells and neutrophils in the walls of the arteries in the context of risk factors.

Fat and cholesterol, stress, inflammation, smoking, and periodontal disease are associated with increased risk of atherosclerosis. Moderate-consumption alcohol is associated with a lower risk of atherosclerosis. The garlic-laced Mediterranean diet is associated with lower risk, as is exercise and treatment with aspirin and statins. But making the logical step from risk to causation often raises paradoxes, which indicate that thinking about these diseases as a simple sum of the parts is missing something important. In each case, however, as the analysis is broadened to consider interactions between risk factors, the paradoxes are resolved. This resolution does not mean that we have the correct answer, but it does mean that we have at least a viable explanation.

3.3.3 Cholesterol and Atherosclerosis

Cholesterol has been incriminated in the development of atherosclerosis for more than a century, with detailed quantitative studies of the association between cholesterol and heart disease beginning in the middle of the twentieth century (Endo 2010). This incrimination arose in part because atherosclerotic plaques are laden with cholesterol. If cholesterol levels in the blood are high it seemed reasonable to presume that the cholesterol would be more likely to build up on the lining of arteries and eventually clog them or break off and block an artery further downstream where arterial diameters become progressively smaller as they branch and eventually connect to capillary beds.

Since the 1960s, the hereditary genetic disorder familial hypercholesterolaemia has anchored evidence for a causal association between cholesterol and cardiovascular disease (Endo 2010). In the most common variant of this disorder, a defective receptor for low-density lipoproteins (LDL, also known as 'bad cholesterol') was associated with very high levels of cholesterol in the blood and increased risk of heart attacks. On the other hand, a variant of the apolipoprotein E gene, called epsilon 2, also results in elevated cholesterol but has a more complicated relationship with cardiovascular disease—it is associated with increases in heart attacks in old people but protects against heart disease at younger ages.

When epsilon 4 was discovered as a risk factor for atherosclerosis, it reinforced the focus on cholesterol because epsilon 4 is associated with elevated serum cholesterol (Wilson et al. 1994; Ilveskoski et al. 1999). The mechanism by which epsilon 4 and high cholesterol contribute to cardiovascular disease, however, remains incompletely understood.

During the last quarter of the twentieth century, pathologists realised that the initial damage occurred within the wall of arteries rather than on the luminal surface. This location complicates the conventional argument about the role of cholesterol, because it is unclear why elevated cholesterol in the blood would cause build-up first within the wall rather than on the more directly exposed lining of the artery. But the early damage within the arterial wall involves the build-up of cholesterol deposits, so one way or another it seemed reasonable to presume that higher cholesterol in the blood led to increased deposition within the wall of the artery. One mechanism involves the movement of foam cells between the cells that line arteries into the arterial wall.

Cholesterol-lowering drugs reduce the probably of cardiovascular events to levels that correlate with the degree to which cholesterol is lowered (Silverman et al. 2015). But similar reductions in cholesterol by different drugs have still been associated with different reductions in cardiovascular events (Silverman et al. 2015). The cholesterol lowering by the statin component of the drug Vytorin® (simvastatin), for example, provides greater protection than that of the non-statin component ezetimibe (Cannon et al. 2015). These different effects suggest that the beneficial effects of a cholesterol-lowering drug might result at least in part from correlates of lowered cholesterol in addition to the effects of lowered cholesterol. The recognition that atherosclerosis progresses partly as a result of inflammation raised concerns that the positive effects of statins resulted more from suppression of inflammation than from lowering of cholesterol. A similar argument can be made for cholestyramine, a cholesterol-lowering drug used mainly prior to the introduction of statins. Cholestyramine sequesters toxins and therefore could be reducing toxin-induced inflammation.

Resolving these ambiguities is important because the lowering of cholesterol has been associated with reduced cardiovascular events even in patients with normal or low cholesterol. This broad effectiveness led some to suggest that use of cholesterol-lowering drugs might be widened to include patients with moderate or low cholesterol. This option raises concern because the synthesis of cholesterol is important for normal functions: neuron conduction, intracellular transport, cell signalling, and synthesis of vitamin D and steroid hormones. The negative effects of interfering with these functions might not be apparent in the necessarily small populations that are studied to assess safety prior to drug approval.

A meta-analysis of all cholesterol-lowering drugs revealed a strong statistically significant association between the extent to which cholesterol was lowered and protection, in spite of the variation among studies and different drugs (Silverman et al. 2015). Taken in its entirety, the evidence pertaining to serum cholesterol suggests that it plays an important role in the development of atherosclerosis, but other correlates of cholesterol need to be invoked to explain, for example, why the lowering of cholesterol by ezetimibe has a less beneficial effect on cardiovascular disease than a similar lowering by simvastatin. One part of the answer is that the positive effects of statins probably result partly from effects other than the lowering of cholesterol. The anti-inflammatory and antimicrobial effects of statins are two possibilities.

3.3.4 Fatty Acids, Inflammation, and Epsilon 4

The fat part of the puzzle focuses on triglycerides, lipids with a glycerol backbone to which three long fatty acid molecules are tethered. For decades, researchers have recognised that a common component of triglycerides, omega 6 fatty acids, can be converted into prostaglandin E, which contributes to inflammation. Omega 3 fatty acids, by contrast, are thought to suppress inflammatory responses (Simopoulos 2008). The widespread recognition that the pathology of atherosclerosis results in part from inflammatory damage has led to suspicions that a high ratio of omega 6 fatty acids relative to omega 3 fatty acids could contribute to atherosclerosis by elevating inflammation and, consequently, that addition of dietary supplementation with omega 3 fats could ameliorate the damage from atherosclerosis (Simopoulos 2008; He et al. 2009). Accordingly, dietary supplementation with omega 3 does lower key indicators of inflammation (Li and Zhang 2014).

People who harbour the epsilon 4 allele have a higher omega 6:3 ratio (Dang et al. 2015). It was therefore thought that epsilon 4 could contribute to atherosclerosis by elevating this ratio, as a result of inflammation. Experts who address this issue from an evolutionary perspective point out that the contemporary diet of humans has a higher omega 6:3 ratio than the traditional diet of hunter-gatherers (Eaton et al. 1998; Simopoulos 2008). (For further discussion, see Chapter 6: Nutrition, Energy Expenditure, Physical Activity, and Body Composition.) Putting all of this together, it is reasonable to propose that epsilon 4 could be contributing to atherosclerosis because humans are not adapted to processing fats with a high omega 6 content, the result being greater inflammation and more rapid progression of atherosclerosis. People who harbour epsilon 4, however, do not respond differently in the short run to enhancement of omega 3 in the diet (Conway et al. 2014; Dang et al. 2015). Studies have not confirmed a preventive effect of omega 3 supplementation for people without illness (Kwak et al. 2012; Rizos et al. 2012). But for atherosclerosis patients, long-term supplementation does appear to offer some protection against heart attacks (Casula et al. 2013).

Although incomplete, the emerging picture therefore is that the negative effects of epsilon 4 do not result simply from a reduced ability to protect against the development of atherosclerosis when diets are low in omega 3 fatty acids. The associations of epsilon 4 with a pro-inflammatory ratio of omega fatty acids and atherosclerosis may therefore reflect inflammatory damage once atherosclerosis has developed. In other words, something other than the omega ratio and epsilon 4 is causing the inflammation that contributes to arterial damage in atherosclerosis; the calming effect of omega 3 on the inflammation may lessen the damage.

3.3.5 Infection, Epsilon 4, and Cholesterol

Infectious agents are prime suspects for instigators of inflammation, because inflammatory responses have evolved largely to control infection. But natural selection is acting on both sides of the conflict, favouring both inflammatory defences against pathogens and pathogens that escape control by inflammation. When a coevolutionary stand-off occurs, the immune system may continue using ineffective inflammation in response to the presence of the pathogen, causing peripheral damage by friendly fire.

High on the list of suspected pathogens is *Chlamydia pneumoniae* (also known as *Chlamydophila pneumoniae*), which has been correlated with coronary artery disease and found in atherosclerotic lesions (Filardo et al. 2015). When it infects macrophages it can transform them into foam cells (He et al. 2009; Mei et al. 2009), a tell-tale sign of early atherosclerosis. People who carry the epsilon 4 allele are especially vulnerable to damage caused by *C. pneumoniae.* This vulnerability was first documented by a study of arthritis patients (Gerard et al. 1999), who often have bacteria in their joint tissues. When knee cartilage was tested for the presence of bacteria, about 70% of the patients who tested positive for *C. pneumoniae* carried the epsilon 4 allele. Patients who tested positive for other bacteria or negative for all bacteria carried the epsilon 4 allele at frequencies ranging from about 15% to 25%, not significantly different from the frequency in the general population. A follow-up study discovered a reason for the increased vulnerability: *C. pneumoniae* enhances its invasion of cells by hitching a ride on the epsilon 4 protein (Gerard et al. 2008). It stimulates inflammation, and is therefore a plausible explanation for inflammatory damage in atherosclerosis (Filardo et al. 2015).

Infection also offers an explanation for the association of elevated serum lipids with atherosclerosis. The lipopolysaccharide (LPS) component of bacterial cell walls leads to elevated serum cholesterol and triglycerides (Feingold et al. 1992, 1993; Memon et al. 1993; Urosevic and Martins 2008), an effect that occurs in response to intracellular bacteria (Samanta et al. 2017). This effect seems to occur in part because the elevation in lipids has a protective effect against bacteria systemically (Urosevic and Martins 2008) and intracellularly (Mulye et al. 2017); however, pathogens often evolve resistance to such defences, in which case the stimulation of the response may persist because it is not terminated by bacterial clearance. This is probably the case with *C. pneumoniae,* which tends to be lipophilic and survives accumulation of lipids within foam cells.

The enhancement of lipid concentrations is mediated by pro-inflammatory cytokines. Accordingly, periodontal disease, which is characterised by inflammation and caused by pathogens that are also linked to atherosclerosis, is strongly associated with the same lipid changes that characterise atherosclerosis: high LDL, low high-density lipoproteins (HDL), and high triglycerides (Penumarthy et al. 2013; Sandi et al. 2014, Nepomuceno et al. 2017).

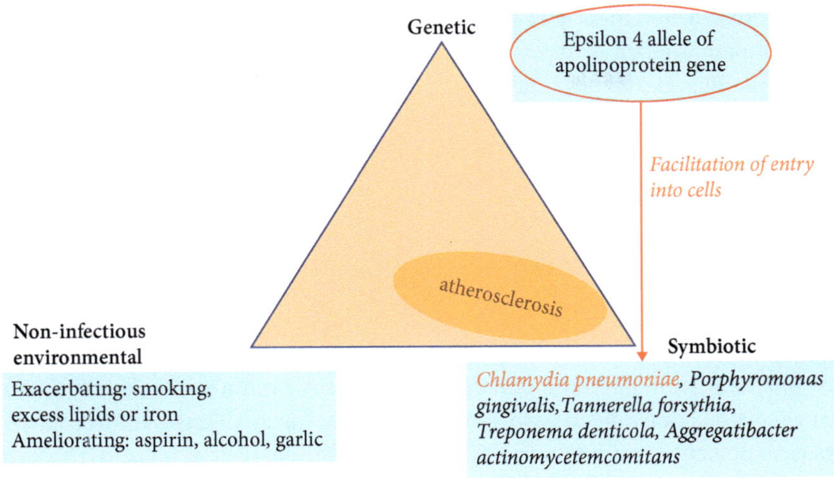

FIGURE 3.2 Representation of atherosclerosis within the triad of disease causation. Although knowledge about causation is still incomplete, atherosclerosis is placed closest to the symbiosis vertex to illustrate the explanatory power of infectious agents as essential causes and other risk factors as exacerbating causes. Infectious agents listed in the symbiotic category have all be found in atherosclerotic lesions. Risk factors in the non-infectious environmental category have all been associated, presumably causally, with atherosclerosis.

To integrate the risk factors of atherosclerosis into a causal framework they need to be considered in the context of essential and exacerbating causes. For each contributing cause it is important to ask whether it could initiate the pathogenesis without input from the other contributing causes and whether the pathogenic process can occur without it. For atherosclerosis this initiation would involve transformation of macrophages into foam cells, initiation of pathogenesis within the arterial wall, the initiation of inflammatory response including attraction of phagocytic neutrophils to the site of pathogenesis within the arterial wall, and lipid dysregulation. Although knowledge about the pathogenesis of atherosclerosis is still incomplete, the only category that seems able to account for these early events is infectious causation. This tentative conclusion is represented in Figure 3.2 by placement of atherosclerosis closest to the symbiotic vertex.

3.3.6 Alzheimer's Disease, Epsilon 4, and Infection

If this explanation is correct we would expect the other epsilon 4-associated diseases also to be associated with *C. pneumoniae*. Accordingly, Alzheimer's disease has been associated with *C. pneumoniae*. The first study used immunofluorescent antibodies and polymerase chain reaction (PCR) to show that *C. pneumoniae* was present in 22 of the 23 brains of people who had died from sporadic Alzheimer's disease but in only one of the 25 brains of people who did not have Alzheimer's at the time of death (Balin et al. 1998; A. Hudson, personal communication, for some additional cases tested after publication). Several authors did not find an association between *C. pneumoniae* and Alzheimer's disease, but a meta-analysis found that the reported association of *C. pneumoniae* and Alzheimer's is statistically significant (Maheshwari and Eslick 2015). *C. pneumoniae* has been shown in mice to stimulate beta amyloid deposition, a presumed cause of Alzheimer's disease (Little et al. 2004; Boelen et al. 2007). In accordance with the documentation of epsilon 4 and *C. pneumoniae* as risk factors for both Alzheimer's disease and atherosclerosis, a meta-analysis confirmed that Alzheimer's disease is significantly associated with stroke, a major consequence of atherosclerosis (Zhou et al. 2015).

The increased vulnerability to *C. pneumoniae* that is associated with the epsilon 4 allele helps integrate the accepted genetic association of epsilon 4 with the candidate infectious causes of these diseases. This integration, however, does not preclude a causal role for other infectious agents any more than it excludes other genetic influences. In Alzheimer's disease, the epsilon 4 allele increases vulnerability to another leading candidate pathogen, human herpes simplex 1 (HHSV-1), although the exact mechanism is unclear (Itzhaki 2017).

Spirochetes have also been associated with Alzheimer's disease according to several studies and a recent meta-analysis (Miklossy 1993; Miklossy et al. 1994; Maheshwari and Eslick 2015). Other pathogens (*Helicobacter pylori, Porphyromonas gingivalis*, and other periodontal pathogens and herpes viruses) have been associated individually and collectively (Kamer et al. 2009; Bu et al. 2015; Harris and Harris 2015). Each of these pathogens is pro-inflammatory, but they have not been assessed for associations with the epsilon 4 allele. Like *C. pneumoniae*, spirochetes and herpes viruses have been found in brain tissue at the sites of Alzheimer's damage; their contribution to Alzheimer's may therefore involve direct damage in addition to inflammation-induced pathology.

Resolving the causation of Alzheimer's disease has been difficult because its insidious development masks obvious assignment of cause and effect and opens the door to contributions from many different factors. The overall trend in interpreting the associations between infectious agents and Alzheimer's disease over the past quarter century, however, has been that infectious agents are contributing causes rather than just being spurious bystanders or chance correlates.

In concert with the gradual shift towards acceptance of infectious causation of Alzheimer's disease, a string of interventions have inhibited amyloid plaques but have not generated the therapeutic benefits that would be predicted if the amyloid plaques were the cause of brain dysfunction (Panza et al. 2014). These results, together with the presence of beta amyloid plaques in healthy individuals, and reports of weak or absent associations between plaque density and brain dysfunction suggest that the pathology of late-onset Alzheimer's disease may involve other causes (Alagiakrishnan et al. 2012). If infectious agents are contributing causes, the plaques and tangles may be downstream in the pathological process or side-effects. The recognition that beta amyloid is antimicrobial (Kagan et al. 2012) lends credence to the idea that pathogens are more centrally involved in the causal process. (For further discussion, see Chapter 5: Senescence and Ageing.)

Geographic distributions also implicate vulnerability to infectious environmental causes. Island–mainland comparisons can be particularly useful because island populations tend to have less continuous exposure to pathogens and may therefore have more intense disease when the pathogens are introduced or reintroduced. Island–mainland comparisons for Alzheimer's disease have not been extensively studied, but some evidence suggests a difference consistent with genetic vulnerabilities to pathogens. The epsilon 4 allele was almost six times more strongly associated with Alzheimer's disease on the island of Sicily than on the Italian mainland (Bosco et al. 2005).

In light of these considerations, the emerging picture of pathogenesis mirrors that of atherosclerosis. Infections are the most parsimonious essential cause of Alzheimer's disease, because infectious causation can encompass all of the other risk factors in a causal framework, with epsilon 4 exacerbating infections with two infectious causes: *C. pneumoniae* and *HSV-1*. The presence of *C. pneumoniae, HSV-1*, and spirochetes in brain tissue suggests direct damage. The other pathogens may be involved indirectly as exacerbating causes of the pathological process by generating inflammation.

3.3.7 Garlic and Epsilon 4 Diseases

Garlic and its constituents have protective effects on various markers of cardiovascular disease event rates and indicators of risk, such as C-reactive protein, total cholesterol, plaque volume, and immune activity (Koscielny et al. 1999; Ried 2016; Varshney and Budoff 2016). It has similarly been associated with protection against Alzheimer's disease (Chauhan 2005; Borek 2006). Direct effects of garlic have been proposed. In some cases, research has generated supportive evidence for a direct effect on atherosclerosis risk factors. Most of the measured benefits, however, could arise indirectly through antimicrobial effects. This possibility has been largely overlooked in spite of the fact that the antibacterial activity of garlic is one of the most powerful of all food additives (Billing and Sherman 1998; Ankri and Mirilman 1999; Harris et al. 2001).

Antimicrobial effects of garlic are attributed to the action of allicin and its derivatives, which are produced by the garlic bulb to defend against organisms that would otherwise exploit it. When a bulb is physically damaged it rapidly produces allicin from a precursor, and then modifies some of the allicin to produce other compounds such as diallyl sulfide. Garlic extracts, allicin, and diallyl sulfide strongly inhibit one of the periodontal pathogens associated with atherosclerosis: *Aggregatibacter actinomycetemcomitans* (Bachrach et al. 2011; Velliyagounder et al. 2012; Shetty et al. 2013). The activity of allicin was degraded after heating at boiling temperature for 20 minutes, but diallyl sulfide was still active after this treatment and when *A. actinomycetemcomitans* was in a biofilm (Velliyagounder et al. 2012). The concentrations of allicin needed to inhibit *P. gingivalis* were higher than for *A. actinomycetemcomitans* (Shetty et al. 2013). The action against *P. gingivalis* neutralised protein-destroying enzymes, such as gingipains, which contribute to the virulence of *P. gingivalis* but also confer resistance to allicin (Bachrach et al. 2011; Shetty et al. 2013).

Overall, the evidence is consistent with the idea that garlic generates its beneficial effects at least in part by suppression of microbial causes of epsilon 4 diseases, but the available evidence is not yet sufficient to distinguish such indirect effects from direct effects.

3.3.8 Smoking, Epsilon 4, and Infection

Genetic causation must take into account not only vulnerabilities to non-living hazards but also the possible interaction between these non-living causes and infectious causes. The possible interplay between epsilon 4, tobacco smoke, and infectious causation of atherosclerosis provides an illustration. Smoking about two packs of cigarettes per day is associated with doubling of the risk of cardiovascular events (Lubin et al. 2016). It is often presumed that toxic components of tobacco smoke damage artery linings, leading to the development of atherosclerosis. One problem with this argument is that the early stages of atherosclerosis involve pathology within the wall of arteries rather than on the surface, as was once thought. Any model for direct toxic effects of smoke components on arterial tissue must therefore explain why these toxic effects occur on cells that are less exposed to the compounds than on the more exposed cells that line the arteries.

An alternative explanation is that the smoke could exacerbate atherosclerosis by suppressing immune defences. Cigarette smoke suppresses immunity in ways that can help explain the association of smoking with lung infections. However, smoke constituents tend to be anti-inflammatory, whereas atherosclerosis is characterised by an inflammatory pathology (Stämpfli and Anderson 2009). This apparent paradox is explainable if the immune-suppressive effect of smoking exacerbates persistent infections that then trigger inflammatory damage over longer periods of time. Accordingly, the main infectious candidates for atherosclerosis, *C. pneumoniae* and periodontal pathogens, are all persistent infectious agents that can cause chronic inflammatory damage. Smoking is associated with exacerbations of infections with these pathogens and suppression of immune defences against them at the primary sites of infection: the lungs and periodontal areas, respectively (Stämpfli and Anderson 2009; Guglielmetti et al. 2014; Souto et al. 2014a, b).

In the case of cigarette smoking, comparison of heart attacks among smokers, second-hand smokers, and non-smokers raises another paradox. People who smoke two packs per day are twice as likely as non-smokers to experience heart attacks. But second-hand smokers—people who live with smokers—have an increased risk that is about one-third of this amount, even though they inhale less than one-one-hundredth the amount of smoke. The paradox vanishes when infection is considered. Smokers have more florid lung infections, which can be transmitted through coughing to the people who live with the smokers (Arnold et al. 1993). The combination of *C. pneumoniae* and exposure to second-hand smoke is associated with acceleration of atherosclerosis (Zhou et al. 2012). The disproportionate increase in heart attacks linked to living with smokers may therefore result from pathogens the non-smokers inhale rather than from the smoke itself.

This explanation raises the possibility that lung infections in smokers may contribute to their doubled risk of heart attacks. Significantly, a study of associated chronic illnesses revealed that the association between smoking and heart attacks occurred only among smokers who had chronic obstructive pulmonary disease (COPD), chronic bronchitis, or periodontitis (Kiechl et al. 2002a). These three diseases are known to be caused, or strongly suspected of being caused, by candidate atherosclerosis pathogens. *C. pneumoniae* contributes to chronic bronchitis and has been identified as a candidate risk factor for COPD (Erkan et al. 2008; Papaetis et al. 2009; Choroszy-Król et al. 2014; Muro et al. 2016); *P. gingivalis* contributes to periodontitis (Mysak et al. 2014). If smoking caused atherosclerosis directly, the association between smoking and heart attacks should persist even in people who do not have COPD, bronchitis, or periodontitis.

One of the strongest correlates of impending heart attacks is elevated levels of the inflammatory marker C-reactive protein. Although this association is useful for clinicians who want to be able to determine at-risk patients, it does not resolve the more basic issue of why inflammation is so elevated even though it is potentially lethal. C-reactive protein is produced in response to a variety of insults, including infection with *C. pneumoniae* and *P. gingivalis*. A study of atherosclerosis of the main artery in the thigh showed that C-reactive protein was strongly correlated with the presence and severity of disease; but antibodies to *C. pneumoniae* were even more strongly associated (Kaperonis et al. 2006).

3.3.9 Evolutionary Decline in Epsilon 4

The vulnerability to infection associated with epsilon 4 generates a hypothesis for the decline in the epsilon 4 allele during the evolutionary history of *H. sapiens*: increased severity of infections among epsilon 4 individuals disfavoured the epsilon 4 allele relative to the other apolipoprotein E alleles, especially epsilon 3. The disadvantage of epsilon 4 would have been relatively small because the negative effects of atherosclerosis and Alzheimer's disease on fitness tend to occur late in life. Reassessments of the longevity of humans in pre-agricultural societies, however, indicate that humans probably lived regularly into their sixties (Gurven and Kaplan 2007), at which age males can still reproduce and both sexes can enhance their inclusive fitness by helping children and other relatives. Atherosclerosis and Alzheimer's disease could therefore have reduced the success of the epsilon 4 allele slightly in addition to any increased severity of illnesses at younger ages that were associated with epsilon 4 exacerbated infections (e.g. respiratory infections for *C. pneumoniae*). A slight fitness disadvantage is all that would be needed to decrease the frequency of epsilon 4 from near fixation to a minority allele during the time period over which it declined, estimated to be about 50,000 years (Fullerton et al. 2000).

An alternative hypothesis argues that epsilon 4 was selected against when human ancestors began eating meat and concomitantly began living longer (Finch and Stanford 2004). According to this hypothesis, meat diets increase lipids in the blood and longer life allowed the main epsilon 4 diseases which occur later in life to exert a selection pressure favouring epsilon 3 (Finch and Stanford 2004). The more rapid selection against epsilon 4 in populations with a long history of agriculture could be explained by this hypothesis for agricultural settings that involve increased consumption of domestic mammals (with their high omega 6 to omega 3 ratio) and longer life. Much evidence, however, does not accord with this hypothesis. People with Mayan ancestry, for example, have a lower frequency of epsilon 4 than Amerindians who were recently hunter-gatherers (Corbo and Scacchi 1999), even though the ancient Mayan diet was based on plants supplemented with undomesticated animals (White 1999). Other evidence mustered in support of this meat diet hypothesis, such as the association of epsilon 4 with high serum lipids and inflammation, accords with the infectious causation hypothesis as well, as discussed in Section 3.3.5. Epsilon 4 vulnerability to infection therefore explains the entire spectrum of evidence implicating epsilon 4 as an exacerbating cause, better than the switch to a meat diet, though the hypotheses are not mutually exclusive.

3.4 Multiple Genetic Contributions to Disease

The rapid expansion of genetic information about diseases makes it clear that most diseases are influenced by more than one genetic cause. Each genetic cause needs to be assessed in the context of interactions between genetic, environmental, and infectious causes. The epsilon 4 diseases illustrate this point. The epsilon 4 allele makes the most significant genetic contribution to atherosclerosis and late-onset Alzheimer's disease, but other genetic contributions exist. An allele for toll-like receptor 4 (TLR4) contributes to atherosclerosis by ratcheting up inflammation in response to LPS, a ubiquitous component of bacterial cell walls. This response makes sense because inflammation is an important part of immunological defences against infection. Two alleles that encode for less-responsive TLR4 molecules are referred to as 299gly and 399Ile because the TLR protein has substituted a glycine or isoleucine in amino acid positions 299 or 399 (Arbour et al. 2000). Individuals who harbour these alleles are more vulnerable to severe bacterial infections but have less atherosclerosis (Kiechl et al. 2002b). They are less vulnerable to periodontal disease (James et al. 2007; Ozturk and Vieira 2009) and less immunologically responsive to *P. gingivalis* (Kinane et al. 2006). This concordance probably arises because *P. gingivalis* can modify its LPS molecules to manipulate the TLR4 receptor (Hajishengallis and Lamont 2014). These modifications can enhance inflammatory responses in a way that benefits *P. gingivalis* (Hajishengallis 2014; Hajishengallis and Lamont 2014). The less-responsive TLR4 alleles neutralise this manipulation and thus protect the person from *P. gingivalis* while reducing long-term damage from inflammation.

The TLR4 alleles are not, however, less responsive to *C. pneumoniae* (Rupp et al. 2004). To turn on inflammation, TLR4 interacts with another membrane protein, called CD14 (Kiechl et al. 2002b). Like the TLR4 variants, a variant of the CD14 gene (called 159C>T) is associated with blunting of the inflammatory response to bacteria and decreased risk of atherosclerosis. This CD14 allele is not, however, associated with decreased risk of periodontitis (James et al. 2007), but is associated with decreased risk of chronic *C. pneumoniae* infection (Rupp et al. 2004). If *C. pneumoniae* contributes to atherosclerosis, this result suggests that enhanced vulnerability to persistent *C. pneumoniae* infection is associated with low risk of atherosclerosis so long as inflammation is toned down by the CD14 variant.

In accordance with their role in atherosclerosis, TLR4 and CD14 also contribute to Alzheimer's disease, apparently through beta amyloid-induced inflammation (Fassbender et al. 2004; Walter et al. 2007; Reed-Geaghan et al. 2010). The 299gly mutation in TLR4, which blocks inflammation induction by *P. gingivalis* and is associated with protection from atherosclerosis (see above), also confers protection from Alzheimer's disease (Minoretti et al. 2006). Whether the CD14 allele that suppresses *C. pneumoniae*-induced inflammation in atherosclerosis also reduces inflammation in Alzheimer's disease still needs to be assessed.

Essential genetic causes can also interact with each other, generating a synergistic pathological effect. Individuals who are heterozygous for the sickle cell allele generally have few if any adverse health effects, so long as the other alleles for haemoglobin are normal. The genetic bases for other prevalent haemoglobinopathies—thalassaemias, haemoglobin C, and haemoglobin E—are also virtually silent when paired with a normal haemoglobin allele. The thalassaemias are due to a mutation in either the A or the B subunit of the haemoglobin gene, and are therefore categorised as alpha and beta thalassaemias. Haemoglobins C and E result from mutations in the A and B subunits, respectively.

These altered haemoglobin alleles are often favoured in the same population because they are thought to provide protection against malaria. This commonality means that individuals who are heterozygous for the sickle cell allele may also be heterozygous for one of the other haemoglobinopathy alleles. The negative effects of the sickle cell in homozygous individuals have generally kept sickle cell allele frequencies below 20%, even where *falciparum* malaria has been prevalent and endemic (Grosse et al. 2011). The vulnerability of individuals who do not have a sickle cell allele has favoured the evolution of these other alleles. As is the case with sickle cell anaemia, most of these alleles can cause life-threatening disease when individuals are homozygous. If, however, a person has one copy of the allele for any of these haemoglobinopathies and one copy of the sickle cell allele, the blood cells will be vulnerable to sickling and the person is categorised as having 'sickle cell disease', a general category that includes sickle cell anaemia. The presence of sickle cell disease in individuals who are heterozygous for the sickle cell anaemia allele and one or more of the other haemoglobin mutations illustrates the interaction between essential genetic causes of different diseases. Sickle cell disease tends to be less severe than sickle cell anaemia and occurs in older ages (Manci et al. 2003). Infection is also an exacerbating cause of the sickle cell disease that arises from joint heterozygotes. In an autopsy study, death was attributed to infection in 43% of individuals heterozygous for the sickle cell allele and haemoglobin C and in 33% of individuals heterozygous for the sickle cell allele and beta thalassaemia, compared with 48% for sickle cell anaemia (Manci et al. 2003). Largely through efforts to control exacerbating infectious contributions to sickle cell disease, lifespan has increased dramatically. In high-income countries, since 1970, life expectancy has increased from about 14 to 40–60 years (https://www.nhlbi.nih.gov/health/health-topics/topics/sca).

3.5 Evaluating Genetic Causation in a 'Complex Genetic Disease': Schizophrenia

3.5.1 Schizophrenia as a Complex Genetic Disease

As mentioned in the Introduction, when geneticists began studying diseases that appeared to have a genetic basis, but low to moderate penetrance, they often applied the term 'complex genetic diseases'. These diseases can be revisited using the organisation presented in this chapter. Schizophrenia provides an illustration.

Schizophrenia is probably an umbrella category of disorders characterised by a mix of 'positive symptoms', which include hallucinations, delusions, and thought disorder; and 'negative symptoms', which include withdrawal, flattened emotional responses, and impairment of memory. Patients with schizophrenia have a high suicide rate, few children, and a high rate of abnormality in their children (Bassett et al. 1996; Osby et al. 2000). Schizophrenic mothers are more likely than non-schizophrenics to have stillborn babies and children with congenital malformations (Sobel 1961; Rieder et al. 1975; Bennedsen et al. 2001). Later in life, individuals with schizophrenia have a high incidence of immune and metabolic disorders, which are frequently accompanied by vision impairments, brain damage, and seizures (Maricle et al. 1987; Müller et al. 2000). Parents of schizophrenics are more likely to have a variety of medical conditions, including various types of cancer and thyroid disease, and first-degree relatives of schizophrenics are at a higher risk of developing personality disorders and insulin-dependent diabetes mellitus (Guggenheim et al. 1969; Onstad et al. 1991; Schwarz et al. 2001).

Studies of schizophrenia generated concordances among identical (i.e. monozygotic) twins of about 40–60%. Fraternal (dizygotic) twins had about half this level of concordance. These concordances led to the conclusion that schizophrenia is largely genetically based.

This conclusion leads to an evolutionary paradox, namely, how these genes could be maintained at high frequency in spite of their fitness costs. Attempts to resolve this paradox have invoked compensating benefits for the costly alleles. These evolutionary explanations, however, falter on various accounts. They do not explain the diversity of the manifestations (Brüne 2004). Nor do they identify fitness benefits that could outweigh the fitness costs associated with these manifestations. Geneticists have not found any alleles that could explain individually any more than a small percentage of schizophrenia. They therefore have presumed that the genetic basis involves a large number of alleles each with a small effect, or alternatively rare mutations each with a large effect (McClellan et al. 2007). Either way, it becomes important to understand whether proposed genetic mechanisms are sufficient to explain the substantial monozygotic twin concordance. This assessment also needs to evaluate whether the interpretations of the evidence for genetic causation are correct.

The argument for genetic causation without compensating benefits often invokes the idea that schizophrenia rates are the same throughout the world (e.g. Jablensky 2000). This apparent worldwide uniformity of schizophrenia prevalence contributes to the paradox of schizophrenia because it tends to negate the possibility that schizophrenia could be attributable to geographically variable environmental influences favouring the influence of background mutation rates (Brüne 2004). The worldwide prevalence of schizophrenia averages about 1% and actually varies considerably geographically: 4% in western Ireland, 2.2% in an isolated Finnish population, 0.1–0.2% in the Inuit, about 0.1% in the Hutterites and aboriginal Australians, and 0.0–1.0% among Micronesians on islands with similar climates (Sampath 1974; Kiloh 1975; Torrey 1980; Dale 1981; Haukka et al. 2001; reviewed by Ledgerwood et al. 2003). A comparison of incidence using uniform methods generated a range, with the highest incidence being 5.6 times the lowest (McGrath et al. 2004; McGrath 2006). Recognition of this substantial variation in prevalence and incidence instantly resolves one of the paradoxes of schizophrenia, by weakening the hypothesis that the genetic basis for schizophrenia results largely from the continued generation of uncompensated mutations. These considerations draw attention to the need to search for environmental (infectious and non-infectious) causes (Torrey 1980; Fañanás and Bertranpetit 1995; Ledgerwood et al. 2003) and the interaction between environmental and genetic contributions (McGrath 2006).

3.5.2 Familial Associations and Infection

Monozygotic twin *discordance* sets a minimum figure for environmental causation. Several infectious agents have been associated with schizophrenia (Arias et al. 2012). Prenatal infections are particularly relevant candidates for environmental causation because their effects may be strongly associated with familial correlations that have been advanced as proof of genetic causation. Identical twins, for example, may be more likely than fraternal twins to be exposed to the same environmental causes of disease. As fetuses, fraternal twins are more likely to be separated from each other by both an amnion and a chorion. Identical twins are more likely to be separated by neither. Approximately 70% of monozygotic twins share a chorion and 5% share an amnion; almost no dizygotic twins are monochorionic and none shares an amnion (Bourne 1962; Danforth 1986). Monochorionic twins are also more likely to share a placenta, a state that has been associated with increasing concordance for infection (Phung et al. 2002).

These findings indicate that prenatal exposure of twins to infection will be correlated with genetic relatedness. Any disease that arises from prenatal infection could therefore be more concordant in identical twins than in fraternal twins. Interpretations about genetic causation could be similarly confounded if the environmental exposure occurs during childhood, when children have more contact with other children of the same sex than the opposite sex (because identical twins are always of the same sex). Identical twin concordances therefore indicate a ceiling of genetic causation rather than an estimate of it. To determine how far below this ceiling the actual level of genetic causation lies requires evaluation of alternative hypotheses such as prenatal infection.

Concordance for schizophrenia occurred in 60% of monozygotic twins with markers of monochorionicity but in only 11% of those who did not have the indicators (Davis et al. 1995). Because monochorionic monozygotic twins are not more closely related genetically than other monozygotic twins, this difference in concordance for schizophrenia implicates some prenatal environmental influence, such as prenatal infection.

In accordance with the hypothesis of prenatal infectious exposure, several studies now identify maternal infections during pregnancy as a risk factor for developing schizophrenia. Women who had particularly active or intense infections with *T. gondii* during pregnancy were more likely to have babies who eventually developed schizophrenia (Buka et al. 2001; Blömstrom et al. 2012, 2015). Influenza and herpes infections during pregnancy have also been associated with development of schizophrenia in offspring (Buka et al. 2001, 2008; Brown et al. 2000, 2005, 2009; Mortensen et al. 2010).

More generally, many studies have correlated schizophrenia with positivity for *T. gondii* infections (see preceding references; Hamidinejat et al. 2010; and older studies reviewed by Torrey and Yolken 2003 and Ledgerwood et al. 2003). This association also occurs between long-standing infections and first-episode schizophrenics (Yolken et al. 2001). This finding together with the maternal associations mentioned above negate the possibility of the reverse direction of causality, namely that the association of *T. gondii* with schizophrenia occurs because schizophrenia increases the probability of acquiring *T. gondii* infections (e.g. if schizophrenia increased the tendency to own *Toxoplasma*-infected cats).

Comparison of dizygotic twins with full siblings can provide insight into the potential significance of prenatal infection. If genetic causation were the sole explanation of the dizygotic twin concordance, the concordance values for these two categories of twins should be the same, because dizygotic twins and non-twin full siblings have the same genetic relatedness. Contributions of prenatal infection to schizophrenia will tend to lower the full sibling concordance relative to the dizygotic twin concordance because the mother's state of infection during pregnancy will be identical for dizygotic twins but not for non-twin full siblings. The extent to which full sibling concordances are lower that dizygotic twin concordances therefore corresponds to the minimal extent to which environmental causation is contributing to dizygotic twin concordance. It indicates the minimal effect, because a prenatal infection during one pregnancy could still affect a second pregnancy, if, for example, the infection was persistent or if it contributed through epigenetic effects. Concordance values for full siblings are generally about half those of dizygotic twins, roughly 3–12% versus 8–27%, respectively (Essen-Möller and Fischer 1979; Kringlen 2000).

The greater concordance with biological parents than with adoptive parents has also been interpreted as evidence that schizophrenia is genetically based (Ingraham and Kety 2000; Tsuang et al. 2001), without adequate consideration of infectious causation. Prenatal and perinatal infections are inherited from biological parents rather than from adoptive parents. The concordance of schizophrenia with biological parents could therefore result from a genetic and/or an infectious cause that they inherited from their biological parents. Similarly, having older siblings is associated with an increased risk of developing schizophrenia (Sham et al. 1993). Older siblings may transmit pathogens to their pregnant mothers, thereby increasing the fetus's risk of developing schizophrenia (Sham et al. 1993).

Increased parental age is a risk factor for schizophrenia (Kelemen 1977; Dalén 1988; Malaspina et al. 2001). This association is generally assumed to be caused by genetic damage that increases with age, but it is also consistent with infectious causation, because the cumulative exposure of parents to pathogens increases with age. The two factors could also be related because infectious agents are known to cause genetic damage in humans (Livezey and Simon 1997; Shima et al. 2000; Gualandi et al. 2001; McDougall 2001; Nevels et al. 2001).

The association of schizophrenia with having older siblings does not, however, appear to be simply a result of increased parental age, because the increased risk is associated with having siblings who are a few years older, but not with having siblings who are many years older and who would tend to have even older parents. These associations therefore are not readily explained by genetic causation but can be explained by infectious causation.

Prenatal or perinatal infections are the most common known cause of severe congenital malformations such as those associated with schizophrenia (Rawlins 2001). Allelic associations with schizophrenia are known to confer susceptibility to pathogens and to cause birth defects and miscarriages. Several pathogens may be involved: cytomegalovirus, influenza, herpes simplex, rubella, and *T. gondii* (Carter 2009). A strong effect of a gene for a cytokine receptor also implicates prenatal infection (Lencz et al. 2007).

3.5.3 Season of Birth

In temperate climates, the frequency of late winter and early spring births is generally 5–15% greater among babies that eventually develop schizophrenia than among controls (Videbech et al. 1974; Torrey et al. 1977; Pallast et al. 1994; Castrogiovanni et al. 1998). The association was not found in Singapore, where distinct warm and cold seasons are absent (Parker et al. 2000; de Messias et al. 2001). This set of findings implicates an environmental cause that is associated with spending time indoors, such as increased exposure to infection during the second or third trimester of pregnancy.

3.5.4 Geographic Associations

The prevalence of schizophrenia tends to be greater among people born in urban areas than in nearby rural areas with the same ethnic composition (Torrey et al. 1997; Haukka et al. 2001; Torrey et al. 2001). The pattern is consistent with infectious causation because exposure to infectious diseases is often more prevalent in more densely populated urban environments (Yolken and Torrey 1995; Torrey et al. 1997). To explain the urban–rural correlations by genetic causation one would need to assume that a greater frequency of alleles coding for schizophrenia could persist in urban environments despite high rates of immigration and emigration from urban areas.

Characteristics of schizophrenia also vary with latitude. Season of birth associations tend to dissipate in lower latitudes, as would be expected if schizophrenia is sometimes caused by an infection or some other environmental factor that is acquired indoors (Ledgerwood et al. 2003). Age of onset in urban populations also varies directly with latitude (Shaner et al. 2007). The reasons for this association are unclear but could involve greater overall exposure to parasitism in lower latitudes (Shaner et al. 2007).

3.5.5 *Toxoplasma gondii* as a Candidate Pathogen

Schizophrenia, being an umbrella category of psychotic disorders, could be caused by a variety of infections. The documented associations with *T. gondii* have been and remain the strongest associations between schizophrenia and any particular agent (Ledgerwood et al. 2003). Mothers of schizophrenia patients had elevated titres of immunoglobulin M (IgM) against *T. gondii*, but not IgG, during pregnancy (Buka et al. 2001), suggesting that new *T. gondii* infections tend to be more important than 'slow-burning' or reactivated chronic infection, with the caveat that elevated IgM sometimes occurs in chronic *T. gondii* infections (Contreras et al. 2000; Gomez-Marin et al. 2000). If the association between *T. gondii* and schizophrenia is indeed causal, the elevated positivity in first-episode schizophrenia (e.g. Yolken et al. 2001) suggests that about one-third of cases of schizophrenia would be attributable to *T. gondii*.

If schizophrenia is caused by prenatal infection, the process must involve long delays between the onset of infection and the onset of schizophrenia. Mothers who are first infected with *T. gondii* during pregnancy have about a 60% chance of transmitting the infection to the fetus during the third trimester (Remington et al. 1976; Ho-Yen and Joss 1992). Manifestations of congenital *T. gondii* infection are sometimes delayed until the second or third decade of life (Remington et al. 1976; Jones et al. 2001). Extended follow-up of subclinical congenital toxoplasmosis, for example, revealed that eye lesions and neurological symptoms of *T. gondii* infection often develop over a decade after the initial infection, sometimes even when no such symptoms occurred during the first decade of life (de Roever-Bonnet et al. 1979; Koppe et al. 1986; Koppe and Rothova 1989; Lappalainen et al. 1995).

Discordance for *T. gondii* infection is rare in monochorial twins but common in bichorial twins (Couvreur et al. 1976; Remington et al. 1976; Tjalma et al. 1998). This difference offers empirical support for the presumed difference between monozygotic and dizygotic concordances for infection and hence for schizophrenia caused by prenatal infection. *T. gondii* infections, therefore, have characteristics that are consistent with the higher schizophrenia concordances of monozygotic twins relative to dizygotic twins, of dizygotic twins relative to full siblings, and of full siblings relative to less closely related individuals.

This broad integrative perspective on schizophrenia suggests how a disease that has been considered largely genetic is increasingly looking like a collection of largely infectious illnesses with exacerbating genetic vulnerabilities. The component diseases for which alleles may be essential causes probably result from rare mutations that are lost from the population soon after they occur as a result of the greatly reduced reproduction associated with schizophrenia.

Major histocompatibility complex (MHC) allelic variations feature prominently in genetic associations with schizophrenia (Duan et al. 2010). Genetic variants have been associated with the main candidate pathogens perhaps because of effects on immunological responses (Anders and Kinney 2015; Avramopoulos et al. 2015; Blömstrom et al. 2015). Genome scans also tend to reveal rare alleles associated with neurodevelopment and neuron function (e.g. Loohuis et al. 2015); these alleles could have effects independent of infection if they alter neurobiology directly, or they could be related to infection if they alter neurobiological responses to infection. Although much is unknown, the current evidence suggests that genetic bases for schizophrenia involve alleles related to protection against infection, immune-related pathology, and neural development, with substantial evidence favouring major roles for infection, but without evidence demonstrating a major role for alleles that have effects that are entirely independent from infection.

3.6 Epigenetics in Health and Disease

The genomic plasticity afforded by epigenetics—alterations in gene expression that do not involve changes to the DNA sequence—is almost certainly an evolved mechanism for dealing with changes in the individual's (macro- and micro-) environment. DNA methylation and histone modifications are key epigenetic mechanisms (Figure 3.3). Epigenetic modifications may help level the generational advantage regarding genetic adaptation of parasites, viruses, and pathogenic bacteria relative to more long-lived organisms in host–pathogen arms races. These mechanisms, however, may become a target for manipulation by pathogens. Cancer, infectious disease, cognition, and mental health are among the areas

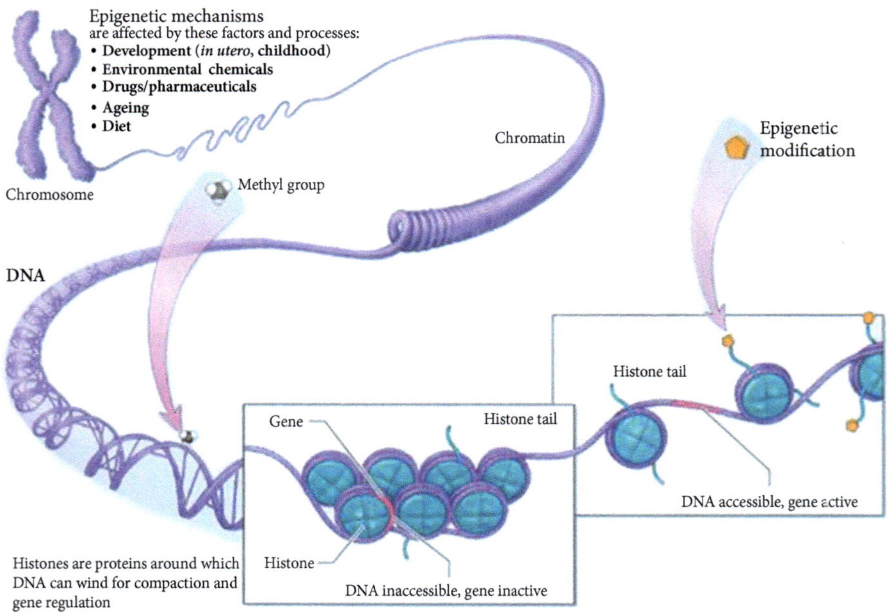

FIGURE 3.3 Two epigenetic mechanisms: DNA methylation and histone acetylation.
Source: https://commonfund.nih.gov/sites/default/files/epigeneticmechanisms.pdf.

associated with epigenetic changes. We will discuss these topics as they relate to evolution and medicine.

3.6.1 DNA Methylation

The epigenetic modification of the cytosine base of DNA by addition of a methyl group can result in gene silencing, and this process, referred to as DNA methylation, is, among other things, crucial to cell differentiation, often dysregulated in cancer, and a target of cancer therapies (Allis and Jenuwein 2016). DNA methylation occurs primarily where cytosine is located directly next to guanine (CpG) in the strand. Overall, the amount of CpG in the human genome is lower than expected, but repeats called CpG islands are often found within genes and gene promoters. DNA methylation marks can be preserved during DNA replication and thus are heritable across cell divisions. Originally thought of strictly as a mechanism for suppressing gene expression, it is now understood that while gene promotor methylation is associated with gene silencing, methylation within the gene is correlated with expression and may play a role in alternative splicing (Jones 2012), resulting in different protein products from a given gene. DNA methylation is essential for many normal cellular processes, including silencing of pluripotent genes to allow cell differentiation during development, X-chromosome inactivation in females which prevents a double gene dose through silencing of one X gene, and gene imprinting (Cedar and Bergman 2009; Li and Zhang 2014).

DNA methylation is frequently altered in cancer and metastasis—these changes can include reduced methylation of oncogenes and increased methylation of tumour suppressor genes (Tam and Weinberg 2013). In acute lymphoblastic leukaemia (ALL) the promotor for the critical tumour suppressor protein p53 has been shown to be hypermethylated in over 30% of the samples reviewed (Agirre et al. 2003). Additionally, aberrant methylation of many genes in the p53 regulatory pathway has been identified (Vilas-Zornoza et al. 2011). This suggests a significant role for epigenetic modifications in ALL because the p53 gene is seldom mutated in this form of leukaemia. Relatedly, this cancer is highly responsive to treatment, possibly suggesting a role for reversion to a normal methylation state in the p53 gene and thus normal p53 protein levels within the cell. Investigation of p53 methylation by many research groups has implicated this epigenetic process in cases of breast, brain, and liver cancers, for example (Saldana-Meyer and Recillas-Targa 2011).

Additionally, in childhood ALL it has been suggested that cancer-promoting environmental exposures may be mediated via methylation changes (Timms et al. 2016). *In-utero* exposure to smoking, alcohol, and iron and postnatal exposure to infection are among the proposed environmental candidates where evidence exists for a possible role in related changes in DNA methylation (Timms et al. 2016). Of note, an infectious aetiology has been proposed for ALL (Greaves 2006); infection with the Epstein-Barr virus (EBV) has been associated with some cases of childhood ALL (Sehgal et al. 2010), and EBV is a known disruptor of host epigenetics (Niller et al. 2009). (For further discussion, see Chapter 9: Senescence and Ageing.)

3.6.1.1 *Genomic Imprinting*

Genomic imprinting involves DNA methylation that generally results in monoallelic expression (via silencing of one of the parental alleles). In humans, the majority of the roughly

300 imprinted genes appear to involve regulation of fetal growth and escape the removal of methylation responsible for silencing that normally occurs shortly after fertilisation. Paternal genes that remain active in imprinting generally increase fetal growth and maternal genes that are expressed suppress growth (for an overview of imprinting see Barlow and Bartolomei 2014). Wilkins and Haig (2003) suggest that a parental conflict may explain why a fundamental advantage of sexual reproduction (that expression from two alleles of a given gene may compensate for a weakness, i.e. recessive mutation, in either of the genes) would be discarded at these parentally imprinted alleles. The selective advantage from imprinting, then, might be to counterbalance the genetic interest of paternal alleles that promote fetal growth and thus use of maternal resources and the maternal alleles that restrict growth, protecting the mother's own health and potential future offspring (Wilkins and Haig 2003). Diseases associated with imprinted genes have helped reveal mechanistic details of this process as well as its potential cost and significance to the organism (Butler 2009).

3.6.2 Histone Modifications

The structure of chromatin, DNA plus histone proteins, plays a critical role in gene expression by restricting or relaxing access to DNA regions. The chromosomal DNA of multicellular (and some unicellular) organisms is wrapped, roughly 147 base pairs at a time, around histone proteins in units of eight, forming nucleosomes. These repeating DNA–histone associations can appear as 'beads on a string' microscopically and can proximately influence one another's structure. This overall organisation of chromatin is highly conserved among organisms with nuclei, suggesting its importance in DNA regulation (Alberts et al. 2015). Epigenetic modifications of histone proteins—for example, through methylation or acetylation of specific amino acids—change how tightly DNA associates with these chromatin proteins. Alterations in DNA–histone avidity as well as direct influences of modified histones not only appear to play a role in regulating gene expression (influencing cellular activity, fate, and environmental response) but also are involved in other essential processes such as DNA repair and replication (see Bannister and Kuozarides 2011 for a broad overview). While there are numerous types of histone modifications, including phosphorylation, sumoylation, and ubiquitination, here we focus on the increasingly characterised role of acetylation, which is almost always activational, and methylation which can facilitate or hinder gene expression.

3.6.2.1 *Histone Acetylation*

Acetylation of the amino acid lysine in the nucleosomal histone shifts its charge from positive to neutral, which reduces the attraction between the histone and DNA (Bannister and Kuozarides 2011); it is thus considered an activating modification. Certain histone core lysines (versus ones located in the histone tail portions that extend out from the nucleosome) are in close contact with the wrapped DNA. It has been shown that when acetylated, these core lysines can be directly involved in stimulating transcription of genes whose promotors are located in the proximate DNA. Additionally, work in model organisms indicates that mutation of these lysines can lead to perturbation of transcription (Tropberger et al. 2013). Acetylation of histone tail lysines appears to be involved in nucleosome cross-talk, facilitating access to large portions of DNA (Allis et al. 2015). The process of histone acetylation and its removal is highly dynamic, allowing for rapid changes in access to DNA by DNA polymerases (Zentner and Henikoff 2013).

In many cancers, cells revert to a more primordial state and are thus less like their fated type—that is, they dedifferentiate. This process involves epigenetic changes. In the mid-1970s, researchers found that cervical cancer HeLa cells exposed to n-butyrate rapidly (re)differentiated, reversing many of the cancer-like properties such as excessive proliferation (Riggs et al. 1977). In other words, they began to act more in accordance with normal cervical cells. This cellular differentiation was thought to be caused by rapid acetylation of histones brought about through n-butyrate's ability to block histone acetylation erasers (histone deacetylases (HDACs)) (Verdin and Ott 2015). When HDACs are blocked, histone acetyltransferases can rapidly acetylate histones (Waterborg 2002), opening the door for gene transcription.

A review of the role of butyrate in brain health suggests that dietary sources of butyrate may decrease blood–brain barrier permeability, enhance immunological protection in the brain, reduce anxiety, and improve cognitive function (Bourassa et al. 2016). Also discussed were findings supporting a role for butyrate as a treatment for many neurologic disorders. A primary mechanism for its effectiveness appears to result from increased acetylation via HDAC inhibition.

A recent large-scale study found that vegetarians had lower rates of colon cancer (Orlich et al. 2015)—this may suggest an intriguing area for consideration. Vegetarians are more likely to consume a high-fibre diet and gut microbes use this fibre to synthesise butyrate (Sonnenburg and Sonnenburg 2014). Could the microbiome be playing a role in reducing the risk of cancer? Could this be in part through the production of butyrate and its effects on acetylation (Fung et al. 2012)? Exploring the role of the microbiome in epigenetic regulation can involve evaluating the mutually beneficial/cooperative interactions between humans and microbial communities and the potential direct applications to health and disease. (For further discussion, see Chapter 13: Digestive System.)

3.6.2.2 *Histone Methylation*

Methylation of histone lysines can be either supportive or suppressive of gene expression. Oddly, methylation of one specific histone lysine that is viewed as an active mark appears primarily to prevent enzymes from methylating lysines at other locations that are associated with gene silencing (Zentner and Henikoff 2013). Histone methylation associated with gene silencing appears to facilitate the activity of enzymes that remove acetylation and increase the histone binding affinity of multiprotein silencing complexes (Zentner and Henikoff 2013).

Not surprisingly, there appears to be cross-talk between histone and DNA methylation which is bi-directional regarding methylation outcome (Cedar and Bergman 2009). Histone methylation may result in gene expression changes that are more easily modified, while changes associated with DNA methylation may be more durable (Cedar and Bergman 2009). Like DNA methylation, changes in histone methylation are associated with cancer and may be involved in treatment failure (Weissman and Knudsen 2009). Mutations in specific nucleosomal histones at amino acids that are typically methylated have been found in a number of cancers, and these 'oncohistones' appear to be able to affect the methylation state of proximate non-mutated histones, leading to large-scale alterations in chromatin methylation and thus associated gene expression (Weinberg et al. 2017).

3.6.3 Non-Coding RNAs: A Mechanism for Specificity in Epigenetics

An overview of epigenetic mechanisms reveals their importance in gene regulation and disease. But how do you get specificity? Non-coding RNAs appear to play a role in directing epigenetic modifiers to specific sequences. Small non-coding RNAs can direct epigenetic modifying complexes, which include HDACs and histone and DNA methylases, to gene promoters with complementary sequences, thereby targeting specific genes for silencing (Weinberg and Morris 2016). Like all DNA methylation, this targeted silencing can be maintained across cell generations by methyltransferases that recognise the hemi-methylated DNA that results from replication, and then methylate the new strand. Long non-coding RNAs also play a role in sequence-specific targeting. X-chromosome inactivation is regulated by the long non-coding RNA Xist (X-inactive specific transcript); this process occurs early in embryonic development and leads to variation in which one of the parental X alleles is silenced from cell to cell (Cerase et al. 2015). It is early days in understanding the breadth of long non-coding RNA activity regarding epigenetically based regulation, but research encompasses the corresponding genes related to cellular regulations, and specific tumour suppressors as well as other aspects of oncogenesis (Weinberg and Morris 2016).

Researchers have found piRNAs (approximately thirty-nucleotide small RNAs that interact with piwi proteins) in neurons in brain regions associated with memory. These small non-coding RNAs have sequences complementary to genes involved in memory consolidation, specifically encoding memory suppressors, and are able to direct epigenetic modifiers to their target gene in response to transient synaptic (serotonin) signals. The resulting silencing of these memory suppressor genes through DNA methylation contributes to the development of long-term memory (Landry and Kandel 2013). Once this target-specific gene silencing has occurred it is maintained through general epigenetic mechanisms of DNA methylation. These exciting findings suggest a (epigenetic) mechanism by which repeated stimulus involving learning circuits can be extended from short- into long-term memories (Kandel et al. 2014). This is an excellent illustration of how external environmental stimuli can be translated by epigenetic mechanisms into altered expression of specific genes in a long-lasting way.

3.6.4 The Fetal to Adult Haemoglobin Switch: An Epigenetic Model

During fetal and early postnatal development, haemoglobin peptide subunit composition changes—this process involves epigenetic mechanisms and results in expression of haemoglobin molecules that are suited to oxygen availability. At the time of birth, a transition occurs from fetal to adult haemoglobin which requires silencing of the fetal γ-globin gene and concurrent activation of the adult haemoglobin subunit gene β-globin. The role of DNA methylation in this switch from fetal to adult haemoglobin production was one of the earliest studied epigenetic modifications (Ginder 2015). Histone methylation in proximity to the fetal haemoglobin gene appears to facilitate DNA methylation of the promotor, resulting in gene silencing (Ginder 2015). The methylation state of the β-globin gene correlates with its expression during development, thus it becomes demethylated when switched on near birth and throughout adulthood (Kiefer et al. 2008). Like DNA methylation, histone acetylation levels correlate with gene expression, and the fetal to adult haemoglobin switch involves removal of acetylation from the active γ-globin gene and addition to β-globin (Kiefer et al. 2008). The ultimate (or why) explanation for the change in fetal to adult haemoglobin composition is the environmental change that occurs at birth—air breathing. (For further discussion, see Chapter 9: Senescence and Ageing.) How dynamic is this switch?

In many individuals with sickle cell anaemia , low levels of fetal haemoglobin continue to be stably produced along with adult haemoglobin (Odenheimer et al. 1984). Within the sickle cell anaemia patient population, having a relatively increased ratio of fetal to adult haemoglobin is correlated with a reduction in disease sequelae and delayed mortality (Platt et al. 1994; Serjeant 2013). In a Jamaican study, individuals homozygous for the sickle cell mutation who lived to the age of 60 or older (lifespans are normally into the 40s or considerably shorter) had 5–10% greater percentages of fetal haemoglobin than individuals from the same patient population who fell within the normal life expectancy for sickle cell anaemia (Serjeant et al. 2007). The association between increased fetal haemoglobin and reduced symptoms and extended survival has led to great interest in non-toxic approaches to stimulating fetal haemoglobin expression (Bertles 1974). A search for epigenetic modifiers—for example, inhibitors of enzymes that would remove acetylation from or methylate fetal haemoglobin genes—has led to clinical trials with mixed success. The transcription factor BCL11A appears to be essential to the process of fetal haemoglobin inactivation and therefore is of interest as a target for sickle cell anaemia therapy (Sankaran and Orkin 2013). While inhibition of BCL11A would be a more surgical approach than previous inhibitors of global-acting epigenetic enzymes (HDACs and DNA methylators), many of the same concerns would remain from an evolutionary medicine standpoint. Efforts to target a transcription factor are very likely to have significant side-effects—BCL11A is thought to be important in neurogenesis and B cell development, for example (Sankaran and Orkin 2013).

An evolutionarily informed approach would suggest that it is difficult to interfere with signals/actors embedded in highly complex pathways that have been adapted to resist perturbation (for example, from pathogens or other environmental or intrinsic effectors) (Ewald and Swain Ewald 2011). On the other hand, an evolutionary approach would suggest that broad epigenetic mechanisms such as methylation and acetylation must have very specific regulators at the individual gene—non-coding RNAs which have DNA sequence specificity appear to be an example of this. Finding such an intermediary between BCL11A transcriptional activity, for example, and fetal haemoglobin DNA for targeting might result in safer therapy.

3.6.5 Epigenetics and the Environment

The mechanisms by which many individuals with sickle cell anaemia continue to produce fetal haemoglobin are not well understood. In some families, it appears to involve mutations in regulatory genes associated with fetal haemoglobin expression (Sankaran and Orkin 2013), but it is also worth considering that an environmental stimulus (possibly driven by reduced oxygen transport) might alter or reverse normal epigenetic silencing of fetal haemoglobin. Interestingly, individuals with β-thalassemia are also found to have an increased ratio of fetal to adult (β-globin) haemoglobin, with a higher percentage of fetal haemoglobin molecules correlating with reduced symptoms (Musallam et al. 2012). This suggests that diseases with different mechanisms of anaemia may create similar environmental conditions leading to continued expression of fetal haemoglobin into adulthood.

Additionally, production of fetal haemoglobin appears to be associated with some cancers (Bertles 1974) and possibly correlates with more aggressive variants (Wolk et al. 2004). It will be interesting to see if research into the mechanism(s) of this unusual fetal haemoglobin expression reveals that this is associated with hypoxic conditions within the tumour microenvironment. Along these lines, researchers have found that inhibition of reactive oxygen species, which are commonly overexpressed during cellular stress, interferes with reactivation of fetal haemoglobin (Mabaera et al. 2008). To resolve the paradox generated by finding that multiple agents—including epigenetic modifiers and cell cytotoxic drugs—can induce fetal haemoglobin expression in patients with haemoglobinopathies and *in vitro* in adult erythrocyte precursors, Mabaera and colleagues (2008) suggest induction via cell stress as the common thread. From an evolutionary perspective, this supports the idea that this fetal to adult switch is facultative and can thus respond to environmental change and may be subject to manipulation by tumour cells or pathogens. Of course, this plasticity is exactly what motivates research into epigenetic modifier drugs designed to ameliorate disease symptoms associated with haemoglobinopathies.

Investigating whether fetal haemoglobin is induced by oncogenic pathogens and any correlations within a cancer type between fetal haemoglobin production and infection might prove informative. In other words, this might provide a signature for pathogen involvement (or the lack of a role) in specific instances of oncogenesis. In general, evolutionary principles provide the insight that disease processes will often act by redirection of existing adaptations—in this case, towards unusual expression of a peptide, γ-globin, that normally is turned on in the more oxygen-depleted environment of the developing fetus and is switched off (in exchange for expression of β-globin with its lower oxygen affinity) after birth.

Like a number of other cancers, the childhood leukaemia juvenile myelomonocytic leukaemia is associated with high levels of fetal haemoglobin production (Fluhr et al. 2017). Oncogenic viruses including EBV have been associated with juvenile myelomonocytic leukaemia (Manabe et al. 2004). Correlations between increased fetal haemoglobin production, haemoglobinopathies, and cancers—including some like non-Hodgkin's lymphoma (Wolk and Newlands 2004) which is also associated with viral infection—suggest complex interactions. For example, there appears to be an increased risk of haematological cancers among individuals with β-thalassaemia (Chung et al. 2015).

In general, research emphasis is often on 'how' questions—What mechanisms are involved in controlling fetal haemoglobin expression? for example. 'Why' questions are of great interest from an evolutionary standpoint—Why is fetal haemoglobin regulation (epigenetically) altered in a number of diseases? Are erythroid cancer cells more successful if they express fetal haemoglobin? Are tumours better oxygenated if tumour cells send out signals to erythroid cells to shift from adult to fetal haemoglobin production? Would oncogenic viruses benefit from manipulating this switch? Efforts to answer why questions will benefit greatly from the wealth of mechanistic information available, including, by way of illustration, detailed information on the role of epigenetics in the fetal to adult switch during normal development and under stressed/disease conditions. Further research in these areas will no doubt shed light on epigenetic plasticity in response to environmental change.

From an evolutionary standpoint, epigenetic mechanisms provide an opportunity to (rapidly) respond to changes in the individual's internal or external environment in a stable, reversible, low-cost way. This is in contrast to genetics which require essentially irreversible changes in DNA bases—which may have different implications if they occur in a somatic cell or via germline mutations. Epigenetics then can serve as an environmental response system that does not rely on adaptations across generations of progeny. Environmental effects on phenotype have been shown to be modulated through DNA methylation and histone modifications (Feil and Fraga 2012).

Can socioeconomic-related environmental factors such as poverty and stress affect brain health through epigenetic changes at the cellular level? Socioeconomic status (SES) appears to impact learning, memory, anxiety, and other measures of cognitive function and mental health. Lower SES is associated with too few resources—for example, reduced access to healthcare and educational enrichment, and decreased positive parent–child interactions. Additionally, low SES correlates with an excess of toxin exposure, negative parent–child interactions, and stress. Studies have shown that effects of SES on cognitive measures are malleable; for example, interventions such as economic improvements and parenting education can reverse differences in cognition and improve mental health (Hackman et al. 2010). Stressors are known to cause changes in DNA methylation in genes regulating the hypothalamic–pituitary–adrenal axis and in specific relevant brain areas such as the hippocampus and prefrontal cortex (Zannas and West 2014). In humans, epigenetic alterations in DNA methylation have been identified in genes associated with stress (affecting glucocorticoid receptor expression). In animal models, changes in environmental stress were correlated with changes in glucocorticoid receptor methylation status (see review by Hackman et al. 2010). Cross-generational epigenetic effects have been reported as well. Much further work will be necessary to determine the role of epigenetics in response to high stress, low resource conditions created by SES. Epigenetics suggest the kind of mechanism that would evolve to respond to relatively stable (difficult) environmental conditions that might none-the-less change in an individual's lifetime. Exploring epigenetic mechanisms in the context of SES seems critical to aiding in developing useful interventions and general policy regarding health and well-being.

Epigenetics as a form of environmental responsiveness can operate at the macro to micro level—for example, through high stress experiences and concordant impact on epigenetic markers, via alterations in the makeup of the gut microbiome, at the memory synapse, and through changes associated with cellular stress. Environment responsiveness is arguably the heart of what epigenetics is about—Why and how does a cell become and stay a neuron, or a liver cell? Why did a system evolve to imprint parental genes? How does the individual or cell respond effectively (or ineffectively) to changes in the environment with the same DNA across a lifetime? Among answers yet to be determined is the role of cross-generational epigenetic changes—can the environment of the parent influence durable epigenetic change in offspring? (Grossniklaus et al. 2013)—in this regard, was Lamarck onto something?

3.6.6 Host–Pathogen Interactions and Epigenetics

DNA and histone methylation can serve as a flexible lockdown system of gene access for which the owner of the DNA has the main key. This might prevent pathogen access to host genes that are methylated and silence invading pathogen or endogenous retroviral

DNA, but will favour evolution of pathogen interference with (de)methylators necessary to alter, initiate, or maintain gene silence. This same construct of a host–pathogen arms race will apply to other histone modifications as well. In fact, many pathogens are expert at taking control of and subverting host epigenetic machinery (Silmon de Monerri and Kim 2014). Furthermore, recognition of unmethylated pathogen DNA is part of the host immune surveillance system and can activate the host immune response (Hoelzer et al. 2008). We present the human papillomavirus (HPV) as a model of viral infection associated with epigenetic changes to the host and the viral genomes.

HPV is the causative agent of cervical, additional anogenital, and some oropharyngeal cancers (Soto et al. 2017). As is true for many oncogenic viruses, infection with HPV is associated with epigenetic changes within the host cell. Examples of aberrant epigenetic modifications include methylation of tumour suppressor promotors in HPV-associated cancers and HPV alteration of cell–cell adhesion by mediating host gene methylation, as well as widespread alterations in histone acetylation and methylation (Soto et al. 2017). Evasion of the host immune system by HPV appears to involve viral-induced epigenetic alterations resulting in infected cells with a reduced capacity to activate innate immunity and to attract key immune system cells (Westrich et al. 2017).

Methylation of viral DNA in oncogenic strains of HPV is found after infection and to differing degrees across cellular disease states (Hoelzer et al. 2008). During asymptomatic carriage, a study showed there is little methylation of genes involved in virion production, and this lack of methylation correlates with gene expression of the viral protein that suppresses the HPV oncogenic proteins E6 and E7 (Fernandez et al. 2009). The same research group found that methylation of the HPV serotype 16 E6/E7 control region may prevent viral suppression of these genes and this region was methylated in 94% of cervical cancer cells analysed. There appear to be differences in the methylation patterns of HPV strains—this is not surprising when considering the process of adaptation. Further exploration of pattern differences may be important for anticipating responses to epigenetically targeted therapy or evaluating mixed results from existing treatment trials. The HPV genome does not include proteins that can alter methylation, but E6 and E7 are able to interact with host methylation modifiers (Soto et al. 2017). Host cells are able to methylate integrated pathogen DNA as a defence against infection (Shalginskikh et al. 2013). So, the question remains, is viral methylation a host viral silencing strategy or a virally hijacked host capability, or when is it both?

HPV, like most known oncogenic viruses, has evolved the ability to have either a short game (virion production) or a long game (replicating its DNA along with the host cell). DNA methylation is likely a significant part of HPV commitment to one or the other. A critical question is what environmental factors trigger the switch in strategy. For example, if the virus can evade the host immune system *and* conditions are not ideal for virion production, then a latent (but potentially oncogenic) infection strategy may be favoured. A better understanding of environmental influences on viral strategy will help determine which epigenetic changes are pathogen driven and which are mediated by the host—this may be critical to developing safe epigenetic-based therapies.

Hepatitis B virus (HBV) provides a final example of virally orchestrated epigenetic modifications. HBV is the causative agent of over 50% of liver cancer. Its virally encoded protein HBx is able to recruit host epigenetic modifiers to specific host genes. This results in targeted silencing of IL4 (as part of immune suppression) and tumour suppressors, and activation of oncogenes (Tian et al. 2013). Like HPV, there is an effective vaccine for HBV, so prevention of infection will lead to prevention of cervical cancers and the majority of liver cancers. It would be expected that all oncogenic viruses have evolved surgical abilities to manipulate the host cell to immortalise, proliferate, and not activate immune responses (Ewald and Swain Ewald 2013). Developing vaccines and antivirals as well as identifying all pathogen-related cancers should be a high priority endeavour.

Unlike genetic changes, epigenetic modifications are alterable and thus make appealing targets in cancer therapy development (Allis and Jeuwein 2016; Soto et al. 2017). For pathogen-driven cancers, it will be essential to determine what epigenetic alterations are pathogen manipulations versus host defences. An evolutionary approach will be critical to not just assuming that all cellular changes as measured experimentally are deleterious and emphasising that interfering with some host processes, for example immunological defences, could be dangerous.

3.6.7 Epigenetics and the Brain: Alzheimer's Disease, Neuropsychiatric Diseases, and Neurodegenerative Diseases

Memory of significant duration requires converting briefly held information into lasting storage and, as discussed earlier, epigenetics appear to play a role in this process, referred to as long-term consolidation (see Kandel et al. 2014 for a detailed overview). Alzheimer's disease is characterised by the loss of memory from brain regions possessing neurons whose memory phenotype is epigenetically modified during the conversion from short-term to lasting memory. It is logical to consider that epigenetic mechanisms may be disturbed in the disease process. A recent study found a difference in small non-coding piRNAs regulation in Alzheimer relative to normal brains (Roy et al. 2017). Large epigenome studies have found significant epigenetic irregularities in DNA methylation and histone modifications, but it is too early to tell whether they play a role in disease aetiology or are a consequence of Alzheimer's pathology (Klein et al. 2016). Overall, these changes appear to involve gene silencing and it has been suggested that this may be a mechanism of decreased neuronal plasticity (Sanchez-Mut and Graff 2015).

In schizophrenia, changes in DNA methylation and epigenetic mechanisms have been identified, but it will be difficult to distinguish whether the origin of dysregulation is the disease or the effect of antipsychotic therapy (Ibi and Gonzalez-Maeso 2015). Dysregulation of the N-methyl-D-aspartate (NMDA) receptor has been suggested as a major factor in the aetiology of schizophrenia, and NMDA antagonism, via drug use or in animal models appears to recreate schizophrenia-like psychosis (Olney et al. 1999; Moghaddam and Jackson 2003). Altered NMDA subunit components have been identified in schizophrenia. An epigenetic change involving a faulty developmental 'switch' mediated by the RE1-silencing transcription factor has been proposed to explain these changes in the NMDA subunit make-up. This switch mechanism appears to be sensitive to environmental insult (Tamminga and Zukin 2015).

A decrease in the RE1-silencing transcription factor has been found in Alzheimer's patients (Lu et al. 2014). Many Alzheimer's patients experience behavioural manifestations, including psychotic-like episodes (Lyketsos et al. 2011), and memory perturbations are common in schizophrenia (Stone and Hsi 2011). So there may be shared epigenetic mechanisms dysregulated in both diseases.

Other neurodegenerative diseases appear to involve epigenetic changes as well, and the use of epigenetic modifiers has shown promise in disease models (Sleiman et al. 2009). Studies using a butyrate analogue as treatment in a mouse model of Huntington's disease have shown reduced cell death and increased longevity (Ferrante et al. 2003; Bourassa et al. 2016). Another study found that butyrate (and other HDAC inhibitors) reduced loss of dopamine-producing neurons relevant to Parkinson's disease (Kidd and Schneider 2010). The effects of these treatments are presumably through blocking removal of acetylation from histones. Problems with the use of HDAC inhibitors in humans is anticipated as they currently lack specificity (Sleiman et al. 2009).

A high-fibre diet that promotes butyrate production may be closer to our ancestral diet than the current western diet, suggesting that we may be adapted to eating foods that might ameliorate brain disorders and improve normal brain health, possibly via epigenetic mechanisms. The fact that colon epithelial cells use butyrate as a primary energy source (Roediger 1980) supports the idea that they have adapted to a long-standing evolutionary relationship with gut microbes and to the high-fibre diet that enriches this microbial community. It is clear that there is a need for extensive further study into the epigenetic role in brain health. The use of evolutionary theory should provide a lens to interpret findings, and to evaluate and devise (safe) interventions.

3.7 Implications of an Integrated Approach

3.7.1 Implications for Categorisation of Disease

In some cases, the diversity of genetic influences on an accepted category of disease is so great that it serves as a basis for dividing a single disease category into new categories. In Alzheimer's disease, for example, the growing base of knowledge has solidified the fission of Alzheimer's disease into late-onset Alzheimer's disease and three categories of early-onset Alzheimer's, which are increasingly considered different diseases. Current evidence accords with the hypothesis that late-onset Alzheimer's disease is a chronic infectious disease, with epsilon 4 and other genetic causes exacerbating the infectious pathology.

Early-onset Alzheimer's disease accounts for about 3–5% of all Alzheimer's disease (Harvey et al. 2003; Alagiakrishnan et al. 2012). Mutations in genes encoding the proteins presenilin 1, presenilin 2, and beta amyloid are accepted genetic causes (Table 3.2) (Rademakers et al. 2003). The Mendelian concordance between these alleles and the occurrence of early-onset Alzheimer's disease has led to consensus that each is necessary for, and therefore an essential cause of, its associated variant.

The epsilon 4 allele has been associated with the presenilin 1 variant of early-onset Alzheimer's disease in a large study, though not in smaller and hence less-reliable studies. It has also been associated with the variant ascribed to a mutation in the beta amyloid percursor gene, but has not been assessed for association with the rarer presenilin 2

Table 3.2 Genetic Causes of Early-Onset Alzheimer's Disease

Mutated protein (gene)	Prevalence in early onset Alzheimer's disease	Proposed pathological mechanism
Beta amyloid precursor protein (APP)	5%[1]	Mutant amyloid protein forms toxic plaques[2]
Presenilin-1 (PSEN1)	40–70%[2]	Increases toxic:benign beta amyloid ratio[2]
Presenilin-2 (PSEN2)	< 1%[1]	Increases toxic:benign beta amyloid ratio[2]

[1] Rademakers et al. 2003
[2] Alagiakrishnan et al. 2012

variant. Epsilon 4 may therefore be considered an exacerbating genetic cause for more than half of early-onset Alzheimer's disease (for relative prevalence see Table 3.2). If the epsilon 4 allele contributes to sporadic Alzheimer's disease by increasing vulnerability to *C. pneumoniae* and HHSV-1, then these pathogens may also be exacerbating causes of early-onset Alzheimer's disease. We therefore appear to be in the midst of the separation of Alzheimer's disease into at least three distinct diseases, which will probably be renamed, especially if the amyloid plaques and tangles that have been used to define Alzheimer's disease are effects rather than causes of the critical damage.

As knowledge becomes more complete, a similar division of schizophrenia according to causation will likely occur. Rare psychoses for which particular alleles are essential causes may be placed in separate categories. The different infectious causes of schizophrenia will probably be associated with at least slightly different combinations or intensities of manifestations. Individuals with schizophrenia who are positive for *T. gondii*, for example, are more prone to suicide (Okusaga et al. 2011). As knowledge on causation develops, we can envision formation of different categories based on different infectious causes. A category, called toxoplasmal psychosis, for example, may be formed incorporating about one-third of what is now considered schizophrenia, and with *T. gondii* as its essential cause.

3.7.2 Implications for Choice of Treatments

The most obvious practical implication of the integrative approach suggested in this chapter is the effect on prevention and therapy. Distinguishing essential from exacerbating causes allows a focus on the actions that will prevent or cure disease rather than ameliorate it—knowing that the human papillomavirus is an essential cause of cervical cancer draws attention to the possibility of preventing the cancer by preventing the infection.

Understanding the interaction among causes may similarly improve the effectiveness of interventions. Guidelines to reduce cholesterol and inflammation will be more effective if the causes of elevated cholesterol and inflammation are better understood. If infections are a major cause of these manifestations, then treating or preventing infection may allow for more safe and effective efforts to reduce damage from cardiovascular disease and late-onset Alzheimer's disease. Controlling infectious causes of these manifestations may be safer because cholesterol and inflammation are often beneficial to states of health. Lowering cholesterol to very low levels may compromise important processes. Inflammation also has positive and negative effects—natural selection has favoured inflammatory defences against pathogens, but pathogens can escape control by inflammation. When a coevolutionary stand-off occurs, the immune system may continue using ineffective inflammation in response to the presence of the pathogen, causing peripheral damage by friendly fire. (For further discussion, see Chapter 10: Immune System.) Using anti-inflammatories will suppress both beneficial and harmful aspects of inflammation, creating uncertainty in whether they should be used in any particular case. Eliminating the causal pathogen resolves the quandary.

3.7.3 Implications of Compensating Benefits

Understanding compensating benefits is important for medical practice and public health policy. People who have an inherent resistance to a particular pathogen may be less prone to damage and therefore benefit less from preventive or therapeutic interventions. The net benefit of a medical action may therefore be negative for such patients. For example, males who carry a copy of the mutated X-linked gene encoding glucose-6-phosphate dehydrogenase, and who therefore have glucose-6-phosphate dehydrogenase deficiency, are resistant to malaria but are particularly vulnerable to damage from quinoline antimalarials. This vulnerability occurs because this enzyme normally regenerates reduced glutathione, which is used both to neutralise toxins and as an energy source for the malaria organism. The mutation terminates both processes. Prophylactic and therapeutic treatment of these individuals with chloroquine or other quinoline antimalarials is therefore contraindicated.

Other applications may arise from more complete knowledge about compensating benefits. As discussed in Sections 3.2.2 and 3.2.3, people who are heterozygous for the cystic fibrosis or Tay-Sachs alleles may be inherently resistant to *M. tuberculosis*. Knowledge about such resistance may influence choices regarding potential exposures to *M. tuberculosis* (e.g. through work in homeless shelters or hospitals) or the use of the Bacille Calmette Guerin (BCG) tuberculosis vaccine, which has low and variable efficacy and significant side-effects (Grode et al. 2013; Moliva et al. 2015).

3.7.4 Implications for Medical Ethics

An integrated understanding of the causes of disease also provides more subtle benefits related to the ethics of alternative policy options. Ever since the discovery of the associations of epsilon 4 with Alzheimer's disease and cardiovascular disease, experts have not encouraged actions that would provide a patient with knowledge about his or her epsilon 4 positivity (Farlow 1997). It has been felt that little could be done to compensate for the negative psychological effects that might stem from knowledge about a genetic predisposition to Alzheimer's disease and atherosclerosis. The integrative approach presented in this chapter indicates that the withholding of this information may often be misguided.

A recent consideration of providing patients with unrequested information on epsilon 4 status provides an illustration. In light of the 12- to 15-fold increased risk for Alzheimer's disease among people who are homozygous for the epsilon 4 allele and a three-fold increased risk for heterozygotes, five committees of ethicists and neurologists were asked to recommend whether patients who were found to be epsilon 4 positive through routine testing should be informed of this status (New York Times 2010). Every committee concluded that these individuals should not be told. Their rationale was that this knowledge would not have a positive overall effect because the knowledge could not lead to effective therapy or prevention.

This conclusion is reasonable when genetic causation is considered in isolation. But when the problem is considered in the context of vulnerability to other causes, it is apparent that individuals who carry the allele should benefit disproportionately from protection against these other causes. This point was illustrated in this chapter for individuals who have the allele for sickle cell anaemia or berylliosis. But it may apply even more strongly to individuals who harbour an exacerbating genetic cause if they can eliminate their risk entirely by eliminating the essential non-genetic cause. If infections are the essential cause of sporadic Alzheimer's disease, for example, then preventing or curing the culpable pathogens may prevent Alzheimer's disease. Even if the medical establishment does not share a consensus on the role of these pathogens, knowledge about their possible causal role may clarify potentially beneficial options; for example, epsilon 4 individuals may be more conscientious about treating *C. pneumoniae* infections and reducing life-style risks, thereby avoiding epsilon 4-associated diseases later in life. By the time consensus on this matter is reached it may be too late to take preventive action because the pathogenesis may proceed for decades before Alzheimer's disease develops. A similar benefit of knowing one's epsilon 4 status would arise for infectious causes that are discovered or accepted at a later date. Those who know that they are epsilon 4 positive can be poised to act upon discovery of a preventable cause to which epsilon 4 increases vulnerability. Moreover, they may be more motivated to keep abreast of new knowledge about such vulnerabilities.

Acknowledgements

Discussions with G. M. Cochran and M. Brüne contributed to the development of ideas in this chapter.

References

Agirre, X., Novo, F. J., Calasanz, M. J., et al. (2003). TP53 is frequently altered by methylation, mutation, and/or deletion in acute lymphoblastic leukaemia. *Mol Carcinog* 38, 201–8. doi: 10.1002/mc.10159.

Alagiakrishnan, K., Gill, S. S., and Fagarasanu, A. (2012). Genetics and epigenetics of Alzheimer's disease. *Postgrad Med J* 88, 522–9. doi: 10.1136/postgradmedj-2011-130363.

Allis, C. D., Kaparros, M., Jenuwein, T., et al. (2015). Overview and concepts. In: Allis, C. D., Kaparros, M., Jenuwein, T., et al. (eds) *Epigenetics*, 2nd ed. Cold Spring Harbor, NY: Cold Spring Harbor Laboratory Press, pp. 47–115.

Alberts, B., Johnson, A., Lewis, J., et al. (2015). *Molecular Biology of the Cell*, 6th ed. New York: Garland.

Allis, C. D. and Jenuwein, T. (2016). The molecular hallmarks of epigenetic control. *Nat Rev Genet* 17, 487–500. doi: 10.1038/nrg.2016.59.

Anders, S. and Kinney, D. K. (2015). Abnormal immune system development and function in schizophrenia helps reconcile diverse findings and suggests new treatment and prevention strategies. *Brain Res* 1617, 93–112. doi: 10.1016/j.brainres.2015.02.043.

Ankri, S. and Mirelman, D. (1999). Antimicrobial properties of allicin from garlic. *Microbes Infect* 1, 125–9.

Arbour, N. C., Lorenz, E., Schutte, B. C., et al. (2000). TLR4 mutations are associated with endotoxin hyporesponsiveness in humans. *Nat Genet* 25, 187–91. doi: 10.1038/76048.

Arias, I., Sorlozano, A., Villegas, E., et al. (2012). Infectious agents associated with schizophrenia: a meta-analysis. *Schizophr Res* 136, 128–36. doi: 10.1016/j.schres.2011.10.026.

Arnold, C., Makintube, S., and Istre, G. R. (1993). Day care attendance and other risk factors for invasive *Haemophilus influenzae* type b disease. *Am J Epidemiol* 138, 333–40.

Avramopoulos, D., Pearce, B. D., McGrath, J., et al. (2015). Infection and inflammation in schizophrenia and bipolar disorder: a genome wide study for interactions with genetic variation. *PLoS One* 10, e0116696. doi: 10.1371/journal.pone.0116696.

Bachrach, G., Jamil, A., Naor, R., et al. (2011). Garlic allicin as a potential agent for controlling oral pathogens. *J Med Food* 14, 1338–43. doi: 10.1089/jmf.2010.0165.

Balin, B. J., Gerard, H. C., Arking, E. J., et al. (1998). Identification and localization of *Chlamydia pneumoniae* in the Alzheimer's brain. *Med Microbiol Immunol* 187, 23–42.

Bannister, A. J. and Kouzarides, T. (2011). Regulation of chromatin by histone modifications. *Cell Res* 21, 381–95. doi: 10.1038/cr.2011.22.

Barlow, D. P. and Bartolomei, M. S. (2014). Genomic imprinting in mammals. *Cold Spring Harb Perspect Biol* 6. doi: 10.1101/cshperspect.a018382.

Bassett, A. S., Bury, A., Hodgkinson, K. A., et al. (1996). Reproductive fitness in familial schizophrenia. *Schizophr Res* 21(3), 151–60.

Bennedsen, B. E., Mortensen, P. B., Olesen, A. V., et al. (2001). Congenital malformations, stillbirths, and infant deaths among children of women with schizophrenia. *Arch Gen Psychiatry* 58, 674–9.

Bertles, J. F. (1974). Human fetal hemoglobin: significance in disease. *Ann N Y Acad Sci* 241, 638–52.

Billing, J. and Sherman, P. W. (1998). Antimicrobial functions of spices: why some like it hot. *Q Rev Biol* 73, 3–49.

Blomstrom, A., Karlsson, H., Wicks, S., et al. (2012). Maternal antibodies to infectious agents and risk for non-affective psychoses in the offspring—a matched case-control study. *Schizophr Res* 140, 25–30. doi: 10.1016/j.schres.2012.06.035.

Blomstrom, A., Gardner, R. M., Dalman, C., et al. (2015). Influence of maternal infections on neonatal acute phase proteins and their interaction in the development of non-affective psychosis. *Transl Psychiatry* 5, e502. doi: 10.1038/tp.2014.142.

Boelen, E., Stassen, F. R., Van Der Ven, A. J., et al. (2007). Detection of amyloid beta aggregates in the brain of BALB/c mice after *Chlamydia pneumoniae* infection. *Acta Neuropathol* 114, 255–61. doi: 10.1007/s00401-007-0252-3.

Booth, C., Inusa, B., and Obaro, S. K. (2010). Infection in sickle cell disease: a review. *Int J Infect Dis* 14, e2–12. doi: 10.1016/j.ijid.2009.03.010.

Borek, C. (2006). Garlic reduces dementia and heart-disease risk. *J Nutr* 136, 810S–12S.

Bosco, P., Gueant-Rodriguez, R. M., Anello, G., et al. (2005). Allele epsilon 4 of APOE is a stronger predictor of Alzheimer risk in Sicily than in continental South Italy. *Neurosci Lett* 388, 168–72. doi: 10.1016/j.neulet.2005.06.056.

Bourassa, M. W., Alim, I., Bultman, S. J., et al. (2016). Butyrate, neuroepigenetics and the gut microbiome: can a high fiber diet improve brain health? *Neurosci Lett*, 625, 56–63. doi: 10.1016/j.neulet.2016.02.009.

Bourne, G. (1962). *The Human Amnion and Chorion*. London: Lloyd-Luke.

Brown, A. S., Schaefer, C. A., Wyatt, R. J., et al. (2000). Maternal exposure to respiratory infections and adult schizophrenia spectrum disorders: a prospective birth cohort study. *Schizophr Bull* 26(2), 287–95.

Brown, A. S., Schaefer, C. A., Quesenberry, C. P., Jr, et al. (2005). Maternal exposure to toxoplasmosis and risk of schizophrenia in adult offspring. *Am J Psychiatry* 162, 767–73. doi: 10.1176/appi.ajp.162.4.767.

Brown, A. S., Vinogradov, S., Kremen, W. S., et al. (2009). Prenatal exposure to maternal infection and executive dysfunction in adult schizophrenia. *Am J Psychiatry* 166, 683–90. doi: 10.1176/appi.ajp.2008.08010089.

Brüne, M. (2004). Schizophrenia—an evolutionary enigma? *Neurosci Biobehav Rev* 28, 41–53. doi: 10.1016/j.neubiorev.2003.10.002.

Bu, X. L., Yao, X. Q., Jiao, S. S., et al. (2015). A study on the association between infectious burden and Alzheimer's disease. *Eur J Neurol* 22, 1519–25. doi: 10.1111/ene.12477.

Buka, S. L., Tsuang, M. T., Torrey, E. F., et al. (2001). Maternal cytokine levels during pregnancy and adult psychosis. *Brain Behav Immun* 15, 411–20. doi: 10.1006/brbi.2001.0644.

Buka, S. L., Cannon, T. D., Torrey, E. F., et al. (2008). Maternal exposure to herpes simplex virus and risk of psychosis among adult offspring. *Biol Psychiatry* 63, 809–15. doi: 10.1016/j.biopsych.2007.09.022.

Butler, M. G. (2009). Genomic imprinting disorders in humans: a mini-review. *J Assist Reprod Genet* 26, 477–86. doi: 10.1007/s10815-009-9353-3.

Cannon, C. P., Blazing, M. A., Giugliano, R. P., et al. (2015). Ezetimibe added to statin therapy after acute coronary syndromes. *N Engl J Med* 372, 2387–97. doi: 10.1056/NEJMoa1410489.

Carter, C. J. (2009). Schizophrenia susceptibility genes directly implicated in the life cycles of pathogens: cytomegalovirus, influenza, herpes simplex, rubella, and *Toxoplasma gondii*. *Schizophr Bull* 35, 1163–82. doi: 10.1093/schbul/sbn054.

Castrogiovanni, P., Iapichino, S., Pacchierotti, C., et al. (1998). Season of birth in psychiatry. A review. *Neuropsychobiology* 37, 175–81.

Casula, M., Soranna, D., Catapano, A. L., et al. (2013). Long-term effect of high dose omega-3 fatty acid supplementation for secondary prevention of cardiovascular outcomes: a meta-analysis of randomized, placebo controlled trials [corrected]. *Atheroscler Suppl* 14, 243–51. doi: 10.1016/S1567-5688(13)70005-9.

Cedar, H. and Bergman, Y. (2009). Linking DNA methylation and histone modification: patterns and paradigms. *Nat Rev Genet* 10, 295–304. doi: 10.1038/nrg2540.

Cerase, A., Pintacuda, G., Tattermusch, A., et al. (2015). Xist localization and function: new insights from multiple levels. *Genome Biol* 16, 166. doi: 10.1186/s13059-015-0733-y.

Chauhan, N. B. (2005). Multiplicity of garlic health effects and Alzheimer's disease. *J Nutr Health Aging* 9, 421–32.

Choroszy-Krol, I., Frej-Madrzak, M., Hober, M., et al. (2014). Infections caused by *Chlamydophila pneumoniae*. *Adv Clin Exp Med* 23, 123–6.

Chung, W. S., Lin, C. L., Lin, C. L., et al. (2015). Thalassaemia and risk of cancer: a population-based cohort study. *J Epidemiol Community Health* 69, 1066–70. doi: 10.1136/jech-2014-205075.

Contreras, M. C., Sandoval, L., Salinas, P., et al. (2000). Diagnostic use of ELISA, IgG, IgM, IgA and ELISA IgG avidity in recent and chronic toxoplasmosis. *Bol Chil Parasitol* 55, 17–24.

Conway, V., Allard, M. J., Minihane, A. M., et al. (2014). Postprandial enrichment of triacylglycerol-rich lipoproteins with omega-3 fatty acids: lack of an interaction with apolipoprotein E genotype? *Lipids Health Dis* 13, 148. doi: 10.1186/1476-511X-13-148.

Corbo, R. M. and Scacchi, R. (1999). Apolipoprotein E (APOE) allele distribution in the world. Is APOE*4 a 'thrifty' allele? *Ann Hum Genet* 63, 301–10.

Couvreur, J., Desmonts, G., and Girre, J. Y. (1976). Congenital toxoplasmosis in twins: a series of 14 pairs of twins: absence of infection in one twin in two pairs. *J Pediatr* 89, 235–40.

Dai, S., Falta, M. T., Bowerman, N. A., et al. (2013). T cell recognition of beryllium. *Curr Opin Immunol* 25, 775–80. doi: 10.1016/j.coi.2013.07.012.

Dale, P. W. (1981). Prevalence of schizophrenia in the Pacific Island populations of Micronesia. *J Psychiatr Res* 16, 103–11.

Dalen, P. (1988). Schizophrenia, season of birth, and maternal age. *Br J Psychiatry* 153, 727–33.

Danforth, D. N. (1986). *Danforth's Obstetrics and Gynecology*. Philadelphia: Lippincott Williams & Wilkins.

Dang, T. M., Conway, V., and Plourde, M. (2015). Disrupted fatty acid distribution in HDL and LDL according to apolipoprotein E allele. *Nutrition* 31, 807–12. doi: 10.1016/j.nut.2014.11.019.

Davis, J. O., Phelps, J. A., and Bracha, H. S. (1995). Prenatal development of monozygotic twins and concordance for schizophrenia. *Schizophr Bull* 21, 357–66.

De Messias, E. L., Cordeiro, N. F., Sampaio, J. J., et al. (2001). Schizophrenia and season of birth in a tropical region: relationship to rainfall. *Schizophr Res* 48, 227–34.

De Roever-Bonnet, H., Koppe, J. G., and Loewer-Sieger, D. H. (1979). Follow-up of children with congenital toxoplasmosis infections and children who became serologically negative after 1 year of age, all born in 1964–1965. In: Thalhammer, O., Baumgarten, K., and Pollak, A. (eds.) *Perinatal Medicine: Sixth European Congress, Vienna*. Stuttgart: Georg Thieme.

Dekeuwer, C. (2015). Defining genetic disease. In: Huneman, P., Lambert, G., and Silberstein, M. (eds.) *Classification, Disease and Evidence*. Dordrecht: Springer Science + Business Media.

Diamond, J. (1988). Tay-Sachs carriers and tuberculosis resistance. *Nature* 331, 666.

Duan, J., Sanders, A. R., and Gejman, P. V. (2010). Genome-wide approaches to schizophrenia. *Brain Res Bull* 83, 93–102. doi: 10.1016/j.brainresbull.2010.04.009.

Eaton, S. B., Eaton, S. B., 3rd, Sinclair, A. J., et al. (1998). Dietary intake of long-chain polyunsaturated fatty acids during the Paleolithic. *World Rev Nutr Diet* 83, 12–23.

Endo, A. (2010). A historical perspective on the discovery of statins. *Proc Jpn Acad Ser B Phys Biol Sci* 86, 484–93.

Erkan, L., Uzun, O., Findik, S., et al. (2008). Role of bacteria in acute exacerbations of chronic obstructive pulmonary disease. *Int J Chron Obstruct Pulmon Dis* 3, 463–7.

Essen-Moller, E. and Fischer, M. (1979). Do the partners of dizygotic schizophrenic twins run a greater risk of schizophrenia than ordinary siblings? *Hum Hered* 29, 161–5.

Ewald, P. W. and Swain Ewald, H. A. (2011). Evolutionary insights for immunological interventions. In: Poiani, A. (ed.) *Pragmatic Evolution*. Cambridge: Cambridge University Press.

Ewald, P. W. and Swain Ewald, H. A. (2013). Toward a general evolutionary theory of oncogenesis. *Evol Appl* 6, 70–81. doi: 10.1111/eva.12023.

Falta, M. T., Pinilla, C., Mack, D. G., et al. (2013). Identification of beryllium-dependent peptides recognized by CD4+ T cells in chronic beryllium disease. *J Exp Med* 210, 1403–18. doi: 10.1084/jem.20122426.

Fananas, L. and Bertranpetit, J. (1995). Reproductive rates in families of schizophrenic patients in a case-control study. *Acta Psychiatr Scand* 91, 202–4.

Farlow, M. R. (1997). Alzheimer's disease: clinical implications of the apolipoprotein E genotype. *Neurology* 48, S30–4.

Fassbender, K., Walter, S., Kuhl, S., et al. (2004). The LPS receptor (CD14) links innate immunity with Alzheimer's disease. *FASEB J* 18, 203–5. doi: 10.1096/fj.03-0364fje.

Feil, R. and Fraga, M. F. (2012). Epigenetics and the environment: emerging patterns and implications. *Nat Rev Genet* 13, 97–109. doi: 10.1038/nrg3142.

Feingold, K. R., Staprans, I., Memon, R. A., et al. (1992). Endotoxin rapidly induces changes in lipid metabolism that produce hypertriglyceridemia: low doses stimulate hepatic triglyceride production while high doses inhibit clearance. *J Lipid Res* 33, 1765–76.

Feingold, K. R., Hardardottir, I., Memon, R., et al. (1993). Effect of endotoxin on cholesterol biosynthesis and distribution in serum lipoproteins in Syrian hamsters. *J Lipid Res* 34, 2147–58.

Fernandes Filho, J. A. and Shapiro, B. E. (2004). Tay-Sachs disease. *Arch Neurol* 61, 1466–8. doi: 10.1001/archneur.61.9.1466.

Fernandez, A. F., Rosales, C., Lopez-Nieva, P., et al. (2009). The dynamic DNA methylomes of double-stranded DNA viruses associated with human cancer. *Genome Res* 19, 438–51. doi: 10.1101/gr.083550.108.

Ferrante, R. J., Kubilus, J. K., Lee, J., et al. (2003). Histone deacetylase inhibition by sodium butyrate chemotherapy ameliorates the neurodegenerative phenotype in Huntington's disease mice. *J Neurosci* 23, 9418–27.

Filardo, S., Di Pietro, M., Farcomeni, A., et al. (2015). *Chlamydia pneumoniae*-mediated inflammation in atherosclerosis: a meta-analysis. *Mediators Inflamm* 2015, 378658. doi: 10.1155/2015/378658.

Finch, C. E. and Stanford, C. B. (2004). Meat-adaptive genes and the evolution of slower aging in humans. *Q Rev Biol* 79(1), 3–50.

Fluhr, S., Krombholz, C. F., Meier, A., et al. (2017). Epigenetic dysregulation of the erythropoietic transcription factor KLF1 and the beta-like globin locus in juvenile myelomonocytic leukemia. *Epigenetics* 12(8), 1–9. doi: 10.1080/15592294.2017.1356959.

Freiman, D. G. and Hardy, H. L. (1970). Beryllium disease. The relation of pulmonary pathology to clinical course and prognosis based on a study of 130 cases from the U.S. beryllium case registry. *Hum Pathol* 1, 25–44.

Fullerton, S. M., Clark, A. G., Weiss, K. M., et al. (2000). Apolipoprotein E variation at the sequence haplotype level: implications for the origin and maintenance of a major human polymorphism. *Am J Hum Genet* 67, 881–900. doi: 10.1086/303070.

Fung, K. Y., Cosgrove, L., Lockett, T., et al. (2012). A review of the potential mechanisms for the lowering of colorectal oncogenesis by butyrate. *Br J Nutr* 108, 820–31. doi: 10.1017/S0007114512001948.

Gerard, H. C., Wang, G. F., Balin, B. J., et al. (1999). Frequency of apolipoprotein E (APOE) allele types in patients with Chlamydia-associated arthritis and other arthritides. *Microb Pathog* 26, 35–43. doi: 10.1006/mpat.1998.0242.

Gerard, H. C., Fomicheva, E., Whittum-Hudson, J. A., et al. (2008). Apolipoprotein E4 enhances attachment of *Chlamydophila (Chlamydia) pneumoniae* elementary bodies to host cells. *Microb Pathol* 44(4), 279–85. doi: 10.1016/j.micpath.2007.10.002.

Ginder, G. D. (2015). Epigenetic regulation of fetal globin gene expression in adult erythroid cells. *Transl Res* 165, 115–25. doi: 10.1016/j.trsl.2014.05.002.

Gomez-Marin, J. E., Montoya-De-Londono, M. T., Castano-Osorio, J. C., et al. (2000). Frequency of specific anti-*Toxoplasma gondii* IgM, IgA and IgE in Colombian patients with acute and chronic ocular toxoplasmosis. *Mem Inst Oswaldo Cruz* 95, 89–94.

Greaves, M. (2006). Infection, immune responses and the aetiology of childhood leukaemia. *Nat Rev Cancer* 6, 193–203. doi: 10.1038/nrc1816.

Grode, L., Ganoza, C. A., Brohm, C., Weiner, J., 3rd, et al. (2013). Safety and immunogenicity of the recombinant BCG vaccine VPM1002 in a phase 1 open-label randomized clinical trial. *Vaccine* 31(9), 1340–8. doi: 10.1016/j.vaccine.2012.12.053.

Grosse, S. D., Odame, I., Atrash, H. K., et al. (2011). Sickle cell disease in Africa: a neglected cause of early childhood mortality. *Am J Prev Med* 41, S398–405. doi: 10.1016/j.amepre.2011.09.013.

Grossniklaus, U., Kelly, W. G., Kelly, B., et al. (2013). Transgenerational epigenetic inheritance: how important is it? *Nat Rev Genet* 14, 228–35. doi: 10.1038/nrg3435.

Gualandi, G., Giselico, L., Carloni, M., et al. (2001). Enhancement of genetic instability in human B cells by Epstein-Barr virus latent infection. *Mutagenesis* 16, 203–8.

Guggenheim, F. G., Pollin, W., Stabenau, J. R., et al. (1969). Prevalence of physical illness in parents of identical twins discordant for schizophrenia. *Psychosom Med* 31, 288–99.

Guglielmetti, M. R., Rosa, E. F., Lourencao, D. S., et al. (2014). Detection and quantification of periodontal pathogens in smokers and never-smokers with chronic periodontitis by real-time polymerase chain reaction. *J Periodontol* 85, 1450–7. doi: 10.1902/jop.2014.140048.

Gurven, M. and Kaplan, H. (2007). Longevity among hunter-gatherers: a cross-cultural examination. *Popul Dev Rev* 33, 321–65.

Hackman, D. A., Farah, M. J., and Meaney, M. J. (2010). Socioeconomic status and the brain: mechanistic insights from human and animal research. *Nat Rev Neurosci* 11, 651–9. doi: 10.1038/nrn2897.

Hajishengallis, G. (2014). Immunomicrobial pathogenesis of periodontitis: keystones, pathobionts, and host response. *Trends Immunol* 35, 3–11. doi: 10.1016/j.it.2013.09.001.

Hajishengallis, G. and Lamont, R. J. (2014). Breaking bad: manipulation of the host response by *Porphyromonas gingivalis*. *Eur J Immunol* 44, 328–38. doi: 10.1002/eji.201344202.

Hamidinejat, H., Ghorbanpoor, M., Hosseini, H., et al. (2010). *Toxoplasma gondii* infection in first-episode and inpatient individuals with schizophrenia. *Int J Infect Dis* 14, e978–81. doi: 10.1016/j.ijid.2010.05.018.

Handa, T., Nagai, S., Kitaichi, M., et al. (2009). Long-term complications and prognosis of chronic beryllium disease. *Sarcoidosis Vasc Diffuse Lung Dis* 26, 24–31.

Harris, J. C., Cottrell, S. L., Plummer, S., et al. (2001). Antimicrobial properties of *Allium sativum* (garlic). *Appl Microbiol Biotechnol* 57, 282–6.

Harris, S. A. and Harris, E. A. (2015). Herpes simplex virus type 1 and other pathogens are key causative factors in sporadic Alzheimer's disease. *J Alzheimer's Dis* 48, 319–53. doi: 10.3233/JAD-142853.

Harvey, R. J., Skelton-Robinson, M., and Rossor, M. N. (2003). The prevalence and causes of dementia in people under the age of 65 years. *J Neurol Neurosurg Psychiatry* 74, 1206–9.

Haukka, J., Suvisaari, J., Varilo, T., et al. (2001). Regional variation in the incidence of schizophrenia in Finland: a study of birth cohorts born from 1950 to 1969. *Psychol Med* 31, 1045–53.

He, P., Mei, C., Cheng, B., et al. (2009). *Chlamydia pneumoniae* induces macrophage-derived foam cell formation by up-regulating acyl-coenzyme A: cholesterol acyltransferase 1. *Microbes Infect* 11, 157–63. doi: 10.1016/j.micinf.2008.11.001.

Hoelzer, K., Shackelton, L. A., and Parrish, C. R. (2008). Presence and role of cytosine methylation in DNA viruses of animals. *Nucleic Acids Res* 36, 2825–37. doi: 10.1093/nar/gkn121.

Ho-Yen, D. O. and Joss, A. W. L. (1992). *Human Toxoplasmosis*. Oxford: Oxford University Press.

Hussels, I. E. and Morton, N. E. (1972). Pingelap and Mokil Atolls: achromatopsia. *Am J Hum Genet* 24, 304–9.

Hyman, B. T., Gomez-Isla, T., Rebeck, G. W., et al. (1996). Epidemiological, clinical, and neuropathological study of apolipoprotein E genotype in Alzheimer's disease. *Ann N Y Acad Sci* 802, 1–5.

Ibi, D. and Gonzalez-Maeso, J. (2015). Epigenetic signaling in schizophrenia. *Cell Signal* 27, 2131–6. doi: 10.1016/j.cellsig.2015.06.003.

Ilveskoski, E., Perola, M., Lehtimaki, T., et al. (1999). Age-dependent association of apolipoprotein E genotype with coronary and aortic atherosclerosis in middle-aged men: an autopsy study. *Circulation* 100, 608–13.

Ingraham, L. J. and Kety, S. S. (2000). Adoption studies of schizophrenia. *Am J Med Genet* 97, 18–22.

Itzhaki, R. F. (2017). Herpes simplex virus type 1 and Alzheimer's disease: possible mechanisms and signposts. *FASEB J* 31, 3216–26. doi: 10.1096/fj.201700360.

Jablensky, A. (2000). Epidemiology of schizophrenia: the global burden of disease and disability. *Eur Arch Psychiatry Clin Neurosci* 250, 274–85.

James, J. A., Poulton, K. V., Haworth, S. E., et al. (2007). Polymorphisms of TLR4 but not CD14 are associated with a decreased risk of aggressive periodontitis. *J Clin Periodontol* 34, 111–17. doi: 10.1111/j.1600-051X.2006.01030.x.

Jarvik, G. P., Wijsman, E. M., Kukull, W. A., et al. (1995). Interactions of apolipoprotein E genotype, total cholesterol level, age, and sex in prediction of Alzheimer's disease: a case-control study. *Neurology* 45, 1092–6.

Jones, J. L., Lopez, A., Wilson, M., et al. (2001). Congenital toxoplasmosis: a review. *Obstet Gynecol Surv* 56, 296–305.

Jones, P. A. (2012). Functions of DNA methylation: islands, start sites, gene bodies and beyond. *Nat Rev Genet* 13, 484–92. doi: 10.1038/nrg3230.

Juengst, E. T. (2004). Concept of disease after the Human Genome Project. In: Caplan, A. L., McCartney, J. J., and Sisti, D. A. (eds.) *Health, Disease, and Illness. Concepts in Medicine*. Washington, DC: Georgetown University Press.

Kagan, B. L., Jang, H., Capone, R., et al. (2012). Antimicrobial properties of amyloid peptides. *Mol Pharm* 9, 708–17. doi: 10.1021/mp200419b.

Kamer, A. R., Craig, R. G., Pirraglia, E., et al. (2009). TNF-alpha and antibodies to periodontal bacteria discriminate between Alzheimer's disease patients and normal subjects. *J Neuroimmunol* 216, 92–7. doi: 10.1016/j.jneuroim.2009.08.013.

Kandel, E. R., Dudai, Y., and Mayford, M. R. (2014). The molecular and systems biology of memory. *Cell* 157, 163–86. doi: 10.1016/j.cell.2014.03.001.

Kaperonis, E. A., Liapis, C. D., Kakisis, J. D., et al. (2006). Inflammation and *Chlamydia pneumoniae* infection correlate with the severity of peripheral arterial disease. *Eur J Vasc Endovasc Surg* 31, 509–15. doi: 10.1016/j.ejvs.2005.11.022.

Kavanagh, P. L., Sprinz, P. G., Vinci, S. R., et al. (2011). Management of children with sickle cell disease: a comprehensive review of the literature. *Pediatrics* 128, e1552-74. doi: 10.1542/peds.2010-3686.

Kelemen, A. (1977). Maternal age at conception of children who became schizophrenic in their adult lives. *Act Nerv Super (Praha)* 1977 Jul (19 Suppl 2), 381.

Kidd, S. K. and Schneider, J. S. (2010). Protection of dopaminergic cells from MPP+-mediated toxicity by histone deacetylase inhibition. *Brain Res* 1354, 172–8. doi: 10.1016/j.brainres.2010.07.041.

Kiechl, S., Lorenz, E., Reindl, M., et al. (2002a). Toll-like receptor 4 polymorphisms and atherogenesis. *N Engl J Med* 347, 185–92. doi: 10.1056/NEJMoa012673.

Kiechl, S., Werner, P., Egger, G., et al. (2002b). Active and passive smoking, chronic infections, and the risk of carotid atherosclerosis: prospective results from the Bruneck Study. *Stroke* 33, 2170–6.

Kiefer, C. M., Hou, C., Little, J. A., et al. (2008). Epigenetics of beta-globin gene regulation. *Mutat Res* 647, 68–76. doi: 10.1016/j.mrfmmm.2008.07.014.

Kiloh, L. G. (1975). Psychiatry amongst the Australian Aborigines. *Br J Psychiatry* 126, 1–10.

Kinane, D. F., Shiba, H., Stathopoulou, P. G., et al. (2006). Gingival epithelial cells heterozygous for Toll-like receptor 4 polymorphisms Asp299Gly and Thr399ile are hypo-responsive to *Porphyromonas gingivalis*. *Genes Immun* 7, 190–200. doi: 10.1038/sj.gene.6364282.

Klein, H. U., Bennett, D. A., and De Jager, P. L. (2016). The epigenome in Alzheimer's disease: current state and approaches for a new path to gene discovery and understanding disease mechanism. *Acta Neuropathol* 132, 503–14. doi: 10.1007/s00401-016-1612-7.

Knudsen, A. G. (1973). Founder effect in Tay-Sachs. *Amer J Hum Genet* 25, 108–111.

Koo, I. C., Ohol, Y. M., Wu, P., et al. (2008). Role for lysosomal enzyme beta-hexosaminidase in the control of mycobacteria infection. *Proc Natl Acad Sci U S A* 105, 710–15. doi: 10.1073/pnas.0708110105.

Koppe, J. G. and Rothova, A. (1989). Congenital toxoplasmosis. A long-term follow-up of 20 years. *Int Ophthalmol* 13, 387–90.

Koppe, J. G., Loewer-Sieger, D. H., and De Roever-Bonnet, H. (1986). Results of 20-year follow-up of congenital toxoplasmosis. *Lancet* 1, 254–6.

Koscielny, J., Klussendorf, D., Latza, R., et al. (1999). The antiatherosclerotic effect of *Allium sativum*. *Atherosclerosis* 144, 237–49.

Kringlen, E. (2000). Twin studies in schizophrenia with special emphasis on concordance figures. *Am J Med Genet* 97, 4–11.

Kwak, S. M., Myung, S. K., Lee, Y. J., et al (2012). Efficacy of omega-3 fatty acid supplements (eicosapentaenoic acid and docosahexaenoic acid) in the secondary prevention of cardiovascular disease: a meta-analysis of randomized, double-blind, placebo-controlled trials. *Arch Intern Med* 172, 686–94. doi: 10.1001/archinternmed.2012.262.

Landry, C. D., Kandel, E. R., and Rajasethupathy, P. (2013). New mechanisms in memory storage: piRNAs and epigenetics. *Trends Neurosci* 36, 535–42. doi: 10.1016/j.tins.2013.05.004.

Lang, L. (1994). Beryllium: a chronic problem. *Environ Health Perspect* 102, 526–31.

Lappalainen, M., Koskiniemi, M., Hiilesmaa, V., et al. (1995). Outcome of children after maternal primary *Toxoplasma* infection during pregnancy with emphasis on avidity of specific IgG. The Study Group. *Pediatr Infect Dis J* 14, 354–61.

Ledgerwood, L. G., Ewald, P. W., and Cochran, G. M. (2003). Genes, germs, and schizophrenia: an evolutionary perspective. *Perspect Biol Med* 46, 317–48.

Lencz, T., Morgan, T. V., Athanasiou, M., et al. (2007). Converging evidence for a pseudoautosomal cytokine receptor gene locus in schizophrenia. *Mol Psychiatry* 12, 572–80. doi: 10.1038/sj.mp.4001983.

Li, E. and Zhang, Y. (2014). DNA methylation in mammals. *Cold Spring Harb Perspect Biol* 6, a019133. doi: 10.1101/cshperspect.a019133.

Little, C. S., Hammond, C. J., Macintyre, A., et al. (2004). *Chlamydia pneumoniae* induces Alzheimer-like amyloid plaques in brains of BALB/c mice. *Neurobiol Aging* 25, 419–29. doi: 10.1016/S0197-4580(03)00127-1.

Livezey, K. W. and Simon, D. (1997). Accumulation of genetic alterations in a human hepatoma cell line transfected with hepatitis B virus. *Mutat Res* 377, 187–98.

Loohuis, L. M., Vorstman, J. A., Ori, A. P., et al. (2015). Genome-wide burden of deleterious coding variants increased in schizophrenia. *Nat Commun* 6, 7501. doi: 10.1038/ncomms8501.

Lu, T., Aron, L., Zullo, J., et al. (2014). REST and stress resistance in ageing and Alzheimer's disease. *Nature* 507, 448–54. doi: 10.1038/nature13163.

Lubin, J. H., Couper, D., Lutsey, P. L., et al. (2016). Risk of cardiovascular disease from cumulative cigarette use and the impact of smoking intensity. *Epidemiology* 27, 395–404. doi: 10.1097/EDE.0000000000000437.

Lyketsos, C. G., Carrillo, M. C., Ryan, J. M., et al. (2011). Neuropsychiatric symptoms in Alzheimer's disease. *Alzheimer's Dement* 7, 532–9. doi: 10.1016/j.jalz.2011.05.2410.

Mabaera, R., West, R. J., Conine, S. J., et al. (2008). A cell stress signaling model of fetal hemoglobin induction: what doesn't kill red blood cells may make them stronger. *Exp Hematol* 36, 1057–72. doi: 10.1016/j.exphem.2008.06.014.

Magnus, D. (2004). The concept of genetic disease. In: Caplan, A. L., McCartney, J. J., and Sisti, D. A. (eds.) *Health, Disease, and Illness. Concepts in Medicine.* Washington, DC: Georgetown University Press.

Maheshwari, P. and Eslick, G. D. (2015). Bacterial infection and Alzheimer's disease: a meta-analysis. *J Alzheimer's Dis* 43, 957–66. doi: 10.3233/JAD-140621.

Mahley, R. W. (1988). Apolipoprotein E: cholesterol transport protein with expanding role in cell biology. *Science* 240, 622–30.

Malaspina, D., Harlap, S., Fennig, S., et al. (2001). Advancing paternal age and the risk of schizophrenia. *Arch Gen Psychiatry* 58, 361–7.

Manabe, A., Yoshimasu, T., Ebihara, Y., et al. (2004). Viral infections in juvenile myelomonocytic leukemia: prevalence and clinical implications. *J Pediatr Hematol Oncol* 26, 636–41.

Manci, E. A., Culberson, D. E., Yang, Y. M., et al. (2003). Causes of death in sickle cell disease: an autopsy study. *Br J Haematol* 123, 359–65.

Maricle, R. A., Hoffman, W. F., Bloom, J. D., et al. (1987). The prevalence and significance of medical illness among chronically mentally ill outpatients. *Community Ment Health J* 23, 81–90.

Matzaraki, V., Kumar, V., Wijmenga, C., et al. (2017). The MHC locus and genetic susceptibility to autoimmune and infectious diseases. *Genome Biol* 18, 76. doi: 10.1186/s13059-017-1207-1.

McCavit, T. L., Quinn, C. T., Techasaensiri, C., et al. (2011). Increase in invasive *Streptococcus pneumoniae* infections in children with sickle cell disease since pneumococcal conjugate vaccine licensure. *J Pediatr* 158, 505–7. doi: 10.1016/j.jpeds.2010.11.025.

McClellan, J. M., Susser, E., and King, M. C. (2007). Schizophrenia: a common disease caused by multiple rare alleles. *Br J Psychiatry* 190, 194–9. doi: 10.1192/bjp.bp.106.025585.

McDougall, J. K. (2001). 'Hit and run' transformation leading to carcinogenesis. *Dev Biol (Basel)* 106, 77–82; discussion 82–3, 143–60.

McGrath, J. J. (2006). Variations in the incidence of schizophrenia: data versus dogma. *Schizophr Bull* 32(1), 195–7. doi: 10.1093/schbul/sbi052.

McGrath, J., Saha, S., Welham, J., et al. (2004). A systematic review of the incidence of schizophrenia: the distribution of rates and the influence of sex, urbanicity, migrant status and methodology. *BMC Med* 2, 13. doi: 10.1186/1741-7015-2-13.

Mei, C. L., He, P., Cheng, B., et al. (2009). *Chlamydia pneumoniae* induces macrophage-derived foam cell formation via PPAR alpha and PPAR gamma-dependent pathways. *Cell Biol Int* 33, 301–8. doi: 10.1016/j.cellbi.2008.12.002.

Memon, R. A., Grunfeld, C., Moser, A. H., et al. (1993). Tumor necrosis factor mediates the effects of endotoxin on cholesterol and triglyceride metabolism in mice. *Endocrinology* 132, 2246–53. doi: 10.1210/endo.132.5.8477669.

Miklossy, J. (1993). Alzheimer's disease—a spirochetosis? *Neuroreport* 4, 1069.

Miklossy, J., Kasas, S., Janzer, R. C., et al. (1994). Further ultrastructural evidence that spirochaetes may play a role in the aetiology of Alzheimer's disease. *Neuroreport* 5, 1201–4.

Minoretti, P., Gazzaruso, C., Vito, C. D., et al. (2006). Effect of the functional toll-like receptor 4 Asp299Gly polymorphism on susceptibility to late-onset Alzheimer's disease. *Neurosci Lett* 391, 147–9. doi: 10.1016/j.neulet.2005.08.047.

Moghaddam, B. and Jackson, M. E. (2003). Glutamatergic animal models of schizophrenia. *Ann N Y Acad Sci* 1003, 131–7.

Moliva, J. I., Turner, J., and Torrelles, J. B. (2015). Prospects in *Mycobacterium bovis* Bacille Calmette et Guerin (BCG) vaccine diversity and delivery: why does BCG fail to protect against tuberculosis? *Vaccine* 33(39), 5035–41. doi: 10.1016/j.vaccine.2015.08.033.

Mortensen, P. B., Pedersen, C. B., Hougaard, D. M., et al. (2010). A Danish National Birth Cohort study of maternal HSV-2 antibodies as a risk factor for schizophrenia in their offspring. *Schizophr Res* 122, 257–63. doi: 10.1016/j.schres.2010.06.010.

Morton, N. E., Lew, R., Hussels, I. E., et al. (1972). Pingelap and Mokil Atolls: historical genetics. *Am J Hum Genet* 24, 277–89.

Muller, N., Riedel, M., Gruber, R., et al. (2000). The immune system and schizophrenia. An integrative view. *Ann N Y Acad Sci* 917, 456–67.

Mulye, M., Samanta, D., Winfree, S., et al. (2017). Elevated cholesterol in the *Coxiella burnetii* intracellular niche is bacteriolytic. *MBio* 8(1), pii: e02313-16. doi: 10.1128/mBio.02313-16.

Muro, S., Tabara, Y., Matsumoto, H., et al. (2016). Relationship among *Chlamydia* and *Mycoplasma pneumoniae* seropositivity, IKZF1 genotype and chronic obstructive pulmonary disease in a general Japanese population: the Nagahama study. *Medicine (Baltimore)* 95, e3371. doi: 10.1097/MD.0000000000003371.

Musallam, K. M., Sankaran, V. G., Cappellini, M. D., et al. (2012). Fetal hemoglobin levels and morbidity in untransfused patients with beta-thalassemia intermedia. *Blood* 119, 364–7. doi: 10.1182/blood-2011-09-382408.

Mysak, J., Podzimek, S., Sommerova, P., et al. (2014). *Porphyromonas gingivalis*: major periodontopathic pathogen overview. *J Immunol Res* 2014, 476068. doi: 10.1155/2014/476068.

Nepomuceno, R., Pigossi, S. C., Finoti, L. S., et al. (2017). Serum lipid levels in patients with periodontal disease. A meta-analysis and meta-regression. *J Clin Periodontol* 44(12), 1192–1207. doi: 10.1111/jcpe.12792.

Nevels, M., Tauber, B., Spruss, T., et al. (2001). 'Hit-and-run' transformation by adenovirus oncogenes. *J Virol* 75, 3089–94. doi: 10.1128/JVI.75.7.3089-3094.2001.

Niller, H. H., Wolf, H., and Minarovits, J. (2009). Epigenetic dysregulation of the host cell genome in Epstein-Barr virus-associated neoplasia. *Semin Cancer Biol* 19, 158–64. doi: 10.1016/j.semcancer.2009.02.012.

Odenheimer, D. J., Whitten, C. F., Rucknagel, D. A., et al. (1984). Stability over time of hematological variables in 197 children with sickle cell anemia. *Am J Med Genet* 18, 461–70. doi: 10.1002/ajmg.1320180316.

Okusaga, O., Langenberg, P., Sleemi, A., et al. (2011). *Toxoplasma gondii* antibody titers and history of suicide attempts in patients with schizophrenia. *Schizophr Res* 133, 150–5. doi: 10.1016/j.schres.2011.08.006.

Olney, J. W., Newcomer, J. W., and Farber, N. B. (1999). NMDA receptor hypofunction model of schizophrenia. *J Psychiatr Res* 33, 523–33.

Onstad, S., Skre, I., Edvardsen, J., et al. (1991). Mental disorders in first-degree relatives of schizophrenics. *Acta Psychiatr Scand* 83, 463–7.

Orlich, M. J., Singh, P. N., Sabate, J., et al. (2015). Vegetarian dietary patterns and the risk of colorectal cancers. *JAMA Intern Med* 175, 767–76. doi: 10.1001/jamainternmed.2015.59.

Osby, U., Correia, N., Brandt, L., et al. (2000). Mortality and causes of death in schizophrenia in Stockholm county, Sweden. *Schizophr Res* 45, 21–8.

Ozturk, A. and Vieira, A. R. (2009). TLR4 as a risk factor for periodontal disease: a reappraisal. *J Clin Periodontol* 36, 279–86. doi: 10.1111/j.1600-051X.2009.01370.x.

Pallast, E. G., Jongbloet, P. H., et al. (1994). Excess seasonality of births among patients with schizophrenia and seasonal ovopathy. *Schizophr Bull* 20, 269–76.

Panza, F., Logroscino, G., Imbimbo, B. P., et al. (2014). Is there still any hope for amyloid-based immunotherapy for Alzheimer's disease? *Curr Opin Psychiatry* 27, 128–37. doi: 10.1097/YCO.0000000000000041.

Papaetis, G. S., Anastasakou, E., and Orphanidou, D. (2009). *Chlamydophila pneumoniae* infection and COPD: more evidence for lack of evidence? *Eur J Intern Med* 20, 579–85. doi: 10.1016/j.ejim.2009.05.006.

Parker, G., Mahendran, R., Koh, E. S., et al. (2000). Season of birth in schizophrenia: no latitude at the equator. *Br J Psychiatry* 176, 68–71.

Penumarthy, S., Penmetsa, G. S., and Mannem, S. (2013). Assessment of serum levels of triglycerides, total cholesterol, high-density lipoprotein cholesterol, and low-density lipoprotein cholesterol in periodontitis patients. *J Indian Soc Periodontol* 17, 30–5. doi: 10.4103/0972-124X.107471.

Phung, D. T., Blickstein, I., Goldman, R. D., et al. (2002). The Northwestern Twin Chorionicity Study: I. Discordant inflammatory findings that are related to chorionicity in presenting versus nonpresenting twins. *Am J Obstet Gynecol* 186, 1041–5.

Pier, G. B., Grout, M., Zaidi, T., et al. (1998). *Salmonella typhi* uses CFTR to enter intestinal epithelial cells. *Nature* 393, 79–82. doi: 10.1038/30006.

Platt, O. S., Brambilla, D. J., Rosse, W. F., et al. (1994). Mortality in sickle cell disease. Life expectancy and risk factors for early death. *N Engl J Med* 330, 1639–44. doi: 10.1056/NEJM199406093302303.

Poolman, E. M. and Galvani, A. P. (2007). Evaluating candidate agents of selective pressure for cystic fibrosis. *J R Soc Interface* 4, 91–8. doi: 10.1098/rsif.2006.0154.

Quinn, C. T., Rogers, Z. R., McCavit, T. L., et al. (2010). Improved survival of children and adolescents with sickle cell disease. *Blood* 115, 3447–52. doi: 10.1182/blood-2009-07-233700.

Rademakers, R., Cruts, M., and Van Broeckhoven, C. (2003). Genetics of early-onset Alzheimer dementia. *Scientific World Journal* 3, 497–519. doi: 10.1100/tsw.2003.39.

Rawlins, S. (2001). Nonviral sexually transmitted infections. *J Obstet Gynecol Neonatal Nurs* 30, 324–31.

Reed-Geaghan, E. G., Reed, Q. W., Cramer, P. E., et al. (2010). Deletion of CD14 attenuates Alzheimer's disease pathology by influencing the brain's inflammatory milieu. *J Neurosci* 30, 15369–73. doi: 10.1523/JNEUROSCI.2637-10.2010.

Remington, J. S., Klein, J. O., Wilson, C. B., et al. (1976). *Infectious Diseases of the Fetus and Newborn Infant*. Philadelphia: W. B. Saunders.

Ried, K. (2016). Garlic lowers blood pressure in hypertensive individuals, regulates serum cholesterol, and stimulates immunity: an updated meta-analysis and review. *J Nutr* 146, 389S–96S. doi: 10.3945/jn.114.202192.

Rieder, R. O., Rosenthal, D., Wender, P., et al. (1975). The offspring of schizophrenics. Fetal and neonatal deaths. *Arch Gen Psychiatry* 32, 200–11.

Riggs, M. G., Whittaker, R. G., Neumann, J. R., et al. (1977). n-Butyrate causes histone modification in HeLa and Friend erythroleukaemia cells. *Nature* 268, 462–4.

Risch, N., Tang, H., Katzenstein, H., et al. (2003). Geographic distribution of disease mutations in the Ashkenazi Jewish population supports genetic drift over selection. *Am J Hum Genet* 72, 812–22. doi: 10.1086/373882.

Rizos, E. C., Ntzani, E. E., Bika, E., et al. (2012). Association between omega-3 fatty acid supplementation and risk of major cardiovascular disease events: a systematic review and meta-analysis. *JAMA* 308, 1024–33. doi: 10.1001/2012.jama.11374.

Roediger, W. E. (1980). Role of anaerobic bacteria in the metabolic welfare of the colonic mucosa in man. *Gut* 21, 793–8.

Roy, J., Sarkar, A., Parida, S., Ghosh, Z., et al. (2017). Small RNA sequencing revealed dysregulated piRNAs in Alzheimer's disease and their probable role in pathogenesis. *Mol Biosyst* 13, 565–76. doi: 10.1039/c6mb00699j.

Rupp, J., Goepel, W., Kramme, E., et al. (2004). CD14 promoter polymorphism -159C>T is associated with susceptibility to chronic *Chlamydia pneumoniae* infection in peripheral blood monocytes. *Genes Immun* 5, 435–8. doi: 10.1038/sj.gene.6364112.

Saldana-Meyer, R. and Recillas-Targa, F. (2011). Transcriptional and epigenetic regulation of the p53 tumor suppressor gene. *Epigenetics* 6, 1068–77. doi: 10.4161/epi.6.9.16683.

Saltini, C., Amicosante, M., Franchi, A., et al. (1998). Immunogenetic basis of environmental lung disease: lessons from the berylliosis model. *Eur Respir J* 12, 1463–75.

Samanta, D., Mulye, M., Clemente, T. M., et al. (2017). Manipulation of host cholesterol by obligate intracellular bacteria. *Front Cell Infect Microbiol* 7, 165. doi: 10.3389/fcimb.2017.00165.

Sampath, H. M. (1974). Prevalence of psychiatric disorders in a southern Baffin Island Eskimo settlement. *Can Psychiatr Assoc J* 19, 363–7.

Sanchez-Mut, J. V. and Graff, J. (2015). Epigenetic alterations in Alzheimer's disease. *Front Behav Neurosci* 9, 347. doi: 10.3389/fnbeh.2015.00347.

Sandi, R. M., Pol, K. G., Basavaraj, P., et al. (2014). Association of serum cholesterol, triglyceride, high and low density lipoprotein (HDL and LDL) levels in chronic periodontitis subjects with risk for cardiovascular disease (CVD): a cross sectional study. *J Clin Diagn Res* 8, 214–16. doi: 10.7860/JCDR/2014/6686.3927.

Sankaran, V. G. and Orkin, S. H. (2013). The switch from fetal to adult hemoglobin. *Cold Spring Harb Perspect Med* 3, a011643. doi: 10.1101/cshperspect.a011643.

Saunders, A. M., Strittmatter, W. J., Schmechel, D., et al. (1993). Association of apolipoprotein E allele epsilon 4 with late-onset familial and sporadic Alzheimer's disease. *Neurology* 43, 1467–72.

Schwarz, M. J., Muller, N., Riedel, M., et al. (2001). The Th2-hypothesis of schizophrenia: a strategy to identify a subgroup of schizophrenia caused by immune mechanisms. *Med Hypotheses* 56, 483–6. doi: 10.1054/mehy.2000.1203.

Sehgal, S., Mujtaba, S., Gupta, D., et al. (2010). High incidence of Epstein Barr virus infection in childhood acute lymphocytic leukemia: a preliminary study. *Indian J Pathol Microbiol* 53, 63–7. doi: 10.4103/0377-4929.59186.

Serjeant, G. R. (2013). The natural history of sickle cell disease. *Cold Spring Harb Perspect Med* 3, a011783. doi: 10.1101/cshperspect.a011783.

Serjeant, G. R., Higgs, D. R., and Hambleton, I. R. (2007). Elderly survivors with homozygous sickle cell disease. *N Engl J Med* 356, 642–3. doi: 10.1056/NEJMc066547.

Shalginskikh, N., Poleshko, A., Skalka, A. M., et al. (2013). Retroviral DNA methylation and epigenetic repression are mediated by the antiviral host protein Daxx. *J Virol* 87, 2137–50. doi: 10.1128/JVI.02026-12.

Sham, P. C., Maclean, C. J., and Kendler, K. S. (1993). Risk of schizophrenia and age difference with older siblings. Evidence for a maternal viral infection hypothesis? *Br J Psychiatry* 163, 627–33.

Shaner, A., Miller, G., and Mintz, J. (2007). Evidence of a latitudinal gradient in the age at onset of schizophrenia. *Schizophr Res* 94(1–3), 58–63. doi: 10.1016/j.schres.2007.04.001.

Shetty, S., Thomas, B., Shetty, V., Bhandary, R., et al. (2013). An in-vitro evaluation of the efficacy of garlic extract as an antimicrobial agent on periodontal pathogens: a microbiological study. *Ayu* 34, 445–51. doi: 10.4103/0974-8520.127732.

Shima, K., Kobayashi, I., Saito, I., et al. (2000). Incidence of human papillomavirus 16 and 18 infection and p53 mutation in patients with oral squamous cell carcinoma in Japan. *Br J Oral Maxillofac Surg* 38, 445–50. doi: 10.1054/bjom.2000.0162.

Silmon De Monerri, N. C., and Kim, K. (2014). Pathogens hijack the epigenome: a new twist on host–pathogen interactions. *Am J Pathol* 184, 897–911. doi: 10.1016/j.ajpath.2013.12.022.

Silverman, M. G., Ference, B. A., Im, K., et al. (2016). Association between lowering LDL-C and cardiovascular risk reduction among different therapeutic interventions: a systematic review and meta-analysis. *JAMA* 316(12), 1289–97. doi: 10.1001/jama.2016.13985.

Simopoulos, A. P. (2008). The importance of the omega-6/omega-3 fatty acid ratio in cardiovascular disease and other chronic diseases. *Exp Biol Med (Maywood)* 233(6), 674–88. doi: 10.3181/0711-MR-311.

Slatkin, M. (2004). A population-genetic test of founder effects and implications for Ashkenazi Jewish diseases. *Am J Hum Genet* 75, 282–93. doi: 10.1086/423146.

Sleiman, S. F., Basso, M., Mahishi, L., et al. (2009). Putting the 'HAT' back on survival signalling: the promises and challenges of HDAC inhibition in the treatment of neurological conditions. *Expert Opin Investig Drugs* 18, 573–84. doi: 10.1517/13543780902810345.

Sobel, D. E. (1961). Children of schizophrenic patients: preliminary observations on early development. *Am J Psychiatry* 118, 512–7. doi: 10.1176/ajp.118.6.512.

Sonnenburg, E. D. and Sonnenburg, J. L. (2014). Starving our microbial self: the deleterious consequences of a diet deficient in microbiota-accessible carbohydrates. *Cell Metab* 20, 779–86. doi: 10.1016/j.cmet.2014.07.003.

Soto, D., Song, C., and McLaughlin-Drubin, M. E. (2017). Epigenetic alterations in human papillomavirus-associated cancers. *Viruses* 9(9), pii: E248. doi: 10.3390/v9090248.

Souto, G. R., Queiroz-Junior, C. M., Costa, F. O., et al. (2014a). Effect of smoking on immunity in human chronic periodontitis. *Immunobiology* 219, 909–15. doi: 10.1016/j.imbio.2014.08.003.

Souto, G. R., Queiroz-Junior, C. M., Costa, F. O., et al. (2014b). Smoking effect on chemokines of the human chronic periodontitis. *Immunobiology* 219, 633–6. doi: 10.1016/j.imbio.2014.03.014.

Stampfli, M. R. and Anderson, G. P. (2009). How cigarette smoke skews immune responses to promote infection, lung disease and cancer. *Nat Rev Immunol* 9, 377–84. doi: 10.1038/nri2530.

Steel, A. J. and Eslick, G. D. (2015). Herpes viruses increase the risk of Alzheimer's disease: a meta-analysis. *J Alzheimer's Dis* 47, 351–64. doi: 10.3233/JAD-140822.

Stone, W. S. and Hsi, X. (2011). Declarative memory deficits and schizophrenia: problems and prospects. *Neurobiol Learn Mem* 96, 544–52. doi: 10.1016/j.nlm.2011.04.006.

Strittmatter, W. J., Saunders, A. M., Schmechel, D., et al. (1993). Apolipoprotein E: high-avidity binding to beta-amyloid and increased frequency of type 4 allele in late-onset familial Alzheimer disease. *Proc Natl Acad Sci U S A* 90, 1977–81.

Sundin, O. H., Yang, J. M., Li, Y., et al. (2000). Genetic basis of total colour blindness among the Pingelapese islanders. *Nat Genet* 25, 289–93. doi: 10.1038/77162.

Tam, W. L. and Weinberg, R. A. (2013). The epigenetics of epithelial–mesenchymal plasticity in cancer. *Nat Med* 19, 1438–49. doi: 10.1038/nm.3336.

Tamminga, C. A. and Zukin, R. S. (2015). Schizophrenia: evidence implicating hippocampal GluN2B protein and REST epigenetics in psychosis pathophysiology. *Neuroscience* 309, 233–42. doi: 10.1016/j.neuroscience.2015.07.038.

Tian, Y., Yang, W., Song, J., et al. (2013). Hepatitis B virus X protein-induced aberrant epigenetic modifications contributing to human hepatocellular carcinoma pathogenesis. *Mol Cell Biol* 33, 2810–6. doi: 10.1128/MCB.00205-13.

Timms, J. A., Relton, C. L., Rankin, J., et al. (2016). DNA methylation as a potential mediator of environmental risks in the development of childhood acute lymphoblastic leukemia. *Epigenomics* 8, 519–36. doi: 10.2217/epi-2015-0011.

Tjalma, W., Vanderheyden, T., Naessens, A., et al. (1998). Discordant prenatal diagnosis of congenital toxoplasmosis in a dizygotic pregnancy. *Eur J Obstet Gynecol Reprod Biol* 79, 107–8.

Torrey, E. F. (1980). *Schizophrenia and Civilization*. New York: Jason Aronson.

Torrey, E. F. and Yolken, R. H. (2003). *Toxoplasma gondii* and schizophrenia. *Emerg Infect Dis* 9, 1375–80. doi: 10.3201/eid0911.030143.

Torrey, E. F., Torrey, B. B., and Peterson, M. R. (1977). Seasonality of schizophrenic births in the United States. *Arch Gen Psychiatry* 34(9), 1065–70.

Torrey, E. F., Bowler, A. E., and Clark, K. (1997). Urban birth and residence as risk factors for psychoses: an analysis of 1880 data. *Schizophr Res* 25, 169–76.

Torrey, E. F., Mortensen, P. B., Pedersen, C. B., et al. (2001). Risk factors and confounders in the geographical clustering of schizophrenia. *Schizophr Res* 49, 295–9.

Tropberger, P., Pott, S., Keller, C., et al. (2013). Regulation of transcription through acetylation of H3K122 on the lateral surface of the histone octamer. *Cell* 152, 859–72. doi: 10.1016/j.cell.2013.01.032.

Trumble, B. C., Stieglit, J., Blackwell, A. D., et al. (2017). Apolipoprotein E4 is associated with improved cognitive function in Amazonian forager-horticulturalists with a high parasite burden. *FASEB J* 31, 1508–15. doi: 10.1096/fj.201601084R.

Tsuang, M. T., Stone, W. S., and Faraone, S. V. (2001). Genes, environment and schizophrenia. *Br J Psychiatry Suppl* 40, s18–24.

Urosevic, N. and Martins, R. N. (2008). Infection and Alzheimer's disease: the APOE epsilon 4 connection and lipid metabolism. *J Alzheimer's Dis* 13, 421–35.

van Exel, E., Koopman, J. J. E., Bodegom, D. V., et al. (2017). Effect of APOE epsilon 4 allele on survival and fertility in an adverse environment. *PLoS One* 12, e0179497. doi: 10.1371/journal.pone.0179497.

Varshney, R. and Budoff, M. J. (2016). Garlic and heart disease. *J Nutr* 146, 416S–21S. doi: 10.3945/jn.114.202333.

Velliyagounder, K., Ganeshnarayan, K., Velusamy, S. K., et al. (2012). In vitro efficacy of diallyl sulfides against the periodontopathogen *Aggregatibacter actinomycetemcomitans*. *Antimicrob Agents Chemother* 56, 2397–407. doi: 10.1128/AAC.00020-12.

Verdin, E. and Ott, M. (2015). 50 years of protein acetylation: from gene regulation to epigenetics, metabolism and beyond. *Nat Rev Mol Cell Biol* 16, 258–64. doi: 10.1038/nrm3931.

Videbech, T., Weeke, A., and Dupont, A. (1974). Endogenous psychoses and season of birth. *Acta Psychiatr Scand* 50, 202–18.

Vilas-Zornoza, A., Agirre, X., Martin-Palanco, V., et al. (2011). Frequent and simultaneous epigenetic inactivation of TP53 pathway genes in acute lymphoblastic leukemia. *PLoS One* 6, e17012. doi: 10.1371/journal.pone.0017012.

Walter, S., Letiembre, M., Liu, Y., et al. (2007). Role of the toll-like receptor 4 in neuroinflammation in Alzheimer's disease. *Cell Physiol Biochem* 20, 947–56. doi: 10.1159/000110455.

Walters, M. C. (2015). Update of hematopoietic cell transplantation for sickle cell disease. *Curr Opin Hematol* 22, 227–33. doi: 10.1097/MOH.0000000000000136.

Waterborg, J. H. (2002). Dynamics of histone acetylation in vivo. A function for acetylation turnover? *Biochem Cell Biol* 80, 363–78.

Weinberg, D. N., Allis, C. D., and Lu, C. (2017). Oncogenic mechanisms of histone H3 mutations. *Cold Spring Harb Perspect Med* 7(1), pii: a026443. doi: 10.1101/cshperspect.a026443.

Weinberg, M. S. and Morris, K. V. (2016). Transcriptional gene silencing in humans. *Nucleic Acids Res* 44, 6505–17. doi: 10.1093/nar/gkw139.

Weissman, B. and Knudsen, K. E. (2009). Hijacking the chromatin remodeling machinery: impact of SWI/SNF perturbations in cancer. *Cancer Res* 69, 8223–30. doi: 10.1158/0008-5472.CAN-09-2166.

Westrich, J. A., Warren, C. J., and Pyeon, D. (2017). Evasion of host immune defenses by human papillomavirus. *Virus Res* 231, 21–33. doi: 10.1016/j.virusres.2016.11.023.

White, C. (1999). *Reconstructing Ancient Maya Diet*. Salt Lake City: University of Utah Press.

Wilkins, J. F. and Haig, D. (2003). What good is genomic imprinting: the function of parent-specific gene expression. *Nat Rev Genet* 4, 359–68. doi: 10.1038/nrg1062.

Wilson, P. W., Myers, R. H., Larson, M. G., et al. (1994). Apolipoprotein E alleles, dyslipidemia, and coronary heart disease. The Framingham Offspring Study. *JAMA* 272, 1666–71.

Wilson, P. W., Schaefer, E. J., Larson, M. G., et al. (1996). Apolipoprotein E alleles and risk of coronary disease. A meta-analysis. *Arterioscler Thromb Vasc Biol* 16, 1250–5.

Withrock, I. C., Anderson, S. J., Jefferson, M. A., et al. (2015). Genetic diseases conferring resistance to infectious diseases. *Genes and Diseases* 2, 247–54.

Wolk, M. and Newlands, E. S. (2004). The correlation between fetal hemoglobin and outcome in non-Hodgkin's lymphoma. *Int J Biol Markers* 19, 168–9.

Wolk, M., Martin, J. E., and Constantin, R. (2004). Blood cells with fetal haemoglobin (F-cells) detected by immunohistochemistry as indicators of solid tumours. *J Clin Pathol* 57, 740–5. doi: 10.1136/jcp.2003.013938.

Yolken, R. H. and Torrey, E. F. (1995). Viruses, schizophrenia, and bipolar disorder. *Clin Microbiol Rev* 8, 131–45.

Yolken, R. H., Bachmann, S., Ruslanova, I., et al. (2001). Antibodies to *Toxoplasma gondii* in individuals with first-episode schizophrenia. *Clin Infect Dis* 32, 842–4. doi: 10.1086/319221.

Zannas, A. S. and West, A. E. (2014). Epigenetics and the regulation of stress vulnerability and resilience. *Neuroscience* 264, 157–70. doi: 10.1016/j.neuroscience.2013.12.003.

Zentner, G. E. and Henikoff, S. (2013). Regulation of nucleosome dynamics by histone modifications. *Nat Struct Mol Biol* 20, 259–66. doi: 10.1038/nsmb.2470.

Zhao, X., Bu, D. X., Hayfron, K., et al. (2012). A combination of secondhand cigarette smoke and *Chlamydia pneumoniae* accelerates atherosclerosis. *Atherosclerosis* 222, 59–66. doi: 10.1016/j.atherosclerosis.2012.02.005.

Zhou, J., Yu, J. T., Wang, H. F., et al. (2015). Association between stroke and Alzheimer's disease: systematic review and meta-analysis. *J Alzheimers Dis* 43, 479–89. doi: 10.3233/JAD-140666.

CHAPTER 4

GROWTH AND DEVELOPMENT

ROBIN M. BERNSTEIN AND BARRY BOGIN

Abstract

An evolutionary and biocultural approach is taken to the study of human growth and development. The evolutionary perspective focuses on the unusual process of human postnatal growth and development, a process that takes two decades to complete and traverses the stages of infancy, childhood, juvenility, and adolescence. Human childhood and adolescence are highly unusual even compared to our closest living relatives, perhaps unique. The biocultural perspective of human development focuses on the constant interaction taking place during all phases of human development between genes and hormones within the body and the sociocultural environment that surrounds the body. While humans are often considered to be cooperative breeders, depending on social group helpers to successfully rear offspring, it may be more accurate to understand humans as practising biocultural reproduction as an adaptation to minimise risks to health.

Keywords

adolescence, biocultural reproduction, childhood, chimpanzee, cooperative breeding, evolution, medicine, infancy, juvenility, risk

4.1 Introduction

In the broadest sense, the task of a species' or an individual's developmental trajectory is to survive to reproduction, to reproduce, and to help ensure that offspring survive and, in turn, reproduce. *Growth* and *development*, in their broadest sense, are multigenerational and encompass one's own lifespan and the lifespans of past and future generations. A multigenerational perspective seems obvious when we consider the flow of energy required for organisms to grow and develop. The main components of energy flow in the traditional *life*

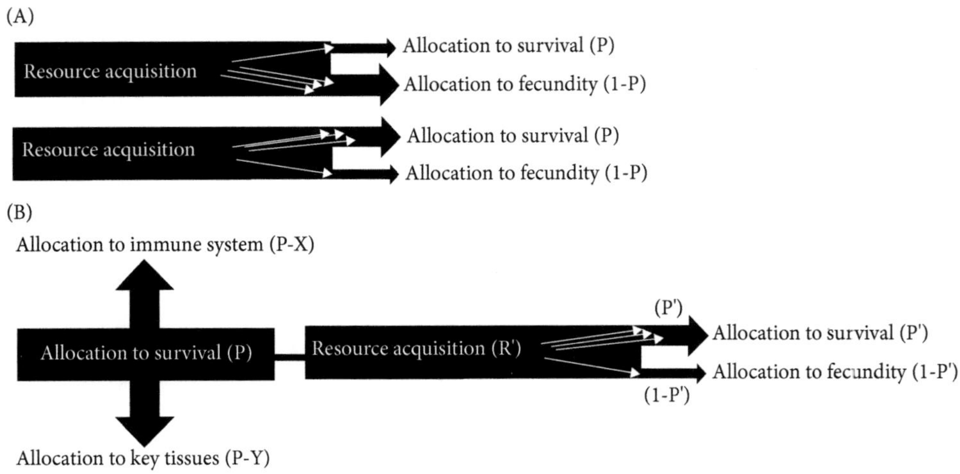

FIGURE 4.1 Life history trade-offs. (A) Classic Y-model illustrating the mechanism for a near-ubiquitous life history trade-off between survival and fecundity via differential allocation of resources to each when resource acquisition is similar. Width of arrow represents relative proportion of resource availability and allocation. (B) Modified Y-model showing allocation of resources to survival (P) via investment in immunity (P-X) and tissues important to whole-animal performance (P-Y). The model also shows how allocation of resources to survival via performance can lead to further acquisition of resources, which can later be allocated to future survival (P′) or fecundity (1-P′).

Source: Reproduced from Simon P. Lailvaux and Jerry F. Husak, The Life History of Whole-Organism Performance, *The Quarterly Review of Biology*, 89 (4), pp. 285–318, doi.org/10.1086/678567. Copyright © 2014, The University of Chicago Press.

History 'Y' model of energy allocation (Figure 4.1) indicate fundamental shifts at varying times during development in all organisms. From the perspective of the individual organism, the primary shifts are from allocating the majority of resources to one's own growth and development, to allocating energy for repair (from trauma, infection, etc.), or to allocating the majority of available energy to reproduction. The pathway to *maturation*, with various milestones along the way (e.g. skeletal and dental development, nervous system development), can vary significantly both among species and within species between males and females.

The key adaptive role of the growth trajectory means that it has been subject to extensive selective pressures throughout evolutionary time. The human growth pattern, with both similarities and differences to the growth patterns of our closest living ape relatives, is no exception. Compared with the apes, we have larger brains relative to our bodies and we grow those brains quickly; we wean our infants from mother's milk relatively early; we have clear and extended developmental stages of childhood and adolescence; we postpone reproductive maturation and first pregnancy, but keep alive more of our offspring; and we double the years of healthy lifespan. From the perspective of proximate causation, we can investigate genetic, endocrine, and other mechanistic pathways for evidence of change. In terms of ultimate causation, we can look to social and environmental factors that influence age-specific mortality risk and other key life history variables. This chapter reviews the evolutionary and comparative aspects of growth; the hormonal and epigenetic influences on growth; and the ways in which the human developmental trajectory creates risks for disease both in the short term and across the lifespan in the context of the *Developmental Origins of Health and Disease (DOHaD)* paradigm.

4.2 Evolution of Development

For multicellular organisms, most major evolutionary change proceeds by alterations in life cycles, that is, the patterns of growth, development, and maturation (Bonner 1965). The human species is no exception, which means that the biological and behavioural characteristics of human beings, including those shared with other mammals and those that set our species apart from all others, are derived from the features of our life cycle. Features of growth, development, and maturation among mammals can be *homologous*, that is, due to a common evolutionary origin. An area of debate is whether a common developmental trajectory for similar traits in closely related organisms is required for those traits to be considered homologous (Lieberman 1999). Different species may also share traits due to a process called *homoplasy*, also called convergence, an example being convergence towards a streamlined body that minimises water resistance in both fish and marine mammals.

One of the most persistent themes in the older literature concerning growth and development is the belief that 'ontogeny recapitulates phylogeny'. This phrase was coined by Ernst Haeckel, a nineteenth-century biologist, and in its strictest sense means that the entire evolutionary history of an organism is revealed by its embryonic development. This idea is related to the concept of the *scala naturae*, that successive adult ancestral stages were added to development as evolution proceeded from its lower to its highest form, humanity. Both recapitulation and the *scala naturae*, while important proposals in their time, are now understood to be incorrect (Gould 1977). Evolutionary change does occur through modifications to development, but not through the stacking of compressed ancestral developmental stages into the ontogeny of descendants. Broad similarities across all mammalian species, and even between mammals, insects, fish, and reptiles, do exist. Some of the similarities are due to shared regulatory sequences of DNA, for example, *Hox* genes that produce body segments or eyes. Other similarities are due to shared phylotypic stages (conserved stages of embryogenesis between species, such as the organogenesis stage of early vertebrate development that results in morphological similarities). These types of shared evolutionary homologies and homoplasies usually occur early in development (Raff 1996).

Transformations to development can occur in many ways. *Heterochrony*—shifts in rates or timing of developmental stages or events, relative to ancestral patterns—has provided one valuable framework for understanding some of the ways in which human development has been shaped (Shea 1989). Often, when relying on fossil evidence, with either static snapshots of developmental processes or adult outcomes of development, the actual processes behind heterochronic transformations cannot be reconstructed but are hypothesised based on evidence of some heterochronic shift. These shifts are described by two main categories of change: (1) paedomorphosis, where descendants possess ancestral juvenile features at later stages of development; and (2) peramorphosis, where ancestral trajectories are extended beyond their original endpoints to produce new morphologies (Klingenberg 1998). Various combinations of shifts in age at sexual maturity, rate of growth, and duration of growth period can result in paedomorphosis (time hypomorphosis, rate hypomorphosis, neoteny) or peramorphosis (time hypermorphosis, rate hypomorphosis, acceleration). Brief definitions of these terms are given here (and more detail may be found in the references cited in this section).

Hypermorphosis is a process that can be produced by either extending the growth period of an ancestral species such that descendants reach maturation at later ages (i.e. time hypermorphosis) or maintaining the age at maturation and increasing growth rate in the period leading up to it (i.e. rate hypermorphosis). Hypomorphosis is a process that results in descendant adults that appear like ancestral juveniles. This can occur through shortening growth duration (i.e. time hypomorphosis) by growing for the same amount of time at a slower rate (i.e. rate hypomorphosis), or by moving maturation earlier in the growth trajectory (i.e. neoteny). Historically, neoteny has often been invoked as being particularly important in human evolution, and scholars have pointed to traits including hairlessness, short and flat faces, and relatively large brains (traits found in a juvenile chimpanzee) as evidence of a global neotenic shift in modern humans (McKinney and McNamara 1991). However, attributing all key human traits to one heterochronic process overlooks traits that do not conform to this role; for example, peramorphosis probably best describes modern human lower limb development (Tardieu 1998). In addition, in neoteny the developmental period overall is truncated relative to the ancestral trajectory, by moving up reproductive maturation, which is the opposite of what has happened in humans. (For further discussion, see Chapter 17: Brain, Spinal Cord, and Sensory Systems.)

However, not all developmental change occurs through heterochrony. Closely related taxa can diverge in ontogeny by skipping over whole stages of development (e.g. direct- versus indirect-developing species; Raff 1992). In addition, the variation in adult form of any given species is shaped by factors encountered during a lifetime, such as seasonality and day length, food and resource availability, and predator pressure. Phenotypic plasticity, or the potential for a given genotype to produce a range of phenotypes in response to environmental differences, is responsible for facultative and adaptive shifts in form that arise from transformations to growth rate and timing, as well as the presence or absence of very specific morphological features at an individual level. The timing of menarche is suggested to represent an example of phenotypic plasticity in humans, in response to environmental predictability (Stearns and Koella 1986).

4.3 Modern Human Growth in Comparative Context

4.3.1 Stages of Growth (Including Intrauterine Growth)

Human growth proceeds through several stages, each corresponding to specific markers of development and maturation of both the body and behaviour (Bogin 1999; see Bogin and Smith (2012) for details of the following brief descriptions). These stages and the markers are listed in Table 4.1). Detailed description of each stage could form a book of its own. We have space here to describe only the most salient features.

4.3.2 Fertilisation and Prenatal Stage

The prenatal stage of the mammalian life cycle encompasses greater amounts of growth, development, and maturation than any other stage. Prenatal life is, in fact, the most critical

Table 4.1 Stages in the Human Life Cycle

Stage	Growth events/duration (approximate or average)
Prenatal life	
Fertilisation	
First trimester	Fertilisation to 12th week: embryogenesis
Second trimester	Fourth through sixth lunar month: rapid growth in length
Third trimester	Seventh lunar month to birth: rapid growth in weight and organ maturation
Birth	
Postnatal life	
Neonatal period	Birth to 28 days: extrauterine adaptation; most rapid rate of postnatal growth and maturation
Infancy	Second month to end of lactation, usually by 36 months: rapid growth velocity, but with steep deceleration in growth rate; feeding by lactation; deciduous tooth eruption; many developmental milestones in physiology, behaviour, and cognition
Childhood	Years 3–7: moderate growth rate; dependency on older people for care and feeding; midgrowth spurt; eruption of first permanent molar and incisor; cessation of brain growth by end of stage
Juvenile	Years 7–10 for girls, 7–12 for boys: slower growth rate; capable of self-feeding; cognitive transition leading to learning of economic and social skills
Puberty	Occurs at end of juvenile stage and is an event of short duration (days or a few weeks): reactivation of central nervous system of sexual development; dramatic increase in secretion of sex hormones
Adolescence	The stage of development that lasts for 5–10 years after the onset of puberty: growth spurt in height and weight; permanent tooth eruption almost complete; development of secondary sexual characteristics; sociosexual maturation; intensification of interest in and practice of adult social, economic, and sexual activities
Adulthood	
Prime and transition	From 20 years old to end of child-bearing years: homeostasis in physiology, behaviour, and cognition; menopause for women
Old age and senescence	From end of child-bearing years to death: decline in function of many body tissues or systems

period for the formation of most of the body's anatomical and physiological systems. *Critical periods* in developmental biology are times during the life cycle when one or more properties of the organism must grow or develop or when this property develops most rapidly. Human prenatal life has many critical periods, for example, maternal exposure to 'German measles' (rubella) in the first 3 months of pregnancy is associated with an elevated risk of birth defects. Teratology is the study of developmental abnormalities, and the analytical approach is based in part on the fact that the earlier in life an insult is introduced, generally the more severe the consequences (Pauli and Feldman 1986). The concept of the 'developmental hourglass' has been used to describe critical windows during development due to phylogenetic constraints or canalisation that force phenotypes to 'squeeze' through developmental bottlenecks that have a low threshold for variance and result in a finite number of body plans (Kalinka et al. 2010).

The prenatal gestation of all the apes, including the human ape, takes a longer amount of time compared with non-primate mammals, particularly long for our body size. Compare the average gestation length in days of humans = 266, chimpanzees = 240, gorillas = 240, and orangutan = 260, with, for example, the African lion = 110, which is less than half that of the apes. Female lions have an adult body mass of about 150 kg, which is more than twice the 68-kg average of female body weight for humans and chimpanzees. Even the longest gestation in mammals, the African elephant = 640 days, is not extremely long given the immense size of a female African elephant, weighing about 5000 kg. In proportion to adult female body weight, human and chimpanzee gestation is about five times longer than that of lions (chimpanzee 240/68 = 3.5; lion 110/150 = 0.73) and about 27 times longer than that of elephants (640/5000 = 0.13).

The fertilised ovum grows at first by cell division and each division takes from 12 to 24 hours. By day 4 there are 16 cells and these cells have become differentiated, that is, specialised in their development. Some of these cells will form the embryo and others will form the amniotic sac and placenta. Several changes take place, such as cavitation, gastrulation, and implantation in the wall of the mother's uterus. Distinct layers of cells differentiate that will eventually form the skin and teeth, muscle and bone, and internal organs. This stage of prenatal life is often called embryogenesis and occurs during the first trimester of pregnancy (fertilisation to end of week 12).

Rapid growth in length occurs during the second trimester, so that at 24 weeks the fetus measures about 30 cm and about 300 g. By the end of the second trimester, all body parts are formed but not fully functional. Significant growth in weight occurs during the third trimester, so that at 36 weeks the fetus averages 47.4 cm in length and 2622 g in weight. This is a gain of 1.6 times in length and 8.7 times in weight. There is also rapid development of the organs, so that birth and survival of the new-born without extraordinary medical care is possible. Normal 'term' births may occur between weeks 37 and 42, with a typical birth length of 48–52 cm and birth weight between 3000 and 4000 g—the mean birth weight for native white English and Welsh women in 2005 was 3393 g (Moser 2008).

4.3.3 Birth, New-born, and Infancy Stages

Birth is a critical transition between life *in utero* and life independent of the support systems provided by the uterine environment. We discuss here only births within the clinically normal range of gestation length and birth weight.

Preterm neonates are at elevated risk of mortality (Goldenberg et al. 2008). The medical technology needed to sustain preterm neonates was unlikely to be available for most of human evolutionary history. The neonate moves from a fluid to a gaseous environment, from a nearly constant external temperature to one with potentially great volatility. The new-born is also removed from a supply of oxygen and nutrients (provided by the mother's blood and passed through the placenta, which also handles the elimination of fetal waste products) to reliance on his or her own systems for digestion, respiration, and elimination. Any failure by the fetus to grow, develop, and mature these systems will cause illness or death. Worldwide, in the year 2015, 45% of deaths to infants and children under 5 years old occurred during the neonatal stage (birth to day 28) and most of these deaths occurred in the first few days after birth (http://www.who.int/gho/child_health/mortality/neonatal/en/).

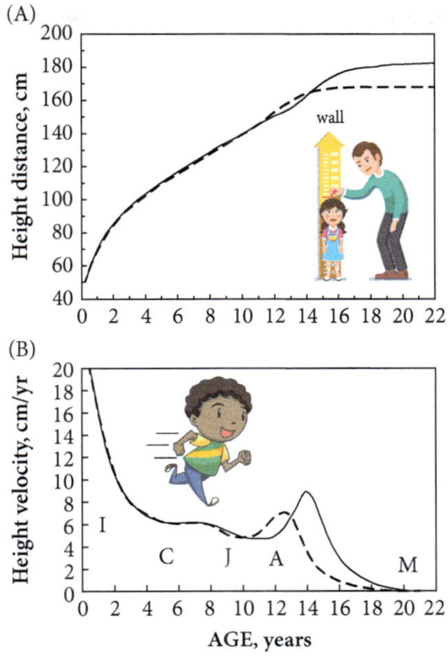

FIGURE 4.2 Average distance (A) and velocity (B) curves of growth in height for healthy girls (dashed lines) and boys (solid lines). Distance is the amount of height achieved at a given age. (A) shows a child's height being measured. Velocity is the rate of growth at a given time, in this case shown as centimetres per year. In (B), the running figure represents 'velocity'. Velocity curves show postnatal periods of the pattern of human growth. Note spurts in growth rate at mid-childhood and adolescence for both girls and boys. Postnatal periods: I, infancy; C, childhood; J, juvenile; A, adolescence; M, mature adult.

Source: (A) Maxim Popov © 123RF.com; (B) iStock.com/Victor_Brave.

During the neonatal stage and much of the infancy stage, growth rates are more rapid than they will ever be again in postnatal life (Figure 4.2). The infant stage lasts from the second postnatal month to about 2.99 years of age; during this stage, growth remains rapid, although growth rate constantly decelerates. As for all mammals, human infancy is the period when the mother provides all or some nourishment to her offspring via lactation or some culturally derived imitation of lactation. For most of human evolution and in many contemporary human populations, infants are/were mainly fed with mother's milk and varying types of transitional foods (Sellen 2007). Complementary feeding may begin between the ages of 0 and 9 months, depending on infant growth rates and energy needs that are not met by breastfeeding, as well as cultural practices relating to infant care. Complementary foods encompass a wide range of liquids and solids, based on fruits, vegetables, carbohydrates, and animals. Sellen (2001) surveyed ethnographic and demographic reports published between 1873 and 1998 and found that the modal age of introduction of both liquids and solids was 6 months, but varied from birth to 12 months in particular societies. In some societies, insects are an especially important source of complementary food, as reported by Schiefenhövel and Blum (2007). Per unit weight, insects provide protein and fat comparable to vertebrates (Dufour 1987; Fink 2013; http://www.the-scientist.com/?articles.view/articleno/34172/title/why-insects-should-be-in-your-diet/). During infancy, the deciduous dentition (the so-called milk teeth) erupts through the gums. Human infancy ends when the child is weaned from the breast, which in pre-industrialised societies occurs between 24 and 36 months of age. By this age, all the deciduous teeth have erupted, even for very late-maturing infants. Once the deciduous dentition has erupted the infant is capable of processing more and different kinds of foods.

Motor skills (i.e. what a baby can do physically) develop rapidly during infancy (http://www.who.int/childgrowth/standards/motor_milestones/en/). Cognitive and learning skills also advance rapidly. Accounting for all of these advancements is the development of the skeletal, muscular, and nervous systems, especially brain growth and development. The human brain grows rapidly during infancy, much more rapidly than almost any other tissue or organ of the body. All parts of the brain seem to take part in this fast pace of infant growth and maturation, including the hypothalamus, a centre of neurological and endocrine control. The endocrine system interacts in complex ways with the environment and the genome to direct the course of growth, development, and maturation by release of hormones. In addition, changes in the operation of the endocrine system seem to play a primary role in the evolution of life cycles (Finch and Rose 1995). (For further discussion, see Chapter 15: Endocrinology.)

Human infancy includes what is often called the 'critical first 1000 days' of life, and insults to growth during the first years of life (reviewed in Victora et al. 2008) have been linked to later-life health outcomes (e.g. DOHaD (http://www.dohadsoc.org/); Barker and Osmond 1986). In cases where an infant is born prematurely, or a child experiences prolonged nutritional deficiency, disease, or severe psychological distress, normal growth patterns may be disrupted and growth in various aspects (e.g. linear, organ growth) can be affected. If these stressors are ameliorated, a marked increase in growth velocity can allow the child to return to age-appropriate size; pronounced catch-up growth has been documented in length/height, weight, head circumference, and adiposity. If the deficit was severe or persisted over a significant period of time, then catch-up growth may or may not serve to bring all systems back to a normal size or growth trajectory; this is especially true for brain growth. Catch-up growth can also have significant long-term consequences: failure to catch up can result in reduced stature or compromised cognitive development, whereas rapid catch-up growth can increase the risk of obesity, insulin resistance, and other metabolic and cardiovascular diseases later in life.

4.3.4 Childhood

The childhood life history stage spans from the end of infancy (~ 3.00 years) to approximately 6.99 years of age. The primary definition of childhood in terms of feeding behaviour is that the youngster is weaned (i.e. is no longer receiving breast milk), but remains dependent on the provisioning and protection of people other than the mother. Those other people may include older siblings, the parents' siblings, or grandparents (especially mother's sister and mother's mother), other genetic kin, non-genetically related kin (such as sister-in-laws and adoptive parents), and fathers. In some societies, childcare is provided by paid caretakers, slaves, and governmental institutions.

Human children, especially just after weaning at age 3 years, often require specially prepared foods because of the immaturity of their dentition. Deciduous teeth have a thinner enamel and shallower roots than do the permanent teeth. This means that children with only deciduous teeth, that is, before age 5–6 years old, cannot mechanically masticate some of the tougher foods that older people can eat. In addition, the small size of their stomachs and intestines, and the rapid growth of their brain, mean that human children need a diet that is easy to chew and swallow, low in total volume, and more energy dense than foods that older people eat.

The child's relatively large and active brain, almost three times the size of an adult chimpanzee's brain, requires that the low-volume diet be dense in energy, lipids, and proteins. An energy-nutrient dense diet is required to support the metabolic activity of the body and, especially, the brain. The human new-born uses 87% of its resting metabolic rate (RMR) energy for brain growth and function. By the age of 5 years, the percentage RMR usage is still high at 44%, and higher than any other mammal, whereas in the adult human, the figure is between 20% and 25% of RMR. At comparable stages of development, the chimpanzee devotes about 45%, 20%, and 9%, respectively, of its RMR to brain growth. Children do not yet have the motor and cognitive skills to prepare an energy-nutrient-rich diet for themselves. Accidents and disease are often risks and thus children require protection. In some hunter-gather environments, such as the Ache of Paraguay and the !Kung of Namibia, children are also vulnerable to predation. Interviews of !Kung adults report that a common concern is that children will wander off into the bush and become lost and then die. Eight cases of lost children were reported, with two deaths, one from a leopard attack (Blurton Jones et al. 1994). !Kung people claim that leopards, lions, and wild dogs will hunt children and !Kung women are reported to carry children up to the age of 6 years when they are in the bush foraging for food. Children will not survive in any society if deprived of the care provided by older individuals. The so-called wolf children and even street children, who are sometimes alleged to have lived on their own, are either myths or not children at all. A search of the literature finds no case of a child (i.e. a youngster under the age of 6 years) living alone, either in the wild or on urban streets (Bogin and Smith 2012). So-called street children are more appropriately named 'street juveniles' (see Section 4.3.5).

The pattern of growth and development also defines human childhood. The rapid growth deceleration of infancy begins to moderate as childhood begins. Growth rate declines modestly from about 7–8 cm/year at age 3 years to 5–6 cm/year by age 4.5 years, and levels off at this velocity until about age 6.99 years. This moderation of growth rate decline and the levelling-off in growth rate are unusual for mammals, because almost all other species continue a pattern of relatively rapid deceleration after infancy. Dental traits, digestive systems, and brains add to the features of human childhood. Two of the important physical developmental milestones of childhood are the eruption of the first permanent teeth/replacement of the deciduous teeth, and completion of brain growth (in weight). First molar eruption takes place, on average, between the ages of 5.5 and 6.5 years in most human populations. Eruption of the central incisor quickly follows, or sometimes precedes, the eruption of the first molar. By the end of childhood, usually at the age of 7 years, most children have erupted the four first molars, and permanent incisors have begun to replace 'milk' incisors. Along with growth in size and strength of the jaws and the muscles for chewing, these new teeth provide sufficient capabilities to eat a diet similar to that of adults.

Another feature of the childhood phase of growth associated with these developmental changes, seen in some populations but not others, is a modest acceleration in growth velocity at about 6–8 years, called the midgrowth spurt (shown in Figure 4.2). Some studies note the presence of the midgrowth spurt in the velocity curve of boys but not girls, while other studies find that up to two-thirds of boys and girls have midgrowth spurts.

The midgrowth spurt is linked with an endocrine event found in all populations called adrenarche, defined as the postnatal onset of secretion of the androgen hormones dehydroepiandrosterone (DHEA) and DHEA sulfate (DHEA-S) from the zona reticularis of the adrenal gland. In humans and chimpanzees, adrenarche occurs between the ages of 6 and 10 years, with some evidence for hormonal signatures of adrenarche in the orangutan and gorilla (Bernstein et al. 2012; Bernstein 2017). In some other primates, such as the rhesus monkey, the up-regulation of DHEA and DHEA-S begins perinatally along with early development of the adrenal gland (Conley et al. 2012).

In humans, it is unclear if adrenal androgens produce the midgrowth spurt in height, but these androgens seem to cause the appearance of a small amount of axillary and pubic hair.

Prior research in some populations, but not others, has found a correspondence of adrenarche with the so-called adiposity rebound seen at the transition between the childhood and juvenile stages of the life cycle. The adiposity rebound describes the increase in the body fatness that takes place between the ages of 5 and 7 years, usually measured as the body mass index (BMI), calculated as weight (in kilograms) divided by height (in metres). In many human populations, the BMI is a reasonable proxy measure of body fatness, but it is a measure of total mass and could be influenced by lean body tissue as well as fatness (Bogin and Varela-Silva 2012). In addition, body proportions of leg length to trunk length are in flux during stages of growth and development and this is likely to produce changes in BMI that are unrelated to fatness. Additional research is needed to demonstrate the presence and function of the adiposity rebound in human populations.

The combination of adrenarche and, possibly, the midgrowth spurt may be unique to human beings, but additional longitudinal studies of non-human primate growth and development are needed. The physical changes induced by adrenarche are accompanied by a change in cognitive function, called the '5- to 7-year-old shift' by some psychologists, or the shift from the preoperational to concrete operational stage, using the terminology of Piaget. This shift leads to new learning and work capabilities in the juvenile. It is proposed that adrenarche and DHEA-S may play a role in hominin evolution in terms of extended brain development and prolonged lifespan compared with other primates (Campbell 2011). Human brain growth in weight is virtually complete at a mean age of 7.0 years, but brain maturation, cognitive and motor skills, neural network growth, pruning, and myelination continue for several decades (Fuhrmann et al. 2015). By the end of the childhood stage, the slow rate of growth in brain mass means that the nutrient requirements diminish, which means less dependence on specially prepared, nutrient-dense foods from older people. Moreover, cognitive and emotional capacities mature to new levels of self-sufficiency. Language and symbolic thinking skills mature rapidly, social interaction in play and learning become common, and the 7-year-old individual can perform many basic tasks, including food preparation with little or no supervision (Locke and Bogin 2006).

In summary, human childhood is defined by the following traits: (1) a slower rate of growth than during infancy and relatively small body size; (2) a large, fast-growing brain; (3) higher brain RMRs than any other mammalian species; (4) immature dentition; (5) motor immaturity; (6) cognitive immaturity; and (7) adrenarche. Some aspects of human childhood are shared with other species, including our non-human primate kin, but we know of no other mammalian species with this entire suite of features. Human-like childhood seems to have evolved sometime in the past 2 million years or so and the search for its species of origin is an active area of research (Cameron et al. 2017).

4.3.5 Juvenile Stage

Juvenile mammals may be defined as prepubertal individuals who are largely responsible for their own feeding and protection. The juvenile stage is most commonly found in some social mammals, such as wolves, hyenas, lions, wild dogs, most primates, especially monkeys and apes, meerkats, mole rats, and possibly some of the cetaceans (porpoises and whales) (Table 4.2). The definition of 'juvenile' may apply to juvenile humans, some of whom are forced to care and feed themselves, such as street juveniles (the proper name for 'street children'), but it is not their usual state of living. Juvenile humans are prepubertal, but usually live in families of several types or social groups consisting of people ranging in age from infancy to old adults. Juveniles in these groups may provide significant amounts of labour, for example in the form of child care or collection of water and firewood. Juveniles may produce some of their own food and even share food with others, but the total energy production of human juveniles is usually less than their own needs. In most human societies, juveniles are net food/energy consumers.

The human juvenile stage begins at about 7.0 years of age. In girls, the juvenile period ends, on average, at about the age of 10 years, 2 years before it usually ends in boys, the difference reflecting the earlier onset of adolescence in girls. The juvenile stage is characterised by the slowest rate of growth since birth.

The evolution of the juvenile stage of primates is associated with both social complexity, especially larger social groups, and diet complexity, including foraging for fruits and seeds.

Table 4.2 Mammalian Life History Stages According to Social Organisation and Cooperative Breeding Status and Human Biocultural Reproduction. Examples of mammalian species (see Lukas and Clutton-Brock 2012, Supplementary Table 1 for complete list): *Non-social and most social mammals*—cattle, bison, horses, zebras, red deer, yellow and banded mongoose, most rodents, prairie dogs, domestic cats, kangaroos, elephant seals, killer whale, pilot whale; *Social primates, carnivores, and elephants, no cooperative breeding*—non-human apes, most monkeys, hyenas, lions, cheetahs; *Social, with cooperative breeding*—wolves, meerkats, wild dogs, naked mole rats, some mice, mongoose, beaver, tamarins, and marmosets

Life history stage	Non-social and most social mammals	Social primates, carnivores, and elephants, no cooperative breeding	Social, with cooperative breeding	Human, with biocultural reproduction
Infancy	X	X	X	X
Childhood		?		X
Juvenile		X	X	X
Adolescence				X
Adult	X	X	X	X
Post-reproductive females				X Also killer and pilot whales

At least three hypotheses have been offered to account for the evolution of juvenility: (1) a 'learning hypothesis' has often been proposed, based on studies of juvenile primates and human juveniles in many cultures indicating that much social learning takes place during this stage; (2) a strategy to avoid death from competition with older individuals while living in a social group; (3) a consequence of older, more socially dominant individuals repressing the sexual development of younger social group members. Any of these three reasons, or all of them, may have played a role.

4.3.6 Adolescence

Human adolescence is the stage of life when social and sexual maturation takes place. Adolescence begins with puberty, or more technically with gonadarche, which is the reactivation of the production and release of gonadotropin releasing hormone (GnRH) from the hypothalamus (Figure 4.3). Old World monkeys, apes, and humans have an 'on–off–on' pattern of GnRH activity during postnatal development. Non-primate mammals, such as rodents, do not show this pattern; instead, these animals have a progressive and uninterrupted increase in GnRH production from birth, through infancy, to sexual maturation (adulthood). The steady, progressive sexual maturation of these mammals reflects the fact that they have only the infancy and adulthood stages of postnatal development. The insertion of childhood and juvenile stages into human life history necessitated a 'turning-off' of GnRH production.

How the 'on–off–on' pattern is regulated has been an active area of research since 1975, when first described by Melvin Grumbach and colleagues (Plant 2015). The current understanding is that one, or perhaps a few, centres of the brain change their pattern of neurological and endocrinological activity and their influence on the hypothalamus. The hypothalamus becomes, basically, inactive in terms of sexual development by about age 2–3 years. The 'inhibitor' has not been identified but likely is located in the brain and certainly not in the gonads. Human children born without gonads, as well as rhesus monkeys and other primates whose gonads have been surgically removed at birth, still undergo both GnRH inhibition in infancy and hypothalamus reactivation at the age when puberty would occur normally. There is evidence that kisspeptin, a protein hormone made in the hypothalamus, may be part of the puberty regulation system. Similar to GnRH, kisspeptin is produced in a rapid pulsatile manner until about age 2 years in humans, then the pulses become less frequent until puberty, when they again become more frequent. An increase in another excitatory factor, the amino acid glutamate, is also associated with the onset of puberty. It is not known if kisspeptin and glutamate activity are causes or consequences of GnRH activity. Tony Plant (2015, p. 85), one of the leading researchers on puberty, writes that, 'From an evolutionary perspective it seems reasonable to argue that in primates with extensive development of the neocortex, the evolution of a mechanism to effect a prolonged delay in the onset of puberty (the neurobiological brake on GnRH pulse generator prepubertally) would be adaptive . . .'. Plant's words may support any or all of the three hypotheses for primate juvenility mentioned above.

Adolescence lasts for between 5 and 10 years following puberty (Bogin 2011). The adolescent stage includes the development of secondary sexual characteristics, such as development of the external genitalia, sexual dimorphism in body size, and the composition of muscle (more in boys) and fatness (more in girls). There is often an abrupt increase in the density

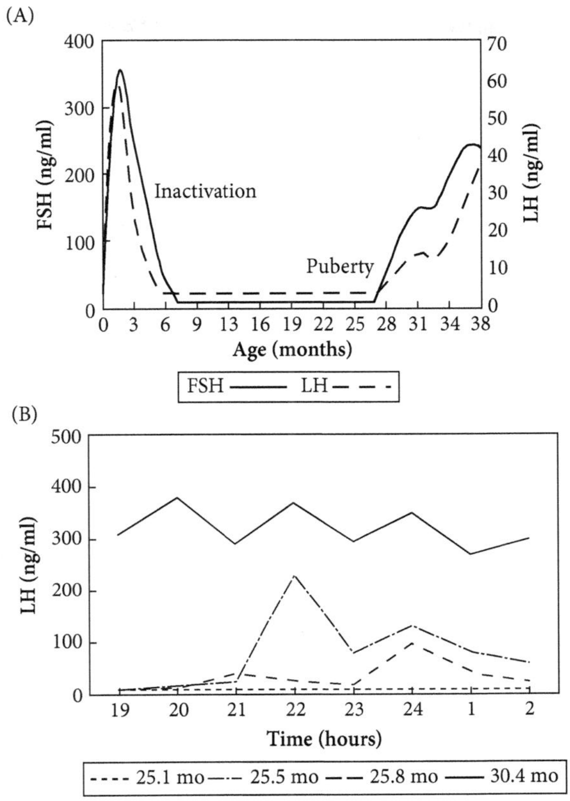

FIGURE 4.3 (A) Pattern of secretion of FSH and LH in a male rhesus monkey (genus *Macaca*). Testes of the monkey were removed surgically at birth. Curves for FSH and LH indicate production and release of GnRH from the hypothalamus. After age 3 months (i.e. during infancy), the hypothalamus is inactivated. Puberty takes place at 27 months, and the hypothalamus is reactivated. (B) Development of hypothalamic release of GnRH during puberty measured as luteinising hormone (LH) in serum in a male rhesus monkey with testes surgically removed. The X-axis is time in hours of a 24-hour clock. At 25.1 months (mo) of age, the hypothalamus remains inactivated. At 25.5 and 25.8 months, modest hypothalamic activity is observed, indicating the onset of puberty. By 30.4 months, the adult pattern of LH release is nearly achieved. This pattern shows increases in both the number of pulses of release and the amplitude of release. In human beings, a very similar pattern of infant inactivation and late juvenile reactivation of the hypothalamus takes place.

Source for Part(A): Adapted from Tony M. Plant, Ei Terasawa, and Selma Feldman Witchel, 'Puberty in Non-Human Primates and Man', in E. Knobil and J. D. Neill (eds.), *Knobil and Neill's Physiology of Reproduction*, 4e, pp. 1487–1536, Figure 32.5. Copyright © 2015 Elsevier Inc.

Source for Part(B): Adapted from Tony M. Plant, 'Puberty in primates', in E. Knobil and J. D. Neill (eds.), *The Physiology of Reproduction*, 2e., pp. 453–85. Copyright © 1994 Raven Press.

of pubic hair (indeed, the term 'puberty' is derived from the Latin *pubescere*, 'to grow hairy'). In boys, the increased density and darkening of facial hair and the deepening of the voice (voice 'cracking') are other signs of puberty. In girls, a visible sign is the development of the breast bud, the first stage of breast development. The pubescent boy or girl, his or her parents, and relatives, friends, and sometimes everyone else in the social group can observe one or more of these signs of early adolescence.

The most notable physical growth event of adolescence is the growth spurt (Figure 4.2). During this life stage, both boys and girls experience a rapid acceleration in the growth velocity of almost all skeletal tissue. Puberty and the onset of the growth spurt may occur as early as 8 years of age or as late as 17 years of age in any individual. Similar age variation is seen in the age at menarche (Garnier et al. 2005; Leenstra et al. 2005; Belachew et al. 2011), but menarche almost always occurs about 1 year after the peak of the adolescent growth spurt. This variation in growth timing is ascribed to various 'environmental' influences (e.g. food availability, disease burden, stress), all of which exert influence on growth through several key hormone systems (reviewed in Section 4.4.2).

The magnitude of this acceleration in growth varies quite a bit between individuals and populations. In some studies, up to 10% of seemingly healthy girls showed no discernible adolescent growth spurt. The absence of an adolescent growth spurt in healthy, well-nourished boys is rare. The near ubiquity of the adolescent growth spurt is, in fact, used as a biomarker for pathology. The absence of a growth spurt may indicate underlying physical disease, chronic malnutrition, or major psychological trauma. Examples of typical adolescent growth spurts are found in samples of healthy Swiss and Guatemalan boys and girls measured annually between the ages of 5 and 18 years (Bogin 1999). Despite differences in geography, culture, and (possibly) genetics, both samples at the peak of their adolescent growth spurt have average velocity of growth in height of +9.0 cm/year for boys and +7.1 cm/year (Swiss) to +7.9 cm/year (Guatemala) for girls.

No other primate species, not even chimpanzees, exhibits linear growth spurts as humans do. Most primate species have rapid growth in length and body weight during infancy and then a declining rate of growth from weaning to adulthood. Some primate species may show a rapid acceleration in soft tissue growth at puberty, especially of muscle mass in male monkeys and apes. Sexually maturing non-human primates may have skeletal spurts in the face, (e.g. due to the eruption of large canine teeth in male baboons). However, unlike humans, other primate species have either no pubertal acceleration in total skeletal growth or a very small increase in growth rate of the long bones.

Human adolescence is also characterised by changes in brain maturation that are less outwardly visible but equally important. Some of the brain changes are easily noted in behaviour, especially the intensification of interest and practice in adult social, economic, and sexual activities. Adolescents learn and integrate these adult roles into their daily behaviour, sometimes with the support of older adults and sometimes in competition with older adults. Human adolescence is, in this sense, an apprenticeship for the productive and reproductive roles that will be assumed at adulthood. Many of these behavioural changes of human adolescence usually occur with puberty in many species of social mammals (e.g. monkeys, apes, wolves, lions, meerkats). What makes human adolescence unusual among the primates is the length of time between age at puberty and age at first birth. Humans take, on average, some 10 years for this transition. On a worldwide basis, and throughout history, the average age for the first external manifestations of puberty for healthy girls is 9 years and first birth is at 19 years. On the same basis, boys show external signs of puberty, on average, at 11 years and fatherhood at 21–25 years. Monkeys and apes typically take less than 3 years to make the transition from puberty to parenthood.

4.3.7 Why Did Human Childhood and Adolescence Evolve?

The universality of the anatomical, developmental, physiological, neurological, and behavioural characteristics of the human childhood and adolescent stages is strong evidence that this combination of traits is species-specific and must ultimately have been shaped by natural selection. Childhood could be selected because it allowed hominin (human ancestor) females to give birth at shorter intervals compared with apes (see Bogin and Smith 2012 for details). In the wild, the interval between successful births for chimpanzee females averages nearly 5 years, and for orangutan mothers, who may be under nutritional stress, it averages 6–8 years. In contrast, human women in traditional societies (foragers, horticulturalists, pastoralists) wait, on average, 3 years between successful births. Humans are able to do this by reducing the length of the infancy stage of life (i.e. lactation) and by the evolution of the special features of the human childhood that allow other people, called allo-parents, to feed and protect children. The evolution of childhood provided humans with the potential for greater lifetime fertility than any ape. But producing offspring is only a small part of reproductive fitness. Rearing the young to their own reproductive maturity is a surer indicator of success. Adolescence may be a key in helping the next generation rear its own young successfully. The adolescence stage may provide boys and girls with a life history strategy to survive to adulthood via the cooperative work they do to support their social group (this is explained in Section 4.7). Survival to adulthood is the immediate benefit to the adolescent and there is also a longer-term benefit, the learning and practice of complex social and economic skills required for effective mating and parenting.

The biological mechanisms for the prolonged human adolescent delay between puberty and first birth or fatherhood may have evolved by the processes of natural selection and sexual selection. Both types of selection were identified by Charles Darwin. Natural selection operates to increase the frequency of genotypes and phenotypes, which confer reproductive advantages on the individuals possessing them, and sexual selection 'depends on the advantage which certain individuals have over other individuals of the same sex and species, in exclusive relation to reproduction' (Darwin 1871, vol. 1, p. 276). Today, biologists replace Darwin's word 'reproduction' with 'mating'. In many mammalian species, a good deal of mating is more about social relations rather than a reproductive event, such as fertilisation. It is known today that sexual selection also works for females, meaning that female-specific physical and behavioural traits may evolve via competition between females for mating opportunities with males.

Sexual selection in human evolutionary history may have led to cooperative child care, also called cooperative breeding (Hrdy 1999; Burkart et al. 2014). Key features of human cooperation are food sharing and division of labour by age and sex. The contribution of food and labour may be quantified in terms of energy (kilocalories). Another way is to quantify the food and labour sharing in terms of reproductive outcomes, such as birth and survival of offspring to adulthood. Reiches et al. (2009) did just this by proposing a 'pooled energy budget' hypothesis. They defined the pooled energy budget as the combined energetic allocations of all members of a reproductive community that might result in direct or indirect reproductive effort. 'These transactions can take many forms at one time and vary across the life course.... Individuals draw on the pooled energy budget by consuming calories and by diverting time and energy [to] reproduction. They contribute by diminishing their own energetic costs and by contributing to the energy budgets of others. The output of the pooled energy budget is the production of new individuals' (p. 424).

According to Reiches and colleagues, the pooled energy budget allows human women to sustain a higher fertility and greater survival of their offspring than would be possible for energetically isolated individuals. Energetic isolation, relative to the human condition, is the case for adult female chimpanzees, bonobos, gorillas, and orangutans. Females in these ape species are, generally, competitors. In terms of survival, about 50–60% of infants and children in human forager societies reach adulthood (Kaplan et al. 2000). In the great apes, survival averages less than 40%.

No non-human ape species is a cooperative breeder; neither is any Old World monkey (Table 4.2). The only primate species with cooperative breeding systems are the New World tamarin and marmoset monkeys. It is most likely that the cooperation shown by humans evolved after the evolutionary split from more ape-like species, perhaps after 6 million years ago (Hrdy 2011). Several types of cooperative breeding, meaning care of offspring by non-parents, are found in some species of birds and other mammals (e.g. wolves, meerkats, naked mole rats; see Table 4.2), and it works to increase net reproductive output. In those species, and in many but not all human groups, the cooperative breeders are close genetic relatives of the mother (Clutton-Brock 2002). By assisting the mother to care for her offspring, the helpers increase their own inclusive fitness, meaning that they help to ensure that their genetic kin survive to reproductive age.

Human societies transcend the limits of genetically based inclusive fitness and cooperate in terms of a wide network of social relationships defined by kinship and friendship relations. Human kinship itself is constructed on the basis of genetic and social ties. Humans are the only species to use language and the cultural institution of marriage to define kinship categories. The overarching importance of kinship for the human species is that in traditional societies (foragers, horticulturalist, pastoralists, and pre-industrial agriculturalists), kinship is the central organising principle for economic production, social organisation, and ideology (e.g. moral codes, religious behaviour). Many societies make use of fictive kinship, the application of kinship names to people unrelated by marriage or descent, to enhance social relations, including rights and responsibilities towards each other's offspring. An example is calling the close friend of one's mother by the name 'Aunt Maria' instead of Mrs Smith. 'Aunt Maria' may provide food, supervision, protection, gifts, and other types of parental investment to her 'niece', and the 'niece' is expected to behave in accordance with the rules of interaction between family members.

Another feature of human cooperation is intergenerational transfers of goods, property, wealth, and social status (Kaplan and Robson 2002; Kohli and Künemund 2003). These types of intergenerational transfers, which are highly structured by kinship relations, political affiliations, and economic rules, do not usually occur in non-human species. An exception may be that some chimpanzee daughters seem to 'inherit' their mother's social rank within the community, but not any material property or items of value.

Human cooperative breeding, therefore, is biocultural in nature—explained by both genetic and fictive kinship. The non-genetic and complex sociocultural nature of human cooperation in production and reproduction is, it seems, quantitatively and qualitative distinct from that of any other species. The term *biocultural reproduction* has been proposed for the human system (Bogin et al. 2014a). Biocultural reproduction describes a suite of biological and sociocultural adaptations that include (based on Bogin and Varea 2017):

- early weaning of infants by mothers, leading to an increased rate of reproduction compared with apes;
- a life-cycle period of childhood, characterised by the absence of nursing but considerable ongoing nutritional dependence, creating extended opportunities and needs for the provision of care by individuals other than the parents;

- cognitive capacities that allow a shift from genetically based prosocial behaviour to hyper-cooperation based on non-genetic marriage and kinship relationships;
- reproductive strategies that provide for culturally universal alloparental care of offspring that is demographically and ecologically flexible to local environments; and
- decreased lifetime reproductive effort for women (meaning a reduced energy investment per pregnancy and infancy per woman) compared with ape females.

Biocultural reproduction is correlated with the special human life history stages of childhood and adolescence (Table 4.2). Biocultural reproduction helps to explain why the human population will exceed 7.6 billion by the time this chapter is published. The total population of all great ape species (chimpanzee, bonobo, orangutan, and gorilla) is less than 500,000, as estimated by the World Wildlife Fund in 2011.

4.3.8 Adulthood

Adolescence ends and early adulthood begins with the completion of the growth spurt, the attainment of adult stature, and the achievement of full reproductive maturity, meaning both physical and psychosocial maturity. Height growth stops when the long bones of the skeleton (e.g. femur and tibia) and the vertebral bodies of the backbone lose their ability to increase in length. In the long bones, this usually occurs when the epiphysis, the growing end of the bone, fuses with the diaphysis, the shaft of the bone. The fusion of epiphysis and diaphysis is stimulated by the gonadal hormones, the androgens and oestrogens. However, it is not merely the fusion of epiphysis and diaphysis that stops growth, because children without gonads or whose gonads are not functional do not have epiphyseal fusion, even though they stop growing. Rather, it is a change in the sensitivity to growth stimuli of cartilage and bone tissue in the growth plate region that causes these cells to lose their hyperplastic growth potential.

Reproductive maturity is another hallmark of adulthood. The production of viable spermatozoa in boys and viable oocytes in girls is achieved during adolescence, but these events mark only the early stages, not the completion, of reproductive maturation. Socioeconomic and psychobehavioural maturation must accompany physiological development. As described in Section 4.3.7, all of these developments coincide, on average, by about 19 years of age in women and 21–25 years in men.

The course of growth and development during the prime reproductive years of adulthood is relatively uneventful. Most tissues of the body lose the ability to grow by hyperplasia, but many may grow by hypertrophy. Exercise training can increase the size of skeletal muscles, and caloric oversufficiency certainly will increase the size of adipose tissue. However, the most striking feature of the prime adult stage of life is its stability, or homeostasis, and its resistance to pathological influences, such as disease-promoting organisms and psychological stress. It contrasts with the preceding stages of life, which were characterised by change and susceptibility to pathology.

4.3.9 Late Life Stage

Old age and senescence follow the prime years of adulthood. Senescence is defined as a gradual or sometimes rapid decline in the ability to adapt to environmental stress and maintain homeostasis. The pattern of decline varies greatly between individuals. Although specific molecular, cellular, and organismic changes can be measured and described, not all changes occur in all people. Unlike the biological regulation of growth and development prior to adulthood, senescence appears to follow no species-specific uniform plan. Menopause may be the only event of the later adult years that is experienced universally by women who live past 50 years of age; men have no similar event. The only non-human species with verified menopause in the wild are killer and pilot whales, species in which females also commonly live past 50 years of age (Brent et al. 2015). Compared with human women, these female whales have a menopausal phase that is rather short in duration prior to their death. No ape species has a median life expectancy over 50 years. African and Asian elephants, the largest land-living mammal, have median life expectancies of 36 years and 43 years, respectively (Clubb et al. 2008).

The biology and possible value of menopause are topics of much interest, empirical research, and speculation. The most parsimonious explanation for menopause is that by 50 years of age women, and the two species of whales, have out lived their supply of primary oocytes, which was established during their prenatal development. If this is the cause, then there is no need for an evolutionary explanation for menopause. Rather, the need is for an evolutionary explanation for the human and whale capacity to live healthy, vigorous lives for decades past menopause. Various versions of 'mother', 'grandmother', and 'grandfather' hypotheses to explain latter life vigour have been proposed (Voland et al. 2005; Hawkes and Coxworth 2013). The most plausible hypothesis seems to be a version of biocultural reproduction. The few species able to out-live their supply of oocytes are large-bodied and large-brained, and acquire a good deal of socially based learned behaviour during their lives. In these species (humans and whales), post-menopausal females shift their reproductive investment to assisting their adult offspring to reproduce. This may be explained by a combination of kin selection (Darwin 1859; Hamilton 1964) and the value of post-reproductive females, and older males, as repositories of ecological knowledge and contributors of food and other material and social resources that buffer kin against environmental hardships (Hawkes and Coxworth 2013; Brent et al. 2015).

There are many hypotheses about senescence and why it occurs at all (e.g. antagonistic pleiotropy, mutation accumulation—see Kirkwood and Rose 1991). Some more recent discussions focus on the mechanistic roles of oxidative damage, telomere shortening, and how directed interventions (e.g. Niki 2014) or lifestyle changes (e.g. caloric restriction) might modify their effects. Discussion of these is beyond the scope of this chapter, but senescence is certainly a multifaceted process that includes genetic and environmental determinants. (For further discussion, see Chapter 5: Senescence and Ageing.)

4.4 Physiological Regulation of Growth in Comparative Context

4.4.1 Epigenetics

Epigenetics refers to modifications in gene expression caused by mechanisms other than changes in the underlying DNA sequence. Epigenetic mechanisms such as DNA methylation, histone acetylation, and micro RNA interference (see Figure 3.3 in Chapter 3) can affect gene activation and inactivation; methylation, for example, inactivates or represses gene expression. Epigenetic mechanisms may be activated by exposure to temperature extremes, exposure to disease, excess or lack of dietary factors (e.g. minerals, vitamins, amino acids), and many behavioural practices including physical activity, smoking, and alcohol consumption. More recent understanding of epigenetic regulation comes from studies of methylated adenine bases in both DNA and messenger RNA (mRNA). More than a hundred different types of chemical marks have been identified on mRNA that regulate genetic expression. In the early days of epigenetics, prior to about 2009, the focus of research was on chemical methylation of cytosine–guanine DNA base pairs, as it was believed generally that cytosine was the target of epigenetic marking. In 2015, it was shown that adenine is also methylated, especially the adenine of mRNA (Willyard 2017). Researchers in the field now write of both epigenomics, the study of the chemical marks on DNA and RNA, as well as epitranscriptomics, the multiple ways in which a DNA sequence may be modified during its expression from the genotype to the phenotype. (For further discussion, see Chapter 3: Genetics and Epigenetics.)

From identical DNA sequences, epigenetic mechanisms can produce profound alterations between individuals in growth, development, appearance, behaviour, risk of congenital abnormality, and lifetime risk of diseases such as obesity and cancer.

Epigenetic expression in the phenotype may be a heritable change in biology or behaviour, but a change that does not alter DNA sequence. In this sense, epigenetic biology is a departure from the traditional genetic dogma of DNA:

DNA → amino acid → polypeptide chain → protein.

The flow of information in epigenetic biology may begin with a social factor, such as the decision of families to migrate from a poorer country to a richer country or the choice by a woman to deliver an infant via caesarean section (C-section). In the first case, second-generation Bangladeshi women, that is, the daughters of women who migrated to the United Kingdom, have higher levels of salivary progesterone and higher ovarian function than first-generation migrants. The difference in progesterone level is in part due to greater methylation of the progesterone receptor protein in the first-generation migrants, who grew up in Bangladesh (O'Connor et al. 2009). Why this is so is not known, but a consequence of elevated progesterone levels is greater risk of breast cancer in the second-generation women (Nuñez-De La Mora et al. 2008). In the second case, women giving birth by C-section deliver infants with greater DNA methylation in general (Schlinzig et al. 2009). Why this is so and how it influences later health is unknown, but infants delivered by C-section have an 'increased risk for allergy, diabetes and leukaemia' (Schlinzig et al. 2009, p. 1096). In a more general sense, the traditional 'gene-to-protein' dogma is changing to a perspective of greater environmental control of genomic programming and DNA expression. Several nutrients, such as vitamins A, C, niacin, and D, are known to regulate DNA activity and be related to diseases such as diabetes, atherosclerosis, and cancer (Willyard 2017).

A socioeconomic factor such as poverty can influence the availability of vitamin D due to limited food choices. Vitamin D is found in a small number of foods, such as expensive oily fish, for example, salmon, in liver, and in eggs, which poor families may not eat on a regular basis. Low socioeconomic status (SES) may also lead to a lack of exposure to sunlight due to the need to work at low-paid indoor jobs. Humans get most of their vitamin D from sunlight (ultraviolet radiation) striking the skin and converting cholesterol-based substances into precursors of vitamin D.

In this case, the flow of epigenetic information is as follows:

> *Social–economic–political forces producing poverty → inability to purchase vitamin D-containing foods/low sunlight exposure → low bioavailability of vitamin D → low transactivation of DNA expression → low amino acid production → insufficient protein → possible harm to health.*

An important human example is risk of the disease multiple sclerosis (MS). Many studies show that people living in northern latitudes, with low exposure to sunlight, low vitamin D intake, and with a specific genetic variant of the major histocompatibility complex (MHC) on chromosome 6 are at greater risk of MS. People with the same MHC genetic variant but with adequate vitamin D bioavailability have significantly lower risk of MS (Dankers et al. 2017). Other nutrients such as methionine and vitamins B6, B12, and B9 (folate) are known to be related to DNA methylation, and the availability of these nutrients during fetal development may influence susceptibility to complex diseases, such as diabetes and obesity. Via this nutrient route, there is a connection between intergenerational effects, developmental origins, and epigenetic events.

There are other epigenetic mechanisms such as genomic imprinting that restricts gene expression to only the allele inherited from the mother or the father (also called parental imprinting). This is discussed in Chapter 3. More detail on imprinting syndromes and other aspects of human epigenetics may be found at https://embryology.med.unsw.edu.au/embryology/index.php/molecular_development_-_epigenetics, an educational website on the epigenetics of human development.

4.4.2 Hormones, Nutrition, Infection, and Growth

Hormones regulate and coordinate critical developmental processes, integrating across several systems (e.g. the central nervous system and digestive system), and are influenced by nutritional status and infections. Therefore, hormones provide a mechanism by which 'real-time' information about a body's health is communicated to the brain and processed by its regulatory centre in the hypothalamus, through which growth is affected accordingly (the hypothalamus plays a central role in many of the central hormone feedback loops involved in growth and development). Growth is disrupted in individuals with nutritional deficiencies, disease burden, and congenital hormone abnormalities (e.g. thyroid hormone deficiency and growth hormone (GH) insufficiency or insensitivity). The severity of the growth disruption and the potential for recovery to 'normal' growth patterns, or to achieve a particular size, depend on a number of factors.

The first formalised model to tie hormones together with human growth throughout its different stages was developed by Johan Karlberg (1987). Karlberg proposed that human growth could best be understood as proceeding through three distinct yet overlapping trajectories, each of which had specific hormonal underpinnings. This model specified differences in the infancy, childhood, and puberty (ICP) phases of growth. He proposed that the infancy growth stage began, in fact, in mid-gestation, and was driven by insulin- and the insulin-like growth factor (IGF) axis; namely, growth was driven by hormones that were dependent on and affected by maternal nutritional status *in utero* and by infant nutritional status following birth. In infancy, nutrition is a key stimulator of IGF-I; specifically, protein and energy intake stimulate IGF-I secretion in the earliest years of life (Larnkjaer et al. 2012). For example, compared to breastfed infants, IGF-I concentrations are higher in formula-fed infants at 3 months of age; these same formula-fed infants are taller, heavier, and have higher BMI values by 1 year of age (Ong et al. 2009). Across several populations, IGF-I appears to increase from birth throughout the first few months of life, followed by a gradual decrease until approximately 9 months of age, at which point levels again increase (Larnkjaer et al. 2009).

At some point around 9–12 months following birth, Karlberg proposed that the main hormonal regulation of growth switched over from external, nutritional control to endogenous rhythmic hormone secretion. The evidence for increased IGF-I secretion at about 9 months of age overlaps with this suggestion, indicating a more constant regulatory process governing IGF-I secretion. Karlberg suggested that the childhood stage of growth was best defined as starting when endogenous GH production takes over as the main driver of somatic growth. Karlberg also proposed that the GH regulation of growth is further modified during puberty (what he called the 'puberty' stage, and what we describe as adolescence) by both the independent contributions of sex steroids and their additive effect on GH pulsatile secretion.

The full duration of the childhood stage of Karlberg includes several of our previously described stages: approximately the last 2 years of infancy plus all of the childhood and juvenile stages. The differences in definition of the growth stages is mostly semantic. We all agree that the changes in amount and rate of growth over time are rooted in the physiological regulation of each growth period. Both our stage system and the Karlberg model provide an understanding of how individual growth strategies can be best matched to available energy and nutrient supply, and other environmental cues (Bogin 1999, 2012; Bernstein 2010; Hochberg et al. 2011). Paediatric clinicians have applied these growth systems and models to account for height variation; for example, a delayed infancy to childhood transition has been proposed to feed forward to reduced growth outcomes (Hochberg and Albertson-Wikland 2008).

Beyond nutrition, disease states can also affect IGF-I secretion in developing infants and children. Children with chronic inflammatory diseases, or recurrent infections, often show growth stunting. While factors such as malabsorption have been implicated in their growth failure, it has also been demonstrated that inflammation can down-regulate IGF-I production. Animal models have shown that this down-regulation can occur independently of effects on GH production, and that markers of systemic inflammation (e.g. C-reactive protein) are inversely correlated with IGF-I levels (de Benedetti et al. 1997). This down-regulation of IGF-I may play a role in the high frequency of lower bone mass and high rate of bone fracture in children with chronic inflammatory diseases (Burnham 2012).

Leptin and ghrelin are both involved in the regulation of energy metabolism and food intake, playing key roles in signalling hunger, satiety, and satiation in postnatal life. They also have important effects on immune function, inflammation, and growth. Many investigations of paediatric leptin concentrations are situated within the context of the development of overweight and obesity (e.g. Granado et al. 2012); however, considering that overnutrition is a recent phenomenon, it is more likely that leptin acts as a starvation hormone, signalling an energy-deficient state with its own downstream consequences (Prentice et al. 2002). Leptin is produced in several different tissues, and acts to regulate energy balance over the long term by signalling to long form receptor sites within the central nervous system (Gautron and

Elmquist 2011). In addition to exerting its actions through binding to leptin receptors within the brain, leptin likely influences brain development itself (especially in the hypothalamus; Bouret et al. 2004). Leptin has been proposed to mediate the predisposition of small-for-gestational-age infants to preferentially accumulate fat mass during catch-up growth, via altered adipose tissue metabolism (Lukaszewski et al. 2013). In children with chronic undernutrition, leptin production is suppressed through the combined effects of decreased fat mass and energy intake, as well as lowered insulin and IGF-I concentrations—all working to divert energy from growth to maintaining metabolic homeostasis (Soliman et al. 2000).

Ghrelin, a gastrointestinal peptide, often demonstrates reciprocal effects to those of leptin. Its secretion is regulated by signals integrated by the central nervous system (e.g. food cues, stress, insulin, leptin) that can feed back to inhibit or enhance release of ghrelin. As such, ghrelin production increases not only in the presence of food cues but also in response to chronic or acute stress. In addition to its orexigenic effects, ghrelin can stimulate the release of GH, and modulate digestive properties, sleep, and muscle integrity (Müller et al. 2015). In healthy children, ghrelin concentrations peak at approximately 2 years of age, then gradually decrease until puberty, at which point low concentrations of ghrelin secretion are patterned differently between the sexes, likely to ensure sustained growth via energy intake (King et al. 2010). The sex differences in patterns of ghrelin secretion are influenced by the sex-specific patterns of GH secretion that appear with puberty. Any effect of ghrelin on linear growth is likely to be indirect (Chanoine et al. 2009); however, most studies have measured total ghrelin rather than the acetylated (active) form, and patterns described so far may be obscured by this methodological choice. For further discussion see Chapter 13.

4.5 DOHaD and Human Growth and Development in Changing Environments

The relationship between growth, infection and inflammation, and nutritional status is complex, and dependent on the developmental trajectory of the individual. Different systems (e.g. reproductive, nervous, gastrointestinal) mature along different trajectories, and the interrelationships among these systems also do not remain static throughout life. It has been proposed that certain 'critical periods' during development define the limits of response for challenges experienced during these periods. For example, the period from birth to 2 years of age has been proposed as one such critical period for brain growth, since the majority of postnatal increase in absolute brain size, and the highest velocity in brain growth, is achieved during this time frame (Dobbing 1968). As such, any insults experienced during this time are proposed to have especially dire consequences for fulfilling the 'genetic potential' of that developmental trajectory. However, when one considers the developmental trajectory of the different components and processes within the brain, rather than just growth in absolute brain size, the previously sharply defined borders of this critical period grow fuzzy (Morgane et al. 1993). Other authors (e.g. Wachs et al. 2014) have pointed to a distinction between critical periods (during which effects from exposures are irreversible) and sensitive periods (while sensitive to exposures, plasticity in subsequent development may allow reversibility of effects). Moving from intellectual argument to application, this idea of critical periods has been used to identify ages when efforts should be directed to interventions: the 'first 1000 days' approach has emphasised the time period from early fetal development throughout the first 2 years of postnatal life as particularly critical for interventions aimed to prevent/ameliorate stunting. The evidence to support this approach comes largely from several studies that examine the nutritional correlates of stunting in infancy and childhood (reviewed in Christian et al. 2015).

A very active area of research related to critical periods is the DOHaD (http://www.dohadsoc.org/). As early as 1927, researchers noted that adult mortality in England and Wales depended on the year of birth (see Smith and Kuh 2001). A few years later, Kermack et al. (1934) confirmed the association between year of birth and rates of adult mortality for England, Scotland, and Sweden. Even more to the point, Kermack et al. found that 'infantile mortality is dependent in large measure on improvement in maternal health' (2001 reprint, p. 683). Kermack and colleagues also suggested that the environmental conditions up to age 15 years were key to later health and mortality risk. By implication, improvements in maternal health would have to take place before age 15 of the mother to have an impact on infant mortality a generation later. Fifty-two years later, Emanuel et al. (1986) formalised these findings into the 'intergenerational effects hypothesis' (IEH), defined as 'those factors, conditions, exposures and environments experienced by one generation that relate to the health, growth and development of the next generation'. Working with British data, Emanuel and colleagues (1992) found that the birth weight of a woman, her health history during infancy and childhood, and her adult stature (which reflects the total history of her growth and development) are strong predictors of the birth weight of her offspring. Many human epidemiological and anthropological studies support the IEH (Drake and Walker 2004; Varela-Silva et al. 2009) and experimental studies confirm the power of intergenerational effects (Drake et al. 2007; Benyshek et al. 2008).

Other DOHaD research focuses on the 'fetal origins' hypothesis (Barker 1990; Benyshek 2007; Kuzawa 2007; Gluckman et al. 2009) to explain the origins of several adult chronic illnesses such as coronary heart disease and diabetes. As stated by Kuzawa (2004, p. 194), 'The fetal origins hypothesis proposes that intrauterine nutrition influences the development of various hormonal systems and organs, with lasting effects on adult risk for cardiovascular disease.' The hypothesis has been expanded to other adult diseases, such as diabetes and obesity. Postnatal environments during infancy and childhood are also proving to be important for the development of health or the risks of disease in later life (Bailey and Schell 2007; Bogin and Varela-Silva 2010; McDade et al. 2010). Human biologists and biological anthropologists are at the forefront of this research, due to their cross-cultural and evolutionary perspectives.

4.6 Maya of Guatemala and Mexico as a Living Laboratory for Growth Research and Evolutionary Medicine

In this section, we describe research on the growth and development of a living group of people, the Maya of Mexico and Central America. We do so to show some of the ways in which human growth and development, which evolved over the past two million years, are influenced by recent human behaviour and environments. In today's world, evolutionary adaptations may have both beneficial and deleterious consequences for health. New environments brought on by European colonisation, industrialisation, urbanisation, and globalisation in the past 500 years have appeared more rapidly than the pace of evolutionary change. Even today, humans retain the basic anatomy and physiology, including brains, emotions, and cognition, of our ancestors of 20,000 years ago. A conflict between 'stone-age' phenotypes and 'industrial-age' environments may push human development beyond its limits for health. This conflict lies at the heart of evolutionary medicine.

At all times in our evolutionary history, human growth and development reflect the biocultural environment in which groups of people live. The growth and health of Maya children and adults is one well-researched example. The Maya are one of the native peoples of the Americas. There are an estimated 7–8 million Maya living in Guatemala, the Yucatan Peninsula, and a few other areas of southern Mexico, Belize, El Salvador, and western Honduras. The majority, about 6.5 million, live in the highlands of Guatemala (Bogin 2012). These demographic statistics make the Maya the most populous Native American ethnic group.

The human biocultural environment is comprised of the *technology* people use to feed and protect themselves, the *sociology* people devise to order their lives, and the *ideology* people imagine to give meaning to their lives. The Maya are 'people of corn', created by Xmucane, the Divine Grandmother of the gods (ideology), when she took dried corn kernels (*Zea maize*) and ground them on a stone pestle with a cylindrical mortar (technology). She mixed the corn meal with water and fashioned human shapes, from which arose the Maya people and Maya social order (sociology). This creation story was written some time before 250 CE when the Maya people began to practice intensive agriculture (more technology) of corn and other crops to feed a large population, including many non-food producers such as soldiers, artists, crafts people, and architects (more sociology), as well as politicians and religious leaders (more ideology). Corn was the basic matter of life, as the kilocalories from the corn that people consumed provided the energy to run Maya civilisation.

The biocultural nature of the Maya people is reflected not only in their 'people of corn' creation story but also in the physical growth of the people. Analysis of skeletons recovered from Maya graves at the site of Tikal allowed for the estimation of the height of the grave occupants. Tikal, dating to the Classic Period (250–900 CE), was one of several large civil and religious centres noted for monumental architecture. During the Early Classic period, skeletons from tomb burials average about 170 cm in total height. The tomb burials are those of high-status individuals, often Maya royalty. These individuals would have enjoyed the best possible conditions for life that Maya society could have provided. For comparison, the average stature of Mexican-American men 18–24 years old, a relatively healthy age, living in the United States was 171.2 cm in 1984. The Classic Maya buried lower-ranking people under the floor of their homes. The size of home is an indication of the wealth and social status of both the occupants and the burials. Individuals buried in small-sized, lower-class Maya homes averaged 163 cm in stature. The difference between the elites and the lower social class is an indication of a lower-quality environment for growth for the general population.

By the late Classic Period (~ 700–900 CE), the mean stature of both tomb and non-tomb burials declined, by about 6.0 cm to 164 cm for tombs and 157 cm for small-sized homes. In late Classic times, we have burials from mid-size homes that average 159 cm. The decrease in estimated stature with smaller-size homes shows how closely stature follows social rank.

The decline in stature occurred during a time of increasing warfare between Maya city-states, increased investment in militarisation (larger armies, weapons production, construction of fortifications, etc.), and declines in food production and public building (Bogin 2001). The material and moral conditions of Maya society were directed away from the diet and health factors that would promote growth and towards those factors that would inhibit growth.

The living Maya are the biological and cultural descendants of people inhabiting the same culture area prior to European Conquest in the year 1500 CE. Following the Conquest, the Guatemala Maya population, which totalled about two million people in 1520, was reduced to 128,000 people by 1625 (Lovell and Lutz 1996). It is estimated that about 1.8 million people died in Guatemala due to warfare, starvation, and disease, all of which destroyed the biocultural basis of Maya society (Bogin et al. 2014b). Perhaps closer to three million people died throughout Mexico and Central America in the century after European contact. Any consideration of Maya health and nutrition must begin with an understanding of that biocultural catastrophe which still reverberates in Mexico and Central America.

Today, Maya adults are the shortest population in the world. This statement excludes so-called pygmy populations, whose short average stature is due to some combination of genetic mutations, hormone and receptor variants, and missing carrier proteins for hormones, among other things (Perry and Dominy 2009). None of these causes for short stature is known to afflict the Maya. In our most recent study of Maya women living in the city of Merida, Mexico the mean height of 109 women was 147.91 cm (standard deviation (SD) = 4.84). The mean age of these women was 32.75 years (SD = 5.67), so all height growth had been completed and age-associated declines in stature were not present (Azcorra et al. 2013). In 2001, the mean height for all adult Maya women from Guatemala, most of whom lived in rural areas, was 145.3 cm (Rios and Bogin 2010). The mean height of adult Maya men was 160.7 cm. These values are about 10 cm less than those for the Maya skeletons from the tombs at Tikal. It is important to note that about 50% of women and men are shorter than these mean values.

An adult mean height of women less than 150 cm or of men less than 162 cm may be considered as evidence of stunting during the years of growth. Stunting in the infancy, childhood, juvenile, or adolescence stages of growth is a height < −2.0 SD scores (SDS, or Z-scores) based on a growth standard or reference for a healthy, adequately nourished population. The World Health Organization (WHO) published such references (http://www.who.int/childgrowth/en/). The Maya women of Merida we measured had an average height of −2.01 SDS. Children of these same women, aged 6.0–8.99 years old, averaged −0.66 SDS (Azcorra et al. 2013). Other research we carried out in Mexico and Guatemala indicates that as Maya children age they continue to decrease in height SDS. The reason for this is the cumulative effects of poverty and poor living conditions. For the past 500 years, most Maya have lived in poverty, with little access to education, gainful employment, and medical care. Their impoverished environment persistently exposes them to disease vectors via contaminated water and food. At the same time, an environment of poverty denies them sufficient nutrients, material resources, social support, and technology for healthy growth throughout their lives. The Maya have a rich tradition of ideology, but that is not sufficient for healthy growth. The Maya families of our most recent study had an average income of US$8.91/day. With a median family size of six people, this equates to US$1.49/day/person—which is well below the official Mexican poverty line of at least US$4.00/day/person. In Guatemala, the World Bank estimates that between 70 and 100% of rural Maya families live in poverty as of

2016, which is a per person income of less than US$2.00/day (http://www.worldbank.org/en/topic/poverty/overview).

Nutrition surveys carried out in Mexico and Guatemala between the 1960s and 1990s found that Maya consume only 75–80% of the total food energy needed for adequate physical growth (Bogin 2012). The most recent Mexican nutrition survey found outright deficiencies for several essential nutrients, including vitamin A, folate, iron, zinc, and calcium. A deficiency in any one of these essential nutrients will stunt growth. At the same time, Maya are consuming ever-more highly processed, industrialised foods, such as pasta, breads, and cakes, along with bottled, boxed, and canned foods. Food frequency surveys in Yucatan carried out since the year 2000 found that tortillas and beans remain staples of the diet, but are now purchased as industrially prepared products. Store-bought tortillas are not at all like those made by Xmucane when she created the Maya people from ground corn flour. The factory-made tortillas often have a lower content of vitamins and minerals, such as calcium. Highly processed factory foods have less fibre, but more added sugars and fats to make them more palatable and to allow them to be stored for lengthy periods and transported long distances. In general, the food we all eat is increasingly globalised, being produced in one part of the world, processed in another part, and then distributed for purchase and consumption to other parts of the world. These globalised foods are relatively inexpensive compared with fresh vegetables, fruit, and animal foods. During our measurement sessions in Mexico, we gave the children an apple and a carton of milk. These foods were consumed immediately, with no refusals. Many children could not tell us when they last ate fresh fruit or milk.

One of the most ubiquitous globalised foods is carbonated sugary drinks. Containers of these 'soda-pops' are often less expensive, and more available, than clean drinking water. Globalised foods, especially soda-pop, fill the stomach but leave people hungry for the missing nutrients. The result is a persistence of short stature and an excess of energy intake. Our direct measures of body composition, by skinfolds and bio-electric impedance, showed that Maya children and their mothers have an excess of body fat. The combination of stunted height with excess body fatness are expressions of both under- and overnutrition in the same population, and this is called the nutritional dual-burden. This burden places the sufferers at risk of both physical and mental illness (WHO 2006; Huang et al. 2013). The Maya are not unique in terms of the nutritional dual-burden, but they are an extreme example. Twenty years ago, the term 'nutritional dual-burden' did not exist. Human biologists thought it was physically impossible to suffer from both stunting due to diet inadequacy and overfatness due to excess energy intake. It took new globalised forms of technology, social organisation, and ideology to create food environments that never existed in human history. Biologically, the human species is not adapted to the globalised food environment. It was estimated that in 2013 there were 161 million stunted children aged < 5 years globally (de Onis and Branca 2016). This was 22.9% of all infants and children of that age. More than 98% of stunted infants and children were living in lower- and middle-income countries. Guatemala and Mexico are considered middle-income (http://www.who.int/nutgrowthdb/publications/stunting1990_2020/en/). The number of stunted Maya in Guatemala is estimated to have increased since 2010, despite many intervention programs to reduce stunting.

Worldwide obesity has more than doubled since 1980, including in the low- and middle-income nations. In 2014, the global estimate was that 41 million infants and children under the age of 5 were overweight or obese. The persistence of stunting in so many parts of the world, along with a global obesity epidemic, is unambiguous evidence of a nutritional maladaptation. It is simplistic to ascribe the dual-burden to poor diet and inadequate physical activity, as is so often done by governments, and ask that the people affected change their behaviour. Insecurity imposed by economic inequality, the persistence of preventable infectious diseases and parasites, unsafe water, globalised food production, and unbearable emotional stress from violence and inequitable political systems are major drivers of the nutritional dual-burden (Bogin et al. 2014b, 2017; Subramanian et al. 2016). The children, juveniles, and adolescents subjected to these upstream causes of malnutrition grow up to be adults with elevated risks of several diseases, including diabetes, hypertension, cardiovascular disease, and dementias. Today, even infants show evidence of the dual-burden, and the risk of these diseases is elevated from childhood onwards (Wells 2017). (For further discussion, see Chapter 6: Nutrition, Energy Expenditure, Physical Activity, and Body Composition.)

4.7 Community Effects

In Section 4.4.1 we reviewed the influence of epigenetic factors as one important mechanism for the regulation of gene-by-environment interactions and development of the phenotype. Still unknown is how the epigenome and transcriptome are marked and regulated to promote a target body size and body composition for a population or group of people. Growth in height has long been understood to be target-seeking. James Tanner proposed that the growth of all mammals is 'target seeking' (Tanner 1963), that is, following a trajectory 'governed by the control systems of their genetical constitution and powered by energy absorbed from the natural environment' (p. 845). Tanner hypothesised that the locus of the control system was in the hypothalamus, and much research since 1963 supports this idea but rejects the notion of a simple genetic control (i.e. a gene or genes for height). A question still unanswered since 1963 is: What or who sets the target?

One proposal is that in otherwise healthy, well-nourished populations it is social status and social competition that are key players in the setting of target height (Bogin et al. 2015; Hermanussen and Scheffler 2016). This proposal is grounded in social network theory, a field of inquiry that studies how people relate to each other, interact with each other, and affect each other. Applications of social network theory have shown clusterings of social–behavioural–health properties such as happiness, loneliness, depression, sleep, drug use, divorce, sexuality and sexual orientation, political orientation, and tastes in movies, music, and books. These types of networks are not surprising, but less expected are findings of networks of social contagion in the 'spread' of food consumption, obesity, smoking, alcohol consumption, likelihood of health screening, cooperative behaviour, and influenza (Christakis and Fowler 2013). These associations are social in nature—they do not depend on genetic relatedness between people—but are of biological importance in that they produce phenotypic variation that leads to elevated risk of morbidity and mortality.

A social network theory approach to the hypothesis of an adult target height differs from traditional explanations. The conventional wisdom assumes that it is the attributes of individuals, via their gene–environment interactions, which act on growth. Social network theory views biological outcomes as due, in large part, to the social relationships and ties with other actors within the network. During the years of growth in height after birth, an individual is enmeshed in a multilayered hierarchy of social networks, as shown in Figure 4.4. Within each layer, the social network may be illustrated as a map of all the relevant ties between the individuals being studied. The maps delineate communities of people who interact with each other more than with members of other communities.

Membership within and between communities may set targets for adult height via contention for status. Height is a social signal (Hermanussen and Scheffler 2016) and in human societies height is positively correlated with social status, reproductive success, educational attainment, economic gains, upward social mobility, and political success (Bogin 1999, pp. 304–28). The 'tall tend to rise in SES and the short tend to sink' (ibid, p. 325). There is increasing evidence that competition for higher social status leads to competitive growth in body size. It is well known that socially dominant individuals of many species (and their offspring), from fish to primates, are often larger in body size than subordinates, and get there by growing more quickly (e.g. Altmann and Alberts 2005). Less well known is how this growth differential is regulated.

A recent study with free-living meerkats (*Suricata suricatta*) may help to provide clarity on how social rank relates to size targeting. The authors report that social dominance itself may be a strong stimulus for growth (Huchard et al. 2016). As for many species of community-living animals, meerkats that grow less, or grow more slowly, may lose their position in the social hierarchy to challengers growing faster. Because social position determines access to mates and reproductive success, it may be expected that evolution would favour competitive growth strategies. Such strategies were found in an experiment with a community of wild South Africa meerkats. Pairs of same-sex juvenile (i.e. feeding independent) littermates were divided into a food-supplemented group comprised of the lower body weight individuals, labelled 'challengers', and an unsupplemented group comprised of the higher body weight individuals, labelled 'challenged'. The challenged individuals would normally be more likely to become socially dominant adults. Changes in their body weight were measured over 3 months. Animals were trained to climb onto a laboratory balance to receive drops of water or food. As the challengers increased in body size due to receiving the food supplements, so did their unsupplemented challenged siblings. The increase in body size required the challenged to eat more, but their growth rate increased first and then food consumption followed.

The growth of newly dominant challenged individuals was followed for the next 5 months. A secondary phase of accelerated growth occurred and was most 'pronounced when the heaviest same-sex subordinate was closer to their own weight at the time of dominance acquisition' (Huchard et al. 2016, p. 534). The authors conclude that meerkat juveniles can strategically adjust their growth in relation to the size of close competitors. Once dominant, meerkats further adjust growth, as if they do not want to be displaced. It is known that dominant, breeding meerkats of both sexes have higher plasma levels of sex steroids and cortisol than subordinates. It is also known that these hormones interact and regulate insulin/IGF-1 pathways and, therefore, may be linked to the changes in growth (Bogin et al. 2015). The authors conclude that a new perspective on social competition is needed, one that replaces 'greater-size-leads-to-dominance' with 'dominance-leads to-greater-size'.

Meerkat-like studies of human growth have long been available, but never, to our knowledge, interpreted in terms of social network theory and its association with the regulation of cortisol, GH/IGF-1, and a community effect on growth. The human data come from family studies of number of siblings and the birth order of each sibling. One British

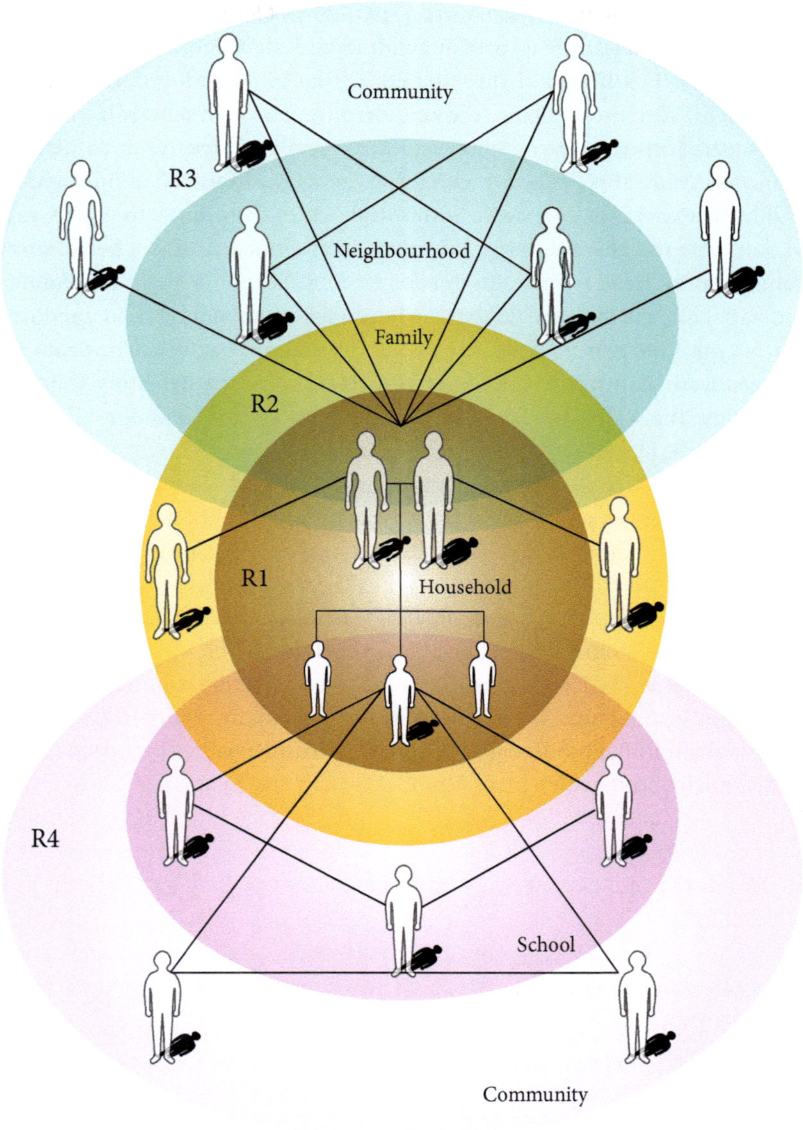

FIGURE 4.4 Levels of social networks for children, juveniles, and adolescents. Levels include the household, family, neighbourhood, school, and their associated communities (from Koehly and Loscalzo 2009). This figure was originally designed for an adolescent obesity prevention programme. Recommended levels for intervention are labelled R1–R4. The same levels may provide the biosocial context that, in part, regulates target height.

Source: Reproduced from Laura M. Koehly and Aunchalee Loscalzo, Adolescent Obesity and Social Networks. *Preventing Chronic Disease. Public Health Research, Practice, and Policy*, 6 (3), A99, http://www.cdc.gov/pcd/issues/2009/jul/08_0265.htm. © United States Department of Health and Human Services, 2009.

study is especially relevant, the Avon Longitudinal Study of Parents and Children (ALSPAC), an on-going birth cohort study of British children and their families. The participants are mostly white native British and of varying SES, similar to the SES variation of the UK as a whole. Growth measurements of the study children were taken regularly from birth onward and families were interviewed and observed up to three times per year. Lawson and Mace (2008) analysed more than 12,000 records for the associations of sibling number, birth order, and birth spacing with growth of the study child from birth to age 10 years. They reported that there is a near monotonic decrease in amount and rate of height growth with more siblings in the family. Adjusting for family SES did not change the findings. When compared with children without siblings, 'children with four siblings had a reduced rate of growth by 2.3 mm per year (95% CI: 3.8 to 0.8), leading to a deficit of 31.5 mm by age 10'. Older siblings were associated with larger deficits of height in the study child than were younger siblings. The sex of the older sibling did not influence the findings.

The authors interpreted their findings in terms of parental investment and resource allocation by parents, that is, with more mouths to feed, each additional sibling receives a smaller share and younger siblings fare worse. This is the conventional interpretation offered by other such studies in the literature. The ALSPAC families are relatively wealthy, the children are well fed, and British society has universal healthcare and education, with extra benefits to those in need. It is difficult to understand how the sibling effect on height can be explained by resource allocation from parents. Social competition may be a better explanation. With or without intentional malice, older siblings are known to inflict physical and emotional stress on younger brothers and sisters (Buist et al. 2013). Sibling rivalry likely has impacts on the hypothalamic–pituitary–adrenal (HPA) axis, cortisol reactivity, and GH/IGF-1 levels (Fey and Trillmich 2008), which can all impact growth and final size.

4.8 Conclusion and Future Directions

Human growth progresses through several distinct stages, each in turn bounded by behavioural and physical milestones. Human growth takes a more leisurely path than many of our closest living relatives, and this is particularly true for pre-adolescent growth. Human growth takes a long time, comparatively speaking, and shows a fair degree of plasticity for responding to environmental conditions. These factors help to explain how perturbations to 'normal' growth, chronic nutritional imbalances, stress, or several other factors can contribute to poor outcomes (e.g. short stature and obesity). Growth and development are the result of multiple systems interacting within the total biocultural environment of each individual. Measurable outcomes of the growth process reflect the quality of growth regulatory systems. This opens multiple downstream possibilities for health in later life, as conceived within the DOHaD paradigm. In this case, a visible outcome such as small birth size can carry with it a risk of several 'hidden' diseases that only become apparent well after the growth outcome has been noted. In this way, growth patterns and size-for-age measurements can act as harbingers of health issues to come. Intergenerational and community-based perspectives also help us to understand population growth patterns, size outcomes, and how these change over time. In the context of a rapidly globalising world, where transitions to new economies and diets happen very quickly, we are still teasing apart how these shifts are reflected in growth—and the implications these shifts have for generations to come.

The future of growth research and its relationship to the rapidly changing landscape of human sickness and health will rely a great deal on techniques available for monitoring growth. Researchers need more precise means to measure important physiological regulators and correlates of growth. For example, blood samples are still required to measure many of the key growth regulatory hormones that are implicated in some of the models discussed above, but repeated blood sampling in paediatric populations is problematic in many settings. Within the past decade, collection of dried blood spots has permitted measurement of many biomarkers related to growth and metabolism (McDade et al. 2007); while this decreases the volume of blood needed for analysis, it still requires finger- or foot-pricking. The collection of minimally invasive samples (i.e. saliva, urine/stool, hair) for regular, longitudinal monitoring of growth biomarkers will be key for future studies. Automated measurements (i.e. photogrammetry) have been used in investigations of the development of facial features (Ramanthan and Chellappa 2006), and this also presents an opportunity for accurate and unbiased measurement of height or length in certain areas of the world, where teams of trained anthropometrists may not be available to monitor growth on a year-round basis.

Also needed is a greater interdisciplinary integration of methods and ideologies about the ways in which evolutionary biology relates to the biocultural environment and variation in growth. Our discussion of the application of social network theory to strategic body growth is just one example. In general, the future of growth research is likely to benefit from major technological and theoretical advances in methods of measuring growth and understanding its relation to health.

References

Altmann, J. and Alberts, S. C. (2005). Growth rates in a wild primate population: ecological influences and maternal effects. *Behav Ecol Sociobiol* 57, 490–501.

Azcorra, H., Varela-Silva, M. I., Bogin, B., et al. (2013). Nutritional status of Maya children, their mothers, and their grandmothers residing in the City of Merida, Mexico: revisiting the leg-length hypothesis. *Am J Hum Biol* 25, 659–65.

Bailey, S. M., and Schell, L. (2007). Symposium introduction. AAPA symposium: Is adaptation healthy? Interpreting growth patterns in adverse environments. *Am J Hum Biol* 19, 603–5.

Barker, D. (1990). The fetal and infant origins of adult disease. *BMJ* 301, 1111–17.

Barker, D. J. and Osmond, C. (1986). Infant mortality, childhood nutrition, and ischaemic heart disease in England and Wales. *Lancet* 1, 1077–81.

Belachew, T., Hadley, C., Lindstrom, D., et al. (2011). Food insecurity and age at menarche among adolescent girls in Jimma Zone Southwest Ethiopia: a longitudinal study. *Reprod Biol Endocrinol* 9, 125.

Benyshek, D. C. (2007). The developmental origins of obesity and related health disorders—prenatal and perinatal factors. *Coll Antropol* 31, 11–17.

Benyshek, D. C., Johnston, C. S., Martin, J. F., et al. (2008). Insulin sensitivity is normalized in the third generation (F3) offspring of developmentally programmed insulin resistant (F2) rats fed an energy-restricted diet. *Nutr Metab* 5, 26–31.

Bernstein, R. M. (2010). The big and small of it: how body size evolves. *Yearb Phys Anthropol* 143, 43–62.

Bernstein, R. M. (2017). Hormones and human and nonhuman primate growth. *Horm Res Paediatr* 88, 15–21.

Bernstein, R. M., Sterner, K. N., and Wildman, D. E. (2012). Adrenal androgen production in catarrhine primates and the evolution of adrenarche. *Am J Phys Anthropol* 147, 389–400.

Blurton Jones, N. G., Hawkes, K., and Draper, P. (1994). Differences between Hadza and !Kung children's work: original affluence or practical reason? In: Burch, E. S. and Ellana, L. (eds) *Key Issues in Hunter-Gatherer Research*. Oxford: Berg, pp. 189–215.

Bogin, B. (1999). *Patterns of Human Growth*, 2nd ed. Cambridge: Cambridge University Press.

Bogin, B. (2001). *The Growth of Humanity*. New York: Wiley-Liss.

Bogin, B. (2011). Puberty and adolescence: an evolutionary perspective. In: Brown, B.B. and Prinstein, M. J. (eds) *Encyclopedia of Adolescence*, Vol. 1. San Diego: Academic Press, pp. 275–86.

Bogin, B. (2012). The Maya in Disneyland: child growth as a marker of nutritional, economic, and political ecology. In: Dufour, D. L., Goodman, A. H., and Pelto, G. H. (eds) *Nutritional Anthropology: Biocultural Perspectives on Food and Nutrition*, 2nd ed. Oxford: Oxford University Press, pp. 231–44.

Bogin, B. and Smith, B. H. (2012). Evolution of the human life cycle. In: Stinson, S., Bogin, B., and O'Rourke, D. (eds) *Human Biology: An Evolutionary and Biocultural Perspective*, 2nd ed. New York: John Wiley and Sons, pp. 515–86.

Bogin, B. and Varea, C. (2017). Evolution of human life history. In: Kaas, J. (ed.) *Evolution of Nervous Systems*, Vol 4, 2nd ed. Oxford: Elsevier, pp. 37–50.

Bogin, B. and Varela-Silva, I. (2010). Leg length, body proportion, and health: a review with a note on beauty. *Int J Environ Res Public Health* 7, 1047–75.

Bogin, B., Bragg, J., and Kuzawa, C. (2014a). Humans are not cooperative breeders but practice biocultural reproduction. *Ann Hum Biol* 41, 368–80.

Bogin, B., Azcorra, H., Wilson, H., et al. (2014b). Globalization and children's diets: the case of Maya of Mexico and Central America. *Antropol Rev* 77, 11–32.

Bogin, B., Hermanussen, M., Blum, W. F., et al. (2015). Sex, sport, IGF-1 and the community effect in height hypothesis. *Int J Environ Res Pub Health* 12, 4816–32.

Bogin, B., Scheffler, C., and Hermanussen, M. (2017). Global effects of income and income inequality on adult height and sexual dimorphism in height. *Am J Hum Biol* 29(2), e22980. doi: 10.1002/ajhb.22980.

Bonner, J. T. (1965). *Size and Cycle*. Princeton, NJ: Princeton University Press.

Bouret, S. G., Draper, S. J., and Simerly, R. B. (2004). Trophic action of leptin on hypothalamic neurons that regulate feeding. *Science* 304, 108–10.

Brent, L. J. N., Franks, D. W., Foster, E. A., et al. (2015). Ecological knowledge, leadership, and the evolution of menopause in killer whales. *Curr Biol* 25, 746–50.

Buist, K. L., Deković, M., and Prinzie, P. (2013). Sibling relationship quality and psychopathology of children and adolescents: a meta-analysis. *Clin Psychol Rev* 33, 97–106.

Burkart, J. M., Allon, O., Amici, F., et al. (2014). The evolutionary origin of human hyper-cooperation. *Nature Comm* 5, 4747.

Burnham, J. M. (2012). Inflammatory diseases and bone health in children. *Curr Opin Rheumatol* 24, 548–53.

Cameron, N., Bogin, B., Bolter, D., et al. (2017). The post-cranial skeletal maturation of *Australopithecus sediba*. *Am J Phys Anthropol* 163, 633–40.

Campbell, B. C. (2011). Adrenarche and middle childhood. *Hum Nat* 22, 327–49.

Chanoine, J.-P., De Waele, K., and Walia, P. (2009). Ghrelin and the growth hormone secretagogue receptor in growth and development. *Int J Obes* 33, S48–52.

Christian, P., Mullany, L. C., Hurley, K. M., et al. (2015). Nutrition and maternal, neonatal, and child health. *Semin Perinatol* 39, 361–72.

Clubb, R., Rowcliffe, M., Lee, P., et al. (2008). Compromised survivorship in zoo elephants. *Science* 322,1649.

Clutton-Brock, T. (2002). Breeding together: kin selection and mutualism in cooperative vertebrates. *Science* 296, 69–72.

Conley, A. J., Bernstein, R. M., and Nguyen, A. D. (2012). Adrenarche in non-human primates: the evidence for it and the need to re-define it. *J Endocrinol* 214, 121–31.

Christakis, N., and Fowler, J. H. (2013). Social contagion theory: examining dynamic social networks and human behavior. *Stat Med* 32: 556–77.

Dankers, W., Colin, E., van Hamburg, J. P., et al. (2017). Vitamin D in autoimmunity: molecular mechanisms and therapeutic potential. *Front Immunol* 7, 697.

Darwin, C. (1859). *On the Origin of Species*. London: John Murray.

Darwin, C. (1871). *The Descent of Man and Selection in Relation to Sex*. London: John Murray.

de Benedetti, F., Alonzi, T., Moretta, A., et al. (1997). Interleukin 6 causes growth impairment in transgenic mice through a decrease in insulin-like growth factor-I. *J Clin Invest* 99, 643–50.

de Onis, M. and Branca, F. (2016). Childhood stunting: a global perspective. *Matern Child Nutr* 12, 12–26.

Dobbing, J. (1968). Vulnerable periods in developing brain. In: Davison, A. N. and Dobbing, J. (eds) *Applied Neurochemistry*. Philadelphia: FA Davis, pp. 287–316.

Drake, A. J. and Walker, B. (2004). The intergenerational effects of fetal programming: non-genomic mechanisms for the inheritance of low birth weight and cardiovascular risk. *J Endocrinol* 180, 1–16.

Drake, A. J., Tang, J. I., and Nyirenda, M. J. (2007). Mechanisms underlying the role of glucocorticoids in the early life programming of adult disease. *Clin Sci* 113, 219–32.

Dufour, D. L. (1987). Insects as food: a case study from the northwest Amazon. *Am Anthropol* 89, 383–97.

Emanuel, I. (1986). Maternal health during childhood and later reproductive performance. *Ann N Y Acad Sci* 477, 27–39.

Emanuel, I., Filakti, H., Alberman, E., et al. (1992). Intergenerational studies of human birthweight from the 1958 birth cohort. 1. Evidence for a multigenerational effect. *Br J Obstet* 99, 67–74.

Fey, K. and Trillmich, F. (2008). Sibling competition in guinea pigs (*Cavia aperea* f. *porcellus*): scrambling for mother's teats is stressful. *Behav Ecol Sociobiol* 62, 321–9.

Finch, C. E. and Rose, M. R. (1995). Hormones and the physiological architecture of life history evolution. *Q Rev Biol* 70, 1–52.

Fuhrmann, D., Knoll, L. J., and Blakemore, S.-J. (2015). Adolescence as a sensitive period of brain development. *Trends Cogn Sci* 19, 558–66.

Garnier, D., Simondon, K. B., and Bénéfice, E. (2005). Longitudinal estimates of puberty timing in Senegalese adolescent girls. *Am J Hum Biol* 17, 718–30.

Gautron, L. and Elmquist, J. K. (2011). Sixteen years and counting: an update on leptin in energy balance. *J Clin Invest* 121, 2087–93.

Gluckman, P. D., Hanson, M. A., Bateson, P., et al. (2009). Towards a new developmental synthesis: adaptive developmental plasticity and human disease. *Lancet* 9, 1654–7.

Goldenberg, R. L., Culhane J. F., Iams, J. D., et al. (2008). Epidemiology and causes of preterm birth. *Lancet* 371, 75–84.

Gould, S. J. (1977). *Ontogeny and Phylogeny*. Cambridge, MA: Harvard University Press.

Granado, M., Fuente-Martín, E., García-Cáceres, C., et al. (2012). Leptin in early life: a key factor for the development of the adult metabolic profile. *Obes Facts* 5, 138–50.

Hamilton, W. D. (1964). The genetical evolution of social behaviour. *J Theor Biol* 7, 1–16.

Hawkes, K. and Coxworth, J. E. (2013). Grandmothers and the evolution of human longevity: a review of findings and future directions. *Evol Anthropol* 22, 294–302.

Hermanussen, M. and Scheffler, C. (2016). Stature signals status: the association of stature, status and perceived dominance—a thought experiment. *Anthropol Anzeiger* 73, 265–74.

Hrdy, S. B. (1999). *Mother Nature: A History of Mothers, Infants, and natural Selection*. New York: Pantheon, Random House.

Hrdy, S. B. (2011). *Mothers and Others*. Cambridge, MA: Harvard University Press.

Huang, C., Phillips, M. R., Zhang, Y., et al. (2013). Malnutrition in early life and adult mental health: evidence from a natural experiment. *Soc Sci Med* 97, 259–66. doi: 10.1016/j.socscimed.2012.09.051.

Huchard, E., English S., Bell, M. B. V., et al. (2016). Competitive growth in a cooperative mammal. *Nature* 533, 532–4.

Hochberg, Z., and Albertsson-Wikland, K. (2008). Evo-devo of infantile and childhood growth. *Pediatr Res* 64, 2–7.

Hochberg, Z., Feil, R., Constancia, M., et al. (2011). Child health, developmental plasticity, and epigenetic programming. *Endocr Rev* 32, 159–224.

Kaplan, H. S. and Robson, A. J. (2002). The emergence of humans: the coevolution of intelligence and longevity with intergenerational transfers. *Proc Natl Acad Sci U S A*, 99, 10221–6.

Kalinka, A. T., Varga, K. M., Gerrard, D. T., et al. (2010). Gene expression divergence recapitulates the developmental hourglass model. *Nature* 468, 811–14.

Kaplan, H., Hill, K., Lancaster, J., and Hurtado, A. M. (2000). A theory of human life history evolution: diet, intelligence, and longevity. *Evol Anthropol* 9, 156–85.

Karlberg, J. (1987). On the modeling of human growth. *Stat Med* 6, 185–92.

Kermack, W. O., McKendrick, A. G., and McKinlay P. L. (1934). Death-rates in Great Britain and Sweden: some general regularities and their significance. *Lancet* 223, 698–703. Reprinted in 2001, *Int J Epidemiol* 30, 678–83.

King, N. A., Gibbons, C. H., and Martins, C. (2010). Ghrelin and obestatin concentrations during puberty: relationships with adiposity, nutrition and physical activity. *Med Sport Sci* 55, 69–81.

Kirkwood, T. B. L. and Rose, M. R. (1991). Evolution of senescence: late survival sacrificed for reproduction. *Phil Trans Biol Sci* 332, 15–24.

Klingenberg, C. P. (1998). Heterochrony and allometry: the analysis of evolutionary change in ontogeny. *Biol Rev* 73, 79–123.

Koehly, L. M. and Loscalzo, A. (2009). Adolescent obesity and social networks. *Prev Chronic Dis* 6, A99.

Kohli, M. and Künemund, H. (2003). Intergenerational transfers in the family: what motivates giving. In: Bengtson, V. L. and Lowenstein, A. (eds) *Global Aging and Challenges to Families*. New York: Aldine de Gruyter, pp. 123–42.

Kuzawa, C. W. (2004). Modeling fetal adaptation to nutrient restriction: testing the fetal origins hypothesis with a supply–demand model. *J Nutr* 134, 194–200.

Kuzawa, C. W. (2007). Developmental origins of life history: growth, productivity, and reproduction. *Am J Hum Biol* 19, 654–61.

Larnkjaer, A., Ingstrup, H. K., Schack-Nielsen, L., et al. (2009). Early programming of the IGF-I axis: negative association between IGF-I in infancy and late adolescence in a 17-year longitudinal follow-up study of healthy subjects. *Growth Horm IGF Res* 19, 82–6.

Larnkjaer, A., Mølgaard, C., and Michaelsen, K. F. (2012). Early nutrition impact on the insulin-like growth factor axis and later health consequences. *Curr Opin Clin Nutr Metabol Care* 15, 285–92.

Lawson, D. W. and Mace, R. (2008). Sibling configuration and childhood growth in contemporary British families. *Int J Epidemiol* 37, 1408–21.

Leenstra, T., Petersen, L. T., Kariuki, S. K., et al. (2005). Prevalence and severity of malnutrition and age at menarche; cross-sectional studies in adolescent schoolgirls in western Kenya. *Eur J Clin Nutr* 59, 41–8.

Lieberman, D. E. (1999). Homology and hominid phylogeny: problems and potential solutions. *Evol Anthropol* 7, 142–51.

Locke, J. L. and Bogin, B. (2006). Language and life history: a new perspective on the development and evolution of human language. *Behav Brain Sci* 29, 259–325.

Lovell, G. W. and Lutz, C. H. (1996). 'A Dark Obverse': Maya survival in Guatemala, 1520–1994. *Geographic Review* 86, 398–407.

Lukas, D. and Clutton-Brock, T. (2012). Cooperative breeding and monogamy in mammalian societies. *Proc Biol Sci* 25 Jan. doi: 10.1098/rspb.2011.2468.

Lukaszewski, M.-A., Eberlé, D., Vieau, D., et al. (2013). Nutritional manipulations in the perinatal period program adipose tissue in offspring. *Am J Physiol Endocrinol Metab* 305, E1195–207.

McDade, T. W., Williams, S., and Snodgrass, J. J. (2007). What a drop can do: dried blood spots as a minimally invasive method for integrating biomarkers into population-based research. *Demography* 44, 899–925.

McDade, T. W., Rutherford, J., Adair, L., et al. (2010). Early origins of inflammation: microbial exposures in infancy predict lower levels of C-reactive protein in adulthood. *Proc Biol Sci* 277, 1129–37.

McKinney, M. L. and McNamara, K. J. (1991). *Heterochrony*. Berlin: Springer.

Morgane, P. J., Austin-LaFrance, R., Bronzino, J., et al. (1993). Prenatal malnutrition and development of the brain. *Neurosci Biobehav Rev* 17, 91–128.

Moser, K. (2008). Birthweight and gestational age by ethnic group, England and Wales 2005: introducing new data on births. *Health Stat Q* 39, 22–31.

Müller, T. D., Nogueiras, R., Andermann, M. L., et al. (2015). Ghrelin. *Mol Metab* 4, 437–60.

Niki, E. (2014). Role of vitamin E as a lipid-soluble peroxyl radical scavenger: in vitro and in vivo evidence. *J Free Radic Biol Med* 66, 3–12.

Nuñez-De La Mora, A., Bentley, G. R., Choudhury, O. A., et al. (2008). The impact of developmental conditions on adult salivary estradiol levels: why this differs from progesterone? *Am J Hum Biol* 21, 263.

O'Connor, C. E., Bentley, G. R., Apostolidou, S., et al. (2009). Differential methylation in PGR may explain varying progesterone levels in migrant Bangladeshi women. *Am J Hum Biol* 21, 263.

Ong, K. K., Langkamp, M., Ranke, M. B., et al. 2009. Insulin-like growth factor I concentrations in infancy predict differential gains in body length and adiposity: the Cambridge Baby Growth Study. *Am J Clin Nutr* 90, 156–61.

Pauli, R. M. and Feldman, P. F. (1986). Major limb malformations following intrauterine exposure to ethanol: two additional cases and literature review. *Teratology* 33, 273–80.

Perry, G. H. and Dominy, N. J. (2009). Evolution of the human pygmy phenotype. *Trends Ecol Evol* 24, 218–25.

Plant, T. M. (2015). Neuroendocrine control of the onset of puberty. *Front Neuroendocrinol* 38, 73–88.

Prentice, A. M., Moore, S. E., Collinson, A. C., et al. (2002). Leptin and undernutrition. *Nutr Rev* 60, S56–67.

Raff, R. A. (1992). Direct-developing sea urchins and the evolutionary reorganization of early development. *BioEssays* 14, 211–18.

Raff, R. A. (1996.) *The Shape of Life*. Chicago: University of Chicago Press.

Ramanthan, N. and Chellappa, R. (2006). Modeling age progression in young faces. *Computer Vision and Pattern Recognition, 2006 IEEE Computer Society Conference*, pp. 387–94. New York: IEEE.

Reiches, M. W., Ellison, P. T., Lipson, S. F., et al. (2009). Pooled energy budget and human life history. *Am J Hum Biol* 21, 421–9.

Rios, L. and Bogin, B. (2010). An anthropometric perspective on Guatemalan ancient and modern history. In: Salvatore, R. D., Coatsworth, J. H., and Challú, A. E. (eds) *Living Standards in Latin American History: Height, Welfare, and Development*, 1750–2000. Cambridge, MA: Harvard University David Rockefeller Center for Latin American Studies, pp. 273–309.

Schiefenhövel, W. and Blum, P. (2007). Insects: forgotten and rediscovered as food. Entomophagy among the Eipo, highlands of West New Guinea, and in other traditional societies. In: MacClancy, J., Henry, C. J., and Macbeth, H. (eds) *Consuming the Inedible: Neglected Dimensions of Food Choice*. Oxford: Berghahn, pp. 163–76.

Schlinzig, T., Johannson, G., Gunnar, A., et al. (2009). Epigenetic modulation at birth—altered DNA-methylation in white blood cells after Caesarean section. *Acta Paediatr* 98, 1096–9.

Sellen, D. W. (2001). Comparison of infant feeding patterns reported for nonindustrial populations with current recommendations. *J Nutr* 131, 2707–15.

Sellen, D. W. (2007). Evolution of infant and young child feeding: implications for contemporary public health. *Annu Rev Nutr* 27, 123–48.

Shea, B. T. (1989). Heterochrony in human evolution: the case for neoteny reconsidered. *Am J Phys Anthropol* 32, 69–101.

Smith, G. D. and Kuh, D. (2001). Commentary: William Ogilvy Kermack and the childhood origins of adult health and disease. *Int J Epidemiol* 30, 696–703.

Soliman, A. T., El Zalabany, M. M., Salama, M., et al. (2000). Serum leptin concentrations during severe protein-energy malnutrition: correlation with growth parameters and endocrine function. *Metabolism* 49, 819–25.

Stearns, S. C. and Koella, J. C. (1986). The evolution of phenotypic plasticity in life-history traits: predictions of reaction norms for age and size at maturity. *Evolution* 40(5), 893–913.

Subramanian, S. V., Mejía-Guevara, I., and Krishna, A. (2016). Rethinking policy perspectives on childhood stunting: time to formulate a structural and multifactorial strategy. *Matern Child Nutr Suppl* 1, 219–36.

Tanner, J. M. (1963). Regulation of growth in size in mammals. *Nature* 199, 845–50.

Tardieu, C. (1998). Short adolescence in early hominids: infantile and adolescent growth of the human femur. *Am J Phys Anthropol* 107, 163–78.

Varela-Silva, M. I., Azcorra, H., Dickinson, F., et al. (2009). Influence of maternal stature, pregnancy age, and infant birth weight on growth during childhood in Yucatan, Mexico: a test of the intergenerational effects hypothesis. *Am J Hum Biol* 21, 657–63.

Victora, C. G., Adair, L., Fall, C., et al. (2008). Maternal and child undernutrition: consequences for adult health and human capital. *Lancet* 371, 340–57.

Voland, E., Chasiotis, A., and Schiefenhovel, W. (eds) (2005). *Grandmotherhood: The Evolutionary Significance of the Second Half of Female Life*. New Brunswick: Rutgers University Press.

Wachs, T. D., Georgieff, M., Cusick, S., et al. (2014). Issues in the timing of integrated early interventions: contributions from nutrition, neuroscience, and psychological research. *Ann N Y Acad Sci* 1308, 89–106.

Wells, J. C. K. (2017). Body composition and susceptibility to type 2 diabetes: an evolutionary perspective. *Eur J Clin Nutr* 71, 881–9.

WHO (2006). Mental Health and Psychosocial Well-being among Children in Severe Food Shortage Situations. WHO/MSD/MER/06.1. Geneva: World Health Organization. http://www.who.int/nmh/publications/msd_mhchildfss9.pdf.

Willyard, C. (2017). A new twist on epigenetics. *Nature* 542, 406–8.

CHAPTER 5

SENESCENCE AND AGEING

XIAQING ZHAO
AND DANIEL E. L. PROMISLOW

Abstract

Ageing, or senescence, is the decline in fitness components caused by the progressive deterioration of virtually all physiological functions. Given the direct fitness costs of ageing, and the rich history of evolutionary and biomedical studies on the topic, it fits well within the framework of evolutionary medicine. This chapter explores the history of evolutionary studies of ageing, but attempts to integrate this rich literature with proximate, molecular studies of ageing. Evolutionary biologists have long seen ageing not as an adaptive process, but rather as the inevitable outcome of natural selection's inability to eliminate alleles that are associated with deleterious effects if these effects occur at later ages. This principle underlies two complementary theories. First, germline mutations that do not affect fitness early in life but show deleterious effects at late ages will accumulate over evolutionary time, leading to increased ageing (the 'mutation accumulation' theory). Second, selection would actually favour late-acting deleterious mutations if they conferred early-acting beneficial effects (the 'antagonistic pleiotropy' theory). This chapter discusses the rationale behind these theories and the challenges facing efforts to test them. It discusses a variety of proximate mechanisms thought to underlie ageing, including environmental, genetic, and epigenetic factors, placing these in an evolutionary context and exploring the possibility that proximate mechanisms might provide us with a biological clock. Finally, it explores three of the most common pathophysiological processes associated with ageing—cancer, cardiovascular disease, and neurodegenerative disease—and the potential to decrease or delay onset of these major age-related drivers of ageing.

Keywords

ageing, senescence, mutation accumulation, antagonistic pleiotropy, biological clock, cancer, cardiovascular disease, neurodegenerative disease, evolution, medicine

5.1 Defining and Measuring Ageing

In the animal world, ageing is the rule rather than the exception. We define ageing (or 'senescence'—here we use the two words interchangeably) as the intrinsic, progressive, and irreversible deterioration of virtually every bodily function. This deterioration begins around the age of reproductive maturation, and results in progressively higher risks of disease and death (Rose 1994; Austad 1999). In most species, ageing is also usually accompanied by a decline in fertility (Partridge and Barton 1996).

Despite its straightforward definition, it is far less straightforward to define and implement meaningful and accurate measures of variation in the rate of ageing among individuals or populations. Chronological age is an important determinant of health and survival, yet it is not a good measure of ageing, as rates of ageing differ among species, and even within human populations (Shock 1984). Similarly, although comparative evolutionary studies have often assumed that maximum lifespan is inversely associated with ageing, it is not actually a measure of ageing, as it does not indicate age-related decline.

Even within the same individual, different tissues and organ systems show different rates of functional decline (Martin 2002). In addition to chronological age, we speak of *biological age* as capturing the functional status of an individual relative to his/her chronological age peers, and thus as presumably a more reliable indicator of one's relative fitness and risk of mortality (Borkan and Norris 1980).

This concept of 'biological age' led to a high-profile and well-funded search for ageing biomarkers—one or more molecular or physiological parameters that would describe physical vitality, and would predict the future onset of age-related diseases and risk of mortality (so-called biomarkers of ageing) more accurately than one's chronological age (Baker and Sprott 1988; Karasik et al. 2005). Despite tremendous scientific effort and a large number of suggested physical, biochemical, molecular, and functional parameters, there was little progress in this attempt to define functional predictors of ageing (Johnson 2006; Wagner et al. 2016), with some researchers doubting that biomarkers of ageing could ever be found. With the advent of high-throughput, high-dimensional '-omic' technologies, there is renewed promise that we might identify meaningful biomarkers (e.g. Hannum et al. 2013; Horvath 2013). We discuss this later in this chapter.

From an evolutionary perspective, the forces that shape patterns of ageing are defined at the population level (Rose 1994), and aggregate measures of ageing at this level are generally more informative than individual-level metrics. Historically, most evolutionary studies of ageing have focused not on physiological processes but on a single demographic measure: age at death. But what we do with that point estimate is critical. A vast comparative literature has used mean or maximum age at death, which is unambiguous and easy to measure, to test hypotheses about why some species live longer than others (e.g. Promislow and Harvey 1990). But these average values tell us little about rates of ageing. As illustrated in Figure 5.1, two species can differ in their mean lifespan while not differing at all in rates of ageing, and vice versa. Within humans, the life expectancy in developed countries has almost doubled since the early nineteenth century, yet no evidence suggests that we are ageing at a slower rate. Much of the increase in mean lifespan is due to lower childhood mortality, thanks to antibiotics, advances in medicine, and improved healthcare, and among

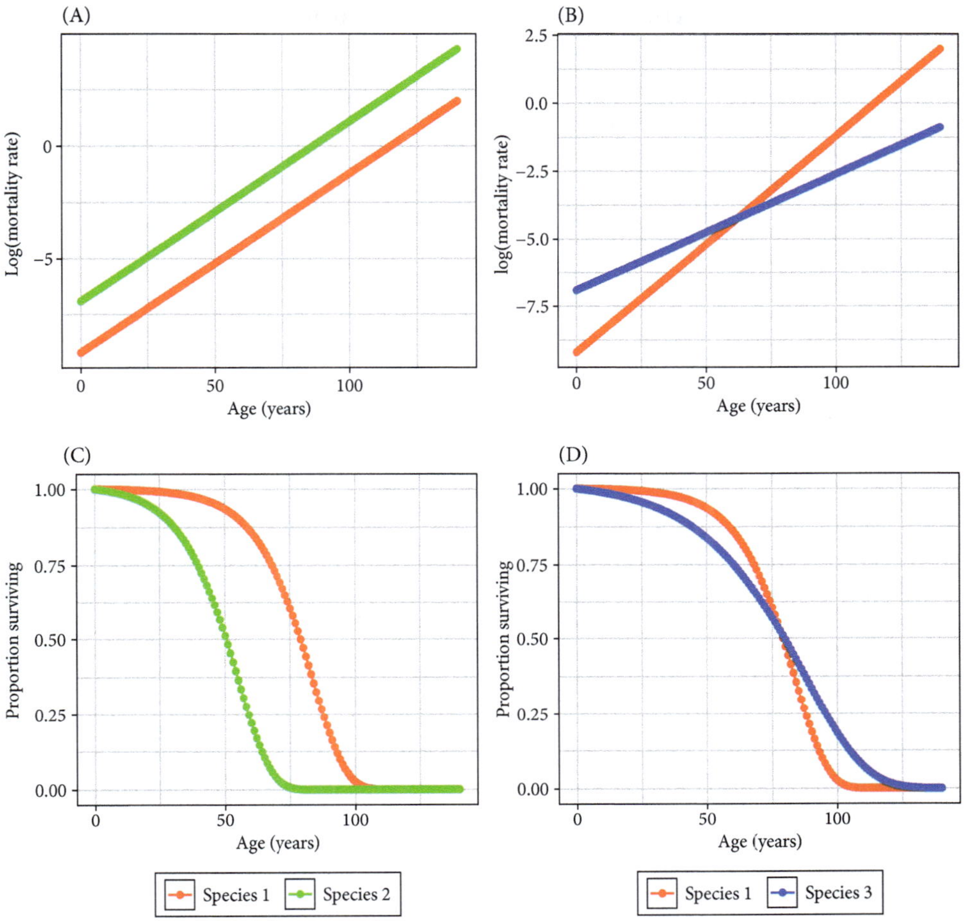

FIGURE 5.1 Mortality curves and survival curves of three hypothetical species. In (A) and (C), species 1 (shown in red) has a longer mean lifespan than species 2 (shown in green), but these two species have the same rate of ageing (i.e. their mortality risks increase at the same rate on a log-linear scale). In (B) and (D), species 1 (shown in red) and species 3 (shown in blue) have the same mean lifespan yet different rates of ageing. Species 1 ages faster than species 3 (i.e. the mortality rate of species 1 increases faster than that of species 3).

females, due to similar advances that keep women alive at childbirth. Analysing lifespan by itself does not distinguish the intrinsic deterioration of bodily function from other causes of death, and differences in lifespan do not necessarily indicate differences in ageing.

While mean lifespan contains little information about age-specific changes (which is exactly how ageing is defined), age-specific mortality rate in a population provides an informative measure of the ageing process. The rationale is that at any particular age, the level of mortality of a cohort reflects the risk of death of each individual. As physiological systems decline with advancing age, the ability of each individual to avoid death decreases. Thus, in aggregate, we see an increase in mortality rate with age. Most (but not all) researchers define the rate of ageing as the rate at which age-specific mortality rate increases.

So how *should* we define ageing rates? The answer dates back to a study of human mortality published almost 200 years ago, and turns out to apply not only to humans, but also to an impressive range of animals both in the laboratory (e.g. fruit flies, nematode worms, mice) and in the wild (e.g. Promislow 1991; Bronikowski et al. 2011). In a study of mortality rates from annuity tables published in 1825, Benjamin Gompertz found that age-specific mortality rate in adults increased exponentially with age. Plotted on a logarithmic axis, mortality rates will appear to increase linearly with age, and in modern human populations, age-specific rate of mortality doubles approximately every 8 years. The Gompertz equation is defined as

$$\mu(x) = \alpha e^{\beta x},$$

or, equivalently,

$$ln(\mu(x)) = ln(\alpha) + \beta x,$$

where $\mu(x)$ is the *instantaneous mortality rate* (also called the *force of mortality*) at age x, α denotes the baseline mortality, and β is the Gompertz slope, which measures the rate of mortality increase over age (Gompertz 1825). The parameter β is an appropriate measure of ageing. Some have also used a transformation of β to define a mortality rate doubling time, $MRDT = ln(2)/\beta$. Standard measures of μ are provided in Box 5.1.

If we compare the patterns of age-at-death of the modern Swedish population (Figure 5.2), men consistently show higher mortality rate (greater α as in the Gompertz model), yet the mortality rates of men and women increase in parallel over most of their adult lifespan (equivalent β as in the model). It seems that the longer lifespan of women is not due to slower ageing, but a lower baseline mortality rate.

Box 5.1 Standard Measures of Mortality Rate

In practice, patterns of age-at-death can be analysed with either cross-sectional or longitudinal data. Cross-sectional studies record age-at-death of individuals dying within a defined time interval regardless of their birth time. Longitudinal studies follow a cohort of individuals born within a small time interval, periodically recording the number of deaths for the lifetime of the cohort. In either case, survivorship (denoted as l_x) is defined as the proportion of the original population that is still alive at age x. Age-specific survivorship p_x is the probability that an individual who has made it to age x will survive to the next age, and can be calculated from the survivorship data as $p_x = l_{x+1}/l_x$. Age-specific mortality rate q_x, which is the probability that an individual who has survived to age x will die before the next age, can be easily calculated as $q_x = 1 - p_x = 1 - (l_{x+1}/l_x)$. Since the value of age-specific mortality rate is sensitive to the size of the age interval used to construct it, demographers prefer to use the parameter instantaneous mortality rate, also called force of mortality, instead. Denoted as μ, the force of mortality is the continuous version of age-specific mortality rate as the age interval is made infinitely small, and is approximated by $\mu_x \approx -ln(p_x)$.

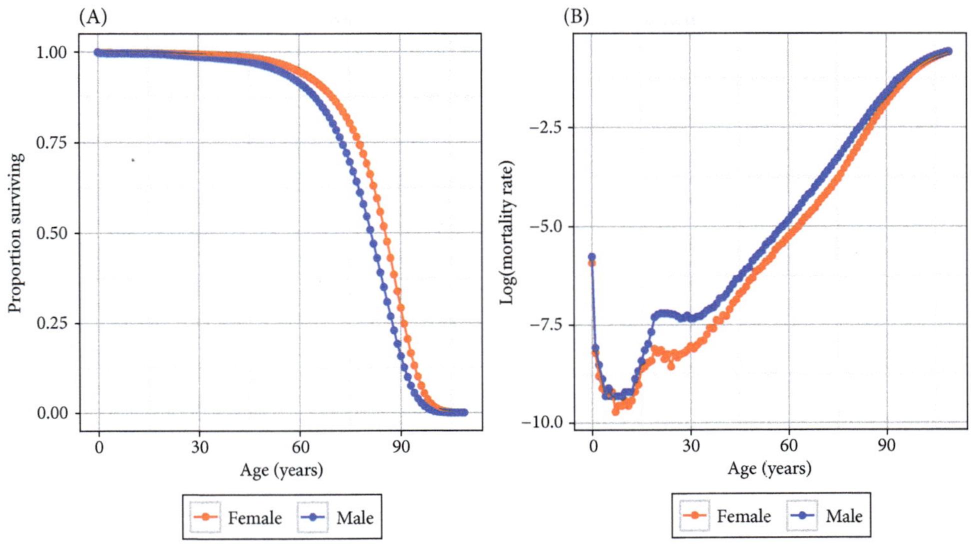

FIGURE 5.2 Age-at-death data of the Swedish population from the year 2000. (A) Survivorship over age. Note that females show longer lifespan than males. (B) Logarithm of mortality rate over age. Note that men show higher baseline mortality rate than women, but mortality rates of men and women increase in a parallel fashion.

Source: Data are from the Human Mortality Database (http//www.mortality.org/).

5.2 Evolutionary Theories of Ageing

With few exceptions (Martínez 1998), ageing appears to be ubiquitous throughout the kingdom Animalia. The deleterious effects of age on survival and reproduction, the two components of fitness, pose a conundrum. If ageing decreases fitness, why has it not been eliminated by natural selection? Evolutionary theories of ageing attempt to explain why and how ageing evolved despite its negative impact on Darwinian fitness.

5.2.1 Adaptive Theories of Ageing

The earliest explanations for the evolution of ageing suggested that natural selection actively favours ageing because it removes old and worn-out individuals (Weismann 1899). Weismann argued that while ageing would be bad for individuals, it would benefit species. Ageing would prevent overcrowding and release more space and resources for young and vigorous individuals better able to sustain the population. This theory, known as the *programmed death* hypothesis, has independently and repeatedly re-emerged in various forms since Weismann's time.

There are serious problems with the programmed death hypothesis. First, Weismann's argument is circular: in order to explain why this age-related decline exists it assumes that older individuals are worn out and reproductively exhausted. If old individuals are as valuable to the fitness of the population as younger ones, eliminating the old via ageing would not increase the collective fitness of the population. Therefore, this theory is not able to explain the *de novo* evolution of ageing from a non-ageing state. Second, while selection for traits that benefit the group, rather than the individual, can be effective under some circumstances (Wade 2015), individual-level selection typically supersedes group selection. Ageing is obviously deleterious to the individual organism (Abrams 1991). If selection favoured ageing for the sake of the group, an individual that carried a mutation deactivating the ageing programme (a 'cheater') would enjoy a fitness advantage over its senescent competitors. Such mutations would be expected to spread and fix. Although it is not impossible that the advantage to the group can overwhelm the disadvantage for individuals, such a scenario requires very restrictive conditions and rarely occurs in natural populations (Goodnight and Stevens 1997; Wade 2015).

The past few decades have seen a resurgence of programmed ageing theories. These new models argue that there exists an optimal age beyond which all individuals should die to achieve maximum individual fitness or inclusive fitness, either because it facilitates the turnover of generations so that organisms are better at adapting to changing environments (Libertini 1988; Skulachev 1997; Goldsmith 2008; Martins 2011; Joshua Mitteldorf and Martins 2014), or because in spatially structured populations with localized dispersal, programmed death reduces the spread of diseases (Mitteldorf and Pepper 2009) or makes room for kin with higher fecundity (Travis 2004; Werfel et al. 2015). However, these models either suffer from conceptual caveats such as the circularity argument mentioned above, or rely on unrealistic parameters such as excessively fast-changing environment and overly high influx of positive mutations (Kowald and Kirkwood 2016). At least thus far, we have yet to see a rigorous mathematical argument to support the idea that ageing is adaptive (Kirkwood and Melov 2011; de Grey 2015).

In addition to lack of solid theoretical arguments for programmed ageing, empirical studies do not appear to support these theories. For a self-destruction programme to be maintained by natural selection, it must have a non-negligible opportunity to be expressed in natural conditions. Therefore, the age at which the death programme switches on cannot be too high. However, it has been shown in many taxa that animals living in protected environments achieve much longer lifespan than the vast majority of the wild populations of the same species. A 'death programme', if it existed, would only guarantee the loss of an exceedingly small number of chance survivors in the wild, and thus be unlikely to provide sufficient force of selection favouring senescence and death. In addition, in most species (except for semelparous species, such as Pacific salmon and bamboo), the age-specific mortality rates increase smoothly even in genetically homogeneous populations living under a uniform environment, inconsistent with the prediction of a programmed death scenario, under which one would expect a 'wall of death' at some point. Furthermore, from a genetic perspective, both traditional and more recent adaptive theories of ageing predict the existence of genetic mechanisms whose specific purpose is to cause ageing and death. If such 'pro-ageing' genes or pathways exist, they would be susceptible to inactivation by mutation. Many mutations have been found in model organisms to increase lifespan, yet these life-extending mutants do not slow down or eliminate ageing—they generally delay its onset. Moreover, knocking out these 'lifespan-shortening' genes typically exacts reproductive costs (Jenkins et al. 2004), consistent with non-adaptive theories of ageing (see Section 5.2.2).

5.2.2 Non-Adaptive Theories of Ageing

The classical evolutionary theory of ageing, supported by mathematical models (Hamilton 1966; Charlesworth 2000) and a wealth of empirical evidences, points to the age-related decline in the force of natural selection as the cause of ageing. This theory argues that young parents make a greater relative genetic contribution to the next generation than older parents. Mutations that negatively affect the survival or fertility of young individuals will be effectively selected against because they severely undermine fitness. Mutations that impact fitness traits only at later ages, however, are more likely to remain in the population, with individuals bearing these mutations already having passed them on to the next generation before they are even expressed phenotypically. Natural selection has uniformly high power until reproductive maturity, after which it progressively declines as the fraction of reproduction that remains in the future monotonically decreases. This idea was first elucidated by Haldane (Haldane 1941), then developed by Medawar (Medawar 1952; Partridge and Gems 2002a), and the quantitative details of the argument derived by Hamilton (Hamilton 1966).

Based on this theory, ageing can evolve as a consequence of the waning power of natural selection via two potential genetic mechanisms. First, of all the mutations that randomly arise in the genome, a considerable fraction do not have any effect early in life but are exclusively deleterious at late ages. According to the argument above, natural selection does not have sufficient power to select against these mutations: they are effectively neutral in terms of their fitness consequences, and will be subject primarily to genetic drift. The frequencies of these mutations will randomly fluctuate in the population, and although most of them will eventually go extinct, a small fraction will reach high frequencies or even be fixed in the population by chance alone. As these late-acting mutations slowly accumulate in the genome over evolutionary time, organisms show compromised performance as they get old. The argument that ageing is a result of evolutionary neglect is known as the *mutation accumulation* (MA) hypothesis (Medawar 1952).

Inspired by Medawar's theory, Williams (1957) suggested that genes associated with ageing could actually be favoured due to opposing fitness consequences of mutations at early versus late ages. The majority of all mutations uniformly impair fitness of both young and old organisms, and thus are likely to be eliminated by purifying selection. A small fraction of mutations consistently enhance fitness of both young and old individuals and will spread under positive selection. There are also mutations that reduce survival and reproduction early in life but enhance fitness at late ages. Although these mutations may extend the lifespan of old organisms, they will be selected against, because strong selection against early disadvantage overwhelms weak selection for late benefit. In contrast, mutations with early benefit and late cost will be favoured because natural selection cares more about early life fitness, even if these mutations cause senescence and death at old ages. The theory that ageing is a by-product of genetic trade-offs between early- and late-life fitness is known as *antagonistic pleiotropy* (AP) (Williams 1957).

One potential physiological mechanism that could explain how genes could have opposite effects at different ages was suggested by Kirkwood's *disposable soma* (DS) theory (Kirkwood 1977). Kirkwood posited that ageing would evolve due to the trade-off between reproduction versus somatic maintenance and repair (such as antioxidant systems and DNA repair). Metabolic resources are limited. Any investment in reproduction takes away the resources that could otherwise be used on somatic maintenance, and vice versa.

To achieve the highest fitness, the best strategy is not to keep the organism in sound physiological condition infinitely, but only to maintain the somatic tissues for as long as the organism has a reasonably good chance to survive in the wild, and invest the remaining resources in reproduction. Ageing is therefore a result of the gradual accumulation of unrepaired damage.

Note that the mutation accumulation and antagonistic pleiotropy theories are not mutually exclusive, and the two genetic mechanisms may operate simultaneously. The relative contribution of each evolutionary mechanism to ageing is not yet resolved, despite decades of debate and empirical research.

5.3 Assumptions and Predictions of Evolutionary Theories of Ageing

The mutation accumulation and antagonistic pleiotropy theories gave us the conceptual and theoretical framework for how to think about the evolution of senescence, and Hamilton provided the mathematical tools. These theories have led to a set of predictions (Williams 1957; Hughes and Charlesworth 1994; Charlesworth and Hughes 1996), which have in turn inspired a large body of empirical studies both in laboratory systems and in natural populations. These studies evaluate the validity of the theories or assess the relative importance of different models, and will ultimately provide a framework for our understanding of the genetic, physiological, and proximate mechanisms of ageing.

5.3.1 Age-Specific Mutational Effects

Both MA and AP theories assume that spontaneous germline mutations with late-acting deleterious effects exist to account for the observed age-related decrease of fitness components. The difference between MA and AP with respect to mutational effects is that MA assumes that late-acting deleterious mutations which linger in the genome are effectively neutral at early ages, while AP argues that late-acting deleterious mutations could have beneficial effects in early life. While both types of mutations may exist and work together to shape patterns of ageing, it is possible to evaluate which pattern of age-specific effect is more likely seen in novel spontaneous mutations.

Several studies have tested the effects of MA on ageing by allowing novel mutations to accumulate within a population. Mutation accumulation experiments create specific regimes to maintain populations so that the force of selection is effectively removed, and spontaneous mutations are allowed to accumulate in the genome. Analysing fitness consequences of the accumulated mutations across ages provides an opportunity to evaluate the age-specificity of *de novo* mutations. A few studies in *Drosophila* have shown that novel mutations are pleiotropic across ages with respect to fitness, but the directions of effects are the same. Specifically, novel mutations tend to be more detrimental to survival at earlier ages than at later ages (Pletcher et al. 1998; Pletcher et al. 1999; Yampolsky et al. 2000; Gong et al. 2006). Note that alleles expected to accumulate under MA would confer greater detrimental effects at later ages.

Subsequent theoretical work extended the models of Medawar and Williams, and generated predictions that are closer in line with observation. With the diminishing strength of selection with age, novel mutations are expected to show more early-acting effects: in the starting population, powerful selection at early ages has already pushed the population near the age-specific fitness optima. Thus, new mutations are likely to take the population away from the fitness peak. In contrast, since selection is weak at late ages, fitness may be far from the optimum, and additional mutations are likely less detrimental to age-specific fitness (Moorad and Promislow 2008). Since in practice we are not able to start the mutation accumulation experiment from a non-ageing population, the phenotypic effects of the accumulated mutations depend upon the existing genetic architecture of age-specific fitness components in the starting population.

Quantitative trait locus (QTL) analyses of lifespan (Nuzhdin et al. 1997) and fecundity (Leips et al. 2006) with recombinant inbred lines have shown that genes contributing to variation in early life fitness are distinct from those producing variation in late life, indirectly suggesting that the effect of these genes is age-specific.

5.3.2 Genetic Variation

Genetic variation arises from naturally occurring genetic differences among individuals within a population. One prediction of the MA model is that genetic variance of fitness traits will increase with age. Since purifying selection of deleterious alleles becomes increasingly inefficient with age under MA, the equilibrium frequencies of deleterious alleles at mutation–selection balance increase with age. The additive genetic variance, dominance genetic variance, homozygote variance (i.e. variance in inbred line means), and inbreeding depression are all proportional to equilibrium allele frequencies, and thus are all predicted to increase with age under the MA model (Charlesworth and Hughes 1996). The AP model does not make explicit predictions about genetic variance: under specific assumptions, AP may produce a wide range of genetic variance, from no pattern to a pattern resembling MA (Moorad and Promislow 2009).

Historically, the genetic variance partitioning approach was widely used to determine the relative importance of AP and MA to the evolution of senescence, both in model systems in the laboratory (Hughes and Charlesworth 1994; Charlesworth and Hughes 1996) and in natural populations across a broad taxonomic range (e.g. Fox et al. 2006; Brommer et al. 2007; Escobar et al. 2008). With the later realisation that any age-specific pattern of genetic variation that is consistent with MA could also be consistent with AP (Moorad and Promislow 2009), this strategy was considered of limited value for this particular purpose. We need clear and mutually exclusive predictions to diagnose the two putative mechanisms of ageing.

5.3.3 Trade-offs

Both AP and DS predict a general trade-off between survival and reproduction. These two models differ in the specific mechanisms of trade-off: AP suggests that the trade-off results from single genes or genetic pathways that have opposite effects on early- and late-life fitness, while DS suggests the trade-off is a result of the allocation of finite metabolic resources between components of fitness. MA does not make an explicit prediction on trade-offs.

One of Williams' original deductions from AP was that 'successful selection for increased longevity should result in decreased vigour in youth'. Many experimental evolution studies, especially in flies, have successfully created long-lived populations by selecting on late reproduction, such that only eggs laid by older females were allowed to form the next generation. Artificial selection experiments of this type have generally been effective in extending lifespan and enhancing late-life fecundity. In some studies these changes were associated with a decrease of early-life fecundity (Luckinbill et al. 1984; Rose 1984), suggesting a negative genetic correlation between early- and late-life fitness as predicted by AP. This result needs to be interpreted with caution, though, as the particular selective regime is not able to separate selection for increased lifespan from selection for increased late-life fecundity, so the decreased early-life fecundity could be a result of negative genetic correlation between early and late fecundity, instead of increased lifespan.

In another selection experiment on late reproduction, the increased longevity was not accompanied by a depression of early-life fecundity (Partridge and Fowler 1992). This result was interpreted as consistent with MA, on the basis that the artificial selection worked against late-acting deleterious genes that harm both survival and reproduction, without affecting early-life fitness. This result again needs to be interpreted with caution, as in practice it is difficult to control for every aspect of the rearing scheme of selected and control populations, leaving possibilities for inadvertent selection for other traits that could confound the result (Zwaan 1999). One study took advantage of family selection to apply selection directly on lifespan without selecting simultaneously on fecundity, and showed that female flies selected for long lifespan not only live longer than control females, but also produce fewer eggs than control females at all ages (Zwaan et al. 1995).

More recent studies in *Drosophila melanogaster* have tested the pleiotropic nature of genes underlying the trade-off between longevity and reproduction (Khazaeli and Curtsinger 2010, 2013). These studies developed recombinant inbred lines with parental lines that were artificially selected for delayed reproduction compared with the baseline control. If pleiotropic genes are primarily responsible for the trade-off, since recombination cannot break the opposing effects of a single pleiotropic gene, the negative correlation between early- and late-life fitness should be seen in most of the recombinant inbred lines. However, these studies found recombinant genotypes of 'superflies', exhibiting longer life as well as elevated early-life fecundity, demonstrating that longevity and reproduction can be genetically uncoupled, and indicating that AP did not fully explain the observed trade-off between survival and reproduction, at least in this population.

Besides observations of phenotypic and genetic correlations, specific genes with opposing effects on fitness at early and late life provide direct evidence for the pleiotropic nature of trade-offs. These genes are further discussed in Section 5.5.

5.3.4 Extrinsic Mortality

One of the most widely tested hypotheses in evolutionary studies of ageing is Williams' (1957) conjecture that high extrinsic mortality would select for high rates of ageing. It turns out that this is also one of the most widespread misconceptions. This misconception arises from the original qualitative argument of the waning power of selection with age. Medawar originally claimed that the intensity of selection declines because of accidental deaths: organisms inevitably die from stochastic environmental hazards such as predators, infectious diseases, and lack of resources, such that progressively fewer individuals will be alive at progressively later ages. Based on this argument, Williams made the prediction that faster ageing will evolve under conditions of higher external, age-independent mortality. Hamilton's mathematical model has shown that selection gradients do decline with age, but this is not due to survivorship by itself, but the combined effects of the relative abundance of adults in each age class, and their respective reproductive values. But in this case, verbal arguments have overshadowed mathematical ones. Williams' argument on extrinsic mortality was so influential that it has been repeatedly tested in laboratory and wild populations, between protected and unprotected populations of the same species, and comparatively among different species that are presumably exposed to higher and lower levels of external risks (Austad 1993; Shattuck and Williams 2010). This body of empirical work, not surprisingly, has found contradictory results (Reznick et al. 2004).

It is true that increased extrinsic hazards result in more rapid death, but high *age-independent* extrinsic mortality does not necessarily mean faster declining strengths of selection. Selection gradients depend on the stable age distribution of individuals belonging to each age group. Additional age-independent mortality reduces survivorship and fitness, but, *ceteris paribus*, does not change the selection gradients, and thus does not alter the rate of ageing (Caswell 2007; Wensink et al. 2016). In fact, an age-structured population with zero mortality (i.e. *immortality*) would still experience a decline in the intensity of selection with age.

There are scenarios where external hazard will affect the evolution of ageing, although in directions more complicated than simply accelerating or decelerating it. If the externally imposed hazard is a positive function of an organism's age or internal condition, namely, older individuals suffer more severely from the hazard than younger individuals, or individuals in poorer condition experience higher susceptibility compared to those in better condition, then rates of deterioration at different life stages could be modified. In this case, however, the external risk is not truly 'extrinsic' as it depends on an organism's age or condition (Williams and Day 2003). Another scenario where additional external mortality risks could modify the rate of ageing is when populations are unstable and population growth (i.e. survival and reproduction) is density-dependent. In this case, additional extrinsic mortality will reduce population density, and the population will subsequently exhibit an altered age structure, which will in turn change the selective gradient and thus patterns of ageing. But again, depending on specific assumptions on how extrinsic hazards modify population growth, patterns of ageing can change in various directions (Abrams 1993).

5.4 Proximate Mechanisms of Ageing

Evolutionary theories of ageing explain *why* senescence exists, but do not specify the biological mechanisms of *how* it occurs. Due to the complex nature of the ageing phenotype, many processes have been proposed to be the physiological sources of age-related damage, and many others are considered compensatory responses to restore homeostasis. For example, because the enzymes that duplicate DNA during cell division lack the capacity to replicate all the way to the end of the chromosome, in each cell division the ends of the chromosomes are shortened. Telomeres are a region of repetitive DNA sequence at the ends of chromosomes, and their existence protects functional sequences from being truncated. The enzyme tolemerase, which can lengthen telomeres, is expressed in all mammals during development, but, at least in large species, telomerase expression is lost in adults (Gomes et al. 2011). In the absence of telomerase expression, telomeres deteriorate with age as more rounds of cell division occur. While short telomeres are thought to protect organisms against cancer, these same short telomeres can increase the risk of cellular senescence (di Fagagna et al. 2003). Such cost–benefit trade-offs are a hallmark of the 'antagonistic pleiotropy' theory for the evolution of senescence, which we discuss in Section 5.5.

For those interested in the molecular mechanisms that explain the *how* of ageing, we encourage them to refer to an excellent review that proposed nine candidate hallmarks that are generally considered to contribute to the ageing process and determine the ageing phenotype, including not only telomere attrition, but also genomic instability, epigenetic alterations, loss of proteostasis, stem cell exhaustion, and altered intercellular communication (López-Otín et al. 2013).

Keeping our focus on the *why* of ageing, in this section, we will restrict our discussion to environmental, genetic, and epigenetic components of ageing as well as their interactions. We will also consider the significance of these proximate mechanisms in the light of evolutionary theories of ageing.

It is worth noting that, as discussed in Section 5.1, ageing is measured by the rate of mortality risk increase; in practice, however, since measuring age-specific mortality rates in a large population is not always feasible, lifespan is usually used as a proxy for ageing, with the assumption that genetic or environmental factors that substantially increase mean or maximal lifespan (under protected environments) are likely to have done so by postponing or retarding ageing. Research seeking to identify these mechanisms typically focuses on perturbations that increase lifespan, since shortening of lifespan is likely a consequence of novel pathologies that may or may not be causally related to ageing. An alternative way of studying modifiers of ageing, especially in long-lived organisms like humans, is to leverage a secondary phenotype (such as a biomarker of ageing) that correlates with risk of mortality to test whether certain genetic variants or environmental manipulations are associated with altered rate of ageing.

5.4.1 Environmental Modulation of Ageing

Several methods of environmental manipulation have been shown to impact lifespan, such as acute heat shock, long-term changes in environmental temperature, and reduced oxygen concentration. Dietary restriction (DR), a reduction in nutrient availability without malnutrition, was first shown to extend the lifespan of rats and is now known as the most consistent and powerful environmental intervention of ageing, with at least some support in species as diverse as yeast, nematode worms, fruit flies, fish, mice, hamsters, dogs, and primates (Masoro 2005; Fontana et al. 2010). In vertebrates, dietary restriction has also been shown to protect against age-related diseases and functional decline. For instance, DR reduces the risk for diabetes, cardiovascular disease, and cancer (Fontana and Klein 2007; Anderson et al. 2009; Colman et al. 2009), and protects against age-related muscle loss and brain atrophy (Anderson et al. 2009). There has been a tremendous amount of effort in identifying genetic pathways that interact with DR, in hopes of finding drugs that mimic DR to promote healthy ageing and extend lifespan.

Despite the apparent conservative effects of DR on ageing, the outcome of DR depends upon a number of factors, including the degree of restriction, composition of food, specific feeding regimen, age at onset, and genetic background of the subject population. In worms, flies, and mice, as food intake decreases, lifespan rises to a maximum but then rapidly declines through starvation if food availability is further reduced (Partridge et al. 2005; Piper and Partridge 2007; Fontana et al. 2010). The optimal food level with respect to longevity may be genotype-dependent (Metaxakis and Partridge 2013), and since traditional DR assays are usually carried out at two nutrition levels, the perceived varying degrees of responses to DR could be due to some genotypes being closer than others to their optimal dietary intake when fed ad libitum. An ideal DR experiment should therefore include multiple nutritional levels to examine the entire dietary dose response, rather than comparing just two food levels (Figure 5.3).

DR is often referred to as caloric restriction or calorie restriction because work with rodents has shown that reduction of calories in diet, rather than the macronutrient composition of the diet, modulates lifespan (Masoro 2005). However, in other work with rodents and also in *Drosophila*, reduction in protein levels or even concentrations of specific amino acids have been shown to play an important role in extending lifespan (Zimmerman 2003; Miller et al. 2005; Grandison et al. 2009; Sun et al. 2009). Dietary restriction can be administered in various ways, such as limiting the quantity of available food, diluting the concentration of food, or restricting the time that animals have access to food. In worms and mice it has been shown that certain genotypes only respond to specific DR regimens, leading to the hypothesis that different modes of dietary restriction influence lifespan via distinct nutrient sensors (Greer and Brunet 2009; Kenyon 2010). In mice and humans, a periodic diet that mimics fasting improved metabolic and cognitive function, and reduced multiple risk factors of ageing (Brandhorst et al. 2015). The age at which DR is initiated can affect the outcome of lifespan extension in rodents (Goodrick et al. 1990). In contrast, in *Drosophila*, studies have found that the effect of DR on mortality is acute: within 2 days of DR initiation, the instantaneous mortality rate of previously fully fed flies fell to the same level as that of flies of the same age that had been subject to long-term DR, supporting the hypothesis that DR extends lifespan by reducing the short-term risk of death (Good and Tatar 2001; Mair et al. 2003).

Genetic background adds a layer of complexity. Genetic analyses of lifespan and DR have identified various nutrient-sensing pathways that apparently regulate responses to DR (reviewed in Fontana et al. 2010; Kenyon 2010), which are discussed in more detail in Section 5.4.2. Analyses of recombinant inbred strains of mice (Liao et al. 2010) and single-gene deletion strains of yeast (Schleit et al. 2013) both found significant variability in response to dietary restriction. Meta-analysis of DR in mice and rats found that outbred populations benefit more from DR compared to inbred strains (Swindell 2012). These findings suggest that individual responses to DR are strongly genotype-dependent.

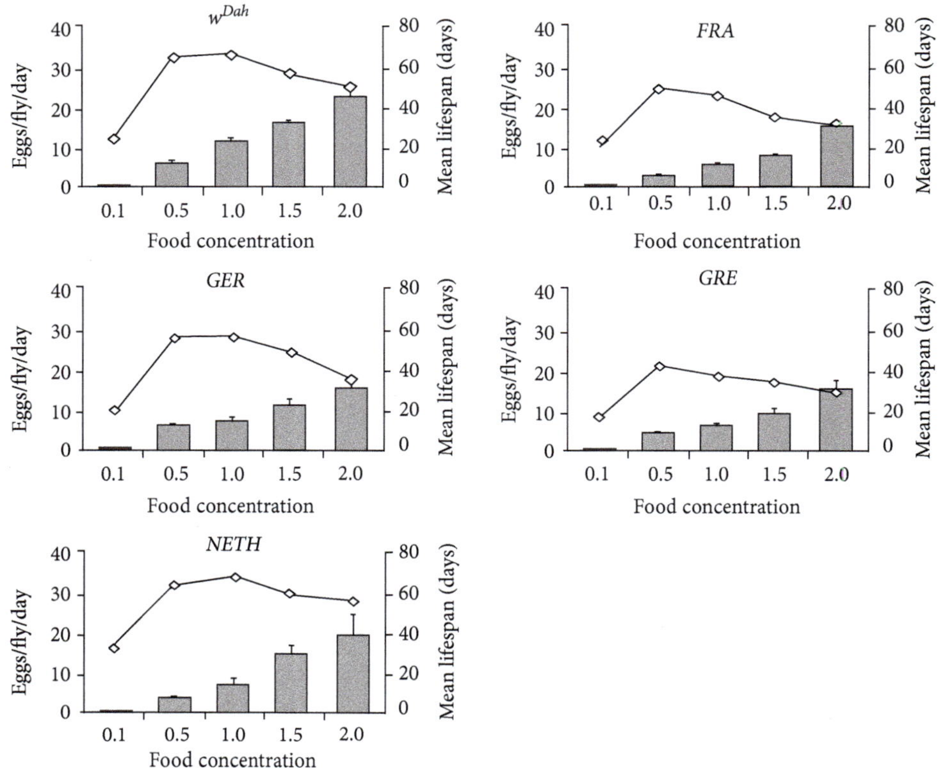

FIGURE 5.3 Lifespan and fecundity of wild-derived *Drosophila melanogaster* strains in response to DR. Five concentrations of yeast (0.1, 0.5, 1.0, 1.5, and 2.0) were used, where 1 corresponds to 100 g of brewer's yeast per liter of food. Mean lifespan of all wild-derived strains (FRA, France; GER, Germany; GRE, Greece; NETH, the Netherlands) and the laboratory strain (W^{Dah}) exhibit a tent-shaped response to DR. Highest mean lifespan values are found at different food levels in different strains. Female fecundity showed a monotonic increase with yeast concentration in all strains. Points indicate mean lifespan. Bars estimate mean number of eggs laid per fly per day ± standard error.

Source: Reproduced from A. Metaxakis and L. Partridge, Dietary Restriction Extends Lifespan in Wild-Derived Populations of *Drosophila melanogaster*, *PLoS ONE* 8(9), e74681, doi.org/10.1371/journal.pone.0074681.

5.4.2 Genetics of Ageing

Genetic analyses in yeast, worms, flies, and mice point to at least three evolutionarily conserved pathways associated with age—insulin signalling, target of rapamycin (TOR) signalling, and sirtuins. The first single-gene mutations that promoted extended lifespan were identified in the nematode worm *Caenorhabditis elegans* as a result of systematic mutagenesis. These initial mutations were all involved in the insulin/insulin-like growth factor 1 (IGF-1) signalling (IIS) pathway (Johnson 1990; Kenyon et al. 1993; Apfeld and Kenyon 1998), which connects nutrient levels to metabolism and growth. Mutations in *daf-2*, a gene encoding an insulin/IGF-1 receptor orthologue, double the lifespan of the worm. This effect requires the activity of *daf-16*, which encodes a forkhead transcription factor (FOXO) that activates suites of downstream genes that promote somatic maintenance and stress resistance. It has subsequently been shown in the fruit fly *D. melanogaster* that elements of the IIS pathway affect ageing: hypomorph mutations of the gene *insulin-like receptor* (*InR*), which is homologous to *daf-2* as well as to mammalian insulin receptors, increase the lifespan of *Drosophila* by 85% (Tatar et al. 2001). Hypomorph mutations of *chico*, the insulin-receptor substrate homologue, increase lifespan by approximately 40% (Clancy et al. 2001) and decelerate mortality risk (Tu et al. 2002). Increasing the activity of dFOXO (the *Drosophila* homologue of DAF-16) specifically in adult adipose tissues increases *Drosophila* lifespan (Giannakou et al. 2004; Hwangbo et al. 2004).

In mammals, ageing also appears to be under neuroendocrine modulation. Mammals have separate receptors for insulin and IGF-1, with the insulin receptor primarily controlling metabolism and the IGF-1 receptor controlling growth. Both receptors retain the ability to influence lifespan: in mice, mutations that reduce the function of both receptors (Blüher et al. 2003; Holzenberger et al. 2003), and their upstream regulators (Coschigano et al. 2003; Bartke 2008) and downstream effectors (Bartke 2008; Selman et al. 2008) can all extend lifespan. Small dogs, which have reduced levels of IGF-1, live much longer than large dogs (Eigenmann et al. 1984). In humans, mutations known to reduce IGF-1 receptor activity are overrepresented in Ashkenazi Jewish centenarians (Suh et al. 2008); sequence variation in the insulin receptor *INSR* is associated with exceptional longevity in a Japanese cohort (Kojima et al. 2004); and variants of *FOXO3A* and *FOXO1* have been linked to longevity in multiple populations (reviewed in Kenyon 2010). The IIS pathway senses nutrients, and is therefore a good candidate for mediating the longevity response to DR. Indeed, in worms and mice, DR appears to extend lifespan by inhibiting IIS (Arum et al. 2009; Honjoh et al. 2009).

Like effects of IIS, effects of the TOR pathway on ageing are conserved across a broad evolutionary spectrum. The TOR kinase is a nutrient sensor that integrates signals from amino acid availability, growth factors, energy, and stress to regulate cell growth and proliferation. When nutrients are abundant, the TOR kinase is activated and up-regulates translation and inhibits autophagy (Kaeberlein et al. 2005; Kenyon 2010). Loss-of-function or hypomorph mutations in components of the TOR pathway extend lifespan in yeast (Fabrizio et al. 2001; Kaeberlein et al. 2005; Bonawitz et al. 2007), worms (Vellai et al. 2003; Hansen et al. 2007), flies (Kapahi et al. 2004; Luong et al. 2006), and mice (Selman et al. 2009; Lamming et al. 2012; Wu et al. 2013). Lifespan extension modulated by TOR inhibition is usually accompanied by increased stress resistance (Hansen et al. 2007) as well as improved healthspan and physiological function (Luong et al. 2006; Selman et al. 2009; Wu et al. 2013). TOR signalling has multiple interactions and feedback loops with IIS signalling, but its downstream effectors appear to be distinct, as TOR inhibition does not require DAF-16/FOXO to extend lifespan (Vellai et al. 2003; Jia et al. 2004; Hansen et al. 2007). The TOR pathway is consistently linked to dietary restriction: TOR activity is inhibited by DR (Kaeberlein and Kennedy 2011), and lifespan extension conferred by TOR inhibition is not further increased by DR (Kapahi et al. 2004; Kaeberlein et al. 2005; Hansen et al. 2007), supporting the hypothesis that DR promotes longevity by reducing TOR activity (Kenyon 2005).

Finally, sirtuins, a group of nicotinamide adenine dinucleotide (NAD^+)-dependent histone deacetylases, are another set of genes that modulate lifespan in an evolutionarily conserved manner. Sirtuin overexpression is shown to extend lifespan in yeast (Kaeberlein et al. 1999), worms (Tissenbaum and Guarente 2001), flies (Rogina and Helfand 2004), and mice (Kanfi et al. 2012; Satoh et al. 2013). In humans, genetic variation in sirtuins is shown to be associated with survival (Rose et al. 2003; Bellizzi et al. 2005; Zhang et al. 2010; Figarska et al. 2013; Albani et al. 2013) and age-related diseases (Figarska et al. 2013), but not in all studies (Flachsbart et al. 2006; Lescai et al. 2009). It is not yet clear how sirtuins influence lifespan. In yeast, it is argued that Sir2 regulates ageing by maintaining chromatin silencing and genome stability (Guarente 2000; Dang et al. 2009). In higher animals, sirtuin overexpression appears to work in parallel with IIS inhibition to trigger the same downstream cascade that extends lifespan, that is, by activating DAF-16/FOXO (Berdichevsky et al. 2006; Kenyon 2010; Kanfi et al. 2012).

Since sirtuin catalytic activity is NAD^+-dependent and NAD^+ level is a good indicator of nutrient availability in certain tissues, it has been hypothesized that sirtuins respond to dietary restriction to confer beneficial effects on healthspan and lifespan (Guarente 2000). Evidence that sirtuin activity is required for DR to extend lifespan has been found in some studies (Lin et al. 2000; Rogina and Helfand 2004; Narasimhan et al. 2009; Banerjee et al. 2012) but not in others (Kaeberlein and Powers 2007; Burnett et al. 2011). The discordant findings may be explained by the fact that multiple nutrient-sensing pathways exist in metazoa, such that variability in laboratory conditions might favour signalling through different subsets of pathways that affect lifespan (Guarente 2013). It remains to be revealed how genetic background and specific modes of DR affect the modulation of ageing through sirtuins.

Except for the pioneering mutagenesis-based studies and deletion screens in yeast, the above-mentioned results all come from the candidate gene approach, where researchers select candidate ageing genes based on *a priori* knowledge of their functions and test whether they significantly impact ageing in model organisms, assuming evolutionary conservation in gene function. However, despite the parallel effects on ageing between these major nutrient-sensing pathways in invertebrates and mammals, it is unclear how important these conserved pathways are in the genetic architecture of ageing in nature.

Systematic screens in short-lived model organisms and gene-mapping/genome-wide association (GWA) studies in higher animals and humans present a more comprehensive picture of the genetic basis of ageing. In *C. elegans*, multiple RNAi screens have identified over 200 genes for which a reduction in expression results in extended lifespan (Hamilton et al. 2005; Hansen et al. 2005; Samuelson et al. 2007). These genes are involved in a broad range of biological processes, including metabolism, signal transduction, mitochondrial function, genome stability, and the stress response. Single-gene knockout screens in yeast have identified over 100 pro-ageing genes (Powers et al. 2006; Matecic et al. 2010; Burtner et al. 2011) with functions such as protein homeostasis, metabolism, and stress response. Quantitative genetic mapping in mice (de Haan et al. 1998; Miller et al. 1998; Guo et al. 2000; Lang et al. 2010; Rikke et al. 2010) has identified dozens of chromosome regions associated with lifespan. In humans, linkage association studies (Boyden and Kunkel 2010; Kerber et al. 2012; Beekman et al. 2013) reported ten chromosome loci for longevity; initial GWA studies with large groups of long-lived and control populations detected a few suggestive single nucleotide polymorphisms (SNPs) for association with longevity, though few reached genome-wide significance due to lack of power (Newman et al. 2010; Deelen et al. 2011, 2014; Nebel et al. 2011; Sebastiani et al. 2012, 2013). These results demonstrate that lifespan is under complex genetic control, likely involving a large number of genes, each with small effect.

It is worth noting that in worms and yeast there is little overlap in the specific genes identified between screens (Smith et al. 2007; Yanos et al. 2012); in mice and humans similarly, no loci are identified in more than one association study, with the exception of *APOE*, which we discuss in more detail in Section 5.5.2. This small degree of overlap may be due to the complex genetic architecture of lifespan, involving dominance and epistasis, subtle differences in experiment design and conditions, or the high false-positive rate inherent in the methodology.

Human twin studies suggest that the genetic contribution to human lifespan is around 25–30% (Herskind et al. 1996). However, SNPs detected via GWA studies can only explain a small proportion of the genetic variance, the so-called missing heritability problem seen in GWA studies on the vast majority of complex traits. Researchers have suggested a variety of causes for this missing heritability, including alleles with effect sizes too small to capture, alleles that are too rare to capture in all but the largest studies, or effects due to structural variants such as insertions, deletions, inversions, and translocations (Manolio et al. 2009). Genetic control of normal ageing in humans is likely determined by subtle variations in many genes belonging to multiple biological pathways. It remains a challenge to collect more inclusive data and develop novel methodologies to comprehensively characterize the genetic basis of ageing-related phenotypes. (For further discussion, see Chapter 9: Haematopoetic System.)

5.4.3 Epigenetics of Ageing

We have seen that ageing can be modulated by genetics and environment. However, even cohorts of isogenic individuals housed in identical and controlled environments exhibit substantial variation in lifespan (Khazaeli et al. 1998; Libert et al. 2012). Although this non-genetic and non-environmental variation may come from somatic mutations, *de novo* germline mutations, and intrinsic stochasticity of the developmental processes, it is likely that a considerable proportion results from epigenetic factors (Benayoun et al. 2015).

Epigenetics is broadly defined as gene expression regulation that is not directly encoded in DNA sequences (Van Speybroeck 2002). Common epigenetic mechanisms include DNA methylation, post-translational histone modifications, chromosomal remodelling, and non-coding regulatory RNAs. These mechanisms work together to regulate chromatin structure, which contributes to transcriptional control of genes and the maintenance of genome stability. Epigenetic state is stable in that it persists in cells between cell divisions; on the other hand, it changes as a function of physiological state and can be altered by environmental signals. The majority of epigenetic signals are reset between generations, but some escape reprogramming, resulting in transgenerational epigenetic memory.

The epigenetic machinery becomes gradually deregulated as organisms age, a process termed 'epigenetic drift' (Issa 2014). Cohort studies of identical twins of various ages have shown that epigenetic markings of young identical twins are nearly indistinguishable, but they become increasingly distinct as a function of age (Fraga et al. 2005; Tan et al. 2016). The reported epigenetic shift in ageing identical twins could be a result of endogenous stochastic epigenetic errors accumulated over time, or it could result from differences in lifetime environmental exposure, such as chronic inflammation, dietary factors, and exposure to chemicals (Martin 2005; Hannum et al. 2013).

Epigenetic changes that occur during ageing are extremely complex, in that specific types of marks are either gained or lost in different parts of the genome, age-associated epigenetic changes are both tissue-specific and common across tissues, and depending on specific genes affected, age-related changes in the epigenome lead to both activation and repression of transcription. Despite the high level of complexity, some epigenetic changes do occur relatively consistently during ageing. For instance, CpG islands near promoters, which are generally unmethylated and open for transcription factors in young cells, tend to become hypermethylated in various ageing mammalian tissues (Maegawa et al. 2010; Day et al. 2013). Methylation of repetitive regions, which is important to maintain the silence of these 'junk sequences' and thus maintain genetic integrity, tends to decrease as animals age (Day et al. 2013). Histone H4 Lys 16 acetylation (H4K16ac), which is normally involved in regulation of telomere silencing and the DNA damage response, tends to decrease in ageing mammalian cells (Krishnan et al. 2011; Shah et al. 2013). Epigenetic changes that occur in ageing cells are highly similar to those observed in cancer cells, although it remains unclear whether age-related chromatin modifications contribute to tumorigenesis and tumour susceptibility (Zane et al. 2014).

Because DNA methylation is mitotically heritable and has the tendency to drift away from normal state as animals age, it represents a good candidate for a 'biological clock', where the initial methylation landscape reflects cell identity and youthfulness, and is potentially correlated with organismal chronological age or biological age (Figure 5.4). Recent studies suggest that the methylation status of a defined set of CpG dinucleotides in specific human tissues is a highly accurate predictor of organismal age (Bocklandt et al. 2011; Hannum et al. 2013; Horvath 2013; Weidner et al. 2014). While DNA methylation age is highly correlated with chronological age, individuals with a range of illnesses tend to have methylomes that look older than average (Weidner et al. 2014; Horvath et al. 2015). Moreover, cancer cells generally exhibit increased DNA methylation age (Horvath 2013), liver tissue from obese individuals appears epigenetically older than muscle or blood tissue from the same donor (Horvath et al. 2014), and DNA methylation age of blood cells predicts all-cause mortality in later life (Marioni et al. 2015). The ability of DNA methylation to act as a biological clock appears to outperform previously proposed biomarkers of ageing, including telomere attrition rate (Epel et al. 2004). It remains to be determined why some sites are better predictors than others, and what the underlying mechanisms linking biological age and methylation status might be.

Environmental factors that modify organismal longevity appear to have done so partly by modulating epigenetic states. In flies, dietary restriction delays the age-related loss of repressive heterochromatin, and diet switch experiments show that this effect works rapidly and is reversible (Jiang et al. 2013). In humans, high nutrient intake generates 'aged-like' DNA methylation profiles in the liver (Horvath et al. 2014). Nutrient sensing pathways, besides their well-known non-chromatin substrates, may partly mediate dietary-induced longevity through chromatin. For instance, mammalian insulin-like growth factor II (*IGF-2*) is maternally imprinted via DNA methylation in differentially methylated regions of the gene (Smith et al. 2006). SIRT6, a mammalian sirtuin whose overexpression extends lifespan (Kanfi et al. 2012), recruits chromatin remodellers and deacetylates histone H3K56 to prevent DNA damage (Toiber et al. 2013). Intermediate metabolites in the Krebs cycle, whose levels fluctuate as a function of nutrient intake, affect global levels of histone acetylation (Wellen et al. 2009). It remains to be established, though, whether chromatin remodelling is required in dietary restriction-mediated lifespan modulation.

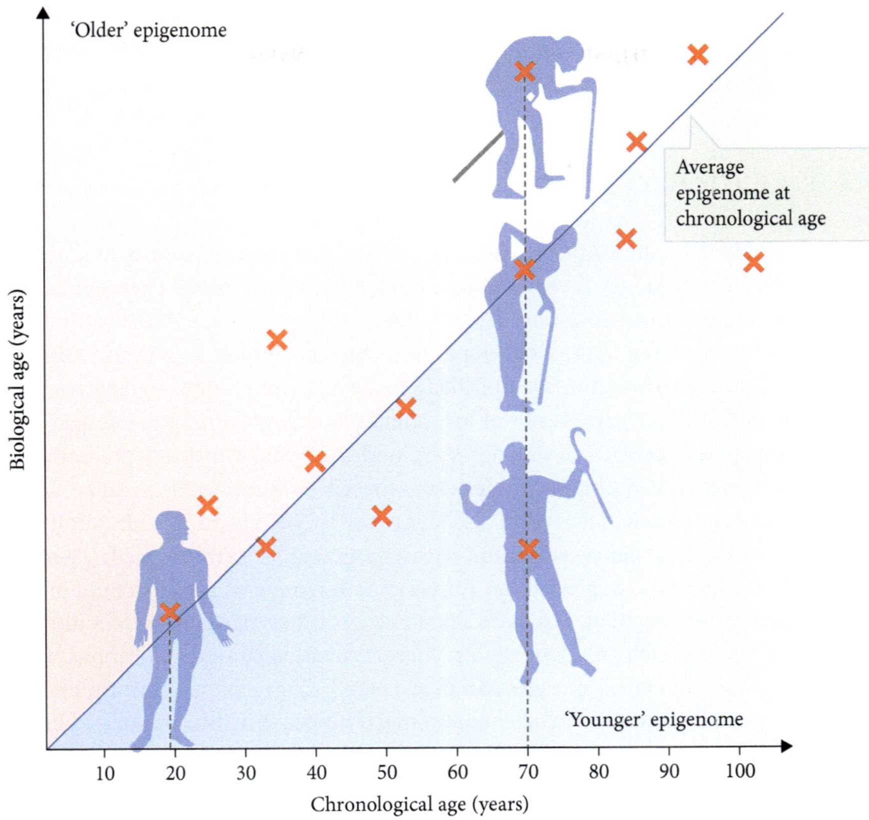

FIGURE 5.4 Changes in DNA methylation have been shown to be an accurate predictor of chronological age and biological age in humans. In specific tissues such as saliva and peripheral blood cells, the methylation status of a defined number of CpGs (termed DNA methylation age) is a highly accurate predictor of chronological age. Although DNA methylation age is highly correlated with chronological age, there is significant deviation from the linear fit, suggesting that methylation patterns may also reflect biological age. People with an apparently 'younger' epigenome are biologically younger than people with the same chronological age, and vice versa.

Source: Reprinted by permission from Macmillan Publishers Ltd, *Nature Reviews Molecular Cell Biology*, 16 (10), Bérénice A. Benayoun, Elizabeth A. Pollina, and Anne Brunet, Epigenetic regulation of ageing linking environmental inputs to genomic stability, pp. 593–610, Box 2, doi.10.1038/nrm4048. Copyright © 2015, Nature Publishing Group.

Altered epigenetic states induced by environmental or nutritional factors can even escape reprogramming and be transmitted between generations. In *C. elegans*, exposure to starvation in the great grandparental generation leads to lifespan extension in F3 offspring (Rechavi et al. 2014). In humans, food availability during specific growth periods of grandparents was shown to affect risk of death from cardiovascular disease and diabetes in grandchildren in a small isolated Swedish population (Kaati et al. 2002; Bygren et al. 2014). Women who lived through the Dutch Hunger Winter during the first trimester of pregnancy gave birth to children with increased risk of metabolic and cardiovascular disease late in life, as well as neuropsychiatric disorders such as schizophrenia, compared to children not exposed to famine *in utero* (Susser and Lin 1992; Painter et al. 2005). Hypomethylation of the nutrient sensor *IGF-2* was found to be associated with this exposure (Heijmans et al. 2008), supporting the hypothesis that transgenerational memory of past conditions is passed on via environmental modulation of chromatin states.

5.4.4 Evolutionary Theories of Ageing Revisited

Now that we have a wealth of information about the molecular mechanisms of ageing, we can ask whether evolutionary theories of *why* ageing occurs are in good agreement with the proximate mechanisms of *how* it occurs.

Unlike mechanisms that control development, which are highly conservative across taxa, mechanisms that control ageing are not likely to be as conserved, as ageing is not a programmed process, but merely a result of inefficient natural selection at older ages. The two evolutionary genetic theories of ageing—MA and AP—make distinct predictions as to whether mechanisms of ageing are public (i.e. shared between evolutionary lineages) or private (i.e. lineage-specific) (Partridge and Gems 2002b; Martin 2006). Mutation accumulation predicts that most genes associated with ageing will be lineage specific, while under antagonistic pleiotropy, strong selection for early-acting beneficial traits could give rise to fixation of ageing-causing alleles that are shared across different lineages. MA also predicts that many genes with small effects contribute to accumulated late-life deterioration, and the mechanisms of ageing within lineages can be extremely complex. In contrast, antagonistic pleiotropy theory allows the existence of conservative mechanisms of ageing, as life history trade-offs, such as that between somatic maintenance and reproduction, could occur by similar mechanisms across different taxa. Furthermore, there may be a manageable number of general processes modulating ageing rates. Thus, untangling the mechanisms of ageing may turn out to be a more soluble problem than some researchers have supposed.

From our discussion on proximate mechanisms of ageing, we can see that both public and private, simple and complex mechanisms exist. The fact that simple dietary and genetic alterations can substantially increase lifespan across broad evolutionary lineages demonstrates that ageing could be modulated through a small number of genes and pathways. On the other hand, screens for life-extending mutations, as well as genetic association and mapping studies of longevity, identified large numbers of longevity genes within species, and little overlap of longevity genes across different species. These results suggest that both processes are likely at play, and the truth may lie somewhere between the extreme predictions.

In fact, it may turn out to be more difficult than previously thought to distinguish MA from AP, not only in quantitative genetic studies (Moorad and Promislow 2009) but also in molecular genetic ones. We have seen that the effect of dietary restriction (Liao et al. 2010) and the effects of longevity genes themselves depend upon genetic background, physiological state, and environment (Spencer et al. 2003; Kapahi et al. 2004; Kaeberlein et al. 2005; Hansen et al. 2007; Kenyon 2010; Guarente 2013). The early-life effects of a lifespan-shortening novel mutation that are beneficial (consistent with AP) may be neutral (consistent with MA) or even deleterious in another environment or genetic background (e.g. Jenkins et al. 2004). Although researchers in the field of evolutionary theories of ageing have made substantial efforts to evaluate the relative importance of MA and AP (with inconclusive results), the distinction between these two mechanisms may in the end be largely artificial and of little practical value.

5.4.5 Mechanisms of Ageing—Conserved or Convergent?

As discussed in Section 5.4.1 and 5.4.2, dietary restriction and genetic alterations of the nutrient sensing pathways consistently yield lifespan extension across taxa. Does this make evolutionary sense? We have seen that ageing is not likely to be selected for. It is reasonable, however, to argue that ageing may be selected against, for example by situations that favour current survival and future reproductive output. Limited nutrient availability may be a good indicator of a harsh environment, conditions that could favour reducing reproduction and increasing survival rate until conditions improve, thus yielding higher fitness. Nutrient sensing pathways may be the mediator of such life history shifts. Note that the overall fitness effects of shifting resource allocation in response to food restriction depend upon numerous parameters, such as the effects of a period of food shortage on organismal survival and fecundity, as well as the frequency, duration, and severity of food shortage. Existing models do not show consistent conclusions in terms of whether resource reallocation between survival and fecundity in response to food shortage is sufficient to explain why organisms appear to age slower under dietary restriction (Shanley and Kirkwood 2000; Hou et al. 2011). In fact, classic life history theory suggests that in sufficiently harsh conditions, one might expect an *increase* in reproductive effort as a terminal investment strategy (Pianka and Parker 1975). Despite decades of experimental work on diet restriction and ageing, we still need more data and theory to fully understand how and why reproduction and survival rates respond over a full range of nutrient limitation.

Another possibility is that the observed life-extending effects of dietary restriction and mutant nutrient sensing genes are exclusively associated with laboratory strains, and have arisen as a by-product of laboratory life as animals are unintentionally subjected to selection for early breeding in an artificially rich nutritional environment. When exposed to conditions more closely resembling a natural environment, namely cycles of starvation, worms with mutant *daf-2* exhibit dramatically reduced fitness as compared to wild type (Jenkins et al. 2004). Mice that are recently derived from wild populations did not show a typical DR response (Harper et al. 2006). The finding that insulin signalling is associated with lifespan in species that diverged hundreds of millions of years ago suggests ancient convergence. But perhaps this convergence occurred over the few decades that yeast, worms, flies, and mice have been in the laboratory. At this point, the suggestion that apparently evolutionarily conserved ageing pathways are an artefact of convergent artificial selection is unproved, but worth testing. Moreover, much work remains to determine the extent to which candidate genes with major effects on longevity in model systems show segregating allelic variation associated with lifespan and/or healthspan in human populations.

5.5 Age-Related Pathology

Within human populations, we observe enormous variation not only in age at death, but also in the *quality* of ageing. While some live long lives in good health and normal cognitive and physical functionality, ageing leads others to suffer impaired physical and cognitive capabilities that can lead to disability, dependency, and relatively high risk of mortality. Even in a healthy cohort of people, age is the single greatest risk factor for a wide range of diseases, including cancer, Alzheimer's disease, cardiovascular disease, hypertension, type 2 diabetes, arthritis, osteoporosis, and cataracts—diseases that are among the leading causes of death in modern societies. The many hallmarks of ageing—among them, genomic instability, shortened telomeres, loss of proteostasis, mitochondria dysfunction, and stem cell exhaustion (López-Otín et al. 2013)—are likely to underlie age-related increases in disease risk. The diseases themselves are not how we define the ageing process. Unlike these age-related diseases, the ageing process itself will eventually reduce fitness components in all individuals. Nevertheless, by understanding the evolutionary forces that have shaped specific diseases, we can better understand the evolution of ageing. In this section, we focus on specific types of cancer, cardiovascular disease, and neurodegenerative disease to illustrate the genetic and environmental forces that have shaped the evolution of ageing and age-related disease risk in modern society. Why does senescence increase disease risk, and why are some individuals particularly resistant or vulnerable to certain age-related diseases?

We note here that many social scientists have suggested that the age-related increase in disease risk could be due to accumulated wear and tear, associated with increases in what they have called 'allostatic load'. Seeman and others (Seeman et al. 1997; McEwen and Seeman 1999; Crimmins et al. 2003) developed the notion of 'allostatic load' as a way to quantify the degree to which age-related increases in disease risk might be due in part to the cumulative deleterious effects of the stressors—not only biological, but also social—that we experience throughout our lives. It is highly likely that part of the variation among individuals in age-related risk is due to variation in this 'allostatic load'. However, most of the literature on allostatic load and ageing has fallen outside of the scope of classic evolutionary ideas about ageing. Readers interested in this environmental component of ageing are encouraged to explore this extensive literature.

As noted in Section 5.2, the primary factor leading to the evolution of ageing is the inability of natural selection to eliminate mutations with detrimental effects only at late life. But at least two additional factors are required to fully understand the prevalence of age-associated diseases in modern society.

First, human life expectancy has increased dramatically over the past two centuries (Figure 5.5). This increase has exposed the deleterious effects of genes that were hitherto solely beneficial. As discussed in Section 5.2, genes or genetic variants with beneficial fitness effects early in life but deleterious effects late in life are subject to positive selection (antagonistic pleiotropy). With a substantial increase in lifespan, people in modern societies start to express genetic costs that were previously not often paid because people died earlier for other reasons.

Second, rapid changes in social and cultural environments pose challenges to bodies adapted to our ancestral environment, which differs dramatically from the modern, industrialised environment in which most of us live (see Chapter 1 about the mismatch between our modern and ancestral environments). Agriculture, industrialisation, and post-industrial globalisation have profoundly changed our environment in the recent past: post-industrial diets contain more calories but less diversity (Demment et al. 2003); the automobile and automated technologies have promoted a sedentary lifestyle (see Chapter 6: Nutrition, Energy Expenditure, Physical Activity, and Body Composition); improved hygiene has reduced exposure not only to deleterious parasites but also to beneficial symbionts (see Chapter 10: Immune System); and the elevated socioeconomic status of women and birth control methods have remodelled patterns of reproduction and menstrual cycles

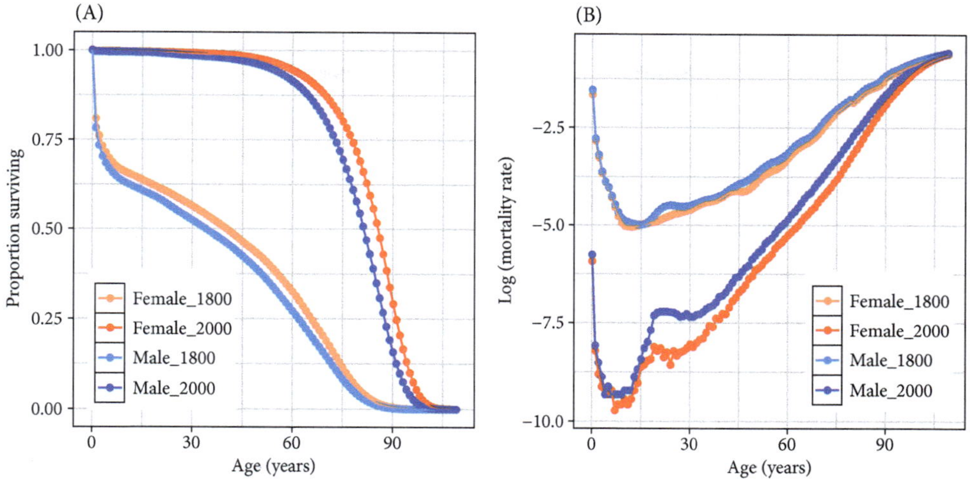

FIGURE 5.5 Age-at-death data of the Swedish population from years 1800 and 2000. (A) Survivorship over age. Note that in 1800, when infectious diseases were the major cause of death, the survivorship curve does not show a pronounced right shoulder as in the curves of 2000. The lack of a flat right shoulder usually indicates that environmental hostility, rather than ageing, is dominating patterns of age-at-death. (B) Logarithms of mortality rate over age. Note that the baseline mortality rate of 1800 is significantly higher than that of 2000.

Source: Data are from the Human Mortality Database (http//www.mortality.org/).

(see Chapter 16: Sexuality, Reproduction, and Birth). In all these cases, one can identify diseases that are associated, at least in part, with our having evolved in a quite different environment and so being mismatched to this new and dramatically different environment. In this section, we consider how our modern environment, while allowing us to live healthier and longer lives in so many ways, has also led to increased frequency of certain age-related morbidities, including cancer, cardiovascular disease, and neurodegenerative disease.

5.5.1 Cancer

Cancers are a group of diseases characterised by inappropriate and uncontrolled cell growth and movement. Genetic changes associated with cancer can be inherited, or they can arise during the organism's lifetime as a result of somatic mutations as cells divide. Cancer-causing somatic mutations can arise spontaneously, or as a result of environmental exposure to chemicals, such as nicotine from tobacco, or ultraviolet (UV) radiation from the sun.

Clinical incidence of cancer increases exponentially with age (Figure 5.6) (Armitage and Doll 1954; Doll 1978; Frank 2004; Frank 2005). Cancer prevalence is particularly high in older people for at least three reasons. First, each cycle of mitosis is an opportunity for spontaneous mutation, and decades of continued stem cell proliferation and tissue turnover provide opportunities for somatic mutations to accumulate in cell lineages, a small proportion of which eventually gain the 'right' combination to become cancerous (Greaves 2007). Second, chronic mutagenic exposure over the course of a lifetime exacerbates somatic mutation accumulation as cells divide. These explanations are consistent with the multistage theory

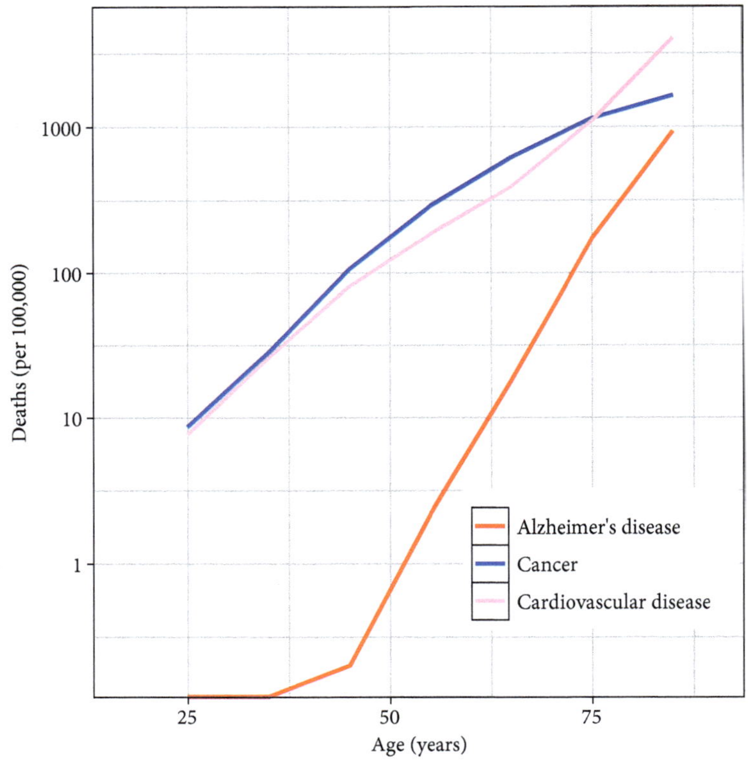

FIGURE 5.6 Annual deaths per 100,000 by cancer, cardiovascular disease, and Alzheimer's disease.
Source: Data are from the 2013 CDC National Vital Statistics Report (https://www.cdc.gov/nchs/products/nvsr.htm).

of carcinogenesis (Armitage and Doll 1954). Third, age-associated deterioration in the tissue microenvironment (Campisi 1997; Campisi 2013) or in immunosurveillance (Dunn et al. 2004) increases permissiveness to cancer initiation and progression.

In many cases, we can think of cancer as an extreme example of the evolutionary trade-offs that shape all organisms. Many adaptive biological processes in multicellular organisms contribute to the risk of cancer (Greaves 2008). The intrinsic capability of DNA to mutate and recombine generates potentially beneficial genetic variation in the germline (Sniegowski 1997), with the trade-off of deleterious mutations in the germline that decrease offspring fitness, and, germane to this discussion, cancer-initiating mutations in somatic cells. Early embryogenesis is precisely regulated by a suite of genes mediating cell proliferation, survival, adhesion, and migration, but this developmental programme is available for co-option or hijack by cancer cells (Thiery 2002; Topczewska et al. 2006). Stem cells are essential for tissue regeneration and repair, and thus are critical for sustained tissue function and longevity in multicellular organisms; however, most malignant cancers originate from the transformation of normal stem cells (Reya et al. 2001), presumably because stem cells are 'pre-adapted' for cancerous lifestyle. For example, telomerase expressed in germ cells protects the integrity of germline DNA, and telomerase expressed in normal stem cells ensures cells' extensive regenerative ability. In humans, normal somatic cells have turned off telomerase. Cancer cells take advantage of this gene, acquiring persistent telomerase activity via unknown mutational mechanisms and thereby escaping the restraint on cell proliferation conferred by telomere attrition (Weinberg 2013). (For further discussion, see Chapter 9: Haematopoetic System.)

Cellular senescence and the tumour suppressor protein p53 represent an example of evolutionary antagonistic pleiotropy. Cellular senescence refers to the permanent arrest of cell proliferation that occurs when cells experience potentially oncogenic stimuli, such as DNA double-strand breaks, strong and chronic mitogenic signals, epigenomic perturbations, and tumour suppressor gene activation (Campisi and di Fagagna 2007; Campisi 2013). In addition to arrested proliferation, senescent cells exhibit significant changes in gene expression, including the secretion of numerous pro-inflammatory cytokines, growth factors, and proteases, a feature termed the senescence-associated secretory phenotype (SASP) (Campisi 2013). Cell senescence suppresses the development of cancer, not only because it halts cell division (Prieur and Peeper 2008; Collado and Serrano 2010), but also because SASP promotes repair of damaged tissue (Adams 2009) and the pro-inflammatory nature of SASP attracts immune cells to clear pre-malignant cells (Kang et al. 2011). However, with increasing age, accumulated senescent cells lead to a loss of tissue structure and function, contributing to degenerative diseases of ageing (Faragher 2000). The chronic and universal presence of SASP accelerates age-associated wear and tear, and creates a tissue microenvironment which, ironically, promotes the development of cancer late in life (Campisi 2005, 2013).

The genetic basis of cell senescence also suggests that there may be a cost to innate cancer prevention. Cell senescence is mediated, in part, by the tumour suppressor p53. People who have inherited only one functional copy of the gene *TP53* (which encodes p53) are at high risk of developing cancer in early adulthood (Malkin 2011). However, mice with enhanced function of p53 exhibit progeroid syndromes and reduced lifespan (Tyner et al. 2002; Maier 2004), presumably because these mice are more susceptible to p53-mediated cell senescence (Rufini et al. 2013). Apparently, elephants, which have twenty copies of *TP53*, have figured out one way to avoid such a cost (Abegglen et al. 2015). It appears that at least some mechanisms that protect against cancer in early life are antagonistically pleiotropic, in that they also promote ageing and late-life cancer in old organisms.

While some cancers have likely been shaped by evolutionary trade-offs, others might be due to the mismatch between the environment in which humans evolved and our current one. One example of this is provided by the high risk of skin cancer in people with pale skin living in sunny climates. Several hypotheses attempt to explain the known correlation between skin colour and latitude. Epidermal pigmentation is presumably favoured by selection to filter excess sunlight to prevent UVB-induced skin cancers (Yamaguchi et al. 2007; Osborne and Hames 2014), to prevent folic acid degradation (Jablonski 1999), or to enhance the permeability and antimicrobial function of the skin (Elias and Williams 2016). In high latitudes there is selection for dilution of epidermal pigmentation, presumably because fair skin allows maximal penetrance of UV so that sufficient vitamin D can be synthesised at low intensities of sunlight (Loomis 1967), or because reduced pigmentation conserves metabolic resources (Elias and Williams 2013).

Nevertheless, given the distribution of pigmentation, it is likely that variation in the skin colour of humans is a result of adaptation to the environment in which the population evolved. But regardless of the specific selective forces that have shaped pigmentation, mismatches can arise when individuals adapted to lower levels of UV intermittently or permanently migrate to sunnier regions and fail to protect themselves from excess UV exposure. Indeed, Caucasians exhibit the highest incidence of skin cancer, and the lifetime risk has continued to increase over the past few decades (Chang et al. 1998); Japanese have much lower incidence of skin cancer than Caucasians, but Japanese-Americans in Hawaii, who are exposed to high levels of UV radiation, show dramatic increase in skin cancer incidence (Woodhead et al. 1999).

It is further hypothesised that an unnatural ratio of UVA and UVB in indoor environments, presumably created when sunlight passes through glass windows in offices and cars, is responsible for the steady increase in skin cancer incidence (Godar et al. 2009; Godar 2011). The depletion of the ozone layer raises further concerns that skin cancer could become an important health issue in the future. (For further discussion, see Chapter 8: Skin and Integument.)

A third class of cancers suggests a role for the joint effects of both environmental shift and antagonistic pleiotropy. As discussed earlier in this section, in every cell division there is a risk of mutation that could lead to cancer. Cell division in the female reproductive tract, particularly in breasts and ovaries, takes place during every menstrual cycle. After the first full-term pregnancy, cells lining a woman's milk ducts differentiate so that they can produce milk, and remain essentially undivided thereafter (Russo et al. 2005). In our evolutionary past, women started giving birth not long after puberty, and spent most of their reproductive lives pregnant or nursing, both processes that repress ovulation and menstruation. In modern society, however, with early onset of menses, increased age at first childbirth, decreased number of childbirths, and shorter duration of lactation, women are experiencing about three times as many menstrual cycles as female hunter-gatherers (Strassmann 1999), and cells lining the milk duct now go through many more cycles of division before differentiation. These mismatches may explain the current high incidence of breast and ovarian cancer (MacMahon et al. 1970; Lambe et al. 1996; Titus-Ernstoff et al. 2001; Britt et al. 2007). Oral contraceptives and obesity, which are mostly seen in modern society, also affect the risk of reproductive cancers (Marchbanks et al. 2002; Morimoto et al. 2002; Collaborative Group on Epidemiological Studies of Ovarian Cancer 2008).

Germline mutations in the DNA repair genes *BRCA1* and *BRCA2* are associated with familial breast and ovarian cancer (Friedman et al. 1994; Miki et al. 1994). Some sequence analyses show that particular regions of *BRCA1* are subject to positive selection (Huttley et al. 2000; Yang and Nielsen 2002; Fleming et al. 2003; Pavlicek 2004), while others show evidence of purifying selection (Pavard and Metcalf 2007). Nevertheless, mutations in *BRCA1* and *BRCA2* are relatively prevalent in human populations, raising the possibility that they may have beneficial early-life effects coupled with elevated cancer risk. Indeed, a recent study of human population data suggests that female *BRCA1/2* mutation carriers have significantly more children than non-carriers (Smith et al. 2011). One hypothetical mechanism for this phenomenon is that mate choice by males and natural selection for fat reserves have selected for alleles promoting breast development, with the cost of increased cancer risk (Crespi and Summers 2006). It is also worth mentioning that cancer susceptibility of some *BRCA* mutations may have been buffered by reproductive patterns in the more distant past, when women started giving birth early and experienced fewer menstrual cycles throughout their lives (Smith et al. 2011). (For further discussion, see Chapter 16: Sexuality, Reproduction, and Birth.)

5.5.2 Cardiovascular Diseases

Cardiovascular diseases are a class of diseases involving the heart or blood vessels and are the leading cause of death in all continents except Africa (Mendis and Puska 2011; Santulli 2013). A large proportion of cardiovascular diseases are due to a pathological process called atherosclerosis ('hardening of the arteries'). Fatty material and cholesterol are deposited inside the lumen of arterial walls, forming plaques that narrow the arteries and decrease the elasticity of blood vessels. Eventually, the plaque can rupture, forming blood clots in arteries. If clots develop in a coronary artery, they can cause a heart attack; if clots develop in the brain, they can cause a stroke.

Age is the most important risk factor of cardiovascular disease (Figure 5.6) (Finegold et al. 2013). Plaque formation can start as early as childhood (Vanhecke et al. 2006), but it takes decades to unfold to produce full-blown disease. General changes in the mechanical and structural properties of the vascular wall over time increase the tendency to rupture (Jani and Rajkumar 2006). Other age-related diseases such as diabetes (Howard and Wylie-Rosett 2002; Micha et al. 2012) and hypertension (Santulli 2013) are risk factors of atherosclerosis. Serum cholesterol level, which increases with age, may be another reason why most cardiovascular disease affects older adults (Jousilahti et al. 1999). (For further discussion, see Chapter 11: Cardiovascular System.)

Atherosclerosis-associated cardiovascular disease is thought to be a disease of environmental mismatch. Low-density lipoproteins (LDL) transport cholesterol from the liver to the rest of the body, and high-density lipoproteins (HDL) transport excess cholesterol from tissues back to the liver so that it can be converted and excreted. Excessive LDL cholesterol is the key contributor to plaque build-up in the arteries, while HDLs help get rid of excessive cholesterol so that they do not end up in the arteries. High levels of triglycerides combined with high blood levels of LDL cholesterol and low blood levels of HDL also contribute to fatty build-up in the artery wall.

The environmental factors that elevate the risks of cardiovascular diseases by increasing the blood levels of LDL and triglycerides, and decreasing those of HDL, include poor diet, physical inactivity, obesity, alcohol consumption, tobacco use, and emotional stress (Mendis and Puska 2011), all of which are commonly seen in post-industrial society, yet at least some are novel to the human body. This argument is supported by a recent study of heart disease among the Tsimane, a Bolivian tribe with a forager-horticulturalist lifestyle, in which atherosclerosis is extraordinarily rare, even among those in their 70s and 80s (Kaplan et al. 2017). The Tsimane are notable for a lifelong practice of high levels of low-intensity daily exercise and a diet very high in complex carbohydrates and low in saturated fats. (For further discussion, see Chapter 6: Nutrition, Energy Expenditure, Physical Activity, and Body Composition.)

Variants of the gene *APOE* contribute to differential susceptibility to atherosclerosis-associated cardiovascular disease. *APOE* encodes apolipoprotein E, which is responsible for lipid metabolism and cholesterol transport (Davignon et al. 1988; Mahley and Huang 1999). In human populations, *APOE* has three major alleles: ε2, ε3, and ε4. Carriers of the ε4 allele show higher levels of total cholesterol than ε2 and ε3 carriers, presumably because the apoE4 isoform has high binding affinity for LDLs (Eichner et al. 2002). Carrying the ε4 allele increases the risk of atherosclerosis-associated cardiovascular disease by up to 40% (Stengård et al. 1998; Eichner et al. 2002), and sometimes manifests early in life (Ilveskoski et al. 1999).

Given its effects on cardiovascular health and the fact that cardiovascular disease is the most important cause of death in the industrialised world, it is not surprising that *APOE* repeatedly shows up in GWA studies of human lifespan (Deelen et al. 2014).

It is worth noting that although atherosclerosis has a genetic component, the genetic risks are expressed through environmental interactions. Recent work suggests that *APOE* genotype interacts with exercise and physical activity to mediate cardiovascular function (Raichlen and Alexander 2014). Athletic ε4 carriers have similar lipid profiles as ε2 and ε3 carriers, while sedentary ε4 carriers show significantly higher levels of LDLs compared with physically active ε4 carriers, and compared with ε3 carriers regardless of activity level (Bernstein et al. 2002; Pisciotta et al. 2003). Thus, exercise appears to buffer the risk of cardiovascular disease among ε4 carriers, while sedentary lifestyle exacerbates the risk. (For further discussion, see Chapter 6: Nutrition, Energy Expenditure, Physical Activity, and Body Composition.)

5.5.3 Neurodegenerative Diseases

The global prevalence of age-related neurodegenerative diseases such as Alzheimer's disease and Parkinson's disease has been increasing as populations get older (Reitz et al. 2011; Reeve et al. 2014). Interestingly, among leading causes of mortality in industrialised countries, the age-related rate at which the major neurodegenerative diseases increase is about twice that of other causes (Figure 5.6). In this section, we use the most common age-related neurodegenerative disease, Alzheimer's disease (AD), as an example to discuss from an evolutionary perspective why some people are susceptible to this disease at old age.

AD is an irreversible, progressive brain disorder that is the most common cause of dementia among older adults. It is not yet clear what causes AD, but the key pathological changes observed in AD brain tissues are accumulations of plaques and tangles in particular regions of the brain. These plaques and tangles lead to neurons that function less efficiently and eventually are unable to function or communicate with other neurons. Increased levels of the gluey amyloid-β (Aβ) peptide lead to Aβ clumps that are deposited extracellularly to form neuritic plaques. Tau protein, which normally forms microtubules in the cell, when hyperphosphorylated, accumulates intracellularly to form neurofibrillary tangles (Finder 2010). Although most people develop some plaques and tangles as they age, this process is significantly accelerated in AD patients. AD is a disease of the elderly (Figure 5.6), and the pathological process can go on for many years without symptoms. Certain aspects of the normal ageing process such as mitochondrial dysfunction (Moreira et al. 2010) and chronic inflammation (Akiyama et al. 2000) appear to trigger or promote AD. Other age-related diseases such as stroke (Wen et al. 2007), hypertension (Launer et al. 1995; Kivipelto et al. 2001; Whitmer et al. 2005), and type 2 diabetes (Leibson et al. 1997; Ott et al. 1999; Luchsinger et al. 2001) increase the risk of AD.

Susceptibility to AD may involve several evolutionary trade-offs. Epidemiological studies have found that individuals diagnosed with AD show reduced risk of cancer and vice versa (Ma et al. 2014; Shi et al. 2014; Zhang et al. 2015). This inverse correlation may reflect opposite ends of the spectrum of cellular homeostasis: in AD there is a propensity of excessive cell degeneration and reduced cell regeneration/survival, while in cancer there is uncontrolled cell proliferation and resistance to cell death (Staropoli 2008; Li et al. 2014).

The shared genetic pathways of AD and cancer remains to be established, but some overlap has been implicated, especially in genes involving DNA repair, cell cycle control, and kinase signalling (Plun-Favreau et al. 2010; Ganguli 2015).

The major components of plaques and tangles of the AD brain—Aβ and tau—are present in normal brains in alternative forms, and may confer beneficial effects in individuals without AD. Aβ is generated as a normal product of amyloid precursor protein (APP) metabolism. Although some animal studies show that Aβ is not required for normal physiological function (Luo et al. 2003), recent studies indicate that it may be involved in protection against oxidative stress (Zou et al. 2002; Baruch-Suchodolsky and Fischer 2009), regulation of cholesterol transportation (Yao and Papadopoulos 2002; Igbavboa et al. 2009), and function as a transcription factor (Bailey et al. 2011; Maloney and Lahiri 2011). What is particularly interesting is that Aβ peptides show potent antimicrobial activities against several clinically important pathogens that can cause brain infection (Soscia et al. 2010), making it a good candidate for antagonistic pleiotropy.

The polymorphisms of the *APOE* gene also influence susceptibility to common late-onset AD. *APOE* is the only gene that appears repeatedly in GWA studies of AD, with carriers of the ε4 allele showing significantly increased risk (Bertram and Tanzi 2009; Kamboh et al. 2012; Tanzi 2012). The specific mechanisms by which the ε4 allele increases AD risk are not yet clear, but may have to do with Aβ transportation (Cedazo-Mínguez 2007). The maintenance of polymorphism in *APOE* may be due to antagonistic pleiotropy. Female ε4 carriers have elevated levels of luteal progesterone, indicating higher potential fertility (Jasienska et al. 2015). As a major risk factor of late-onset AD, the effects of the ε4 allele on cognitive ability appear to be strongly age dependent: younger ε4 carriers outperform non-carriers in various aspects of cognitive performance (Yu et al. 2000; Wright et al. 2003; Jochemsen et al. 2012; Rusted et al. 2013).

5.6 Conclusion

The age-related decline in fitness components that we refer to as ageing or senescence has been observed in a tremendous diversity of species, and a careful body of theoretical, comparative, and experimental studies has attempted to understand both the how and why of this ubiquitous phenomenon. Despite enormous scientific effort, the field is full of unanswered questions, making it an exciting area for continued exploration. For lack of space, even here we have left off some critical and exciting questions: Why do women but not men exhibit menopause, and why do women live longer than men? Will the dramatic demographic shifts that human populations have experienced very recently in our evolutionary history lead to long-term evolutionary changes in patterns of ageing? (For further discussion, see Chapter 4: Growth and Development.) And what, in the end, will laboratory studies teach us about the causes of human ageing, and our ability to modify it through pharmacological interventions? Just in the past few years we have seen a concerted effort to create a translational geroscience, with hopes for improved healthspan pinned on a variety of treatments, including rapamycin in dogs (Urfer et al. 2017) and humans (Mannick et al. 2014), so-called senolytics that remove senescent cells (Kirkland et al. 2017), and efforts to initiate the first-ever clinical trial on healthspan using the diabetes drug metformin (Barzilai et al. 2016). We look forward to evolutionary perspectives on this attempt to reverse what many regard as an evolutionarily inevitable process.

This is an exciting time to be studying the evolution of ageing. We have learned a tremendous amount about the basic mechanisms of ageing over the past decades, but these discoveries have also raised new and fascinating questions. The future of this discipline will depend in large part on intellectually stimulating collaborations between researchers representing diverse disciplines, from demography, evolutionary theory, epidemiology, and mathematics, to biochemistry, genetics, systems biology, and pathology. Such interdisciplinary approaches hold out the promise of a unified theory that tells us both how *and* why ageing occurs. And finally, for those hoping to delay or decrease the deleterious effects of age, the best bet for now appears to be maintaining a healthy weight through nutritious diet and moderate exercise. (For further discussion, see Chapter 6: Nutrition, Energy Expenditure, Physical Activity, and Body Composition.)

References

Abegglen, L. M., Caulin, A. F., Chan, A., et al. (2015). Potential mechanisms for cancer resistance in elephants and comparative cellular response to DNA damage in humans. *JAMA* 314(17), 1850–60.

Abrams, P. A. (1991). The fitness costs of senescence: the evolutionary importance of events in early adult life. *Evol Ecol* 5(4), 343–60.

Abrams, P. A. (1993). Does increased mortality favor the evolution of more rapid senescence? *Evolution* 47(3), 877.

Adams, P. D. (2009). Healing and hurting: molecular mechanisms, functions, and pathologies of cellular senescence. *Mol Cell* 36(1), 2–14.

Akiyama, H., Barger, S., Barnum, S., et al. (2000). Inflammation and Alzheimer's disease. *Neurobiol Aging* 21(3), 383–421.

Albani, D., Ateri, E., Mazzuco, S., et al. (2013). Modulation of human longevity by SIRT3 single nucleotide polymorphisms in the prospective study 'Treviso Longeva (TRELONG)'. *Age* 36(1), 469–78.

Anderson, R. M., Shanmuganayagam, D., and Weindruch, R. (2009). Caloric restriction and aging: studies in mice and monkeys. *Toxicol Pathol* 37(1), 47–51.

Apfeld, J. and Kenyon, C. (1998). Cell nonautonomy of *C. elegans* daf-2 function in the regulation of diapause and life span. *Cell* 95(2), 199–210.

Armitage, P. and Doll, R. (1954). The age distribution of cancer and a multi-stage theory of carcinogenesis. *Br J Cancer* 8(1), 1–12.

Arum, O., Bonkowski, M. S., Rocha, J. S., et al. (2009). The growth hormone receptor gene-disrupted mouse fails to respond to an intermittent fasting diet. *Aging Cell* 8(6), 756–60.

Austad, S. N. (1993). Retarded senescence in an insular population of Virginia opossums (*Didelphis virginiana*). *J Zool* 229(4), 695–708.

Austad, S. N. (1999). *Why We Age: What Science Is Discovering about the Body's Journey Through Life*. Chichester: Wiley.

Bailey, J. A., Maloney, B., Ge, Y.-W., et al. (2011). Functional activity of the novel Alzheimer's amyloid β-peptide interacting domain (AβID) in the APP and BACE1 promoter sequences and implications in activating apoptotic genes and in amyloidogenesis. *Gene* 488(1–2), 13–22.

Baker, G. T., III and Sprott, R. L. (1988). Biomarkers of aging. *Exp Gerontol* 23(4–5), 223–39.

Banerjee, K. K., Ayyub, C., Ali, S. Z., et al. (2012). dSir2 in the adult fat body, but not in muscles, regulates life span in a diet-dependent manner. *Cell Rep* 2(6), 1485–91.

Bartke, A. (2008). Insulin and aging. *Cell Cycle* 7(21), 3338–43.

Baruch-Suchodolsky, R. and Fischer, B. (2009). Aβ40, either soluble or aggregated, is a remarkably potent antioxidant in cell-free oxidative systems. *Biochemistry* 48(20), 4354–70.

Barzilai, N., Crandall, J. P., Kritchevsky, S. B., et al. (2016). Metformin as a tool to target aging. *Cell Metabol* 23(6), 1060–5.

Beekman, M., Blanché, H., Perola, M., et al. (2013). Genome-wide linkage analysis for human longevity: Genetics of Healthy Aging Study. *Aging Cell* 12(2), 184–93.

Bellizzi, D., Rose, G., Cavalcante, P., et al. (2005). A novel VNTR enhancer within the SIRT3 gene, a human homologue of SIR2, is associated with survival at oldest ages. *Genomics* 85(2), 258–63.

Benayoun, B. A., Pollina, E. A., and Brunet, A. (2015). Epigenetic regulation of ageing: linking environmental inputs to genomic stability. *Nat Rev Mol Cell Biol* 16(10), 593–610.

Berdichevsky, A., Viswanathan, M., Horvitz, H. R., et al. (2006). *C. elegans* SIR-2.1 interacts with 14-3-3 proteins to activate DAF-16 and extend life span. *Cell* 125(6), 1165–77.

Bernstein, M. S., Costanza, M. C., James, R. W., et al. (2002). Physical activity may modulate effects of ApoE genotype on lipid profile. *Arterioscler Thromb Vasc Biol* 22(1), 133–40.

Bertram, L. and Tanzi, R. E. (2009). Genome-wide association studies in Alzheimer's disease. *Hum Mol Genet* 18(R2), R137–45.

Blüher, M., Kahn, B. B., and Kahn, C. R. (2003). Extended longevity in mice lacking the insulin receptor in adipose tissue. *Science* 299(5606), 572–4.

Bocklandt, S., Lin, W., Sehl, M. E., et al. (2011). Epigenetic predictor of age. *PLoS One* 6(6), e14821.

Bonawitz, N. D., Chatenay-Lapointe, M., Pan, Y., et al. (2007). Reduced TOR signaling extends chronological life span via increased respiration and upregulation of mitochondrial gene expression. *Cell Metabol* 5(4), 265–77.

Borkan, G. A. and Norris, A. H. (1980). Assessment of biological age using a profile of physical parameters. *J Gerontol* 35(2), 177–84.

Boyden, S. E. and Kunkel, L. M. (2010). High-density genomewide linkage analysis of exceptional human longevity identifies multiple novel loci. *PLoS One* 5(8), e12432.

Brandhorst, S., Choi, I. Y., Wei, M., et al. (2015). A periodic diet that mimics fasting promotes multi-system regeneration, enhanced cognitive performance, and healthspan. *Cell Metabol* 22(1), 86–99.

Britt, K., Ashworth, A., Smalley, M. et al. (2007). Pregnancy and the risk of breast cancer. *Endocr Relat Cancer* 14(4), 907–33.

Brommer, J. E., Wilson, A. J., Gustafsson, L. (2007). Exploring the genetics of aging in a wild passerine bird. *Am Nat* 170(4), 643–50.

Bronikowski, A. M., Altmann, J., Brockman, D. K., et al. (2011). Aging in the natural world: comparative data reveal similar mortality patterns across primates. *Science* 331(6022), 1325–8.

Burnett, C., Valentini, S., Cabreiro, F., et al. (2011). Absence of effects of Sir2 overexpression on lifespan in *C. elegans* and *Drosophila*. *Nature* 477(7365), 482–5.

Burtner, C. R., Murakami, C. J., Olsen, B., et al. (2011). A genomic analysis of chronological longevity factors in budding yeast. *Cell Cycle* 10(9), 1385–96.

Bygren, L. O., Tinghög, P., Carstensen, J., et al. (2014). Change in paternal grandmothers' early food supply influenced cardiovascular mortality of the female grandchildren. *BMC Genet* 15(1), 12.

Campisi, J. (1997). Aging and cancer: the double-edged sword of replicative senescence. *J Am Geriatr Soc* 45(4), 482–8.

Campisi, J. (2005). Aging, tumor suppression and cancer: high wire-act! *Mech Ageing Dev* 126(1), 51–8.

Campisi, J. (2013). Aging, cellular senescence, and cancer. *Annu Rev Physiol* 75(1), 685–705.

Campisi, J. and di Fagagna, F. D. (2007). Cellular senescence: when bad things happen to good cells. *Nat Rev Mol Cell Biol* 8(9), 729–40.

Caswell, H. (2007). Extrinsic mortality and the evolution of senescence. *Trends Ecol Evol* 22(4), 173–4.

Cedazo-Mínguez, A. (2007). Apolipoprotein E and Alzheimer's disease: molecular mechanisms and therapeutic opportunities. *J Cell Mol Med* 11(6), 1227–38.

Chang, A. E., Karnell, L. H., Menck, H. R. (1998). The National Cancer Data Base report on cutaneous and noncutaneous melanoma. *Cancer* 83(8), 1664–78.

Charlesworth, B. (2000). Fisher, Medawar, Hamilton and the evolution of aging. *Genetics* 156(3), 927–31.

Charlesworth, B. and Hughes, K. A. (1996). Age-specific inbreeding depression and components of genetic variance in relation to the evolution of senescence. *Proc Natl Acad Sci U S A* 93(12), 6140–5.

Clancy, D. J., Gems, D., Harshman, L. G., et al. (2001). Extension of life-span by loss of CHICO, a *Drosophila* insulin receptor substrate protein. *Science* 292(5514), 104–6.
Collaborative Group on Epidemiological Studies of Ovarian Cancer (2008). Ovarian cancer and oral contraceptives: collaborative reanalysis of data from 45 epidemiological studies including 23 257 women with ovarian cancer and 87 303 controls. *Lancet* 371(9609), 303–14.
Collado, M. and Serrano, M. (2010). Senescence in tumours: evidence from mice and humans. *Nat Rev Cancer* 10(1), 51–7.
Colman, R. J., Anderson, R. M., Johnson, S. C., et al. (2009). Caloric restriction delays disease onset and mortality in rhesus monkeys. *Science* 325(5937), 201–4.
Coschigano, K. T., Holland, A. N., Riders, M. E., et al. (2003). Deletion, but not antagonism, of the mouse growth hormone receptor results in severely decreased body weights, insulin, and insulin-like growth factor I levels and increased life span. *Endocrinology* 144(9), 3799–810.
Crespi, B. J. and Summers, K. (2006). Positive selection in the evolution of cancer. *Biol Rev* 81(3), 407–24.
Crimmins, E. M., Johnston, M., Hayward, M., et al. (2003). Age differences in allostatic load: an index of physiological dysregulation. *Exp Gerontol* 38(7), 731–4.
Dang, W., Steffen, K. K., Perry, R., et al. (2009). Histone H4 lysine 16 acetylation regulates cellular lifespan. *Nature* 459(7248), 802–7.
Davignon, J., Gregg, R. E., and Sing, C. F. (1988). Apolipoprotein E polymorphism and atherosclerosis. *Arteriosclerosis* 8(1), 1–21.
Day, K., Waite, L. L., Thalacker-Mercer, A., et al. (2013. Differential DNA methylation with age displays both common and dynamic features across human tissues that are influenced by CpG landscape. *Genome Biol* 14(9), R102.
Deelen, J., Beekman, M., Uh, H.-W., et al. (2011). Genome-wide association study identifies a single major locus contributing to survival into old age; the APOE locus revisited. *Aging Cell* 10(4), 686–98.
Deelen, J., Beekman, M., Uh, H.-W., et al. (2014). Genome-wide association meta-analysis of human longevity identifies a novel locus conferring survival beyond 90 years of age. *Hum Mol Genet* 23(16), 4420–32.
de Grey, A. D. N. J. (2015). Do we have genes that exist to hasten aging? New data, new arguments, but the answer is still no. *Curr Aging Sci* 8(1), 24–33.
de Haan, G., Gelman, R., Watson, A., et al. (1998). A putative gene causes variability in lifespan among genotypically identical mice. *Nat Genet* 19(2), 114–16.
Demment, M. W., Young, M. M., and Sensenig, R. L. (2003). Providing micronutrients through food-based solutions: a key to human and national development. *J Nutr* 133(11 Suppl 2), 3879S–85S.
di Fagagna, F. D., Reaper, P. M., Clay-Farrace, L., et al. (2003). A DNA damage checkpoint response in telomere-initiated senescence. *Nature* 426(6963), 194–8.
Doll, R. (1978). An epidemiological perspective of the biology of cancer. *Cancer Res* 38(11 Pt 1), 3573–83.
Dunn, G. P., Old, L. J., and Schreiber, R. D. (2004). The immunobiology of cancer immunosurveillance and immunoediting. *Immunity* 21(2), 137–48.
Eichner, J. E., Dunn, S. T., Perveen, G., et al. (2002). Apolipoprotein E polymorphism and cardiovascular disease: a HuGE review. *Am J Epidemiol* 155(6), 487–95.
Eigenmann, J. E., Patterson, D. F., Froesch, E. R. (1984). Body size parallels insulin-like growth factor I levels but not growth hormone secretory capacity. *Acta Endocrinol* 106(4), 448–53.
Elias, P. M. and Williams, M. L. (2013). Re-appraisal of current theories for the development and loss of epidermal pigmentation in hominins and modern humans. *J Hum Evol* 64(6), 687–92.
Elias, P. M. and Williams, M. L. (2016). Basis for the gain and subsequent dilution of epidermal pigmentation during human evolution: the barrier and metabolic conservation hypotheses revisited. *Am J Phys Anthropol* 161(2), 189–207.
Epel, E. S., Blackburn, E. H., Lin, J., Dhabhar, et al. (2004). Accelerated telomere shortening in response to life stress. *Proc Natl Acad Sci U S A* 101(49), 17312–15.
Escobar, J. S., Jarne, P., Charmantier, A., et al. (2008). Outbreeding alleviates senescence in hermaphroditic snails as expected from the mutation-accumulation theory. *Current Biology* 18(12), 906–10.
Fabrizio, P., Pozza, F., Pletcher, S. D., et al. (2001). Regulation of longevity and stress resistance by Sch9 in yeast. *Science* 292(5515), 288–90.

Faragher, R. G. (2000). Cell senescence and human aging: where's the link? *Biochem Soc Trans* 28(2), 221–6.
Figarska, S. M., Vonk, J. M., Boezen, H. M. (2013). SIRT1 polymorphism, long-term survival and glucose tolerance in the general population. *PLoS One* 8(3), e58636.
Finder, V. H. (2010). Alzheimer's disease: a general introduction and pathomechanism. *J Alzheimer's Dis* 22(S3), S5–19.
Finegold, J. A., Asaria, P., Francis, D. P. (2013). Mortality from ischaemic heart disease by country, region, and age: statistics from World Health Organisation and United Nations. *Int J Cardiol* 168(2), 934–45.
Flachsbart, F., Croucher, P., Nikolaus, S., et al. (2006). Sirtuin 1 (SIRT1) sequence variation is not associated with exceptional human longevity. *Exp Gerontol* 41(1), 98–102.
Fleming, M. A., Potter, J. D., Ramirez, C. J., et al. (2003). Understanding missense mutations in the BRCA1 gene: an evolutionary approach. *Proc Natl Acad Sci U S A* 100(3), 1151–6.
Fontana, L. and Klein, S. (2007). Aging, adiposity, and calorie restriction. *JAMA* 297(9), 986–94.
Fontana, L., Partridge, L., and Longo, V. D. (2010). Extending healthy life span—from yeast to humans. *Science* 328(5976), 321–6.
Fox, C. W., Scheibly, K. L., Wallin, W. G., et al. (2006). The genetic architecture of life span and mortality rates: gender and species differences in inbreeding load of two seed-feeding beetles. *Genetics* 174(2), 763–73.
Fraga, M. F., Ballestar, E., Paz, M. F., et al. (2005). Epigenetic differences arise during the lifetime of monozygotic twins. *Proc Natl Acad Sci U S A* 102(30), 10604–9.
Frank, S. A. (2004). Age-specific acceleration of cancer. *Curr Biol* 14(3), 242–6.
Frank, S. A. (2005). Age-specific incidence of inherited versus sporadic cancers: a test of the multistage theory of carcinogenesis. *Proc Natl Acad Sci U S A* 102(4), 1071–5.
Friedman, L. S., Ostermeyer, E. A., Szabo, C. I., et al. (1994). Confirmation of BRCA1 by analysis of germline mutations linked to breast and ovarian cancer in ten families. *Nat Genet* 8(4), 399–404.
Ganguli, M. (2015). Cancer and dementia: it's complicated. *Alzheimer Dis Assoc Disord* 29(2), 177–82.
Giannakou, M. E., Goss, M., Jünger, M. A., et al. (2004). Long-lived *Drosophila* with overexpressed dFOXO in adult fat body. *Science* 305(5682), 361–1.
Godar, D. E. (2011). Worldwide increasing incidences of cutaneous malignant melanoma. *J Skin Cancer* 2011(12), 1–6.
Godar, D. E., Landry, R. J., and Lucas, A. D. (2009). Increased UVA exposures and decreased cutaneous vitamin D3 levels may be responsible for the increasing incidence of melanoma. *Med Hypotheses* 72(4), 434–43.
Goldsmith, T. C. (2008). Aging, evolvability, and the individual benefit requirement; medical implications of aging theory controversies. *J Theor Biol* 252(4), 764–8.
Gomes, N. M. V., Ryder, O. A., Houck, M. L., et al. (2011). The comparative biology of mammalian telomeres: hypotheses on ancestral states and the roles of telomeres in longevity determination. *Aging Cell* 10(5), 761–8.
Gompertz, B. (1825). On the nature of the function expressive of the law of human mortality, and on a new mode of determining the value of life contingencies. *Philos Trans R Soc Lond* 115, 513–83.
Gong, Y., Thompson, J. N., and Woodruff, R. C. (2006). Effect of deleterious mutations on life span in *Drosophila melanogaster*. *J Gerontol A Biol Sci Med Sci* 61(12), 1246–52.
Good, T. P. and Tatar, M. (2001). Age-specific mortality and reproduction respond to adult dietary restriction in *Drosophila melanogaster*. *J Insect Physiol* 47(12), 1467–73.
Goodnight, C. J. and Stevens, L. (1997). Experimental studies of group selection: what do they tell us about group selection in nature? *Am Nat* 150(S1), S59–79.
Goodrick, C. L., Ingram, D. K., Reynolds, M. A., et al. (1990). Effects of intermittent feeding upon body weight and lifespan in inbred mice: interaction of genotype and age. *Mech Ageing Dev* 55(1), 69–87.
Grandison, R. C., Piper, M. D. W., Partridge, L. (2009). Amino-acid imbalance explains extension of lifespan by dietary restriction in *Drosophila*. *Nature* 462(7276), 1061–4.
Greaves, M. (2007). Darwinian medicine: a case for cancer. *Nature Reviews Cancer* 7(3), 213–21.
Greaves, M. (2008). Cancer: evolutionary origins of vulnerability. In: Stearns, S. C. and Koella, J. C. (eds) *Evolution in Health and Disease*, 2nd ed. Oxford: Oxford University Press, pp. 277–88.

Greer, E. L. and Brunet, A. (2009). Different dietary restriction regimens extend lifespan by both independent and overlapping genetic pathways in *C. elegans*. *Aging Cell* 8(2), 113–27.
Guarente, L. (2000). Sir2 links chromatin silencing, metabolism, and aging. *Genes Dev* 14(9), 1021–6.
Guarente, L. (2013). Calorie restriction and sirtuins revisited. *Genes Dev* 27(19), 2072–85.
Guo, Z., Toichi, E., Hosono, M., et al. (2000). Genetic analysis of lifespan in hybrid progeny derived from the SAMP1 mouse strain with accelerated senescence. *Mech Ageing Dev* 118(1–2), 35–44.
Haldane, J. B. S. (1941). *New Paths in Genetics*. London: George Allen & Unwin.
Hamilton, B., Dong, Y., Shindo, M., et al. (2005). A systematic RNAi screen for longevity genes in *C. elegans*. *Genes Dev* 19(13), 1544–55.
Hamilton, W. D. (1966). The moulding of senescence by natural selection. *J Theor Biol* 12(1), 12–45.
Hannum, G., Guinney, J., Zhao, L., et al. (2013). Genome-wide methylation profiles reveal quantitative views of human aging rates. *Mol Cell* 49(2), 359–67.
Hansen, M., Hsu, A.-L., Dillin, A., et al. (2005). New genes tied to endocrine, metabolic, and dietary regulation of lifespan from a *Caenorhabditis elegans* genomic RNAi screen. *PLoS Genetics* 1(1), e17.
Hansen, M., Taubert, S., Crawford, D., et al. (2007). Lifespan extension by conditions that inhibit translation in *Caenorhabditis elegans*. *Aging Cell* 6(1), 95–110.
Harper, J. M., Leathers, C. W., Austad, S. N. (2006). Does caloric restriction extend life in wild mice? *Aging Cell* 5(6), 441–9.
Heijmans, B.T., Tobi, E. W., Stein, A. D., et al. (2008). Persistent epigenetic differences associated with prenatal exposure to famine in humans. *Proc Natl Acad Sci U S A* 105(44), 17046–9.
Herskind, A. M., McGue, M., Holm, N., et al. (1996). The heritability of human longevity: a population-based study of 2872 Danish twin pairs born 1870–1900. *Hum Genet* 97(3), 319–23.
Holzenberger, M., Dupont, J., Ducos, B., et al. (2003). IGF-1 receptor regulates lifespan and resistance to oxidative stress in mice. *Nature* 421(6919), 182–7.
Honjoh, S., Yamamoto, T., Uno, M., et al. (2009). Signalling through RHEB-1 mediates intermittent fasting-induced longevity in *C. elegans*. *Nature* 457(7230), 726–30.
Horvath, S. (2013). DNA methylation age of human tissues and cell types. *Genome Biol* 14(10), 3156.
Horvath, S., Erhart, W., Brosch, M., et al. (2014). Obesity accelerates epigenetic aging of human liver. *Proc Natl Acad Sci U S A* 111(43), 15538–43.
Horvath, S., Garagnani, P., Bacalini, M. G., et al. (2015). Accelerated epigenetic aging in Down syndrome. *Aging Cell* 14(3), 491–5.
Hou, C., Bolt, K., Bergman, A. (2011). A general life history theory for effects of caloric restriction on health maintenance. *BMC Sys Biol* 5(1), 78.
Howard, B. V. and Wylie-Rosett, J. (2002). Sugar and cardiovascular disease. *Circulation* 106(4), 523–7.
Hughes, K. A. and Charlesworth, B. (1994). A genetic analysis of senescence in *Drosophila*. *Nature* 367(6458), 64–6.
Huttley, G. A., Easteal, S., Southey, M. C., et al. (2000). Adaptive evolution of the tumour suppressor BRCA1 in humans and chimpanzees. *Nat Genet* 25(4), 410–13.
Hwangbo, D. S., Gersham, B., Tu, M.-P., et al. (2004). *Drosophila* dFOXO controls lifespan and regulates insulin signaling in brain and fat body. *Nature* 429(6991), 562–6.
Igbavboa, U., Sun, G. Y., Weisman, G. A., et al. (2009). Amyloid β-protein stimulates trafficking of cholesterol and caveolin-1 from the plasma membrane to the Golgi complex in mouse primary astrocytes. *Neuroscience* 162(2), 328–38.
Ilveskoski, E., Perola, M., Lehtimäki, T., et al. (1999). Age-dependent association of apolipoprotein E genotype with coronary and aortic atherosclerosis in middle-aged men: an autopsy study. *Circulation* 100(6), 608–13.
Issa, J.-P. (2014). Aging and epigenetic drift: a vicious cycle. *J Clin Invest* 124(1), 24–9.
Jablonski, N. G. (1999). A possible link between neural tube defects and ultraviolet light exposure. *Med Hypotheses* 52(6), 581–2.
Jani, B. and Rajkumar, C. (2006). Ageing and vascular ageing. *Postgrad Med J* 82(968), 357–62.
Jasienska, G., Ellison, P. T., Galbarczyk, A., et al. (2015). Apolipoprotein E (ApoE) polymorphism is related to differences in potential fertility in women: a case of antagonistic pleiotropy? *Proc Biol Sci* 282(1803), 20142395.

Jenkins, N. L., McColl, G., Lithgow, G. J. (2004). Fitness cost of extended lifespan in *Caenorhabditis elegans*. *Proc R Soc Lon B Biol Sci* 271(1556), 2523–6.
Jia, K., Chen, D., Riddle, D. L. (2004). The TOR pathway interacts with the insulin signaling pathway to regulate *C. elegans* larval development, metabolism and life span. *Development* 131(16), 3897–906.
Jiang, N., Du, G., Tobias, E., et al. (2013). Dietary and genetic effects on age-related loss of gene silencing reveal epigenetic plasticity of chromatin repression during aging. *Aging* 5(11), 813–24.
Jochemsen, H. M., Muller, M., van der Graaf, Y., et al. (2012). APOE ε4 differentially influences change in memory performance depending on age. The SMART-MR study. *Neurobiol Aging* 33(4), 832.e15–22.
Johnson, T. E. (1990). Increased life-span of age-1 mutants in *Caenorhabditis elegans* and lower Gompertz rate of aging. *Science* 249(4971), 908–12.
Johnson, T. E. (2006). Recent results: biomarkers of aging. *Exp Gerontol* 41(12), 1243–6.
Jousilahti, P., Vartiainen, E., Tuomilehto, J., et al. (1999). Sex, age, cardiovascular risk factors, and coronary heart disease. *Circulation* 99(9), 1165–72.
Kaati, G., Bygren, L. O., Edvinsson, S. (2002). Cardiovascular and diabetes mortality determined by nutrition during parents' and grandparents' slow growth period. *Eur J Hum Genet* 10(11), 682–8.
Kaeberlein, M. and Kennedy, B. K. (2011). Hot topics in aging research: protein translation and TOR signaling, 2010. *Aging Cell* 10(2), 185–90.
Kaeberlein, M. and Powers, R. W., III. (2007). Sir2 and calorie restriction in yeast: a skeptical perspective. *Ageing Research Reviews* 6(2), 128–40.
Kaeberlein, M., McVey, M., Guarente, L. (1999). The SIR2/3/4 complex and SIR2 alone promote longevity in *Saccharomyces cerevisiae* by two different mechanisms. *Genes Dev* 13(19), 2570–80.
Kaeberlein, M., Powers, R. W., Steffen, K. K., et al. (2005). Regulation of yeast replicative life span by TOR and Sch9 in response to nutrients. *Science* 310(5751), 1193–6.
Kamboh, M.I., Demirci, F. Y., Wang, X., et al. (2012). Genome-wide association study of Alzheimer's disease. *Transl Psychiatry* 2(5), e117.
Kanfi, Y., Naiman, S., Amir, G., et al. (2012). The sirtuin SIRT6 regulates lifespan in male mice. *Nature* 483(7388), 218–21.
Kang, T.-W., Yevsa, T., Woller, N., et al. (2011). Senescence surveillance of pre-malignant hepatocytes limits liver cancer development. *Nature* 479(7374), 547–51.
Kapahi, P., Zid, B. M., Harper, T., et al. (2004). Regulation of lifespan in *Drosophila* by modulation of genes in the TOR signaling pathway. *Curr Biol* 14(10), 885–90.
Kaplan, H., Thompson, R. C., Trumble, B. C., et al. (2017). Coronary atherosclerosis in indigenous South American Tsimane: a cross-sectional cohort study. *Lancet* 389(10080), 1730–9.
Karasik, D., Demissie, S., Cupples, L. A., et al. (2005). Disentangling the genetic determinants of human aging: biological age as an alternative to the use of survival measures. *J Gerontol A Biol Sci Med Sci* 60(5), 574–87.
Kenyon, C. (2005). The plasticity of aging: insights from long-lived mutants. *Cell* 120(4), 449–60.
Kenyon, C. J. (2010). The genetics of ageing. *Nature* 464(7288), 504–12.
Kenyon, C., Chang, J., Gensch, E., et al. (1993). A *C. elegans* mutant that lives twice as long as wild type. *Nature* 366(6454), 461–4.
Kerber, R. A., O'Brien, E., Boucher, K. M., et al. (2012). A genome-wide study replicates linkage of 3p22-24 to extreme longevity in humans and identifies possible additional loci. *PLoS One* 7(4), e34746.
Khazaeli, A. A. and Curtsinger, J. W. (2010). Life history variation in an artificially selected population of *Drosophila melanogaster*: pleiotropy, superflies, and age-specific adaptation. *Evolution* 64(12), 3409–16.
Khazaeli, A. A. and Curtsinger, J. W. (2013). Pleiotropy and life history evolution in *Drosophila melanogaster*: uncoupling life span and early fecundity. *J Gerontol A Biol Sci Med Sci* 68(5), 546–53.
Khazaeli, A. A., Pletcher, S. D., Curtsinger, J. W. (1998). The fractionation experiment: reducing heterogeneity to investigate age-specific mortality in *Drosophila*. *Mech Ageing Dev* 105(3), 301–17.
Kirkland, J. L., Tchkonia, T., Zhu, Y., et al. (2017). The clinical potential of senolytic drugs. *J Am Geriatr Soc* 6, a025908.
Kirkwood, T. B. L. (1977). Evolution of ageing. *Nature* 270(5635), 301–4.
Kirkwood, T. B. L. and Melov, S. (2011). On the programmed/non-programmed nature of ageing within the life history. *Curr Biol* 21(18), R701–7.

Kivipelto, M., Helkala, E. L., Hänninen, T., et al. (2001). Midlife vascular risk factors and late-life mild cognitive impairment: a population-based study. *Neurology* 56(12), 1683–9.

Kojima, T., Kamei, H., Aizu, T., et al. (2004). Association analysis between longevity in the Japanese population and polymorphic variants of genes involved in insulin and insulin-like growth factor 1 signaling pathways. *Exp Gerontol* 39(11–12), 1595–8.

Kowald, A. and Kirkwood, T. B. L. (2016). Can aging be programmed? A critical literature review. *Aging Cell* 15(6), 986–98.

Krishnan, V., Chow, M. Z. Y., Wang, Z., et al. (2011). Histone H4 lysine 16 hypoacetylation is associated with defective DNA repair and premature senescence in Zmpste24-deficient mice. *Proc Natl Acad Sci* 108(30), 12325–30.

Lambe, M., Hsieh, C.-C., Chan, H.-W., et al. (1996). Parity, age at first and last birth, and risk of breast cancer: a population-based study in Sweden. *Breast Cancer Res Treat* 38(3), 305–11.

Lamming, D. W., Ye, L., Katajisto, P., et al. (2012). Rapamycin-induced insulin resistance is mediated by mTORC2 loss and uncoupled from longevity. *Science* 335(6076), 1638–43.

Lang, D. H., Gerhard, G. S., Griffith, J. W., et al. (2010). Quantitative trait loci (QTL) analysis of longevity in C57BL/6J by DBA/2J (BXD) recombinant inbred mice. *Aging Clin Exp Res* 22(1), 8–19.

Launer, L. J., Masaki, K., Petrovitch, H., et al. (1995). The association between midlife blood pressure levels and late-life cognitive function: the Honolulu–Asia Aging Study. *JAMA* 274(23), 1846–51.

Leibson, C. L., Rocca, W. A., Hanson, V. A., et al. (1997). The risk of dementia among persons with diabetes mellitus: a population-based cohort study. *Ann N Y Acad Sci* 826, 422–7.

Leips, J., Gilligan, P., and Mackay, T. F. C. (2006). Quantitative trait loci with age-specific effects on fecundity in *Drosophila melanogaster*. *Genetics* 172(3), 1595–605.

Lescai, F., Blanché, H., Nebel, A., et al. (2009). Human longevity and 11p15.5: a study in 1321 centenarians. *Eur J Hum Genet* 17(11), 1515–19.

Li, J.-M., Liu, C., Hu, X., et al. (2014). Inverse correlation between Alzheimer's disease and cancer: implication for a strong impact of regenerative propensity on neurodegeneration? *BMC Neurol* 14(1), e47.

Liao, C.-Y., Rikke, B. A., Johnson, T. E., et al. (2010). Genetic variation in the murine lifespan response to dietary restriction: from life extension to life shortening. *Aging Cell* 9(1), 92–5.

Libert, S., Bonkowski, M. S., Pointer, K., et al. (2012). Deviation of innate circadian period from 24 h reduces longevity in mice. *Aging Cell* 11(5), 794–800.

Libertini, G. (1988). An adaptive theory of the increasing mortality with increasing chronological age in populations in the wild. *J Theoret Biol* 132(2), 145–62.

Lin, S.-J., Defossez, P.-A., Guarente, L. (2000). Requirement of NAD and SIR2 for life-span extension by calorie restriction in *Saccharomyces cerevisiae*. *Science* 289(5487), 2126–8.

Loomis, W. F. (1967). Skin-pigment regulation of vitamin-D biosynthesis in man. *Science* 157(3788), 501–6.

López-Otín, C., Blasco, M. A., Partridge, L., et al. (2013). The hallmarks of aging. *Cell* 153(6), 1194–217.

Luchsinger, J. A., Tang, M. X., Stern, Y., et al. (2001). Diabetes mellitus and risk of Alzheimer's disease and dementia with stroke in a multiethnic cohort. *Am J Epidemiol* 154(7), 635–41.

Luckinbill, L. S., Arking, R., Clare, M. J., et al. (1984). Selection for delayed senescence in *Drosophila melanogaster*. *Evolution* 38(5), 996.

Luo, Y., Boton, B., Damore, M. A., et al. (2003). BACE1 (β-secretase) knockout mice do not acquire compensatory gene expression changes or develop neural lesions over time. *Neurobiol Dis* 14(1), 81–8.

Luong, N., Davies, C. R., Wessells, R. J., et al. (2006). Activated FOXO-mediated insulin resistance is blocked by reduction of TOR activity. *Cell Metabol* 4(2), 133–42.

Ma, L.-L., Yu, J.-T., Wang, H.-F., et al. (2014). Association between cancer and Alzheimer's disease: systematic review and meta-analysis. *J Alzheimer's Dis* 42(2), 565–73.

MacMahon, B., Cole, P., Lin, T. M., et al. (1970). Age at first birth and breast cancer risk. *Bull World Health Org* 43(2), 209–21.

Maegawa, S., Hinkal, G., Kim, H. S., et al. (2010). Widespread and tissue specific age-related DNA methylation changes in mice. *Genome Res* 20(3), 332–40.

Mahley, R. W. and Huang, Y. (1999). Apolipoprotein E: from atherosclerosis to Alzheimer's disease and beyond. *Curr Opin Lipidol* 10(3), 207–17.

Maier, B. (2004). Modulation of mammalian life span by the short isoform of p53. *Genes Dev* 18(3), 306–19.

Mair, W., Goymer, P., Pletcher, S. D., et al. (2003). Demography of dietary restriction and death in *Drosophila*. *Science* 301(5640), 1731–3.

Malkin, D. (2011). Li-Fraumeni syndrome. *Genes Cancer* 2(4), 475–84.

Maloney, B. and Lahiri, D. K. (2011). The Alzheimer's amyloid β-peptide (Aβ) binds a specific DNA Aβ-interacting domain (AβID) in the APP, BACE1, and APOE promoters in a sequence-specific manner: characterizing a new regulatory motif. *Gene* 488(1–2), 1–12.

Mannick, J. B., Del Giudice, G., Lattanzi, M., et al. (2014). mTOR inhibition improves immune function in the elderly. *Sci Transl Med* 6(268), 268ra179.

Manolio, T. A., Collins, F. S., Cox, N. J., et al. (2009). Finding the missing heritability of complex diseases. *Nature* 461(7265), 747–53.

Marchbanks, P. A., McDonald, J. A., Wilson, H. G., et al. (2002). Oral contraceptives and the risk of breast cancer. *N Engl J Med* 346(26), 2025–32.

Marioni, R. E., Shah, S., McRae, A. F., et al. (2015). DNA methylation age of blood predicts all-cause mortality in later life. *Genome Biol* 16(1), 25.

Martin, G. M. (2002). Help wanted: physiologists for research on aging. *Sci Aging Knowledge Environ* 2002(9), vp2.

Martin, G. M. (2005). Epigenetic drift in aging identical twins. *Proc Natl Acad Sci U S A* 102(30), 10413–14.

Martin, G. (2006). Keynote lecture: an update on the what, why and how questions of ageing. *Exp Gerontol* 41(5), 460–3.

Martins, A. C. R. (2011). Change and aging senescence as an adaptation. *PLoS One* 6(9), e24328.

Martínez, D. E. (1998). Mortality patterns suggest lack of senescence in hydra. *Exp Gerontol* 33(3), 217–25.

Masoro, E. J. (2005). Overview of caloric restriction and ageing. *Mech Ageing Dev* 126(9), 913–22.

Matecic, M., Smith, D. L., Jr, Pan, X., et al. (2010). A microarray-based genetic screen for yeast chronological aging factors. *PLoS Genet* 6(4), e1000921.

McEwen, B. S. and Seeman, T. (1999). Protective and damaging effects of mediators of stress: elaborating and testing the concepts of allostasis and allostatic load. *Ann N Y Acad Sci* 896(1), 30–47.

Medawar, P. B. (1952). *An Unsolved Problem of Biology*. London: H. K. Lewis.

Mendis, S. and Puska, P. (2011). *WHO Global Atlas on Cardiovascular Disease Prevention and Control*. Geneva: World Health Organization.

Metaxakis, A. and Partridge, L. (2013). Dietary restriction extends lifespan in wild-derived populations of *Drosophila melanogaster*. *PLoS One* 8(9), e74681.

Micha, R., Michas, G., Mozaffarian, D. (2012). Unprocessed red and processed meats and risk of coronary artery disease and type 2 diabetes—an updated review of the evidence. *Curr Atheroscler Rep* 14(6), 515–24.

Miki, Y., Swensen, J., Shattuck-Eidens, D., et al. (1994). A strong candidate for the breast and ovarian cancer susceptibility gene BRCA1. *Science* 266(5182), 66–71.

Miller, R. A., Chrisp, C., Jackson, A. U., et al. (1998). Marker loci associated with life span in genetically heterogeneous mice. *J Gerontol A Biol Sci Med Sci* 53(4), M257–63.

Miller, R. A., Buehner, G., Chang, Y., et al. (2005). Methionine-deficient diet extends mouse lifespan, slows immune and lens aging, alters glucose, T4, IGF-I and insulin levels, and increases hepatocyte MIF levels and stress resistance. *Aging Cell* 4(3), 119–25.

Mitteldorf, J. and Martins, A. C. R. (2014). Programmed life span in the context of evolvability. *Am Nat* 184(3), 289–302.

Mitteldorf, J. and Pepper, J. (2009). Senescence as an adaptation to limit the spread of disease. *J Theor Biol* 260(2), 186–95.

Moorad, J. A. and Promislow, D. E. L. (2008). A theory of age-dependent mutation and senescence. *Genetics* 179(4), 2061–73.

Moorad, J. A. and Promislow, D. E. L. (2009). What can genetic variation tell us about the evolution of senescence? *Proc Biol Sci* 276(1665), 2271–8.

Moreira, P. I., Carvalho, C., Zhu, X., et al. (2010). Mitochondrial dysfunction is a trigger of Alzheimer's disease pathophysiology. *Biochim Biophys Acta* 1802(1), 2–10.

Morimoto, L. M., White, E., Chen, Z., et al. (2002). Obesity, body size, and risk of postmenopausal breast cancer: the Women's Health Initiative (United States). *Cancer Causes Control* 13(8), 741–51.

Narasimhan, S. D., Yen, K., Tissenbaum, H. A. (2009). Converging pathways in lifespan regulation. *Curr Biol* 19(15), R657–66.

Nebel, A., Kleindorp, R., Caliebe, A., et al. (2011). A genome-wide association study confirms APOE as the major gene influencing survival in long-lived individuals. *Mech Ageing Dev* 132(6–7), 324–30.

Newman, A. B., Walter, S., Lunetta, K. L., et al. (2010). A meta-analysis of four genome-wide association studies of survival to age 90 years or older: the Cohorts for Heart and Aging Research in Genomic Epidemiology Consortium. *J Gerontol A Biol Sci Med Sci* 65(5), 478–87.

Nuzhdin, S. V., Pasyukova, E. G., Dilda, C. L., et al. (1997). Sex-specific quantitative trait loci affecting longevity in *Drosophila melanogaster*. *Proc Natl Acad Sci U S A* 94(18), 9734–9.

Osborne, D. L. and Hames, R. (2014). A life history perspective on skin cancer and the evolution of skin pigmentation. *Am J Phys Anthropol* 153(1), 1–8.

Ott, A., Stolk, R. P., van Harskamp, F., et al. (1999). Diabetes mellitus and the risk of dementia: the Rotterdam Study. *Neurology* 53(9), 1937–42.

Painter, R. C., Roseboom, T. J., Bleker, O. P. (2005). Prenatal exposure to the Dutch famine and disease in later life: an overview. *Reprod Toxicol* 20(3), 345–52.

Partridge, L. and Barton, N. H. (1996). On measuring the rate of ageing. *Proc Biol Sci* 263(1375), 1365–71.

Partridge, L. and Fowler, K. (1992). Direct and correlated responses to selection on age at reproduction in *Drosophila melanogaster*. *Evolution* 46(1), 76.

Partridge, L. and Gems, D. (2002a). Mechanisms of aging: public or private? *Nat Rev Genet* 3(3), 165–75.

Partridge, L. and Gems, D. (2002b). Mechanisms of aging: public or private? *Nat Rev Genet* 3(3), 165–75.

Partridge, L., Pletcher, S. D., and Mair, W. (2005). Dietary restriction, mortality trajectories, risk and damage. *Mech Ageing Dev* 126(1), 35–41.

Pavard, S. and Metcalf, C. J. E. (2007). Negative selection on BRCA1 susceptibility alleles sheds light on the population genetics of late-onset diseases and aging theory. *PLoS One* 2(11), e1206.

Pavlicek, A. (2004). Evolution of the tumor suppressor BRCA1 locus in primates: implications for cancer predisposition. *Hum Mol Genet* 13(22), 2737–51.

Pianka, E. R. and Parker, W. S. (1975). Age-specific reproductive tactics. *Am Nat* 109(968), 453–64.

Piper, M. D. W. and Partridge, L. (2007). Dietary restriction in *Drosophila*: delayed aging or experimental artefact? *PLoS Genet* 3(4), e57.

Pisciotta, L., Cantafora, A., Piana, A., et al. (2003). Physical activity modulates effects of some genetic polymorphisms affecting cardiovascular risk in men aged over 40 years. *Nutr Metab Cardiovasc Dis* 13(4), 202–10.

Pletcher, S. D., Houle, D., Curtsinger, J. W. (1998). Age-specific properties of spontaneous mutations affecting mortality in *Drosophila melanogaster*. *Genetics* 148(1), 287–303.

Pletcher, S. D., Houle, D., Curtsinger, J. W. (1999). The evolution of age-specific mortality rates in *Drosophila melanogaster*: genetic divergence among unselected lines. *Genetics* 153(2), 813–23.

Plun-Favreau, H., Lewis, P. A., Hardy, J., et al. (2010). Cancer and neurodegeneration: between the devil and the deep blue sea. *PLoS Genet* 6(12), e1001257.

Powers, R. W., Kaeberlein, M., Caldwell, S. D., et al. (2006). Extension of chronological life span in yeast by decreased TOR pathway signaling. *Genes Dev* 20(2), 174–84.

Prieur, A. and Peeper, D. S. (2008). Cellular senescence in vivo: a barrier to tumorigenesis. *Curr Opin Cell Biol* 20(2), 150–5.

Promislow, D. E. L. (1991). Senescence in natural populations of mammals: a comparative study. *Evolution* 45(8), 1869.

Promislow, D. E. L. and Harvey, P. H. (1990). Living fast and dying young: a comparative analysis of life-history variation among mammals. *J Zool* 220(3), 417–37.

Raichlen, D. A. and Alexander, G. E. (2014). Exercise, APOE genotype, and the evolution of the human lifespan. *Trends Neurosci* 37(5), 247–55.

Rechavi, O., Houri-Ze'evi, L., Anava, S., et al. (2014). Starvation-induced transgenerational inheritance of small RNAs in *C. elegans*. *Cell* 158(2), 277–87.

Reeve, A., Simcox, E., Turnbull, D. (2014). Ageing and Parkinson's disease: why is advancing age the biggest risk factor? *Ageing Res Rev* 14, 19–30.

Reitz, C., Brayne, C., Mayeux, R. (2011). Epidemiology of Alzheimer disease. *Nat Rev Neurol* 7(3), 137–52.

Reya, T., Morrison, S. J., Clarke, M. F., et al. (2001). Stem cells, cancer, and cancer stem cells. *Nature* 414(6859), 105–11.

Reznick, D. N., Bryant, M. J., Roff, D., et al. (2004). Effect of extrinsic mortality on the evolution of senescence in guppies. *Nature* 431(7012), 1095–9.

Rikke, B. A., Liao, C.-Y., McQueen, M. B., et al. (2010). Genetic dissection of dietary restriction in mice supports the metabolic efficiency model of life extension. *Exp Gerontol* 45(9), 691–701.

Rogina, B. and Helfand, S. L. (2004). Sir2 mediates longevity in the fly through a pathway related to calorie restriction. *Proc Natl Acad Sci U S A* 101(45), 15998–6003.

Rose, G., Dato, S., Altomare, K., et al. (2003). Variability of the SIRT3 gene, human silent information regulator Sir2 homologue, and survivorship in the elderly. *Exp Gerontol* 38(10), 1065–70.

Rose, M. R. (1984). Laboratory evolution of postponed senescence in *Drosophila melanogaster*. *Evolution* 38(5), 1004.

Rose, M. R. (1994). *Evolutionary Biology of Aging*. Oxford: Oxford University Press.

Rufini, A., Tucci, P., Celardo, I., et al. (2013). Senescence and aging: the critical roles of p53. *Oncogene* 32(43), 5129–43.

Russo, J., Moral, R., Balogh, G. A., et al. (2005). The protective role of pregnancy in breast cancer. *Breast Cancer Res* 7(3), 131–42.

Rusted, J. M., Evans, S. L., King, S. L., et al. (2013). APOE e4 polymorphism in young adults is associated with improved attention and indexed by distinct neural signatures. *NeuroImage* 65, 364–73.

Samuelson, A. V., Klimczak, R. R., Thompson, D. B., et al. (2007). Identification of *Caenorhabditis elegans* genes regulating longevity using enhanced RNAi-sensitive strains. *Cold Spring Harb Symp Quant Biol* 72(1), 489–97.

Santulli, G. (2013). Epidemiology of cardiovascular disease in the 21st century: updated numbers and updated facts. *J Cardiovasc Dis* 1(1), 1.

Satoh, A., Brace, C. S., Rensing, N., et al. (2013). Sirt1 extends life span and delays aging in mice through the regulation of Nk2 homeobox 1 in the DMH and LH. *Cell Metabolism* 18(3), 416–30.

Schleit, J., Johnson, S. C., Bennett, C. F., et al. (2013). Molecular mechanisms underlying genotype-dependent responses to dietary restriction. *Aging Cell* 12(6), 1050–61.

Sebastiani, P., Solovieff, N., DeWan, A. T., et al. (2012). Genetic signatures of exceptional longevity in humans. *PLoS One* 7(1), e29848.

Sebastiani, P., Bae1, H., Sun, F. X., et al. (2013). Meta-analysis of genetic variants associated with human exceptional longevity. *Aging (Albany NY)* 5(9), 653–61.

Seeman, T. E., Singer, B. H., Rowe, J. W., et al. (1997). Price of adaptation—allostatic load and its health consequences. MacArthur studies of successful aging. *Arch Intern Med* 157(19), 2259–68.

Selman, C., Lingard, S., Choudhury, A. I., et al. (2008). Evidence for lifespan extension and delayed age-related biomarkers in insulin receptor substrate 1 null mice. *FASEB J* 22(3), 807–18.

Selman, C., Tullet, J. M. A., Wieser, D., et al. (2009). Ribosomal protein S6 kinase 1 signaling regulates mammalian life span. *Science* 326(5949), 140–4.

Shah, P. P., Donahue, G., Otte, G. L., et al. (2013). Lamin B1 depletion in senescent cells triggers large-scale changes in gene expression and the chromatin landscape. *Genes Dev* 27(16), 1787–99.

Shanley, D. P. and Kirkwood, T. B. (2000). Calorie restriction and aging: a life-history analysis. *Evolution* 54(3), 740–50.

Shattuck, M. R. and Williams, S. A. (2010). Arboreality has allowed for the evolution of increased longevity in mammals. *Proc Natl Acad Sci U S A* 107(10), 4635–9.

Shi, H.-B., Tang, B., Liu, Y.-W., et al. (2014). Alzheimer disease and cancer risk: a meta-analysis. *J Cancer Res Clin Oncol* 141(3), 485–94.

Shock, N. (1984). *Normal Human Aging: The Baltimore Longitudinal Study of Aging*. Washington, DC: NIH Publication.

Skulachev, V. P. (1997). Aging is a specific biological function rather than the result of a disorder in complex living systems: biochemical evidence in support of Weismann's hypothesis. *Biochem Biokhim* 62(11), 1191–5.

Smith, E. D., Kennedy, B. K., Kaeberlein, M. (2007). Genome-wide identification of conserved longevity genes in yeast and worms. *Mech Ageing Dev* 128(1), 106–11.

Smith, F. M., Garfield, A. S., Ward, A. (2006). Regulation of growth and metabolism by imprinted genes. *Cytogenet Genome Res* 113(1–4), 279–91.

Smith, K. R., Hanson, H. A., Mineau, G. P., et al. (2011). Effects of BRCA1 and BRCA2 mutations on female fertility. *Proc Biol Sci* 279(1732), 1389–95.

Sniegowski, P. (1997). Evolution: setting the mutation rate. *Curr Biol* 7(8), R487–88.

Soscia, S. J., Kirby, J. E., Washicosky, K. J., et al. (2010). The Alzheimer's disease-associated amyloid beta-protein is an antimicrobial peptide. *PLoS One* 5(3), e9505.

Spencer, C. C., Howell, C. E., Wright, A. R., et al. (2003). Testing an 'aging gene' in long-lived *Drosophila* strains: increased longevity depends on sex and genetic background. *Aging Cell* 2(2), 123–30.

Staropoli, J. F. (2008). Tumorigenesis and neurodegeneration: two sides of the same coin? *BioEssays* 30(8), 719–27.

Stengård, J. H., Weiss, K. M., Sing, C. F. (1998). An ecological study of association between coronary heart disease mortality rates in men and the relative frequencies of common allelic variations in the gene coding for apolipoprotein E. *Hum Genet* 103(2), 234–41.

Strassmann, B. I. (1999). Menstrual cycling and breast cancer: an evolutionary perspective. *J Women's Health* 8(2), 193–202.

Suh, Y., Atzmon, G., Cho, M.-O., (2008). Functionally significant insulin-like growth factor I receptor mutations in centenarians. *Proc Natl Acad Sci U S A* 105(9), 3438–42.

Sun, L., Sadighi Akha, A. A., Miller, R. A., (2009). Life-span extension in mice by preweaning food restriction and by methionine restriction in middle age. *J Gerontol A Biol Sci Med Sci* 64A(7), 711–22.

Susser, E. S. and Lin, S. P. (1992). Schizophrenia after prenatal exposure to the Dutch hunger winter of 1944–1945. *Arch Gen Psychiatry* 49(12), 983–8.

Swindell, W. R. (2012). Dietary restriction in rats and mice: a meta-analysis and review of the evidence for genotype-dependent effects on lifespan. *Ageing Res Rev* 11(2), 254–70.

Tan, Q., Heijmans, B. T., Hjelmborg, J. V. B., et al. (2016). Epigenetic drift in the aging genome: a ten-year follow-up in an elderly twin cohort. *Int J Epidemiol* 45(4), 1146–58.

Tanzi, R. E. (2012). The genetics of Alzheimer disease. *Cold Spring Harb Perspect Med* 2(10), a006296.

Tatar, M., Kopelman, A., Epstein, D., et al. (2001). A mutant *Drosophila* insulin receptor homolog that extends life-span and impairs neuroendocrine function. *Science* 292(5514), 107–10.

Thiery, J. P. (2002). Epithelial–mesenchymal transitions in tumour progression. *Nat Rev Cancer* 2(6), 442–54.

Tissenbaum, H. A. and Guarente, L. (2001). Increased dosage of a sir-2 gene extends lifespan in *Caenorhabditis elegans*. *Nature* 410(6825), 227–30.

Titus-Ernstoff, L., Perez, K., Cramer, D. W., et al. (2001). Menstrual and reproductive factors in relation to ovarian cancer risk. *Br J Cancer* 84(5), 714–21.

Toiber, D., Erdel, F., Bouazoune, K., et al. (2013). SIRT6 recruits SNF2H to DNA break sites, preventing genomic instability through chromatin remodeling. *Mol Cell* 51(4), 454–68.

Topczewska, J. M., Postovit, L.-M., Margaryan, N. V., et al. (2006). Embryonic and tumorigenic pathways converge via nodal signaling: role in melanoma aggressiveness. *Nat Med* 12(8), 925–32.

Travis, J. M. J. (2004). The evolution of programmed death in a spatially structured population. *J Gerontol A Biol Sci Med Sci* 59(4), 301–5.

Tu, M.-P., Epstein, D., Tatar, M. (2002). The demography of slow aging in male and female *Drosophila* mutant for the insulin-receptor substrate homologue chico. *Aging Cell* 1(1), 75–80.

Tyner, S. D., Venkatachalam, S., Choi, J., et al. (2002). p53 mutant mice that display early ageing-associated phenotypes. *Nature* 415(6867), 45–53.

Urfer, S. R., Kaeberlein, T. L., Mailheau, S., et al. (2017). A randomized controlled trial to establish effects of short-term rapamycin treatment in 24 middle-aged companion dogs. *GeroScience* 39(2), 117–27.

Vanhecke, T. E., Miller, W. M., Franklin, B. A., et al. (2006). Awareness, knowledge, and perception of heart disease among adolescents. *Eur J Prev Cardiol* 13(5), 718–23.

Van Speybroeck, L. (2002). From epigenesis to epigenetics. *Ann N Y Acad Sci* 981(1), 61–81.

Vellai, T., Takacs-Vellai, K., Zhang, Y., et al. (2003). Genetics: influence of TOR kinase on lifespan in *C. elegans*. *Nature* 426(6967), 620.

Wade, M. J. (2015). A critical review of the models of group selection. *Q Rev Biol* 53(2), 101–14.

Wagner, K.-H., Cameron-Smith, D., Wessner, B., et al. (2016). Biomarkers of aging: from function to molecular biology. *Nutrients* 8(6), 338.

Weidner, C. I., Lin, Q., Koch, C. M., et al. (2014). Aging of blood can be tracked by DNA methylation changes at just three CpG sites. *Genome Biol* 15(2), R24.

Weinberg, R. A. (2013). *The Biology of Cancer*, 2nd ed. Oxford: Garland Science.

Weismann, A. (1899). *Essays upon Heredity and Kindred Biological Problems*. Oxford: Clarendon Press.

Wellen, K. E., Hatzivassiliou, G., Sachdeva, U. M., et al. (2009). ATP-citrate lyase links cellular metabolism to histone acetylation. *Science* 324(5930), 1076–80.

Wen, Y., Yang, S.-H., Liu, R., et al. (2007). Cdk5 is involved in NFT-like tauopathy induced by transient cerebral ischemia in female rats. *Biochim Biophys Acta* 1772(4), 473–83.

Wensink, M. J., Caswell, H., Baudisch, A. (2016). The rarity of survival to old age does not drive the evolution of senescence. *Evol Biol* 44(1), 5–10.

Werfel, J., Ingber, D. E., Bar-Yam, Y. (2015). Programed death is favored by natural selection in spatial systems. *Phys Rev Lett* 114(23), 238103.

Whitmer, R. A., Sidney, S., Selby, J., et al. (2005). Midlife cardiovascular risk factors and risk of dementia in late life. *Neurology* 64(2), 277–81.

Williams, G. C. (1957). Pleiotropy, natural selection, and the evolution of senescence. *Evolution* 11(4), 398.

Williams, P. D. and Day, T. (2003). Antagonistic pleiotropy, mortality source interactions, and the evolutionary theory of senescence. *Evolution* 57(7), 1478–88.

Woodhead, A. D., Setlow, R. B., Tanaka, M. (1999). Environmental factors in nonmelanoma and melanoma skin cancer. *J Epidemiol* 9(6 Suppl), S102–14.

Wright, R. O., Hu, H., Silverman, E. K., et al. (2003). Apolipoprotein E genotype predicts 24-month Bayley Scales infant development score. *Pediatr Res* 54(6), 819–25.

Wu, J. J., Liu, J., Chen, E. B., et al. (2013). Increased mammalian lifespan and a segmental and tissue-specific slowing of aging after genetic reduction of mTOR expression. *Cell Reports* 4(5), 913–20.

Yamaguchi, Y., Brenner, M., Hearing, V. J. (2007). The regulation of skin pigmentation. *J Biol Chem* 282(38), 27557–61.

Yampolsky, L., Pearse, L. E., Promislow, D. E. (2000). Age-specific effects of novel mutations in *Drosophila melanogaster* I. Mortality. *Genetica* 110(1), 11–29.

Yang, Z. and Nielsen, R. (2002). Codon-substitution models for detecting molecular adaptation at individual sites along specific lineages. *Mol Biol Evol* 19(6), 908–17.

Yanos, M. E., Bennett, C. F., Kaeberlein, M. (2012). Genome-wide RNAi longevity screens in *Caenorhabditis elegans*. *Curr Genomics* 13(7), 508–18.

Yao, Z.-X. and Papadopoulos, V. (2002). Function of b-amyloid in cholesterol transport: a lead to neurotoxicity. *FASEB J* 16(12), 1677–9.

Yu, Y. W. Y., Lin, C.-H., Chen, S.-P., et al. (2000). Intelligence and event-related potentials for young female human volunteer apolipoprotein E ε4 and non-ε4 carriers. *Neurosci Lett* 294(3), 179–81.

Zane, L., Sharma, V., Misteli, T. (2014). Common features of chromatin in aging and cancer: cause or coincidence? *Trends Cell Biol* 24(11), 686–94.

Zhang, Q., Guo, S., Zhang, X., et al. (2015). Inverse relationship between cancer and Alzheimer's disease: a systemic review meta-analysis. *Neurol Sci* 36(11), 1987–94.

Zhang, W.-G., Bai, X.-J., Chen, X.-M. (2010). SIRT1 variants are associated with aging in a healthy Han Chinese population. *Clin Chim Acta* 411(21–22), 1679–83.

Zimmerman, J. (2003). Nutritional control of aging. *Exp Gerontol* 38(1–2), 47–52.

Zou, K., Gong, J.-S., Yanagisawa, K., et al. (2002). A novel function of monomeric amyloid β-protein serving as an antioxidant molecule against metal-induced oxidative damage. *J Neurosci* 22(12), 4833–41.

Zwaan, B. J. (1999). The evolutionary genetics of ageing and longevity. *Heredity* 82(6), 589.

Zwaan, B., Bijlsma, R., Hoekstra, R. F. (1995). Direct selection on life span in *Drosophila melanogaster*. *Evolution* 49(4), 649.

CHAPTER 6

NUTRITION, ENERGY EXPENDITURE, PHYSICAL ACTIVITY, AND BODY COMPOSITION

ANN E. CALDWELL, STANLEY BOYD EATON, AND MELVIN KONNER

Abstract

An imbalance of energy intake, expenditure, and storage substantially increases the risk of chronic, deadly conditions, including heart disease, stroke, type 2 diabetes mellitus (T2DM), certain cancers, and depression. Our understanding of the physiological, lifestyle, and environmental factors that contribute to poor energy regulation and balance can be informed in meaningful ways in the context of evolutionary medicine. There is a substantial mismatch, or discordance, between ancestral environments (environments of evolutionary adaptedness (EEAs)) and most modern environments with regards to the availability and accessibility of food, and the connection between physical activity energy expenditure and energy acquisition. Evolutionary medicine can inform research and approaches to reduce the disease burden associated with an imbalance of energy based on the discordance model for diet and activity, and further applying life history theory to understand physical activity. Challenges to the evolutionary approach to chronic disease are thought-provoking, but they do not negate its value. Novel approaches incorporating insights that consider human evolutionary history and integrate across Tinbergen's levels of analysis can lead to the development of interventions that are more compatible with evolved physiology and psychology and environments that are more conducive to lifestyles that reduce the risks of chronic diseases.

Keywords

evolutionary medicine, Palaeolithic diets, modern diets, physical activity, energy balance, hunter-gatherers, discordance model, fetal programming, thrifty phenotype hypothesis, mismatch

6.1 Introduction

Four out of the five leading risk factors for mortality worldwide are related to energy intake, expenditure, and storage (WHO 2009; Hall et al. 2012). An imbalance of these factors is known to increase the risk of at least 22 chronic conditions, including heart disease, stroke, type 2 diabetes mellitus (T2DM), certain cancers, and depression (Booth et al. 2012). These conditions affect people in every country, across racial, ethnic, and socioeconomic lines. They are increasingly prevalent, often fatal, and can be prevented or delayed through lifestyle modifications that influence energy regulation and balance. Our understanding of the physiological, lifestyle, and environmental factors that contribute to poor energy regulation and balance can be informed in meaningful ways in the context of evolutionary medicine.

One emphasis of evolutionary medicine is the mismatch, or discordance, between ancestral environments (environments of evolutionary adaptedness (EEAs)) and most modern environments. Our genome has changed little since the beginnings of agriculture, and is therefore largely selected for pre-agricultural or at least preindustrial conditions (see Sections 6.5.2 and 6.5.7 for notable exceptions). These did not change substantially during 99% of our species' history, but differ markedly from our current circumstances, particularly in the availability and accessibility of food (Eaton and Konner 1988; Egger and Dixon 2009). Agriculture and the Industrial Revolution changed what foods were available, while mechanisation and urbanisation changed patterns of physical activity and eating, and a growing disconnect between energy expenditure and acquisition, largely outpacing genetic evolution. While technological advances have improved health overall through sanitation, public health measures and medical advances, a lack of physical activity is a now considered a key risk factor for two of the leading causes of disease burden worldwide, cardiovascular disease (CVD) and depression (WHO 2008). Sitting is often described as 'the new smoking'.

In a seminal paper, Neel (1962) put forward the thrifty gene hypothesis to explain why mismatches between past and current environments would lead to greater disease prevalence. He suggested that natural selection favoured gene-based mechanisms and traits that promoted survival during times of resource scarcity. However, when paired with abundant access to foods, these formerly adaptive genes increase the risk of overweight, obesity, T2DM, CVD, and the metabolic syndrome. Neel's hypothesis was later expanded to the thrifty phenotype hypothesis by Hales and Barker (1992) to include not only genes, but the ways genes are expressed as phenotypes through gene × environment interactions, which influence insulin signalling and the onset of T2DM. This framework remains particularly strong for understanding the fetal origins of hypertension and diabetes (Vaag et al. 2012). Although genetic evolution is slow, there is a range of phenotypes that can result from the same genome in different environments. Genes, environments, development, and behaviour all shape phenotypes. Genetic changes are not necessary for an organism to develop a very different phenotype when developing and living in an environment with very different energy intake and expenditure than prevailed across human history.

6.1.1 Integrating Tinbergen's Levels of Analysis

Tinbergen proposed four categories of analysis—phylogenetic, functional, developmental, and mechanistic (Bateson and Laland 2013)—and considering these can enhance understanding of human energy metabolism. (For further discussion, see Chapter 1: Core Principles for Evolutionary Medicine.)

A phylogenetic or comparative perspective leads us to examine the ways humans compare with other mammals in physiological traits that influence nutrition, energy expenditure, and body composition. Many of our nutritional requirements, most of our intermediary metabolism, and our digestive physiology have been largely conserved over time (Milton 2000). Primates, and humans in particular, stand out because of our extraordinarily large brains. Brains are more energetically costly than other types of tissue, and about 20% of the body's oxygen and 25% of the glucose is required for human cerebral function. Humans are able to accommodate the higher metabolic demands of a large brain by consuming a nutrient-dense diet, but this, in turn, necessitated high levels of physical activity, as evidenced by the relatively high day ranges exhibited by humans living in preindustrial, subsistence-based economies. (For further discussion, see Chapter 17: Brain, Spinal Cord, and Sensory Systems.) Animal flesh has much more protein per gram than almost all plant matter, and consuming it played an important role in the emergence of *Homo*. As detailed in Chapter 13 (Digestive System), the human gut is substantially smaller than what is predicted based on regressions of gut and body size among primates (Aiello 1997). It is particularly well suited for digesting small volumes of nutritionally dense cooked or preprocessed food that can be absorbed in the small intestine (Milton 2000). Thus humans are unique from a comparative perspective in our large brains, small guts, nutrient-dense diet, and, after industrialisation, the low levels of energy expenditure required to obtain our nutrients.

Tinbergen's analysis stresses the interaction of genes, environment, and development in shaping phenotypes throughout development. This interaction confers a flexibility, or plasticity, that has been favoured by natural selection. An organism's genotype can give rise to a range of phenotypes depending on the environmental conditions experienced throughout development (West-Eberhard 2003). Factors that influence energy metabolism include resource availability, the energy required to extract resources, exposure to pathogens and infections, and sociocultural factors. The resulting phenotypes need not be associated with variation in gene frequency, and are not necessarily heritable, although epigenetic modifications of genes may sometimes be (Giuliani et al. 2015). Rather, they represent outcomes of a system that tended to be adaptive in the environments in which it evolved. Physiological and psychological factors that influence diet and physical activity are affected by environments and experiences during development, as well as those that shaped these mechanisms throughout our evolutionary history.

The uricase gene provides one example of integration across Tinbergen's levels of analysis and environmental mismatch. During the Miocene, there were a series of mutations in the great ape (and human) lineage beginning approximately 24 million years ago (mya) that led to progressive reductions in uricase activity, and the eventual silencing of the uricase gene (about 15 mya; Johnson and Andrews 2010). Silencing of the uricase gene leads to accumulation of uric acid in the blood, and apes are unique among mammals in having become knockouts for this gene. A parallel silencing of uricase also occurred in the lesser apes around 10–12 mya. This parallel silencing suggests that accumulation of uric acid may have conferred adaptive advantages, including promoting fat storage after fructose consumption, and preserving blood pressure in a low-salt state (Watanabe et al. 2002; Johnson and Andrews 2010; Johnson et al. 2017). The earth was experiencing global cooling during this period; thus seasonal fruits would have been more difficult to come by, and extra fat stores may have provided a buffer for longer periods between seasons. In modern, industrialised environments with diets rich in sugar and salt, however, blood urate accumulation increases susceptibility to kidney disease, hypertension, gout, overweight, and obesity. It has therefore been proposed that the uricase gene, or its deletion, is a candidate for what Neel envisioned as a 'thrifty gene'. While all great apes (and some lesser apes) have the same genetic predisposition, the interactions of (no) gene × environment (sugar and salt availability) × behaviour (intake of large amounts of sugar and salt) result in poor health outcomes. This integration is central to evolutionary medicine, and key to understanding the following sections examining nutrition, energy expenditure, and body composition. (For further discussion, see Chapter 14: Excretory System.)

6.2 Nutrition

Suboptimal dietary intake, specifically excess sodium, high intake of processed meat and sugar-sweetened beverages, and insufficient intake of nuts/seeds, seafood, omega-3 fats, vegetables, fibre, and fruits, is related to nearly half (45%) of all cardiometabolic deaths (Micha et al. 2017). Since the first dietary recommendations based on a characterisation of the diet retrodicted for the vast majority of human history were put forward (Eaton and Konner 1985), dietary recommendations made by the American Heart Association have become increasingly and now remarkably similar. (For further discussion, see Chapter 11: Cardiovascular System.)

6.2.1 Palaeolithic Diets: How Can We Characterise Them?

Early estimates of ancestral nutrition were based on then available proximate analyses of foods consumed by modern hunter-gatherers and on data about the diets of current populations who live in pre-industrial conditions with subsistence-based economies (Table 6.1). Subsequent research has improved these characterisations, notably emphasising the importance of plant material during human evolution (Konner and Eaton 2010), while affirming that meat and seafood were major dietary components (Marlowe 2010; Suzman 2017), and detailing a range of diets consumed by modern populations living in pre-industrial conditions. For the purposes of reconstructing ancestral diets, there has been increasing recognition that the broad range of diets consumed by recent foragers living in varying environments includes nutritional 'noise'. Extra-African, pre-agricultural diets demonstrate that humans can survive, grow, and reproduce, over a broad dietary spectrum (as do contemporary humans in affluent societies), some of which is relevant to human evolution (see Section 6.5). However, tolerable is not optimal and, increasingly, the nutrition retrodicted for Africans before the exodus of behaviourally modern humans, say between 100,000 and 50,000 years ago (kya), is being considered the Palaeolithic diet of greatest significance for contemporary dietary recommendations.

One widely accepted aspect of ancestral diets—in Africa and elsewhere—concerns what they lacked: added sugars, highly refined cereal grains, trans fats, and commercial salt. These are major components of contemporary diets increasingly linked to obesity, T2DM, CVD, and hypertension. The antibiotics, hormones, and additives commonly fed to domestic livestock are not found in wild game, and there was little or no heavy-metal

Table 6.1 Dietary Recommendations Compared with Estimates of Ancestral Dietary Intakes for Hunter-Gatherers

Nutrients	Dietary recommendations	Estimated ancestral dietary intake
Carbohydrate (% daily energy)	45–65	35–40
Added sugar (% daily energy)	< 10	2
Fibre (g/day)	38 male; 25 female	> 70
Protein (% daily energy)	10–35	25–30
Fat (% daily energy)	20–35	20–35
Saturated fat (% daily energy)	< 10	7.5–12
Cholesterol (mg/day)	< 300	500+
Eicosapentaenoic acid and docosahexaenoic acid (g/day)	0.65	0.7–6.0
Vitamin C (mg/day)	90 male; 75 female	500
Vitamin D (IU/day)	1000	4000 (sunlight)
Calcium (mg/day)	1000	1000–1500
Sodium (mg/day)	1500	< 1000
Potassium (mg/day)	4700	7000

Source: Reproduced from Melvin Konner and S. Boyd Eaton, Paleolithic Nutrition, Nutrition in Clinical Practice, 25 (6), p. 600, Table 2, doi.org/10.1177/0884533610385702, Copyright © 2010, SAGE Publications. Reprinted by Permission of SAGE Publications, Inc.

contamination of fish and other aquatic foods. Modern cultivars have been bred over centuries to be less fibrous than are their wild relatives; consequently, fibre intake in developed nations is much below that during the African Stone Age.

Retrodictions of the Palaeolithic diet remain imperfect, but are becoming ever more defensible. Evolutionary considerations imply that such understanding is critical for formulating dietary recommendations. Closely approximating what human physiology is best designed for may reduce risks for the 'diseases of civilisation' while promoting optimal health and vigour.

6.2.2 Fat and Saturated Fat

The proportion of total fat in American diets averages approximately 34% (US Department of Agriculture 2016). This is similar to total fat intake for recent African hunter-gatherers (Cordain et al. 1998, 2005; Kuipers et al. 2010). Total fat intake is unrelated to heart disease, obesity, diabetes, and cancer (Oh et al. 2005; Chowdhury 2014). Epidemiological studies have consistently shown an association between trans fats and coronary heart disease (Chowdhury 2014), but the relationship between saturated fat and coronary disease is less clear. Some respected authorities express reservations (Jakobsen et al. 2009; Willett 2014); still, recent meta-analyses appear to show no connection (Chowdhury 2014; de Souza et al. 2015; Malhotra et al. 2017).

In post-industrial diets, the primary sources of saturated fat are red meat together with full-fat dairy and bakery products (according to US Health and Human Services dietary guidelines). Wild game has less saturated fat but more mono- and polyunsaturated fatty acids (PUFAs) than does meat from domesticated animals (Eaton and Konner 1985; Cordain et al. 2002, 2005).

Foragers in natural environments rich in marine plants and animals have PUFA intake nearly twice that observed in post-industrial diets. For them, the ratio of ω-6 to ω-3 fatty acids is 2:1 (Broadhurst et al. 2002; Marean et al. 2007; Kuipers et al. 2010), which is significantly lower than the estimated modern ratio of 20:1 (Simopoulos 2016) or even the currently *recommended* ratio of 8:1 (Trumbo et al. 2002). This implies that Stone Agers in coastal and riverine locales consumed much more ω-3 PUFA relative to ω-6 PUFA than do people in industrialised countries. Before hydrogenation of vegetable oils became common in the 1950s, foods contained little trans fat; intake was negligible for foragers (and everyone else).

6.2.3 Dietary Cholesterol

Recent hunter-gatherers have higher total dietary cholesterol intake (~ 480 mg/day) than is currently recommended (< 300 mg/day) (Konner and Eaton 2010). Dietary cholesterol intake was once thought related to coronary heart disease risk, but recent research has shown that this is not the case. There is a subset of 'responders', about 25% of Americans, in whom dietary cholesterol tends to increase serum cholesterol. However, neither in responders nor in others does dietary cholesterol intake affect risk of coronary heart disease (Fernandez and Calle 2010). The 2015 Dietary Guidelines (US Department of Health and Human Services and US Department of Agriculture 2015) no longer recommend a limit for dietary cholesterol consumption. Instead, cholesterol-raising fatty acids and dietary refined carbohydrates are now considered what should be limited in order to keep serum cholesterol and the low-density lipoprotein (LDL)/high-density lipoprotein (HDL) ratio low (Mensink et al. 2003; Forsythe et al. 2008; Hu et al. 2012).

6.2.4 Protein

As a proportion of total energy, protein intake in the United States (~ 16%; US Department of Agriculture 2016) is at the low end of the range observed among populations currently living in pre-industrial conditions (14–50%; Cordain et al. 2000a, 2000b, 2002; Martin et al. 2012) and well below that observed among many recent hunter-gatherer populations: San 30%, Australian Aborigines 77%, Ache 70%, and Hadza 48% (Suzman 2017). Isotopic analyses of pre-agricultural human skeletal remains indicate similarly high consumption of meat and fish (Hu et al. 2009; Richards 2009). The acceptable range for protein intake has been set at 10–35% (Trumbo et al. 2002), overlapping forager intake levels. The topic of meat consumption raises complex issues: (1) the proportion of high-quality protein, especially animal protein, is the most important nutritional factor positively affecting adult height (Grasgruber et al. 2014); (2) protein intake may be important for regulating appetite and intake of other macronutrients (Simpson and Raubenheimer 2005; Raubenheimer and Simpson 2016); however, (3) there is evidence that red meat is 'probably' carcinogenic—increasing lifetime risk of colorectal cancer to 6% from the background rate of 5%

(Bouvard et al. 2015); and (4) red meat consumption at the forager level is environmentally unsustainable. Red meat, especially beef, production requires far more water and energy input and causes much greater carbon dioxide (CO_2) emission than does production of vegetable protein (Pimentel and Pimentel 2003).

6.2.5 Carbohydrates

Americans and people living in industrialised countries obtain about 50% of their daily energy input from carbohydrate, predominantly refined grains (20%) and added sugar (15%). Fruits and vegetables contribute only about 14%. Carbohydrates provided only about 35% of daily energy for African Stone Agers, and nearly all (~ 33%) came from fruits and vegetables. Current recommendations are that carbohydrates provide 45–65% of daily caloric intake, and that up to 25% can be added sugar (Trumbo et al. 2002; Kaplan et al. 2017). Recommendations for fruits (1.5–2.0 cups/day) and vegetables (2–3 cups/day) are expressed in volumes rather than percentages. Even so, Americans are failing to consume even these meagre quantities: median fruit intake for adults is ½ cup per day and for vegetables 1½ cups per day (Moore and Thompson 2015). For some foragers, honey is a major dietary constituent. For example, honey makes up 10–15% of the Hadza diet, and, consequently, they suffer a high rate (36%) of dental caries (Crittenden et al. 2017). For pre-agricultural foragers, caries prevalence varied from 0.0 to 5.3% (Lofranco 2012), suggesting that honey (and sugar) consumption was typically much less(see also Crittenden 2011; Marlowe et al. 2014). Compared with Americans, Palaeolithic foragers ate twice the fruit and vegetables, far less added sugar, and no highly refined grains. Accordingly, the overall glycaemic index of the Palaeolithic diet was substantially lower than that of the average post-industrial diet.

6.2.6 Fibre

Current recommendations for fibre intake are 20–35 g/day for adults, but actual intake averages ~ 15 g/day (US Department of Agriculture 2016). Both recommendations and actual intake are far below estimates for pre-agricultural hunter-gatherers, which average around 100 g/day (Konner and Eaton 2010). In industrialised nations, cereal grains provide most dietary fibre, but for Palaeolithic foragers, fibre came almost exclusively from uncultivated fruits and vegetables. These are markedly more fibrous (13.3 g fibre/100 g) than are commercial varieties (4.2 g/100 g) (Jenike 2001). Because fibre from grains tends to be more insoluble than that from fruits and vegetables, current fibre intake averages about 75% insoluble and 25% soluble. For Palaeolithic foragers, the ratio was probably closer to 50:50. Both forms of dietary fibre have health benefits. Soluble fibre tends to lower serum cholesterol levels, both reduce insulin demand in diabetics, and insoluble fibre aids prevention and treatment of diverticulosis (Anderson et al. 2009). In one 9-year study of over 500,000 Americans, those consuming the most fibre (25–30 g/day) had a 22% lower all-cause mortality rate than those consuming the least (14–16 g/day) (Park et al. 2011).

6.2.7 Sodium and the Sodium : Potassium Ratio

Among recent hunter-gatherers, sodium (Na^+) intake averages about 1 g/day or less and potassium (K^+) intake much exceeds that of Na^+. The same pattern is seen in most pre-industrial societies that lack commercial salt and in all free-living, non-marine mammals. Given this near-universal electrolyte intake pattern, that of humans in developed nations seems extreme (but see Section 6.5.4 for notable exceptions). On average, Americans consume nearly 3500 g of sodium and 2650 g of potassium each day (Gunn et al. 2010). In industrialised countries, increasing blood pressure is a near-inevitable part of ageing (O'Rourke and Nichols 2005) and, worldwide, over one billion individuals are hypertensive. However, such rises do not occur in forager societies, nor in other groups without access to commercial salt (Eaton et al. 1988; Intersalt Cooperative Research Group 1988; Gurven et al. 2012). In the medical literature, the relation between Na^+, K^+, hypertension, and CVD is controversial (Oparil 2014). So few people consume electrolytes at the forager level that available data do not allow for evaluating a proposed Palaeolithic diet. Surveys have generally focused on Na^+ intake, but a minority have analysed dietary Na^+: K^+ ratios (O'Donnell et al. 2014). These demonstrate that increased K^+ intake, resulting in an Na^+: K^+ ratio closer to the ancestral pattern, reduced cardiovascular mortality.

6.2.8 Electrolyte and Acid Base Balances

Before the 'out of Africa exodus', humans are thought to have obtained about half their energy from plant sources and half from meat and fish (Eaton et al. 2010), as do recent warm-climate foragers (Marlowe 2010). Consequently, their diets were net alkaline-producing. In contrast, the acid-base impact of post-industrial diets is acidic (Sebastian et al. 2002). Cereal and dairy products that characterise the bulk of post-industrial carbohydrate intake are net acid yielding, which over time results in metabolic measures that cause urinary calcium loss, accelerated skeletal calcium depletion, calcific urolithiasis, age-related muscle wasting, and deteriorating renal function.

6.2.9 Experimental Clinical Studies

In the first study to test the effects of a more Palaeolithic-type diet among people with T2DM, now-sedentary Australian former hunter-gatherers were returned to their traditional diet and lifestyle for 7 weeks. This resulted in lowered fasting and postprandial glucose, increased insulin response, and markedly lowered fasting plasma triglycerides (O'Dea 1984). Since that time, several randomised and well-controlled clinical studies have been conducted to test the effects of diets based on Palaeolithic nutrition guidelines compared to standard dietary guidelines on metabolic health among modern populations. The results of these studies demonstrate promising improvements in metabolic parameters, even in as little as 2 weeks.

There have been four studies to date completely fulfilling the criteria for randomised controlled trials (RCTs) that compare the effects of diets based on Palaeolithic nutrition guidelines to control diets based on current dietary recommendations on the components of the metabolic syndrome (Lindeberg et al. 2007; Jönsson et al. 2009; Boers et al. 2014;

Mellberg et al. 2014). In each of these studies, Palaeolithic nutrition was broadly defined as consisting only of unprocessed meat, fish, eggs, vegetables, fruit, and nuts in variable proportions, and no dairy, cereal grains, or added sugar or salt. Control diets were based on international or national dietary guidelines. A meta-analysis of pooled effects from these studies demonstrated that the Palaeolithic nutrition recommendations led to significantly greater short-term improvements in waist circumference, triglycerides, and blood pressure across the studies (Mannheimer et al. 2015). The pooled effects of improvements in HDL cholesterol and fasting blood sugar were greater in those eating diets based on Palaeolithic nutrition recommendations, but the pooled differences were not statistically significant. Interestingly, in one RCT (Boers et al. 2014), isocaloric meals were provided for 2 weeks to participants who met two criteria for the metabolic syndrome (n = 32) in order to determine the effect of the diets on metabolic syndrome independent of weight loss. Despite these efforts, participants who were fed a 'Palaeolithic-type diet' lost weight and reported decreased hunger. The improvements in blood pressure, cholesterol, and triglycerides observed among those on the Palaeolithic-type diet remained significant when statistically controlling for weight loss. In another trial (Lindeberg et al. 2007), 29 patients with ischaemic heart disease and either glucose intolerance or T2DM were randomised to 12 weeks of a 'Palaeolithic' diet (n = 14) or a Mediterranean-like 'consensus' diet (n = 15). The Palaeolithic group showed a 26% reduction in area under the curve (AUC) glucose in a glucose tolerance test compared to a 7% reduction in the consensus group, and greater reductions in waist circumference. Importantly, changes in glucose were independent of the reductions in waist circumference.

An RCT of 39 healthy women comparing a Palaeolithic (n = 22) and a diet based on Australian dietary guidelines (n = 17) over 4 weeks demonstrated greater weight loss in the Palaeolithic diet group (−1.99 kg, 95% CI −2.9, −1.0), $p < 0.001$), but no differences in cardiovascular or metabolic risk factors (Genoni et al. 2016). Another RCT examined a Palaeolithic diet with and without exercise in terms of fat mass, insulin sensitivity, and glycaemic control among individuals with T2DM (Otten et al. 2016). All outcomes improved in both groups with the Palaeolithic diet, and the diet + exercise group lost less lean mass and had greater improvements in cardiovascular fitness.

In a randomised cross-over pilot study (Jönsson et al. 2009), 13 patients (3 women) with T2DM were sequentially placed on a Palaeolithic diet and an American Diabetes Association (ADA) guideline-based diet (emphasising balanced meals and low total and saturated fat) for 3 months each. Compared to the ADA diet, the Palaeolithic diet produced lower mean levels of HbA1c, triacylglycerol, diastolic blood pressure, weight, body mass index (BMI), and waist circumference, and higher mean HDL. The Palaeolithic diet also resulted in lower fasting plasma concentrations of leptin, but no significant differences in glucagon, insulin, incretins, ghrelin, C-peptide, or adipokines (Fontes-Villalba et al. 2016).

A similar, but non-randomised, two-phase diet intervention trial compared the American Heart Association (AHA) heart-healthy diet (4 months, phase 1) to a Palaeolithic nutrition diet (subsequent 4 months) with regard to cholesterol and triglycerides among 20 patients diagnosed with hypercholesterolaemia (Pastore et al. 2015). The AHA guidelines emphasise a diet rich in fruits and vegetables, whole grains, high fibre, little to no salt, eating fish at least twice per week, and minimising foods and beverages with added sugars (Eckel et al. 2014). The Palaeolithic diet was similar in many ways, but emphasised vegetables, lean animal protein, eggs, nuts, and fruit and excluded dairy, grains, and legumes (Jönsson et al. 2009). The Palaeolithic diet phase resulted in significantly lower triglycerides, total cholesterol, and LDL, and significant increases in HDL compared to baseline and after 4 months of the AHA diet. Although it was not an aim of the study, participants also lost significantly more weight during the Palaeolithic diet phase, and the improvements in cholesterol were independent of weight loss.

Two small, non-controlled studies demonstrated significant metabolic improvements, one in as little as 10 days. Ryberg et al. (2013) fed 10 overweight, post-menopausal women prepared meals based on Palaeolithic dietary recommendations of macronutrient composition (30% protein, 40% fat, mainly monounsaturated fatty acids (MUFAs), and 30% carbohydrate). In addition to metabolic measures, high-resolution spectroscopy measured liver fat. After 5 weeks, liver triglycerides actually decreased by 49%, and although not an aim, participants lost weight and demonstrated significant decreases in overall intake from baseline to end (−22%). In a non-controlled challenge study, Frassetto et al. (2009) had nine obese, sedentary, but otherwise healthy volunteers consume their usual diets for 3 days, then consume three 'ramp-up' diets with increasing fibre and K^+intake for 7 days each, and finally consume a Palaeolithic-type diet for 10 days. Participants experienced modest but significant reductions in blood pressure with improved arterial distention; decreased insulin secretion (AUC) in a 2-hour oral glucose tolerance test; marked reductions in insulin/glucose ratio; and reductions in total serum and LDL cholesterol. Given the short duration of this study, these results are noteworthy.

In summary, since the first recommendation that adopting a diet that more closely resembles the diet and eating patterns experienced by humans for the majority of human history (based on palaeontology, archaeology, and ethnographic studies of modern humans living in pre-industrial conditions) would be beneficial for human health, the framework has gained traction. In general, existing studies broadly define Palaeolithic nutrition recommendations as consisting only of unprocessed meat, fish, eggs, vegetables, fruit, and nuts, with no dairy, cereal grains, or added sugar or salt. These studies have demonstrated promising improvements in health, particularly decreases in hypertension, waist circumference, weight, and triglycerides. In addition, dietary recommendations by the AHA have moved closer to Palaeolithic nutrition recommendations, with small differences: the AHA recommendations include an emphasis on whole grains and Palaeolithic nutrition had much higher fibre, no dairy, and no added salt or sugar. Although the differences are small, they may be clinically important, because RCTs demonstrate significantly improved metabolism with Palaeolithic compared to other healthy diet recommendations, even over short time periods. However, all of these studies were done with relatively small sample sizes, which can limit the generalisability of the results. Studies with large sample sizes are warranted.

6.3 Energy Expenditure and Physical Activity

Physical inactivity is one of the leading risk factors for 22 chronic diseases including cardiometabolic disorders such as CVD and T2DM, as well as colon and breast cancer (Booth et al. 2012). Physical activity improves both physical and mental health, including mental acuity and brain function (Colcombe and Kramer 2003). Despite these benefits, current levels of physical activity are very low. The American College of Sports Medicine recommends 150 minutes of moderate to vigorous intensity physical activity per week, but very few people meet this recommendation (in the United States only 5% of adults meet this criterion when measured objectively) (Troiano et al. 2008). It is therefore critical to understand how to increase physical activity to improve health and prevent or at least delay the onset of chronic diseases.

Applying an evolutionary framework to understand energy expenditure and physical activity sheds new light on why and how physical activity is important for health, and how human physiology and behaviour have both changed and remained the same across evolutionary history. Traditionally, researchers in health psychology, public health, and epidemiology examine the proximate predictors of physical activity or the ways physical activity is related to chronic disease. In contrast, an evolutionary framework allows for a more thorough understanding by integrating Tinbergen's four levels of analysis. Existing approaches are limited to proximate predictors of physical activity, without considering how they are directly or indirectly linked to ultimate/functional factors. An ultimate level of analysis makes predictions about how the physiological and psychological systems that underlie physical activity and energy expenditure have been (and continue to be) shaped by natural selection in ways that improve survival and reproductive success. This can lead to predictions about factors that influence energy expenditure and physical activity that may not otherwise be considered. A comparative or phylogenetic approach is informed by how modern humans compare to our ancestors and other hominins, primates, and mammals in physiology, physical activity, and energy expenditure. A developmental level of analysis leads to an examination of how energy allocation changes across the lifespan, including the shaping effects of environments. Integrating these four levels of analysis provides a broader perspective that can offer novel insights into the evolved barriers (i.e. costs) and benefits of physical activity.

It can also lead to novel approaches to encourage physical activity in increasingly sedentary modern environments. Understanding the barriers to physical activity and exercise that are functionally adaptive—or were in the environments where they evolved—and identifying which factors are more and less modifiable than others can aid in the development of interventions and environments that are more conducive to physical activity. It can also lead to hypotheses about why or when certain proximate predictors of physical activity are more or less likely to influence behaviour at a given time or in a given environment. Interventions can be targeted to focus on more malleable traits or psychological processes, or encourage physical activity in ways that are more consistent with evolved psychological and physiological mechanisms. Including an ultimate or functional level of explanation has advantages, but requires consideration of how natural selection operates, and how it can lead to seemingly maladaptive physiology and behaviour.

Throughout this section the term fitness will be used in two distinct ways. When discussing fitness in terms of physical prowess, including cardiorespiratory and muscular endurance, muscle strength, flexibility, balance, agility, and body composition, the term physical fitness will be used. When discussing fitness in an evolutionary sense, meaning the relative success of passing genes on to a subsequent generation, the terms reproductive or Darwinian fitness will be used.

6.3.1 Mismatch and Life History Theory

As with diet, when environmental conditions are very different from those under which selection was shaping a behaviour (or underlying mechanism), this leads to a discordance or mismatch between the EEAs and current lifestyles (Eaton et al. 1988, 2002a; Lieberman 2013). In the novel environment, the behaviours or mechanisms may no longer increase reproductive success, and can even be detrimental to reproduction or survival. (For further discussion, see Chapter 1: Core Principles of Evolutionary Medicine.) To understand the evolution of psychological and physiological factors that influence physical activity, we must consider the mechanisms and behaviours that would have been adaptive in the environments faced by humans during the majority of our history, when selection was acting to shape them.

Physical activity levels in industrialised populations contrast starkly with physical activity estimated based on palaeontological research of ancestral remains, or measured in current populations living in pre-industrial conditions with subsistence-based economies. This evidence suggests that high levels of physical activity were an integral part of our species lifestyle for the vast majority of our existence. Fossil remains suggest humans were highly muscular (Booth et al. 2002). Furthermore, humans have several anatomical and physiological adaptations, such as spring-like, collagen-rich tendons and ligaments, arches in the feet, long legs, and other traits that help stabilise and correctly move our upright torso and head, which are unique among primates and suggest that endurance running was an important part of human evolution (Bramble and Lieberman 2004). Endurance running also necessitated the evolution of brain cooling mechanisms to prevent seizures when the body heats up. (For further discussion, see Chapter 17: Brain, Spinal Cord, and Sensory Systems.) Endurance running is not only possible because of these adaptations, but also associated with neurobiological rewards, evidenced by increases in endocannabinoid signalling in response to high-intensity exercise (Raichlen et al. 2012). There is evidence for a 'runner's high' dependent on natural opioids (Boecker et al. 2008), and animal models suggest involvement of dopamine reward pathways (Ruegsegger et al. 2016). Recently, researchers have hypothesised that selection for endurance running (and an increased aerobic capacity) in humans also played a role in the evolution of our uniquely large brains. According to this hypothesis, endurance exercise increased the signalling of compounds responsible for brain growth and metabolic regulation during exercise (neurotrophin and growth factor), which led to increased overall brain growth and development (Raichlen and Polk 2013). (For further discussion, see Chapter 17: Brain, Spinal Cord, and Sensory Systems.)

The view that human physiology has evolved in response to high levels of obligatory physical activity helps explain why low levels of physical activity lead to deleterious health outcomes, and higher levels result in mental and physical health benefits. A large body of evidence supports the claim that human physiology is not designed to function properly under conditions of unlimited caloric intake and low physical activity (Booth et al. 2002a, 2002b, 2012; Chakravarthy and Booth 2004). But this perspective sometimes leads to the assumption that natural selection has *only* shaped human physiology and psychology to support high levels of physical activity and exercise. This is unlikely, however, given the low levels of physical activity observed in modern, industrialised populations.

Levels and patterns of physical activity have changed drastically over a very short evolutionary time (about 200 years). Beyond acknowledging these differences, considering *why* there are differences leads to a better understanding of how to increase physical activity in current environments. There is a stark contrast between the physical effort necessary to extract resources needed to fulfil our biological needs in current environments compared to those faced by our ancestors (Eaton et al. 1988; Cordain et al. 1998; Eaton and Eaton 2003). However, it has been hypothesised that, over the course of human evolutionary history, a willingness to *selectively* exert oneself physically would have generated survival and reproductive benefits through the ability to produce more resources, protect oneself and one's family from danger, compete with rivals, defend resources, migrate over long distances, meet the energetic demands of pregnancy, childbirth, and lactation, care and provide for children, ward off sexual predators, and display one's physical health and condition to gain mates and social partners (Hönekopp et al. 2004, 2007; Caldwell Hooper 2013). The Agricultural Revolution allowed humans to produce and store food to make it more widely and consistently available. The Industrial Revolution led to more rapid changes in mechanisation and automation of work.

This level of technological innovation is unique to humans, but a drive to efficiently allocate energy is a fundamental feature of life. The way organisms allocate energy between competing demands across the lifespan is key to understanding energy expenditure and physical activity, and is a central premise of life history theory (LHT). LHT leverages the observation that all organisms face finite budgets of both time and energy, and each must therefore be strategically allocated. (For further discussion, see Chapter 4: Growth and Development.) Humans have been so successful at creating ways to improve the efficiency of food production and the consistency of availability that this has resulted in environments that promote chronic diseases. Food is available at such low and separable energetic costs that our inherent drive to be strategic and conservative with energy expenditure leads to deleterious health outcomes.

From an adaptive, functional level of analysis, how energy is prioritised and allocated has been shaped by natural selection over millions of years to maximise the probability of expected reproductive success in predictable ecological contexts. The optimal way to partition energy and time is balanced between competing demands of somatic maintenance, growth, and reproduction. The optimal allocation strategy varies across species and environments, and among individuals within species, depending on environmental and individual factors. Energy devoted to physical activity is fundamentally linked to these other forms of energy expenditure. As with diet composition, natural selection has shaped a flexible system of energy allocation that responds to the environmental and individual costs and benefits of expending energy on physical activity relative to the other trade-offs across the lifespan. Natural selection has not yet had the opportunity to act on the phenotypes that result from environments with few demands but many resources, where formerly adaptive strategies are actually harmful. This nuanced application of evolutionary theory leads to a more sophisticated understanding of the circumstances under which natural selection favours *both* sedentary and active behaviours in predictable ways.

Energy allocation strategies (or life history strategies) of humans can also be compared to other primates, and are informative because they define the unique timing and degree of energetic trade-offs competing with physical activity. For example, chimpanzees are one of our closest primate relatives, having diverged from a common ancestor 6–8 mya. But humans and chimpanzees take different approaches to reproduction and child rearing. Chimpanzee mothers care for and provide food for themselves and one offspring at a time, without the help of fathers or close kin, for approximately 5 years (Sugiyama 2004). At around age 5, offspring can provide for themselves, and the mother can shift energy to her next offspring. In contrast, human females have a much shorter interbirth interval (2–3 years in natural fertility populations) and, due to our unique, learning-intensive foraging niche, offspring cannot provide for themselves for decades (Gurven et al. 2006; Howell 2010; Marlowe, 2010; Hooper et al. 2015). Thus human females care for and help provision multiple dependent offspring simultaneously, and can be pregnant, lactating, and raising older, dependent children at the same time. (For further discussion, see Chapter 16: Sexuality, Reproduction, and Birth.) All of these activities are energetically costly and decrease the energy that women can devote to producing food, while increasing the demands on food production. It would not be possible for human females to meet these energetic demands without the ability to store and metabolise fat stores during lactation, and without the help of male partners, grandparents, and larger kin and social networks. The differences in human patterns of reproduction and providing for offspring are intricately linked to male and female patterns of activity throughout the lifespan, as well as body mass differences in fat and muscle distribution. These differences in parental investment and body composition also lead to differential costs and benefits of physical activity over human history, and continue to influence patterns of physical activity in modern contexts. Mismatch is particularly relevant in this instance, because selection in the human EEAs primarily shaped the optimal energy allocation strategies for humans based on probabilities of reproducing in resource-constrained environments without unlimited access to energy-dense foods. These environments drove selective pressures that favour energy conservation in order to buffer reproduction and provision offspring during food shortages or seasonal availability to a greater degree than strategies that increase activity in response to caloric excesses, particularly in women. When physical activity is detached from fulfilling biological needs and producing food, the benefits of being physically active that would have been favoured by natural selection historically are also removed.

6.3.2 Measuring Energy Expenditure and Physical Activity

Energy expenditure and physical activity are related, but distinct, and are measured in different ways. Physical activity is any bodily movement produced by skeletal muscles, while energy expenditure is the amount of energy, or calories, that are required to carry out bodily functions including physical activity, breathing, digesting, and circulating blood. Both energy expenditure and physical activity are important for health, energy balance, and body composition, but must be considered separately.

Doubly labelled water (DLW) is considered the gold standard measure of energy expenditure. This technique uses ingested stable isotopes of deuterium (^2H) and oxygen (^{18}O) to trace the flow of water and CO_2 through the body, and computes the rate of oxygen consumption (Butler et al. 2004). The primary advantage of DLW is that it is a relatively precise and accurate way to objectively measure total energy expenditure (TEE). TEE encompasses all energy expended, including basal metabolic rate (BMR), the thermogenic effect of food (the energy it takes to digest food, typically estimated to be 10%), and the remainder is labelled activity energy expenditure (AEE). AEE and physical activity level (PAL) are computed with either measured resting energy expenditure (REE), which is as close an approximation of BMR as possible, or a predicted value of BMR based on one of several calculations that consider individual characteristics (such as sex, age, height, weight, body composition, and organ size). AEE and PAL are computed with the following equations:

$$\text{AEE} = \text{TEE} - (0.10 \times \text{REE or predicted BMR})$$
$$\text{PAL} = \text{TEE/REE or predicted BMR}$$

Although it is considered the gold standard, the DLW technique has limitations, particularly for estimating physical activity. First, when predicted BMR is used instead of measured REE, the calculations of AEE and PAL include non-negligible error; and since several equations can be used to estimate BMR, it is difficult to compare between studies. This detracts from the primary advantage of this method as being precise and accurate. It is also much more accurate to measure REE, because it is the primary component of TEE and is shown to change due to several factors. It is affected by minor respiratory infections (Muehlenbein et al. 2010), reduced caloric intake (Martin et al. 2007), pregnancy and lactation (Melzer et al. 2009), and acute and long-term bouts of physical activity (Speakman and Selman 2003). Moreover, the respiratory quotient (RQ) measured by REE is used to estimate TEE from expired CO_2, and when not measured must be estimated, which contributes another source of error in TEE estimations. In addition, BMR is often estimated using weight, but is also affected by body composition, with fat-free mass explaining over 81% of the variance (Sparti et al. 1997). Large amounts of adipose tissue can also contribute meaningfully to TEE, and are important to consider, particularly when comparing two samples with markedly different body composition (Kaiyala and Schwartz 2011). Finally, when interested in understanding the proportion of energy allocated across competing demands (i.e. physical activity and reproduction), the DLW method alone does not provide the right level of detail needed to understand differences across populations or between individuals. TEE does not provide any information about the frequency, intensity, or duration of physical activity, which are critical for understanding the health benefits of physical activity and exercise.

6.3.3 Comparing Physical Activity Across Ecological and Economic Contexts

Physical activity varies widely across populations. Figure 6.1 depicts measures of PAL in hunter-gatherer/forager populations from around the world and highlights the great variation in PAL between men and women, and across a range of ecologies and modes of food production (Norgan et al. 1974; Pearson 1990; Katzmarzyk et al. 1994; Leonard et al. 1995; Panter-Brick 1996; Yamauchi 2000; Yamauchi et al. 2001; Dufour and Piperata 2008; Leonard 2008; Pontzer et al. 2012; Gurven et al. 2013). A clear, predictable pattern does not emerge when comparing PAL across these dimensions and populations. Men are sometimes, but not always more active than women, and differences in physical activity are even observed within the same mode of food production. PAL estimates were made using different methods of data collection, which can contribute to these differences, but similar variability is also seen within PAL estimates from studies using the DLW method.

The variation in physical activity highlighted in Figure 6.1 exemplifies that there is not simply one level of physical activity that humans have evolved to perform. Instead, the

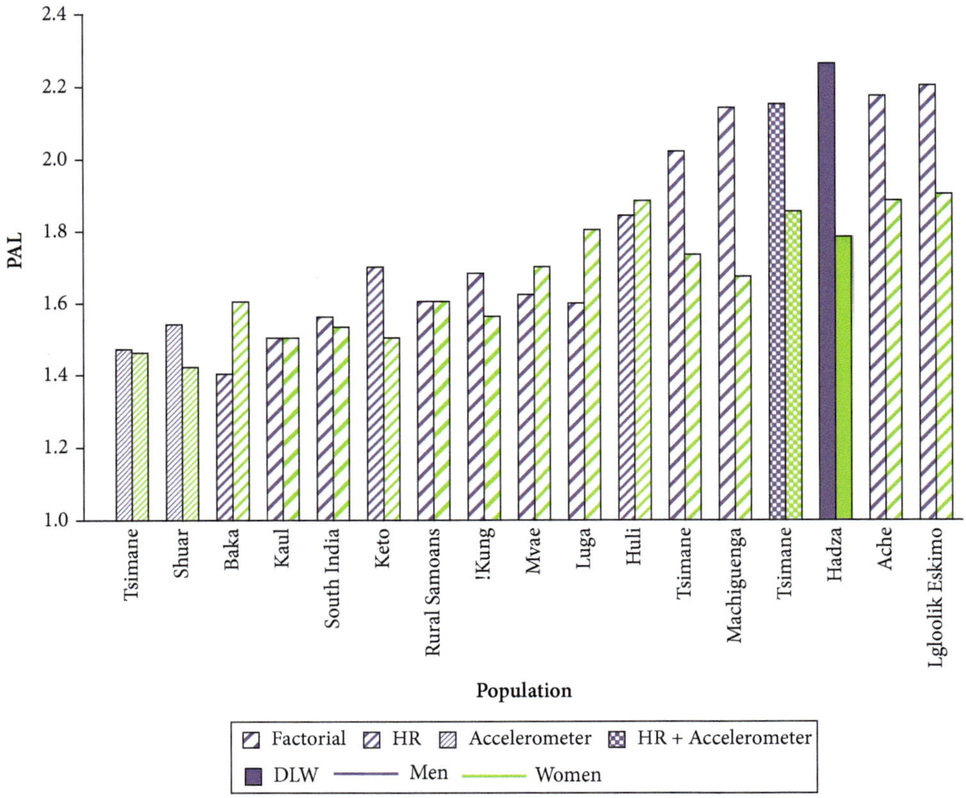

FIGURE 6.1 Variation in physical activity level (PAL) by sex among hunter-gather/forager populations. PAL was estimated using a variety of data collection modalities: Factorial, factorial method; HR, heart rate; Accelerometer, heart rate and accelerometer together; DLW, doubly labelled water.

Source: Reproduced from Ann E. Caldwell, *Human Physical Fitness and Activity: An Evolutionary and Life History Perspective*, p. 32, Fig 4.1. © The Author, 2016.

variation suggests that humans have also evolved to flexibly respond to competing energetic demands across environments, across seasons, and across the lifespan to selectively exert energy when the adaptive benefits to doing so outweigh the costs, and otherwise conserve and/or redirect energy. Section 6.3.5 expands on life history and epidemiological research to help explain this variation in the costs and benefits of physical activity across the lifespan.

6.3.4 Comparative Approach with Other Primates and Mammals

This section focuses on the unique morphological, physiological, and metabolic adaptations that influence energy expenditure and physical activity in humans compared to non-human primates and other mammals. Next, it examines the evolution of human life history patterns, and how the human pattern compares to other primates and mammals. Finally, it compares physical activity patterns and energy expenditure in primates and humans. The focus is on the ways the unique aspects of human anatomy, energetics, and life history are hypothesised to influence the costs and benefits of physical activity, as well as developing and maintaining physical fitness and strength.

6.3.4.1 *Overall Energy Expenditure among Primates*

Humans and other primates are similar in having TEE that is 50% lower than predicted from body size based on scaling relationships among mammals (Pontzer et al. 2014). However, primates devote a greater relative proportion of their energy budget to BMR, with their BMR in line with predictions based on body size. Preliminary findings among orangutans suggest that a lower TEE evolved in orangutans to reduce the risk of starvation in unpredictable environments (Sibly and Brown 2007; Pontzer et al. 2014).

6.3.4.2 *Human Morphological, Physiological, and Metabolic Adaptations*

The hominin line diverged from the last common ancestor with other apes between 6 and 8 mya. Ecological changes led to two major shifts in behaviour after this divergence: a descent from the trees and a shift to bipedal locomotion. At the end of the Miocene, forested habitats were becoming less common, and vast open landscapes with a cooler, drier climate set the scene for hominin ancestors, who increased roaming distances to find more distantly spaced food patches (Cerling et al. 1997, 1998). Many of the details of hominin evolution, including the evolution of bipedalism, are not conclusive, but current research suggests that hominin ancestors have morphological adaptations like sideways-facing pelvic bones that stabilise the torso, long hind limbs, and feet with arches and toe joints that allow for propulsion that made locomotion, especially bipedal locomotion, more efficient by the mid-Pliocene (Sockol et al. 2007; Pontzer et al. 2009, 2014). Researchers have estimated that the earliest evidence for efficient bipedalism is in *Australopithecus afarensis*, about 3.4 mya (Pontzer et al. 2014). Modern subsistence humans exhibit much larger ranging distances (9–15 km/day among the Hadza) (Marlowe 2010) compared to chimpanzees (2–5 km/day) (Pontzer and Wrangham 2004, 2006).

6.3.4.3 *Human Adaptations for Endurance Activities and Energy Storage*

Around 2 mya, *Homo erectus* emerged, with a much larger brain, long legs, and generally humanlike body (Lieberman 2013, 2015; Lieberman 2013). Since the divergence from the last common ancestor with chimpanzees and bonobos, human ancestors became morphologically adapted to cover long distances by walking and running. As described in Section 6.3.1, humans have evolved anatomical and physiological adaptations that make endurance activity more efficient (Bramble and Lieberman 2004), which decreases the costs, and rewarding (Boecker et al. 2008; Raichlen et al. 2012; Ruegsegger et al. 2016), which increases the benefits. This may help to encourage intense aerobic activity despite the high energetic costs and risks associated with it.

On the other hand, humans have not simply evolved in ways that favour cardiovascular and muscular fitness. Instead, humans seem unique in needing consistent and repeated activity to *maintain* physical fitness. Palaeontological research suggests that human ancestors were substantially more muscular than humans living in modern, sedentary contexts (F.W. Booth et al. 2002). But when we compare humans to other primates, we actually have much less muscle tissue and more adipose tissue, which some researchers have suggested has helped humans balance the increased metabolic costs of a large brain (Leonard and Robertson 1997; Leonard et al. 2007; Zihlman and Bolter 2015). Humans seem to rapidly shift energy from maintaining muscular fitness to storing fat in response to changes in energy expenditure. One study found decreases in fat-free mass and increased abdominal fat after only one 1 week of reduced physical activity (Olsen et al. 2008). These rapid declines in allocation of effort to muscle tissue when not being used may be unique to humans. Captive bonobos have high muscle mass, despite having lower levels of activity than bonobos in the wild, which suggests that high levels of activity are not necessary for maintaining muscle mass in one of our closest primate relatives (Zihlman and Bolter 2015). Humans rapidly lose muscle tissue in the absence of activity, and it appears that human physiology is driven to avoid the metabolic costs of building and maintaining skeletal muscle, particularly in the upper body, when not necessary.

Humans are unusual among primates in their propensity for storing energy outside of the body, and to a lesser degree as fat. San hunter-gatherers dry strips of meat in the sun to save for longer-term use (Lee and Barnard 1980). On an evolutionary timescale, this is still a relatively recent development. Over longer time spans, humans have had complex networks for food sharing, which are unique in the animal kingdom and have allowed us to buffer fluctuations in resource availability and competing demands of energy allocation that interfere with resource production (Kaplan and Gurven 2005). Humans have larger fat deposits (14–26% in healthy adults, even among modern foragers) than wild or captive chimpanzees and bonobos (3–10%) (Zihlman 1984; Zihlman and Bolter 2015) and wild baboons (Altmann et al. 1993). However, the body fat of other captive primates (lemurs, baboons, and Western lowland gorillas) tends to fall within the human range and is much higher than wild counterparts (summarised in Dufour and Sauther 2002). Female primates also tend to have more fat than males.

6.3.5 Life History Trade-offs and Physical Activity across the Lifespan

From an LHT perspective, the ways energy is allocated across the lifespan determine pre- and postnatal growth rate, gestation length, size at birth, age of reproductive maturation, length of reproductive age, rate of reproduction, and lifespan, which are all considered to be life history parameters. They also influence the relevant trade-offs between energy allocated to physical activity and energetic demands.

Life history parameters are sensitive to energetic condition, and comparing wild and captive chimpanzees and orangutans provides evidence for how energy allocation systems have evolved to maximise the probability of reproductive fitness in different environments. Wild chimpanzees typically reach menarche at age 10, and have their first offspring at around 13 years old. Captive chimpanzees, on the other hand, reach menarche earlier, at around 8 years, and have their first offspring between ages 10 and 11 (Sugiyama 2004). Among captive chimpanzees, where food is consistently available and extrinsic mortality is low, the reproductive lifespan is extended, demonstrating that while life history parameters have been shaped at a species level, they are still sensitive to changes in environment. Similarly, wild orangutans reproduce at 15 years and show a mean interbirth interval of

9.3 years. In captivity, however, age of first reproduction is halved and interbirth intervals are much shorter (less than 1 year when infants are removed from mothers and a median of 5.5 years when they are kept together) (Shumaker et al. 2008). In captivity, food availability is consistent and is not associated with foraging, analogous to the ways modern environments contrast with ancestral ones.

One way to utilise a life history approach to understanding physical activity is to consider each stage independently to examine the energetic trade-offs theoretically expected to influence the optimisation of energy allocation. This optimisation has been shaped by natural selection to increase predicted reproductive success, and considering this can lead to hypotheses about what factors influence the costs and benefits of energy expenditure in the form of physical activity. We can also take a cumulative approach, because LHT predicts that experiences and environments at earlier stages of life, particularly during development, will influence later energy optimisation strategies. It is therefore important to look at each stage of life to know what the salient costs and benefits are within that stage, how it may influence behaviour in later stages, and how earlier experiences and environments may have shaped responses to current environmental cues.

6.3.5.1 *Fetal and Early-Life Development*

Several lines of research suggest that poor nutrition or energetic stress during prenatal and early life environments has a profound impact on health, development, and even physical activity. During fetal development and in infancy, all energy is focused on somatic development of brains and bodies, growing, accumulating fat, and maintaining tissue. Epidemiological evidence has consistently shown that early-life undernutrition, indexed by low birth weight (alone and when paired with postnatal overnutrition), is related to high blood pressure, T2DM, CVD, and stroke in adulthood (Barker 1989, 1994, 2004; Curhan et al. 1996). After observing these patterns, researchers termed this phenomenon the fetal or developmental origins of adult disease (reviewed in Barker 2004), or the Barker hypothesis. Hales and Barker (1992) hypothesised that poor fetal and postnatal nutrition can lead to a 'thrifty phenotype' which results in impaired development of the pancreas and endocrine disruptions that lead to increased risk of T2DM.

Some evolutionary researchers have proposed that when faced with energetic stress *in utero* and during infancy, developmentally plastic energy allocation systems have predictive adaptive responses (PARs) that shape more thrifty or conservative energy allocation systems later in life (Ellison 1990; Lipson 2001; Gluckman et al. 2005; Kuzawa 2005; Ellison and Jasienska 2007; Kuzawa et al. 2007). It is hypothesised that maternal cues of constrained resource availability were likely to signal future environments with unstable resource availability, and therefore it was adaptive to shift reaction norms to develop a suite of metabolic characteristics that help buffer against hard times with increases in fat accumulation, lower muscle mass, insulin resistance, lower vascular density, compact body shape, and earlier reproductive age. Proponents of the PAR hypothesis contend that over the course of hominin evolution, later environments predictably resembled prenatal environments, and selection would favour the development of anticipatory energy-saving mechanisms (Gluckman et al. 2005; Kuzawa 2007). They argue that the changes in metabolic physiology that buffer against hard times would only lead to harmful health outcomes when energetic stress during fetal and infant development is followed by later-life environments with abundant resources and much lower levels of obligatory physical activity.

Other researchers have argued that rather than being an adaptive response, correlations between later-life health or life history traits and early-life nutritional stressors merely reflect individuals 'making the best of a bad start' (e.g. Bogin et al. 2007), or results from a high metabolic load (large tissue masses, rich diet, and sedentary lifestyle) relative to metabolic capacity (physiological traits contingent on development) (Wells 2011). PAR theorists agree that some environmental factors merely disrupt development, rather than serving an adaptive function. It remains an open question whether or not these energetic strategies reflect an adaptive shift in reaction norms to a more conservative energy allocation strategy, or merely a disruption during development that leads to poorer health outcomes. But the energy costs of performing physical activity and developing and maintaining physical fitness and strength in humans are high. Part of a thrifty phenotype may be a conservative energy allocation strategy that leads to lower levels of non-obligatory and even obligatory forms of physical activity. (For further discussion, see Chapter 4: Growth and Development.)

Studies on humans and rats are consistent with this perspective. A meta-analysis of 13 Scandinavian birth cohorts demonstrated that lower than normal birth weight was related to less leisure time physical activity later in life (Andersen et al. 2009). Another study found low birth weight was related to higher energy costs of running (Baraldi et al. 1991). In another, small size at birth was associated with lower muscle efficiency (Workman et al. 2015). Lower running and muscle efficiency may be one mechanism through which physical activity levels are lower among those with lower birth weight, because it would increase the metabolic costs and decrease the likelihood of doing it. Lower muscle efficiency and higher energy costs of running among those with low birth weight may reflect a system that is not fully or optimally developed in response to energetic stress early in life, or may reflect an adaptive, energy-conserving shift.

Similar patterns are observed in animal models. In an experimental study of Wistar rats, offspring from mothers who were undernourished had offspring with lower levels of physical activity irrespective of postnatal feeding (Vickers et al. 2003). Surprisingly, but consistent with the PAR hypothesis, the least active rats were offspring of malnourished mothers who were fed a *hypercaloric* diet after birth. Taken together, in light of LHT, these results suggest that reaction norms for energy allocation strategies are sensitive to cues of resource availability early in development. This may result from poorer development (as in the pancreas), or from adaptive shifts in energy away from physical activity similar to the shifts observed in metabolism and immune function that prioritise growth and reproduction. This shift may lead to higher costs for performing physical activity (e.g. if it is relatively less efficient or more difficult to develop muscle mass) or more psychological barriers (e.g. more displeasure in response to activity, and thus lower motivation).

6.3.5.2 *Childhood and Pubertal Development*

Extended childhood is a unique characteristic of the human lifespan. In contrast with other primates, human offspring remain primarily dependent on their parents to produce food for well over a decade after weaning. Most energy during childhood is focused on somatic growth and maintenance, including substantial continued investments in brain growth, learning, and immune system development. The benefits of activity are also inherently high because physical activity can help to build the coordination, strength, and skills necessary to survive in a given ecology, learn the energetic demands of the environment, test individual ability to meet those demands, and foster play. Play provides opportunities for children to interact with others, developing critical skills necessary for living in our extraordinarily social species (Gray 2009; Konner 2010). However, hunter-gatherer children also contribute significantly to their own subsistence (helping to produce food) and caring for younger siblings (Crittenden et al. 2013; Konner 2016).

Overall, a life history framework predicts that physical activity should be high among children, particularly those who are healthy, well fed, and safe. Children's productivity varies between populations and depends in part on the type of provisioning activity and the safety of performing this provisioning (Jones et al. 1996). In less safe settings, children produce less food and tend to stay close to home. In addition to environmental cues of resource availability, childhood environments vary in exposure to infectious diseases and parasites, which both affect mortality risks and thus are likely to have lasting influences on optimal energy allocation strategies. The energetic budget necessary to fight parasites or infections affects how much energy can be put towards other endeavours: immune functions, growth, reproduction, and physical activity. In addition, where mortality risks are higher, it is predicted that energy allocation will favour reproducing earlier over investing in longevity-related somatic investment and longer growth. (For further discussion, see Chapter 4: Growth and Development.)

Physical activity should theoretically be reduced during times when somatic and brain growth are higher. Humans' slow growth rate is likely due to the high energy costs of our brains, which peak in childhood, at 66% of resting metabolism and 43% of the overall energy budget around age 4 (Kuzawa et al. 2014). Consistent with theory, Kuzawa et al. (2014) demonstrated that growth is slower and physical activity is lower at this age, when glucose uptake in the brain is higher, suggesting a trade-off between allocating energy between brain and body growth and physical activity.

Once children reach puberty, the opportunity costs of physical activity increase because adolescents start allocating energy to support reproductive maturation, and growth velocity increases during the adolescent growth spurt (Baxter-Jones et al. 2005; Walker et al. 2006). A life history perspective predicts a decrease in physical activity at this stage of life in any environment because of these trade-offs, but particularly among those who are in poorer energetic condition and those in environments where physical activity is not related to food production, survival, or reproduction.

Decreases in physical activity and increases in sedentary behaviour during pubertal maturation have been observed in at least two subsistence-based populations, the !Kung and the Tsimane. In a classic ethnographic documentary that details the life of a !Kung woman named N!ai, she describes her son by saying, 'my son was almost a man. He was of the age, the lazy, difficult age which we call "owners of the shade"' (Miesmer and Marshall 1980). A cross-sectional study of sedentary and physical activity measured by accelerometers among the Tsimane demonstrates that sedentary time increases linearly and physical activity decreases linearly as puberty progresses (Caldwell 2016).

This pattern is also consistently seen in large-scale epidemiological studies in post-industrial populations. In the United States, when physical activity is estimated objectively with accelerometers and strict cut-offs are used to define moderate physical activity, less than half (42%) of children get the recommended amount of activity, but only 8% of adolescents get the recommended amount (Troiano et al. 2008). In a European sample, 97% of children at age 9 get the recommended amount of physical activity, whereas only 82% of boys and 62% of girls get the recommended amount at age 15 (Riddoch et al. 2004).

Taken together, the allocation of energy throughout childhood is tightly and predictably coordinated to prioritise brains, growth, and reproductive capacity over running around playing when appropriate. This perspective suggests these metabolic trade-offs are barriers to physical activity that are not typically considered by traditional public health approaches to increase physical activity.

6.3.5.3 *Adulthood in the Reproductive Years*

In adulthood, the energetic demands of reproduction increase and vary between the sexes. The energetic costs of parental investment are much greater for humans compared to other primates, because of the huge investment in neurological development, altriciality at birth, and caloric dependency at least into the second decade of life (Kaplan et al. 2000; Aiello and Key 2002; Hooper et al. 2015). Humans have babies that are bigger than predicted by maternal body size and have significantly more body fat (15%) than other primates (Zihlman 1997; Kuzawa 1998). The energy cost of fat deposition in human babies increases the energy demands on human mothers compared to other primates (Dufour and Sauther 2002). At birth, the human brain utilises a large proportion of resting metabolic rate and total energy budget, which has historically been produced completely through lactation (Lieberman 2013). Women pay a much higher price than men for the initial metabolic investment in reproduction from gametes through pregnancy, particularly in the last trimester when fetuses store fat, increasing maternal energy expenditure by an estimated 465 kcal/day over pre-pregnancy levels (Ellison 2001; Butte and King 2005). Lactation is estimated to necessitate an extra 625–700 calories/day for nursing mothers (Butte and King 2005). The high energy costs of reproduction for women, and the returns to sex- and age-specific division of labour, combine to favour increased investment from fathers and other close kin in food production, while pregnant and lactating women *decrease* energy devoted to food production when possible (Hurtado et al. 1992; Marlowe 2010; Madimenos et al. 2011; Hooper et al. 2015). Because women pay a much higher energetic cost to reproduce, their energetic trade-offs to devote energy to physical activity are steeper than men's, and the benefits of storing fat are higher (Mauvais-Jarvis 2015).

In addition to the substantial energetic costs human females face to reproduce, studies have consistently shown that both negative energy balance alone and high levels of energy expenditure coupled with neutral or negative energy balance can adversely affect ovarian function and fecundity, or the biological capacity to reproduce, among women. Female fecundity is sensitive to transient shifts in energy availability in both industrialised and pre-industrial populations. Low gonadotropin and ovarian steroid levels, for example, have been observed among women in industrial populations who have low energy availability, are losing weight, are restraining their intake below appetite, have less stored fat and lower lipid profiles, and have high energy expenditure through exercise or workloads with neutral or negative energy balance (Loucks et al. 1994, 1998; Loucks and Thuma 2003; Ellison 2008; Vitzthum 2009). The combination of high energy expenditure and low energy intake has a more pronounced negative affect on fecundity (Ellison 2001). It is interesting to note, however, that increased workload or physical activity/exercise has been shown to decrease steroid profiles even when food availability is not limited (e.g. Jasienska and Ellison 2004; Jasienska et al. 2006; Williams et al. 2015).

In a traditional setting, the Tamang in Nepal provide a natural experiment for examining the impact of energetics on reproductive function because they subsist on small-scale agriculture and pastoralism and have seasonal variation in physical workloads—from moderately heavy in the winter to very heavy during the monsoon season. Overall, Tamang women exhibit weight loss and decreased ovulation between seasons (from 71% of women ovulating in winter to 38% during the monsoon). Notably, however, young women aged 17–23 exhibited negligible seasonal changes in ovarian function and actually gained weight during the heavy work season (Panter-Brick et al. 1993).

A study by Jasienska and colleagues (2006) also demonstrated that energetic stress during fetal development may affect the sensitivity of women's reproductive systems to increases in physical activity or workload. Among a sample of rural Polish women, size at birth, measured by ponderal index (PI, in kilograms per cubic metre), moderated the relationship between physical activity and ovarian suppression (measured by mid-follicular salivary oestradiol) such that those who had high PI at birth were buffered from ovarian suppression at moderate levels of physical activity. On the other hand, women who experienced a constrained fetal environment, reflected by low or moderate PI at birth, showed a heightened sensitivity of fecundity to increased energetic demands. These results support the PAR hypothesis that fetal environments with constrained energetic resources may lead to adaptive shifts in metabolism or energy allocation that favour energy conservation, in this case developing a more thrifty reproductive phenotype that is more sensitive to fluctuations in energy availability. Given the high metabolic costs of gestation and lactation for women, researchers have hypothesised that this heightened responsiveness to energetic condition and energetic stress would be adaptive in order to decrease the likelihood of conception when it is questionable whether the energetic demands of pregnancy and/or lactation can be met (e.g. Ellison 2001; Jasienska 2003; Vitzthum 2009). (For further discussion, see Chapter 16: Sexuality, Reproduction, and Birth.)

Energy balance also influences female reproduction in terms of offspring birth weight and length of gestation, with smaller birth weight and shorter length of gestation among women facing undernutrition (Kline et al. 1989; Ellison 2003). Energetics also modulate the length of time that breastfeeding women do not menstruate, termed lactational amenorrhea. Lactation is extremely energetically costly for women, and typically leads to ovarian suppression. The duration of lactational amenorrhoea varies greatly between and within populations, and is thought to be directly affected by energetic demands and fat stores available for producing breast milk. The direct role of physical activity in the duration of lactational ovarian suppression is not well understood and is difficult to tease apart from other factors that influence lactation, such as frequency and duration of breastfeeding. Women who have heavy workloads often breastfeed less frequently and begin to supplement children's diet with alternative food sources earlier, decreasing maternal levels of prolactin and reducing the drive for milk production (Panter-Brick and Pollard 1999).

Evidence suggests that women catabolise more fat than men, during physical activity and at rest (Blaak 2001), in spite of having higher returns to stored adipose tissues, since these are utilised during pregnancy and lactation. Furthermore, in a study of previously sedentary men and women, exercise changed metabolism-regulating hormones in a direction associated with increased appetite (i.e. increased ghrelin and decreased insulin), but only in women (Hagobian et al. 2009). Physical activity may therefore increase required food intake, or perceptions of required food intake, specifically in women. In an environment of limited food resources, as was experienced for humans over the majority of human history, this would have been a substantial cost to all physical activity, but particularly any inessential physical activity. In modern environments, this may also make it more difficult for women who are using physical activity or exercise as a means to lose weight, as they are also likely to experience increases in appetite and have a more difficult or unpleasant experience when reducing caloric intake.

Thus, numerous lines of evidence suggest that, theoretically and empirically, it is advantageous for human females to be more conservative with physical activity whenever possible during reproductive years to divert more energy to pregnancy and lactation. This coevolved with humans' unique pattern of sociality involving pair-bonding, sex-specific divisions of labour, extended family food sharing, and cooperative breeding social networks.

Not surprisingly, this pattern of physical activity is observed in modern contexts as well. Physical activity decreases as women get married, become pregnant, or are postpartum (Brown and Trost 2003; Fell et al. 2009). A life history perspective helps us to understand why this is the case, and provides novel ways to help women resist becoming sedentary during this critical period with higher costs of physical activity. One study noted that while moderate and vigorous activity decreased during pregnancy and postpartum, walking did not (Pereira et al. 2007). Walking and other activities of a lower intensity may be preferable for women at this time when competing energetic demands and opportunity costs for physical activity are so high. Consistent with LHT, the women in this study who reported having other children in the home, having long work hours, and lacking child care during physical activity became less active during and after pregnancy.

Activities that are compatible with childcare or more childcare opportunities that women are comfortable with can help reduce the costs of physical activity for reproducing women. Among recent hunter-gatherers, most subsistence activity performed by women (mainly gathering plant foods and home-based tasks) was compatible with infant and child care; infants were always carried in slings on gathering trips and children often walked along (Konner 2010, 2016). A recent study held focus groups to talk to working mothers and fathers about the barriers and facilitators of physical activity. The barrier listed most frequently was family responsibilities, while what facilitated exercise the most was participating in an activity that was consistent with childcare responsibilities (Mailey et al. 2014). This perspective can inform even very inexpensive approaches for helping make physical activity more likely among reproductive age women by changing the way messages are framed. Parents (particularly mothers) may need tailored and more manageable recommendations that are lower than overall recommendations, and recommended activities consistent with pregnancy, lactation, and parenting.

Men's reproductive physiology, in contrast to women's, is generally less energetically costly and more robust in the face of reductions in energy availability or increases in energy expenditure. The quantity and quality of sperm production appear to be insensitive to short-term changes in energy availability (Bribiescas 2006). Moreover, physical activity is largely compatible with men's reproductive effort, increasing their ability to compete with other males for access to mates, and in environments with strenuous demands for resource production, provisioning wives and children.

In addition, research in modern contexts shows that investment in fitness and muscle mass actually *increases* men's access to mates (Frederick and Haselton 2007; Gallup et al. 2007; Hönekopp et al. 2007; Lassek and Gaulin 2009). While men do not appear to face nearly the same trade-offs between physical activity and reproductive investment as women, variation in physical activity may still be affected by resource availability and individual condition. Lassek and Gaulin (2009) examined the costs of musculature in terms of increased energy intake and trade-offs with immune function in the United States. They found that fat-free mass and limb muscle volume were positively related to food intake and inversely related to two proxies of immune function: C-reactive protein (CRP) and white blood cell counts.

Testosterone (T)—which modulates muscle mass, as well as allocations to mating effort (Burnham et al. 2003; Gray et al. 2007; Ellison and Gray 2009; Kuzawa et al. 2009)—appears to be unresponsive to short-term changes in energy availability, but sensitive to longer-term energetic deficit (Ellison 2003). T has also been shown to be negatively related to immune function (e.g. Muehlenbein and Bribiescas 2005; Muehlenbein et al. 2010). Some researchers have argued that investment in muscle mass can be thought of primarily as somatic effort aimed at reproduction by increasing mating opportunities (Bribiescas 2001, 2006). More recently, others have pointed out that somatic investment in musculature is required for paternal investment through provisioning and protecting partner and offspring from external threats (Lancaster and Kaplan 2009). Long-term energetic deficits, environments with constrained resources, heavy pathogen or infection risk, or a burdened immune system are all predicted to result in lower levels of T, which may therefore hinder men's reproductive effort in both mating and provisioning for offspring.

However, Alvarado and colleagues (2015) demonstrated that T was not correlated with strength or musculature among a sample of rural Polish subsistence agriculturalists who regularly engage in strenuous manual labour. The authors propose the paternal provisioning hypothesis which predicts that in humans, muscle is more sensitive to physical workloads than androgens, allowing for lower T levels to foster pair-bonding and paternal investment while simultaneously maintaining muscle mass through physically demanding resource production. In this case, energetic stress would not necessarily influence men's mating or paternal investment significantly. This may also influence humans' ability to maintain muscle in the absence of activity.

The opportunity costs of physical activity for men increase substantially, however, in environments where provisioning for offspring requires sedentary rather than strenuous physical labour, as is the case in many modern, technologically advanced environments. Physical activity is therefore predicted to be lower among committed, mated men with children in these environments, but men focused on mating rather than parenting effort are still expected to favour somatic investment in muscle mass in order to attract mates. As with mothers, it may be possible to increase physical activity among mated fathers in these more sedentary economies by reframing it to be more compatible with providing for children.

6.3.5.4 *Adulthood in the Post-Reproductive Years*

Another unique characteristic of the human lifespan is that humans have a long period of life after reproduction ceases, especially women (i.e. after menopause). In contrast, most other animals continue reproducing until close to the end of their lives. Some men have children with other partners after their partners reach menopause, but it is far more common for men to have their last child at the same time their partners do, and thus similarly have an extended post-reproductive life. This is thought to have evolved because of late-life returns to production and concentrated physiological demands of maternal investment.

Help from partners, parents, and in-laws allows human females to decrease production during pregnancy and lactation when the demands for resources go up (Marlowe 2010; Hooper et al. 2015). (For further discussion, see Chapter 4: Growth and Development.) A life history perspective would predict that decreases in trade-offs for reproduction might allow for increases in energy available for physical activity, but physical activity tends to decrease at older ages and particularly after menopause (Sallis 2000). The costs and benefits of physical activity in post-reproductive individuals are theoretically predicted to be influenced by the following: (1) how much grandparental investment in offspring and grand-offspring is dependent on physical activity, and (2) how much grandparental investment significantly contributes to offspring and grand-offspring survival and reproduction. Messages about the long-term health effects of physical activity are likely to be much more salient and effective at increasing physical activity during this stage of life. At earlier stages, these messages are unlikely to have much of an effect, particularly among those with a more energetically conservative life history strategy that does not favour investments in longevity.

Taken together, this research suggests that physiological and psychological mechanisms that influence physical activity have been shaped by natural selection to flexibly respond to energetic demands and resource availability in order to optimise survival and reproduction within a given environment. This system was, for the vast majority of human existence, consistently constrained by resource scarcity, and thus mechanisms have evolved to decrease energy expenditure in physical activity adaptively in response to decreased intake, early cues of nutritional constraints, or fewer adaptive benefits of expenditure (Shetty 1999; Vickers et al. 2003). However, there is little evidence that selection has favoured mechanisms to increase physical activity in order to match the resource abundance that has arisen in recent history.

A life history perspective can help us to understand how patterns of physical activity across the lifespan have been shaped to strategically allocate energy in ways that can optimise reproductive fitness. It can predict stages of life where allocating energy to physical activity will be more likely to occur (e.g. play among children, hunting, gathering, or competing for access to mates), as well as stages where it is less likely to occur (e.g. during puberty, pregnancy, and lactation, and when—as in post-hunting-gathering societies—offspring care is incompatible with physical activity). From there, approaches to increase voluntary physical activity can be tailored according to sex and life stage, and specific to individual condition and developmental experiences.

This perspective builds on evolutionary approaches that have been used thus far that highlight the ways humans are physically and mentally adapted for high levels of physical activity. This research adds to the evidence demonstrating that high levels of physical activity have been an important part of the life of humans for most of our evolutionary history. But considering the evolutionarily relevant benefits and costs of engaging in physical activity and investing in physical fitness across the lifespan is important for understanding the ways humans have also evolved to conserve energy and strategically allocate it to physical activity when the benefits outweigh the costs.

Furthermore, environments are important for shaping the costs and benefits of physical activity. Among horticulturalists with seasonal resource availability, physical activity is reduced when not required for harvesting resources, particularly among women. Human females also demonstrate reproductive function that is acutely responsive to increased energetic demands (Ellison and Jasienska 2007), and female foragers show greater variation than males in production according to the intensity and compatibility of childcare demands

(Hurtado et al. 1992; Dufour and Piperata 2008; Lancaster and Kaplan 2009); however, in many foraging societies, women's subsistence activities were largely compatible with infant and child care (Konner 2010). Changes to the built environment that increase opportunities for obligatory physical activity (transportation, etc.), as well as economic incentives or reduced trade-offs between provisioning, childcare, and physical activity are potential avenues for public and private policy initiatives that may help increase physical activity and overall health.

6.4 Body Composition

Prevailing palaeoanthropological opinion is that throughout evolution our pre-human and human progenitors had body compositions similar to those in contemporary naturally living primate species. For wild lemurs, thought to resemble our Eocene ancestors of 50 mya, percentage body fat varies seasonally from 5.2 to 20.5% (Simmen et al. 2010). For free-living baboons, whose size and lifestyles to some extent resemble those of Australopithecines, body fat ranges from 1.9 to 23.2% (Altmann et al. 1993). Ancestral hominins are calculated to have carried 11–14% fat (males) and 15–24% fat (females) primarily stored in adipose tissue (Ruff et al. 2005; Squyres and Ruff 2015). These proportions are believed to have prevailed from the time of *A. afarensis*, about 4 mya, to the appearance of behaviourally modern *H. sapiens* between 300 and 100 kya (McBrearty and Brooks 2000).

Anthropological studies of recent hunter-gatherers have included analyses of their BMIs. For 11 different forager groups from five continents, male BMI averaged 21.4 (Jenike 2001). Measurements of weight and height are available for British army recruits between 1860 and 1864. For these men, height averaged 169.7 cm (66.8 inches, or just under 5 ft 7 inches), mean weight was 61.9 kg (136.3 lb), and, accordingly, BMI averaged 21.6 (Rosenbaum and Crowdy 1992). In the nineteenth century, the BMIs of West Point cadets averaged 21.3 (Hiermeyer 2010). These data suggest that, at least for young male adults, BMIs during the 1800s were essentially similar to those during the pre-agricultural Palaeolithic era. In contrast, the BMIs of contemporary American male college students now average 24.1 (Behrens and Dinger 2003; Pribis et al. 2010), a 12% increase compared to either American or British youths of the 1860s.

Anthropological data on skinfold thickness are available for recently studied hunter-gatherers. For males and females (averaged), mean triceps skinfold thickness determined for multiple forager groups was 4.7 mm (Eaton et al. 1988). The figure for Americans (females and males averaged) in the National Health and Nutrition Examination Survey (NHANES) survey (2005–2006) was 19.5 mm (Fryar et al. 2008). This suggests that much of the BMI difference between hunter-gatherers and contemporary westerners is the result of added adipose tissue.

Physical activity is no longer a requirement for daily living, so that the immemorial tie between eating and physical work has been severed. For many, this 'new' circumstance results in energy surplus. Basal metabolic needs have, presumably, changed little and, accordingly, energy for disease resistance, growth, and reproduction has become more available. Secular trends in height (increasing), age at menarche (decreasing), and average life expectancy (increasing) are the result. All these processes may now be asymptotically approaching the phenotypic limits imposed by our underlying genetics.

In contrast, there seems to be almost no limit to our capacity for energy storage; at least the trend towards increasing obesity shows little sign of slowing. One anatomical consequence is that average body composition has been changing and, in turn, this has important physiological implications. Throughout evolution, our pre-human and human progenitors exhibited body composition similar to that observed in other naturally living primate species. However, in recent years, proportions of muscle and fat have deviated from the ancestral pattern conducive to normal carbohydrate metabolism.

The bony remains of pre-agricultural humans indicate that their muscularity was comparable to that of contemporary superior athletes (Frenkl et al. 2001; Ruff et al. 2005). For today's highly trained males, the ratio of skeletal muscle to adipose tissue approximates 5:1 (about 50% body weight as muscle and 10% as fat) (Kyle et al. 2001; Janssen et al. 2004). For current female athletes, the ratio can be as high as 3:1 (about 45% muscle and 15% fat) or more (Kyle et al. 2001; Janssen et al. 2004).

Today, in the United States and increasingly all over the world, people with excess fat (men > 25%; women > 35%) and reduced muscle (< 40% and < 30%, respectively) have become common (Kyle et al. 2001). These are the individuals at high risk for insulin resistance and T2DM. Their evolutionarily novel but increasingly widespread body compositional pattern alters the physiological conditions from those for which mammalian glucoregulatory mechanisms were originally selected. For such individuals, the proportion of adipocyte insulin receptors is increased relative to those of myocytes (Figure 6.2).

6.4.1 Adipocyte–Myocyte Competition for Circulating Insulin

Myocyte and adipocyte insulin receptors are identical. There are two insulin receptor isoforms which differ in their affinity for insulin, but the proportions present in skeletal muscle

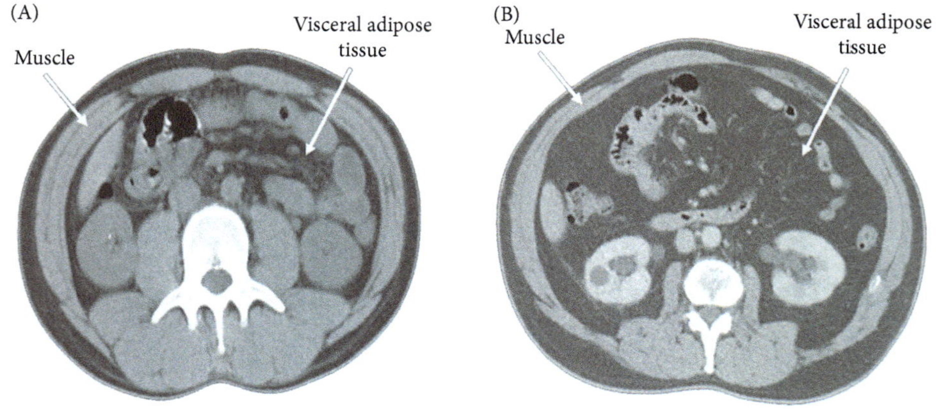

FIGURE 6.2 (A) CT scan of the abdomen of a lean individual, with little fat relative to muscle, analogous to the muscle to adipose ratio in humans for the vast majority of history. (B) CT scan of the abdomen of an overweight individual with much higher fat relative to muscle.

Source: Reproduced from S. Boyd Eaton and Stanley B. Eaton, Physical Inactivity, Obesity, and Type 2 Diabetes: An Evolutionary Perspective, *Research Quarterly for Exercise and Sport*, 88 (1), p. 4, Figure 2 a and b, doi.org/10.1080/02701367.2016.1268519. © 2017, Taylor & Francis.

and adipose tissue are similar (Benecke et al. 1992). Accordingly, fatty tissue and skeletal muscle engage in a whole-body competition for circulating insulin: insulin molecules that dock on adipocyte receptors are unavailable to muscle insulin receptors and vice versa (Eaton et al. 2002a; Wells 2010). For this reason, relative tissue proportions play a major role in determining how the insulin molecules released during any given pancreatic secretory pulse are distributed. Another important determinant is the proportion of cardiac output reaching various body regions—a factor responsible for the well-known apple–pear dichotomy, in which the predominant fat distribution is upper or lower body, respectively (Williams and Leggett 1989).

While the insulin receptors of adipocytes and myocytes are structurally identical, their biochemical impact differs. In humans, an insulin molecule activating a muscle insulin receptor induces 2.1 to 3.1 times more glucose clearance than does one reacting with an adipose tissue receptor (Virtanen et al. 2002). Thus, insulin molecules interacting with fatty tissue are being utilised inefficiently. Other factors being equal, this means that lean, muscular individuals exhibit greater whole-body insulin sensitivity than those who are sarcopenic and overfat. Insulin receptor imbalance—too many adipose tissue receptors relative to myocyte receptors—necessitates extra insulin secretion (hyperinsulinaemia) for any given carbohydrate load.

6.4.2 Hyperinsulinaemia and Intrinsic Insulin Resistance

Insulin, like other hormones, stimulates its target cells, but also resets their responsiveness, down-regulating the effect of subsequent hormone exposure. This process, known as 'homologous desensitisation', is widespread. In the case of insulin, persistently elevated levels desensitise target cells (myocytes, adipocytes, hepatocytes, etc.) through multiple intracellular mechanisms (Shanik et al. 2008). These multiple regulatory responses to elevated insulin levels typically lead to generalised intrinsic insulin resistance (Kahn and Flier 2000). Intrinsically resistant tissues (muscle, fat, liver) take up less glucose than normal, despite similar tissue mass, insulin stimulation, and glucose concentration. This is the kind of insulin resistance ordinarily thought of in connection with T2DM: an inappropriately reduced response of target cells to insulin stimulation. For example, (intrinsic) skeletal muscle insulin resistance has been called 'the primary defect in Type 2 Diabetes' (DeFronzo and Tripathy 2009). However, insulin resistance could be more broadly defined as 'a state in which a normal amount of insulin produces a subnormal biologic response', or reduced responsiveness of the whole organism to the insulin concentration to which it is exposed (Shanik et al. 2008). This whole organismal, or systemic, reduced responsiveness, which occurs earlier, is readily explained by adipocyte–myocyte imbalance.

As a 'primary defect', intrinsic cellular resistance begs the question as to how it comes about and how it relates to the secular increase in T2DM; extraordinarily rapid genetic evolution seems unlikely. On the other hand, secular changes in body composition as 'the primary cause' can be readily understood (although they, in turn, indicate sedentary behaviour and nutritional overload—both evolutionary novelties—as the true initiating factors). Because adipocyte–myocyte imbalance leads, ultimately, to intrinsic cellular insulin resistance, development of T2DM can be understood as a two-phase process. The phases are not sequentially discrete, they overlap, but the initiating phenomenon is abnormal body composition and associated insulin receptor imbalance. Intrinsic insulin resistance, myocytic or otherwise, is the secondary, reactive phase.

6.4.3 Effect of Adipose Tissue Distribution

The anatomic location of adipocytes affects the likelihood that they will capture circulating insulin molecules, and the apple–pear dichotomy arises from this principle. The intra-abdominal (visceral) compartment receives 25% of cardiac output at rest and 35% during digestion. Skin and subcutaneous tissue receive only 5% (Williams and Leggett 1989). Accordingly, adipocyte insulin receptors within the abdominal cavity (as in apple body type) are favourably placed to compete for insulin molecules because they are exposed to five to seven times as many such molecules as are adipocyte insulin receptors in subcutaneous adipose tissue (as in pear body type). Even individuals whose BMI is 40 or more may exhibit normal insulin sensitivity if they have relatively little visceral fat and marked muscularity—Sumo wrestlers being an extreme example (Matsuzawa 1997; Yamauchi et al. 2004).

In summary, throughout the entire evolution of animal life on earth, energy acquisition and energy expenditure have been inextricably linked and humanity's recent uncoupling of these vital functions must surely be a major contributor to, and perhaps the ultimate driver of, both the obesity and the diabetes (T2DM) epidemics. Distortion of human body composition—excessive adipose tissue relative to muscle—has reordered the proportions of myocyte and adipocyte insulin receptors. Insulin acting on adipocyte receptors produces less glucose uptake than does comparable interaction with myocyte receptors, Accordingly, in individuals with disproportionate muscle–fat composition, any given glucose load must require greater than normal pancreatic insulin secretion for adequate disposal. This hyperinsulinaemia then becomes T2DM's proximate cause as insulin-sensitive tissues become desensitised—producing intrinsic insulin resistance.

6.5 Questions and Challenges

Since two of us (Eaton and Konner 1985; Eaton et al. 1988) first proposed our version of the mismatch (discordance) hypothesis, many challenges have arisen. Principal among these are evidence suggesting that: (1) hunter-gatherer diets had far *higher* proportions of animal products than our original estimate of 35%; (2) saturated fat is not atherogenic or unnatural; (3) healthy diets should have a much *lower* proportion of animal products; (4) salt restriction does not prevent hypertension as we claimed, and furthermore marine ecosystems important for some hunter-gatherers provided high salt levels; (5) atherosclerosis (AS) was common in ancient populations and is part of the human condition; (6) energy throughput corrected for body size does not differ between current hunter-gatherers and sedentary Westerners; and (7) genes have evolved so much in the past 10,000 years that the theoretical basis of the discordance concept is voided. We have taken up some of these challenges before (Eaton et al. 1997, 2002b; Konner and Eaton 2010), and so have others (Kuipers et al. 2012), but here we address their current status systematically.

6.5.1 Claim: Hunter-Gatherers are Basically Carnivores

'Palaeo' diets on the internet and in recent books vary, but one popular version promotes a low or very low intake of plant products (Wolf 2010). Some versions restrict intake of or eliminate grains, dairy products, legumes, and alcohol in order to hew faithfully to their concept of our ancestral diet. Some internet versions of 'palaeo' are more moderate and scientifically informed (Cordain 2017; Noakes and Windt 2017). The more flesh-dominated versions have something in common with some weight-loss diets, for example the very-low-carbohydrate, high-protein, high-fat Atkins diet (Atkins 2002; Atkins et al. 2003) and the low-carbohydrate, low-saturated fat, high-protein 'South Beach' diet (Agatson 2003). Like previous and still popular high-carbohydrate diets restricting refined carbohydrates (Ornish et al. 1998; Esselstyn et al. 2014), these varied diets result in weight loss if adhered to (Johnston et al. 2014). However, most also claim to be advisable for health maintenance, a more controversial matter.

Low-carbohydrate diets labelled 'palaeo' have their origins in estimates of hunter-gatherer diets published by a group including one of us (SBE); (Cordain et al. 2000a, 2002). The key study showed the distribution of the animal and plant contributions to the diets of 229 hunter-gatherer groups (Cordain et al. 2000a), concluding that the modal diets were 55–65% animal. However, as pointed out in a commentary accompanying the paper (Milton 2000), the sample included equestrian and arctic hunter-gatherers, who consume very high proportions of flesh, which is not relevant to ancestral human diets; also, humans do not have gastrointestinal tracts consistent with carnivory. Some hunter-gatherers living in tropical forests, where paradoxically plant foods may be scarce due to lack of sunlight on the forest floor, also have high proportions of meat consumption (Hill et al. 1996), but the question of how much of modern human evolution could have taken place in forests as opposed to savannas remains controversial (Roberts et al. 2016).

A sample of 179 warm-climate, non-equestrian hunter-gatherer groups resulted in a median estimate for dietary composition of 53% plants, 26% meat, and 21% fish (Marlowe 2010). This estimate of the contribution of animal products (47%) is below the lower end of the 55–65% range proposed by Cordain's group (Cordain et al. 2000a), but above the 35% estimate two of us originally provided (Eaton and Konner 1985). Many, though not all, hunter-gatherer diets included substantial amounts of insects, reflecting a deep evolutionary history of insectivory in primates (Raubenheimer et al. 2014). Almost all living primates, including all great apes, eat insects (Fox et al. 2004; Rothman et al. 2014; Hamad et al. 2015), and they almost certainly figured in the long course of hominin evolution (Lesnik 2014).

However, the ultimate arbiter is archaeology (Warinner et al. 2013; Lin and Epstein 2014). The prosimian line from which monkeys and apes originated was significantly insectivorous as well as herbivorous, but all monkeys and apes are primarily plant eaters. Our common ancestor with chimpanzees and bonobos may have consumed some meat (chimp diets are 3–10% flesh, bonobos less), but meat became important early in hominin evolution (Mann 2000; McPherron et al. 2010). However, there is strong evidence that plant foods continued to be important throughout the emergence of modern *Homo sapiens* (Eoin 2016). Both starches and grass seeds were important in hominin diets in Mozambique in the Middle Stone Age (Mercader et al. 2008; Mercader 2009) and the Middle Palaeolithic of Israel (Lev et al. 2005). Geophytes among other plants provided a stable base of carbohydrates for humans in Western Cape Province, South Africa, roughly 164 kya (Marean 2010).

Plants were important in human diets in the Middle Pleistocene of Germany and the Late Pleistocene of Spain (Allué et al. 2010; Bigga et al. 2014). Exploitation of plant foods in mangrove forests near maritime ecozones may have been crucial in enabling humans to leave Africa and spread along coastal routes across Asia (Erlandson and Braje 2015).

Even Neanderthals, who are not our direct ancestors, were not as dependent on meat as previously thought (Lin and Epstein 2014; Fiorenza et al. 2015; Estalrrich et al. 2017). Microfossils in dental calculus show that they were consuming starches and ate something like barley broth (Henry et al. 2011). Neanderthals from open habitats were more meat dependent, but those living in forested areas had substantial amounts of plant food (Lin and Epstein 2014; Krueger et al. 2017). Contemporary and post-Neanderthal modern humans had very varied plant and animal diets and this may have helped them outcompete Neanderthals (Holst 2010; Henry et al. 2014; Janz 2016). Notably, at least 10,000–20,000 years before routine domestication of grains, grindstones were used to process wild grass seeds in widely separate sites (Piperno et al. 2004; Revedin et al. 2010, 2015; Mariotti Lippi et al. 2015; Groman-Yaroslavski et al. 2016), and the consumption of wild grasses goes back much earlier, disproving the hypothesis that grains were not part of ancestral diets (Mercader 2009; Eoin 2016). The claim that legumes were not part of those diets can also be rejected (Lev et al. 2005). Cooking was an important part of the dietary adaptation of *H. sapiens* and perhaps the genus *Homo*, making plant as well as animal food sources more digestible (Conklin-Brittain 2003; Wobber et al. 2008; Wrangham et al. 2009; Wrangham and Carmody et al. 2016).

6.5.2 Claim: Saturated Fat is Healthy and Ancestral

In the 1980s, there was a consensus that saturated fat in post-industrial diets was the main cause of atherosclerotic disease, but it is now clear that this was wrong (Taubes 2008, 2011, 2016). Dietary carbohydrate drives circulating levels of saturated fatty acids (Volk et al. 2014). Saturated fat intake is unrelated to ischaemic heart disease (de Souza et al. 2015; Praagman et al. 2016), and it does not predict cardiac events or mortality even in patients with established heart disease (Puaschitz et al. 2015). Dietary Saturated fat increases LDL cholesterol in some studies (Mansoor et al. 2016), and improvement in other risk factors outweighs this added risk (Wood et al. 2016). Low-carbohydrate high-fat (LCHF) diets are effective and safe in weight loss both generally and in the context of metabolic syndrome and T2DM (Hu et al. 2012; Saslow et al. 2014; Feinman et al. 2015; Noakes and Windt 2017). Most studies of LCHF diets do not stratify results by fat type, but those that do have not produced a consensus (Kaplan et al. 2010; Ramsden et al. 2013; de Souza et al. 2015; Hooper et al. 2016).

Carbohydrates, especially highly refined (manufactured, high-glycaemic-index) carbohydrates, are considered drivers of the diabetes epidemic (Imamura et al. 2015; Popkin 2015; Lustig 2016), and are strongly implicated in non-alcoholic fatty liver disease (Neuschwander-Tetri 2013) and CVD (Johnson et al. 2009). High-fructose corn syrup causes metabolic syndrome and/or T2DM in rats and monkeys (Bremer et al. 2011; Toop and Gentili 2016). Arguments are still made defending sugars (Sievenpiper 2016), but meta-analyses show that financial bias influences research that tends to exonerate them (Bes-Rastrollo et al. 2013). Sugar addiction is real and probably transgenerational (Bayol et al. 2007; Avena et al. 2008). Manufactured carbohydrates were non-existent in ancestral diets, although non-negligible energy was obtained from honey (Crittenden 2011; Marlowe et al. 2014).

How much did fat in general, and saturated fat in particular, contribute to ancestral diets? Our original estimates suggested that 20% of calories came from dietary fat, with a much higher PUFA to saturated fatty acid (P:S) ratio than in modern diets (Eaton and Konner 1985). This was later challenged based on a sample of hunter-gatherers that inappropriately (see Section 6.5.1) included polar and equestrian groups (Cordain et al. 2000a; Eaton et al. 2002b). However, it was also pointed out that we had omitted depot fat from our calculations. This fat would certainly have been eaten by hunter-gatherers throughout the world. (No one who has seen the ethnographic film classic *Nanook of the North* will forget the Eskimo boy eagerly eating a hunk of blubber.) We now consider the 20% estimate to have been substantially too low, although the overall P:S ratio remains uncertain.

Mammalian fat is necessary and adaptive (Young 1976; Lindstedt and Boyce 1985; Pond et al. 1993). The P:S ratio is demonstrably low in game animal muscle fat (Davidson et al. 2011; Bartoň et al. 2014; Bureš et al. 2015; Hoffman et al. 2015), but not in depot fat (Eaton et al. 2002b), except in aquatic mammals (Arnould et al. 2005; Guerrero et al. 2016; Guerrero and Rogers 2017). However, among terrestrial mammals, bodies become larger (Bergmann's rule) and rounder (Allen's rule) as latitude increases, but in addition the proportion of body mass devoted to fat increases (Lindstedt and Boyce 1985; Mugaas and Seidensticker 1993). This means that fat content is highest in the cold climates that were not formative during the evolution of modern humans (McBrearty and Brooks 2000). To the extent that fish (including shellfish) were important in modern human evolution, fats with a high P:S ratio and n-3 fatty acids were also important.

It has been suggested that 'fat hunting' capabilities were important in the replacement of *H. erectus* by *H. sapiens* (Ben-Dor et al. 2011). Hunter-gatherers certainly like meat (Biesele 1993); no doubt fat was desired, but the amount and fatty acid partitioning of fat in meat as well as in fat depots remains controversial (Mann 2000; Li et al. 2005). The Masai, cattle herders of Kenya who have high dietary levels of whole milk and blood, were found early on to be basically free of CVD risks (Mann et al. 1964), and were recently shown to have specific genetic adaptations for serum cholesterol regulation and lactase persistence (Wagh et al. 2012). Despite the palatability of saturated fat, its possible role in satiety, and its proven role in weight-loss diets, long-term negative effects on health cannot be ruled out. However, the older recommendations making it the chief culprit in CVD were clearly wrong. Evidence suggests that replacing dietary saturated fat with PUFAs may decrease CVD risk, while replacing saturated fat with carbohydrate, especially sugar, increases it (Siri-Tarino et al. 2015).

6.5.3 Claim: High-Carbohydrate, Low-Fat Diets are Healthiest

This claim, if true, would imply that the substantial flesh content of ancestral diets has nothing to say about health. While unlikely, this is what we would have to conclude if epidemiology, laboratory studies, and RCTs supported it. Do they? The idea of 'meat as a condiment' was made famous by Nathan Pritikin, a layman who reversed some of his own heart disease symptoms with a near-vegetarian diet and exercise plan (Pritikin and McGrady 1979). Physician-scientists like Dean Ornish and Caldwell Esselstyn created similar diets (combined with exercise and stress reduction), doing research that established the safety of these diets. Radiologically proven reversal of atherosclerotic arterial stenosis (Ornish et al. 1998) and reductions in major cardiovascular events (Esselstyn et al. 2014) were achieved. Importantly, these diets are very low in refined carbohydrates.

Epidemiologically, the China Study, suggesting that in 69 rural Chinese counties vegetarianism is associated with many positive measures of health, has been widely influential (Campbell and Campbell 2006; Chen et al. 2007). Since its recommendations are diametrically opposed to those of some versions of 'palaeo' diets, as well as of LCHF and LCHP diets, vigorous debates ensued (Atkins et al. 2003; Cordain and Campbell 2008). Claims are often made about how easy or difficult different diets are to adhere to and about how they differ in subjective feelings of satiety, but these things are difficult to measure. However, in one RCT a Palaeolithic-type diet (more or less as we define it) was found to be more satiating per calorie (i.e. equally satiating at a lower caloric intake) than a Mediterranean-type diet (Jönsson et al. 2010). The China Study has been strongly criticised on statistical and methodological grounds (Chen et al. 2007; Minger 2010). Overall, HCLF diets that exclude highly refined carbohydrates will, if adhered to, produce weight loss (like LCHF and LCHP diets) and improved cardiovascular health, but as they are not the only diets that do these things, there is no basis for recommending that everyone follow them.

6.5.4 Claim: Salt Restriction is Not Necessary and Perhaps Not Ancestral

Daily intake of sodium (Na^+) for hunter-gatherers (~ 1.0 g/day) is far less than for those living in post-industrial nations (3–4 g/day) (Eaton and Konner 1985; Elliott and Brown 2007). The Intersalt Study (Intersalt Cooperative Research Group 1988) correlated 24-hour urinary Na^+ excretion and blood pressure in men and women from 32 countries. For three of the groups studied (Brazilian Xingu Indians, Papua New Guinea highlanders, and Venezuelan Yanomamo Indians), Na^+ excretion was less than 1g/day—similar to that estimated for ancestral humans. None of these three groups had ready access to commercial salt, their average blood pressure was 103/63 mmHg, and none had hypertension. For the other groups, modal Na^+ excretion was 4 g/day for men and 2.9 g/day for women and over 30% were hypertensive. All the remote groups consumed substantially less Na^+ than K^+, typically less than one-fourth as much. In contrast, the groups with access to commercial salt consumed anywhere from two to seven times more Na^+ than K^+.

To date, the Intersalt Study is the only epidemiological investigation that has included subjects whose Na^+ and K^+ intakes were comparable to those of typical hunter-gatherers. From an evolutionary perspective, epidemiologists have been studying a subgroup, humans whose electrolyte intake patterns differ from those of not only their ancestors, but also most other free-living, terrestrial mammals. Even so, many studies have shown a linear increase in blood pressure with salt intake or salt excretion within the range examined, and there is a dose-response effect in RCTs (Graudal et al. 2017). Also, there are negative effects of Na^+ in CVD and all-cause mortality independent of blood pressure (Konerman and Hummel 2016), linear down to the lowest levels observed, in some studies (Mozaffarian et al. 2014; Cook et al. 2016). In one global modelling study, '1.65 million deaths from cardiovascular causes that occurred in 2010 were attributed to Na^+ consumption above a reference level of 2.0 g per day' (Mozaffarian et al. 2014). Hypertension is considered pandemic worldwide (Rossier et al. 2017).

However, some studies have found a non-linear (U- or J-shaped) relationship between Na^+ intake and outcomes, with suggested optimal intakes being between 3 and 6 g/day (Mente et al. 2014; O'Donnell et al. 2014; Kong et al. 2016; Lamelas et al. 2016). Guidelines currently (2018) recommend lowering salt intake throughout the world, but some experts think that most of the world is in the optimal range and the guidelines are too strict (Graudal 2016). However, there is much less evidence for increased risk at low intake levels (< 3 g/day) than at high levels (> 6 g/day), and the lower limit of this range warrants further justification. There is little or no argument that in the post-industrial West intake is too high. Ongoing research is exploring possible risks of low Na^+ intake. Part of the discrepancy may be in methodology, notably the frequency of urine sampling. We have also argued that, partly due to higher K^+ intake from varied sources, hunter-gatherer K^+/Na^+ ratios were much higher; the protective effect of K^+ is well known (Mente et al. 2014; O'Donnell et al. 2014; Rodrigues et al. 2014).

Did we underestimate Na^+ intake in hunter-gatherers? Recent scenarios for the evolution of modern humans suggest that maritime, riverine, and lacustrine habitats played an important role (Marean 2016). Maritime locations may have been important in the spread of modern humans from South Africa to Asia and Australia (Erlandson and Braje 2015; De Vynck et al. 2016; Marean 2016). Such populations would have had unlimited access to sea salt and could perhaps have traded it inland; one of the oldest towns in Europe, although agricultural, was a centre of salt trade 7 kya (Nikolov 2012). Compared with table salt, sea salt is lower in Na^+, rich in minerals, and higher in K^+ (Greenfield et al. 1984), but the K^+ is not sufficient to balance Na^+, so K^+ has to come from other sources to produce an effect that buffers Na^+ effects on blood pressure (O'Donnell et al. 2014; Rodrigues et al. 2014). However, sea salt is less hypertensive than table salt for laboratory rats, probably because of its lower Na^+ content by weight. We conclude that it is possible that some ancestral diets had high Na^+ intake from sea salt, and that more evidence is needed to determine whether Na^+ intake should be reduced to less than 2 g/day. (For further discussion, see Chapter 14: Excretory System.)

6.5.5 Claim: Atherosclerosis was Always Common and is Part of the Human Condition

An important and surprising report in 2013, extending earlier findings, found AS in mummies over a period of 4000 years (Finch 2011; Thompson et al. 2013). Known as the Horus Study, it stated that 'Probable or definite atherosclerosis was noted in 47 (34%) of 137 mummies and in all four geographical populations: 29 (38%) of 76 ancient Egyptians, 13 (25%) of 51 ancient Peruvians, two (40%) of five Ancestral Puebloans, and three (60%) of five Unangan hunter gatherers' (Thompson et al. 2013, p. 1211). The Horus Study authors conclude: 'The presence of atherosclerosis in premodern human beings suggests that the disease is an inherent component of human ageing and not characteristic of any specific diet or lifestyle' (Thompson et al. 2013, p. 1221), and others have reached similar conclusions (Clarke et al. 2014). However, a co-author of the Horus Study wrote separately: 'The relatively small samples of elite Egyptian individuals who were mummified preclude generalisations about atherosclerosis at the population level' (Finch 2011). The Peruvian sample is also likely biased toward elites, as any deliberately mummified sample would be. Only the five Unangan individuals were hunter-gatherers randomly mummified by their very cold climates; none of these populations is really relevant to human evolution. The Horus Study has been criticised and the criticisms replied to (Fornaciari et al. 2013; Thompson et al. 2013).

There is no reason to doubt that elite individuals in the ancient world could have had atherogenic diets and lifestyles, and there is equally little doubt that AS became much more prevalent in the West in the nineteenth and twentieth centuries and is now endemic in most of the world (Cervellin and Lippi 2014). For example, in the Rotterdam Study of community-living adults in a Dutch suburb, only around 3% of individuals (mean age early 70s) were free of arterial calcifications, compared with almost 50% of the Horus Study mummies in their 40s and 60% of those over 50 (Thompson et al. 2013). In asymptomatic subjects at a preventive medicine clinic in San Diego (Allison et al. 2004), 'Approximately one third of subjects younger than 50 were free of calcified disease, whereas all subjects older than 70 were found to have some calcium' (p. 331).

Some arterial beds are more involved in particular risks than others. Specifically, with respect to coronary artery calcification (CAC), which predicts myocardial infarction, in an analysis of three major studies, the US-based Multi-Ethnic Study of Atherosclerosis (MESA, n = 6726), the Dallas Heart Study (DHS, n = 1080), and the German Heinz Nixdorf Recall study (HNR, n = 3692), the respective percentages of zero scores for CAC were 50.1 (mean age 62), 33.4 (mean age 53), and 30.8 (mean age 60). In the Horus study, only 4% of individuals showed CAC; that is, 96% were free of CAC (Thompson et al. 2013). Of 18 Egyptian and Peruvian mummies in the study, three (17%) had calcifications in the internal carotid arteries, which predicts stroke risk; the comparable percentages for modern Greeks and Brazilians in that age range were 46 and 55, respectively (Da Silveira et al. 2016).

Atherosclerotic arterial calcifications have been found in other mummified remains since the Horus Study (Binder and Roberts 2014; Piombino-Mascali et al. 2014, 2017; Gabrovsky et al. 2016). While it is interesting to know that certain ancient populations, particularly elite ones, had significant prevalence of AS, their prevalence does not compare with the modern range in many countries encompassing whole populations. All samples studied show increases in prevalence with age, but in modern industrial and post-industrial populations AS is endemic or even pandemic in unprecedented ways related to modern lifestyles.

The best study to date of a population that does not have a modern lifestyle, the Tsimane forager-horticulturalists of the Bolivian Amazon, shows very low levels of cardiac risk factors, including arterial calcifications, and much less steep increases with age (Gurven et al. 2017; Kaplan et al. 2017). Of 705 Tsimane individuals over 40 years of age (mean = 58), 85% had zero CAC, and for those over 75 years old (n = 48) the corresponding percentage was 65 (Kaplan et al. 2017). Only four in that age group had CAC scores over 100, proportionately a highly significant five-fold difference from the MESA population in a direct comparison (Kaplan et al. 2017). More research on arterial calcification in traditional peoples is needed, but their much lower levels of serum cholesterol, blood pressure, and other CVD risk factors have long been clear (Eaton et al. 1988; Konner and Eaton 2010).

6.5.6 Claim: Dietary Intake is Much More Important than Physical Activity for the Recent Rise in Obesity

Recent observational studies examining TEE measured by DLW across time and different populations have led to unexpected findings. Although obesity has increased dramatically since the 1980s, and physical activity is considered an important part of maintaining a healthy weight (Jakicic and Otto 2005), TEE has not changed significantly (Westerterp and Speakman 2008). A meta-analysis compared TEE in countries with a high development index to those with a low or medium development index and did not find differences between the two in TEE (Dugas et al. 2011), despite presumably different lifestyles based on economic development. Another study compared TEE (adjusted for body size) between the Hadza in Africa and Westerners and found no significant differences between these populations, despite extremely different PAL (2.26 and 1.81, respectively) (Pontzer et al. 2012) and objectively measured activity patterns (Raichlen et al. 2016). In light of the counterintuitive observations of similar TEE over time and across different lifestyles, Pontzer et al. (2012) interpret this finding to suggest that energy intake is responsible for recent increases in overweight and obesity, but not expenditure, and by extension physical activity/exercise (Pontzer 2012b).

It is paradoxical that physical activity does not translate into increases in TEE (controlling for body size) in the linear way scientists predicted but do not fully understand currently (Westerterp and Plasqui 2004; Westerterp 2013). However, in the case of the Hadza, the conclusion that physical activity energy expenditure does not influence body composition and increases in obesity does not accurately reflect the analysis that was done. This conclusion is framed as if body size was the outcome variable, when in actuality it was a covariate. In other words, because body size was treated as a covariate, the relationship between body size and TEE was statistically removed, and conclusions about the effect of TEE on body size cannot be drawn. A more appropriate statistical approach to draw that conclusion would be to demonstrate a lack of relationship between PAL (or TEE) and body composition, with body size/composition (or fat mass) as the outcome variable. The extension of these findings to imply that the recent epidemic rise in overweight and obesity is due primarily to energy intake and *not* expenditure or physical activity does not accurately represent the analyses conducted in the study of the Hadza, nor the complexity of the relationship between physical activity and TEE.

The relative contributions of body size (i.e. the energy costs of a large body mass) and physical activity to TEE and the development of overweight/obesity require further investigation. The Hadza example demonstrates that TEE levels (controlling for body size) do not help us to understand the myriad health benefits of physical activity on health, well-being, and body size that are consistently observed. Simultaneous, objective measures of physical activity with accelerometers and appropriate analytical techniques will add meaningfully to our understanding of the effects of physical activity on TEE and obesity. For example, high levels of TEE (measured with DLW) were observed in a sample of individuals with obesity, which masked the very low physical activity (measured with accelerometers) observed in these individuals (DeLany et al. 2013). We interpret these findings and those of Pontzer, Raichlen, and colleagues to suggest that a similar level of TEE can be achieved through high levels of physical activity and a small body size, or through a large body size and low levels of physical activity. Although this distinction cannot be observed in analyses that statistically remove the effect of body size on TEE, it is likely to have meaningful implications for weight management and disease risk. The physically active Hadza are in fact leaner and have no evidence of CVD risk (Raichlen et al. 2016). Teasing apart the proportion of TEE 'spent' on physical activity versus body size will lead to clearer understanding of the role of physical activity on energy balance, weight management, and related chronic disease risk.

6.5.7 Claim: Genetic Evolution in the Past 10,000 Years Negates the Mismatch Model

The mismatch or discordance hypothesis holds that our genomes evolved over millions of years, but particularly during the emergence of modern *H. sapiens* 300 to 100 kya. This formative period of the human species in our EEAs left us with a genome that is adapted to environmental conditions then, and cultural changes since the hunting-gathering period (the past 10–12 kya, also known as the Holocene) have been proposed to have been too rapid for the genes to catch up. Hence the mismatch and all the arguments above. But some anthropological geneticists have argued that there has been so much evolutionary change in human genes that very recent adaptations have neutralised the mismatch, calling Holocene evolution 'the ten thousand year explosion' (Cochran and Harpending 2010).

Certainly there is growing evidence of recent genetic evolution (Fu et al. 2012, 2016; Field et al. 2016). We have known since the mid-twentieth century of the life-course extension of lactose tolerance through gene changes in dairying populations during the Holocene, and we now understand the details of this convergent evolution affecting around 30% of the human species in different locations around the globe (Enattah et al. 2002; Tishkoff et al. 2007; Ranciaro et al. 2014). Remains of Basque individuals in 5000-year-old burials have much lower levels of lactase persistence than do Basque populations today, due to post-Neolithic adoption of dairy cattle and fresh milk consumption (Plantinga et al. 2012).

Evidence of other new dietary adaptations is accumulating. Copy number variation in the salivary amylase gene (AMY1) is directly correlated with salivary amylase proteins and people from populations with high-starch diets have more AMY1 copies than those whose traditional diets were low-starch (Perry et al. 2007). Furthermore, a selective sweep led to fixation of multiple AMY1 copies in modern humans after the modern human/Neanderthal split, giving us a genetic signature unique among primates (Inchley et al. 2016)—and incidentally confirming the importance of starch in the diet of early modern humans. Reduced copy number in populations with very low starch intake (as in northeastern Siberia) was a *secondary* recent adaptation.

Many were impressed too by the demonstration that Greenland Inuit (Eskimo) have been under strong selection for a suite of fatty acid desaturase (FADS) genes specific to metabolising PUFAs, which are plentiful in their marine-animal-heavy diet (Fumagalli et al. 2015). It can be pointed out that the Inuit have experienced an approximately 20,000-year bottleneck due to isolation that has resulted in a unique genome (Pedersen et al. 2017). However, selection on FADS genes in relation to diet has also been shown in Europeans (Buckley et al. 2017). Specifically, a comparison of present-day to Bronze Age Europeans indicates that selection has favoured a decrease in linoleic acid and an increase in arachidonic and eicosapentaenoic acids, the opposite of the pattern seen in Greenland. The authors hypothesise that after the transition to agriculture, European diets were lower in arachidonic acid and eicosapentaenoic acids, but higher in linoleic and alpha-linolenic acids.

Furthermore, one of the new alleles interacts with dietary PUFAs such that carriers have lower LDL cholesterol levels with higher PUFA intake, but non-carriers do not. This suggests a post-Neolithic adaptation to dietary change affecting serum cholesterol metabolism, a major influence on CVD risk. There has also been recent and local selection in relation to toxic dietary chemicals. A population in the Argentine Andes has a genetic adaptation that allows more efficient metabolism of arsenic, compared with a closely related Peruvian population with no arsenic exposure in its history, due to high frequency of an arsenic methylase gene (Schlebusch et al. 2015).

These findings are a few examples focused on dietary adaptation, but there are many more outside the realm of diet, including recent genetic evolution of adaptations to altitude (Bigham et al. 2010; Yi et al. 2010; Huerta-Sánchez et al. 2014), pigmentation of skin, hair, and eyes (Jablonski and Chaplin 2010; Wilde et al. 2014), and immune responses (Kwiatkowski 2005; Fumagalli et al. 2011; Mathieson et al. 2015). Between 15 and 2 kya, Europeans experienced an increase in mutation rate, an important contributor to the variation that is the substrate of evolutionary change (Harris and Pritchard 2017). A bottleneck occurred in the Y-chromosome around 10 kya, coincident with an increase in polygyny and other measures of variance in male reproductive success (Karmin et al. 2015), and many millions of men carry Y haplotypes belonging to two individual males who lived less than 1 kya (Zerjal et al. 2003; Moore et al. 2006). Signals of substantial selection between 2 and 3 kya are found in UK residents today (Field et al. 2016), and the United Kingdom has also seen recent evolution of genetic variants related to blood pressure (Galinsky et al. 2016). Many genetic changes throughout the world have been recent and local (Fan et al. 2016).

It is clear that the 10–12 kya since the emergence of agriculture have allowed time for some genetic adaptation to diet and lifestyle. However, there are many reasons to doubt that all or even most of the needed evolutionary genetic changes have occurred. First, it has been known for decades and recently confirmed that there was a decrease in body size and health at the time of the emergence of agriculture, whenever it occurred, throughout the world (Cohen and Armelagos 1984; Mummert et al. 2011; Macintosh et al. 2016). Second, despite the instances of genetic adaptation to diets discussed above (and more to be discovered), it is evident from the emerging pandemic of chronic diet-related disease that whatever genetic adaptations have evolved are not sufficient to protect us. Specifically, dietary and activity changes in the past two centuries, with their huge increase in manufactured foods (including sugars and other refined carbohydrates), have outpaced any possible evolutionary gene changes (Lieberman 2003, 2006; Wolf et al. 2008; Babbitt et al. 2011; Armelagos 2014). Third, evolution is ongoing but has clearly not caught up, as specific cases show.

For example, in an indigenous population of Oaxaca, Mexico, there is evidence of ongoing natural selection against individuals with T2DM (Little et al. 2017). The 'thrifty genotype' hypothesis, proposed by James Neel (Neel 1962), has not been found to be generally applicable in recent genetic studies (Speakman 2013; Ayub et al. 2014), but appears to apply in particular situations, such as among Samoans and Mexicans (Minster et al. 2016; Sánchez-Pozos and Menjívar 2016). It may seem circular to argue that because there is a growing pandemic of chronic diseases, evolution cannot have caught up with lifestyle changes, but it is a fact that our genes, at the species level, are not protecting us against obesity, T2DM, hypertension, and CVD, among other diseases (Armelagos 2014; Feigin et al. 2016). A summary of these counterarguments can be found in Box 6.1.

Part of the uncertainty reflected in these controversies is due to the great variability evident in hunter-gatherer lifestyles, both ancient and current (Marlowe 2010; Henry et al. 2014). Rick Potts proposed decades ago that the most important human adaptation has been flexibility in the face of environment variability in time and space (Potts 1996, 1998), a formulation that has held up well (Kingston 2007). Thus it is not surprising that a wide range of dietary sources and macronutrients are compatible with health. However, based on all we know today about

Box 6.1 Summary of Counterarguments for Questions and Challenges

1. Hunter-gatherer diets relevant to modern human evolution did not consist predominantly of flesh. A diet of half animal and half plant sources is and probably was modal. However, variability was great.
2. Saturated fat intake was not as high in ancestral diets as is recommended in some 'palaeo' and ketogenic diets, and probably not as high as in typical western diets today. Overall calories from fat were high, but PUFAs predominated, including omega-3 fatty acids.
3. High-carbohydrate diets may be one way to achieve health, provided that sugars and manufactured starches are low, but ancestral diets were about half animal flesh, and growing evidence supports the claim that this can be healthy too. Refined sugars and starches in large amounts cause obesity, T2DM, and CVD, among other chronic diseases.
4. Na^+ intake varied in ancestral populations, with maritime foragers having potentially large amounts. K^+ intake was higher. Epidemiological and experimental research continues to support a linear increase in blood pressure with increases of Na^+ above 2 g/day, and there are CVD and other consequences independent of blood pressure. Further research is needed to determine how low Na^+ intake should go.
5. AS existed in elites of some ancient civilisations, and in some cold-climate hunter-gatherers, but not at levels like those pervasive in modern populations. Studies of recent populations with traditional lifestyles show it is possible for humans to be largely free of AS and CVD.
6. The conclusion that diet, but not physical activity, is responsible for recent increases in obesity because energy expenditure (controlling for body size) is comparable in the Hadza and sedentary westerners does not accurately reflect the methodology or analyses conducted. Physical activity was indeed higher among the Hadza, who were also significantly leaner. Because body size is the strongest predictor of energy expenditure, it was statistically controlled for in the analyses. However this conclusion is framed as if body size was the dependent variable of interest (rather than TEE). The Hadza achieve the same level of TEE as a sedentary western sample through high levels of physical activity, while the westerners achieved the same TEE through a larger body size and low physical activity. The relative proportions of TEE derived from body size versus physical activity are likely to have important implications for obesity and health. Palaeontological research has demonstrated that ancestral activity levels were much higher than modern ones, and body size and composition were also significantly different. Both diet and physical activity are clearly important for health, body size, and composition.
7. New genetic and genomic studies show much more recent evolution than we thought, but not enough to negate the mismatch between our lagging ancestral genome and modern lifestyle, especially the accelerating mismatch of the past two centuries, too short a period for genetic evolution.

ancestral diets and, more importantly, epidemiological and experimental (both laboratory and clinical) research testing such diets, if you have an average western industrial or post-industrial lifestyle you should decrease your dietary level of sugars and simple starches, decrease your sodium and increase your potassium intake, increase complex carbohydrates (fibre), probably replace saturated with polyunsaturated fats, and increase your activity level—that is, if you want to minimise conflict between your genome and your lifestyle and avoid obesity, diabetes, CVD, and stroke, and other chronic degenerative diseases of civilisation. Despite controversy over details, hunter-gatherer diets and lifestyles remain a valid source of hypotheses about what is likely to improve and maintain health (Ruiz-Núñez et al. 2013; Turner and Thompson 2013, 2014; Muskiet and Carrera-Bastos 2014; Crittenden and Schnorr 2017).

6.6 Conclusion

Diet, physical activity, and body composition are all associated with chronic and deadly diseases of civilisation that were not widespread causes of mortality until very recently in evolutionary timescales. Evolutionary medicine can inform research and approaches to reduce the disease burden associated with an imbalance of energy, based on the discordance model and applying LHT for understanding physical activity across the lifespan. Novel approaches incorporating insights that consider human evolutionary theory and integrating across Tinbergen's levels of analysis can lead to the development of interventions and environments that are more compatible with evolved physiology and psychology and conducive to lifestyles that reduce the risks of chronic diseases.

References

Agatson, A. (2003). *The South Beach Diet*. New York: St. Martins.

Aiello, L. C. (1997). Brains and guts in human evolution: the expensive tissue hypothesis. *Braz J Genet* 20(1), 14025.

Aiello, L. C. and Key, C. (2002). Energetic consequences of being a *Homo erectus* female. *Am J Hum Biol* 14, 551–65.

Allison, M. A., Criqui, M. H., and Wright, C. M. (2004). Patterns and risk factors for systemic calcified atherosclerosis. *Arterioscler Thromb Vasc Biol* 24(2), 331–6.

Allué, E., Ibáñez, N., Saladié, P., et al. (2010). Small preys and plant exploitation by Late Pleistocene hunter-gatherers. A case study from the Northeast of the Iberian Peninsula. *Archaeol Anthropol Sci* 2(1), 11–24.

Altmann, J., Schoeller, D., Altmann, S. A., et al. (1993). Body size and fatness of free-living baboons reflect food availability and activity levels. *Am J Primatol* 30, 148–61.

Alvarado, L. C., Muller, M. N., Emery Thompson, M., et al. (2015). The paternal provisioning hypothesis: effects of workload and testosterone production on men's musculature. *Am J Phys Anthropol* 158(1), 19–35.

Andersen, L. G., Angquist, L., Gamborg, M., et al. (2009). Birth weight in relation to leisure time physical activity in adolescence and adulthood: meta-analysis of results from 13 Nordic cohorts. *PLoS One* 4(12), e8192.

Anderson, J. W., Baird, P., Davis, R. H., et al. (2009). Health benefits of dietary fiber. *Nutr Rev* 67(4), 188–205.

Armelagos, G. J. (2014). Brain evolution, the determinates of food choice, and the omnivore's dilemma. *Crit Rev Food Sci Nutr* 54(10), 1330–41.

Arnould, J. P. Y., Nelson, M. M., Nichols, P. D., et al. (2005). Variation in the fatty acid composition of blubber in Cape fur seals (*Arctocephalus pusillus pusillus*) and the implications for dietary interpretation. *J Comp Physiol B* 175(4), 285–95.

Atkins, R. C. (2002). *Dr Atkins' New Diet Revolution*. New York: HarperCollins.

Atkins, R. C., Ornish, D., and Wadden, T. (2003). Low-carb, low-fat diet gurus face off. Interview by Joan Stephenson. *JAMA* 289(14), 1767–8, 1773.
Avena, N. M., Rada, P., and Hoebel, B. G. (2008). Evidence for sugar addiction: behavioral and neurochemical effects of intermittent, excessive sugar intake. *Neurosci Biobehav Rev* 32(1), 20–39.
Ayub, Q., Moutsianas, L., Chen, Y., et al. (2014). Revisiting the thrifty gene hypothesis via 65 loci associated with susceptibility to type 2 diabetes. *Am J Hum Genet* 94(2), 176–85.
Babbitt, C. C., Warner, L. R., Fedrigo, O., et al. (2011). Genomic signatures of diet-related shifts during human origins. *Proc R Soc Lond B Biol Sci* 278(1708), 961–9.
Baraldi, E., Zanconato, S., Zorzi, C., et al. (1991). Exercise performance in very low birth weight children at the age of 7–12 years. *Eur J Pediatr* 150(10), 713–16.
Barker, D. J. P. (1989). Growth in utero, blood pressure in childhood and adult life, and mortality from cardiovascular disease. *BMJ* 298, 564–7.
Barker, D. J. P. (1994). *Mothers, Babies, and Disease in Later Life*. London: British Medical Journal Group.
Barker, D. J. P. (2004). The developmental origins of adult disease. *J Am Coll Nutr* 23(Suppl 6), 588S–95S.
Bartoň, L., Bureš, D., Kotrba, R., et al. (2014). Comparison of meat quality between eland (*Taurotragus oryx*) and cattle (*Bos taurus*) raised under similar conditions. *Meat Sci* 96(1), 346–52.
Bateson, P. and Laland, K. N. (2013). Tinbergen's four questions: an appreciation and an update. *Trends Ecol Evol*, 28 (12), 712–718.
Baxter-Jones, A., Eisenmann, J., Baxter-Jones, A. D. G., et al. (2005). Controlling for maturation in pediatric exercise science. *Pediatr Exerc Sci* 17, 18–30.
Bayol, S. A., Farrington, S. J., and Stickland, N. C. (2007). A maternal 'junk food' diet in pregnancy and lactation promotes an exacerbated taste for 'junk food' and a greater propensity for obesity in rat offspring. *Br J Nutr* 98(4).
Behrens, T. K. and Dinger, M. K. (2003). A preliminary investigation of college students' physical activity patterns. *Am J Health Stud* 18(2/3), 169–72.
Ben-Dor, M., Gopher, A., Hershkovitz, I., et al. (2011). Man the fat hunter: the demise of *Homo erectus* and the emergence of a new hominin lineage in the Middle Pleistocene (ca. 400 kyr) Levant. *PLoS One* 6(12), e28689.
Benecke, H., Flier, J. S., and Moller, D. E. (1992). Alternatively spliced variants of the insulin receptor protein. Expression in normal and diabetic human tissues. *J Clin Invest* 89(6), 2066–70.
Bes-Rastrollo, M., Schulze, M. B., Ruiz-Canela, M., et al. (2013). Financial conflicts of interest and reporting bias regarding the association between sugar-sweetened beverages and weight gain: a systematic review of systematic reviews. *PLoS Med* 10(12), 1–9.
Biesele, M. (1993). *Women Like Meat: The Folklore and Foraging Ideology of the Kalahari Ju/'hoan*. Johannesburg: Witwatersrand University Press, pp. 124–38.
Bigga, G., Schoch, W. H., and Urban, B. (2014). Paleoenvironment and possibilities of plant exploitation in the Middle Pleistocene of Schöningen (Germany). Insights from botanical macro-remains and pollen. *J Hum Evol* 89, 92–104.
Bigham, A., Bauchet, M., Pinto, D., et al. (2010). Identifying signatures of natural selection in Tibetan and Andean populations using dense genome scan data. *PLoS Genet* 6(9), e1001116.
Binder, M. and Roberts, C. A. (2014). Calcified structures associated with human skeletal remains: possible atherosclerosis affecting the population buried at Amara West, Sudan (1300–800 BC). *Int J Paleopathol* 6(1), 20–9.
Blaak, E. (2001). Gender differences in fat metabolism. *Curr Opin Clin Nutr Metab Care* 4(6), 499–502.
Boecker, H., Sprenger, T., Spilker, M. E., et al. (2008). The runner's high: opioidergic mechanisms in the human brain. *Cereb Cortex* 18(11), 2523–31.
Boers, I., Muskiet, F. A., Berkelaar, E., et al. (2014). Favourable effects of consuming a Palaeolithic-type diet on characteristics of the metabolic syndrome: a randomized controlled pilot-study. *Lipids Health Dis* 13, 160.
Bogin, B., Varela Silva, M. I., and Rios, L. (2007). Life history trade-offs in human growth: adaptation or pathology? *Am J Hum Biol* 19, 631–42.
Booth, F. W., Chakravarthy, M. V., Gordon, S. E., et al. (2002a). Waging war on physical inactivity: using modern molecular ammunition against an ancient enemy. *J Appl Physiol* 93, 3–30.

Booth, F. W., Chakravarthy, M. V., and Spangenburg, E. E. (2002b). Exercise and gene expression: physiological regulation of the human genome through physical activity. *J Appl Physiol* 543, 399–411.

Booth, F. W., Roberts, C. K., and Laye, M. J. (2012). Lack of exercise is a major cause of chronic diseases. *Compr Physiol* 2(2), 1143–211.

Bouvard, V., Loomis, D., Guyton, K. Z., et al. (2015). Carcinogenicity of consumption of red and processed meat. *Lancet Oncol* 16(16), 1599–600.

Bramble, D. M. and Lieberman, D. (2004). Endurance running and the evolution of *Homo*. *Nature* 432, 345–52.

Bremer, A. A., Stanhope, K. L., Graham, J. L. et al. (2011). Fructose-fed rhesus monkeys: a nonhuman primate model of insulin resistance, metabolic syndrome, and type 2 diabetes. *Clin Transl Sci* 4(4), 243–52.

Bribiescas, R. G. (2001). Reproductive ecology and life history of the human male. *Am J Phys Anthropol* 33(Suppl), 148–76.

Bribiescas, R. G. (2006). *Men: Evolutionary and Life History*. Cambridge, MA: Harvard University Press.

Broadhurst, C. L., Wang, Y., Crawford, M. A., et al. (2002). Brain-specific lipids from marine, lacustrine, or terrestrial food resources: potential impact on early African *Homo sapiens*. *Comp Biochem Physiol Biochem Mol Biol* 131(4), 653–73.

Brown, W. J. and Trost, S. G. (2003). Life transitions and changing physical activity patterns in young women. *Am J Prev Med* 25(2), 140–3.

Buckley, M. T., Racimo, F., Allentoft, M. E., et al. (2017). Selection in Europeans on fatty acid desaturases associated with dietary changes. *Mol Biol Evol* 34(6), 1307–18.

Bureš, D., Bartoň, L., Kotrba, R., et al. (2015). Quality attributes and composition of meat from red deer (*Cervus elaphus*), fallow deer (*Dama dama*) and Aberdeen Angus and Holstein cattle (*Bos taurus*). *J Sci Food Agric* 95(11), 2299–306.

Burnham, T. C., Flynn Hapman, J., Gray, P. B., et al. (2003). Men in committed romantic relationships have lower testosterone. *Horm Behav* 44, 1–4.

Butler, P. J., Green, J. A., Boyd, I. L., et al. (2004). Measuring metabolic rate in the fields: the pros and cons of the doubly labeled water and heart rate methods. *Funct Ecol* 18, 168–83.

Butte, N. F. and King, J. C. (2005). Energy requirements during pregnancy and lactation. *Public Health Nutr* 8(7a), 1010–27.

Caldwell, A. E. (2016). *Human Physical Fitness and Activity: An Evolutionary and Life History Perspective*. Basel: Springer International.

Caldwell Hooper, A. E. (2013). *An evolutionary approach to understanding social facilitation: energy exertion and exercise motivation*. Doctoral Dissertation. Albuquerque: University of New Mexico. http://hdl.handle.net/1928/23333.

Campbell, T. C. and Campbell, T. M. (2006). *The China Study*. Dallas: Benbella Books.

Carmody, R. N., Dannemann, M., Briggs, A. W., et al. (2016). Genetic evidence of human adaptation to a cooked diet. *Genome Biol Evol* 8(4), 1091–103.

Cerling, T. E., Harris, J. M., Macfadden, B. J., et al. (1997). Global vegetation change through the Miocene/Pliocene boundary. *Nature* 389(6647), 153.

Cerling, T. E., Ehleringer, J. R., and Harris, J. M. (1998). Carbon dioxide starvation, the development of C4 ecosystems, and mammalian evolution. *Proc R Soc Lond B Biol Sci* 353, 159–71.

Cervellin, G. and Lippi, G. (2014). Of MIs and men—a historical perspective on the diagnostics of acute myocardial infarction. *Semin Thromb Hemost* 40(5), 535–43.

Chakravarthy, M. V. and Booth, F. W. (2004). Eating, exercise, and 'thrifty' genotypes: connecting the dots toward an evolutionary understanding of modern chronic diseases. *J Appl Physiol* 96, 3–10.

Chen, J., Peto, R., Pan, W., et al. (2007). Mortality, biochemistry, diet and lifestyle in rural China. Geographical study of the characteristics of 69 counties in mainland China and 16 areas in Taiwan. *J Epidemiol Community Health* 61(3), 271.

Chowdhury, R. (2014). Association of dietary, circulating, and supplement fatty acids with coronary risk: a systematic review and meta-analysis. *Ann Inter Med* 160(6), 398–406.

Clarke, E. M., Thompson, R. C., Allam, A. H., et al. (2014). Is atherosclerosis fundamental to human aging? Lessons from ancient mummies. *J Cardiol* 63(5), 329–34.

Cochran, G. and Harpending, H. (2010). *The Ten Thousand Year Explosion: How Civilization Accelerated Human Evolution*. New York: Basic Books.

Cohen, M. N. and Armelagos, G. J. (1984). *Paleopathology at the Origins of Agriculture*. New York: Academic Press.

Colcombe, S. and Kramer, A. F. (2003). Fitness effects on the cognitive function of older adults: a meta-analytic study. *Psychol Sci* 14(2), 125–30.

Cook, N. R., Appel, L. J., and Whelton, P. K. (2016). Sodium intake and all-cause mortality over 20 years in the trials of hypertension prevention. *J Am Coll Cardiol* 68(15), 1609–17.

Cordain, L. (2017). The paleo diet. https://thepaleodiet.com/tag/the-paleo-diet/.

Cordain, L. and Campbell, T. C. (2008). The protein debate. *Performance Menu: Journal of Nutrition and Athletic Excellence*. https://www.catalystathletics.com/articles/dayownloads/proteindebate.pdf.

Cordain, L., Gotshall, R. W., Eaton, S. B., et al. (1998). Physical activity, energy expenditure and fitness: an evolutionary perspective. *J Sports Med* 19, 328–35.

Cordain, L., Miller, J. B., Eaton, S. B., et al. (2000a). Macronutrient estimations in hunter-gatherer diets. *Am J Clin Nutr* 72(6), 1589–92.

Cordain, L., Miller, J. B., Eaton, S. B., et al. (2000b). Plant–animal subsistence ratios and macronutrient energy estimations in worldwide hunter-gatherer diets. *Am J Clin Nutr* 71(3), 682–92.

Cordain, L., Eaton, S., Miller, J. B., et al. (2002). The paradoxical nature of hunter-gatherer diets: meat-based, yet non-atherogenic. *Eur J Clin Nutr* 56, 42–52.

Cordain, L., Eaton, S. B., Sebastian, A., et al. (2005). Origins and evolution of the western diet: health implications for the 21st century. *Am J Clin Nutr* 81(2), 341–54.

Crittenden, A. N. (2011). The importance of honey consumption in human evolution. *Food Foodways* 19(4), 257–73.

Crittenden, A. N. and Schnorr, S. L. (2017). Current views on hunter-gatherer nutrition and the evolution of the human diet. *Am J Phys Anthropol* 162, 84–109.

Crittenden, A. N., Conklin-Brittain, N. L., Zes, D. A., et al. (2013). Juvenile foraging among the Hadza: implications for human life history. *Evolution and Human Behavior* 34(4), 299–304.

Crittenden, A. N., Sorrentino, J., Moonie, S. A., et al. (2017). Oral health in transition: the Hadza foragers of Tanzania. *PLoS One* 12(3), e0172197.

Curhan, G. C., Chertow, G. M., Willett, W. C., et al. (1996). Birth weight and adult hypertension and obesity in women. *Circulation* 94, 1310–15.

Da Silveira, H. L. D., Damaskos, S., Arús, N. A., et al. (2016). The presence of calcifications along the course of internal carotid artery in Greek and Brazilian populations: a comparative and retrospective cone beam CT data analysis. *Oral Surg Oral Med Oral Pathol Oral Radiol* 121(1), 81–90.

Davidson, B., Maciver, J., Lessard, E., et al. (2011). Meat lipid profiles: a comparison of meat from domesticated and wild southern African animals. *In Vivo* 25(2), 197–202.

DeFronzo, R. A. and Tripathy, D. (2009). Skeletal muscle insulin resistance is the primary defect in type 2 diabetes. *Diabetes Care* 32(Suppl 2), S157–63.

DeLany, J. P., Kelley, D. E., Hames, K. C., et al. (2013). High energy expenditure masks low physical activity in obesity. *Int J Obes* 37(7), 1006–11.

de Souza, R. J., Mente, A., Maroleanu, A., et al. (2015). Intake of saturated and trans unsaturated fatty acids and risk of all cause mortality, cardiovascular disease, and type 2 diabetes: systematic review and meta-analysis of observational studies. *BMJ (Clin Res Ed)* 351, h3978.

De Vynck, J. C., Anderson, R., Atwater, C., et al. (2016). Return rates from intertidal foraging from Blombos Cave to Pinnacle Point: understanding early human economies. *J Hum Evol* 92, 101–15.

Dufour, D. L. and Piperata, B. A. (2008). Energy expenditure among farmers in developing countries: what do we know? *Am J Hum Biol* 20(3), 249–58.

Dufour, D. L. and Sauther, M. L. (2002). Comparative and evolutionary dimensions of the energetics of human pregnancy and lactation. *Am J Hum Biol* 14(5), 584–602.

Dugas, L. R., Harders, R., Merrill, S., et al. (2011). Energy expenditure in adults living in developing compared with industrialized countries: a meta-analysis of doubly labeled water studies. *Am J Clin Nutr* 93(2), 427–41.

Eaton, S. B. and Eaton, S. B. (2003). An evolutionary perspective on human physical activity: implications for health. *Comp Biochem Physiol* 136, 153–9.

Eaton, S. B. and Konner, M. (1985). Paleolithic nutrition. A consideration of its nature and current implications. *N Eng J Med* 312(5), 283–9.

Eaton, S. B., Konner, M., and Shostak, M. (1988). Stone agers in the fast lane: chronic degenerative diseases in evolutionary perspective. *Am J Med* 84(4), 739–49.

Eaton, S. B., Eaton, S. B., and Konner, M. J. (1997). Paleolithic nutrition revisited: a twelve-year retrospective on its nature and implications. *Eur J Clin Nutr* 51(4), 207–16.

Eaton, S. B., Konner, M. J., and Cordain, J. (2010). Diet-dependent acid load, Paleolithic nutrition, and evolutionary health promotion. *Am J Clin Nutr* 91(4), 295–7.

Eaton, S. B., Strassman, B. I., Nesse, R.M., et al. (2002a). Evolutionary health promotion. *Prev Med* 34(2), 109–18.

Eaton, S. B., Cordain, L., and Lindeberg, S. (2002b). Evolutionary health promotion: a consideration of common counterarguments. *Prev Med* 34(2), 119–23.

Eckel, R. H., Jakicic, J. M., Ard, J. D., et al. (2014). 2013 AHA/ACC guideline on lifestyle management to reduce cardiovascular risk: a report of the American College of Cardiology/American Heart Association Task Force on practice guidelines. *Circulation* 129(25 Suppl 1).

Egger, G. and Dixon, J. (2009). Should obesity be the main game? Or do we need an environmental makeover to combat the inflammatory and chronic disease epidemics? Viewpoint. *Obes Rev* 10(2), 237–49.

Elliott, P. and Brown, I. (2007). *Sodium Intakes Around the World*. Geneva: WHO.

Ellison, P. T. (1990). Human ovarian function and reproductive ecology. *Am Anthropol* 92, 933–52.

Ellison, P. T. (2001). *On Fertile Ground*. Cambridge, MA: Harvard Press.

Ellison, P. T. (2003). Energetics and reproductive effort. *Am J Hum Biol* 15, 342–51.

Ellison, P. T. (2008). Energetics, reproductive ecology, and human evolution. *PaleoAnthropology* 2008, 172–200.

Ellison, P. T. and Gray, P. B. (2009). *Endocrinology of Social Relationships*. Cambridge, MA: Harvard University Press.

Ellison, P. T. and Jasienska, G. (2007). Constraint, pathology, and adaptation: how can we tell them apart? *Am J Hum Biol* 19, 622–30.

Enattah, N. S., Sahi, T., Savilahti, E., et al. (2002). Identification of a variant associated with adult-type hypolactasia. *Nat Genet* 30(2), 233–7.

Eoin, L. N. (2016). Geophytes, grasses and grindstones: replanting ideas of gathering in Southern Africa's Middle and Later Stone Ages. *SAAB* 71(203), 36–45.

Erlandson, J. M. and Braje, T. J. (2015). Coasting out of Africa: the potential of mangrove forests and marine habitats to facilitate human coastal expansion via the Southern Dispersal Route. *Q Int* 382, 31–41.

Esselstyn, C. B., Gendy, G., Doyle, J., et al. (2014). A way to reverse CAD? *J Family Pract* 63(7), 356–64.

Estalrrich, A., El Zaatari, S. and Rosas, A. (2017). Dietary reconstruction of the El Sidrón Neandertal familial group (Spain) in the context of other Neandertal and modern hunter-gatherer groups. A molar microwear texture analysis. *J Hum Evol* 104, 13–22.

Fan, S., Lo, Y., and Tishkoff, S. A. (2016). Going global by adapting local: a review of recent human adaptation. *Science* 354(6308), 54–9.

Feigin, V. L., Roth, G. A., Naghavi, M., et al. (2016). Global burden of stroke and risk factors in 188 countries, during 1990–2013: a systematic analysis for the Global Burden of Disease Study 2013. *Lancet Neurol* 15(9), 913–24.

Feinman, R. D., Pogozelski, W. K., Astrup, A., et al. (2015). Dietary carbohydrate restriction as the first approach in diabetes management: critical review and evidence base. *Nutrition* 31(1), 1–13.

Fell, D. B., Joseph, K. S., Armson, B. A., et al. (2009). The impact of pregnancy on physical activity level. *Matern Child Health J* 13(5), 597–603.

Fernandez, M. L. and Calle, M. (2010). Revisiting dietary cholesterol recommendations: does the evidence support a limit of 300 mg/day? *Curr Atheroscler Rep* 12(6), 377–83.

Field, Y., Boyle, E. A., Telis, N., et al. (2016). Detection of human adaptation during the past 2000 years. *Science* 354(6313), 760–4.
Finch, C. E. (2011). Atherosclerosis is an old disease: summary of the Ruffer Centenary Symposium. The paleocardiology of Ancient Egypt, a meeting report of the Horus Study team. *Exp Gerontol* 46(11), 843–6.
Fiorenza, L., Benazzi, S., Henry, A. G., et al. (2015). To meat or not to meat? New perspectives on Neanderthal ecology. *Am J Phys Anthropol* 156(Suppl 59), 43–71.
Fontes-Villalba, M., Lindeberg, S., Granfeldt, Y., et al. (2016). Palaeolithic diet decreases fasting plasma leptin concentrations more than a diabetes diet in patients with type 2 diabetes: a randomised cross-over trial. *Cardiovasc Diabetol* 15(1), 80.
Fornaciari, G., Gaeta, R., Thompson, R. C., et al. (2013). Atherosclerosis in ancient populations (critique and reply). *Lancet* 382(9887), 123–4.
Forsythe, C. E., Phinney, S. D., Fernandez, M. L., et al. (2008). Comparison of low fat and low carbohydrate diets on circulating fatty acid composition and markers of inflammation. *Lipids* 43(1), 65–77.
Fox, E. B. A., Van Schaik, C. P., Sitompul, A., et al. (2004). Intra- and interpopulational differences in orangutan (*Pongo pygmaeus*) activity and diet: implications for the invention of tool use. *Am J Phys Anthropol* 125(2), 162–74.
Frassetto, L. A., Schloetter, M., Mietus-Synder, M., et al. (2009). Metabolic and physiologic improvements from consuming a Paleolithic, hunter-gatherer type diet. *Eur J Clin Nutr* 63(8), 947–55.
Frederick, D. A. and Haselton, M. G. (2007). Why is muscularity sexy? Tests of the fitness indicator hypothesis. *Pers Soc Psychol Bull* 33(8), 1167–83.
Frenkl, R., Mészáros, J., Soliman, Y. A., et al. (2001). Body composition and peak aerobic power in male international level Hungarian athletes. *Acta Physiol Hung* 88(3–4), 251–8.
Fryar, C. D., Gu, Q., and Ogden, C. L. (2012). Anthropometric reference data for children and adults: United States, 2007–2010. National Center for Health Statistics. *Vital Health Stat* 11(252).
Fu, W., O'Connor, T. D., Jun, G., et al. (2012). Analysis of 6,515 exomes reveals the recent origin of most human protein-coding variants. *Nature* 493(7431), 216–20.
Fumagalli, M., Sironi, M., Pozzoli, U., et al. (2011). Signatures of environmental genetic adaptation pinpoint pathogens as the main selective pressure through human evolution. *PLoS Genet* 7(11).
Fumagalli, M., Moltke, I., Grarup, N., et al. (2015). Greenlandic Inuit show genetic signatures of diet and climate adaptation. *Science* 349(6254), 1343–7.
Gabrovsky, A. N., O'Neill, K. D., and Gerszten, E. (2016). Paleopathology of cardiovascular diseases in South American mummies. *Int J Cardiol* 223, 101–7.
Galinsky, K. J., Loh, P. R., Mallick, S., et al. (2016). Population structure of UK biobank and ancient Eurasians reveals adaptation at genes influencing blood pressure. *Am J Hum Genet* 99(5), 1130–9.
Gallup, A. C., White, D. D., and Gallup, G. G. (2007). Handgrip strength predicts sexual behavior, body morphology, and aggression in male college students. *Evol Hum Behav* 28(6), 423–9.
Genoni, A., Lyons-Wall, P., Lo, J., et al. (2016). Cardiovascular, metabolic effects and dietary composition of ad-libitum Paleolithic vs. Australian guide to healthy eating diets: a 4-week randomised trial. *Nutrients* 8(5).
Giuliani, C., Bacalini, M. G., Sazzini, M., et al. (2015). The epigenetic side of human adaptation: hypotheses, evidences and theories. *Ann Hum Biol* 42(1), 1–9.
Gluckman, P. D., Hanson, M. A., and Spencer, H. G. (2005). Predictive adaptive responses and human evolution. *Trends Ecol Evol* 20, 527–33.
Grasgruber, P., Cacek, J., Kalina, T., et al. (2014). The role of nutrition and genetics as key determinants of the positive height trend. *Econ Hum Biol* 15, 81–100.
Graudal, N. (2016). A radical sodium reduction policy is not supported by randomized controlled trials or observational studies: grading the evidence. *Am J Hypertens* 29(5), 543–8.
Graudal, N. A., Hubeck-Graudal, T., and Jurgens, G. (2017). Effects of low sodium diet versus high sodium diet on blood pressure, renin, aldosterone, catecholamines, cholesterol, and triglyceride. *Cochrane Database Syst Rev* 2017(4).
Gray, P. B. (2009). Play as a foundation for hunter-gatherer social existence. *Am J Play* 1, 476–522.

Gray, P. B., Ellison, P. T., and Campbell, B. C. (2007). Testosterone and marriage among Ariaal men of northern Kenya. *Curr Anthropol* 48(5), 750–5.

Greenfield, H., McCullum, D., and Wills, R. B. H. (1984). Sodium and potassium contents of salts, salt substitutes, and other seasonings. *Med J Aust* 140(8), 460462.

Groman-Yaroslavski, I., Weiss, E., and Nadel, D. (2016). Composite sickles and cereal harvesting methods at 23,000-years-old Ohalo II, Israel. *PLoS One* 11(11), e0167151.

Guerrero, A. I. and Rogers, T. L. (2017). Blubber fatty acid composition and stratification in the crabeater seal, *Lobodon carcinophaga*. *J Exp Mar Bio Ecol* 491, 51–7.

Guerrero, A. I., Negrete, J., Márquez, M. E. I., et al. (2016). Vertical fatty acid composition in the blubber of leopard seals and the implications for dietary analysis. *J Exp Mar Bio Ecol* 478, 54–61.

Gunn, J. P., Kuklina, E. V., Keenan, N. L., et al. (2010). Sodium intake among adults—United States, 2005–2006. *Morb Mortal Wkly Rep* 59(24), 746–9.

Gurven, M., Kaplan, H., and Gutierrez, M. (2006). How long does it take to become a proficient hunter? Implications for the evolution of extended development and long life span. *J Hum Evol* 51(5), 454–70.

Gurven, M., Blackwell, A. D., Rodríguez, D. E., et al. (2012). Does blood pressure inevitably rise with age? Longitudinal evidence among forager-horticulturalists. *Hypertension* 60(1), 25–33.

Gurven, M., Jaeggi, A. V., Kaplan, H. S., et al. (2013). Physical activity and modernization among Bolivian Amerindians. *PLoS One* 8(1), e55679.

Gurven, M., Stieglitz, J., Trumble, B., et al. (2017). The Tsimane Health and Life History Project: Integrating anthropology and biomedicine. *Evol Anthropol* 26(2), 54–73.

Hagobian, T. A., Sharoff, C. G., Stephens, B. R., et al. (2009). Effects of exercise on energy-regulating hormones and appetite in men and women. *Am J Physiol Regul Integr Comp Physiol* 296, R233–42.

Hales, C. N. and Barker, D. J. P. (1992). Type 2 (non-insulin-dependent) diabetes mellitus: the thrifty phenotype hypothesis. *Int J Epidemiol* 35, 595–601.

Hall, K. D., Heymsfield, S. B., Kemnitz, J. W., et al. (2012). Energy balance and its components: implications for body weight regulation. *Am J Clin Nutr* 95(4), 989–94.

Hamad, I., Delaporte, E., Raoult, D., et al. (2015). Detection of termites and other insects consumed by African great apes using molecular fecal analysis. *Sci Rep* 4(1), 4478.

Harris, K. and Pritchard, J. K., (2017). Rapid evolution of the human mutation spectrum. *eLife*, 6.

Henry, A. G., Brooks, A. S., and Piperno, D. R. (2011). Microfossils in calculus demonstrate consumption of plants and cooked foods in Neanderthal diets (Shanidar III, Iraq; Spy I and II, Belgium). *Proc Natl Acad Sci U S A* 108(2), 486–91.

Henry, A. G., Brooks, A. S., and Piperno, D. R. (2014). Plant foods and the dietary ecology of Neanderthals and early modern humans. *J Hum Evol* 69(1), 44–54.

Hiermeyer, M. (2010). The height and BMI values of West Point cadets after the Civil War. *Econ Hum Biol* 8(1), 127–33.

Hill, K., Hurtado, A., and Low, B. (1996). *Ache Life History: The Ecology and Demography of a Foraging People*. Piscataway: Transaction Publishers.

Hoffman, L. C., Schalkwyk, D. L. Van, McMillin, K. W., et al. (2015). Intramuscular fat characteristics of Namibian common eland (*Tragelaphus oryx*). *Afr J Wildl Res* 45(October), 312–20.

Holst, D. (2010). Hazelnut economy of early Holocene hunter-gatherers: a case study from Mesolithic Duvensee, northern Germany. *Journal of Archaeological Science* 37(11), 2871–80.

Hönekopp, J., Bartholomé, T., and Jansen, G. (2004). Facial attractiveness, symmetry, and physical fitness in young women. *Hum Nat* 15(2), 147–67.

Hönekopp, J., Rudolph, U., Beier, L., et al. (2007). Physical attractiveness of face and body as indicators of physical fitness in men. *Evol Hum Behav* 28, 106–11.

Hooper, L., Summerbell, C. D., Thompson, R., et al. (2016). Reduced or modified dietary fat for preventing cardiovascular disease. *Sao Paulo Med J* 134(2), 182.

Hooper, P. L., Gurven, M., Winking, J., et al. (2015). Inclusive fitness and differential productivity across the life course determine intergenerational transfers in a small-scale human society. *Proc R Soc Lond B Biol Sci* 282(1803), 20142808.

Howell, N. (2010). *Life Histories of the Dobe! Kung: Food, Fatness, and Well-Being Over the Life Span.* Berkeley: University of California Press.

Hu, Y., Shang, H., Tong, H., et al. (2009). Stable isotope dietary analysis of the Tianyuan 1 early modern human. *Proc Natl Acad Sci U S A* 106(27), 10971–4.

Hu, T., Mills, K. T., Yao, L., et al. (2012). Effects of low-carbohydrate diets versus low-fat diets on metabolic risk factors: a meta-analysis of randomized controlled clinical trials. *Am J Epidemiol* 176(Suppl 7).

Huerta-Sánchez, E., Jin, X., Asan Bianba, Z., et al. (2014). Altitude adaptation in Tibetans caused by introgression of Denisovan-like DNA. *Nature* 512(7513), 194–7.

Hurtado, A. M., Hill, K., Kaplan, H., et al. (1992). Trade-offs between female food acquisition and child care among Hiwi and Ache foragers. *Hum Nat* 3(185–216).

Imamura, F., O'Connor, L., Ye, Z., et al. (2015). Consumption of sugar sweetened beverages, artificially sweetened beverages, and fruit juice and incidence of type 2 diabetes: systematic review, meta-analysis, and estimation of population attributable fraction. *BMJ (Clin Res Ed)* 351, h3576.

Inchley, C. E., Larbey, C. D. A., Shwan, N. A. A., et al. (2016). Selective sweep on human amylase genes postdates the split with Neanderthals. *Sci Rep* 6(1), 37198.

Intersalt Cooperative Research Group (1988). Intersalt: an international study of electrolyte excretion and blood. Results for 24 hour urinary sodium and potassium excretion. *BMJ* 297, 319–28.

Jablonski, N. G. and Chaplin, G. (2010). Human skin pigmentation as an adaptation to UV radiation. *Proceedings of the National Academy of Sciences of the United States of America* 107, 8962–8.

Jakicic, J. M. and Otto, A. D. (2005). Physical activity considerations for the treatment and prevention of obesity. *Am J Clin Nutr* 82(Suppl 1).

Jakobsen, M. U., O'Reilly, E. J., Heitmann, B. L., et al. (2009). Major types of dietary fat and risk of coronary heart disease: a pooled analysis of 11 cohort studies. *Am J Clin Nutr* 89(5), 1425–32.

Janssen, I., Katzmarzyk, P. T., Ross, R., et al. (2004). Fitness alters the associations of BMI and waist circumference with total and abdominal fat. *Obes Res* 12(3), 525–37.

Janz, L. (2016). Fragmented landscapes and economies of abundance: the broad-spectrum revolution in arid East Asia. *Curr Anthropol* 57(5), 537–64.

Jasienska, G. (2003). Energy metabolism and the evolution of reproductive suppression in the human female. *Acta Bioitheor* 51, 1–18.

Jasienska, G. and Ellison, P. T. (2004). Energetic factors and seasonal changes in ovarian function in women from rural Poland. *Am J Hum Biol* 16, 563–80.

Jasienska, G., Ziomkiewicz, A., Thune, I., et al. (2006). Habitual physical activity and estradiol levels in women of reproductive age. *Eur J Cancer Prevention* 15, 439–45.

Jenike, M. (2001). Nutritional ecology: diet, physical activity and body size. In: Panter-Brick, C., Layton, R. H., and Owley-Conwy, P. (eds) *Hunter-Gatherers: An Interdisciplinary Perspective.* Cambridge: Cambridge University Press, 205–38.

Johnson, R. J. and Andrews, P. (2010). Fructose, uricase, and the Back-to-Africa hypothesis. *Evol Anthropol* 19(6), 250–7.

Johnson, R. K., Appel, L. J., Brands, M., et al. (2009). Dietary sugars intake and cardiovascular health a scientific statement from the American Heart Association. *Circulation* 120(11), 1011–20.

Johnson, R. J., Appel, L. J., Brands, M., et al. (2017). Perspective: a historical and scientific perspective of sugar and its relation with obesity and diabetes. *Adv Nutr Int Rev J* 8(3), 412–22.

Johnston, B. C., Kanters, S., Bandayrel, K., et al. (2014). Comparison of weight loss among named diet programs in overweight and obese adults. *JAMA* 312(9), 923.

Jones, N. B., Hawkes, K., and O'Connell, J. F. (1996). The global process and local ecology: how should we explain differences between the Hadza and the !Kung? In: Kent, S. (ed.) *Cultural Diversity among Twentieth-Century Foragers: An African Perspective.* Cambridge: Cambridge University Press, pp. 159–87.

Jönsson, T., Granfeldt, Y., Ahrén, B., et al. (2009). Beneficial effects of a Paleolithic diet on cardiovascular risk factors in type 2 diabetes: a randomized cross-over pilot study. *Cardiovasc Diabetol* 8, 35.

Jönsson, T., Granfeldt, Y., Erlanson-Albertsson, C., et al. (2010). A Paleolithic diet is more satiating per calorie than a Mediterranean-like diet in individuals with ischemic heart disease. *Nutr Metabol* 7(1), 85.

Kahn, B. and Flier, J. (2000). Obesity and insulin resistance. *J Clin Invest* 106(4), 473–81.

Kaiyala, K. J. and Schwartz, M. W. (2011). Toward a more complete (and less controversial) understanding of energy expenditure and its role in obesity pathogenesis. *Diabetes* 60(1), 17–23.

Kaplan, H. S. and Gurven, M. (2005). The natural history of human food sharing and cooperation: a review and a new multi-individual approach to the negotiation of norms. In: Gintis, H., Bowles, S., Boyd, R., and Fehr, E. (eds) *Moral Sentiments and Material Interests: The Foundations of Cooperation in Economic Life*. Cambridge, MA: MIT Press, pp. 75–113.

Kaplan, H. S., Hill, K. R., Lancaster, J. B., et al. (2000). A theory of human life history evolution: diet, intelligence, and longevity. *Evol Anthropol* 9, 156–85.

Kaplan, H. S., Gurven, M., Winking, J., et al. (2010). Learning, menopause, and the human adaptive complex. *Ann N Y Acad Sci* 1204, 30–42.

Kaplan, H. S, Thompson, R. C., Trumble, B. C., et al. (2017). Coronary atherosclerosis in indigenous South American Tsimane: a cross-sectional cohort study. *Lancet* 389(10080), 1730–9.

Karmin, M., Saag, L., Vicente, M., et al. (2015). A recent bottleneck of Y chromosome diversity coincides with a global change in culture. *Genome Res* 25(4), 459–66.

Katzmarzyk, P. T., Leonard, W. R., Crawford, M. H., et al. (1994). Resting metabolic rate and daily energy expenditure among two indigenous Siberian populations. *Am J Hum Biol* 6(6), 719–30.

Kingston, J. D. (2007). Shifting adaptive landscapes: progress and challenges in reconstructing early hominid environments. *Yearb Phys Anthropol* 50, 20–58.

Kline, J., Stein, Z., and Susser, M. (1989). *Conception to Birth: Epidemiology of Prenatal Development*. New York: Oxford University Press.

Konerman, M. C. and Hummel, S. L. (2016). Does limiting salt intake prevent heart failure? A critical appraisal. *Curr Cardiovasc Risk Rep* 10(2).

Kong, Y. W., Baqar, S., Jerums, G., et al. (2016). Sodium and its role in cardiovascular disease—the debate continues. *Frontiers in Endocrinology* 7(Dec).

Konner, M. (2010). *The Evolution of Childhood: Relationships, Emotion, Mind*. Cambridge, MA: Belknap Press.

Konner, M. (2016). Hunter-gatherer infancy and childhood in the context of human evolution. In: Meehan, C. L. and Crittenden, A. (eds) *Childhood: Origins, Evolution, and Implications*. Santa Fe and Albuquerque, NM: School for Advanced Research (SAR) Press and University of New Mexico Press, pp. 123–54.

Konner, M. and Eaton, S. B. (2010). Paleolithic nutrition: twenty-five years later. *Nutr Clin Prac* 25(6), 594–602.

Krueger, K. L., Ungar, P. S., Guatelli-Steinberg, D., et al. (2017). Anterior dental microwear textures show habitat-driven variability in Neandertal behavior. *J Hum Evol* 105, 13–23.

Kuipers, R. S., Luxwolda, M. F., Dijck-Brouwer, D. A. J., et al. (2010). Estimated macronutrient and fatty acid intakes from an East African Paleolithic diet. *Br J Nutr* 104(11), 1666–87.

Kuipers, R. S., Joordens, J. C. A., and Muskiet, F. A. J. (2012). A multidisciplinary reconstruction of Palaeolithic nutrition that holds promise for the prevention and treatment of diseases of civilisation. *Nutr Res Rev* 25(1), 96–129.

Kuzawa, C. W. (1998). Adipose tissue in human infancy and childhood: an evolutionary perspective. *Am J Phys Anthropol* 107(Suppl 27), 177–209.

Kuzawa, C. W. (2005). Fetal origins of developmental plasticity: are fetal cues reliable predictors of future nutritional environments? *Am J Hum Biol* 17, 5–21.

Kuzawa, C. W. (2007). Developmental origins of life history: growth, productivity, and reproduction. *Am J Hum Biol* 19, 654–61.

Kuzawa, C. W., Gluckman, P. D., Hanson, M. A., et al. (2007). Evolution, developmental plasticity, and metabolic disease. In: Stearns, S. C. and Koella, J. C. (eds) *Evolution in Health and Disease*. Oxford: Oxford University Press.

Kuzawa, C. W., Gettler, L. T., Muller, M. N., et al. (2009). Fatherhood, pairbonding and testosterone in the Philippines. *Horm Behav* 56(4), 429–35.

Kuzawa, C. W., Chugani, H. T., Grossman, L. I., et al. (2014). Metabolic costs and evolutionary implications of human brain development. *Proc Natl Acad Sci U S A* 111(36), 13010–15.

Kwiatkowski, D. P. (2005). How malaria has affected the human genome and what human genetics can teach us about malaria. *Am J Hum Genet* 77(2), 171–92.

Kyle, U. G., Genton, L., Slosman, D. O., et al. (2001). Fat-free and fat mass percentiles in 5225 healthy subjects aged 15 to 98 years. *Nutrition* 17(7–8), 534–41.

Lamelas, P. M., Mente, A., Diaz, R., et al. (2016). Association of urinary sodium excretion with blood pressure and cardiovascular clinical events in 17,033 Latin Americans. *Am J Hypertens* 29(7), 796–805.

Lancaster, J. B. and Kaplan, H. S. (2009). The endocrinology of the human adaptive complex. In: Ellison, P. T. and Gray, P. B. (eds) *Endocrinology of Social Relationships*. Cambridge, MA: Harvard University Press, 95–119.

Lassek, W. D. and Gaulin, S. J. C. (2009). Costs and benefits of fat-free muscle mass in men: relationship to mating success, dietary requirements, and native immunity. *Evol Hum Behav* 30(5), 322–8.

Lee, R. B. (1980). *The !Kung San: Men, Women, and Work in a Foraging Society*. Cambridge: Cambridge University Press.

Leonard, W. R. (2008). Lifestyle, diet, and disease: comparative perspectives on the determinants of chronic health risks. In: Stenrs, S. C. and Koella, J. C. (eds) *Evolution in Health and Disease*. New York: Oxford University Press.

Leonard, W. R. and Robertson, M. L. (1997). Comparative primate energetics and hominid evolution. *Am J Phys Anthropol* 102, 265–81.

Leonard, W. R., Katzmarzyk, P. T., Stephen, M. A., et al. (1995). Comparison of the heart rate-monitoring and factorial methods: assessment of energy expenditure in highland and coastal Ecuadoreans. *Am J Clin Nutr* 61, 1146–52.

Leonard, W. R., Snodgrass, J. J., and Robertson, M. L. (2007). Effects of brain evolution on human nutrition and metabolism. *Annu Rev Nutr* 27, 311–27.

Lesnik, J. J. (2014). Termites in the hominin diet: a meta-analysis of termite genera, species and castes as a dietary supplement for South African robust australopithecines. *J Hum Evol* 71, 94–104.

Lev, E., Kislev, M. E., and Bar-Yosef, O. (2005). Mousterian vegetal food in Kebara Cave, Mt. Carmel. *J Archaeol Sci* 32(3), 475–84.

Li, D., Siriamornpun, S., Wahlqvist, M. L., et al. (2005). Lean meat and heart health. *Asia Pacific J Clin Nutr* 14(2), 113–19.

Lieberman, L. S. (2003). Dietary, evolutionary, and modernizing influences on the prevalence of type 2 diabetes. *Annu Rev Nutr* 23(1), 345–77.

Lieberman, L. S. (2006). Evolutionary and anthropological perspectives on optimal foraging in obesogenic environments. *Appetite* 47(1), 3–9.

Lieberman, D. E. (2013). *The Story of the Human Body: Evolution, Health, and Disease*. New York: Pantheon Books.

Lieberman, D. E. (2015). Is exercise really medicine? An evolutionary perspective. *Curr Sports Med Rep* 14(4), 313–19.

Lin, H. and Epstein, L. H. (2014). Living in the moment: effects of time perspective and emotional valence of episodic thinking on delay discounting. *Behav Neurosci* 128(1), 12–19.

Lindeberg, S., Jönsson, T., Granfeldt, Y., et al. (2007). A Palaeolithic diet improves glucose tolerance more than a Mediterranean-like diet in individuals with ischaemic heart disease. *Diabetologia* 50(9), 1795–807.

Lindstedt, S. L. and Boyce, M. S. (1985). Seasonality, fasting endurance, and body size in mammals. *Am Nat* 125(6), 873.

Lipson, S. F. (2001). Metabolism, maturation, and ovarian function. In: Ellison, P. T. (ed.) *Reproductive Ecology and Human Evolution*. New York: Aldine de Gruyter.

Little, B. B., Peña Reyes, M. E., and Malina, R. M. (2017). Natural selection and type 2 diabetes-associated mortality in an isolated indigenous community in the valley of Oaxaca, southern Mexico. *Am J Phys Anthropol* 162(3), 561–72.

Lofranco, L. P. (2012). Caries through time. In: Ming-Yu, L. (ed.) *Contemporary Approach to Dental Caries*. Shanghai: Shanghai Jiao Tong University.

Loucks, A. B. and Thuma, J. R. (2003). Luteinizing hormone pulsatility is disrupted at a threshold of energy availability in regularly menstruating women. *J Clin Endocrinol Metabol* 88(1), 297–311.

Loucks, A. B., Heath, E. M., Law, T., et al. (1994). Dietary restriction reduces luteinizing hormone (LH) pulse frequency during waking hours and increases LH pulse amplitude during sleep in young menstruating women. *J Clin Endocrinol Metabol* 78(4), 910–15.

Loucks, A. B., Verdun, M., and Heath, E. M. (1998). Low energy availability, not stress of exercise, alters LH pulsatility in exercising women. *J Appl Physiol* 84(1), 37–46.

Lustig, R. H. (2016). Sickeningly sweet: does sugar cause type 2 diabetes? Yes. *Can J Diabetes* 40(4), 282–6.

Macintosh, A. A., Pinhasi, R., and Stock, J. T. (2016). Early life conditions and physiological stress following the transition to farming in central/southeast Europe: skeletal growth impairment and 6000 years of gradual recovery. *PLoS One* 11(2), e0148468.

Madimenos, F. C., Snodgrass, J. J., Blackwell, A. D., et al. (2011). Physical activity in an indigenous Ecuadorian forager-horticulturalist population as measured using accelerometry. *Am J Hum Biol* 23(4), 488–97.

Mailey, E. L., Huberty, J., Dinkel, D., et al. (2014). Physical activity barriers and facilitators among working mothers and fathers. *BMC Public Health* 14(1), 657.

Malhotra, A., Redberg, R. F., and Meier, P. (2017). Saturated fat does not clog the arteries: coronary heart disease is a chronic inflammatory condition, the risk of which can be effectively reduced from healthy lifestyle interventions. *Br J Sports Med* 51(15), 1111–12.

Mannheimer, E., Zuuren, E., Fedorowicz, Z., et al. (2015). Paleolithic nutrition for metabolic syndrome: systematic review and meta-analysis. *Am J Clin Nutr* 102(C), 922–32.

Mann, G. V., Shaffer, R. D., Anderson, R. S., et al. (1964). Cardiovascular disease in the Masai. *J Atheroscler Res* 4(4), 289–312.

Mann, N. (2000). Dietary lean red meat and human evolution. *Eur J Nutr* 39, 71–9.

Mansoor, N., Vinknes, K. J., Veierød, M. B., et al. (2016). Effects of low-carbohydrate diets v. low-fat diets on body weight and cardiovascular risk factors: a meta-analysis of randomised controlled trials. *Br J Nutr* 115(3), 466–79.

Marean, C. W. (2010). Pinnacle Point Cave 13B (Western Cape Province, South Africa) in context: the Cape floral kingdom, shellfish, and modern human origins. *J Hum Evol* 59(3–4), 425–43.

Marean, C. W. (2016). The transition to foraging for dense and predictable resources and its impact on the evolution of modern humans. *Philos Trans R Soc Lond B Biol Sci* 371(1698), 20150239.

Marean, C. W., Bar-Matthews, M., Bernatchez, J., et al. (2007). Early human use of marine resources and pigment in South Africa during the Middle Pleistocene. *Nature* 449(7164), 905–8.

Mariotti Lippi, M., Foggi, B., Aranguren, B., et al. (2015). Multistep food plant processing at Grotta Paglicci (southern Italy) around 32,600 cal B.P. *Proc Natl Acad Sci U S A* 112(39), 12075–80.

Marlowe, F. W. (2010). *The Hadza: Hunter-Gatherers of Tanzania*. Berkeley: University of California Press.

Marlowe, F. W., Berbesque, J. C., Wood, B., et al. (2014). Honey, Hadza, hunter-gatherers, and human evolution. *J Hum Evol* 71, 119–28.

Martin, C. K., Heilbronn, L. K., de Jonge, L., et al. (2007). Effect of calorie restriction on resting metabolic rate and spontaneous physical activity. *Obesity* 15(12), 2964–73.

Martin, M. A., Lassek, W. D., Gaulin, S. J. C., et al. (2012). Fatty acid composition in the mature milk of Bolivian forager-horticulturalists: controlled comparisons with a US sample. *Matern Child Nutr* 8(3), 404–18.

Mathieson, I., Lazaridis, I., Rohland, N., et al. (2015). Genome-wide patterns of selection in 230 ancient Eurasians. *Nature* 528(7583), 499–503.

Matsuzawa, Y. (1997). Pathophysiology and molecular mechanisms of visceral fat syndrome: the Japanese experience. *Diabetes Metab Res Rev* 13, 3–13.

Mauvais-Jarvis, F. (2015). Sex differences in metabolic homeostasis, diabetes, and obesity. *Biol Sex Differ* 6, 14.

McBrearty, S. and Brooks, A. S. (2000). The revolution that wasn't: a new interpretation of the origin of modern human behavior. *J Hum Evol* 39(5), 453–563.

McPherron, S. P., Alemseged, Z., Marean, C. W., et al. (2010). Evidence for stone-tool-assisted consumption of animal tissues before 3.39 million years ago at Dikika, Ethiopia. *Nature* 466(7308), 857–60.

Mellberg, C., Sandberg, S., Ryberg, M., et al. (2014). Long-term effects of a Palaeolithic-type diet in obese postmenopausal women: a 2-year randomized trial. *Eur J Clin Nutr* 68(3), 350–7.

Melzer, K., Schutz, Y., Boulvain, M., and Kayser, B. (2009). Pregnancy-related changes in activity energy expenditure and resting metabolic rate in Switzerland. *Eur J Clin Nutr* 63(10), 1185–91.

Mensink, R. P., Zock, P. L., Kester, A. D. M., et al. (2003). Effects of dietary fatty acids and carbohydrates on the ratio of serum total to HDL cholesterol and on serum lipids and apolipoproteins: a meta-analysis of 60 controlled trials. *Am J Clin Nutr* 77(5), 1146–55.

Mente, A., O'Donnell, M. J., Rangarajan, S., et al. (2014). Association of urinary sodium and potassium excretion with blood pressure. *N Eng J Med* 371(7), 601–11.

Mercader, J. (2009). Mozambican grass seed consumption during the Middle Stone Age. *Science* 326(5960), 1680–3.

Mercader, J., Bennett, T., and Raja, M. (2008). Middle Stone Age starch acquisition in the Niassa Rift, Mozambique. *Quart Res* 70(2), 283–300.

Micha, R., Peñalvo, J. L., Cudhea, F. I., et al. (2017). Association between dietary factors and mortality from heart disease, stroke, and type 2 diabetes in the United States. *JAMA* 317(9), 912.

Miesmer, A. and Marshall, J. (1980). *N!ai, The Story of !Kung Woman*. Watertown, MA: Documentary Educational Resources. https://archive.org/dayetails/naithestoryofakungwoman.

Milton, K. (2000). Hunter-gatherer diets—a different perspective. *Am J Clin Nutr* 71(3), 665–7.

Minger, D. (2010). 'The China Study': a formal analysis and response. https://dayeniseminger.com/2010/8/06/final-china-study-response-html/.

Minster, R. L., Hawley, N. L., Su, C.-T., et al. (2016). A thrifty variant in CREBRF strongly influences body mass index in Samoans. *Nat Genet* 48(9), 1049–54.

Moore, L. V. and Thompson, F. E. (2015). Adults meeting fruit and vegetable intake recommendations—United States (2013). *Morb Mortal Wkly Rep* 64(26), 709–13.

Moore, L. T., McEvoy, B., Cape, E., et al. (2006). A Y-chromosome signature of hegemony in Gaelic Ireland. *Am J Hum Genet* 78(2), 334–8.

Mozaffarian, D., Fahimi, S., Singh, G. M., et al. (2014). Global sodium consumption and death from cardiovascular causes. *N Eng J Med* 371(7), 624–34.

Muehlenbein, M. P. and Bribiescas, R. G. (2005). Testosterone-mediated immune functions and male life histories. *Am J Hum Biol* 17(5), 527–58.

Muehlenbein, M. P., Hirschtick, J. L., Bonner, J. Z., et al. (2010). Toward quantifying the usage costs of human immunity: altered metabolic rates and hormone levels during acute immune activation in men. *Am J Hum Biol* 22(4), 546–56.

Mugaas, J. N. and Seidensticker, J. (1993). Geographic variation of lean body mass and a model of its effect on the capacity of the raccoon to fatten and fast. *Bull Florida Mus Nat Hist Biol Sci* 36(3), 85–107.

Mummert, A., Esche, E., Robinson, J., et al. (2011). Stature and robusticity during the agricultural transition: evidence from the bioarchaeological record. *Econ Hum Biol* 9(3), 284–301.

Muskiet, F. A. J. and Carrera-Bastos, P. (2014). Beyond the Paleolithic prescription: commentary. *Nutr Rev* 72(4), 285–6.

Neel, J. V. (1962). Diabetes mellitus: a 'thrifty' genotype rendered detrimental by 'progress.' *Am J Hum Genet* 14, 353–62.

Neuschwander-Tetri, B. A. (2013). Carbohydrate intake and nonalcoholic fatty liver disease. *Curr Opin Clin Nutr Metabol Care* 16(4), 446–52.

Nikolov, V. (2012). Salt, early complex society, urbanization: Provadia-Solnitsata (5500–4200 BC). *Bulgarian National Institute of Archaeology*.

Noakes, T. D. and Windt, J. (2017). Evidence that supports the prescription of low-carbohydrate high-fat diets: a narrative review. *Br J Sports Med* 51(2), 133–9.

Norgan, N. G., Ferro-Luzzi, A., and Durning, J. V. (1974). The energy and nutrient intake and the energy expenditure of 204 New Guinean adults. *Philos Trans R Soc Lond B Biol Sci* 268(893), 309–48.

O'Dea, K. (1984). Marked improvement in carbohydrate and lipid metabolism in diabetic Australian Aborigines after temporary reversion to traditional lifestyle. *Diabetes* 33(6), 596–603.

O'Donnell, M., Mente, A., Rangarajan, S., et al. (2014). Urinary sodium and potassium excretion, mortality, and cardiovascular events. *N Eng J Med* 371(7), 612–23.

O'Rourke, M. F. and Nichols, W. W. (2005). Aortic diameter, aortic stiffness, and wave reflection increase with age and isolated systolic hypertension. *Hypertension* 45(Part 2), 652–8.

Oh, K., Hu, F. B., Manson, J. E., et al. (2005). Dietary fat intake and risk of coronary heart disease in women: 20 years of follow-up of the Nurses' Health Study. *Am J Epidemiol* 161(7), 672–9.

Olsen, R. H., Krogh-Madsen, R., Thomsen, C., et al. (2008). Metabolic responses to reduced daily steps. *JAMA* 299(11), 1261–3.

Oparil, S. (2014). Low sodium intake—cardiovascular health benefit or risk? *N Eng J Med* 371(7), 677–9.

Ornish, D., Scherwitz, L. W., Billings, J. H., et al. (1998). Intensive lifestyle changes for reversal of coronary heart disease. *JAMA* 280(23), 2001–7.

Otten, J., Stomby, A., Waling, M., et al. (2016). Effects of a Paleolithic diet with and without supervised exercise on fat mass, insulin sensitivity, and glycemic control: a randomized controlled trial in individuals with type 2 diabetes. *Diabetes/Metabol Res Rev* 33(1).

Panter-Brick, C. (1996). Seasonal and sex variation in physical activity levels among agro-pastoralists in Nepal. *Am J Phys Anthropol* 100(1), 7–21.

Panter-Brick, C. and Pollard, T. M. (1999). Work and hormonal variation in subsistence and industrial contexts. In: Panter-Brick, C. and Worthman, C. M. (eds) *Hormones, Health and Behavior*. Cambridge: Cambridge University Press.

Panter-Brick, C., Lotstein, D. S., and Ellison, P. T. (1993). Seasonality of reproductive function and weight loss in rural Nepali women. *Hum Reprod* 8(5), 684–90.

Park, Y., Subar, A. F., Hollenbeck, A., et al. (2011). Dietary fiber intake and mortality in the NIH-AARP Diet and Health Study. *Arch Intern Med* 171(12).

Pastore, R. L., Brooks, J. T., and Carbone, J. W. (2015). Paleolithic nutrition improves plasma lipid concentrations of hypercholesterolemic adults to a greater extent than traditional heart-healthy dietary recommendations. *Nutr Res* 35(6), 474–9.

Pearson, J. A. Y. D. (1990). Estimation of energy expenditure in Western Samoa, American Samoa, and Honolulu by recall interviews and direct observation. *Am J Hum Biol* 326, 313–26.

Pedersen, C. E. T., Lohmueller, K. E., Grarup, N., et al. (2017). The effect of an extreme and prolonged population bottleneck on patterns of deleterious variation: insights from the Greenlandic Inuit. *Genetics* 205(2), 787–801.

Pereira, M. A., Rifas-Shiman, S. L., Kleinman, K. P., et al. (2007). Predictors of change in physical activity during and after pregnancy project Viva. *Am J Prev Med* 32(4), 312–19.

Perry, G. H., Dominy, N. J., Claw, K. G., et al. (2007). Diet and the evolution of human amylase gene copy number variation. *Nat Genet* 39(10), 1256–60.

Pimentel, D. and Pimentel, M. (2003). Sustainability of meat-based and plant-based diets and the environment. *Am J Clin Nutr* 78(3 Suppl), 660S–3S.

Piombino-Mascali, D., Jankauskas, R., Tamošiunas, A., et al. (2014). Atherosclerosis in mummified human remains from Vilnius, Lithuania (eighteenth–nineteenth centuries AD): a computed tomographic investigation. *Am J Hum Biol* 26(5), 676–81.

Piombino-Mascali, D., Zink, A. R., and Panzer, S. (2017). Paleopathology in the Piraino mummies as illustrated by X-rays. *Anthropol Sci* 125(1), 25–33.

Piperno, D. R., Weiss, E., Holst, I., et al. (2004). Processing of wild cereal grains in the Upper Palaeolithic revealed by starch grain analysis. *Nature* 430(7000), 670–3.

Plantinga, T. S., Alonso, S., Izagirre, N., et al. (2012). Low prevalence of lactase persistence in Neolithic South-West Europe. *Eur J Hum Genet* 20(7), 778–82.

Pond, C. M., Mattacks, C. A., Calder, P. C., et al. (1993). Site-specific properties of human adipose depots homologous to those of other mammals. *Comp Biochem Physiol A Physiol* 104(4), 819–24.

Pontzer, H. (2012). Debunking the hunter-gatherer workout. Gray Matter. *New York Times Sunday Review* 26 August. https://www.nytimes.com/2012/08/26/opinion/sunday/debunking-the-hunter-gatherer-workout.html.

Pontzer, H. and Wrangham, R. W. (2004). Climbing and the daily energy cost of locomotion in wild chimpanzees: implications for hominoid locomotor evolution. *J Hum Evol* 46(3), 317–35.
Pontzer, H. and Wrangham, R. W. (2006). Ontogeny of ranging in wild chimpanzees. *Int J Primatol* 27(1), 295–309.
Pontzer, H., Raichlen, D. A., and Sockol, M. D. (2009). The metabolic cost of walking in humans, chimpanzees, and early hominins. *J Hum Evol* 56(1), 43–54.
Pontzer, H., Raichlen, D. A., Wood, B. M., et al. (2012). Hunter-gatherer energetics and human obesity. *PLoS One* 7(7), 1–8.
Pontzer, H., Raichlen, D. A., Gordon, A. D., et al. (2014). Primate energy expenditure and life history. *Proc Natl Acad Sci U S A* 111(4), 1433–7.
Popkin, B. M. (2015). Nutrition transition and the global diabetes epidemic. *Curr Diabetes Rep* 15(9).
Potts, R. (1996). *Humanity's Descent: The Consequences of Ecological Instability*. New York: William Morrow and Company.
Potts, R. (1998). Variability selection in hominid evolution. *Evol Anthropol* 7(3), 81–96.
Praagman, J., Beulens, J. W. J., Alssema, M., et al. (2016). The association between dietary saturated fatty acids and ischemic heart disease depends on the type and source of fatty acid in the European Prospective Investigation into Cancer and Nutrition—Netherlands cohort. *Am J Clin Nutr* 103(2), 356–65.
Pribis, P., Burtnack, C. A., Mckenzie, S. O., et al. (2010). Trends in body fat, body mass index and physical fitness among male and female college students. *Nutrients* 2(10), 1075–85.
Pritkin, N. and McGrady, P. M. (1979). *The Pritkin Program for Diet and Exercise*. New York: Gosset and Dunlap.
Puaschitz, N. G., Strand, E., Norekvål, T. M., et al. (2015). Dietary intake of saturated fat is not associated with risk of coronary events or mortality in patients with established coronary artery disease. *J Nutr* 145(2), 299–305.
Raichlen, D. A. and Polk, J. D. (2013). Linking brains and brawn: exercise and the evolution of human neurobiology. *Proc R Soc Lond Biol Sci* 280(1750), 20122250.
Raichlen, D. A., Foster, A. D., Gerdeman, G. L., et al. (2012). Wired to run: exercise-induced endocannabinoid signaling in humans and cursorial mammals with implications for the 'runners high'. *J Exp Biol* 215(8), 1331–6.
Raichlen, D. A., Pontzer, H., Harris, J. A., et al. (2016). Physical activity patterns and biomarkers of cardiovascular disease risk in hunter-gatherers. *Am J Hum Biol* (July), 1–13.
Ramsden, C. E., Zamora, D., Leelarthaepin, B., et al. (2013). Use of dietary linoleic acid for secondary prevention of coronary heart disease and death: evaluation of recovered data from the Sydney Diet Heart Study and updated meta-analysis. *BMJ* 346, e8707.
Ranciaro, A., Campbell, M. C., Hirbo, J. B., et al. (2014). Genetic origins of lactase persistence and the spread of pastoralism in Africa. *Am J Hum Genet* 94(4), 496–510.
Raubenheimer, D. and Simpson, S. J. (2016). Nutritional ecology and human health. *Annu Rev Nutr* 36(1), 603–26.
Raubenheimer, D., Rothman, J. M., Pontzer, H., et al. (2014). Macronutrient contributions of insects to the diets of hunter-gatherers: a geometric analysis. *J Hum Evol* 71, 70–6.
Revedin, A., Aranguren, B., Becattini, R., et al. (2010). Thirty thousand-year-old evidence of plant food processing. *Proc Natl Acad Sci U S A* 107(44), 18815–19.
Revedin, A., Aranguren, B., Becattini, R., et al. (2015). New technologies for plant food processing in the Gravettian. *Quat Int* 359, 77–88.
Richards, M. P. (2009). Stable isotope evidence for European Upper Paleolithic human diets. In: Hublin, J.-J. and Richards, M. P. (eds) *The Evolution of Hominin Diets*. Berlin: Springer Science + Business Media, pp. 251–7.
Riddoch, C. J., Bo Andersen, L., Wedderkopp, N., et al. (2004). Physical activity levels and patterns of 9- and 15-yr-old European children. *Med Sci Sports Exerc* 36(1), 86–92.
Roberts, P., Boivin, N., Lee-Thorp, J., et al. (2016). Tropical forests and the genus *Homo*. *Evol Anthropol* 25(6), 306–17.
Rodrigues, S. L., Baldo, M. P., Machado, R. C., et al. (2014). High potassium intake blunts the effect of elevated sodium intake on blood pressure levels. *J Am Soc Hypertens* 8(4), 232–8.

Rosenbaum, S. and Crowdy, J. P. (1992). British Army recruits: 100 years of heights and weights. *J R Army Med Corps* 138(2), 81–6.

Rossier, B. C., Bochud, M., and Devuyst, O. (2017). The hypertension pandemic: an evolutionary perspective. *Physiology* 32(2), 112–25.

Rothman, J. M., Raubenheimer, D., Bryer, M. A. H., et al. (2014). Nutritional contributions of insects to primate diets: implications for primate evolution. *J Hum Evol* 71, 59–69.

Ruegsegger, G. N., Brown, J. D., Kovarik, M. C., et al. (2016). Mu-opioid receptor inhibition decreases voluntary wheel running in a dopamine-dependent manner in rats bred for high voluntary running. *Neuroscience* 339, 525–37.

Ruff, C., Niskanen, M., Junno, J. A., et al. (2005). Body mass prediction from stature and bi-iliac breadth in two high latitude populations, with application to earlier higher latitude humans. *J Hum Evol* 48(4), 381–92.

Ruiz-Núñez, B., Pruimboom, L., Dijck-Brouwer, D. A. J., et al. (2013). Lifestyle and nutritional imbalances associated with western diseases: causes and consequences of chronic systemic low-grade inflammation in an evolutionary context. *J Nutr Biochem* 24(7), 1183–201.

Ryberg, M., Sandberg, S., Mellberg, C., et al. (2013). A Palaeolithic-type diet causes strong tissue-specific effects on ectopic fat deposition in obese postmenopausal women. *J Intern Med* 274(1), 67–76.

Sallis, J. F. (2000). Age-related decline in physical activity: a synthesis of human and animal studies. *Med Sci Sports Exerc* 32(9), 1598–600.

Sánchez-Pozos, K. and Menjívar, M. (2016). Genetic component of type 2 diabetes in a Mexican population. *Arch Med Res* 47(7), 496–505.

Saslow, L. R., Kim, S., Daubenmier, J. J., et al. (2014). A randomized pilot trial of a moderate carbohydrate diet compared to a very low carbohydrate diet in overweight or obese individuals with type 2 diabetes mellitus or prediabetes. *PLoS One* 9(4), e91027.

Schlebusch, C. M., Gattepaille, L. M., Engström, K., et al. (2015). Human adaptation to arsenic-rich environments. *Mol Biol Evol* 32(6), 1544–55.

Sebastian, A., Frassetto, L. A., Sellmeyer, D. E., et al. (2002). Estimation of the net acid load of the diet of ancestral preagricultural *Homo sapiens* and their hominid ancestors. *Am J Clin Nutr* 76(6), 1308–16.

Shanik, M. H., Xu, Y., Skrha, J., et al. (2008). Insulin resistance and hyperinsulinemia: is hyperinsulinemia the cart or the horse? *Diabetes Care* 31 (Suppl 2).

Shetty, P. S. (1999). Adaptation to low energy intakes: the responses and limits to low intakes in infants, children and adults. *Eur J Clin Nutr* 53, S14–33.

Shumaker, R. W., Wich, S. A., and Perkins, L. (2008). Reproductive life history traits of female orangutans (Pongos). *Interdiscip Top Gerontol* 36, 147–61.

Sibly, R. M. and Brown, J. H. (2007). Effects of body size and lifestyle on evolution of mammal life histories. *Proc Natl Acad Sci U S A* 104(45), 17707–12.

Sievenpiper, J. L. (2016). Sickeningly sweet: does sugar cause chronic disease? No. *Can J Diabetes* 40(4), 287–95.

Simmen, B., Bayart, F., Rasamimanana, H., et al. (2010). Total energy expenditure and body composition in two free-living sympatric lemurs. *PLoS One* 5(3), e9860.

Simopoulos, A. P. (2016). An increase in the omega-6/omega-3 fatty acid ratio increases the risk for obesity. *Nutrients* 8(3), 128.

Simpson, S. J. and Raubenheimer, D. (2005). Obesity: the protein leverage hypothesis. *Obes Rev* 6(2), 133–42.

Siri-Tarino, P. W., Chiu, S., Bergeron, N., et al. (2015). Saturated fats versus polyunsaturated fats versus carbohydrates for cardiovascular disease prevention and treatment. *Annu Rev Nutr* 35(1), 517–43.

Sockol, M. D., Raichlen, D. A., and Pontzer, H. (2007). Chimpanzee locomotor energetics and the origin of human bipedalism. *Proc Natl Acad Sci U S A* 104(30), 12265–9.

Sparti, A., Delany, J. P., De La Bretonne, J. A., et al. (1997). Relationship between resting metabolic rate and the composition of the fat-free mass. *Metabolism: Clin Exp* 46(10), 1225–30.

Speakman, J. R. (2013). Evolutionary perspectives on the obesity epidemic: adaptive, maladaptive, and neutral viewpoints. *Annu Rev Nutr* 33(1), 289–317.

Speakman, J. R. and Selman, C. (2003). Physical activity and resting metabolic rate. *Proc Nutr Soc* 62(3), 621–34.

Squyres, N. and Ruff, C. B. (2015). Body mass estimation from knee breadth, with application to early hominins. *Am J Phys Anthropol* 158(2), 198–208.

Sugiyama, Y. (2004). Demographic parameters and life history of chimpanzees at Bossou, Guinea. *American J Phys Anthropol* 124(2), 154–65.

Suzman, J. (2017). *Affluence Without Abundance*. London: Bloomsbury.

Taubes, G. (2008). *Good Calories, Bad Calories: Fats, Carbs, and the Controversial Science of Diet and Health*. New York: Anchor.

Taubes, G. (2011). *Why We Get Fat: And What to Do About It*. New York: Anchor.

Taubes, G. (2016). *The Case Against Sugar*. New York: Knopf.

Thompson, R. C., Allam, A. H., Lombardi, G. P., et al. (2013). Atherosclerosis across 4000 years of human history: the Horus study of four ancient populations. *Lancet* 381(9873), 1211–22.

Tishkoff, S. A., Reed, F. A., Ranciaro, A., et al. (2007). Convergent adaptation of human lactase persistence in Africa and Europe. *Nat Genet* 39(1), 31–40.

Toop, C. R. and Gentili, S. (2016). Fructose beverage consumption induces a metabolic syndrome phenotype in the rat: a systematic review and meta-analysis. *Nutrients* 8(9).

Troiano, R. P., Berrigan, D., Dodd, K. W., et al. (2008). Physical activity in the United States measured by accelerometer. *Med Sci Sports Exerc* 40(1), 181–8.

Trumbo, P., Schlicker, S., Yates, A. A., et al. (2002). Dietary reference intakes for energy, carbohydrate, fiber, fat, fatty acids, cholesterol, protein and amino acids. *J Am Diet Assoc* 102(11), 1621–30.

Turner, B. L. and Thompson, A. L. (2013). Beyond the Paleolithic prescription: incorporating diversity and flexibility in the study of human diet evolution. *Nutr Rev* 71(8), 501–10.

Turner, B. L. and Thompson, A. L. (2014). Beyond the Paleolithic prescription: authors' reply to Commentary. *Nutr Rev* 72(4), 287–8.

US Department of Agriculture (2016). *What We Eat in America*. Agricultural Research Service. Food Surveys Research Group. http://www.ars.usda.gov/nea/bhnrc/fsrg.

US Department of Health and Human Services and US Department of Agriculture (2015). *2015–2020 Dietary Guidelines for Americans*, 8th ed. https://health.gov/dietaryguidelines/2015/guidelines/.

Vaag, A. A., Grunnet, L. G., Arora, G. P., et al. (2012). The thrifty phenotype hypothesis revisited. *Diabetologia* 55(8), 2085–8.

Vickers, M. H., Breier, B. H., McCarthy, D., et al. (2003). Sedentary behavior during postnatal life is determined by the prenatal environment and exacerbated by postnatal hypercaloric nutrition. *Am J Physiol Regul Integr Comp Physiol* 285, 271–3.

Virtanen, K. A., Lönnroth, P., Parkkola, R., et al. (2002). Glucose uptake and perfusion in subcutaneous and visceral adipose tissue during insulin stimulation in nonobese and obese humans. *J Clin Endocrinol Metabol* 87(8), 3902–10.

Vitzthum, V. J. (2009). The ecology and evolutionary endocrinology of reproduction in the human female. *Am J Phys Anthropol* 52, 95–136.

Volk, B. M., Kunces, L. J., Freidenreich, D. J., et al. (2014). Effects of step-wise increases in dietary carbohydrate on circulating saturated fatty acids and palmitoleic acid in adults with metabolic syndrome. *PLoS One* 9(11), e113605.

Wagh, K., Bhatia, A., Alexe, G., et al. (2012). Lactase persistence and lipid pathway selection in the Maasai. *PLoS One* 7(9), e44751.

Walker, R., Hill, K., Burger, O., et al. (2006). Life in the slow lane revisited: ontogenetic separation between chimpanzees and humans. *Am J Phys Anthropol* 129(4), 577–83.

Warinner, C., Garcia, N. R., and Tuross, N. (2013). Maize, beans and the floral isotopic diversity of highland Oaxaca, Mexico. *J Archaeol Sci* 40(2), 868–73.

Watanabe, S., Kang, D. H., Feng, L., et al. (2002). Uric acid, hominoid evolution, and the pathogenesis of salt-sensitivity. *Hypertension* 40(3), 355–60.

Wells, J. C. K. (2010). The fitness value of fat. In: *The Evolutionary Biology of Human Body Fatness: Thrift and Control*. Cambridge: Cambridge University Press, pp. 153–94.

Wells, J. C. K. (2011). The thrifty phenotype: an adaptation in growth or metabolism? *Am J Hum Biol* 23(1), 65–75.

West-Eberhard, M. J. (2003). *Developmental Plasticity and Evolution*. Oxford: Oxford University Press.

Westerterp, K. R. (2013). Physical activity and physical activity induced energy expenditure in humans: measurement, determinants, and effects. *Front Physiol* 4(April), 1–11.

Westerterp, K. R. and Plasqui, G. (2004). Physical activity and human energy expenditure. *Curr Opin Clin Nutr Metab Care* 7, 607–13.

Westerterp, K. R. and Plasqui, G. (2009). Physically active lifestyle does not decrease the risk of fattening. *PLoS One* 4(3), e4745.

Westerterp, K. R. and Speakman, J. R. (2008). Physical activity energy expenditure has not declined since the 1980s and matches energy expenditures of wild mammals. *Int J Obes* 32(8), 1256–63.

WHO (2008). *The Global Burden of Disease: 2004 Update*. Geneva: World Health Organization.

WHO (2009). Global health risks: mortality and burden of disease attributable to selected major risks. *Bull World Health Org* 87, 646–6.

Wilde, S., Timpson, A., Kirsanow, K., et al. (2014). Direct evidence for positive selection of skin, hair, and eye pigmentation in Europeans during the last 5,000 y. *Proc Natl Acad Sci* 111(13), 4832–7.

Willett, W. C. (2014). *Dietary Fat and Heart Disease*. Boston: Harvard School of Public Health.

Williams, L. R. and Leggett, R. W. (1989). Reference values for resting blood flow to organs of man. *Clin Phys Physiol Meas* 10(3), 187–217.

Williams, N. I., Leidy, H. J., Hill, B. R., et al. (2015). Magnitude of daily energy deficit predicts frequency but not severity of menstrual disturbances associated with exercise and caloric restriction. *Am J Physiol Endocrinol Metabol* 308(1), E29–39.

Wobber, V., Hare, B., and Wrangham, R. (2008). Great apes prefer cooked food. *J Hum Evol* 55(2), 340–8.

Wolf, A., Bray, G. A., and Popkin, B. M. (2008). A short history of beverages and how our body treats them. *Obes Rev* 9(2), 151–64.

Wood, T. R., Hansen, R., Sigurdsson, A. F., et al. (2016). The cardiovascular risk reduction benefits of a low-carbohydrate diet outweigh the potential increase in LDL-cholesterol. *Br J Nutr* 115(6), 1126–8.

Workman, M., Baker, J., Lancaster, J. B., et al. (2015). Birth weight predicted baseline muscular efficiency, but not response of energy expenditure to calorie restriction: an empirical test of the predictive adaptive response hypothesis. *American J Hum Biol* 28(4), 484–92.

Wrangham, R. and Conklin-Brittain, N. L. (2003). Cooking as a biological trait. *Comp Biochem Physiol A Mol Integr Physiol* 136(1), 35–46.

Wrangham, R., Cheney, D., Seyfarth, R., et al. (2009). Shallow-water habitats as sources of fallback foods for hominins. *Am J Phys Anthropol* 140(4), 630–42.

Yamauchi, T. (2000). Nutritional status, activity pattern, and dietary intake among the Baka hunter-gatherers in the village camps in Cameroon. *Afr Study Monogr* 21(2), 67–82.

Yamauchi, T., Umezaki, M., and Ohtsuka, R. (2001). Physical activity and subsistence pattern of the Huli, a Papua New Guinea Highland population. *Am J Phys Anthropol* 114(3), 258–68.

Yamauchi, T., Abe, T., Midorikawa, T., et al. (2004). Body composition and resting metabolic rate of Japanese college Sumo wrestlers and nonathlete students. Are Sumo wrestlers obese? *Anthropol Sci* 112, 79–85.

Yi, X., Liang, Y., Huerta-Sanchez, E., et al. (2010). Sequencing of 50 human exomes reveals adaptation to high altitude. *Science* 329(5987), 75–8.

Young, R. A. (1976). Fat, energy and mammalian survival. *Integr Comp Biol* 16(4), 699–710.

Zerjal, T., Xue, Y., Bertorelle, G., et al. (2003). The genetic legacy of the Mongols. *Am J Hum Genet* 72(3), 717–21.

Zihlman, A. L. (1984). Body build and tissue composition in *Pan paniscus* and *Pan troglodytes* with comparison to other hominoids. In: Susman, R. L. (ed.) *The Pygmy Chimpanzee*. Berlin: Springer, pp. 179–200.

Zihlman, A. L. (1997). Natural history of apes: life-history features in females and males. In: Morbeck, M. E., Galloway, A., and Zihlman, A. (eds) *The Evolving Female: A Life-History Perspective*. Princeton, NJ: Princeton University Press, pp. 86–106.

Zihlman, A. L. and Bolter, D. R. (2015). s. *Proc Natl Acad Sci U S A* 112(24), 7466–71.

PART II

SPECIFIC SYSTEMS

CHAPTER 7

MUSCULOSKELETAL SYSTEM

MARTIN HÄUSLER, NICOLE BENDER, LAFI ALDAKAK, FRANCESCO M. GALASSI, PATRICK EPPENBERGER, MACIEJ HENNEBERG, AND FRANK RÜHLI

Abstract

This chapter on the musculoskeletal system aims to give an overview of the evolutionary implications for health and diseases of this body system. Using specific examples, the chapter shows how musculoskeletal diseases develop from an evolutionary point of view, with a specific emphasis on mismatches between the evolved human body and the modern environment, as well as on consequences of human bipedal gait. The chapter discusses the evolutionary aetiology of common musculoskeletal disorders as a consequence of mismatch to the modern world, such as carpal tunnel syndrome, flat feet, osteoporosis, and impacted third molars. As a consequence of the passage to bipedalism, the evolutionary background of low back pain, and knee and hip arthritis is examined. The chapter concludes with a discussion of implications for treatment and prevention of musculoskeletal diseases, focusing on the life course approach to health. In fact, every life stage shows specific needs and asks for specific interventions in order to optimally maintain musculoskeletal health up to old age.

Keywords

bipedalism, low back pain, osteoporosis, osteoarthritis, relaxed natural selection, physical activity, palaeopathology, evolution, medicine, life course approach

7.1 Introduction

ACTIVE body movement and locomotion are fundamental functions defining an animal and differentiating it from other forms of life. These functions in animals are effected by muscle tissue—a specialised kind of organic material consisting of cells capable of fast mechanical actions. Muscle cells translate energy stored in chemical bonds of molecules of the body into shifts in structures of specific proteins. These shifts change dimensions of cells releasing kinetic energy. In short, muscles translate energy of chemical bonds into kinetic energy. Here only somatic voluntary muscles—skeletal muscles—that move parts of the body as a result of control by the central nervous system are discussed. Visceral (smooth) involuntary muscles that serve various vegetative functions of internal organs and systems are not described, since they do not belong to the musculoskeletal system.

Skeletal muscles (mostly muscles moving bones against each other) consist of large tubular multinuclear cells—myocytes—some centimetres long and of about 50 μm in diameter. These multinuclear cells contain chains of myofibrils containing linear protein molecules. Among those are thin molecules of actin and thick molecules of myosin that, using adenosine triphosphate (ATP) in their 'heads' to obtain energy, move against actin chains, shortening the fibres when a nerve signal reaching the cell at the neuromuscular junction causes release of Ca^{2+} ions. When the release of energy from the decomposition of ATP into adenosine diphosphate (ADP) and phosphoric acid is completed and no further signalling comes in, the titin (spring-like) molecules maintain elasticity of muscle fibres. To maintain muscle contraction, a steady flow of nerve signals is needed many times a second. This comes from the central nervous system. This process constantly uses energy by transforming ATP. The strength of the muscle is proportional to the number of contractile fibres and the efficiency and constancy of nerve signalling. A muscle with many fibres may not produce expected strength if the nerve signalling is faulty. An optimal signalling of muscles requires training of the nerve system to deliver fast and coordinated signals to an appropriate set of muscle fibres. A number of muscle fibres is supplied by a single nerve. Such a set is called 'motor unit'. All fibres of the unit always contract in unison, reacting to the same nerve signal. It follows that a muscle containing only one motor unit can act in only one way—contract or relax in response to the signal of a single nerve. If a muscle contains numerous motor units, each of them can be signalled separately by a different nerve, so that actions of the muscle can be varied depending on signals from the central nervous system.

Muscles can only produce force when they contract; the single muscle fibre exerts a force only in one direction. Thus, for the locomotion of the whole body, groups of muscles must operate in a coordinated manner—they either support each other's function (agonists) or oppose it (antagonists). In animals without a skeleton, such as worms, rhythmic contractions of contralateral arrays of muscles provide propulsion by weaving the body. In chordates, segments of muscle tissue are lined on both sides of the axial skeleton—notochord or vertebral column organised around the primordial notochord—and by rhythmic contralateral contractions can weave the body and by bilateral contraction can stabilise the body, making it into a strong trunk that can be held vertically, supporting the head on its upper end, or horizontally, when it can itself be supported by front and hind limbs. The weaving action achieved by shortening and lengthening of trunk muscles is useful, for instance, in fish for swimming and reptiles for sliding on the ground. The stabilisation of the trunk is achieved by contraction of muscle fibres that do not change their length, but oppose outside forces that try to change the shape of the trunk—isometric muscle contraction. The isometric contraction, though not releasing kinetic energy, consumes energy to oppose the force of gravity, or some other force trying to displace an element of the body. It is, for example, important for providing stability to the trunk, so that actions of the head and limbs may be executed from a firm basis. Isometric contraction uses ATP stored energy the same as kinetic contractions. There is an energy cost to keeping the body in a particular, non-relaxed position. When the entire animal moves, the muscles of the trunk adjust the body's position by slight movements and stabilise it by isometric contractions. Coordination of these functions involves abundant signalling, orchestrated by the central nervous system, especially when the trunk is in a vertical position supported only on its lower end by the legs or in sitting.

Chemical reactions occurring during muscle actions are exothermic; thus heat is a by-product of muscle work. One simple way to heat the body exposed to cold is to start quick rhythmic muscle contractions—shivering. This heats the body well, but uses lots of energy. During their normal actions, muscles produce substantial amounts of heat that must be dissipated from the body to prevent overheating. Several physiological mechanisms serve this function: blood flow to the skin, radiation from the body surface, convection against air/water flow, and sweating. Also body anatomical structure—surface area to volume ratio and length of appendages—and its covering—such as hair—are important for heat dissipation. (For further discussion, see Chapter 8: Skin and Integument and Chapter 17 about Brain, Spinal Cord, and Sensory Systems.) Another by-product of muscle function are electric potentials. Though they are weak, they can be detected and measured by electromyography.

ATP that provides energy for muscle actions is produced in the body from catabolic processes breaking down sugars, fats, and proteins. This breakdown can happen without involvement of oxygen—an anaerobic process that is fast, but inefficient—or an aerobic process, which is slower but more efficient. When a quick and strong muscle action is required, anaerobic processes provide energy. Their by-product is lactic acid, which accumulates and after a short time lowers the efficiency of muscle work. Thus, for prolonged muscle action, aerobic processes (the citric acid cycle) must be used. This involves gas exchange through respiration, which becomes increased with increased muscle function. (For further discussion, see Chapter 12: Respiratory System.) Increasing respiratory movements in itself involves greater work of the muscles of respiration—primarily the diaphragm, and intercostal and trunk muscles—stabilising and moving the chest. Muscle metabolism and maintenance of the protein structure of muscles eventually involves nutrition, and hence the workings of the gastrointestinal system and urinary system; in short, actions of the musculoskeletal apparatus are a part of the complex systemic operation of the whole body.

Sets of muscle fibres are covered by connective tissue sheets. These come together to form tendons where muscle fibres end. Tendons are non-elastic connective tissue structures that attach to other muscle fibre sets or to bones and cartilages. The relation of muscle fibres to tendons directs forces with which sets of muscle fibres act and influences movements of structures they attach to. When a single set of muscle fibres forms tendons at both ends, a fusiform muscle is formed. It has a muscular belly and two tendons that provide attachments to skeletal elements. Some muscles attach directly to bone without forming a rope-like tendon, others form tendons between bellies (e.g. the rectus abdominis in humans), while others have sets of fibres running into a tendon at an angle—unipennate or bipennate muscles, or even multipennate muscles where there are many intermediate tendons separating bellies running in various directions—an example is the masseter muscle attaching to the ramus of the mandible and the zygomatic arch. Its bellies are short, but their simultaneous contraction produces a large chewing force. The muscle is palpably hard at the side of the mandible when contracted in forcible closing of the jaws. Interestingly, in humans a mutation led to the weakening of the masseter muscle compared to apes (Stedman et al. 2004).

There are some muscles whose fibres run in circular fashion around openings of the body—the mouth, eyes, and anus—and are not attached to a bone. Their contraction closes the opening without moving any skeletal elements. Some other muscles attach to the skin, providing its movements. Since the human face plays a key role in the communication of emotions (Darwin 1872), humans have complex sets of small muscles attached to the skin capable of changing facial expressions in a multitude of subtle ways. Yet other muscles enable vocal communication by moving the larynx, hyoid bone, and tongue. Small muscles adjusting leverage between auditory ossicles transmitting vibrations of tympanic membrane into the inner ear are also important in vocal communication, as they may widely change the amplification of sounds reaching the inner ear. Some skeletal muscles move just one bone over another—they act on one joint between bones. Others have bellies and tendons extending over multiple joints—flexors of the digits are a simple example. Long flexors of fingers run from the forearm over the wrist joints and then metacarpal and interphalangeal joints. Depending on the coordinated action of other muscles attaching to the hand, they can flex any of these joints separately or together in many combinations. Individual muscles and muscle groups are adjusted in the process of evolution to efficiently produce movements, ensuring good mechanical adaptation to specific living conditions. Muscles moving the digits of a horse have been reduced and simplified as only one digit remaining in the horse's limbs used almost exclusively for propulsion and support, while muscles of the human hand are many, ensuring precise manipulation of objects between digits.

Effective action of most muscles occurs when their ends are attached to elastic or rigid structures that can passively move. Such structures comprise the skeleton. The skeleton can be external, like in arthropods, or internal, like in vertebrates. It is made of organic molecules (chitin in invertebrates, collagen in vertebrates) that can be strengthened by inclusion of inorganic compounds, typically hydroxyapatite in vertebrate bones.

In vertebrates, the main types of tissue in skeletons are cartilage and bone. Cartilage is entirely made of organic collagenous matrix containing proteoglycans and elastin that give it rigidity and elasticity, while bone contains hydroxyapatite crystals held together by collagen fibres. The basic unit of the structure of compact bone is an osteon—a column of lamellae organised around the central canal containing blood vessels (Haversian canal). Living cells—osteoblasts and osteocytes—are present on seams connecting osteons and in lacunae in the bone matrix. These cells communicate with each other by thin processes. Thus, bone is a living substance. Its surface is covered by well-innervated tissue called periosteum comprising irregular connective tissue, fibroblasts, and osteoblasts. Osteoblasts produce osteoid, an organic material that turns to bone when permeated by inorganic salts. On the surface of the bone there are large cells—osteoclasts that break down the bone. The bone is a dynamic, living system, continuously remodelled by osteoclasts removing and osteoblasts adding tissue, as required to respond to mechanical demands and to pathological influences. Therefore, bones of the skeleton are undergoing a continuous process of adaptability—the evolved ability of an organism to respond to direct environmental stimuli. Macroscopically, bone tissue can be organised into continuous layers (compact bone) or occur in small units called trabeculae that produce a network (spongy bone). Cortical and spongy types of bone are arranged so as to provide the best mechanical support for a given part of the body at the minimum expense of material; thus bone structure follows the basic law of natural selection: maximisation of reproduction (function) at the minimum cost of energy (and material) expenditure. Depending on the body part, and thus the function, bones can have different shapes. The simplest classification of bone shapes is that into:

- flat
- short
- irregular
- long.

Long bones are large bones of the limbs. They contain a shaft (diaphysis) in the middle and ends that articulate with other bones (epiphyses). Parts between the shaft and epiphyses are called metaphyses. In young animals, epiphyses are separated from the shaft by plates of cartilage—physeal plates. Cartilage cells in those plates divide only in the vertical direction, increasing the thickness of the plate, while parts of the plate adjacent to bone tissue of the diaphysis and epiphysis ossify or mineralise. This way the bone grows in length, while thickness of the physeal plate does not change much. Physeal plates have growth hormone receptors that respond with increased cell proliferation to greater levels of the hormone. At reaching adulthood, under the influence of sex hormones, plates completely ossify, thus ending bone growth in length.

Individual bones are connected by joints. Depending on the function, joints may be permanently immovable (synostoses), allow for small degree of movement while holding bones firmly together (fibrous syndesmoses), allow limited movement while elastically cushioning compressive forces (cartilaginous, or synchondroses), and finally those joints that allow wide ranges of movement without cushioning compressive forces and thus preserve dimensions of body parts supported by jointed bones (synovial joints). The synovial joints surround the articulating ends of bones covered by thin compact cartilage (hyaline cartilage) with a capsule containing lubricating (synovial) fluid. Blood vessels in the capsule walls remove gases from the inside of the joint space, producing negative gas pressure inside the capsule. Thus, the atmospheric pressure, together with ligaments joining bones on the outside of the capsule, and forces of muscles extending over the joint hold articulating bones together. Synovial joints, depending on the shapes of articulating surfaces of bones, allow a wide range of movements including flexion and extension, abduction and adduction, rotation, circumduction, and sliding.

Structure of bones, the way they are connected by joints, and arrangements of muscles acting on them together determine the mechanical properties of the body. These properties reflect adaptations to particular ways of life. Basic structural arrangements are determined during ontogeny by genetic endowment shaped over generations of natural selection, but developmental plasticity and ability to remodel during life allows a substantial range of adjustments to occur in the musculoskeletal system due to individual activities and circumstances.

The musculoskeletal system develops mostly from the embryonic mesoderm that forms on both sides of the notochord, producing metameric (linearly organised sections) somites at the end or the third week of gestation. Somites will differentiate into bones of the axial skeleton and trunk muscles. The first somites form at the rostral (anterior) end of the body and then 38–44 somites develop consecutively towards the caudal end over 20–30 days of gestation. Limb buds—first anterior, then posterior—appear about the fourth week of gestation. The apical ectodermal ridge forms at the tip of the bud. Mesoderm from the lateral plate of the nearest sclerotomes enters the bud. The apical ridge prompts growth of limbs that is regulated by Hox genes inducing development of myotome, giving rise to muscles and chondrogenesis, forming templates of bones. In both forelimbs and hindlimbs the part closest to the body (stylopod) forms one bone, the more distal part (zeugopod) two bones, and the end (autopod) usually has five parallel bone rays. In the fetal period, already formed miniature limbs undergo growth and differentiation of tissues, separate muscles form, and blood vessels and nerves establish their distribution pattern that will persist for the rest of life. Bone and muscle differentiation and growth continue until adulthood. In the whole process of individual ontogeny (growth and development) a general sequence of growth is from the front (rostral end) towards the posterior end. Therefore, the fetus and new-born have very large heads in proportion to the rest of the body, the trunk grows next, then the upper limbs and lower limbs. Last growing lower limbs are mostly responsible for the 'pubertal growth spurt'. Thus individuals who had a smaller growth spurt usually have shorter legs in proportion to the rest of the body. (For further discussion, see Chapter 4: Growth and Development.) Human upper and lower limbs are very complex structures adapted to a wide range of object manipulations and locomotion. Although we do not use our upper limbs for walking, their position and movements during walking and running are important for balancing the body and ensuring energetic efficiency of locomotion.

Embryonic development of the musculoskeletal system is very fast and involves a series of precise differentiations of complex structures. Sometimes this evolved developmental homeostasis becomes slightly altered, producing anatomical variations. Numbers of fingers and toes may be greater than five (polydactyly), and digits may sometimes be fused together (syndactyly) or shortened (bradydactyly). The vertebral arches may not completely close, leaving parts of the spinal canal opened. Some muscles may not develop—the simplest example is the M. palmaris longus—a small muscle in the forearm that is missing in 10–40% of people. Frequencies of those developmental variations seem to have increased in recent generations, reflecting microevolutionary changes (Solomon et al. 2009). Ultimately, the shape and function of the musculoskeletal system are a reflection of the regulation of its embryonic development by gene complexes established by the evolution of developmental homeostasis and its adaptability providing responses to environmental stimuli during the fetal and postnatal ontogeny.

The musculoskeletal system, the largest system of the human body, is a result of millions of years of vertebrate evolution, which in the last 5–6 million run in a particular way in the hominin lineage, being shaped by the event of collective parenting that favoured erect posture, freeing hands for manipulatory actions of increasing complexity and demanding lower limb changes that allowed for locomotion of the erect trunk, with upper limbs capable of carrying objects and manipulating them (Preuschoft 2004; Clark and Henneberg 2015). As with all evolutionary adaptations, these structural and functional features are working 'just so', not being ideally formed. In the fast-changing modern human-made environment, long-term evolutionary musculoskeletal adaptations, even with their substantial range of adaptability, are not capable of providing optimal functions of the bodies, encountering long periods of immobility, often in poorly designed workspaces and living spaces. These spaces, and the lifestyles they imply, were not constructed with the hindsight of an understanding of human evolutionary history. (For further discussion, see Chapter 1: Core Principles for Evolutionary Medicine.) Such understanding may lead to the redesign of our living and working spaces and activity schedules.

Due to the relaxation of natural selection in the last few generations of humanity as a result of improvements in public health and medical treatments (Rühli and Henneberg 2013, 2016; Rühli et al. 2016b; You and Henneberg 2016, 2017), an accumulation of mutations, destabilising human evolved developmental homeostasis and resistance to pathologies, has occurred. This phenomenon alters development of the musculoskeletal system, affecting skeletal growth (gigantism or dwarfism, changing body proportions, and skeletal reductions such as spina bifida occulta); formation of muscles, including a changed number of muscle heads and their attachments, or disappearance of some muscles (e.g. M. palmaris longus); and skeletal robusticity and resistance of muscle, bone, and connective tissues to pathological processes, especially non-infectious degenerative processes. Muscle physiology and the quality of muscle signalling by the central nervous system may also be affected.

In Section 7.2 we address the evolutionary origins of musculoskeletal diseases and discuss the role of bipedalism.

7.2 Evolution, Functions, and Mechanisms of the Musculoskeletal System

Musculoskeletal problems are among the major complaints of modern humans. Disorders of the back, knee, hip, and shoulder constitute an increasing public health issue with immense socioeconomic costs. According to a recent study on healthcare spending in the United States, musculoskeletal disorders rank among the top five aggregated conditions (Dieleman et al. 2016). If individual conditions are analysed separately, low back and neck pain is estimated to be the third-largest contributor to healthcare costs after diabetes and ischaemic heart disease.

Musculoskeletal disorders are often conjectured to be a trade-off to the evolution of bipedalism (e.g. Krogman 1951) or a result of a mismatch with the modern environment. In the following sections we use an evolutionary approach and discuss the relevant fossil record of the hominids, our closest extinct relatives on the human lineage since the chimpanzee–human split 4–7 million years ago (mya) (Scally et al. 2012). Another approach to test this hypothesis involves a comparative morphological–pathological analysis of our closest relatives among other mammals, the great apes. Indeed, non-human primates including chimpanzees and gorillas show a significantly lower prevalence of osteoarthritis (OA) (degenerative joint disease) than modern humans (Jurmain 2000). On the other hand, the loading pattern of the joints is similar in quadrupedal animals and modern humans. This particularly includes the spine, which not only in bipeds but also in quadrupeds is mainly axially loaded (Smit 2002). Contrary to common belief, it has even been demonstrated that the vertebrae of quadrupeds sustain significantly higher compressive stresses than those of humans (Smit 2002). Putz and Müller-Gerbl (1996) therefore suggested that our spine is an optimised compromise between stability and mobility, and that occasional failures of this system should primarily be attributed to our increased lifespan and a changed lifestyle compared to that of our ancestors. First, we briefly discuss the evolution of bipedalism and the early hominid fossil record before we explore the evolutionary background of major musculoskeletal disorders.

7.2.1 The Fossil Record of Hominid Bipedalism

The common ancestor of humans and chimpanzees that lived some 4–7 mya was adapted to an arboreal environment. The earliest fossils that are attributed to the hominid lineage, that is, the lineage leading to modern humans, suggest that the common ancestor lacked the specialisations for suspension and knuckle-walking such as a long and broad pelvis seen in extant African great apes (Lovejoy et al. 2009). As in most other non-human primates, the very diverse locomotor repertoire of our first ancestors included occasional bipedalism. The percentage of terrestrial bipedalism then gradually increased in early hominids in relation to a more open environment, together with concomitant musculoskeletal adaptations. Rather than bipedalism per se, it is therefore these adaptations to upright locomotion that represent the initial step during human evolution and that distinguish our fossil ancestors and us. Only much later followed the production of stone tools, the increase in brain size, the loss of body hair, and the gradual abandonment of an arboreal habitat.

The diverse hominid fossil record with multiple lineages living contemporaneously that variously became extinct or reunified suggests that different pathways of bipedalism have been developed and that there was no gradual transition to a modern human-like gait pattern. Accordingly, a surprising variety of mosaics of derived and primitive morphologies in the feet of different *Australopithecus* species has been described, suggesting a corresponding diversity in modes of bipedalism (Clarke and Tobias 1995; Harcourt-Smith and Aiello 2004; Zipfel et al. 2011; Haile-Selassie et al. 2012; DeSilva et al. 2013). Nevertheless, their pelves uniformly demonstrate a remarkably derived morphology. The pelvis therefore seems to have played a crucial role in the adaptation to upright bipedalism. It is essential for balance of the trunk on the hip joints and for efficient bipedal locomotion (Tardieu et al. 2017).

Often, also the evolution of a human-like spinal curvature and particularly a lumbar lordosis is considered as the hallmark evolutionary adaptation to bipedal locomotion. New studies indicate, however, that this characteristic evolved secondary to the modification of the pelvis. The double S-shape of the vertebral column became then functionally linked to the pelvis and they evolved together (Tardieu and Haeusler in press).

The earliest claimed hominid fossil is *Sahelanthropus*, a 7 million years (Ma) old crushed skull from Chad (Brunet et al. 2002). Based on derived morphologies, including an anteriorly placed and downwards oriented foramen magnum and a downwards oriented nuchal plane, it was inferred that *Sahelanthropus* walked bipedally upright (Zollikofer et al. 2005). However, the skull needed extensive reconstruction and the orientation of the foramen magnum might not be useful for inferring bipedal posture, as it would imply that later hominids such as the australopithecines were quadrupeds (Wolpoff et al. 2006). Moreover, the age of *Sahelanthropus* is incompatible with its hominid status based on simple genetic models of the last common ancestor of chimpanzees and humans, unless a slowdown of the mutation rate is postulated (Scally et al. 2012).

The second oldest fossil attributed to the hominid lineage is *Orrorin* from Kenya with an age of 6 Ma. It is mainly known from skull and teeth fragments and isolated postcranial elements. Characteristics of the femora such as an obturator externus groove, an elongated femoral neck, and an asymmetric cortical thickness of the femoral neck suggested that *Orrorin* was bipedal (Senut et al. 2001; but see Ohman et al. 2005).

A more complete skeleton is known for the 4.4-Ma-old *Ardipithecus ramidus* from Ethiopia (White et al. 2009). With limb proportions similar to monkeys rather than to great apes and a grasping big toe, it was a palmigrade tree climber that may have facultatively walked upright when on the ground. The relatively short and narrow hipbone but wide sacrum suggests a more mobile lumbar spine than that of extant great apes. This implies that the great ape morphologies are derived rather than primitive. It is, however, also possible that *A. ramidus* was an extinct ape not closely related to the human lineage (Sarmiento 2010).

Better known is the skeleton of the australopithecines that lived between 4.1 and about 1 mya. It shows a mosaic of primitive and more derived characteristics, suggesting that the locomotor repertoire of this diverse group of early hominids was still very polyvalent and more versatile than that of modern humans. *Australopithecus* was therefore called a 'compromise biped' (Rose 1991) that still spent a significant amount of time in the trees (Schmid 1991; Ward 2013). The ilium already demonstrated a human-like reduced length, which was associated with changes in the birth mechanism (Häusler and Schmid 1995). The frontally oriented and laterally flaring iliac blades provided extensive origin for the latissimus dorsi, an important climbing muscle, while the lesser gluteal muscles functioned as extensors of the hip rather than as abductors as in humans (Haeusler 2002). These early hominids therefore possessed a different spino-pelvic balance and gait pattern than modern humans. In addition, the large, cone-shaped thorax, powerful arms, and a skull with a still-prominent face suggest a more ventrally located centre of body weight (Schmid 1983; McHenry 1991).

Representatives of the genus *Homo* evolved slightly more than 2 mya. They were the earliest 'committed' bipeds that possessed near-modern human-like body proportions (Haeusler and McHenry 2004, 2007). This implies adaptation to long-distance walking and running (Bramble and Lieberman 2004), allowing them to follow the game that has become their new staple diet, and to migrate out of Africa to populate Asia and Europe. Best known is *Homo erectus* from an almost complete 1.5-Ma-old skeleton from Nariokotome, Kenya, that belongs to an approximately 9-year-old boy (Walker and Leakey 1993). Its axial skeleton demonstrates a modern bauplan (Haeusler et al. 2002, 2011) and its locomotion is considered to be virtually indistinguishable from modern humans, although a new reconstruction of its pelvis shows remarkable similarities to the australopithecine morphology and their different lateral pelvic balance (Haeusler et al. 2017a).

Neanderthals were successors of *H. erectus* in Europe and in large parts of Asia after about 400,000 years ago. They show a robust, stocky body build, with markedly curved femora and radii, and thick cortical bone. This has variously been interpreted as evidence for high activity levels and climatic adaptations. Their extinction ca. 40,000 years ago may at least partially be explained by their genetic assimilation with early modern humans. Successive migration waves of anatomically modern humans from the Levant may then have diluted the Neanderthal contribution to the gene pool of Europeans to the present-day 1.5–4%, but many Neanderthal alleles still have a significant effect on our susceptibility to various diseases (Simonti et al. 2016).

In Section 7.3 we present an illustrative example of a hormonally caused skeletal alteration, acromegaly. We present this example from a palaeopathological and palaeopathographical point of view, before discussing the most common musculoskeletal diseases from an evolutionary point of view in Section 7.4.

7.3 Palaeopathology and Palaeopathography of the Musculoskeletal System: Acromegaly as an Illustrative Example

In order to properly explain the evolutionary course of diseases still affecting mankind in this century of supreme technological advancement, a leap into the past is perceived as a first necessary step to describe the historical presentation of pathological processes. While this goal can partially be achieved through the study of the development of the medical art (traditional history of medicine), over the past one hundred years the discipline of palaeopathology, by applying modern scientific techniques to the assessment of osteological and mummified ancient human remains, has tremendously enriched our understanding of the history of diseases. In particular, the major and statistically more relevant successes of palaeopathological research were achieved by the study of skeletons, the material arguably more prone to preservation, on a par with the muscle under special environmental conditions or as a result of apposite artificial embalming procedures, unlike visceral parts mostly unpreserved (with some notable exception, like viscera in Egyptian canopic jars). Although—as one of the pioneers of palaeopathology Walter G. J. Putschar (1904–1987) used to remark—skeletal diseases only reflect the condition of the skeletal system at the time of death and that such osteological alterations do not obviously always indicate the cause of death, though they can tell us much about the impact of internal conditions which can be interpreted in the light of evolutionary rationale (Ortner et al 2012). In this section, we analyse the effects of an endocrinological entity, acromegaly, on the musculoskeletal system, not only focusing on the traces left by disease on this anatomical district, but also adding notions of palaeosymptomatology as derived through the application of the newly introduced branch of palaeopathology, referred to as palaeopathography (Rühli et al. 2016a).

Acromegaly is a rare disease with a prevalence of 2.8–13.7 cases/100,000 people and with an annual incidence of 0.2–1.1 cases/100,000 people. It affects men more commonly than women, with a median age at diagnosis generally more than 40 years. The underlying aetiology is an excess of growth hormone (GH) after completed closure of the epiphyseal gaps. Among its systemic complications, cardiac and respiratory disorders, tumours, diabetes, and arthropathy pay a prominent role. In particular, acromegalic arthropathy, characterised by joints showing thicker cartilage and alterations of its biochemical composition, has a prominent impact on patients' health in spite of biochemical control of the disease. The whole skeleton is affected by new bone deposits (e.g. on vertebral body surfaces), but its most eye-catching deformations occur in the facial skeleton, with the mandibular bone (prognathism) and malar structures being mainly affected (Aufderheide and Rodiguez-Martin 2005). For a long time this condition was considered a relatively recent one, its first description dating back to 1886, when the French physician Pierre Marie (1853–1940) provided a full clinical classification, although he failed to understand the cause of the disease, paradoxically maintaining that the pituitary gland was hypofunctional in the process. Hyperfunctionality of the gland was indeed proposed by R. Massalongo (1857–1920) and A. Tamburini (1848–1919) and ultimately later proved by neurosurgeon Harvey Williams Cushing (1869–1939), who found an enlarged sella turcica in the Irish giant Charles Byrne (1761–1783) preserved in the London John Hunter Museum. However, contrary to this assumption, historical research has allowed scientists to understand the antiquity of the condition. The disease was first described in the medical literature by Johann Wier (1515–1588) in 1567, while several more cases were at the time considered merely osteological manifestations (hyperostosis) with prominent facial components ('prosopoectasia') (de Herder 2009). Interestingly, multidisciplinary research has finally backdated the discovery of the association between skeletal changes and enlargement of visceral organs to 1808, the year in which the Bolognese anatomist Alessandro Moreschi (1771–1826) holistically investigated the case and mortal remains of an acromegalic known as 'Il Bottaro' (Armocida et al. 2017) (Figure 7.1). This study was possible through a combination of archival, historical, ceroplastic—the skull is matched by the first wax representation of acromegaly—and

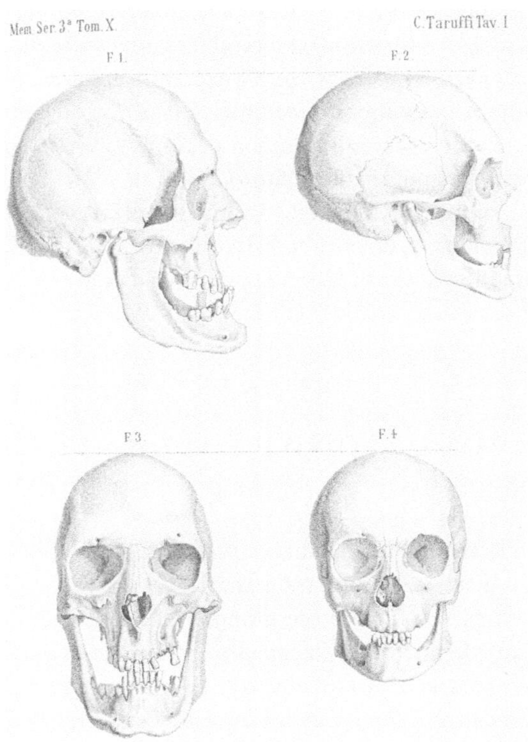

FIGURE 7.1 The Bottaro's skull (left) compared to a normal skull (right).

Source: Cesare Taruffi, 'Scheletro con prosopectasia e tredici vertebre dorsali' in Memorie della Reale Accademia delle Scienze dell'Istituto di Bologna, Ser III, Tom. X (1879), pp. 63–108. Image courtesy of UCL Library Services.

palaeopathological lines of evidence, in order to optimise the global understanding of the palaeo-phenotype of this endocrinological entity.

From a purely palaeopathological perspective, the oldest case of acromegaly appears to be that of a skeleton dated 9,500–11,500 years ago discovered in New Mexico in 1923, while in the Old World an acromegaloid habitus has been described in the mortal remains of history's oldest case of gigantism, that of the presumed remains of Pharaoh Sa-Nakht (ca. 2700 BCE) (Brauer 1991; Galassi et al. 2017). Furthermore, this hard biological evidence appears to be strongly corroborated by indirect ancient artistic and literary representations and descriptions highly suggestive of acromegalic pathology. The Greek mythological character Polyphemus has been suggested to show acromegalic traits and his monocular vision has been interpreted as the result of the loss of peripheral vision, a known outcome when the exuberant growth of the pituitary gland ends up compressing the optic chiasm (Papavramidou et al. 2007). Sticking to the Old World, other potential cases have been identified in Northern sagas, as well as in a large number of statuettes dating back to the Hellenistic period which show the typical coarse features resulting from facial bony alterations. In addition, even folk masks such as Mr Punch, originating from the Roman god Maccus, have been proposed to have been born out of not only cultural elements, but also pathological ones, including acromegaly (Armocida et al. 2016).

As previously astutely observed by some authors (Johnson 1912; Toni 2016), like other endocrinopathies, acromegaly's (and of its juvenile counterpart, gigantism) genetic background could be characterised by ancestral molecular pathways, whose alleles are currently silenced in adult *Homo sapiens*, while they exerted a prominent role in most ancient vertebrates (e.g. dinosaurs) for whom a gigantic size seemingly offered an advantage. Since these pathways have, however, been preserved in human genomes, even their partial reactivation may lead to pathological outcomes (Toni 2016). Such an evolutionary approach to acromegaly appears of substantial relevance not only in retrospective diagnostics and theoretical considerations of the condition, but also to better understand the increased incidence and prevalence of this condition in the modern western world (Dal et al. 2016; Toni 2016).

7.4 Clinical Diseases as Expressions of Evolutionary Constraints

Natural selection operates by giving individuals that are the most fit for their environment a better chance to reproduce. As a result, traits that are adaptive for a specific environment become more frequent and possibly more pronounced in the population over time in a process known as adaptation. However, adaptation is specific to a certain environment, so a trait that is adaptive in one environment may be maladaptive in another. As modern humans, we have evolved from hunter-gatherer ancestors who lived in physical environments that were very different from ours. Hunter-gatherers lived in small groups, mainly ate fruits and vegetables and lean meat, and performed a lot of physical activities. In contrast to this, modern humans mostly show poor diets, sedentary lifestyles, and high levels of stress and sleep deprivation, especially in the western world. (For further discussion, see Chapter 6: Nutrition, Energy Expenditure, Physical Activity, and Body Composition.) However, evolution has shaped our bodies to fit the environment where our ancestors evolved in a process that took hundreds of thousands of years. There is little doubt, therefore, that we are, to some degree, mismatched to our modern environment which has been extensively altered by the agricultural and industrial revolutions. The term mismatch has been coined to describe the discrepancy between the environment where an organism evolved and the environment where it currently lives. An organism that is adapted to an environment different to the one in which it lives is hence mismatched. We continue to change our environment and lifestyles today at an ever-increasing rate. However, genetic adaptation depends on slow processes that require the succession of multiple generations and cannot keep up with the rapid human-induced environmental changes. Moreover, culture as a product of human psychosocial interactions has become a major component of the modern human's environment. The problem is that culture is not a physical entity and unlike genotypic change it is not bound to physical processes that can only happen at slower rates. In fact, human culture has known major transitions, with an increasing magnitude ever since the agricultural revolution. Hence, in theory, we can never be optimally adapted to such rapidly changing environments. (For further discussion, see Chapter 1: Core Principles for Evolutionary Medicine.)

Mismatch can be temporal when the environment changes relatively quickly (mismatch in time) or spatial when a species emigrates to an environment different to the one to where it is most adapted (mismatch in space). Since the rate of biological change and the rate of change in the environment and living conditions are fundamentally different in modern humans, it must be expected that humans continuously suffer from temporal and spatial mismatches. Such mismatches have been blamed for many diseases of modern life including obesity, diabetes, coronary heart disease, and low back pain.

Evolutionary mismatch is therefore considered as one of the main aetiologies of modern human diseases and conditions, but other mechanisms are equally relevant, such as evolutionary trade-offs (like the obstetrical dilemma), pathogen–host interactions, or the persistence of disease-related gene variants. Such variants can persist because of several reasons including genetic heterogeneity (as in the case of HLA B 27 variants and psoriasis arthritis), pleiotropy, recessive mode of heredity, or disease outbreak beyond the reproductive age.

In this section, we discuss the aetiology of several diseases of the musculoskeletal system due to mismatch or other evolutionary causes. We start with diseases affecting the vertebral column as the most frequent medical problems of the musculoskeletal system. We then continue to describe diseases due to a mainly genetic predisposition (impacted or missing third molars), a combination of genetic predisposition and environmental influences (spina bifida, osteoporosis, OA), or mainly due to environmental influences (carpal tunnel syndrome (CTS), flat feet).

7.4.1 Back Problems

Back problems are the foremost musculoskeletal disorder of modern humans, with 60–80% of all people being affected by low back pain at some point in their life. The Global Burden of Disease 2010 study demonstrated that the prevalence and burden of low back pain is very high throughout the world and causes more disability globally than any other condition (Hoy et al. 2014). It is now acknowledged that low back pain is a multifactorial condition with various factors contributing in some way, including internal beliefs and expectations, life experience, family issues, and socioeconomic aspects (Battié et al. 1995; Katz 2006; Rubin 2007; Elfering et al. 2015). This agrees with historical analyses of sickness certifications for back pain that found a dramatic increase of chronic low back disability since the 1960s (Waddell 1996; Wynne-Jones and Dunn 2016). Yet, there is no indication that low back disorders have changed, or have become more common or more severe than they always have been, and the prevalence is globally similar independent of the occupation, behaviour, environment, or culture of the patient (Volinn 1997). In fact, it is mainly our attitude towards low back disorders that has changed in the past decades and what we do about them (Waddell 1996).

An important pathogenetic factor for low back disorders is spinal degeneration. As radiological features of spinal degeneration increase with age and because they are also common in asymptomatic patients, their relevance for low back conditions has been questioned (Brinjikji et al. 2015). Yet, there is a clear association of all types of low back pain with signs of disc degeneration (Luoma et al. 2000). Twin studies demonstrated that spinal degeneration is determined to a great part by genetic influences, thus rejecting the once commonly held view that back problems are primarily the result of ageing and environmental factors leading to 'wear and tear' from mechanical insults (Battié et al. 1995). However, the major part of the variability of spinal degenerative disease still remained unexplained in these studies and might be attributable to our everyday exposure to bipedal locomotion.

The importance of biomechanical factors for degenerative changes of the intervertebral discs is also demonstrated by studies of bipedal rats. They suggest that prolonged upright posture induces degenerative changes in the intervertebral discs of the rat lumbar spine (Liang et al. 2008). Moreover, disc herniation, isthmic spondylolysis and spondylolisthesis, facet joint OA, and also spinal stenosis have a predilection for the caudalmost segments of the lumbar curve (Takatalo et al. 2009). Such a predilection for the caudalmost spinal levels cannot be observed in quadrupedal monkeys (Nuckley et al. 2008).

Scheuermann's disease is another spinal pathology where disc lesions and mechanical overloading at the peaks of the spinal curves during the adolescent growth spurt seem to play a role, leading to thoracic hyperkyphosis or lumbar hypolordosis (Sørensen 1964). And also the development of adolescent idiopathic scoliosis seems to be related to the thoracic and lumbar curves of the spinal column and to upright bipedalism. Like Scheuermann's disease, adolescent idiopathic scoliosis is restricted to the species *Homo*, although its aetiology is still unknown (Kouwenhoven and Castelein 2008).

7.4.1.1 *Back Problems in the Hominid Fossil Record*

The hominid fossil record for the vertebral column is remarkably sparse. From the entire time period between 50,000 and 7 mya, less than a dozen partial skeletons have been discovered so far that preserve significant elements of the spinal column as well as a few isolated thoracolumbar vertebrae. Despite this small sample size, a surprising variety of vertebral pathologies have been described. This includes Scheuermann's disease in the 3.1-Ma-old Lucy skeleton (*Australopithecus afarensis*) (Cook et al. 1983). Similar bony alterations have been recognised in several additional isolated *A. afarensis* vertebrae, as well as in a partial *Australopithecus africanus* skeleton from South Africa (Haeusler et al. 2015b). Another *Australopithecus* partial skeleton shows facet joint OA of the cervical spine

(Meyer 2016), and a further *A. africanus* partial skeleton suffered from lumbar spondylosis and facet joint arthritis in addition to possible brucellosis (D'Anastasio et al. 2009). The best-preserved early hominid skeleton discovered to date, that of a 1.5-Ma-old *H. erectus* juvenile from Nariokotome, Kenya (Walker and Leakey 1993; Haeusler et al. 2011), was said to have suffered from severe axial skeletal dysplasia with disproportionate dwarfism, spinal stenosis, and scoliosis (Latimer and Ohman 2001; Lovejoy 2005). A thorough reanalysis refuted these claims (Schiess and Haeusler 2013; Schiess et al. 2014; Meyer and Haeusler 2015). Rather, evidence for telescoping facet joint subluxation was found in the lower lumbar region, most likely as the result of juvenile traumatic disc herniation at L4–5, which is very rare in modern humans (Figure 7.2) (Haeusler et al. 2013). The formation of a nearthrosis and extensive bony remodelling of the articular processes suggests that the disc herniation occurred several months before death. Disabling backache and recurrent sciatica might have, at least, temporarily restricted the daily activities of this *H. erectus* youth, which raises interesting questions for social care and nursing in early *Homo*.

The number of fossils with at least a partially preserved vertebral column increases dramatically in the Upper Palaeolithic. Many Neanderthals demonstrate advanced spondylosis and facet joint OA. Several specimens also have a hypolordotic or even kyphotic lumbar spine. This has already been observed in a 430,000-year-old Neanderthal ancestor from Spain who also suffered from spondylolisthesis (Bonmatí et al. 2010). As a flat spinal curve is associated with significant back problems in modern humans, this raised a discussion regarding a different spinal posture in Neanderthals (Been et al. 2017; but see Haeusler et al. 2017b).

This brief overview suggests an extraordinarily high prevalence of vertebral pathologies related to biomechanical factors in this small sample size of early hominids despite the young adult or even juvenile age of most of these specimens. More than a third of the early hominid vertebral columns suffered from Scheuermann's disease, implying a considerably higher prevalence of this condition in the Plio-Pleistocene than today where about 2.8% of

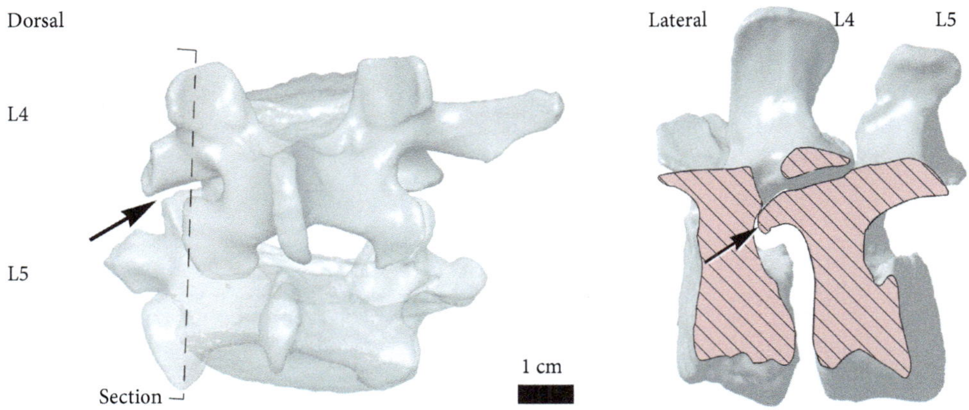

FIGURE 7.2 Subluxation of caudalmost lumbar vertebrae of *H. erectus* juvenile from Nariokotome, Kenya, probably due to juvenile disc herniation. Osteophytic remodelling of facet joints (arrow) suggests that the disc herniation was survived for a long time.

Source: Modified from Martin Haeusler, Regula Schiess, and Thomas Boeni, Evidence for Juvenile Disc Herniation in a *Homo erectus* Boy Skeleton, *Spine*, 38 (3), E123–8, doi.10.1097/BRS.0b013e31827cd245. Copyright © 2013 Lippincott Williams.

the general population are affected (Damborg et al. 2006). Scheuermann's disease and juvenile disc herniation observed in the Nariokotome *H. erectus* juvenile have many aspects in common. They have therefore been argued to be different manifestations of the same phenomenon (Parisini et al. 2001). Both entities occur during the increased vulnerability phase of the pubertal growth spurt, result from displacement of disc material, and have a higher incidence following trauma. In this context, the disc herniation of the Nariokotome juvenile loses some of its singularity despite being rare among adolescents today. These disc lesions most likely testify to the massive mechanical strains the axial skeleton was exposed to in our ancestors.

A possible explanation for the high prevalence of these signs of biomechanical failure might be the relatively small cross-sectional area of the early hominid vertebrae compared to modern humans (Figure 7.3) (McHenry 1992; Schiess and Haeusler 2013). Because stress is directly proportional to the force over the loaded area, a small cross-section makes vertebrae and discs more susceptible to injury. Conversely, this suggests that our spinal column has been shaped by a long process of natural selection to become less vulnerable than that

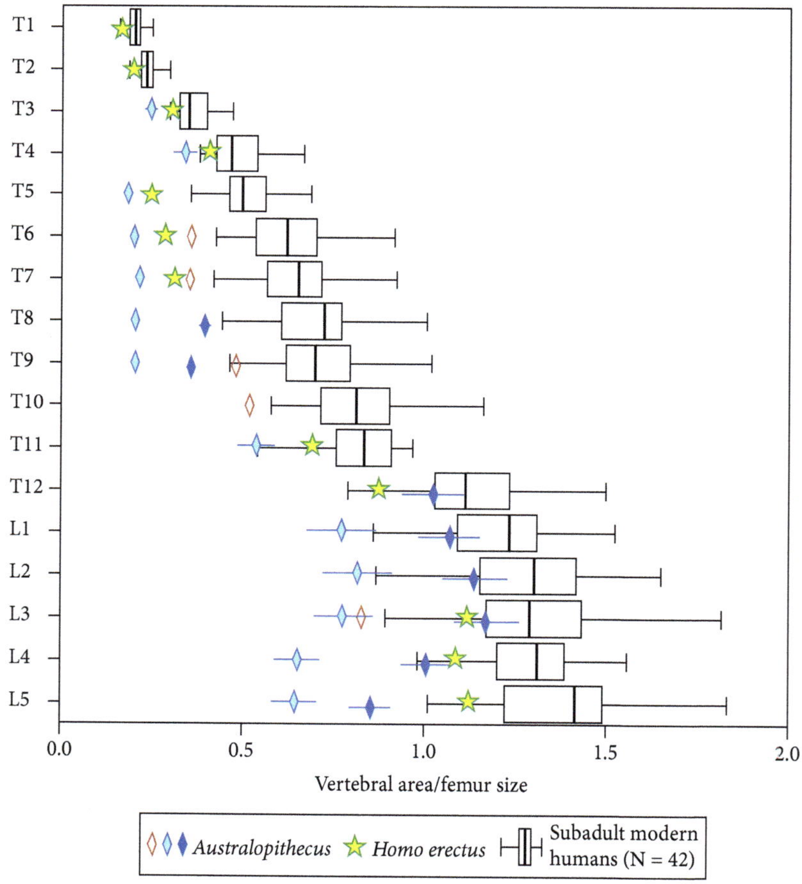

FIGURE 7.3 Cross-sectional area of vertebrae relative to proximal femur size as a measure for body size in *Australopithecus* (open diamonds, Lucy; light blue diamonds, Sts 14; dark blue diamonds, Stw 431), *H. erectus* (stars, Nariokotome boy), and subadult modern humans (box-plots).

of our ancestors. Increased vertebral cross-sectional area has therefore been considered as one of the most important vertebral adaptations to habitual bipedal locomotion. It is likely that this makes our vertebral column better adapted to vertical loading than that of early hominins, and we seem to have less problems with our spine than our ancestors.

These adaptations considerably pre-date industrialisation and the spread of agriculture. The hominid fossil record therefore does not support the hypothesis that the changed lifestyle of contemporary humans or a mismatch of our body to modern-day environments is related to our widespread back problems. Yet, the interaction of biomechanical factors with back problems may not be a simple linear relationship. In fact, both excessive physical activity as well as inactivity presents a risk for developing back problems (Heneweer et al. 2009; Adams et al. 2015). Perhaps, then, today's predominance of low back conditions relate to more than external influences, such as our society's perception of back pain (Waddell 1996).

A notable absent pathology is osteoporotic fractures of the vertebrae, not only from the hominid fossil record, but also during most of the human past. Likewise, osteoporotic fractures are not observed in great apes, although female chimpanzees can develop severe age-related bone loss (Gunji et al. 2003). A decline in activity levels associated with the adoption of agriculture and industrialisation significantly contributed to the reduction in bone volume fraction in recent modern humans (Chirchir et al. 2015).

7.4.2 Osteoarthritis

OA is one of the most common joint diseases and causes of pain in developed countries. This is often attributed to an imperfect adaptation of humans to bipedalism and vertical posture, and that we are not adapted to prolonged sitting, leading to low back pain and other musculoskeletal diseases (but see also Section 7.4.1).

7.4.2.1 *Knee Joint*

The knee is the largest joint in our body and supports nearly the entire body weight. This exposes it to enormous stress during locomotion. It is the site of the appendicular skeleton most affected by joint pain in older adults, where it is usually attributed to knee OA. Thus, the general US population has a lifetime risk of 45% to be affected by symptomatic disorders of the knee joint (Murphy et al. 2008). This contrasts to great apes with their more mobile and less tightly guided knee, where knee OA is almost unknown (Jurmain 2000).

Major risk factors for knee OA in humans include obesity and previous knee trauma, particularly meniscus lesions. Most often the medial meniscus is injured. This meniscus is relatively immobile due to its widely separated insertions on the tibial plateau and the attachment to the tibial collateral ligament, while in apes the medial meniscus is separated from this ligament by a bursa. Also the lateral meniscus with two closely adjacent insertions is less mobile than in apes, which have a single insertion. This primitive configuration was also exhibited by *Australopithecus*, while early *Homo* evolved a human-like double insertion of the lateral meniscus (Tardieu 2010). This supplementary insertion prevents an excessive anterior displacement of the meniscus in full extension and is thus an adaptation to bipedal walking with a fully extended knee joint. Yet, these specialisations restrict the capacity for rotation of the knee, making it more susceptible to injury and OA compared to that of apes.

Another trade-off to the evolution of bipedalism is patella luxation, which is principally a consequence of the valgum axis of our knee. The *genu valgum* is one of the most striking differences between humans and apes. It is necessary for an efficient bipedalism as it brings the knee joint below the body's centre of weight. However, it causes the quadriceps muscles to pull the patella laterally as the knee extends, thus favouring patella luxation. As a response, during hominid evolution the trochlea became deepened and its lateral lip elevated (Tardieu 2010). In modern humans, a flat trochlea can occur as an anatomic variant. Because it predisposes to habitual patella luxation and OA; operative trochleoplasty is indicated in these cases.

7.4.2.2 *Knee Osteoarthritis in the Human Past*

Knee OA is almost absent from the hominid fossil record. It can only be observed in a 1.8-Ma-old *H. erectus* knee joint from Dmanisi, Georgia, and the *sapiens*-like Femur I from Trinil, Java, that belongs to the holotype of *H. erectus* (Dubois 1892; but see Ruff et al. 2015). In prehistoric and historic modern human populations, remarkably low levels of knee OA have been reported (Rogers and Dieppe 1994).

In a recent study, knee OA has been found to have doubled in prevalence in a post-industrial sample from the United States (late twentieth to early twenty-first century) compared to an early industrial sample (nineteenth to early twentieth century) (Wallace et al. 2017). They used cadavers of people who died 50 years old or older and who had their body mass index (BMI) at death documented. They also analysed indicators of knee OA in prehistoric hunter-gatherers and early farmers (6000–300 years BP) estimated to be ≥ 50 years old at death. Knee OA was found to be more prevalent in the post-industrial individuals compared to the other two samples. Even after controlling for age and BMI, knee OA prevalence was more than two-folds higher in the post-industrial sample compared to the early industrial one. These results suggest that although high load due to increased BMI is more prevalent today, it is not sufficient to explain the higher incidence of knee OA. Potential culprits for the increased prevalence of knee OA today include physical inactivity, which has become epidemic in modern times. Physical inactivity may lead to thinner and softer joint cartilage and to weaker muscles that stabilise and protect the joints. Inactivity in conjunction with modern diets rich in processed carbohydrates and excessive adiposity may also exacerbate systemic chronic low-grade inflammation that further aggravates knee OA (Wallace et al. 2017).

7.4.2.3 *Hip Joint*

Symptomatic OA of the hip joint is less prevalent than knee OA but associated with a higher socioeconomic burden. Idiopathic or primary OA is the most common form of OA in western populations. It is attributable to subtle anatomical variations of the proximal femur and acetabulum leading to femoroacetabular impingement (Ganz et al. 2008). This contrasts to other parts of the world where hip joint OA often results secondary to other conditions such as developmental hip dysplasia. This condition is associated with certain swaddling practices of the babies, but is rarely observed in populations where the mothers carry their babies on the back. On the other hand, femoroacetabular impingement is virtually absent in non-European populations, suggesting that a genetic predisposition is necessary to express this form of OA (Hoaglund 2013). Alternatively, it has been speculated that femoroacetabular impingement in males is a trade-off to their adaptation to running, while in females it should be associated with the obstetrical dilemma (Hogervorst et al. 2011), a reasoning that contradicts basic evolutionary principles.

7.4.2.4 *Hip Osteoarthritis in the Human Past*

Hip joint OA is remarkably rare in ancient humans. Particularly compared to OA of other joints, hip OA is infrequent in prehistoric people of North America (Bridges 1991) and Africa (Jurmain and Kilgore 1995). In even earlier times, it has only been described in a 50,000-year-old Neanderthal skeleton, the 'old man' from La Chapelle-aux-Saints (Trinkaus 1985), and a 2.6-Ma-old *A. africanus* femoral fragment (Reed et al. 1993). The hip pathology of the skeleton from La Chapelle-aux-Saints may, however, have resulted from an injury (Trinkaus 1985), and from a haematological condition in the *Australopithecus* specimen (Haeusler et al. 2015a). The fossil and archaeological record thus fits the notion that the high prevalence of hip OA in modern western populations is a recent phenomenon related to genetic or behavioural shifts.

7.4.2.5 *Shoulder Joint*

Shoulder pain and dysfunction are the third most common musculoskeletal complaint in the general population after back and knee pain. The majority of these disorders are related to structures of the subacromial space that includes the subacromial bursa, the supraspinatus tendon, the capsule of the humeral joint, and the long head of the biceps brachii tendon. These different pathologies are often subsumed under the umbrella term 'shoulder impingement syndrome'. Its multifactorial causes include morphological variants of the scapula that might compromise the subacromial space, and muscular dysbalance.

Degenerative shoulder disorders and rotator cuff tears are surprisingly uncommon among wild-living non-human primates (Potau et al. 2009). In contrast to apes, humans have a narrower supraspinous fossa and a reduced size of the supraspinatus muscle. Together with a more laterally facing glenoid cavity and a more horizontal scapular spine, this was linked to the habitually lowered posture of the human arm. Fossil evidence suggests that the most important morphological changes occurred during the transition from *Australopithecus* to the genus *Homo* around 2 mya. The australopithecine scapula shows similarity to the shoulder girdle of apes in the orientation of both its spine and glenoid joint and in the proportions of its supraspinous and infraspinous fossa heights. The *H. erectus* scapula, on the other hand, closely resembles that of modern humans (Larson 2013).

It has been proposed that this evolutionary reorganisation of the human supraspinatus muscle and muscular imbalance made it susceptible to overuse (Lewis et al. 2001). Accordingly, the high prevalence of shoulder disorders related to the subacromial space in humans has been hypothesised to be an evolutionary trade-off of bipedalism and the associated freeing of the upper limb from locomotor function (Lewis et al. 2001; Potau et al. 2007; Bello-Hellegouarch et al. 2013; Craik et al. 2014). The proportions of the rotator cuff muscles are reflected by the supraspinous and infraspinous fossa areas. Using area measurements instead of diameters, the supraspinatus muscle and the subacromial space are, however, similarly sized in modern humans, early hominins, and great apes relative to shoulder joint size. This suggests that the reorganisation of the shoulder joint during the evolution of bipedalism is not relevant for the high prevalence of shoulder disorders in humans (Mathews and Haeusler 2015).

7.4.3 Osteoporosis

Osteoporosis has been qualitatively defined in a consensus development conference (Anonymous 1993) as 'a disease characterised by low bone mass and microarchitectural deterioration of bone tissue, leading to enhanced bone fragility and a consequent increase in fracture risk'. At the age of 50 years, 15% of the population is affected; at age 80 and more, the prevalence reaches 70%. Women are four to five times more affected than men. The main risk factor for osteoporosis is vitamin D deficiency. Vitamin D is essential for calcium, phosphate, and magnesium absorption, as well as mineralisation of bone tissues, and it can be photochemically synthesised in the skin upon exposure to UV light. Vitamin D deficiency is defined as a serum concentration of less than 20 ng/ml (50 nmol/l). The overall prevalence of vitamin D deficiency in western countries is estimated at 40–50%, depending on the population and season.

There are many potential evolutionary culprits for the increasing risk of osteoporosis. These include changes in diet and physical activity, changes in behaviours related to sun exposure, and increased life expectancy in most human societies. The modern western diet contains high amounts of processed animal meat rich in saturated fatty acids, and lower amounts of fruits and vegetables. This misbalance creates a chronic low-grade acidosis that worsens with age due to declining kidney function. As a result, the osteoclast resorptive activity is increased and the osteoblast deposition of alkaline minerals in bone is reduced to buffer the excessive protons. Furthermore, biomechanical loading is an essential factor to trigger the building of bone tissue. Thus, it comes with no surprise that the incidence of osteoporosis is increasing due to lack of physical activity associated with a sedentary lifestyle. The argument has also been made that the risk of vitamin D deficiency has increased with the emigration of *H. sapiens* outside Africa. *Homo sapiens* evolved in the African plains and highlands where natural sunlight was abundant and there was probably some selection for protection against excessive exposure rather than lack of it. The problem of not getting enough sun exposure probably started with emigration to Europe through the Middle East and was only exacerbated by the need of covering the body with fabrics to protect against the colder climate. The evolution of fair skin colour in Europeans and convergently in North Asians was partially explained by the need to synthesise more vitamin D in northern countries with less sunlight. (For further discussion, see Chapter 8: Skin and Integument.) In modern times, the abundant use of sun creams may play an additional role in increasing the prevalence of vitamin D deficiency and thus osteoporosis.

Lastly, the increased prevalence of OA, related to the increased life expectancy of the modern human population, increases the risk of bone fractures, especially in combination with osteoporosis. The bone-maintaining effect of oestrogen and progesterone has been the focus of a substantial number of studies. The levels of these two hormones decline precipitously with the cessation of the menstrual cycle in women after menopause. An increased life expectancy, combined with a high prevalence in osteoporosis and OA due to low levels of sexual hormones after menopause, increases the accumulative risk of OA-related fractures in women. Moreover, this effect might be aggravated by the fact that OA-related fractures usually happen past the reproductive age, which means that natural selection cannot select out risk gene variants for osteoporosis and OA (Purdie 1996; Brandao-Burch et al. 2005; Karasik 2008).

7.4.4 Impaction or Lack of Third Molars as an Example of a Genetically Predisposed Condition

Impacted wisdom teeth occur when the third molar does not emerge into the mouth due to occlusion (impaction) by other molars. When this happens, it might cause pain, inflammation, infection, or damage of other teeth. The impaction of at least one wisdom tooth is estimated to reach a prevalence of 70–90%, and 10–30% of the population show an M3 agenesis, a condition in which at least one wisdom tooth did not develop at all.

The aetiology of impacted wisdom teeth has been linked to the wider and smaller jaws of modern humans that evolved in connection to skulls that must accommodate larger brains. In fact, human brains have tripled in size since the split from the last common ancestor of humans and chimpanzees, while the dentition and the facial prognathism decreased compared to apes. As the jaw decreased in size and the chewing muscles became weaker, the pressure on the cranial sutures lowered, and encephalisation was facilitated. Early hominins like australopithecines showed smaller molars and canines than apes, but their teeth were still larger than those of modern humans. In early *H. erectus* the teeth were larger than in modern humans, but smaller than in archaic *H. sapiens*. The incisors began to show the 'shovel' shape, while the molars showed a reduction in size that was linked to the consumption of softer foods, including cooked foods as well as more meat.

7.4.5 Spina Bifida as an Example of Combined Genetic and Environmental Disorders

Spina bifida comes from the Latin words for spine and cleft. It refers to a medical condition where there is incomplete closure of the spinal canal around the spinal cord, mostly in the sacral bone. It is a congenital disease with mixed genetic and environmental causes and reaches a prevalence of 0.1–5 per 1000 births in western populations. One of the main causes is folate deficiency, and pregnant women are recommended to supplement their diet with folate. Spina bifida can cause paralysis of the legs and other neurological symptoms.

The mildest form of this disease is called spina bifida occulta, with a prevalence of 5–20% in the general population. *Occulta* is the Latin word for hidden. This form is usually considered harmless as it does not cause paralysis, even though one study found that low back pain can be more severe if spina bifida occulta is present.

There are two evolutionary mechanisms that may have affected the prevalence of spina bifida occulta in modern human populations. The first one is related to mismatch, particularly changes in diet, and the second was discussed in the context of the filtration effect of natural selection. One of the major risk factors of spina bifida and other congenital neural tube anomalies is vitamin B9 (also known as folate) deficiency before and during pregnancy. Folic acid is abundant in vegetables, particularly dark leafy ones. Shifting to a diet rich in carbohydrates and poor in green leaves increases the risk of folic acid deficiency. Our modern diet contains less green leaves and nuts than that of our hunter-gatherer ancestors and might therefore represent a risk factor for folate deficiency (Eaton and Cordain 1997). (For further discussion, see Chapter 6: Nutrition, Energy Expenditure, Physical Activity, and Body Composition.)

The other evolutionary mechanism that might have increased the incidence rate of spina bifida is related to the effect of spina bifida occulta on survival and reproductive fitness, or, more precisely, the absence of this effect. Individuals with spina bifida occulta might experience some discomfort, but their chances of reproductive success are not compromised. Therefore, we should not expect natural selection to eliminate the gene variants and behaviours that increase the risk of this condition. This fact could explain why the relatively high prevalence of this condition was maintained through evolutionary time (Lachmann 1982; Murphy 1993). Moreover, studies have shown that within a few generations the prevalence of spina bifida occulta has increased in the last decades. This increase was explained by a general effect of relaxed natural selection and genetic drift in the last century due to modern medicine and improved life conditions (Solomon et al. 2009). As this effect will persist and even increase with the further development of modern medicine, it is expected that variability in many human traits will continue to increase in future.

7.4.6 Carpal Tunnel Syndrome as an Example of an Environmentally Determined Condition

Carpal tunnel syndrome (CTS) is a fairly common medical condition among clerks of all kinds. It reaches a prevalence of 3–6% in the adult population in western countries. It typically causes pain, numbness, and tingling in the fingers innervated by the median nerve. CTS is caused by compression of the median nerve as it travels through the carpal tunnel together with the flexor tendons of the thumb and finger muscles. The carpal tunnel is a tight space formed between the carpal arch and the flexor retinaculum. Inflammation of the tendons and tissues in the carpal tunnel creates oedema and excessive fluid in the carpal tunnel, and therefore compression of the median nerve. Any structure innervated by the median nerve downstream of the compression can be affected and the sensory symptoms can be accompanied by wasting of the thenar muscles. Risk factors include any condition that can cause or exacerbate inflammation and tissue swelling, such as obesity, pregnancy, hypothyroidism, and chronic inflammatory diseases such as rheumatoid arthritis. CTS is assumed to be triggered by repetitive physical stress like typing for lengthy periods of time. For this reason, CTS is usually bilateral and worsens at night after day use.

The evolutionary perspectives of CTS are related to the evolution of bipedalism and the mode of use of the hands. Perez et al. (2017) argue that a reduction in size and functional orientation of the pisiform bone caused by the loss of locomotor function of the forelimbs in humans reduced the cross-sectional area of the carpal tunnel (Figure 7.4). In apes, the pisiform bone retains a growth plate and a secondary centre of ossification, making the pisiform longer (Figure 7.5). The described changes in human anatomy lead to a narrower carpal tunnel in humans. The decreased amount of space available for the median nerve and the adjacent structures increases the risk of compression. The narrower carpal tunnel, together with the increased likelihood of repetitive overuse in modern activities, render humans particularly susceptible to CTS. Due to the increasing dependence on computers and smart phones in modern times, we are at risk of overusing our hands by excessive typing or similar activities, and until keyboards are completely replaced by dictaphones that efficiently convert voice into text, and other similar cultural adaptations, our clerks are expected to suffer a high prevalence of CTS (Perez and Simpson 2017).

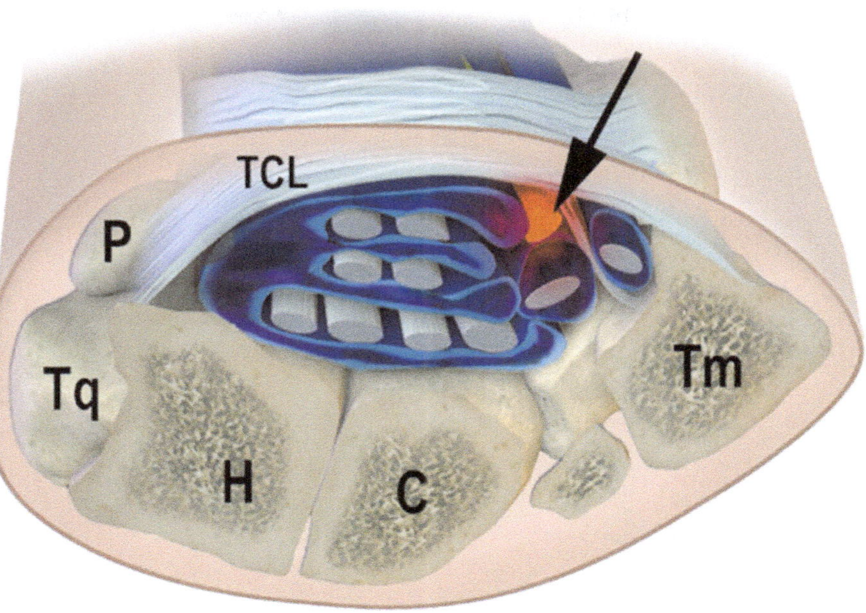

FIGURE 7.4 Cross section of a right proximal wrist, palmar side up. Arrow points to median nerve. P, Pisiform; Tq, triquetral; H, hamate; C, capitate; Tm, trapezium; TCL, transverse carpal ligament.

Source: Reproduced from Althea Anne D. Perez and Scott W. Simpson, Evolutionary medicine carpal tunnel syndrome, *Evolution, Medicine, and Public Health*, 17 (1), pp. 51–2, Figure 1, doi.org/10.1093/emph/eox006.

7.4.7 Flat and Splayed Feet

Another important evolutionary principle that influences our susceptibility to many musculoskeletal disorders is the fact that evolution most often works to modify existing structures rather than creating them *de novo*. Exaptation, a term suggested by Stephen Jay Gould (1982), describes the process when a certain trait serving a particular function later evolved to serve another. We evolved from quadrupedal ancestors whose grasping feet were adapted for tree climbing. We then gradually changed posture and became to be bipedal. In the course of adaptation to bipedalism, the arrangement of the feet bones gradually changed and eventually formed the arches in the modern human foot. The use of hard pavements and modern footwear, however, might have contributed to a weakening of the musculoskeletal apparatus of the foot, increasing the risk of a collapse of the foot arches, and leading to what is known as flat and splayed feet (DeSilva and Throckmorton 2010). In flat feet, the head of the talus bone is shifted medially and distal from the navicular bone, leading to a collapse of the medial longitudinal and the transversal arches. This condition reaches a prevalence of 20–30% in the general western population and can lead to chronic foot pain. Furthermore, because of a functional relationship between foot architecture and leg biomechanics, flat feet can also cause pain and malfunction in the knees and hips.

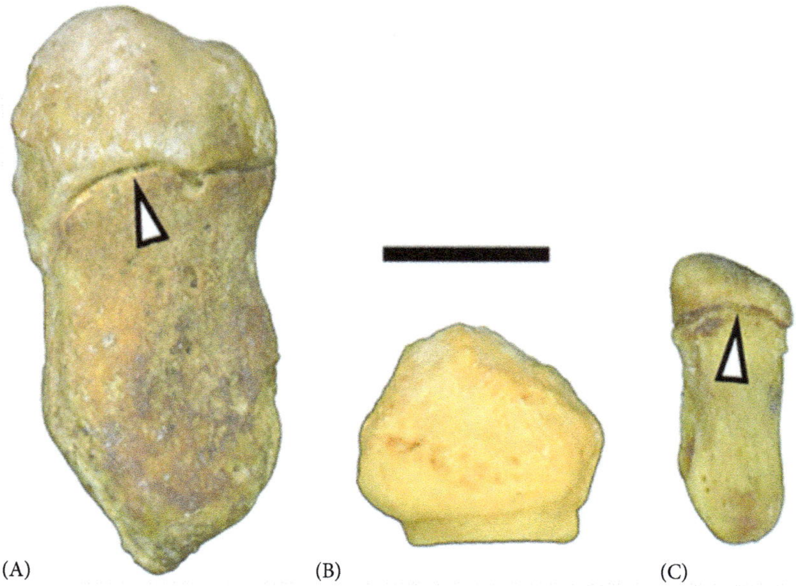

FIGURE 7.5 Pisiform bones: (A) gorilla, (B) human, and (C) chimpanzee. Arrows indicate secondary ossification centres. Scale bar 10 mm.

Source: Althea Anne D. Perez and Scott W. Simpson 2017. Published by Oxford University Press on behalf of the Foundation for Evolution, Medicine, and Public Health. Reproduced under the terms of the Creative Commons Attribution-ShareAlike 4.0 International license. https://creativecommons.org/licenses/by-sa/4.0/

7.4.8 Conclusion

As we have shown with several examples, many musculoskeletal diseases and conditions can be explained by evolutionary means. A common explanation of several conditions is the fact that the human musculoskeletal system was strongly influenced by the passage to the bipedal gait, leading to a number of anatomical trade-offs, exaptations, and incomplete adaptations. Furthermore, the modern human environment led to a number of conditions caused by a combination of mechanical overuse (as in the case of CTS and certain types of arthritis), underuse (as in the case of flat feet and wisdom teeth), or of wrong use (i.e. mismatch, as in the case of prolonged sitting and low back pain). Aetiologically we showed examples of genetically determined diseases, environmentally determined diseases, and combinations thereof. In Section 7.5 we discuss consequences for prevention and treatment of the most common musculoskeletal diseases.

7.5 Consequences for Prevention and Treatment of Musculoskeletal Diseases

As previously discussed, evolutionary medicine emphasises the interactions between the slowly evolving nature of the human body and a rapidly changing man-made environment. Knowledge about this mismatch not only contributes to explain the aetiology of many modern diseases, but also can be used to develop new approaches to prevention and treatment of such diseases.

Humans have the longest lifespan of all primates. Even under high mortality circumstances, such as those of pre-industrialised hunter-gatherer populations, human life expectancy at birth (LEo) was twice that of closely related extant great apes. For modern western populations, human LEo has doubled once more in the course of the last 200 years of industrialisation (Oeppen 2002; Finch and Crimmins 2004). In historical populations, up to the twentieth century, high early mortality and shorter life expectancy were principally due to highly infectious environments. (For further discussion, see Chapter 10: Immune System.) At the same time, most non-communicable diseases in modern western societies, such as obesity, musculoskeletal diseases, cardiovascular disease, cancer, diabetes, and Alzheimer's disease, at least partly coincide with an obvious lack of exercise. As in most mammalian species, mortality across the human lifespan is distributed in a J-shaped curve, with a high early-age mortality declining to a minimum at the approach of adulthood, followed by exponential acceleration of mortality after midlife, along with an increasing incidence of chronic degenerative alterations and dysfunctions, which essentially define senescence. Our pre-industrialised hunter-gatherer ancestors had to run up to 15 km a day to survive. Thanks to their ability to cross long distances in constant running, they were superior in hunting their prey. Today's adults are advised to do at least 150 minutes of more or less intense sports per week. (For further discussion, see Chapter 6: Nutrition, Energy Expenditure, Physical Activity, and Body Composition.) This corresponds to 30 minutes 5 days a week, during which the heart rate should increase by 50%.

In this context, it must be considered that humans continue to evolve (Stearns et al. 2010) and that microevolutionary changes can be observed over a few generations, including changes that are brought about by improved living conditions, hygiene, and medical care (Rühli and Henneberg 2013). This has consequences for the treatment and prevention of most civilisational and degenerative diseases, including musculoskeletal conditions. The importance of manifest musculoskeletal conditions, which can be grossly divided into the five categories osteoarthritis, inflammatory arthritis, soft tissue rheumatism, osteoporosis, and back pain, has steadily increased in western industrialised nations over the last decades, despite the increasing automation of industrial production processes and far-reaching improvements in workplace ergonomics. Musculoskeletal diseases are today the most important cause of disability among women and the second most significant one among men. Lifetime prevalence of back pain, for instance, in modern western industrialised countries ranges between 58% and 85%. However, since the underlying aetiology of back pain is a multidimensional process of formation and chronification, individual factors for preventive measures are difficult to identify.

At first sight, musculoskeletal diseases appear to be less related to increased mortality in comparison with other chronic diseases. While that does apply for osteoarthritis or back pain, various forms of inflammatory arthritis and osteoporosis are indeed associated with increased mortality. Some cardiovascular risk factors are almost twice as frequent in rheumatoid arthritis patients as in an adjusted normal population, as a recent evaluation of the US-Medicare database illustrates. The incidence of arterial hypertension, hyperlipidaemia, and diabetes was 65.3%, 53.0%, and 29.2%, respectively, in rheumatoid arthritis patients, compared to 37.0%, 29.6%, and 18.9%, respectively, in an adjusted comparative group (Kaeley and Dodani 2016).

Evolutionary pressures during the Miocene and Pleistocene periods likely selected for individuals with a proficient inflammatory response and innate immunity in order to counteract infection and ensure for survival of traumata, which might explain why some musculoskeletal conditions and cardiovascular diseases share similar aetiological pathways. Future directions in clinical practice will need to take the common underlying molecular mechanisms into account to tackle these highly prevalent, costly, and disabling morbidities.

The increased risk for impairment and mortality due to osteoporosis follows a different pattern and results from complications of osteoporotic fractures. Consequently, for osteoporosis and fracture prophylaxis, physical activity and nutrition-related measures are recommended, namely a sufficient intake of calcium and vitamin D, besides a general prevention of underweight for the elderly. In general, the treatment and prevention of musculoskeletal disorders at all ages must be based, apart from the sporadic use of specific drugs (e.g. anti-inflammatory drugs), on a combination of dietary measures and physical activity.

Preventive measures follow a life course approach, since at every life stage the musculoskeletal system shows specific needs and metabolic properties. The World Health Organization (WHO) promotes a life course approach to maintain independence and prevent disability from a large number of diseases, including musculoskeletal conditions (WHO 1999). Preventive measures for the musculoskeletal system should therefore start in childhood and adolescence, as the maximal bone mass is reached in early adulthood. In many countries, new-borns are supplemented with vitamin D to prevent rickets and ensure an optimal start to skeletal growth. Sufficient and age-adjusted physical activity, combined with a diet rich in all essential minerals and vitamins, is essential for optimal skeletal growth during childhood. Improved life conditions, better hygiene, and modern medicine have permitted children in western societies to develop optimally since several decades. Indeed, the secular increase of body height in western countries has plateaued, indicating that the genetic potential may already have been reached (Staub et al. 2015). (For further discussion, see Chapter 4: Growth and Development.)

During adulthood, a diet sufficient in calcium and vitamin D, as well as enough physical activity and proper work ergonomics are essential to maintain bone mass for as long as possible. In the elderly, as mentioned, sufficient nutrition in general, combined with sufficient but adapted physical activity, contributes to maintain musculoskeletal health.

References

Adams, M. A., Lama, P., Zehra, U., et al. (2015). Why do some intervertebral discs degenerate, when others (in the same spine) do not? *Clin Anat* 28, 195–204.

Anonymous (1993). Consensus development conference: diagnosis, prophylaxis, and treatment of osteoporosis. *Am J Med* 94(6), 646–50.

Armocida, E., Böni, T., Rühli, F. J., et al. (2016). Does acromegaly suffice to explain the origin of Pulcinella? A novel interpretation. *Eur J Intern Med* 28, e16–17.

Armocida, E., Galassi, F. M., Ballestriero, R., et al. (2017). The 'Bottaro': when the study acromegaly went beyond osteology. Proceedings of 3rd Meeting of Italian Paleopathology Group (GIPaleo), Pisa, Italy, May 2017.

Aufderheide, A. C. and Rodiguez-Martin, C. (2005). Acromegaly. In: *The Cambridge Encyclopedia of Human Paleopathology*. Cambridge: Cambridge University Press, pp. 326–8.

Battié, M. C., Videman, T., Gibbons, L. E., et al. (1995). Volvo Award in Clinical Sciences. Determinants of lumbar disc degeneration. A study relating lifetime exposures and magnetic resonance imaging findings in identical twins. *Spine* 20, 2601–12.

Been, E., Gomez-Olivencia, A., Shefi, S., et al. (2017). Evolution of spinopelvic alignment in hominins. *Anat Rec* 300, 900–11.

Bello-Hellegouarch, G., Potau, J. M., Arias-Martorell, J., et al. (2013). A comparison of qualitative and quantitative methodological approaches to characterizing the dorsal side of the scapula in Hominoidea and its relationship to locomotion. *Int J Primatol* 34, 315–36.

Bogdan, S., Ruehli, F. J., Lee, Y. C., et al. (2009). Secular trend in the opening of the sacral canal: an Australian study. *Spine* 34(3), 244–8.

Bonmatí, A., Gómez-Olivencia, A., Arsuaga, J. L., et al. (2010). Middle Pleistocene lower back and pelvis from an aged human individual from the Sima de los Huesos site, Spain. *Proc Natl Acad Sci U S A* 107, 18386–91.

Bramble, D. M. and Lieberman, D. E. (2004). Endurance running and the evolution of *Homo*. *Nature* 432, 345–52.

Brandao-Burch, A., Utting, J. C., Orriss, I. R., et al. (2005). Acidosis inhibits bone formation by osteoblasts in vitro by preventing mineralization. *Calcif Tissue Int* 77(3), 167–74. doi: 10.1007/s00223-004-0285-8.

Brauer, J. L. (1991). A case of acromegaly in a prehistoric skeleton from the San Cristobal Ruins, New Mexico. *Am J Phys Anthropol* 43, 53.

Bridges, P. S. (1991). Degenerative joint disease in hunter-gatherers and agriculturalists from the southeastern United States. *Am J Phys Anthropol* 85, 379–92.

Brinjikji, W., Luetmer, P. H., Comstock, B., et al. (2015). Systematic literature review of imaging features of spinal degeneration in asymptomatic populations. *Am J Neuroradiol* 36, 811–16.

Brunet, M., Guy, F., Pilbeam, D., et al. (2002). A new hominid from the Upper Miocene of Chad, Central Africa. *Nature* 418, 145–51.

Chirchir, H., Kivell, T. L., Ruff, C. B., et al. (2015). Recent origin of low trabecular bone density in modern humans. *Proc Natl Acad Sci U S A* 112, 366–71.

Clark, G. and Henneberg, M. (2015). The life history of *Ardipithecus ramidus*: a heterochronic model of sexual and social maturation. *Anthropol Rev* 78(2), 109–32.

Clarke, R. J. and Tobias, P. V. (1995). Sterkfontein Member 2 foot bones of the oldest South African hominid. *Science* 269, 521–4.

Cook, D. C., Buikstra, J. E., DeRousseau, C. J., et al. (1983). Vertebral pathology in the Afar australopithecines. *Am J Phys Anthropol* 60, 83–101.

Craik, J. D., Mallina, R., Ramasamy, V., et al. (2014). Human evolution and tears of the rotator cuff. *Int Orthop* 38, 547–52.

Dal, J., Feldt-Rasmussen, U., Andersen, M., et al. (2016). Acromegaly incidence, prevalence, complications and long-term prognosis: a nationwide cohort study. *Eur J Endocrinol* 175, 181–90.

Damborg, F., Engell, V., Andersen, M., et al. (2006). Prevalence, concordance, and heritability of Scheuermann kyphosis based on a study of twins. *J Bone Joint Surg Am* 88, 2133–6.

D'Anastasio, R., Zipfel, B., Moggi-Cecchi, J., et al. (2009). Possible brucellosis in an early hominin skeleton from Sterkfontein, South Africa. *PLoS One* 4, e6439.

Darwin, C. (1872). *The Expression of the Emotions in Man and Animals*. London: John Murray.

de Herder, W. W. (2009). Acromegaly and gigantism in the medical literature. Case descriptions in the era before and the early years after the initial publication of Pierre Marie (1886). *Pituitary* 12(3), 236–44.

DeSilva, J. M. and Throckmorton, Z. J. (2010). Lucy's flat feet: the relationship between the ankle and rearfoot arching in early hominins. *PLoS One* 5(12), e14432. doi:10.1371/journal.pone.0014432.

DeSilva, J. M., Holt, K. G., Churchill, S. E., et al. (2013). The lower limb and mechanics of walking in *Australopithecus sediba*. *Science* 340, 1232999.

Dieleman, J. L., Baral, R., Birger, M., et al. (2016). US spending on personal health care and public health, 1996–2013. *JAMA* 316, 2627–46.

Dubois, E. (1892). Paleontologische onderzoekingen op Java. *Verslagen van het Mijnwezen* 3, 10–14.

Eaton, S. B. and Cordain, L. (1997). Evolutionary aspects of diet: old genes, new fuels—nutritional changes since agriculture. *Nutrition and Fitness: Evolutionary Aspects, Children's Health, Programs and Policies* 81, 26–37.

Elfering, A., Müller, U., Rolli Salathé, C., et al. (2015). Pessimistic back beliefs and lack of exercise: a longitudinal risk study in relation to shoulder, neck, and back pain. *Psychol Health Med* 20, 767–80.

Finch, C. E. and Crimmins, E. M. (2004). Inflammatory exposure and historical changes in human life-spans. *Science* 305, 1736–9.

Galassi, F. M., Henneberg, M., de Herder, W., et al. (2017). Oldest case of gigantism? Assessment of the alleged remains of Sa-Nakht, king of ancient Egypt. *Lancet Diabetes Endocrinol* 5(8), 580–58.

Ganz, R., Leunig, M., Leunig-Ganz, K., et al. (2008). The etiology of osteoarthritis of the hip: an integrated mechanical concept. *Clin Orthop Relat Res* 466, 264–72.

Gould, S. J. and Vrba, E. S. (1982). 'Exaptation'—a missing term in the science of form. *Paleobiology* 8(1), 4–15.

Gunji, H., Hosaka, K., Huffman, M. A., et al. (2003). Extraordinarily low bone mineral density in an old female chimpanzee (*Pan troglodytes schweinfurthii*) from the Mahale Mountains National Park. *Primates* 44, 145–9.

Haeusler, M. (2002). New insights into the locomotion of *Australopithecus africanus* based on the pelvis. *Evol Anthropol* 11, 53–7.

Haeusler, M. and McHenry, H. M. (2004). Body proportions of *Homo habilis* reviewed. *J Hum Evol* 46, 433–65.

Haeusler, M. and McHenry, H. M. (2007). Evolutionary reversals of limb proportions in early hominids? Evidence from KNM-ER 3735 (*Homo habilis*). *J Hum Evol* 53, 383–405.

Haeusler, M., Martelli, S., and Boeni, T. (2002). Vertebrae numbers of the early hominid lumbar spine. *J Hum Evol* 43, 621–43.

Haeusler, M., Schiess, R., and Boeni, T. (2011). New vertebral and rib material point to modern bauplan of the Nariokotome *Homo erectus* skeleton. *J Hum Evol* 61, 575–82.

Haeusler, M., Schiess, R., and Boeni, T. (2013). Evidence for juvenile disc herniation in a *Homo erectus* boy skeleton. *Spine* 38, 123–8.

Haeusler, M., Landis, S., and Zipfel, B. (2015a). Hip joint osteoarthritis in the MLD 46 (*Australopithecus africanus*) proximal femur. *Am J Phys Anthropol* S60, 157.

Haeusler, M., Nakita, F., Hangartner, R., et al. (2015b). Are our back problems evolutionary trade-offs of bipedalism? *J Evol Med* 3, 235924.

Haeusler, M., Bonneau, N., and Fornai, C. (2017a). The morphology of the *Homo erectus* pelvis. *Proc Eur Soc Hum Evol* 6, 82.

Haeusler, M., Fornai, C., Frater, N., et al. (2017b). The vertebral column of La Chapelle-aux Saints: the evidence of spinal osteoarthritis for Neanderthal spinal curvature. *Am J Phys Anthropol* 162(Suppl 64), 206.

Haile-Selassie, Y., Saylor, B. Z., Deino, A., et al. (2012). A new hominin foot from Ethiopia shows multiple Pliocene bipedal adaptations. *Nature* 483, 565–9.

Harcourt-Smith, W. E. H. and Aiello, L. C. (2004). Fossils, feet and the evolution of human bipedal locomotion. *J Anat* 204, 403–16.

Häusler, M. and Schmid, P. (1995). Comparison of the pelves of Sts 14 and AL 288–1: implications for birth and sexual dimorphism in australopithecines. *J Hum Evol* 29, 363–83.

Heneweer, H., Vanhees, L., and Picavet, H. S. (2009). Physical activity and low back pain: a U-shaped relation? *Pain* 143, 21–5.

Hoaglund, F. T. (2013). Primary osteoarthritis of the hip: a genetic disease caused by European genetic variants. *J Bone Joint Surg Am* 95, 463–8.

Hogervorst, T., Bouma, H., de Boer, S. F., et al. (2011). Human hip impingement morphology: an evolutionary explanation. *J Bone Joint Surg Br* 93, 769–76.

Hoy, D., March, L., Brooks, P., et al. (2014). The global burden of low back pain: estimates from the Global Burden of Disease 2010 study. *Ann Rheum Dis* 73, 968–74.

Johnson, R. H. (1912). The analysis of natural selection. *Science* 36, 750–60.

Jurmain, R. D. (2000). Degenerative joint disease in African great apes: an evolutionary perspective. *J Hum Evol* 39, 185–203.

Jurmain, R. D. and Kilgore, L. (1995). Skeletal evidence of osteoarthritis: a palaeopathological perspective. *Ann Rheum Dis* 54, 443–50.

Kaeley, G. S. and Dodani, S. (2016). Call to action: cardiovascular comorbidities in medicare rheumatology beneficiaries [abstract]. *Arthritis Rheumatol* 68(Suppl 10).

Karasik, D. (2008). Osteoporosis: an evolutionary perspective. *Hum Genet* 124(4), 349–56. doi: 10.1007/s00439-008-0559-8.

Katz, J. N. (2006). Lumbar disc disorders and low-back pain: socioeconomic factors and consequences. *J Bone Joint Surg Am* 88A, 21–4.

Kouwenhoven, J. W. and Castelein, R. M. (2008). The pathogenesis of adolescent idiopathic scoliosis: review of the literature. *Spine* 33, 2898–908.

Krogman, W. M. (1951). The scars of human evolution. *Sci Am* 185, 54–7.

Lachmann, S. M. (1982). Spina-bifida occulta is of significance in back injuries in sport. *Ann Rheum Dis* 41(6), 625.

Larson, S. (2013). Shoulder morphology in early hominin evolution. In: Reed, K. E., Fleagle, J. G., and Leakey, R. E. (eds) *The Paleobiology of* Australopithecus. Dordrecht: Springer, pp. 247–61.

Latimer, B. and Ohman, J. C. (2001). Axial dysplasia in *Homo erectus*. *J Hum Evol* 40, A12.

Lewis, J., Green, A., Yizhat, Z., et al. (2001). Subacromial impingement syndrome. Has evolution failed us? *Physiotherapy* 87, 191–8.

Liang, Q. Q., Zhou, Q., Zhang, M., et al. (2008). Prolonged upright posture induces degenerative changes in intervertebral discs in rat lumbar spine. *Spine* 33, 2052–8.

Lovejoy, C. O. (2005). The natural history of human gait and posture. Part 1. Spine and pelvis. *Gait Posture* 21, 95–112.

Lovejoy, C. O., Suwa, G., Simpson, S. W., et al. (2009). The great divides: *Ardipithecus ramidus* reveals the postcrania of our last common ancestors with African apes. *Science* 326, 100–6.

Luoma, K., Riihimaki, H., Luukkonen, R., et al. (2000). Low back pain in relation to lumbar disc degeneration. *Spine* 25, 487–92.

Mathews, S. and Haeusler, M. (2015). Are rotator cuff lesions of the human shoulder joint related to the evolution of bipedalism? *Proc Eur Soc Study Hum Evol* 4, 154.

McHenry, H. M. (1991). First steps? Analyses of the postcranium of early hominids. In: Coppens, Y. and Senut, B. (eds) *Origine(s) de la Bipédie Chez les Hominidés*. Paris: CNRS, pp. 133–41.

Meyer, M. R. (2016). The cervical vertebrae of KSD-VP-1/1. In: Haile-Selassie, Y. and Su, D. (eds) *The Postcranial Anatomy of* Australopithecus afarensis*: New Insights from KSD-VP-1/1*. New York: Springer, pp. 63–111.

Meyer, M. R. and Haeusler, M. (2015). Spinal cord evolution in early *Homo*. *J Hum Evol* 88, 43–53.

McHenry, H. M. (1992). Body size and proportions in early hominids. *Am J Phys Anthropol* 87, 407–31.

Murphy, L., Schwartz, T. A., Helmick, C. G., et al. (2008). Lifetime risk of symptomatic knee osteoarthritis. *Arthritis Rheum* 59, 1207–13.

Murphy, P. A. (1993). Beneficial effects of folic acid supplementation in the prevention of neural tube defects (NTDs). *J Nurse Midwifery* 38(1), 54.

Nuckley, D. J., Kramer, P. A., Del Rosario, A., et al. (2008). Intervertebral disc degeneration in a naturally occurring primate model: radiographic and biomechanical evidence. *J Orthop Res* 26, 1283–8.

Oeppen, J. and Vaupel, J. W. (2002). Demography. Broken limits to life expectancy. *Science*. 296, 1029–31.

Ohman, J. C., Lovejoy, C. O., and White, T. D. (2005). Questions about *Orrorin* femur. *Science* 307, 845.

Ortner, D. J., Ragsdale, B. D., Walter, G. J. Putschar, MD. (2012). Pathologist, paleopathologist, teacher. In: Buikstra, J. E. and Roberts, C. (eds) *The Global History of Paleopathology: Pioneers and Prospects*. New York: Oxford University Press, pp. 104–5.

Papavramidou, N., Christopoulou-Aletra, H., and Pozzilli, P. (2007). Endocrinology and art. *J Endocrinol Invest* 30, 83.

Parisini, P., Di Silvestre, M., Greggi, T., et al. (2001). Lumbar disc excision in children and adolescents. *Spine* 26, 1997–2000.

Perez, A. A. and Simpson, S. W. (2017). Evolutionary medicine. Carpal tunnel syndrome. *Evol Med Public Health* 2017(1), 51–2. doi: 10.1093/emph/eox006.

Potau, J. M., Bardina, X., and Ciurana, N. (2007). Subacromial space in African great apes and subacromial impingement syndrome in humans. *Int J Primatol* 28, 865–80.

Potau, J. M., Bardina, X., Ciurana, N., et al. (2009). Quantitative analysis of the deltoid and rotator cuff muscles in humans and great apes. *Int J Primatol* 30, 697–708.

Preuschoft, H. (2004). Mechanisms for the acquisition of habitual bipedality: are there biomechanical reasons for the acquisition of upright bipedal posture? *J Anat* 204, 363–84.

Purdie, D. W. (1996). Evolution of osteoporosis. *Ann Rheum Dis* 55(6), 335–7.

Putz, R. L. V. and Müller-Gerbl, M. (1996). The vertebral column—a phylogenetic failure? A theory explaining the function and vulnerability of the human spine. *Clin Anat* 9, 205–12.

Reed, K. E., Kitching, J. W., Grine, F. E., et al. (1993). Proximal femur of *Australopithecus africanus* from Member 4, Makapansgat, South Africa. *Am J Phys Anthropol* 92, 1–15.

Rogers, J. and Dieppe, P. (1994). Is tibiofemoral osteoarthritis in the knee joint a new disease? *Ann Rheum Dis* 53, 612–13.

Rose, M. D. (1991). The process of bipedalization in hominids. In: Coppens, Y. and Senut, B. (eds) *Origine(s) de la Bipédie Chez les Hominidés*. Paris: CNRS, pp. 37–48.

Rubin, D. I. (2007). Epidemiology and risk factors for spine pain. *Neurol Clin* 25, 353–71.

Rühli, F. and Henneberg, M. (2013). New perspectives on evolutionary medicine: the relevance of microevolution for human health and disease. *BMC Medicine* 11, 115.

Ruff, C. B., Puymerail, L., Macchiarelli, R., et al. (2015). Structure and composition of the Trinil femora: functional and taxonomic implications. *J Hum Evol* 80, 147–58.

Rühli, F. and Henneberg, M. (2016). Biological future of humankind: ongoing evolution and the impact of recognition of human biological variation. In: Tibayrenc, M. and Ayala, F. J. (eds) *On Human Nature*. Amsterdam: Academic Press, pp. 263–75.

Rühli, F. J., Galassi, F. M., and Haeusler, M. (2016a). Palaeopathology: current challenges and medical impact. *Clin Anat* 29, 816–22.

Rühli, F., van Schaik, K., and Henneberg, M. (2016b). Evolutionary medicine: the ongoing evolution of human physiology and metabolism. *Physiology* 31(6), 392–7.

Sarmiento, E. E. (2010). Comment on the paleobiology and classification of *Ardipithecus ramidus*. *Science* 328, 1105; author reply.

Scally, A., Dutheil, J. Y., Hillier, L. W., et al. (2012). Insights into hominid evolution from the gorilla genome sequence. *Nature* 483, 169–75.

Schiess, R. and Haeusler, M. (2013). No skeletal dysplasia in the Nariokotome boy KNM-WT 15000 (*Homo erectus*): a reassessment of congenital pathologies of the vertebral column. *Am J Phys Anthropol* 150, 365–74.

Schiess, R., Boeni, T., Rühli, F., et al. (2014). Revisiting scoliosis in the KNM-WT 15000 *Homo erectus* skeleton. *J Hum Evol* 67, 48–59.

Schmid, P. (1983). Eine Rekonstruktion des Skelettes von AL 288-1 (Hadar) und deren Konsequenzen. *Folia Primatol* 40, 283–306.

Schmid, P. (1991). The trunk of the australopithecines. In: Coppens, Y. and Senut, B. (eds) *Origine(s) de la Bipédie Chez les Hominidés*. Paris: CNRS, pp. 225–34.

Senut, B., Pickford, M., Gommery, D., et al. (2001). First hominid from the Miocene (Lukeino Formation, Kenya). *C R Acad Sci* 332, 137–44.

Simonti, C. N., Vernot, B., Bastarache, L., et al. (2016). The phenotypic legacy of admixture between modern humans and Neandertals. *Science* 351, 737–41.

Smit, T. H. (2002). The use of a quadruped as an in vivo model for the study of the spine—biomechanical considerations. *Eur Spine J* 11, 137–44.

Solomon, L. B., Rühli, F. J., Lee, Y. C., et al. (2009). Secular trend in the opening of the sacral canal: an Australian study. *Spine* 34(3), 244–8.

Sørensen, K. H. (1964). *Scheuermann's Juvenile Kyphosis. Clinical Appearances, Radiography, Aetiology, and Prognosis*. København: Munksgaard.

Staub, K., Floris, J., Woitek, U., and Rühli, F. (2015). From left-skewness to symmetry: how body-height distribution among Swiss conscripts has changed shape since the late nineteenth century. *Ann Hum Biol* 42(3), 262–9.

Stearns, S. C., Byars, S. G., Govindaraju, D. R., et al. (2010). Measuring selection in contemporary human populations. *Nat Rev Genet* 11, 611–22.

Stedman, H. H., Kozyak, B. W., Nelson, A., et al. (2004). Myosin gene mutation correlates with anatomical changes in the human lineage. *Nature* 428(6981), 415–18.

Takatalo, J., Karppinen, J., Niinimaki, J., et al. (2009). Prevalence of degenerative imaging findings in lumbar magnetic resonance imaging among young adults. *Spine* 34, 1716–21.

Tardieu, C. (2010). Development of the human hind limb and its importance for the evolution of bipedalism. *Evol Anthropol* 19, 174–86.

Tardieu, C. and Haeusler, M. (in press). The acquisition of human verticality with an emphasis on sagittal balance. In: Roussouly, P., Pinheiro-Franco, J. L., Labelle, H., and Gehrchen, M. (eds) *Sagittal Balance of the Spine*. New York: Thieme.

Tardieu, C., Hasegawa, K., and Haeusler, M. (2017). How the pelvis and vertebral column became a functional unit in human evolution during the transition from occasional to permanent bipedalism? *Anat Rec* 300, 912–31.

Toni, R. (2016). Selezione naturale positiva, gigantismo, acromegalia e interferenti endocrini. *L'Endocrinologo* 17, 262–5.

Trinkaus, E. (1985). Pathology and the posture of the La Chapelle-aux-Saints Neandertal. *Am J Phys Anthropol* 67, 19–41.

Volinn, E. (1997). The epidemiology of low back pain in the rest of the world. A review of surveys in low- and middle-income countries. *Spine* 22, 1747–54.

Waddell, G. (1996). Low back pain: a twentieth century health care enigma. *Spine* 21, 2820–5.

Walker, A. and Leakey, R. (eds) (1993). *The Nariokotome* Homo erectus *Skeleton*. Berlin: Springer.

Wallace, I. J., Worthington, S., Felson, D. T., et al. (2017). Knee osteoarthritis has doubled in prevalence since the mid-twentieth century. *Proc Natl Acad Sci U S A* 114(35), 9332–6. doi: 10.1073/pnas.1703856114.

Ward, C. V. (2013). Postural and locomotor adaptations of *Australopithecus* species. In: Reed, K. E., Fleagle, J. G., and Leakey, R. E. (eds) *The Paleobiology of* Australopithecus. Dordrecht: Springer, pp. 235–45.

White, T. D., Asfaw, B., Beyene, Y., et al. (2009). *Ardipithecus ramidus* and the paleobiology of early hominids. *Science* 326, 75–86.

WHO (1999). *A Life Course Perspective of Maintaining Independence in Older Age*. Geneva: World Health Organization.

Wolpoff, M. H., Hawks, J., Senut, B., et al. (2006). An ape or the ape: is the Toumaï cranium TM 266 a hominid? *Paleoanthropology* 2006, 36–50.

Wynne-Jones, G. and Dunn, K. M. (2016). Has there been a change in the rates of UK sickness certification for back pain over time? An examination of historical data from 2000 to 2010. *BMJ Open* 6, e009634.

You, W. and Henneberg, M. (2016). Type 1 diabetes prevalence increasing globally and regionally: the role of natural selection and life expectancy at birth. *BMJ Open Diabetes Res Care* 4(1), e000161.

You, W. and Henneberg, M. (2017). Cancer incidence increasing globally: the role of relaxed natural selection. *Evol Appl* 11(2), 140–52.

Zipfel, B., DeSilva, J. M., Kidd, R. S., et al. (2011). The foot and ankle of *Australopithecus sediba*. *Science* 333, 1417–20.

Zollikofer, C. P. E., Ponce de León, M., Lieberman, D. E., et al. (2005). Virtual cranial reconstruction of *Sahelanthropus tchadensis*. *Nature* 434, 755–9.

CHAPTER 8

SKIN AND INTEGUMENT

MARK A. HILL

Abstract

The skin and its associated specialisations together form the integumentary system, the barrier and sensory structure between the body and our environment. Human skin compared to other mammalian species has undergone significant and specialised evolutionary changes reflected in both development and mature structure. In development, the vernix caseosa protects the skin during prolonged uterine incubation. A lengthy childhood then leaves secondary sex characteristic development for many years, and the cyclic changes of female mammary glands have exposed some key cancer genes. In mature structure, we are comparatively 'hairless' and uniquely endowed with copious sweat glands for thermoregulation. The now-exposed skin needs to be protected from UV damage by melanocyte-sourced pigments that are both genetically and environmentally regulated; these cells are today a major cancer cell population. Conversely, we have evolved a UV requirement to generate key vitamin D precursors utilised in many different physiological systems. Evolutionary local modification of the skin, nails, and sensory receptors in the hands has supported fine grip and touch sensitivity. The patterning on the hand surface not only is a unique identification of an individual, but also may be diagnostically associated with some abnormalities. Therapeutically, we can replace damaged skin, have identified stem cells that hold great potential, and now interface with medical devices for monitoring and drug delivery. The future challenges for the skin, as a barrier, are changing environmental conditions, technology assistance and interaction, and how the many other unique functions are dynamically rebalanced in health and disease.

Keywords

skin evolution, colour, hairlessness, vitamin D production, integumentary specialisations, diseases and infections of the skin, evolution, medicine

8.1 Skin: The Largest Organ of the Body

The integument (Latin *integumentum* meaning 'a covering') is the term used to describe the covering of an animal, comprising the skin and its appendages (hair and nails in the case of humans); it is the outer cover between us and our environment. The lay term 'skin' is generally used to describe this interface and it includes many specialisations associated with its functions. We now know that this interface has evolved to be much more than a simple barrier between the wet internal environment of the body and the terrestrial air that surrounds us. Evolution from an animal solely within a water environment (fish), to a water/land environment (amphibia), and finally to a wide terrestrial environment (reptiles, birds, and mammals, including humans) involved a large range of climatic conditions. Cornification is a specialised type of cell death that has evolved to protect terrestrial animals against water loss, as well as providing other barrier functions. It is possible that both the central and peripheral nervous systems evolved originally as a form of 'integumentary specialisation' that allowed the interpretation of, and responses to, the external environment. Along with brain evolutionary development, the skin has evolved its own complex signalling molecules and neuroendocrine systems. The human brain receives information about the surrounding environment from the specialised sensory systems (vision, hearing, balance, taste, and smell), but also from a large array of specialised sensory structures throughout the skin.

This chapter describes how the mammalian integumentary system has evolved, and this continues as our environment changes, from a simple barrier function to the many roles we know today. This barrier function is generally measured by the parameter of transepidermal water loss (TEWL). The skin protects the organism against external physical, chemical, and biological insults, including wounding.

The basic structure of the skin is shared by all mammals, but some features are unique to primates and humans. In humans, compared to other primates there has been a significant loss of hair and gain of epidermal pigmentation. The development of prehensile limbs has allowed fine motor development and increased sensory receptor density and specialisation. Integumentary and other human evolutionary changes have allowed us to colonise almost the entire planet, compared to the restricted distribution of our nearest primate relatives. This chapter addresses integumentary issues such as human embryological development, specialised regional appendages, pigmentation, hair loss, metabolic roles, stem cells, ageing, diseases, and abnormalities. Note that genetic variation in the human population results in differential susceptibility to a broad range of disease agents and environmental effects.

Within this chapter, both original research articles and reviews are given for the reader's reference, and the Online Mendelian Inheritance in Man (OMIM) gene database entries are shown within parentheses by their associated entry number and can be added to the address https://www.omim.org to see the online entry.

8.2 Skin: Introducing the Basic Structure

The epidermis is histologically classified as a stratified squamous epithelium (Figure 8.1). The stratum lucidum is a sub-layer only present in the thick skin of the palms and soles of the feet; both these regions lack hair and are described as glabrous skin. The remainder of the body is covered in thin hairy (long or vellus) skin. The thick/thin histological classification does not refer to the actual overall skin thickness. The human palms have evolved as highly

(A)

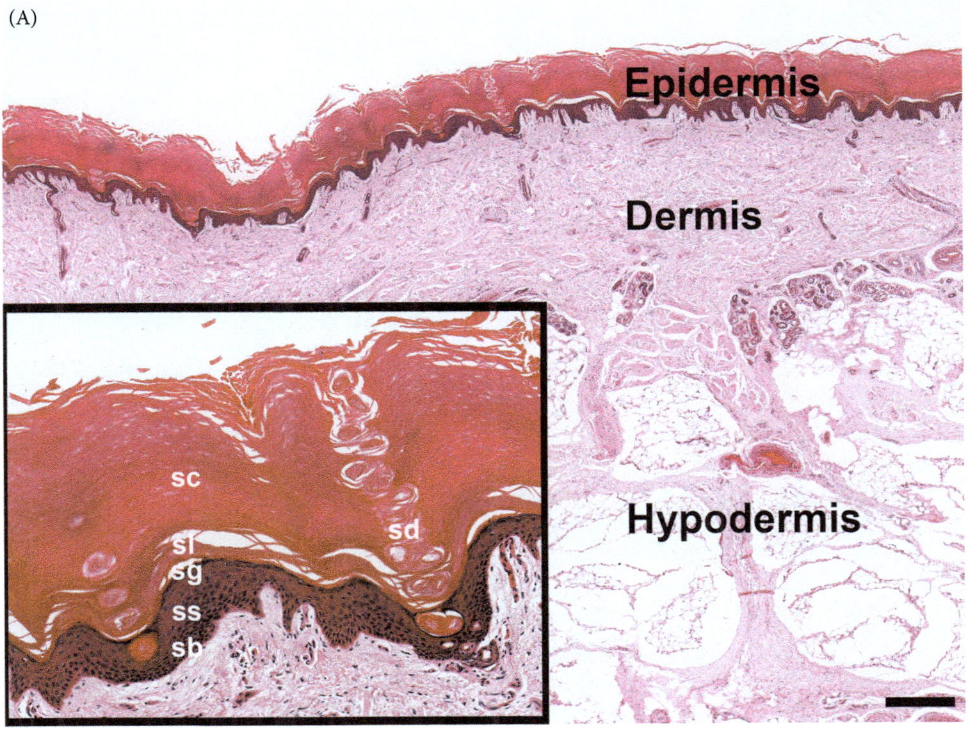

FIGURE 8.1 Human skin histology of thick skin from the palm (A) and thin hairy skin from the scalp (B). Both panels show the three main layers of skin: epidermis, dermis, and hypodermis. (A) Inset shows a detailed view of epidermis layers: stratum basale (sb) containing keratinocyte stem cells of the skin; stratum spinosum (ss) containing differentiating keratinocytes; stratum granulosum (sg) keratinocytes which accumulates keratohyalin granules; stratum lucidum (sl) keratinocytes are a pale cell layer and only found in thick skin; stratum corneum (sc), the dead keratinocyte outermost barrier layer of the skin. (B) Thin skin of the scalp showing hair bulb (hb) at base of hair follicle containing hair dermal papilla: hair (h) extends from the skin surface; sebaceous gland (seb) that secretes oily sebum as a lubricant. Inset shows a labelled overview of this region.

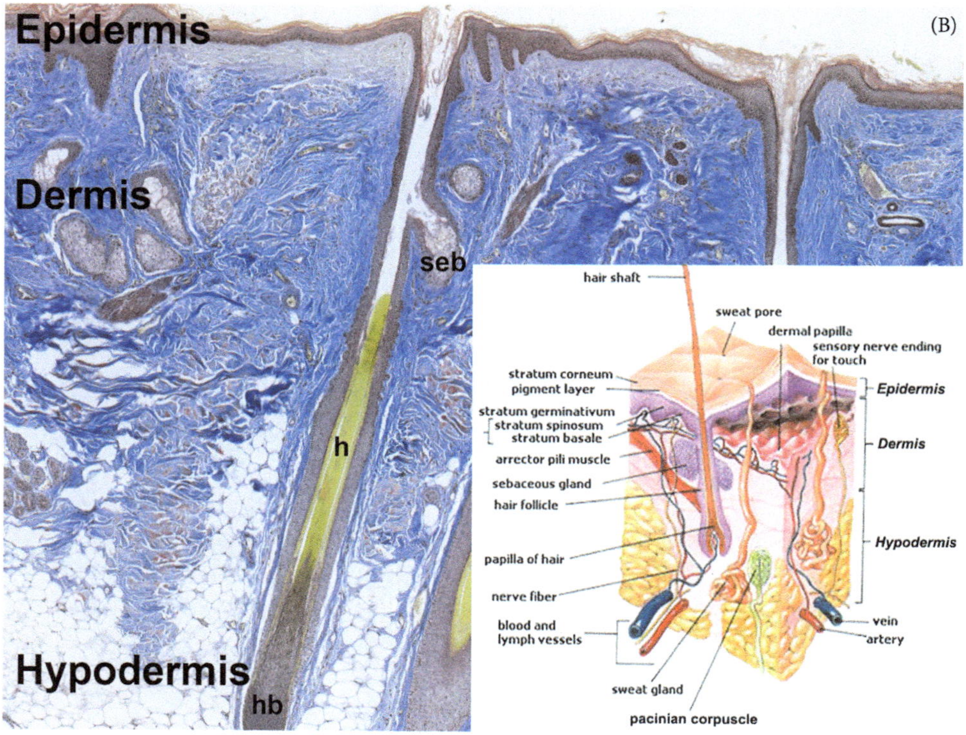

FIGURE 8.1 Continued

Table 8.1 Components of Human Skin

Layers	Sub-layers	Cells	Extrace lular matrix
Epidermis	Stratum corneum, stratum granulosum, stratum lucidum (thick skin), stratum spinosum, basal layer	Keratinocytes, keratinocyte stem cell melanocytes	Basement membrane
Dermis	Papillary layer, reticular layer	Fibroblasts, Langerhans blood and immune cells	Collagen fibres, elastic fibres
Hypodermis	Adipose tissue	Adipocytes, fibroblasts	Collagen fibres

specialised mechanical and sensory structures, with dorsal nails, dermal ridges (prints), and a higher density of sensory receptors.

The underlying dermis is classified as a dense irregular connective tissue containing blood vessels, lymphatic vessels, immune cells, nerve fibres, specialised sensory end-organs, and bare nerve endings. Beneath the dermis, the hypodermis is the chief repository of adipose (fat) cells as an energy store and provides the 'cushioning' that internal organs and tissues require from the external environment and during mechanical movement of the skeleton. Each layer of the skin has its own sub-layers, cells, and specialised extracellular matrix (see Table 8.1).

8.3 The Skin: Ongoing Development

An understanding of the adult human integumentary system and its clinical features is best begun by studying an overview of embryological–morphological development. For detailed aspects of development in an online resource see (http://tiny.cc/integumentary_development).

8.3.1 Embryology of the Skin

In humans, skin development occurs over the entire prenatal period (Table 8.2), protected in the latter stages by a specialised coating, and finally undergoing dramatic changes postnatally with the shift from a watery environment to a dry air environment. Furthermore, the skin now at birth becomes populated with a microbiota that can influence the skin in health and disease. Many of the specialised structures of the skin do not complete their development until puberty, years later.

Development in humans is uniquely measured (clinically) from the last time the mother was not pregnant, using the first day of the last menstrual period as a 'gestational age' (GA). This timing corresponds to approximately 2 weeks before the actual fertilisation event, about mid-point in the next cycle. Therefore week 3 of development becomes week 5 GA. Development in all other animals is measured from the time of fertilisation. Within this chapter I provide both the fertilisation date and GA together.

In week 3 (GA week 5), the three human germ layers (ectoderm, mesoderm, and endoderm) form their initial contributions to the skin from the ectoderm and mesoderm (Table 8.3). Note the third layer, the endoderm, contributes to the human internal barrier associated with the gastrointestinal and respiratory tract. (For further discussion, see Chapter 12: Respiratory System and Chapter 13: Digestive System) Interestingly, where the external integumentary and internal mucosal environment meet (oral cavity, nasal cavity, anus, cornea) are located unique anatomical and functional junctions.

The process of formation of the three germ layers, gastrulation, is shared by most animal species, and the adult structures formed from these layers has been completely evolutionarily

Table 8.2 Time-Course of Human Epidermis Development

Stages	Days
Single layer ectoderm	21–55
Periderm develops	40–65
Intermediate layers	66–95
Early terminal differentiation	96–160
Late terminal differentiation	160+
Postnatal epidermis	266+

Table 8.3 Human Embryo Germ Cell Layers and their Contribution to Adult Skin Layers. Note: these layers and their cell contributions are the same for all mammals

Ectoderm		Mesoderm
Surface ectoderm	Neural crest	Adipocytes
Keratinocytes	Melanocytes	Arrector pili muscle
Merkel cells		Fibroblasts
		Endothelial and blood cells
		Langerhans cells

conserved (Figure 8.1). Key differences lie in development of the head region, where the neural crest has evolved as a major contributor to human head structure. During development, the ectoderm- and mesoderm-derived tissues form intermediate structures that interact through complex signalling mechanisms.

The ectoderm layer forms both the integumentary epithelium and the nervous system. The central portion of the ectoderm differentiates as the neural plate that will fold to form the entire central nervous system (brain and spinal cord). The edges of this plate will form the neural crest which contributes many different structures, including the entire peripheral nervous system (sensory, sympathetic, and enteric ganglia) as well as components of the integument. (For further discussion, see Chapter 17: Brain, Spinal Cord, and Sensory Systems).

In week 4 (GA week 5), the integument consists of a simple ectoderm epithelium overlying mesoderm forming an embryonic connective tissue (mesenchyme). Epithelial–mesenchymal interactions (signalling) between the ectoderm and mesoderm underlie the development for all the appendages: hair, glands, nails, and teeth (Rendl et al. 2005). These interactions are required for many of the specialised skin cells, including melanocytes (Yamaguchi et al. 2004) and stem cells (Yang et al. 2017).

In week 8 (GA week 10), the epidermis is initially formed as two cell layers: peridermal cells (outer layer) and basal cells (underlying layer) (Figure 8.2). The periderm is a unique transient embryological layer (Akiyama et al. 1999) that is initially replaced by an underlying layer of keratinocyte stem cells. The underlying connective tissue layer is far more extensive and is vascularised (Figure 8.2). During this week, in the human head region a superficial scalp vascular plexus is clearly visible through the overlying epidermis. As this plexus develops it regularly changes position, shifting downwards on the scalp.

At the time of embryonic to fetal transition, periderm cells undergo a series of changes in morphology that continue into the second trimester fetal period and these changes are similar to keratinisation. The periderm also contributes to formation of the vernix caseosa. Other mammals, like the mouse, which is often used as a genetic disease model, develop a periderm layer (days 9–12), but this does not function to protect against amniotic fluid or contribute to formation of a vernix caseosa (M'Boneko and Merker 1988). The periderm therefore has evolved as a unique structure and function for human skin development.

By the end of the first trimester, the ectoderm layer now develops into a germinative basal cell layer, which through repeated rounds of cell division generates the first stratified epithelium. This basal cell layer will develop into the location of stem cells for the continuous

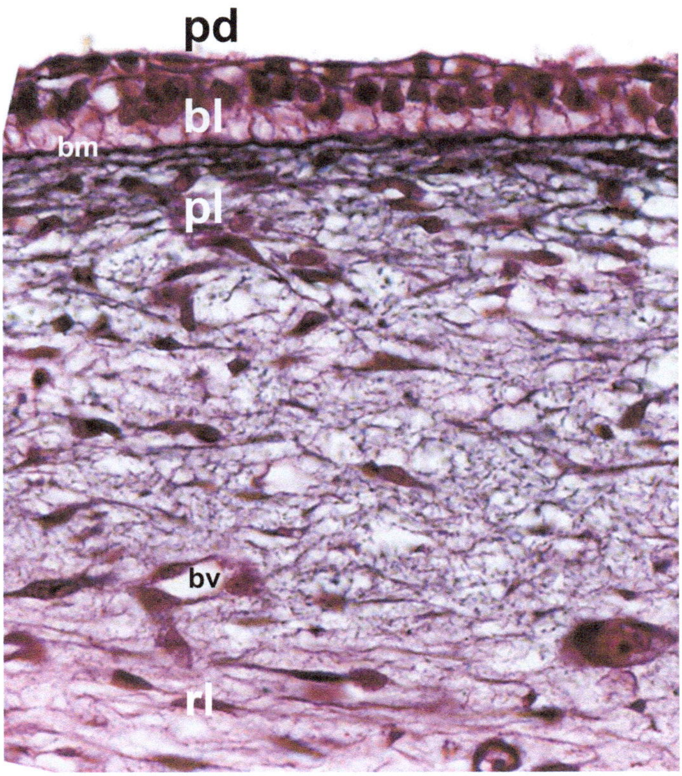

FIGURE 8.2 Histological section of developing human embryo skin in week 8 (GA week 10), with skin surface at top of image. Epidermis consists of just two layers: the upper flattened (squamous) periderm (pd) and a lower basal layer (bl) that sits on top of a basement membrane (bm). The basal stem cell layer will in the fetal period generate keratinocyte cell layers of skin stratified epithelium. The basal layer will also be continuously replaced as stem cells throughout life. The early dermis is more extensive, consisting of fibroblasts and a loose network of collagen fibres. The dermis early differentiation of the upper papillary layer (pl) and underlying reticular layer (rl) can be identified; note presence of a blood vessel (bv) within the dermis. There is no hypodermis layer present.

replacement of the epidermis throughout postnatal life. The basal cell layer will also sit upon a unique extracellular matrix, the basement membrane, which polarises these basal cells through adhesive interactions and allows the transfer of nutrients to the overlying epidermis. All mammals have a similar basal cell layer and basement membrane structure, but these develop at different times within each species embryo.

The underlying mesoderm begins to differentiate to form connective tissue and blood vessels of the dermis. By week 11 (GA week 13), blood vessels are now visible in the early fetal skin, small blood vessels in the upper papillary region, and larger vessels in the deep reticular dermis. This is the beginning of the extensive vascularisation of the dermis that will later bring nutrients and immune mediator cells and remove waste and cellular debris.

On the digits of the upper and lower limbs the nails begin to form, first as digital ectodermal placodes in the embryonic period. Like the other skin specialisations, their development involves a series of epithelial–mesenchymal (epidermis–dermis) signalling interactions (Cai and Ma 2011). The nail field then forms a fold at the very digital tip, and later development will see the nail move to the dorsum of the digit. The nail regions are macroscopically apparent by week 10 on the fingers and week 14 on the toes. This rostro–caudal timing difference reflects the general timing difference between the entire upper and lower limb development. Humans and some primates are unique in having dorsal flattened nails rather than claws or hoofs on their limb digits. The morphological change in human nails allows for development of a precision grip, the nails being utilised for fine motor functions; they also provide a source of increased dorsal physical resistance which can contribute mechanically and touch sensitivity of the fingertips.

By 4 months, the ectoderm-derived basal cell layer continues to proliferate and generates folds in basement membrane. Ectoderm will contribute to nails, hair follicles, and glands. Neural crest cells have previously migrated into regions near the epithelium as precursor melanoblasts. The melanoblast differentiates either in migration, in situ, or both to form the future melanocyte pigment cell of the skin and hair. In mammals, these pigmented cells can be found in other tissues such as the eyes, ears, heart, and central nervous system meninges. The pattern of pigmentation abnormalities identified in humans, Blaschko's lines (lines of Blaschko), first described in 1901, suggests an epidermal cell and neural crest migration pattern not present in other mammals. Skin diseases such as a type of lupus erythematosus show this same distinctive pattern. It is not clear whether this additional population is specific to these pigmentation abnormalities (Kinsler and Larue 2017). A similar patterning of the underlying dermis connective tissue, Langer's lines (skin cleavage lines, cleavage lines), refers to an orientation of reticular dermis collagen bundles causing tensions on skin and subcutaneous tissues. Human Langer's lines tend to be horizontal in the trunk and neck and longitudinal in the skin and limbs and have surgical significance (Byard et al. 2005).

Embryonic connective tissue differentiates into dermis, a loose connective tissue layer over a dense connective tissue layer. Beneath the dense connective tissue layer is another loose connective tissue layer that will form the subcutaneous layer.

Nail matrix begins to develop within the proximal nail fold which then undergoes localised cornification. The nail plate grows from this nail matrix as keratinised cells, flattening and compacting into dense nail tissue. Cords of epidermis cells now extend into mesoderm-forming epithelial columns; these contribute to hair follicle, and sebaceous and sweat gland development.

By 5 months, hair growth is initiated at the base of these cell cords, and lateral outgrowths form associated sebaceous glands. Other cords continue to elongate and coil to form eccrine sweat glands. Uniquely in humans, we have evolved to develop far fewer hair follicles and significantly more sweat glands than any other mammal. This gives us a 'hairless' appearance and thermoregulation, and other properties, due to these additional sweat glands.

Epithelial cords in the mammary region thicken to form a mammary ridge that branches as they elongate to form primordial mammary glands. These glands will not complete development until puberty and only in females. Functional maturity only occurs in late pregnancy. Uniquely to humans and in some primates, only a pair of mammary glands develops, while in other species the mammary ridge along the ventral surface gives rise to several pairs of mammary glands. This is only seen in some human developmental abnormalities such as an accessory or supernumerary nipple. Supernumerary nipples have been shown to have a higher prevalence for the left side and male gender and more recently with BRCA2 cancer mutations (McKay and Coad 2017).

By 7 months, nail matrix at the nail root becomes differentiated, with active growth and the terminal margin of the nails becoming free. The nail then grows over the corneous layer which covers the terminal row of papillae of the nail bed. The finger nails, then the toe nails by week 32–36 reach the tips of their respective digits. A key skin external feature now appears in the third trimester—the secretion and attachment of the vernix caseosa.

8.3.2 The Vernix Caseosa

The vernix caseosa (Latin: *vernix* meaning 'to varnish' and *caseous* meaning 'cheese-like'), also simply called vernix, has uniquely evolved with humans. The naming refers to its general covering and its appearance. Its chief function is protective for skin development and from infections of both the fetal and new-born skin. Preterm infants may lack the same coverage as term infants, and the coverage and composition are variable. A study has shown significant differences in composition between boys and girls, with new-born girls having a higher proportion of wax esters and triacylglycerols with longer hydrocarbon chains (Míková et al. 2014). It is not clear whether this has evolved as a sexual dimorphism with a specific function.

The vernix begins to cover the fetal skin in a craniocaudal direction during the second trimester from about week 17, and its protective role is mainly in the third trimester. The antimicrobial function has been shown to be mediated through defensins and cathelicidin LL-37 (Yoshio et al. 2004). Vernix production coincides *in utero* with terminal differentiation of the epidermis and formation of the stratum corneum.

Glandular sebum production has been shown to be dependent upon both the gland and endocrine development. Secretion of sebum that generates the lipid film appears to be regulated by the hypothalamic–pituitary–adrenal (HPA) axis. This would therefore involve development of the 'adult' adrenal cortex from the 'fetal' cortex. Postnatally, disruption to the HPA axis and its circadian cycling has been identified as being involved with development of stress-related skin diseases, mediated through corticotrophin-releasing hormone (CRH) from the paraventricular nuclei of the hypothalamus (Kim et al. 2013).

The developing skin is protected by vernix from the *in utero* amniotic fluid environment by provision of a waterproofing film and regulating TEWL. The barrier function protects the skin from the high levels of alpha-chymotrypsin found in amniotic fluid (Tansirikongkol et al. 2007). Towards term, fragments of vernix can mix with the amniotic fluid, resulting in (normal) turbidity. Fetal swallowing of amniotic fluid then may provide a positive role for vernix in both gastrointestinal tract and respiratory tract protection and possible development (Dasgupta et al. 2016).

During stage two of labour, this slippery layer may aid passage of the infant through the birth canal. After birth, the hairless neonatal skin is aided in adapting to the dry air environment and provides a possible thermoregulatory role. Human vernix has evolved to perform many antibacterial and antifungal functions, such as an opsonising capacity, protease inhibitors, and parasite inactivation, in addition to the host defence active molecules cystatin A, UGRP-1, and carbanion A, B, and C (Tollin et al. 2005).

In some societies, the protective nature of the vernix is prematurely lost through extensive neonatal washing (Adejuyigbe et al. 2015; Sacks et al. 2015). The vernix also contains relatively high levels of vitamin E as an antioxidant, and these levels can be reduced by ultraviolet B (UVB) irradiation (Visscher et al. 2005). (For further discussion, see Chapter 16: Sexuality, Reproduction, and Birth) This new-born protective function of vernix has been investigated for manufacture of therapeutic artificial substitutes in infants and potentially for damaged barrier-deficient skin (Rissmann et al. 2009).

There are several known pathological conditions related to the vernix and therefore unique to humans. First is the infant with vernix caseosa aspiration syndrome (Nakwan et al. 2017). This condition results from a negative effect of the fetus ingesting the vernix during normal fetal swallowing and respiratory movements. Second is the mother with vernix caseosa peritonitis (Selo-Ojeme 2007; Chambers et al. 2012) and associated vernix caseosa granuloma (Boothby et al. 1985).

The developing epidermis sheds the initial periderm cell layer that is then mixed with sebaceous gland sebum secretions within the epithelial walls. Vernix is composed of sebum, cells that have sloughed off the fetus's skin, and lanugo hair that binds it to the fetal skin. There is a significant water content (80%), largely compartmentalised within fetal corneocytes (the cells forming the stratum corneum), surrounded by a matrix of lipids (10%) and proteins (10%) (Hoeger et al. 2002).

8.3.3 Postnatal Period

Postnatally, the epidermis stratum corneum continues to develop and matures over the first weeks to months. Premature infants lack the well-formed barrier of term infants, and both have initial microbiota sourced from the mother during birth. During the initial postnatal period, the stratum corneum changes hydration and TEWL is quite high and also regionally different. The skin pH is changing to become more acidic, establishing the protective acid mantle (Kanti et al. 2014). This shift is due to keratinocytes in the stratum corneum and glandular secretion products. From birth to about 2 years of age, the mechanical properties of the skin change with increasing elasticity and decreasing viscoelasticity (Visscher et al. 2017). Viscoelasticity, as the name suggests, is when the skin undergoes deformation of both viscous and elastic properties. Many of these mechanical properties reflect the dermis extracellular matrix collagen and elastic fibre composition, organisation, and expression.

8.3.4 Puberty and Sex Differences

Sexual maturation occurs quite late in humans compared to all other mammals. There is also a significant difference in the timing of this event between girls (10–14 years) and boys (12–16 years). While genital development is covered elsewhere (For further discussion, see Chapter 4: Growth and Development), this chapter discusses puberty's hormonal impact upon the skin. The increase in sex hormones can be recognised externally by the appearance of secondary sex characteristics that include breast development and axillary, genital and facial hair. This may actually be preceded by sexual dimorphic changes in skin pigmentation (Sitek et al. 2013). Such a functional dimorphism has been shown in mice for collagen expression under the influence of testosterone (Arai et al. 2017).

Testosterone is a key regulator of skin changes at puberty, resulting in increasing gland sebum production, while axillary apoeccrine glands develop from eccrine glands and increase sweat production (Sato et al. 1987). Testosterone is also responsible for stimulating pubic and axillary hair growth. Another male-related steroid, androstadienone, is found both in the skin and with axillary hair. Interestingly, androstadienone does not appear to have the same androgenic action as testosterone, but has a strong pheromone-like activity in humans (Lundström et al. 2003). This potential pheromone function acting through the vomeronasal organ can have neurological effects related to sexuality. Female menstrual cycle hormonal changes cause premenstrual exacerbation of psoriasis, atopic eczema, and lupus erythematosus.

Acne vulgaris is an inflammatory disease affecting about 80% of adolescents and young adults and worldwide affecting 650 million people (Tuchayi et al. 2015). Pathogenesis has been described as increased sebum production and *Propionibacterium acnes*, present on normal skin, colonising the duct of the sebaceous follicle, stimulating an immune response. Other hairy mammals do not tend to get acne and it is also significantly lower in human non-westernised and non-modernised societies (Cordain et al. 2002). This has suggested more than just a genetic correlation, but also an association with other environmental factors. Some examples include changing western diets, clothing, lifestyle, drugs, and environmental changes.

8.4 Skin Structure

8.4.1 Epidermis

Epidermis in the adult is continuously being replaced and the underlying dermis extracellular matrix remodelled. The dermis contains many different residing and migrating (into and out of) cell types. These processes are similar to those in all other mammal skin. A basement membrane underlies the entire epidermis and is located at the epidermal–dermal junction. The basement membrane is a specialised extracellular matrix required to organise the basal layer of the epidermis and associated cells, consisting of the glycoproteins laminin and type VI collagen.

Through the entire human life, epidermis keratinocytes differentiate directionally from inside to outside, expressing different genes, histological features, and differentiation processes—from the stratum basale keratinocyte stem cells, into the stratum spinosus forming many desmosome junctions, then the stratum granulosum forming keratohyalin granules, and finally the stratum corneum where they undergo cornification and are continually shed as flattened dehydrated squames. Within the stratum corneum, lipids extruded from the cells form a water-repelling envelope around the cornified envelope.

Keratinocyte turnover occurs regularly, taking between 28 and 30 days in young adults and later 45–50 days in elderly adults. Compare this to the mouse keratinocyte turnover of between 8 and 10 days; this means that individual human keratinocytes live longer than the animal models we use to study skin. This continuous lifelong turnover requires the presence of stem cells within the basal layer.

8.4.2 Dermis

Dermis in the skin can be further divided histologically into the upper papillary region, which is vascular and nutritionally supporting the epidermis, and the lower reticular region, containing connective tissue, hair follicles, and glands. The papillary fibroblasts may represent an undifferentiated phenotype, while reticular fibroblasts represent a more differentiated population (Janson et al. 2013). Mixed within the dermis are fibroblasts, adipocytes, blood and immune cells, lymph and blood vessels, specialised sensory structures, bare sensory nerves, and immune cells. The dermis key feature though is the large amount of dense irregular connective tissue formed from collagen bundles.

A study has shown that there are at least two separate fibroblast lineages within the dermis (Driskell et al. 2013). The upper dermis lineage regulates hair follicle formation, growth, and the erector pili muscle. The lower dermis lineage forms the reticular fibroblasts and is involved in initial wound repair. Dermal fibroblasts interact with immune cells during infections by forming collagen 'walls' around polymorph nuclear cells to prevent microbial spreading.

The dermis contains a unique member of the immune dendritic cell family called the Langerhans cells, also found in the cornea and internal mucosa. Langerhans cells are myeloid cells generated in the embryo and remain as extremely long-living cells (Collin and Milne 2016). They appear to exist as distinct subsets that regulate immunity and tolerance within the skin and regulate T lymphocyte function. Once presented with antigen, the cells can migrate to the regional lymph nodes, present to naive T cells, and initiate the adaptive immune response. Langerhans cells can coordinate with other immune cells, natural killer cells, to impair skin squamous cell carcinoma development caused by chemical carcinogens (Atmatzidis et al. 2017). Langerhans cells can be associated with skin abnormalities of contact hypersensitivity and chronic inflammatory diseases of the skin.

The dermis contains different collagens. Human fetal skin contains mainly type III collagen (30–60% of total collagen), compared to adult skin (10–20%) (Smith 1994). The collagen isoform ratio of type III/type I changes with age, with type I increasing with age. Later the skin displays sexual dimorphism under the influence of sex hormones (see Section 8.3.4). The dermis and hypodermis (subcutis) of children and adolescents undergo an age-related thickening that is associated with increasing body mass index (BMI). The current first-world trend in increasing BMI is leading to significant changes in this layer's composition and health effects. The dermal thickening process begins to decrease with ageing in adults, differing in body regions between the sexes (Derraik et al. 2014).

8.4.3 Hypodermis

Hypodermis, also called the subcutaneous adipose or subcutis, lies beneath the dermis. This layer contains similar though less collagens and is a second connective tissue layer, described as skin-associated adipose tissue (Driskell et al. 2014). The adipose cells in this region, white adipose tissue (WAT), appear to have the same embryonic origin as the overlying dermis fibroblasts. This WAT develops in specific anatomical locations and is deposited in a sex-specific manner. These adipocytes serve several different functions: energy storage as fatty acids that can be metabolised into ATP; and endocrine functions related to sex differences, food intake, glucose homeostasis, lipid metabolism, inflammation, and angiogenesis. As humans are hairless, this layer provides both thermal insulation and a mechanical cushion to the body surface. The hypodermis layer can also, in different anatomical locations, differ in overall thickness between the two sexes. The ability to store energy in specific tissues evolved along with multicellular organisms. In recent years, we have seen significant increases in the amount of adipose tissue within skin and other tissues. Increasing adipose tissue contributes to obesity, which has evolved as one of the most significant worldwide health problem. This is a multifactorial problem and relates to diet, lifestyle, and potentially both genetic and epigenetic changes in the human population. (For further discussion, see Chapter 6: Nutrition, Energy Expenditure, Physical Activity, and Body Composition).

8.4.4 Keratinocytes Inside and Outside

In an early evolutionary aquatic environment, both the external and internal animal surfaces were protected from an aqueous medium with specialised outer and inner surface tissues. This environment is similar to our current prenatal environment. Transiently living between the aquatic and air environments, for example like amphibians, led to differentiation of a modified outer skin that could resist water loss and drying as well as modified renal mechanisms for dealing with waste. (For further discussion, see Chapter 14: Excretory System.) Permanent movement to the land by animals led to differentiation of a specialised skin.

The internal gastrointestinal, renal, and genital mucosal connective tissue and epithelial barrier is covered in other chapters. The mucosa differs in fibroblast organisation and keratinocyte turnover and is 'non-keratinised'. The major protein within keratinocytes is keratin or cytokeratin, an intermediate filament protein that has evolved into many different isoforms. The cytokeratin isoform profile of both the oral and the surface epithelium in humans is similar except for CK20 (Barrett et al. 2000), though there are isoform differences found in other species (Barrett et al. 1998 2000). The process of wound repair is similar, though there are major differences in the rate of epithelial healing and associated scarring. For example, oral mucosal wounds heal more rapidly than skin wounds, and with minimal or no scar formation (Turabelidze et al. 2014; Evans 2017). This difference in healing is associated with several factors, including intrinsic keratinocyte differences (Turabelidze et al. 2014); the expression of antimicrobials in the saliva (Mohanty et al. 2017); and a lower inflammatory response with less neutrophil, macrophage, and T-cell infiltration (Szpaderska et al. 2003).

There exists a specialised 'border region' between the internal wet environment and the outer dry environment. This mucocutaneous junctional region is seen at the upper and lower gastrointestinal tract, eyelid, and urogenital openings in all mammalian species (Lanfranchi and de Rey 1978). The eyelids and the lips have been shown to share common cellular markers (Barrett et al. 2005) and capillary loops (Wolfram-Gabel and Sick 2002) at their transitional zones. The oral mucosa is persistently exposed to environmental effects at the lips, also called the vermilion border of the lips. In humans, these regions are richly vascularised with dermal capillary loops, generating the colouration, as well as increasing the surface temperature compared to the face. This red colouration has been cosmetically enhanced in many cultures throughout human history. The lips are hairless and have no sweat glands, until after puberty, when about half the population express sebaceous glands in this region (Kobayashi and Tagami 2004).

8.4.5 Structure and Transepidermal Water Loss

TEWL is used as the non-invasive measurement of the skin barrier function and permeability. A number of different techniques have been developed to measure this parameter (Farahmand et al. 2009). Humans over evolutionary time have decreased their hair covering and increased their eccrine sweat gland number. The greater number of eccrine sweat glands for cooling have a much higher TEWL contribution from this source than all other mammals. Water loss is also affected by the external environment parameters temperature, humidity, and clothing. A review of the published literature showed TEWL varied by region over the body surface, with the lowest for breast skin and the highest for the axilla (Kottner et al. 2013). Infant TEWL is much higher than in adults (Kelleher et al. 2013; Ludriksone et al. 2014), and adult values are similar for most genetic groups (Mack et al. 2016), though much lower in older (> 65 years old) humans (Kottner et al. 2013).

8.5 Skin Gene Regulation

8.5.1 Epidermal Differentiation Complex

The epidermal differentiation complex (EDC) is an evolutionarily conserved genomic region required for late stages of epidermal differentiation and barrier formation (Oh and de Guzman Strong 2017). The EDC locus is located on human chromosome arm 1q21 and mouse chromosome arm 3q and comprises a dense cluster of genes whose protein products are the major molecular markers for terminal differentiation in the stratified epidermis. Loricrin and involucrin were the first EDC genes to be discovered and are key proteins in keratinocyte terminal differentiation in the stratum corneum.

Loricrin is a structural protein forming more than 70% of the cornified envelope, contributing to the protective barrier function of the stratum corneum (Nithya et al. 2015). Involucrin binds to loricrin within the stratum corneum. Involucrin expression begins within the early spinous layer and continues in the granular layer. Involucrin expression is regulated by a unique signalling cascade (Eckert et al. 2004) and cross-linked by transglutaminases to membrane proteins during keratinocyte terminal differentiation.

8.5.2 Ectodysplasin Signalling

The ectodysplasin (EDA) pathway is a skin signalling molecule that has been strongly conserved in all vertebrates. EDA is involved in the development of ectodermal appendages: teeth, hairs, and mammary glands. In developing skin, EDA is critical for regulation of the early epithelial–mesenchymal interactions (Botchkarev and Fessing 2005; Mikkola and

Thesleff 2013). In the adult skin, the signalling role continues in regulating hair follicle cycling. The EDA family includes two trimeric type II membrane proteins (Eda A1 and Eda A2) differing by only two amino acids (OMIM 300451). A novel member of the tumour necrosis factor (TNF)-related ligand family is located on lateral and apical surfaces of cells that can be cleaved by furin to produce a secreted form that interacts with the Eda receptor (Edar). A single Edar independent mutation more than 10,000 years ago in both the Asian and American Indian populations resulted in thicker hair and tooth differences (Bryk et al. 2008). In these individuals, this mutation enhances another transcription factor, NF-κB (nuclear factor kappa light chain enhancer of activated B cells), whose activity directly affects signal transduction during hair development in the skin (Thesleff and Mikkola 2002). Mutations in the EDA gene Xq13.1 cause the X-linked disease hypohidrotic ectodermal dysplasia and tooth agenesis.

8.5.3 Cornification

The epithelial layer of the skin in all animals undergoes a unique form of programmed cell death called 'cornification'. Cornification developed evolutionarily to prevent water loss as animals became terrestrial. The other main form of programmed cell death, apoptosis, occurs in most other body tissues. Apoptosis results in the destruction of all cell organelles and encapsulation of the small cellular fragments within parts of the original plasma membrane; these are in turn removed by macrophages and other cells. Unlike apoptosis, cornification preserves the (modified) cell membrane, but does remove all the internal cell organelles. These two forms of cell death appear to be related to the key family of mediator enzymes, the caspases. Caspases are a family of evolutionarily conserved cysteinyl aspartate-specific proteinases (Table 8.4).

CASP12 in most humans is an inactive pseudogene, though active in about 20% of humans of African descent. CASP11 is a mouse homologue of human CASP4.

Table 8.4 The Caspase Enzyme Family

Caspase	Group	OMIM No.	Gene location
CASP1	I–inflammatory	147678	11q22.3
CASP2	II–apoptosis initiator	600639	7q34
CASP3	III–apoptosis effector	600636	4q35.1
CASP4	I–inflammatory	602664	11q22.3
CASP5	I–inflammatory	602665	11q22.3
CASP6	III–apoptosis effector	601532	4q25
CASP7	III–apoptosis effector	601761	10q25.3
CASP8	II–apoptosis initiator	601763	2q33.1
CASP9	II–apoptosis initiator	602234	1p36.21
CASP10	II–apoptosis initiator	601762	2q33.1
CASP11	I–inflammatory	602664	Mouse
CASP12	I–inflammatory	608633	11q22.3
CASP13	I–inflammatory	N/A	Bovine
CASP14	I–inflammatory	605848	19p13.12

CASP14 was first identified in 1998 (Van de Craen et al. 1998) and is required for the processing of the cornifying protein filaggrin (OMIM 135940). Mutations in CASP14 result in several different types of cornification disorders (ichthyosis). Profilaggrin forms a component of the keratohyalin granules, in the granulosa layer, of the epidermis and when processed to filaggrin it acts as a keratin-aggregating protein for the terminal differentiation of the stratum corneum. Interestingly, filaggrin protein demonstrates wide species variations.

During skin development, CASP14 expression occurs differentially in each trimester (Gkegkes et al. 2013). In the first trimester, nuclear expression is seen in the periderm, basal layer, and all layers present. In the second trimester, cytoplasmic expression occurs mainly in the more differentiated upper epidermal layers. In the third trimester, expression occurs only in the cornified layer.

In adult skin, inactive CASP14 expression occurs only in differentiating and not in proliferating keratinocytes (Figure 8.3). Expression of inactive CASP14 starts in the spinous layer; then at the transition from the granular to cornified layer it is cleaved into two components (p20 and p10). In the cornified layer with dehydration, CASP14 is active and proteolytically cleaves filaggrin, which is required for maintaining epidermal hydration. CASP14 expression also occurs in the more differentiated layers of the hair inner root sheath (Gkegkes et al. 2013).

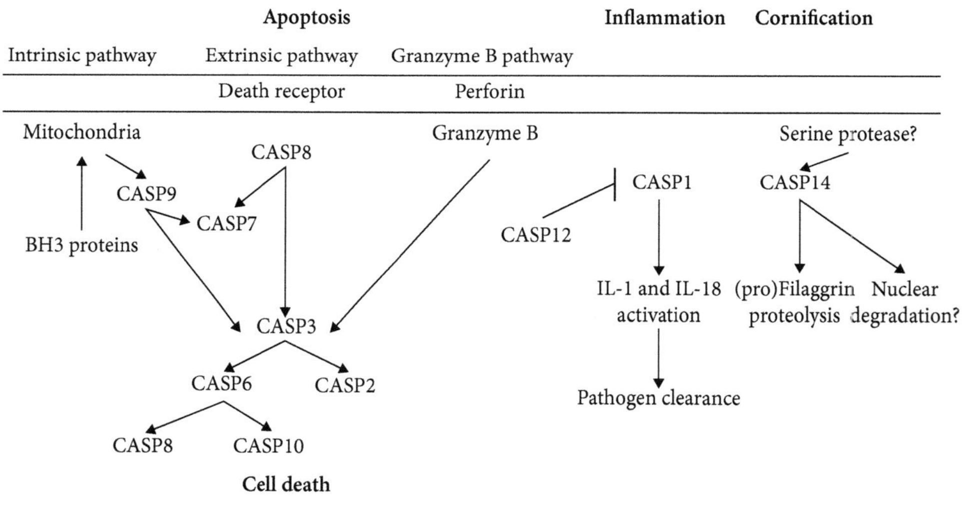

FIGURE 8.3 Simplified overview of caspase pathways. The three main processes of apoptosis, inflammation, and cornification involve different members of the caspase enzyme family. Within cells, all caspases generally exist in an inactive pro-caspase form, and cleavage activates their enzymatic activity, the basis of the naming (Cysteinyl ASPartate-specific proteinASE). Many caspases are activated by other caspases in a cascading mechanism. Apoptosis has three main pathways; intrinsic, extrinsic, and granzyme B. The extrinsic pathway has a membrane receptor (death receptor), the granzyme B pathway has a membrane pore (perforin), and the intrinsic pathway is mediated through a pore forming within mitochondria. All pathways are activated by many mechanisms, including UV irradiation, and utilise many caspases, with CASP3 key in all three pathways. Inflammation associated with infection is mediated by CASP1 that can be inhibited by CASP12. Cornification uniquely uses CASP14, which has evolved and is expressed only in the skin of terrestrial mammals and not in birds or reptiles. The enzyme is involved in cleavage of (pro)filaggrin, hydration, and protection against UVB. CASP14 is activated only in upper layers of epithelium (stratum granulosum and corneum; see also Figure 8.4) and not by other caspases.

Profilaggrin undergoes many post-translational modifications, eventually leading to release from the keratin intermediate filaments. Within the lower stratum corneum, the dehydration changes cause stepwise proteolytic degradation of filaggrin monomers into free hygroscopic amino acids that maintain epidermal hydration. Mutations in the human gene encoding profilaggrin and filaggrin are the cause of the common skin condition ichthyosis vulgaris. These mutations have been shown to be a strong genetic predisposing factor for other skin-related diseases such as atopic eczema, asthma, and allergies (Sandilands et al. 2009).

8.5.4 Keratin

Keratins are a large family of intermediate filament proteins that are differentially expressed in different tissues. These intermediate filaments form strong structural bundles that provide the cell's mechanical properties within these tissues. Keratins along with epidermal differentiation genes appear to be under evolutionary selection based upon environmental conditions (Gautam et al. 2015). For example, α-keratins are expressed in all vertebrates (hair, horns, nails, and epidermis), while β-keratins are only found in birds (feathers, claws, and beaks) and reptiles (scales, claws, and shells) (Wu et al. 2015).

Keratin-19 is an epithelial marker and a potential marker of epidermal stem cells that is expressed during human fetal skin development (Gkegkes et al. 2013). Keratins K5 and K14 (OMIM 148066) are mainly expressed in the basal layer proliferating keratinocytes, while the interfollicular differentiating keratinocytes express K1 and K10 (Figure 8.4). During keratinocyte differentiation within the epidermis, K14 is downregulated by the transcription factor p53 (Cai et al. 2012). An inverse relationship is shown by another transcription factor p63 (OMIM 603273) in keratin expression (Figure 8.4), as well as maintenance of keratinocyte progenitor-cell population and other keratinocyte differentiation processes (Koster and Roop 2004, 2007).

8.5.5 Animal Skin Models

Pig skin has been used to model human skin (Jacobi et al. 2007; Debeer et al. 2013) and for tissue engineering purposes using porcine skin extracellular matrix, including xenogenous dermal matrices for hernia repair and breast reconstruction (Mirastschijski et al. 2013). However, it is still not an ideal skin model as there have been differences shown in basic structure (Debeer et al. 2013), for example in dermis type I and type III collagen distribution (Turner et al. 2015).

8.6 Skin Specialisations

8.6.1 Glands

Humans have evolved the highest density of eccrine sweat glands of any mammal. Other mammals do not appear to use these glands for thermoregulation, but they do have them

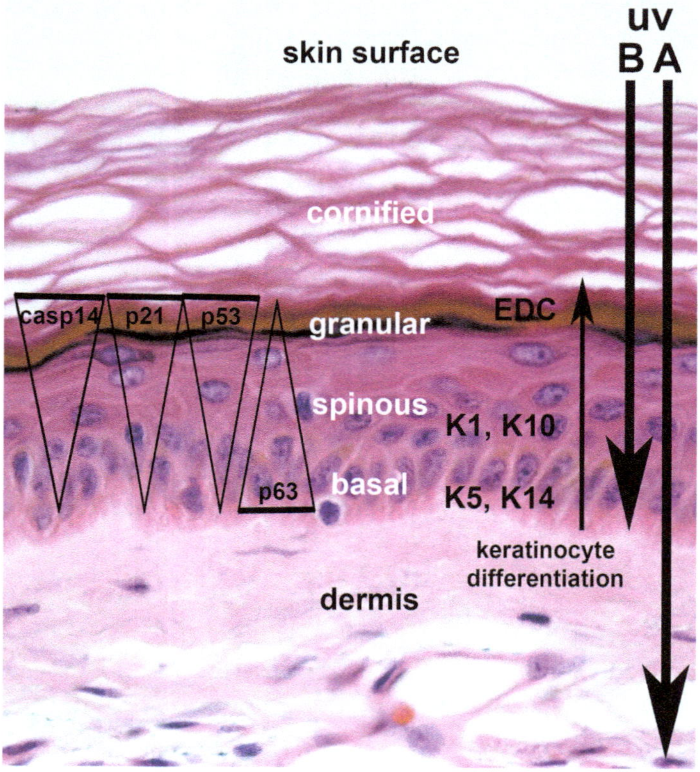

FIGURE 8.4 Patterns of transcription factor expression and keratin in layers of epidermis of thin skin. The skin surface is shown at the top and basal (stem cell) layer at the bottom. CASP14 is a key regulator of the cornification process (see Figure 8.3), increases in expression above the basal layer, and is activated in upper layers. The basal stem cell layer expresses p63 which decreases with keratinocyte differentiation, and p53 and its downstream target p21 conversely both increase in expression level. Keratin isoforms also change with differentiation; basal stem cells express K14 and K5, while differentiated keratinocytes express K1 and K10 isoforms. In upper skin layers, proteins of EDC are expressed, cell organelles are lost, and lamellar bodies are produced and extruded. On the right, large arrows show approximate level of skin penetration of solar UV light. UVA (320–400 nm) penetrates through to the deeper dermis layer. UVB (280–320 nm) penetrates mainly only the superficial epithelial layer. UV light can increase melanin production, help produce vitamin D, damage DNA, generate free radicals, and break down folate in the skin.

for traction in footpad regions. This means that most research animal skin models for this function do not apply to humans. Sweating contributes to TEWL, allowing effective body cooling by evapotranspiration of water on the skin's surface. This cooling effect is thought to allow humans periods of extreme prolonged activity and higher metabolic activity.

A second role for human eccrine sweat glands could relate to wound healing. Skin wound re-epithelialisation is supported by keratinocytes derived and recruited from nearby eccrine sweat glands (Rittié et al. 2013). The third role for human eccrine sweat glands is their contribution to the 'acid mantle', the acidic film that covers the surface of the skin and contributes to the barrier to microbiota. Rather than 'hairless', perhaps we should be referred to as the 'sweaty' mammals.

8.6.2 Breast

The breast evolved as the defining feature of the vertebrate class mammalia (Latin: *mamma* meaning 'breast'), which functions to provide postnatal nutrition to the young (Lefèvre et al. 2010). Recent studies have shown that mammary buds developed evolutionarily from 'reuse' of the Hox signalling pathway involving members Hoxd8 and Hoxd9, which are normally used in embryonic limb development (Schep et al. 2016).

In humans, the female breast, unlike many other mammals, remains developed even during non-pregnant periods and has evolved an additional neurological role in relation to sexual attraction. Neurologically, following birth, lactation of the infant has an important role in establishing mother–infant bonding. (For further discussion, see Chapter 16: Sexuality, Reproduction, and Birth).

In the embryo, the mammary gland develops from a placode, like all other skin specialisations. It is a highly modified sexually dimorphic gland that does not complete functional development in females until puberty. Prior to puberty the male and female glands are the same.

In 1976 Tanner and Whitehouse established a series of descriptive stages for the development of primary and secondary sexual characteristics at puberty (Tanner and Whitehouse 1976). The female secondary sex characteristics of non-pregnant breast development were divided into five numbered stages (I–V) that could be classified by approximate age and external appearance (Table 8.5).

During the menstrual cycle and in pregnancy, raised oestrogens and progesterone stimulate gland development. During the non-pregnant menstrual cycles, mammary glands go through repeated rounds of development and regression (involution). After an infant ceases breastfeeding (weaning), the mammary gland milk-producing epithelial cells undergo involution. Involution requires cell death by apoptosis and involves the caspase family (CASP3) (Watson 2006) (see Section 8.5.3). Note that the hair follicle goes though similar cycles of growth, resting, and regression.

For some time, the focus on breast function was upon lactation and the nutritional role of milk, but more recently there has been recognition of the many other roles the breast can play in development of the young. In recent history, the clinical focus has been upon the hormonal responsiveness and genetics that can lead to breast cancer. Both BRCA1 (OMIM 113705) and BRCA 2 (OMIM 600185) are tumour suppressor genes that have critical roles in DNA repair, cell cycle checkpoint control, and maintenance of genomic stability. Mutations in these two genes contribute up to 25% of all breast and 15% of all ovarian cancers and are currently the targets of genetic testing. The mutation rate in these genes varies across racial, ethnic, and age groups. Mutations in BRCA1 and BRCA2 in humans and primates (Figure 8.5) probably relate to maintaining genomic integrity (Huttley et al. 2000), and in both species the amino acid substitutions evolved rapidly under positive selection (Huttley et al. 2000; Lou et al. 2014). A population study has shown that BRCA1 mutations are more common in white and Jewish women

Table 8.5 Tanner Stages of Breast Development

Tanner stage	Age (approximate years)	Breast feature
I	10 and younger (pre-pubertal)	No glandular tissue, areola follows skin contours of chest
II	10– 11.5	Breast bud forms with small area of surrounding glandular tissue, areola begins to widen
III	11.5–13	Breast begins to become more elevated and extends beyond borders of the areola, which continues to widen but remains in contour with surrounding breast
IV	13–15	Increased breast size and elevation of areola and papilla form a secondary mound projecting from contour of surrounding breast
V	15+	Breast reaches final adult size, areola returns to contour of surrounding breast with a projecting central papilla

than in black women; and BRCA2 mutations are slightly more frequent in black than white women (Malone et al. 2006). A full explanation for these differences has not yet been provided.

8.6.3 Hair

This section describes hair development. Section 8.9 discusses 'hairlessness' that is specific to humans and some primates and is a relative term.

In humans, hair follicle development begins as an epithelial–mesenchymal interaction (ectoderm–mesoderm) at weeks 9–12 (GA weeks 11–14). This initial hair formed over the entire body is described as lanugo hair (Latin: *lana* meaning 'wool') and it has an important role in binding the vernix to the fetal skin. Lanugo hair is later shed and replaced in the late fetal or early neonate period by vellus and terminal hairs. New-born infants have therefore two different types of hair.

Vellus hairs are short hairs, 1–2 cm long, and contain little or no pigment. The follicles that produce these hairs do not have associated sebaceous glands and never produce any other kind of hairs. Terminal hairs are the long hairs that grow on the head and other parts of the body. The follicles that produce these hairs are associated with sebaceous glands. The hairs in many of these follicles gradually become thinner and shorter until they appear similar to vellus hairs.

On the new-born infant head, there are two periods of hair development in which hair growth begins at the forehead and then extends to the back of the neck. Then at 2–3 months old, the first hairs may be shed naturally over an area on the back of the head. This is often mistakenly thought to be due to head rubbing.

At puberty, another round of hair development occurs, under the influence of steroidal hormones, associated with secondary sexual characteristics. Pigmentation is supplied to the hair by melanocytes, with the follicle and terminal hair colour genetically linked.

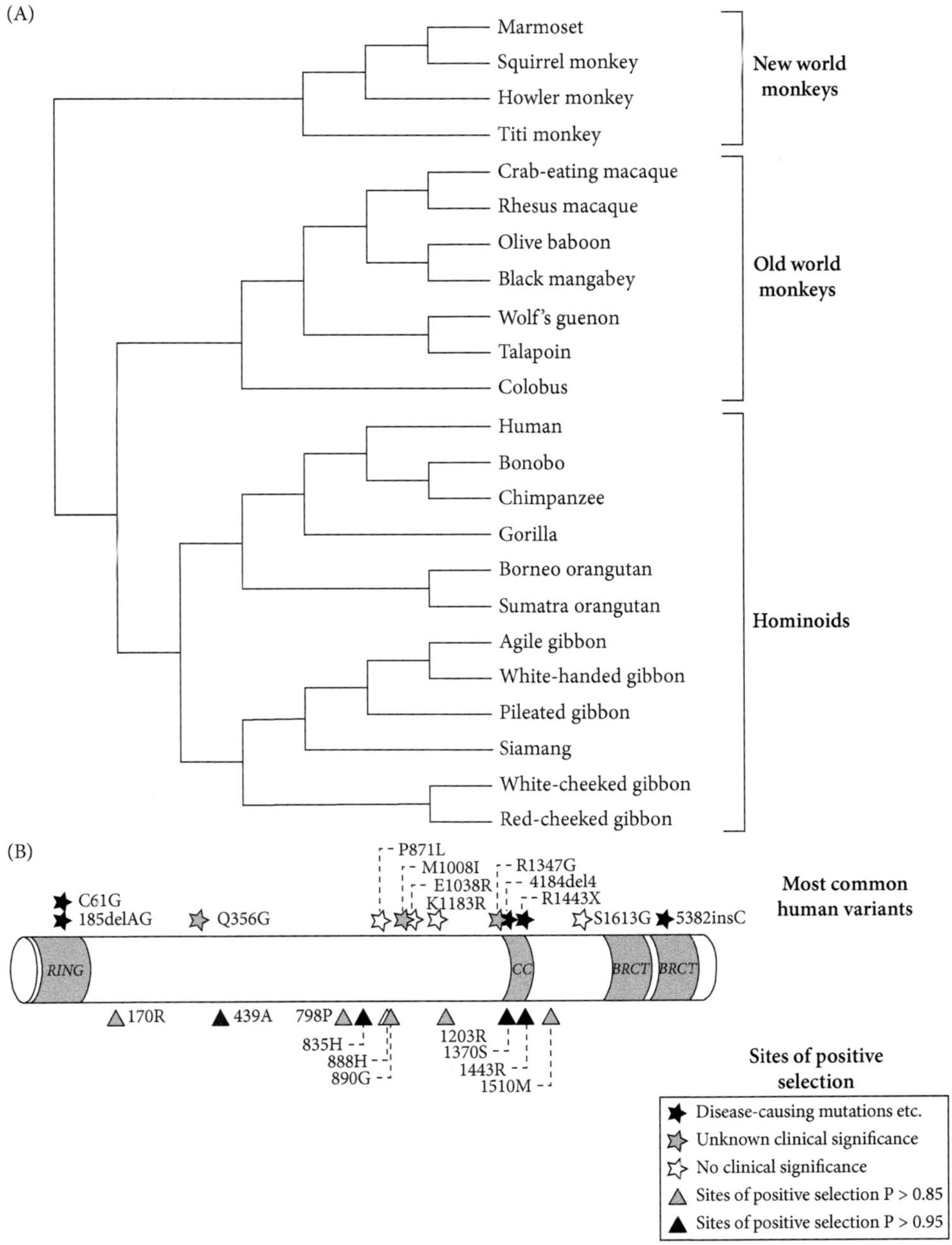

FIGURE 8.5 BRCA1 evolution and mutations based on data from Lou et al. (2014). (A) Evolution of BRCA1 during primate speciation, based on the free-ratio model of non-synonymous/synonymous (dN/dS) values for each primate phylogeny branch. (B) Human BRAC1 protein domains. Stars show sites of most common human variations: black stars, disease-causing mutations; white stars, variants with no known clinical significance; grey stars, those with unknown significance. Triangles show sites of evolutionary selection. Numbers indicate position and letters the single-letter amino acid code.

Source: Reproduced from Dianne I. Lou, Ross M. McBee, Uyen Q. Le, Anne C. Stone, Gregory K. Wilkerson, Ann M. Demogines, and Sara L. Sawyer, Rapid evolution of BRCA1 and BRCA2 in humans and other primates, *BMC Evolutionary Biology*, 14 (1), p. 155, Figures 1a and 3b, doi.org/10.1186/1471-2148-14-155.

Adult hair follicles go through several well-described phases of hair growth:

- anagen—active growing phase;
- catagen—apoptosis-driven involution, the end of the active growing phase of the life cycle of the hair;
- telogen—hair follicle resting phase of hair growth cycle.

In human old age, hairs tend to decrease in colouration (greying) and hair loss occurs. Many other factors can lead to hair loss and in modern times it has been the source of a growth industry in 'hair replacement'. Androgenetic alopecia can occur in both males and females, with distinct patterns of hair loss in each sex. At least two major genetic risk loci have been identified: the X-chromosome AR/EDA2R locus and the chromosome 20p11 locus (Lolli et al. 2017). Telogen effluvium describes the alteration of the normal hair cycle due to different stress stimuli, including severe stress, chemotherapy, childbirth, major surgery, severe chronic illness, and a rare occurrence in vaccination (Harrison and Sinclair 2002). Alopecia areata is a common autoimmune disease where antibodies are generated against some of the person's own hair follicles, leading to a distinct circular pattern of hair loss (Gilhar et al. 2016).

The hair follicle has been identified as a site for stem cells, allowing continual replacement of the follicle cells. These stem cells have recently been investigated for potential therapeutic applications.

8.6.4 Nails

Humans and higher primates, with their evolved acquisition of manual dexterity and opposable thumb, have developed from primitive claws the current flattened nails that lie on the dorsum of the terminal phalanges (Dawber 1980). This fingertip feature is said to distinguish us from other mammals (Figure 8.6). This location allows protection of the fingertips, the mechanical advantage of a hard tissue for fine grasping, and an increase in sensory structures within the fingertips. The nail itself contains no sensory structures, though there are many sensory endings within the nail bed, as well as being richly vascularised.

The evolution of these flattened ectodermal appendages has led to several developmental and clinical issues specific to humans. Interestingly, abnormalities in the nail structure are indicative of a number of known syndromic and non-syndromic genetic disorders (OMIM 148360, 161200, 161050, 189500, 129200, 153300, 614149).

Nails develop initially as placodes on the dorsum of the terminal phalanges (GA week 9). These placodes interact with the underlying mesenchyme through bone morphogenic proteins (BMPs) and Wnt signalling. The BMP downstream factor Msx2 is required during onychocyte (nail cell) terminal differentiation (Cai and Ma 2011). The differentiated nail begins in the proximal portion of the nail bed area and its growth is by the addition of new keratinised cells to the proximal edge and lower surface. New-born infant nail abnormalities can be associated with hypertrophy, leading to an incomplete development of the hallux nail, which grows triangular to trapezoidal in shape (Milano et al. 2010). Postnatally, nail growth changes with age; in children, growth of about 150 μ/day is greater than in adults, and decreases further in old age to less than 60 μ/day.

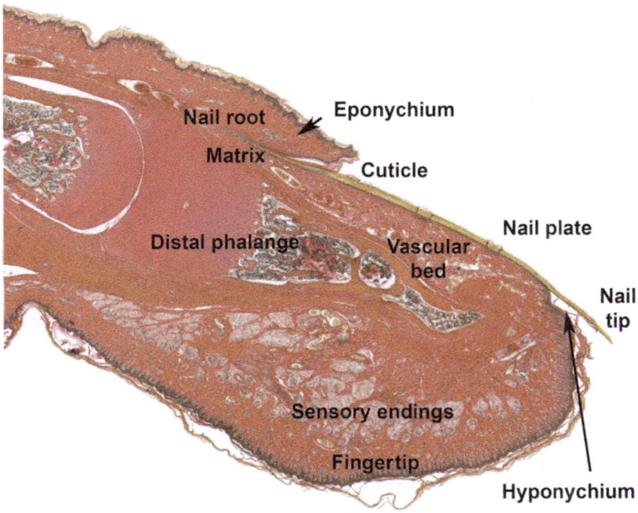

FIGURE 8.6 The human nail histological structure. This section through a neonatal digit shows the developing nail, surrounding skin, fingertip, and sensory endings within the dermis. The eponychium is the fold of skin located at the base and above the nail, while the hyponychium is a similar skinfold beneath the free end of the nail tip. The nail matrix forming the nail plate consists of a range of keratin type 1 and 2 isoforms (1, 5, 10, 14, 17, 31, 34, 81, 85, 86). Note the structure of the developing fingertip skin and underlying sensory region. Bone is still replacing cartilage within the fingertip as the distal phalange develops.

8.6.5 Fingerprints

Epidermal ridges on the fingers are also known as 'fingerprints' and reflect the histological structure of the underlying dermis interdigitating with the overlying epithelium. The study of fingerprints is called dermatoglyphics and each person's set of fingerprints and palm prints are unique and also remain stable over the individual's lifetime. This fact has been used in our modern history from the last century for specifically identifying individuals in relation to crimes, and more recently as a unique identifier for unlocking electronic devices such as mobile phones. Specific named fingerprint ridge features identified include ridge ending, bifurcation, and short ridge (or dot) (Figure 8.7).

Fingerprints are genetically related within human populations and allow the identification of individual ethnic groups (Zhang et al. 2010). Researchers have also linked specific 'patterns' to a range of genetic disorders and diseases, including Klinefelter's syndrome, trisomy 13/18/21, Turner syndrome, and schizophrenia (Golembo-Smith et al. 2012). Today, dermatoglyphic alteration for many known genetic abnormalities is more relevant to understanding developmental alterations rather than a clinical diagnostic tool, as more powerful molecular tools are now available.

Individuals with schizophrenia have been shown to have reductions in palmar a–b ridge counts (ABRCs) (Bramon et al. 2005). The neural association has been suggested to be based upon the original shared embryonic origin (ectoderm) of the central nervous system and the epidermis, or other developmental insults/effects. In terms of the neural relationship,

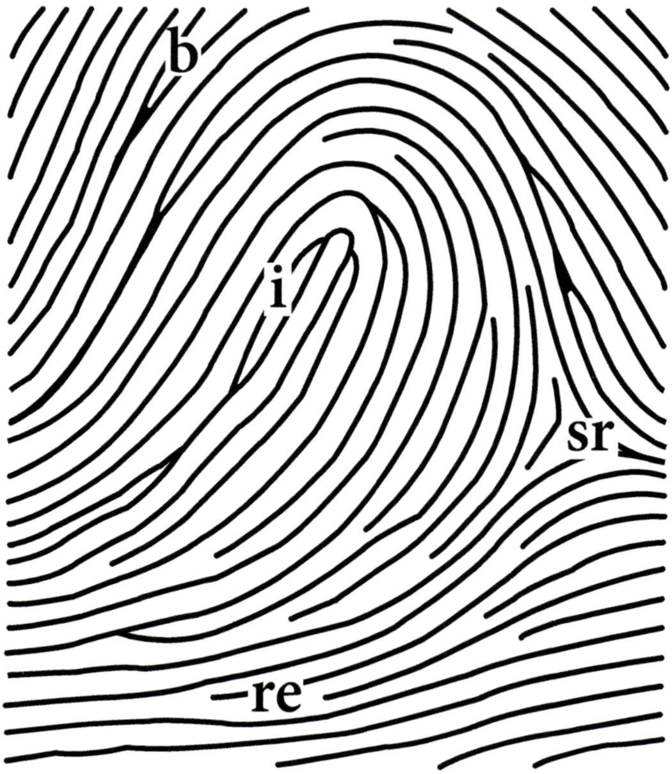

FIGURE 8.7 Example of a fingerprint (left loop) formed by dermal ridges of the skin fingertip. There are three basic fingerprint ridge patterns classified as arch, loop (left or rift), and whorl. In this classification, minutia features can be identified, shown by letters: b, bifurcation; i, island; sr, short ridge; re, ridge ending. No two human prints have been shown to be identical. Palm print analysis is made by palmar (interdigital) a–b ridge count; this is done by counting the dermal ridges that intersect a line drawn joining the triradii located on the distal palm at the base of the second and third fingers.

other primates have fingerprints and have shown a relationship to grip and hand preferences based upon specific patterns (Hopkins et al. 2005).

While fingerprints have traditionally been analysed as a two-dimensional (2D) print, current research technologies now allow us to look at three-dimensional (3D) differences in these prints; this form of analysis may in the future provide additional genetic information and linkages, as well as a potential diagnostic tool (Liu et al. 2017).

8.6.6 Sensory Receptors and Dermatomes

The skin has many different sensory receptors in three main classes: mechanoreceptors, nociceptors, and thermoreceptors. Nociceptor and thermoreceptors are mainly associated free nerve endings. Mechanoreceptors can be classified by their specialised skin sensory structures, each with a specific role. The sensory innervation pattern of the skin can be mapped to 'dermatomes' that represent the skin area innervated by a single spinal nerve. This mapping is used clinically as a diagnostic tool for identifying specific nerve damage or other neural degenerative disorders. Interestingly, the 'map' is established in the embryo before limb rotation occurs, leading to the limbs having a rotated innervation map. (For further discussion, see Chapter 17: Brain, Spinal Cord, and Sensory Systems).

The grasping and fine motor skills of the human hand have evolved along with other unique features such as the opposable thumb, thick skin, nails, fingerprints, and dense sensory innervation. This dense sensory innervation and specialist sensory receptors contribute the sensory feedback required for the evolution of the human hand for grasping and fine motor skills. This has led to an evolutionary increase and change in both the somatosensory and motor cortex mapping of the hand in the human brain compared to other mammals (Eickhoff et al. 2008). Interestingly, the other human body region to gain an evolutionary increase in its cortical map space is the face.

In humans, the hand is innervated by three mixed nerves arising from many spinal segmental levels (shown in parentheses): the median (C5 to C8), radial (C6 to C8) and ulnar (C8 to T1) nerves all pattern in a topographical manner (Bas and Kleinert 1999). Within the hand are concentrations of all the pain, temperature, and mechanosensory structures. These endings are found in other body skin regions, but not at the same high density as in the hand and finger region. Mechanosensory information reaches the central nervous system through the dorsal root ganglion, to brainstem nuclei neurons, then the thalamus ventral posterior nuclear complex, finally reaching the somatic sensory cortex. (For further discussion, see Chapter 17: Brain, Spinal Cord, and Sensory Systems).

Meissner corpuscles are a specialised rapidly adapting (RA) sensory touch receptor, type I RA mechanoreceptor, located in the dermis superficial region. It extends within the dermal papillae (peg), bringing it close to the overlying epidermis, to be activated by its own distortion. These receptors are abundant within the fingertip region. This conversion of mechanical energy into electrical signals is mediated by stretch-gated ion channels (Vega et al. 2009). While males and females do not appear to differ in the number of Meissner corpuscles within their fingers, the generally larger fingers of males leads to a lower overall density (Dillon et al. 2001). Clinically, these corpuscles, along with Pacinian corpuscles, have been used for pathological diagnosis of some peripheral neuropathies and neurodegenerative diseases (Vega et al. 2012).

Pacinian corpuscles (Vater-Pacini) are a specialised RA sensory pressure receptor, type II RA mechanoreceptor, located in the dermis deep region and sensitive to vibration and pressure (Biswas et al. 2015). Primates show a similar distribution of these sensory structures within their hands (Kumamoto et al. 1993). These sensory structures have been found in non-skin locations throughout the human body such as the joints, mesentery, mesocolon, urinary bladder, and lymph nodes (Feito et al. 2017). In the ageing human fingers, these sensory structures undergo morphological changes and decrease significantly in number, suggesting a contribution to the loss of sensitivity with ageing (Kobayashi et al. 2017).

Merkel cells are a specialised slowly adapting (SA) sensory light touch receptor, type I SA mechanoreceptor, and can also be the cellular origin of a rare human skin cancer, Merkel cell carcinoma. This carcinoma develops in sun-exposed skin (Amaral et al. 2017). Merkel cells have a fascinating origin and a still unfolding role in skin function. They were thought to be neural crest in origin, though research now shows that embryonically they are derived from skin lineage and in the adult are replaced from the epidermal stem cell population (Morrison et al. 2009; Van Keymeulen et al. 2009). Furthermore, there is evidence that suggests that they participate in other processes in skin disorders, react to histamine, and release vasoactive intestinal peptide (Boulais et al. 2009a). In rodent species, these cells release met-enkephalin (Hartschuh et al. 1983) and act separately as osmoreceptors, with hypo-osmolarity activating Merkel cells (Boulais et al. 2009b).

Ruffini's corpuscles are a SA sensory ending receptor, type II SA mechanoreceptor, sensitive to pressure from skin stretch, involved with the control of finger position. Structurally similar to the Golgi tendon organs, several Ruffini-like structures, including an association with hair follicles, have been identified within human skin. In skin regions of different species, the distribution of these receptors is still being elucidated (Fleming and Luo 2013).

In hairy human skin, the hair follicles are associated with lanceolate endings, which are also found in animals with a broad range of hair follicle types for fur and whiskers (Li and Ginty 2014). Lanceolate endings are associated with dorsal root ganglia low-threshold mechanosensory neurons (Aβ, Aδ, and C). Each may have a different role. For example, Aδ fibres appear to act as mechanosensory nociceptors, acting in rapid response to hair pulling (Ghitani et al. 2017). While mechanosensory input from hairs is still relevant in humans, the decrease in body hair density compared with fur-covered mammals implies a comparative decrease in this form of sensory input compared with other species. This may have relevance to changes in somatosensory cortex mapping from these receptors.

The sensation of skin pain (nociception) has evolved as a defensive/protective mechanism to prevent ongoing tissue damage. However, this mechanism can be inappropriate in humans. Overactive pain can be seen with chronic pain states and is a significant clinical problem (Sommer 2016). Phantom pain and phantom limbs occur following limb amputations. Underactive pain can be seen with diabetes and occurs as diabetic foot with a lack of pain sensation (Chantelau 2015). The central sensory pathways for all (mechanosensory, pain, and temperature) begin at the dorsal root ganglia, but after that diverge. Pain and temperature take the same pathway through the spinothalamic system. The face is unique, having its own trigeminal central pathway, perhaps reflecting the evolutionary importance of the face to human development.

8.6.7 Cornea

The cornea is a vision-specific, specialised sensory epithelium differentiating mainly in the postnatal period. It arises initially from cranial ectoderm adjacent to the lens placode and forms a presumptive corneal epithelium. Later, neural crest cells migrate between the lens and presumptive structure to form both the corneal endothelium and the stromal fibroblasts (keratocytes) (Lwigale 2015). Neural crest development in humans, reptiles, and birds differs from that seen in other species such as rodents, cats, rabbits, and cattle. The human cornea also differs mechanically from the most commonly used model porcine cornea (Zeng et al. 2001). The human cornea can be damaged by many environmental insults, including UV radiation, and UV-blocking glasses have been shown to prevent this type of damage and the associated loss of visual performance (Liou et al. 2015).

8.7 The Evolved Function of Skin Colour

Skin colouring is thought to have evolved in hairless humans as a protective mechanism against UV damage to skin cell DNA (Jablonski and Chaplin 2010). This may be an oversimplification; latitude and climate were important but so too was diet, lifestyle, and shifting metabolic priorities (Barsh 2003; Elias and Williams 2016).

Human skin colour shows a wide genetic variation, from dark (black) in Africans to light (white) in Celtics. Such a broad variation is skin colour is not seen in any other species and may call into question application of some animal models for human skin disorders. Note that in most individuals the palms of the hands and the soles of the feet are hypo-pigmented compared with other body regions (Yamaguchi et al. 2004).

In general, the more melanin that is present, the lower the percentage of light reflected from the skin. Colour can be influenced endogenously by other skin properties such as hormonal and neural stimulation, erythema, haemoglobin levels, and other chromophores, as well as the level of blood oxygenation (Hajizadeh-Saffar et al. 1990). Colour can be influenced exogenously mainly by UV radiation. Melanin levels can be altered physiologically during pregnancy, in ageing, and by certain drugs (Costin and Hearing 2007). A recent genetic study has used the wide range in pigmentation levels in different African populations to identify pigmentation-associated loci near SLC24A5 (OMIM 609802), MFSD12, DDB1 (OMIM 600045), TMEM138 (OMIM 614459), OCA2 (OMIM 611409), and HERC2 (OMIM 605837) (Crawford et al. 2017). This study also shows the light pigmentation variant at SLC24A5, a transmembrane potassium-dependent sodium/calcium exchanger, which was introduced into East Africa by gene flow from non-Africans. Evolutionarily significant mutations occurred within the melanocyte-specific regulatory regions near DDB1/TMEM138 correlated with expression of UV response genes under selection in Eurasians (Crawford et al. 2017).

There have been several subjective tools used to classify and distinguish natural skin colour/pigmentation in health and disease. These include the Felix von Luschan chromatic scale (0–36), Fitzpatrick scale (I–VI), Taylor hyperpigmentation scale (1–15) (Taylor et al. 2005, 2006), Skin Tone Colour Scale (Konishi et al. 2007), and a number of proprietary scales. More recently, narrow-band and visible light reflectance spectrophotometry has developed as the most objective quantification of human skin colour (Bjerring and Andersen 1987) and can be used to measure skin post-mortem changes (Sterzik et al. 2014).

Within the skin, melanocytes interdigitate with the basal layer of the epithelium and transfer melanin to the keratinocytes. The major role of this transferred pigment is to protect the (basal) cells within the epithelium from DNA damage due to UV radiation, a known mutagen. This protection can be either directly through melanin or through its precursor molecules (Schmitz et al. 1995).

Melanocytes produce melanin as an oxidative derivative of the non-essential amino acid tyrosine (4-hydroxyphenylalanine). It is also the pigment that colours both hair and the eyes. Two types of melanin can be formed by processing tyrosine: pheomelanin, a cysteine-rich red–yellow form; or eumelanin, the black-brown form (Figure 8.8). Pheomelanin synthesis occurs as a loss-of-function allele of the melanocortin 1 receptor.

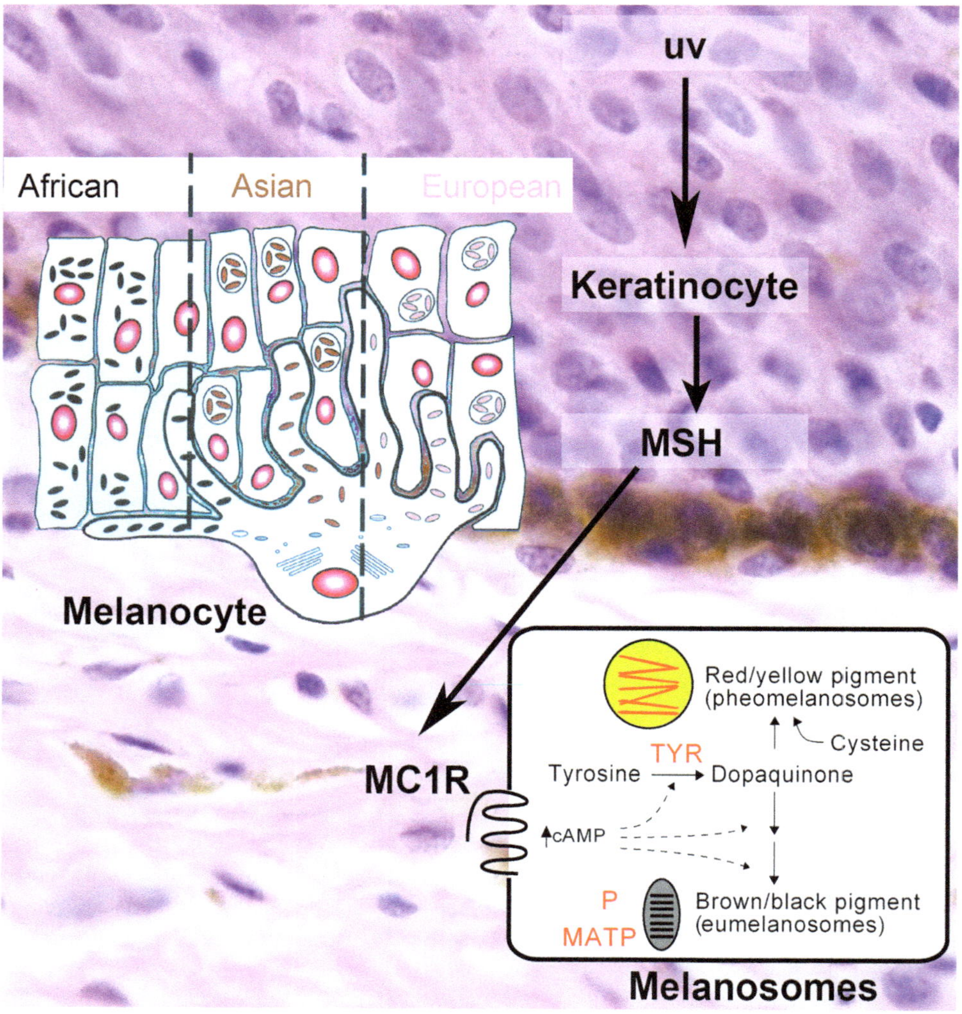

FIGURE 8.8 Melanin biosynthesis and the major forms generated. UV light stimulates keratinocytes to produce melanocyte-stimulating hormone (MSH), which binds with melanocytes melanocortin 1 receptor (MC1R). Activation of MC1R then promotes pigment synthesis of eumelanin (brown/black) at the expense of pheomelanin (red/yellow). Oxidation of tyrosine by the tyrosinase enzyme (TYR) is required for synthesis of both these pigment types. Melanosome pigment synthesis levels can be affected by membrane-associated transport protein (MATP) and the pink-eyed dilution protein (P). These melanosomes are then transported and exchanged back to keratinocytes. Skin colour is thus dependent upon melanosome size, number, and dispersion within the keratinocytes. Finally, melanin within keratinocytes, seen as brown in the background histology image, acts to block the damaging effects of UV radiation upon cells of the skin. Note that a similar pigment transfer process occurs within the hair follicle to determine hair colour.

Source: Modified from Gregory S. Barsh, What Controls Variation in Human Skin Color? *PLoS Biology*, 1(3) e91, Figure 1, doi.org/10.1371/journal.pbio.0000091.

Mutations in genes encoding melanosome transport-related molecules, such as MYO5A, RAB27A, and SLAC-2A, have also been reported to cause a rare human pigmentary disease known as Griscelli syndrome (OMIM 214450, 607624, 609227). The disorder shows both hair and skin pigment dilution, with the presence of hair shaft pigment clumps and melanosome accumulation in melanocytes (Thomas et al. 2009).

There are many social mores associated with skin colour, both lighter and darker within different cultures. In many Asian cultures, there has been an historic and modern focus on development of pharmaceutical 'skin whitening agents'. These are generally designed to decrease melanin production by inhibition of tyrosinase (Pillaiyar et al. 2017). Conversely, in some western cultures darkening of the skin by 'tanning' has been considered cosmetically attractive (Miyamura et al. 2011). Finally, both ancient and modern cultures continue to colour the skin for various (decorative) purposes by ochres, tattooing, and cosmetics. Note the cosmetic (beauty) industry worldwide today is one of the largest, with an estimated US$ 445 billion in annual sales. The long-term biological effects of these cosmetics, which include many newly developed nano-materials and chemicals, have yet to be established (Choksi et al. 2010).

8.7.1 Geographic and Seasonal Gradations of UV Exposure

Human skin pigmentation may have evolved as a balanced adaptation to high UV radiation, geographic gradations, and an opposing requirement for vitamin D3 in low-UVB geographic regions (Jablonski and Chaplin 2010). Near the equator we see high levels of both UVA and UVB, leading to dark constitutive pigmentation. Away from the equator, near the poles, lower and seasonal UV radiation leads to lighter constitutive pigmentation. Between the two geographic extremes, in the intermediate latitudes there are seasonally high loads of UVB, which lead to moderate constitutive pigmentation with the ability for facultative pigmentation or tanning (Jablonski and Chaplin 2010). There has been historic and ongoing controversy as to whether skin cancer has contributed as the main selective agent (Greaves 2014; Jablonski and Chaplin 2014) or whether other factors such as compromising the skin barrier function contributed to the evolution of pigmentation (Elias et al. 2010). There have been other evolutionary theories, for example UV degradation of folate in the skin. A trial exposing individuals to UVA radiation over an extended period and then measuring blood folate levels showed no differences in folate levels to those of control individuals (Gambichler et al. 2001).

8.7.2 Protecting Exposed Skin from UV Damage

Prolonged skin sun exposure can cause photo-ageing or dermatoheliosis (Jackson 2001) which can affect all layers and components of the skin:

- epidermis—solar keratosis (actinic keratosis);
- dermis—solar elastosis (actinic elastosis);
- blood vessels—telangiectasia;
- sebaceous glands—solar comedones (senile comedones);
- melanocytes—brown patches, diffuse or mottled.

Our current ageing population has experienced sun exposure during their lifetime and we are seeing an increased prevalence of these photo-ageing effects, combined with the general ageing properties of the skin. With the recent development of new effective sunscreens and a changing awareness, and thus exposure to sun, it will be interesting to see the differences that occur in this new ageing population.

8.7.3 Tanning Response to Exposure

As a suggested biological role for facultative skin pigmentation, 'tanning or sun tanning is to increase skin protection against subsequent UV radiation exposure by increasing melanin levels' (Miyamura et al. 2011). In some western cultures, darkening of the skin is considered cosmetically attractive. Historically, tanning was considered associated with outdoor manual work and had associated lower socioeconomic connotations. During the last century, beginning in the 1920s and 1930s, tanning began to be associated with fashion and an increase in leisure time (Martin et al. 2009). In the 1980s, first-world countries began using indoor 'tanning beds' where artificial UV radiation can be up to ten times more powerful than sunlight. This tanning trend, along with changes in the ozone layer, has led to a significant increase in UV-associated skin cancers (Karagas et al. 2002). In the 2000s, with awareness of the carcinogenic effects of sun exposure, a new pharmaceutical sunscreen industry has developed. There is a growing search for sunscreens that have the beneficial effect of blocking UV related to skin damage but not that associated with vitamin D production.

Animal (Fell et al. 2014) and now human (Fell et al. 2014; Jussila et al. 2016) studies have shown an increase in circulating β-endorphins, known hormones associated with addictive behaviours, following UV exposure. UV radiation stimulates keratinocytes to produce β-endorphin (Wintzen et al. 2001), an endogenous opioid neuropeptide, along with other hormonal compounds, such as pro-opiomelanocortin (POMC) and α-melanocyte-stimulating hormone (α-MSH) (Jussila et al. 2016). In this context, β-endorphin can be seen as enhancing the neurological 'pleasure' or sun-seeking response and perhaps the concept of sunlight exposure as being 'good for the health'. It is also possible that the endorphin effect may have enhanced evolutionary vitamin D biosynthesis (Fell et al. 2014).

The longer wavelength of UVA allows deeper penetration into the dermis, while the shorter wavelength of UVB penetrates the epidermis. The shortest wavelength UVC is absorbed entirely by atmospheric ozone, though this may also change with ozone depletion. Both UVA and UVB have tanning effects, and their levels vary both geographically and seasonally. The UVA tanning effect oxidises pre-existing melanin or melanogenic precursors, inducing an immediate tanning effect and persistent pigment darkening. The UVB tanning effect requires activation of melanocytes, delaying tanning by several days or longer. Contrasting these effects, several studies have shown that UVA tanning appears to provide no beneficial skin photo-protective effect (Miyamura et al. 2011; Coelho et al. 2015).

UV radiation-stimulated keratinocyte production and secretion of α-MSH is regulated by p53 (TP53) (Cui et al. 2007; Chang et al. 2017). This is a novel role for p53 which is generally associated with tumour suppression and cell cycle regulation in all other tissues. This p53 effect appears to be mediated specifically by narrow band UVB (NB-UVB) radiation (Jussila et al. 2016).

Sensitivity to UV is generally more pronounced in the fair-skinned, with less eumelanin, than darker-skinned people. Both types have about the same level of pheomelanin. Sensitivity to UV radiation can lead to overexposure in the form of sunburn, a transient erythema associated with inflammation and neutrophil infiltration (Andersen et al. 1992). The sensitivity to UV radiation is quantified by the minimal erythemal dose (MED), defined as the dose of UVB radiation that produces a perceptible erythema 24 hours after administration (Heckman et al. 2013). The MED dose will also depend on skin type (I–V). In countries like Australia on a summer's day with high UV levels, sunburn can occur in some skin types in as little as 15 minutes. The associated inflammatory response is complex, with some contribution from cyclo-oxygenase-derived prostaglandin E2 (PGE2) stimulating dermal vasodilatation (Nicolaou et al. 2012). A recent study has shown that oral vitamin D supplementation can decrease inflammation associated with sunburn (Scott et al. 2017).

8.8 Functions of the Skin in Hormone Production

8.8.1 Vitamin D

Vitamin D is the principal regulator of calcium homoeostasis, required for bone development and many other essential physiological processes. The evolutionary movement from a marine to a terrestrial (sun) environment required both more body support and increased mobility through the development in animals of a significant musculoskeletal system. There was also less available environmental calcium. The skin evolved as a key component in the production of vitamin D for calcium regulation and this broadened further into many other physiological functions. Placental animals require vitamin D also for many developmental roles.

Calcium is the key to bone development, and without it humans develop rickets. The link was first identified in the nineteenth century in the search for a cure for rickets with cod-liver oil, which is rich in vitamin D. The regulation of calcium levels by vitamin D is now described as the 'classical' role of this vitamin. A small proportion of our total vitamin D was derived from what we eat in plants and yeast as vitamin D_2 (ergocalciferol) as well as in fatty fish, beef liver, and egg yolk as vitamin D_3. The modern diet also gets vitamin D from fortified dairy and dietary supplements.

There is a growing understanding of the relevance of vitamin D for placentation and pregnancy (Pérez-López 2007; Liu and Hewison 2012). For a placental animal, vitamin D has a role in trophoblast function for implantation and maternal decidualisation (Tamblyn et al. 2015). A study has shown that the current recommended vitamin D intake guideline for diet and supplements during pregnancy is not adequate to achieve vitamin D sufficiency in many pregnancies (Aghajafari et al. 2016). The vitamin D developmental role extends beyond placentation, with potential key roles in neural, bone, and renal development.

The vast majority is formed as vitamin D_3 (cholecalciferol) by the action of sunlight on the skin (Holick et al. 1980). The skin, liver, and kidney organ axis is vital to the production of active vitamin D. Within both the epidermis and dermis, cells exposed to UVB radiation (between 290 and 315 nm) cleave the B-ring 7-dehydrocholesterol to previtamin D_3 (ergosterol). Vitamin D is described as a 'secosteroid'—a type of steroid with a 'broken' ring. This unstable molecule is rearranged at the cell surface to the more stable form as vitamin D_3. This form is then released by the cells, and within the dermis enters the capillary bed to be both bound and transported by the vitamin D binding protein (DBP).

The DBP transport first target organ for this form of vitamin D is the liver. Within the liver it is converted to 25-hydroxyvitamin D [25(OH)D], still as an inactive form but the form clinically measured in humans to assess vitamin D status.

The DBP next target organ after the liver is the kidney, where 25(OH)D is converted to 1,25-dihydroxyvitamin D [1,25$(OH)_2$D], the biologically active form. The enzyme that produces the active 1,25$(OH)_2$D has been identified in other tissues and cells such as the skin, colon, prostate, brain, lungs, monocytes, macrophages, and placenta (Tamblyn et al. 2015).

It is the 1,25$(OH)_2$D form that can bind the sole vitamin D receptor (VDR). The VRD is a nuclear receptor belonging to the class II steroid hormone family, related to both the retinoic acid and the thyroid hormone receptors. 1,25$(OH)_2$D binding to VDR displaces a repressor, leading to receptor conformational change, binding retinoid X receptor, an activator, and several other proteins to form a transcription complex. The transcription complex binds vitamin D-responsive elements (VDREs) found upstream in the many target genes.

This single VDR is expressed in many different tissues (Jones et al. 1998; DeLuca 2004; Holick 2011). In the gastrointestinal mucosa of the intestines, epithelial cell VDR binding increases the absorption of calcium and expression of calcium regulatory proteins. In these cells, it also stimulates absorption of phosphorus. In the skeletal bone, osteoblast VDR binding increases receptor activator of NF-κB ligand expression, leading to osteoclast release of calcium from the bone matrix. In the endocrine system, both the parathyroid glands and pancreas respond. In the parathyroid glands, VDR binding negatively feeds back and regulates parathyroid hormone (PTH) production. In the pancreas, islet β-cells are stimulated to secrete insulin. In the kidneys, distal renal tubule VDR binding increases the tubular reabsorption of calcium. In skeletal muscles, vitamin D supplementation can be useful in preventing injuries in the aged, improving muscle performance and therefore reducing falls in vitamin D deficient older adults (Ceglia 2009). In the immune system, vitamin D has been shown in studies to have an anti-inflammatory effect (Calton et al. 2015).

In the skin, VDR mutations affect hair formation and result in alopecia, impacting on keratinocyte stem cells (Cianferotti et al. 2007). In melanocytes, population polymorphisms in both VDR and pigmentation genes are associated with an increased risk of developing cutaneous malignant melanoma (Kosiniak-Kamysz et al. 2014).

What about the neurological relationship? There should be more research than there currently is on the linkage of vitamin D production in humans to the diurnal rhythm (Jones et al. 2017). There is a potential neural relationship, linked to the pituitary and hypothalamus, yet to be fully explored. Vitamin D levels have been observed in depressed patients and those with other psychiatric disorders (Grudet et al. 2014). VDREs have been identified within genes that regulate serotonin and tryptophan hydroxylase. A potential association with higher-latitude seasonal affective disorder has also been identified (Stewart et al. 2014; Kerr et al. 2015).

Mutations in VDR in children with autism have been recently described (Cieślińska et al. 2017; Cui et al. 2017). Treatment with vitamin D of mouse models for autism during pregnancy shows improvements in the offspring (Vuillermot et al. 2017). Rat models low in vitamin D during development have poor spatial learning (Al-Harbi et al. 2017). Rats have previously also been shown to require vitamin D for the normal development of their dopaminergic systems (Kesby et al. 2013).

Vitamin D deficiency has been implicated in several skin diseases, including atopic dermatitis and psoriasis, due to its important role in both the innate and adaptive immune system, though a UK study has questioned the atopic dermatitis association in their study population (Manousaki et al. 2017).

8.8.2 Melatonin

The pineal gland is the main source of melatonin, a hormone that is the body's diurnal clock. Melatonin levels are low during daylight hours and increase at night. It has many different actions in different tissues; related to this are cyclic changes, and antioxidant and free radical scavenging properties. Recent therapeutic uses have been for the treatment of asynchrony associated with 'jet lag' sleep disturbances and depression. In animal skin, but not human, the hormone can lighten the pigmentation and regulate hair growth.

In humans, melatonin and its metabolites can be produced within the epidermis (Kim et al. 2015) and scalp hair follicle (Fischer et al. 2007). Skin melatonin production may affect the barrier function, and levels vary genetically with race and gender, and also with age (Kim et al. 2015). The hair follicle production may show some correlation to the hair cycle.

8.9 Hairlessness in Humans

The hair follicle is a skin organ composed of epidermal keratinocytes and mesodermal papilla cells. Postnatally, each hair follicle goes through a dynamic cycling of activity. The cycle consists of periods of active growth and hair fibre formation (anagen), followed by apoptosis-driven involution (catagen), relative resting, and hair shedding (telogen) (Botchkarev and Fessing 2005).

Humans are often referred to as 'hairless' mammals, referring to the lack of terminal hairs, while other mammals are often covered in a thick layer of these hairs, though, to be accurate, our bodies are still covered in the small unpigmented vellus hairs. These are found in all body regions, except for on palms and soles, and at a range of different regional densities. Terminal hairs are most visible on the adult scalp and with the secondary sex characteristics of armpits, groin, and male face.

In other fur-covered species, such as the mouse, there are three separate prenatal waves of prenatal hair development (Duverger and Morasso 2009). This hair in other species is generally associated with protective, sensory, and thermal insulation functions.

Anthropologists continue today to argue the evolutionary benefit of hairlessness. One theory even suggests that our brain development was predisposed upon this evolutionary hair loss, due to early hominids' dietary requirements for these two tissues (Dror and Hopp 2014). Historically, other theories have ranged widely from sexual selection, an aquatic evolutionary period, bipedality, ectoparasites, photobiomodulation (Mathewson 2015), acid mantle formation, wound healing, vitamin D production, and thermoregulation. The most accepted theory is that of developing a thermoregulatory advantage (Schwartz and Rosenblum 1981; Sandel 2013); that is, there is a reciprocal relationship between the amount of hair and body mass and the number of eccrine sweat glands. This would be required to support strenuous activity and the high metabolic rate of humans to support neurological and other functions.

8.10 Social Changes and Clothing

Through most of human history, clothing was quite different between cultures and eras. One common feature was that most of the body was covered, except for the face and hands. The exceptions were cultures where social, work, or environmental conditions, such as heat, meant that clothing was minimised. Historically, tanned skin was associated with manual labour and therefore both social and economic status. This began to change in western culture in the early twentieth century and it has not been as prevalent in other cultures (Chang et al. 2014). First, medicine began to link the lack of sunlight as a cause of rickets and sunlight inhibited the growth of microorganisms and was used as treatment for a range of diseases (tuberculosis, anaemia, Hodgkin's disease, dermatological disorders, and wound healing). Later, an increase in leisure time, travel, and more revealing fashions in turn led to more sun exposure and tanning (facultative tans). Tanning became a trend and clothing could both aid exposure as well as 'display' the tan. The increase in skin cancer paralleled these changes in exposure (Chang et al. 2014).

Modern clothing materials also differ from those in the past. The twentieth century saw the development of synthetic materials and new types of fabrics. Patients with dermatological disorders had previously been recommended to use natural fibres like cotton, rather than synthetics, due to its low irritability and breathable structure (Brambilla et al. 2016).

8.11 Environmental and Genetic Impacts on the Skin

There are far too many dermatological conditions that have evolved specific to humans to cover in any depth within the scope of this chapter. Some selected topics are covered in relation to association with immunodeficiency, skin cancers, and ageing. For a compendium of disease, the reader is therefore referred to the extensive online resources of DermNet NZ (https://www.dermnetnz.org) and for reviews of clinical trials relating to skin conditions to Cochrane Skin (http://skin.cochrane.org).

8.11.1 Modern Diseases, HIV/AIDS, and the Skin

The skin surface is normally populated by a large variety of bacteria, archaea, fungi, and viruses. Some of these species are infective, while others are generally commensal, deriving food or other benefits without hurting or helping us. In contrast, several skin-related infections have increased in severity since the 1980s with the occurrence of human immunodeficiency virus/ acquired immune deficiency syndrome (HIV/AIDS) infections, though these do not directly affect the skin. This changed the landscape of what had been the typical historic outcomes of these skin infections. The common initial feature of most of these viral and fungal infections is a compromised outer layer of the skin (stratum corneum) or mucosa (stratified squamous epithelium). A few examples of these viral infections are described in this section.

8.11.1.1 *Herpes Simplex Virus*

Viruses are classified by the International Committee on Taxonomy of Viruses (ICTV) (https://talk.ictvonline.org). Herpes simplex virus belongs to the order *Herpes virales*, family *Alpha herpes virinae*, and genus *simplex virus*. Herpesviruses have evolved alongside animals and have their specific species targets; the skin is the barrier they need to overcome. Three groupings of herpesvirus exist: viruses with mammals, birds, and reptiles as natural hosts; viruses of amphibians and fish; and a single invertebrate herpesvirus. Mammalian and avian herpesviruses were descended from a common ancestor (McGeoch et al. 2006).

According to World Health Organization (WHO) 2015 retrospective estimates, 67% of the world's population are infected with herpes simplex virus type 1 (HSV-1) (Looker et al. 2015). Herpes simplex virus is categorized into two types: HSV-1 and HSV-2. Both are highly infectious and incurable. HSV-1 is primarily transmitted by oral–oral contact and causes orolabial herpes or 'cold sores' around the mouth. HSV-2 is primarily transmitted sexually through skin–skin contact, causing genital herpes. Recently, though, HSV-1 has been identified as an increasing cause of genital herpes (Looker et al. 2015). With immunocompromised people, such as those with advanced HIV infection, HSV-1 can have both more severe symptoms and more frequent recurrences. HSV-1 was identified as one of the first HIV-related opportunistic infections (Wald et al. 1997). A rare form of neonatal herpes can occur when the infant is exposed to HSV in the mother's genital tract during delivery.

HSV-1 is a double-stranded DNA virus with linear DNA and unpaired, complementary nucleotides at each terminus. Initially the virus infects the skin; subsequently it spreads to the nervous system and the peripheral sensory system, establishing a latent infection. The major physical barrier to any viral invasion is the skin via intact stratum corneum, and therefore minor injuries can aid transmission. Reactivation can then occur with sun exposure, minor trauma, surgery, upper respiratory tract infections, hormonal factors, and emotional stress. Rhesus macaques can be infected by the same HSV-1 and display similar symptoms and viral latency in nervous tissues (Fan et al. 2017). In comparison, mouse models of HSV-1 infection have shown a much wider tissue distribution of the virus and lethality (Khoury-Hanold et al. 2016).

8.11.1.2 *Kaposi Sarcoma*

Kaposi sarcoma herpesvirus (KSHV), described historically by Moritz Kaposi in 1872 as an 'idiopathic multiple pigmented sarcoma of the skin', is more common in men than women. The virus is taxonomically human gamma-herpesvirus 8, which can induce tumours and is classified as a class I carcinogen, an old human virus that has coevolved with human populations (Mariggiò et al. 2017). KSVH in the general population is transmitted mainly during childhood, and became a dominant infection with the rise of immunocompromised HIV/AIDS infections.

8.11.1.3 *Molluscum Contagiosum*

Molluscum contagiosum (MC) is a fairly common skin infection. It is an infectious dermatosis spread by skin–skin contact that presents in children and immunocompromised individuals (Forbat et al. 2017). The MC virus is generally harmless and in healthy people will disappear on its own over a long period of time. In immunocompromised HIV/AIDS people, the infection can spread over the entire body and last for a much longer time.

8.11.1.4 Candida *Fungus*

The human skin microbiota has a large variety of fungi, some species of which tend to be topographically located (Findley et al. 2013). Some species can become pathogenic, for example, *Candida tropicalis*, *Candida parapsilosis*, and *Candida orthopsilosis*, while symptomatic skin infections are mainly due to *Candida albicans*.

Hippocrates first described oral candidiasis in 400 BCE, and we now know this fungal infection is caused by more than twenty yeasts that belong to the genus *Candida*. Infections are more common in internal or genital environments and the most common yeast is *C. albicans*. Candidiasis in the mouth or throat is called oropharyngeal candidiasis or 'thrush'. Candidiasis in the vagina is commonly referred to as a 'yeast infection'. The skin has resident fungal species including *C. albicans*. The dry skin barrier is more effective and has efficient defence mechanisms, with both residential non-immune and immune cells and immune cells rapidly recruited to sites of infection (Kühbacher et al. 2017). When they do occur, *Candida* skin infections cause skin thickening (hyperkeratosis) and redness (erythema). Candidiasis associated mainly with the mucus membranes was identified very early as the most common opportunistic fungal infection in HIV-positive patients (Imam et al. 1990).

8.11.1.5 *Photodermatitis and Prurigo Nodularis*

Dermatitis that is provoked by electromagnetic radiation within mainly UV and occasionally the visible spectra is described as photodermatitis. The two main subtypes are immunologically mediated and the phototoxic or photoallergic (exogenous) type (Coffin et al. 2017).

A common immunological form of primary photosensitivity that mainly occurs in young adult women in temperate climates during spring and summer is described as polymorphic light eruption, polymorphous light eruption, or prurigo aestivalis. 'Polymorphic' refers to the topographical appearance of many different rash forms but always located in the same place on an individual. UVA is the trigger and leads to impaired skin T-cell function, altered production of cytokines, and a reduced immune suppression in the skin. Several contributory factors have been implicated such as vitamin D insufficiency, hormonal (oestrogen) levels, and changes in either microbiome and/or antimicrobial factors. Oral vitamin D supplementation has been used therapeutically in patients to avoid sun exposure. The rash appears hours to days after exposure and rarely affects the face. The rash appears on newly light-exposed areas such as the low neckline, forearms, backs of hands, lower legs, and feet (Gruber-Wackernagel et al. 2009). With climate change and alteration worldwide in UV radiation levels, this form of photosensitivity is likely to increase.

Prurigo nodularis (PN) is the most severe form of prurigo, with intensely itchy spots. It is of unknown cause, affects adults equally of both sexes, and follows sensitisation of the spinal nerves. Nodularis refers to the thickened (lichenified) nodules. There is a genetic linkage, with a family history of atopic dermatitis, asthma, or hay fever.

8.11.2 Now-Common Infections of the Skin

The prevalence of common non-malignant skin conditions in humans varies by age, genetic grouping, reproductive status, geography, and climatic conditions. The non-genetic conditions can relate to skin microbiota (bacterial, viral, and fungal) that have evolved with and been selected by human evolution. Note that the term 'flora' used to be used for this mixture of microorganisms and viruses. This microbiota resides on the skin usually without any effect, but damage or alteration to the skin barrier function can lead to opportunistic pathogens and a variety of skin conditions. The skin microbiota barrier function is also contributed to by the film of antimicrobial peptides, enzymes, and lipid matrix present in the acid mantle (Coates et al. 2014).

Following barrier breakdown, there is often a significant contribution from the human immune response, which itself can cause recurring common skin conditions. (For further discussion, see Chapter 10: Immune System) There has also emerged in the late-twentieth century many infectious diseases caused by microorganisms that would rarely cause disease in normal, healthy immunocompetent hosts (Casadevall and Pirofski 2003), demonstrating that changes within the human population have in turn offered 'opportunities' for previously docile microbiota. Many new materials and industrial contaminants developed within the twentieth century may also have impacted upon our immune response and interaction with these new pathogens (Ge et al. 2017; Honda et al. 2017; Hong et al. 2017; Tsai et al. 2017). This leads to the question: Do these twentieth-century industrial-age materials, products, and pollutants, as they enter our environment and food chain, become not only our microbiota, but also our twenty-first-century Darwinian selection agents?

The twentieth century has seen huge advances in medical technology, a by-product being the increased use of catheters for the delivery of drugs and homoeostasis for patients. Catheters have broken down the skin barrier in a new way and allowed direct transmission across the skin barrier, introducing skin and other microbiota—for example, with fungaemia (fungi or yeasts in the blood) and bacteriuria (bacterial infections in the urinary tract) (Nicolle 2014). The most serious though are the central-venous-catheter-related bloodstream infections caused by mainly Gram-positive bacteria (Gahlot et al. 2014). A 2011 Indian hospital study identified the catheter infection prevalence by mainly skin microbiota as: *Staphylococcus aureus* 40%, *Pseudomonas aeruginosa* 16%, *coagulase negative staphylococci* 8%, *Escherichia coli* 8%, *Klebsiella pneumoniae* 8%, and *Acinetobacter baumanii* 4% (Parameswaran et al. 2011).

In a United States survey of dermatological conditions (Wilmer et al. 2014), dermatologists identified the top ten skin conditions as: acne, actinic keratosis (solar keratosis), non-melanomic skin cancers, benign tumours, contact dermatitis, seborrheic dermatitis, viral warts, psoriasis, rosacea, and epidermoid cyst.

Table 8.6 Self-Reported Common Skin Conditions in a Rural Australian Community

Skin condition	Description	Prevalence (%)
Seborrheic keratosis (basal cell papilloma)	Harmless warty spot, with ageing	58.2
Cherry angiomas	Campbell de Morgan spots, benign tumours with ageing	54.4
Acne	Prevalent in adolescents and young adults, all races and ethnicities	12.8
Culture-positive tinea	Fungus (ringworm) infection	12.0
Seborrheic dermatitis	Relapsing form of eczema, affecting the scalp and face	9.7
Asteatotic dermatitis (eczema craquelé)	Very dry skin affecting the limbs (mainly lower) and trunk	8.6
Human papillomavirus (HPV)	Warts	7.1
Atopic dermatitis	Eczema, more common in children	6.9
Psoriasis	Immune-mediated inflammatory disease	6.6

Source: Reproduced from Anne Plunkett, Kate Merlin, David Gill, Yeqin Zuo, Damien Jolley, and Robin Marks, The frequency of common nonmalignant skin conditions in adults in central Victoria, Australia, *International Journal of Dermatology*, 38 (12), 901–8, doi: 10.1046/j.1365-4362.1999.00856.x. Copyright © 2001, John Wiley and Sons.

A smaller Australian survey study of an adult rural community (Plunkett et al. 1999) allowed self-identification of some common non-malignant skin conditions (Table 8.6). Many of these conditions are associated with both an ageing population and ageing skin. With the current ageing population, these conditions are thus increasing. Changing UV exposure means that actinic keratosis (solar keratosis) is increasing, associated with sun-damaged skin, and is considered a precancerous form of cutaneous squamous cell carcinoma.

Some skin conditions are more common in black skin, in children (tinea capitis), and in adults (acne keloidalis nuchae), and others are less common (skin cancer) (Child et al. 1999; Dlova et al. 2015; Ogunbiyi 2016). Acne keloidalis nuchae is also known as folliculitis keloidalis nuchae, a chronic form of hypertrophic or keloid-like scarring seen mostly in men of African descent (Ogunbiyi 2016).

8.11.3 Skin Cancers

There are many research articles, reviews, and books written on the topic of skin cancers. This chapter refers to the topic in relation to the changing environment and human population genetic background. In human evolution, decreased hair coverage/protection has resulted in an increased likelihood of UV radiation causing mutations in the skin cell DNA. In the case of melanoma, the most common UV-induced DNA nucleotide mutational change is from cytosine (C) to thymine (T). The important recent evolving correlation is the increasing incidence with increasing UV radiation in the environment. Countries with a high UV index and an outdoor lifestyle have shown high prevalence of most skin cancers. However, the same sunscreens that are used to prevent carcinomas and sun damage can also impact negatively upon skin vitamin D production.

Worldwide, cancer outcomes differ significantly, with first-world countries having better early cancer identification and treatment regimes. On the other hand, developing regions in Africa, Asia, Central America, and South America now contribute more than 60% of cancer cases and 70% of cancer deaths (McGuire 2016).

The non-melanoma skin cancers are generally non-lethal and are classified as either a basal cell carcinoma or squamous cell carcinoma. These carcinomas do not generally spread to other tissues, though they use the same tumour, nodes, and metastasis (TNM) staging of other cancers (stage I– IV). Evidence suggests a predisposition to these non-lethal cancers may correlate with a higher risk for other types of cancers, particularly in the young (Small et al. 2016).

Basal cell carcinoma accounts for about 70% of non-melanoma skin cancers and is higher among Caucasians. These carcinomas occur on the body on parts that receive high or intermittent sun exposure such as the head, face, neck, shoulders, and back. The mutated cells are generated from the basal keratinocyte stem cell population that has not been shielded enough from UV radiation by the melanin within the skin, leading to DNA damage.

Squamous cell carcinoma accounts for the other 30% of non-melanoma skin cancers and is the second most common cancer among Caucasians in the United States. This cancer is formed from the keratinocytes in the more superficial layers of the epidermis. It is also due to sun damage and develops over weeks to months. An evolutionary relationship to immune genetics has been identified. Human leukocyte antigen (HLA) gene family associates with the initial carcinoma development, and there is differing ongoing tumour development between immunocompetent and immunosuppressed patients (Yesantharao et al. 2017).

Melanoma is a type of skin cancer that usually occurs on the parts of the body that have been overexposed to the sun, and rarely on non-exposed skin. The mutated cells in this cancer are derived from the skin melanocytes. Unlike the previous two cancers, this is an aggressive, metastasising, and lethal cancer. When metastases do occur, studies show a survival rate of less than 20% (Sandru et al. 2014). Secondary melanoma tumours can be found in many different tissue locations throughout the body. A recent large database study has shown that prognosis with these regional metastases is influenced mainly by the actual tumour stroma immunobiology (Akbani et al. 2015).

Malignant melanoma was historically classified into four major histopathological subtypes: superficial spreading melanoma, nodular melanoma, acral lentiginous melanoma, and lentigo maligna melanoma. Superficial spreading melanoma occurs on intermittently sun-exposed skin and is the most common subtype in populations of European descent. Since the 1970s, differences in melanoma incidence rates between racial origins have been identified. White skin populations, from different genetic backgrounds, showed a wide range of melanoma incidences and these were mostly higher in females, while non-white populations generally had a lower incidence of melanoma, though African descent had a higher sole-of-the-foot rate, with no sex difference (Crombie 1979). More recent studies have provided a much more detailed analysis of genetic correlation and mutational analysis (Ossio et al. 2017) and have shown a risk associated with skin colour (Hulur et al. 2017).

A problem still with many cancer population analyses is that DNA databanks are overrepresented by western white population contributions. For example, the Cancer Genome Atlas Network has just 2.4% of samples of admixed Hispanic origin and less than 5% of samples from patients of Asian or African descent (McGuire 2016). Since the late 1980s, this cancer has been increasing significantly and is currently thought to be doubling in incidence every 10–20 years. While some of these changes are due to better reporting rates, some increases must also be attributed to lifestyle changes and the increases in UV radiation levels.

Merkel cell carcinoma is a rare type of skin cancer, though similarly dangerous. This cancer is associated with polyomavirus infection and the tumour has a neuroendocrine phenotype (Schadendorf et al. 2017). Cancer incidence has been shown to vary according to geographic region (Amaral et al. 2017) and in recent years has been increasing, particularly in regions with high exposure to sunlight, a predominantly white population, and an outdoor lifestyle (Youlden et al. 2014).

8.11.4 Evolving Antibiotic and Fungicide Resistance

Medical antibacterial therapy developed in the mid-twentieth century and contributed to the major improvements in human health we have today. In the hospital setting with more than a hundred antibiotics now available, the gradual selection of new multidrug-resistant Gram-negative bacteria has become a huge public health issue (Cerceo et al. 2016; Friedman et al. 2016). Innocuous skin-borne microbiota, such as *Staphylococcus aureus* (golden staph), when they become a bloodstream infection have a 12-month death rate of between 20 and 35%. Antibiotic-resistant strains of the same bacteria, methicillin-resistant *Staphylococcus aureus* (MRSA), have an even higher death rate. One key contributor is the loss of the skin barrier function through either disease or medical procedure.

Another contributor may be the more common use of antibiotics and fungicides in agriculture, aquaculture, and human medicine which have led to the 'evolution' (selection) of new microbiota. There is now good evidence that antibiotic use in animals for food production and antibiotic resistance in humans is linked (Aitken et al. 2016). Since the 1980s, aquaculture for food production has grown significantly, from just 7% of global fisheries to today over 40%. Though we do not yet have good separate data on the potential changing antibiotic resistance effects (Done et al. 2015), agriculture and aquaculture have been shown to share the same antibiotic resistance mechanisms.

8.11.5 Mosquitoes and Malaria

Malaria caused by the *Plasmodium* parasite may have been historically limited to just a few populations associated with the tropical environments and with the appropriate female *Anopheles* mosquito hosts. Worldwide there are about 400 different *Anopheles* species, of which 30 are significant malaria vectors. During the last few centuries, human exploration and travel allowed the spread of this disease to new environs and communities. WHO reported in 2016 that in 91 countries there were 212 million cases of malaria worldwide and in 2015 more than 430,000 deaths. Genetics plays its part; the blood disease sickle cell anaemia has shown a differential outcome with malaria (For further discussion, see Chapter 9: Haematopoetic System). Heterozygotes for the sickle gene are protected to a degree against the danger of dying from malaria, while those homozygous for the same gene suffer sickle cell anaemia and are highly susceptible to its lethal effects (Luzzatto 2012).

Transmission of *Plasmodium* requires the female mosquito to break the skin barrier and inject the parasite as a by-product of feeding on blood from dermis blood vessels. Mosquitoes rely heavily upon olfaction to find their targets and show an attraction to individuals' skin surface populated by specific microbiota. The 'attractive' individuals have a higher abundance, but lower diversity, of bacteria on their skin than 'poorly attractive' individuals (Verhulst et al. 2011). The human immune system also has a role in this mosquito attraction. The HLA gene family encodes the HLA complex present on nearly all cells, which distinguishes the body's own proteins from proteins made by viruses and bacteria. HLA is also involved in the regulation of human body odour (Jacob et al. 2002) and people carrying the HLA gene Cw*07 have been shown to be more attractive to mosquitoes (Figure 8.9), though the different odours from different body regions do not appear to have a role (Verhulst et al. 2016).

8.11.6 Ectoparasites

Nearly all animals, including humans, have ectoparasites, parasites that live on or outside their body. The human skin parasites can be classified as either arachnids or insects. The arachnid class includes ticks and mites. The insect class consists of flies, mosquitoes, fleas, and lice. There are also some that become subdermal parasites.

The human head louse is an extremely common worldwide problem, infesting millions of school-aged children every year, and is only found on the human head or hair. Head lice do not live on furniture, hats, bedding, carpet, or anywhere else in the environment. Human lice appear to have originated in Africa and spread along with humans (Kittler et al. 2003). The human head louse (*Pediculus humanus capitis*), as its name suggests, lives on the scalp and feeds only on human blood. The other anatomical location, the body, has been occupied by the human body louse (*Pediculus humanus corporis* or *Pediculus humanus humanus*). The body louse attaches its eggs to clothing rather than the base of hairs. There is still taxonomic disagreement on whether the head/body morphotypes should be considered as separate species (Light et al. 2008), with the difference of only one gene absent in the head louse. In terms of modern human anatomical locations, the human body louse apparently arose from the head louse only recently with the advent and widespread use of clothing (Kittler et al. 2003). Future changes in clothing material and fashion may also lead to changes in the body louse.

The myiasis-causing flies are geographically distributed and infest human skin by laying eggs that develop into maggots, the developing fly larvae. The two myiasis-causing fly species are the human botfly (*Dermatobia hominis*) and tumbu fly (*Cordylobia anthropophaga*). The human botfly is found in the Americas. Using the mosquito as an egg carrier, bites from the mosquito transfer botfly eggs to the human host. The tumbu fly is found in sub-Saharan Africa and the hatched eggs are found in the soil; the larvae can penetrate unbroken skin (Mathison and Pritt 2014).

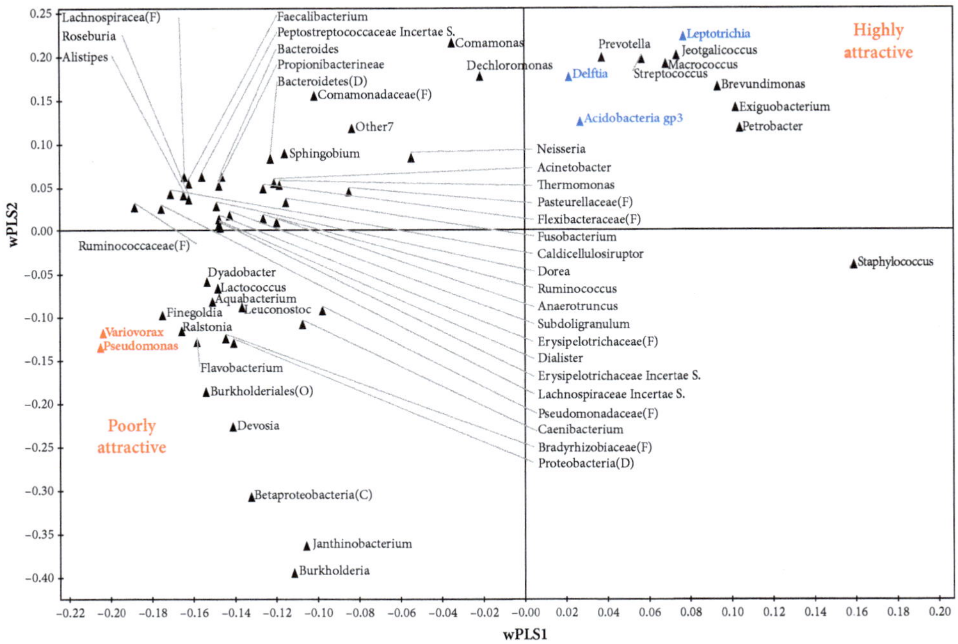

FIGURE 8.9 Multivariate data analysis of bacterial profiles of poorly attractive (red) and highly attractive (blue) individuals. Partial least squares discriminant analysis (PLS-DA) loading plot based on relative abundance of bacterial genera in microbiota profiles of poorly attractive and highly attractive individuals. Some sequences are only identified by division (D), class (C), order (O), or family (F); axes refer to comparative sequence analysis and further details are available from the original source.

Source: Adapted from Niels O. Verhulst, Yu Tong Qiu, Hans Beijleveld, Chris Maliepaard, Dan Knights, Stefan Schulz, Donna Berg-Lyons, Christian L. Lauber, Willem Verduijn, Geert W. Haasnoot, Roland Mumm, Harro J. Bouwmeester, Frans H. J. Claas, Marcel Dicke, Joop J. A. van Loon, Willem Takken, Rob Knight, and Renate C. Smallegange, Composition of Human Skin Microbiota Affects Attractiveness to Malaria Mosquitoes, *PLoS ONE*, 6(12), Figure 4, e28991, https//doi.org/10.1371/journal.pone.0028991. © 2011 Verhulst et al. This is an open-access article distributed under the terms of the Creative Commons Attribution License, which permits unrestricted use, distribution, and reproduction in any medium, provided the original author and source are credited.

8.12 Skin Ageing

Ageing skin undergoes changes in the vascular network, reduction in melanocyte and Langerhans cell number, decreased overall thickness of the epidermis, and reduction in specific collagen types within the extracellular matrix. During skin ageing, the dermis papillary layer decreases in volume.

Aged skin, both dermal fibroblasts and some terminally differentiated keratinocytes, display an increase in a lamin A splicing event that occurs in Hutchinson–Gilford progeria syndrome (OMIM 176670) (McClintock et al. 2007). A mutation within exon 11 of the lamin A gene leads to activation of a splice donor site that results in production of a dominant negative form of lamin, progerin. This type of genetic change appears unique to humans with their extended longevity.

8.13 The Future of Skin

8.13.1 Stem Cells of the Skin

There are several sources of stem cells within the skin, initially identified by historic skin grafting and currently by stem cell research. The normal physiological function of these stem cells is to replace cells continuously lost, in the case of epidermis; cycling, in the case of hair follicles and associated sebaceous glands; and remodelling after injury or disease, in the case of the dermis. There are unique changes specific to placental animals; during pregnancy, specific female abdominal skin areas expand from a unique subpopulation of stem cells (Ichijo et al. 2017).

The skin is continuously replaced throughout our lives by stem cells residing in the basal layer of the epidermis (Blanpain and Fuchs 2006). Hair follicles are also cycling continuously through growth and degeneration, with both the dermal papilla (bulge) and sebaceous glands replaced by their own resident stem cells (Tiede et al. 2007). There has been extensive research (20,000 papers) on possible replacement of hair follicles lost with male pattern baldness and alopecia and stem cells are the new frontier for this work. There is even the possibility of stem cell-like neural crest precursors within the skin (Fernandes et al. 2008). Dermal 'stem cells' are also involved in the process of wound healing from skin injury and disease (Lee et al. 2016).

While the physiological role of these stem cells is important, clinical medicine has now begun to look at these cells as a source for many therapeutic applications. Yamanaka in 2006 developed 'inducible' stem cells by reprogramming fibroblasts using just four genes: OCT4, SOX2, KLF4, and cMyc (Takahashi and Yamanaka 2006). This has led to a technical expansion of potential pluripotential stem cells that can be sourced. Other induced pluripotent stem cells have been reprogrammed by specific growth factors to differentiate specifically into melanocytes, with potential for use in pigmentation disorders (Ohta et al. 2011; Kawakami et al. 2017).

The skin is a clinically readily accessible source of stem cells, and it is now being looked at as a source of many cell types, reprogramming these cells into a broad range of cell types (neuron, muscle, glia, connective tissue) for autologous therapy to replace or repair tissues. Being from the same patient, it overcomes the problem of graft versus host immune response with other types of transplantation. These stem cells can also be used to develop, along with other recent technologies (skin-on-chip), *in vitro* skin models (van den Broek et al. 2017). Another potential *in vitro* stem cell application is for the development of new models of disease using cells from the same diseased patients (Jungverdorben et al. 2017; Zheng et al. 2017).

8.13.2 The Ageing Skin

Until this last century, few humans survived to old age and the current calculated oldest human age achievable is about 115 years. In the United States, data from the National Center for Health Statistics show that during the 1800s average life expectancy of all races was only

39.4 years. Average life expectancy by the 1930s was 60 years, and by 2010 men reached 76.2 years and women 81.1 years. The 1940s to 1970s saw the greatest overall increase in life expectancy; that increase in age has subsequently plateaued. The vast majority of these life expectancy changes come from improvements in both medicine and food quality.

Historically, the human skin only had to last about 40 years. With increasing ages achieved for the general population there is a huge increase in age-related changes and diseases of the skin (dermatoses), including conditions such as dry skin, wrinkles, dyspigmentation, ulcers, itching, and fungal infections, as well as benign and malignant tumours (Blume-Peytavi et al. 2016). Pigmentation changes in the ageing skin are related to a decrease in total number of melanocytes and an increase in size of the remaining melanocytes. Some of these conditions would have been seen in the past in only a few surviving individuals.

There are also twenty-first-century environmental and lifestyle factors that can negatively impact upon the ageing skin. Negative environmental factors include increasing UV radiation, climate change, pollution, and newly synthesised materials. This environmental impact has been demonstrated in monozygotic twin studies (Ichibori et al. 2014). Negative lifestyle factors include new drugs, smoking, sunbed use, and clothing leading to skin exposure. The dermis in particular undergoes significant age-related changes that affect its structure, thickness, mechanical properties, and repair (Quan and Fisher 2015). These dermal changes lead to a decline in aged human skin function, including wound healing. The way the skin ages is also clearly affected by the individual's genetic background and we are now identifying epigenetic changes associated with a reduction in DNA methylation patterning (Bormann et al. 2016).

8.13.3 Predicting the Future of Skin in the Twenty-First Century

We look to basic research and medicine to enhance the quality and function of our skin, prevent its ageing decline, aid its repair following wounding or operative procedures, and prevent skin diseases. There are new technologies for the delivery of pharmaceutical hydrophilic and lipophilic macromolecules using novel dermal and transdermal methods (Münch et al. 2017). Nicotine patches are now a common-place example of transdermal delivery of a drug (nicotine) as a therapy to prevent smoking-related disease. Microneedles also hold great promise for the delivery of vaccines, as well as the added bonus of not requiring refrigeration and allowing prolonged storage. These properties are all useful for delivery of vaccines to third-world locations. Other new clinical techniques include iontophoresis (electrical current), sonophoresis (absorption of semisolid topical compound), and ultrasound, as well as a variety of permeator and carrier molecules such as nanoparticles (Amjadi et al. 2017).

Artificial 'skin' is being developed for prosthetic devices that researchers hope one day will provide sensory feedback (heat, cold, touch) to the user. Technological medical applications currently include the transdermal measurement of heart rate and blood pressure, while future sensor development may allow measurement and monitoring of other diagnostic parameters.

The largest impact on the future of skin will still relate to its barrier function. We remain the same on the inside, but the world outside is changing in many ways and the question is can the human skin adapt to these changing requirements?

References

Adejuyigbe, E., Bee, M., Amare, Y., et al. (2015). Why not bathe the baby today? A qualitative study of thermal care beliefs and practices in four African sites. *BMC Pediatr* 15(1), 156. doi: 10.1186/s12887-015-0470-0.

Aghajafari, F., Field, C., Kaplan, B., et al. (2016). The current recommended vitamin D intake guideline for diet and supplements during pregnancy is not adequate to achieve vitamin D sufficiency for most pregnant women. *PLoS One* 11(7), e0157262. doi: 10.1371/journal.pone.0157262.

Aitken, S., Dilworth, T., Heil, E., et al. (2016). Agricultural applications for antimicrobials. A danger to human health: an official position statement of the Society of Infectious Diseases Pharmacists. *Pharmacotherapy* 36(4), 422–32. doi: 10.1002/phar.1737.

Akbani, R., Akdemir, K., Aksoy, B., et al. (2015). Genomic classification of cutaneous melanoma. *Cell* 161(7), 1681–96. doi: 10.1016/j.cell.2015.05.044.

Akiyama, M., Smith, L., Yoneda, K., et al. (1999). Periderm cells form cornified cell envelope in their regression process during human epidermal development. *J Invest Dermatol* 112(6), 903–9. doi: 10.1046/j.1523-1747.1999.00592.x.

Al-Harbi, A. N., Khan, K. M., and Rahman, A. (2017). Developmental vitamin D deficiency affects spatial learning in Wistar rats. *Journal of Nutrition* 147(9), 1795–805. doi: 10.3945/jn.117.249953.

Amaral, T., Leiter, U., and Garbe, C. (2017). Merkel cell carcinoma: epidemiology, pathogenesis, diagnosis and therapy. *Rev Endocr Metab Disord* 18(4), 517–32. doi: 10.1007/s11154-017-9433-0.

Amjadi, M., Mostaghaci, B., and Sitti, M. (2017). Recent advances in skin penetration enhancers for transdermal gene and drug delivery. *Curr Gene Ther* 17(2), 139–46. doi: 10.2174/1566523217666170510151540.

Andersen, P. H., Abrams, K., and Maibach, H. (1992). Ultraviolet B dose-dependent inflammation in humans: a reflectance spectroscopic and laser Doppler flowmetric study using topical pharmacologic antagonists on irradiated skin. *Photodermatol Photoimmunol Photomed* 9(1), 17–23.

Arai, K., Hara, T., Nagatsuka, T., et al. (2017). Postnatal changes and sexual dimorphism in collagen expression in mouse skin. *PLoS One* 12(5), e0177534. doi: 10.1371/journal.pone.0177534.

Atmatzidis, D. H., Lambert, W. C., and Lambert, M. W. (2017). Langerhans cell: exciting developments in health and disease. *J Eur Acad Dermatol Venereol* 31(11), 1817–24. doi: 10.1111/jdv.14522.

Barrett, A., Selvarajah, S., Franey, S., et al. (1998). Interspecies variations in oral epithelial cytokeratin expression. *J Anat* 193(2), 185–93.

Barrett, A., Cort, E., Patel, P., et al. (2000). An immunohistological study of cytokeratin 20 in human and mammalian oral epithelium. *Arch Oral Biol* 45(10), 879–87.

Barrett, A., Morgan, M., Nwaeze, G., et al. (2005). The differentiation profile of the epithelium of the human lip. *Arch Oral Biol* 50(4), 431–8. doi: 10.1016/j.archoralbio.2004.09.012.

Barsh, G. S. (2003). What controls variation in human skin color? *PLoS Biol* 1(1), E27. doi: 10.1371/journal.pbio.0000027.

Bas, H. and Kleinert, J. M. (1999). Anatomic variations in sensory innervation of the hand and digits. *J Hand Surg* 24(6), 1171–84. doi: 10.1053/jhsu.1999.1171.

Biswas, A., Manivannan, M., and Srinivasan, M. A. (2015). Multiscale layered biomechanical model of the Pacinian corpuscle. *IEEE Trans Haptics* 8(1), 31–42. doi: 10.1109/TOH.2014.2369416.

Bjerring, P. and Andersen, P. H. (1987). Skin reflectance spectrophotometry. *Photodermatol* 4(3), 167–71.

Blanpain, C. and Fuchs, E. (2006). Epidermal stem cells of the skin. *Annu Rev Cell Dev Biol* 22(1), 339–73. doi: 10.1146/annurev.cellbio.22.010305.104357.

Blume-Peytavi, U., Kottner, J., Sterry, W., et al. (2016). Age-associated skin conditions and diseases: current perspectives and future options. *Gerontologist* 56(Suppl 2), S230–42. doi: 10.1093/geront/gnw003.

Boothby, R., Lammert, N., Benrubi, G., et al. (1985). Vernix caseosa granuloma: a rare complication of cesarean section. *South Med J* 78(11), 1395–6.

Bormann, F., Rodríguez-Paredes, M., Hagemann, S., et al. (2016). Reduced DNA methylation patterning and transcriptional connectivity define human skin aging. *Aging Cell* 15(3), 563–71. doi: 10.1111/acel.12470.

Botchkarev, V. A. and Fessing, M. Y. (2005). Edar signaling in the control of hair follicle development. *J Invest Dermatol* 10(3), 247–51. doi: 10.1111/j.1087-0024.2005.10129.x.

Boulais, N., Pereira, U., Lebonvallet, N., et al. (2009a). Merkel cells as putative regulatory cells in skin disorders: an in vitro study. *PLoS One* 4(8), e6528. doi: 10.1371/journal.pone.0006528.

Boulais, N., Pennec, J., Lebonvallet, N., et al. (2009b). Rat Merkel cells are mechanoreceptors and osmoreceptors *PLoS One* 4(11), e7759. doi: 10.1371/journal.pone.0007759.

Brambilla, L., Brena, M., and Tourlaki, A. (2016). Textiles in dermatology: our experience and literature review. *Giorn Ital Dermatol Venereol* 151(3), 266–74.

Bramon, E., Walshe, M., McDonald, C., et al. (2005). Dermatoglyphics and schizophrenia: a meta-analysis and investigation of the impact of obstetric complications upon a–b ridge count. *Schizophr Res* 75(2–3), 399–404. doi: 10.1016/j.schres.2004.08.022.

Bryk, J., Hardouin, E., Pugach, I., et al. (2008). Positive selection in East Asians for an EDAR allele that enhances NF-kappaB activation. *PLoS One* 3(5), e2209. doi: 10.1371/journal.pone.0002209.

Byard, R. W., Gehl, A., and Tsokos, M. (2005). Skin tension and cleavage lines (Langers lines) causing distortion of ante- and postmortem wound morphology. *Int J Legal Med* 119(4), 226–30. doi: 10.1007/s00414-005-0539-7.

Cai, B., Hsu, P., Hsin, I., et al. (2012). p53 acts as a co-repressor to regulate keratin 14 expression during epidermal cell differentiation. *PLoS One* 7(7), e41742. doi: 10.1371/journal.pone.0041742.

Cai, J. and Ma, L. (2011). Msx2 and Foxn1 regulate nail homeostasis. *Genesis* 49(6), 449–59. doi: 10.1002/dvg.20744.

Calton, E., Keane, K., Newsholme, P., et al. (2015). The impact of vitamin D levels on inflammatory status: a systematic review of immune cell studies. *PLoS One* 10(11), e0141770. doi: 10.1371/journal.pone.0141770.

Casadevall, A. and Pirofski, L. (2003). The damage-response framework of microbial pathogenesis. *Nat Rev Microbiol* 1(1), 17–24. doi: 10.1038/nrmicro732.

Ceglia, L. (2009). Vitamin D and its role in skeletal muscle. *Curr Opin Clin Nutr Metab Care* 12(6), 628–33. doi: 10.1097/MCO.0b013e328331c707.

Cerceo, E., Deitelzweig, S., Sherman, B., et al. (2016). Multidrug-resistant Gram-negative bacterial infections in the hospital setting: overview, implications for clinical practice, and emerging treatment options. *Microb Drug Resist* 22(5), 412–31. doi: 10.1089/mdr.2015.0220.

Chambers, A., Patil, A., Alves, R., et al. (2012). Delayed presentation of vernix caseosa peritonitis. *Ann R Coll Surg Engl* 94(8), 548–51. doi: 10.1308/003588412X13373405385296.

Chang, C., Murzaku, E., Penn, L., et al. (2014). More skin, more sun, more tan, more melanoma. *Am J Public Health* 104(11), e92–9. doi: 10.2105/AJPH.2014.302185.

Chang, C., Kuo, C., Ito, T., et al. (2017). CK1α ablation in keratinocytes induces p53-dependent, sunburn-protective skin hyperpigmentation. *Proc Natl Acad Sci U S A* 114(38), E8035–44. doi: 10.1073/pnas.1702763114.

Chantelau, E. A. (2015). Nociception at the diabetic foot, an uncharted territory. *World J Diabetes* 6(3), 391. doi: 10.4239/wjd.v6.i3.391.

Child, F., Fuller, L., Higgins, E., et al. (1999). A study of the spectrum of skin disease occurring in a black population in south-east London. *Br J Dermatol* 141(3), 512–17.

Choksi, A. N., Poonawalla, T., and Wilkerson, M. G. (2010). Nanoparticles: a closer look at their dermal effects. *J Drugs Dermatol* 9(5), 475–81.

Cianferotti, L., Cox, M., Skorija, K., et al. (2007). Vitamin D receptor is essential for normal keratinocyte stem cell function. *Proc Natl Acad Sci U S A* 104(22), 9428–33. doi: 10.1073/pnas.0702884104.

Cieślińska, A., Kostyra, E., Chwała, B., et al. (2017). Vitamin D receptor gene polymorphisms associated with childhood autism. *Brain Sci* 7(9), 115. doi: 10.3390/brainsci7090115.

Coates, R., Moran, J., and Horsburgh, M. J. (2014). Staphylococci: colonizers and pathogens of human skin. *Future Microbiol* 9(1), 75–91. doi: 10.2217/fmb.13.145.

Coelho, S., Yin, L., Smuda, C., et al. (2015). Photobiological implications of melanin photoprotection after UVB-induced tanning of human skin but not UVA-induced tanning. *Pigment Cell Melanoma Res* 28(2), 210–16. doi: 10.1111/pcmr.12331.

Coffin, S. L., Turrentine, J. E., and Cruz, P. D. (2017). Photodermatitis for the allergist. *Curr Allergy Asthma Rep* 17(6), 36. doi: 10.1007/s11882-017-0705-2.

Collin, M. and Milne, P. (2016). Langerhans cell origin and regulation. *Curr Opin Hematol* 23(1), 28–35. doi: 10.1097/MOH.0000000000000202.

Cordain, L., Lindeberg, S., Hurtado, M., et al. (2002). Acne vulgaris: a disease of western civilization. *Arch Dermatol* 138(12), 1584–90.

Costin, G.-E. and Hearing, V. J. (2007). Human skin pigmentation: melanocytes modulate skin color in response to stress. *FASEB J* 21(4), 976–94. doi: 10.1096/fj.06-6649rev.

Crawford, N., Kelly, D., Hansen, M., et al. (2017). Loci associated with skin pigmentation identified in African populations. *Science* 358(6365), pii, eaan8433. doi: 10.1126/science.aan8433.

Crombie, I. K. (1979). Racial differences in melanoma incidence. *Br J Cancer* 40(2), 185–93.

Cui, R., Widlund, H., Feige, E., et al. (2007). Central role of p53 in the suntan response and pathologic hyperpigmentation. *Cell* 128(5), 853–64. doi: 10.1016/j.cell.2006.12.045.

Cui, X., Gooch, H., Petty, A., et al. (2017). Vitamin D and the brain: genomic and non-genomic actions. *Mol Cell Endocrinol* 453, 131–43. doi: 10.1016/j.mce.2017.05.035.

Dasgupta, S., Arya, S., Choudhary, S., et al. (2016). Amniotic fluid: source of trophic factors for the developing intestine. *World J Gastrointest Pathophysiol* 7(1), 38. doi: 10.4291/wjgp.v7.i1.38.

Dawber, R. P. (1980). The ultrastructure and growth of human nails. *Arch Dermatol Res* 269(2), 197–204.

Debeer, S., Le Luduec, J., Kaiserlian, D., et al. (2013). Comparative histology and immunohistochemistry of porcine versus human skin. *John Libbey Eurotext* 23(4), 456–66. doi: 10.1684/ejd.2013.2060.

DeLuca, H. F. (2004). Overview of general physiologic features and functions of vitamin D. *Am J Clin Nutr* 80(Suppl 6), 1689S–96S.

Derraik, J., Rademaker, M., Cutfield, W., et al. (2014). Effects of age, gender, BMI, and anatomical site on skin thickness in children and adults with diabetes. *PLoS One* 9(1), e86637. doi: 10.1371/journal.pone.0086637.

Dillon, Y. K., Haynes, J., and Henneberg, M. (2001). The relationship of the number of Meissners corpuscles to dermatoglyphic characters and finger size. *J Anat* 199(Pt 5), 577–84.

Dlova, N., Mankahla, A., Madala, N., et al. (2015). The spectrum of skin diseases in a black population in Durban, KwaZulu-Natal, South Africa. *Int J Dermatol* 54(3), 279–85. doi: 10.1111/ijd.12589.

Done, H. Y., Venkatesan, A. K., and Halden, R. U. (2015). Does the recent growth of aquaculture create antibiotic resistance threats different from those associated with land animal production in agriculture? *AAPS J* 17(3), 513–24. doi: 10.1208/s12248-015-9722-z.

Driskell, R., Lichtenberger, B., Hoste, E., et al. (2013). Distinct fibroblast lineages determine dermal architecture in skin development and repair. *Nature* 504(7479), 277–81. doi: 10.1038/nature12783.

Driskell, R., Jahoda, C., Chuong, C., et al. (2014). Defining dermal adipose tissue. *Exp Dermatol* 23(9), 629–31. doi: 10.1111/exd.12450.

Dror, Y. and Hopp, M. (2014). Hair for brain trade-off, a metabolic bypass for encephalization *Springerplus* 3(1), 562. doi: 10.1186/2193-1801-3-562.

Duverger, O. and Morasso, M. I. (2009). Epidermal patterning and induction of different hair types during mouse embryonic development. *Birth Defects Res C Embryo Today* 87(3), 263–72. doi: 10.1002/bdrc.20158.

Eckert, R., Crish, J., Efimova, T., et al. (2004). Regulation of involucrin gene expression. *J Invest Dermatol* 123(1), 13–22. doi: 10.1111/j.0022-202X.2004.22723.x.

Eickhoff, S. B., Grefkes, C., Fink, G. R., et al. (2008). Functional lateralization of face, hand, and trunk representation in anatomically defined human somatosensory areas. *Cereb Cortex* 18(12), 2820–30. doi: 10.1093/cercor/bhn039.

Elias, P. M. and Williams, M. L. (2016). Basis for the gain and subsequent dilution of epidermal pigmentation during human evolution: the barrier and metabolic conservation hypotheses revisited. *Am J Phys Anthropol* 161(2), 189–207. doi: 10.1002/ajpa.23030.

Elias, P. M., Menon, G., Wetzel, B. K., et al. (2010). Barrier requirements as the evolutionary 'driver' of epidermal pigmentation in humans. *Am J Hum Biol* 22(4), 526–37. doi: 10.1002/ajhb.21043.

Evans, E. W. (2017). Treating scars on the oral mucosa. *Facial Plast Surg Clin North Am* 25(1), 89–97. doi: 10.1016/j.fsc.2016.08.008.

Fan, S., Cai, H., Xu, X., et al. (2017). The characteristics of herpes simplex virus type 1 infection in rhesus macaques and the associated pathological features. *Viruses* 9(2), 26. doi: 10.3390/v9020026.

Farahmand, S., Tien, L., Hui, X., et al. (2009). Measuring transepidermal water loss: a comparative *in vivo* study of condenser-chamber, unventilated-chamber and open-chamber systems. *Skin Res Technol* 15(4), 392–8. doi: 10.1111/j.1600-0846.2009.00376.x.

Feito, J., Cobo, J. L., Santos-Briz, A., et al. (2017). Pacinian corpuscles in human lymph nodes. *Anat Rec* 300(12), 2233–8. doi: 10.1002/ar.23679.

Fell, G. L., Robinson, K. C., Mao, J., et al. (2014). Skin β-endorphin mediates addiction to UV light. *Cell* 157(7), 1527–34. doi: 10.1016/j.cell.2014.04.032.

Fernandes, K. J., Toma, J. G., and Miller, F. D. (2008). Multipotent skin-derived precursors: adult neural crest-related precursors with therapeutic potential. *Philos Trans R Soc Lond B Biol Sci* 363(1489), 185–98. doi: 10.1098/rstb.2006.2020.

Findley, K., Oh, J., Yang, J., et al. (2013). Topographic diversity of fungal and bacterial communities in human skin. *Nature* 498(7454), 367–70. doi: 10.1038/nature12171.

Fischer, T. W., Slominski, A., Tobin, D. J., et al. (2007). Melatonin and the hair follicle. *J Pineal Res* 44(1), 1–15. doi: 10.1111/j.1600-079X.2007.00512.x.

Fleming, M. S. and Luo, W. (2013). The anatomy, function, and development of mammalian Aβ low-threshold mechanoreceptors. *Front Biol* 8(4), 408–20. doi: 10.1007/s11515-013-1271-1.

Forbat, E., Al-Niaimi, F., and Ali, F. R. (2017). Molluscum contagiosum: review and update on management. *Pediatr Dermatol* 34(5), 504–15. doi: 10.1111/pde.13228.

Friedman, N. D., Temkin, E., and Carmeli, Y. (2016). The negative impact of antibiotic resistance. *Clin Microbiol Infect* 22(5), 416–22. doi: 10.1016/j.cmi.2015.12.002.

Gahlot, R., Nigam, C., Kumar, V., et al. (2014). Catheter-related bloodstream infections. *Int J Crit Illn Inj Sci* 4(2), 162–7. doi: 10.4103/2229-5151.134184.

Gambichler, T., Bader, A., Sauermann, K., et al. (2001). Serum folate levels after UVA exposure: a two-group parallel randomised controlled trial. *BMC Dermatol* 1, 8.

Gautam, P., Chaurasia, A., Bhattacharya, A., et al. (2015). Population diversity and adaptive evolution in keratinization genes: impact of environment in shaping skin phenotypes. *Mol Biol Evol* 32(3), 555–73. doi: 10.1093/molbev/msu342.

Ge, T., Han, J., Qi, Y., et al. (2017). The toxic effects of chlorophenols and associated mechanisms in fish. *Aquatic Toxicol* 184, 78–93. doi: 10.1016/j.aquatox.2017.01.005.

Ghitani, N., Barik, A., Szczot, M., et al. (2017). Specialized mechanosensory nociceptors mediating rapid responses to hair pull. *Neuron* 95(4), 944–54. doi: 10.1016/j.neuron.2017.07.024.

Gilhar, A., Schrum, A. G., Etzioni, A., et al. (2016). Alopecia areata: animal models illuminate autoimmune pathogenesis and novel immunotherapeutic strategies. *Autoimmun Rev* 15(7), 726–35. doi: 10.1016/j.autrev.2016.03.008.

Gkegkes, I. D., et al. (2013). Expression of caspase-14 and keratin-19 in the human epidermis and appendages during fetal skin development. *Arch Dermatol Res* 305(5), 379–87. doi: 10.1007/s00403-013-1319-8.

Golembo-Smith, S., Walder, D. J., Daly, M. P., et al. (2012). The presentation of dermatoglyphic abnormalities in schizophrenia: a meta-analytic review. *Schizophr Res* 142(1–3), 1–11. doi: 10.1016/j.schres.2012.10.002.

Greaves, M. (2014). Response to Jablonski and Chaplin. *Proc Biol Sci* 281(1789), 20140940. doi: 10.1098/rspb.2014.0940.

Gruber-Wackernagel, A., Byrne, S. N., and Wolf, P. (2009). Pathogenic mechanisms of polymorphic light eruption. *Front Biosci (Elite edn)* 1, 341–54.

Grudet, C., Malm, J., Westrin, A., et al. (2014). Suicidal patients are deficient in vitamin D, associated with a pro-inflammatory status in the blood. *Psychoneuroendocrinology* 50, 210–19. doi: 10.1016/j.psyneuen.2014.08.016.

Hajizadeh-Saffar, M., Feather, J. W., and Dawson, J. B. (1990). An investigation of factors affecting the accuracy of in vivo measurements of skin pigments by reflectance spectrophotometry. *Phys Med Biol* 35(9), 1301–15.

Harrison, S. and Sinclair, R. (2002). Telogen effluvium. *Clin Exp Dermatol* 27(5), 389–5.

Hartschuh, W., Weihe, E., Yanaihara, N., et al. (1983). Immunohistochemical localization of vasoactive intestinal polypeptide (VIP) in Merkel cells of various mammals: evidence for a neuromodulator function of the Merkel cell. *J Invest Dermatol* 81(4), 361–4.

Heckman, C. J., Chandler, R., Kloss, J., et al. (2013). Minimal erythema dose (MED) testing. *J Vis Exp* (75), e50175. doi: 10.3791/50175.

Hoeger, P. H., Schreiner, V., Klaassen, I. A., et al. (2002). Epidermal barrier lipids in human vernix caseosa: corresponding ceramide pattern in vernix and fetal skin. *Br J Dermatol* 146(2), 194–201.

Holick, M. F. (2011). Vitamin D: evolutionary, physiological and health perspectives. *Curr Drug Targets* 12(1), 4–18.

Holick, M. F., MacLaughlin, J. A., Clark, M. B., et al. (1980). Photosynthesis of previtamin D3 in human skin and the physiologic consequences. *Science* 210(4466), 203–5.

Honda, A., Fukushima, W., Oishi, M., et al. (2017). Effects of components of PM2.5 collected in Japan on the respiratory and immune systems. *Int J Toxicol* 36(2), 153–64. doi: 10.1177/1091581816682224.

Hong, F., Zhou, Y., Zhou, Y., et al. (2017). Immunotoxic effects of thymus in mice following exposure to nanoparticulate TiO_2. *Environ Toxicol* 32(10), 2234–43. doi: 10.1002/tox.22439.

Hopkins, W. D., Russell, J. L., Hostetter, A., et al. (2005). Grip preference, dermatoglyphics, and hand use in captive chimpanzees (*Pan troglodytes*). *Am J Phys Anthropol* 128(1), 57–62. doi: 10.1002/ajpa.20093.

Hulur, I., Skol, A. D., Gamazon, E. R., et al. (2017). Integrative genetic analysis suggests that skin color modifies the genetic architecture of melanoma. *PLoS One* 12(10), e0185730. doi: 10.1371/journal.pone.0185730.

Huttley, G. A., Easteal, S., Southey, M. C., et al. (2000). Adaptive evolution of the tumour suppressor BRCA1 in humans and chimpanzees. Australian Breast Cancer Family Study. *Nat Genet* 25(4), 410–13. doi: 10.1038/78092.

Ichibori, R., Fujiwara, T., Tanigawa, T., et al. (2014). Objective assessment of facial skin aging and the associated environmental factors in Japanese monozygotic twins. *J Cosmet Dermat* 13(2), 158–63. doi: 10.1111/jocd.12081.

Ichijo, R., Kobayashi, H., Yoneda, S., et al. (2017). Tbx3-dependent amplifying stem cell progeny drives interfollicular epidermal expansion during pregnancy and regeneration. *Nat Commun* 8(1), 508. doi: 10.1038/s41467-017-00433-7.

Imam, N., Carpenter, C. C., Mayer, K. H., et al. (1990). Hierarchical pattern of mucosal candida infections in HIV-seropositive women. *Am J Med* 89(2), 142–6.

Jablonski, N. G. and Chaplin, G. (2010). Colloquium paper: human skin pigmentation as an adaptation to UV radiation. *Proc Natl Acad Sci U S A* 107(Suppl 2), 8962–8. doi: 10.1073/pnas.0914628107.

Jablonski, N. G. and Chaplin, G. (2014). Skin cancer was not a potent selective force in the evolution of protective pigmentation in early hominins. *Proc Biol Sci* 281(1789), 20140517. doi: 10.1098/rspb.2014.0517.

Jackson, R. (2001). Elderly and sun-affected skin. Distinguishing between changes caused by aging and changes caused by habitual exposure to sun. *Can Fam Physician* 47, 1236–43.

Jacob, S., McClintock, M. K., Zelano, B., et al. (2002). Paternally inherited HLA alleles are associated with women's choice of male odor. *Nat Genet* 30(2), 175–9. doi: 10.1038/ng830.

Jacobi, U., Kaiser, M., Toll, R., et al. (2007). Porcine ear skin: an in vitro model for human skin. *Skin Res Technol* 13(1), 19–24. doi: 10.1111/j.1600-0846.2006.00179.x.

Janson, D., Saintigny, G., Mahé, C., et al. (2013). Papillary fibroblasts differentiate into reticular fibroblasts after prolonged in vitro culture. *Exp Dermatol* 22(1), 48–53. doi: 10.1111/exd.12069.

Jones, G., Strugnell, S. A., and DeLuca, H. F. (1998). Current understanding of the molecular actions of vitamin D. *Physiol Rev* 78(4), 1193–231.

Jones, K. S., Redmond, J., Fulford, A. J., et al. (2017). Diurnal rhythms of vitamin D binding protein and total and free vitamin D metabolites. *J Steroid Biochem Mol Biol* 172, 130–5. doi: 10.1016/j.jsbmb.2017.07.015.

Jungverdorben, J., Till, A., and Brüstle, O. (2017). Induced pluripotent stem cell-based modeling of neurodegenerative diseases: a focus on autophagy. *J Mol Med (Berl)*, 95(7), 705–18. doi: 10.1007/s00109-017-1533-5.

Jussila, A., Huotari-Orava, R., Ylianttila, L., et al. (2016). Narrow-band ultraviolet B radiation induces the expression of β-endorphin in human skin in vivo. *J Photochem Photobiol B* 155, 104–8. doi: 10.1016/j.jphotobiol.2016.01.007.

Kanti, V., Bonzel, A., Stroux, A., et al. (2014). Postnatal maturation of skin barrier function in premature infants. *Skin Pharmacol Physiol* 27(5), 234–41. doi: 10.1159/000354923.

Karagas, M. R., Stannard, V. A., Mott, L. A., et al. (2002). Use of tanning devices and risk of basal cell and squamous cell skin cancers. *J Natl Cancer Inst* 94(3), 224–6.

Kawakami, T., Okano, T., Takeuchi, S., et al. (2017). New approach for the derivation of melanocytes from induced pluripotent stem (Ips) cells. *J Invest Dermatol* 138(1), 150–8. doi: 10.1016/j.jid.2017.07.849.

Kelleher, M. M., O'Carroll, M., et al. (2013). Newborn transepidermal water loss values: a reference dataset. *Pediatr Dermatol* 30(6), 712–16. doi: 10.1111/pde.12106.

Kerr, D. C., Zava, D. T., Piper, W. T., et al. (2015). Associations between vitamin D levels and depressive symptoms in healthy young adult women. *Psychiatry Res* 227(1), 46–51. doi: 10.1016/j.psychres.2015.02.016.

Kesby, J. P., Cui, X., Burne, T. H., et al. (2013). Altered dopamine ontogeny in the developmentally vitamin D deficient rat and its relevance to schizophrenia. Front Cell Neurosci. 17(7) 111, 1–13. doi: 10.3389/fncel.2013.00111.

Khoury-Hanold, W., Yordy, B., Kong, P., et al. (2016). Viral spread to enteric neurons links genital HSV-1 infection to toxic megacolon and lethality. *Cell Host Microbe* 19(6), 788–99. doi: 10.1016/j.chom.2016.05.008.

Kim, J. E., Cho, B. K., Cho, D. H., et al. (2013). Expression of hypothalamic–pituitary–adrenal axis in common skin diseases: evidence of its association with stress-related disease activity. *Acta Derm Venereol* 93(4), 387–93. doi: 10.2340/00015555-1557.

Kim, T. K., Lin, Z., Tidwell, W. J., et al. (2015). Melatonin and its metabolites accumulate in the human epidermis in vivo and inhibit proliferation and tyrosinase activity in epidermal melanocytes in vitro. *Mol Cell Endocrinol* 404, 1–8. doi: 10.1016/j.mce.2014.07.024.

Kinsler, V. A. and Larue, L. (2017). The patterns of birthmarks suggest a novel population of melanocyte precursors arising around the time of gastrulation. *Pigment Cell Melanoma Res* 31(1), 95–109. doi: 10.1111/pcmr.12645.

Kittler, R., Kayser, M., and Stoneking, M. (2003). Molecular evolution of *Pediculus humanus* and the origin of clothing. *Curr Biol* 13(16), 1414–17.

Kobayashi, H. and Tagami, H. (2004). Functional properties of the surface of the vermilion border of the lips are distinct from those of the facial skin. *Br J Dermatol* 150(3), 563–7. doi: 10.1046/j.1365-2133.2003.05741.x.

Kobayashi, K., Cho, K. H., Yamamoto, M., et al. (2018). Tree of Vater–Pacinian corpuscles in the human finger and thumb: a comparison between the late fetal stage and old age. *Surg Radiol Anat* 40(3), 243–57. doi: 10.1007/s00276-017-1894-z.

Konishi, N., Kawada, A., Morimoto, Y., et al. (2007). New approach to the evaluation of skin color of pigmentary lesions using Skin Tone Color Scale. *J Dermatol* 34(7), 441–6. doi: 10.1111/j.1346-8138.2007.00307.x.

Kosiniak-Kamysz, A., Marczakiewicz-Lustig, A., Marcińska, M., et al. (2014). Increased risk of developing cutaneous malignant melanoma is associated with variation in pigmentation genes and VDR, and may involve epistatic effects *Melanoma Res* 24(4), 388–96. doi: 10.1097/CMR.0000000000000095.

Koster, M. I. and Roop, D. R. (2004). The role of p63 in development and differentiation of the epidermis. *J Dermatol Sci* 34(1), 3–9.

Koster, M. I. and Roop, D. R. (2007). Mechanisms regulating epithelial stratification. *Annu Rev Cell Dev Biol* 23(1), 93–113. doi: 10.1146/annurev.cellbio.23.090506.123357.

Kottner, J., Lichterfeld, A., and Blume-Peytavi, U. (2013). Transepidermal water loss in young and aged healthy humans: a systematic review and meta-analysis. *Arch Dermatol Res* 305(4), 315–23. doi: 10.1007/s00403-012-1313-6.

Kühbacher, A., Burger-Kentischer, A. and Rupp, S. (2017). Interaction of *Candida* species with the skin. *Microorganisms* 5(2), pii: E32. doi: 10.3390/microorganisms5020032.

Kumamoto, K., Senuma, H., Ebara, S., et al. (1993). Distribution of Pacinian corpuscles in the hand of the monkey, *Macaca fuscata*. *J Anat* 183 (Pt 1), 149–54.

Lanfranchi, H. E. and de Rey, B. M. (1978). Comparative morphometric analysis of vermilion border epithelium and lip epidermis. *Acta Anatomica* 101(2), 187–91.

Lee, S. B., Shim, S., Kim, M. J., et al. (2016). Identification of a distinct subpopulation of fibroblasts from murine dermis: CD73(–) CD105(+) as potential marker of dermal fibroblasts subset with multipotency. *Cell Biol Int* 40(9), 1008–16. doi: 10.1002/cbin.10623.

Lefèvre, C. M., Sharp, J. A., and Nicholas, K. R. (2010). Evolution of lactation: ancient origin and extreme adaptations of the lactation system. *Annu Rev Genomics Hum Genet* 11(1), 219–38. doi: 10.1146/annurev-genom-082509-141806.

Li, L. and Ginty, D. D. (2014). The structure and organization of lanceolate mechanosensory complexes at mouse hair follicles. *eLife* 3, e01901. doi: 10.7554/eLife.01901.

Light, J. E., Toups, M. A., and Reed, D. L. (2008). What's in a name: the taxonomic status of human head and body lice. *Mol Phylogenetics Evol* 47(3), 1203–16. doi: 10.1016/j.ympev.2008.03.014.

Liou, J. C., Teng, M. C., Tsai, Y. S., et al. (2015). UV-blocking spectacle lens protects against UV-induced decline of visual performance. *Mol Vis* 21, 846–56.

Liu, F., Liang, J., Shen, L., et al. (2017). Case study of 3D fingerprints applications. *PLoS One* 12(4), e0175261. doi: 10.1371/journal.pone.0175261.

Liu, N. Q. and Hewison, M. (2012). Vitamin D, the placenta and pregnancy. *Arch Biochem Biophys* 523(1), 37–47. doi: 10.1016/j.abb.2011.11.018.

Lolli, F., Pallotti, F., Rossi, A., et al. (2017). Androgenetic alopecia: a review. *Endocrine* 57(1), 9–17. doi: 10.1007/s12020-017-1280-y.

Looker, K. J., Magaret, A. S., May, M. T., et al. (2015). Global and regional estimates of prevalent and incident herpes simplex virus type 1 infections in 2012. *PLoS One* 10(10), e0140765. doi: 10.1371/journal.pone.0140765.

Lou, D. I., McBee, R. M., Le, U. Q., et al. (2014). Rapid evolution of BRCA1 and BRCA2 in humans and other primates. *BMC Evol Biol* 14(1), 155. doi: 10.1186/1471-2148-14-155.

Ludriksone, L., Garcia Bartels, N., Kanti, V., et al. (2014). Skin barrier function in infancy: a systematic review. *Arch Dermatol Res* 306(7), 591–9. doi: 10.1007/s00403-014-1458-6.

Lundström, J. N., Gonçalves, M., Esteves, F., et al. (2003). Psychological effects of subthreshold exposure to the putative human pheromone 4, 16-androstadien-3-one. *Horm Behav* 44(5), 395–401.

Luzzatto, L. (2012). Sickle cell anaemia and malaria. *Mediterr J Hematol Infect Dis* 4(1), e2012065. doi: 10.4084/MJHID.2012.065.

Lwigale, P. Y. (2015). Corneal development: different cells from a common progenitor. *Prog Mol Biol Transl Sci* 134, 43–59. doi: 10.1016/bs.pmbts.2015.04.003.
M'Boneko, V. and Merker, H. J. (1988). Development and morphology of the periderm of mouse embryos (days 9–12 of gestation). *Acta Anatomica* 133(4), 325–36.
Mack, M. C., Chu, M. R., Tierney, N. K., et al. (2016). Water-holding and transport properties of skin stratum corneum of infants and toddlers are different from those of adults: studies in three geographical regions and four ethnic groups. *Pediatr Dermatol* 33(3), 275–82. doi: 10.1111/pde.12798.
Malone, K. E., Daling, J. R., Doody, D. R., et al. (2006). Prevalence and predictors of *BRCA1* and *BRCA2* mutations in a population-based study of breast cancer in white and black American women ages 35 to 64 years. *Cancer Res* 66(16), 8297–308. doi: 10.1158/0008-5472.CAN-06-0503.
Manousaki, D., Paternoster, L., Standl, M., et al. (2017). Vitamin D levels and susceptibility to asthma, elevated immunoglobulin E levels, and atopic dermatitis: a Mendelian randomization study. *PLoS Med* 14(5), e1002294. doi: 10.1371/journal.pmed.1002294.
Mariggiò, G., Koch, S., and Schulz, T. F. (2017). Kaposi sarcoma herpesvirus pathogenesis. *Philos Trans R Soc Lond B Biol Sci* 372(1732), pii: 20160275. doi: 10.1098/rstb.2016.0275.
Martin, J. M., Ghaferi, J. M., Cummins, D. L., et al. (2009). Changes in skin tanning attitudes. Fashion articles and advertisements in the early twentieth century. *Am J Public Health* 99(12), 2140–6. doi: 10.2105/AJPH.2008.144352.
Mathewson, I. (2015). Did human hairlessness allow natural photobiomodulation 2 million years ago and enable photobiomodulation therapy today? This can explain the rapid expansion of our genus's brain. *Med Hypotheses* 84(5), 421–8. doi: 10.1016/j.mehy.2015.01.032.
Mathison, B. A. and Pritt, B. S. (2014). Laboratory identification of arthropod ectoparasites. *Clin Microbiol Rev* 27(1), 48–67. doi: 10.1128/CMR.00008-13.
McClintock, D., Ratner, D., Lokuge, M., et al. (2007). The mutant form of lamin A that causes Hutchinson–Gilford progeria is a biomarker of cellular aging in human skin. *PLoS One* 2(12), e1269. doi: 10.1371/journal.pone.0001269.
McGeoch, D. J., Rixon, F. J., and Davison, A. J. (2006). Topics in herpesvirus genomics and evolution. *Virus Res* 117(1), 90–104. doi: 10.1016/j.virusres.2006.01.002.
McGuire, S. (2016). World Cancer Report 2014. Geneva, Switzerland: World Health Organization, International Agency for Research on Cancer, WHO Press, 2015. *Adv Nutr* 7(2), 418–19. doi: 10.3945/an.116.012211.
McKay, M. and Coad, R. (2017). A brother and sister with breast cancer, BRCA2 mutations and bilateral supernumerary nipples. *Ann Transl Med* 5(5), 106. doi: 10.21037/atm.2017.03.02.
Mikkola, M. L. and Thesleff, I. (2013). Ectodysplasin signaling in development. *Cytokine Growth Factor Rev* 14(3–4), 211–24.
Míková, R., Vrkoslav, V., Hanus, R., et al. (2014). Newborn boys and girls differ in the lipid composition of vernix caseosa. *PLoS One* 9(6), e99173. doi: 10.1371/journal.pone.0099173.
Milano, A., Cutrone, M., Laforgia, N., et al. (2010). Incomplete development of the nail of the hallux in the newborn. *Dermatol Online J* 16(6), 1.
Mirastschijski, U., Kerzel, C., Schnabel, R., et al. (2013). Complete horizontal skin cell resurfacing and delayed vertical cell infiltration into porcine reconstructive tissue matrix compared to bovine collagen matrix and human dermis. *Plast Reconstr Surg* 132(4), 861–9. doi: 10.1097/PRS.0b013e31829fe461.
Miyamura, Y., Coelho, S. G., Schlenz, K., et al. (2011). The deceptive nature of UVA tanning versus the modest protective effects of UVB tanning on human skin. *Pigment Cell Melanoma Res* 24(1), 136–47. doi: 10.1111/j.1755-148X.2010.00764.x.
Mohanty, T., Alberius, P., Schmidtchen, A., et al. (2017). Saliva induces expression of antimicrobial peptides and promotes intracellular killing of bacteria in keratinocytes by epidermal growth factor receptor transactivation. *Br J Dermatol* 176(2), 403–12. doi: 10.1111/bjd.14883.
Morrison, K. M., Miesegaes, G. R., Lumpkin, E. A., et al. (2009). Mammalian Merkel cells are descended from the epidermal lineage. *Dev Biol* 336(1), 76–83. doi: 10.1016/j.ydbio.2009.09.032.

Münch, S., Wohlrab, J., and Neubert, R. H. H. (2017). Dermal and transdermal delivery of pharmaceutically relevant macromolecules. *Eur J Pharm Biopharm* 119, 235–42. doi: 10.1016/j.ejpb.2017.06.019.

Nakwan, N., Kamolvisit, W., Napapongsuriya, C., et al. (2017). Fatal vernix caseosa aspiration associated with persistent pulmonary hypertension of the newborn. *Pediatr Dev Pathol* 20(2), 168–71. doi: 10.1177/1093526616686243.

Nicolaou, A., Masoodi, M., Gledhill, K., et al. (2012). The eicosanoid response to high dose UVR exposure of individuals prone and resistant to sunburn. *Photochem Photobiol Sci* 11(2), 371–80. doi: 10.1039/c1pp05272a.

Nicolle, L. E. (2014). Catheter associated urinary tract infections. *Antimicrob Resist Infect Control* 3(1), 23. doi: 10.1186/2047-2994-3-23.

Nithya, S., Radhika, T., and Jeddy, N. (2015). Loricrin—an overview. *J Oral Maxillofac Pathol* 19(1), 64–8. doi: 10.4103/0973-029X.157204.

Ogunbiyi, A. (2016). Acne keloidalis nuchae: prevalence, impact, and management challenges. *Clin Cosmet Investig Dermatol* 9, 483–9. doi: 10.2147/CCID.S99225.

Oh, I. Y. and de Guzman Strong, C. (2017). The molecular revolution in cutaneous biology: EDC and locus control. *J Invest Dermatol* 137(5), e101–4. doi: 10.1016/j.jid.2016.03.046.

Ohta, S., Imaizumi, Y., Okada, Y., et al. (2011). Generation of human melanocytes from induced pluripotent stem cells. *PLoS One* 6(1), e16182. doi: 10.1371/journal.pone.0016182.

Ossio, R., Roldán-Marín, R., Martínez-Said, H., et al. (2017). Melanoma: a global perspective. *Nat Rev Cancer* 17(7), 393–4. doi: 10.1038/nrc.2017.43.

Parameswaran, R., Sherchan, J. B., Varma, D. M., et al. (2011). Intravascular catheter-related infections in an Indian tertiary care hospital. *J Infect Dev Ctries* 5(6), 452–8.

Pérez-López, F. R. (2007). Vitamin D: the secosteroid hormone and human reproduction. *Gynecol Endocrinol* 23(1), 13–24.

Pillaiyar, T., Manickam, M., and Namasivayam, V. (2017). Skin whitening agents: medicinal chemistry perspective of tyrosinase inhibitors. *J Enzyme Inhib Med Chem* 32(1), 403–25. doi: 10.1080/14756366.2016.1256882.

Plunkett, A., Merlin, K., Gill, D., et al. (1999). The frequency of common nonmalignant skin conditions in adults in central Victoria, Australia. *Int J Dermatol* 38(12), 901–8.

Quan, T. and Fisher, G. J. (2015). Role of age-associated alterations of the dermal extracellular matrix microenvironment in human skin aging: a mini-review. *Gerontology* 61(5), 427–34. doi: 10.1159/000371708.

Rendl, M., Lewis, L., and Fuchs, E. (2005). Molecular dissection of mesenchymal–epithelial interactions in the hair follicle. *PLoS Biol* 3(11), e331. doi: 10.1371/journal.pbio.0030331.

Rissmann, R., Oudshoorn, M. H., Zwier, R., et al. (2009). Mimicking vernix caseosa—preparation and characterization of synthetic biofilms. *Int J Pharm* 372(1–2), 59–65. doi: 10.1016/j.ijpharm.2009.01.013.

Rittié, L., Sachs, D. L., Orringer, J. S., et al. (2013). Eccrine sweat glands are major contributors to reepithelialization of human wounds. *Am J Pathol* 182(1), 163–71. doi: 10.1016/j.ajpath.2012.09.019.

Sacks, E., Moss, W. J., Winch, P. J., et al. (2015). Skin, thermal and umbilical cord care practices for neonates in southern, rural Zambia: a qualitative study. *BMC Pregnancy Childbirth* 15(1), 149. doi: 10.1186/s12884-015-0584-2.

Sandel, A. A. (2013). Brief communication: hair density and body mass in mammals and the evolution of human hairlessness *Am J Phys Anthropol* 152(1), 145–50. doi: 10.1002/ajpa.22333.

Sandilands, A., Sutherland, C., Irvine, A. D., et al. (2009). Filaggrin in the frontline: role in skin barrier function and disease. *J Cell Sci* 122(9), 1285–94. doi: 10.1242/jcs.033969.

Sandru, A., Voinea, S., Panaitescu, E., et al. (2014). Survival rates of patients with metastatic malignant melanoma. *J Med Life* 7(4), 572–6.

Sato, K., Leidal, R., and Sato, F. (1987). Morphology and development of an apoeccrine sweat gland in human axillae. *Am J Physiol* 252(1 Pt 2), R166–80.

Schadendorf, D., Lebbé, C., zur Hausen, A., et al. (2017). Merkel cell carcinoma: epidemiology, prognosis, therapy and unmet medical needs. *Eur J Cancer* 71, 53–69. doi: 10.1016/j.ejca.2016.10.022.

Schep, R., Necsulea, A., Rodríguez-Carballo, E., et al. (2016). Control of *Hoxd* gene transcription in the mammary bud by hijacking a preexisting regulatory landscape. *Proc Natl Acad Sci U S A* 113(48), E7720–9. doi: 10.1073/pnas.1617141113.

Schmitz, S., Thomas, P. D., Allen, T. M., et al. (1995). Dual role of melanins and melanin precursors as photoprotective and phototoxic agents: inhibition of ultraviolet radiation-induced lipid peroxidation. *Photochem Photobiol* 61(6), 650–5.

Schwartz, G. G. and Rosenblum, L. A. (1981). Allometry of primate hair density and the evolution of human hairlessness. *Am J Phys Anthropol* 55(1), 9–12. doi: 10.1002/ajpa.1330550103.

Scott, J. F., Das, L. M., Ahsanuddin, S., et al. (2017). Oral vitamin D rapidly attenuates inflammation from sunburn: an interventional study. *J Invest Dermatol* 137(10), 2078–86. doi: 10.1016/j.jid.2017.04.040.

Selo-Ojeme, D. (2007). Vernix caseosa peritonitis. *J Obstet Gynaecol* 27(7), 660–3. doi: 10.1080/01443610701582792.

Sitek, A., Żądzińska, E., Rosset, I., et al. (2013). Is increased constitutive skin and hair pigmentation an early sign of puberty? *Homo* 64(3), 205–14. doi: 10.1016/j.jchb.2013.03.003.

Small, J., Barton V., Peterson B., et al. (2016). Keratinocyte carcinoma as a marker of a high cancer-risk phenotype. *Adv Cancer Res Treat* 130, 257–91. doi: 10.1016/bs.acr.2016.01.003.

Smith, L. T. (1994). Patterns of type VI collagen compared to types I, III and V collagen in human embryonic and fetal skin and in fetal skin-derived cell cultures. *Matrix Biol* 14(2), 159–70.

Sommer, C. (2016). Exploring pain pathophysiology in patients. *Science* 354(6312), 588–92. doi: 10.1126/science.aaf8935.

Sterzik, V., Belenkaia, L., Liehr, A. W., et al. (2014). Spectrometric evaluation of post-mortem optical skin changes. *Int J Legal Med* 128(2), 361–7. doi: 10.1007/s00414-013-0855-2.

Stewart, A. E., Roecklein, K. A. Tanner, S., et al. (2014). Possible contributions of skin pigmentation and vitamin D in a polyfactorial model of seasonal affective disorder. *Med Hypotheses* 83(5), 517–25. doi: 10.1016/j.mehy.2014.09.010.

Szpaderska, A. M., Zuckerman, J. D., and DiPietro, L. A. (2003). Differential injury responses in oral mucosal and cutaneous wounds. *J Dent Res* 82(8), 621–6. doi: 10.1177/154405910308200810.

Takahashi, K. and Yamanaka, S. (2006). Induction of pluripotent stem cells from mouse embryonic and adult fibroblast cultures by defined factors. *Cell* 126(4), 663–76. doi: 10.1016/j.cell.2006.07.024.

Tamblyn, J. A., et al. (2015). Immunological role of vitamin D at the maternal–fetal interface. *J Endocrinol* 224(3), R107–21. doi: 10.1530/JOE-14-0642.

Tanner, J. M. and Whitehouse, R. H. (1976). Clinical longitudinal standards for height, weight, height velocity, weight velocity, and stages of puberty. *Arch Dis Child* 51(3), 170–9.

Tansirikongkol, A., Wickett, R. R., Visscher, M. O., et al. (2007). Effect of vernix caseosa on the penetration of chymotryptic enzyme: potential role in epidermal barrier development. *Pediatr Res* 62(1), 49–53. doi: 10.1203/PDR.0b013e318067b442.

Taylor, S. C., Arsonnaud, S., Czernielewski, J., et al. (2005). The Taylor Hyperpigmentation Scale: a new visual assessment tool for the evaluation of skin color and pigmentation. *Cutis* 76(4), 270–4.

Taylor, S. C., Westerhof, W., Im, S., et al. (2006). Noninvasive techniques for the evaluation of skin color. *J Am Acad Dermatol* 54(5 Suppl 2), S282–90. doi: 10.1016/j.jaad.2005.12.041.

Thesleff, I. and Mikkola, M. L. (2002). Death receptor signaling giving life to ectodermal organs. *Sci STKE* 131, pe22. doi: 10.1126/stke.2002.131.pe22.

Thomas, E. R., Walker, L. J., Pullaperuma, S., et al. (2009). Griscelli syndrome type 1: a report of two cases and review of the literature. *Clin Dysmorphol* 18(3), 145–8. doi: 10.1097/MCD.0b013e328317b870.

Tiede, S., Kloepper, J., Bodò, E., et al. (2007). Hair follicle stem cells: walking the maze. *Eur J Cell Biol* 86(7), 355–76. doi: 10.1016/j.ejcb.2007.03.006.

Tollin, M., Bergsson, G., Kai-Larsen, Y., et al. (2005). Vernix caseosa as a multi-component defence system based on polypeptides, lipids and their interactions. *Cell Mol Life Sci* 62(19–20), 2390–9. doi: 10.1007/s00018-005-5260-7.

Tsai, M. S., Chen, M. H., Lin, C. C. et al. (2017). Children's environmental health based on birth cohort studies of Asia. *Sci Total Environ* 609, 396–409. doi: 10.1016/j.scitotenv.2017.07.081.

Tuchayi, M. S., Makrantonaki, E., Ganceviciene, R., et al. (2015). Acne vulgaris. *Nat Rev Dis Primers* 1, 15029. doi: 10.1038/nrdp.2015.29.

Turabelidze, A., Guo, S., Chung, A. Y., et al. (2014). Intrinsic differences between oral and skin keratinocytes. *PLoS One* 9(9), e101480. doi: 10.1371/journal.pone.0101480.

Turner, N. J., Pezzone, D., and Badylak, S. F. (2015). Regional variations in the histology of porcine skin. *Tissue Eng Part C Methods* 21(4), 373–84. doi: 10.1089/ten.TEC.2014.0246.

Van de Craen, M., et al. (1998). Identification of a new caspase homologue: caspase-14. *Cell Death Differ* 5(10), 838–46. doi: 10.1038/sj.cdd.4400444.

van den Broek, L. J., et al. (2017). Progress and future prospectives in skin-on-chip development with emphasis on the use of different cell types and technical challenges. *Stem Cell Rev* 13(3), 418–29. doi: 10.1007/s12015-017-9737-1.

Van Keymeulen, A., Mascre, G. Youseff, K., et al. (2009). Epidermal progenitors give rise to Merkel cells during embryonic development and adult homeostasis. *J Cell Biol* 187(1), 91–100. doi: 10.1083/jcb.200907080.

Vega, J. A., García-Suárez, O., Montaño, J. A., et al. (2009). The Meissner and Pacinian sensory corpuscles revisited new data from the last decade. *Microsc Res Tech* 72(4), 299–309. doi: 10.1002/jemt.20651.

Vega, J. A., López-Muñiz, A., Calavia, M. G., et al. (2012). Clinical implication of Meissner's corpuscles. *CNS Neurol Disord Drug Targets* 11(7), 856–68.

Verhulst, N. O., Qiu, Y. T., Beijleveld, H., et al. (2011). Composition of human skin microbiota affects attractiveness to malaria mosquitoes. *PLoS One* 6(12), e28991. doi: 10.1371/journal.pone.0028991.

Verhulst, N. O., Weldegergis, B. T., Menger, D., et al. (2016). Attractiveness of volatiles from different body parts to the malaria mosquito *Anopheles coluzzii* is affected by deodorant compounds. *Sci Rep* 6(1), 27141. doi: 10.1038/srep27141.

Visscher, M. O., et al. (2005). Vernix caseosa in neonatal adaptation. *J Perinatol* 25(7), 440–6. doi: 10.1038/sj.jp.7211305.

Visscher, M. O., Burkes, S. A., Adams, D. M., et al. (2017). Infant skin maturation: preliminary outcomes for color and biomechanical properties. *Skin Res Technol* 23(4), 545–51. doi: 10.1111/srt.12369.

Vuillermot, S., Luan, W., Meyer, U., et al. (2017). Vitamin D treatment during pregnancy prevents autism-related phenotypes in a mouse model of maternal immune activation. *Mol Autism* 8(1), 9. doi: 10.1186/s13229-017-0125-0.

Wald, A., Schacker, T., and Corey, L. (1997). HSV-2 and HIV: consequences of an endemic opportunistic infection. *STEP Perspect* 9(3), 2–4.

Watson, C. J. (2006). Involution: apoptosis and tissue remodelling that convert the mammary gland from milk factory to a quiescent organ. *Breast Cancer Res* 8(2), 203. doi: 10.1186/bcr1401.

Wilmer, E. N., Gustafson, C. J., Ahn, C. S., et al. (2014). Most common dermatologic conditions encountered by dermatologists and nondermatologists. *Cutis* 94(6), 285–92.

Wintzen, M., de Winter, S., Out-Luiting, J. J., et al. (2001). Presence of immunoreactive beta-endorphin in human skin. *Exp Dermatol* 10(5), 305–11.

Wolfram-Gabel, R. and Sick, H. (2002). Microvascularization of the mucocutaneous junction of the eyelid in fetuses and neonates. *Surg Radiol Anat* 24(2), 97–101.

Wu, P., Ng, C. S., Yan, J., et al. (2015). Topographical mapping of α- and β-keratins on developing chicken skin integuments: functional interaction and evolutionary perspectives. *Proc Natl Acad Sci U S A* 112(49), E6770-9. doi: 10.1073/pnas.1520566112.

Yamaguchi, Y., Itami, S., Watabe, H., et al. (2004). Mesenchymal–epithelial interactions in the skin: increased expression of dickkopf1 by palmoplantar fibroblasts inhibits melanocyte growth and differentiation. *J Cell Biol* 165(2), 275–85. doi: 10.1083/jcb.200311122.

Yang, H., Adam, R. C., Ge, Y., et al. (2017). Epithelial–mesenchymal micro-niches govern stem cell lineage choices. *Cell* 169(3), 483–96.e13. doi: 10.1016/j.cell.2017.03.038.

Yesantharao, P., Wang, W., Ioannidis, N. M., et al. (2017). Cutaneous squamous cell cancer (cSCC) risk and the human leukocyte antigen (HLA) system. *Hum Immunol* 78(4), 327–35. doi: 10.1016/j.humimm.2017.02.002.

Yoshio, H., Lagercrantz, H., Gudmundsson, G. H., et al. (2004). First line of defense in early human life. *Semin Perinatol* 28(4), 304–11.

Youlden, D. R., Soyer, H. P., Youl, P. H., et al. (2014). Incidence and survival for Merkel cell carcinoma in Queensland, Australia, 1993–2010. *JAMA Dermatol* 150(8), 864. doi: 10.1001/jamadermatol.2014.124.

Zeng, Y., Yang, J., Huang, K., et al. (2001). A comparison of biomechanical properties between human and porcine cornea. *J Biomech* 34(4), 533–7.

Zhang, H. G., Chen, Y. F., Ding, M., et al. (2010). Dermatoglyphics from all Chinese ethnic groups reveal geographic patterning. *PLoS One* 5(1), e8783. doi: 10.1371/journal.pone.0008783.

Zheng, X., Zhu, Z., Chen, K., et al. (2017). Establishment of an induced pluripotent stem cell line ZZUi003-A from a 65-year-old male with sporadic Parkinson's disease. *Stem Cell Res* 23, 119–21. doi: 10.1016/j.scr.2017.07.004.

CHAPTER 9

HAEMATOPOIETIC SYSTEM

ERIC M. PIETRAS AND JAMES DEGREGORI

Abstract

The haematopoietic system provides numerous essential functions for animals, including transport of gases and nutrients, wound repair, and host defence. Given the fundamental importance of the blood system, these roles are conserved across animals, with specific features shaped by the unique needs and adaptations of different organisms. While even the simplest organisms have haematopoietic systems, increasing size and complexity of organisms has necessitated the evolution of more efficient clotting and oxygen transport systems, more complex circulatory systems, and more diverse blood cell lineages for immune defence. Evolution has sculpted haematopoietic systems for different animals by modification of previously existing programmes and developmental systems, with striking examples of conservation and convergent evolution in the blood systems of distantly related organisms suggesting common adaptive solutions to a range of selective pressures. Notably, our own haematopoietic system recapitulates many features found in ancestral organisms. This chapter discusses how blood and vascular systems have evolved together and share common endothelial heritage, as well as how different blood lineages are produced and how they have evolved to meet new challenges from pathogens. Moreover, it examines how pathogenic threats to blood cells have influenced modern population genetics for humans and in turn impact our susceptibility to various disorders of the blood system. Finally, the chapter suggests how evolved life histories to maximise reproductive success have influenced ageing and disease patterns, such as for blood cancers.

Keywords

clotting, oxygen transport, immunity, ABO blood groups, interferon, ageing, haematopoietic stem cell, sickle cell anaemia, leukaemia, evolutionary medicine

9.1 The Evolutionary Biology of Haematopoiesis

Science and medicine have spent thousands of years trying to understand the workings of blood, with a more functional and detailed appreciation of the development, regulation, and diverse functions of haematopoietic cells arising over the last hundred or so years. This proximate understanding of haematopoiesis and blood cell functions is clearly important, and has contributed to numerous interventions, from stem cell transplants to treatments of various inherited and acquired blood disorders to immune modulators. The blood system, with its uses and associated disorders, is covered in any respectable medical school.

What are much less well appreciated are the evolutionary origins of this complex system. Understanding the evolution of blood cell production is important at a higher level, in that we must know where we come from—our evolutionary origins. This chapter discusses when and how different blood circulation systems evolved, facilitating oxygen transport, clotting, wound healing, tissue remodelling, and immunity. What was the adaptive value of these processes, and the selective pressures that led to their evolution? Beyond the importance of appreciating how evolution led to such an important and complex system, we will discuss how these origins can lead to constraints and vulnerabilities that have health and clinical relevance. We will further learn how human evolution in response to pathogens has shaped population genetics, with inherent trade-offs resulting from pathogen resistance. Throughout this discussion, it is important to *not* think of the human or mammalian systems as better or more evolved, as each organism has evolved the system that maximises its own survival and reproductive success—its fitness. So, while we might present the evolution of blood systems in what seems like progression to the 'optimal form'—ours, this simply reflects our anthropocentric perspective.

9.1.1 Increasing Complexity and Sizes of Animals Created the Need for Circulatory Systems

Circulatory systems, either open or closed, are present in all metazoa, with the likely origin of the first vascular systems 600–700 million years ago (mya), thought to have evolved to deal with the limits of diffusion (Monahan-Earley et al. 2013). Circulatory systems allowed for larger body sizes. In addition, circulation supports greater organisation and communication between tissues. The last common ancestor of vertebrates, annelids (worms), and mollusks is thought to have been a segmented bilaterian. The blood system may have evolved in all of these lineages as a means to overcome the barriers created by segmentation. Scientists have learned about the evolution of circulatory systems from comparative biology (Figure 9.1), as blood vessels and blood cells leave scant fossil records. Comparisons of such systems in modern (extant) species, together with an understanding of phylogenetic relationships (evolutionary trees) connecting these species, can allow researchers to estimate when and in what common ancestors different blood systems evolved. Using these methods, it is likely that the endothelium (lining blood vessels) evolved 540–510 mya in ancestral

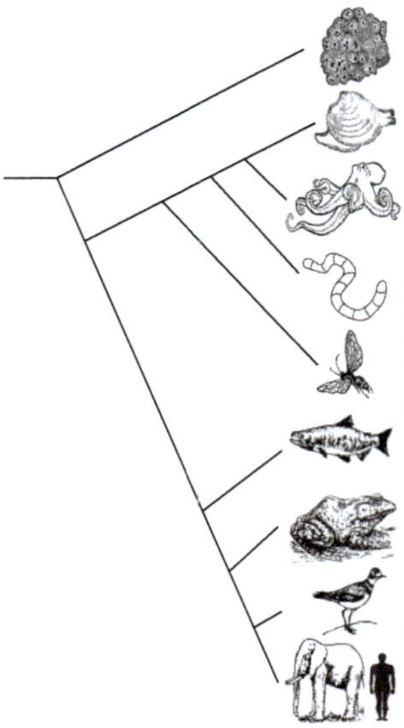

Organism	Circulation	Pump	Blood pressure	Vasculature	Erythrocytes
Porifera	Diffusion	None	None	None	No
Mollusk	Open	Multi-chambered heart	Low	ECM	No
Cephalapod mollusk	Closed	Brachial and central hearts	Low	ECM	No
Annelid	Closed	Contractile blood vessel	Low	ECM	No
Arthropod	Open	Single -chambered heart	Low	ECM	No
Telost fish	Closed	Two-chambered heart	Intermediate	Endothelium	Nucleated
Amphibian	Closed	Three-chambered heart	Intermediate	Endothelium	Nucleated
Bird	Closed	Four-chambered heart	Intermediate	Endothelium	Nucleated
Mammal	Closed	Four-chambered heart	High	Endothelium	Enuclated

FIGURE 9.1 Evolution of circulatory systems. Key characteristics that differentiate blood systems throughout evolution include whether blood circulates through an open or closed circulatory system, mechanisms for pumping blood or haemolymph, type of vasculature, and whether oxygen transport occurs via specialised cells. Note potential features that may suggest convergent evolution between various circulatory systems (such as whether systems are open or closed, or complexity of the heart). ECM, Extracellular matrix.

vertebrates, probably to optimise blood flow, barrier functions, and/or immunity (Monahan-Earley et al. 2013). Only vertebrates possess a true polarised endothelium.

Circulation is required for nutrient/gas delivery, given time-distance constraints of diffusion. Larger body sizes, in particular, require internal systems for transport and exchange of gases, nutrients, and wastes. Such circulatory systems are also critical for immunity and hormonal communication. In diploblasts like sponges and jellyfish, circulation is not internal, but involves passage of seawater through a body cavity that is open to the environment. In triploblasts (from worms to flies to mammals), circulation is internal, with extracellular fluids actively pumped through vessels and/or sinuses. Vessels can be surrounded by smooth muscle, which pumps fluids unidirectionally (perhaps a primitive precursor of the more complex heart). Some small animals like flatworms rely on diffusion to transport oxygen and nutrients across the skin and gut. Most triploblasts rely on a coelom, a fluid-filled cavity, with cilia or pumps to move the fluid.

Invertebrates can have either closed or open systems. Arthropods (insects and crustaceans) possess open systems. Organs are directly bathed in the body fluid, instead of being innervated with blood vessels. Circulation of these fluids is achieved through what is considered a true heart. Interestingly, while non-cephalopod mollusks (such as clams and slugs) have an open circulatory system, cephalopod mollusks (such as the octopus, nautilus, and squid) have closed systems, with hearts possessing both atrial and ventricular compartments. Such hearts resemble those of vertebrates, and represent an example of convergent evolution.

Open systems have high fluid volume with low pressure, while closed circulatory systems exhibit lower volume and higher pressures. Unlike the open systems, closed capillary-based systems allow for much more even distribution of oxygen and other contents to tissues and organs, and this distribution can be regulated. The open circulatory system of insects is not responsible for oxygen delivery, which is instead mediated by a highly branched tracheal system. For closed systems, exchange of gases, nutrients, wastes, etc. occurs in capillary beds, where the walls of the vessels are thin enough to facilitate diffusion. Closed systems are likely important for more active animals (whether cephalopods or vertebrates), allowing for more efficient and regulated pumping of fluids to different regions of the body.

Note that circulatory systems can be hybrids of open and closed. In many mollusks and crustaceans, a network of vessels distributes fluids towards a central coelom (the haemocoel). Even the vascular beds of vertebrates, such as in sinusoids of the liver, spleen, and bone marrow, resemble open systems where there is direct contact of blood with interstitial space. A similar 'open-like' system is found in the primate placenta, where maternal arteries release fluids directly into the placental labyrinth, bathing the chorion.

Vertebrates have evolved endothelium, lining blood vessels, to greatly facilitate the transport of cells and fluid, with additional roles in clotting, tissue repair, pathogen surveillance, and host defence. The evolution of the endothelium from mesoderm in vertebrates may have facilitated blood movement, improved barrier function, and allowed for higher pressure systems. The endothelium may have been necessary for the restricted localisation of novel coagulation and immune functions that also evolved in a common ancestral vertebrate. Vertebrates also possess lymphatic systems that recycle interstitial fluids back into circulation (and serve as conduits for immune cells).

Increasing animal size, particularly in warmer climates, engenders a need to deliver oxygen for respiration. Haemoglobin is used across life, including in bacteria, to sequester oxygen. Haemoglobin can be cell-bound or cell-free. Other respiratory pigments can be employed, such as haemocyanins commonly occurring in mollusks and arthropods. Air breathing led to changes in circulation, impacting red blood cell (RBC) and vessel design (Snyder and Sheafor 1999). As seen in lung fishes and amphibians, the first air-breathers are thought to have evolved large blood vessels with low systemic pressure to accommodate the low pressures needed in lungs and the limitations of the early heart. Separation of the pulmonary vasculature system with division of the ventricle allowed for a return to high-pressure circulation in reptiles, which itself facilitated an extensive microvasculature. High-pressure systemic circulation evolved in birds and mammals led to additional adaptations in RBC and vessel structures (both getting smaller). Snyder and Sheafor propose that smaller RBC in mammals and birds evolved as an adaptation to smaller blood vessels, which increase gas diffusion into tissues. Squeezing of RBC through vessels with diameters smaller than the RBC may have further facilitated oxygen diffusion (Snyder and Sheafor 1999). (For further discussion, see Chapter 12: Respiratory System.)

All vertebrates have red blood cells—with one known exception. The cold-water ice fish of Antarctica, of the family Channichthyidae, lack red blood cells and functional haemoglobin proteins. These fish do possess the 'remains' of ancestral haemoglobin genes. These genes are riddled with mutations including stop codons, thus rendering them non-functional (Carroll 2006). Given that oxygen solubility is high at cold temperatures, haemoglobin is unnecessary. The ice fish likely lost their haemoglobin genes and RBCs due to a lack of purifying selection, the selective pressure important for eliminating deleterious mutations from genomes. Weakened purifying selection on genes that lack adaptive value in the current environment will lead to the irreversible degradation of such genes. Moreover, the elimination of RBCs may actually have provided an adaptive value, leading to positive selection for RBC loss. The absence of RBCs makes for thinner blood, which is easier to pump at low temperatures. In addition, ice fish have evolved an antifreeze glycoprotein in blood and bodily fluids, and larger hearts and blood volume to improve circulation in the face of cold temperatures.

It is important to consider how these different systems in diverse animals evolved as adaptations to differential requirements based on life histories (lifespan, body size, etc.) and environments (temperature, gas availability, predation, etc.). The mammalian system does not represent the apex of evolution, but instead a different evolved solution for a particular life strategy that facilitates adaptation to the environment. Still, these systems are not 'design from scratch'—there are evolutionary constraints. As we learn in Chapter 11, the evolved design of the mammalian circulatory system leads to disease vulnerabilities, such as cardiovascular disease and stroke. We must appreciate that our circulatory and blood systems evolved in mammals and our hominid ancestors under circumstances very different from modern ones, with substantial changes in diet, longevity, habits like smoking, salt intake, and so on. The 'resulting gene–environment mismatch has led to an epidemic of hypertension, hypercholesterolemia, diabetes, and atherosclerosis' (Monahan-Earley et al. 2013). (For further discussion, see Chapter 11: Cardiovascular System.)

9.1.2 Clotting and Tissue Repair

The evolution of a circulatory system requires a mechanism to avoid fluid loss. Invertebrates like arthropods can use 'sticky cells', while vertebrates and crustaceans have independently and additionally evolved fibrin clots (Doolittle 2009). Unlike the crustacean clotting system, vertebrate clotting involves proteolytic cascades. Notably, arachnids also use a completely unrelated proteolytic cascade, another great example of convergent evolution. For vertebrates, fibrinogen makes up about 3% of plasma protein. Protease (thrombin) cleavage of fibrinogen releases fibrin, which assembles into a tight mesh to facilitate clot formation. Thrombin is also activated by cleavage, by Factor X, which is itself activated by Factor IX-mediated cleavage. Factor IX is similarly activated by cleavage, by Factor XI, which is activated by Factor XII. This process is activated by cell damage, such as from exposure to air. Alternatively, tissue damage results in release of tissue-resident Tissue Factor, which activates Factor VII, which then cleaves Factor X. Thus, we can understand how a chain of protease cleavages leads eventually to clot formation (Miller 1999). Based on deriving phylogenetic trees (relatedness based on gene sequence), we can deduce the order with which these factors appeared via gene duplication, and their relationships to each other.

These proteases appear to have evolved via gene duplication, eventually leading to a division of labour. In fact, prothrombin and Factors X, IX, XI, and VII are all highly homologous. Tissue Factor further acquired an exon from an ancestral epidermal growth factor (EGF) gene, and since many cells have EGF receptors, this directs clot formation to cell surfaces. The whole cascade appears to have evolved in the last common ancestor of vertebrates. It evolved over about 50 million years and has remained relatively unchanged since (450 million years) (Doolittle 2009).

The protease cascade provides a key advantage—amplification. Since enzymes like proteases can catalyse many events, a protease like Factor VII can cleave and activate many Factor X molecules. While an invertebrate such as a lobster has a simpler system with only two factors (a protease that activates a fibrinogen-like protein to form the clot), the evolution of a more complex protease chain allows for a much more rapid mobilisation of a clot, which would be particularly important in high-pressure and high-volume circulatory systems (particularly for mammals and birds).

A complex cascade provides numerous points where the system can be controlled, which is important for the clotting system. Inappropriate clots can contribute to strokes and heart disease. More importantly, given that a minor stimulus could lead to a clot cascade that spreads throughout the circulatory system, mammals have evolved mechanisms to buffer proteolytic activation. In particular, antithrombin is a thrombin inhibitor that prevents spurious clotting protease activation. Antithrombin evolved from a gene duplication of an ancestral thrombin-like gene. Similarly, plasminogen evolved from a similar protease gene, and its cleavage to plasmin allows for clot breakdown. In all, gene duplication has led to a complex and highly regulated system that can rapidly form a clot in response to tissue damage, preventing blood loss, while at the same time the system is well controlled enough to prevent spurious clot formation.

Thrombocytes in most animals, and platelets in mammals, further promote clot formation by forming a cellular barrier in concert with the fibrin clot. Platelets are unique to mammals—thousands form from the cytoplasm of each megakaryocyte, creating more surface area. An analogous system functions in insects using cells called coagulocytes. Both the cells themselves and released factors can contribute to clot formation. The platelets and thrombocytes of vertebrates appear to be more specialised towards haemostasis (clotting and blood vessel repair) than more multifunctional thrombocyte-like haemocytes of invertebrates, such as the horseshoe crab amoebocytes. These circulating haemocytes in their coelomic fluid (haemolymph—equivalent to blood) appear critical for haemostasis, aggregating to seal wounds (Levin 1997). Circulating haemocytes perform varied functions beyond coagulation, including reacting to bacterial endotoxins with associated phagocytosis and granule release. Notably, mammalian platelets also have some antibacterial and phagocytic functions, and can contribute to inflammation, so they appear to retain some multifunctionality. Platelets can even facilitate the expulsion of *Schistosoma* and their eggs into the intestinal lumen to mediate their elimination. These antipathogen activities may be ancestral functions maintained from the more multifunctional haemocytes (Figure 9.2).

While wound repair in mammals relies on platelets, non-mammalian vertebrates use nucleated thrombocytes (often spindle-shaped), which can also form clots. Birds have this system, despite having a high-pressure arterial system like mammals, and birds are less susceptible to clot formation (Schmaier et al. 2011). Avian thrombocytes lack fibrinogen receptor and the adenosine diphosphate receptor required for the formation of occlusive arterial clots, which are also critical targets of anticlotting drugs. These studies raise interesting and important questions. Are platelets unique to mammals due to drift, a characteristic of our last common ancestor, but not a necessary feature of being mammalian? Or was the evolution of platelets a key adaptation necessary for being mammal? Platelets do not appear to be required for homeothermia (warm-bloodedness) or for high-pressure circulation, which mammals share with birds. Nor are platelets a feature specific for viviparous births, as monotremes have platelets, but lay eggs.

(A)

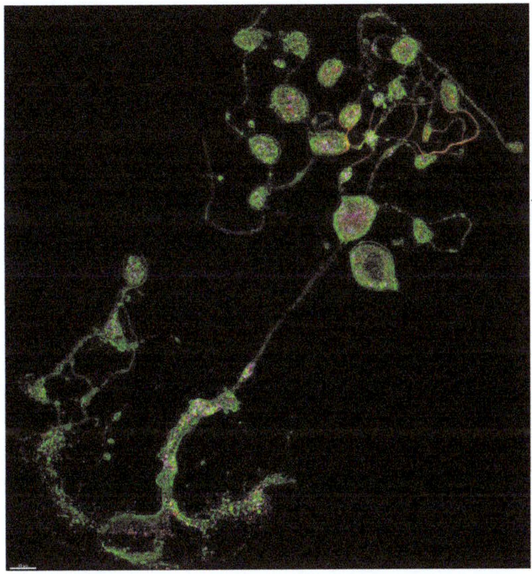

(B)

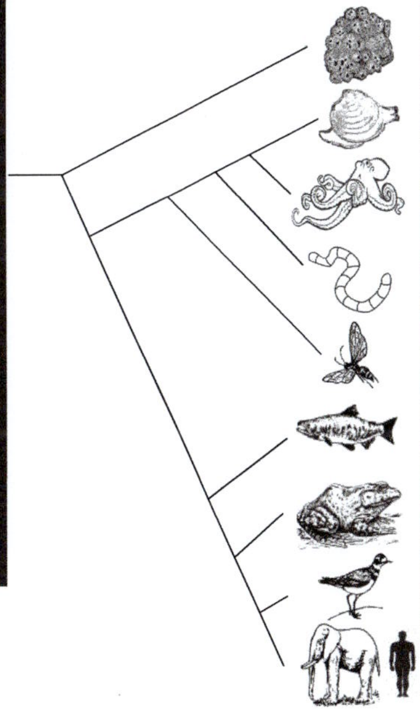

Organism	Blood pressure	Haemostatic cell	Clotting cascade
Porifera	Diffusion	None	No
Mollusk	Low	Haemocytes	No
Cephalapod mollusk	Low	Haemocytes	No
Annelid	Low	Haemocytes	No
Arthropod	Low	Haemocytes	Yes
Telost fish	Intermediate	Thrombocytes	Yes
Amphibian	Intermediate	Thrombocytes	Yes
Bird	Intermediate	Thrombocytes	Yes
Mammal	High	Platelets	Yes

FIGURE 9.2 Evolution of haemostasis systems. (A) Mouse megakaryocyte generating platelets—the main cell body (bottom left) has ejected its nucleus, and most of the cell has resolved into proplatelet extensions, including several platelet-sized terminal buds evident at top right. Immunostaining for CD41 (green), alpha granule membrane (P-selectin; red), and granule cargo (von Willebrand factor; magenta) is shown. Bar, 10 μM. (B) Key characteristics that differentiate blood clotting systems throughout evolution, including emergence of specialised clotting cells (thrombocytes) as well as presence of a proteolytic clotting cascade. Note that platelets are a relatively recent evolutionary feature associated with high-pressure mammalian circulatory systems.

Source: Photo micrograph kindly provided by Dr. Walter H. A. Kahr, University of Toronto.

Researchers have proposed that the platelet system improves the rapidity with which wounds can be healed, sealing blood vessels and preventing blood loss (Schmaier et al. 2011). They showed that mice, but not similarly sized birds, developed flow-impairing clots after arterial injury. While this system should have a clear adaptive advantage, allowing mammals to better survive injury, this system may increase the risk of cardiovascular disease and strokes. However, these costs are mostly apparent in humans at older ages, when the odds of reproductive success (for most of our evolutionary history) were minimal. Thus, the fitness costs were very low, while the benefits, particularly in ancestral times, were large. This system represents a good example of two evolutionary concepts: antagonistic pleiotropy and maladaptation to the modern world. Antagonistic pleiotropy in relation to ageing was a concept originally proposed by George Williams (Williams 1957), whereby a phenotype that contributes to ageing-associated decline or disease is offset by the selective advantage conferred in younger ages. As noted by Williams, 'natural selection may be said to be biased in favor of youth over old age whenever a conflict of interests arises'. The costs of this system have been amplified in the modern world, due not only to substantially longer lives for humans living in industrialised societies, but also to modern lifestyles and diets. So, if clotting underlies heart disease and strokes, and we have drugs to block clotting, can we fix the problem? Simple solutions belie a fundamental reality—clotting systems evolved for a reason. Just as these systems have costs and benefits, the costs and benefits of any intervention must be carefully considered.

At least in vertebrates, the immune system plays an important role in orchestrating wound healing. A type 2 immune response, mediated by cytokines like interleukin (IL)-4 and IL-13, helper T cells type 2 (Th2), basophiles (and their immunoglobulin E (IgE)), mast cells, and M2 macrophages (among other players), is critical for effective wound healing, and appears to have evolved as a defence against metazoan invaders like helminth worms (Gause et al. 2013). The response suppresses inflammation while promoting tissue repair and the expulsion of the invader. The lack of endemic worms in modern developed societies has removed the normal suppressive effects of their presence on inflammation (driven by coevolution of humans and worms), increasing the incidence of diseases like inflammatory bowel disease and type 1 diabetes. The absence of this tuning by helminths may also contribute to increased allergies in developed societies (a disconnect between modern environments and ancient genetics), and also provides a rationale for why we have IgE-driven allergic responses in the first place: to eliminate invaders like worms and other foreign substances. In all, we can appreciate how the adaptive value of a system can nonetheless contribute to disease susceptibility, and why medical interventions that disrupt these systems need to consider their evolved contributions to fitness. (For further discussion, see Chapter 10: Immune System.)

9.1.3 Host Defence

A key function for circulating cells is host defence. Animal survival is constantly threatened by a diverse array of pathogens, including viruses, bacteria, fungi, and parasites like worms. The evolution of the immune system is also covered in Chapter 10. Immune functions are classified as either innate or adaptive. Innate immunity refers to the germline-encoded ability of cells and molecules to eliminate pathogens or other entities that threaten the host (including cancers) (Janeway et al. 2001; Flajnik and Du Pasquier 2004). Innate immunity includes cells with phagocytic or other killing abilities, typically guided by receptor-mediated recognition of pathogen-associated molecular patterns (PAMPs). These PAMP receptors recognise certain classes of molecules specific to groups of pathogens, from endotoxins and flagella proteins of bacteria to double-stranded (ds) RNA of certain viruses. Many studied metazoans have phagocytic cells (from haemocytes to macrophages) that can engulf or otherwise eliminate invaders. These cells must possess the ability to distinguish self from non-self—otherwise, they would engulf and destroy other host cells.

Innate immune pathways have been studied extensively in model organisms like *Drosophila melanogaster* (fruit flies), *Caenorhabditis elegans* (soil nematodes), and *Mus musculus* (mice; with corollary studies in humans). The TOLL pathway for innate immune recognition was first described in *Drosophila*, but has very ancient origins, as it is present across animals and even in plants (Figure 9.3). In mammals, the key recognition receptors are referred to as TOLL-like receptors (TLR). Notably, different TOLL receptors and TLRs recognise different PAMPs. Loss of individual receptors, as studied in spontaneous mutant individuals or in engineered mutants, engenders defective responses to *specific* pathogens. TOLL receptors and TLR, upon engagement of pathogenic ligands, all activate nuclear factor kappa light chain enhancer of activated B cells (NF-κB) transcription factors. NF-κB

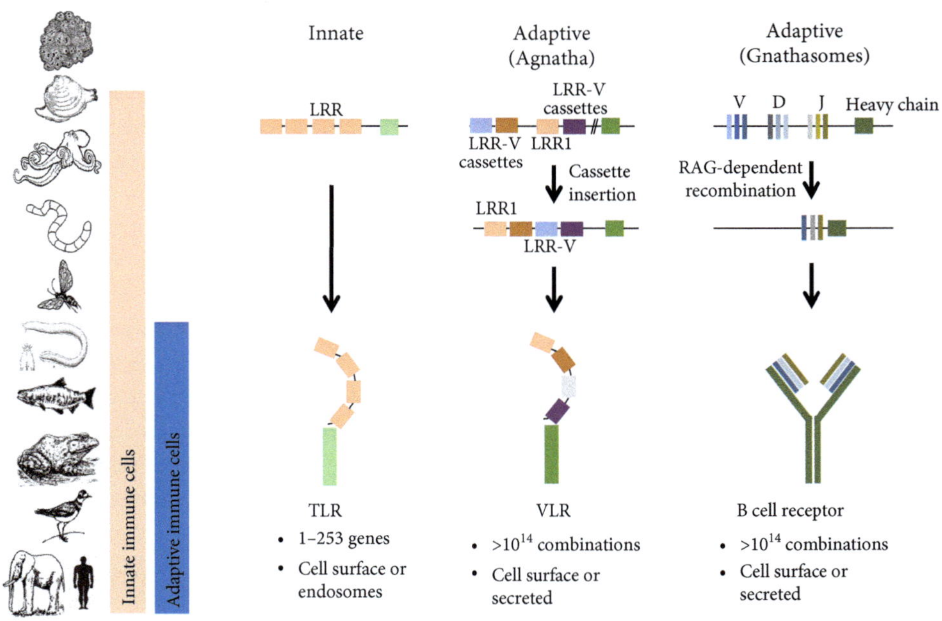

FIGURE 9.3 Evolution of innate and adaptive blood system components. Innate immune cells, commonly myeloid-lineage cells, are found in most organisms and recognise pathogens in large part via different numbers and combinations of conserved leucine-rich repeat (LRR)-containing Toll-like receptor (TLR) genes. Lymphocytes first appear in agnathans (jawless fishes), with antigen specificity and clonal diversity generated through the recombination of multiple LRR cassettes in the variable lymphocyte receptor (VLR) gene. T and B cells using recombination-activating gene (RAG)-mediated recombination of V, D, and J regions of B and T cell receptor genes first appear in bony fishes and are conserved down through mammals.

activation leads to the expression of numerous immune and inflammatory genes in order to mount an orchestrated response. Genes in the TOLL pathway, from the receptors to the MyD88 adaptor to NF-κB factors, are highly conserved at the sequence level across animals. This high conservation indicates the extreme importance of this recognition system—avoiding pathogen-mediated death is quite obviously very important for fitness. The basic mechanisms for recognition and response are very similar across animals, further validating the importance of research on model organisms, even when no medically translatable endpoint is evident.

Adaptive immunity refers to immune systems that 'adapt' to recognise foreign antigens. Germline-encoded genetic modules are modified, by recombination and/or mutation, to provide recognition that is tailored for specific antigens (Cooper and Alder 2006; Litman and Cooper 2007). Consequently, adaptive immunity can also provide immune memory—improved protection from subsequent challenge from the same pathogen. Jawed vertebrates (gnathostomes; sharks to humans) use RAG enzyme-mediated recombination of large families of variable antigen-recognition segments (V, D, and J segments in genomic DNA), which when combined create a diverse repertoire of T-cell and B-cell antigen receptors encoding for over 10^{14} possible sequences (including via junctional diversity). The resulting T-cell receptors (TCR) recognise small peptides presented on major histocompatibility complexes (MHC), displayed either by specialised antigen-presenting cells (for class II MHC) or by any cell (for class I MHC). Presentation by class II MHC can alert the adaptive immune system to the presence of foreign invaders engulfed by phagocytes like macrophages, key players in innate immunity. Presentation by class I MHC of viral antigens (for example) on a cell can lead to its elimination by recognising T cells. Finally, B-cell receptors (BCR) and the secreted immunoglobulins (BCR lacking the transmembrane segment) by B cells are critical for immune clearance of pathogens. The persistence of memory B and T cells after an infection can facilitate lifelong memory, protecting from future challenges.

Jawless vertebrates (agnathans) like the lamprey also generate a diverse array of antigen receptors, but through a very different process (*not* involving RAG enzymes): a diverse array of leucine-rich repeat modules are recombined, likely using gene conversion, to generate lymphocyte antigen receptors (Pancer et al. 2004) (Figure 9.3). That jawed and jawless vertebrates convergently evolved independent but analogous adaptive immune systems speaks to the importance for organismal fitness of the ability to respond to a diverse array of foreign antigens. The RAG1 and RAG2 recombinases appear to have been derived from a transposon of the Transib family that inserted in a common ancestor of deuterostomes (includes echinoderms and vertebrates)—we can credit our adaptive immune systems to 'selfish DNA' that turned out to be useful! Mobile genomic elements like transposons are '*naturally evolved tools for genome engineering*' which have been adapted by organisms ranging from bacteria to vertebrates (Koonin and Krupovic 2015). Over half of the human genome is derived from selfish DNA, transposons, and viral-derived sequences, and substantial fractions of genomes across all life are constituted by such sequences. While much of this genetic material likely does not contribute significantly to host fitness, some transposon or viral relics can serve as substrates for adaptive evolution by providing 'pre-evolved' functional attributes (such as the discussed recombinase activity of the ancestor of RAG1 and telomerase reverse transcriptases from yeast to vertebrates).

While the RAG enzyme (together with non-homologous end-joining machinery) was only leveraged for immune recognition diversification in jawed vertebrates (Litman and

Cooper 2007), non-vertebrates have capitalised on analogous systems. It was previously thought that adaptive immunity was limited to vertebrates, but there is evidence for somatic diversification of antigen receptors in invertebrates, such as through combinatorial RNA processing (flies) or somatic hypermutation (snails) even though they lack lymphocytes. These antigen receptor genes are also in the immunoglobulin superfamily (like BCR and TCR genes). Capitalising on the versatility of immunoglobulin domains for foreign antigen recognition seems to be a common theme across animals (Cooper and Alder 2006). Thus, some form of adaptive, protective immunity appears essential across animals to fend off microbial threats and, for some invertebrates, invasion by totipotent stem cells from other members of the species (Laird et al. 2005). Even archaea and bacteria use the clustered regularly interspaced short palindromic repeat (CRISPR) system to insert viral DNA into their genomes to generate a form of protective immunity mediated by nucleic acid recognition and cleavage (Koonin and Krupovic 2015). The CRISPR system proves *transgenerational* immunity to archaea and bacteria, and represents a form of 'Lamarckian evolution'.

The innate and adaptive immune systems are not separable defence mechanisms. They work together to coordinate the recognition, response, and elimination of pathogens. Innate immune functions such as phagocytosis lead to the presentation of pathogen antigens on class II MHC on phagocytic granulocytes and macrophages, which can then alert lymphocytes to the invasion through recognition via pathogen-specific TCRs and BCRs. The lymphocytes in turn become activated, rapidly proliferating to generate armies of T and B cells. In addition, these activated lymphocytes secrete cytokines that signal back to cells of the innate immune system.

A key mediator of cross-talk between the innate and adaptive immune systems is the interferon (IFN) response pathway. The activation of TOLL receptors and TLRs by PAMPs, such as from intracellular pathogens, leads to potent stimulation of IFN production. These receptors activate interferon regulatory factors (IRFs), increasing the expression of multiple IFN genes. IFNs then activate IFN receptors on both the infected cell and other cells in the tissue, signalling through Janus kinase (JAK) and signal transducer and activator of transcription (STAT) factors to orchestrate responses to limit pathogen propagation (including production of more IFNs, amplifying the response). The IFN response can lead to shutdown of cellular systems (such as cell cycle or RNA transcription) or cellular suicide via apoptosis of the infected cell (an attempt to prevent virus production), danger signalling to other cells, and mobilisation and activation of immune effector cells. In particular, IFNs augment the ability of CD8 T cells to kill infected cells, increase antigen presentation by MHC on immune cells, and activate innate immune cells (such as macrophages and natural killer cells). Type I IFNs, like IFNα and IFNβ, and type III IFNs (called IFNλs) are critical for antiviral responses. Type II IFN responses, via IFNγ, primarily resist bacterial, fungal, and parasitic infections. Germline mutations in IFN genes confer susceptibility to viral (for type I and III genes) and bacterial (for type II) infections, and other gene variants can increase the incidence of autoimmune disease.

Genes encoding multiple IFNs, their receptors, and signalling mediators are highly conserved across all jawed vertebrates (Secombes and Zou 2017). The diversity of IFN genes appears to have been created by a combination of genome duplication, which has happened multiple times in vertebrate evolution, and retrotransposition (Secombes and Zou 2017). The latter has resulted in the lack of introns in many IFN genes in amniotes (reptiles, birds, and mammals). Following the sequencing of the genomes of many vertebrates, researchers have acquired a good understanding of the sequence conservation of IFN genes across jawed vertebrates, from sharks to mammals (Manry et al. 2011). Genetic conservation, whereby gene sequences can show minimal changes over millions of years, can provide a good idea of functional importance, as changes that reduce function are eliminated by purifying selection. These analyses have shown that multiple type I IFN genes and the one type II IFNγ gene are under very strong purifying selection, indicative of their high value for eliminating infections. Examination of IFN sequences within the human population has substantiated this strong conservation, with many of these IFN genes showing no or almost no variation across humans. In contrast, other IFNs show dramatic changes and even mutations that result in production of a truncated protein (which should eliminate activity). Therefore, some IFN genes appear to be under very weak selection in humans, likely related to their redundancy with other IFN gene family members.

Some of the type III IFN genes show signs of positive selection in individuals from particular geographic locations, which could reflect selection for resistance to particular endemic viruses. At least one of these alleles common among Asians has been shown to enhance the clearance of hepatitis C virus (HCV) (Secombes and Zou 2017). The challenge will now be to determine why a particular change in a type III IFN gene enhances antiviral responses, which could provide a framework for manipulating this pathway to improve immunity. Different selective pressures experienced by our ancestors have influenced allele distributions for IFN pathway genes, which can have implications for disease susceptibility—increased risk of infections for alleles contributing to reduced activity and autoimmunity for alleles that potentiate activity. Both too much and too little IFN activity can contribute to morbidity, with the fitness impact of IFN allele combinations highly dependent on the environmental context experienced by our ancestors.

In all, we can appreciate the incredible power of coordinated actions of innate and adaptive immunity to recognise and eliminate pathogenic threats. Nonetheless, humans and all other animals still suffer microbial infections and parasitic invasions, leading to negative impacts on our health and sometimes to death. Evolutionary battles between organisms with competing interests result in strong selection for adaptations by both host *and* pathogen to improve their outcomes (and thus their fitness). Viruses have evolved mechanisms to counter IFN responses. Pathogens from viruses to parasites have evolved antigen shuffling that prevents efficient immune-mediated elimination—their own recombination-mediated diversification method to counter the vertebrate adaptive immune repertoire. The bacteria in our guts, as well as resident nematodes, have also evolved mechanisms to dampen immune responses, such as by enhancing T-regulatory cell activity (which suppresses adaptive immune responses). For this last case, you will learn in Chapter 10 how changes to our microbiome have unleashed immune responses in some individuals in industrialised societies. We have not had sufficient time to evolutionarily adapt to modern conditions. (For further discussion, see Chapter 10: Immune System.)

While we have discussed the deep evolutionary history of our immune systems, the more recent changes in hominid environments have stimulated rapid evolutionary changes in the DNA encoding immune functions. When humans migrated out of Africa in the second great wave around 100,000 years ago, they encountered new pathogenic threats in the climates of Europe and Asia. They also encountered, and mated with, Neanderthals and Denisovans, our evolutionary cousins, having evolved from the first wave of hominid migration out of Africa (thought to have been *Homo heidelbergensis*). When humans arrived, these Neanderthals and Denisovans were already adapted to the environments of Europe and Asia, including to pathogens present at the time. The introgression of Neanderthal and Denisovan DNA into the human lineage likely conferred adaptive advantages to these early humans. Indeed, Neanderthal DNA that remains in individuals of European decent (approximately 2% of DNA) is enriched for genes associated with responses to infections and immunity, suggesting that introgression of these immune genes allowed these humans to better counter infections in Europe (Houldcroft and Underdown 2016). Moreover, recent studies comparing immune responses against bacteria and viruses using macrophages from individuals of African or European descent reveal key differences in these responses, with gene expression differences enriched for recently introgressed Neanderthal genes and regulatory sequences in the Europeans (Nédélec et al. 2016; Quach et al. 2016). Of note, African ancestry is associated with stronger inflammatory responses to bacteria, which could contribute to substantially higher incidences of autoimmunity in women of African descent. We can now appreciate how differential contributions of Neanderthal and Denisovan DNA in humans, often dependent on the geographic origins of their ancestors, can impact disease risk, such as for autoimmunity.

Finally, in the last 10,000 or so years, the advent of agriculture, livestock rearing, and crowded living led to exposures to new pathogens (Wolfe et al. 2007). For example, many common viruses that now infect humans were transferred from domesticated animals. In addition, crowded living will select for more aggressive pathogens, as competition for hosts favours a more rapid disease course (while in small populations a pathogen will either evolve to be more commensal or likely eliminate its host population). The rate of substitutions in DNA that change coding sequence (non-synonymous changes) can be used as a proxy for evolutionary change. When this rate suddenly accelerates in a species, it can indicate adaptive change. Such is the case for immune genes, which show accelerated change in recent millennia (Vallender and Lahn 2004).

Human adaptation to pathogens has clearly affected our genetics and traits, including disease susceptibility. The ability to mount effective responses against pathogens was strongly selected for, even if the same system can contribute to autoimmune and inflammatory disease. These latter negative impacts would have minimally affected human fitness if they either manifested at older ages (when reproductive success was unlikely) or are only really prevalent in the modern world (as our immune systems are evolutionarily 'tuned' to our ancestral environments).

9.1.4 Impacts of Environmental Pressures and Disease on Human Genes Controlling Haematopoietic Function

The environment, and changes in it, is the major determinant of evolutionary change, or lack thereof. Likewise, environmental conditions have substantially contributed to genetic and phenotypic divergence among humans. Climate, food sources, predators, and pathogens exert strong selective pressures on all organisms, including humans, selecting for phenotypes that maximise survival and reproduction within each environment. As a dramatic example for the blood cell system, malaria has driven natural selection on polymorphisms in β-globin/haemoglobin and other key RBC genes, as well as in genes important in immunity (Wellems et al. 2009). Malaria is caused by the parasite *Plasmodium falciparum* or related species such as *Plasmodium vivax* and is spread by mosquitos. Large population size and strong selective pressure lead to genetic adaptation, and such conditions have been clearly present for affected humans. Hundreds of millions of humans are infected with malaria globally. In previous centuries, up to 20% of African children died in their first decade from malaria (even though most infections are relatively mild).

Various genetic polymorphisms confer resistance from the lethal impact of malaria. The HbS allele of β-globin (HBB gene) confers such resistance, but also leads to sickle cell anaemia, which is lethal in the homozygous state. Similar HBB mutations have arisen at least five independent times globally. This adaptation appears to be relatively recent, as genetic evidence indicates that malarial parasites only made the jump to humans (from gorillas) in the last 10,000 years in Africa (Liu et al. 2010). The malarial resistance conferred by heterozygosity for HbS is balanced in the population by the lethality of HbS homozygosity, with this trade-off limiting the prevalence of this allele in the population (Figure 9.4). For individuals living in areas with endemic malaria, the HbS allele in the heterozygous state confers greater fitness than homozygosity for either the parental or HbS allele, referred to as *heterozygous advantage*. In countries where malaria is rare, the HbS allele now only confers a disadvantage, another example for how changes in the environment (such as movement of humans from Africa to Europe or the United States) can lead to maladaptation in the new environment.

HBB mutations (such as HbS) result in reduced sequestration of malaria parasite-infected erythrocytes in microvessels of critical organs (like brain, heart, liver, and lung), which normally contributes to disease severity (Wellems et al. 2009). These alleles also improve fetal fitness in infected mothers. While protecting from lethal malaria, these polymorphisms

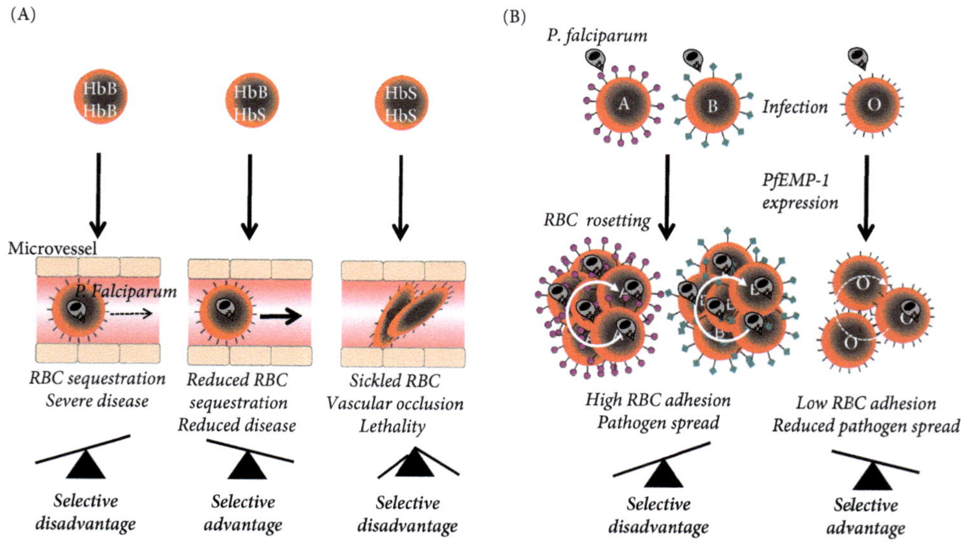

FIGURE 9.4 Erythrocyte haemoglobin and ABO blood groups in pathogen resistance. (A) Impact of HbS heterozygosity and homozygosity on RBC sequestration and malaria severity. Note significant selective disadvantage to HbS homozygosity. (B) Impact of ABO blood group on RBC rosetting and *P. falciparum* spread.

confer less protection against uncomplicated malaria or from infection in general, which may explain why *P. falciparum* has not evolved resistance to these defence mechanisms. In this case, HbS alleles really represent a tolerance mechanism, rather than a resistance mechanism, as the former reduces the negative impact of the disease on host fitness without necessarily affecting pathogen fitness. Similarly, survivors of childhood infections with *P. falciparum* develop sufficient immunity to typically protect from lethal disease upon subsequent challenges, but this does not prevent parasite infection and propagation. Based on understanding the evolutionary battle between humans and malarial parasites, we can speculate as to whether one could design a drug or vaccine that mimics either what HBB polymorphisms do or what acquired immunity does. If the therapy or vaccine does *not* target the parasite for elimination, but just lessens disease severity, there should not be selective pressure for the development of parasite resistance.

In general, we should be able to use population genetics and evolutionary theory to understand the presence and persistence of disease-predisposing alleles in the human population. In some cases, disease-causing alleles arise infrequently enough to be explained by the mutation–selection balance. Even if the mutation only confers a fitness disadvantage, depending on the severity of this effect it may take multiple generations to eliminate the allele from the population. This balance is particularly evident for alleles that cause disease only in the homozygous state, such as for Fanconi anaemia. If the allele is infrequent enough in the population, and is mostly maladaptive in the homozygous state, then the elimination of this allele from the population could be balanced by the new occurrence of these gene mutations. On the other hand, as is the case for HBB polymorphisms, the persistence of an allele that promotes disease in the homozygous state with an allele frequency greater than expected based on mutation–selection balance could indicate a heterozygous advantage yet to be discovered.

Malaria highlights an evolutionary battle between pathogen and host. Even within a host, somatic evolution of immune cells is countered by the evolved ability of the parasite to shuffle through different displayed antigens. Immunity still plays a role, as prior exposures greatly reduce the severity of subsequent exposures to malaria (without preventing them). Humans have developed multiple drugs to treat malaria, targeting the *P. falciparum* parasites for elimination. However, large numbers of parasites plus strong drug-conferred selective pressures have rapidly led to the evolution of drug resistance by these parasites in each case (White 2004) . Notably, when drug application is removed from a region, drug sensitivity of the disease returns (Laufer et al. 2006).

The drug-resistant phenotype is only adaptive for *P. falciparum* with the drug, as the genetic change conferring this resistance clearly came with a cost to the parasite's fitness. Each gene in a well-adapted organism should be optimised to maximise fitness in a particular environment. While a mutation that confers drug resistance can clearly improve parasite fitness in the *presence of the drug*, the underlying phenotypic change is likely to be disadvantageous in the absence of the drug, as the original phenotype was optimal in this environment. Consequently, we can speculate that public policy that mandates periodic changes in the antimalarial drug used for a particular region could improve the ability to maintain a treatable disease. This policy would need to leverage evolutionary theory, and consider the fitness costs of the malarial resistance, whether the resistance mechanisms are distinct for different drugs, and the rapidity of changes in allelic frequencies (for resistance) among malarial parasites.

Other genetic alleles conferring resistance to malaria can result in thalassaemias and glucose-6-phosphate dehydrogenase (G6PD) deficiency in homozygous states, eliciting similar trade-offs as seen with HBB mutations (Wellems et al. 2009). α-Thalassaemia results from deletions in the HBA1/HBA2 genes, leading to reduced expression of complement receptor, which normally mediates binding of non-infected erythrocytes to parasite-infected erythrocytes, reducing severe anaemia. Mutations in the G6PD gene that lead to G6PD deficiency enhance the phagocytosis of parasite-infected erythrocytes. Moreover, some recently described polymorphisms in glycophorin (GYP) genes may reduce *P. falciparum* entry into erythrocytes, and associate with reduced rosetting (Wassmer and Carlton 2016). However, these do not appear to be associated with overt defects in the blood system. Additional protective polymorphisms have been described for the FLT1, TLR genes, and MyD88 genes regulating inflammation, and the IRF1 gene that is key for T-cell activity. The protective MyD88 polymorphism has been shown to dampen TLR2 signalling. Interestingly, tumour necrosis factor-α (TNFα) promoter polymorphisms that increase expression enhance parasite clearance but also increase severity of disease (by increasing inflammation). In all, it appears that 'overreaction' to the parasite infection by the host in terms of inflammation can enhance disease severity.

In a similar vein as HBB polymorphisms, ABO blood group evolution in humans is closely linked to the species' relationship with malaria, as well as other pathogens. Independently described by Karl Landsteiner and Jan Jansky in the early twentieth century, the ABO blood type classification scheme is of critical importance in transfusion medicine. The *ABO* gene encodes a glycosyltransferase expressed primarily on erythrocytes, but also on endothelial cells and kidney parenchyma. The ABO glycosyltransferase decorates the H antigen sugar (also termed substance H) with either an N-acetylgalactosamine (GalNAc), a galactose (Gal), or no additional sugar moiety (Yamamoto et al. 2012). Strikingly, three separate alleles of the *ABO* gene exist in humans and many other primate species, resulting in differing enzymatic activities of the glycosyltransferase. The A allele, which is considered the ancestral ABO variant in humans and primates, leads to deposition of GalNAc on the H antigen, whereas the B allele, in humans differing by four amino acids, leads to Gal deposition on H. On the other hand, the O allele, generated by a frameshift at amino acid 261, is enzymatically non-functional, leaving H with no additional sugar decorations.

Early in life, the immune system develops tolerance towards the ABO antigen expressed in the individual, and antibodies against the antigens are not present. Hence individuals whose erythrocytes express only A antigen develop antibodies against B, B against A, and O against A and B, with AB individuals expressing both antigens producing no antibodies. Most of these 'naturally occurring' antibodies are of an IgM subtype, with IgG and IgA species also occasionally present. How is the immune system educated to produce these antibodies? Intriguingly, this process is thought to result from sampling of gut microbiota and/or exposure to pathogens (Makivuokko et al. 2012), which have similar moieties present on their surfaces, thereby having potential roles in downstream immunity against specific pathogens, a point to which we will return in subsequent paragraphs. Such an immunological mechanism is discussed in more detail in Chapter 10, and further underscores the role of the microbiota in educating the immune system at an early age in humans and other animals.

The clinical importance of the ABO system is well documented, particularly in the transfusion and organ transplantation fields. Transfusion of erythrocytes with a specific antigen into individuals expressing antibodies against that antigen—for example, type A erythrocytes transfused into a type B individual—leads to a rapid and potentially fatal rejection process termed acute haemolytic transfusion reaction, in which anti-A antibodies bind to the donor cells and induce a rapid activation of the complement system and other immune components (Savage 2016). This results in highly toxic haemolysis and haemagglutination reactions. Similar rejections are known to occur in response to transplantation of ABO-incompatible (ABOi) tissues such as the kidney. Clinical approaches that mitigate rejection of ABOi tissues include anti-A/B antibody depletion via apheresis and B-cell immunomodulation using depletion drugs such as Rituximab (Bohmig et al. 2015). Importantly and more evolutionarily relevant, as blood transfusion or organ transplantation are extremely recent developments in the arc of human existence, ABO incompatibility between mother and fetus can result in haemolytic disease of the new-born. This alloimmune condition is potentially fatal to the fetus, and is particularly common when type O mothers carry a type A or B fetus, due to a higher proportion of anti-A/B IgG, which, unlike IgM, can cross the placental barrier. So why do humans exhibit polymorphism for these blood groups?

The ABO blood groups exhibit a unique geographical distribution, related quite strikingly to patterns of malaria incidence, with the O allele predominant in Africa, as well as in Central and South America where malaria is common. Such differing allele frequencies suggest ABO alleles are under balancing selection, offering an advantage under certain circumstances. The A and B alleles arose at least 20 million years ago (mya) and are present in many other primate species (Segurel et al. 2013). The maintenance of genetic polymorphisms for such a long period would be extremely rare by chance alone, thus supporting a role for balancing selection for the persistence of ABO types. Notably, while distinct loss-of-function alleles have independently arisen in multiple primate species, single nucleotide polymorphism (SNP) genotyping analyses of human populations reveal that the frameshift mutation giving rise to the O allele has arisen in the human population independently at least twice, indicating a strong selective pressure for its emergence during human history. Interestingly, the human O alleles are evolutionarily more recent, with two emergences of the two most prevalent O frameshifts at 2.5 and 1.15 mya, well before the emergence of modern *Homo sapiens* (Calafell et al. 2008).

How does the O allele impact parasitisation of erythrocytes by *P. falciparum*? The key lies in erythrocyte rosetting, the pathogenic mechanism by which infected erythrocytes are induced to express the protein PfEMP-1, a *Plasmodium*-encoded gene, on their surface, allowing them to bind multiple surface antigens on uninfected erythrocytes, platelets, and endothelium, and generating rosette-like formations that provide easy access to new cells for *Plasmodium* to parasitise (Cserti and Dzik 2007) (Figure 9.4). Erythrocytes with the O antigen display reduced levels of rosetting, likely in part due to decreased adhesion with other cells (Rowe et al. 2007). Interestingly, the O allele is also associated with reduced expression of von Willebrand factor (vWF), a cytokine secreted by platelets and endothelium that drives blood clotting and may also promote erythrocyte adhesion during malaria infection. Still, since ABO antigens are not expressed on red blood cells (RBCs) from both New World and Old World monkeys, and yet these species maintain ABO polymorphisms, balancing selection likely did not arise from selective pressure exerted by *P. falciparum* infection of RBC (Segurel et al. 2013), although such pressure could have contributed to the maintenance of polymorphism in apes. Notably, polymorphism at the ABO glycosyltransferase in primates extends well beyond the A, B, and O alleles, which have been defined by haemagglutination patterns of blood, and could be very relevant for fitness given the expression of ABO antigens in other tissues susceptible to pathogens. In fact, the ABO glycosyltransferase is one of the most polymorphic genes in humans.

Further contributing to the advantage of the O allele, individuals expressing anti-A/B antibodies may have greater resistance to viral and bacterial pathogens such as influenza and *Escherichia coli*, which in some cases possess similar moieties (Yang and Boettcher 1992). Different alleles of the ABO glycosyltransferase influence cellular entry of pathogens that use the transferred carbohydrate for cell binding, which could provide resistance to O homozygotes (Calafell et al. 2008). The expression of ABO antigens in the gut epithelium of all examined primates suggests that selective pressure mediated by gut pathogens may have been key. It is particularly interesting to note that reduced expression of vWF and Factor VIII, and therefore reduced efficiency of blood clotting following a wound, could constitute a significant disadvantage for the survival of ancestral humans during conflicts or following accidents. In this sense, in tropical areas one might anticipate that child mortality due to cerebral malaria is likely a stronger selective pressure than injuries due to accident or fighting. Interestingly, in modern-day life, humans with O alleles are likely to have reduced susceptibility to venous thromboembolism (VTE) than non-O individuals (Dentali et al. 2012), which may provide resistance to the negative effects of high-fat western diets on health and lifespan relative to populations with A and/or B alleles. It is also interesting to note that the B allele is particularly common in southern Asia, and is likely selected for due to increased resistance to enteric pathogens such as *Vibrio cholera* and *Helicobacter pylori*, likely by virtue of ABO antigen expression in gut epithelia. Conversely, the O allele appears to confer susceptibility, emphasising the evolutionary impact of ABO antigens in other tissues aside from the blood (Anstee 2010).

In sum, individuals that are O homozygotes would appear to have greater protection from pathogens due to the production of anti-A and anti-B antibodies, impaired pathogen binding to cells via the glycosyltransferase-generated cell surface carbohydrates, and reduced RBC rosetting during malarial infection, which is balanced by the cost of immune incompatibility between mother and fetus (for O mothers carrying fetuses with A and/or B genotypes). Since pathogens may use either A and B glycosylation marks for infection, there could be both selection against AB individuals and frequency-dependent selection preventing dominance by either A or B in the population. For the former, contrasting with heterozygous advantage for HBB polymorphisms that provide protection from malaria, AB heterozygosity may be disadvantageous in terms of pathogen resistance. In addition, frequency-dependent selection can arise from the selection for pathogens that target the most common genotype in a population, which can lead to fluctuation in allele dominance (and the maintenance of variation) in the targeted population as both pathogen and host adapt in an ongoing evolutionary battle.

Finally, similar to ABO, variation in the human Rh blood group at the RHD and RHCE genes is common in human populations. Until recently, RhD-positive children of RhD-negative women were at substantial risk of haemolytic anaemia of the new-born, if the mother produced anti-D antibodies generated from prior gestation of an RhD-positive child (Perry et al. 2012). Haemolytic anaemia in the new-born results from anti-D IgG, produced by a D^{neg} mother, crossing the placental barrier. This risk appears to have been as high as approximately 1 in 50 births of Rh- positive children to -negative mothers. And yet this polymorphism is highly prevalent in humans. First discovered by Landsteiner and

Wiener based on the ability of sera from the Rhesus macaque to haemagglutinate human blood, the Rh genes RHD and RHCE encode for a large, multi-subunit ion transporter of unknown biological significance that is expressed on the surface of erythrocytes (Perry et al. 2012). Interestingly, the D^{neg} genotype is uniquely enriched in Europe, particularly in the Iberian Peninsula, but also in the Balkans and the Ukraine. Is this a result of a selective advantage for the D^{neg} allele that counterbalances the cost of fetal mortality due to mother–fetal incompatibility, or is there another evolutionary mechanism in play? Strikingly, these areas overlay with known human refuges during the last Ice Age. Analyses of Y-chromosome markers by Semino and others has suggested that most modern Europeans are descended from relatively small founder populations in the Palaeolithic period who expanded in response to a warming climate, and intermixed with Neolithic farmers migrating to Europe from the Middle East (Semino et al. 2000).

Well before these studies, it was speculated by Mourant that prevalence of the D^{neg} allele may result from intermixing of D^{neg} and D^{+} populations during this time, with the founder effect resulting in increased frequency of the D^{neg} allele (Anstee 2010). Supporting this theory, most Europeans carry the same form of D^{neg} (a deletion), whereas D^{neg} populations in Africa and Asia carry a 37-basepair (bp) insertion or silenced expression, respectively. These studies would indicate that the prevalence of the D^{neg} allele is more the result of a founder effect rather than ongoing positive selective pressure. Consistent with this role for genetic drift, analyses of haplotype genetic data from many individuals did not reveal any signals of positive selection on RHD deletion (Perry et al. 2012). Perhaps the negative fitness effects of the D^{neg} allele could have been less strong through ancestral times than previously thought, consistent with epidemiological studies that often fail to observe differences in numbers of surviving offspring for D^{pos} and D^{neg} mothers.

9.2 Haematopoietic Development

On any given day, the blood system of an adult human produces anywhere between 10^{11} and 10^{12} new blood cells in different lineages, including erythrocytes, megakaryocytes, lymphocytes, and myeloid cells, in order to maintain normal peripheral blood levels. Such a scale of production is particularly noteworthy given estimates that indicate the number of blood-forming haematopoietic stem cells (HSCs) in the bone marrow of an adult human is a vanishingly small pool of 10,000–300,000 cells, or approximately 1 per 10^{8} to 1 per 10^{6} nucleated cells in the marrow (Wang et al. 1997; Abkowitz et al. 2002). Even more remarkably, this estimate is no greater than the number of HSCs in a mouse, despite the fact that humans live longer and have much larger blood systems to generate. That such a small pool of stem cells, first formed from an even smaller number of fetal HSCs emerging during embryogenesis, can function as a lifelong genetic and functional reservoir that builds and subsequently maintains such a complex, high-turnover liquid tissue seems on its face counterintuitive. How could evolution favour a solution to the problem of tissue building and maintenance that requires so much output from such a rare cell population, with little apparent redundancy in numbers to support lifelong production? Just as importantly, how can such a system adapt to both short- and long-term needs related to physiology or environment? In this section, we will explore the evolution and development of the human haematopoietic system, its unique hierarchical structure and adaptability to change. We will come to understand, if incompletely, how such a system could in fact be an evolutionarily robust solution to the problem of tissue maintenance, at least through the reproductive years of an organism.

9.2.1 What is (and isn't) a Haematopoietic Stem Cell?

Before understanding the developmental and evolutionary history of HSCs, it is critical to understand the functional properties by which these cells are defined. In the most basic sense, an HSC is not simply a phenotypic definition based on surface proteins or other markers, but instead is the basic functional unit of blood production, from which all mature lineages of blood cells can arise. Thus, multipotency is a key defining feature of an HSC, but is by no means the only defining feature (Figure 9.5). In addition to its capacity to generate all mature lineages of blood cell, an HSC, by definition, is a self-renewing cell, capable of undergoing asymmetrical divisions in which one daughter cell is a differentiated progeny that loses its self-renewal capacity while activating molecular programmes that define commitment to a particular lineage(s), whereas the second daughter cell retains its stem cell properties and is in essence a 'copy' of the mother cell. These features were defined based on

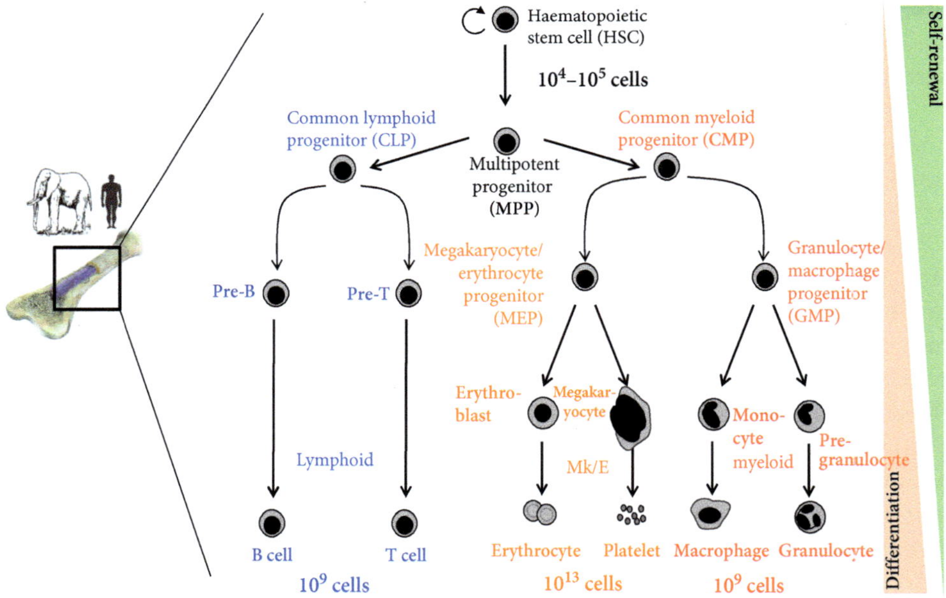

FIGURE 9.5 Mammalian haematopoiesis. Mammals possess a hierarchically organised blood system that is generated in the bone marrow by a rare population of HSCs. This relatively small number of largely quiescent, self-renewing stem cells gives rise to progressively more lineage-committed progenitor cells such as common myeloid and lymphoid progenitors, that in turn specify into lymphoid, megakaryocyte/erythrocyte (Mk/E), and myeloid-lineage cells. Note the degree of amplification between the relatively small number of HSCs in human bone marrow and mature cells in peripheral blood of an adult human.

seminal experiments conducted in the mid-twentieth century by McCullough and Till, in which different numbers of bone marrow cells from a donor mouse were transplanted into a lethally irradiated recipient animal, hence defining a dose curve in which a specific number of transplanted marrow cells could provide radioprotection and reconstitute normal blood production in the recipient (Till and McCulloch 1961).

Subsequent transplantation studies in mice and other vertebrate model organisms defined HSC self-renewal capacity as the ability of donor marrow cells, and in fact even a single cell, to migrate to the bone marrow and serially reconstitute the blood system of multiple irradiated recipients, thereby differentiating HSCs from non-self-renewing haematopoietic progenitors in which the blood-forming activity is extinguished following a single transplantation (Keller and Snodgrass 1990). Following these experiments, the advent of flow cytometry has allowed investigators to identify panels of cell-surface proteins and functional markers that enrich for HSCs (Spangrude et al. 1988; Kiel et al. 2005); notably, none of these enrichment approaches produces a functionally pure HSC population in which every cell is a serially repopulating long-term haematopoietic stem cell, due both to technical limitations and the emergence of functionally distinct subsets of cells with differing reconstitution activity within phenotypic HSC compartments (Wilson et al. 2008). Nonetheless, these phenotypic definitions do allow unprecedented access to highly enriched functional populations for further analyses, and the relative ease with which these populations are isolated makes the haematopoietic system a compelling model in which to understand the evolution and life cycle of this complex tissue.

9.2.2 Waves of Haematopoietic Development

The vertebrate haematopoietic system develops during embryogenesis in two successive waves: primitive, or embryonic, haematopoiesis, which functions to transiently provide a limited array of blood cells to the developing embryo prior to the emergence of a fully formed vascular system; and definitive haematopoiesis, in which true self-renewing, long-term HSCs emerge from the haemogenic endothelium and subsequently expand to form the foundation for lifelong, multilineage blood production. Primitive haematopoiesis in mammals occurs primarily in blood islands located in the yolk sac, giving rise to erythroid cells, as well as myeloid cells such as macrophages, that subsequently colonise the embryo (Medvinsky et al. 2011). Indeed, myeloid cells, particularly macrophages, are a common and likely ancient denominator in haematopoietic system evolution, perhaps tracing their origin to the phagocytic cells present in *Drosophila* larvae well prior to the evolution of erythroid and lymphoid lineages, and possibly even to the phagocytic haemocytes present in mollusks lacking a closed circulatory system. Along these lines, tissue macrophages in mammals have been shown to develop essentially independently from definitive haematopoietic cells, which require the transcription factor Myb for their emergence (Schulz et al. 2012) during development. This suggests that these macrophages arise during primitive haematopoiesis. Some of these macrophages, particularly microglia, appear to persist even through adulthood (Ginhoux et al. 2010). In addition, primitive macrophages have been shown to play a key role in 'assisting' the generation of definitive HSCs from the haemogenic endothelium of zebrafish embryos (Travnickova et al. 2015). Hence, primitive myeloid cells can serve multiple functions, including host defence, removal of apoptotic

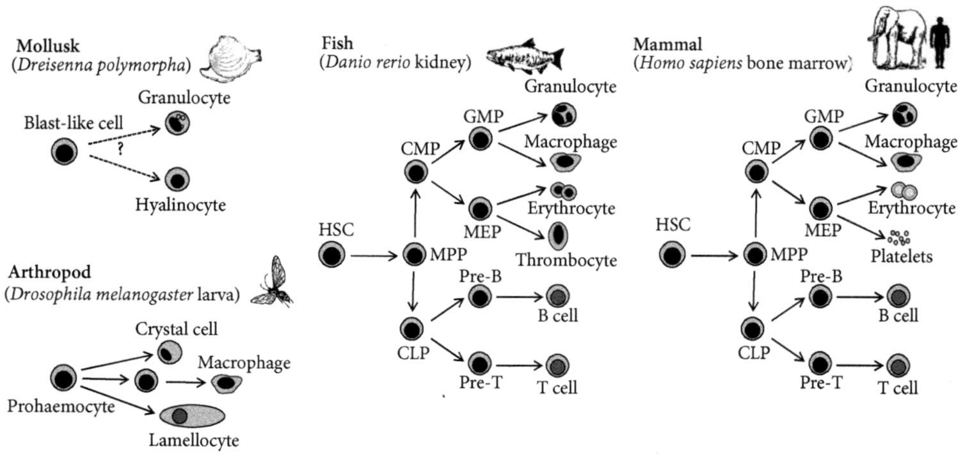

FIGURE 9.6 Comparison of haematopoiesis across evolution. Representative blood system hierarchies in non-cephalopod mollusks, arthropods, teleost fishes, and mammals, depicting organisation of haematopoietic progenitors as well as lineage output. Note that myeloid-lineage cells, particularly macrophages, are common to all species across evolution.

cells, and other duties that are critical for efficient development. Based on their common functional properties and similar cellular compartments, we might be able to draw an ontogenic connection between early, myeloid-based haematopoietic systems and primitive haematopoiesis in vertebrates (Figure 9.6). In this sense, retaining the primitive myeloid system, particularly as a facilitator of tissue development and remodelling, likely constitutes an evolutionary advantage for complex organisms such as fish or mammals.

Importantly, the primitive haematopoietic wave generates short-term erythromyeloid progenitors (EMP); it does not generate long-term HSCs that form the lifelong foundation of the vertebrate haematopoietic system. Instead, this process occurs in the second, or definitive, haematopoietic wave, where HSCs are specified from haemogenic endothelial cells localised in the aorta–gonad–mesonephros (AGM) region of the embryo, which is derived from the mesodermal layer of the blastocyst. This process, which requires Notch signalling and expression of the transcription factors Myb and Runx 1, results in the specification and budding of HSCs from the floor of the AGM (Medvinsky et al. 2011). Interestingly, HSCs retain at least some residual evidence of their endothelial origin well after their emergence, and continue to express endothelial surface markers such as CD31, EPCR, and ESAM (Baumann et al. 2004; Yokota et al. 2009; Iwasaki et al. 2010). This shared heritage with vascular endothelium is probably ancient; in mollusks, such as the snail *Biomphalaria glabrata*, haemocytes emerge from the anterior pericardial wall (Zhang et al. 2016). Likewise, in *Drosophila* larvae, haematopoietic cells emerge from the lymph gland, with the AGM a key source of definitive HSCs in vertebrates. Why would haematopoietic emergence from vasculature be advantageous for fitness? On the one hand, it makes logistical sense. As circulatory systems evolve into hybrid or closed systems, the most proximal site for subsequent development of cells that move through the vasculature would be the vascular endothelium itself. Interestingly, the biophysical properties of the vascular system also have an impact—specifically, shear stress and resulting nitric oxide (NO) signalling, an angiogenic signal, is produced by endothelial cells in response to mechanosensation of blood flow (Adamo et al. 2009). Hence, active blood circulation leads to upregulation of Myb and Runx1, and blockade of this mechanism by exposing embryos to the NO inhibitor L-NAME impairs HSC specification (North et al. 2009). Therefore, definitive haematopoietic development can take advantage of pre-existing factors such as macrophages and shear forces to ensure correct timing of specification.

A final point worth noting is the emerging role of inflammatory signalling in HSC specification during definitive haematopoiesis. In a number of vertebrate models from fish to humans, a growing body of work has identified a role for inflammatory cytokines normally associated with host defence against pathogens, such as TNF, IFN-γ, and signalling via TLR4 and its adapter MyD88, in promoting efficient haematopoietic specification via activation of Notch ligand expression (Espin-Palazon et al. 2014; Li et al. 2014). These factors are produced by myeloid cells, likely formed during primitive haematopoiesis, and this again highlights the role of primitive haematopoiesis in establishing conditions for the definitive wave. Interestingly, genetic knockout of these inflammatory factors does not prevent HSC specification, but instead simply slows it. From an evolutionary standpoint, many inflammatory signalling pathways (such as the TLR and TNF pathways) are ancient mechanisms that among other functions can induce cell proliferation and survival, along with tissue repair or regeneration pathways. It is no stretch of the imagination for us to see how appropriately timed induction of such a host defence system can have a dual role in 'speeding up' the HSC specification process by promoting the recruitment of macrophages, production of regenerative and angiogenic signals, and supporting the proliferation and survival of nascent HSCs.

9.2.3 A Nomadic Journey during Fetal and Adult Haematopoiesis

Once HSCs are specified in the AGM, HSCs colonise a variety of sites in the fetus and begin producing mature blood cells (Figure 9.7). In mammals, definitive HSCs can be found in the AGM, placenta, and yolk sac early in development. These cells subsequently migrate to niches in the fetal liver and spleen, and once the bones develop, migrate to the bone marrow (Orkin and Zon 2008). Likewise, the bone marrow is the final home of choice for haematopoiesis in birds (Godin and Cumano 2005). In fishes, however, the kidney, rather than the bones, is colonised by HSCs following their emergence from the AGM (Jing and Zon 2011). What is the selective advantage to colonising these different sites? A key common denominator is vascularity. Kidneys, spleen, bone marrow, and liver are all highly vascularised organs harbouring large numbers of sinusoids, large blood vessels characterised by low flow rate and high permeability, facilitating transport of cells across the endothelial barrier, as well as easier entry and egress of HSCs and their progeny cells to and from the bloodstream. Such characteristics are also present in the lung, which has been recently described as a site of haematopoiesis in mammals (Lefrancais et al. 2017).

In addition, timing clearly plays a role, with locations such as the fetal liver playing intermediate host until the lifelong niche in the bone marrow is prepared. The bone marrow itself would be a particularly attractive location due to not only its vascularity, but also protection from physical damage offered by such a hardened location. Interestingly, the

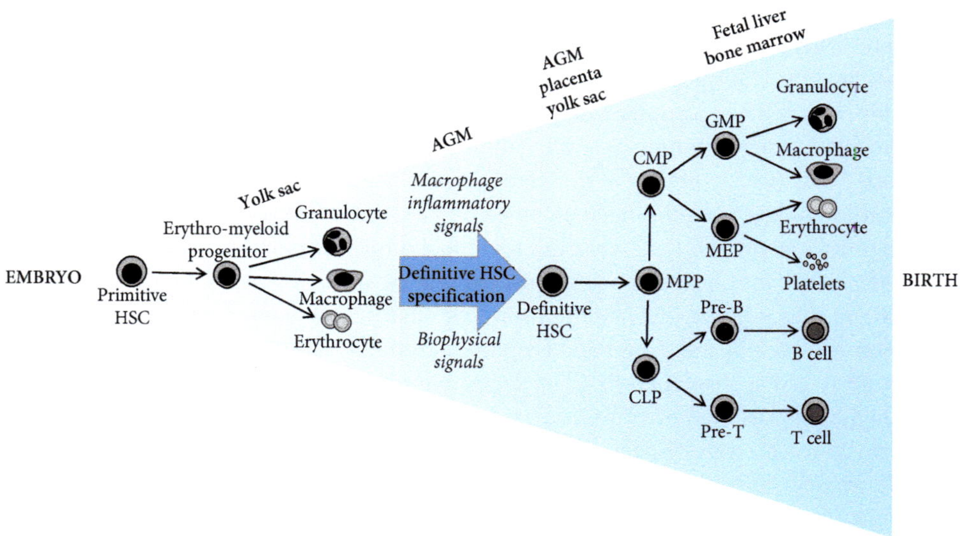

FIGURE 9.7 Timeline of haematopoietic development and colonisation in mammals.

marrow itself is a dynamic organ, and haematopoiesis continues to shift locations through adulthood as the marrow becomes more adipose. While HSCs colonise most bones at birth, in humans haematopoiesis rapidly retreats from the long bones of the arms and legs, followed by the sternum and ribs in mid-life, with the vertebral column serving as a remaining site of haematopoiesis in older individuals (Fernandez and de Alarcon 2013); this gradual retreat is not observed in mice, suggesting it is not a common property of mammalian haematopoiesis, but perhaps species specific. One possibility is that this progressive shutdown improves fitness by conserving energy for other functions such as hunting or mating/child rearing; at the same time, the number of phenotypic HSCs expands in older individuals (which we discuss more thoroughly in Section 9.3), perhaps in some way compensating for the reduction in available bones to colonise.

Once localised in the bone marrow, HSCs expand significantly in number and actively proliferate to build all lineages of blood cells, with mature, functional B and T lymphocytes generally appearing last (though their progenitors are produced in a similar timescale as myeloid cells during definitive haematopoiesis), likely due to distinct requirements for their development in the bone marrow and thymus, respectively (Dorshkind and Montecino-Rodriguez 2007). In this way, the haematopoietic system appears to build itself in a manner reflecting its evolutionary history, beginning with ancient myeloid lineages and concluding with the development of functional lymphocytes, which first appear in a more primitive form in jawless fishes, and in their contemporary form, featuring RAG-mediated somatic hypermutation, in jawed fishes. We thus might that argue the haematopoietic system maximises use of available resources, producing in sequence only immediately needed cell types, and relying on *in utero* protection by the mother's immune system, as well as acquisition of maternal immunoglobulins after birth in mammals, when the adaptive immune system remains immature.

In the first years of life in humans, HSCs undergo a significant reprogramming event, shifting from a highly proliferative phenotype geared towards building the blood system, to a dormant, or quiescent, phenotype with more limited self-renewal capacity geared towards HSC maintenance (Pietras et al. 2011). This molecular switch is driven by a number of cell-intrinsic factors, including the transcription factor Sox17, the cell cycle inhibitor CDKN2C (p18), and the Lin28B-let7 axis (Copley et al. 2013), which has been shown to regulate the function of multiple stem cell types. This shift to quiescence may facilitate the function of HSCs as a lifelong genetic reservoir for blood maintenance by limiting proliferation-associated stressors including reactive oxygen and nitrogen species, as well as potential DNA damage and mutations resulting from genome replication. Indeed, studies using molecular barcoding approaches to track HSC clonal activity have suggested that HSCs generally do not participate in day-to-day blood production, with certain highly dormant HSC subsets entering the cell cycle once every 40 weeks in humans (Pietras et al. 2011; J. Sun et al. 2014b). In this way, quiescence could limit the potential for leukaemic transformation or haematopoietic failure, at least up to and during the reproductive phase of life.

9.2.4 Haematopoietic Hierarchy and Fate Control

If HSCs are not involved in day-to-day maintenance of the blood system, how do these exceedingly rare cells meet the daily demand for 10^{11}–10^{12} new blood cells in a healthy adult? Likewise, how are different lineages of blood cells generated from multipotent HSCs? The answers lie in understanding the complex hierarchical nature of haematopoiesis, and how cell fate decisions are regulated in this context. As pointed out in Section 9.2, HSCs are rare, as many tissue stem cells are, which likely evolved as a way to limit energy expenditure on maintaining self-renewing stem cell populations, which, as we will see in the case of HSCs, require complex niches with multiple signalling outputs, to maintain (Figure 9.5).

The production of mature myeloid, erythroid, lymphoid, and platelet populations is the result of a continuum of successive fate decisions that lead to lineage commitment and, eventually, differentiation into cells of that lineage. This continuum of fate decisions is often depicted as an inverted hierarchical tree, with HSCs at the apex, branching into different lineage-committed progenitor cells and ultimately mature blood cells (Orkin and Zon 2008). Across evolution, this process varies in terms of the number of isolatable intermediate populations, that is, progenitor cell types representing specific lineage decision points. Thus, *Drosophila* appears to have fewer known intermediate stages between its HSC equivalent and the mature cell populations it gives rise to, relative to mammals. Perhaps this is related to the fact that fewer lineage decisions need to be made in the differentiation process. Intermediates could also allow for greater amplification of mature cell numbers while maintaining tight control of decisions regarding production. Nonetheless, many of the key molecular regulators required for haematopoiesis, such as RUNX, are highly conserved between *Drosophila* and humans, suggesting common evolutionary heritage of the overall mechanism.

In mammals, production of mature blood cells is thought to begin with an HSC undergoing an asymmetric division that generates a non-self-renewing, actively dividing multipotent progenitor (MPP). Myeloid cells arise from the MPP compartment via differentiation into lineage-committed granulocyte–macrophage progenitors, followed by immature granulocyte progenitors, and finally granulocytes. Similar successive differentiation processes occur for other lineages, with common lymphoid progenitors giving rise to pre-B cells or early thymic progenitors that migrate and differentiate into T cells in the thymus. Erythrocytes and platelets undergo unique end-stage differentiation processes, culminating in the enucleation of erythrocyte progenitors to generate the mature erythrocyte, and successive rounds of endomitosis in megakaryocyte progenitors, leading to polyploid (≥ 2N DNA content during G_0/G_1 phases of the cell cycle) megakaryocytes that send cytoplasmic processes termed proplatelets into the vascular lumen, where they are sheared into platelets.

Importantly, the haematopoietic differentiation process is also a proliferative process, and the actively dividing intermediate progenitor populations serve as transit-amplifying compartments, numerically expanding while differentiating (Figure 9.5). In this way, HSCs generate significantly more numerous progenitor cells, which in turn generate significantly more numerous mature cell populations (Orkin and Zon 2008). Hence, these relatively abundant progenitors contribute to the day-to-day maintenance of bone marrow and peripheral blood cell numbers, allowing HSCs to remain quiescent. Along these lines, despite the potential for replication-associated DNA damage in proliferating cells, mutation or genomic damage in a non-self-renewing progenitor cell would only result in extinction of that particular cell or transmission of the mutation to a limited number of progeny cells, which is significantly less harmful to the fitness of the organism than acquisition of such a mutation in the self-renewing HSC compartment. Hence, while leukaemias do occur in progenitor populations that acquire aberrant self-renewal activity, the founding mutation(s) can often be traced to the HSC compartment (see Section 9.4).

While dormant HSCs may not be heavily involved in day-to-day blood production, careful regulation of their fate decisions is required. HSCs can follow a number of cell fate pathways: they can enter the cell cycle and proliferate, with the division resulting in differentiation (generation of a progenitor) or self-renewal (generation of another HSC) or, during asymmetrical divisions, both. In addition, damaged or defective HSCs can either activate a pro-apoptotic programme and die, or enter a senescent state (Seita and Weissman 2010). Lastly, HSCs can mobilise, exiting the bone marrow into peripheral blood and lymphatic circulation, and subsequently return to the marrow without losing their self-renewal capacity. This behaviour was elegantly demonstrated by joining syngeneic (same haplotype) mice with different CD45 allotypes that can be detected with unique antibodies (termed CD45.1 and CD45.2), via skin on the flanks to establish a common circulation, an approach known as parabiosis (Wright et al. 2001). Strikingly, upon separation of the mice, functional CD45.1-expressing HSCs were found in the bone marrow of CD45.2 mice, and vice versa. As HSCs themselves express multiple pathogen recognition receptors including TLRs, this is thought to have evolved as a form of immunosurveillance, as activated HSCs and progenitors are capable of generating myeloid cells rapidly in response to inflammatory signals, even in extramedullary tissues (Mazo et al. 2011). As these fate decisions can have profound impacts on the short- and long-term functionality of the blood system, they must be tightly regulated. This regulation occurs via collaboration between transcriptional and epigenetic regulators in HSC, as well as inputs from the surrounding environment, particularly the specialised niches in which HSCs reside (Asking and Gjorstrup 1987).

The bone marrow itself is highly complex, with multiple stromal and haematopoietic cell types contributing to the lifelong maintenance of HSCs. It is also an excellent example of how multiple physiological systems can intersect to regulate a tissue. The precise character of the bone marrow niche and whether it represents a defined, finite number of specific 'places' for HSCs to engraft versus an emergent property of the marrow based on the pantheon of cell types and structural components within the bone is currently debated. However, most studies now agree that HSCs closely associate with vascular endothelium, as well as mesenchymal stem cells (MSCs), which produce critical HSC survival factors such as stem cell factor (SCF) and the chemokine CXCL12, which binds to CXCR4 expressed on the HSC surface to retain them in the marrow (Schepers et al. 2015). Likewise, other haematopoietic cells play regulatory roles for HSCs; notably, tissue macrophages localised in the bone marrow (often referred to as osteomacs) play critical roles in maintaining HSC localisation in the niche, and megakaryocytes produce the chemokine CXCL4, which maintains HSC quiescence. Physical connections, in the form of integrin interactions with extracellular matrix proteins in the marrow, serve to anchor HSCs in place in the niche.

Perhaps most intriguingly, the bone marrow is highly enervated, and multiple lines of evidence now point to the nervous system as a key controller of HSC fate. Non-myelinating Schwann cells lining sympathetic nerves have been shown to produce TGF-β, which maintains HSC quiescence. Sympathetic nerve cells also transmit circadian rhythm information to the bone marrow, modulating production of CXCL12 and in turn regulating HSC mobilisation. In this way, HSCs and progenitors are released into the blood during the organism's resting period, that is, evening for diurnal organisms like humans and daytime for nocturnal organisms like mice. What is the function of this oscillation? While a definitive answer is not known, from an evolutionary perspective, release of HSCs and progenitors could potentially function as a repair or surveillance mechanism, or allow for regeneration of the bone marrow niche during resting periods when energy requirements for other activities is minimal (Mendez-Ferrer et al. 2009). Such oscillations can be reversed based on altering the light–dark cycle, indicating a direct connection between sensation of light (presumably via the eye) and HSC mobilisation. This raises the interesting question of how and whether this mechanism functions in individuals who are blind, or organisms without eyes (such as blind cave fish), and whether alternative inputs (such as activity) can substitute for the light cycle. It also raises the important point of how closely the evolution of our physiological systems is tied to circadian cycles, and the potential for dysregulation of key tissues if the cycle is disrupted, reversed, or broken. Indeed, we know that rates of leukaemia and other cancers are higher in night workers (Fu and Kettner 2013)—could these malignancies somehow represent a modern disconnect from evolved systems?

An HSC placed into a simple *ex vivo* liquid culture system with basic survival cytokines such as SCF will not self-renew, but instead will deterministically differentiate along the myeloid lineage in the absence of additional instructive signals that 'substitute' for the presence of the bone marrow niche (Ema et al. 2000). That myeloid output appears to be the 'default' setting of HSCs in the absence of the complex series of signals from the niche can be inferred to reflect the evolutionary history of haematopoiesis, with a minimum of additional energy or external input required to generate this ancestral cell lineage even in mammalian cells that, at the genetic level, can generate numerous blood lineages. As a corollary, this suggests that the evolution of new blood lineages has in turn required the evolution of increasingly complex and energy-intensive niches to maintain HSCs in a self-renewing state and, when they differentiate, provide key signals to guide their lineage output. Hence, it seems reasonable that the HSC pool is relatively small in humans and other organisms; on the one hand, the cells are kept dormant and a transit-amplifying compartment maintains blood levels; on the other hand, the capacity of an organism to maintain complex functional niches required for HSC maintenance in an economical manner may be self-limiting. Lastly, the number of individual HSC clones, while small relative to the mature progeny they generate, may already be in sufficient excess to provide functional redundancy; indeed, recent work suggests that only a small fraction (~ 10%) of the HSC pool is required to efficiently produce blood and respond to acute needs (Schoedel et al. 2016). From an evolutionary standpoint, such redundancy (and no more) is probably advantageous, by preventing excessive replication of individual HSCs (which carries the risk of stress and genome replication errors that lead to failure or malignancy), while not burdening the organism with the energetic requirements of maintaining an overabundant HSC pool that will likely be disadvantageous to the fitness of the organism.

9.2.5 Adapting Blood Production to Physiological Needs

Our lives are rarely conducted in a 'room-temperature' environment, at sea level, in germ-free conditions, where all of the surfaces we touch are soft. Instead, as a species we must regularly contend with injuries and other physiological insults. We encounter and defend ourselves from infection by pathogenic organisms. We become pregnant, endure emotional stress, eat fatty (or healthy) diets, and go hiking in the mountains or play strenuous sports. Hence, living requires our blood systems to adapt, in some cases very quickly, to maintain or re-establish physiological homeostasis in response to problems that are ancient (infection and injury) or sufficiently modern that evolution has not yet caught up (psychological stress, high-fat diets), leaving our blood systems to perform from an outdated script. In Section 9.2.4, we learned how our blood systems are regulated under homeostatic conditions, in best-case scenarios. In this section, we discuss how our blood systems have evolved to respond to situations beyond the steady state, or in some cases to mount a maladaptive response to them.

Each lineage of blood cell has one or more cytokine signals capable of instructing HSCs and certain haematopoietic progenitors to differentiate and generate that lineage. The liver and kidney (which are haematopoietic locations in mammalian fetuses and fish, respectively) produce erythropoietin (EPO) and thrombopoietin (TPO) to signal production of erythrocytes and megakaryocytes, respectively. Mature immune cells and stromal cells in the bone marrow produce granulocyte- and macrophage-colony stimulating factor (G-CSF and M-CSF) that activate production of myeloid lineages, namely granulocytes and monocytes/macrophages, respectively. Other stromal populations in the bone marrow produce interleukin (IL)-7, which promotes lymphoid specification (Schepers et al. 2015). These cytokines are all produced at low tonic levels to support normal haematopoiesis, and their levels can be increased to respond to need. Low platelet numbers (such as due to injury) trigger TPO production to promote megakaryocyte production and maturation. Hypoxic stress in response to high altitude or physical exertion triggers EPO production to increase oxygen carrying capacity via production of additional erythrocytes, which has also led to the abuse of EPO by some athletes. Likewise, infection or injury can trigger G-CSF production, leading to an 'emergency' granulopoiesis response that leads to rapid production of myeloid cells that assist in host defence (Manz and Boettcher 2014). Each of these signals acts on the haematopoietic system at multiple levels, to instruct lineage choice in multipotent cells and accelerate differentiation in progenitors already committed to that lineage. In essence, they are also 'dual-purpose' signals, as they govern blood production under homeostatic conditions, as well as non-homeostatic 'emergency' conditions, with the levels of each cytokine dictating their effect. For this reason, TPO, EPO, and G-CSF are used clinically to boost production of each cell lineage in patients with deficient production; G-CSF also mobilises haematopoietic progenitors into the peripheral blood, facilitating the collection of these cells for transplantation or scientific study.

Although HSCs are maintained in a largely dormant state, they are responsive to a striking array of systemic signals for which haematopoietic cells are not the sole targets. Indeed, HSCs have been shown to express the oestrogen receptor (ER) and are directly responsive to the female hormone oestrogen. Interestingly, increased oestrogen levels during pregnancy induce elevated HSC self-renewal activity and production of additional erythrocytes (Nakada et al. 2014). This increases the oxygen carrying capacity of the blood system, likely an evolutionary adaptation that maintains sufficient oxygenation to mother and fetus. In a similar vein, glucocorticoid hormones were recently shown to signal directly to HSCs and contribute to the mobilising effect of G-CSF (Pierce et al. 2017). Glucocorticoids are typically produced as a feedback mechanism in response to stress and can turn down inflammatory responses. It is tempting to consider whether the role of glucocorticoids in enhancing the G-CSF mobilisation response evolved as a mechanism to initiate acute HSC-mediated surveillance or repair outside of the normal circadian mechanism described above. Along the same lines, social stress induces β-adrenergic hormones that in turn activate increased monocyte production from the bone marrow, in essence generating a 'stress myelopoiesis' phenotype that likely evolved to 'prime' the blood system to respond to potential injuries resulting from either negative social interactions, such as fighting, or other danger situations activating the 'fight or flight' response (Powell et al. 2013). (For further discussion, see Chapter 17: Brain, Spinal Cord, and Sensory Systems.) In modern human society, where social stresses (career, social or family relationships) are not resolvable via a 'fight or flight' response and thus can become chronic in nature, it stands to reason that continuous activation of this pathway could lead to a toxic, chronic inflammatory state in which continued overproduction of myeloid cells promotes damage instead of repair. Such a scenario is a critical reminder of how the complexity of modern human society can outstrip our evolutionary programming in the blood system and elsewhere.

HSCs and progenitor cells boast a surprisingly comprehensive array of receptors to detect the presence of pathogens and/or inflammatory signals. These include TLR, receptors for pro-inflammatory cytokines such as IL-1, TNFα, and IFNs, and receptors for complement, alarmins, and other 'emergency' signals (Baldridge et al. 2011). In response to these stimuli, HSCs can rapidly enter the cell cycle and proliferate, typically resulting in increased production of myeloid cells or megakaryocytes, the two ancient immune cell types with critical 'first responder' roles in host defence (Manz and Boettcher 2014; Haas et al. 2015) (Figure 9.8).

Interestingly, recent work has suggested that HSCs use distinct 'emergency' differentiation pathways that presumably allow for faster production of megakaryocytes and myeloid cells. As inflammatory cytokines like IL-1 have been shown to activate more rapid proliferation and upregulation of myeloid transcription factors such as PU.1 in HSCs (Pietras et al. 2016), it appears feasible that HSCs upregulating such factors will proceed more rapidly through the lineage commitment process. Interestingly, similar megakaryocyte and myeloid-biased 'emergency' pathways are invoked during haematopoietic regeneration in response to injury to the bone marrow or transplantation of HSCs into a recipient (Pietras et al. 2015). In some respects, inflammation-driven haematopoiesis mirrors fetal haematopoiesis,

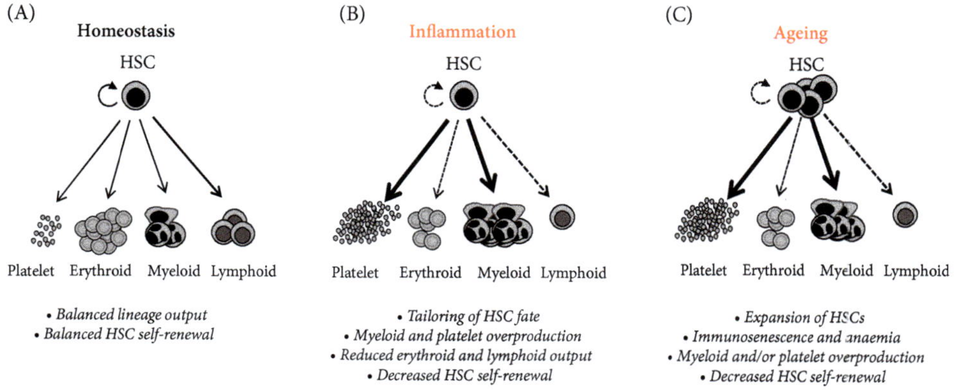

FIGURE 9.8 Common and distinct features of inflamed and aged adult mammalian blood systems. Under homeostatic conditions (A), HSCs and progenitors generate a balanced blood output, yielding the relative abundances of each lineage specified in Figure 9.5. During acute and chronic inflammation (B), blood lineage output is increased in the myeloid and platelet lineages, while in chronic conditions both erythroid and lymphoid lineage output is often suppressed. Moreover, HSC self-renewal can be impaired by inflammatory signals. Aged blood systems (C) share some features of inflamed blood systems, including lineage imbalance due to immunosenescence and loss of erythroid potential, while platelets and sometimes myeloid cells are overproduced at least in part due to ageing-related inflammation. Notably, the aged HSC compartment expands considerably in number and exhibits decreased fitness, which may contribute to the increased propensity toward haematological malignancy in aged individuals.

wherein a limited array of possible lineages is produced by actively proliferating HSCs, essentially in order of priority to ensure the survival of the organism. In the context of inflammation, this can be viewed as a specific adaptation to maximise resource use to produce only those cell types needed. On the other hand, in the context of bone marrow injury and transplantation (as well as inflammation), the bone marrow niche becomes disorganised, and HSCs are likely freed from their regulatory environment, at which point—much like the HSCs in *ex vivo* culture—they simply revert their default myeloid programme until the niche is re-established. These models are not mutually exclusive, and may represent an evolutionary reason why bone marrow niche control over an otherwise actively proliferating, lineage-restricted HSC would be selected for, versus an HSC that defaults to a dormant state or continues to expend resources producing less acutely important lineages (i.e. lymphoid) under 'emergency' conditions.

While the above evolutionary playbook works well for acute injuries or insults, it may be somewhat maladaptive in response to chronic inflammatory stresses or modern-day habits. In this sense, we have just discussed the impairment of lymphoid output being potentially advantageous during an acute inflammatory response. Indeed, B and T cells take considerably longer to produce than a myeloid cell. An organism infected with a pathogen must make do with the complement of T and B lymphocytes already present. Production of new cells is necessarily focused on controlling parasite/pathogen burden and presenting antigen to T and B cells, functions best suited to the myeloid and platelet lineages. On the other hand, long-term production of inflammatory signals due to autoimmunity and other forms of chronic inflammation, including obesity and intake of a high-fat diet, can create a maladaptive state for the blood system, in essence 'tricking' it into maintaining the 'emergency' output indefinitely.

Aside from the numerous physiological deregulations induced by chronic inflammatory diseases, individuals with such conditions sometimes exhibit reduced production of new lymphocytes as well as anaemia of chronic disease, while at the same time exhibiting thrombocytosis and/or overabundant myeloid cell output. While the HSC compartment is thought to maintain a quiescent state during chronic inflammation (Pietras et al. 2014), haematopoietic progenitors from patients with autoimmune rheumatoid arthritis show attrition of telomeres (Colmegna and Weyand 2011), and patients with chronic inflammatory diseases show increased propensity towards clonal haematopoiesis and haematological malignancy, which we will address in Section 9.4.

Altogether, this suggests that despite the remarkable evolutionary adaptations present in the blood system to deal with stress conditions such as injury and infection, the response to chronic diseases that are either largely modern in origin, such as obesity, or found primarily in ageing individuals, such as rheumatoid arthritis, is largely maladaptive. Such cases are prime examples of George C. Williams' antagonistic pleiotropy theory (Williams 1957), in which a gene that provides an advantageous trait in one context is maladaptive in another, but is nonetheless positively selected. Indeed, it is likely that there has been insufficient selective pressure for the blood system 'emergency' response to be modified on account of these diseases, due to either their recent emergence as common health problems or their higher frequency in post-reproductive demographics. In Section 9.3, we discuss a particularly significant example of antagonistic pleiotropy that is perhaps inevitable in the blood systems of all humans: ageing.

9.3 Alterations in the Haematopoietic System with Age

9.3.1 What is Ageing?

Ageing is a complex physiological process of change over time, in most organisms culminating in functional decline and failure of tissue systems and (in the absence of other causes) the death of the organism. The question of why an organism ages, and how the ageing process is regulated, if at all, by evolutionary or other processes remains a complex question that is open to debate. Interestingly, ageing is a heterogeneous process. While most metazoan species do age, some organisms such as *Hydra* have sufficient regenerative capacity to be effectively immortal (Martinez 1998). Among organisms that do age, there is considerable variability in life expectancy (defined as the mean time an organism could be expected to live), and within each species there is likewise variability in lifespan (defined as the amount of time any one individual lives). To illustrate this point, the worm *C. elegans*, laboratory mice, and humans all undergo an ageing process. Yet their life expectancies are drastically different, ranging from weeks (*C. elegans*) to 1 or 3 years (laboratory mice), to several decades, approaching a century (for modern humans). Embedded within these observations is the difference between life expectancy and longevity; the latter typically refers to the maximal age an organism can reach; however, the former can be impacted by environmental factors such as exposure to predators and other dangers. Hence, a mouse may have longevity of approximately 2 years, but due to predation and other environmental conditions, the life expectancy of a mouse in nature is less than 6 months.

Likewise, there are significant distinctions between chronological and biological age that are important to any discussion of evolution and ageing. Chronological age refers to the length of time an organism has been alive, whereas biological age is how old the organism seems; this is sometimes referred to as 'age versus frailty', or other similar dichotomies. Particularly in the field of regenerative medicine, we often consider this dichotomy as lifespan versus 'health-span', or the length of time an individual remains healthy despite undergoing an ageing process (Niedernhofer et al. 2017). In this sense, certain modern human lifestyle traits such as smoking, improper diet, inactivity, and emotional stress level can exacerbate the rate of biological ageing, resulting in individuals showing advanced features of biological ageing at a much earlier chronological age. Thus, it is important to consider biological and chronological ageing as separate phenotypic elements that, alongside environmental factors, directly influence each other in dictating the lifespan of an individual organism. (For further discussion, see Chapter 5: Senescence and Ageing.)

9.3.2 Theories of Evolution and Ageing

The fact that worms, mice, and humans have incredibly different longevities and yet all undergo an ageing process begs the question: why? A number of evolutionary theories have been developed to try to unravel this conundrum (Kowald and Kirkwood 2016). Most modern theories reject the idea that ageing is a programmed trait. Peter Medawar and other evolutionary biologists have posited that ageing largely falls outside of the purview of natural selection. In this sense, the process of natural selection favours maximising the reproductive success of an organism. Following the reproductive phase, the capacity of natural selection to influence the physiology of an organism declines rapidly as reproductive fitness ceases to act as a selective factor. Since then, multiple hypotheses as to *how* organisms age within this paradigm have emerged. The first, described by Medawar himself, identifies the progressive accumulation of germline mutations in organisms with costs in post-reproductive periods as drivers of ageing, which promote progressive tissue dysfunction. The second, Williams' pleiotropic antagonism theory discussed in Section 9.1.2, suggests in the context of ageing that traits favouring survival and reproduction early in life can be maladaptive later in life, hence driving ageing phenotypes. In 1977, Thomas Kirkwood proposed the *disposable soma* theory, which stipulates that organisms have finite resources to allocate between reproductive and regenerative goals (Kirkwood 2005). A balance between the two processes is required, which ultimately sacrifices regenerative capacity in favour of reproductive output.

Amidst these three theories, a number of other considerations arise. First, different animals have evolved different strategies to maximise reproductive success. In some cases, these strategies do not necessarily result in longevity overlapping directly with the reproductive period. For instance, mice are short-lived and encounter significant extrinsic danger in the wild due to predation and other hazards. They produce large litters of pups that are born and weaned every 3 weeks and require minimal parental training before the young are independent. On the other hand, long-lived organisms with few predators, such as humans, elephants, dolphins, and whales, have small broods (usually individual young) which nonetheless require a significant investment in time and energy (sometimes decades) to rear before they are considered independent. In this way, it is possible that other factors, including parenting behaviours, fertility windows, social behaviours, and culture (such as group parenting versus individual parenting) can impact on longevity and biological ageing in the post-reproductive period while nonetheless favouring reproductive success and, therefore, falling under the influence of natural selection. (For further discussion, see Chapter 4: Growth and Development.) Thus, the degree to which specific tissue ageing processes are similar between species with different longevities and reproductive strategies can provide a window into the selective forces (or lack thereof) driving the ageing phenotype in that tissue, as well as the theoretical framework(s) in which that ageing phenotype may fall (Kowald and Kirkwood 2015).

In this context, it is well worth noting that much of our empirical understanding of the aged blood system has been generated in mice, which in the wild typically do not live long enough to undergo an ageing phenotype in that tissue. Strikingly, many of our observations in mice have been validated in humans, who pursue distinctly different evolutionary strategies for maximising reproductive fitness and live on average many decades longer than a mouse. Nevertheless, ageing phenotypes in each organism's blood system do not typically emerge until after the period where reproductive success was most likely (note that modern humans can now live past this period, but selection acted on historical probabilities). This suggests that in mice and humans, blood system ageing is likely unrelated to reproductive success. Instead, like blood system responses to chronic inflammation, blood system ageing may be the result of antagonistic pleiotropy. At the same time, some aspects of blood system ageing could perhaps pose a selective advantage in the context of organisms with extended post-reproductive longevity tied to cultural or parenting duties. In Sections 9.3.3 and 9.3.4, we discuss the features of aged blood systems, the mechanisms driving them, and their evolutionary implications.

9.3.3 Features of the Aged Blood System

The haematopoietic system undergoes a dynamic series of changes in the context of ageing, in some ways mirroring the phenotype of a chronically inflamed haematopoietic system. Old blood systems are characterised by anaemia and loss of lymphoid potential, with concurrent increases in platelets and myeloid cells. Initially described in the mouse, similar alterations have been verified in humans (Pang et al. 2011; Akunuru and Geiger 2016). A hallmark of ageing is immune senescence, a loss of lymphoid potential accompanied by decreased output of T- and B-cell progenitors, and an overall reduction in the generation of new lymphoid clones alongside expansion of memory-type clones (Figure 9.8). In essence, the aged individual is left to rely solely on the existing complement of lymphoid cells for immunological protection, leading to the potential for reduced adaptive immunity. Why does this occur? In the context of ageing, these changes accompany increased systemic levels of pro-inflammatory cytokines associated with tissue senescence, also known as the senescence-associated secretory phenotype (van Deursen 2014). These cytokines can directly inhibit the production of lymphoid progenitors by simultaneously interfering with

IL-7 signalling and activating an inflammatory myelopoiesis programme in HSCs and MPPs (Muller-Sieburg et al. 2004; Henry et al. 2015).

Second, age-associated bone remodelling, typified by conditions including osteoporosis, could deprive haematopoietic progenitors of key niche components required for lymphopoiesis; in this context, laboratory investigations have described an IL-7-rich 'osteoblastic niche' where lymphoid progenitors are generated and maintained (Ding and Morrison 2013). Ongoing bone remodelling and loss of this niche could be a contributing factor to the reduction in lymphoid potential in aged individuals. Interestingly, laboratory experiments comparing old mice with transgenic counterparts overexpressing anti-inflammatory factors demonstrate a less severe reduction in lymphopoiesis (Henry et al. 2015), suggesting that the inflammatory phenotype that accompanies ageing does impact lymphopoiesis, albeit incompletely.

This suggests that other mechanisms may be at play to shut down lymphoid output. Indeed, from an evolutionary standpoint, there may be certain selective advantages to the reduction in lymphoid output. For example, lymphopoiesis is reduced in individuals well before other signs of physiological decline, with thymic involution starting in the teenage years for humans (Berent-Maoz et al. 2012). Older individuals will have already encountered and developed immunological memory to a broad array of pathogens throughout their life; the likelihood that such an individual will encounter a novel pathogen concurrently may decrease. Thus, it may be less advantageous to commit physiological resources to continued production of new lymphoid clones that are statistically unlikely to contribute to host defence, relative to maintaining the pre-existing memory compartment to respond to the array of 'known' pathogens. In addition, it has been proposed that decreased lymphocyte production reduces the likelihood of errors during somatic hypermutation that could lead to cellular transformation and lymphoid leukaemia development (Signer et al. 2007). Nonetheless, the trade-off is clear: old individuals are much more susceptible to 'novel' pathogens, such as new strains of influenza, to which they do not have prior immunity (Weinberger and Grubeck-Loebenstein 2012). While vaccinations against influenza can offer protection, the reduction in new lymphocyte clones may nonetheless impair the ability of the immune system to generate a robust memory compartment following the inoculation.

In contrast, the aged haematopoietic system is often described as 'myeloid-biased' based on the relative increase in myeloid output observed following transplantation of aged HSCs (Weinberger and Grubeck-Loebenstein 2012). Once again, such a myeloid bias could be interpreted as both a response to systemic inflammatory signals produced by senescent tissues, and potential degradation of the bone marrow niche, which allows HSCs to revert to their 'default' programme of myeloid output in the absence of contravening instructions. More recently, laboratory work using mice has suggested that the aged blood system is not myeloid-biased per se, but more specifically is platelet-biased (Grover et al. 2016), with an expansion of vWF-expressing HSCs that fuel the chronic overproduction of megakaryocytes and platelets (Figure 9.8). Such experiments fall in line with clinical observations of increased incidence of thrombosis, including myocardial infarct (heart attack) and deep-vein thrombosis (DVT) in aged individuals (Costantino et al. 2016). While some thrombosis events are likely linked to (or at least exacerbated by) cumulative effects of a high-fat diet, smoking, and other unhealthy lifestyle habits resulting in atherosclerosis, these negative trade-offs are novel in the evolutionary timescale and would have not provided a basis for selection. On the other hand, the platelet-biased haematopoietic system could hold some advantage for aged individuals. Because platelets are critical for clotting and wound repair, one possibility is that increased platelet numbers ensure a robust restoration of haemostasis in response to wounds or tissue dysfunction, perhaps in turn compensating for the reduced regenerative function of tissue stem cells throughout the body. Still, for selection to have favoured this programme, there would need to have been sufficient individuals at older ages increasing the contributions of their genes to subsequent generations.

9.3.4 Mechanisms and Impact of HSC Ageing

The HSC compartment itself undergoes significant change during ageing. First and foremost, the number of phenotypic HSCs expands by up to five- to ten-fold the number of cells in the bone marrow of a young individual. The HSC compartment also exhibits reduced capacity to reconstitute the blood system on a per-cell basis, suggesting a significant reduction in regenerative potential (Dykstra et al. 2011). Such old-age-dependent reductions in regenerative potential are also observed for other tissue stem cells such as epidermis (Doles and Keyes 2013), muscle, and brain (Brack et al. 2005; Renault et al. 2009), suggesting a potentially common ageing mechanism. In addition, as discussed above, the HSC compartment gives rise to a skewed haematopoietic output, with expansion of myeloid- and platelet-producing clones and a concurrent reduction in lymphoid-biased cells present in the bone marrow. Altogether, these observations point to significantly reduced fitness of individual HSCs with age.

What are the mechanisms driving this phenotype? These defects appear to be at least in part related to the replicative history of the HSC compartment; *in vivo* tracking experiments in mice have shown that HSC fitness is successively reduced with each cell division (Bernitz et al. 2016), accompanied by a loss of cell polarity while undergoing division (Florian et al. 2013), which could in part underlie the aberrant expansion of the HSC compartment, due to failure to asymmetrically divide. In addition, aged HSCs show significant changes to their epigenetic regulation that include expansion of repressive histone and DNA methylation marks on numerous genes, including lymphoid gene targets, that serve to repress their expression (Sun et al. 2014a). This may well be an evolutionary strategy to limit spurious gene expression, thereby delaying or limiting tissue deregulation and/or cellular transformation in the ageing organism. Interestingly, reprogramming of old haematopoietic stem and progenitor cells using induced pluripotent stem cell (iPS) technology appears to reverse some epigenetic changes and restore normal function (Wahlestedt et al. 2013), raising the question of how much the reductions in HSC fitness with age are 'hard-wired'. Indeed, changes in signals from the bone marrow niche also appear to play a role in the ageing phenotype. As a notable example, mouse experiments have shown that the switch from canonical Wnt signals (such as Wnt3a) to non-canonical (such as Wnt5a) in aged animals drives the HSC loss-of-polarity phenotype, and blockade of Wnt5a can restore lymphoid potential and reduce the numerical expansion of old HSCs (Florian et al. 2013). Similar findings in intestinal stem cells (Nalapareddy et al. 2017) suggest that changes in niche signals (in addition to the inflammatory ageing phenotype) may play a cooperative role in altering HSC lineage output in the context of ageing. Altogether, these findings raise the interesting question as to whether any aspects of HSC ageing are in essence programmed and therefore selected by evolutionary pressures, as opposed to a result or even a bystander effect of functional exhaustion and senescence arising in post-reproductive individuals, for which little selective pressure exists.

In this sense, it is interesting to note the interplay between the maintenance of genome integrity and ageing in the haematopoietic system. Aged HSCs exhibit an increased frequency of mutations in their genome (Moehrle et al. 2015), though this two- to three-fold increase is well below the mutational burden required to generate a *de novo* leukaemia (Geiger et al. 2006), likely due to retention of functional DNA repair machinery and damage checkpoints in aged cells. Such mechanisms essentially function to suppress somatic evolution that could promote leukaemic transformation, with HSC quiescence also protecting HSCs from undue proliferative stress by limiting their replicative history. Quiescence is particularly important in the context of ageing, as aged HSCs undergo replicative stress and exhibit an increased number of DNA breaks when they do proliferate (Flach et al. 2014), likely increasing the chance of acquiring a genetic mutation. Perhaps the numerical expansion of HSCs in aged individuals could serve as a mechanism to suppress somatic evolution, by creating a larger pool of individual clones and limiting the chance that any one cell will undergo a sufficient number of replicative events and associated stresses to generate a leukaemia. Of course, this strategy only works provided that pre-existing mutations are not amplified during expansion of the HSC pool during ageing. Recent studies of blood cells from aged individuals has revealed the expansion of specific HSC clones carrying point mutations in key epigenetic regulatory genes such as *DNMT3A* and *TET2* (Genovese et al. 2014; Jaiswal et al. 2014; Xie et al. 2014; McKerrell et al. 2015), suggesting that the aged bone marrow may provide these clones with a selective advantage, allowing them to expand and increase the odds of acquiring secondary mutations that lead to haematological malignancy. We discuss these concepts in further detail in Section 9.4.

9.4 Haematopoietic Malignancies

Haematopoietic malignancies represent almost 10% of cancers in humans, and include leukaemias of all major blood cell lineages (e.g. lymphoid, myeloid, megakaryocyte, and erythroid) and lymphomas from the lineages making T and B cells. While the vast majority of these malignancies develop in old age, a separate peak of incidence for leukaemias is present in the first few years of life. In this section we discuss the evolutionary forces that have contributed to the age-dependent incidence of cancers, with a focus on haematopoietic malignancies. We will learn how the haematopoietic system can produce more than 10^{11} blood cells per day (mostly red blood cells), and yet despite the enormous numbers of cell divisions (and resulting mutations) required for this feat, blood cancers remain rare (particularly through youth). This section describes some of the mechanisms that we and other animals have evolved to limit cancer through periods of likely reproductive success.

9.4.1 Why is Cancer Mostly a Disease of Old Age?

Just as natural selection limits physiological ageing until beyond periods of likely reproduction, it will also limit the incidence of cancer so as to maximise individual fitness. From the theories developed by Medawar, Williams, and Kirkwood (see Section 9.3.2), we can understand how the force of natural selection to maintain tissues and limit disease wanes in concert with reduced odds of reproductive success (Medawar 1952; Williams 1957; Kirkwood 2005). Investments in maintenance of the soma will only be made to the extent that it maximises reproductive success. Moreover, even programmes that contribute to ageing phenotypes and disease in late life could be positively selected if this programme increased reproductive success in youth (such that, in balance, overall reproductive success was improved).

Natural selection has acted to greatly limit cancer incidence through 'youth', the period when reproductive success is most likely (DeGregori 2011). As mentioned in Section 9.3.1, mice can survive for up to 2 years or possibly even 3 years as pets or in laboratories, but most mice will live only weeks or a few months in the wild. Given that mice produce eight to ten progeny every few weeks, we would otherwise be up to our noses in mice (millions of mice would be produced within 1 year starting with a single pair, assuming unlimited resources and no external hazards). In the laboratory, most cancers develop in mice after 1 year of age. Similarly, cancer is generally a disease of ageing for most mammals where it has been studied. For these reasons, cancer is relatively rare 'in the wild' (Hochberg and Noble 2017), as external hazards, from limited food to disease to predation, typically limit survival to ages where cancer is more prevalent (Figure 9.9).

Different animals have evolved different life histories, which describe parameters such as body size and the characteristics of maturation, reproduction, and ageing. These parameters evolved to maximise adaptation to a particular environment. Mice and similar short-lived animals like rabbits evolved a 'fast life strategy', maturing early and producing large numbers of offspring per reproductive cycle. As external hazards are high, mice do not invest in long-term maintenance of tissues—instead, investment is made in early reproduction. Humans, like other long-lived vertebrates such as whales, elephants, and Galapagos tortoises, have evolved 'slow' life strategies, with slower maturation and greater investment in fewer offspring. (For further discussion, see Chapter 4: Growth and Development.) Given reductions in external hazards (through size and improved defences), natural selection invested in longer lives to provide for additional opportunities for reproduction and time for successful rearing of offspring (Figure 9.9).

Still, even for humans, investments to maintain tissues and avoid ageing and cancer have been limited. For most of human evolutionary history, survival to what we now consider old age (more than 60 years of age) was rare, and odds of reproduction and successful child rearing declined progressively post-maturation (Kaplan et al. 2000; Gurven and Kaplan 2007). In industrialised nations, more than 90% of cancers occur in humans over 50 years of age. Blood cancers are no exceptions. Most cases of acute myeloid leukaemia (AML), chronic myeloid leukaemia (CML), and chronic lymphocytic leukaemia (CLL) are diagnosed in the elderly. Given the enormous number of cell divisions required each day to produce blood cells, how are these haematopoietic malignancies relegated to late life?

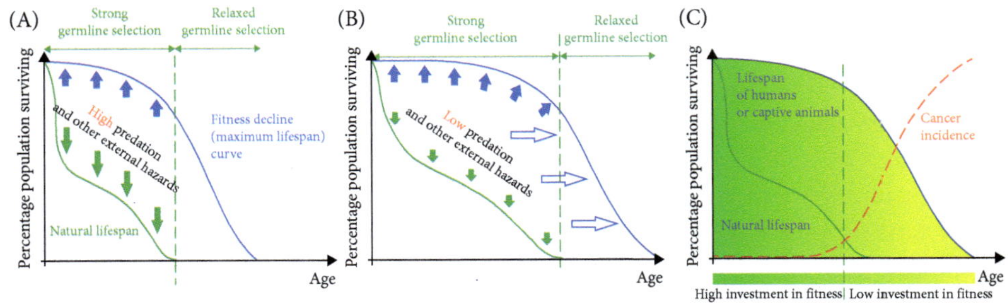

FIGURE 9.9 Impact of age on cellular fitness and cancer incidence. (A) Lifespan of animals in the wild (dark green curve) is usually defined by external hazards, such as predators, food availability, and infectious diseases (dark green arrows). Maximum lifespan, limited by their rates of physiological ageing (blue line), can only be observed in captivity or in modern humans. Germline selection acts to maintain high body fitness (solid blue arrows) for the duration of likely survival and reproduction in the wild. Physiological ageing in humans and captive animals accelerates after the time of their likely survival in the wild promoted by progressively relaxing germline selection for body fitness. (B) If external hazards are reduced, animals survive into more advanced ages in the wild and trigger germline selection to extend (hollow blue arrows) their maximum lifespan, pushing the curve of physiological ageing rightwards on the age axis and leading to evolution of a longer potential lifespan. (C) Cancer is rare in wild animals, as well as in humans and captive animals during the period of their likely survival in the wild (natural lifespan), with carcinogenesis being suppressed by higher tissue and stem cell fitness relative to ages past natural lifespan. A dramatic increase in cancer incidence coincides with higher rates of physiological decline.

Source: Reprinted from *Trends in Cancer*, 2 (10), Andrii I. Rozhok and James DeGregori, The Evolution of Lifespan and Age-Dependent Cancer Risk, pp. 552–60, Figure 2, doi.10.1016/j.trecan.2016.09.004. Copyright © 2016 Elsevier Inc., with permission from Elsevier.

Above, we discussed the evolutionary reason—there has been strong selective pressure to avoid such malignancies during periods where cancers would interfere with reproductive success. We next discuss the proximate mechanisms to explain how we limit cancer through youth—the pathways, processes, and systems evolved to suppress cancer.

9.4.2 Evolved Tumour Suppressive Strategies

From more than 3.5 billion years ago to about 700 million years ago, unicellular organisms appear to have represented all life on earth. The development of multicellular animals (metazoan) created a new problem—avoiding somatic evolution that can lead to function-disrupting malignancies. Cancer biologists often propose that somatic cells surrendered their cellular fitness, forming a covenant of selfless cells with minimal fitness (proliferation and survival are limited). In this scenario, somatic cells are envisioned in a fitness valley, and progression to cancer involves the sequential acquisition of mutations that increase fitness (Fortunato et al. 2017). However, experimental studies, such as with chimeric animals, have shown that normal somatic cells are very competitive, and thus possess somatic fitness (DeGregori 2011). Here, we argue that high somatic cell fitness maximises tissue and organismal fitness, and limits carcinogenesis. Note that we define somatic cell fitness as the ability of a cell of a particular genotype (or epigenotype) to pass this type on to other cells. This measure of *cellular* reproductive success will be highly dependent on the tissue environment (the microenvironment) and competing cells.

In the development and maintenance of animal bodies, many mutations occur. Selection acts on somatic cells to change the frequency of cells with particular genotypes, depending on their somatic fitness, in a process that we will refer to as *somatic evolution*. Cancer is one form of somatic evolution. As it disrupts normal tissue and organ function, typically leading to death in the absence of intervention, carcinogenesis is the somatic evolution that concerns us the most. Given its destructive consequences, natural selection has led to effective mechanisms to limit cancer's impact on organismal fitness, which we will categorise as mechanisms that are cell intrinsic, cell extrinsic, and integral to the system (DeGregori 2011).

Cell intrinsic mechanisms that limit cancer include DNA repair systems that limit mutation accumulation. However, in the development and maintenance of animal bodies composed of billions or trillions of cells, many mutations will nonetheless accumulate, given that somatic mutation rates are in the order of 10^{-8} to 10^{-9}/nucleotide/cell division (at least for mammals). For the production of 2×10^{11} blood cells each day (not including platelets), at this mutation rate, every possible position in the human genome will be mutated hundreds of times over, including every possible oncogenic mutation. Fortunately, we and other animals have evolved mechanisms to eliminate mutated cells, particularly when such mutations lead to potentially oncogenic changes. Oncogenic changes include mutations that activate oncogenes as well as mutations that inactivate tumour suppressor genes. Animals have evolved signalling pathways that lead to programmed cell death, such as apoptosis, in response to DNA damage (Lowe et al. 2004). Additional pathways lead to cell senescence, the permanent withdrawal of a cell from the cell cycle, providing an additional mechanism to prevent the propagation of cells with DNA damage (or other cellular damage). These mechanisms not only are tumour suppressive, by eliminating cells that could possess oncogenic mutations, but also help maintain the health of tissues by purging damaged cells. Importantly, these pathways leading to cellular apoptosis or senescence are also activated by oncogenic signalling or the loss of tumour suppressor genes (Lowe et al. 2004). Oncogenic signalling can also lead to cellular differentiation, removing a cell from the replicative cell pool (such as stem cell pools). Consequently, oncogenic mutations will occur in cells in our bodies, but cells with these mutations will be permanently removed through cell death, senescence, or differentiation, eliminating the risk of further progression towards cancer.

Cell extrinsic mechanisms function outside of the cells receiving oncogenic mutations. The immune system, from innate immunity to adaptive immunity, limits the persistence of malignant clones. Natural killer cells appear particularly important, as mutations in mice that prevent the development of these cells lead to substantially increased cancer rates (Waldhauer and Steinle 2008). These cells seek out and destroy tumour cells that have downregulated MHC. MHC present peptide antigens to the immune system and function as cellular identification cards. Presenting abnormal peptides can lead to activation of adaptive immune cells, specifically T cells, which can destroy the presenting cell. In response, cancers evolve somatically to eliminate cells presenting antigens recognised as foreign by the immune system. This immune editing can lead to tumour cell escape from immune surveillance. Like all evolved tumour suppressive mechanisms, immune-mediated elimination of malignant cells is not perfect.

An additional cell extrinsic mechanism that limits cancer is cell competition. First described in fruit flies, and since shown in mice, cells with function-impairing mutations have been shown to be actively eliminated from a tissue, in some cases via engulfment by other more fit cells (Merino et al. 2016). This process thus maintains a fit stem cell pool, and we will learn in Section 9.4.3 how a fit stem cell pool can be tumour suppressive. Importantly, cells with oncogenic mutations are eliminated from tissues through similar pathways (Menendez et al. 2010; Ballesteros-Arias et al. 2014). These cells can be either engulfed by other cells, or actively extruded from a monolayer into the lumen (leading to their elimination) (Kajita and Fujita 2015). So, if an oncogenically mutated cell escapes cell intrinsic mechanisms like apoptosis, the immune system and cell competition will further limit the odds of malignant cell survival.

There are additional critical mechanisms that are *integral to evolved tissue systems*. A key mechanism is the hierarchical organisation of tissues, described in Section 9.2.4. While 2×10^{11} cells are generated each day in our blood, mutations occurring in different cells in the haematopoietic system do not possess equal chances of contributing to blood cancer. In fact, most blood cell malignancies initiate in HSCs. It is estimated that a human will possess from 11,000 to 300,000 HSCs (Wang et al. 1997; Abkowitz et al. 2002), protected within the bone marrow. This represents a much smaller target size for oncogenic mutations than the billions of more committed blood progenitor cells and the hundreds of billions of mature blood cells. In addition, these HSCs divide very infrequently, as little as once every 1.5 years. Since most mutations happen during cell divisions, a small stem cell pool that divides infrequently will greatly reduce cancer risk. Moreover, studies have shown how normal tissue architecture can suppress transformed phenotypes (Bissell and Hines 2011). Even cancer cells can behave relatively normally when placed in a normal tissue environment.

A final mechanism of integral tumour suppression will be discussed in Section 9.4.3: stabilising selection in young stem cell pools, which eliminates most mutations (including oncogenic ones) that alter cellular phenotype. We will describe how the evolution of highly fit stem cell pools, with strict dependency on their tissue niche, will lead to the elimination of phenotype-altering mutations.

9.4.3 Evolutionary and Proximate Explanations for Late-Life Incidence of Most Haematopoietic Malignancies

Most leukaemias and lymphomas occur in old age. We learned in Section 9.4.1 the evolutionary reason—there is minimal selection against cancer and other diseases of ageing beyond periods of likely reproductive success. Natural selection acted to maximise the utility of the haematopoietic system, from oxygen transport to clotting to immunity, while at the same time minimising fitness-reducing complications that can arise from this system. These complications include cardiovascular disease, autoimmunity, inflammatory disorders, and of course haematopoietic malignancies. When we recognise that selection has acted to maximise fitness (reproductive success) and that evolution has largely acted during the millennia of pre-industrial human history, we can understand how the costs of these complications have been very small from an evolutionary perspective. These costs are experienced to a much greater extent today in industrialised nations, due to longer lives and modern lifestyles. These diseases and disorders also take very large tolls on families, our society, and us. Therefore, we need to understand their origins in order to better prevent and treat them.

For cancers, the predominant paradigm posits that cancers arise late in life due to the requirement for the age-dependent accumulation of oncogenic mutations (Hoeijmakers 2009; Beerman et al. 2010; Tomasetti and Vogelstein 2015). However, a number of observations are inconsistent with the mutation accumulation model for cancer with age (see DeGregori 2012 and Rozhok et al. 2015 for details), including that about half of mutations accumulate in most tissues by maturity, that oncogenic events in young HSCs have been shown to frequently *reduce* stem cell self-renewal (which should reduce somatic fitness by causing loss from the stem cell pool), the frequent detection of oncogenic mutations in healthy tissues (described below for blood), and that current models ignore a well-known tenet of evolutionary biology: the fitness effects of most mutations are highly context (environment)-dependent. The role of the tissue environment in determining the course of somatic evolution is largely ignored. And yet we know that the evolution of species is largely driven by environmental changes that alter selective pressures (Eldredge 1999). Major periods of speciation throughout Earth's history have been driven by dramatic changes in environments, resulting from alterations in atmospheric gases, plate tectonics, huge fluctuations in temperature, and collisions with extraterrestrial meteors (Ward and Kirschvink 2015). Still, cancer evolution is commonly deemed to be limited by mutation occurrence.

An alternative theory, adaptive oncogenesis, proposes that changes in tissues in old age or with other carcinogenic contexts promote selection for adaptive oncogenic mutations, which would *not* be selected for in a healthy tissue landscape (DeGregori 2012; Rozhok et al. 2015). This theory has two critical components: (1) natural selection at the organismal level acts to maximise (albeit imperfectly) stem and progenitor cell adaptation to tissue microenvironments in order to optimise tissue function, which promotes stabilising selection (eliminating most phenotype-altering mutations, including oncogenic mutations); and (2) ageing and insults like smoking alter tissue microenvironments, such that stem/progenitor cells are no longer well adapted (thus reducing their cellular fitness), which promotes selection for mutations that can be adaptive within the altered tissue landscape. Some of these mutations are oncogenic. Expansion of an oncogenically initiated clone consequently increases the risk of further cancer progression. In total, this new theory proposes that the dominant effect of carcinogenic contexts like ageing in terms of cancer risk is to change the adaptive landscape in which cells with potentially oncogenic mutations exist. Adaptive oncogenesis theory connects cancer to its causes (ageing, smoking, etc.) through impacts on microenvironments, which alter the strength and directionality of selection on oncogenic mutations.

The relevance of this theory to haematopoietic malignancies has been tested in mouse models. As described in Section 9.3.3, ageing is associated with a decline in HSC fitness (Henry et al. 2011). Ageing is also associated with chronic inflammation, which has been implicated in many ageing-associated disease processes (Caruso et al. 2004; Krabbe et al. 2004; Shaw et al. 2010; Green et al. 2011). HSC and haematopoiesis are highly responsive to inflammation (Mirantes et al. 2014). Studies demonstrated that particular oncogenic mutations are selectively adaptive in aged B-cell progenitor pools (*not* in young), and this adaptation involves the restoration of key somatic cell fitness determinants like STAT5 signalling and MYC expression (Henry et al. 2010, 2015). This selective adaptation significantly promoted leukaemogenesis in aged *backgrounds*, even when progenitors bearing oncogenic mutations were young. Importantly, transgenic expression of anti-inflammatory mediators in old mice prevents ageing-associated increases in IL-1, TNF-α, and IL-6, prevents ageing-associated defects in signalling and gene expression in B-cell progenitors, and abrogates selection for oncogenic events normally adaptive in old progenitor pools (Henry et al. 2015). Thus, these studies show that chronic, inflammatory microenvironments in old age lead to reductions in the fitness of B-cell progenitor populations. This reduced progenitor pool fitness engenders selection for cells harbouring oncogenic mutations due to their ability to correct ageing-associated functional defects, which is prevented by blocking inflammation. Thus, modulating inflammation, a common feature of ageing, can limit ageing-associated oncogenesis.

Similar evolutionary dynamics are likely at play in HSC pools. Computational modelling of dynamic microenvironment-dependent fitness alterations in HSC pools demonstrated that ageing-imposed reductions in HSC fitness and increases in somatic evolution leading to leukaemia are unlikely to result from cell-autonomous processes, and require considerable contribution from age-altered microenvironmental factors (Rozhok et al. 2014). Interestingly, several studies used deep sequencing to detect somatic mutations in peripheral blood cells across thousands of adult humans of various ages (Genovese et al. 2014; Jaiswal et al. 2014; Xie et al. 2014; McKerrell et al. 2015). These studies revealed that clonal expansions exceeding approximately 5% of cells were rare in those under 40 years of age, but quite common in the elderly, detectable in up to 20% of individuals over 70. The clonal expansions resembled the late-life increases in the frequency of haematopoietic malignancies. Most but not all clones possessed oncogenic mutations, typically only a single oncogenic mutation. Thus, multiple driver mutations do not appear to be required for the age-dependent pattern of clonal expansions. In all, we can envision a model whereby mutations that are purged by stabilising selection in young, healthy tissues can become positively selected late in life or in damaged tissue, due to adaptation to altered tissue environments (Figure 9.10).

Late-life clonal haematopoiesis coincides with a greater risk for haematopoietic malignancies (Jacobs et al. 2012; Laurie et al. 2012; Aghili et al. 2014; Genovese et al. 2014; Jaiswal et al. 2014), although it is notable that most people with oncogenic clonal expansions do not go on to develop leukaemia. Thus, these expansions confer greater risk, but other factors are required (including a role for chance) for progression. Since roughly half of mutations and epigenetic changes accumulate in the haematopoietic system by maturity (Vijg et al. 2005; Horvath 2013), these studies reveal how selection for oncogenic mutations can be suppressed in youth but promoted in old age, likely due to microenvironmental changes. Interestingly, clonal haematopoiesis was also associated with all-cause mortality, particularly for cardiovascular diseases, which could either indicate roles for altered haematopoiesis or common underlying systemic causes. For the latter, systemic changes such as inflammation could impact multiple systems, increasing risks for multiple diseases, from cancer to cardiovascular disease. Clonal haematopoiesis could represent a 'symptom' of overall physiological decline, resulting from both proximate causes (lifestyles) and evolutionary causes (the waning of the force of selection to maintain the soma beyond reproductive years).

We can consider whether modern lifestyles and exposures have influenced blood cancer risk (Zeeb and Blettner 1998; Ilhan et al. 2006). Certainly, exposure to radiation is highly associated with increased cancer risk, primarily for acute leukaemias. Similarly, smoking is associated with increased risk of haematopoietic malignancies. Moreover, treatment with

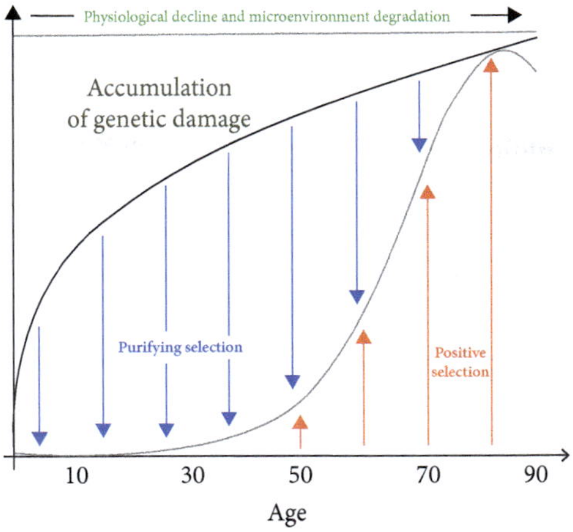

FIGURE 9.10 Adaptive oncogenesis in cancer development. Mutations and other genetic damage accumulate rapidly in tissues cells early in life, coinciding with rapid body growth, and then more slowly after maturation. Through periods of likely reproduction, purifying selection is dominant in the larger stem cell pools, which leads to suppression of oncogenic clonal expansions. Late in life, the degradation of the tissue microenvironment leads to positive selection for adaptive oncogenic mutations, which in turn leads to greater risk of oncogenic clonal expansions. Expansion of HSC clones bearing oncogenes increases the risk of cancer.

Source: Reprinted from *Cancer Research*, 77 (22), James DeGregori, Connecting Cancer to its Causes Requires Incorporation of Effects on Tissue Microenvironments, pp. 1–4, Figure 2, doi.10.1158/0008-5472.CAN-17-1207. Copyright © 2017 AACR.

chemo and/or radiation therapy is associated with a several-fold increased risk of developing acute myeloid leukaemia. Obesity is also associated with increased leukaemia risk. We did not evolve to deal with whole body radiation exposure, cigarette smoking, anticancer therapies, or modern diets. Evolution has not had time to change adaptive strategies in response to recent alterations in human lifestyles and exposures. Moreover, since these leukaemias are primarily relegated to later years of life, there will be minimal selective pressure to evolve additional tumour suppressive mechanisms.

Interestingly, several lymphomas show earlier age-dependence, and thus should disrupt reproductive success. For example, Hodgkin's lymphoma and Burkitt's lymphoma occur in young adults and children, respectively. Why has evolution not acted to limit these cancers? The answer may be that these cancers may have viral origins, and we are in evolutionary battles with pathogens like viruses. The Epstein-Barr virus (EBV) is thought to contribute to the aetiology of both lymphomas (Flavell and Murray 2000; Grywalska and Rolinski 2015). While over 90% of humans carry EBV, the vast majority of us do not develop associated cancers. In the case of Burkitt's lymphoma, infection with malarial parasites appears to substantially contribute to disease risk. These infections can reduce immune function, and also promote inflammation. We have learned how chronic inflammation can alter selective pressures, promoting the selection for adaptive oncogenic mutations. Just as evolution has not been able to eliminate our susceptibility to various pathogens, which have their own evolutionary agendas, natural selection has not eliminated the risk of cancers caused by viruses and other pathogens. See Chapter 10 for more in-depth discussion of viruses and cancers.

9.4.4 Evolutionary Explanations for Childhood Leukaemias

Young children (0–5 years of age) are at greater risk of cancers than in the following years of youth and young adulthood (years 5–35). Through the impartial lens of evolutionary theory, childhood cancers represent a quandary, as natural selection should strongly act to limit diseases that prevent an individual from successfully reproducing. Yet until around 50 years ago, childhood cancers were almost invariably lethal, reducing individual fitness to zero. In addition, these cancers are hard to understand through the lens of the current cancer paradigm, as stem cells should accumulate fewer mutations in younger children than in older children or young adults. Childhood cancers also are difficult to explain through adaptive oncogenesis, which proposes that the high fitness of youthful stem cell pools should limit oncogenic adaptation. We will discuss several models developed to explain the small peak in childhood cancer incidence in young children, noting that these explanations are not necessarily mutually exclusive.

We first need to consider relative risks. Evolution does not eliminate risk, but only limits it to the point that maximises fitness. If the cost to fitness of further risk reduction is greater than the gain, natural selection will not favour it. Leukaemias represent the most common childhood cancers, with acute lymphocytic leukaemia (ALL) representing the majority of leukaemias. Still, the risk of developing childhood ALL is only about 1 in 2000, although the impact on the children and their families is immense. Regardless, we need to consider the impact of this risk relative to other risks such as infections, which contributed to high infant mortality before the twentieth century (and continue to have this impact in the developing world). In addition, drift plays a significant role in human genetics, and given genetic bottlenecks in human evolutionary history (Eller et al. 2004), it is estimated that a mutation having an impact of less than about 0.01% will be governed by drift rather than selection. Selection may be relatively blind to at least some fraction of childhood cancers.

Additional explanations have been offered. Some have suggested that selection on immune function, bone growth, and brain development in recent millennia has come with a cost in terms of associated malignancies (Leroi et al. 2003). Humans have experienced strong selective pressures spurring evolutionary changes in the last 10,000–100,000 years, and these adaptive changes could have come with costs. For example, in the last 10,000 years, humans have experienced many new pathogens, mainly introduced following the domestication of animals. These threats have likely contributed to rapid evolutionary change in the immune system. If such a change decreased the risk of death or impairment from some infectious challenge, while leading to a much smaller increased risk of the development of lymphocytic leukaemia, natural selection would have favoured this change.

Mel Greaves has proposed that the early peak in childhood ALL incidence may be explained by delayed exposures to pathogens (Greaves 2006). He notes that childhood ALL is far more prevalent in industrialised societies than in the developing world. Moreover, having an older sibling, early entry into daycare, and other contexts that increase exposure to pathogen antigens are associated with lower ALL risk. He proposes that an immune system that has not been tuned to inflammatory insults like infections early in life becomes hyperactivated in the third to fifth year of life (such as upon entry into preschool), leading to hyperinflammation and leukaemia promotion. Natural selection has tuned our immune systems to the conditions present for the vast majority of human history, which included ubiquitous exposure to pathogens and other microbes in early life. Thus, childhood leukaemias may in part represent a disconnect between our evolved genomes and modern conditions.

Some leukaemias are caused by inheritance of genetic alleles that greatly increase risk. Fanconi anaemia results from inheritance of loss-of-function alleles of one or more than a dozen different genes whose products function in DNA repair. Individuals with Fanconi anaemia suffer from severe anaemia due to impaired haematopoiesis, with greatly reduced function of HSC and other blood cell progenitors. These individuals also exhibit a 700-fold increase in the risk of myeloid leukaemias, primarily in childhood. Grover Bagby has provided a proximate explanation, founded in an evolutionary understanding of somatic evolution (Bagby and Fleischman 2011). He has theorised that HSC and haematopoietic progenitor cells from Fanconi anaemia individuals are defective in survival, which leads to strong selection for mutations that provide resistance to this enhanced cell death. Bagby et al.'s experimental models in mice support this interpretation. Increased production of the inflammatory cytokine TNFα is observed in these individuals, leading to impairment of HSC and progenitors. Since certain oncogenic mutations can bypass this TNFα-mediated inhibition, these mutations can be adaptive selectively in Fanconi anaemia stem cell pools, leading to their selective expansion. Thus, while most investigators have focused on the increases in mutation frequency engendered by this syndrome, Bagby and colleagues have shown a critical role for alterations in selective pressures in promoting leukaemias.

Finally, computational modelling revealed another possible explanation for childhood leukaemias. A stochastic model of HSC clonal dynamics demonstrated that somatic evolution in HSC pools is differentially affected by drift and selection at different points in life (Rozhok et al. 2016). Early-life HSC pools (which are small) were permissive to clonal evolution driven by drift, with greater mutation accumulation potentiated by the rapid HSC cycling during ontogeny. Conversely, clonal evolution was suppressed by stabilising selection in larger young adult pools, and driven by positive selection in the elderly in the presence of bone marrow microenvironmental decline. These results indicate that leukaemia risk in adults is driven by the context-dependent selective advantage conferred by particular oncogenic mutations, but that such risk in early childhood HSC pools is less dependent on the selective value of mutations. Leukaemia risk at different phases in life is governed by distinct evolutionary forces acting on somatic stem cell pools.

9.4.5 Moving Evolutionary Theory into the Oncology Clinic

At this point, you should appreciate that cancer risk is dictated by evolutionary dynamics at two levels. At the organismal level, natural selection has acted to limit cancer incidence in order to maximise reproductive success, with risk with age dependent on evolved life histories. At the somatic level, we can appreciate how changes in tissues due to ontogeny, ageing, exposures, or inherited genetics can affect the evolutionary forces of mutation, selection, and drift. An evolutionary perspective may also offer solutions for the prevention and treatment of cancer.

Recent studies have highlighted the substantial heterogeneity present within cancers, both genetic and phenotypic (Marusyk and Polyak 2010; Lipinski et al. 2016). This heterogeneity results from accumulating mutations (including epigenetic changes), acted on by the forces of drift and selection. For the latter, cancers can present diverse ecological niches that promote adaptation by cells of different phenotypes. For example, selection will act on cells on the periphery of the tumour and areas farther from blood vasculature to increase glycolysis, leading to more aggressive phenotypes (Ibrahim-Hashim et al. 2017). Tumour heterogeneity will present more phenotypes upon which drug-induced selection can act, increasing the odds of cancer resistance to therapy. Leukaemias have been shown to possess multiple distinct clones with different combinations of driver mutations (Anderson et al. 2011). These clones likely represent differential adaptation to separate niches in the patient (or insufficient time has passed for one genotype to sweep through the population). Importantly, this clonal diversity can increase the odds that some cancer genotype will be resistant to therapy, and a previously minor leukaemic clone can dominate after therapy.

A well-described example of cancer evolution is the acquisition of drug resistance, which mirrors the development of antibiotic resistance in bacteria. Large tumour populations, with as many as a trillion cells, together with the strong selective pressures engendered by therapies, create ideal ingredients for the selection of resistant clones. Evolutionary theory can be used to try to limit this evolution of resistance. Analogous to the triple antiretroviral therapies for acquired immune deficiency syndrome (AIDS), targeting multiple cancer dependencies simultaneously should limit the development of resistance, assuming that there is not a common mechanism of resistance. Unfortunately, there are often common resistance mechanisms, such as through expression of multidrug resistance drug pumps or through activation of a general survival pathway.

Another strategy is to reduce the pressure for development of resistance. We can learn from integrated pest management, whereby farmers only apply pesticide or herbicide when the pests/weeds reach some threshold level, leading to greatly reduced chemical application while maintaining yields. The reason that this strategy works is that evolution of resistance typically comes with a cost (as the pest was previously optimised by evolution). Robert Gatenby and colleagues have exploited this concept to design 'adaptive therapy' regimens that involve only treating patients when the tumour size exceeds some threshold (Enriquez-Navas et al. 2016). In theory, this approach should better maintain drug-sensitive cancer cells in the population, which should rebound in between therapeutic treatments (given the reduced somatic fitness of the drug-resistant cells in the absence of therapy). Results from mouse models look promising, and there are ongoing clinical trials.

Finally, we should be able to target the microenvironment to influence the fitness value of oncogenic mutations. As we have discussed, the somatic fitness conferred by any mutation is highly dependent on microenvironmental context. For example, if we know that a leukaemia possesses an oncogenic KRAS mutation, and that previous studies have demonstrated that the leukaemia is 'addicted' to this oncogene, we can try to determine what the microenvironmental context is that promoted the selection of this oncogene in the first place. Knowing this could facilitate the design of interventions that reverse this microenvironmental factor, in order to disfavour the oncogenic-driven phenotype. Recent studies have shown how subtle alterations in the microenvironment (in this case, pH) can dampen the selective value of certain aggressive cancer phenotypes (Ibrahim-Hashim et al. 2017). Recognising that cancer is not striving towards any goal, but simply adapting to its current environment, we should be able to alter environments to favour more benign phenotypes.

REFERENCES

Abkowitz, J. L., Catlin, S. N., McCallie, M. T., et al. (2002). Evidence that the number of hematopoietic stem cells per animal is conserved in mammals. *Blood* 100(7), 2665–7.

Adamo, L., Naveiras, O., Wenzel, P. L., et al. (2009). Biomechanical forces promote embryonic haematopoiesis. *Nature* 459(7250), 1131–5.

Aghili, L., Foo, J., DeGregori, J., et al. (2014). Patterns of somatically acquired amplifications and deletions in apparently normal tissues of ovarian cancer patients. *Cell Rep* 7(4), 1310–19.

Akunuru, S. and Geiger, H. (2016). Aging, clonality, and rejuvenation of hematopoietic stem cells. *Trends Mol Med* 22(8), 701–12.

Anderson, K., Lutz, C., van Delft, F. W., et al. (2011). Genetic variegation of clonal architecture and propagating cells in leukaemia. *Nature* 469(7330), 356–61.

Anstee, D. J. (2010). The relationship between blood groups and disease. *Blood* 115(23), 4635–43.

Asking, B. and Gjorstrup, P. (1987). Synthesis and secretion of amylase in the rat parotid gland following autonomic nerve stimulation in vivo. *Acta Physiol Scand* 130(3), 439–45.

Bagby, G. C. and Fleischman, A. G. (2011). The stem cell fitness landscape and pathways of molecular leukemogenesis. *Front Biosci (Schol Ed)* 3, 487–500.

Baldridge, M. T., King, K. Y., and Goodell, M. A. (2011). Inflammatory signals regulate hematopoietic stem cells. *Trends Immunol* 32(2), 57–65.

Ballesteros-Arias, L., Saavedra, V., and Morata, G. (2014). Cell competition may function either as tumour-suppressing or as tumour-stimulating factor in *Drosophila*. *Oncogene* 33(35), 4377–84.

Baumann, C. I., Bailey, A. S., Li, W., et al. (2004). PECAM-1 is expressed on hematopoietic stem cells throughout ontogeny and identifies a population of erythroid progenitors. *Blood* 104(4), 1010–16.

Beerman, I., Maloney, W. J., Weissmann, I. L., et al. (2010). Stem cells and the aging hematopoietic system. *Curr Opin Immunol* 22(4), 500–6.

Berent-Maoz, B., Montecino-Rodriguez, E., and Dorshkind, K. (2012). Genetic regulation of thymocyte progenitor aging. *Semin Immunol* 24(5), 303–8.

Bernitz, J. M., Kim, H. S., MacArthur, B., et al. (2016). Hematopoietic stem cells count and remember self-renewal divisions. *Cell* 167(5), 1296–309.

Bissell, M. J. and Hines, W. C. (2011). Why don't we get more cancer? A proposed role of the microenvironment in restraining cancer progression. *Nat Med* 17(3), 320–9.

Bohmig, G. A., Farkas, A. M., Eskandary, F., et al. (2015). Strategies to overcome the ABO barrier in kidney transplantation. *Nat Rev Nephrol* 11(12), 732–47.

Brack, A. S., Bildsoe, H., and Hughes, S. M. (2005). Evidence that satellite cell decrement contributes to preferential decline in nuclear number from large fibres during murine age-related muscle atrophy. *J Cell Sci* 118(Pt 20), 4813–21.

Busque, L., Patel, J. P., Figueroa, M. E., Vasanthakumar, A., Provost, S., Hamilou, Z., Mollica, L., Li, J., Viale, A., Heguy, A., et al. (2012). Recurrent somatic TET2 mutations in normal elderly individuals with clonal hematopoiesis. *Nature Genetics* 44, 1179–81.

Calafell, F., Roubinet, F., Ramirez-Soriano, A., et al. (2008). Evolutionary dynamics of the human ABO gene. *Hum Genet* 124(2), 123–35.

Carroll, S. B. (2006). *The Making of the Fittest: DNA and the Ultimate Forensic Record of Evolution.* New York: W. W. Norton & Co.

Caruso, C., Lio, D., Cavallone, L., et al. (2004). Aging, longevity, inflammation, and cancer. *Ann N Y Acad Sci* 1028, 1–13.

Colmegna, I. and Weyand, C. M. (2011). Haematopoietic stem and progenitor cells in rheumatoid arthritis. *Rheumatol (Oxford)* 50(2), 252–60.

Cooper, M. D. and Alder, M. N. (2006). The evolution of adaptive immune systems. *Cell* 124(4), 815–22.

Copley, M. R., Babovic, S., Benz, C., et al. (2013). The Lin28b-let-7-Hmga2 axis determines the higher self-renewal potential of fetal haematopoietic stem cells. *Nat Cell Biol* 15(8), 916–25.

Costantino, S., Paneni, F., and Cosentino, F. (2016). Ageing, metabolism and cardiovascular disease. *J Physiol* 594(8), 2061–73.

Cserti, C. M. and Dzik, W. H. (2007). The ABO blood group system and *Plasmodium falciparum* malaria. *Blood* 110(7), 2250–8.

DeGregori, J. (2011). Evolved tumor suppression: why are we so good at not getting cancer? *Cancer Res* 71(11), 3739–44.

DeGregori, J. (2012). Challenging the axiom: does the occurrence of oncogenic mutations truly limit cancer development with age? *Oncogene* 32(15), 1869–75.

Dentali, F., Sironi, A. P., Ageno, W., et al. (2012). Non-O blood type is the commonest genetic risk factor for VTE: results from a meta-analysis of the literature. *Semin Thromb Hemost* 38(5), 535–48.

Ding, L. and Morrison, S. J. (2013). Haematopoietic stem cells and early lymphoid progenitors occupy distinct bone marrow niches. *Nature* 495(7440), 231–5.

Doles, J. and Keyes, W. M. (2013). Epidermal stem cells undergo age-associated changes. *Aging* 5(1), 1–2.

Doolittle, R. F. (2009). Step-by-step evolution of vertebrate blood coagulation. *Cold Spring Harb Symp Quant Biol* 74, 35–40.

Dorshkind, K. and Montecino-Rodriguez, E. (2007). Fetal B-cell lymphopoiesis and the emergence of B-1-cell potential. *Nat Rev Immunol* 7(3), 213–19.

Dykstra, B., Olthof, S., Schreuder, J., et al. (2011). Clonal analysis reveals multiple functional defects of aged murine hematopoietic stem cells. *J Exp Med* 208(13), 2691–703.

Eldredge, N. (1999). *The Pattern of Evolution*. New York: W. H. Freeman.

Eller, E., Hawks, J., and Relethford, J. H. (2004). Local extinction and recolonization, species effective population size, and modern human origins. *Hum Biol* 76(5), 689–709.

Ema, H., Takano, H., Sudo, K., et al. (2000). In vitro self-renewal division of hematopoietic stem cells. *J Exp Med* 192(9), 1281–8.

Enriquez-Navas, P. M., Kam, Y., Das, T., et al. (2016). Exploiting evolutionary principles to prolong tumor control in preclinical models of breast cancer. *Sci Transl Med* 8(327), 327ra24.

Espin-Palazon, R., Stachura, D. L., Campbell, C. A., et al. (2014). Proinflammatory signaling regulates hematopoietic stem cell emergence. *Cell* 159(5), 1070–85.

Fernandez, K. S. and de Alarcon, P. A. (2013). Development of the hematopoietic system and disorders of hematopoiesis that present during infancy and early childhood. *Pediatr Clin North Am* 60(6), 1273–89.

Flach, J., Bakker, S. T., Mohrin, M., et al. (2014). Replication stress is a potent driver of functional decline in ageing haematopoietic stem cells. *Nature* 512(7513), 198–202.

Flajnik, M. F. and Du Pasquier, L. (2004). Evolution of innate and adaptive immunity: can we draw a line? *Trends Immunol* 25(12), 640–44.

Flavell, K. J. and Murray, P. G. (2000). Hodgkins disease and the Epstein-Barr virus. *Mol Pathol* 53(5), 262–9.

Florian, M. C., Nattamai, K. J., Dorr, K., et al. (2013). A canonical to non-canonical Wnt signalling switch in haematopoietic stem-cell ageing. *Nature* 503(7476), 392–6.

Fortunato, A., Boddy, A., Mallo, D., et al. (2017). Natural selection in cancer biology: from molecular snowflakes to trait hallmarks. *Cold Spring Harb Perspect Med* 7(2), pii: a029652.

Fu, L. and Kettner, N. M. (2013). The circadian clock in cancer development and therapy. *Prog Mol Biol Transl Sci* 119, 221–82.

Gause, W. C., Wynn, T. A., and Allen, J. E. (2013). Type 2 immunity and wound healing: evolutionary refinement of adaptive immunity by helminths. *Nat Rev Immunol* 13(8), 607–14.

Geiger, H., Schleimer, D., Nattamai, K. J., et al. (2006). Mutagenic potential of temozolomide in bone marrow cells in vivo. *Blood* 107(7), 3010–11.

Genovese, G., Kahler, A. K., Handsaker, R. E., et al. (2014). Clonal hematopoiesis and blood-cancer risk inferred from blood DNA sequence. *N Engl J Med* 371(26), 2477–87.

Ginhoux, F., Greter, M., Leboeuf, M., et al. (2010). Fate mapping analysis reveals that adult microglia derive from primitive macrophages, *Science*, 330 (6005), 841–5.

Godin, I. and Cumano, A. (2005). Of birds and mice: hematopoietic stem cell development. *Int J Dev Biol* 49(2–3), 251–7.

Greaves, M. (2006). Infection, immune responses and the aetiology of childhood leukaemia. *Nat Rev Cancer* 6(3), 193–203.
Green, D. R., Galluzzi, L., and Kroemer, G. (2011). Mitochondria and the autophagy–inflammation–cell death axis in organismal aging. *Science* 333(6046), 1109–12.
Grover, A., Sanjuan-Pla, A., Thongjuea, S., et al. (2016). Single-cell RNA sequencing reveals molecular and functional platelet bias of aged haematopoietic stem cells. *Nat Commun* 7, 11075.
Grywalska, E. and Rolinski, J. (2015). Epstein-Barr virus-associated lymphomas. *Semin Oncol* 42(2), 291–303.
Gurven, M. and Kaplan, H. (2007). Longevity among hunter- gatherers: a cross-cultural examination. *Popul Dev Rev* 33(2), 321–65.
Haas, S., Hansson, J., Klimmeck, D., et al. (2015). Inflammation-induced emergency megakaryopoiesis driven by hematopoietic stem cell-like megakaryocyte progenitors. *Cell Stem Cell* 17(4), 422–34.
Henry, C. J., Marusyk, A., Zaberezhnyy, V., et al. (2010). Declining lymphoid progenitor fitness promotes aging-associated leukemogenesis. *Proc Natl Acad Sci U S A* 107(50), 21713–18.
Henry, C. J., Marusyk, A., and DeGregori, J. (2011). Aging-associated changes in hematopoiesis and leukemogenesis: what's the connection? *Aging* 3(6), 643–56.
Henry, C. J., Casas-Selves, M., Kim, J., et al. (2015). Aging-associated inflammation promotes selection for adaptive oncogenic events in B cell progenitors. *J Clin Invest* 125(12), 4666–80.
Hochberg, M. E. and Noble, R. J. (2017). A framework for how environment contributes to cancer risk. *Ecol Lett* 20(2), 117–34.
Hoeijmakers, J. H. (2009). DNA damage, aging, and cancer. *N Engl J Med* 361(15), 1475–85.
Horvath, S. (2013). DNA methylation age of human tissues and cell types. *Genome Biol* 14(10), R115.
Houldcroft, C. J. and Underdown, S. J. (2016). Neanderthal genomics suggests a Pleistocene time frame for the first epidemiologic transition. *Am J Phys Anthropol* 160(3), 379–88.
Ibrahim-Hashim, A., Robertson-Tessi, M., Enrizues-Navas, P., et al. (2017). Defining cancer subpopulations by adaptive strategies rather than molecular properties provides novel insights into intratumoral evolution. *Cancer Res* 77(9), 2242–54.
Ilhan, G., Karakus, S., and Andic, N. (2006). Risk factors and primary prevention of acute leukemia. *Asian Pac J Cancer Prev* 7(4), 515–17.
Iwasaki, H., Arai, F., Kubota, Y., et al. (2010). Endothelial protein C receptor-expressing hematopoietic stem cells reside in the perisinusoidal niche in fetal liver. *Blood* 116(4), 544–53.
Jacobs, K. B., Yeager, M., Zhou, W., et al. (2012). Detectable clonal mosaicism and its relationship to aging and cancer. *Nat Genet* 44(6), 651–8.
Jaiswal, S., Fontanillas, P., Flannick, J., et al. (2014). Age-related clonal hematopoiesis associated with adverse outcomes. *N Engl J Med* 371(26), 2488–98.
Janeway, C. A., et al. (2001). Evolution of the innate immune system. In: Janeway, C. A., Jr, Travers, P., Walport, M., et al. (eds) *Immunobiology: The Immune System in Health and Disease*, 5th ed. New York: Garland Science.
Jing, L. and Zon, L. I. (2011). Zebrafish as a model for normal and malignant hematopoiesis. *Dis Model Mech* 4(4), 433–8.
Kajita, M. and Fujita, Y. (2015). EDAC: epithelial defence against cancer-cell competition between normal and transformed epithelial cells in mammals. *J Biochem* 158(1), 15–23.
Kaplan, H., Hill, K., Lancaster, J., et al. (2000). A theory of human life history evolution: diet, intelligence, and longevity. *Evol Anthropol* 9(4), 156–85.
Keller, G. and Snodgrass, R. (1990). Life span of multipotential hematopoietic stem cells in vivo. *J Exp Med* 171(5), 1407–18.
Kiel, M. J., Yilmaz, O. H., Iwashita, T., et al. (2005). SLAM family receptors distinguish hematopoietic stem and progenitor cells and reveal endothelial niches for stem cells. *Cell* 121(7), 1109–21.
Kirkwood, T. B. (2005). Understanding the odd science of aging. *Cell* 120(4), 437–47.
Koonin, E. V. and Krupovic, M. (2015). Evolution of adaptive immunity from transposable elements combined with innate immune systems. *Nat Rev Genet* 16(3), 184–92.
Kowald, A. and Kirkwood, T. B. (2015). Evolutionary significance of ageing in the wild. *Exp Gerontol* 71, 89–94.

Kowald, A. and Kirkwood, T. B. (2016). Can aging be programmed? A critical literature review. *Aging Cell* doi: 10.1111/acel.12510 [Epub ahead of print].
Krabbe, K. S., Pedersen, M., and Bruunsgaard, H. (2004). Inflammatory mediators in the elderly. *Exp Gerontol* 39(5), 687–99.
Laird, D. J., De Tomaso, A. W., and Weissman, I. L. (2005). Stem cells are units of natural selection in a colonial ascidian. *Cell* 123(7), 1351–60.
Laufer, M. K., Thesing, P. C., Eddington, N. D., et al. (2006). Return of chloroquine antimalarial efficacy in Malawi. *N Engl J Med* 355(19), 1959–66.
Laurie, C. C., Laurie, C. A., Rice, K., et al. (2012). Detectable clonal mosaicism from birth to old age and its relationship to cancer. *Nat Genet* 44(6), 642–50.
Lefrancais, E., Ortiz-Munoz, G., Caudrillier, A., et al. (2017), 'The lung is a site of platelet biogenesis and a reservoir for haematopoietic progenitors', *Nature,* 544 (7648), 105–09.
Leroi, A. M., Koufopanou, V., and Burt, A. (2003). Cancer selection. *Nat Rev Cancer* 3(3), 226–31.
Levin, J. (1997). The evolution of mammalian platelets. In: Kuter, D. J. et al. (eds) *Thrombopoiesis and Thrombopoietins: Molecular, Cellular, Preclinical, and Clinical Biology*. Totowa, NJ: Humana Press, pp. 63–78.
Li, Y., Esain, V., Teng, L., et al. (2014). Inflammatory signaling regulates embryonic hematopoietic stem and progenitor cell production. *Genes Dev* 28(23), 2597–612.
Lipinski, K. A., Barber, L. J., Davies, M. N., et al. (2016). Cancer evolution and the limits of predictability in precision cancer medicine. *Trends Cancer* 2(1), 49–63.
Litman, G. W. and Cooper, M. D. (2007). Why study the evolution of immunity? *Nat Immunol* 8(6), 547–8.
Liu, W., Li, Y., Learn, G. H., et al. (2010). Origin of the human malaria parasite *Plasmodium falciparum* in gorillas. *Nature* 467(7314), 420–5.
Lowe, S. W., Cepero, E., and Evan, G. (2004). Intrinsic tumour suppression. *Nature* 432(7015), 307–15.
Makivuokko, H., Lahtinen, S. J., Wacklin, P., et al. (2012). Association between the ABO blood group and the human intestinal microbiota composition. *BMC Microbiol* 12, 94.
Manry, J., Laval, G., Patin, E., et al. (2011). Evolutionary genetic dissection of human interferons. *J Exper Medicine* 208(13), 2747–59.
Manz, M. G. and Boettcher, S. (2014). Emergency granulopoiesis. *Nat Rev Immunol* 14(5), 302–14.
Martinez, D. E. (1998). Mortality patterns suggest lack of senescence in hydra. *Exp Gerontol* 33(3), 217–25.
Marusyk, A. and Polyak, K. (2010). Tumor heterogeneity: causes and consequences. *Biochim Biophys Acta* 1805(1), 105–17.
Mazo, I. B., Massberg, S., and von Andrian, U. H. (2011). Hematopoietic stem and progenitor cell trafficking. *Trends Immunol* 32(10), 493–503.
McKerrell, T., Park, N., Moreno, T., et al. (2015). Leukemia-associated somatic mutations drive distinct patterns of age-related clonal hemopoiesis. *Cell Reports* 10(8), 1239–45.
Medawar, P. B. (1952). *An Unsolved Problem of Biology*. London: H. K. Lewis.
Medvinsky, A., Rybtsov, S., and Taoudi, S. (2011). Embryonic origin of the adult hematopoietic system: advances and questions. *Development* 138(6), 1017–31.
Mendez-Ferrer, S., Chow, A., Merad, M., et al. (2009). Circadian rhythms influence hematopoietic stem cells. *Curr Opin Hematol* 16(4), 235–42.
Menendez, J., Perez-Garijo, A., Calleja, M., et al. (2010). A tumor-suppressing mechanism in *Drosophila* involving cell competition and the Hippo pathway. *Proc Natl Acad Sci U S A* 107(33), 14651–6.
Merino, M. M., Levayer, R., and Moreno, E. (2016). Survival of the fittest: essential roles of cell competition in development, aging, and cancer. *Trends Cell Biol* 26(10), 776–88.
Miller, K. R. (1999). *Finding Darwin's God: A Scientist's Search for Common Ground Between God and Evolution*. New York: Cliff Street Books.
Mirantes, C., Passegue, E., and Pietras, E. M. (2014). Pro-inflammatory cytokines: emerging players regulating HSC function in normal and diseased hematopoiesis. *Exp Cell Res* 329(2), 248–54.

Moehrle, B. M., Nattamai, K., Brown, A., et al. (2015). Stem cell-specific mechanisms ensure genomic fidelity within HSCs and upon aging of HSCs. *Cell Rep* 13(11), 2412–24.

Monahan-Earley, R., Dvorak, A. M., and Aird, W. C. (2013). Evolutionary origins of the blood vascular system and endothelium. *J Thromb Haemost* 11(Suppl 1), 46–66.

Muller-Sieburg, C. E., Cho, R. H., Karlsson, L., et al. (2004). Myeloid-biased hematopoietic stem cells have extensive self-renewal capacity but generate diminished lymphoid progeny with impaired IL-7 responsiveness. *Blood* 103(11), 4111–18.

Nakada, D., Oguro, H., Levi, B. P., et al. (2014). Oestrogen increases haematopoietic stem-cell self-renewal in females and during pregnancy. *Nature* 505(7484), 555–8.

Nalapareddy, K., Nattamai, K. J., Kumar, R. S., et al. (2017). Canonical Wnt signaling ameliorates aging of intestinal stem cells. *Cell Rep* 18(11), 2608–21.

Nédélec, Y., Sanz, J., Baharian, G., et al. (2016). Genetic ancestry and natural selection drive population differences in immune responses to pathogens. *Cell* 167(3), 657–69.

Niedernhofer, L. J., Kirkland, J. L., and Ladiges, W. (2017). Molecular pathology endpoints useful for aging studies. *Ageing Res Rev* 35, 241–9.

North, T. E., Goessling, W., Peeters, M., et al. (2009). Hematopoietic stem cell development is dependent on blood flow. *Cell* 137(4), 736–48.

Orkin, S. H. and Zon, L. I. (2008). Hematopoiesis: an evolving paradigm for stem cell biology. *Cell* 132(4), 631–44.

Pancer, Z., Amemiya, C. T., Ehrhardt, G. R., et al. (2004). Somatic diversification of variable lymphocyte receptors in the agnathan sea lamprey. *Nature* 430(6996), 174–80.

Pang, W. W., Price, E. A., Sahoo, D., et al. (2011). Human bone marrow hematopoietic stem cells are increased in frequency and myeloid-biased with age. *Proc Natl Acad Sci U S A* 108(50), 20012–17.

Perry, G. H., Xue, Y., Smith, R. S., et al. (2012). Evolutionary genetics of the human Rh blood group system. *Hum Genet* 131(7), 1205–16.

Pierce, H., Zhang, D., Magnon, C., et al. (2017). Cholinergic signals from the CNS regulate G-CSF-mediated HSC mobilization from bone marrow via a glucocorticoid signaling relay. *Cell Stem Cell* 20(5), 648–58.

Pietras, E. M., Warr, M. R., and Passegue, E. (2011). Cell cycle regulation in hematopoietic stem cells. *J Cell Biol* 195(5), 709–20.

Pietras, E. M., Lakshminarasimhan, R., Techner, J. M., et al. (2014). Re-entry into quiescence protects hematopoietic stem cells from the killing effect of chronic exposure to type I interferons. *J Exp Med* 211(2), 245–62.

Pietras, E. M., Reynaud, D., Kang, Y. A., et al. (2015). Functionally distinct subsets of lineage-biased multipotent progenitors control blood production in normal and regenerative conditions. *Cell Stem Cell* 17(1), 35–46.

Pietras, E. M., Mirantes-Barbeito, C., Fong, S., et al. (2016). Chronic interleukin-1 exposure drives haematopoietic stem cells towards precocious myeloid differentiation at the expense of self-renewal. *Nat Cell Biol* 18(6), 607–18.

Powell, N. D., Sloan, E. K., Bailey, M. T., et al. (2013). Social stress up-regulates inflammatory gene expression in the leukocyte transcriptome via beta-adrenergic induction of myelopoiesis. *Proc Natl Acad Sci U S A* 110(41), 16574–9.

Quach, H., Rotival, M., Pothlichet, J., et al. (2016). Genetic adaptation and Neandertal admixture shaped the immune system of human populations. *Cell* 167(3), 643–56.

Renault, V. M., Rafalski, V. A., Morgan, A. A., et al. (2009). FoxO3 regulates neural stem cell homeostasis. *Cell Stem Cell* 5(5), 527–39.

Rowe, J. A., Handel, I. G., Thera, M. A., et al. (2007). Blood group O protects against severe *Plasmodium falciparum* malaria through the mechanism of reduced rosetting. *Proc Natl Acad Sci U S A* 104(44), 17471–6.

Rozhok, A. I., Salstrom, J. L., and DeGregori, J. (2014). Stochastic modeling indicates that aging and somatic evolution in the hematopoietic system are driven by non-cell-autonomous processes. *Aging* 6(12), 1033–48.

Rozhok, A. I., Wahl, G. M., and DeGregori, J. (2015). A critical examination of the bad luck explanation of cancer risk. *Cancer Prev Res* 8(9), 762–4.

Rozhok, A. I., Salstrom, J. L., and DeGregori, J. (2016). Stochastic modeling reveals an evolutionary mechanism underlying elevated rates of childhood leukemia. *Proc Natl Acad Sci U S A* 113(4), 1050–5.

Savage, W. J. (2016). Transfusion reactions. *Hematol Oncol Clin North Am* 30(3), 619–34.

Schepers, K., Campbell, T. B., and Passegue, E. (2015). Normal and leukemic stem cell niches: insights and therapeutic opportunities. *Cell Stem Cell* 16(3), 254–67.

Schmaier, Alec A., Stalker, Timothy, J., Runge, Jeffrey, J., et al. (2011). Occlusive thrombi arise in mammals but not birds in response to arterial injury: evolutionary insight into human cardiovascular disease. *Blood* 118(13), 3661–9.

Schoedel, K. B., Morcos, M. N., Zerjatke, T., et al. (2016). The bulk of the hematopoietic stem cell population is dispensable for murine steady-state and stress hematopoiesis. *Blood* 128(19), 2285–96.

Schulz, C., Gomez Perdiguero, E., Chorro, L., et al. (2012). A lineage of myeloid cells independent of Myb and hematopoietic stem cells. *Science* 336(6077), 86–90.

Secombes, C. J. and Zou, J. (2017). Evolution of interferons and interferon receptors. *Front Immunol* 8, 209.

Segurel, L., Gao, Z., and Przeworski, M. (2013). Ancestry runs deeper than blood: the evolutionary history of ABO points to cryptic variation of functional importance. *Bioessays* 35(10), 862–7.

Seita, J. and Weissman, I. L. (2010). Hematopoietic stem cell: self-renewal versus differentiation. *Wiley Interdiscip Rev Syst Biol Med* 2(6), 640–53.

Semino, O., Passarino, G., Oefner, P. J., et al. (2000). The genetic legacy of Paleolithic *Homo sapiens sapiens* in extant Europeans: a Y chromosome perspective. *Science* 290(5494), 1155–9.

Shaw, A. C., Joshi, S., Greenwood, H., et al. (2010). Aging of the innate immune system. *Curr Opin Immunol* 22(4), 507–13.

Signer, R. A., Montecino-Rodriguez, E., and Dorshkind, K. (2007). Aging, B lymphopoiesis, and patterns of leukemogenesis. *Exp Gerontol* 42(5), 391–5.

Snyder, G. K. and Sheafor, B. A. (1999). Red blood cells: centerpiece in the evolution of the vertebrate circulatory system. *Am Zool* 39(2), 189–98.

Spangrude, G. J., Heimfeld, S., and Weissman, I. L. (1988). Purification and characterization of mouse hematopoietic stem cells. *Science* 241(4861), 58–62.

Sun, D., Luo, M., Jeong, M., et al. (2014a). Epigenomic profiling of young and aged HSCs reveals concerted changes during aging that reinforce self-renewal. *Cell Stem Cell* 14(5), 673–88.

Sun, J., Ramos, A., Chapman, B., et al. (2014b). Clonal dynamics of native haematopoiesis. *Nature* 514(7522), 322–7.

Till, J. E. and McCulloch, E. A. (1961). A direct measurement of the radiation sensitivity of normal mouse bone marrow cells. *Radiat Res* 14, 213–22.

Tomasetti, C. and Vogelstein, B. (2015). Cancer etiology. Variation in cancer risk among tissues can be explained by the number of stem cell divisions. *Science* 347(6217), 78–81.

Travnickova, J., Tran Chau, V., Julien, E., et al. (2015). Primitive macrophages control HSPC mobilization and definitive haematopoiesis. *Nat Commun* 6, 6227.

Vallender, E. J. and Lahn, B. T. (2004). Positive selection on the human genome. *Hum Mol Genet* 13(Spec No. 2), R245–54.

van Deursen, J. M. (2014). The role of senescent cells in ageing. *Nature* 509(7501), 439–46.

Vijg, J., Busuttil, R. A., Bahar, R., et al. (2005). Aging and genome maintenance. *Ann N Y Acad Sci* 1055, 35–47.

Wahlestedt, M., Norddahl, G. L., Sten, G., et al. (2013). An epigenetic component of hematopoietic stem cell aging amenable to reprogramming into a young state. *Blood* 121(21), 4257–64.

Waldhauer, I. and Steinle, A. (2008). NK cells and cancer immunosurveillance. *Oncogene* 27(45), 5932–43.

Wang, J. C., Doedens, M., and Dick, J. E. (1997). Primitive human hematopoietic cells are enriched in cord blood compared with adult bone marrow or mobilized peripheral blood as measured by the quantitative in vivo SCID-repopulating cell assay. *Blood* 89(11), 3919–24.

Ward, P. D. and Kirschvink, J. (2015). *A New History of Life: The Radical New Discoveries About the Origins and Evolution of Life on Earth*. New York: Bloomsbury Press.

Wassmer, S. C. and Carlton, J. M. (2016). Glycophorins, blood groups, and protection from severe malaria. *Trends Parasitol* 32(1), 5–7.

Weinberger, B. and Grubeck-Loebenstein, B. (2012). Vaccines for the elderly. *Clin Microbiol Infect* 18(Suppl 5), 100–8.

Wellems, T. E., Hayton, K., and Fairhurst, R. M. (2009). The impact of malaria parasitism: from corpuscles to communities. *J Clin Invest* 119(9), 2496–505.

White, N. J. (2004). Antimalarial drug resistance. *J Clin Invest* 113(8), 1084–92.

Williams, G. C. (1957). Pleiotropy, natural selection, and the evolution of senescence. *Evolution* 11(4), 398–411.

Wilson, A., Laurenti, E., Oser, G., et al. (2008). Hematopoietic stem cells reversibly switch from dormancy to self-renewal during homeostasis and repair. *Cell* 135(6), 1118–29.

Wolfe, N. D., Dunavan, C. P., and Diamond, J. (2007). Origins of major human infectious diseases. *Nature* 447(7142), 279–83.

Wright, D. E., Wagers, A. J., Gulati, A. P., et al. (2001). Physiological migration of hematopoietic stem and progenitor cells. *Science* 294(5548), 1933–6.

Xie, M., Lu, C., Wang, J., et al. (2014). Age-related mutations associated with clonal hematopoietic expansion and malignancies. *Nat Med* 20(12), 1472–8.

Yamamoto, F., Cid, E., Yamamoto, M., et al. (2012). ABO research in the modern era of genomics. *Transfus Med Rev* 26(2), 103–18.

Yang, N. and Boettcher, B. (1992). Development of human ABO blood group A antigen on *Escherichia coli* Y1089 and Y1090. *Immunol Cell Biol* 70(Pt 6), 411–16.

Yokota, T., Oritani, K., Butz, S., et al. (2009). The endothelial antigen ESAM marks primitive hematopoietic progenitors throughout life in mice. *Blood* 113(13), 2914–23.

Zeeb, H. and Blettner, M. (1998). Adult leukaemia: what is the role of currently known risk factors? *Radiat Environ Biophys* 36(4), 217–28.

Zhang, S. M., Loker, E. S., and Sullivan, J. T. (2016). Pathogen-associated molecular patterns activate expression of genes involved in cell proliferation, immunity and detoxification in the amebocyte-producing organ of the snail *Biomphalaria glabrata*. *Dev Comp Immunol* 56, 25–36.

CHAPTER 10

IMMUNE SYSTEM

GRAHAM A. W. ROOK

Abstract

As humans move from the natural environment in which we evolved into modern urban settings, there are striking increases in chronic inflammatory and psychiatric disorders. To understand and eventually take control of this phenomenon we have to understand how humans, and in particular our immune systems, evolved in partnership with microorganisms in the environment and in our own bodies. Humans are holobionts, composed of human cells containing the human genome passed on via the germline, but also a much larger number of microbial cells acquired from mother, family members, and the environment. This microbiota provides signals involved in the development of essentially all organ systems, including the brain, and provides data and signals that regulate metabolism and the immune system. The immune system evolved to perform the dual functions of managing this microbiota, while simultaneously protecting us from pathogens. By considering the evolution of the immune system and the ways in which lifestyle changes have altered our exposures to, and colonisation by microorganisms, we can identify the crucial factors leading to the modern urban pattern of disease.

Keywords

inflammatory, microbiota, holobiont, lifestyle, urban, psychiatric disorder, metabolism, immune system, natural environment, evolutionary medicine

10.1 Introduction

We tend to think of the immune system as a mechanism for destroying infections invading from outside and for destroying mutated cells inside our bodies that are becoming cancerous. We expect it to do all this while simultaneously censoring any inappropriate attacks on our

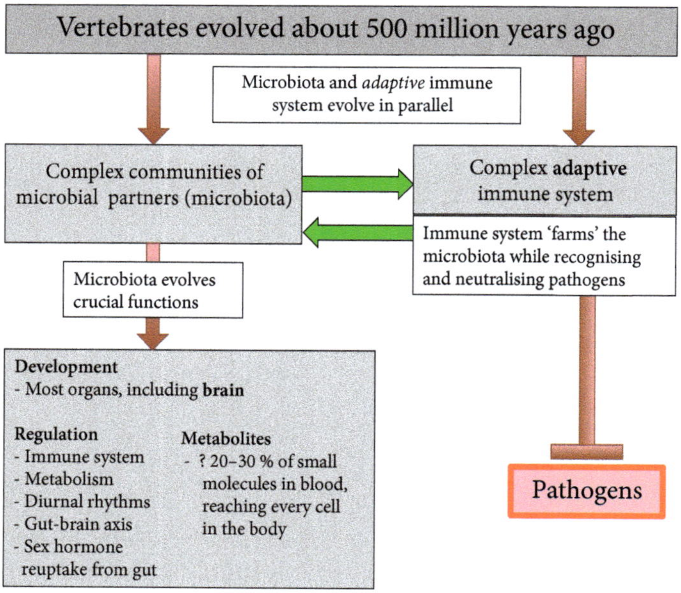

FIGURE 10.1 The adaptive immune system and microbiota evolved in parallel soon after the appearance of vertebrates about 500 million years ago. The microbiota is an essential 'organ', performing numerous essential physiological functions, listed in the Figure. The adaptive immune system works in collaboration with the innate immune system to manage ('farm') the microbiota, while excluding pathogens.

own healthy tissues. Thus, we assume that the immune system maintains our integrity and individuality. However, the realisation that we are not individuals has caused this view of the immune system to become obsolete (McFall-Ngai et al. 2013). All multicellular life forms are ecosystems. Where humans are concerned, at least 50%, perhaps more, of the cells that make up our bodies are microbial, and they contribute far more genes, DNA, and metabolic pathways than are encoded in our human genomes. These organisms, known collectively as the microbiota, are not mere passengers. They contribute to the development of every organ system and to every physiological function including metabolism, diurnal rhythms, the balance of sex hormones, and the development and function of the brain (Figure 10.1).

Thus, we are 'holobionts' composed of multiple different microbes and macrobes in symbiotic relationships. The immune system, therefore, has to perform the protective functions listed above, while simultaneously tolerating the microbiota. Indeed, the immune system does much more than tolerate the microbiota—it actively 'farms' these organisms and helps to determine and maintain their composition. In both vertebrates and invertebrates this involves the innate immune system, of which the various receptors that recognise microorganisms are encoded in the genome. This seems to work well in invertebrates. The invertebrate microbiota is not usually complex, and invertebrates use various strategies to manage their microbiota, such as keeping small numbers of organisms in intracellular locations, or walling them off in chitin containers.

However, shortly after the vertebrates evolved about 500 million years ago (mya), their microbiota became more complex and numerous, and in close contact with host tissues.

For example, the human gut may contain about 2 kg of microorganisms, spread over an intestinal surface of about 32 m^2 (as a conservative estimate) (Helander and Fandriks 2014). It is not surprising that at least 30% of all small molecules in human blood are microbial metabolites, many of them exceedingly biologically active, ranging from toxic to physiologically essential (Wikoff et al. 2009; Hsiao et al. 2013). This point highlights the need to regulate the components of the microbiota very carefully. Thus, the microbiota became physiologically essential, providing flexibility and the rapid evolution of novel physiological functions, but it remained potentially dangerous, so how could it be managed? Most evolutionary biologists now think that vertebrates solved the problem of maintaining this huge complex microbiota by evolving the adaptive immune system that uses mutations to generate novel specificities. This provides an almost infinitely expandable repertoire and the ability to keep up with rapidly evolving organisms. Thus, the problem was to exclude pathogens, while simultaneously tolerating continuous inputs of usually harmless organisms from the natural environment, and maintaining an appropriate, non-toxic, physiologically essential microbiota. This was solved by supplementing the innate immune system with an adaptive system that could generate entirely novel recognition specificities and regulatory feedback pathways.

The resulting immune system in which the innate and adaptive branches work closely together shares many characteristics with the brain (Figure 10.2). Just like the brain, it is a learning system, so, like the brain, it must receive appropriate data inputs, and these must be received early in life and then maintained and updated throughout life. These inputs come from the microbiota, from the natural environment, and from some pathogens with

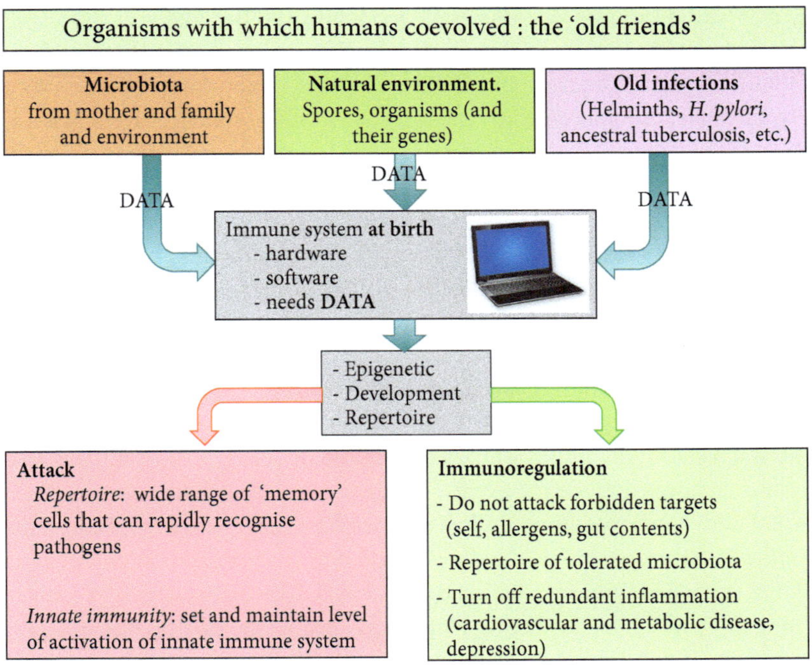

FIGURE 10.2 Organisms with which humans coevolved provide essential data to the immune system. If deprived of these microbial inputs, the development, epigenetic priming, repertoire, and immunoregulatory functions of the immune system are not set up normally and can malfunction.

which we coevolved. Deprivation or corruption of these inputs, for example by depletion or distortion of the microbiota, is now known to have widespread physiological consequences. Thus, if we can identify lifestyle changes that disrupt the data inputs and underlie the poor regulation of our immune responses, we can expect to identify prophylactic and therapeutic strategies.

10.2 Evolution of the Immune System

In high-income urban settings, we are seeing striking increases in a range of inflammatory, metabolic, and psychiatric disorders. Faulty regulation of the immune system plays a significant role in these increases. The purpose of this chapter is to use an 'evolutionary' or 'Darwinian' approach to these medical problems. The chapter does not contain a detailed account of how the immune system evolved: there have been numerous excellent reviews of this topic (Flajnik and Kasahara 2010; Boehm 2012). Rather, the chapter will show how an understanding of the broad principles of that evolution, of the environments in which the immune system evolved, and of the sources of the 'data' that it evolved to anticipate enables us to explain many of the malfunctions of the immune system that we see and to identify lifestyle changes in high-income settings that deprive the immune system of essential inputs.

The innate immune system has its origins in early eukaryotes such as amoebae. It uses germline-encoded pattern-recognition receptors (PRRs) such as the Toll-like receptors (TLRs), nucleotide-binding oligomerisation domain (NOD)-like receptors (nLRs), and scavenger receptors to recognise conserved microbial components. These PRRs are found throughout the animal kingdom and there are similar molecules in plants. This system is relatively inflexible, though some invertebrates use forms of somatic diversification, such as alternative splicing, in order to generate thousands of isoforms, while in some plants and invertebrates there has been expansion of families of PRRs to increase the repertoire. Clearly this strategy has the disadvantage that it increases genetic complexity.

The development of the adaptive immune system in vertebrates provided a way to create a very large repertoire of different receptors with minimal increase in genetic complexity. A transposon of the *Transib* superfamily that had invaded non-vertebrate species multiple times invaded a crucial site in the ancestral jawed vertebrate genome about 500 mya. It is suggested that the transposon invaded an immunoglobulin superfamily exon, so that the gene could not be expressed unless its separate parts were reassembled through the action of a recombinase (Flajnik and Kasahara 2010; Boehm 2012). This event was followed by several whole genome duplication events. The genes encoding antibodies and B-cell receptors (BCRs) and T-cell receptors (TCRs) are members of the duplicated immunoglobulin superfamily. The transposon eventually evolved into the recombination-activating genes (RAGs) that encode enzymes mediating ligation of TCR or BCR variable (V), diversity (D), and joining (J) gene segments during lymphocyte ontogeny. During this process, somatic hypermutation and variable–diversity–joining (VDJ) rearrangements occur. Random sequences are generated in the third complementarity-determining regions (CDR3) of immunoglobulins, BCRs, and TCRs due to an error-prone non-homologous end-joining process, initiated by the RAG family proteins. This process creates large numbers of distinct TCRs and BCRs. It is also important that these diversified receptors generated by mutation are expressed clonally. That is to say, each cell clone expresses only one receptor, so that if the receptor turns out to be autoreactive, the relevant cell line can be eliminated.

10.2.1 The Censoring Role of the Thymus

The evolution of these mechanisms that generate a diverse range of lymphocyte receptors by somatic hypermutation and VDJ rearrangements could not have occurred without simultaneous evolution of mechanisms to censor and eliminate any of these mutated cells that would be either useless (and so waste resources) or damaging to the host because of anti-self activity (autoimmunity) (Boehm 2008; Franchini and Ottaviani 2017; Kondo et al. 2017). The generation of receptor diversity described in Section 10.2 and these subsequent censoring functions occur in the thymus, which was previously thought to be the location of the soul. (The name *thymus* is derived from the Greek word for soul.) By the 1950s, the thymus was recognised as a site of production of lymphocytes, but it was not until the 1960s that the major role of the thymus began to be understood (reviewed in Ribatti et al. 2006). Thus, the thymus plays a crucial role in generating essential lymphocyte populations in early life and in censoring these cells before they are released into the periphery. After these censored lymphocyte populations have been created, the thymus decreases in size, as had been noted by Galen in the second century. In mice, this involution is triggered at least in part by sex hormones at puberty, but in humans the thymus is fully developed at birth and begins to involute in the first year of life before hormone levels rise, so other mechanisms must be involved.

10.2.1.1 *Evolutionary Development of the Thymus*

The thymus is present in all jawed vertebrates (Gnathostomata) and arises from the endoderm in the pharyngeal area. In unjawed vertebrates such as lampreys (Agnatha), primitive structures that are the site of development of lymphocyte-like cells have been noted and named thymoids. However, the oldest jawed vertebrates (cartilaginous fish), such as sharks, skates, and rays, have structures that resemble a true thymus both in organisation and function (Boehm 2008; Ge and Zhao 2013). The mechanisms that operate to generate and then censor the receptor repertoire are of interest from an evolutionary point of view.

10.2.1.2 *Thymic Cortex*

The thymic cortex attracts haematopoietic T-lymphoid progenitors and induces their differentiation into TCR-expressing CD4+CD8+ double-positive (DP) thymocytes. Peptides associated with class I or class II major histocompatibility (MHC) molecules are then presented to these DP thymocytes by a complex and varied community of cells known as cortical thymic epithelial cells (cTECs). These cTECs express a unique spectrum of lysosomal and endosomal proteases that generate an unusual set of peptides derived from self-proteins. It is probable that some novel variants of self-peptides are generated by splicing (Liepe et al. 2016). DP thymocytes with receptors that have no affinity or dangerously high affinity for the peptide-MHC complexes displayed by cTEC are eliminated, but those with low affinity are induced to survive and differentiate into CD4+CD8− or CD4−CD8+ single-positive (SP) thymocytes (Kondo et al. 2017). These positively selected cells are then attracted to the thymic medulla for further selection processes. To understand why this mechanism evolved it is important to remember that about 65% of our genes originated with bacteria, archaea, and unicellular eukaryotes (Rook et al. 2017). Or to put it the other way round, all life forms are based on similar building blocks. Thus, to generate a repertoire of TCRs likely to recognise structures that have evolved in life forms on planet earth, it 'makes sense' to use peptides derived from self, especially when these are derived by unusual enzyme systems and splicing, so as to create variants that might resemble those of other organisms.

10.2.1.3 *Thymic Medulla*

In addition to becoming SP CD4+ or CD8+ lymphocytes, the cells positively selected in the thymic cortex as described in Section 10.2.1.2 are induced to express chemokine receptor 7 (CCR7), so they migrate to the thymic medulla where CCR7 ligands are expressed at high levels. Here they are again exposed to peptides derived from self, but this time presented by *medullary* thymic epithelial cells (mTECs), which are different from those found in the cortex, and by dendritic cells. Some mTECs express a nuclear protein known as Aire (autoimmune regulator) or a zinc finger transcription factor known as Fezf2. These two molecules promote 'promiscuous' gene expression. That is to say that between them they cause expression of essentially the entire cellular self-protein repertoire, including proteins usually confined to specific peripheral organs (Boehm 2008; Kondo et al. 2017). Thus, the cells that were positively selected in the cortex and that have become SP CD4+ or CD8+ are screened in the medulla for excessive affinity for self-peptides presented by MHC molecules. Those SP lymphocytes that have such excessive affinity are destroyed or become regulatory T cells. However, cells bearing receptors that have intermediate affinity for self-peptides are allowed to leave the thymus to populate the periphery. There they will fail to respond to self-peptides, but will sometimes have a high affinity for subtly different versions of the peptide derived from other organisms or created by mutations associated with neoplasia or by chemical changes secondary to inflammation.

10.2.2 Modern Man–Neanderthal/Denisovan Exchange of Genes

Recent studies have shown that up to 4–6% of the modern Eurasian genome is derived from archaic human populations such as Neanderthals or Denisovans (Green et al. 2010). Introgression of archaic hominin genetic material occurred many times in many different non-African populations (Vernot et al. 2016). These hominin groups had existed outside Africa for hundreds of thousands of years, and so were presumably well adapted to the infectious challenges outside Africa and to modifications of the microbiota driven by the diets available in these regions, which might explain why much of the genetic material involved in these introgressions (or that has been retained during subsequent evolution) is concerned with the immune system. There is evidence of a Neanderthal or Denisovan origin for several alleles of the human leucocyte antigen (HLA) system that are involved in adaptive immunity (Abi-Rached et al. 2011). But now there is also evidence for introgression of archaic genes into the TLR6–TLR1–TLR10 cluster of the innate immune system

(Dannemann et al. 2016; Deschamps et al. 2016). However, we cannot simply assume that the archaic versions of these genes were selected because they were more pro-inflammatory and protective. It has been revealed recently that TLR10 is unique among the TLRs in that it is anti-inflammatory when it forms heterodimers with TLR2 (Oosting et al. 2014). Thus, it is equally possible that the recent arrivals from Africa were suffering from inflammation triggered by novel microbiota or infections and that the anti-inflammatory roles of TLR10 were in some way beneficial. This argument is supported by the fact that introgression from Neanderthal or Denisovan genomes also includes *TNFAIP3* which encodes the immunoregulatory protein A20. This protein is critical for limiting and terminating nuclear factor kappa-light-chain-enhancer of activated B cells (NF-κB)-mediated inflammatory responses, and is discussed later in relation to endotoxin tolerance (Gittelman et al. 2016).

10.3 Evolution of the Microbiota

Since the evolving immune system plays a crucial role in 'farming' the microbiota, the next issue to consider is what we know about the evolution of the microbiota. Did the microbiota and the immune system evolve in parallel? A study of sequence divergence of gut bacteria from humans, chimpanzees, bonobos, and gorillas suggested that the split between human microbiota and chimpanzee microbiota occurred about 5.3 mya, which is close to estimates of the human–chimpanzee split based on host mitochondrial and nuclear genomes. Similarly, the microbiota of humans seems to have split from that of gorillas about 15.6 mya, which is within the range of estimates based on nuclear genomes, though older than estimates based on mitochondrial genomes (Moeller et al. 2016).

Interestingly, recent studies suggest that the appendix (together with the caecum and associated lymphoid tissue), far from being vestigial, has evolved many times, particularly in species that live in large communities, so that gut infections are easily spread (Smith et al. 2017). The current view is that these structures maintain, guard, and 'farm' a reservoir of beneficial gut organisms that are not subject to the main flow of gut contents and that can help reconstitute the microbiota after it has been disturbed. This explains why appendectomy is associated with an increased risk of infection with *Clostridium difficile* after exposure to broad-spectrum antibiotics (Smith et al. 2017).

10.3.1 Innate and Adaptive Immunity Regulate the Microbiota

Both the innate and adaptive branches of the immune system play crucial roles in regulating the composition of the microbiota. Some of this influence is germline encoded. Thus, the microbiomes of monozygotic (MZ) twins are more similar than those of dizygotic (DZ) twins. Using a large cohort of MZ and DZ twins it was possible to identify many microbial taxa of which the abundance was influenced by host genetics, including one (*Christensenella minuta*) that correlated with reduced obesity in the human subjects and was shown to oppose obesity in an animal model (Goodrich et al. 2014). This important observation indicated that host genes can influence phenotype by controlling the organisms present in the microbiota (Goodrich et al. 2014, 2016).

Some of this regulation of the microbiota is mediated via the innate immune system, of which TLR and inflammasomes are essential components. If the gene encoding TLR5 is knocked out, mice develop the metabolic syndrome (hyperlipidaemia, hypertension, insulin resistance, and adiposity) and altered microbiota. This altered microbiota will induce similar physiological changes when transferred to wild-type germ-free mice (Vijay-Kumar et al. 2010). Similar changes occur if components of inflammasomes are disabled. Inflammasomes are multiprotein oligomers of variable composition expressed in myeloid cells. They promote the maturation of the inflammatory cytokines IL-1β and IL-18 and play a major role in activating inflammatory responses. If various components of inflammasomes are knocked out, there are metabolic (Henao-Mejia et al. 2012) or inflammatory consequences (Elinav et al. 2011), and as in the TLR5 knockouts, these consequences are mediated by a changed microbiota. Several other genes have been shown to regulate the microbiota in a variety of animal models (Kostic et al. 2013). So once again, it is evident that host genes within the immune system can influence phenotype by controlling the organisms present in the microbiota.

These experiments prove the role of the innate immune system, but other experiments make it clear that the adaptive immune system is just as crucial. For example, in the absence of luminal immunoglobulin A (IgA), there is increased gut leakiness which will lead to more translocation of intestinal bacteria (Johansen et al. 1999). More recently, it has been shown that expression of MHC class II on conventional dendritic cells (cDCs) is needed for effective control of the gut microbiota (Loschko et al. 2016). When MHC II was not expressed on cDCs in a mouse model, there was chronic intestinal inflammation, which could be reduced by antibiotic treatment. This inflammation did not occur in germ-free animals. Since the role of MHC II is to present antigens to the T cells of the adaptive system, this is formal proof of an essential role. Similar evidence comes from mice that lack both the transcription factor Tbet and the RAG2 gene. These animals cannot develop T cells or generate adaptive receptor diversity. Such mice develop a severe colitis, which can be treated with antibiotics or by infusion of regulatory T cells (Garrett and Glimcher 2009). These observations confirm that both the innate and the adaptive immune systems are involved in maintaining and controlling the gut microbiota, and so indirectly modulate the phenotype.

10.3.2 Diet, Evolution of Microbiota, and the Immune System

We know that the microbiota is profoundly affected by diet. Adults who consume a high proportion of fruit and vegetables but little meat have a diverse microbiota, in which *Prevotella* is more abundant than *Bacteroides*, while the reverse is seen in individuals whose diet is dominated by meat (Jeffery and O'Toole 2013). The genus *Homo* is thought to have evolved from ancestors that spent much of their lives in trees eating fruit and leaves. The evolution of modern man will not be reviewed in detail here, but certain aspects of our evolutionary history and the accompanying dietary changes will have had massive effects on human microbiota and therefore on the human immune system.

A crucial feature of humans is our large brain, which has expanded three-fold over the last 2 or 3 million years. But this large brain has specific nutritional requirements and uses large quantities of energy. In new-borns the brain uses 70–80% of all energy consumption, though this falls to 20–30% in adults. The nutritional requirements for building this large brain include omega-3 fatty acids, particularly docosahexaenoic acid (DHA), minerals such as iodine, iron, zinc, copper, and selenium, and vitamins A and D (Cunnane and Crawford 2014).

A series of adaptations and behavioural changes allowed *Homo* to access high-energy food sources and to develop an accelerated metabolic rate (Pontzer et al. 2016). These adaptations included efficient walking and foraging, and adoption of stone tools with which to process meat. Access to meat as a high-energy source (together with cooking, discussed in Section 10.3.2.1) is thought to have facilitated the larger brain size and reduced gut size. (For further discussion, see Chapter 17: Brain, Spinal Cord, and Sensory Systems.) In order to obtain the brain-specific nutrients, particularly the fats, in sufficient quantities from wild animal sources it would have been necessary to eat the livers and brains as hunter-gatherers do to this day (Schnorr et al. 2014). Wild animals are usually lean, and an exclusively high-protein diet is toxic (Bilsborough and Mann 2006). Some suggest that in addition to eating offal, there was extensive consumption of fish and shellfish which are rich in DHA and minerals. During human evolution, East Africa was intermittently dominated by large lakes and waterways, and early humans tended to spread along coastlines (Maslin et al. 2015). Either way, these dietary changes will have profoundly affected the microbiota and, therefore, the immune system.

10.3.2.1 *Cooking*

The adoption of fire and cooking will have had profound effects on the immune system and microbiota. Recently, mice were fed raw or cooked meat in order to identify host genes that were differentially expressed when food was cooked (Carmody et al. 2016). Interestingly, the genes identified correlated with those differentially expressed in liver between humans and other primates. Moreover, sequence changes in these genes appear to have occurred before the split between modern humans and Neanderthals or Denisovans, suggesting that adaptation to cooked food was already underway at least 275,000 years ago (Carmody et al. 2016).

Indeed, cooking was probably common by 400,000–500,000 years BP and was definitely in use 300,000 years ago (Shahack-Gross et al. 2014). Six of seven putatively selected cooking-related genes were involved in functions of the immune system, and the authors suggest that cooking reduces immune up-regulation triggered by food and so conserves energy for other purposes. So, the adoption of cooking will have affected our immune systems. (For further discussion, see Chapter 13: Digestive System.)

Cooking also helps to make starch available. Mammals and birds lack cellulases, so whole grains are indigestible, unless the husk and the cell walls are broken down mechanically or by bacteria, or weakened by cooking. Humans mechanically opened grains from about 30,000 years ago, but amyloid gene copy number variation (increases) might have occurred earlier, presumably following the cooking of grains that helped to render the starch available (Perry et al. 2007).

10.3.2.2 *Fermented Foods*

In the absence of refrigerators, fermentation was inevitable, and when scavenging meat or fallen fruit, the process would have begun before the food was gathered. Learning to control these processes involving bacteria (e.g. lactic fermentation), yeast (alcoholic beverages), or molds (oncom, tempeh) so as to develop palatable non-toxic storage options was a crucial advance. Lactic fermentation of vegetables (sauerkraut, kimchi, gundruk, khalpi, sinki, etc.) and olives (Benitez-Cabello et al. 2016) or fermentation of meat (for example, the Eskimo fermented fish, walrus, sea lion, and whale flippers, beaver tails, animal oils, and birds) adds nutritional and microbiological diversity to the diet (Breidt et al. 2013; Leroy et al. 2013; Selhub et al. 2014; Swain et al. 2014). Fermentation increases the content of vitamins, lactoferrin, bioactive peptides, and newly formed phytochemicals such as flavonoids, which may in turn modulate the intestinal microbiota (Lu et al. 2013).

Fermented beverages containing alcohol (McGovern et al. 2004) were an important source of human intake of microbiota from the environment. Chemical analyses of residues in pottery reveal that some of these methods were in use at least 9000 years ago (McGovern et al. 2004) and probably a great deal earlier (McGovern 2009). A mutation in alcohol dehydrogenase 4 that increased the efficiency of alcohol metabolism appears to have arisen in distant ancestors of mankind about 10 mya, perhaps triggered by consumption of fruit that had fermented after falling to the ground (Carrigan et al. 2014).

Interestingly, it is clear that microbiota from foods transiently colonises the gut (David et al. 2014), and therefore when fermented foods and beverages are consumed daily this must cause long-term influences on the microbiota, both directly and via the immune system. Even conventional diets used in the United States provide microbial inputs in the range 10^6–10^9 organisms/day (Lang et al. 2014). But if fermented foods are being consumed, this can easily rise to 10^{12}/day (Derrien et al. 2015). Moreover, since there are relatively few organisms in the ileum, ingested organisms can form a significant percentage of the organisms present at this site, or even transiently outnumber the resident ones (Derrien et al. 2015). This is important because the ileum is where mucosal dendritic cells sample the gut contents and 'inform' the immune system about the antigenic repertoire of gut contents (Schulz and Pabst 2013). Therefore, organisms that are constituents of food, or merely contaminating it, contribute to the *data* supplied to the immune system even if they do not colonise.

10.4 Organisms on which the Immune System is Dependent

Which organisms does the immune system need to encounter in order to function correctly? The inappropriately named 'hygiene hypothesis' was first suggested in 1989, following the observation that children who had been brought up with older siblings, especially older brothers, were less likely to suffer from hay-fever (Strachan 1989). At first there was a tendency to assume that the effect was due to exposure to the common infections of childhood, but epidemiological studies rapidly revealed that the childhood virus infections do not protect against allergic disorders (Bremner et al. 2008), and often actually trigger them (Yoo et al. 2007). We now understand that the 'crowd infections' have only relatively recently started to plague humans (Wolfe et al. 2007). A virus such as measles will either kill those affected or induce solid sterilising immunity, so measles, for example, dies out from small communities (Black 1966) and cannot persist in small hunter-gatherer groups. A study suggests that measles might have not entered the human population until the first millennium of the Common Era (Wertheim and Kosakovsky Pond 2011), so it is not an organism with which our immune system coevolved, except to the extent that it has eliminated some susceptible individuals. So, although modern air travel and huge population sizes will make crowd infections an increasing threat, they are not the topic of this chapter (except to the extent that the existence of this threat makes talk of abandoning hygiene totally absurd, even dangerous, as will be discussed in Section 10.6.3).

Since all complex multicellular animals evolved from and in the continuous presence of microorganisms, there must be some organisms that perform functions for the immune system that were never encoded in the mammalian genome, because they did not need to be. This we can call 'initial dependence', because these organisms were present as the immune system evolved and they became essential components of that system. For example, background levels of microbial polysaccharides, lipopolysaccharides, phospholipids, and peptidoglycans are required signals for establishment of the immune system (McFall-Ngai et al. 2013).

In other cases, there will be 'evolved dependence'. This term refers to situations where an organism has become adapted to the presence of a partner through loss of genetic material and can no longer function without that partner (De Mazancourt et al. 2005). A classic example was seen in the laboratory environment when a strain of *Amoeba discoides* became infected with a bacterium (Jeon 1972). Initially this infection compromised the growth of both species, so it was not a case of mutualism. However, after 5 years neither organism could survive without the other. This indicates genetic changes leading to dependence. For instance, an enzyme that is encoded in the genome of both species might be dropped from the genome of one of them. Access to that gene is now 'entrusted' to the other species. This idea is at first somewhat alien to immunologists, but is in fact rather commonplace. For instance, most mammals can synthesise vitamin C, but large primates and guinea pigs have lost the relevant pathways. Man and guinea pig are now in a state of evolved dependence on fruit and vegetables; we had the genes in the past, but we do not any more. Perhaps the best example of evolved dependence is the mitochondrion of which the genome is derived from an endosymbiotic event involving an organism resembling the aerobic alpha-proteobacteria.

Logically, there might also have been situations where a newly evolved (or newly encountered) microorganism offered a function or molecule that allowed the immune system to evolve a new capability, on which animals came to rely. One might call this 'exploitative dependence'. This is not likely to be common.

There are multiple categories of organism to which humans need to be exposed in order to set up appropriate regulation of the immune system (and of the metabolic system that is often regulated in parallel, at least partly because of competition for energy). As a close approximation, these categories can be regarded as the 'Old Infections', the microbiota and organisms from the natural environment. Pathogens such as malaria to which humans have been forced to adapt, but on which we are clearly not dependent, are discussed briefly in Section 10.4.1.6.

10.4.1 'Old Infections'

The term 'old infections' refers to organisms that established life-long carrier states or subclinical infections and so were able to survive within small hunter-gatherer groups (Wolfe et al. 2007) (Figure 10.3). *Helicobacter pylori*, *Mycobacterium tuberculosis*, gut

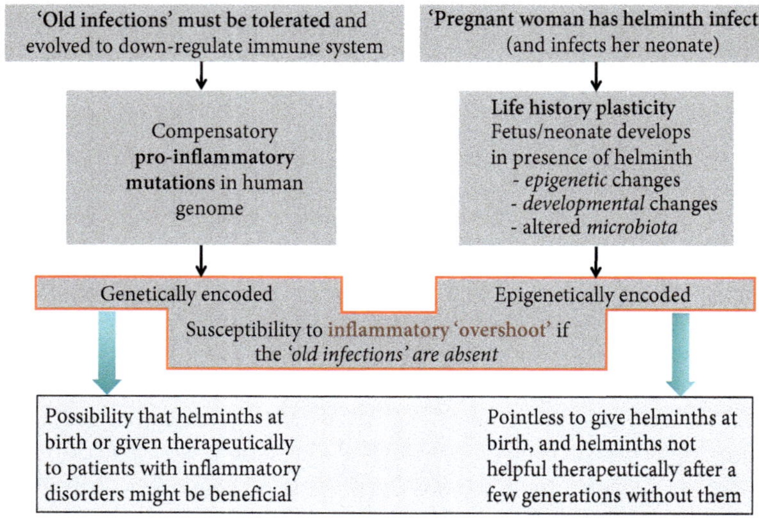

FIGURE 10.3 'Old infections' have had both genetic and epigenetic impacts on the immune system. It is not yet clear whether we need to compensate for the fact that modern medicine rapidly causes removal of old infections. If their impact is largely epigenetic (right-hand-side of the figure), then after a few generations without them they might become irrelevant.

helminths, and blood nematodes fall into this category and were present in human populations before the migrations out of Africa. Analysis of their phylogenetic trees and comparison with the human phylogenetic tree reveal how these old infections coevolved and spread over the globe with human populations (Linz et al. 2007; Wolfe et al. 2007; Galagan 2014).

Once established, these organisms were constantly present and had to be tolerated, and so coevolved roles in setting up immunoregulatory pathways. It is therefore possible that immunoregulation becomes deficient when the old infections are eliminated by modern medicine. In high-income, developed countries, most of these infections are now rare. For example, it is estimated that in 1947 about 36% of the population of Europe carried helminths such as *Enterobius vermicularis*, *Trichuris trichiura*, and *Ascaris lumbricoides*, but these have almost totally disappeared (Stoll 1947). Since it is standard practice in many developing countries to deworm pregnant women, the consequences for the child of removing these organisms from the mother can be monitored. Deworming in pregnancy increases the risk of eczema and wheeze in the resulting infant (Mpairwe et al. 2011). Latent tuberculosis provides another example. Tuberculin-positive children are less likely to have allergic rhinitis or positive allergen skin prick tests (Obihara et al. 2005), and individuals carrying *H. pylori* are also somewhat protected from allergic disorders (Hussain et al. 2016). These infections are being progressively eliminated from high-income urban populations. In Sections 10.4.1.1–10.4.1.5 we consider some of the immunoregulatory strategies used by these organisms.

10.4.1.1 *Helminths*

Since helminths are large organisms and could not be eliminated, the immune responses needed to be modulated, to reduce immunopathology, such as destruction of the lymphatic system leading to elephantiasis, and unproductive use of energy resources. Numerous immunoregulatory strategies are used by different helminths. Some authors suggest that the

Th2 pattern of response itself, which is accompanied by activation of M2 macrophages and repair mechanisms, evolved in order to cope with chronic metazoan infections such as helminths by limiting the inflammation and repairing the damage (Gause et al. 2013). Helminths use a range of mechanisms to modulate the immune system, and a selection of these is outlined below.

Helminth infections in the gut alter the composition of the microbiota and can lead to an increase in anti-inflammatory lactobacilli (Walk et al. 2010). Another example is the ability of helminth infection to protect mice deficient in NOD2 from intestinal abnormalities (Ramanan et al. 2016). This is interesting because some NOD2 variants are associated with Crohn's disease. Nod2-deficient mice harbouring the helminths developed a Th2 response that reduced colonisation with an inflammatory *Bacteroides* species and promoted a protective microbiota enriched in *Clostridiales*. There is evidence that individuals in helminth-endemic regions harbour a similar protective microbiota (Ramanan et al. 2016). *Bacteroidales* might be pathogenic only in the small subset of Crohn's disease patients who carry NOD2 variants, so trials of helminth therapy should perhaps target this subgroup.

Some helminth products alter the phenotype of dendritic cells (DC) so that they are more likely to induce an anti-inflammatory response (Hang et al. 2010). The soluble egg antigen (SEA) from *Shistosoma mansoni* also drives Treg development (Zaccone et al. 2009), probably because it imitates the Lewis X trisaccharide motif and thus mimics lacto-N-fucopentaose III (LNFPIII), an immunomodulatory glycan found in human milk. Fucosylated glycans, which are uncommon in vertebrates and bacteria, bind DC-SIGN and induce expansion of Th2 and Treg responses (Van Liempt et al. 2006; Lowry et al. 2016). Other helminth products also drive Treg, though the identity of the molecules and pathways responsible are not always clear. In a mouse model, infection with *Heligmosomoides polygyrus* induces expansion of the Foxp3+ Treg population via the transforming growth factor beta (TGF-β) receptor (Grainger et al. 2010). The hookworm (*Necator americanus*), which protects Ethiopians from allergic wheeze (Scrivener et al. 2001), releases a protein that resembles a tissue inhibitor of metalloproteases (TIMP), and drives increased Treg activity, probably by modulating CD103+ dendritic cells (Navarro et al. 2016).

10.4.1.2 Helicobacter pylori

H. pylori is an inducer of Treg, both locally in the stomach and duodenum (Lundgren et al. 2005; Robinson et al. 2008) and systemically (Lundgren et al. 2005). Interestingly, epidemiological studies indicate that infection with *H. pylori* protects against allergic disorders (Matricardi et al. 2000; Reibman et al. 2008), and similar effects are seen in laboratory models (Codolo et al. 2008; D'Elios et al. 2009). *H. pylori*-infected patients had higher frequencies of IL-10-secreting CD4+CD25hi Tregs, and lower allergen-specific IgE concentrations when there was a strong Treg response (Hussain et al. 2016). Not all the mechanisms are known, but some *H. pylori* strains, like the helminths discussed in Section 10.4.1.1, express the Lewis antigen on their lipopolysaccharides, and this material binds DC-SIGN and reduces Th1 polarisation (Bergman et al. 2004).

10.4.1.3 *Interactions between* H. pylori *and Helminths*

Since helminths and *H. pylori* coevolved with humans, it is possible that the presence of both infections in Palaeolithic man provided an immunological equilibrium, where the helminths suppressed the excessive inflammatory response to *H. pylori*. Both infections 'survived' the first 'Epidemiological Transition' (agriculture/husbandry about 13,000 years ago), but both were progressively lost after the Second Epidemiological Transition, starting in the mid-nineteenth century with urbanisation, concrete, tarmac, and loss of contact with animals, soil, and faeces. The helminths were lost first, and this might have allowed more inflammatory responses to *H. pylori*. Severe inflammation induced by *H. pylori* can drive ulceration and gastric cancer, and this timing fits the birth-cohort event in the incidence of peptic ulcer, with the peak of disease occurring in those *born* during the late nineteenth century (Sonnenberg 2006). An experimental example of protection from *Helicobacter*-induced oncogenesis by helminth-induced Treg has been demonstrated in C57BL/6 mice infected with *Helicobacter felis*. This organism induces Th1- and IFN-γ-dependent premalignant lesion reminiscent of that which can occur in humans infected with *H. pylori*. However, when *H. felis*-infected mice were co-infected with *Heligmosomoides polygyrus*, a natural murine nematode parasite with a strictly enteric life cycle, there was minimal gastritis, and the premalignant lesions did not appear despite an increased load of *Helicobacter* (Fox et al. 2000). Moreover, the inflammation in the co-infected mice showed reduced IFN-γ and increased expression of IL-10, TGF-β, and some Th2 cytokines (Fox et al. 2000). Thus it is reasonable to suppose that *Helicobacter* plus helminth results in immunoregulatory equilibrium, which does not lead to cancer, but elimination of the helminth results in a more Th1 pattern, and increased oncogenesis.

More recently, the story has changed in an interesting way. We often eliminate *H. pylori* with antibiotics in order to reduce the risk of peptic ulceration, but this reveals another problem. Since over the course of a lifetime *H. pylori* reduces gastric acid production, elimination of *H. pylori* by antibiotics might indeed have led to the waning of the 'epidemic' of peptic ulcers, but it has led to the appearance of oesophagitis, short- and long-segment Barrett's oesophagus, and its malignant complications (Vaezi et al. 2000; Cover and Blaser 2009; Rook and Dalgleish 2011).

10.4.1.4 *Do we Need Helminths at Birth?*

Some authors suggest that, like the microbiota, these helminth-mediated immunoregulatory mechanisms might have evolved to become a physiological necessity (Fumagalli et al. 2009). However, greatly varying nature, prevalence and intensity of infection (Montresor 1987), and the diverse range of immunoregulatory mechanisms exerted by the helminths (Babu et al. 2006; Zaccone et al. 2009; Grainger et al. 2010; Walk et al. 2010; Reynolds et al. 2015; Ramanan et al. 2016; Zaiss and Harris 2016) make it unlikely that there is a helminth-driven germline-encoded requirement for helminths. Interestingly, there is some evidence that helminth exposure (Correale and Farez 2011), by driving immunoregulation (Correale and Farez 2013), is effective in patients with multiple sclerosis in Argentina where helminth infections are common, but results have so far been disappointing using *Trichuris suis* in a high-income setting where the subjects might have been helminth-free for several generations (Fleming 2013). This might simply mean that *T. suis* is inappropriate, but it is equally possible that a requirement for the presence of helminths can be epigenetically programmed if the mother and infant are helminth-infected, but that after a few generations without helminths in a high-income setting their presence is no longer required (Figure 10.3). Only clinical trials can resolve this debate.

10.4.1.5 *Tuberculosis*

It used to be thought that tuberculosis was a recent arrival in human populations, but that is no longer believed. Human-infecting organisms resembling *Mycobacterium canettii* probably evolved in Africa from environmental soil mycobacteria as much as 2.8 mya, in which case they might have infected human ancestors as far back as *Homo habilis* (Galagan 2014). The *M. tuberculosis* complex evolved from these strains in Africa at least 70,000 years ago and accompanied the out-of-Africa human migrations. The earliest archaeological records yield only skeletal evidence of tuberculous bone lesions. However, ancient DNA and lipid biomarkers have confirmed human tuberculosis in a woman and child who lived in what is now Israel about 9000 years BP (Hershkovitz et al. 2008), and there is both archaeological and molecular evidence that the disease was present in the Americas before historical contacts with Europeans (Galagan 2014).

An interesting evolutionary hypothesis has been inspired by this long association between humans and *M. tuberculosis*. Meat provides nicotinamide (the amide of niacin; vitamin B_3) and also tryptophan which can be converted to nicotinamide in the liver. The nicotinamide is incorporated into nicotinamide adenine dinucleotide (NAD) and nicotinamide adenine dinucleotide hydride (NADH), essential for development and function of the big human brain and of the immune system. Nicotinamide deficiency leads to pellagra, which is accompanied by inflammatory lesions, mental disturbances, and eventually dementia. Nicotinamide deficiency is often seen with diets dominated by maize, or with severe lack of protein. But *M. tuberculosis* secretes nicotinamide, and it has been suggested that subclinical infection with early variants of *M. tuberculosis* provided a backup source of this essential nutritional factor (Williams and Dunbar 2014). These early tuberculosis strains were necessarily of low virulence and so able to persist subclinically in small hunter-gatherer groups like the other 'old infections' (Wolfe et al. 2007). More recently, tuberculosis has been evolving into a 'crowd infection', with increasing virulence and detrimental effects on lifespan, but the early strains probably derived from *M. canettii* were very different.

Modern strains of *M. tuberculosis* still show evidence of mechanisms that regulate the host immune system and help to reduce immunopathology. They produce a range of molecules that act via the mannose receptor (MR), dectin-2, and TLR2 (Lowry et al. 2016). They also release lipoarabinomannan, lipoarabinomannan carrier protein, glyceraldehyde-3 phosphate dehydrogenase, and Hsp60 and Hsp70 which signal via DC-SIGN, like the products of helminths and of *H. pylori* discussed in Sections 10.4.1.1 and 10.4.1.2 (Lowry et al. 2016). Thus, subclinical mycobacterial infections might play a useful role by activating appropriate background levels of innate and adaptive immunity, while simultaneously driving some aspects of immunoregulation. This concept has been called 'trained immunity' and might explain the evidence that vaccination with an attenuated strain of a related species, Bacille Calmette Guérin (BCG), protects not only against tuberculosis but also against other infections, allergies, and some cancers (Netea and Van Crevel 2014).

10.4.1.6 *Malaria*

Most human malaria is caused by *Plasmodium falciparum*. Malaria might not deserve to be included within the 'old infections' because recent genetic analyses indicate that this organism jumped from gorillas to humans in Africa within the last 10,000 years (Liu et al. 2010; Loy et al. 2016). Thus, *falciparum* malaria left Africa long after the dispersal of modern humans, spreading via the movement of people and mosquitos through pre-existing human populations. However, malaria clearly exerted a rapid selection pressure, and multiple different genetic adaptations occurred in different populations and geographical regions (Wellems et al. 2009). The adaptations are remarkable because they cause significant health problems for the host (Wellems et al. 2009). Evidently these penalties were offset by increased resistance to malaria. Most of the adaptations, such as sickle cell anaemia, the thalassaemias, other haemoglobin variants, and glucose-6-phosphate dehydrogenase deficiency (G6PD; favism), are more relevant to the chapter on blood and are discussed there. (For further discussion, see Chapter 9: Haematopoetic System.) However, an additional adaptation that occurred in Sardinia might be of relevance to the theme of this chapter. There is a high incidence of multiple sclerosis (MS) in Sardinia and this has been rising since the elimination of malaria in 1951, whereas most increases in MS have occurred in northern Europe. It has emerged that Sardinians (or genetically Sardinian mainland Italians) with MS show exaggerated inflammatory responses *in vitro* to *P. falciparum*, possibly linked to polymorphisms of MHC class II and tumour necrosis factor (TNF) (Sotgiu et al. 2007). It is tentatively suggested that in the absence of malaria, these hyper-responsive adaptations predispose to MS.

10.4.2 Organisms from the Natural Environment

Large epidemiological studies demonstrate that living close to the natural rural or coastal environment, often denoted 'green space' or 'blue space', respectively, reduces overall mortality, cardiovascular disease, and depressive symptoms and increases subjective feelings of well-being (Maas et al. 2006; Mitchell and Popham 2008; Wheeler et al. 2012). It used to be assumed that these effects are explained by psychological mechanisms, but this view is untenable and supported only by experiments that lack relevant controls (Rook 2013). While there undoubtedly are health benefits attributable to relaxation induced by exposure to the delights of nature and benefits from accompanying exercise and sunlight, there is solid and rapidly mounting evidence for biological effects on the immune system mediated by exposures to environmental microorganisms. This is the topic discussed here.

The beneficial effects of exposure to green and blue space are particularly prominent in urban individuals of low socioeconomic status who tend to be most severely deprived of green space (Maas et al. 2006; Mitchell and Popham 2008; Dadvand et al. 2012; Wheeler et al. 2012). The natural environment also protects from inflammatory disorders. Exposure of pregnant mothers or infants to the farming environment protects the child against allergic disorders and juvenile forms of inflammatory bowel disease (Riedler et al. 2001; Radon et al. 2007). This protection appears to be attributable to airborne microbial biodiversity that can be assayed in children's bedrooms (Ege et al. 2011). Similarly, in a study performed in Finland, mere proximity to agricultural land rather than to urban agglomerations increased the biodiversity of skin microbiota, reduced atopic (allergic) sensitisation, and increased release by blood cells of IL-10, an *anti*-inflammatory mediator (Hanski et al. 2012). It is important to note that in this study, hygiene was a constant, not a variable. The effect of the environment was seen in the presence of universally high levels of home hygiene.

Some of the relevant microbiota comes from animals (Figure 10.4). Contact with cows and pigs protects against allergic disorders (Riedler et al. 2001; Sozanska et al. 2013). Contact

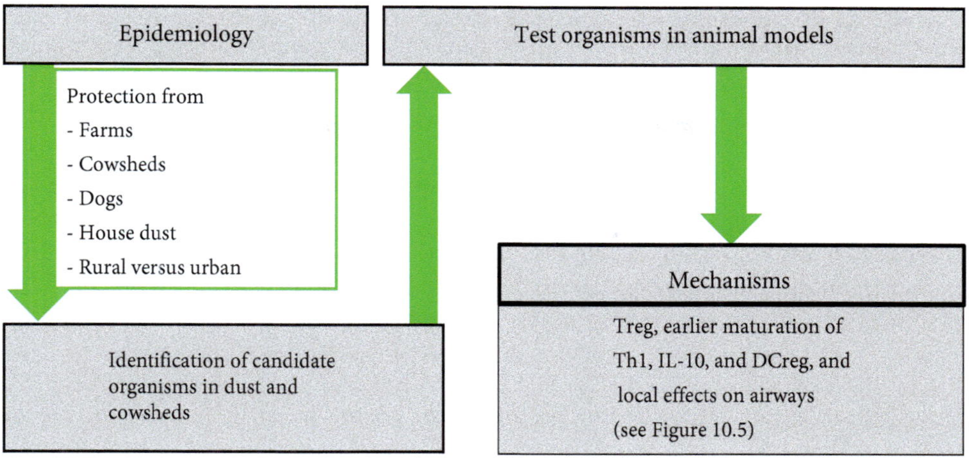

FIGURE 10.4 Exposure to farms, cowsheds, dogs, and house dust protects from allergic disorders. Candidate organism isolated from these sites can protect in animal models too, via various immunoregulatory pathways. See Figure 10.5 which illustrates additional pathways locally within the airways.

with dogs, with which humans have coevolved for many millennia (Axelsson et al. 2013; Thalmann et al. 2013), also protects from allergic disorders (Ownby et al. 2002; Aichbhaumik et al. 2008). Dogs greatly increase the microbial biodiversity of the home (Fujimura et al. 2010; Dunn et al. 2013). In a developing country, the presence of animal faeces in the home correlated with better ability to control background inflammation in adulthood (McDade et al. 2012), and in Russian Karelia (where the prevalence of childhood atopy is four times lower and type 1 diabetes is six times lower than in Finnish Karelia), house dust contained a seven-fold higher number of clones of animal-associated species than was present in Finnish Karelian house dust (Pakarinen et al. 2008). A recent study investigated 10,201 participants aged 26–54 years from 14 countries and generated a 'biodiversity score' based on reported childhood exposures to farms, rural versus urban environment, cats, dogs, day care, bedroom sharing, and older siblings. It emerged that a high biodiversity score correlated with reduced allergic sensitisation and improved lung health (Campbell et al. 2016). Unfortunately, the increasing use of industrial agricultural methods and monoculture might be reducing the microbial biodiversity of the countryside in developed countries.

10.4.2.1 *The Natural Environment and the Immune System*

How then does contact with the natural environment modulate the immune system? The effects are mediated both via the airways and via the gut, though the latter is discussed in Section 10.4.3.1 and its subsections in relation to the gut microbiota. In this section, we discuss mechanisms operating via the airways (Figure 10.5). An adult human breathes in about 10,000 litres of air every 24 hours. The intake of airborne organisms and of other biogenic particulates such as leaf fragments rich in plant polyphenols, pollen, and microbe-bearing soil particles depends on the degree of exertion and the environment. With exertion in a natural environment, the figures could be as high as 10^{10} organisms, 500 mg of pollen, and 10 mg of plant polyphenols in 24 hours. While not all of this material will be retained in the airways, the quantities that are retained clearly fall within the biologically relevant range. Some of this material will be trapped in mucus, carried up the airways, and swallowed, ending up in the gut.

10.4.2.1.1 THE NATURAL ENVIRONMENT, AIRWAYS, AND ASTHMA

The airways contain a number of cellular sensor systems that can monitor the content of inhaled air (Figure 10.5). One of these involves the PI3K/Akt/mTORC1 signalling system that plays a role in inflammatory pathways via NF-κB. Overactive Akt and mTOR (mTORC1) contribute to many pathological processes including cancers, diabetes, inflammation, and neurodegenerative diseases (Moore 2015). The evidence that many natural products from bacteria, algae, fungi, and higher plants can inhibit the activities of these protein kinases was reviewed recently, and the overall effect is thought to be anti-inflammatory (Moore 2015). Plant polyphenols can also exert anti-inflammatory effects via the Aryl hydrocarbon receptor (AhR). Thus, quercetin, resveratrol, and curcumin interfere with metabolic degradation of the endogenous AhR ligand 6-formylindolo[3,2-b]carbazole (FICZ). This increases FICZ levels and indirectly activates the AhR (Mohammadi-Bardbori et al. 2012). Microbial metabolites can also trigger the AhR. Tryptophan can be metabolised to produce AhR ligands that drive production of IL-22 by activated DC, T cells, and innate lymphoid cells (ILC) that are abundant at mucosal surfaces. IL-22 regulates reactions to microbial pathogens especially in respiratory and gut epithelial cells, and plays a major role

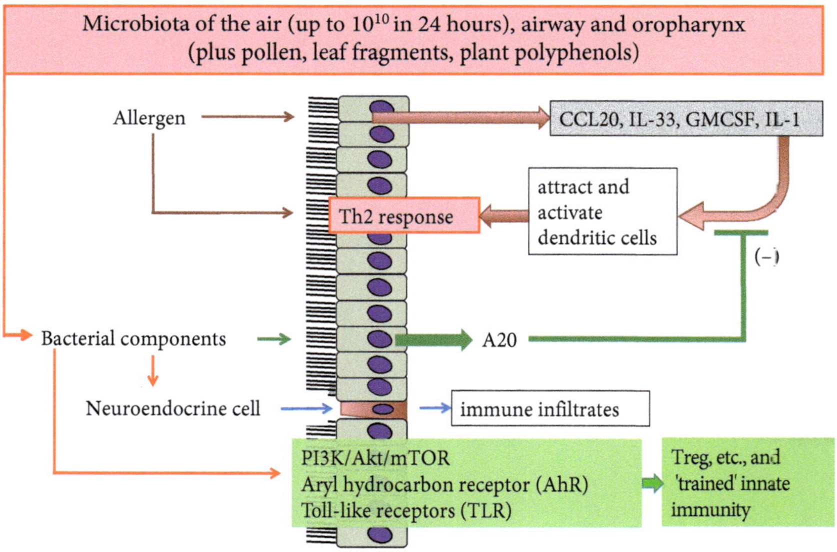

FIGURE 10.5 Immunoregulation by microorganisms in the airways. Allergens trigger mild inflammation in the airways that recruits dendritic cells (DC). These then initiate the Th2 allergic response. However, microbial components, including repeated low doses of endotoxin (lipopolysaccharides), can cause release of protein A20 which inhibits recruitment of DC. Other microbial components can increase cellular infiltration the neuroendocrine cells, or be sensed by the other receptors shown. Akt, Serine/threonine kinase Akt, or protein kinase B (PKB); CCL20, C-C motif chemokine ligand 20; GMCSF, granulocyte-macrophage colony-stimulating factor; IL-1, interleukin 1; IL-33, interleukin 33; mTOR, mammalian target of rapamycin; NF-κB, nuclear factor kappa B; PI3K, phosphatidylinositol-3 kinase.

in resistance to colonisation by fungi and by some bacterial species (Zelante et al. 2014). But this pathway also activates host indoleamine-2,3-dioxygenase 1. This enzyme generates further tryptophan-derived AhR agonists that drive production of TGF-β (Bessede et al. 2014) and Treg (Quintana et al. 2008).

A major microbial component in inhaled air is endotoxin (lipopolysaccharide (LPS)), and the phenomenon known as 'endotoxin tolerance' is important in the gut and the airways. Allergens induce a mild inflammatory response in the airways. The epithelial cells then release mediators that attract and activate DCs. The DCs can process inhaled allergens and initiate the Th2 allergic response. In a mouse model, however, frequent low doses of LPS trigger epithelial cells via TLR4 and NF-κB, and in addition to a range of inflammatory mediators, drive increased production of A20 (encoded by *Tnfaip3*). A20 is a ubiquitin-modifying enzyme that attenuates NF-κB activation and therefore reduces influx and activation of DC in the airways (Schuijs et al. 2015). Thus, exposure to LPS, or to farm dust, blocked a mouse model of allergic asthma induced by house dust mite, and this effect was attributable to A20 (Schuijs et al. 2015). LPS did not block induction of asthma in animals that did not express A20 in their lung epithelial cells (Schuijs et al. 2015). Interestingly, a previously unrecognised genetic disorder has been described in human families with an early-onset systemic inflammatory disorder. The disease is caused by germline mutations in the gene which encodes A20 (Zhou et al. 2016).

The relevance of these observations to humans is strongly suggested by a recent study of two culturally isolated farming communities in the United States. The Amish use traditional farming methods, while the Hutterites are industrialised. The Amish have much lower levels of asthma. The study revealed that the traditional farming methods of the Amish expose them to higher levels of LPS in dust. The Amish children have fewer eosinophils in their peripheral blood and more neutrophils. Peripheral blood leukocytes from the Amish have differences in gene expression and release lower levels of cytokines when exposed to LPS *in vitro*. Notably, cells from the Amish children express more *Tnfaip3* (Stein et al. 2016). Thus, it is suggested that chronic exposure of the airways to low-dose environmental microbiota sets up regulatory pathways within the airways. (For further discussion, see Chapter 12: Respiratory System.) Endotoxin tolerance is likely to be an important part of this phenomenon. (This point and its relationship to induction of Treg and other anti-inflammatory pathways is discussed further in Section 10.4.3.1 and its subsections on gut microbiota.)

Pulmonary neuroendocrine cells constitute another airway sensory system that is inevitably involved in the conditioning of the airway and its immune system. What is known is that stimulating the pulmonary neuroendocrine cells causes release of neuropeptides that increase immune cell infiltrates (Branchfield et al. 2016).

10.4.2.1.2 HORIZONTAL GENE TRANSFER

The natural environment can provide genes as well as whole microorganisms. There is some controversial evidence that horizontal gene transfer (HGT) from bacteria and protists to vertebrates has occurred, usually involving genes encoding metabolic enzymes (Crisp et al. 2015). However, what is definitely known is that the gut microbiota can acquire such genes by HGT from environmental organisms (Hehemann et al. 2010; Smillie et al. 2011). One such example is that enzymes acquired by HGT from marine bacteria enable the gut microbiota of Japanese individuals to metabolise seaweed carbohydrates

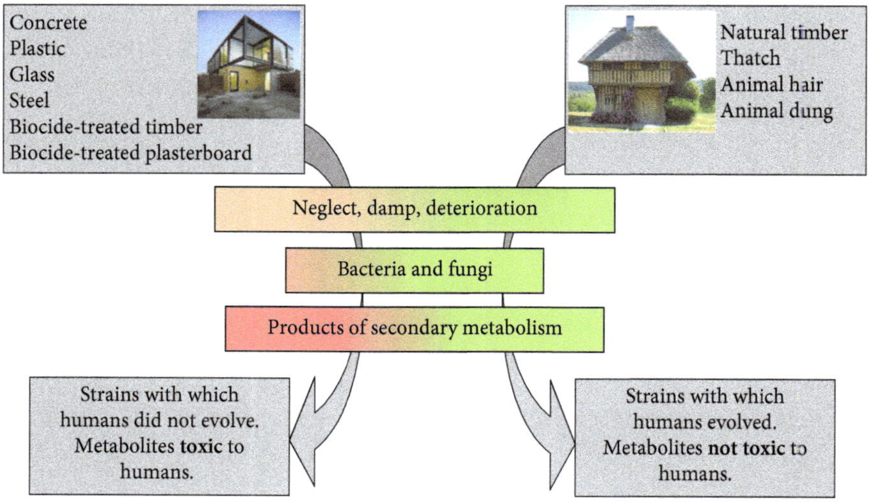

FIGURE 10.6 Evolving humans lived with modern materials, and so did modern humans while we built homes with timber, thatch, animal hair, and dung. Modern buildings, especially if damp and deteriorating, can harbour toxic microbial strains with which we did not evolve and that are toxic to us.

(Hehemann et al. 2010). The natural environment thus constitutes a resource of genetic diversity for the microbiota, though the timescale of such events is unclear (Smillie et al. 2011; Forsberg et al. 2012).

10.4.2.1.3 THE URBAN ENVIRONMENT

Humans evolved in a natural environment and in contact with animals. Until recently, even our homes were constructed with timber, mud, animal hair, animal dung, thatch, and other natural products, and ventilated by outside air. By contrast, modern buildings are constructed with synthetic materials, plastics, and concrete, while the timber and cardboard are treated with adhesives and biocides, and ventilated by air-conditioning systems (Figure 10.6). When these modern structures degrade, or become damp, or accumulate condensation in cavity walls, they do not become colonised with the bacterial strains with which we coevolved. They harbour a low biodiversity and become habitats for unusual strains that we did not encounter during our evolutionary history, some of which synthesise toxic molecules that we are unable to inactivate (Andersson et al. 1998; Sahlberg et al. 2010). Some examples of 'sick building syndrome' have been tentatively attributed to prolonged exposure to these inappropriate airborne microbiota (Andersson et al. 1998; Sahlberg et al. 2010).

10.4.2.1.4 SPORES

The issue of spores has been neglected. Spores are remarkably resistant and can remain viable for thousands, possibly millions of years (reviewed in Nicholson 2002). They are relevant in two contexts. First, about 1/3 of the organisms in the gut microbiota are spore-forming, and spores are readily demonstrable in human faeces (Hong et al. 2009a). Human faeces contain up to 10^4 spores/g, while soil contains approximately 10^6 spores/g (Hong et al. 2009b). Wherever humans have lived, the natural environment is inevitably seeded with human gut-adapted bacterial strains (Figure 10.7). A recent study revealed

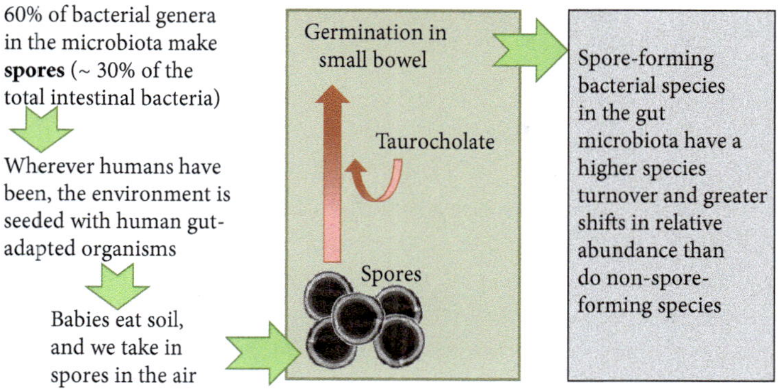

FIGURE 10.7 Humans seed the natural environment with spores of human gut-adapted bacteria. These may play a role in establishing microbiota in early life and re-establishing it after dietary or antibiotic abuse.

that the spore-forming strains within the human microbiota are more diverse than non-spore-forming bacteria and show a higher species turnover or a greater shift in relative abundance over the course of a year (Browne et al. 2016). Therefore, it is possible that when a gut organism becomes extinct as a result of dietary inadequacy or antibiotic misuse (Sonnenburg et al. 2016), it can be 'reinstalled' via spores from the environment.

Other spore-forming organisms from the environment might also be important despite not being definite components of the human microbiota. Spores in soil have tended to be studied by environmental microbiologists and ecologists, and the soil has been regarded as the natural habitat of the spore-forming organisms such as *Bacillus* spp., despite awareness of the fact that many of them can germinate and replicate in the intestinal tracts of insects and other animals (Nicholson 2002). It has been reported that spores of *Bacillus subtilis* can germinate in the small bowels of mice and rabbits (Casula and Cutting 2002; Tam et al. 2006). Moreover, after germination they replicated in the small bowel and then resporulated as they entered the colon. The same was observed in humans. *Bacillus subtilis* strains were obtained from biopsies of human ileum and from faecal samples (Hong et al. 2009a). Most of these strains could form biofilms, sporulate anaerobically, and secrete antimicrobials, properties that could be advantageous for survival within the gut (Hong et al. 2009a). There is therefore a growing view that *B. subtilis* and other environmental spore-forming species are gut commensals rather than soil microorganisms (Hong et al. 2009b). This might be very relevant to the 'old friends' mechanism, particularly to the clear importance of exposure to animals, agricultural land, and green spaces. For example, *B. subtilis* is an important stimulus for development of the gut-associated lymphoid tissue (GALT) in rabbits, and sporulation of live bacilli within the GALT was considered critical to this process (Rhee et al. 2004). At the very least, after germinating in the small bowel these organisms will provide data to the immune system in the ileum where dendritic cells sample gut contents, and where recently ingested organisms can constitute a significant proportion of the microbes present (Schulz and Pabst 2013). Thus, the presence of spores of gut-adapted strains in the natural environment leads us on to the microbiota.

10.4.3 Microbiota

Modern lifestyles cause our gut microbiota to deviate from the pattern seen in contemporary hunter-gatherers. For example, microbiota from hunter-gatherers from Africa and South America contain *Treponema* species (Obregon-Tito et al. 2015; Rampelli et al. 2015; Gomez et al. 2016), whereas these organisms are absent from European microbiota (De Filippo et al. 2010). Such changes, together with the loss of 'old infections' and diminished exposure to the natural environment, might play a role in the increases in chronic inflammatory disorders in high-income countries. We now consider contemporary lifestyle changes that: (1) impair transfer of microbiota from mother to infant, and 2) disrupt the composition of the microbiota. These lifestyle changes are considered in relation to health problems that are thought to be caused or exacerbated by dysbiosis.

10.4.3.1 *Microbiota and the Immune System before Birth*

It used to be thought that the fetus was sterile before birth, but we now know that some transfer of maternal microbiota to the placenta and fetus starts *in utero* (Funkhouser and Bordenstein 2013; Meropol and Edwards 2015). Interestingly, the microbiota undergoes complex changes during pregnancy (Nuriel-Ohayon et al. 2016). Gut microbiota from the third trimester shows an increase in *Proteobacteria* and *Actinobacteria* and reduced richness. Moreover, when third-trimester microbiota was transferred into germ-free mice, the recipients developed greater adiposity and insulin insensitivity compared to recipients of first-trimester microbiota (Koren et al. 2012), suggesting an adaptation for energy extraction and conservation that might benefit the neonate. Animal experiments suggest that molecules derived from the maternal microbiota, some bound to maternal antibodies, cross the placenta and influence development of intestinal group 3 innate lymphoid cells and intestinal transcriptional profiles (Gomez De Aguero et al. 2016). These effects, some of which are mediated via the AhR, help to limit inflammatory responses to microbial molecules and translocation of intestinal microbes across the gut wall (Gomez De Aguero et al. 2016).

Aspects of modern high-income life, such as stress and obesity, will modify the maternal microbiota, and so change the composition of the organisms transferred to the infant. When human mothers are stressed during pregnancy the infants are found to carry higher relative abundances of *Proteobacteria*, some of which might be pathogens, and lower relative abundances of lactic acid bacteria and *Bifidobacteria* (Zijlmans et al. 2015). Similarly, it is extremely probable that the transfer of a dysfunctional microbiota from an obese mother to her offspring explains the fact that maternal obesity is a major risk factor for obesity in the child (Galley et al. 2014; Soderborg et al. 2016).

10.4.3.2 *Mechanisms of Immunoregulation by Microbiota*

How does the microbiota regulate the immune system? Germ-free animals show defects on the development of the immune system and of the gut itself. In particular, the intestinal epithelial cells have reduced turnover and altered patterns of microvilli and glycosylation, perhaps attributable in part to their requirement for short chain fatty acids (SCFA) as energy source. In animals with a gut microbiota these SCFA are generated by fermentation of polysaccharides. As far as the immune system is concerned, it fails to develop normally in the absence of microbiota. This is particularly true of GALT, Peyer's patches, and mesenteric lymph nodes (Round and Mazmanian 2009). Bacterial strains that drive expansion of components of the immune system are beginning to be identified. In mouse models, certain members of the *Clostridia* (Atarashi et al. 2011) or *Bacteroides fragilis* (Round and Mazmanian 2010) will drive Treg formation, while segmented filamentous bacteria (SFB; provisionally known as *Candidatus savagella*) will expand Th17 cells. There may be an evolutionary relationship between these cell types, and some Th17 cells (which are only found in vertebrates) can be derived from Treg (Lee and Mazmanian 2010). It is suggested that Treg developed to allow the microbiota to be tolerated, while Th17 cells evolved later to provide a balancing level of protective background inflammation in the gut (Lee and Mazmanian 2010). There is a distinct subset of RORγ+ Treg in the colon and their formation is driven by a range of gut organisms from various different genera (Sefik et al. 2015). Interestingly, the RORγ+ transcription factor is also involved in driving Th17 cells in the small intestine, but this involves a different subset of gut organisms (Sefik et al. 2015).

The microbiota also drives the development of intestinal innate lymphoid cells (ILC1, ILC2, and ILC3) (Gury-Benari et al. 2016). In general, ILC1 cells make IFN-γ, ILC2 cells secrete IL-5 and IL-13, and ILC3 cells secrete IL-22, but the transcriptomes of these cells are profoundly altered by antibiotic treatments that disrupt the microbiota (Gury-Benari et al. 2016).

Progress is being made towards the identification of specific signals from the microbiota to the immune system, and these include tryptophan metabolites that are AhR ligands, SCFA, microbial molecules that signal via NOD or RIG-1 (Gomez De Aguero et al. 2016), histamine, spermine, and taurine (Levy et al. 2016). For example, the composition of the gut microbiota has strong influences on the release of cytokines by the donor's peripheral blood cells *in vitro* in the presence of bacterial and fungal stimuli. The strongest influences were on production of IFNγ and TNF, and the effects appeared to involve the tryptophan metabolite tryptophol, which has strong inhibitory effects on the TNF response, while metabolism of palmitoleic acid was important for the IFNγ response (Schirmer et al. 2016). No doubt many other molecular signals remain to be discovered, and this topic is not reviewed in detail here.

Other ways in which the immune system is regulated by the microbiota are considered in Section 10.4.3.4 in relation to diet and Section 10.5 in relation to specific immune-system-associated disorders. (For further discussion, see Chapter 13: Digestive System.)

10.4.3.3 *Endotoxin (LPS) Tolerance and Immunoregulation*

Endotoxin (LPS) tolerance, mentioned in Section 10.4.2.1.1 in relation to the airways, has been studied for many decades, but only recently has it become recognised as fundamental to immunoregulation by microbes. Animals can survive a potentially lethal dose of endotoxin if they have previously received one or more sublethal doses, and *in vitro*, macrophages that have been exposed to endotoxin respond differently when challenged again, with less release of TNF and reduced NF-κB translocation (Biswas and Lopez-Collazo 2009). Further studies show that repeated low-dose endotoxin administration *in vivo* leads eventually to increased Treg activity, which can, for example, block induction of type 1 diabetes in the NOD mouse (Caramalho et al. 2011; Wang et al. 2015). These facts have become of striking relevance to human pathology. The prevalence of childhood atopy is four-fold higher in Finland than in a bordering area of Russia with a genetically similar population, while the prevalence of type 1 diabetes is six-fold higher (Kondrashova et al. 2005; Pakarinen et al. 2008). The lifestyle on the Finnish side is very modern, while development has been retarded on the Russian side. The microbiota of Russian homes is quite different from that of Finnish homes, with much higher levels of Gram-positive organisms and of animal-derived strains (Pakarinen et al. 2008). A recent study has revealed that the endotoxin in the guts of Russian infants is mostly derived from *Escherichia coli*, which can protect NOD mice and drive endotoxin tolerance. But the endotoxin in the guts of Finnish infants is overwhelmingly derived from a *Bacteroides* species. This endotoxin fails to protect NOD mice and acts as an *inhibitor* of the agonist effects of *E. coli* endotoxin (Vatanen et al. 2016). Thus, it seems likely that one major problem with the 'modern' microbiota in Finland is a failure to induce endotoxin tolerance. Consequently, there is reduced immunoregulation, which manifests itself as increases in allergic disorders and in autoimmune destruction of pancreatic β cells, leading to type 1 diabetes. Indeed, it is clear that the gut microbiota plays an important role in mouse models of type 1 diabetes (Costa et al. 2016), and the same is probably true in the human disorder (Knip and Siljander 2016).

10.4.3.4 *Modern Diets, Microbiota, and the Immune System*

Although the microbiota is initially established by transfer from the mother and the environment, diet plays a crucial role in maintaining it in a state that is compatible with good health (Wu et al. 2011). But the ability to switch microbiota to cope with a different diet is clearly adaptive. Early humans were often exposed to major seasonal changes in the availability of different foods such as fruits, and successful hunting was also often seasonal. So, the microbiota will have changed with the seasons. Therefore, the question we must confront is just how far the human microbiota can be changed by modern diets before it starts to become incompatible with human physiology? Is it possible that the high-income urban diet is beyond our 'environment of evolutionary adaptedness (EEA)' and is pushing our microbiota into configurations to which we have not had time to adapt genetically?

10.4.3.4.1 FIBRE AND SCFA

The modern high-income urban diet clearly deviates from the diet with which humans coevolved, and 'modern' microbiota has lower microbial richness and biodiversity and different composition from that of traditional hunter-gatherers (Schnorr et al. 2014).

The microorganisms in the human gut are mainly from two phyla: Firmicutes and Bacteroidetes. Diets rich in fat and protein raise levels of bacteria belonging to the *Bacteroides* (a genus within the phylum Bacteroidetes), whereas diets rich in fibre increase the abundance of *Prevotella*, which is another genus within the same phylum (De Filippo et al. 2010; Wu et al. 2011). *Prevotella* made up 53% of the gut bacteria of children exposed to a traditional village lifestyle in Africa (Burkina Faso), whereas this genus was absent from the guts of age-matched European children (De Filippo et al. 2010; Wu et al. 2011). The same was true for *Treponema* species which were present in microbiota from hunter-gatherers from Africa and South America but absent from Europeans (De Filippo et al. 2010; Schnorr et al. 2014; Obregon-Tito et al. 2015; Rampelli et al. 2015; Gomez et al. 2016. *Treponema* species can hydrolyse cellulose and xylan, which probably explains their abundance in hunter-gatherer guts (Schnorr et al. 2014). The same authors noted an absence of *Bifidobacter* in hunter-gatherer adults, whereas this organism was routinely found in Europeans, perhaps driven by consumption of dairy products (Schnorr et al. 2014).

Crucial factors in maintaining the microbiota seem to be dietary fibre (polysaccharides that are metabolised by gut microorganisms to yield SCFA) and polyphenols (considered in Section 10.4.3.4.2). The importance of fibre is illustrated by experiments in which mice were given a low-fibre diet (Sonnenburg et al. 2016). If this was done for short periods, despite some apparent loss of biodiversity, an essentially normal microbiota could be recovered by reintroducing fibre, implying that the diminished strains were still present, even if difficult to detect. However, if the diet was deficient in fibre over several generations, there were true extinctions of important organisms, so that restoring fibre to the diet failed to restore the microbiota (Sonnenburg et al. 2016). This might be the situation in many modern human populations. A crucial immunoregulatory role of fibre has been elucidated in a mouse model (Tan et al. 2016). Fibre is fermented by the microbiota to generate SCFA that then act on G-protein-coupled receptors (GPR43, GPR109a, and GPR41) which are expressed on epithelial cells and cells of the immune system. These effects, together with other signals from the microbiota that require MyD88, result in enhanced retinaldehyde dehydrogenase-2 activity in CD103+ DC. These DC convert vitamin A to retinoic acid, and then, together with this retinoic acid, they enhance differentiation of naïve T cells into Treg and so inhibit food allergy in this model (Tan et al. 2016).

10.4.3.4.2 POLYPHENOLS

Another major dietary factor that stabilises the gut microbiota is the intake of polyphenols (Vanamala et al. 2015). These are a heterogeneous group of phytochemicals found in plant-based foods such as fruits, wine, tea, and coffee. The polyphenols include phenolic acids, flavonoids, stilbenoids, resveratrol, proanthocyanidins, curcuminoids, tannins, and lignans. Dietary polyphenols are poorly absorbed, but they are metabolised by organisms in the colon that might provide some absorbable derivatives. Nevertheless, it now seems that the major action of the polyphenols is on the microbiota rather than the host. Consuming red wine polyphenols (with or without the alcohol) inhibited the growth of the potential pathogen *Clostridium perfringens* and increased the growth of beneficial *Bifidobacterium, Eggerthella lenta,* and *Bacteroides uniformis* (Queipo-Ortuno et al. 2012). We can hypothesise that our ancestors' diets were rich in polyphenol-containing plant foods, and therefore the altered microbiota that will result from a modern polyphenol-deficient diet might be another example of a gene–environment mismatch that predisposes us to metabolic and inflammatory clinical problems. Intake of polyphenols is epidemiologically linked to patterns of faecal microbiota in allergic patients (Cuervo et al. 2016). Similarly, dietary polyphenols have been shown to attenuate glycaemic responses to carbohydrate intake, to improve insulin sensitivity, and to limit fasting hyperglycaemia (Hanhineva et al. 2010).

10.4.3.4.3 FATS AND REFINED SUGARS

The fats consumed in the diet also have effects on the composition of the microbiota, and such changes can lead to excessive absorption of endotoxin, and to persistent background inflammation which predisposes to a multitude of inflammatory, metabolic, and psychiatric disorders. Endotoxaemia is seen following a high-fat diet in humans (Pendyala et al. 2012), and also in mice where the effect can be shown to be secondary to dysbiosis

(Cani et al. 2008). Saturated fats have this effect, but so also does a high ratio of omega-6 to omega-3 polyunsaturated fatty acids. It is estimated that during human evolution we consumed roughly equal quantities of omega-6 and omega-3 fatty acids, but thanks to recent dietary changes we now consume 10× to 50× more omega-6 than omega-3 (Blasbalg et al. 2011; Kaliannan et al. 2015). In a mouse model, feeding a diet high in omega-6 fatty acids resulted in high levels of metabolic endotoxaemia and systemic low-grade inflammation that could be blocked by antibiotic treatment, implying that the effect was secondary to changed microbiota (Kaliannan et al. 2015).

The modern tendency to obesity may also be encouraged by easy access to refined sugars, which would not have been available to human ancestors. Similarly, the ability to consume the fructose from ten or more oranges as fruit juice in a few minutes, without simultaneous intake of the fibre contained within the fruit itself, is thought by some to be a metabolic shock that we have not evolved to handle. Whether these refined sugars are inherently detrimental or merely too easily quantitatively abused is far from clear, and hotly debated (Rippe and Angelopoulos 2016). Meanwhile, the need for more research in this area is emphasised by the recent discovery that artificial sweeteners also drive obesity despite containing negligible calories (Suez et al. 2015).

10.4.3.5 *Other Behavioural Changes that Compromise the Microbiota*

In addition to dietary changes, there are several other behavioural changes that coincide with economic development and urbanisation and that lead to striking changes in our microbiota. Most of these operate by impeding transfer of maternal microbiota to the offspring.

10.4.3.5.1 CAESAREAN SECTION

Caesarean section results in delayed transfer of maternal microbiota and delayed maturation of the infant microbiota (Dominguez-Bello et al. 2016). This can be partially corrected by exposing the neonate to maternal vaginal fluids at birth (Dominguez-Bello et al. 2016). There is strong evidence from human studies that caesarean delivery increases the risk of obesity in later life (Blustein et al. 2013; Yuan et al. 2016). There is also controversial evidence that it increases the risk of allergic disorders (Thavagnanam et al. 2008), but not all studies confirm this (Almqvist and Oberg 2014). However, there is certainly a relationship between the composition of the infant gut microbiota and the risk of atopic sensitisation, so caesarean delivery is likely to contribute to the relevant changes in some situations (Fujimura et al. 2016).

10.4.3.5.2 BREASTFEEDING

Breast milk contains oligosaccharides that cannot be metabolised by the infant and that serve as nutrients for *Bifidobacteria* (Zivkovic et al. 2011; Garrido et al. 2012). The composition of these oligosaccharides is dependent upon the mother's Lewis secretor status (Smith-Brown et al. 2016). The breast milk of non-secretor mothers lacks α1,2 fucosyl-oligosaccharides and their infants have lower relative counts of *Bifidobacterium* and *Bacteroides* in their first 4 months (Smith-Brown et al. 2016). Milk also provides *Bifidobacterium* and *Lactobacillus* species that seem to be transferred from the maternal gut to the breast (Jost et al. 2013; Melnik et al. 2016).

Nutritional factors in breast milk include omega-3 fatty acids, which are important for development of the brain. Finally, milk contains exosomes that carry microRNAs and TGF-β.

These constituents are thought to drive FoxP3 expression and long-lasting Treg differentiation (Saarinen et al. 1999; Melnik et al. 2016). In addition to influencing Treg development, recent work in animals suggests that maternal T lymphocytes might enter the thymus of the offspring and help to instruct the neonate's Th1 cell repertoire (Ghosh et al. 2016). It is therefore not surprising that breastfeeding modulates the microbiota (Stark and Lee 1982), and this might explain why breastfeeding protects against eczema, and possibly against other allergic disorders (Kramer 2011). However, much of the research in this area has failed to take the maternal Lewis secretor status into account, and once this is done more disease associations with failure of breast-feeding might emerge (Smith-Brown et al. 2016).

Moreover, whether via immunoregulatory pathways, via microbiota-induced effects, or via other mechanisms, breastfeeding is important for the brain. Duration of breastfeeding is related to verbal and non-verbal intelligence later in life (Belfort et al. 2013), to better cognitive and motor development (Bernard et al. 2013), and to greater social mobility (Sacker et al. 2013). (For further discussion, see Chapter 16: Sexuality, Reproduction, and Birth.)

10.4.3.5.3 *ANTIBIOTICS*

Antibiotics represent a particularly clear cause of dysbiosis. The relevance of this to human health is most obvious in allergic disorders, which have become very common and so have received a lot of attention. For example, the microbiota of human infants of median age 5 weeks fell into three distinct categories, one of which was associated with a more atopic sensitisation (Fujimura et al. 2016). In mouse models, antibiotic treatment can increase susceptibility to food allergy by eliminating immunoregulation-inducing *clostridia* (Stefka et al. 2014). Studies in human infants in Europe show that administration of antibiotics during the second or third trimesters of pregnancy can increase the risk of allergies in the infant (McKeever et al. 2002; Metsala et al. 2013; Lapin et al. 2015), and a Chinese study found that antibiotic use in the first trimester increased the incidence of asthma in the offspring (Chu et al. 2015). While not all studies agree, there is strong evidence that administering antibiotics to infants during the first months after birth can increase the risk of asthma (Korpela et al. 2016).

Infants are also exposed to antimicrobial consumer products containing, for example, Triclosan (5-chloro-2-(2,4-dichlorophenoxy)phenol). This works by inhibiting the final step in bacterial fatty acid synthesis. It is uncertain whether triclosan in toothpaste, hard and liquid soap, and products used in dish-washers alters the microbiota (Yee and Gilbert 2016), but it does encourage antibiotic resistance (Hartmann et al. 2016) and has been declared useless by the US Food and Drug Administration (FDA) (Food and Drug Administration 2016).

10.4.3.5.4 ANTIBIOTICS, OBESITY, AND TYPE 2 DIABETES

Animal studies suggest that in addition to allergic disorders, other health problems might be related to the dysbiosis that follows administration of antibiotics in the periconceptual period (Figure 10.8). In mice, periconceptual antibiotics can lead to weight gain later in life (Cox et al. 2014) and to behavioural abnormalities (Degroote et al. 2016). Epidemiological studies suggest that the same might be true for obesity in humans (Trasande et al. 2013; Azad et al. 2014; Korpela et al. 2016). This would be in good agreement with studies showing

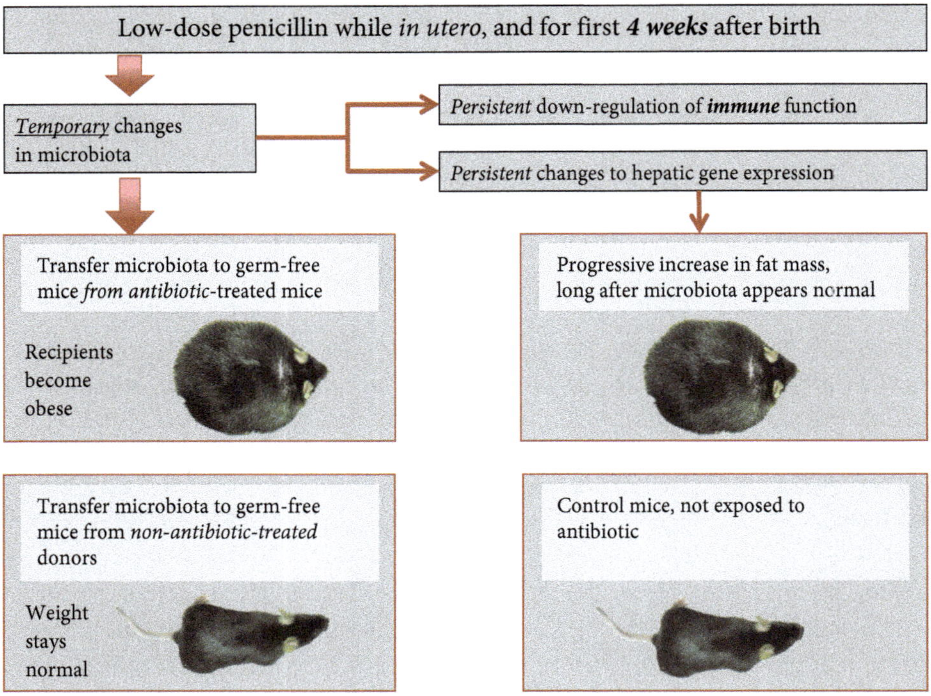

FIGURE 10.8 The early life 'window' of development that requires an appropriate microbiota. Low-dose penicillin in the perinatal period (during pregnancy and in the early post-natal period) can cause long-lasting defects in regulation of metabolism and of the immune system. There is evidence that the same is true in humans.

that obesity (Turnbaugh et al. 2006) and type 2 diabetes (Qin et al. 2012) are routinely associated with altered microbiota.

10.5 Other Inflammatory Disorders Associated with Dysbiosis

Numerous other disorders have been tentatively linked to changes in the gut microbiota. Metabolic disorders, mentioned briefly in Sections 10.4.3.4.3 and 10.4.3.5.4, are discussed in Chapter 6 and will not be discussed further here. (For further discussion, see Chapter 6: Nutrition, Energy Expenditure, Physical Activity, and Body Composition.)

The role of the microbiota and of dysbiosis is usually easy to demonstrate in animal models, but attempts to define the relevant changes in human disease are often contradictory. For example, a subpopulation of patients with early rheumatoid arthritis (RA) had high intestinal *Prevotella copri* (Maeda et al. 2016). When microbiota from such patients was transferred to a mouse model, it induced increased intestinal Th17 cells and increased susceptibility to severe arthritis. However, others find dysbiosis in RA, but not increased *Prevotella copri* (Chen et al. 2016b). Similarly, there have been several attempts to link multiple sclerosis (MS) to dysbiosis, and abnormalities are reported, though not necessarily showing the same changes (Chen et al. 2016a; Jangi et al. 2016). This contrasts with the clear role of gut microbiota in animal models of MS (Lee et al. 2011).

As far as intestinal disorders are concerned, overgrowth of *C. difficile*, usually a consequence of antibiotic administration, is unquestionably the cause of a gastrointestinal disorder, and it also provides the one reproducible example of a condition that can be treated by faecal microbial transplantation from a healthy donor (Cammarota et al. 2014). The situation is not so clear for irritable bowel syndrome (IBS) or inflammatory bowel disease (IBD). IBS, an extremely common disorder, is associated with variable changes in the microbiota and might be a disorder of the gut–microbiota–brain axis (Kennedy et al. 2014), but there is no consistent link with particular identified species.

The data for IBD are also in a state of flux and emphasise that Crohn's disease (CD) and ulcerative colitis (UC) might represent a number of different disorders, with different aetiologies, involving different types of organism within the gut. Dysbiosis of the small intestinal microbiota (depletion of Firmicutes and Bacteroidetes) was reported in a subset of patients (Frank et al. 2007). A recent study suggested that *Candida tropicalis* (i.e. a yeast), *E. coli*, and *Serratia marcescens* are more abundant in dysbiotic CD microbiota and might work together to create a pro-inflammatory biofilm with fungal hyphae that are associated with pathogenicity (Hoarau et al. 2016). So, this study implicated a yeast. Meanwhile it is known that the gut contains viruses that can influence the local immune system and, therefore, the microbiota. Following a norovirus infection, some individuals develop a long-term distortion of the microbiota with reduced Bacteroidetes and increased *E. coli* (Pfeiffer and Virgin 2016). In a mouse model, some persistent norovirus strains infect Paneth cells bearing a human CD susceptibility gene (*ATG16L1*) and drive an abnormal gene transcription pattern that in turn leads to susceptibility to inflammation driven by the microbiota (Cadwell et al. 2010). This study therefore suggests a triple aetiology involving a virus, a genetic risk factor, and a dysbiotic microbiota (Cadwell et al. 2010). If this is correct it is not surprising that current attempts to establish links between microbiota and disease are yielding encouraging but erratic results. There is therefore a problem of overinterpretation of results.

10.5.1 Cancer

Between a fifth and a quarter of all human cancers are associated with chronic inflammation caused by infections, irritants (such as asbestos, alcohol, or tobacco smoke), or chronic inflammatory disorders such as IBD (Kundu and Surh 2008; Pesic and Greten 2016). Colitis-associated colorectal cancer is a particularly clear example of the latter, but inflammatory environments caused by *H. pylori*, prostatitis, chronic hepatitis, cholecystitis, or pancreatitis have been implicated in cancers of the stomach, prostate, liver, gallbladder, and pancreas, respectively. Inflammation has two types of role in cancer development. First, inflammation can promote oncogenesis, by inducing DNA damage and defective DNA repair and by causing chronic stimulation of cancer-susceptible cell types, including the lymphocytes that mediate the immune response. Second, inflammation releases growth factors and angiogenic factors that can facilitate growth, vascularisation, and spread of the tumour (Mantovani et al. 2008).

Therefore, the issue of immunoregulation in cancer initiation and cancer progression is complex but of great interest. Any of the inflammatory stimuli listed above could become more oncogenic if control of inflammation were impaired. Several inflammation-related cancers, notably prostate, breast, colon, and pancreas, increase as countries develop economically or when a defined population moves from a developing to a high-income environment (Rastogi et al. 2008). The same is true for cancers of the immune system such as Hodgkin's lymphoma and acute lymphocytic leukaemia (reviewed in Rook and Dalgleish 2011). (For further discussion, see Chapter 9: Haematopoetic System.) Moreover, these increases parallel the increases in chronic inflammatory disorders (Bach 2002; Rook and Dalgleish 2011). The parallel between the increased incidence of cancer and the increases in chronic inflammatory disorders is supported by a number of other observations. First, colorectal cancer is associated with dysbiosis, and with the modern high-income pattern of gut microbiota rather than with that seen in rural Africans (Ou et al. 2013). Second, frequent use of antibiotics is associated with increased risk of colorectal cancer (Dik et al. 2016). Third, regular use of aspirin, an anti-inflammatory agent, significantly reduced the risk of death from cancer when all cancer types were considered together (Rothwell et al. 2012). Most cancer types were too infrequent in the sample to be analysed separately, but the protection from colorectal cancer and lymphomas was significant (Rothwell et al. 2012).

The findings summarised in the preceding paragraph indicate that failing immunoregulation in high-income settings associated with diminished exposure to immunoregulation-inducing 'old friends' might also play a role in the increased incidences of some cancers. This hypothesis has been discussed in detail elsewhere (Rook and Dalgleish 2011; Von Hertzen et al. 2011). This might seem an odd assertion in an era when immunotherapy of cancer is increasingly successful. For example, immune checkpoint inhibitors block the physiological negative feedback mechanisms that control the immune system and so intensify immune attack, and these agents are proving effective (Naidoo et al. 2014). How can this beneficial effect of blocking immunoregulation be reconciled with the view that failing immunoregulation is increasing the risk of cancer? There are two crucial points to make here. First, well-functioning suppression of chronic inflammation can help to stop inflammation-induced *initiation* of oncogenesis. Second, animal and human studies reveal that Treg are heterogeneous and various subtypes differ in their roles in cancer, possibly because of differences in secretion of IL-10 or conversion to Th17 cells, and differences in homing. Animal experiments in models of breast and colorectal cancer revealed that tumour progression was associated with Foxp3$^+$ cells in lymph nodes and in tissue *surrounding* the tumour, where they were likely to oppose development of the immune response (Erdman et al. 2010). On the other hand, remission was associated with Foxp3$^+$ cells located in the tumour itself, where they might reduce local release of factors that induce growth and angiogenesis (Erdman et al. 2010). Interestingly, in cell transfer experiments, Treg that were protective and that localised within the tumour could be derived from *Helicobacter*-infected donors, but not from very clean uninfected animals (Erdman et al. 2010). This finding has clear parallels with the 'hygiene' or 'old friends' hypothesis. A similar observation was made in a human cancer. A study that recorded the precise location of the Treg in breast cancer concluded that Treg in surrounding lymphoid-enriched areas correlated with a higher risk of relapse and reduced overall survival, whereas the presence of Foxp3$^+$ Treg within the tumour bed itself did not affect the patients' clinical evolution (Menetrier-Caux et al. 2009).

10.5.2 Psychiatric Disorders

Major depressive disorder (MDD) is rapidly becoming the major cause of human disability, so it deserves emphasis here (Ustun et al. 2004). Psychiatric disorders are frequently comorbid with the chronic inflammatory disorders that were discussed in relation to immune dysregulation and to lack of exposure to the 'old friends' (O'Donovan et al. 2014; Zerbo et al. 2015). Moreover, the evidence associating several psychiatric disorders with inflammation is now overwhelming. But because psychiatric disorders are heterogeneous, difficult to classify with precision, and affected by external stressors, it is difficult to prove that they are increasing in high-income urban settings in parallel with the other disorders of immunoregulation. Nevertheless, a powerful case can be made by examining urban–rural differences in disease prevalence, and the effects of migration from developing to rich urban environments. Urbanisation and migration both lead to loss of exposure to 'old friends', and both have been shown to correlate with increases in chronic inflammatory disorders (reviewed and referenced in Rook et al. 2014b). It is generally assumed that urbanisation and immigration modulate the prevalence and expression of psychiatric disorders via psychosocial stress, and no doubt stress plays a role, but evidence for other effects mediated via the immune system is now compelling. (For further discussion, see Chapter 17: the Brain, Spinal Cord, and Sensory Systems.)

10.5.2.1 *Depression*

A meta-analysis of high-quality studies performed in high-income countries since 1985 found that the prevalence of depression in urban areas was 39% higher than in rural areas. Similarly, the prevalence of anxiety disorders was 21% higher in urban than in rural areas (Peen et al. 2010), though a small minority of studies fails to find this urban–rural difference (Kovess-Masfety et al. 2005). Peen and colleagues also noted an increased urban prevalence of psychiatric disorders in general (38% more in urban communities) (Peen et al. 2010).

Depression in immigrants is also very revealing (Vega et al. 2004; Breslau et al. 2009). Mexicans, Cubans, and African/Caribbean peoples have a two- to three-fold increase in the prevalence of depression if immigration to the United States occurred when the individual was less than 13 years old or was born in the United States, compared to the prevalence in those who migrated after the age of 13 (Vega et al. 2004; Breslau et al. 2009). This implies that there is a protective effect of early environmental influences, as has been shown for autoimmune disorders and IBD (reviewed and referenced in Rook et al. 2014b).

These epidemiological findings are therefore compatible with the view that early exposure to immunoregulation-inducing organisms is protective against depression. The failure of immunoregulation in depression is suggested by data showing that raised C-reactive protein (CRP) or interleukin-6 (IL-6) can *predict* later depression in children (Khandaker et al. 2014) and adults in the United Kingdom (Gimeno et al. 2009), and also *predict* later susceptibility to post-traumatic stress disorder (PTSD) in army recruits (Eraly et al. 2014), while non-invasive positron emission tomography (PET) scans reveal the presence of inflammation in the brains of depressed individuals (Setiawan et al. 2015). As expected therefore, a large subset of depressed individuals has persistently raised levels of pro-inflammatory cytokines and other downstream inflammatory markers (Maes et al. 1992;

Miller et al. 2009), together with a relative deficit in anti-inflammatory mediators and regulatory T cells (Chen et al. 2011; fully referenced in Raison et al. 2010). Finally, depression can be induced by therapeutic use of a pro-inflammatory cytokine (Capuron et al. 2002), and those cases of depression that have raised background inflammation can be treated with a neutralising antibody to TNF (Raison et al. 2013). Interestingly, depressed individuals also show exaggerated release of inflammatory mediators in response to psychosocial stressors (Pace et al. 2006), which provides further evidence for defective regulation of cytokine release (Figure 10.9).

These human findings are supported by a mass of animal data that will not be reviewed here, but two animal studies are of direct relevance. First, spleen cells from individual mice were tested *in vitro* for IL-6 output in the presence of endotoxin. Individual animals could then be classified as high or low IL-6 releasers. When the spleen cell donors were subsequently subjected to stress (Figure 10.10), only the high IL-6 releasers showed depression-like behavioural changes (Hodes et al. 2014). This is analogous to the human study mentioned in the previous paragraph (Pace et al. 2006). A still more relevant study exploited a model in which exposure to stress induces changes in the gut microbiota accompanied by an inflammatory colitis, inflammatory changes in the brain, and depression-like behavioural changes (Reber et al. 2016). In this model, injections of a heat-killed environmental *Mycobacterium* were able to block the colitis, the brain inflammation, and the behavioural changes, and all these effects were shown to be attributable to induction of Treg (Reber et al. 2016).

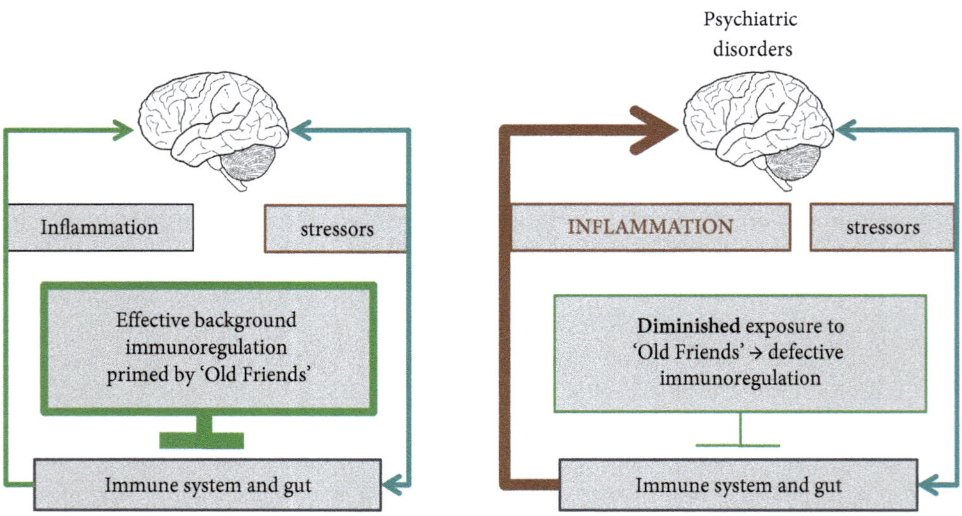

FIGURE 10.9 Stress-induced inflammation is subject to immunoregulation. When immunoregulatory pathways are compromised because early life inputs to the immune system were deficient, a given level of stress induces higher levels of inflammatory mediators that persist for longer. This can promote psychiatric disorders.

Source: Adapted from Graham A. W. Rook, Christopher A. Lowry, and Charles L. Raison, Microbial 'old friends', immunoregulation and stress resilience, *Evolution, Medicine, and Public Health*, 7 (1), pp. 46–64, Figure 2, doi.org/10.1093/emph/eot004.

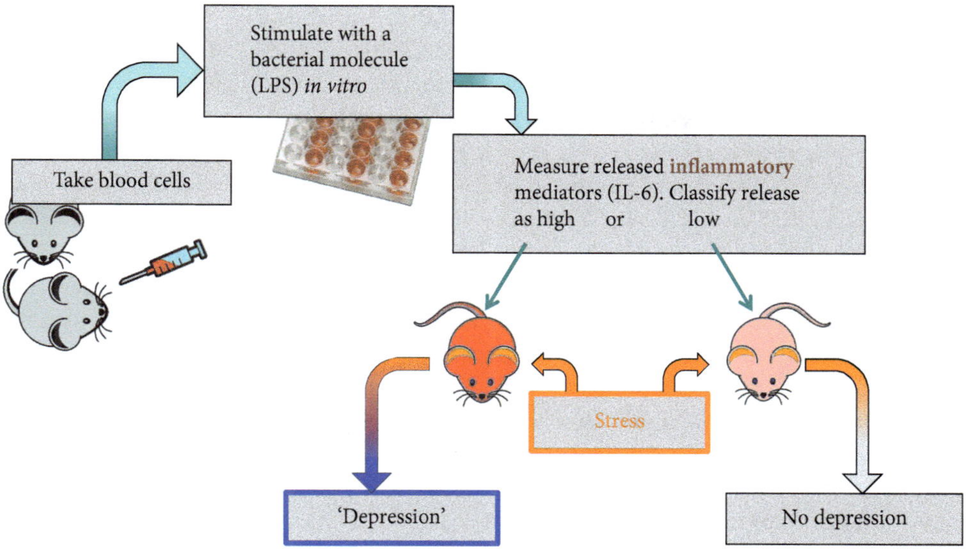

FIGURE 10.10 Experimental demonstration of the link between behavioural response to stress and efficiency of control of release of inflammatory mediators.

This simultaneous modulation of a peripheral inflammatory disorder (colitis) and a behavioural disorder by microbe-induced Treg is a striking finding.

So, is there any evidence of changes to the microbiota in human depression? There are few studies, and they are not conclusive. Patients suffering from MDD had reduced levels of *Faecalibacterium* which showed a negative correlation with the severity of depressive symptoms (Jiang et al. 2015). Patients also had increased levels of Enterobacteriaceae and Alistipes (Jiang et al. 2015). A second study found some increases in members of the Alistipes group in the faecal microbiota of depressed individuals (Naseribafrouei et al. 2014). There is some evidence that transfer of faecal microbiota from depressed patients to germ-free mice results in depression-like behavioural changes in the latter (Zheng et al. 2016).

However, there is now good evidence that human behaviour can be changed by modulation of the microbiota. In one study, a probiotic-rich fermented milk product or matching placebo were given to women for 4 weeks. Both before and after this regimen the women were exposed to emotive images of faces while undergoing functional magnetic resonance (fMRI) imaging of their brains. Consumption of the probiotic fermented milk product was shown to have affected the activity of brain regions involved in central processing of emotion and sensation (Tillisch et al. 2013). Another randomised and blinded study used 20 healthy participants without current mood disorder who received a complex probiotic food supplement or placebo for 4 weeks. Assessment before and after the intervention indicated that the active preparation reduced negative thoughts associated with sad mood (Steenbergen et al. 2015).

10.5.2.2 *Autism and Schizophrenia*

Similar arguments apply to autism and schizophrenia. A study of all children born in Denmark between 1 January 1984 and 31 December 1998 found that the degree of urbanisation of place of birth was very significantly correlated to risk of autism ($p < 0.0001$) (Lauritsen et al. 2005). Moreover, autism spectrum disorders are markedly more common in the offspring of immigrants. This risk was greatest if the mother had migrated from a developing to a high-income country (Keen et al. 2010; Becerra et al. 2014) and had done so within 1 year of parturition (Magnusson et al. 2012). The increased risk was no longer seen if the migration took place 20 years before (Magnusson et al. 2012). Interestingly, inflammation during pregnancy, from any cause, increases the risk of autism and schizophrenia in the child (Meyer et al. 2011; Brown et al. 2014). It is hypothesised that diminished immunoregulation can increase this risk further, and that this could occur in high-income urban settings when women from developing countries, whose immune systems are epigenetically programmed to receive heavy contact with immunoregulatory 'old friends', are subjected to modern medicine and deworming (Rook et al. 2014a).

If a deficit in immunoregulation were involved, we would expect comorbidity with other disorders of immunoregulation. Allergies and autoimmune diseases were diagnosed significantly more often among children with autism than among controls (Zerbo et al. 2015). There is also an association with gastrointestinal symptoms and IBDs, and numerous studies have indicated abnormalities in the gut microbiota of autistic children (reviewed in Vuong and Hsiao 2016).

Finally, diet, a major regulator of the microbiota as described in Sections 10.3.2 and 10.4.3.4, is also implicated. Mothers with obesity were 1.5 times more likely to have a child with autism spectrum disorder, and the risk was increased still further if the mother had both obesity and gestational diabetes (Connolly et al. 2016). Similarly, feeding a high-fat diet to pregnant mice caused dysbiosis in the offspring with striking depletion of certain bacterial strains from the gut microbiota. These offspring had autism-like behavioural and social deficits that could be reversed by administering *Lactobacillus reuteri* or by co-housing with control mice (Buffington et al. 2016).

Similar remarks can be made about schizophrenia. A large meta-analysis found a significantly raised prevalence of schizophrenia in urban communities (McGrath et al. 2004), and immigration into Denmark when less than 4 years old was associated with an increased risk for psychotic disorders, whereas the increased risk gradually decreased with older age at migration and disappeared in those immigrating when more than 29 years old (Veling et al. 2011). And again, a large meta-analysis confirmed that schizophrenia was increased among first-generation immigrants, and further increased among second-generation immigrants, particularly when the country of origin was a developing one (Cantor-Graae and Selten 2005). Moreover, schizophrenia, like autism, is associated with increased risk of autoimmune disorders (Benros et al. 2014). Therefore, as described for depression in Section 10.5.2.1, early events and the immune system seem to play significant roles in autism and schizophrenia.

10.5.2.3 *Microbial Metabolites and Psychiatric Disorders*

So far, this review has almost ignored the fact that 20–30% of all small molecules in the blood are microbial products, but they are discussed here as a separate topic. Many of these molecules affect development and function of multiple organs. Some are required for normal function, while others are toxic. However, the tasks of identifying these molecules and their effects and then relating the production of these molecules to particular families, genera, species, or strains of bacteria are immense and have barely begun.

Several amino acids are precursors of neurotransmitters within the central nervous system. The metabolic pathways used by humans to generate these compounds are in fact mediated by genes that entered the genomes of our ancestors by horizontal gene transfer from bacteria (Iyer et al. 2004). Microorganisms routinely synthesise the same molecules, and also a range of variants that not surprisingly have important physiological effects. Tyrosine (the precursor of dopamine, noradrenaline, and adrenaline) and tryptophan (the precursor of serotonin and melatonin) are two examples. It was observed that when pregnant mice were exposed to poly(I:C) which mimicked inflammation induced by virus infections, the offspring exhibited behavioural impairments reminiscent of autism spectrum disorder (Hsiao et al. 2013). This was accompanied by dysbiosis and changes in blood levels of gut microbiota-derived metabolites of tyrosine (4-ethylphenyl sulfate) and tryptophan (indolepyruvate). Interestingly, administering 4-ethylphenyl sulfate to normal mice was able to elicit the autism-like symptoms. Moreover, a probiotic preparation was able to normalise the blood levels of these metabolites. Other tryptophan metabolites are agonists for AhR, and several of these (indole, indoxyl-3-sulfate, indole-3-propionic acid, and indole-3-aldehyde) were found to exert anti-inflammatory effects in the CNS via AhR expressed on astrocytes (Rothhammer et al. 2016). These observations provide potential mechanisms for the association between inflammation of any cause in pregnancy and the prevalence of autism in the offspring.

These gut-derived tryptophan metabolites are accompanied by SCFA, derived by the fermentation of fibre by organisms in the colon. SCFA have been shown to regulate the development of microglia (Erny et al. 2015) and to be necessary for development of an intact blood–brain barrier (Braniste et al. 2014).

It should be noted that the microbe-derived amino acid metabolites have systemic roles in the regulation of inflammation, and are not only active in the central nervous system. For example, another tryptophan metabolite, indole-3-acetic acid, was shown to have anti-inflammatory effects in the gut via AhR expressed on many cell types including DC (Lamas et al. 2016), and excessive production of histamine from histidine by dysbiotic gut microbiota can inhibit activation of the NLRP6 inflammasome (Levy et al. 2015).

10.6 Future Attempts to Reconcile Our Environment with Our Evolution

Humans no longer exist in the human 'environment of evolutionary adaptedness' (Bowlby 1971 (first published by Hogarth Press in 1969)). Changes to our environment, child rearing, medical practices, and diet have been rapid, particularly since the early twentieth century, and there has not been time for genetic adaptation to these changes. Many of the consequences of this gene–environment mismatch are associated with malfunction of the immune system (Figure 10.11). Thus, there have been huge increases in the incidences of a range of chronic inflammatory diseases, where the immune system is targeting things that it should not attack, such as self, harmless allergens, and gut contents. There has also been a simultaneous increase in diseases associated with a failure to terminate background inflammation, including cardiovascular, psychiatric, and metabolic disorders. How then can we try to mitigate the effects of modern lifestyles? This review has attempted to identify

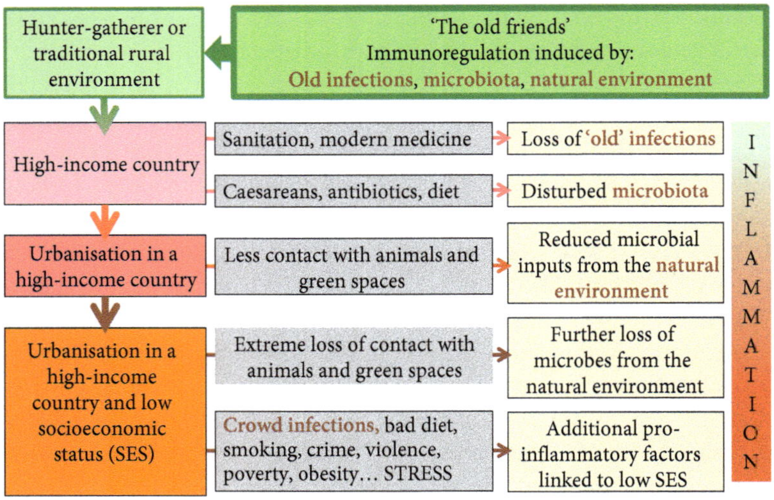

FIGURE 10.11 Progressive loss of microbial inputs to the immune system as humans move from traditional to modern lifestyles. This results in progressive loss of immunoregulation.

Source: Adapted from G. A. W. Rook, C. L. Raison, and C. A. Lowry, Microbial 'old friends', immunoregulation and socioeconomic status, *Clinical and Experimental Immunology*, 177 (1) pp. 1–12, Figure 3, doi.10.1111/cei.12269. © 2014 The Authors. Clinical and Experimental Immunology published by John Wiley & Sons Ltd on behalf of British Society for Immunology.

the crucial environmental changes, and this last section suggests how we can try to avoid the consequences of the current state of gene–environment mismatch.

10.6.1 Old Infections

There are authors who think that some of the 'old infections' lost as a result of modern medicine might need to be replaced, and name this concept 'biome reconstitution' (Parker and Ollerton 2013). For example, some kind of 'domesticated' helminth could be given to every infant at birth (Parker and Ollerton 2013). However, as discussed in Section 10.4.1.4, the extraordinary variability of helminth loads and the diversity of helminth-mediated immunoregulatory mechanisms make it unlikely that we have a genetically encoded requirement for these organisms, even if people born in developing countries have an *epigenetically* encoded need for the species with which they developed. Moreover, from a regulatory and pharmaceutical point of view, a domesticated helminth is not a practical objective.

Of the other 'old infections', *H. pylori* has also been considered as a means of 'biome reconstitution', but this is equally impractical, especially if, as suggested, it needs to be accompanied by a simultaneous load of helminths (see Section 10.4.1.3), though it is possible that a strain could be developed that does not cause peptic ulceration.

However, tuberculosis provides an interesting case. There is evidence that vaccination with BCG, which is an attenuated strain of *Mycobacterium bovis*, a related mycobacterial species, protects against not only tuberculosis, but also other infections, allergies, and some cancers (Netea and Van Crevel 2014). This phenomenon has been called 'trained immunity', and might conceivably be related to the long association of humans with the mycobacteria.

However, unlike the concept of biome reconstitution with a living persistent organism, exposure to BCG is transient and extraordinarily safe.

10.6.2 Microbiota

Control and modulation of the microbiota clearly has huge potential, but although we know the microbiota is crucial, in the context of most disorders, we do not know why. The results of different studies are too inconsistent. We also know that the composition of the microbiota is influenced by the genetic background of the individual (Goodrich et al. 2014). For example, the optimum diet for an individual, in terms of limiting the postprandial glycaemic response, depends on their genetics and their microbiota (Zeevi et al. 2015). In other words, even within a single modern population there is no such thing as a 'correct' microbiota. Therefore, matching dysbiotic microbiotas with specific clinical entities is difficult. This difficulty is exacerbated by the fact that the clinical entities may also be heterogeneous in aetiology.

The reasons for the inconsistent results are relatively easily identified, but will be more difficult to resolve. First, the material sampled is most often the free-flowing colonic material recovered from faeces, so that organisms associated with the mucosa or with the small bowel (where DC sample the antigenic repertoire) tend to be underrepresented (Schulz and Pabst 2013). Second, few papers discuss the fungal populations or the virome, although viruses (both prokaryotic and eukaryotic) are the most abundant organisms in the gut (Pfeiffer and Virgin 2016). Prokaryotic viruses such as bacteriophages affect turnover of their target bacteria. Some *abundant* organisms may have low turnover, so little replication or death. But some apparently *less abundant* organisms may have huge turnover due to bacteriophages, so they potentially have a greater biological impact on the host as dying organisms release biologically potent molecules (Korem et al. 2015).

There are also important technical differences in different papers. The ease with which DNA can be extracted from bacteria is very variable, and extraction from spores and from mycobacteria is particularly difficult. For example, mycobacteria are hardly ever reported, though when appropriate methods are used they are abundant, as expected for a group of organisms that is ubiquitous in the environment (Macovei et al. 2015). Another crucial technical variant is the use of cut-offs, so that organisms are ignored if their abundance is below an arbitrary threshold. This is technically necessary, but irrational from a microbiological point of view; the most clinically important organisms might be present in low numbers. Interestingly, a recent study used culture techniques in novel media made from sterile cow rumen fluid. Using only three faecal samples, this study found 174 species never previously reported in the human gut (Lagier et al. 2012). A further problem is that an organism will not be reported if its sequence is not yet in the relevant databases, and even if the organism is in the database, the identification rarely extends to the strain level. This is a problem because different strains of the same species can have quite different effects (notably amongst *Lactobacilli*), so the identification of a 'taxonomic unit' does not necessarily provide adequate information.

Finally, the emphasis is switching away from individual organisms, towards overall metabolic pathways. It may be the end metabolites of the entire gut ecosystem that matter, rather than individual organisms considered in isolation (Schirmer et al. 2016).

Despite this background of uncertainty, there is much that can be done. First, we need to ensure that maternal microbiota is transmitted to the infant. Natural birth and breastfeeding can help, and it is possible to mitigate the effect of caesarean section by transferring maternal vaginal secretions to the baby after caesarean delivery (Dominguez-Bello et al. 2016). But we can also try to introduce targeted hygiene that does not block transmission of maternal microbiota to the infant. For example, sucking the baby's pacifier clean or sharing a cup or spoon should not be regarded as unhygienic (Hesselmar et al. 2013).

The most damaging period for exposure to antibiotics is the perinatal period, in pregnancy and early infancy. Clearly, antibiotics will sometimes be essential, but we can anticipate a move towards shorter courses with very specific antibiotics that target only the relevant organism. New molecular technologies should allow rapid, accurate identification of the infecting organism, and drug discoveries will eventually allow development of 'narrow spectrum' antibiotics that cause minimal damage to the microbiota while targeting only the organism that is causing the infection. It might also be possible to bank an infant's microbiota and then restore it as soon as possible after termination of the treatment. Some restoration of microbiota may be possible from the natural environment, since this is inevitably seeded with spores of human gut-adapted strains. Finally, we might be able to combat bacterial infections by administering specific bacteriophages, or harmless bacteria that compete for the same ecological niche within the host.

The other fundamental issue for maintenance of the microbiota is diet. The public has been subjected to absurd swings in the nature of the dietary advice offered, with recommendations that they eat less fat, more carbohydrate, less carbohydrate, more protein, less protein, or a 'palaeolithic diet', whatever that might be. Even the latter has been grossly misrepresented. The average 'palaeolithic diet' enthusiast when eating meat would probably not like to be told to eat the offal and stomach contents. We can now start to give less-complicated advice based on the concepts that the diet should approximate to the varied diet of the human evolutionary past, which approximates to what has been named 'the Mediterranean diet' (Martinez-Gonzalez et al. 2015). But the optimal diet will not be the same for everyone (Zeevi et al. 2015), so, in the future, more specific individualised dietary advice might be based on analysis of the patient's genetics and microbiota.

10.6.3 Natural Environment

The realisation that the natural environment provides important inputs of microbial diversity to the immune system provides another medical reason for promoting the provision of green space within the urban environment (Rook 2013). In addition to this role, it has become clear recently that trees strikingly reduce levels of nitrogen dioxide (NO_2) and pollutant particulate matter (Pugh et al. 2012). So, these two powerful medical arguments can be added to the traditional view that inserting the natural environment into cities encourages exercise, relaxation, and exposure to sunlight. Unfortunately, we are tending to degrade some green spaces.

Antibiotics are signalling molecules within microbial ecosystems that establish the nature and balance of organisms present (Martinez 2009). Widespread use of antibiotics in agriculture, and reuse in city parks of partially purified water supplies, are causing pollution of the environment with antibiotics, together with selection of antibiotic resistance genes in environmental microbial communities (Wang et al. 2014). This will inevitably alter the microbial composition and biodiversity of these ecosystems (Martinez 2009). We do not yet know what impact this is having on the benefits to humans of exposure to the microbiota from these environments, but such effects are likely to be adding to depletion of microbial biodiversity already taking place as a result of agricultural practices and monoculture. Meanwhile, we can increase the microbial biodiversity of our urban environments by creating urban green spaces, by placing vegetation on flat roofs, and by keeping plants in homes (Mahnert et al. 2015).

Finally, we need to switch public awareness away from the notion that hygiene is to blame for the rising scourge of inflammatory disorders. Indeed, the term 'hygiene hypothesis' should perhaps be regarded as obsolete. The media often seize upon the word 'hygiene' and suggest to the public that we have become 'too clean for our own good'. This is quite wrong, even dangerous. Hygiene remains the most useful outcome of medical research and is increasingly important as new crowd infections evolve (Bloomfield et al. 2016). The immune system does indeed require data inputs from microbes, and the system is increasingly deprived of those inputs, but this deprivation is not mainly due to domestic hygiene, but rather to major lifestyle changes, diet, and antibiotics.

References

Abi-Rached, L., Jobin, M. J., Kulkarni, S., et al. (2011). The shaping of modern human immune systems by multiregional admixture with archaic humans. *Science* 334, 89–94.

Aichbhaumik, N., Zoratti, E. M., Strickler, R., et al. (2008). Prenatal exposure to household pets influences fetal immunoglobulin E production. *Clin Exp Allergy* 38, 1787–94.

Almqvist, C. and Oberg, A. S. (2014). The association between caesarean section and asthma or allergic disease continues to challenge. *Acta Paediatr* 103, 349–51.

Andersson, M. A., Mikkola, R., Kroppenstedt, R. M., et al. (1998). The mitochondrial toxin produced by *Streptomyces griseus* strains isolated from an indoor environment is valinomycin. *Appl Environ Microbiol* 64, 4767–73.

Atarashi, K., Tanoue, T., Shima, T., et al. (2011). Induction of colonic regulatory T cells by indigenous *Clostridium* species. *Science* 331, 337–41.

Axelsson, E., Ratnakumar, A., Arendt, M. L., et al. (2013). The genomic signature of dog domestication reveals adaptation to a starch-rich diet. *Nature* 495, 360–4.

Azad, M. B., Bridgman, S. L., Becker, A. B., et al. (2014). Infant antibiotic exposure and the development of childhood overweight and central adiposity. *Int J Obes (Lond)* 38, 1290–8.

Babu, S., Blauvelt, C. P., Kumaraswami, V., et al. (2006). Regulatory networks induced by live parasites impair both Th1 and Th2 pathways in patent lymphatic filariasis: implications for parasite persistence. *J Immunol* 176, 3248–56.

Bach, J. F. (2002). The effect of infections on susceptibility to autoimmune and allergic diseases. *N Engl J Med* 347, 911–20.

Becerra, T. A., Von Ehrenstein, O. S., Heck, J. E., et al. (2014). Autism spectrum disorders and race, ethnicity, and nativity: a population-based study. *Pediatrics* 134, e63–71.

Belfort, M. B., Rifas-Shiman, S. L., Kleinman, K. P., et al. (2013). Infant feeding and childhood cognition at ages 3 and 7 years: effects of breastfeeding duration and exclusivity. *JAMA Pediatr* 167, 836–44.

Benitez-Cabello, A., Bautista-Gallego, J., Garrido-Fernandez, A., et al. (2016). RT-PCR-DGGE analysis to elucidate the dominant bacterial species of industrial Spanish-style green table olive fermentations. *Front Microbiol* 7, 1291.

Benros, M. E., Pedersen, M. G., Rasmussen, H., et al. (2014). A nationwide study on the risk of autoimmune diseases in individuals with a personal or a family history of schizophrenia and related psychosis. *Am J Psychiatry* 171, 218–26.

Bergman, M. P., Engering, A., Smits, H. H., et al. (2004). *Helicobacter pylori* modulates the T helper cell 1/T helper cell 2 balance through phase-variable interaction between lipopolysaccharide and DC-SIGN. *J Experi Med* 200, 979–90.

Bernard, J. Y., De Agostini, M., Forhan, A., et al. (2013). Breastfeeding duration and cognitive development at 2 and 3 years of age in the EDEN mother–child cohort. *J Pediatr* 163, 36–42.

Bessede, A., Gargaro, M., Pallotta, M. T., et al. (2014). Aryl hydrocarbon receptor control of a disease tolerance defence pathway. *Nature* 511, 184–90.

Bilsborough, S. and Mann, N. (2006). A review of issues of dietary protein intake in humans. *Int J Sport Nutr Exerc Metab* 16, 129–52.

Biswas, S. K. and Lopez-Collazo, E. (2009). Endotoxin tolerance: new mechanisms, molecules and clinical significance. *Trends Immunol* 30, 475–87.

Black, F. L. (1966). Measles endemicity in insular populations: critical community size and its evolutionary implication. *J Theor Biol* 11, 207–11.

Blasbalg, T. L., Hibbeln, J. R., Ramsden, C. E., et al. (2011). Changes in consumption of omega-3 and omega-6 fatty acids in the United States during the twentieth century. *Am J Clin Nutr* 93, 950–62.

Bloomfield, S. F., Rook, G. A., Scott, E. A., et al. (2016). Time to abandon the hygiene hypothesis: new perspectives on allergic disease, the human microbiome, infectious disease prevention and the role of targeted hygiene. *Perspect Public Health* 136, 213–24.

Blustein, J., Attina, T., Liu, M., et al. (2013). Association of caesarean delivery with child adiposity from age 6 weeks to 15 years. *Int J Obes (Lond)* 37, 900–6.

Boehm, T. (2008). Thymus development and function. *Curr Opin Immunol* 20, 178–84.

Boehm, T. (2012). Evolution of vertebrate immunity. *Curr Biol* 22, R722–32.

Bowlby, J. (1971 (first published by Hogarth Press in 1969)). *Attachment and Loss, Volume 1: Attachment*. Harmondsworth: Penguin.

Branchfield, K., Nantie, L., Verheyden, J. M., et al. (2016). Pulmonary neuroendocrine cells function as airway sensors to control lung immune response. *Science* 351, 707–10.

Braniste, V., Al-Asmakh, M., Kowal, C., et al. (2014). The gut microbiota influences blood–brain barrier permeability in mice. *Sci Transl Med* 6, 263ra158.

Breidt, F., Mcfeeters, R. F., Perez-Diaz, I., et al. (2013). Fermented vegetables. In: Doyle, M. P. and Buchanan, R. L. (eds) *Food Microbiology: Fundamentals and Frontiers*, 4th ed. Washington, DC: ASM Press.

Bremner, S. A., Carey, I. M., Dewilde, S., et al. (2008). Infections presenting for clinical care in early life and later risk of hay fever in two UK birth cohorts. *Allergy* 63, 274–83.

Breslau, J., Borges, G., Hagar, Y., et al. (2009). Immigration to the USA and risk for mood and anxiety disorders: variation by origin and age at immigration. *Psychol Med* 39, 1117–27.

Brown, A. S., Sourander, A., Hinkka-Yli-Salomaki, S., et al. (2014). Elevated maternal C-reactive protein and autism in a national birth cohort. *Mol Psychiatry* 19, 259–64.

Browne, H. P., Forster, S. C., Anonye, B. O., et al. (2016). Culturing of 'unculturable' human microbiota reveals novel taxa and extensive sporulation. *Nature* 533, 543–6.

Buffington, S. A., Di Prisco, G. V., Auchtung, T. A., et al. (2016). Microbial reconstitution reverses maternal diet-induced social and synaptic deficits in offspring. *Cell* 165, 1762–75.

Cadwell, K., Patel, K. K., Maloney, N. S., et al. (2010). Virus-plus-susceptibility gene interaction determines Crohn's disease gene Atg16L1 phenotypes in intestine. *Cell* 141, 1135–45.

Cammarota, G., Ianiro, G., and Gasbarrini, A. (2014). Fecal microbiota transplantation for the treatment of *Clostridium difficile* infection: a systematic review. *J Clin Gastroenterol* 48, 693–702.

Campbell, B., Raherison, C., Lodge, C. J., et al. (2017). The effects of growing up on a farm on adult lung function and allergic phenotypes: an international population-based study. *Thorax* 72, 236–44.

Cani, P. D., Bibiloni, R., Knauf, C., et al. (2008). Changes in gut microbiota control metabolic endotoxemia-induced inflammation in high-fat diet-induced obesity and diabetes in mice. *Diabetes* 57, 1470–81.

Cantor-Graae, E. and Selten, J. P. (2005). Schizophrenia and migration: a meta-analysis and review. *Am J Psychiatry* 162, 12–24.

Capuron, L., Gumnick, J. F., Musselman, D. L., et al. (2002). Neurobehavioral effects of interferon-alpha in cancer patients: phenomenology and paroxetine responsiveness of symptom dimensions. *Neuropsychopharmacol* 26, 643–52.

Caramalho, I., Rodrigues-Duarte, L., Perez, A., et al. (2011). Regulatory T cells contribute to diabetes protection in lipopolysaccharide-treated non-obese diabetic mice. *Scand J Immunol* 74, 585–95.

Carmody, R. N., Dannemann, M., Briggs, A. W., et al. (2016). Genetic evidence of human adaptation to a cooked diet. *Genome Biol Evol* 8, 1091–103.

Carrigan, M. A., Uryasev, O., Frye, C. B., et al. (2014). Hominids adapted to metabolize ethanol long before human-directed fermentation. *Proc Natl Acad Sci U S A* 112, 458–63.

Casula, G. and Cutting, S. M. (2002). Bacillus probiotics: spore germination in the gastrointestinal tract. *Appl Environ Microbiol* 68, 2344–52.

Chen, J., Chia, N., Kalari, K. R., et al. (2016a). Multiple sclerosis patients have a distinct gut microbiota compared to healthy controls. *Sci Rep* 6, 28484.

Chen, J., Wright, K., Davis, J. M., et al. (2016b). An expansion of rare lineage intestinal microbes characterizes rheumatoid arthritis. *Genome Med* 8, 43.

Chen, Y., Jiang, T., Chen, P., et al. (2011). Emerging tendency towards autoimmune process in major depressive patients: a novel insight from Th17 cells. *Psychiatry Res* 188, 224–30.

Chu, S., Yu, H., Chen, Y., et al. (2015). Periconceptional and gestational exposure to antibiotics and childhood asthma. *PLoS One* 10, e0140443.

Codolo, G., Mazzi, P., Amedei, A., et al. (2008). The neutrophil-activating protein of *Helicobacter pylori* down-modulates Th2 inflammation in ovalbumin-induced allergic asthma. *Cell Microbiol* 10, 2355–63.

Connolly, N., Anixt, J., Manning, P., et al. (2016). Maternal metabolic risk factors for autism spectrum disorder—an analysis of electronic medical records and linked birth data. *Autism Res* 9, 829–37.

Correale, J. and Farez, M. F. (2011). The impact of parasite infections on the course of multiple sclerosis. *J Neuroimmunol* 233, 6–11.

Correale, J. and Farez, M. F. (2013). Parasite infections in multiple sclerosis modulate immune responses through a retinoic acid-dependent pathway. *J Immunol* 191, 3827–37.

Costa, F. R., Francozo, M. C., De Oliveira, G. G., et al. (2016). Gut microbiota translocation to the pancreatic lymph nodes triggers NOD2 activation and contributes to T1D onset. *J Exp Med* 213, 1223–39.

Cover, T. L. and Blaser, M. J. (2009). *Helicobacter pylori* in health and disease. *Gastroenterology* 136, 1863–73.

Cox, L. M., Yamanishi, S., Sohn, J., et al. (2014). Altering the intestinal microbiota during a critical developmental window has lasting metabolic consequences. *Cell* 158, 705–21.

Crisp, A., Boschetti, C., Perry, M., et al. (2015). Expression of multiple horizontally acquired genes is a hallmark of both vertebrate and invertebrate genomes. *Genome Biol* 16, 50.

Cuervo, A., Hevia, A., Lopez, P., et al. (2016). Phenolic compounds from red wine and coffee are associated with specific intestinal microorganisms in allergic subjects. *Food Funct* 7, 104–9.

Cunnane, S. C. and Crawford, M. A. (2014). Energetic and nutritional constraints on infant brain development: implications for brain expansion during human evolution. *J Hum Evol* 77, 88–98.

Dadvand, P., De Nazelle, A., Figueras, F., et al. (2012). Green space, health inequality and pregnancy. *Environ Int* 40, 110–15.

Dannemann, M., Andres, A. M., and Kelso, J. (2016). Introgression of Neandertal- and Denisovan-like haplotypes contributes to adaptive variation in human Toll-like receptors. *Am J Hum Genet* 98, 22–33.

David, L. A., Maurice, C. F., Carmody, R. N., et al. (2014). Diet rapidly and reproducibly alters the human gut microbiome. *Nature* 505, 559–63.

De Filippo, C., Cavalieri, D., Di Paola, M., et al. (2010). Impact of diet in shaping gut microbiota revealed by a comparative study in children from Europe and rural Africa. *Proc Natl Acad Sci U S A* 107, 14691–6.

Degroote, S., Hunting, D. J., Baccarelli, A. A., et al. (2016). Maternal gut and fetal brain connection: increased anxiety and reduced social interactions in Wistar rat offspring following peri-conceptional antibiotic exposure. *Prog Neuropsychopharmacol Biol Psychiatry* 71, 76–82.

D'Elios, M. M., Codolo, G., Amedei, A., et al. (2009). *Helicobacter pylori*, asthma and allergy. *Fems Immunol Med Microbiol* 56, 1–8.

De Mazancourt, C., Loreau, M., and Dieckmann, U. (2005). Understanding mutualism when there is adaptation to the partner. *J Ecol* 93, 305–14.

Derrien, M., Johan, E. T., and Vlieg, H. (2015). Fate, activity, and impact of ingested bacteria within the human gut microbiota. *Trends Microbiol* 23, 354–66.

Deschamps, M., Laval, G., Fagny, M., et al. (2016). Genomic signatures of selective pressures and introgression from archaic hominins at human innate immunity genes. *Am J Hum Genet* 98, 5–21.

Dik, V. K., Van Oijen, M. G., Smeets, H. M., et al. (2016). Frequent use of antibiotics is associated with colorectal cancer risk: results of a nested case-control study. *Dig Dis Sci* 61, 255–64.

Dominguez-Bello, M. G., De Jesus-Laboy, K. M., Shen, N., et al. (2016). Partial restoration of the microbiota of cesarean-born infants via vaginal microbial transfer. *Nat Med* 22, 250–3.

Dunn, R. R., Fierer, N., Henley, J. B., et al. (2013). Home life: factors structuring the bacterial diversity found within and between homes. *PLoS One* 8, e64133.

Ege, M. J., Mayer, M., Normand, A. C., et al. (2011). Exposure to environmental microorganisms and childhood asthma. *N Engl J Med* 364, 701–9.

Elinav, E., Strowig, T., Kau, A. L., et al. (2011). NLRP6 inflammasome regulates colonic microbial ecology and risk for colitis. *Cell* 145, 745–57.

Eraly, S. A., Nievergelt, C. M., Maihofer, A. X., et al. (2014). Assessment of plasma C-reactive protein as a biomarker of posttraumatic stress disorder risk. *JAMA Psychiatry* 71, 423–31.

Erdman, S. E., Rao, V. P., Olipitz, W., et al. (2010). Unifying roles for regulatory T cells and inflammation in cancer. *Int J Cancer* 126, 1651–65.

Erny, D., Hrabe De Angelis, A. L., Jaitin, D., et al. (2015). Host microbiota constantly control maturation and function of microglia in the CNS. *Nat Neurosci* 18, 965–77.

Flajnik, M. F. and Kasahara, M. (2010). Origin and evolution of the adaptive immune system: genetic events and selective pressures. *Nat Rev Genet* 11, 47–59.

Fleming, J. O. (2013). Helminth therapy and multiple sclerosis. *Int J Parasitol* 43, 259–74.

Food and Drug Administration (2016). Safety and effectiveness of consumer antiseptics; topical antimicrobial drug products for over-the-counter human use. *Fed Reg* 81, 61106–30.

Forsberg, K. J., Reyes, A., Wang, B., et al. (2012). The shared antibiotic resistome of soil bacteria and human pathogens. *Science* 337, 1107–11.

Fox, J. G., Beck, P., Dangler, C. A., et al. (2000). Concurrent enteric helminth infection modulates inflammation and gastric immune responses and reduces *Helicobacter*-induced gastric atrophy. *Nat Med* 6, 536–42.

Franchini, A. and Ottaviani, E. (2017). Thymus: conservation in evolution. *Gen Comp Endocrinol* 246, 46–50.

Frank, D. N., St Amand, A. L., Feldman, R. A., et al. (2007). Molecular-phylogenetic characterization of microbial community imbalances in human inflammatory bowel diseases. *Proc Natl Acad Sci U S A* 104, 13780–5.

Fujimura, K. E., Johnson, C. C., Ownby, D. R., et al. (2010). Man's best friend? The effect of pet ownership on house dust microbial communities. *J Allergy Clin Immunol* 126, 410–2, 412.e1–3.

Fujimura, K. E., Sitarik, A. R., Havstad, S., et al. (2016). Neonatal gut microbiota associates with childhood multisensitized atopy and T cell differentiation. *Nat Med* 22, 1187–91.

Fumagalli, M., Pozzoli, U., Cagliani, R., et al. (2009). Parasites represent a major selective force for interleukin genes and shape the genetic predisposition to autoimmune conditions. *J Exp Med* 206, 1395–408.

Funkhouser, L. J. and Bordenstein, S. R. (2013). Mom knows best: the universality of maternal microbial transmission. *PLoS Biol* 11, e1001631.

Galagan, J. E. (2014). Genomic insights into tuberculosis. *Nat Rev Genet* 15, 307–20.

Galley, J. D., Bailey, M., Kamp Dush, C., et al. (2014). Maternal obesity is associated with alterations in the gut microbiome in toddlers. *PLoS One* 9, e113026.

Garrett, W. S. and Glimcher, L. H. (2009). T-bet-/- RAG2-/- ulcerative colitis: the role of T-bet as a peacekeeper of host–commensal relationships. *Cytokine* 48, 144–7.

Garrido, D., Barile, D., and Mills, D. A. (2012). A molecular basis for bifidobacterial enrichment in the infant gastrointestinal tract. *Adv Nutr* 3, 415S–21S.

Gause, W. C., Wynn, T. A., and Allen, J. E. (2013). Type 2 immunity and wound healing: evolutionary refinement of adaptive immunity by helminths. *Nat Rev Immunol* 13, 607–14.

Ge, Q. and Zhao, Y. (2013). Evolution of thymus organogenesis. *Dev Comp Immunol* 39, 85–90.

Ghosh, M. K., Nguyen, V., Muller, H. K., et al. (2016). Maternal milk T cells drive development of transgenerational Th1 immunity in offspring thymus. *J Immunol* 197, 2290–6.

Gimeno, D., Kivimaki, M., Brunner, E. J., et al. (2009). Associations of C-reactive protein and interleukin-6 with cognitive symptoms of depression: 12-year follow-up of the Whitehall II study. *Psychol Med* 39, 413–23.

Gittelman, R. M., Schraiber, J. G., Vernot, B., et al. (2016). Archaic hominin admixture facilitated adaptation to out-of-Africa environments. *Curr Biol* 26, 3375–82.

Gomez, A., Petrzelkova, K. J., Burns, M. B., et al. (2016). Gut microbiome of coexisting BaAka pygmies and Bantu reflects gradients of traditional subsistence patterns. *Cell Rep* 14, 2142–53.

Gomez De Aguero, M., Ganal-Vonarburg, S. C., Fuhrer, T., et al. (2016). The maternal microbiota drives early postnatal innate immune development. *Science* 351, 1296–302.

Goodrich, J. K., Waters, J. L., Poole, A. C., et al. (2014). Human genetics shape the gut microbiome. *Cell* 159, 789–99.

Goodrich, J. K., Davenport, E. R., Beaumont, M., et al. (2016). Genetic determinants of the gut microbiome in UK twins. *Cell Host Microbe* 19, 731–43.

Grainger, J. R., Smith, K. A., Hewitson, J. P., et al. (2010). Helminth secretions induce *de novo* T cell Foxp3 expression and regulatory function through the TGF-beta pathway. *J Exp Med* 207, 2331–41.

Green, R. E., Krause, J., Briggs, A. W., et al. (2010). A draft sequence of the Neandertal genome. *Science* 328, 710–22.

Gury-Benari, M., Thaiss, C. A., Serafini, N., et al. (2016). The spectrum and regulatory landscape of intestinal innate lymphoid cells are shaped by the microbiome. *Cell* 166, 1231–46.

Hang, L., Setiawan, T., Blum, A. M., et al. (2010). *Heligmosomoides polygyrus* infection can inhibit colitis through direct interaction with innate immunity. *J Immunol* 185, 3184–9.

Hanhineva, K., Torronen, R., Bondia-Pons, I., et al. (2010). Impact of dietary polyphenols on carbohydrate metabolism. *Int J Mol Sci* 11, 1365–402.

Hanski, I., Von Hertzen, L., Fyhrquist, N., et al. (2012). Environmental biodiversity, human microbiota, and allergy are interrelated. *Proc Natl Acad Sci U S A* 109, 8334–9.

Hartmann, E. M., Hickey, R., Hsu, T., et al. (2016). Antimicrobial chemicals are associated with elevated antibiotic resistance genes in the indoor dust microbiome. *Environ Sci Technol* 50, 9807–15.

Hehemann, J. H., Correc, G., Barbeyron, T., et al. (2010). Transfer of carbohydrate-active enzymes from marine bacteria to Japanese gut microbiota. *Nature* 464, 908–12.

Helander, H. F. and Fandriks, L. (2014). Surface area of the digestive tract—revisited. *Scand J Gastroenterol* 49, 681–9.

Henao-Mejia, J., Elinav, E., Jin, C., et al. (2012). Inflammasome-mediated dysbiosis regulates progression of NAFLD and obesity. *Nature* 482, 179–85.

Hershkovitz, I., Donoghue, H. D., Minnikin, D. E., et al. (2008). Detection and molecular characterization of 9,000-year-old *Mycobacterium tuberculosis* from a Neolithic settlement in the Eastern Mediterranean. *PLoS One* 3, e3426.

Hesselmar, B., Sjoberg, F., Saalman, R., et al. (2013). Pacifier cleaning practices and risk of allergy development. *Pediatrics* 131, e1829–37.

Hoarau, G., Mukherjee, P. K., Gower-Rousseau, C., et al. (2016). Bacteriome and mycobiome interactions underscore microbial dysbiosis in familial Crohn's disease. *MBio* 7, pii: e01250-16.

Hodes, G. E., Pfau, M. L., Leboeuf, M., et al. (2014). Individual differences in the peripheral immune system promote resilience versus susceptibility to social stress. *Proc Natl Acad Sci U S A* 111, 16136–41.

Hong, H. A., Khaneja, R., Tam, N. M., et al. (2009a). *Bacillus subtilis* isolated from the human gastrointestinal tract. *Res Microbiol* 160, 134–43.

Hong, H. A., To, E., Fakhry, S., et al. (2009b). Defining the natural habitat of *Bacillus* spore-formers. *Res Microbiol* 160, 375–9.

Hsiao, E. Y., McBride, S. W., Hsien, S., et al. (2013). Microbiota modulate behavioral and physiological abnormalities associated with neurodevelopmental disorders. *Cell* 155, 1451–63.

Hussain, K., Letley, D. P., Greenaway, A. B., et al. (2016). *Helicobacter pylori*-mediated protection from allergy is associated with IL-10-secreting peripheral blood regulatory T cells. *Front Immunol* 7, 71.

Iyer, L. M., Aravind, L., Coon, S. L., et al. (2004). Evolution of cell–cell signaling in animals: did late horizontal gene transfer from bacteria have a role? *Trends Genet* 20, 292–9.

Jangi, S., Gandhi, R., Cox, L. M., et al. (2016). Alterations of the human gut microbiome in multiple sclerosis. *Nat Commun* 7, 12015.

Jeffery, I. B. and O'Toole, P. W. (2013). Diet–microbiota interactions and their implications for healthy living. *Nutrients* 5, 234–52.

Jeon, K. W. (1972). Development of cellular dependence on infective organisms: microsurgical studies in amoebas. *Science* 176, 1122–3.

Jiang, H., Ling, Z., Zhang, Y., et al. (2015). Altered fecal microbiota composition in patients with major depressive disorder. *Brain Behav Immun* 48, 186–94.

Johansen, F. E., Pekna, M., Norderhaug, I. N., et al. (1999). Absence of epithelial immunoglobulin A transport, with increased mucosal leakiness, in polymeric immunoglobulin receptor/secretory component-deficient mice. *J Exp Med* 190, 915–22.

Jost, T., Lacroix, C., Braegger, C. P., et al. (2014). Vertical mother–neonate transfer of maternal gut bacteria via breastfeeding. *Environ Microbiol* 16, 2891–904.

Kaliannan, K., Wang, B., Li, X. Y., et al. (2015). A host–microbiome interaction mediates the opposing effects of omega-6 and omega-3 fatty acids on metabolic endotoxemia. *Sci Rep* 5, 11276.

Keen, D. V., Reid, F. D., and Arnone, D. (2010). Autism, ethnicity and maternal immigration. *Br J Psychiatry* 196, 274–81.

Kennedy, P. J., Cryan, J. F., Dinan, T. G., et al. (2014). Irritable bowel syndrome: a microbiome–gut–brain axis disorder? *World J Gastroenterol* 20, 14105–25.

Khandaker, G. M., Pearson, R. M., Zammit, S., et al. (2014). Association of serum interleukin 6 and C-reactive protein in childhood with depression and psychosis in young adult life: a population-based longitudinal study. *JAMA Psychiatry* 71, 1121–8.

Knip, M. and Siljander, H. (2016). The role of the intestinal microbiota in type 1 diabetes mellitus. *Nat Rev Endocrinol* 12, 154–67.

Kondo, K., Takada, K., and Takahama, Y. (2017). Antigen processing and presentation in the thymus: implications for T cell repertoire selection. *Curr Opin Immunol* 46, 53–7.

Kondrashova, A., Reunanen, A., Romanov, A., et al. (2005). A six-fold gradient in the incidence of type 1 diabetes at the eastern border of Finland. *Ann Med* 37, 67–72.

Korem, T., Zeevi, D., Suez, J., et al. (2015). Growth dynamics of gut microbiota in health and disease inferred from single metagenomic samples. *Science* 349, 1101–6.

Koren, O., Goodrich, J. K., Cullender, T. C., et al. (2012). Host remodeling of the gut microbiome and metabolic changes during pregnancy. *Cell* 150, 470–80.

Korpela, K., Salonen, A., Virta, L. J., et al. (2016). Intestinal microbiome is related to lifetime antibiotic use in Finnish pre-school children. *Nat Commun* 7, 10410.

Kostic, A. D., Howitt, M. R., and Garrett, W. S. (2013). Exploring host–microbiota interactions in animal models and humans. *Genes Dev* 27, 701–18.

Kovess-Masfety, V., Lecoutour, X., and Delavelle, S. (2005). Mood disorders and urban/rural settings: comparisons between two French regions. *Soc Psychiatry Psychiatr Epidemiol* 40, 613–18.

Kramer, M. S. (2011). Breastfeeding and allergy: the evidence. *Ann Nutr Metab* 59(Suppl 1), 20–6.

Kundu, J. K. and Surh, Y. J. (2008). Inflammation: gearing the journey to cancer. *Mutat Res* 659, 15–30.

Lagier, J. C., Armougom, F., Million, M., et al. (2012). Microbial culturomics: paradigm shift in the human gut microbiome study. *Clin Microbiol Infect* 18, 1185–93.

Lamas, B., Richard, M. L., Leducq, V., et al. (2016). CARD9 impacts colitis by altering gut microbiota metabolism of tryptophan into aryl hydrocarbon receptor ligands. *Nat Med* 22, 598–605.

Lang, J. M., Eisen, J. A., and Zivkovic, A. M. (2014). The microbes we eat: abundance and taxonomy of microbes consumed in a day's worth of meals for three diet types. *PeerJ* 2, e659.

Lapin, B., Piorkowski, J., Ownby, D., et al. (2015). Relationship between prenatal antibiotic use and asthma in at-risk children. *Ann Allergy Asthma Immunol* 114, 203–7.

Lauritsen, M. B., Pedersen, C. B., and Mortensen, P. B. (2005). Effects of familial risk factors and place of birth on the risk of autism: a nationwide register-based study. *J Child Psychol Psychiatry* 46, 963–71.

Lee, Y. K. and Mazmanian, S. K. (2010). Has the microbiota played a critical role in the evolution of the adaptive immune system? *Science* 330, 1768–73.

Lee, Y. K., Menezes, J. S., Umesaki, Y., et al. (2011). Proinflammatory T-cell responses to gut microbiota promote experimental autoimmune encephalomyelitis. *Proc Natl Acad Sci U S A*, 108(Suppl 1), 4615–22.

Leroy, F., Geyzen, A., Janssens, M., et al. (2013). Meat fermentation at the crossroads of innovation and tradition: a historical outlook. *Trends Food Sci Technol* 31, 130–7.

Levy, M., Thaiss, C. A., Zeevi, D., et al. (2015). Microbiota-modulated metabolites shape the intestinal microenvironment by regulating NLRP6 inflammasome signaling. *Cell* 163, 1428–43.

Levy, M., Thaiss, C. A., and Elinav, E. (2016). Metabolites: messengers between the microbiota and the immune system. *Genes Dev* 30, 1589–97.

Liepe, J., Marino, F., Sidney, J., et al. (2016). A large fraction of HLA class I ligands are proteasome-generated spliced peptides. *Science* 354–8.

Linz, B., Balloux, F., Moodley, Y., et al. (2007). An African origin for the intimate association between humans and *Helicobacter pylori*. *Nature* 445, 915–98.

Liu, W., Li, Y., Learn, G. H., et al. (2010). Origin of the human malaria parasite *Plasmodium falciparum* in gorillas. *Nature* 467, 420–5.

Loschko, J., Schreiber, H. A., Rieke, G. J., et al. (2016). Absence of MHC class II on cDCs results in microbial-dependent intestinal inflammation. *J Exp Med* 213, 517–34.

Lowry, C. A., Smith, D. G., Siebler, P. H., et al. (2016). The microbiota, immunoregulation, and mental health: implications for public health. *Curr Environ Health Rep* 3, 270–86.

Loy, D. E., Liu, W., Li, Y., et al. (2016). Out of Africa: origins and evolution of the human malaria parasites *Plasmodium falciparum* and *Plasmodium vivax*. *Int J Parasitol* 47, 87–97.

Lu, M. F., Xiao, Z. T., and Zhang, H. Y. (2013). Where do health benefits of flavonoids come from? Insights from flavonoid targets and their evolutionary history. *Biochem Biophys Res Commun* 434, 701–4.

Lundgren, A., Stromberg, E., Sjoling, A., et al. (2005). Mucosal FOXP3-expressing CD4+ CD25 high regulatory T cells in *Helicobacter pylori*-infected patients. *Infect Immun* 73, 523–31.

Maas, J., Verheij, R. A., Groenewegen, P. P., et al. (2006). Green space, urbanity, and health: how strong is the relation? *J Epidemiol Community Health* 60, 587–92.

Macovei, L., Mccafferty, J., Chen, T., et al. (2015). The hidden 'mycobacteriome' of the human healthy oral cavity and upper respiratory tract. *J Oral Microbiol* 7, 26094.

Maeda, Y., Kurakawa, T., Umemoto, E., et al. (2016). Dysbiosis contributes to arthritis development via activation of autoreactive T cells in the intestine. *Arthritis Rheumatol* 68, 2646–61.

Maes, M., Scharpe, S., Van Grootel, L., et al. (1992). Higher alpha 1-antitrypsin, haptoglobin, ceruloplasmin and lower retinol binding protein plasma levels during depression: further evidence for the existence of an inflammatory response during that illness. *J Affect Disord* 24, 183–92.

Magnusson, C., Rai, D., Goodman, A., et al. (2012). Migration and autism spectrum disorder: population-based study. *Br J Psychiatry* 201, 109–15.

Mahnert, A., Moissl-Eichinger, C., and Berg, G. (2015). Microbiome interplay: plants alter microbial abundance and diversity within the built environment. *Front Microbiol* 6, 887.

Mantovani, A., Allavena, P., Sica, A., et al. (2008). Cancer-related inflammation. *Nature* 454, 436–44.

Martinez, J. L. (2009). Environmental pollution by antibiotics and by antibiotic resistance determinants. *Environ Pollut* 157, 2893–902.

Martinez-Gonzalez, M. A., Salas-Salvado, J., Estruch, R., et al. (2015). Benefits of the Mediterranean diet: insights from the PREDIMED Study. *Prog Cardiovasc Dis* 58, 50–60.

Maslin, M. A., Shultz, S., and Trauth, M. H. (2015). A synthesis of the theories and concepts of early human evolution. *Philos Trans R Soc Lond B Biol Sci* 370, 20140064.

Matricardi, P. M., Rosmini, F., Riondino, S., et al. (2000). Exposure to foodborne and orofecal microbes versus airborne viruses in relation to atopy and allergic asthma; epidemiological study. *BMJ* 320, 412–17.

McDade, T. W., Tallman, P. S., Madimenos, F. C., et al. (2012). Analysis of variability of high sensitivity C-reactive protein in lowland Ecuador reveals no evidence of chronic low-grade inflammation. *Am J Hum Biol* 24, 675–81.

McFall-Ngai, M., Hadfield, M. G., Bosch, T. C., et al. (2013). Animals in a bacterial world, a new imperative for the life sciences. *Proc Natl Acad Sci U S A* 110, 3229–36.

McGovern, P. E. (2009). *Uncorking the Past: The Quest for Wine, Beer, and Other Alcoholic Beverages.* Berkeley: University of California Press.

McGovern, P. E., Zhang, J., Tang, J., et al. (2004). Fermented beverages of pre- and proto-historic China. *Proc Natl Acad Sci U S A* 101, 17593–8.

McGrath, J., Saha, S., Welham, J., et al. (2004). A systematic review of the incidence of schizophrenia: the distribution of rates and the influence of sex, urbanicity, migrant status and methodology. *BMC Med* 2, 13.

McKeever, T. M., Lewis, S. A., Smith, C., et al. (2002). The importance of prenatal exposures on the development of allergic disease: a birth cohort study using the West Midlands General Practice Database. *Am J Resp Crit Care Medicine* 166, 827–32.

Melnik, B. C., John, S. M., Carrera-Bastos, P., et al. (2016). Milk: a postnatal imprinting system stabilizing FoxP3 expression and regulatory T cell differentiation. *Clin Transl Allergy* 6, 18.

Menetrier-Caux, C., Gobert, M., and Caux, C. (2009). Differences in tumor regulatory T-cell localization and activation status impact patient outcome. *Cancer Res* 69, 7895–8.

Meropol, S. B. and Edwards, A. (2015). Development of the infant intestinal microbiome: a bird's eye view of a complex process. *Birth Defects Res C Embryo Today* 105, 228–39.

Metsala, J., Lundqvist, A., Virta, L. J., et al. (2013). Mother's and offspring's use of antibiotics and infant allergy to cow's milk. *Epidemiology* 24, 303–9.

Meyer, U., Feldon, J., and Dammann, O. (2011). Schizophrenia and autism: both shared and disorder-specific pathogenesis via perinatal inflammation? *Pediatr Res* 69, 26R–33.

Miller, A. H., Maletic, V., and Raison, C. L. (2009). Inflammation and its discontents: the role of cytokines in the pathophysiology of major depression. *Biol Psychiatry* 65, 732–41.

Mitchell, R. and Popham, F. (2008). Effect of exposure to natural environment on health inequalities: an observational population study. *Lancet* 372, 1655–60.

Moeller, A. H., Caro-Quintero, A., Mjungu, D., et al. (2016). Cospeciation of gut microbiota with hominids. *Science* 353, 380–2.

Mohammadi-Bardbori, A., Bengtsson, J., Rannug, U., et al. (2012). Quercetin, resveratrol, and curcumin are indirect activators of the aryl hydrocarbon receptor (AHR). *Chem Res Toxicol* 25, 1878–84.

Montresor, A. (ed.) (1987). *Prevention and Control of Intestinal Parasitic Infections. WHO Technical Report Series 749.* Geneva: World Health Organisation.

Moore, M. N. (2015). Do airborne biogenic chemicals interact with the PI3K/Akt/mTOR cell signalling pathway to benefit human health and wellbeing in rural and coastal environments? *Environ Res* 140, 65–75.

Mpairwe, H., Webb, E. L., Muhangi, L., et al. (2011). Anthelminthic treatment during pregnancy is associated with increased risk of infantile eczema: randomised-controlled trial results. *Pediatr Allergy Immunol* 22, 305–12.

Naidoo, J., Page, D. B., and Wolchok, J. D. (2014). Immune checkpoint blockade. *Hematol Oncol Clin North Am* 28, 585–600.

Naseribafrouei, A., Hestad, K., Avershina, E., et al. (2014). Correlation between the human fecal microbiota and depression. *Neurogastroenterol Motil* 26, 1155–62.

Navarro, S., Pickering, D. A., Ferreira, I. B., et al. (2016). Hookworm recombinant protein promotes regulatory T cell responses that suppress experimental asthma. *Sci Transl Med* 8, 362ra143.

Netea, M. G. and Van Crevel, R. (2014). BCG-induced protection: effects on innate immune memory. *Semin Immunol* 26, 512–17.

Nicholson, W. L. (2002). Roles of *Bacillus* endospores in the environment. *Cell Mol Life Sci* 59, 410–16.

Nuriel-Ohayon, M., Neuman, H., and Koren, O. (2016). Microbial changes during pregnancy, birth, and infancy. *Front Microbiol* 7, 1031.

Obihara, C. C., Beyers, N., Gie, R. P., et al. (2005). Inverse association between *Mycobacterium tuberculosis* infection and atopic rhinitis in children. *Allergy* 60, 1121–5.

Obregon-Tito, A. J., Tito, R. Y., Metcalf, J., et al. (2015). Subsistence strategies in traditional societies distinguish gut microbiomes. *Nat Commun* 6, 6505.

O'Donovan, A., Cohen, B. E., Seal, K. H., et al. (2014). Elevated risk for autoimmune disorders in Iraq and Afghanistan veterans with posttraumatic stress disorder. *Biol Psychiatry* 77, 365–74.

Oosting, M., Cheng, S. C., Bolscher, J. M., et al. (2014). Human TLR10 is an anti-inflammatory pattern-recognition receptor. *Proc Natl Acad Sci U S A* 111, E4478–84.

Ou, J., Carbonero, F., Zoetendal, E. G., et al. (2013). Diet, microbiota, and microbial metabolites in colon cancer risk in rural Africans and African Americans. *Am J Clin Nutr* 98, 111–20.

Ownby, D. R., Johnson, C. C., and Peterson, E. L. (2002). Exposure to dogs and cats in the first year of life and risk of allergic sensitization at 6 to 7 years of age. *JAMA* 288, 963–72.

Pace, T. W., Mletzko, T. C., Alagbe, O., et al. (2006). Increased stress-induced inflammatory responses in male patients with major depression and increased early life stress. *Am J Psychiatry* 163, 1630–3.

Pakarinen, J., Hyvarinen, A., Salkinoja-Salonen, M., et al. (2008). Predominance of Gram-positive bacteria in house dust in the low-allergy risk Russian Karelia. *Environ Microbiol* 10, 3317–25.

Parker, W. and Ollerton, J. (2013). Evolutionary biology and anthropology suggest biome reconstitution as a necessary approach toward dealing with immune disorders. *Evol Med Public Health* 2013, 89–103.

Peen, J., Schoevers, R. A., Beekman, A. T., et al. (2010). The current status of urban–rural differences in psychiatric disorders. *Acta Psychiatr Scand* 121, 84–93.

Pendyala, S., Walker, J. M., and Holt, P. R. (2012). A high-fat diet is associated with endotoxemia that originates from the gut. *Gastroenterology* 142, 1100–1.

Perry, G. H., Dominy, N. J., Claw, K. G., et al. (2007). Diet and the evolution of human amylase gene copy number variation. *Nat Genet* 39, 1256–60.

Pesic, M. and Greten, F. R. (2016). Inflammation and cancer: tissue regeneration gone awry. *Curr Opin Cell Biol* 43, 55–61.

Pfeiffer, J. K. and Virgin, H. W. (2016). Viral immunity. Transkingdom control of viral infection and immunity in the mammalian intestine. *Science* 351(6270), pii: aad5872.

Pontzer, H., Brown, M. H., Raichlen, D. A., et al. (2016). Metabolic acceleration and the evolution of human brain size and life history. *Nature* 533, 390–2.

Pugh, T. A., Mackenzie, A. R., Whyatt, J. D., et al. (2012). Effectiveness of green infrastructure for improvement of air quality in urban street canyons. *Environ Sci Technol* 46, 7692–9.

Qin, J., Li, Y., Cai, Z., et al. (2012). A metagenome-wide association study of gut microbiota in type 2 diabetes. *Nature* 490, 55–60.

Queipo-Ortuno, M. I., Boto-Ordonez, M., Murri, M., et al. (2012). Influence of red wine polyphenols and ethanol on the gut microbiota ecology and biochemical biomarkers. *Am J Clin Nutr* 95, 1323–34.

Quintana, F. J., Basso, A. S., Iglesias, A. H., et al. (2008). Control of T(reg) and T(H)17 cell differentiation by the aryl hydrocarbon receptor. *Nature* 453, 65–71.

Radon, K., Windstetter, D., Poluda, A. L., et al. (2007). Contact with farm animals in early life and juvenile inflammatory bowel disease: a case-control study. *Pediatrics* 120, 354–61.

Raison, C. L., Lowry, C. A., and Rook, G. A. W. (2010). Inflammation, sanitation and consternation: loss of contact with co-evolved, tolerogenic micro-organisms and the pathophysiology and treatment of major depression. *Arch Gen Psychiatry* 67, 1211–24.

Raison, C. L., Rutherford, R. E., Woolwine, B. J., et al. (2013). A randomized controlled trial of the tumor necrosis factor antagonist infliximab for treatment-resistant depression. *JAMA Psychiatry* 70, 31–41.

Ramanan, D., Bowcutt, R., Lee, S. C., et al. (2016). Helminth infection promotes colonization resistance via type 2 immunity. *Science* 352, 608–12.

Rampelli, S., Schnorr, S. L., Consolandi, C., et al. (2015). Metagenome sequencing of the Hadza hunter-gatherer gut microbiota. *Curr Biol* 25, 1682–93.

Rastogi, T., Devesa, S., Mangtani, P., et al. (2008). Cancer incidence rates among South Asians in four geographic regions: India, Singapore, UK and US. *Int J Epidemiol* 37, 147–60.

Reber, S. O., Siebler, P. H., Donner, N. C., et al. (2016). Immunization with a heat-killed preparation of the environmental bacterium *Mycobacterium vaccae* promotes stress resilience in mice. *Proc Natl Acad Sci U S A* 113, E3130–9.

Reibman, J., Marmor, M., Filner, J., et al. (2008). Asthma is inversely associated with *Helicobacter pylori* status in an urban population. *PLoS One* 3(12), e4060.

Reynolds, L. A., Finlay, B. B., and Maizels, R. M. (2015). Cohabitation in the intestine: interactions among helminth parasites, bacterial microbiota, and host immunity. *J Immunol* 195, 4059–66.

Rhee, K. J., Sethupathi, P., Driks, A., et al. (2004). Role of commensal bacteria in development of gut-associated lymphoid tissues and preimmune antibody repertoire. *J Immunol* 172, 1118–24.

Ribatti, D., Crivellato, E., and Vacca, A. (2006). Miller's seminal studies on the role of thymus in immunity. *Clin Exp Immunol* 144, 371–5.

Riedler, J., Braun-Fahrlander, C., Eder, W., et al. (2001). Exposure to farming in early life and development of asthma and allergy: a cross-sectional survey. *Lancet* 358, 1129–33.

Rippe, J. M. and Angelopoulos, T. J. (2016). Added sugars and risk factors for obesity, diabetes and heart disease. *Int J Obes (Lond)* 40(Suppl 1), S22–7.

Robinson, K., Kenefeck, R., Pidgeon, E. L., et al. (2008). *Helicobacter pylori*-induced peptic ulcer disease is associated with inadequate regulatory T cell responses. *Gut* 57, 1375–85.

Rook, G. A. W. (2013). Regulation of the immune system by biodiversity from the natural environment: an ecosystem service essential to health. *Proc Natl Acad Sci U S A* 110, 18360–7.

Rook, G. A. W. and Dalgleish, A. (2011). Infection, immunoregulation and cancer. *Immunol Rev* 240, 141–59.

Rook, G. A., Raison, C. L. and Lowry, C. A. (2014a). Microbiota, immunoregulatory old friends and psychiatric disorders. *Adv Exp Med Biol* 817, 319–56.

Rook, G. A., Raison, C. L. and Lowry, C. A. (2014b). Microbiota, immunoregulatory old friends and psychiatric disorders. In: Lyte, M. and Cryan, J. F. (eds) *Microbial Endocrinology: The Microbiota–Gut–Brain Axis in Health and Disease, Advances in Experimental Medicine and Biology*. New York: Springer, p. 817.

Rook, G., Bäckhed, F., Levin, B. R., et al. (2017). Evolution, human–microbe interactions, and life history plasticity. *Lancet* 390, 521–30.

Rothhammer, V., Mascanfroni, I. D., Bunse, L., et al. (2016). Type I interferons and microbial metabolites of tryptophan modulate astrocyte activity and central nervous system inflammation via the aryl hydrocarbon receptor. *Nat Med* 22, 586–97.

Rothwell, P. M., Price, J. F., Fowkes, F. G., et al. (2012). Short-term effects of daily aspirin on cancer incidence, mortality, and non-vascular death: analysis of the time course of risks and benefits in 51 randomised controlled trials. *Lancet* 379, 1602–12.

Round, J. L. and Mazmanian, S. K. (2009). The gut microbiota shapes intestinal immune responses during health and disease. *Nat Rev Immunol* 9, 313–23.

Round, J. L. and Mazmanian, S. K. (2010). Inducible Foxp3+ regulatory T-cell development by a commensal bacterium of the intestinal microbiota. *Proc Natl Acad Sci U S A* 107, 12204–9.

Saarinen, K. M., Vaarala, O., Klemetti, P., et al. (1999). Transforming growth factor-beta1 in mothers' colostrum and immune responses to cows' milk proteins in infants with cows' milk allergy. *J Allergy Clin Immunol* 104, 1093–8.

Sacker, A., Kelly, Y., Iacovou, M., et al. (2013). Breast feeding and intergenerational social mobility: what are the mechanisms? *Arch Dis Child* 98, 666–71.

Sahlberg, B., Wieslander, G., and Norback, D. (2010). Sick building syndrome in relation to domestic exposure in Sweden—a cohort study from 1991 to 2001. *Scand J Public Health* 38, 232–8.

Schirmer, M., Smeekens, S. P., Vlamakis, H., et al. (2016). Linking the human gut microbiome to inflammatory cytokine production capacity. *Cell* 167, 1125–36.

Schnorr, S. L., Candela, M., Rampelli, S., et al. (2014). Gut microbiome of the Hadza hunter-gatherers. *Nat Commun* 5, 3654.

Schuijs, M. J., Willart, M. A., Vergote, K., et al. (2015). Farm dust and endotoxin protect against allergy through A20 induction in lung epithelial cells. *Science* 349, 1106–10.

Schulz, O. and Pabst, O. (2013). Antigen sampling in the small intestine. *Trends Immunol* 34, 155–61.

Scrivener, S., Yemaneberhan, H., Zebenigus, M., et al. (2001). Independent effects of intestinal parasite infection and domestic allergen exposure on risk of wheeze in Ethiopia: a nested case-control study. *Lancet* 358, 1493–9.

Sefik, E., Geva-Zatorsky, N., Oh, S., et al. (2015). Mucosal immunology. Individual intestinal symbionts induce a distinct population of RORgamma(+) regulatory T cells. *Science* 349, 993–7.

Selhub, E. M., Logan, A. C., and Bested, A. C. (2014). Fermented foods, microbiota, and mental health: ancient practice meets nutritional psychiatry. *J Physiol Anthropol* 33, 2.

Setiawan, E., Wilson, A. A., Mizrahi, R., et al. (2015). Role of translocator protein density, a marker of neuroinflammation, in the brain during major depressive episodes. *JAMA Psychiatry* 72, 268–75.

Shahack-Gross, R., Berna, F., Karkanas, P., et al. (2014). Evidence for the repeated use of a central hearth at Middle Pleistocene (300 ky ago) Qesem Cave, Israel. *J Arch Sci* 44, 12–21.

Smillie, C. S., Smith, M. B., Friedman, J., et al. (2011). Ecology drives a global network of gene exchange connecting the human microbiome. *Nature* 480, 241–4.

Smith, H. F., Parker, W., Kotze, S. H., et al. (2017). Morphological evolution of the mammalian cecum and cecal appendix. *Comptes Rendus Palevol* 16, 39–57.

Smith-Brown, P., Morrison, M., Krause, L., et al. (2016). Mothers secretor status affects development of children's microbiota composition and function: a pilot study. *PLoS One* 11, e0161211.

Soderborg, T. K., Borengasser, S. J., Barbour, L. A., et al. (2016). Microbial transmission from mothers with obesity or diabetes to infants: an innovative opportunity to interrupt a vicious cycle. *Diabetologia* 59, 895–906.

Sonnenberg, A. (2006). Causes underlying the birth-cohort phenomenon of peptic ulcer: analysis of mortality data 1911–2000, England and Wales. *Int J Epidemiol* 35, 1090–7.

Sonnenburg, E. D., Smits, S. A., Tikhonov, M., et al. (2016). Diet-induced extinctions in the gut microbiota compound over generations. *Nature* 529, 212–15.

Sotgiu, S., Sannella, A. R., Conti, B., et al. (2007). Multiple sclerosis and anti-*Plasmodium falciparum* innate immune response. *J Neuroimmunol* 185, 201–7.

Sozanska, B., Blaszczyk, M., Pearce, N., et al. (2013). Atopy and allergic respiratory disease in rural Poland before and after accession to the European Union. *J Allergy Clin Immunol* 133, 1347–53.

Stark, P. L. and Lee, A. (1982). The microbial ecology of the large bowel of breast-fed and formula-fed infants during the first year. *J Med Microbiol* 15, 189–203.

Steenbergen, L., Sellaro, R., Van Hemert, S., et al. (2015). A randomized controlled trial to test the effect of multispecies probiotics on cognitive reactivity to sad mood. *Brain Behav Immun* 48, 258–64.

Stefka, A. T., Feehley, T., Tripathi, P., et al. (2014). Commensal bacteria protect against food allergen sensitization. *Proc Natl Acad Sci U S A* 111, 13145–50.

Stein, M. M., Hrusch, C. L., Gozdz, J., et al. (2016). Innate immunity and asthma risk in Amish and Hutterite farm children. *N Engl J Med* 375, 411–21.

Stoll, N. R. (1947). This wormy world. *J Parasitol* 33, 1–18.

Strachan, D. P. (1989). Hay fever, hygiene, and household size. *BMJ* 299, 1259–60.

Suez, J., Korem, T., Zilberman-Schapira, G., et al. (2015). Non-caloric artificial sweeteners and the microbiome: findings and challenges. *Gut Microbes* 6, 149–55.

Swain, M. R., Anandharaj, M., Ray, R. C., et al. (2014). Fermented fruits and vegetables of Asia: a potential source of probiotics. *Biotechnol Res Int* 2014, 250424.

Tam, N. K., Uyen, N. Q., Hong, H. A., et al. (2006). The intestinal life cycle of *Bacillus subtilis* and close relatives. *J Bacteriol* 188, 2692–700.

Tan, J., Mckenzie, C., Vuillermin, P. J., et al. (2016). Dietary fiber and bacterial SCFA enhance oral tolerance and protect against food allergy through diverse cellular pathways. *Cell Rep* 15, 2809–24.

Thalmann, O., Shapiro, B., Cui, P., et al. (2013). Complete mitochondrial genomes of ancient canids suggest a European origin of domestic dogs. *Science* 342, 871–4.

Thavagnanam, S., Fleming, J., Bromley, A., et al. (2008). A meta-analysis of the association between Caesarean section and childhood asthma. *Clin Exp Allergy* 38, 629–33.

Tillisch, K., Labus, J., Kilpatrick, L., et al. (2013). Consumption of fermented milk product with probiotic modulates brain activity. *Gastroenterology* 144, 1394–401.

Trasande, L., Blustein, J., Liu, M., et al. (2013). Infant antibiotic exposures and early-life body mass. *Int J Obes (Lond)* 37, 16–23.

Turnbaugh, P. J., Ley, R. E., Mahowald, M. A., et al. (2006). An obesity-associated gut microbiome with increased capacity for energy harvest. *Nature* 444, 1027–31.

Ustun, T. B., Ayuso-Mateos, J. L., Chatterji, S., et al. (2004). Global burden of depressive disorders in the year 2000. *Br J Psychiatry* 184, 386–92.

Vaezi, M. F., Falk, G. W., Peek, R. M., et al. (2000). CagA-positive strains of *Helicobacter pylori* may protect against Barrett's esophagus. *Am J Gastroenterol* 95, 2206–11.

Vanamala, J. K., Knight, R., and Spector, T. D. (2015). Can your microbiome tell you what to eat? *Cell Metab* 22, 960–1.

Van Liempt, E., Bank, C. M., Mehta, P., et al. (2006). Specificity of DC-SIGN for mannose- and fucose-containing glycans. *FEBS Lett* 580, 6123–31.

Vatanen, T., Kostic, A. D., D'Hennezel, E., et al. (2016). Variation in microbiome LPS immunogenicity contributes to autoimmunity in humans. *Cell* 165, 842–53.

Vega, W. A., Sribney, W. M., Aguilar-Gaxiola, S., et al. (2004). 12-month prevalence of DSM-III-R psychiatric disorders among Mexican Americans: nativity, social assimilation, and age determinants. *J Nerv Ment Dis* 192, 532–41.

Veling, W., Hoek, H. W., Selten, J. P., et al. (2011). Age at migration and future risk of psychotic disorders among immigrants in the Netherlands: a 7-year incidence study. *Am J Psychiatry* 168, 1278–85.

Vernot, B., Tucci, S., Kelso, J., et al. (2016). Excavating Neandertal and Denisovan DNA from the genomes of Melanesian individuals. *Science* 352, 235–9.

Vijay-Kumar, M., Aitken, J. D., Carvalho, F. A., et al. (2010). Metabolic syndrome and altered gut microbiota in mice lacking Toll-like receptor 5. *Science* 328, 228–31.

Von Hertzen, L. C., Joensuu, H., and Haahtela, T. (2011). Microbial deprivation, inflammation and cancer. *Cancer Metastasis Rev* 30, 211–23.

Vuong, H. E. and Hsiao, E. Y. (2016). Emerging roles for the gut microbiome in autism spectrum disorder. *Biol Psychiatry*, 411–23.

Walk, S. T., Blum, A. M., Ewing, S. A., et al. (2010). Alteration of the murine gut microbiota during infection with the parasitic helminth *Heligmosomoides polygyrus*. *Inflamm Bowel Dis* 16, 1841–9.

Wang, F. H., Qiao, M., Su, J. Q., et al. (2014). High throughput profiling of antibiotic resistance genes in urban park soils with reclaimed water irrigation. *Environ Sci Technol* 48, 9079–85.

Wang, J., Cao, H., Wang, H., et al. (2015). Multiple mechanisms involved in diabetes protection by lipopolysaccharide in non-obese diabetic mice. *Toxicol Appl Pharmacol* 285, 149–58.

Wellems, T. E., Hayton, K., and Fairhurst, R. M. (2009). The impact of malaria parasitism: from corpuscles to communities. *J Clin Invest* 119, 2496–505.

Wertheim, J. O. and Kosakovsky Pond, S. L. (2011). Purifying selection can obscure the ancient age of viral lineages. *Mol Biol Evol* 28, 3355–65.

Wheeler, B. W., White, M., Stahl-Timmins, W., et al. (2012). Does living by the coast improve health and wellbeing? *Health Place* 18, 1198–201.

Wikoff, W. R., Anfora, A. T., Liu, J., et al. (2009). Metabolomics analysis reveals large effects of gut microflora on mammalian blood metabolites. *Proc Natl Acad Sci U S A* 106, 3698–703.

Williams, A. C. and Dunbar, R. I. (2014). Big brains, meat, tuberculosis and the nicotinamide switches: co-evolutionary relationships with modern repercussions on longevity and disease? *Med Hypotheses* 83, 79–87.

Wolfe, N. D., Dunavan, C. P., and Diamond, J. (2007). Origins of major human infectious diseases. *Nature* 447, 279–83.

Wu, G. D., Chen, J., Hoffmann, C., et al. (2011). Linking long-term dietary patterns with gut microbial enterotypes. *Science* 334, 105–8.

Yee, A. L. and Gilbert, J. A. (2016). Is triclosan harming your microbiome? *Science* 353, 348–9.

Yoo, J., Tcheurekdjian, H., Lynch, S. V., et al. (2007). Microbial manipulation of immune function for asthma prevention: inferences from clinical trials. *Proc Am Thorac Soc* 4, 277–82.

Yuan, C., Gaskins, A. J., Blaine, A. I., et al. (2016). Association between cesarean birth and risk of obesity in offspring in childhood, adolescence, and early adulthood. *JAMA Pediatr* 170(11), e162385.

Zaccone, P., Burton, O., Miller, N., et al. (2009). *Schistosoma mansoni* egg antigens induce Treg that participate in diabetes prevention in NOD mice. *Eur J Immunol* 39, 1098–107.

Zaiss, M. M. and Harris, N. L. (2016). Interactions between the intestinal microbiome and helminth parasites. *Parasite Immunol* 38, 5–11.

Zeevi, D., Korem, T., Zmora, N., et al. (2015). Personalized nutrition by prediction of glycemic responses. *Cell* 163, 1079–94.

Zelante, T., Iannitti, R. G., Fallarino, F., et al. (2014). Tryptophan feeding of the IDO1-AhR axis in host–microbial symbiosis. *Front Immunol* 5, 640.

Zerbo, O., Leong, A., Barcellos, L., et al. (2015). Immune mediated conditions in autism spectrum disorders. *Brain Behav Immun* 46, 232–6.

Zheng, P., Zeng, B., Zhou, C., et al. (2016). Gut microbiome remodeling induces depressive-like behaviors through a pathway mediated by the host's metabolism. *Mol Psychiatry* 21, 786–96.

Zhou, Q., Wang, H., Schwartz, D. M., et al. (2016). Loss-of-function mutations in TNFAIP3 leading to A20 haploinsufficiency cause an early-onset autoinflammatory disease. *Nat Genet* 48, 67–73.

Zijlmans, M. A., Korpela, K., Riksen-Walraven, J. M., et al. (2015). Maternal prenatal stress is associated with the infant intestinal microbiota. *Psychoneuroendocrinol* 53, 233–45.

Zivkovic, A. M., German, J. B., Lebrilla, C. B., et al. (2011). Human milk glycobiome and its impact on the infant gastrointestinal microbiota. *Proc Natl Acad Sci U S A* 108(Suppl 1), 4653–8.

CHAPTER 11

CARDIOVASCULAR SYSTEM

KEVIN S. SHAH, KALYANAM SHIVKUMAR, MEHDI NOJOUMI, AND BARBARA NATTERSON-HOROWITZ

Abstract

Cardiovascular (CV) disease is the leading killer of our species. Various evolutionary lenses can be applied to better understand human vulnerability to CV disorders. The evolutionary origins of a healthy human heart—its myocardial, electrophysiologic, valvular and vascular systems—offers a history of the selective pressures, trade-offs and adaptations leading to the normal mammalian CV systems. Beyond these evolutionary-developmental perspectives, the application of a framework based on Tinbergen's four questions offers a novel evolutionary lens for understanding our species' vulnerability to CV pathology. This is done by a consideration of comparative information about non-human animals who spontaneously develop the same CV diseases. This phylogenetic information can then be used to develop trade-off-based adaptive hypotheses to explain the nature and origins of vulnerability to a range of CV pathologies including atherosclerosis, heart failure, valvular heart disease and arrhythmias.

Keywords

cardiovascular, heart, Tinbergen, atherosclerosis, heart failure, aortic stenosis, atrial fibrillation, multicellularity, triploblasty, evolutionary medicine

11.1 Introduction

Structural and physiological origins of the human (mammalian) cardiovascular (CV) system have been characterised at structural, functional, cellular, and molecular levels. Comparative studies of extant, and in some instances extinct, organisms have offered detailed descriptions and depictions of the development of vascular, ventricular, valvular, and electrophysiological systems. Such descriptions typically illustrate changes in anatomy, morphology, and function in sentinel species in which significant modification from previous CV systems has been noted. Less well explored has been an analysis of how these modifications may represent adaptations to changing ecological and environmental conditions.

Like all biological phenomena, the anatomy and physiology of the CV system have emerged through evolutionary processes. The origin of the human heart—in both health and disease—involved a multitude of adaptations, chance occurrences, and 'compromises' occurring over countless generations of pumping hearts. Because contemporary cardiac function and dysfunction have emerged through these evolutionary processes, to be accurate and comprehensively understood, CV physiology and pathophysiology must be considered through this evolutionary perspective.

This requires looking beyond structural analyses and mechanistic conceptualisations to explanations that consider selective pressures, ecological conditions, and adaptive advantages underlying anatomy, physiology, and vulnerability to disease. Such an approach offers a holistic framework through which the evolutionary past can be used to offer insights into modern cardiovascular diseases. This an approach pursues the most foundational and important question: how and why vulnerability to heart disease emerged and persisted.

Moving from mechanistic to more evolutionarily informed perspectives is an approach advanced by Dutch biologist and Nobel laureate Nikolaas Tinbergen in the early part of the twentieth century. Tinbergen recognised that most theories of causation centre on mechanistic or proximate explanations. He conceived of a four-part explanatory system to provide a more holistic conceptualisation of biological processes which included both proximate (mechanistic and developmental/ontological) and evolutionarily informed perspectives (phylogenetic and adaptive). (For further discussion, see Chapter 1: Core Principles for Evolutionary Medicine.)

Tinbergen was formally trained as an ethologist and conceptualised his 'Four Why's' as a method through which animal behaviour could be studied. However, the application of this approach to any biological phenomena, including human medical disorders and diseases, expands and strengthens insights into disease causation. Tinbergen's perspectives encourage a consideration of 'how' and 'why' the vulnerability to disease might have offered adaptive benefits for individuals living in specific ecosystems under unique selective pressures (Alcock 2013).

This chapter offers an evolutionary account of both the origins of human CV systems and its pathologies. In Section 11.2, the evolutionary origins of CV structures and function of the heart from bilaterians to modern mammals is offered and considered specifically through the lens of selective pressures and adaptation. The evolutionary processes will be considered specifically through the lens of selective pressure and adaptation. The emergence of the normal CV system across several major invertebrate and vertebrate taxa will be reviewed. This approach will look at and then beyond the anatomic/physiologic changes across species to the ecological forces and associated adaptations involved in shaping vertebral CV systems.

Section 11.3 will apply a Tinbergen-inspired lens to several forms of high-impact human CV disease including how they manifest across animal species. Moreover, Tinbergen-inspired approach will emphasise how vulnerability to these CV disorders emerged through natural selection and other evolutionary processes.

11.2 Evolutionary Origins of Cardiovascular Systems

While life on Earth in the form of RNA-based undersea organisms first appeared 3.8 billion years ago (bya), the emergence of the eukaryotic cells 2 bya represented the first crucial event facilitating development of the invertebrate and vertebrate CV systems. Symbiosis theory suggests that digestion of prokaryotes without degradation of the digested cell resulted in the appearance of sophisticated organelles including chloroplasts and mitochondria (Roger et al. 1998; McFadden 2001). Cooperation between organelles improved the efficiency and functionality of the cells, offering fitness benefit to these primitive eukaryotes (Bhattacharya et al. 2004). In addition, the capacity for the creation and storage of high-energy molecules offered early cells a significant advantage within their ecological niches. Leading up to the emergence of eukaryotes, prokaryotic cells migrated towards the surface where they had exposure to sunlight. Utilising sunlight via photosynthesis, which probably emerged around 3.5 bya (Schopf et al. 2007), offered essentially an unlimited energy resource. Moreover, exposure to oxygen-rich environments facilitated the emergence of species capable of oxygenic photosynthesis and oxidative phosphorylation.

In addition to eukaryotic development, the emergence of multicellularity (1.2 bya) represented a second crucial development, providing the framework for the subsequent evolution of CV systems (Butterfield 2000). Multicellularity itself emerged as a compromise between competing selective advantages, including the benefits of increased size, and the functional specialisation of cells (Gray 2012). Multicellularity facilitated the evolution of cooperative and altruistic behaviours, division of labour, and the emergence of varying and sometimes competing levels of selection (Nedelcu 2012). Cells began to show cooperative behaviour, with survival of the organism superseding the fitness of individual cells. This shift towards cooperation between cells provided important scaffolding for CV systems which subsequently emerged.

Complex multicellular organisms had increasingly significant energy requirements necessitating the evolution of resource distribution systems. Rudimentary distribution systems emerged during this period to provide resources for these progressively complex and enlarging organisms. Multicellularity, therefore, represented an important driving force behind the emergence of early CV systems because the needs imposed by increasingly multicellular life forms exerted selective pressure, promoting their emergence.

Another important stage in the early evolution of CV systems was the emergence of triploblasty (three differentiated layers of cells). Triploblasty first evolved in the phylum Cnidaria about 500 million years ago (mya) (Boero et al. 2004; Martindale et al. 2004). With the transition from diploblasty to triploblasty and the emergence of a third layer of differentiated cells, namely mesoderm, cells no longer needed to be close to food sources to utilise energy and secreted waste material. This provided an opportunity for development of internal organs and distribution of responsibilities among different organ systems. For instance, differentiation of mesodermal cells led to development of skeletal and circulatory systems which provided locomotion and circulation corresponding to the ecological needs of organisms. Specifically, the emergence of embryonic mesodermal germ-layer cells was crucial for the development of the myocardial cells (Seilacher et al. 1998). Additionally, differentiation of mesothelial germ-layer cells during the gastrulation phase gave rise to epicardial cells. The evolution of a third layer of cells allowed for further specialisation and improved functionality (Burton 2008). Persistence of triploblasty throughout evolutionary history offered important adaptive benefits, including facilitating differentiated tissue within vascular and ventricular structures. Together, the evolution of eukaryotic cells, multicellularity, and triploblasty represented crucial occurrences driving the emergence of vertebral CV systems.

Multicellularity and triploblasty created the need to transport nutrients and waste material from cells that could not simply exchange chemicals due to distance constraints. An array of different mechanisms to transport nutrients and excrete waste evolved in early multicellular organisms. For instance, flat worms (phylum Platyhelminthes) directly diffuse nutrients and waste within the environment. This is possible since they occupy minimal cross-sectional areas that help the diffusion (Halton 1997; Wright and Ahearn 1997). On the other hand, insects have external openings (spiracles) that bring oxygen to the proximity of cells (Lighton 1996). However, the increase in the volume to surface area ratio during evolutionary history made the vascular system the dominant force in satisfying the transportation needs of organisms. The ultimate result of this evolution was the formation of haemangioblasts that differentiate into angioblasts, which migrate and form lumens for vessels around the body in the process of vasculogenesis in vertebrates (Carmeliet 2000).

11.2.1 Evolution of the Conduction System

The conduction system of the heart is a system of fibres that facilitates initiation of impulses and their rapid conduction through myocardium to coordinate synchronised activity. The mammalian system consists of pacemaker and conduction cells—specifically the sinoatrial (SA) node, atrioventricular (AV) node, His bundle and Purkinje fibres. The SA and AV node originated from slow-conducting myocardial tissue, while the remainder of the conducting system developed from fast-conducting ventricular segments. Warm-blooded animals have a formed cardiac conduction system, while cold-blooded animals do not possess an analogously-formed system (Jensen et al. 2012). However, studies have shown that some ectotherms (lizards, frogs, and zebrafish) have a slow-conducting AV canal muscle as well as base-to-apex electrical activation of ventricles.

Across phylogeny, rudimentary endothermic conduction systems are seen in invertebrates, including possible pacemaker activity in *Drosophila melanogaster* (Rizki 1978). Furthermore, chordates such as tunicates have a wandering dominant locus of automaticity. A more well-defined conduction system is seen in fish, including pathways of preferential spread of excitation traversing the atrium towards the AV junction in eel (Solc 2007). Among warm-blooded animals, birds have similar conduction systems as humans; however, the AV node's function may be instituted by an AV ring. It is clear that the most well-defined or 'in-tact' conduction system has been seen in warm-blooded mammals while many ectothermic species have individual components that likely contribute to synchronising cardiac activity.

Notably, concurrent with the evolution of vertebral hearts, other types of CV and energy distribution systems emerged in non-vertebral animals. The features of these CV systems represented organismal adaptations to ecological niches in which these animals existed. Ecological niches are important drivers of adaptations that resulted in genetic and phenotypic changes. Furthermore, evolution does not provide the *best* solution to a problem created by the environment. Rather, evolution works with what is already available; new features are merely 'adjustments' to the old characteristics. This notion exists at both a morphological and molecular level. For instance, most new genes are often modifications of duplicated old genes (Taylor and Raes 2004). Modification in CV systems across the animal kingdom—not exclusively in humans and other mammals—therefore must be framed within the context of changing ecosystems.

11.2.2 Emergence of the First Hearts

Phylogenetic evidence suggests that the first beating, resource-distributing organ appeared in ancient bilaterians (a major group of animals mostly with bilateral symmetry) around 800–500 mya. These primitive hearts were presumably tubular organs with no defined chamber or valves. Electrical impulses stimulated each heart beat which propelled blood. Later on, during the course of the evolution, auxiliary hearts began appearing in annelids (a group of segmented worms). There are five pairs of aortic arches, which have the responsibility of pumping blood into the dorsal and ventral blood vessels. The dorsal and ventral blood vessels are responsible for carrying blood to the front and back of the earthworm's body similar to arteries in a general circulatory system setting.

The emergence of auxiliary hearts in these bilaterian creatures facilitated the propulsion of fluids throughout capillaries (Edwards and Bohlen 1996). Although annelids represent an example of a closed circulatory system (they have capillaries), the open circulatory system (no capillaries) was also emerging around the same evolutionary time in insects and molluscs (407–396 mya). In open systems, blood pours through ostia (pores) from thickened dorsal vessels. The blood then directly bathes the organs with nutrients transferred by diffusion to cells. Pumping in an open circulatory system is mediated through both regulated pulsation of the heart and movement of the body of the organism (Reiber and McGaw 2009). Open systems require more blood, which is energetically costly, placing significant metabolic constraints on the organisms. Efficiency is another compromise in an open circulatory system. Lack of segregation of nutrient-rich and -poor blood significantly decreases the efficiency of transferring resources to their destination cells. Conversely, in closed systems, blood is conserved, targeting organs directly, improving the efficiency and decreasing the metabolic requirement for the organisms (Bourne et al. 1990). It is important to note that relative efficiency of a closed versus open system is not indicative of 'superiority' of one system over the other (McMahon and Burnett 1990). Rather, both systems represent optimising adaptation to the ecosystems in which these animals live.

11.2.3 Fish and Two-Chambered Hearts

The cardiomyocyte-based ventricular heart and evolution of the cardiac valves emerged with early fish 530 mya. Fish have a single atrioventricular system that guides the blood in one direction. Single circulation allows blood to push through two sets of capillaries, gill and systemic capillaries (Greco and Tota 1981). The two-chambered heart emerged to facilitate expanded energetic and metabolic requirements. In addition to better efficiency, counter exchange, which refers to the opposite directionality of blood and water movement in gills, increases efficiency in oxygen intake (Romenskii 1978). Heart development in fish is derived by a fundamental network of transcription factors that regulate the development of cardiac structures. These factors including NK2, MEF2, GATA, Tbx, and Hand dynamically interact and control each other's expression (Olson 2006). MEF2 is the most ancient and evolutionarily conserved gene among the factors that play a central role in the cardiac muscle regulatory network, a basic requirement for the evolution of hearts (Olson 2006). Duplication and subsequent modification of genes to adapt to new roles within the regulatory network created the initial expansion of cardiac transcription factors. These genes facilitated embryonic differentiation and development of the heart, possibly starting with the two-chambered heart of fish.

The fish CV system represented a major change from an auxiliary heart and reflected adaptive change driven in part by dangers and opportunities in novel environmental niches. While fish hearts were the first myocyte-based heart to emerge, they are low-pressure systems with limited speeds of circulation (Greco and Tota 1981). Single AV valves also evolved in fish. These valvular structures prevented the flow of blood from the ventricle to the atrium during systole and provided regulatory functions which underlie the two phases of the cardiac cycle in a two-chambered heart (Santer 1985).

11.2.4 Amphibian and Reptilian Hearts

CV systems with double circulation emerged with amphibians and reptiles around 500 mya. Three-chambered heart systems ensured that oxygenated blood returns to the heart from the lungs through pulmonary circulation into the left atrium, while deoxygenated blood returning from the body passes into the right atrium. The separation of pulmonary and peripheral blood circulation significantly increased the efficiency and blood flow pressure compared to fish, which is an ideal adaptation for more active organisms (Johansen 1959). In modern reptiles, although oxygenated and deoxygenated blood is separate throughout the atria, both pour into the same ventricle, resulting in mixing of the oxygenated blood returning from the lungs with the deoxygenated blood from tissues. While this mixed venous and arterial blood is adaptive for cold-blooded amphibians and reptiles, separation of oxygenated and deoxygenated blood emerged in crocodilian species, presumably providing adaptive metabolic and thermoregulatory effects. Partial separation of ventricles increased the efficiency of the crocodilian heart (Axelsson et al. 1989), and this represents a sentinel species in the evolution of ventricular septation. In this sense, the crocodilian heart is the ancestor of the avian and mammalian four-chambered heart. However, the true four-chambered heart did not emerge until the evolution of birds and mammals.

11.2.5 Avian and Mammalian Hearts

Increasing ventricular septation characterises the transition of the crocodilian three-chambered to avian and mammalian four-chambered hearts. Tbx5 is a transcription factor gene that plays a role in the development of ventricular structure and function. Variable expression of Tbx underlies septation. The expression of Tbx5 in mammals is restricted to left ventricles, while reptilians homogeneously express the gene in a single ventricular chamber and it is gradually restricted to the left chamber, creating a left–right gradient (Koshiba-Takeuchi et al. 2009). This is a strong indication of the refinement of the expression of Tbx5 during evolution (Koshiba-Takeuchi et al. 2009). Septation facilitated the emergence of four-chambered hearts with double circulatory systems closely associated with the emergence of endothermy. Total separation of oxygenated and deoxygenated blood provided enhanced tissue oxygen delivery as compared to non-septated systems (Olson 2006). Oxygen-saturated blood also facilitates muscle movement, and larger oxygen supply allows for thermoregulation and more active behaviour (Challice and Virágh 1974). Figure 11.1 illustrates the evolutionary development of CV systems.

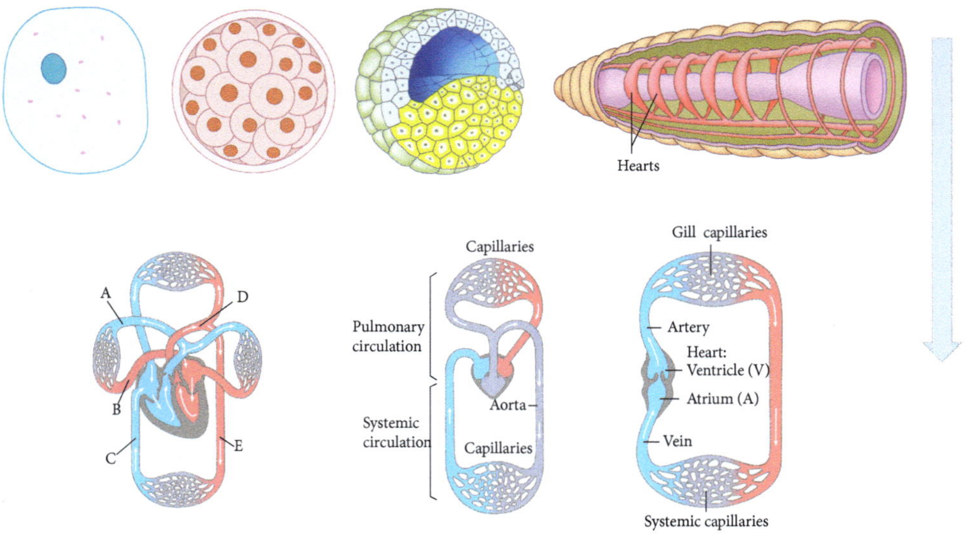

FIGURE 11.1 Evolutionary development of cardiovascular systems, from unicellular organism, multicellular organism, and primitive heart to two, three, and four-chambered heart with extensive vasculature. Although the timeline shows progression of complexity in the CV system, the trend is not completely linear with respect to time. In addition, organisms continuously evolve in their ecological niche according to evolutionary forces such as natural selection. Best fit for the CV system in a particular niche is determined by biological and ecological forces, as well as residual genes of the CV system from closely related organisms. Deoxygenated blood flow from right ventricle to pulmonary circulation (A); oxygenated blood in pulmonary veins returning to left atrium (B); deoxygenated venous blood in inferior vena cava returning to right atrium (C); oxygenated blood in aorta traveling to CNS circulation (D); oxygenated blood in descending aorta traveling to systemic circulation (E).

11.3 Cardiovascular Pathophysiology

Cardiovascular disease (CVD) is the worldwide leading cause of death (WHO 2015; Mozaffarian et al. 2016), claiming 17.5 million lives worldwide in 2012. CV systems are composed of many elements including conduction, valvular, ventricular, and vascular systems, and a vast range of pathological conditions—infectious, congenital, traumatic, and neoplastic—can and do affect each of these systems. Physicians are often asked 'why' heart disease occurs. Most often, the offered responses are mechanistically based and proximate, emphasising acute or hyperacute theories of causation. For example, atherosclerotic ischaemic events are 'caused' by plaque rupture or thrombus formation. Similarly, proximate explanations for arrhythmias point to high catecholamine states or damaged myocardial tissues, and valvular heart disease is mechanistically explained through infection, trauma, or congenital effects.

Each of these proximate explanations is partially accurate and an important piece of a larger puzzle. However, when presented in isolation without the contextualisation provided by evolution, they fail to engage the most foundational of questions: Why does vulnerability to these various forms of heart disease exist at all? Understanding why CV systems are so vulnerable to heart disease is a complex question, requiring explanations which look beyond proximate (molecular/genetic and developmental) explanations to those that consider evolutionary (phylogenetic and adaptive) perspectives as well.

An evolutionary approach to CV disease widens the aperture through which heart disease can be understood. In this chapter, we propose an expanded explanation based on Tinbergen's Four Questions (Tinbergen 1963). For each of the high-impact human CV diseases, we will offer four theories of causation corresponding to Tinbergen's Four Questions. First, common mechanistic explanations will be offered. Developmental (ontogenetic) explanations will follow. A phylogenetic perspective will expose the occurrence or non-occurrence of the disorder in non-human animals. Finally, an adaptive explanation will be offered which will consider why and how vulnerability to the disorder might have offered adaptive benefits in the animal ancestors in whom human hearts evolved (Tinbergen 1963; Bateson and Laland 2013).

Having heart disease has neither adaptive benefit nor survival advantage. However, some aspects of the pathogenesis of CV disease may have provided adaptive benefits to the animal ancestors in whom human hearts evolved. These potential adaptive characteristics are not typically captured in the traditional proximate explanations of causation for heart disease but help expand how we understand why the human CV system is so vulnerable to disease.

Since all aspects of our anatomy and physiology have emerged through selective processes, a consideration of why and how we develop diseases requires this evolutionary perspective. The complexity of evolution and disease makes a comprehensive analysis of the role of adaptation unlikely. However, attempting to understand how vulnerability to heart disease emerged offers an important and underexplored pathway towards a broader understanding of the origins of heart disease.

11.3.1 Atherosclerosis

CVD specifically due to atherosclerosis includes ischaemic heart disease as well as acute coronary syndromes (ACS), cerebrovascular disease (e.g. stroke), and peripheral vascular disease. Ischaemic heart disease is a major contributor to global morbidity and mortality. In 2008, out of the 17+ million CV deaths, heart attacks were responsible for more than 7 million deaths (Mendis et al. 2011). Furthermore, 10% of the global disease burden (disability-adjusted life years) can be attributed to CVD (Mendis et al. 2011). The Framingham Heart Study, a landmark cohort-based study that began in the year 1948, has helped identify many of these as causal contributors to CV disease (Mahmood et al. 2014). Figure 11.2 illustrates Tinbergen's Four Questions framework for atherosclerosis causation.

Many lifestyle-related behaviours and comorbidities have been established as risk factors for development of coronary artery disease (CAD). Advanced age, dyslipidaemia, hypertension, diabetes mellitus, and smoking have all been identified as contributing causes of atherosclerosis.

The genesis of atherosclerosis and the mechanisms through which plaque comes to crack and rupture are complex and multifactorial, and have been exhaustively studied and characterised. The molecular biology, genetics, and physiology contributing to atherosclerosis

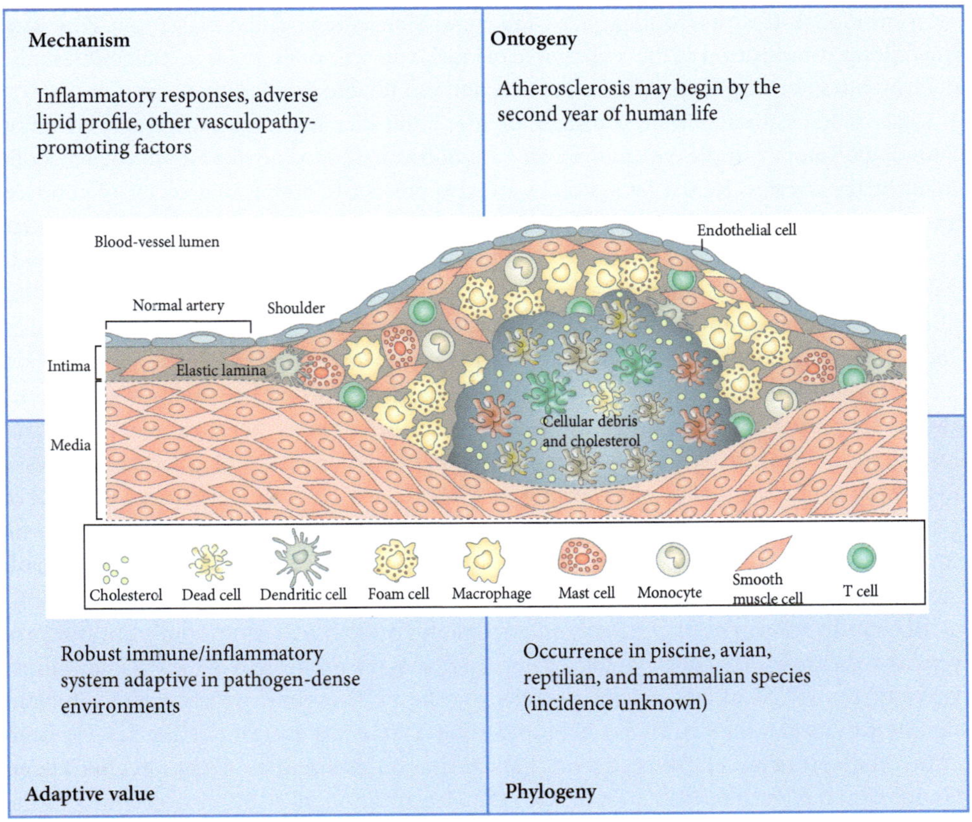

FIGURE 11.2 Example of proximate and evolutionary causation of atherosclerosis.

have been extensively studied and characterised at cellular, organismic, and population-based levels. The first steps in the development of atherosclerosis involve the accumulation of small lipoprotein particles within the intimal layer likely due to injury and attenuated activation of the endothelial cell (Libby et al. 2002; Carmena et al. 2004). An atherogenic diet high in cholesterol and saturated fat is associated with injury and lipoprotein particle accumulation (Kuipers et al. 2011). The next early step in atherogenesis involves the recruitment of leukocytes within the intima to become foam cells (Ley et al. 2011). The mechanistic explanation for variable spatial location of atherosclerosis is complex; the systemic damage from toxins would presumably result in pan-vascular disease. The localising patterns of atherosclerosis may be related to shear stress and laminar flow differences in coronary arteries as compared to other arteries (Stone et al. 2012; Koskinas et al. 2013). Over time, smooth muscle proliferation and matrix deposition with the development of lesion progression is complex and involves a balance between replication and cell death (Geng and Libby 2002).

The role of coronary thrombosis as a cause of myocardial infarction and sudden cardiac death has been established by autopsy and angiographic studies (DeWood et al. 1980; DeWood et al. 1983; Sherman et al. 1986). The mechanisms for plaque rupture and myocardial infarction are related to either a fracture of a plaque fibrous cap or erosion of the intima itself (Jia et al. 2013). Vascular thrombogenicity of the vascular bed is induced by triggers including but not limited to illicit drug use, extreme physical activity, acute infection, and even emotional stressors (Muller 1989; Muller and Mangel 1994; Čulić et al. 2005). However, the cellular composition of the 'vulnerable plaque' is an area of ongoing investigation in the determination of triggers for plaque rupture and modifiable mechanisms.

Atherosclerosis is classically a disease of later adult life. There are exceptions, however. Pathologic autopsy studies of children have demonstrated fatty streaks and fibrous plaques in coronary arteries. Raised fatty streaks are also present in abdominal aortas of children and young adults who undergo autopsies (McGill et al. 2000). These autopsy studies demonstrate an increased prevalence of atherosclerosis (in asymptomatic young individuals) in association with many traditional risk factors, including a high body mass index, blood pressure elevation, and high serum concentrations of total cholesterol, triglycerides, and low-density lipoprotein (LDL) cholesterol (Berenson et al. 1998).

The genetic predisposition to atherosclerosis is an important consideration, as heritability accounts for up to half of the susceptibility to coronary heart disease (Ding and Kullo 2009). The development of technology to perform genome-wide association studies to identify single-nucleotide polymorphisms (SNPs) has significantly advanced the field of genetic studies. The strength of risk associated with any one identified SNP tends to be small and it is difficult to provide additional risk stratification outside of traditional risk factors (Damani and Topol 2011). The greatest promise in genetic studies has been the ability to identify outlier populations and study them for mechanistic innovations towards preventative therapies. The most notable recent example is the proprotein convertase subtilisin/kexin type 9 (PCSK9), an enzyme encoded by the PCSK9 gene, which, when inhibited significantly, decreases serum LDL in human subjects (Gencer et al. 2015). The development of this therapy stemmed from genetic characterisation of a mutation in individuals from France who had very few CV events, with the identification of PCSK9 involvement in the LDL pathway. This approach to the genetic investigation of CV disease continues to evolve as our ability to determine mechanisms of atherogenesis improve.

Much less well studied than the aforementioned mechanistic and developmental explanations relating to atherosclerosis and associated syndromes are the evolutionarily framed questions of phylogeny and adaptation. First, what non-human animals spontaneously develop atherosclerosis and associated syndromes? Second, how might some of the multitude of biological factors ultimately contributing to CV disease have offered *adaptive* benefits in the past and biological benefits in the present. In other words, how might vulnerability of atherosclerosis have provided adaptive benefit to our ancient animal ancestors in the ecosystems in which they evolved?

Atherosclerosis has been identified as spontaneously occurring in animals including walruses, pandas, emus, and many avian species (Tomimura et al. 1970; Gao 1995; Gruber et al. 2002; Beaufrère 2013). Indeed, avian species both wild and captive seem particularly vulnerable to atherosclerosis, with reports of myocardial infarction, stroke, ischaemic cardiomyopathy, and atherosclerotic aortic aneurysms and rupture (Beaufrère 2013). Notably, there is no reliable way to estimate the incidence of prevalence of atherosclerosis in non-human animals given the extremely few numbers of wild animals that ever come into contact with people, let alone have necropsies after death. However, the occurrence of even a single case in a species points to underlying vulnerability in this taxa whether or not the disease emerges (Prescott et al. 1995; Schmaier et al. 2011).

A benefit of looking at atherosclerosis from a phylogenetic perspective is to develop knowledge about the vulnerability (high or low) of various animal groups. Once identified, high- and low-vulnerability groups can serve as natural animal models of disease resilience or susceptibility. Moreover, awareness of the occurrence of the disorder, even in one animal, expands conventional understanding of the contribution of modern environments, practices, and habits. In addition, it casts light on the foundational question of vulnerability.

Vulnerability to atherosclerosis over a wide range of non-human species suggests that some aspects of the biological processes that underlie atherogenesis may participate in foundational and conserved health-promoting functions. Atherogenesis occurs over many decades in humans and some other affected animals, with the involvement of several physiological systems. Two systems that play central roles in the development of atherosclerosis and acute ischaemic events are the immune system and associated inflammatory processes. Clinical atherosclerosis with its significant mortality and morbidity represents a maladaptive 'by-product' (trade-off) associated with a robust immune and inflammatory system.

With respect to inflammation, it is well established that immune cells are readily found in early atherosclerotic lesions. Activation of the inflammatory cascade can trigger ACV (Ross 1999). The fatty streak, a precursor to atherosclerotic lesions, is partly made of macrophages induced by the inflammatory cascade. Multiple contributors to the inflammatory cascade including interleukin-6, C-reactive protein, and tumour necrosis factor are related to the development CV events (Ross 1999). The chronic inflammatory reaction is elevated in response to vascular injuries, which are likely exacerbated by hypertension, smoking, and diabetes (Braunwald 2012). ACS-related biology involves the release of tissue factor, thrombin production, and clot formation. Circulating macrophages are known to be involved in the amplification of the coagulation cascade and the thrombotic process during ACS (Santos-Gallego et al. 2014).

Multiple possible infectious triggers have also been explored, including but not limited to *Chlamydia pneumoniae*. *Chlamydia pneumoniae* is a bacterial pathogen typically affecting the respiratory tract and passes to the endovascular wall via monocytes (Watson and

Alp 2008). In animal models, mice with hypercholesterolaemia had worsened atherosclerosis (lesion area and severity) when infected with *C. pneumoniae* (Hu et al. 1999). Similarly, in humans, *C. pneumoniae* infection is shown to precede both the earliest and more advanced lesions of coronary atherosclerosis (Davidson et al. 1998). The organism may be an innocent bystander or commensal to atherosclerotic tissue, rather than an inciter of chronic inflammation. Given this relationship, multiple studies were attempted to see if CAD and atherosclerotic events could be prevented with antibiotic therapy, which was not shown to be beneficial (Wells et al. 2004). Therefore, it is difficult to determine if *C. pneumoniae* is an early trigger of inflammation in the development of CAD or a witness to the process.

Our animal ancestors relied on inflammation as an adaptive process in dealing with pathogenic and traumatic events. The ability to rapidly recruit inflammatory cells plays an important role in survival for animals living in a pathogen-rich, dangerous, and violent world.

11.3.2 Heart Failure

More than 23 million people worldwide are affected by congestive heart failure (HF), which carries a 50% 5-year and 10% 10-year survival estimate (Roger 2013). The prevalence of HF ranges from 1 to 12%, varying between different populations. It is a syndrome of cardiac dysfunction leading to the body's inability to provide sufficient blood and oxygen to meet its needs. Many symptoms are associated with low cardiac output including fatigue, confusion, and renal dysfunction. Additionally, congestion from elevated cardiac filling pressures leads to an overall state of hypervolaemia including pulmonary oedema, ascites, and peripheral oedema. There are many forms of HF and many causes associated with each of these, including CAD, toxins, viral infections, hypertension, genetic explanations, and even stress.

11.3.2.1 *Takotsubo Cardiomyopathy (Stress Cardiomyopathy)*

A specific form of acute HF in humans offers an example of how Tinbergen's framework can deepen and strengthen our understanding of why it occurs. Stress-induced cardiomyopathy (Takotsubo cardiomyopathy) is a recently recognised phenomenon in humans, first identified in 1990 (Akashi et al. 2015). Approximately 2% of patients (and up to 10% of women) presenting to hospitals with suspected ACS are ultimately diagnosed with Takotsubo syndrome (Akashi et al. 2015). Typically, it affects postmenopausal women and is often associated with a physiologic or emotional stressor (Lyon et al. 2008). Risk factors for the condition include anxiety states, dyslipidaemia, and tobacco and alcohol abuse (Akashi et al. 2015). Takotsubo cardiomyopathy was named by the Japanese cardiologists who discovered it for the similarity in the shape of an affected left ventricle and the pots used by Japanese fishermen to trap octopuses (Lyon et al. 2008). During cardiac systole, the bulging ventricle tends to resemble this same pot.

Patients often present similarly to those experiencing ACV, with abnormal electrocardiograms and elevated cardiac biomarkers indicative of heart strain and myocardial necrosis, respectively. However, angiography always reveals no obstructive coronary lesions as the cause of the syndrome. Typically, the apex of the heart is particularly affected with dyskinesia, which may be related to the concentration of beta-receptor distribution within the heart. Fortunately, humans tend to have a good prognosis when presenting with this unique syndrome. Stress-induced cardiomyopathy in humans often follows a physiologic or emotional stressor and is thought to be driven by excessive sympathetic stimulation (Cocco and Chu 2007; Lyon et al. 2008).

Notably, some consider this disorder to represent a hyperreactivity of the autonomic nervous system. Perturbation in autonomic function is associated with a variety of cardiac arrhythmias and other clinical syndromes in humans. Indeed, balance versus dominance of either arm of the autonomic nervous system underlies many forms of CVD. The evolution of the emergence and integration of both the parasympathetic and sympathetic systems is best understood as a response to constraints and opportunities organisms have faced. Autonomic physiologies evolved to shape neurologic and behavioural responses to optimise safety and survival for the organism. (For further discussion, see Chapter 17: Brain, Spinal Cord, and Sensory Systems.) The phylogeny of autonomic responses to danger spans from purely vagal responses seen in some animals (Porges (2009), which points to defensive freezing behaviour mediated via vagal dominance via sympathetic withdrawal, to the explosive sympathetic responses with high circulating catecholamines seen in others. Figure 11.3 illustrates Tinbergen's Four Questions framework for stress cardiomyopathy causation.

Mechanistic explanations for Takotsubo cardiomyopathy typically centre around an emotionally or physiologically stressful event and focus on the high catecholamine state as the 'cause' for the disorder. Cases of Takotsubo cardiomyopathy have been associated with the death of a loved one, being left at the altar, and other emotionally charged occurrences. It is increasingly recognised that intensive physiologic stress also places patients at risk.

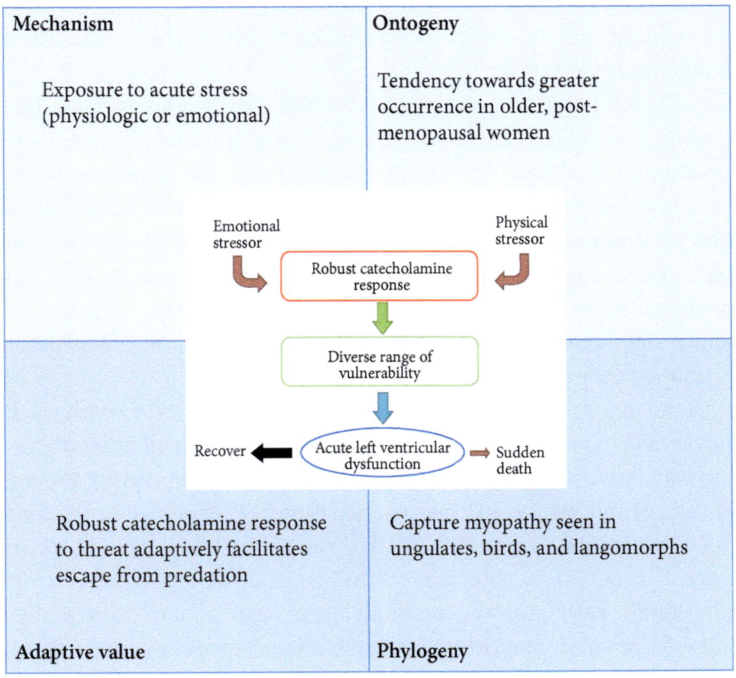

FIGURE 11.3 Example of proximate and evolutionary causation of Takotsubo cardiomyopathy.

Elderly patients are at elevated risk for Takotsubo cardiomyopathy (Lyon et al. 2016) and have the highest risk for complications (i.e. arrhythmias, stroke, shock) and death. The reason why the elderly are much more affected by this condition is unclear.

Looking first phylogenetically, a stress-induced disease entity called capture myopathy is seen in a wide range of species and may represent a natural animal model for Takotsubo cardiomyopathy in humans. The syndrome is seen in association with the pursuit, capture, or handling of an animal, leading to damage of cardiac and skeletal muscle (Beringer et al. 1996; McMahon et al. 2013). It has been demonstrated in many species, including parakeets, dolphins, whales, zebra, buffaloes, and moose (Beringer et al. 1996; Montané et al. 2002; McMahon et al. 2013).

Capture myopathy is thought to be driven by an intense catecholamine-surge as a physiologic attempt to trigger a hyperadrenergic response. Acute left ventricular dysfunction would be highly maladaptive for the CV system of an animal if it resulted in compromised function and escape. However, only a small percentage of animals do succumb to this disorder, suggesting that while surging catecholamines could be problematic for certain individuals within a heterogenetic population, overall the response has protective benefits.

An adaptive explanation for stress cardiomyopathy in humans and animals suggests that while a robust or even explosive catecholamine response to threat may increase vulnerability to the disorder, anti-predation benefits of such a response are extremely strong and most likely extremely old.

HF is a common CV disorder and up-regulation of the neurohormonal system is a fundamental stage in the development and subsequent progression of the illness. Takotsubo cardiomyopathy is a clinical syndrome similarly associated with catecholamines and sympathetic activation. Both disorders are examples of overactivation of the body's response to stressors and are seen across species. An evolutionary perspective of the disorders highlights the commonality of an overactive counter-regulatory system to modern stressors and the possible deleterious effects.

Originally, HF was thought to arise primarily from abnormalities of the heart pumping capacity—truly a 'pump problem'. However, over time, up-regulation of the renin–angiotensin–aldosterone system and various other vasoconstrictive hormones where ultimately recognized as the primary mechanism of myocardial dysfunction. This helped reinforce the concept of the 'neurohormonal hypothesis' as the main factor of the mechanism driving HF (Packer 1992). This change in paradigm was pivotal to the contemporary understanding and discovering of novel treatments for HF (Packer 1992). Up-regulation of hormones including renin, angiotensin, and angiotensin II are what cause an imbalance and myocardial dysfunction over time.

Advances in the management of congestive HF seen in association with the shift towards the 'neurohormonal hypothesis' can be considered a triumph of evolutionary medicine. Endocrine activation, a maladaptive response, has become the primary therapeutic target. This has led to many advances in therapies including beta-blockade, angiotensin-converting enzyme (ACE) inhibition, and aldosterone antagonism to help block the effects of these hormones. In fact, neuroendocrine activation can be highly adaptive and result in protective physiologic responses under certain circumstances. Rapid rise in catecholamines and mineralocorticoids in response to trauma, bleeding, or severe dehydration can be life-saving.

However, what evolved as highly adaptive responses to primary existential threats (dehydration and exsanguination) for our animal ancestors have become maladaptive for modern humans with ventricular damage most often associated with atherosclerosis.

11.3.3 Aortic Stenosis

Aortic valve disease, specifically aortic stenosis (AS), is a common valvular disease in the developed world (Sawaya et al. 2012). It has a very poor prognosis, especially when symptoms in humans begin to develop. Patients typically develop symptoms in their seventh to eight decade of life (Sawaya et al. 2012). These symptoms include, but are not limited to, fatigue, exertional shortness of breath, light-headedness, and angina. Over time, the left ventricle of the heart adapts to a stenotic outflow valve with compensatory left ventricular hypertrophy and dilatation. Figure 11.4 illustrates Tinbergen's Four Questions framework for AS causation.

Calcific AS has been thought to be driven by the 'wear and tear' from the opening and closing of the aortic valve over decades. More recently, basic and translational research has

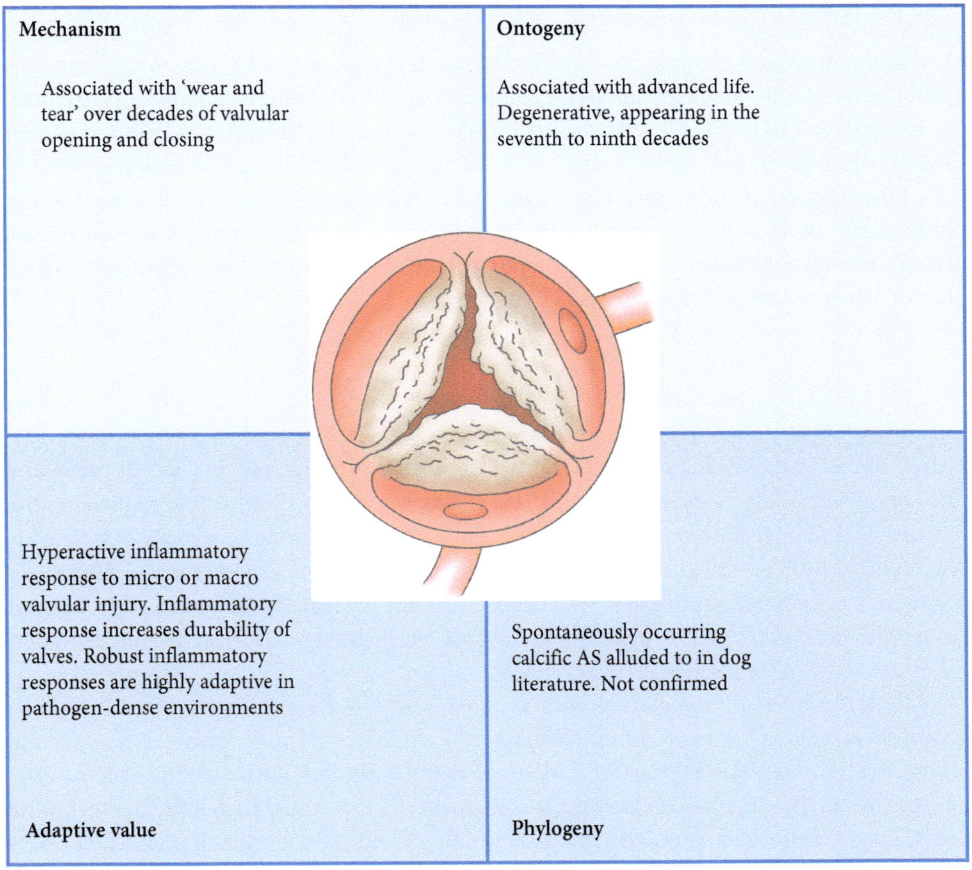

FIGURE 11.4 Example of proximate and evolutionary causation of aortic stenosis.

demonstrated that the development of disease is more complex. Haemodynamic stress affects LDL to promote oxidative effects, leading to inflammation and calcification of the valve leaflets (Yetkin and Waltenberger 2009). The cellular processes that lead to aortic calcification are complex and involve osteoblast up-regulation and procalcific factor secretion (Leopold 2012). As a result of these processes, calcification and fibrosis of the valve itself generates obstruction of blood flow. Additionally, the dysregulation of the metabolic system associated with diabetes mellitus has been shown in mouse models to play a pathogenic role in the development of AS (Le Quang et al. 2014). Thus, our current understanding of the mechanistic cause of AS involves perturbations of metabolic and calcium regulation systems.

Among patients 75 years or older, the prevalence of AS is approximately 4.9 million patients in European countries and 2.7 million in the United States (approximately 12.4% of the US population). Because AS occurs far more frequently with advanced age, the vulnerability to it is related to 'wear and tear'—decades of valvular opening and closing with cumulative degenerative changes. The relative resistance of younger valves to AS argues for some protective physiology which is potentially reduced or lost with advancing age.

Some canine breeds (Golden Retrievers, German Shepherds, Boxers, Rottweilers) are known to have congenital sub-AS (Kienle et al. 1994; Höglund et al. 2007). However, the spontaneous occurrence of calcific degenerative AS in non-human animals has not been confirmed.

Among the many components involved in the development of AS, inflammation plays a prominent role. Notably, inflammatory responses also provide significant benefits to embryonic valves and to necessary remodelling of valves across a lifetime. A dynamic and responsive inflammatory system may increase the durability of developing and ageing valves.

AS is the most common valvular disorder of the elderly and has significant impact on individuals' quality of life and mortality. It is likely that the development of stenosis and calcification is a result of the trade-off between durability during early decades of life and calcification and rigidity in later life.

11.3.4 Atrial Fibrillation

Atrial fibrillation (AF) is an important cause of morbidity and in recent years its prevalence has been increasing worldwide (Chugh et al. 2014). It has received tremendous attention in the fields of cardiology, interventional cardiac electrophysiology, cardiac surgery, and neurology. Besides symptoms, exercise intolerance, and the occurrence of HF, stroke remains a feared complication of AF (Fuster et al. 2006). AF has been described as an emerging 'epidemic' in CVD (Braunwald 1997). Figure 11.5 illustrates Tinbergen's Four Questions framework for AF causation.

AF can be seen in association with multiple comorbid conditions including valvular heart disease, CAD, hypertension, congestive heart failure, hypertrophic cardiomyopathy, lung disease, and diabetes. Certain endocrine states, including untreated hyperthyroidism, as well as undergoing surgery may place some patients at elevated risk. Still the mechanism of AF is not well understood and there are multiple theories regarding its causation (Weiss et al. 2016). Left ventricular dysfunction, for example, is associated with a five-fold increase in the risk of AF (Cha et al. 2004). The prevalence of AF is also related to the extent of the

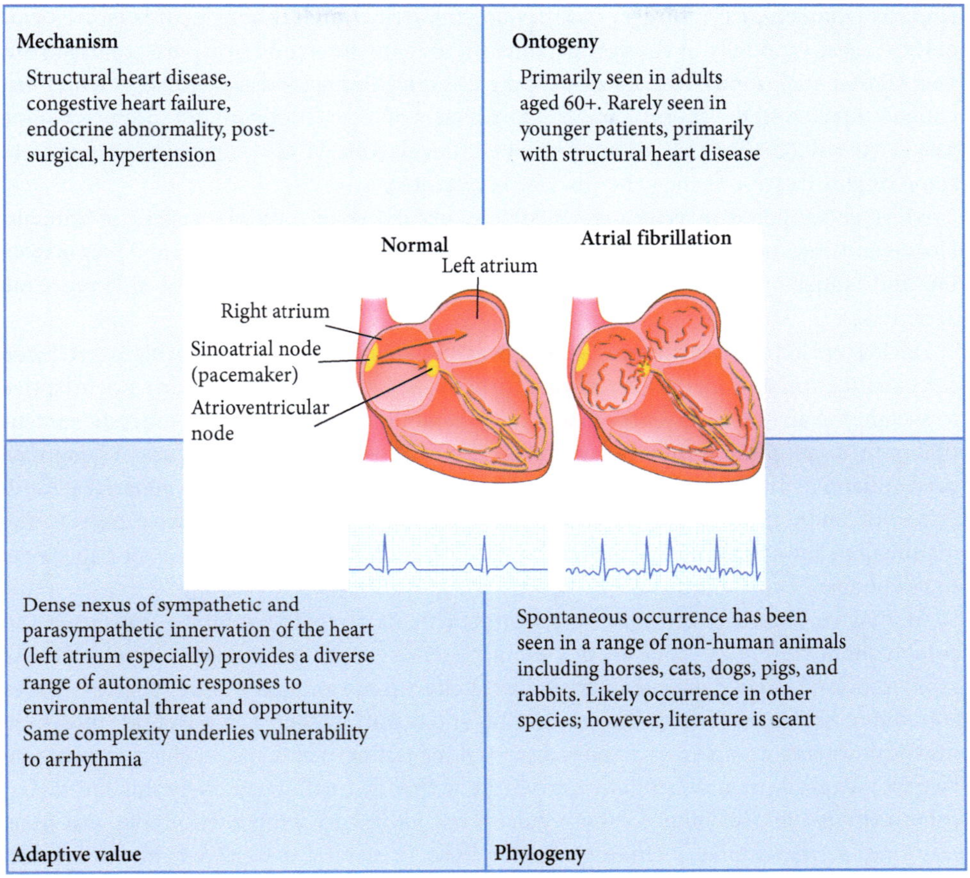

FIGURE 11.5 Example of proximate and evolutionary causation of atrial fibrillation.

left ventricular dysfunction. AF is seen in about10% of New York Heart Association (NYHA) functional class I/II heart failure and in up to 50% of NYHA functional class IV patients in the large congestive heart failure trials. The mechanism underlying this association involves a complex interplay between atrial stretch, neurohormonal activation with AF potentially promoting congestive heart failure via fast irregular ventricular rates, and loss of AV synchrony (Khand et al. 2000; Ehrlich et al. 2002).

In many patients with AF, focal triggers can be identified. Sleeves of atrial myocardium that extend into the pulmonary veins are thought to harbour arrhythmogenic foci in these patients. These patients have been successfully managed using radiofrequency catheter ablation. Other foci that can cause focal AF and are amenable to catheter ablation include the ligament of Marshall and the junction between the superior vena cava and the right atrium. Very little is known about the cellular behaviour of 'triggers' of AF. Why fibrillation persists in some settings and is self-limiting in others is also poorly understood.

AF is the most common cardiac arrhythmia, with a prevalence ranging from 5% in over 60-year-olds to 17% in patients who are 85 years and above (Fuster et al. 2006). It occurs infrequently in infants and children. Younger individuals with structural heart disease or who have undergone CV surgery may develop the arrhythmia, although paroxysmal forms of the disease especially in younger patients without any structural heart disease have been seen (Fuster et al. 2006). Risk for developing this arrhythmia increases with age, with most patients affected in the seventh and eighth decades of life. Genetic inheritance plays some role in the risk of AF. Although most genetic influences on AF are polygenic, in a few cases some single disease-causing genes may be responsible.

AF is encountered in veterinary medicine, occurring in multiple species of animals. Horses and dogs are the animal groups most affected by arrhythmias (Deem and Fregin 1982; Liu and Nattel 1997). Other species include cats and cows (McGuirk et al. 1983; Cote et al. 2004).

Having episodic or persistent AF exposes patients to many risks, bringing reduced cardiac function and increased risk of a thromboembolic event. It is clearly not adaptive to sustain the arrhythmia. Yet the essential vulnerability to AF, the heart's inherent susceptibility to develop the arrhythmia, is connected with selective pressures and adaptations over millennia. The autonomic nervous system plays a central role in the genesis of AF. A review of autonomic anatomy and function, with emphasis on the adaptive aspects of the mammalian autonomic nervous system, may offer an expanded view of why our species is so vulnerable.

AF may be caused by enhanced parasympathetic or sympathetic input. This dense and complex innervation of the heart by both arms of the autonomic nervous system of fish, reptiles, non-primate mammals, and primates offers these animals a wide range of diverse autonomic responses to exploit opportunities and thwart danger. The vertebral autonomic nervous system's complex and interconnected signalling nexus has evolved to promote resource acquisition and facilitate escape. Yet, embedded within precisely this life-saving complexity is the possibility for dysregulation of vagal and sympathetic input and tone, resulting in arrhythmias, including AF. Therefore, vulnerability to AF can be understood as a trade-off between the benefits and liabilities associated with a complex and densely enervated CV system.

11.4 Preventive Medicine and Evolutionary Approaches

Advances in the treatment of CVD in recent decades have led to significant reductions in mortality and morbidity associated with these conditions. Credit for much of these reductions in CV mortality belongs to preventive measures such as improved screening and treatment of hypertension and diabetes, along with smoking cessation efforts. Yet, despite the many scientific and clinical successes, heart disease remains the leading killer of adults. The application of evolutionary perspectives that look beyond mechanistic explanations of causation could improve and amplify preventative efforts. For example, most CVD emerges from a complex interplay of biological factors, some of which may have adaptive benefits for crucial physiological systems. Characterising the complex physiologies responsible for arrhythmias, AS, or HF facilitates the identification of preventative approaches which target maladaptive pathways while strategically avoiding systems with important physiologic benefits. Moreover, phylogenetic review and analyses of CVD across the animal kingdom can help identify possible natural animal models of disease resistance which could serve as the basis for preventive strategies.

11.5 Conclusion

Despite breakthroughs and advances in vascular biology, electrophysiology, CV surgery, and prevention, heart disease remains the leading cause of death in our species. While traditional approaches to the understanding of CVD have resulted in many life-saving innovations, heart disease remains seemingly intractable pointing to the need for new and expanded perspectives. Looking beyond the traditional mechanistic and developmental hypotheses is one such approach. Applying Tinbergen's lens to a diverse range of CVDs offers expanded and evolutionarily informed perspectives, which can generate important insights and generate novel hypotheses. These in turn may spark investigational pathways and even life-saving innovations which could not be possible without an evolutionary lens.

References

Akashi, Y. J., Nef, H. M., and Lyon A. R. (2015). Epidemiology and pathophysiology of Takotsubo syndrome. *Natl Rev Cardiol* 12(7), 387–97.

Alcock, J. (2013). *Animal Behavior: An Evolutionary Approach*, 10th ed. Sunderland, MA: Sinauer Associates.

Axelsson, M., Holm, S., and Nilsson, S. (1989). Flow dynamics of the crocodilian heart. *Am J Physiol Regul Integr Comp Physiol* 256(4), R875–9.

Bateson, P. and Laland, K. N. (2013). Tinbergen's Four Questions: an appreciation and an update. *Trends Ecol Evol* 28(12), 712–18.

Beaufrère, H. (2013). Avian atherosclerosis: parrots and beyond. *J Exot Pet Med* 22(4), 336–47.

Berenson, G. S., Srinivasan, S. R., Bao, W., et al. (1998). Association between multiple cardiovascular risk factors and atherosclerosis in children and young adults. *N Eng J Med* 338(23), 1650–6.

Beringer, J., Hansen, L., Wilding, W., et al. (1996). Factors effecting capture myopathy in white-tailed deer. *J Wildl Manag* 60(2), 373–80.

Bhattacharya, D., Yoon, H. S., and Hackett, J. D. (2004). Photosynthetic eukaryotes unite: endosymbiosis connects the dots. *BioEssays* 26(1), 50–60.

Boero, F., Bouillon, J., and Piraino, S. (2004). The role of *Cnidaria* in evolution and ecology. *Ital J Zool* 72(1), 65–71.

Bourne, G. B., Redmond, J. R., and Jorgensen, D. D. (1990). Dynamics of the molluscan circulatory system: open versus closed. *Physiol Biochem Zool* 63(1), 140–66.

Braunwald, E. (1997). Shattuck lecture: cardiovascular medicine at the turn of the millennium: triumphs, concerns, and opportunities. *N Eng J Med* 337(19), 1360–9.

Braunwald, E. (2012). The treatment of acute myocardial infarction: the past, the present, and the future. *Eur Heart J* 1(1), 9–12.

Burton, P. M. (2008). Insights from diploblasts: the evolution of mesoderm and muscle. *J Exp Zool B Mol Dev Evol* 310(1), 5–14.

Butterfield, N. J. (2000). *Bangiomorpha pubescens* n. gen., n. sp.: implications for the evolution of sex, multicellularity, and the Mesoproterozoic/Neoproterozoic radiation of eukaryotes. *Paleobiology* 26(3), 386–404.

Carmeliet, P. (2000). Mechanisms of angiogenesis and arteriogenesis. *Nat Med* 6(4), 389.

Carmena, R., Duriez, P., and Fruchart, J.C. (2004). Atherogenic lipoprotein particles in atherosclerosis. *Circulation* 109(23), 2–7.

Cha, Y. M., Redfield, M. M., Shen, W. K., et al. (2004). Atrial fibrillation and ventricular dysfunction: a vicious electromechanical cycle. *Circulation* 109(23), 2839–43.

Challice, C. E. and Virágh, S. (1974). The architectural development of the early mammalian heart. *Cell Tissue Res* 6(3), 447–62.

Chugh, S. S., Havmoeller, R., Narayanan, K., et al. (2014). Worldwide epidemiology of atrial fibrillation: a global burden of disease 2010 study. *Circulation* 129(8), 837–47.

Cocco, G. and Chu, D. (2007). Stress-induced cardiomyopathy: a review. *Eur J Intern Med* 18(5), 369–79.

Cote E., Harpster N. K., Laste N. J., et al. (2004). Atrial fibrillation in cats: 50 cases (1979–2002). *J Am Vet Med Assoc* 225(2), 256–60.

Čulić, V., Eterović, D., and Mirić, D. (2005). Meta-analysis of possible external triggers of acute myocardial infarction. *Int J Cardiol* 99(1), 1–8.

Damani, S. B. and Topol, E. J. (2011). Emerging genomic applications in coronary artery disease. *JACC Cardiovasc Interv* 4(5), 473–82.

Davidson, M., Kuo, C. C., Middaugh, J. P., et al. (1998). Confirmed previous infection with *Chlamydia pneumoniae* (TWAR) and its presence in early coronary atherosclerosis. *Circulation* 98(7), 628–33.

Deem, D. A. and Fregin, G. E. (1982). Atrial fibrillation in horses: a review of 106 clinical cases, with consideration of prevalence, clinical signs, and prognosis. *J Am Vet Med Assoc* 180(3), 261–5.

DeWood, M. A., Spores, J., Notske, R., et al. (1980). Prevalence of total coronary occlusion during the early hours of transmural myocardial infarction. *N Eng J Med* 303(16), 897–902.

DeWood, M. A., Spores, J. Hensley, G. R., et al. (1983). Coronary arteriographic findings in acute transmural myocardial infarction. *Circulation* 68(2 Pt 2), I39–49.

Ding, K. and Kullo, I. J. (2009). Evolutionary genetics of coronary heart disease. *Circulation* 119(3), 459–67.

Edwards C. A. and Bohlen, P. J. (1996). *Biology and Ecology of Earthworms*, 3rd ed. Berlin: Springer Science+Business Media.

Ehrlich, J. R., Nattel, S., and Hohnloser, S. H. (2002). Atrial fibrillation and congestive heart failure: specific considerations at the intersection of two common and important cardiac disease sets. *J Cardiovasc Electrophysiol* 13(4), 399–405.

Fuster, V., Rydén, L. E., Cannom, D. S., et al. (2006). ACC/AHA/ESC 2006 guidelines for the management of patients with atrial fibrillation—executive summary: a report of the American College of Cardiology/American Heart Association Task Force on Practice Guidelines and the European Society of Cardiology Committee for Practice Guidelines (Writing Committee to revise the 2001 guidelines for the management of patients with atrial fibrillation): developed in collaboration with the European Heart Rhythm Association, Heart Rhythm Society. *J Am Coll Cardiol* 48(4), 854–906.

Gao, Q. (1995). Pathomorphological observation on atherosclerosis in giant panda. *Acta Vet Zootech Sin* 26(1), 76–80.

Gencer, B., Lambert, G., and Mach, F. (2015). PCSK9 inhibitors. *Swiss Med Wkly* 145, w14094

Geng, Y. J. and Libby, P. (2002). Progression of atheroma: a struggle between death and procreation. *Arterioscler Thromb Vasc Biol* 22(9), 1370–80.

Gray, M. W. (2012). Mitochondrial evolution. *Cold Spring Harb Perspect Biol* 4(9), a011403.

Greco, G. and Tota, B. (1981). The ventricular myocardium of fish: aspects of comparative morphology physiology and pharmacology. II) Coronary circulation. *Boll Soc Ital Biol Sper* 57(20), 2060–6.

Gruber, A. D., Peters, M., Knieriem, A., et al. (2002). Atherosclerosis with multifocal myocardial infarction in a Pacific walrus (*Odobenus rosmarus divergens* Illiger). *J Zoo Wildl Med* 33(22), 139–44.

Halton, D. W. (1997). Nutritional adaptations to parasitism within the platyhelminthes. *Intern J Parasitol* 27(6), 693–704.

Höglund, K., Bussadori, C., Domenech, O., et al. (2007). Contrast echocardiography in Boxer dogs with and without aortic stenosis. *J Vet Cardiol* 9(1), 15–24.

Hu, H., Pierce, G. N., and Zhong, G. (1999). The atherogenic effects of chlamydia are dependent on serum cholesterol and specific to *Chlamydia pneumoniae*. *J Clin Invest* 103(5), 747–53.

Jensen, B., Boukens, B. J. D., Postma, A. V., et al. (2012). Identifying the evolutionary building blocks of the cardiac conduction system. *PLoS One* 7(4), e444231.

Jewell, J. L. and Guan, K. L. (2013). Nutrient signaling to mTOR and cell growth. *Trends Biochem Sci* 38(5), 233–42.

Jia, H., Abtahian, F., Aguirre, A. D., et al. (2013). In vivo diagnosis of plaque erosion and calcified nodule in patients with acute coronary syndrome by intravascular optical coherence tomography. *J Am Coll Cardiol* 62(19), 1748–58.

Johansen, K. (1959). Circulation in the three-chambered snake heart. *Circ Res* 7(6), 828–32.

Khand, A. U., Rankin, A. C., Kaye, G. C., et al. (2000). Systematic review of the management of atrial fibrillation in patients with heart failure. *Eur Heart J* 21(8),614–32.

Kienle, R. D., Thomas, W. P., and Pion, P. D. (1994). The natural clinical history of canine congenital subaortic stenosis. *J Vet Intern Med* 8(6), 423–31.

Kim, D. H., Sarbassov, D. D., Ali, S. M., et al. (2002). mTOR interacts with raptor to form a nutrient-sensitive complex that signals to the cell growth machinery. *Cell* 110(2), 163–75.

Koshiba-Takeuchi, K., Mori, A. D., Kaynak, B. L., et al. (2009). Reptilian heart development and the molecular basis of cardiac chamber evolution. *Nature* 461(7260), 95–8.

Koskinas, K. C., Sukhova, G. K., Baker, A. B., et al. (2013). Thin-capped atheromata with reduced collagen content in pigs develop in coronary arterial regions exposed to persistently low endothelial shear stress. *Arterioscler Thromb Vasc Biol* 33(7), 1494–504.

Kuipers, R. S., de Graaf, D. J., Luxwolda, M. F., et al. (2011). Saturated fat, carbohydrates and cardiovascular disease. *Neth J Med* 69(9), 372–8.

Leopold, J. A. (2012). Cellular mechanisms of aortic valve calcification. *Circ Cardiovasc Interv* 5(4), 605–14.

Le Quang, K., Bouchareb, R., Lachance, D., et al. (2014). Early development of calcific aortic valve disease and left ventricular hypertrophy in a mouse model of combined dyslipidemia and type 2 diabetes mellitus. *Arterioscler Thromb Vasc Biol* 34(10), 2283–91.

Ley, K., Miller, Y. I., and Hedrick, C. C. (2011). Monocyte and macrophage dynamics during atherogenesis. *Arterioscler Thromb Vasc Biol* 31(7), 1506–16.

Libby, P., Ridker, P. M., and Maseri, A. (2002). Inflammation and atherosclerosis. *Circulation* 105(9), 1135–43.

Lighton, J. R. (1996). Discontinuous gas exchange in insects. *Annu Rev Entomol* 41(1), 309–24.

Liu, L. and Nattel, S. (1997). Differing sympathetic and vagal effects on atrial fibrillation in dogs: role of refractoriness heterogeneity. *Am J Physiol* 273(2 Pt 2), H805–16.

Lyon, A. R., Rees, P. S., Prasad, S., et al. (2008). Stress (Takotsubo) cardiomyopathy—a novel pathophysiological hypothesis to explain catecholamine-induced acute myocardial stunning. *Nat Clin Pract Cardiovasc Med* 5(1), 22–9.

Lyon, A. R., Bossone, E., Schneider, B., et al. (2016). Current state of knowledge on Takotsubo syndrome: a position statement from the Taskforce on Takotsubo Syndrome of the Heart Failure Association of the European Society of Cardiology. *Eur J Heart Failure* 18(1), 8–27.

Mahmood, S. S., Levy, D., Vasan, R. S., et al. (2014). The Framingham Heart Study and the epidemiology of cardiovascular disease: a historical perspective. *Lancet* 383(9921), 999–1008.

Martindale, M. Q., Pang, K., and Finnerty, J. R. (2004). Investigating the origins of triploblasty: mesodermal gene expression in a diploblastic animal, the sea anemone *Nematostella vectensis* (phylum, Cnidaria; class, Anthozoa). *Development* 131(10), 2463–74.

McFadden, G. I. (2001). Primary and secondary endosymbiosis and the origin of plastids. *J Phycol* 37(6), 951–9.

McGill, H. C. Jr, McMahan, C. A., Zieske, A. W., et al. (2000). Association of coronary heart disease risk factors with microscopic qualities of coronary atherosclerosis in youth. *Circulation* 102(4), 374–9.

McGuirk, S. M., Muir, W. W., Sams R. A., et al. (1983). Atrial fibrillation in cows: clinical findings and therapeutic considerations. *J Am Vet Med Assoc* 182(12), 1380–6.

McMahon, B. and Burnett, L. (1990). The crustacean open circulatory system: a reexamination. *Physiol Zool* 63(1), 35–71.
McMahon, C. R., Wiggins, N. L., French, V. A., et al. (2013). A report of capture myopathy in the Tasmanian pademelon (*Thylogale billardierii*). *Animal Welfare* 22(1), 1–4.
Mendis, S., Puska, P., and Norrving, B. (2011). *Global Atlas on Cardiovascular Disease Prevention and Control*. Geneva: World Health Organization. http://whqlibdoc.who.int/publications/2011/9789241564373_eng.pdf.
Montané, J., Marco, I., Manteca, X., et al. (2002). Delayed acute capture myopathy in three roe deer. *J Vet Med A Physiol Pathol Clin Med* 49(2), 93–8.
Mozaffarian, D., Benjamin, E. J., Go, A. S., et al. (2016). Heart disease and stroke statistics—2016 update: a report from the American Heart Association. *Circulation* 133(4), e38–60.
Muller, J. E. (1989). Morning increase of onset of myocardial infarction. Implications concerning triggering events. *Cardiology* 76(2), 96–104.
Muller, J. E. and Mangel, B. (1994). Circadian variation and triggers of cardiovascular disease. *Cardiology* 85(Suppl 2), 3–10.
Nedelcu, A. M. (2012). Evolution of multicellularity. *eLS* 15 March.
Olson, E. (2006). Gene regulatory networks in the evolution and development of the heart. *Science* 313(5795), 1922–7.
Packer, M. (1992). The neurohormonal hypothesis: a theory to explain the mechanism of disease progression in heart failure. *J Am Coll Cardiol* 20(1), 248–54.
Porges, S. W. (2009). The polyvagal theory: new insights into adaptive reactions of the autonomic nervous system. *Cleve Clin J Med* 76(Suppl 2), S86–90. doi: 10.3949/ccjm.76.s2.17.
Prescott, M. F., Hasler-Rapacz, J., Von Linden-Reed, J., et al. (1995). Familial hypercholesterolemia associated with coronary atherosclerosis in swine bearing different alleles for apolipoprotein B. *Ann N Y Acad Sci* 748, 283–92.
Reiber, C. L. and McGaw, I. J. (2009). A review of the 'open' and 'closed' circulatory systems: new terminology for complex invertebrate circulatory systems in light of current findings. *Intern J Zool* 2009, 301284. http://dx.doi.org/10.1155/2009/301284.
Rizki, T. M. (1978). The circulatory system and associated cells and tissues. In: Ashburner, M. and Wright, T. R. F. (eds) *The Genetics and Biology of* Drosophila. London: Academic Press, pp. 397–452.
Roger, A. J., Svärd, S. G., Tovar, J., et al. (1998). A mitochondrial-like chaperonin 60 gene in *Giardia lamblia*: evidence that diplomonads once harbored an endosymbiont related to the progenitor of mitochondria. *Proc Natl Acad Sci U S A* 95(1), 229–34.
Roger, V. L. (2013). Epidemiology of heart failure. *Circ Res* 113(6), 646–59.
Romenskii, O. (1978). Blood supply of the compact and spongy myocardium of fish, amphibia and reptiles. *Arkh Anat Gistol Embriol* 75(7), 91–5.
Ross, R. (1999). Inflammation or atherogenesis. *N Eng J Med* 340(2), 115–26.
Santer, R. M. (1985). Morphology and innervation of the fish heart. *Adv Anat Embryol Cell Biol* 89, 1–102.
Santos-Gallego, C. G., Picatoste, B., and Badimón, J. J. (2014). Pathophysiology of acute coronary syndrome. *Curr Atheroscler Rep* 16(4), 401.
Sawaya, F., Liff, D., Stewart, J., Lerakis, S., et al. (2012). Aortic stenosis: a contemporary review. *Am J Med Sci* 343(6), 490–6.
Schmaier, A. A., Stalker, T. J., Runge, J. J., et al. (2011). Occlusive thrombi arise in mammals but not birds in response to arterial injury: evolutionary insight into human cardiovascular disease. *Blood* 118(13), 3661–9.
Schopf, J. W., Kudryavtsev, A. B., Czaja, A. D., et al. (2007). Evidence of Archean life: stromatolites and microfossils. *Precambrian Res* 158(3–5), 141–55.
Seilacher, A., Bose, P., and Pfluger, F. (1998). Triploblastic animals more than 1 billion years ago: trace fossil evidence from India. *Science* 282(5386), 80–3.
Sherman, C. T., Litvack, F., Grundfest, W., et al. (1986). Coronary angioscopy in patients with unstable angina pectoris. *N Eng J Med* 315(15), 913–19.

Solc, D. (2007). The heart and heart conducting system in the kingdom of animals: a comparative approach to its evolution. *Exp Clin Cardiol* 12(3), 113–18.

Stone, P. H., Saito, S., Takahashi, S., et al. (2012) Prediction of progression of coronary artery disease and clinical outcomes using vascular profiling of endothelial shear stress and arterial plaque characteristics: the PREDICTION study. *Circulation* 126(2), 172–81.

Taylor, J. S. and Raes, J. (2004). Duplication and divergence: the evolution of new genes and old ideas. *Annu Rev Genet* 38, 615–43.

Tinbergen, N. (1963). On aims and methods of ethology. *Zeitschr Tierpsychol* 20, 410–33.

Tomimura, T., Kotani, T., Mochizuki, H., et al. (1970). Comparative studies on arteriosclerosis in wild and domestic animals. I. Spontaneous arteriosclerosis in the emu, *Dromiceius novaehollandiae*. *Nihon Juigaku Zasshi* 32(1), 25–34.

Vander Heiden, M. G., Cantley, L. C., and Thompson, C. B. (2009). Understanding the Warburg effect: the metabolic requirements of cell proliferation. *Science* 324(5930), 1029–33.

Watson, C. and Alp, N. J. (2008). Role of *Chlamydia pneumoniae* in atherosclerosis. *Clin Sci* 114(8), 509–31.

Weiss J. N., Qu Z. and Shivkumar K. (2016). Ablating atrial fibrillation: a translational science perspective for clinicians. *Heart Rhythm* 13(9), 1868–77.

Wells, B. J., Mainous, A. G., and Dickerson, L. M. (2004). Antibiotics for the secondary prevention of ischemic heart disease: a meta-analysis of randomized controlled trials. *Arch Intern Med* 164(19), 2156–61.

WHO (2015). *World Health Statistics 2015*. Geneva: World Health Organization.

Wright, S. H. and Ahearn, G. A. (1997). Nutrient absorption in invertebrates. *Compr Physiol* Suppl 30. doi: 10.1002/cphy.cp130216.

Yetkin, E. and Waltenberger, J. (2009). Molecular and cellular mechanisms of aortic stenosis. *Int J Cardiol* 135(1), 4–13.

CHAPTER 12

RESPIRATORY SYSTEM

OLGA CARVALHO AND JOHN N. MAINA

Abstract

The lung is the gas-exchange organ that provides oxygen and removes carbon dioxide from the blood. The environment in which animals live and their metabolic needs determine the evolved design of their gas exchange system. Gills are the primordial respiratory organs that evolved for water 'breathing', while other adaptive solutions evolved for bimodal breathing, that is, the ability to extract oxygen from both water and air. The transition to fully terrestrial life was accompanied by significant changes in dimensions of respiratory units (alveoli) which decreased in size, whereas the number of units and total lung volume increased, leading to more efficient gas exchange and oxygen supply. While the shape and make-up of lungs in humans suggest adaptations to long-distance running and possibly to the exposition of smoke caused by fire, the exposure of the human respiratory system to novel environments has brought about a diverse array of disease patterns, including lung cancer, autoimmune diseases, and infectious diseases.

Keywords

lung, gas exchange organ, lung development, respiratory disease, alveoli, evolution, medicine

12.1 Introduction

During the evolution of animal life, oxygen (O_2) has been an important driver of phenotypic change (Maina 1998a; Dolph 2000; Roux 2002; Losos 2010; Ward et al. 2016). The functional designs of the respiratory organs of animals correlate with factors such as the levels of development, body mass, metabolic rate, environment inhabited, and lifestyle pursued (Meban 1980; Gehr et al. 1981; Hlastala and Berger 1996; Maina 1998a, 2002a, b, c, 2005, 2011). Generally, species that maintain high O_2 to carbon dioxide (CO_2) gas exchange ratios, relative to their body mass, establish high and stable tissue fluid gas concentrations under different environmental conditions and metabolic states, and are biologically and ecologically the most successful taxa. The capacity of efficiently procuring, transporting, and utilising large amounts of O_2 granted significant selective advantages and allowed great advances in animal life (Hammond 1992; May 1992). From the inauguration of life on Earth, the size and the multiplicity of the global biota has increased by sixteen orders of magnitude (Payne et al. 2009, 2011; Bell et al. 2015). Specialisation and refinement of respiratory processes such as incorporation of carrier-mediated gas-transport systems allowed energetically costly lifestyles such as powered flight (Brackenbury 1984, 1991; Maina and King 1984; Maina 1998a, b). Along the same lines, predator avoidance and survival corresponds with the capacity of acquiring O_2 and the expenditure of energy, the type of food consumed, and the habitat occupied (Gaillard et al. 2010).

However, life started in an anaerobic environment, the so-called primordial broth, a mixture of organic molecules in the absence of oxygen (Maina 2000b). Almost 4 billion years ago (bya), the first creatures that inhabited earth were primitive microorganisms living in strictly anaerobic environments, perhaps methanogenic bacteria that still exist in our days and belong to the Archae phyla (Eigen et al. 1989).

Anaerobic fermentation is, however, a quite inefficient metabolic way of extracting energy from organic molecules. So, the transition to an oxygenic environment between 2.3 and 1.8 bya (Owen et al. 1979) was a momentous event in the diversification of life that dramatically shifted from inefficient to sophisticated ecosystems, in which oxygen became an indispensable factor for aerobic metabolism, especially in higher (multicellular) life forms. The change from primitive anaerobic respiration to the more efficient O_2-driven aerobic metabolism thus led to a vast increase in biomass (Davies 1975; Wu 2002). The relatively small size of the O_2 molecule and its high intracellular diffusivity and optimum redox potential (Eh) made it the molecule of choice as the terminal electron acceptor in the process of oxidative phosphorylation that culminates in production of energy in the form of adenosine triphosphate (ATP).

Molecular evidence shows that aerobic respiration evolved prior to oxygenic photosynthesis, suggesting that denitrification (nitric-oxide (NO) reductase) was the probable origin of aerobic respiration, and that aerobic respiration arose from a single evolutionary line, the cyanobacteria (Castresana and Saraste 1995). Practically, all the molecular O_2 that now exists in the biosphere was produced by the blue-green algae (cyanobacteria) between 1 and 3.5 bya (Van Valen 1971; Walker 1977). After green plants invaded land, production of O_2 surpassed that by cyanobacteria two-fold (Knoll 1979). It is envisaged that if all life on Earth ended today, the O_2 in the atmosphere would disappear in less than 4 million years (Greener 2008).

The rise of atmospheric oxygen caused by photosynthetic activity of evolving cyanobacteria must have created a remarkably strong selective pressure on organisms. Adaptations to use the new, chemically superior, and more profitable method of extracting energy, that is, aerobic respiration, led to a dominance of aerobic organisms in the biological community, and probably caused extinction of some anaerobic organisms (Schopf 1978; Alberts et al. 1994).

The first obligatory aerobic eukaryotic cells formed around 2.0–1.3 bya when the atmospheric O_2 level was only a few per cent of the current level (Schopf and Walter 1983; Rasmussen et al. 2008; Donoghue and Antcliffe 2010). The first metazoans (multicellular organisms) appeared about 500–700 million years ago (mya) (Schopf and Walter 1983; Cloud 1993; Canfield 1998; Lyons and Reinhard 2009), possibly even earlier around 2.45 and 2.32 bya (El Albani et al. 2010). In eukaryotic cells, the nucleus and the nuclear membrane formed to reduce the potentially injurious effects of O_2 (Margulis 1981). That is, the nucleus is a fairly anoxic and thus a safer location for DNA (Dyer and Obar 1994).

Together, the evolution of multicellular organisms determined the appearance of a more sophisticated, effective, and specialised system for gas exchange. Progressively, multicellular organisms differentiated morphofunctional groups of cells that became the precursors of tissues, organs, and systems currently present in more complex organisms (Alberts et al. 1994; Grunwald 1996).

More specifically, the evolution of terrestrial vertebrates included several changes: from anaerobic to aerobic life; accretion of single cells into multicellular organs; formation of a closed circulatory system; evolution of metal-based carrier pigments that improved oxygen uptake; formation of invaginated respiratory organs; physical translocation from water to land; development of a double circulation; and progression from ectothermic-heterothermy to endothermic-homoeothermy (Maina 2002a).

Although morphologically different, the respiratory systems from various groups of animals have some characteristics in common: a large capillary network; thin and moist gas exchange surfaces; constant renewal of oxygen-rich fluid (air or water) to provide oxygen and remove carbon dioxide; and free movement of blood within the capillary network (Weichert and Presch 1986; Maina 2002b). According to Atwood (1992), these characteristics are present in external respiration: (1) by cutaneous diffusion in earthworms and some amphibians; (2) by thin tubes called tracheae in some insects; and (3) by gills in fish and respiratory organs (lungs) in amphibians, reptiles, birds, and mammals.

It is puzzling though, that life should become so intricately dependent on O_2 (Tappan 1974; Chang et al. 1983; Howard and Schopf 1983; Schopf 1989; Deamer and Szostak 2010). During its existence on Earth, animal life has undergone periods of massive expansions, contractions, and catastrophic crushes or mass extinctions (Rampino 2010). Many of these events have been associated with lack of O_2 (anoxia) or paucity of it (hypoxia) (Zhuravlev and Wood 1996; Hough et al. 2006; Hurtgen et al. 2009). Together with other environmental events such as a relatively sudden drop in temperature, lowering of the sea level, geochemical changes, and tectonic activities, it is assumed that hypoxia caused the Mid-Palaeozoic Crisis (Robinson 1991; Erwin 1993) when about 90% of fish families succumbed (Tappan 1974; McGhee 1989).

In any event, the ability of utilising oxygen for metabolism must have been tremendously advantageous for multicellular organisms, exceeding the risk of fatal consequences by transient oxygen depletion by orders of magnitude. Yet, organisms had to solve the problem of extracting oxygen from the environment (water or air), while at the same time protecting cells and organs from potentially damaging effects of oxygen. In this chapter, we aim to outline the anatomy and physiology of the respiratory system in relation to its ontogeny and phylogeny, and to embed explanations for certain human diseases affecting the respiratory system in the context of its evolved design.

12.2 Evolutionary Challenges of Gas Exchange Related to the Physico-Chemical Properties of Oxygen

Respiration is a complex process that comprises physical, physiological, biochemical, and behavioural mechanisms and processes by which molecular oxygen is acquired from the external respiratory fluid medium (water or air) and delivered to cells, where it is utilised for production of energy in the form of ATP. No cells, tissues, or structures are unique to respiratory organs and structures like, for example, hepatocytes are to the liver, osteocytes to bone, podocytes to kidneys, and neurons to nervous tissue. Although often stated to be specific to the lung and termed the 'defender of the alveolus' (Mason and Williams 1977), the type II alveolar cells that produce surfactant are not specific to the organ: surfactant-like phospholipids are produced in various tissues and cells in organs like the stomach, the intestines, the swim-bladder, the gas mantle of air-breathing snails (e.g. *Helix aspersa*), the prostate gland, the female reproductive tract, the lacrimal gland, the mesothelial cells of the pleura, pericardium, and peritoneum, and the epithelium of the Eustachian tube (Bernhard et al. 2001; Bourbon and Chailley-Heu 2001; Akiyama et al. 2002; Gowdy et al. 2012; Liu et al. 2013).

The simplest respiratory structure (gas exchanger) is the plain cell membrane of unicellular organisms across which O_2 diffuses along a partial pressure gradient (ΔPO_2). While many animals will survive for weeks without food and days without water, most of them will survive for only a few minutes without O_2. Unlike feeding, thermoregulation, locomotion, and reproduction—activities and processes that can be delayed or even suspended—the importance of O_2 for life comes into sharp personal focus when considering that if you stopped breathing just now, as you read these words, permanent damage, especially of the cells of the most sensitive organs like the brain and the heart, would occur in less than 3 minutes (Piper et al. 1994; Nilsson 2010) and death would be certain in about 5–8 minutes. An adult human being breathes ~ 12,000 l of air per day (Burri 1985) and eliminates ~ 300 l of carbon dioxide (Farhi and Rahn 1955; Farhi 1964). In an adult 70-kg human, only ~ 1.6 l of O_2 are stored in the body: 370 cm^3 in the alveoli, 280 cm^3 in the arterial blood, 600 cm^3 in the capillary and the venous blood, 60 cm^3 dissolved in the body tissues, and 240 cm^3 bound to the myoglobin (Farhi and Rahn 1955; Farhi 1964; Snyder 1983). It is because only ~ 500 cm^3 of O_2 is removed per minute from the ~ 6–7 l of the air inhaled at rest (in addition to the only 1.6 l of the total pool in the body) that injury of cells, especially those of hypoxia/anoxia-intolerant organs, occurs soon after O_2 becomes unavailable (Erecinska and Silver 1994; Lipton 1999; Lutz et al. 2003). The insignificant stores of O_2 in the body explain why the alveolar partial pressure of O_2 (PO_2) changes rapidly with pulmonary circulation during conditions of hypoxia and apnoea. Breathing 100% O_2 produces a large shift in the total O_2 stores, since the functional residual capacity (FRC) rapidly fills up with O_2 (Comroe 1974). Under such a state, the greatest amount of the pool of O_2 in the body exists in the lung and 80% of it can be used without lowering haemoglobin-O_2 saturation. Also, it explains why pre-oxygenation, that is, O_2 supplementation, is highly effective in conditions like hypoxaemia, for example during anaesthesia, and in cases such as obstructive respiratory complications. In a human, breathing pure O_2 at FRC, ~ 3 L of O_2 are found in the lungs, 1 L in blood, and 0.3 L is dissolved or bound to tissues (Comroe 1974). In such cases, the total O_2 store in the body (4.3 L) is about three times the volume that exists when breathing normal air. The body's stores of CO_2 that exist both in solution and in the form of bicarbonate ($-HCO_3^{-1}$) ions considerably surpass those of O_2 (Farhi and Rahn 1955; Farhi 1964; Dejours 1988). In a 70-kg-weight person, there are 35 L of CO_2 in the body, a quantity equivalent to the resting metabolic CO_2 production over a time of ~ 140 minutes (Slonim and Hamilton 1971).

Contrary to metabolic substrates like carbohydrates and fats that can be stored in the body in large quantities (Hochachka 1973), O_2 has to be acquired from air and/or water at rates and measures it is utilised at.

In spite of its extreme importance to life, O_2 is paradoxically also highly injurious to life (Auten and Davis 2009; Cabelli 2010). This emanates from the highly reactive O_2 species (ROS) (also termed 'free radicals') that are produced from leakage of high-energy electrons as they progress down the electron transport chain (Buonocore et al. 2010; Harrison et al. 2010; Maina 2011). Imbalance between production of ROS and the antioxidant defences that protect cells has been associated with pathogenesis of various diseases and conditions such as cancer, asthma, pulmonary hypertension, retinopathy, and ageing (Auten and Davis 2009). Risk of lung injury from O_2 toxicity is particularly high in scuba divers breathing 100% compressed O_2, patients suffering from diseases like chronic obstructive pulmonary disease (COPD), premature babies with respiratory deficiencies who are placed under high concentration of supplemental O_2, and during human space flight (Northway and Rosan 1968; Couroucli et al. 2006). Chronic exposure or acute exposure to high O_2 concentration causes abnormal lung development with poor alveolisation (Jiang et al. 2004; Lin et al. 2005). Even under a lower level of 40% O_2, pulmonary epithelial function deteriorates and collagen breaks down (Aoki et al. 2008). In rats exposed to 60% O_2 for 2 weeks, the interalveolar septa thicken from infiltration and deposition of interstitial collagen fibres (Al-Motabagani 2005). Although in some fish, O_2 held in the swim-bladder at high pressure and concentration can be utilised during hypoxic episodes (Randall and Daxboeck 1984), and sufficient tissue oxygenation can be maintained for several hours, the counter-current gas-concentrating systems that occur in especially pelagic fish mainly serve a hydrostatic and not a respiratory function (Pelster and Scheid 1992; Pelster 2015).

Accordingly, to prevent oxidative cell damage and death, tissue O_2 levels have to be maintained within a very narrow range (Aprelikova et al. 2004). To utilise O_2 safely, mechanisms and processes of sensing it developed very early in the evolution of the amniotes (Nikinmaa 2010; Maina 2011; Hockman et al. 2017) and appear to exist in every cell (Bunn and Poyton 1996; Gorr et al. 2006). Normal O_2 tension and expression of hypoxia-inducible factors (HIFs) is closely regulated in cells (Aprelikova et al. 2004). In human beings, ~ 2–5% of all genes comprise the so-called oxy-genes that are transcriptionally regulated by HIFs (Wenger 2000; Manalo et al. 2005; Wenger et al. 2005).

12.3 Evolutionary Ontogeny of Respiratory Systems

The development of the respiratory organs in vertebrates is closely related to the primitive pharynx, since the gills and the lungs of terrestrial vertebrates and aquatic mammals have pharyngeal embryology origin (Moore and Persaud 1998; Schoenwolf et al. 2009).

In all vertebrates, at a certain stage of their development, a series of diverticula arise bilaterally in craniocaudal direction from the inner side of the pharynx, which evaginate towards the outer surface, forming the pharyngeal pouches (Moore and Persaud 1998; Schoenwolf et al. 2009). The number of pharyngeal pouches is greater in lower vertebrates, reaching fourteen in cyclostomes and only four or five in birds and mammals. The pharyngeal pouches are separated by masses of mesenchyme that have the designation of pharyngeal arches, in which is located an arterial structure, called the aortic arch, which extends from the ventral aorta to the dorsal aorta (Moore and Persaud 1998; Schoenwolf et al. 2009).

During ontogeny of higher vertebrates, the pharyngeal pouches fail to open to the outside, in contrast to what happens in fish and, temporarily, in amphibians. Thus, in terrestrial vertebrates, the pharyngeal pouches just remain during the embryonic period, where they undergo several changes, but very few or none of their initial characteristics are presented in adults. In amniotes, as in humans, only the first pair of pharyngeal pouches remains, giving origin bilaterally to the eustachian tube and middle ear (Moore and Persaud 1998; Schoenwolf et al. 2009). Persistence of pharyngeal clefts may cause health problems such as preauricular cysts or fistulae.

12.3.1 Ontogenetic Development of the Human Respiratory System

In most mammals, lung development from pre- to postnatal life follows five stages: embryonic, pseudoglandular, canalicular, saccular, and alveolar stages (Figure 12.1) (Wert 2004; Joshi and Kotecha 2007).

The embryonic stage (3rd–7th weeks post conception (p.c.)) begins with the emergence of the primordium of the lungs, called the respiratory diverticulum or lung bud, and finishes with a complete pulmonary tree outline, in which right and left budding are visible as well as certain slightly branched bronchial tubules. This bud grows ventrocaudally and undergoes three initial rounds of branching, producing successively the primordia of the two lungs, the lungs lobes, and the bronchopulmonary segments. There is also a complete separation between the trachea and the oesophagus (Moore and Persaud 1998; Wert 2004; Schoenwolf et al. 2009).

The primitive lung buds are constituted by two phenotypically different cell types: the epithelial cell, derived from the primitive gut endoderm, and the mesenchymal cell, emerging from the splanchnopleuric portion of the primitive gut. The epithelial cells will differentiate into specialised cells that line the conducted compartment, including ciliated cells, goblet cells, cells of mucosa glands, Clara cells, and type I and type II alveolar cells. The mesenchymal cells will differentiate into the cellular and extracellular elements of the interstitial compartment, vascular component, cartilaginous tissue of bronchi, and smooth muscle of blood vessels and bronchi (Moore and Persaud 1998).

Several factors are important during these stages of development such as hepatocyte nuclear factor 3β (HNF-3β) and transcription factors GATA6, GATA4, Gli2, and Gli3 that regulate the formation of the foregut endoderm and therefore all its derivatives, including the lungs (Morrisey et al. 1998; Cardoso 2004; Cai et al. 2008). In the development of the lung bud the presence of the sonic hedgehog (SHH) glycoprotein is essential for the formation of the trachea–oesophageal septum and therefore crucial for the separation of the

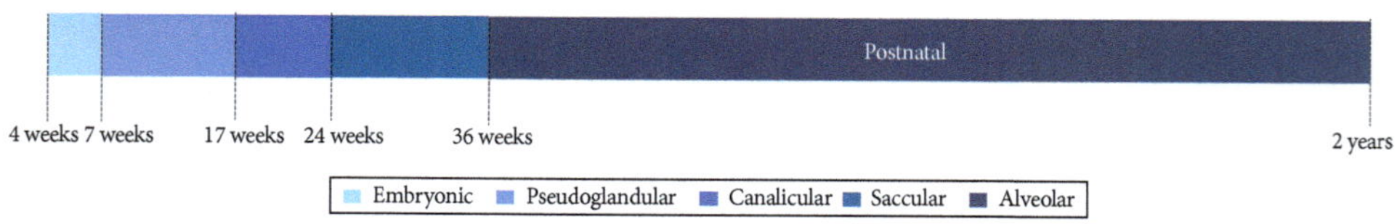

FIGURE 12.1 Stages of human lung development.

digestive and respiratory systems (Litingtung et al. 1998). Fibroblast growth factor-10 (FGF-10) is important in the bud formation and initial branching of the primary bronchi (Min et al. 1998), and retinoids are essential for bud formation and endoderm differentiation (Desai et al. 2004, 2006).

In the pseudoglandular stage (7th–17th weeks p.c.) the lung resembles a gland, and during this stage successive dichotomy branching occurs, such that eventually almost all the conducting portion is established to the terminal bronchioli (Joshi and Kotecha 2007) (Figure 12.2A). The lining epithelium of the air-conducting areas differentiates (Joshi and Kotecha 2007). Bronchial tubules are initially covered by a pseudostratified columnar epithelium, which, as the branch process continues to grow, develops into simple columnar epithelium in the proximal areas and simple cubic epithelium in the distal acinar tubules and buttons (Wert 2004).

Development and branching of the air-conducting portion involves epithelial–mesenchymal interactions through the matrix components (Hilfer 1996), several growth factors such as FGF (Cardoso 2004; Maina 2012a) and transforming growth factors (TGF) TGF-β1, -β2, and -β3 (Cardoso 2001), and retinoic acid (McGowan and Snyder 2004).

In the canalicular stage (17th–27th weeks p.c.) canaliculi branch out of the terminal bronchioli, and the acinar tubules and buttons grow, subdivide, lengthen, and undergo cell differentiation (Wert 2004; Schoenwolf et al. 2009) (Figure 12.2B). All distal portions are remodelled by increasing the capillary network and by the condensation and reduction of mesenchyme thickness between the terminal ducts. The distal epithelium begins to differentiate and one can observe type I and type II alveolar cells in the distal regions (Wert 2004). Type II alveolar cells begin to lose intracellular glycogen, to acquire multilamellar bodies (Chi 1985), and to initiate surfactant excretion (Khoor et al. 1994). Peripheral development is accompanied by the growth and development of intra-acinar capillaries, which are arranged around the airspace, establishing contact with the adjacent epithelium and forming the primitive alveolar–capillary membrane (Wert 2004).

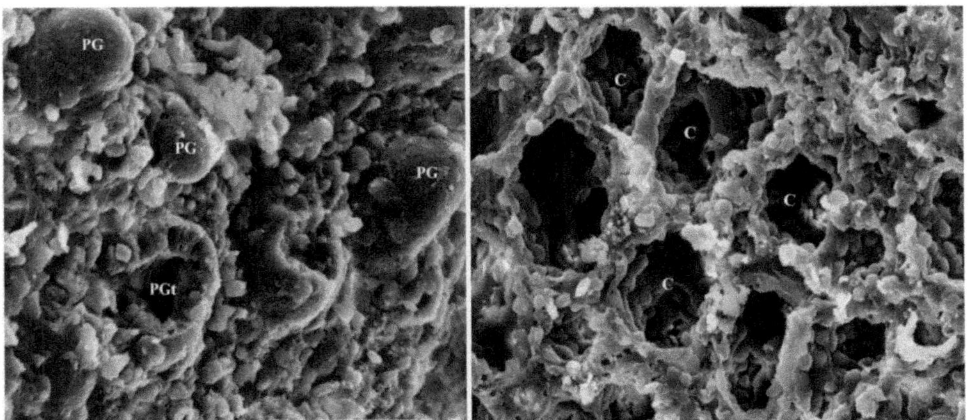

FIGURE 12.2 Scanning electron microscopy images of human lung: (A) pseudoglandular stage, showing 'pseudoglands' (PG) during lung development, some sectioned transversely (PGt); and (B) canalicular stage, with several canaliculi (C) sectioned transversely.

The cell differentiation along the proximal–distal axis is regulated by bone morphogenetic protein 4 (BMP4) (Weaver et al. 1999) and the hepatocyte nuclear factor/forkhead homologue 4 (HFH-4) (Tichelaar et al. 1999).

In the saccular stage (28th–36th weeks p.c.) clusters of acinar tubules and terminal buds begin to expand into alveolar saccules and transitory ducts, which is accompanied by a considerable reduction or condensation of the surrounding mesenchymal tissue (Wert 2004) (Figure 12.3A). Type I alveolar cells continue to mature, reducing the intracellular glycogen content and increasing the size and number of lamellar bodies (Wert 2004). Type II alveolar cells also continue their differentiation, thinning the cell and increasing the distal lung surface area (Joshi and Kotecha 2007). The capillaries are associated more closely with type I cells, decreasing the distance of diffusion between the future air spaces and capillaries.

The alveolar stage (36 weeks p.c. to 2 postnatal years) is characterised by the development of secondary alveolar septa, which divide the terminal ducts and sacs into truly alveolar ducts and alveoli, a process that increases dramatically the pulmonary gas exchange surface area (Wert 2004) (Figure 12.3B). There is also a reduction of the interstitial connective tissue and a remodelling of the capillary bed, which occurs by the fusing of the capillary network from the two adjacent septa (Wert 2004).

The establishment and development of the vascular network is mainly regulated by the vascular endothelial growth factor (VEGF) (Wert 2004), but there are other mediators of angiogenesis and vasculogenesis, whose role in lung development is still scrutinised, such as angiopoietins (Loughna and Sato 2001), ephrins (Kullander and Klein 2002), and semaphorins (Neufeld et al. 2002).

During the septation process, epithelial differentiation and alveolar interstitial rearrangement are regulated by platelet-derived growth factor (PDGF), a potent mitogen that promotes the proliferation of the precursor cells of alveolar myofibroblasts (AFM) (Bostrom et al. 2002) and elastin presence in the alveolar septa (Lindahl et al. 1997). In addition, glucocorticoids are involved in cell maturation, surfactant, and elastin production (Pierce et al. 1995), while retinoic acid contributes to the production of surfactant (Fraslon and Bourbon 1994), induction of the formation of alveolar septa (Massaro and Massaro 1996, 2000), cell differentiation (Costa et al. 2001; Bielsalski and Nohr 2003; Wongtrakool et al. 2003), synthesis and

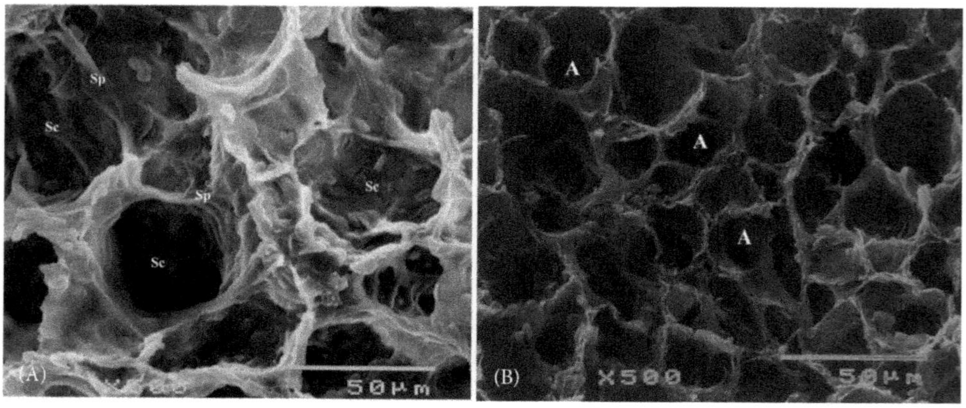

FIGURE 12.3 Scanning electron microscopy images of human lung: (A) saccular stage, with alveolar sacs (Sc) and thick septa (Sp); and (B) alveolar stage, with alveoli (A) and thin inter-alveolar septa.

deposition of elastin (Lui et al. 1993; McGowan et al. 2000), and alveolar regeneration capacity (Hind and Maden 2004; Maden 2006).

Despite the presence of definitive alveoli at birth (36 weeks of development), it is in the first 6 months of postnatal life that 85–90% of all alveoli are formed. From this point onward until the age of 1.5–2 years, more alveoli continue to be formed, though at a slower rate. Lung growth proportionally follows body growth, and until 18 years of age the gas exchange area and diffusion capacity increases linearly with body weight (Wert 2004; Joshi and Kotecha 2007).

12.4 Functions and Mechanisms

12.4.1 Anatomy and Histology

The respiratory system consists of an air-conducting portion for the passage of inhaled and exhaled air, a respiratory portion for gas exchange between air and blood, and a ventilation mechanism. The air-conducting pathway comprises the nasal cavities, pharynx, larynx, trachea, bronchi, and bronchioles, and the respiratory portion consists of respiratory bronchioles, alveolar sacs, and alveoli. The ventilation mechanism involves the thoracic cage with its intercostal muscles, diaphragm, and elastic connective tissue of the lungs.

The nasal cavities are lined by a pseudostratified columnar ciliated epithelium with goblet cells supported by a lamina propria of connective tissue with seromucous glands. In the roof of each nasal cavity exits the specialised olfactory mucosa, which have a pseudostratified epithelium containing olfactory cells (bipolar neurons), supporting cells and basal cells (stem cells that differentiated into olfactory cells), all supported by a lamina propria with the serous glands of Bowman. (For further discussion, see Chapter 17: Brain, Spinal Cord, and Sensory Systems.)

The pharynx is divided into the posterior portion, called the nasopharynx, lined by a pseudostratified columnar ciliated epithelium and abundant lymphoid tissue, and the oropharynx, lined by a stratified non-keratinised squamous epithelium.

The larynx has two functions: one is to produce sound and the other is to close the trachea during swallowing to prevent food and saliva from entering into the airways. The larynx consists of a series of cartilage formations, some made of hyaline cartilage (thyroid, cricoid, and parts of the arytenoid), and others made of elastic cartilage (epiglottis, corniculate, and cuneiform) that are attached to each other and to the hyoid bone. Extrinsic laryngeal muscles, which attach the larynx to the hyoid bone, raise the larynx during swallowing. The larynx can be subdivided into three regions: the supraglottis (epiglottis, false vocal cords, and laryngeal ventricles), the glottis (true vocal cords and anterior and posterior commissures), and the subglottis (from below the true cords until the lower border of the cricoid cartilage). The lingual surface of the epiglottis and the true vocal cords are lined by a stratified squamous epithelium, and the remaining part of the larynx is lined by a pseudostratified ciliated epithelium with goblet cells and seromucous glands in the lamina propria (Kierszenbaum 2011). The human larynx is essential not only for speech, but also for swallowing and respiration. With age, the thickness of the epithelium of the supraglottic region and vocal cord increases, and there is evidence to suggest that there is a significant increase in thickness of these areas in tobacco smokers and people who regularly consume alcohol.

Furthermore, there is a significant difference in tissue thickness of the supraglottic region between heavy and light smokers, with no significant difference between smokers and alcohol drinkers (Hirabayashi et al. 1990). Cigarette smoke and alcohol consumption are associated with an increased risk for cancer of the oral cavity, pharynx, and larynx, whereby smoking is a much stronger risk factor for larynx cancer, while alcohol drinking is a risk factor for oral cavity cancer, with synergistic effects in people who both smoke and drink (Choi and Kahyo 1991).

The trachea, the major segment of the conducting portion, has a mucosa with a pseudostratified columnar ciliated epithelium and a lamina propria containing elastic fibres. The epithelium contains columnar ciliated cells, goblet cells, basal cells, and the neuroendocrine cells of Kulchitsky. The submucosa consists of connective tissue and seromucous glands, followed by the C-shaped hyaline cartilage, with the open ends pointing into the oesophagus, precisely were the trachealis muscle is positioned and attached to the inner ends of the cartilage. The trachea branches to form the right and left primary extrapulmonary bronchi that enter the hilus of each lung in proximity to the pulmonary artery, pulmonary vein, nerves, and lymphatic vessels.

After entering the lung parenchyma, the bronchi divide into intrapulmonary secondary bronchi, three on the right and two in the left lung, according to the number of lobes in each lung. The secondary bronchi subdivide into segmental tertiary bronchi (ten in the right and eight in the left lung), each supplying a bronchopulmonary segment. Further divisions generate progressively smaller subsegmental bronchi, each one continuous with a terminal bronchiole that supplies a pulmonary lobule (Kierszenbaum 2011).

Extrapulmonary bronchi have the same structure of the trachea but with a complete hyaline cartilage ring. The secondary bronchi are lined by a pseudostratified columnar ciliated epithelium with goblet cells supported by a lamina propria containing seromucous glands and several hyaline cartilage plates. As the branching process advances, we observe a reduction in the number and dimensions of cartilage plates, as well as glands and goblet cells, and the epithelium cells shape becomes smaller, with fewer ciliated cells. Smooth muscle fibres develop between the cartilage plates and the mucosa. Bronchi have clusters of neuroepithelial cells, the neuroepithelial bodies, which are chemoreceptors that detect variations of carbon dioxide and oxygen in the air (Kierszenbaum 2011).

The bronchioles are lined by a columnar ciliated epithelium with few goblet cells in the initial portions, and become cuboidal epithelium with few cilia, Clara cells, and absence of goblet cells. The supporting lamina propria is composed of smooth muscle, and elastic and collagenous fibres. The bronchioles divide into five to seven terminal bronchioles that are initially lined by a simple cuboidal epithelium, with few ciliated cells, and the cubic shape of the cell becomes smaller, with non-ciliated epithelium. Some bundles of smooth muscle cells and elastic fibres are observed in their wall. Terminal bronchioles give rise to two respiratory bronchioles which are the transition from conducting to the respiratory portion of the lung. These bronchioles are formed by a simple cuboidal epithelium and already have alveoli in their walls.

The terminal bronchiole and the associated region of pulmonary tissues constitute the pulmonary lobule, which includes the respiratory bronchioles, alveolar ducts, alveolar sacs, and alveoli. The alveolar duct is the distal portion of the respiratory bronchiole with alveoli in the walls and a few groups of cuboidal cells, underneath of which are situated small bundles of collagen, elastic fibres, and smooth muscle cells. The duct is continuous with the alveolar sacs and the sacs consist of several alveoli (Figure 12.4).

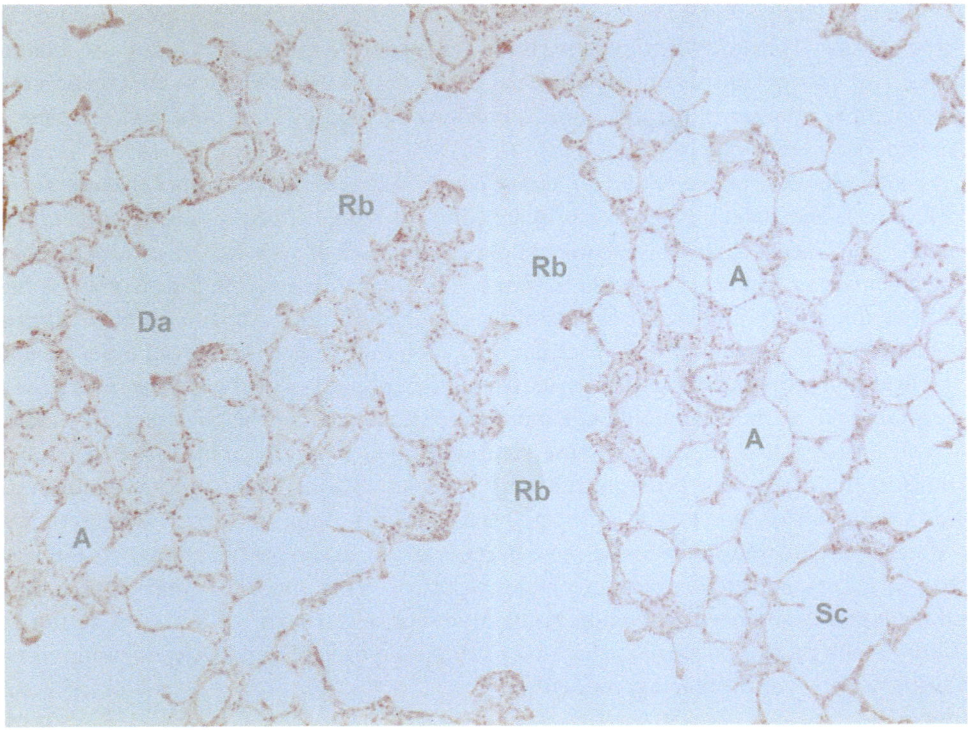

FIGURE 12.4 Light microscopy image of human lung (400×), with respiratory bronchiole (Rb), alveolar duct (Da), alveolar sacs (Sc), and alveoli (A).

The alveolar epithelium consists of two cell types: type I alveolar cells involved in gas exchange and type II alveolar cells that secrete surfactant which reduces the surface tension on the alveolus and prevents alveolar collapsing. Type I are squamous epithelial cells with a flat basal nucleus and a thin cytoplasm that line 95% of the total alveolar surface and can only be identified by specific lectins (McGowan and Snyder 2004) or specific proteins, such as T1-α (Borok et al. 1998) or aquaporins (Sato et al. 2004). Type II are polygonal cells, with a free surface with short microvilli and dense membrane-bound lamellar bodies in the cytoplasm, representing the secretory granules that contain pulmonary surfactant. The surfactant is released by exocytosis and then spreads and forms a thin layer of fluid that lowers the surface tension at the air–fluid interface, reducing the tendency of the alveolus to collapse at the end of expiration. Surfactant contains phospholipids, cholesterol, and specific surfactant proteins (SP-A, SP-B, and SP-C). An additional function for type II cells is the maintenance and repair of the alveolar epithelium when type I are damaged. They can increase their number and differentiate into type I.

The interstitium is the tissue between two adjacent alveoli on the alveolar septa, which contains capillaries, elastic and collagen fibres, fibroblasts, macrophages, and cells that store lipids and vitamin A. The interalveolar septa have small orifices, the alveolar pores or pores of Kohn, which allow communication between adjacent alveoli. The alveolar macrophages (also called dust cells) migrate from the interalveolar septa into the alveoli lumen to phagocyte bacteria and dust, and they are also responsible for the surfactant turnover (Kierszenbaum 2011).

The alveolar capillaries are lined by continuous endothelial cells that are juxtaposed to type I alveolar cells and constitute the blood–gas barrier for the gas exchange passive diffusion. This barrier consists of type I alveolar cells cytoplasmic extensions, endothelial cells cytoplasmic extensions, the dual basal lamina synthesised by both cells, and the plasma membrane of red blood cells.

Each lung is contained within a pleura cavity and covered by the pleura that consist of two layers. The visceral layer, closely attached to the lung, is lined by a simple squamous epithelium (mesothelium). The parietal layer is also lined by mesothelium and supported by a connective tissue with abundant elastic and collagen fibres. The pleural membranes secrete a lubricating fluid, which allows them to move freely against each other during ventilation. The normal pleural space has only a few millilitres of liquid, but fluid, air, and particles can move into the pleural space from different parts of the body because of its low pressure and its ability to hold large amounts of liquid or air. This is clinically relevant, because several conditions may cause pleural effusion (abnormal production of pleural fluid in the cavity that prevents full lung expansion during breathing); haemothorax (accumulation of blood in the pleural space due to trauma or surgery); hydrothorax (water accumulation in the space); pneumothorax (air in the pleural space that can lead to a partial or complete compression of the lung); pleurisy (an infection of the pleural cavity); and pleural tumours (mesothelioma). Pneumothorax can occur following lung injury, chronic obstructive pulmonary disease or other lung disease, tuberculosis, ruptured air blisters (blebs), or mechanical ventilation.

12.5 Phylogeny of the Respiratory System

Generally speaking, respiratory organs develop either by evagination (i.e. they project outwards from the body) (Figure 12.5A) or by invagination (i.e. they extend inwards into the body) (Figure 12.5B) (Maina 1998a, 2015). Gills (Figure 12.5C) fall in the first category, while

FIGURE 12.5 (opposite) Gas exchangers fall into two categories: evaginated (A) and invaginated (B). External gills (A) and so-called internal gills (C) fall into the first category, while lungs (D) represent invaginated ones. In (C), in a teleost fish, operculum has been removed to show internal gills (arrows), while in (D), lung of African tree-frog (*Chiromantis petersi*) is shown, with asterisks depicting respiratory air spaces. (E) Latex corrosion cast of airway (bronchial) system of lung of pig (*Sus scrofa*). Tr, Trachea; TB, tracheal bronchus. (F) Scanning electron micrograph showing an acinus of *S. scrofa* lung. RB, Respiratory bronchiole; asterisks, alveoli. (G) Scanning electron micrograph showing alveoli (asterisks) of lung of greater bush-baby (*Cercopithecus aethiops*). Boxed area, Interalveolar septum; arrows, interalveolar pores (pores of Kohn); circles, blood capillaries. (H) Transmission electron micrograph showing components of blood–gas (tissue) barrier of lung of emu (*Dromaius novehollandiae*). Arrows, Epithelial cell; asterisks, basement lamina; En, endotheial cell; Er, erythrocyte; Al, alveolar air space. (I) Type II cell (asterisk) of lung of epauletted fruit bat (*Epomorphorus wahlbergi*). Arrows, Osmiophilic lamellated bodies, precursors of surfactant; AS, alveolar air spaces; Er, erythrocytes contained in blood capillary. (J) Transmission electron micrograph of alveolar macrophages (asterisk) of lung of baboon (*Papio anubis*). Arrow, Filopodia; BC, blood capillaries; stars, vesicular bodies; Er, erythrocytes contained in blood capillaries.

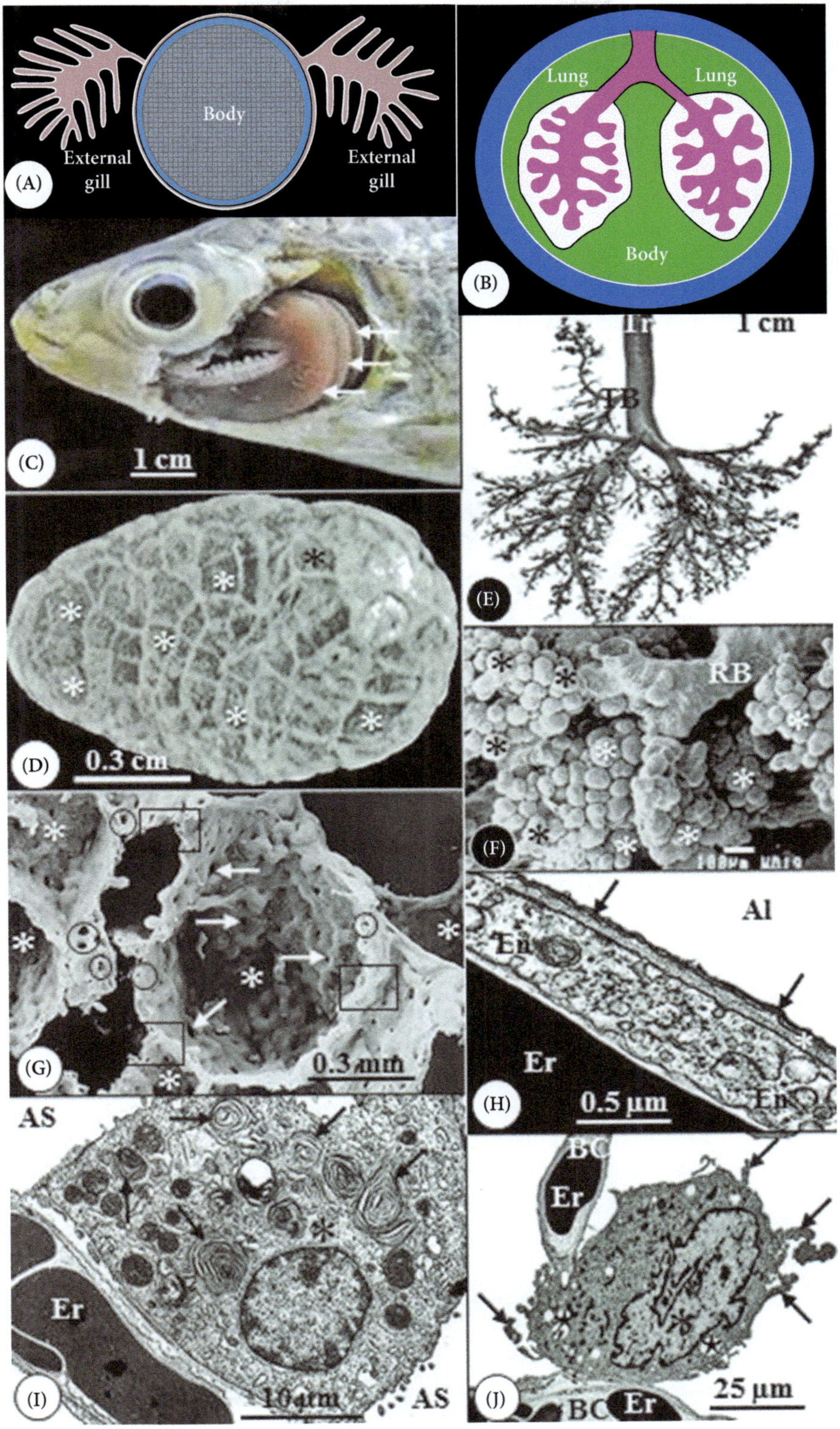
External gill
Body
External gill
(A)
Lung
Lung
Body
(B)
1 cm
(C)
1 cm
TB
(E)
0.3 cm
(D)
RB
(F)
(G)
0.3 mm
Al
En
Er
(H)
0.5 µm
En
AS
Er
(I)
10 µm
AS
BC
Er
(J)
BC
Er
25 µm

lungs (Figure 12.5D) fall in the latter. The physicochemical differences between water and air, especially those relating to how O_2 and CO_2 dissolve or chemically react with the fluid medium, have importantly determined the functional designs of gas exchangers (Dejours 1988; Hlastala and Berger 1996; Maina 1998a, 2002a, b, c, 2011). Accordingly, gills and lungs have developed and specialised so differently and to such extents that a respiratory organ that functions well in a fluid medium works dismally in the other. For example, when a fish is removed from water it dies from lack of O_2 (anoxia), albeit the fact that air contains more O_2 than water.

Invaginated respiratory systems, by contrast, reduce water loss on land. That is, if the adult human lung, which has a respiratory surface area of 142 m^2, a size equivalent to that of a tennis court, and which is crammed in 4.5 L of lung volume (Gehr et al. 1978), was evaginated, that is, the organ protruded outwards from the body into the surrounding air, even in moderately desiccating conditions, water loss would be ~ 500 L per day, a rate that is about 1000 times greater than that which occurs under normal conditions (McCutcheon 1964). Death from loss of water would occur in less than 3 minutes, a time shorter than that from asphyxia.

A particular functional disadvantage that characterises invaginated gas exchangers is that they can only be ventilated bidirectionally, that is, tidally or in-and-out. Unavoidably, dead-space air exists in the airways (Comroe 1974; West 2008). In the human lung, where the volume of the dead space is ~ 150 cm^3, out of the total volume of inhaled air of ~ 500 cm^3, only 70% of the air in the lung is replaced with 'fresh' air. Since the inspired air is 'diluted' by the CO_2-loaded air in the dead space, invaginated gas exchangers cannot fully exploit the high PO_2 in the atmosphere. For example, in the mammalian lung, at sea level the PO_2 drops from 21 kPa in the ambient air to ~ 13 kPa at the alveolar level, a decrease of ~ 40% (Comroe 1974; West 2008). A beneficial attribute of invagination is, however, that local microenvironmental conditions can form within the organ. In the vertebrate lung, the partial pressure of CO_2 (PCO_2) in the intrapulmonary air is higher than that in ambient air. This is physiologically important, because CO_2 is vital in the $-HCO_3^{-1}$ buffering system that regulates the pH of blood. Such microenvironments cannot exist in fish gills that are continuously and unidirectionally perfused with water (Hughes and Shelton 1958, 1962; Lauder 1984).

The primordial respiratory system that evolved for water 'breathing' were the gills, a multifunctional organ responsible for extraction of oxygen from water but also for osmoregulation, acid-base regulation, and excretion of nitrogenous waste (Perry 1997; Evans et al. 2005). It is formed by a large number of filaments spaced out along the gill arches on either side of the pharynx. Each filament has a series of projecting plates, numerous secondary lamellae, where gaseous exchange takes place, forming a fine sieve which ensures that all the water comes into close contact with the blood (Maina 2000b; Evans et al. 2005).

The epithelial surface of a gill arch is structurally and functionally zoned (Maina 2000b). The filaments are covered by two distinct epithelial surfaces: the primary epithelia (or filament epithelia), where non-respiratory functions of the gills take place; and the secondary epithelia (or lamellar), where gas exchange occurs (Laurent and Dunel 1980; Maina 2000b). The exchange of oxygen and carbon dioxide takes place by diffusion from the surrounding water and blood within the capillary network of the gills, and because of this counter-current flow (water/blood), fish can extract 80–90% of dissolved oxygen in the water (Schmidt-Nielsen 1971).

The evolutionary transition from aquatic to terrestrial life led to adaptations in locomotion, breathing, hearing, and other functions, and the first air-breathing vertebrates were fishes. The Devonian air-breathing sarcopterygian (lobefin) occupies the basal position in the lineage that extends from the Palaeozoic fishes to tetrapods (Liem 1986; Clack 1994). The evolution of tetrapods from sarcopterygian fish is one of the major transformations in the history of life and involved numerous structural and functional adaptations (Zhu and Yu 2002). In our days, the lungfish are represented by three genera: the Australian lungfish (*Neoceratodus forsteri*), African lungfish (*Protopterus*), and South American lungfish (*Lepidosiren*) (Bemis and Burgreen 1986). *Neoceratodus* has efficient gills and one lung that is only used to breathe air for short periods and during times of high activity (Kemp 1986; Power et al. 1999). The lung consists of a single elongated chamber compartmentalised by a thick cartilaginous structural scaffold, lined by epithelial cells and with abundant capillaries interspersed (Power et al. 1999). These cells that are the gas-diffusing surface have large numbers of osmiophilic bodies resembling mammalian lamellar bodies and contain a surfactant-like material comprising both SP-A- and SP-B-like proteins. It is possible that these lungfish cells may be the common ancestral cell for the alveolar type I and II found in the mammalian lung (Power et al. 1999).

The other two lungfish genera have paired lungs and much reduced gills (Shelton 1970; Burggren and Johansen 1986) and are considered bimodal breathers that use both gills and lungs for respiratory gas transfer. They are obligatory breathers, because they die if denied access to air (Burggren and Johansen 1986; Greenwood 1986). In *Protopterus*, the gills and skin take up only 10% of the total O_2 and these structures are much more effective in removing CO_2 (Lenfant and Johansen 1968). In *Lepidosiren*, 99.15% of total diffusing capacity lies in the lungs, showing the importance of lungs as gas exchange organs. Oxygen uptake is accomplished by the lungs and carbon dioxide is eliminated by the skin (de Moraes et al. 2005).

The fish–tetrapod transition was one of the most remarkable events in vertebrate evolution. The first tetrapods that appeared in the Late Devonian, about 360 mya, appear to have been primarily aquatic animals (Long and Gordon 2004). The freshwater origin of tetrapods remains the most likely scenario, but several recent findings raise the possibility that the tetrapod land invasion could have come from a marine habitat (Graham and Lee 2004).

The loss of gills brought about several advantages, such as improved head mobility, development of hearing, and the origin of different ventilatory and feeding mechanisms (Graham and Lee 2004; Long and Gordon 2004). The loss of gills is also linked to the development of cutaneous respiration as a site for gaseous exchange which can function in water and land.

Together, the evolution of respiratory mechanisms for terrestrial life originated in ancestral fish adapted for oxygen uptake and CO_2 elimination in an aquatic medium, which, under conditions of low oxygen in the water, adapted to breathe oxygen from the air, while the gills still functioned for CO_2 elimination.

At this stage of evolution, CO_2 was mainly eliminated via the skin. When over time the lung became more efficient in oxygen uptake and elimination of CO_2, the skin became less important in this regard. The need to reduce water loss selected for the evolution of harder scales, which eventually allowed animals to survive outside the water for longer periods of time (Hughes 1966).

The transition from an aquatic to land environment exposed the gas exchange organ to a much richer oxygen ambience, which allowed a drastic reduction in the ventilation requirements, but created problems for the disposal of carbon dioxide, because at 20°C the water solubility of this gas is 28 times greater than that of oxygen (Dejours 1989). To prevent a severe respiratory acidosis, land animals began to use the skin as an important respiratory organ, designed especially for the removal of CO_2 (Howell 1970).

Amphibians have simple and rudimentary lungs that are sufficient for ectothermic and low aerobic metabolism (Maina 2002b, c). Depending on the species, amphibians' lungs differ greatly in size, topographic extension, and exchange surface by the development of interconnected folds with a highly varying number of subdivisions and height of their folds (Duncker 2004). These differences are due to the amount of gas exchange performed via lungs in concert with cutaneous and buccal cavity exchange (Duncker 2004). Furthermore, in the absence of ribs or diaphragm, the pulmonary ventilation is mainly accomplished by swallowing air, which is carried out by the rising oral cavity floor (Carrier and Wake 1995).

The morphological heterogeneity of the amphibian gas exchange matches the environment and lifestyle diversity of amphibians. The skin is the main pathway for gas transfer in aquatic species, while in the terrestrial species it has been relegated or rendered redundant (Maina 2002b, c). The lungs of terrestrial species are highly elaborate, presenting a series of stratified septa that divided the lung, and the orders of *Anura* and *Apoda* have more complex lungs than those of *Urodela* (Maina 2002b, c).

Reptilians were the first vertebrates properly adapted for permanent terrestrial habitation and utilisation of lungs as a sole pathway for oxygen acquisition (Maina 2002b, c). The skin was no longer necessary for gas exchange. Instead, it evolved to protect against dehydration and injury, and in many reptilian species the skin became covered with keratinised epidermal scales or developing dermal bone plates (Howell 1970).

Reptilians exhibit great structural pulmonary heterogeneity, which can be divided into three types: profusely subdivided multicameral lungs (in turtles, monitor lizards, crocodiles, and snakes); simpler lungs (paucicameral) (in chameleons and iguanids); and saccular, smooth-walled, transparent unicameral lungs (the teju lizard) (Maina 2002b, c).

The lungs are localised in the pleuroperitoneal cavity and are not separated from the abdominal cavity by a diaphragm. The presence of ribs and intercostal muscles allows the development of more effective pulmonary ventilation compared to amphibians, which do not bear any one of these anatomical structures.

Generally, the pattern of organisation in the respiratory system of reptiles is identical to mammals, with the lungs coated externally by a serosa (Fleetwood and Munnell 1996), and the air-conducting portions supported by complete cartilaginous rings, which continue through the extra- and intrapulmonary bronchi. The branching of the bronchial intrapulmonary tree is similar to that of mammals; however, they have specific designations (Perry 1983), sequentially designed as bronchus, tubular chambers, niches, and aedicules. The homologous structures in mammals are the respiratory bronchioles, alveolar channels, alveolar sacs, and alveoli, respectively.

The epithelial cells lining the respiratory surface of reptilian lungs are differentiated into type I and type II cells. They also possess multilamellar bodies (Solomon and Purton 1984; Fleetwood and Munnell 1996; Maina 2002b), similar to those present in mammals (Kikkawa et al. 1965; Askin and Kuhn 1971).

Some 300 mya, the ancestors of modern reptiles finally emerged completely from water and made a commitment to air breathing, developing the two great classes of vertebrates with maximal oxygen consumption: mammals and birds.

The respiratory system of birds is the most complex and efficient gas exchanger that has evolved in air-breathing vertebrates (King 1966; Duncker 1971; McLelland 1989; Maina 2005). Their respiratory system allows them to acquire adequate amounts of O_2 under extreme altitudes where the air is hypoxic (Scott 2011; Scott et al. 2014; Bishop et al. 2015; Scott and Dawson 2017), an activity out of reach of any mammal (Tucker 1968; Lasiewski 1972; Maina 1988), including bats, which would excessively lose heat due to their bare skin. The exposure to hypoxia causes an immediate increase in breathing, but birds have the ability to adjust peripheral heat dissipation to facilitate the depression of body temperature, which reduces the metabolic demand, allowing them to fly high and for long periods (Scott et al. 2008). Most importantly, by means of still unclear mechanism(s), unlike mammals, birds can tolerate a high degree of arterial hypoxaemia: they have an unrivalled tolerance of arterial hypocapnia (Grubb et al. 1977; Black and Tenney 1980; Faraci and Fedde 1986; Bernstein 1990; Faraci 1990). During avian flight, the increased ventilatory rate is not accompanied by an increase in the tidal volume, and consequently excessive flush-out of CO_2(Bernstein 1987). By lowering the ΔPO_2 between the arterial blood and that in the inhaled air (Shams and Scheid 1987) and by generating a Bohr effect, in some species of bird, for example, the bar-headed goose (*Anser indicus*) (an extreme high-altitude champion flyer) (Scott et al. 2014; Bishop et al. 2015), low arterial PCO_2 increases blood O_2 content for given PO_2 (Grubb et al. 1978). Furthermore, the hyperventilatory response during high-altitude hypoxia increases blood flow to the brain (Faraci et al. 1984; Pavlov et al. 1987). Interestingly, a hypocapnic bird benefits from increased cerebral O_2 delivery compared to a normoxic one (Grubb et al. 1979).

In total contrast, in humans, lowering of the arterial PCO_2 to 1.3 kPa leads to cerebral vasoconstriction that results in extreme reduction of blood flow to the brain by ~ 50% (Wollman et al. 1968). In *A. indicus*, which tolerates hypoxia at simulated altitude of 11 km (Black and Tenney 1980), cerebral oxygen flow is not a limiting factor to flight activity (Faraci et al. 1984): up to an altitude of 6.1 km, *A. indicus* maintains normal gas exchange without hyperventilating, and at ~ 11 km, when the O_2 concentration is only 1.4 mmol/l, it acquires adequate O_2 by only minimal increase in ventilation (Black and Tenney 1980). Under hypoxia, Fedde et al. (1989) observed that blood supply to muscles and O_2 loading from the muscle blood capillaries, and not ventilation or pulmonary gas transfer, are the limiting parameters for the contraction of the flight muscles of *A. indicus*.

The main morphological and physiological properties that provide the highly efficient respiratory system of birds comprise: a cross-current design and inbuilt multicapillary serial arterialisation system; an auxiliary counter-current system; large tidal volume; large cardiac output; continuous and unidirectional parabronchial ventilation; short pulmonary circulatory time; a particularly large respiratory surface area; and a remarkably thin blood–gas (tissue) barrier (Scheid and Piiper 1972; Maina 1989, 2000a, 2005, 2017a, b; Maina et al. 1989; Maina and West 2005; Hsia et al. 2013).

The respiratory system of birds comprises the lungs (the gas exchanging part) and a series of air sacs: capacious and transparent non-vascularised air chambers that disseminate in the coelomic cavity pneumatising bones (King 1966; Duncker 1971; McLelland 1989; Maina 2005). The air sacs are not directly utilised for gas exchange (Magnussen et al. 1976). The anastomoses of the airways and the air capillaries allow unidirectional air flow in the avian lung, even though like the mammalian one it is tidally ventilated (Schmidt-Nielsen

1971; Scheid 1982; Maina 1988, 2000b, 2003; Duncker 2004). The lack of a diaphragm displaced the lungs in the coelomic cavity, where they come to be attached to the ribs (Maina 2000b); unlike in mammals, in birds the lungs do not surround the heart, it is the liver that stays next to the heart (McLelland 1989). Intercalated between the air sacs, the lungs are largely continuously ventilated back-to-front by a concerted action of the cranial and caudal groups of air sacs (Maina 2000b; Duncker 2004). The pair of lungs are relatively small and non-compliant (Duncker 2004), with little movement between respiratory cycles (Jones et al. 1985). Air is driven through the lung continuously and unidirectionally, by synchronised activity of cranial and caudal air sacs (King 1966; Duncker 1971; Thompson 2007).

The trachea has a cartilaginous or partially ossified support ring lined by a pseudostratified ciliated epithelium with goblet cells (Banks 1986) and bifurcates into two primary bronchi, lined by a similar epithelium but with incomplete cartilaginous rings, which disappear or are reduced when they reach the bronchial lung parenchyma (Banks 1986). The connection between the primary, secondary, and tertiary bronchus is labyrinthic, markedly opposed to the monopodic branch in mammals. Along the inner surface of the tertiary bronchus, small vesicular structures with hexagonal shape emerge, called atria (Maina 1988; Klika et al. 1996; Duncker 2004).

The atria have granular cells (Klika et al. 1996), whose cytoplasm contains multilamellar bodies and are considered analogous to the type II alveolar cells of the mammal's lungs (Lorz and Lopez 1997); and squamous cells, which line the inner surface of the atria (Scheuermann et al. 1997). The blood capillaries are surrounded by extremely small air capillaries, which give an appearance of a dense network, and together they form the gas exchange surface of the bird's lungs.

When comparing the lungs of birds and mammals, the most fundamental difference is the separation of the gas exchange function from the ventilatory function. The bioengineering requirements for these two functions are so dissimilar that it is remarkable that the mammalian lung has combined the two functions and does not encounter more frequent problems (West et al. 2007). In the bird lung, ventilation is carried out by non-vascular air sacs, which are robust, in contrast to the mammalian lungs that have a deformable parenchyma. The blood–gas barrier of birds is much thinner and uniform than that found in mammals (Maina and King 1982), an obvious advantage for gas exchange.

The other fundamental difference is the ventilation flow, which is unidirectional and continuous in bird lungs (Fedde 1980, 1986), while mammalian lungs have a reciprocating pattern of ventilation that results in additional complications. The mammalian lung does not empty completely during expiration and the convective flow alone cannot take the inspired gas to the periphery of the lung where some of the gas-exchanging alveoli are located (West et al. 2007).

Mammals probably evolved from a group of reptile-like vertebrates with a multicameral lung with three rows of lung chambers. The branched conducting bronchial system was originated by further subdivision of these lung chambers that open into respiratory bronchioli and ductus, covered with alveoli (Duncker 2004).

The airways form the so-called respiratory tree which consists of a regularly branching system of passageways (Horsfield and Thurbeck 1981; Weibel 1984; Horsfield and Woldenberg 1986; West et al. 1986) (Figure 12.5E) that pattern the pulmonary vasculature. The trachea bifurcates into left and right principal bronchi, which divide into lobar bronchi, which in turn divide into many generations that end in terminal bronchioles, which give rise to respiratory bronchioles, which terminate in alveolar ducts and sacs and alveoli (Weibel and Gomez 1962; Weibel 1963; Maina and Van Gils 2001) (Figure 12.5F). In gas exchangers, the geometry, size, and branching of the airways and the blood vessels are set by the immutable laws of physics that govern the flow dynamics of the respiratory fluid media, air/water, and blood, granting cost-effective layout and movement in the conduits (Weibel 1991; Metzger and Krasnow 1999; Cardoso 2000; Warburton and Bellusci 2004; Metzger et al. 2008; Warburton 2008; Warburton et al. 2010; Sapoval and Filoche 2013). Developmentally, the airway and the vascular systems of the mammalian lung are space-filling in nature, that is, they are fractal (Mandelbrot 1983; Nelson et al. 1990; Weibel 1991, 1997; Brown et al. 2002; Mancardi et al. 2008). This feature optimises the respiratory surface area and brings air and blood into very close proximity for efficient gas exchange.

In the human lung, the bronchial system consists of 23 generations that give off thousands of air conduits that terminate in millions of alveoli (Weibel 1963, 2000, 2009; Ochs et al. 2004; Kumar et al. 2009). Generations 0–14 comprise bronchi and terminal bronchioles that form the conducting airways, and generations 15–23 form the transition and respiratory zones that consist of respiratory bronchioles, alveolar ducts, alveolar sacs, and alveoli (Haefeli-Bleuer and Weibel 1988; Finlay 2001; Sapoval et al. 2002; Weibel 2009; Sapoval and Filoche 2013). In the human lung, there are ~ 150,000 first-order bronchioles of a diameter of ~ 5 mm (Weibel 1984, 2009). As they bifurcate, the airways decrease in diameter by a constant factor of the cube root of ½ ($2^{-1/3}$ or 0.79) (Weibel and Gomez 1962; Weibel 1963; Horsfield and Woldenberg 1986). The branching of the airways and the blood vessels determine the ventilation and the perfusion of the lung (Hastings and Powell 1986; Hopkins and Powell 1998). It is because the cross-sectional area of the airways increases with the generations of the airways (Weibel 1984; Hou et al. 2010) that in the lung, resistance to air flow decreases with the length (distance) of the airways (Weibel 1984). A terminal bronchiole supplies air to an acinus (Hansen and Ampaya 1975; Weibel 1984; Rodriguez et al. 1987; Haefeli-Bleuer and Weibel 1988; Weibel et al. 2005; Schittny 2017) (Figure 12.5F). In the human lung (Weibel 1984; Haefeli-Bleuer and Weibel 1988; Weibel et al. 2005; Tsuda et al. 2008), as many as 10,000 acini of an average volume of 183 mm^3 exist. In the rat and the rabbit lungs, the volumes of the acini are 1.83 and 3.46 mm^3, respectively (Rodriguez et al. 1987). At 60% total lung capacity, in the human, the rat, and the rabbit lungs, the average lengths of the acini are 8.8, 1.46, and 1.95 mm, respectively (Rodriguez et al. 1987). Termed 'the hallmark of lung structure' by Weibel (2008, p. 483), the alveolus (Figure 12.5G) is the terminal respiratory unit of the mammalian lung. There are ~ 480 million alveoli in an adult human lung (Ochs et al. 2004) compared to ~ 12–13 and 20 million, respectively, in the mouse (Jung et al. 2005; Knust et al. 2009) and rat lungs (Hyde et al. 2004). In the human lung, the mean alveolar volume is 4.2×10^6 μm^3 (Ochs et al. 2004). Interalveolar pores (the pores of Kohn) connect adjacent alveoli (Figure 12.5G): they provide collateral ventilation of the alveoli when respiratory bronchioles are blocked or part of the lung collapses (atelectasis). They may, however, provide a pathway for the spread of pulmonary infections. Pulmonary circulation is a low-pressure, low-resistance circulatory system where the pressure in the pulmonary artery is ~ 2 kPa (15 mm Hg) (Bachofen et al. 1988; Bachofen 2009).

The blood–gas barrier comprises an epithelial cell, a common basement membrane, and an endothelial cell (Figure 12.5H) (Weibel 1973, 1984). The harmonic mean thickness of the blood-gas barrier of the human lung is ~ 0.65 μm (Gehr et al. 1978). By comparison, the corresponding thickness of the barrier in the minute (~ 3 g) Etruscan shrew, *Suncus etruscus*, is 0.270 μm (Gehr et al. 1980, 1981) and 0.120 μm in the phyllostomid bat, *Phyllostomus hastatus* (Maina et al. 1991). Bat lungs generally have relatively thinner blood–gas barriers compared to those of the non-flying mammals (Maina et al. 1982; Maina and King 1984; Maina et al. 1991). Subdivision of the parenchyma giving rise to small terminal respiratory units (alveoli) increases the respiratory surface area (Gehr et al. 1981; Maina et al. 1991). For example, while a geometrical sphere of a volume of 1 cm^3 has a surface area of 4.8 cm^2, in the lung of the shrew, *Sorex minutus*, a respiratory surface area of 2100 cm^2 exists in the ~ 170,000 alveoli that are contained in 1 cm^3 of the volume of the parenchyma (Gehr et al. 1980; Ochs et al. 2004). In mammals, the highest mass-specific respiratory surface area of 138 cm^2/g has been reported in the lung of the epauletted fruit bat, *Epomophorus wahlbergi* (Maina et al. 1982). More than 40 different types of cells exist in the human lung (Breeze and Wheeldon 1977; Pinkerton and Joad 2000; Cardoso and Whitsett 2008; Fine 2009). At the alveolar level, the pneumocytes comprise type I and type II cells. The former are thin and expansive, and are largely devoid of organelles, while the latter are cuboidal in shape (Figure 12.5I) and well endowed with organelles (Crapo et al. 1982; Weibel 1984). In the human lung, a type II cell contains an average of ~ 200–500 osmiophilic lamellated bodies (Ochs et al. 2001) (Figure 12.5I). Type I cells comprise ~ 33% of the total number of pneumocytes and cover as much as 97% of the respiratory surface area. A type I cell covers an area of ~ 5000 μm^2 (Crapo et al. 1982; Weibel 2009). Ochs et al. (2001) estimated that ~ 24 billion type II cells, that is, 50 cells per alveolus, exist in the adult human lung, and in the mouse lung about one type II cell exists in an alveolus (Jung et al. 2005).

The capillaries are located in the alveolar walls which are widely separated from each other. Thus the blood–gas barrier has to withstand the full transmural pressure (West et al. 2006). The capillary is typically polarised, with one side having a very thin blood–gas barrier whereas on the other side the barrier is thicker (West et al. 2006) and contains strands of collagen which provides support for the alveolar wall and maintains the integrity of the alveoli (Weibel 1973). In contrast to a uniform thin blood–gas barrier in birds, in mammals half of the surface area of the capillaries provides inefficient gas exchange due to its increased thickness.

Functionally, in mammals there is no dissociation between locomotion and respiratory movements and both are closely synchronised, especially during exercise. The strong musculature of the diaphragm acts as a forceful inspiratory muscle together with the intercostal musculature, but is also responsible for maintaining a pressure gradient between the pleural and peritoneal activity during strong exercise (Duncker 2001). At rest, expiration is performed by elastic retractile forces of the extended rib cage and by the retraction forces of the lung itself (surface tension of the alveoli together with their extended elastic fibre system) (West et al. 2007). Mammalian lungs do not empty completely during expiration, and the result is that convective flow alone cannot take the inspired gas to the periphery of the lung where some of the gas-exchanging alveoli are located. Instead, the last part of the distance is accomplished by relatively large peripheral airways to allow mixing of the inspired air with that already in the lung, and the resulting large alveoli cause additional problems (West et al. 2006).

12.6 Human Respiratory Evolution, Adaptations, and Evolutionary Challenges

Hominid evolution has involved important changes, particularly in relation to the process of encephalisation and the transition to bipedalism. Some of these changes involved structures related to the respiratory system. The classic understanding of the evolution of hominids has been based on comparison with modern great apes (chimpanzees, bonobos, gorillas, and orangutans) and palaeontological findings of the various species of the genus *Homo* and other related genera.

The transition to an upright position of the body has consequences for our anatomy and physiology and posed an important challenge to the respiratory system. Bipedalism is believed to have developed in the context of the transition from a heavily wooded terrain to a more savannah-like landscape. In this savannah-like environment, the broader view of the surroundings that bipedalism allowed was very useful for catching sight of predators. Bipedalism gave hominids an important competitive advantage by improving their ability to survey the surroundings, allowing them to detect dangers at a greater distance and better identify feeding opportunities. It also frees the upper limbs, which gives hominids the chance to handle instruments that can be used for defence or to obtain food, and carry (furless) babies (Gea 2008).

As far as the respiratory system is concerned, bipedalism has led to some anatomical changes that are relevant, for example, with regard to the size of the anatomical dead space in ventilation. In great apes the pharynx is closely connected to the bones of the skull, while in modern humans only a small portion is in contact (Dean 1985). This, together with the different configuration of the lower part of the skull, has led to a wider and higher nasopharynx (Reznik 1990). Australopithecines, in contrast, seem to have a nasopharynx similar in shape to extant great apes (Reznik 1990), while the nasopharynx of ancestral members of the genus *Homo* was already quite similar to our own (Aiello and Dean 1990).

The larynx provides the source of acoustic energy for vowels and other phonated speech sounds, while the supralaryngeal vocal tract acts as an acoustic filter that determines the phonetic quality of the sounds (Lieberman 2007). However, the vocal tract structures such as the tongue, the hyoid bone, and the larynx are also involved in the process of swallowing. The hyoid supports the larynx, moving upward and forward, opening the oesophagus and placing the larynx in a position that prevents food entering the trachea during swallowing (Ishida et al. 2002). During human ontogenetic development, the tongue gradually descends into the pharynx, changing its shape from relatively long and flat to posteriorly rounded. This uniquely human developmental process is not complete until the age of 6–8 years (Lieberman and McCarthy 1999). As the human tongue descends, it carries the larynx down with it. The human neck gradually lengthens, which is critical for the larynx position (Mahajan and Bharucha 1994). If positioned at the level of the collarbone, it would be impossible to swallow (Palmer et al. 1992; Lieberman et al. 2001).

In chimpanzees, gorillas, and orangutans the larynx is in a more rostral position (near the base of the tongue) and is tilted back more than in modern humans, which makes coordinated vocalisation enormously difficult for apes. In non-human primates, the tongue is positioned almost entirely in the mouth. The larynges of young chimpanzees descend somewhat through elongation of the distance between the hyoid bone and the larynx, but their tongues do not descend (Nishimura et al. 2003). The epiglottis plays an important protective role in aspiration in modern great apes, while it is relatively unimportant in modern humans in this regard. However, in contrast to non-human primates, humans are at risk for inadvertently inhaling food particles both before and after swallowing (Palmer et al. 1992). The risk of choking was evolutionarily outweighed by the ability to produce articulated speech.

An interesting and unexplored question concerns the consequences that the special airway configuration of early hominids had on respiratory diseases, specifically on those related to sleep-disordered breathing. Modern great apes snore, but do not develop sleep apnoea (Barsh 1999). The larynx in adult humans is much lower than that of great apes, and although this position facilitates vocalisation, it also entails the development of a soft oropharynx, making it much easier for the larynx to collapse during sleep. By contrast, the trachea and large airways do not significantly differ between great apes and humans, except for small differences in size (Nakakuki 1991, 1992).

The complexity of mammalian lungs involves the presence of a respiratory tree with multiple branches ending in millions of alveoli (Randall et al. 1998) where gas exchange occurs. In great apes and humans the number of alveoli continues to increase after birth, reaching its maximum during childhood (Núñez and Cosio 2007). Given the fact that modern great apes' lungs are structurally similar to ours, we can deduce that the lungs of earlier *Homo* species were not different from our own.

Another question concerns the distribution of ventilation and perfusion in the lung, since both have a vertical gradient that depends on gravity and respiratory pressures (West 2005). In humans, the upright posture has meant the shift of the ventilation and perfusion gradient to a craniocaudal axis, while in other primates this gradient is mainly dorsoventral (Gea 2008).

In general, the rib cages of modern humans and great apes are similar, having a wide transverse diameter contrasting with a smaller ventrodorsal diameter, which gives rise to a flattened section in the anteroposterior direction. Nevertheless, the human thorax has a characteristic barrel-shaped appearance, while the great apes have a more bell-shaped or inverted funnel appearance (Beckman 1973). This implies a different configuration of rib cage muscle components, especially in the lower intercostal muscles and the diaphragm itself. This resulted in respiratory muscle mechanics that were more efficient for upright posture.

In summary, the evolution of vocalisation and the transition to the upright posture are important features of the genus *Homo* that involved a series of structural and functional changes in our respiratory system. The changes affect the relationship between the skull and the spinal column, together with alterations in laryngeal structure (allowing vocalisation), accompanied by the evolution of a soft and elongated oropharynx with part of the tongue being integrated into its anterior wall. Some of these adaptations are nowadays associated with an increased risk for respiratory diseases such as sleep apnoea (Gea 2008).

12.7 Respiratory System Diseases

The respiratory surface of the human lung is located only ~ 40–50 cm from the external environment (air) (Weibel 1984; Weibel et al. 2005). The close proximity of a large respiratory surface to the commonly polluted atmosphere across a thin blood–gas barrier makes the lung an important portal of entry and assault by pathogens, injurious particulates, and toxic gaseous substances. To counteract it, the respiratory system has developed various cellular defences that are augmented by inflammatory and immune responses (Nguyen et al. 1982; Nicod 2005; Maina 2012b). They include the following: (1) tightly packed epithelial cells that physically prevent entry of harmful agents and substances (Breeze and Wheeldon 1977; Harkema et al. 1991; Godfrey 1997; Nicod 2005); (2) ciliated epithelial cells which generate a mucociliary escalator system that traps and removes foreign agents towards the pharynx, where they are swallowed or expectorated (Kilburn 1968; Lippmann and Schlesinger 1984; Geiser et al. 2003); (3) a surfactant system that directly neutralises harmful biological agents (Gil and Weibel 1971; Gehr et al. 1990); (4) a blood–gas barrier that physically stops pathogenic microorganisms from entering blood and disseminating to the rest of the body (Weibel 1973, 1984); (5) phagocytic cells (i.e. surface macrophages) (Figure 12.5J), which engulf, sequester, and directly destroy harmful particles and pathogens (Geiser et al. 1990; Nicod 2005; Maina 2012b); (6) strategically located mucosal and bronchial lymphoid tissue that filters and destroys foreign agents; and (7) committed antigen-presenting dendritic cells that are highly phagocytic (Dreher et al. 2001; Kiama et al. 2001; Walter et al. 2001; Obregon et al. 2015). The efficiency of the pulmonary defence system is so great that below the larynx, the respiratory system is sterile (Skerret 1994). Albeit elegant defences, respiratory diseases are common, however, especially when polluted air is breathed (Peters et al. 1997).

Changes in human–environment interactions, shifts in subsistence practices and social adaptations, variations in human population health, and microbial adaptations have been associated with outbreaks of new infections (McMichael 2004; Wolfe et al. 2007; Woolhouse and Gaunt 2007; Aoshiba et al. 2015). For example, the start of agriculture and livestock husbandry from ~ 10,000 years ago (Gupta 2004; Thrall et al. 2010) and that of controlled use of fire between 0.2 and 1.7 mya (James 1989; Roebroeks and Villa 2011; Berna et al. 2012; Gowlett 2016), which after the evolution of speech was deemed by Charles Darwin (Darwin 1871) to be the most important human cultural innovation, impacted greatly on human host–pathogen interactions (Wrangham et al. 1999; Weiss 2001; McMichael 2004; Woolhouse and Gaunt 2007; Wolfe et al. 2007). Fire has provided a source of warmth and protection and served the purpose of cooking food. It allowed human geographic dispersal, cultural innovations, and changes to diet and behaviour (Wrangham and Carmody 2010; Attwell et al. 2015). Moreover, human activity could proceed into the dark and colder hours of the night (Wiessner 2014). Chisholm et al. (2016) speculated that by bringing about lifestyle changes (cultural and social) and directly injuring the lung through exposure to smoke, controlled use of fire by early humans during the Pleistocene created favourable conditions for the emergence of tuberculosis as a transmissible disease.

In the last century, the human lifestyle has changed dramatically, especially after the industrialisation period. Sedentariness, pollution, and medical care progress have brought new challenges for our respiratory system.

In Section 12.7.1, we refer to some respiratory diseases caused by this new way of life, such as obesity and asthma, and also to the respiratory challenges and diseases that arise from the great medical care advances, such as increased life expectancy of preterm infants and some genetic disease carriers.

12.7.1 Lifestyle and Respiratory Diseases

Sedentary societies and increased pollution are responsible for the development of some diseases that affect the global population. The metabolic syndrome, a concept that includes obesity, diabetes mellitus, high blood pressure, and hypercholesterolaemia, is one syndrome caused by food abundance and reduced physical activity in developed countries. (For further discussion, see Chapter 6: Nutrition, Energy Expenditure, Physical Activity, and Body Composition.) Other health problems comprise chronic respiratory diseases such as COPD, asthma, obstructive sleep apnoea, obesity hypoventilation syndrome, and other forms of chronic lung disease associated with exposure to particles, such as silicosis (dust from coal mines), asbestosis (asbestos), farmer's lung, silo filler's disease, emphysema, and lung cancer related to tobacco smoking.

12.7.1.1 *Obesity and Respiratory Diseases*

Obesity is an epidemic global problem that has emerged as an important risk factor for respiratory diseases. Obesity plays a key role in the development of obstructive sleep apnoea syndrome (OSAS) and obesity hypoventilation syndrome (OHS), and is strongly linked with respiratory exertional dyspnoea, COPD, asthma, pulmonary embolism, and aspiration pneumonia (Zammit et al. 2010). The influence of obesity in respiratory diseases is complex and goes beyond the obvious mechanical and physical consequences of weight gain.

Two distinct patterns of obesity are recognised in the general population: central obesity (more common in males), with adipose tissue in the anterior chest wall, anterior abdominal wall, and visceral organs; and peripheral obesity (more common in women), with adiposity located peripherally in the subcutaneous tissue. Abdominal obesity is associated with worsening of lung function and respiratory symptoms, and is a significant risk factor for cardiovascular disease, type 2 diabetes, rheumatoid arthritis, cancer, and metabolic syndrome (Poulain et al. 2006). Peripheral obesity is associated with fewer medical complications and better lung function (Zammit et al. 2010).

The thoracic restriction associated with obesity is attributed to the mechanical effects of fat deposited on the diaphragm and the chest wall. The diaphragm excursion is impeded, the chest wall compliance is reduced, and the respiratory muscles are weakened, which increase the work of breathing and airway resistance (Ochs-Balcom et al. 2006). The adipose tissue in the anterior abdominal wall and intra-abdominal visceral tissue hinders diaphragmatic movement, diminishes basal lung expansion during inspiration, and, with the closure of peripheral lung units, causes ventilation–perfusion abnormalities and arterial hypoxaemia. These changes contribute to an increase in prevalence of respiratory problems in obese individuals, particularly during exertion and sleeping in the supine position (Zammit et al. 2010).

There is a clear association between dyspnoea and obesity. The reductions in chest wall compliance and respiratory muscle strength create an imbalance between the demand on the respiratory muscles and their capacity to generate tension, which leads to the perception of increased breathing effort (Poulain et al. 2006). In obesity, the respiratory muscle function deteriorates, which may contribute to the additional oxygen demand required for ventilation and heightened sensation of breathlessness in obese patients. The elevated mechanical workload of obesity can overburden respiratory muscles through a combination of increased work of breathing and apparent reduction in respiratory muscle efficiency (Gibson 2000).

Obesity is a well-recognised risk factor for OSAS that is characterised by intermittent upper airway obstruction due to disturbances in the neuromuscular control of the pharyngeal dilatory muscles that fail to maintain upper airway patency in the presence of alterations in airway shape and diameter. The obstruction causes a decrease in arterial oxygen content, an increase in carbon dioxide levels, and inspiratory efforts, leading to abrupt awakenings as the person struggles to breathe (Poulain et al. 2006). The increase in fat tissue deposition in the neck and pharyngeal region and the reduced operating lung volumes in obesity act together to reduce upper airway calibre, modifying airway configuration and enhancing the collapse of airways, predisposing to repetitive closures (Poulain et al. 2006).

Asthma seems to be more common in the overweight and obese population, and the frequency of self-reported symptoms of breathlessness and wheezing increases with body mass index (BMI), linking obesity with asthma (Beuther and Sutherland 2007). Without demonstrable correlation between obesity and airway inflammation, it has been speculated that the epidemiological association between obesity and asthma is primarily due to ventilatory mechanics rather than an inflammatory cause (Zammit et al. 2010).

The interaction between BMI and asthma is stronger in women than men, suggesting that female sex hormones may play a role in the increased prevalence of asthma among obese women. Oestrogens may modulate the immune response and increase the risk of asthma (Kim and Camargo 2003).

COPD and obesity share a complex interplay of similarities, which seems to compound each condition. Both are associated with deterioration in lung function, hypoxia, and a low-grade systemic inflammation, which predispose to an increase in morbidity and mortality. In COPD and obesity, low-grade inflammation and arterial hypoxaemia have been associated with a reduction in the skeletal muscle tissue, a decrease in muscle fat oxidative capacity, and a loss of respiratory muscle performance (Zammit et al. 2010). Physiological and metabolic factors related to COPD and obesity seem to jeopardise morbidity and mortality further when in association. The combined effects of obesity and COPD add to lung function deterioration, but this interaction is complex and requires additional studies.

12.7.1.2 *Respiratory Diseases Due to Exposure to Novel Environments*

Asbestosis is an interstitial pulmonary fibrosis caused by inhaled asbestos dust. An excess of asbestos, composed of silica, iron, sodium magnesium, and other metals, causes release of macrophages, chemical agents producing alveolitis, and eventual lung fibrosis (increased collagen deposition). The asbestos fibres are heat-resistant fibrous silicates, long and thin (length:diameter ratio > 3), either curved or straight, that penetrate cell membranes. They are incompletely phagocytised and stay in the lungs, setting up cycles of cellular events and the release of cytokines. The initial inflammation occurs in the alveolar bifurcations and is characterised by the influx of alveolar macrophages. Asbestos-activated macrophages produce a variety of growth factors which interact to induce fibroblast proliferation. The oxygen-free radicals (e.g. superoxide anion, hydrogen peroxide, hydroxy radicals) that are released by the macrophages damage proteins and lipid membranes and sustain the inflammatory process. A plasminogen activator, which is also released by macrophages, further damages the interstitium of the lung by degrading matrix glycoproteins.

Patients have breathing difficulties and an increased risk of developing cancers, including lung cancer and pleura mesothelioma. It is regarded as an occupational lung disease of mining, manufacturing workers, and people that handle or remove asbestos.

Emphysema is an abnormal permanent enlargement of distal air spaces and terminal bronchioles, associated with the destruction of their walls. The loss of elasticity and breakdown of elastic fibres, one of the important components of bronchiole and alveolar walls, give rise to a chronic airflow obstruction. As a result, adjacent alveoli become confluent, creating large air spaces or blebs. Terminal and respiratory bronchioles are also affected by the loss of elastic tissue and, as a result, the small airways tend to collapse during expiration, leading to chronic airflow obstruction and secondary infections. In centroacinar emphysema, the respiratory bronchioles are affected, but the alveolar duct and alveoli are intact. In panacinar emphysema, blebs are observed from the respiratory bronchiole down to the alveolar sacs. This type of emphysema is more common in patients with deficiency in the glycoprotein alpha1-antitrypsin (AAT), a major inhibitor of elastase that is secreted by neutrophils during inflammation. If not inactivated by AAT, elastase destroys lung parenchyma connective tissue (elastolysis), leading to emphysema.

Cigarette, cigar, and pipe smoking are the principal risk factors for the development of emphysema, but occupational, inhalational, and environmental exposures to similar toxic particles are also on the list. In the developing countries, women are exposed to biomass cooking of liquids and fuels, including wood, crops, animal dung, and coal, and have an increased risk of developing COPD. Throughout the world, COPD is a disease of occupation and environmental pollutants, and for this reason environmental protection and improvements of air quality are extremely important in the prevention of such diseases (Kierszenbaum 2011).

Asthma is a chronic inflammatory process characterised by the reversible narrowing of the airways (bronchoconstriction) in response to diverse stimuli. The airway hyperresponsiveness is defined by three salient features: the airway wall inflammation that involves activation of mast cells, infiltration of eosinophils, and increased activated T cells, which mediate allergic inflammation through the secretion of an array of cytokines; mucus luminal obstruction caused by the hypersecretion of bronchial mucous glands along with infiltration of inflammatory cells; and vasodilation of bronchial microvasculature with increased vascular permeability and oedema. Chronic inflammation can lead to structural changes in the airways, including airway smooth muscle hypertrophy and hyperplasia (Marandi et al. 2013). Asthma can be triggered by repeated antigen exposure (allergic asthma) or by abnormal autonomic neural regulation of airway function (non-allergic asthma).

In the past, bronchodilators were used for asthma management, but the introduction of corticosteroids has added benefit in the treatment of the underlying inflammatory component of asthma, reducing hospitalisation and mortality (Marandi et al. 2013).

12.7.2 Medical Care Advances and Respiratory Disease

12.7.2.1 *Preterm Births and Respiratory Disease*

In recent years, the rate of preterm births has risen markedly in many industrialised countries. Maternal infections are clinically identified to be responsible for 30–40% of preterm births, but the latter may also be associated with rupture of membranes, pregnancy-induced hypertension, intrauterine fetal growth retardation, low socioeconomic status, increased maternal age, and the impact of fertility treatments (Albertine and Physher 2004).

Advances in neonatology treatments have increased the survival of preterm infants who are at risk of respiratory morbidity and mortality due to a structurally and functionally immature lung (Vollsæter et al. 2013). Preterm infants are born before the lungs are sufficiently mature to sustain adequate gas exchange, and therefore require invasive treatments that include assisted ventilation and long periods of oxygen supplementation (Smith et al. 2010). Because the lungs are still developing, they are vulnerable to injury and disturbances in their fine-tuned programmed development (O'Reilly et al. 2013).

Oxygen administration exposes the immature lung to high concentrations of oxygen, which is the major contributor for lung injury and development of bronchopulmonary dysplasia (BPD), the main significant pulmonary complication of preterm birth. The hyperoxic gas can instigate the generation of oxygen-free radicals, known as ROS, which can directly damage DNA and proteins, cause lipid peroxidation, and contribute to inflammation. In recent years, lower levels of supplemental oxygen have been used, but the risk still continues, due to the immaturity of the new-born's antioxidant system and reduced ability to combat oxidative stress (O'Reilly et al. 2013).

Injury by overexpansion of the lung, continuous opening of the collapsed alveoli, and inflammatory mediators released during ventilatory support are common consequences of positive pressure of mechanical ventilation (MV) in the preterm lung, and all increase the risk of BPD. MV can disrupt elastin deposition in the lung, which can impair the formation of alveoli and lung microvessels. More recently, there has been an increased use of non-invasive ventilatory support techniques in an attempt to reduce the incidence of BPD. The absence of an endotracheal tube is thought to reduce the risk of trauma to the airways, reduce the risk of infection, and cause less acute and chronic lung damage (O'Reilly et al. 2013).

The immediate and long-term impact of preterm birth is inversely related to the gestational age at birth, with the greatest effects being seen in those born very preterm (28–32 weeks) or extremely preterm (24–27 weeks) (O'Reilly et al. 2013). Nevertheless, the majority of preterm births are late preterm, delivered between 32 and 36 weeks of gestation just in the end of the immature saccular stage of lung development.

Very preterm infants are delivered during the initial saccular stage of lung development before the development of definitive alveoli, and extremely preterm birth infants are born during the early canalicular stage, which corresponds to the formation of respiratory bronchioles, precapillaries, bronchi mucous glands, adequate interstitium thinned to form the blood–air barrier, and long before the onset of pulmonary surfactant production, which is vital after birth because it lowers the surface tension in the alveolar air–liquid interface (Smith et al. 2010). The immature pulmonary surfactant system of these infants requires exogenous surfactant administration to prevent alveolar collapse.

An immature pulmonary surfactant system in combination with structurally immature alveoli is the major cause of respiratory distress syndrome, which is common in extremely and very preterm infants that need prolonged respiratory support which causes lung injury and BPD development (O'Reilly et al. 2013).

Although definitions have changed and today the clinical expression is generally less severe, BPD continues to be the major risk to these children (Vollsæter et al. 2013) and this disease has long-term respiratory consequences (O'Reilly et al. 2013).

The original description of BPD includes pulmonary atelectasis, thickened interstitium, interalveolar fibrosis, emphysematous regions with dilated alveolar ducts, and disorganised elastin deposition. The conducting airways are persistently injured, probably due to hyperoxic gas and/or mechanical ventilation. Many bronchioles contain necrotic epithelium, some are occluded by luminal debris, and others show considerable epithelial injury. Hypertrophy of airway smooth muscle is also a common finding, as well as an increased number of mucus-secreting goblet cells (O'Reilly et al. 2013).

BPD has become less severe since the 1970s and 1980s due to care improvements, especially with the use of exogenous surfactant, lower concentrations of administered oxygen, and gentler forms of ventilatory support. In the so-called surfactant era, the conducting airways epithelium is largely intact but thicker than normal, and the airway smooth muscle thickness is increased. The lungs show a more uniform expansion and less fibrosis. The most common feature is the presence of fewer and larger alveoli, resembling emphysema, which is believed to be a result of impaired septation, during the alveolarisation process (O'Reilly et al. 2013).

Recent studies suggest that infants born very preterm that developed BPD have an increased risk of impaired lung function during infancy, childhood, and adulthood, as well as an increased risk of respiratory illness (O'Reilly et al. 2013). During infancy, very and extremely preterm infants, both with and without BPD, show significantly reduced airway function and increased residual lung volume. In childhood, they have an increased risk of respiratory symptoms including cough, wheeze, and asthma (Abe et al. 2010) and their exercise capacity is also compromised (with or without BPD) (O'Reilly et al. 2013).

Very preterm birth and BPD can lead to hyperplasia of airway smooth muscle, impaired alveolarisation, pulmonary inflammation, increase in pulmonary artery muscularisation, and an alveoli deficit that can persist into their adult life (O'Reilly et al. 2013).

It is important to recognise that the studies mentioned above were undertaken with individuals born during the pre-surfactant era. Although adverse pulmonary outcomes have been reported in infants and children born following the introduction of exogenous surfactant, longitudinal follow-up studies are required to understand the long-term pulmonary outcomes of surfactant treatment (O'Reilly et al. 2013).

There are a lack of studies on lung function outcomes for late preterm infants during childhood and adult life, but the available data suggest that they may have an increased risk of short- and long-term respiratory complications and have an increased risk of deficits in lung function in later life (Kotecha et al. 2012).

12.7.2.2 *Genetic Respiratory Diseases*

Cystic fibrosis (CF) is the most common life-shortening genetic disease in the Caucasian population, affecting approximately 75,000 individuals worldwide. The maintenance of such high incidence rates of CF in most European populations can perhaps be explained by a heterozygous advantage. The carriers of one recessive gene may show higher resistance to diarrhoeas (Romeo et al. 1989) and to the human strain of *Mycobacterium tuberculosis* (Meindl 1987).

CF is an autosomal recessive disorder caused by mutations in the cystic fibrosis transmembrane regulator gene (*CFTR*) on chromosome 7 (Riordan et al. 1989). Identification of *CFTR* in the year 1989 opened a new era in understanding the disease, and since then over 2000 mutations have been identified (Fajac and Boeck 2017).

The CFTR protein is an anion channel of primary importance that controls the normal movement of chloride (Cl^-), sodium (Na^+), and water in and out of the cells in different parts of the body. When open or activated, it allows passive diffusion of chloride ions down their electrochemical gradient, followed by water osmosis. It has many other roles such as inhibition of sodium transport through the epithelial sodium channel and regulation of other chloride channels (Fajac and Boeck 2017).

This multisystem disease affects the airways, pancreas, intestines, liver, and reproductive system and is involved in chronic lung infection, malabsorption, salt loss syndrome, infertility, and development of other numerous morbidities. It can impair other organs, such as the hypothalamus and kidney, and bone, and may be implied in growth retardation, delayed pubertal onset, bone density modulation, and susceptibility to renal calculi (Castellani and Assael 2017). The forms of disease connected with CFTR are widely heterogeneous in severity, rate of progression, and body region involved.

In the respiratory system, the CFTR protein plays a major role in determining the airway surface mucus, and therefore impairs the mucociliary clearance that is the primary innate defence mechanism. The epithelial cells lining the airways have two types of channels: the CFRT (releases Cl^-) and the sodium channel (takes up Na^+). When the CFTR protein is defective, there is an imbalance between CFTR-dependent chloride secretion and sodium channel-mediated absorption, leading to low liquid volume and airway surface dehydration. The airways mucus becomes thick and difficult to clear, creating an environment for bacteria and infections (Knowles and Boucher 2002). These defects in the airways' innate defence trigger a chain of events that include mucus stasis and plugging, airway obstruction, infection, and inflammation (Fajac and Boeck 2017).

CF was considered lethal during infancy and childhood, but nowadays individuals with the condition can survive until 40 years of age due to early diagnosis through neonatal screening and an aggressive therapeutic attitude (Farrell 2008).

Current treatments are mainly symptomatic, focusing on compensating the exocrine pancreatic insufficiency with pancreatic enzymes, addition of lipid-soluble vitamins in the diet, respiratory physiotherapy, mucolytics, and aggressive antibiotic therapy. A significant proportion of patients with severe symptoms still require lung or, less frequently, liver transplantation.

New therapies for treatment of CF complications are in development, including novel inhaled antibiotics, anti-inflammatory drugs, agents to enhance mucociliary clearance, and nutritional/pancreatic replacement therapies.

Understanding how mutations translate into synthesis or CFRT protein function disturbance opened a way to 'personalised' treatments, and since 2012, two new drugs called CFTR modulators have become available. The treatment has a CFTR potentiator that augments channel function, and a potentiator that increases CFTR expression at the cell membrane. Other modulators are in development and other mutation type-specific treatments to correct the basic defect at the mRNA level are under clinical investigation (Fajac and Boeck 2017).

Primary ciliary dyskinesia (PCD) is a rare congenital disorder that comprises a range of defects in all cilia of the body, either being immobile or showing an ineffective beating pattern. In 50% of the patients, these syndromes results in a situs inversus, possibly caused by an inability of embryonic cilia to shift the heart to the left side, and then it is called Kartagener's syndrome (Afzelius 1981).

PCD is caused by genetic mutations in the proteins required for the assembly and function of motile cilia. Typically, the disease is inherited in an autosomal recessive manner, although some cases of X-linked inheritance and autosomal dominance have been documented. To date, 21 different disease-causing mutations have been identified, and more are expected to follow (Hosie et al. 2014).

Cilia occur mainly in the respiratory system, and the abnormality in ciliary function results in impaired mucociliary clearance that has different clinical symptoms depending on the stage of development (Hosie et al. 2014). During the neonatal period, patients exhibit respiratory distress, pneumonia, rhinorrhoea, and nasal obstruction immediately after birth. In childhood, patients typically show chronic cough with sputum, rhinosinusitis, secretory otitis media, pneumonia, and bronchiectasis (Takeuchi et al. 2016).

The other system that is affected is the genital tract, and throughout adolescence and adult life fertility problems emerge and females have an increased rate of ectopic pregnancy and males have fertilising problems caused by spermatozoa that are unable to swim progressively (Afzelius 1981).

It is not classified as a life-limiting illness, but some individuals die prematurely because of congenital cardiac abnormalities or respiratory failure. However, for the most part, early diagnosis with effective respiratory therapy can dramatically improve lung conditions. PCD with situs inversus (Kartagener's syndrome) is easy to diagnose, but other cases are probably underdiagnosed (Hosie et al. 2014).

The available tools to diagnose the disease are nasal nitric oxide concentration (extremely low in these patients), nasal or bronchial biopsy/brushing to analyse cilia morphology, and genetic tests to identify the gene mutation (Hosie et al. 2014). An earlier diagnosis is important to prevent bronchiectasis progression and lung function deterioration. The treatment is symptomatic, based on airway clearance techniques (e.g. physiotherapy) and systemic and/or inhaled antibiotics.

12.8 Conclusion

The processes and mechanisms by which O_2 is acquired for production of energy have greatly determined the evolution of animal life and their ecological distribution and diversification. The strategies adopted can be understood by studying the respiratory organs/structures of animals that: (1) occupy various habitats, especially extreme ones; (2) exhibit profound developmental changes, like amphibians which start life in water and end up on land; (3) are at different phylogenetic levels of development, that is, the basal and derived species, and have adopted different respiratory strategies; (4) utilise water, air, or both (bimodal breathers) as source(s) of O_2 and; (5) pursue different lifestyles, for example volant and non-volant taxa. Although gas exchangers occur in different forms, shapes, and sizes, some common structural features and functional mechanisms and processes exist, especially at the gas exchange level. The similarities may be explained by the fact that in nature, at the biological range of temperature, there are only two gases of respiratory significance, O_2 and CO_2, and only two respiratory fluid media, water and air. Gas exchangers evolved to utilise one or the other, and in rare cases both. Most importantly, and probably least appreciated, the functional designs of gas exchangers have occurred under the universal and immutable laws of physics (Thompson 1942). The respiratory fluid media are transported convectionally, while the transfer of O_2 across tissue barriers occurs sorely by diffusion. Through trade-off and compromises, gas exchangers are designed to perform these roles optimally. The foremost structural specialisations that have been employed to promote respiratory efficiency are a thin blood–gas barrier, large respiratory surface area, and large pulmonary capillary blood volume. On the comparative pulmonary morphology of gas exchangers, the observation made by Jürgens and Gros (2002, p. 185) that 'The system that has evolved in each species depends to an impressive extent on environmental conditions, on body build and size, on the animal's patterns of movement and on its energy consumption' captures the message that this chapter has endeavoured to communicate, particularly with regard to the evolution of the human respiratory system and its role in health and disease.

Acknowledgement

The preparation of this chapter was supported by a grant from the National Research Foundation of South Africa awarded to John N. Maina.

References

Abe, K., Shapiro-Mendoza, C. K., Hall, L. R., et al. (2010). Late preterm birth and risk of developing asthma. *J Pediatr* 157, 74–8.

Afzelius, B. A. (1981). Genetical and ultrastructural aspects of the immotile-cilia syndrome. *Am J Hum Genet* 33(6), 852–64.

Aiello, L. and Dean C. (1990). *An Introduction to Human Evolutionary Anatomy*. London: Academic Press.

Akiyama, J., Hoffman, A., Brown, C., et al. (2002). Tissue distribution of surfactant proteins A and D in the mouse. *J Histochem Cytochem* 50, 993–6.

Albertine, K. H. and Physher, T. J. (2004). Pulmonary consequences of preterm birth. In: Harding, R., Pinkerton, K. E., and Plopper, C. G. (eds) *The Lung: Development, Aging and the Environment*. Amsterdam: Elsevier Academic Press, pp. 237–51.

Alberts, B., Bray, D., Lewis, J., et al. (1994). *Molecular Biology of the Cell*, 3rd ed. New York: Garland, pp. 218–34.

Al-Motabagani, M. A. (2005). Histological changes in the alveolar structure of the rat lung after exposure to hyperoxia. *Ital J Anat Embryol* 110, 209–23.

Aoki, T., Yamasawa, F., Kawashiro, T., et al. (2008). Effects of long-term low-dose oxygen supplementation on the epithelial function, collagen metabolism and interstitial fibrogenesis in the guinea pig lung. *Resp Res* 9, 37. doi: org/10.1186/1465-9921-9-37.

Aoshiba, K., Tsuji, T., Itoh, M., et al. (2015). An evolutionary medicine approach to understanding factors that contribute to chronic obstructive pulmonary disease. *Respiration* 89, 243–52.

Aprelikova, O., Chandramouli, G. V. R., Wood, M., et al. (2004). Regulation of HIF prolyl hydroxylases by hypoxia-inducible factors. *J Cell Biochem* 92, 491–501.
Askin, F. B. and Kuhn, C. (1971). The cellular origin of pulmonary surfactant. *Lab Invest* 25, 260–8.
Attwell, L., Kovarovic, K., and Kendal, J. (2015). Fire in the Plio-Pleistocene: the functions of hominin fire use, and the mechanistic, developmental and evolutionary consequences. *J Anthropol Sci* 93, 1–20.
Atwood, W. H. (1992). In: Bahr, L. S. and Johnston, B. (eds) *Collier's Encyclopedia (Volume 6)*. Basingstoke: Macmillan, pp. 137–66.
Auten, R. L. and Davis, J. M. (2009). Oxygen toxicity and reactive oxygen species: the devil is in the details. *Pediatr Res* 66, 121–7.
Bachofen, H. (2009). Why are the lungs dry? *Pneumologie* 63, 346–51.
Bachofen, H., Bachofen, M., and Weibel, E. R. (1988). Ultrastructural aspects of pulmonary edema. *J Thorac Imaging* 3, 1–7.
Banks, W. J. (1986) *Applied Veterinary Histology*, 2nd ed. Baltimore: Williams & Wilkins, pp. 447–66.
Barsh, L. I. (1999). The origin of pharyngeal obstruction during sleep. *Sleep Breath* 3, 17–21.
Beckman, D. L. (1973). Mechanical properties of the primate thorax. *J Med Primatol* 2, 218–22.
Bell, E. A., Boehnke, P., Harrison, T. M., et al. (2015). Potentially biogenic carbon preserved in a 4.1 billion-year old zurcon. *Proc Natl Acad Sci U S A* 112, 14518–21.
Bemis, W. E. and Burggreen, W. W. (1986). The biology and evolution of lungfishes. *J Morphol* 1, 3–4.
Berna, F., Goldberg, P., Kolska Horwitz, L., et al. (2012). Microstratigraphic evidence of *in situ* fire in the Acheulean strata of Wonderwerk Cave, Northern Cape province, South Africa. *Proc Natl Acad Sci U S A* 109, E1215–20.
Bernhard, W., Postle, A. D., Rau, G., et al. (2001). Pulmonary and gastric surfactants. A comparison of the effect of surface requirements on function and phospholipids composition. *Comp Biochem Physiol A Mol Integr Physiol* 129, 173–82.
Bernstein, M. H. (1987). Respiration in flying birds: In: Sutton, T. J. (ed.) *Bird Respiration*, Vol II. Boca Raton: CRC Press, pp. 43–73.
Bernstein, M. H. (1990). Avian respiration and high altitude tolerance. In: Sutton, J. R., Coates, G. C., and Remmers, J. E. (eds) *Hypoxia: The Adaptations*. Burlington, ON: B. C. Decker, pp. 30–40.
Beuther, D. A. and Sutherland, E. R. (2007). Overweight, obesity and incident asthma: a meta-analysis of prospective epidemiologic studies. *Am J Respir Crit Care Med* 175, 661–6.
Bielsalski, H. K. and Nohr, D. (2003). Importance of vitamin-A for lung function and development. *Mol Aspects Med* 24, 431–40.
Bishop, C. M., Spivey, R. J., Hawkes, L. A., et al. (2015). The roller coaster flight strategy of bar-headed geese conserves energy during Himalayan migrations. *Science* 347, 250–4.
Black, C. P. and Tenney, S. M. (1980). Oxygen transport during progressive hypoxia in high altitude and sea level water-fowl. *Respir Physiol* 39, 217–39.
Borok, Z., Danto, S. I., Lubman, R. L., et al. (1998). Modulation of t1alpha expression with alveolar epithelial cell phenotype in vitro. *Am J Physiol* 275(19), L155–64.
Bostrom, H., Gritli-Linde, A., and Betsholtz, C. (2002). PDGF-A/PDGF alpha-receptor signaling is required for lung growth and the formation of alveoli but not for early lung branching morphogenesis. *Dev Dyn* 223, 155–62.
Bourbon, J. B. and Chailley-Heu, B. (2001). Surfactant proteins in the digestive tract, mesentry, and other organs: evolutionary significance. *Comp Biochem Physiol A Mol Integr Physiol* 129, 151–61.
Brackenbury, J. H. (1984). Physiological responses of birds to flight and running. *Biol Rev* 59, 559–75.
Brackenbury, J. H. (1991). Ventilation, gas exchange and oxygen delivery in flying and flightless birds. In: Woakes, A. J., Grieshaber, M. K., and Bridges, C. R. (eds) *Physiological Strategies for Gas Exchange and Metabolism*. Cambridge: Cambridge University Press, pp. 125–47.
Breeze, R. G. and Wheeldon, E. B. (1977) The cells of the pulmonary airways. *Am Rev Respir Dis* 116, 705–17.
Brown, J. H., Gupta, V. K., Li, B. L., et al. (2002). The fractal nature of nature: power laws, ecological complexity and biodiversity. *Philos Trans R Soc Lond B Biol Sci* 357, 619–26.

Bunn, H. F. and Poyton, R. O. (1996). Oxygen sensing and molecular adaptation to hypoxia. *Physiol Rev* 76, 839–85.

Buonocore, G., Perrone, S., and Tataranno, M. L. (2010). Oxygen toxicity: chemistry and biology of oxygen species. *Semin Fetal Neonatal Med* 15, 186–90.

Burggren, W. W. and Johansen, K. (1986). Circulation and respiration in lungfishes (dipnoi). *J Morphol* 1, 217–36.

Burri, P. H. (1985). Morphology and respiratory function of the alveolar unit. *Int Arch Allergy Appl Immunol* 76, 2–12.

Cabelli, D. E. (2010). *Superoxide Dismutases and Reactive Oxygen Species*. Brookhaven National Laboratory, Report No. BNL-93775-2010-BC, pp. 1–32.

Cai, K. Q., Capo-Chichi, C. D., Rula, M. E., et al. (2008). Dynamic GATA 6 expression in primitive endoderm formation and maturation in early mouse embryogenesis. *Dev Dyn* 237, 2820–9.

Canfield, D. E. (1998). A new model for Proterozoic ocean chemistry. *Nature* 396, 450–3.

Cardoso, W. V. (2000). Lung morphogenesis revisited: old facts, current ideas. *Dev Dyn* 219, 121–30.

Cardoso, W. V. (2001). Molecular regulation of lung development. *Annu Rev Physiol* 63, 471–94.

Cardoso, W. V. (2004). Lung morphogenesis, role of growth factors and transcription factors. In: Harding, R., Pinkerton, K. E., and Plopper C.G. (eds) *The Lung: Development, Aging and the Environment*. Amsterdam: Elsevier Academic Press, pp. 3–11.

Cardoso, W. V. and Whitsett, J. A. (2008). Resident cellular components of the lung: developmental aspects. *Proc Am Thorac Soc* 5,767–71.

Carrier. D. R. and Wake, M. H. (1995). Mechanism of lung ventilation in the Caecilian *Dremophis mexicanus*. *J Morphol* 226, 289–95.

Castellani, C. and Assael, B. M. (2017). Cystic fibrosis: a clinical view. *Cell Mol Life Sci* 74 (1), 129–40.

Castresana, J. and Saraste, M. (1995). Evolution of energetic metabolism: the respiration-early hypothesis. *Trends Biochem Sci* 20, 443–8.

Chang, D. M., Mark, R., Miller, S. L., et al. (1983). Prebiotic organic syntheses and the origin of life. In: Schopf, J. W. (ed.) *Earth's Earliest Biosphere: Its Origin and Evolution*. Princeton, NJ: Princeton University Press, pp. 53–92.

Chi, E. Y. (1985). The ultrastructural study of glycogen and lamellar bodies in the development of fetal monkey lung. *Exp Lung Res* 8, 275–89.

Chisholm, R., Trauer, J. M., Curnoe, D., et al. (2016). Controlled fire use in early humans might have triggered the evolutionary emergence of tuberculosis. *Proc Natl Acad Sci U S A* 113, 9051–6.

Choi, S. Y. and Kahyo, H. (1991). Effect of cigarette smoking and alcohol consumption in the aetiology of cancer of the oral cavity, pharynx and larynx, *Int J Epidemiol* 20 (4), 878–85.

Clack, J. A. (1994). Earliest known tetrapod braincase and the evolution of the stapes and fenestra ovalis. *Nature* 369, 392–4.

Cloud, P. (1993). Early biogeologic history: the emergence of a paradigm. In: Schopf, J. W. (ed.) *Earth's Earliest Biosphere: Its Origin and Evolution*. Princeton, NJ: Princeton University Press, pp. 14–31.

Comroe, J. H. (1974). *Physiology of Respiration: An Introductory Text*. Chicago: Year Book Medical Publishers.

Costa, R. H., Kalinichenko, V. V., and Lim, L. (2001). Transcription factors in mouse lung development and function. *Am J Respir Cell Mol Biol* 280, L823–38.

Couroucli, X. I., Liang, Y. W., Jiang, W., et al. (2006). Attenuation of oxygen-induced abnormal lung maturation in rats by retinoic acid: possible role of cytochrome P_{4501A} enzymes. *J Pharmacol Exp Ther* 317, 946–54.

Crapo, J. D., Barry, B. E., Gehr, P., et al. (1982). Cell number and cell characteristics of the normal human lung. *Am Rev Respir Dis* 125, 332–7.

Darwin, C. (1871). *The Descent of Man and Selection in Relation to Sex*. London: John Murray.

Davies, J. C. (1975). Minimal dissolved oxygen requirements of aquatic life with emphasis on Canadian species: a review. *J Fish Res Board Can* 32, 2295–395.

Deamer, D. and Szostak, J. W. (2010). *The Origins of Life*. New York: Cold Spring Harbour Laboratory Press.

Dean, M. C. (1985). Comparative myology of the hominoid cranial base. II. The muscles of the prevertebral and upper pharyngeal region. *Folia Primatol* 44: 40–51.

Dejours, P. (1988). *Respiration in Water and Air: Adaptations, Regulation and Evolution*. New York: Elsevier.

Dejours, P. (1989). From comparative physiology of respiration to several problems of environmental adaptations and to evolution. *J Physiol* 410, 1–19.

de Moraes, M. F., Holler, S., da Costa, O. T., et al. (2005). Morphometric comparison of the respiratory organs in the South American lungfish *Lepidosiren paradoxa* (Dipnoi). *Physiol Biochem Zool* 78, 546–59.

Desai, T. J., Malpel, S., Flentke, G. R., et al. (2004). Retinoic acid selectively regulates Fgf10 expression and maintains cell identity in the prospective lung field of the developing foregut. *Dev Biol* 273, 402–15.

Desai, T. J., Chen, F., Lu, J., et al. (2006). Distinct roles for retinoic acid receptors alpha and beta in early lung morphogenesis. *Dev Biol* 129, 12–24.

Dolph, S. (2000). *The Ecology of Adaptive Radiation*. Oxford: Oxford University Press.

Donoghue, P. C. J. and Antcliffe, J. B. (2010). Origins of multicellularity. *Nature* 466, 41–2.

Dreher, D., Kok, M., Cochand, L., et al. (2001). Genetic background of attenuated *Salmonella typhimurium* has profound influence on infection and cytokine patterns in human dendritic cells. *J Leukoc Biol* 69, 583–9.

Duncker, H. R. (1971). The lung-air sac system of birds. A contribution to the functional anatomy of the respiratory apparatus. *Ergeb Anat Entwicklungsgesch* 45, 1–171.

Duncker, H. R. (2001). The emergence of macroscopic complexity. An outline of the history of the respiratory apparatus of vertebrates from diffusion to language production. *Zoology* 103, 240–59.

Duncker, H. R. (2004). Vertebrate lungs: structure, topography and mechanics. A comparative perspective of the progressive integration of respiratory system, locomotor apparatus and ontogenic development. *Respir Physiol Neurobiol* 144, 111–24.

Dyer, B. D. and Obar, R. A. (1994). *Tracing the History of the Eukryotic Cells*. New York: Columbia University Press.

Eigen, H., Lindemann, B. F., Tietze, M., et al. (1989). How old is the genetic code? Statistical geometry of the tRNA provides an answer. *Science* 244, 673–9.

El-Albani, A., Bengston, S., Canfield, D. E., et al. (2010). Large colonial organisms with co-ordinated growth in oxygenated environments 2.1 Gyr ago. *Nature* 466, 100–4.

Erecinska, M. and Silver, I. A. (1994). Ions and energy in mammalian brain. *Prog Neurobiol* 43, 37–71.

Erwin, D. H. (1993). *The Great Paleozoic Crisis: Life and Death in the Permian*. New York: Columbia University Press.

Evans, D. H., Piermarini, P. M., and Choe, K. P. (2005). The multifunctional fish gill: dominant site of gas exchange, osmoregulation, acid-base regulation, and excretion of nitrogenous waste. *Physiol Rev* 85, 97–177.

Fajac, I. and Boeck, K. (2017). New horizons for cystic fibrosis treatment. *Pharmacol Ther* 70, 205–11.

Faraci, M. F. (1990). Cerebral circulation during hypoxia: is a bird brain better? In: Sutton, J. R., Coates, G., and Remmers, J. E. (eds) *Hypoxia: The Adaptations*. Burlington, ON: B. C. Decker, pp. 26–9.

Faraci, M. F. and Fedde, M. R. (1986). Regional circulatory responses to hypocapnia and hypercapnia in bar-headed geese. *Am J Physiol* 250, R499–504.

Faraci, M. F., Kilgore, D. L., and Fedde, M. R. (1984). Oxygen delivery to the heart and brain during hypoxia: pekin duck vs bar-headed geese. *Am J Physiol* 247, R68–75.

Farhi, L. (1964). Gas stores of the body. In: Fenn, W. O. and Rahn, H. (eds) *Handbook of Physiology, Section 3: Respiration, Vol I*. Washington, DC: American Physiological Society, pp. 873–924.

Farhi, L. and Rahn, H. (1955). Gas stores of the body and the steady state. *J Appl Physiol* 7, 472–84.

Farrell, P. M. (2008). The prevalence of cystic fibrosis in the European Union. *J Cyst Fibros* 7, 450–3.

Fedde, M. R. (1980). The structure and gas flow pattern in the avian lung. *Poult Sci* 59, 2642–53.

Fedde, M. R. (1986). Respiration. In: Sturlie, P. D. (ed.) *Avian Physiology*, 3rd ed. New York: Springer, pp. 191–220.

Fedde, M. R., Orr, J. A., Shams, H., et al. (1989). Cardiopulmonary function in exercising bar-headed geese during normoxia and hypoxia. *Respir Physiol* 77, 239–62.

Fine, A. (2009). Breathing life into the lung stem cell field. *Cell Stem Cell* 4, 468–9.

Finlay, W. H. (2001). *The Mechanics of Pharmaceutical Aerosols: An Introduction.* London: Academic Press.

Fleetwood, J. N. and Munnell, J. F. (1996). Morphology of the airways and lung parenchyma in hatchlings of the loggerhead sea turtle, *Caretta caretta. J Morphol* 227, 289–304.

Fraslon, C. and Bourbon, J. R. (1994). Retinoids control surfactant phospholipid biosynthesis in fetal rat lung. *Am J Physiol* 266, L705–12.

Gaillard, J. M., Hebblewhite, M., Loison, A., et al. (2010). Habitat–performance relationships: finding the right metric at a given spatial scale. *Philos Trans Royal Soc B Biol Sci* 27, 2255–65.

Gea, J. (2008). The evolution of the human species: a long journey for the respiratory system. *Arch Bronconeumol* 44(5), 263–70.

Gehr, P., Bachofen, M., and Weibel, E. R. (1978). The normal human lung: ultrastructure and morphometric estimation of diffusion capacity. *Respir Physiol* 32, 121–40.

Gehr, P., Sehovic, S., Burri, P. H., et al. (1980). The lung of shrews: morphometric estimation of diffusion capacity. *Respir Physiol* 44, 61–86.

Gehr, P., Mwangi, D. K., Amman, A., et al. (1981). Design of the mammalian respiratory system: V. Scaling morphometric diffusing capacity to body mass: wild and domesic animals. *Respir Physiol* 44, 61–86.

Gehr, P., Schürch, S., Berthiaume, Y., et al. (1990). Particle retention in airways by surfactant. *J Aerosol Med* 3, 27–43.

Geiser, M., Baumann, M., Cruz-Orive, L. M., et al. (1990). Assessment of particle retention and clearance in the intrapulmonary conducting airways of hamster lungs with the fractionator. *J Microsc* 160, 75–88.

Geiser, M., Rothen-Rutishauser, B. M., Kapp, N., et al. (2003). Ultrafine particles cross cellular membranes by non-phagocytic mechanisms in lungs and in cultured cells. *Environ Health Perspect* 113, 1555–60.

Gibson, G. J. (2000). Obesity, respiratory function and breathlessness. *Thorax* 55, S41–4.

Gil, J. and Weibel, E. R. (1971). Extracellular lining of bronchioles after perfusion-fixation of rat lungs for electron microscopy. *Anat Rec* 169, 131–45.

Godfrey, R. W. (1997). Human airway epithelial tight junctions. *Microsc Res Tech* 38, 488–99.

Gorr, T. A., Gassmann, M., and Wappner, P. (2006). Sensing and responding to hypoxia via HIF in model invertebrates. *J Insect Physiol* 52, 349–64.

Gowdy, K. M., Cardona, D., Nuggett, J., et al. (2012). Novel role for surfactant protein A in gastrointestinal graft-versus-host disease. *J Immunol* 188, 4897–905.

Gowlett, J. A. J. (2016). The discovery of fire by humans: a long and convoluted process. *Philos Trans Royal Soc B Biol Sci* 371, 20150164.

Graham, J. B. and Lee, H. J. (2004). Breathing air in air: in what ways might extant amphibious fish biology relate to prevailing concepts about early tetrapods, the evolution of vertebrate air breathing, and the vertebrate land transition? *Physiol Biochem Zool* 77, 720–31.

Greener, M. (2008). It's life, but just as we know it. *EMBO Rep* 9, 1067–9.

Greenwood, P. H. (1986). The natural history of African lungfishes. *J Morphol* 1, 163–79.

Grubb, B. R., Mills, C. D., Colacino, J. M., et al. (1977). Effect of arterial carbon dioxide on cerebral blood flow in ducks. *Am J Physiol* 232, H596–601.

Grubb, B. R., Colacino, J. M., and Schmidt-Nielsen, K. (1978). Cerebral blood flow in birds: effects of hypoxia. *Am J Physiol* 234, H230–43.

Grubb, B. R., Jones, J. H., and Schmidt-Nielsen, K. (1979). Avian cerebral blood flow: influence of the Bohr effect on oxygen supply. *Am J Physiol* 236, H744–53.

Grunwald, D. J. (1996). A fin-de siècle achievement: charting new wares in vertebrate biology. *Science* 274, 1634–5.

Gupta, A. K. (2004). Origin of agriculture and domestication of plants and animals linked to early Holocene climate amelioration. *Curr Sci* 87, 54–9.

Haefeli-Bleuer, B. and Weibel, E. R. (1988). Morphometry of the human pulmonary acinus. *Anat Rec* 220, 401–14.

Hammond, P. (1992). Species inventory: In: Groombridge, B. (ed.) *Global Biodiversity: Status of the Earth's Living Resources*. London: Chapman and Hall, pp. 17–39.

Hansen, J. E. and Ampaya, E. P. (1975). Human air space shapes, sizes, areas, and volumes. *J Appl Physiol* 38, 990–5.

Harkema, J. R., Mariassy, A., George, J., et al. (1991). Epithelial cells of the conducting airways: a species comparison. In: Farmer, S. G. and Hay, D. W. P. (eds) *Lung Biology in Health and Disease: The Airway Epithelium*, Vol. 55. New York: Marcel Dekker, pp. 3–39.

Harrison, J. F., Kaiser, A., and VandenBrooks, J. M. (2010). Atmospheric oxygen level and the evolution of insect body size. *Proc Biol Sci* 277, 1937–46.

Hastings, R. H. and Powell, F. L. (1986). Physiological dead space and effective parabronchial ventilation in ducks. *J Appl Physiol* 60, 85–91.

Hilfer, S. R. (1996). Morphogenesis of the lung: control of embryonic and fetal branching. *Annu Rev Physiol* 58, 93–113.

Hind, M. and Maden, M. (2004). Retinoic acid induces alveolar regeneration in the adult mouse lung. *Eur Respir J* 23, 20–7.

Hirabayashi, H., Koshii, K., Uno, K., et al. (1990). Laryngeal epithelial changes on effects of smoking and drinking, *Auris Nasus Larynx* 17(2), 105–14.

Hlastala, M. P. and Berger, A. J. (1996). *Physiology of Respiration*. Oxford: Oxford University Press.

Hochachka, P. W. (1973). Comparative intermediary metabolism. In: Prosser, C. L. (ed.) *Comparative Animal Physiology*, 3rd ed. Philadelphia: W. B. Saunders, pp. 212–78.

Hockman, D., Burns, A. J., Schlosser, G., et al. (2017). Evolution of the hypoxia-sensitive cells involved in amniote respiratory reflexes. *eLife* 6, pii: e21231.

Hopkins, S. R. and Powell, F. L. (1998). Ventilation–perfusion heterogeneity: insights from comparative physiology. In: Hlastala, M. P. and Robertson, H. T. (eds) *Complexity in Structure and Function of the Lung*. New York: Marcel Dekker, pp. 549–70.

Horsfield, K. and Thurbeck, A. (1981). Relation between diameter and flow in the bronchial tree. *Bull Math Biol* 43, 681–91.

Horsfield, K. and Woldenberg, M. J. (1986). Branching ratio and growth of tree-like structures. *Respir Physiol* 63, 97–107.

Hosie, P., Fitzgerald, D. A., Jaffe, A., et al. (2014). Primary ciliary dyskinesia: overlooked and undertreated in children, *J Paediatr Child Health* 50(12), 952–8.

Hou, C., Gheorghiu, S., Huxley, V. H., et al. (2010). Reverse engineering of oxygen transport in the lung: adaptation to changing demands and resources through space-filling networks. *PLoS Comput Biol* 6(8), e1000902.

Hough, M. L., Shields, G. A., Evins, L. Z., et al. (2006). A major sulphur isotope event at c 510 Ma: a possible anoxia–extinction–volcanism connection during the Early–Middle Cambrian transition? *Terra Nova* 18, 257–63.

Howard, G. and Schopf, J. W. (1983). Biochemical evolution of anaerobic energy conversion: the transition from fermentation to anoxygenic photosynthesis. In: Schopf, J. W. (ed.) *Earth's Earliest Biosphere: Its Origin and Evolution*. Princeton, NJ: Princeton University Press, pp. 135–48.

Howell, B. J. (1970). Acid-base balance in transition from water breathing to air breathing. *Fed Proc* 29, 1130–4.

Hsia, C. W., Schmitz, A., Lambertz, M., et al. (2013). Evolution of air breathing: oxygen homeostasis and the transitions from water-to-land and sky. *Compr Physiol* 3, 849–915.

Hughes, G. M. (1966). Species variation in gas exchange. *Proc Royal Soc Med* 59, 494–500.

Hughes, G. M. and Shelton, G. (1958). The mechanism of gill ventilation in three freshwater teleosts. *J Exp Biol* 35, 807–23.

Hughes, G. M. and Shelton, G. (1962). Respiratory mechanisms and their nervous control in fish. *Adv Comp Physiol Biochem* 1, 275–364.

Hurtgen, M. T., Pruss, S. B., and Knoll, A. H. (2009). Evaluating the relationship between the carbon and sulphur cycles in the later Cambrian ocean: an example from the Portal au Port Group, western Newfoundland, Canada. *Earth Planet Sci Lett* 281, 288–97.

Hyde, D. M., Tyler, N. K., Putney, L. F., et al. (2004). Total number and mean size of alveoli in mammalian lung estimated using fractionator sampling and unbiased estimates of the Euler characteristic of alveolar openings. *Anat Rec* 277, 216–26.

Ishida, R., Palmer, J. B., and Hiiemae, K. M. (2002). Hyoid motion during swallowing: factors affecting forward and upward displacement. *Dysphagia* 17, 262–72.

James, S. R. (1989). Hominid use of fire in the lower and middle Pleistocene: a review of the evidence. *Curr Anthropol* 30, 1–26.

Jiang, W., Welty, S. E., Couroucli, X. I., et al. (2004). Disruption of the Ah receptor gene alters the susceptibility of mice to oxygen-mediated regulation of pulmonary and hepatic cytochromes P_{4501A} expression and exacerbates hyperoxic lung injury. *J Pharmacol Exp Ther* 310, 512–19.

Jones, J. H., Effmann, E. L., and Schmidt-Nielsen, K. (1985). Lung volume changes during respiration in ducks. *Respir Physiol* 59, 15–25.

Joshi, S. and Kotecha, S. (2007). Lung growth and development. *Early Hum Dev* 83, 789–94.

Jung, A., Allen, L., Nyengaard, J. R., et al. (2005). Design-based stereological analysis of the lung parenchymal architecture and alveolar type-II cells in surfactant protein A and D double deficient mice. *Anat Rec* 286, 885–90.

Jürgens, D. and Gros, G. (2002). Phylogeny of gas exchange systems. *Anästhesiol Intensivmed Notfallmed Schmerzther* 37, 185–98.

Kemp, A. (1986). The biology of the Australian lungfish, *Neoceratodus forsteri* (Krefft 1870). *J Morphol* 1, 181–98.

Khoor, A., Stahlman, M. T., Gray, M. E., et al. (1994). Temporal–spatial distribution of SP-B and SP-C proteins and mRNAs in developing respiratory epithelium of human lung. *J Histochem Cytochem* 42, 1187–99.

Kiama, S. G., Cochand, L., Karlsson, L. M., et al. (2001). Evaluation of phagocytic activity in human monocyte-derived dendritic cells. *J Aerosol Med* 14, 289–99.

Kierszenbaum, A. L. (2011). *Histology and Cell Biology: An Introduction to Pathology*, 4th ed. Amsterdam: Elsevier, pp. 387–414.

Kikkawa, Y., Motoyama, E. K., and Cook, C. D. (1965). The ultrastructure of the lungs of lambs: the relation of osmiophilic inclusions and alveolar lining layer to fetal maturation and experimentally produced respiratory distress. *Am J Pathol* 47, 877–903.

Kilburn, H. (1968). A hypothesis for pulmonary clearance and its implications. *Am Rev Respir Dis* 98, 449–63

Kim, S. and Camargo, C. A. Jr (2003). Sex–race differences in the relationship between obesity and asthma: the behavioral risk factor surveillance system, 2000. *Ann Epidemiol* 13, 666–73.

King, A. S. (1966). Structural and functional aspects of the avian lung and its air sacs. *Int Rev Gen Exp Zool* 2, 171–267.

Klika, E., Scheurermann, D. W., De Groodt-Lasseel, M. H. A., et al. (1996). Pulmonary macrophages in birds (barn owl, *Tyto tyto alba*), domestic fowl (*Gallus, gallus f. domestica*), quail (*Coturnix coturnix*), and pigeons (*Columbia livia*). *Anat Rec* 246, 87–97.

Knoll, A. H. (1979). Archean photoautotrophy: some alternatives and limits. *Orig Life* 9, 313–27.

Knowles, M. R. and Boucher, R. C. (2002). Mucus clearance as a primary innate defense mechanism for mammalian airways. *J Clin Invest* 109, 571–7.

Knust, J., Ochs, M., Gundersen, J. G., et al. (2009). Stereological estimates of alveolar number and size and capillary length and surface area in mice lungs. *Anat Rec* 292, 113–22.

Kotecha, S. J., Dunstan, F. D., and Kotecha, S. (2012) Long term respiratory outcomes of late preterm-born infants. *Semin Fetal Neonatal Med* 17, 77–81.

Kullander, K. and Klein, R. (2002). Mechanisms and functions of eph and ephrin signalling. *Nat Rev Mol Cell Biol* 3, 475–86.

Kumar, H., Tawhai, M. H., Hoffman, E. A., et al. (2009). The effects of geometry on airflow in the acinar region of the human lung. *J Biomech* 42, 1635–42.

Lasiewski, R. C. (1972). Respiratory function in birds. In: Farner, D. S. and King, J. R. (eds) *Avian Biology*, Vol II. New York: Academic Press, pp. 287–342.

Lauder, G. V. (1984). Pressure and water flow patterns in the respiratory tract of the bass (*Micropterus Salmoides*). *J Exp Biol* 113, 151–64.

Laurent, P. and Dunel, S. (1980). Morphology of gill epithelia in fish. *Am J Physiol* 238, R147–59.

Lenfant, C. and Johansen, K. (1968). Respiration in the African lungfish *Protopterus aethiopicus*. I. Respiratory properties of blood and normal patterns of breathing and gas exchange. *J Exp Biol* 49, 437–52.

Lieberman, D. E. and McCarthy, R. C. (1999). The ontogeny of cranial base angulation in humans and chimpanzees and its implications for reconstructing pharyngeal dimensions. *J Hum Evol* 36, 487–517.

Lieberman, D. E., McCarthy, R. C., Hiiemae, K. M., et al. (2001). Ontogeny of postnatal hyoid and laryngeal descent: implications for deglutition and vocalization. *Arch Oral Biol* 46, 117–28.

Lieberman, P. (2007). The evolution of human speech. *Curr Anthropol* 48, 39–66.

Liem, K. F. (1986). The biology of lungfishes: an epilogue. *J Morphol* 1, 299–303.

Lin, Y. J., Markham, N. E., Balasubramaniam, V., et al. (2005). Inhaled nitric oxide enhances distal lung growth after exposure to hyperoxia in neonatal rats. *Pediatr Res* 58, 22–9.

Lindahl, P., Karlsson, L., Hellstrom, M., et al. (1997). Alveogenesis failure in PDGF-A-deficient mice is coupled to lack of distal spreading of alveolar smooth muscle cell progenitors during lung development. *Development* 124, 3943–53.

Lippmann, M. and Schlesinger, R. B. (1984). Interspecies comparison of particle deposition and mucociliary clearance in tracheobronchial airways. *J Toxicol Environ Health* 13, 441–69.

Lipton, P. (1999). Ischemic cell death in brain neurons. *Physiol Rev* 79:1431–568.

Litingtung, Y., Lei, L., Westphal, H., et al. (1998). Sonic hedgehog is essential to foregut development. *Nat Genet* 20, 58–61.

Liu, J., Hu, F., Wang, G., et al. (2013). Lipopolysaccharide-induced expression of surfactant proteins A1 and A2 in human renal tubular epithelial cells. *J Inflam* December, 10, 2. doi: 10.1186/1476-9255-10-2.

Long, J. A. and Gordon, M. S. (2004). The greatest step in vertebrate history: a paleobiological review of the fish–tetrapod transition. *Physiol Biochem Zool* 77, 700–19.

Lorz, C. and Lopez, J. (1997). Incidence of air pollution in the pulmonary surfactant system of the pigeon (*Columba livia*). *Anat Rec* 249, 206–12.

Losos, J. B. (2010). Adaptive radiation, ecological opportunity and evolutionary determinism. *Am Nat* 175, 623–39.

Loughna, S. and Sato, T. N. (2001). Angiopoietin and Tie signalling pathways in vascular development. *Matrix Biol* 20, 319–25.

Lui, R., Harvey, C. S., and McGowan, S. E. (1993). Retinoic acid increases elastin in neonatal rat lung fibroblast cultures. *Am J Physiol* 265, L430–7.

Lutz, P. L., Nilsson, G. E., and Prentice, H. (2003). *The Brain Without Oxygen*, 3rd ed. Dordrecht: Kluwer Academic Publishers.

Lyons, T. W. and Reinhard, C. T. (2009). An early productive ocean unfit for aerobics. *Proc Natl Acad Sci U S A* 106, 18045–6.

Maden, M. (2006). Retinoids have differing efficacies on alveolar regeneration in dexamethasone-treated mouse. *Am J Respir Cell Mol Biol* 35, 260–7.

Magnussen, H., Willmer, H., and Scheid, P. (1976). Gas exchange in the air sacs: contribution to respiratory gas exchange in ducks. *Respir Physiol* 26, 129–46.

Mahajan, P. V. and Bharucha, B. A. (1994). Evaluation of short neck: percentiles and linear correlations with height and sitting height. *Indian Pediatr* 31, 1193–203.

Maina, J. N. (1988). Scanning electron microscope study of the spatial organization of the air and blood conducting components of the avian lung (*Gallus gallus* variant *domesticus*). *Anat Rec* 222, 145–53.

Maina, J. N. (1989) The morphometry of the avian lung. In: King, A. S. and McLelland, J. (eds) *Form and Function in Birds*, Vol. 4. London: Academic Press, pp. 307–68.

Maina, J. N. (1998a). *The Gas Exchangers: Structure, Function and Evolution of the Respiratory Processes*. Berlin: Springer.

Maina, J. N. (1998b). The lungs of the volant vertebrates—birds and bats: how are they relatively optimized for this elite mode of locomotion? In: Weibel, E. R., Taylor, C. R., and Bolis, L. (eds) *Symmorphosis and Optimization—Fact or Fancy?* Cambridge: Cambridge University Press, pp. 177–85.

Maina, J. N. (2000a). What it takes to fly: the structural and functional respiratory refinements in birds and bats. *J Exp Biol* 203, 3045–64.

Maina, J. N. (2000b). Comparative respiratory morphology: themes and principles in the design and construction of the gas exchangers. *Anat Rec* 261, 25–44.

Maina, J. N. (2002a). Fundamental structure aspects and features in the bioengineering of the gas exchangers: comparatives perspectives. *Adv Anat Embryol Cell Biol* 163, 1–108.

Maina, J. N. (2002b). Structure, function and evolution of the gas exchangers: comparative perspectives. *J Anat* 201, 281–304.

Maina, J. N. (2002c). *Functional Morphology of the Vertebrate Respiratory Organs*. Lebanon, NH: Oxford & IBH Publishing Company.

Maina, J. N. (2003). Developmental dynamics of the bronchial (airway) and air sac system of the avian respiratory system from day 3 to day 26 of life: a scanning electron microscope study of the domestic fowl, *Gallus gallus* variant *domesticus*. *Anat Embryol* 207, 119–34.

Maina, J. N. (2005). *The Lung-Air Sac System of Birds: Development, Structure and Function*. Heidelberg: Springer.

Maina, J. N. (2011). *Bioengineering Aspects in the Design of Gas Exchangers: Comparative Evolutionary, Morphological, Functional and Molecular Perspectives*. Heidelberg: Springer.

Maina, J. N. (2012a). Comparative molecular developmental aspects of the mammalian and the avian lungs, and the insectan tracheal system by branching morphogenesis: recent advances and future directions. *Front Zool* 9, 16.

Maina, J. N. (2012b). Cellular defenses of the lung: comparative perspectives. In: Ghanei, M. (ed.) *Respiratory Diseases*. Rijeka: InTech, pp. 15–56.

Maina, J. N. (2015). Morphological and morphometric properties of the blood–gas barrier: comparative perspectives. In: Makanya, A. N. (ed.) *The Vertebrate Blood–Gas Barrier in Health and Disease: Structure, Development and Remodeling*. Heidelberg: Springer, pp. 15–38.

Maina, J. N. (2017a). Pivotal debates and controversies on the structure and function of the avian respiratory system: setting the record straight. *Biol Rev* 92, 1475–504.

Maina, J. N. (2017b). *The Biology of the Avian Respiratory System: Evolution, Development, Structure and Function*. Heidelberg: Springer.

Maina, J. N. and King, A. S. (1982). The thickness of the avian blood–gas barrier: qualitative and quantitative observations. *J Anat* 134, 553–62.

Maina, J. N. and King, A. S. (1984). The structural–functional correlation in the design of the bat lung. *J Exp Biol* 111, 43–63.

Maina, J. N. and Van Gils, P. (2001). Morphometric characterization of the airway and vascular systems of the lung of the domestic pig, *Sus scrofa*: comparison of the airway, arterial, and venous systems. *Comp Biochem Physiol* 130A, 781–98.

Maina, J. N. and West, J. B. (2005). Thin and strong! The bioengineering dilemma in the structural and functional design of the blood–gas barrier. *Physiol Rev* 85, 811–44.

Maina, J. N., King, A. S., and King, D. Z. (1982). A morphometric analysis of the lung of a species of bat. *Respir Physiol* 50, 1–11.

Maina, J. N., King, A. S., and Settle, G. (1989). An allometric study of the pulmonary morphometric parameters in birds, with mammalian comparison. *Philos Trans Royal Soc B Biol Sci* 326B, 1–57.

Maina, J. N., Thomas, S. P., and Hyde, D. M. (1991). A morphometric study of bats of different size: correlations between structure and function of the chiropteran lung. *Philos Trans Royal Soc B Biol Sci* 333B, 31–50.

Manalo, D. J., Rowan, A., Lavoie, T., et al. (2005). Transcriptional regulation of vascular endothelial cell responses to hypoxia by HIF-1. *Blood* 105, 659–69.

Mandelbrot, B. B. (1983). *The Fractal Geometry of Nature*. San Francisco: W. H. Freeman and Company.

Mancardi, D., Varetto, G., Bucci, E., et al. (2008). Fractal parameters and vascular networks: facts and artifacts. *Theor Biol Med Model* 5, 12. doi: 10.1186/1742-4682-5-12.

Marandi, Y., Farahi, N., and Hashjin, G. S. (2013). Asthma: beyond corticosteroid treatment. *Arch Med Sci* 9(3), 521–6.

Margulis, L. (1981). *Symbiosis in Cell Evolution*. New York: W. H. Freeman and Company.

Mason, R. J. and Williams, M. C. (1977). Type II alveolar cell: defender of the alveolus. *Am Rev Respir Dis* 115, 81–91.

Massaro, G. D. and Massaro, D. (1996). Postnatal treatment with retinoic acid increases the number of pulmonary alveoli in rats. *Am J Physiol* 270(14), L305–10.

May, R. M. (1992). How many species inhabit Earth? *Sci Am* October, 18–24.

McCutcheon, F. H. (1964). Organ systems in adaptation: the respiratory system. In: Dill, D. B., Adolph, E. F., and Wilber, C. G. (eds) *Handbook of Physiology, Section 4. Adaptation to the Environment*. Washington, DC: American Physiological Society, pp. 167–91.

McGhee, G. R. (1989). Frasian–Famennian extinction event. In: Briggs, D. E. G. and Crowther, P. R. (eds) *Paleobiology—A Synthesis*. Oxford: Blackwell Scientific Publications, pp. 97–176.

McGowan, S. E. and Snyder, J. M. (2004). Development of alveoli. In: Harding, R., Pinkerton, K. E., and Plopper, C. G. (eds) *The Lung: Development, Aging and the Environment*. Amsterdam: Elsevier Academic Press, pp. 55–73.

McGowan, S. E., Jackson, S. K., Jenkins-Moore, M., et al. (2000). Mice bearing deletions of retinoic acid receptors demonstrate reduced lung elastin and alveolar numbers. *Am J Respir Cell Mol Biol* 23, 162–7.

McLelland, J. (1989). Anatomy of the lungs and air sacs. In: King, A. S. and McLelland, J. (eds) *Form and Function in Birds, Vol. 4*. London: Academic Press, pp. 221–79.

McMichael, A. J. (2004). Environmental and social influences on emerging infectious diseases: past, present and future. *Philos Trans Royal Soc B Biol Sci* 359, 1049–58.

Meban, C. (1980). Thicknesses of the air–blood barriers in vertebrate lungs. *J Anat* 131, 299–307.

Meindl, R. S. (1987). Hypothesis: a selective advantage for cystic fibrosis heterozygotes. *Am J Phys Anthropol* 74(1), 39–45.

Metzger, R. J. and Krasnow, M. A. (1999). Genetic control of branching morphogenesis. *Science* 284, 1635–59.

Metzger, R. J., Klein, O. D., Martin, G. R., et al. (2008). The branching programme of the mouse lung development. *Nature* 453, 745–50.

Min, H., Danilenko, D. M., Scully, S. A., et al. (1998). Fgf-10 is required for both limb and lung development and exhibits striking functional similarity to *Drosophila* branchless. *Genes Dev* 12, 3156–61.

Moore, K. L. and Persaud, T. V. N. (1998). *Before We Are Born. Essentials of Embryology and Birth Defects*. Philadelphia: W. B. Saunders, pp. 241–54.

Morrisey, E. E., Tang, Z., Sigrist, K., et al. (1998). GATA 6 regulates HNF4 and is required for differentiation of visceral endoderm in the mouse embryo. *Genes Dev* 12, 3579–90.

Nakakuki, S. (1991). The bronchial tree and lobular division of the gorilla lung. *Primates* 32, 403–8.

Nakakuki, S. (1992). The bronchial tree and lobular division of the chimpanzee lung. *Primates* 33, 265–72.

Nelson, T. R., West, B. J., and Goldberger, A. L. (1990). The fractal lung: universal and species related scaling patterns. *Experientia* 46, 251–4.

Neufeld, G., Cohen, T., and Shraga, N. (2002). The neuropilins: multifunctional semaphorin and VEGF receptors that modulate axon guidance and angiogenesis. *Trends Cardiovasc Med* 12, 13–19.

Nguyen, B. T., Peterson, P. K., Verbrigh, H. A., et al. (1982). Differences in phagocytosis and killing by alveolar macrophages from humans, rabbits, rats, and hamsters. *Infect Immun* 36, 504–9.

Nicod, L. (2005). Lung defences: an overview. *Eur Respir Rev* 95, 45–50.

Nikinmaa, M. (2010). Sensing oxygen. In: Nilsson, G. E. (ed.) *Respiratory Physiology of Vertebrates: Life With and Without Oxygen*. Cambridge: Cambridge University Press, pp. 14–48.

Nilsson, G. E. (2010). Introduction: why we need oxygen. In: Nilsson, G. E. (ed.) *Respiratory Physiology of Vertebrates: Life With and Without Oxygen*. Cambridge: Cambridge University Press, pp. 3–13.

Nishimura, T., Mikami, A., Suzuki, J., et al. (2003). Descent of the larynx in chimpanzee infants. *Proc Natl Acad Sci U S A* 100, 6930–3.

Northway, W. H. and Rosan, R. C. (1968). Radiographic features of pulmonary oxygen toxicity in the newborn: bronchopulmonary dysplasia. *Radiology* 91, 49–57.

Núñez, B. and Cosío, B. G. (2007). Estructura y desarrollo del pulmón. In: Casán, P., García-Río, F., and Gea, J. (eds) *Fisiología y Biología Respiratorias*. Madrid: Ergón, pp. 13–21.

Obregon, C., Graf, L., Chung, K. F., et al. (2015). Nitric oxide sustains IL-1β expression in human dendritic cells enhancing their capacity to induce IL-17-producing T-cells. *PLoS One* 10(4), e0120134.

Ochs, M., Nyengaard, J. R., Waizy, H., et al. (2001). Alveolar type II cells and the intracellular surfactant pool in the human lung—a stereological approach. *Am J Respir Crit Care Med* 163, A731.

Ochs, M., Nyengaard, J. R., Jung, A., et al. (2004). The number of alveoli in the human lung. *Am J Respir Crit Care Med* 169, 120–4.

Ochs-Balcom, H. M., Grant, B. J., Muti, P., et al. (2006). Pulmonary function and abdominal adiposity in the general population. *Chest* 129, 853–62.

O'Reilly, M., Sozo, F., and Harding, R. (2013). Impact of preterm birth and bronchopulmonary dysplasia on the developing lung: long-term consequences for respiratory health. *Clin Exp Pharmacol Physiol* 40, 765–73.

Owen, T., Cess, R. D., and Ramanathan, V. (1979). Enhanced CO_2 greenhouse to compensate for reduced solar luminosity on early earth. *Nature* 277, 640–1.

Palmer, J. B., Rudin, N. J., Lara, G., et al. (1992). Coordination of mastication and swallowing. *Dysphagia* 7, 187–200.

Pavlov, N. A., Krivchenko, A. T., Cherepivskaya, E. N., et al. (1987). Reactivity of cerebral vessels in the pigeon, *Columba livia*. *J Evol Biochem Physiol* 23, 447–51.

Payne, J. L., Boyer, A. G., Brown, J. H., et al. (2009). Two phase increase in the maximum size of life over 3.5 billion years reflects biological innovation and environmental opportunity. *Proc Natl Acad Sci U S A* 106, 24–7.

Payne, J. L., McLain, C. R., Boyer, A. G., et al. (2011). The evolutionary consequences of oxygenic photosynthesis: a body size perspective. *Photosynth Res* 107, 37–57.

Pelster, B. (2015). Swimbladder function and the spawning migration of the European eel *Anguilla anguilla*. *Front Physiol* 5, Article 486/1.

Pelster, B. and Scheid, P. (1992). Counter-current concentration and gas secretion in the fish swim bladder. *Physiol Zool* 65, 1–16.

Perry, S. F. (1983). Reptilian lungs: functional anatomy and evolution. *Adv Anat Embryol Cell Biol* 79, 1–81.

Perry, S. F. (1997). The chloride cell: structure and function in the gills of freshwater fishes. *Annu Rev Physiol* 59, 325–47.

Peters, A., Wichmann, H. E., Tuch, T., et al. (1997). Respiratory effects are associated with the number of ultrafine particles. *Am J Respir Crit Care Med* 155, 1376–83.

Pierce, R. A., Mariencheck, W. I., Sandefur, S., et al. (1995). Glucocorticoids upregulate tropoelastin expression during late stages of fetal lung development. *Am J Physiol* 268(12), L491–500.

Pinkerton, K. E. and Joad, J. P. (2000). The mammalian respiratory system and critical windows of exposure for children's health. *Environ Health Perspect* 108, 457–62.

Piper, M. H., Noll, T., and Siegmund, B. (1994). Mitochondrial function in the oxygen depleted and reoxygenated myocardial cell. *Cardiovasc Res* 28, 1–15.

Poulain, M., Doucet, M., Major, G. C., et al. (2006). The effect of obesity on chronic respiratory diseases: pathophysiology and therapeutic strategies. *Can Med Assoc J* 174(9), 1293–9.

Power, J. H. T., Doyle, I. R., Davidson, K., et al. (1999). Ultrastructural and protein analyses of surfactant in the Australian lungfish *Neoceratodus forsteri*: evidence for conservation of composition for 300 million years. *J Exp Biol* 202, 2543–50.

Rampino, M. R. (2010). Mass extinctions of life and catastrophic flood basalt volcanism. *Proc Natl Acad Sci U S A* 107, 6555–6.

Randall, D. J. and Daxboeck, C. (1984). Oxygen and carbon dioxide transfer across fish gills. In: Hoar, W. S. and Randall, D. J. (eds) *Fish Physiology, Vol. 10, Part A*. New York: Academic Press, pp. 263–314.

Randall, D. J., Burggren, W., and French, K. (1998). Intercambio de gases y equilibrio ácido-base. In: Eckert, R. (ed.) *Fisiología Animal. Mecanismos y Adaptaciones*. Madrid: McGraw-Hill, pp. 563–622.

Rasmussen, B., Fletcher, I. R., Brocks, J. J., et al. (2008). Re-assesing the first appearance of eukaryotes and cyanobacteria. *Nature* 455, 1101–4.

Reznik, G. K. (1990). Comparative anatomy, physiology, and function of the upper respiratory tract. *Environ Health Perspect* 85, 171–6.

Riordan, J. R., Rommens, J. M., Kerem, B., et al. (1989). Identification of the cystic fibrosis gene: cloning and characterization of complementary DNA. *Science* 245, 1066–73.

Robinson, J. M. (1991). Phanerozoic atmospheric reconstructions: a terrestrial perspective. *Glob Planet Change* 97, 51–62.

Rodriguez, M., Bur, S., Favre, A., et al. (1987). Pulmonary acinus: geometry and morphometry of the peripheral airway system in rat and rabbit. *Am J Physiol* 180, 143–55.

Roebroeks, W. and Villa, P. (2011). On the earliest evidence for habitual use of fire in Europe. *Proc Natl Acad Sci U S A* 108, 5209–14.

Romeo, G., Devoto, M., and Galietta, L. J. (1989). Why is the cystic fibrosis gene so frequent? *Hum Genet* 84(1), 1–5.

Roux, E. (2002). Origin and evolution of the respiratory tract in vertebrates. *Rev Mal Respir* 19, 601–54.

Sapoval, B. and Filoche, M. (2013). Optimisations and evolution of the mammalian respiratory system: a suggestion of possible gene sharing in evolution. *Eur Phys J E Soft Matter* 36, 105. doi: 10.1140/epje/i2013-13105-1.

Sapoval, B., Filoche, M., and Weibel, E. R. (2002). Smaller is better—but not too small: a physical scale for the design of the mammalian pulmonary acinus. *Proc Natl Acad Sci U S A* 99, 10411–16.

Sato, K., Kobayashi, K., Aida S., et al. (2004). Bronchiolar expression of aquaporin-3 (AQP3) in rat lung and its dynamics in pulmonary oedema. *Pflugers Archiv* 449, 106–14.

Scheid, P. (1982) Respiration and control of breathing. In: Farner, D. S., King, J. R., and Parkes, K. C. (eds) *Avian Biology, Vol. 6*. New York: Academic Press, pp. 405–53.

Scheid, P. and Piiper, J. (1972). Cross-currrent gas exchange in the avian lungs: effects of reversed parabronchial air flow in ducks. *Respir Physiol* 16, 304–12.

Scheuermann, D. W., Klika, E., De Groodt-Lasseel, M. H. A., et al. (1997). An electron microscopic study of the parabronchial epithelium in the mature lung of four bird species. *Anat Rec* 249, 213–25.

Schittny, J. (2017). Development of the lung. *Cell Tissue Res* 367, 427–44.

Schmidt-Nielsen, K. (1971). How birds breathe. *Sci Am* 225, 72–9.

Schoenwolf, G., Bleyl, S., Brauer, P., et al. (2009). *Larsen's Human Embryology*, 5th ed. London: Churchill Livingstone, pp. 319–36.

Schopf, J. W. (1978). The evolution of the earliest cells. *Sci Am* 239, 85–103.

Schopf, J. W. (1989). The evolution of the earliest cells. In: Gould, J. L. and Gould, C. G. (eds) *Life at the Edge: Readings from the Scientific American Magazine*. New York: W. H. Freeman and Company, pp. 7–23.

Schopf, J. W. and Walter, M. R. (1983). Archean microfossils: new evidence of ancient microbes. In: Schopf, J. W. (ed.) *Earth's Earliest Biosphere: Its Origin and Evolution*. Princeton, NJ: Princeton University Press, pp. 214–39.

Scott, G. R. (2011). Elevated performance: the unique physiology of birds that fly at high altitudes. *J Exp Biol* 214, 2455–62.

Scott, G. R. and Dawson, N. J. (2017). Flying high: the unique physiology of birds that fly at high altitudes. In: Maina, J. N. (ed.) *The Biology of the Avian Respiratory System: Evolution, Development, Structure and Function*. Heidelberg: Springer.

Scott, G. R., Cadena, V., Tattersall, G. J., et al. (2008). Body temperature depression and peripheral heat loss accompany the metabolic and ventilatory responses to hypoxia in low and high altitude birds. *J Exp Biol* 211, 1326–35.

Scott, G. R., Hawkes, L. A., Frappel, P. B., et al. (2014). How bar-headed geese fly over the Himalayas. *Physiology* 30, 107–15.

Shams, H. and Scheid, P. (1987). Respiration and blood gases in the duck exposed to normocapnic and hypercapnic hypoxia. *Respir Physiol* 67, 1–12.

Shelton, G. (1970). The regulation of breathing. In: Hoar, W. S. and Randall, D. J. (eds) *Fish Physiology, Vol. 4*. New York: Academic Press, pp. 293–359.

Skerret, S. J. (1994). Host defenses against respiratory infection. *Med Clin N Am* 78, 941–66.

Slonim, N. B. and Hamilton, L. H. (1971). *Respiratory Physiology*, 2nd ed. Saint Louis: C. V. Mosby.

Smith, L. J., McKay, K. O., van Asperen, P. P., et al. (2010). Normal development of the lung and premature birth. *Paediatr Respir Rev* 11, 135–42.

Snyder, G. K. (1983). Respiratory adaptations in diving mammals. *Respir Physiol* 54, 269–94.

Solomon, S. E. and Purton, M. (1984). The respiratory epithelium of the lung in the green turtle (*Chelonia mydas* L.). *J Anat* 139, 353–70.

Takeuchi, K., Kitano, M., Ishinaga, H., et al. (2016). Recent advances in primary ciliary dyskinesia. *Auris Nasus Larynx* 43(3), 229–36.

Tappan, H. (1974). Molecular evolution: In: Hayaishi, O. (ed.) *Molecular Oxygen in Biology*. Amsterdam: Elsevier–North Holland, pp. 81–135.

Thompson, D. A. W. (1942). *On Growth and Form*. Cambridge: Cambridge University Press.

Thompson, M. B. (2007). Comparison of the respiratory transition at birth or hatching in viviparous and oviparous amniote vertebrates. *Comp Biochem Physiol* 148(4), 755–60.

Thrall, T. H., Bever, J. D., and Burdon, J. J. (2010). Evolutionary change in agriculture: the past, present and future. *Evol Appl* 3, 405–8.

Tichelaar, J. W., Lim, L., Costa, R. H., et al. (1999). HNF-3/Forkhead homologue-4 influences lung morphogenesis and respiratory epithelial cell differentiation *in vivo*. *Dev Biol* 213, 405–17.

Tsuda, A., Filipovic, N., Haberthür, D., et al. (2008). Finite element 3-D reconstruction of the pulmonary acinus imaged by synchrotron X-ray tomography. *J Appl Physiol* 105, 964–76.

Tucker, V. A. (1968). Respiratory physiology of house sparrows in relation to high altitude flight. *J Exp Biol* 48, 55–66.

Van Valen, L. (1971). The history and stability of atmospheric oxygen. *Science* 171, 439–43.

Vollsæter, M., Røksund, O. D., Eide, G. E., et al. (2013). Lung function after preterm birth: development from mid-childhood to adulthood. *Thorax* 68, 767–76.

Walker, J. C. G. (1977). *Evolution of the Atmosphere*. New York: Macmillan.

Walter, E., Dreher, D., Kok, M., et al. (2001). Hydrophilic poly (DL-lactide-co-glycolide) microspheres for the delivery of DNA to human-derived macrophages and dendritic cells. *J Control Release* 76, 149–68.

Warburton, D. (2008). Developmental biology: order in the lung. *Nature* 453, 733–5.

Warburton, D. and Bellusci, S. (2004). The molecular genetics of lung morphogenesis and injury repair. *Paediatr Respir Rev* 5, S283–7.

Warburton, D., El-Hashash, A., Carraro, G., et al. (2010). Lung organogenesis. *Curr Top Dev Biol* 90, 73–158.

Ward, L. M., Kirschvink, J. L., and Fischer, W. W. (2016). Timescales of oxygenation following the evolution of oxygenic photosynthesis. *Orig Life Evol Biosph* 46, 51–65.

Weaver, M., Yingling, J. M., Dunn, M. R., et al. (1999). Bmp signalling regulates proximal-distal differentiation of endoderm in mouse lung development. *Development* 126, 4005–15.

Weibel, E. R. (1963). *Morphometry of the Human Lung*. Berlin: Springer.

Weibel, E. R. (1973). Morphological basis of the alveolar–capillary gas exchange. *Physiol Rev* 53, 419–95.

Weibel, E. R. (1984). *The Pathway for Oygen: Structure and Function in the Mammalian Respiratory System*. Cambridge, MA: Harvard University Press.

Weibel, E. R. (1991). Fractal geometry: a design principle for living organisms. *Am J Physiol* 261, L361–9.

Weibel, E. R. (1997). Design of airways and between the organism blood vessels considered as confluent tree. In: Crystal, R. D., West, J. B., Weibel, E. R., et al. (eds) *The Lung: Scientific Foundations*. Philadelphia: Lippincott-Raven, pp. 1061–71.

Weibel, E. R. (2000). *Symmorphosis: On Form and Function in Shaping Life*. Cambridge, MA: Harvard University Press.

Weibel, E. R. (2008). How to make an alveolus. *Eur Respir J* 31, 483–5.

Weibel, E. R. (2009). What makes a good lung? The morphometric basis of lung function. *Swiss Med Wkly* 139, 375–86.

Weibel, E. R. and Gomez, D. M. (1962). Architecture of the human lung. *Science* 137, 577–85.

Weibel, E. R., Sapoval, B., and Filoche, M. (2005). Design of peripheral airways for efficient gas exchange. *Respir Physiol Neurobiol* 148, 3–21.

Weichert, C. K. and Presch, W. (1986). *Elements of Chordate Anatomy*. New York: McGraw-Hill, pp. 209–45.

Weiss, R. A. (2001). The Leeuwenhook Lecture 2001. Animal origins of human infectious disease. *Philos Trans Royal Soc B Biol Sci* 359, 957–77.

Wenger, R. H. (2000). Mammalian oxygen sensing, signaling and gene regulation. *J Exp Biol* 203, 1253–63.

Wenger, R. H., Stiehl, D. P., and Camenisch, G. (2005). Integration of oxygen signaling at the consensus HRE (abstract). *Sci STKE* 306, re12.

Wert, S. E. (2004) Normal and abnormal structural development of the lung. In: Polin, R. A., Fox, W. W., and Abman, S. H. (eds) *Fetal and Neonatal Physiology, Vol.1*, 3rd ed. Philadelphia: W. B. Saunders, pp. 783–94.

West, B. J., Barhava, V., and Goldberger, A. L. (1986). Beyond the principle of similitude: renormalization in the bronchial tree. *J Appl Physiol* 60, 1089–97.

West, J. B. (2005). *Fisiología Respiratoria*, 7th ed. Madrid: Editorial Médica Panamericana.

West, J. B. (2008). *Respiratory Physiology: The Essentials*, 8th ed. Baltimore: Lippincott Williams and Wilkins.

West, J. B., Watson, R. R., and Fu, W. (2006). The honeycomb-like structure of the bird lung allows a uniquely thin blood–gas barrier. *Respir Physiol Neurobiol* 152, 115–18.

West, J. B., Watson, R. R., and Fu, W. (2007). The human lung: did evolution get it wrong? *Eur Respir J* 29, 11–17.

Wiessner, P. W. (2014). Embers of society: firelight talk among the Ju/'hoansi Bushmen. *Proc Natl Acad Sci U S A* 111, 14027–35.

Wolfe, N. D., Dunavan, C. P., and Diamond, J. (2007). Origin of major human infectious diseases. *Nature* 447, 279–83.

Wongtrakool, C., Malpel, S., Gorenstein, J., et al. (2003). Down-regulation of retinoic acid receptor alpha signaling is required for sacculation and type I cell formation in the developing lung. *J Biol Chem* 278, 46911–18.

Wollman, H., Smith, T. C., Stephen, G. W., et al. (1968). Effects of extremes of respiratory and metabolic alkalosis on cerebral blood flow in man. *J Appl Physiol* 24, 60–5.

Woolhouse, M. and Gaunt, E. (2007). Ecological origins of novel human pathogens. *Crit Rev Microbiol* 33, 231–42.

Wrangham, R. W. and Carmody, R. (2010). Human adaptation to the control of fire. *Evol Anthropol* 19, 187–99.

Wrangham, R. W., Jones, J. H., Laden, G., et al. (1999). The raw and the stolen. Cooking and the ecology of human origins. *Curr Anthropol* 40, 567–94.

Wu, R. S. S. (2002). Hypoxia: from molecular responses to ecosystem responses. *Mar Pollut Bull* 45, 35–45.

Zammit, C., Liddicoat, H., Moonsie, I., et al. (2010). Obesity and respiratory diseases. *Int J Gen Med* 3, 335–43.

Zhu, M. and Yu, X. (2002). A primitive fish close to the common ancestor of tetrapods and lungfish. *Nature* 418, 767–70.

Zhuravlev, A. and Wood, R. (1996). Anoxia as the cause of the mid-Early Cambrian (Botomian) extinction event. *Geology* 24, 311–14.

CHAPTER 13

DIGESTIVE SYSTEM

JOHN B. FURNESS, JOSIANE FAKHRY, JOANNA GAJEWSKI, EVE K. BOYLE, AND LINDA J. FOTHERGILL

Abstract

Digestive tracts vary considerably because animals have evolved different processes to convert foods to essential molecular building blocks. Differences in digestive strategies distinguish, for example, foregut and hindgut fermenters, and animals utilising different dominant food types, for example herbivores, carnivores, and folivores. Neither the modern human diet nor the size and proportions of the human gut resemble those of other primates. The human digestive system has evolved and diverged in response to introduction of new food types and food preparation techniques. For example, persistence of lactase activity into adulthood occurred in populations that maintained cattle to harvest milk. Humans have utilised non-thermal food preparation for over 2 million years and cooking for 300,000–400,000 years. For most extant humans, prepared food comprises over 70% of the diet. The modern human digestive system is suited to pre-prepared food because of its smaller volume, relative to other species, and because of differences in dentition and masticatory muscles that results in lower bite strength. Adaptations of human digestion in response to diet involved genetic selection over thousands of years. However, transmissible changes linked to diet occur in a single generation. These are best documented for epigenetic changes related to obesity, and are maladaptive in some cases. Diets for most humans have changed substantially in the last half century, too rapidly for evolutionary change in digestive physiology. The capacity to adapt to recent dramatic dietary changes has proven insufficient to avoid deleterious effects leading to obesity, diabetes, fatty liver disease, and metabolic syndrome.

Keywords

digestive physiology, comparative anatomy, modern foods, digestive enzymes, microbiota, evolution, medicine

13.1 Functions, Physiology, and Structure of Digestive Systems

This chapter focuses on the digestive systems of mammals, although much of what we discuss applies to animals of other classes. All mammals have a tubular digestive tract, utilise similar enzymes to break down food, and possess similar transporters for absorption of nutrient molecules released from the food. They also have similar control of digestive functions through nerves and hormones. However, the organisation of the system differs considerably, with some mammals, such as cattle, utilising bacteria in the stomach in food breakdown, whereas modern humans rely on mechanical, enzymatic, and acid breakdown of foods in the stomach (Figure 13.1). Such differences are considered in more detail later in this chapter. Thus, the building blocks of the digestive system are very similar, but how these are utilised vary considerably.

Apart from sponges, practically all living animals, from hydra to humans, have a digestive tube with a mouth opening. An exception is cestodes, in which the gut tube, which is present in other parasitic helminths, appears to have regressed. Cestodes attach to the gut lining of their host and utilise a tegument to derive and absorb nutrients from the host. Gut tubes are lined with a single layer of epithelial cells that regulates the uptake of nutrients. In all but the simplest aquatic animals with a digestive tube (*Hydra*, *Holothuria*), the contents of the digestive tract are partly or substantially isolated from the environment, for example by sphincters in the foregut and at the anus. This protected space is colonised by microbiota, including bacteria, archaea, fungi, viruses, and protozoa. Commonly also present are small organisms (helminths), including various annelids, nematodes, and platyhelminths. There has been a recent upsurge of interest in gut microbiota, although for digestive physiologists this is an old story (Stevens and Hume 1998). The function of the digestive system is always the same: to gather foodstuff from the environment, process it so that constituent molecules useful to animal maintenance are freed, and transfer the valued molecules to sites where they are stored or used. A second function is to defend the organism against injurious components within or mixed with the food (Furness et al. 2013).

13.1.1 Alimentary Tract Design

The general pattern of the mammalian digestive tract is that there is an oral cavity equipped with teeth and bathed with saliva where the digestive process begins, followed by a conduit, the oesophagus, that takes the macerated food to a reservoir, the stomach, which may have one or several compartments, followed by a long tubular digestive tract which may include reservoirs and diverticula (Figure 13.1). The digestive processes in the gastric reservoir convert the ingested food to a liquid chyme, which is propelled into the small intestine where digestive enzymes (proteases, lipases, and carbohydrases) reduce the food to amino acids, free fatty acids, monoglycerides, and monosaccharides, at the same time liberating inorganic ions (Na, Cl, K, Ca, Mg, Fe, etc.) and vitamins. The nutrients are transported across the epithelium of the small intestine and enter its venous and lymphatic drainage. The residual

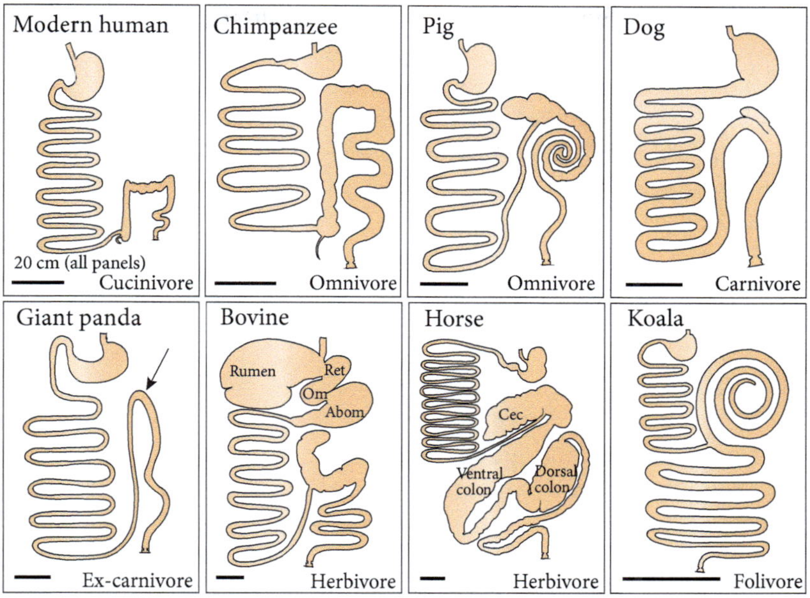

FIGURE 13.1 Comparison of digestive tracts of animals with different diets and digestive strategies. Notional dietary categories (discussed below and in the text) are indicated. The modern human is a cucinivore, having adopted a diet in which prepared foods account for 70–80% of energy intake. The chimpanzee is an omnivore that has a diet dominated by leaves and other vegetable matter. Proportionately it has a substantially larger colon than its close relative, the human. The pig is an omnivore with substantial colonic fermentation supplying energy-rich fatty acids. Note the spiral colon. The dog, a carnivore, has a small caecum and a colon that extracts little energy from food. The giant panda belongs to the order Carnivora and has retained a carnivore gut form, despite having adopted a diet of leaves and shoots. It has no caecum, and from an external view the junction between the small and large intestine (arrow) is not obvious. Bovines and horses are both grazing herbivores, but have adopted different strategies to utilise bacterial breakdown of plant cell walls, the bovine being a foregut fermenter and the horse a hindgut fermenter. The four compartments of the ruminant stomach are labelled rumen, reticulum (Ret), omasum (Om), and abomasum (Abom). The horse has three expansions of the large intestine, the caecum (Cec), right ventral colon, and right dorsal colon. The koala eats eucalyptus leaves, which are rich in tannins and potentially toxic oils. It has an extensive large bowel and very slow transit times. Its large bowel harbours tannin-metabolising bacteria.

contents move into the large intestine, which also has absorptive functions (for example for ions and fatty acids) and a reservoir function that may be enhanced by the presence of a caecum (Figure 13.1). All mammals have some degree of bacterial digestion of material that reaches the large intestine (see Section 13.7.2).

Non-mammalian vertebrates and invertebrates such as worms, insects, and arthropods also have tubular digestive tracts. The most evolutionarily ancient animals with tubular digestive tracts are cnidaria. Sponges have holes or spaces through which water bearing nutrients flows, and they might be considered to have tubular digestive systems, in that digestion occurs within these spaces (Renard et al. 2013). The also primitive placazoans (*Trichoplax*) are flat multicellular organisms that absorb nutrients through their surfaces and truly lack a tubular digestive system. Among the most ancient animals to have a digestive tract with identifiable homologues of the mammalian digestive tract (mouth, oesophagus, stomach, intestine, and anus) are the sea squirts (e.g. *Ciona intestinalis*). Thus, a differentiated gastrointestinal tract whose structure has resemblance to the mammalian gastrointestinal tract can be traced back more than 500 million years (my).

The pancreas, a specialised organ arising as a digestive tract diverticulum, is found in vertebrates, but not in the closely related holothurian or in other invertebrates (Perillo et al. 2016). Digestive enzymes that are released from the pancreas in vertebrates are released from cells of the digestive tract lining in non-vertebrates. Thus, in evolutionary terms, the functional cell type of the vertebrate pancreas predates the emergence of the pancreas as a discrete organ (Perillo et al. 2016).

13.1.2 Building Blocks in Common

In all vertebrates, including the most primitive living vertebrates—the jawless fish, superclass Agnatha (Langille and Youson 1984)—the wall of the tubular gastrointestinal tract from the stomach to the rectum has a similar structure, consisting of an outer musculature that mixes and propels the contents, a single layer of lining epithelium that regulates nutrient absorption, and subepithelial blood vessels that collect absorbed nutrients. Embedded in the wall of the digestive tract, from the upper oesophagus to the internal anal sphincter, is the enteric nervous system that controls, inter alia, contractile patterns in the musculature, local blood flow, and fluid movements between the gut lumen and body compartments.

Digestive enzymes and nutrient transporters are highly conserved. For example, the structures and catalytic properties of amylases are similar from insects to mammals (Payan 2004). Homologues of the mammalian digestive enzymes, alpha-amylase, lipase, phospholipase A2, trypsin, chymotrypsin, and carboxypeptidase, are found in animals as evolutionarily ancient as sea squirts, which originated over 500 million years ago (mya) (Nakayama and Ogasawara 2017). Homologues of mammalian amylases, lipases, and proteases have also been identified in the digestive tracts of echinoderms (Perillo et al. 2016). Likewise, nutrient transporters, such as SGLT1 and GLUT1, are highly conserved, even between vertebrates and invertebrates (Caccia et al. 2007).

All mammals sense food components through olfaction, taste, and texture, and also by specialised chemoreceptors within the stomach and intestines. The same receptors used to detect luminal signals are observed in different species. One well-illustrated example is the taste receptors initially found in the lingual epithelium and later described in enteroendocrine cells of a number of species, including rodents and modern humans (Wu et al. 2002; Rozengurt and Sternini 2007), fish (Latorre et al. 2013), horses (Daly et al. 2012), pigs (Moran et al. 2010), and ruminants (Moran et al. 2014). (For further discussion, see Chapter 17: Brain, Spinal Cord, and Sensory Systems.)

Different species sharing the same molecular and cellular machineries also share physiological responses. For example, the stimulant of appetite, ghrelin, is present in birds, fish, reptiles, and amphibia, as well as in mammals (Kaiya et al. 2008); in most fish species, ghrelin treatment appears to promote food intake and a more positive energy balance (Jönsson 2013), as it does in mammals (Kojima and Kangawa 2005). The enhanced glucose uptake following up-regulation of the glucose transporter SGLT1 in response to glucose or artificial sweeteners appears to share the same molecular series of events in rodents and humans (Margolskee 2002; Margolskee et al. 2007) and horses, piglets, dairy cows, calves, and sheep (Moran et al. 2010; Daly et al. 2012; Nishimura et al. 2013; Moran et al. 2014).

13.1.3 Digestive Tract Design Features that Relate to Disease Vulnerability

In later sections, we discuss how foods have influenced digestive system evolution, leading to diversity between species and limitations for humans to adapt to rapid changes in foods consumed. A design compromise for which evolution has not been able to find an adequate solution is the need for the lining of the intestine to be highly permeable, to facilitate nutrient digestion and absorption, and at the same time to defend against pathogens and toxins. To mount this defence, the intestine has an extensive immune system immediately adjacent to its lining epithelium, with both innate and adaptive components. (For further discussion, see Chapter 10: Immune System.) This defence system is always under challenge, and the intestinal mucosa in healthy individuals exists in a state of mild inflammation. Inflammatory disease may be a manifestation of a failure to balance the needs of the organ to be accessible to luminal contents and to defend against their deleterious components. Inflammatory bowel diseases (IBD; Crohn's disease and ulcerative colitis) are relapsing and remitting, and once established never fully subside (Cosnes et al. 2011). IBD has an annual incidence of 13–17 per 100,000 (Rubin et al. 2014). Another compromise results from the intestine hosting microorganisms. The gut microbiota contributes to the host, for example by digesting plant carbohydrate, and in healthy individuals there is equilibrium between beneficial and pathogenic bacteria. However, disturbance of this equilibrium results in dysbiosis that is associated with disordered digestive function (Cerf-Bensussan and Gaboriau-Routhiau 2010). Vertebrates have gathered together digestive enzyme-producing cells in the pancreas, which makes the compromised organ susceptible to autodigestion. The liver is also vulnerable, as it receives the venous drainage from the gut, and hence any absorbed toxins, bacteria, and nutrients that are in excess of the liver's capacity to manage can cause disease.

13.2 Probing Digestive Tract Evolution

The soft tissues of the digestive system rarely survive in the fossil record, so much of what we can infer has relied on extrapolations from extant species, for example by comparing modern primates and deducing what functions were shared by common ancestors (Carrigan et al. 2015). However, technology to detect and characterise protein remnants and even to sequence genetic material obtained for archaic biological remains (palaeogenetics) is advancing quickly. For example, extensive gene sequences have now been obtained from the remnants of Neanderthal and Denisovan humans dating to 400,000–500,000 years ago (kya) (Perry et al. 2015). This has allowed the evolution of changes in amylase copy numbers to be determined (see Section 13.6.2). In relation to diet, analysis of residues in dental calculus (Henry and Piperno 2008; Adler et al. 2013) and faeces (Battillo and Fisher 2015) has allowed inferences to be drawn about meat and milk consumption, and radiocarbon dating from the same samples provides a temporal context.

13.3 Control Systems for Digestion, the Nervous System and the Enteroendocrine System and How They Evolved

13.3.1 Evolution of the Enteric Nervous System

An enteric nervous system (ENS) that forms a meshwork around and along the gut tube is present in all animals with a gastrointestinal tract (Furness and Stebbing 2018). The most ancient animals to have a nervous system are cnidarians, exemplified by the genus *Hydra*, which evolved approximately 650 mya. The hydra is a tubular animal with an opening to the water in which it lives, and an internal epithelial lining that separates the nutrients in the gut tube from the body tissues. Within the body wall, the cnidarian ENS is distributed as hundreds of neurons that form a synaptically interconnected meshwork around the gut tube, but there is no evidence of the formation of specialised aggregations of neurons (head ganglia) that might be a first stage of brain formation (Bullock and Horridge 1965; Sakaguchi et al. 1996; Hansen et al. 2002; Furness and Stebbing 2018). The neurons of the hydra ENS react to the introduction of food by producing peristaltic movements (Shimizu et al. 2004). Nerve-free polyps, in which neurons are inactivated by colchicine treatment, exhibit only weak segmentation movements, indicating that the cnidarian ENS is necessary for physiologically effective movements of the gut tube, and hydra must be force fed to survive the lack of a nervous system (Galliot et al. 2009). If it is fed, the hydra appears perfectly normal even with the absence of a nervous system.

Other invertebrates, including echinoderms (García-Arrarás et al. 2001), cephalopods (Alexandrowicz 1928), gastropod molluscs (Ito and Kurokawa 2007; Okamoto and Kurokawa 2010), and annelid worms (Ierusalimsky and Balaban 2006) have an ENS, some of which have architectures reminiscent of that of the ENS of vertebrates, including humans. However, the ENS of insects has diverged from that of other invertebrates, in that enteric neurons are in ganglia and along nerve strands on the surface of the gut, rather than being embedded in its wall (Nässel et al. 1998; Copenhaver 2007). In snails, two layers of ganglia, identified as myenteric and submucosal ganglia, as is observed in other invertebrates and in mammals, have been described (Ábrahám 1940; Martinez-Pereira et al. 2013). In mammals, where there are both a well-developed ENS and a central nervous system (CNS), there is a highly developed reciprocal relationship between the ENS and the CNS (Figure 13.2), with pathways taking information in both directions and some organs having inputs from both (Furness et al. 2014).

In all animals, the ENS controls movements of the muscle of the gut tube. Other roles of the ENS have been defined in vertebrates, with the majority of studies having been conducted in mammals. These roles include control of fluid movement across the lining

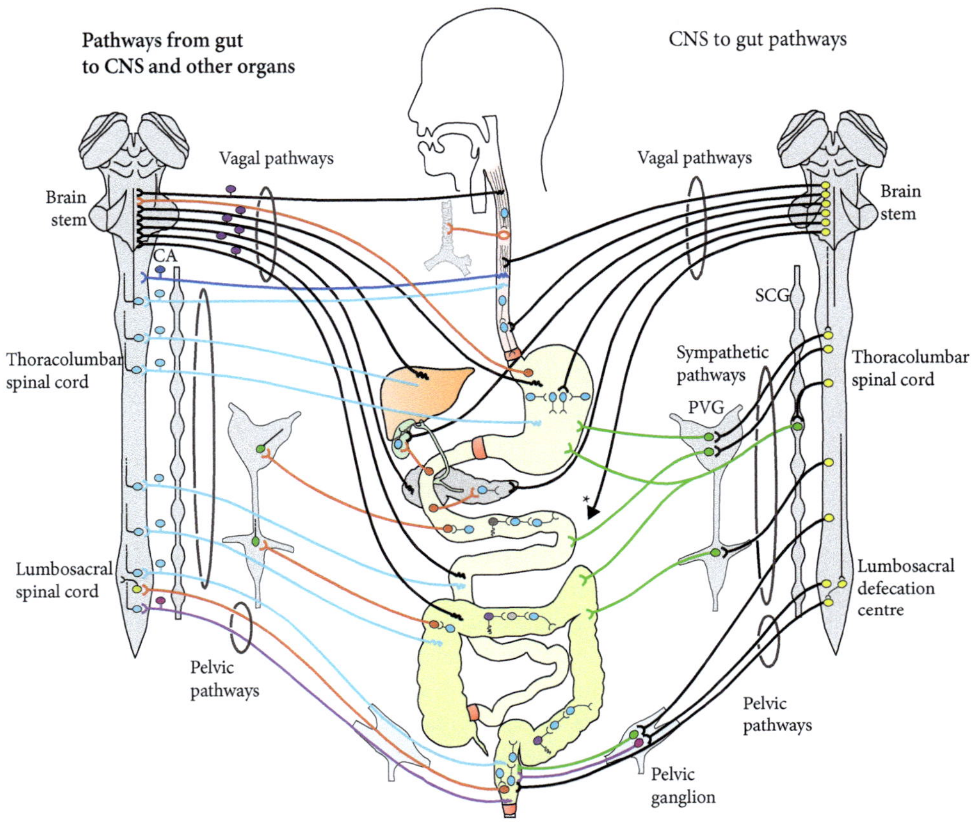

FIGURE 13.2 Reciprocal neural connections between the enteric and central nervous systems occur in mammals, reflecting their separate evolutionary histories. Connections from ENS to other organs and CNS are shown left and connections from CNS are shown right. Small and large intestines (middle of figure) contain full ENS reflex circuits (motor neurons and interneurons in blue, sensory neurons in purple). Pathways from the gastrointestinal tract (left) project outwards, via intestinofugal neurons (red), to CNS, sympathetic chain ganglia (SCG), gallbladder, pancreas, and trachea. Some neurons in prevertebral ganglia (PVG, green neurons) receive both CNS and ENS inputs. Sensory information goes to both ENS, via intrinsic primary afferent (sensory) neurons (purple) within the gut wall, and CNS, via extrinsic primary afferent neurons (left of figure) that follow spinal and vagal nerve connections. Cervical afferents (CA) connect oesophagus to cervical spinal cord. Pathways from CNS reach ENS and gastrointestinal effector tissues through vagal, thoracolumbar, and pelvic pathways (right of figure). Vagal and pelvic nerves include pre-enteric neurons (ending in enteric ganglia) and most gut-projecting sympathetic neurons with cell bodies in PVG are also pre-enteric neurons.

Source: Reproduced from 'Microbial Endocrinology: The Microbiota–Gut–Brain Axis in Health and Disease', in *The Enteric Nervous System and Gastrointestinal Innervation Integrated Local and Central Control*, Advances in Experimental Medicine and Biology, vol 817, pp 39–71. Copyright © 2014, Springer New York. With permission of Springer.

epithelium, control of acid secretion, control of local blood flow, modulation of the gut immune system and tissue defence, and influences on release of gut hormones (Furness et al. 2014). There is extensive communication between the ENS and the CNS in vertebrates (Figure 13.2). (For further discussion, see Chapter 17: Brain, Spinal Cord, and Sensory Systems.)

The chemistries of invertebrate enteric neurons, and, by implication, the chemistries of their neurotransmitters, exhibit some similarities and also major differences from vertebrates (Grimmelikhuijzen and Hauser 2012). In particular, typical invertebrate peptides, such as the molluscan peptide, FMRF amide, the insect peptide, adipokinetic hormone, and invertebrate tachykinin-related peptides (locustatachykinins), are not found in vertebrates or are remote from their vertebrate relatives. The innervation of the gastrointestinal tracts and the architecture of the ENS of all vertebrates is similar, and primary neurotransmitters are well preserved (Olsson and Holmgren 2010; Uyttebroek et al. 2010). Thus, the form and function of the ENS has persisted, but there has been divergence between vertebrates and invertebrates of the chemical messengers through which the neurons communicate.

13.3.2 Evolution of Enteric Hormonal Signalling

The hormones that are stored in and released from enteric endocrine cells (gut hormones) arose early in evolution and have been well preserved across vertebrate species (Rehfeld 2004; Nelson and Sheridan 2006). For example, peptide YY is present in all vertebrates (Larhammar 1996; Conlon 2002). The glucagon-like peptides have been described also in very different vertebrates, including chickens, pigs, sheep, and ruminants (Moran et al. 2010, 2014; Nishimura et al. 2013). Invertebrate enteroendocrine cells (EEC) contain typical invertebrate peptides, which parallels the situation in the nervous system. Surveys of the *Drosophila* gut show that populations of EEC contain other typically invertebrate messenger peptides, allostatins, small neuropeptide F, invertebrate tachykinins (locustatachykinins), and diuretic hormone 31 (Veenstra and Sellami 2008). The cholecystokinin (CCK) peptide family is, however, represented across vertebrates and invertebrates, in which peptides having a C-terminal phenylalanine amide and sulfated tyrosines in the C-terminal region are conserved (Nässel and Williams 2014). The CCK-like peptide, expressed by EEC in *Drosophila*, like CCK in mammals, regulates satiety (Nässel and Williams 2014). A peptide of EEC that belongs to the arthropod messenger peptide family, CCHamide2, has been described to have orexigenic effects in the fruit fly (Ren et al. 2015). In mammals, the EEC peptide, ghrelin, performs this role. Thus, actions of EEC hormones in controlling appetite and digestive functions are well preserved in vertebrates and invertebrates (Nelson and Sheridan 2006; Nässel and Williams 2014).

13.4 Comparisons of Digestive Strategies

Different mammalian species have adapted to consume different types of food. Conversely, adaptations to different foods have constrained the digestive systems of species, which is reflected in their anatomy (Figure 13.1). Such is the degree of specialisation that species are often confined to a narrow range of diets, modern humans being an exception. The diversity of solutions to the problems of nutrient assimilation includes that no mammal has evolved to be a universal digester of foraged or hunted food. In the wild, a modern human could not exist on the diet of a sheep, koala, or panda. And a koala or cat could not exist on the diet of these other species. The differences in the diets that animals prefer or are obliged to eat have led to the classification of animals by their dietary specialisation as carnivore, omnivore, herbivore, folivore, frugivore, and so on.

That digestive physiology and anatomy imposes dietary restriction is exemplified by ruminant dependence on low-protein fibrous plant material, the cat being an obligate carnivore, specialist feeders among birds and mammals being reliant on nectar, and the koala subsisting almost exclusively on eucalyptus leaves. Major differences include ways that bacteria are employed in the digestive process. Mammals lack cellulases and thus bacteria are essential to the breakdown of plant cell walls in the diet. Most herbivores solve the problem of digesting plant cells by holding food in capacious gut reservoirs (Figure 13.1) and enlisting cellulolytic, pectinolytic, and xylanolytic bacteria that are adapted to breakdown of plant cell walls. Foregut fermenters include ruminants, such as cattle and sheep, that have multichambered stomachs, and kangaroos, colobus monkey, and other species that have an enlarged single-chambered forestomach. Hindgut fermenters include horses, most rodents, lagomorphs, and rhinoceros. In addition to ruminants, non-ruminant foregut fermenters and hindgut fermenters, a fourth herbivore specialisation is to possess a very long small intestine in which microbial digestion can proceed, as occurs in the black bear. Sheep have both a ruminant stomach and an exceedingly long small intestine, and bovines have both a rumen and an enlarged colon (Figure 13.1). A cost of microbial reliance for herbivorous species is that anaerobic bacteria within the rumen or other fermentation reservoirs utilise almost all the plant sugars. For ruminants to access sugars, bacterially produced short chain fatty acids (SCFA) must be absorbed across the gut wall and converted enzymatically to glucose and other products, a net energy cost to the host (Stevens and Hume 1995). The food choices and evolutionary history of ruminants means that they are unable to utilise meat protein, other foods rich in protein, or rapidly fermentable fruits or grains as energy sources (Stevens and Hume 1998). High amounts of protein putrify in the rumen and poison the animal. This is the cause of cattle bloat. For humans, it has been suggested that high-protein diets, particularly high-protein/low-carbohydrate diets, can result in excess protein reaching the colon where it ferments to produce injurious sulfides, ammonia, phenols, and indoles (Yao et al. 2016).

Thus, the gastrointestinal tracts of ruminants and hindgut fermenters are very large, and to gain sufficient energy from food, these animals eat almost continuously while awake and digest continuously while asleep. A strategy of granivorous birds, which also lack cellulases, is to swallow small stones into the gizzard, which is lined with a tough, cornified epithelium, which grinds seeds making their starches accessible to digestive enzymes. Modern humans gain access to plant carbohydrate through grinding, pounding, and cooking grains and tubers (see Section 13.5.1).

Foregut fermenters are provided with bacterially produced B vitamins, which are absorbed in the small intestine. Coprophagy (or caecotrophy) allows some hindgut fermenters to deliver vitamin B to the small intestine and also to increase nutrient gain from reprocessing partly digested plant material. Vitamin B_{12} is not available from plant sources, so for humans it is necessary to obtain this vitamin from animal products, although many foods are fortified with vitamin B_{12}.

The interplay between evolution and food choice has influenced several aspects of digestive physiology. For example, cats (*Felidae*) are obligate carnivores that lack functional sweet taste receptors for the detection of carbohydrate (Li et al. 2005), and the increased expression of carbohydrases and monosaccharide transporters that occurs with carbohydrate feeding in other mammals does not occur in cats (Buddington et al. 1991; Kienzle 1993). Consistent with the lack of sweet taste receptors, cats do not exhibit meaningful behavioural responses to sugars in food (Bradshaw et al. 1996). In another example, both the carnivore and the ruminant pancreas lacks pancreatic lipase-related protein 2 (PLRP2), an enzyme that degrades plant triglycerides, whereas the enzyme is present in non-ruminant herbivores and omnivores (De Caro et al. 2008). Lack of PLRP2 relates to diet, as carnivores do not eat plant triglycerides, and gastric bacteria of ruminants hydrolyse plant triglycerides before they can reach the small intestine.

The koala, an arboreal folivore that eats a diet almost exclusively of eucalyptus leaves, has a number of anatomical, physiological, and microbial hosting adaptations (Barker et al. 2013). Its primary digestive organ is the hindgut (Figure 13.1). The koala caecum is the largest of any mammal in relation to body size, and among mammals it has the longest known mean gastrointestinal retention time. Eucalyptus leaves contain high levels of tannins, and the pure eucalyptus leaf diet would be toxic for other mammals. One way the koala has adapted is by an expansion of the expression of detoxifying enzymes, particularly enzymes of the cytochrome P450 monooxygenase (CYP) family, thus enhancing the metabolism of high levels of phenols and terpenes that would be fatal for other species (Johnson et al. 2018). In addition, the koala's large intestine is colonised by tannin-digesting bacteria, including *Lonepinella koalarum*, discovered in the koala and perhaps unique to this species (Osawa et al. 1995; Barker et al. 2013). As young koalas wean, they are fed by the mother a faecal paste (pap), derived from the caecum, that contains tannin-digesting bacteria; this maternal behaviour ensures colonisation with appropriate bacteria as the weanling moves from a milk diet to a diet of tannin-rich eucalyptus leaves (Osawa et al. 1993).

The giant and red pandas diverged separately from otherwise carnivorous clades, to become species that consume an almost exclusively (99%) plant diet (Jin et al. 2007; Nie et al. 2015; Hu et al. 2017). In parallel with the sweet taste receptor being a pseudogene in the cat, an obligate carnivore, in the herbivorous giant and red pandas the umami receptor for the savoury taste of meat is a pseudogene (Zhao et al. 2010; Hu et al. 2017). Despite this diet, the giant panda has a simple stomach and short gastrointestinal tract typical of carnivores (Figure 13.1) (Raven 1936; Dierenfeld et al. 1982), and it needs to eat large amounts of bamboo to survive because of the inefficiency of a carnivore digestive tract design to digest the panda diet. The giant panda alimentary tract clearly has not (yet) adapted to a shift to herbivory during an estimated period of 2–2.4 my of an exclusive bamboo diet (Jin et al. 2007), although there has been evolution of their dentition to a herbivore type (Dierenfeld et al. 1982). The giant and red pandas belong to different carnivore families (Ursidae and Ailuridae, respectively) and it seems certain that the pseudogenes arose independently, as the mutations in the two species are distinct. We can conclude that pseudogenisation of taste receptors is related to disuse in cats and pandas.

Most modern humans have escaped the restrictions they would encounter in the wild, to become 'cucinivores', that is, consumers of cooked or otherwise prepared foods (Furness and Bravo 2015; Furness et al. 2015). Before they were able to prepare and cook foods, the majority of the major nutrient sources used by modern humans were less readily available or unavailable (Cordain et al. 2005). The most obvious example is grains, which only became a common food source for humans about 10–30 kya with the development of grinding to create flour and the use of cooking. Grains, which were formerly poorly accessible food sources, account for up to 70% of dietary intake in modern societies (Copeland et al. 2009). Mechanical break-up and cooking also provided access to the starches of underground storage organs (USOs) such as potatoes and yams. Cooking, food preparation, and storage have contributed to extending the environmental range of human habitation and, it can be argued, have allowed humans the time to develop technologies and led to modern civilisation. It is estimated that 80–90% of food that is cooked in modern societies is preprocessed (Van Boekel et al. 2010).

13.5 History of Food Preparation by Humans and its Evolutionary Influence

From the Early Palaeolithic, at least from 2.6 mya, non-thermal food preparation techniques were used by pre-modern humans. Bovid and equid bones and bone fragments, dating from 2.5–2.6 mya, and associated stone cutting and pounding tools have been found in Gona, Ethiopia (Semaw et al. 2003), indicating non-thermal food preparation by archaic hominins (Wood and Boyle 2016). Bones exhibited cut and percussion marks, showing that very early stone tools were used for processing animal carcasses for meat and bone marrow. Bones with possible cut and percussion marks from another Ethiopian site, Dikika, suggest that non-thermal food preparation may date from as early as 3.4 mya (McPherron et al. 2010). Good evidence that is reliably dated to 1.95 mya indicating that early hominins manufactured and used stone tools for butchery and incorporated into their diets both mammals and aquatic animals, including turtles, crocodiles, and fish, has been found at the Koobi Fora formation in Kenya (Braun et al. 2010). Extensive skeletal remains located in the Olduvai Gorge, Tanzania, and dated to 1.7 mya exhibit signs of skilled butchery, with clusters of cut marks at sites of muscle insertion and percussion marks and fragmentation of long bones, suggesting a stone-on-anvil method to access marrow (Pante et al. 2017). There is no evidence of use of fire to cook the flesh at these early hominin sites. In Europe, there is well-documented evidence of hominin (*Homo erectus*) non-thermal food preparation, for example at Thuringia in central Germany, 1.06 my, where numerous bones with butchery marks and bones broken presumably for marrow access and stone tools have been located (Landeck and Garriga 2016). In the use of non-thermal food preparation, we witness an overlap of hominins with other species. Non-human primates commonly use tools to access foods, including tools to break open nuts and pods, and some birds and other mammals also use tools to gain food.

Humans have consumed cooked foods for at least 300–400 ky (Figure 13.3), and no groups of recent humans who live without cooking have been recorded (Wrangham and Conklin-Brittain 2003; Weaver 2012). Burned seeds, wood, and flint at Gesher Benot Ya'aqov in Israel suggest the controlled use of fire almost 800 ky (Goren-Inbar et al. 2004). At another Israel site, Qesem cave, there is evidence of use of fire for cooking in the period 200–420 ky (Shahack-Gross et al. 2014). A hearth at Qesem cave that has been dated to 300 ky contained charred and calcined bones, microcharcoal, burnt flint, and burnt microscopic clay aggregates in ash layers, identifying it as a cooking hearth that has been repeatedly used.

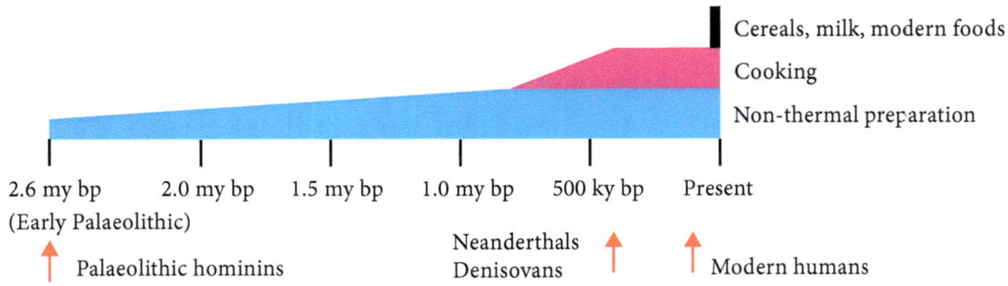

FIGURE 13.3 Timeline of hominin and modern human food preparation and access. Reference to milk is to common ingestion of milk and milk products by adults, assisted by increased incidence of adult lactase-persistence genes.

Micro Fourier transform infrared (FTIR) spectroscopic analysis reveals cooking temperatures of about 500°C. Chipped and cut bones indicate that meat was harvested for cooking. Hearths of similar age (300–400 ky) have been found in Africa, Britain, and France (Gowlett and Wrangham 2013). Remnants at Wonderwerk Cave, South Africa, dated to 1 mya, provide the oldest convincing evidence of hominin use of fire (Berna et al. 2012), although there is speculation that burnt artefacts discovered at several sites in Africa are evidence of controlled use of fire as early as 1.4–1.6 mya (Gowlett and Wrangham 2013).

Early use of fire may have been for warmth and repulsion of insect and animal pests and predators. Thus, we can conservatively estimate that the regular use of fire for cooking dates back 400 ky, about 14,000 generations, and that some use in cooking may have occurred at earlier dates. This is consistent with changes in expression of liver enzymes that are associated with consumption of cooked food (discussed in Section 13.6.6), dating back 270 ky. However, it should be pointed out that anatomically modern humans can be traced back only about 200 ky (Weaver 2012; Wood and Boyle 2016). In other words, fire and cooking were probably the province of pre-modern humans.

13.5.1 Processed Food in the Human Diet: Palaeolithic to Present

The preparation and preservation of foods are important milestones in human evolution (Figure 13.3). Historically, preparation includes grinding, pounding, fermentation, drying, salting, cooling (natural refrigeration), and burying. Modern preservation methods include heat sterilisation, microwaving, mechanical refrigeration, ultrafiltration, irradiation, and vacuum packing. No doubt other methods will appear.

Grinding of grain, presumably for human consumption, dates to at least 30 ky (Revedin et al. 2010). The evidence comes from the discovery of starch grains in the surfaces of grinding stones dated to this period at sites in Russia, the Czech Republic, and Italy. Similar studies in China indicate that grinding of plant starch sources (roots and grains) dates back 19–23 ky (Liu et al. 2014). Starch granules in grinding stones in Israel point to grain processing 23 kya (Snir et al. 2015). In the absence of grinding, pounding, and cooking, grain and USO starches would not be readily available as a human food source, because all mammals lack cellulases, and in animals with only hindgut fermentation, short hindguts, and short hindgut retention times, such as cats, dogs, and humans, cell walls are largely undigested. This is not to say that plants were not eaten; for there is evidence from phytoliths preserved in dental calculus of grain consumption by Neanderthals (Henry et al. 2014). Animals without foregut fermentation, including modern humans, rely on gastric acid, mechanical breakdown, and the enzymes they themselves produce for initial digestion of food. Once our ancestors had secured regular food supplies through agriculture, about 11 kya, food abundance, food storage, and the efficiency of digestion of processed food freed time for the development of technology and culture (Diamond 1997). It has also been suggested that the absence of a need to spend many hours feeding favoured an increase in brain size (Fonseca-Azevedo and Herculano-Houzel 2012), and, by extrapolation, development of human intelligence, technology, and further divergence from other primates. (For further discussion, see Chapter 17: Brain, Spinal Cord, and Sensory Systems.) For example, chimpanzees in the wild spend 4–6 hours/day chewing their food, whereas modern humans typically spend less than 30 minutes to consume a meal (Wrangham and Conklin-Brittain 2003).

Fermentation of fruits and grains, for example to form wines and beers, also preserves foods, by producing a milieu in which bacteria and other pathogens cannot thrive (McGovern et al. 2009). The alcohol itself provides an energy source, and levels of B vitamins of yeast origin (which does not include B_{12}) are increased. Yeast-based fermentation is also used in bread production, the CO_2 that is produced causing the bread to rise, thus increasing its digestibility. The history of the exposure of the human intestine to alcohol, which predates the intentional production of alcoholic beverages, is considered in Section 13.6.4.

13.6 Evidence for Diet-Related Divergence of Digestive Processes in Human Evolution

13.6.1 Adult Lactase Persistence and Domestication of Dairy Animals

Expression of lactase, the small intestinal enzyme necessary for the digestion of lactose, the major sugar of milk, disappears soon after weaning in mammals in general, and in the majority of modern humans. Several lines of evidence show that the development of dairy farming, and the consumption of milk and milk products well beyond weaning, can be dated back for about 8 ky (Itan et al. 2009; Warinner et al. 2014). Direct chemical evidence for extensive use of dairy milk comes from the detection of milk fats in pottery remnants from all 14 sites examined in Britain, dating widespread milk consumption to about 6 kya (Copley et al. 2003). The long history of milk consumption has resulted in the selection of genes for the persistence of lactase expression, and thus adult lactose tolerance (Laland et al. 2010; Perry et al. 2015). The lactase persistence (LP) phenotype is determined by alleles in the promotor region of the lactase gene and is inherited as a dominant trait (Swallow 2003). LP is only common in populations with a long history of herding and milk production, and is presumed to be an advantage only to these populations (Ingram et al. 2009; Itan et al. 2009).

For example, adult lactase expression occurs in up to 75% of individuals in northern European populations that rear cattle and consume cow's milk, but is as low as 5% in hunter-gatherer populations in the same regions (Malmstrom et al. 2010). Another regional comparison, between herders and crop-growing communities in central Asia (Heyer et al. 2011), also found a positive correlation between milk consumption and LP. The data indicated a date of selection for the LP allele of 6–12 kya. The LP allele of some modern human populations is absent in genes isolated from early Neolithic Europeans, 7–8 kya (Burger et al. 2007). It is interesting that times to weaning in Neanderthals and in human hunter-gatherers were most likely similar, in the order of a year or less (Austin et al. 2013), which is consistent with an absence of lactase persistence in populations that do not consume milk after weaning.

13.6.2 Amylase Copy Numbers and Dietary Starch

Amylases, enzymes that hydrolyse glycosidic bonds in polysaccharides to produce glucose and maltose, are necessary for the digestion of grain and USO starches. Salivary amylase gene (*AMY1*) copy number is correlated positively with salivary amylase protein level, and individuals from populations with high-starch diets have, on average, more amylase gene copies than those with traditionally low-starch diets. Individuals from populations whose diets include high amounts of starchy foods had a median *AMY1* copy number of 7, and those on traditionally low starch had a median copy number of 5 (Perry et al. 2007). A further comparison has been made between modern humans consuming high-starch diets (*AMY1* mean copy number 7.34, SD 2.61), Neanderthals (copy number 1.83), and Denisovans from the Siberian Denis cave, about 500 ky (copy number 1.76) (Perry et al. 2015). When significant increases in *AMY1* copy number occurred is uncertain, and it may have been prior to the documented utilisation of grain starches at about 20–30 ky (Perry et al. 2015).

13.6.3 Coeliac Disease

Coeliac disease (CD) is the most common autoimmune intolerance disorder in the world, having a prevalence in western countries of about 1%, although certain regions exhibit a higher incidence (Lionetti and Catassi 2014). Appropriate treatment via a gluten-free diet has only been implemented since the 1950s. Detrimental effects of gluten consumption in CD patients include: malnutrition, increased rate of infection, nutritional deficiencies, weight loss, dehydration, stunted growth, bone fragility, a higher incidence of malignancies, and acute aversive reactions to gluten ingestion (Aaron 2011).

Despite its negative effect on human health, the CD phenotype has persisted; in fact, its incidence is increasing in regions with high levels of gluten consumption. This suggests that the genes underlying the disease have been favoured by natural selection over time, despite the detrimental effects on health. The most important known predisposing genes are human leukocyte antigen (HLA) genes encoding HLA-DQ2 and HLA-DQ8.

Analysis of variations in the prevalence of CD-associated genomic regions suggests that the high prevalence of CD in modern societies may be the by-product of past selection for increased immune responses to combat infections in populations in which agriculture and cereals were introduced early in the post-Neolithic period, that is, in the last 7–10 ky

(Abadie et al. 2011). Zhernakova et al. (2010) demonstrated signs of positive selection for three common loci associated with CD. They suggest that alleles in these regions play a role in protection against bacterial infection. Recent studies provide evidence that whole-grain diets also reduce systemic inflammation (Roager et al. 2017). The HLA-DQ2 CD predisposing allele has also been linked to protection against dental caries, a possible driving selective force in ancient populations (Lionetti and Catassi 2014). The protection is hypothesised to be a product of wheat avoidance, and thus the reduction in the contribution of wheat products to caries.

A high prevalence of CD of 5.6% was found in a large sample of Saharan children when compared to wealthy countries where prevalence is at most about 1% (Catassi et al. 1999). It is proposed by Catassi et al. (1999) that the jejunal mucosa of individuals with CD harbours enterocytes that lack the membrane receptors that are required for adhesion by microorganisms, giving people with CD in the Sahara an advantage by protecting them from intestinal infections and parasites entering the gut lining.

13.6.4 Ethanol Metabolism: Evolution and Population Differences

The majority of ingested ethanol, commonly referred to simply as 'alcohol', is metabolised in two steps. First it is converted to the toxic product acetaldehyde by alcohol dehydrogenase (ADH), and then acetaldehyde is converted to acetate by aldehyde dehydrogenase (ALDH). Acetate is an essential component of many molecules, for example about 85% of proteins are acetylated, and is non-toxic, whereas acetaldehyde has toxic effects. Protein acylation depends on free acetate being bound by coenzyme A to form the acetyl donor, acetyl coenzyme A. The form of ADH that first encounters ingested ethanol is alcohol dehydrogenase 4 (ADH4) in the intestine. It is apparent that enzymes for metabolising ethanol were present long before humans intentionally produced alcohol for consumption about 8–9 kya. A maximum-likelihood analysis of the relatedness of primate ADH4 isoforms indicates that an isoform that efficiently metabolises ethanol emerged in the primate lineage that led to humans about 10 mya (Carrigan et al. 2015). This was preceded by the emergence of plants that bear fleshy fruits that, when infected by suitable yeasts, produce ethanol via fermentation (Thomson et al. 2005), about 12 mya. Modern apes have been observed to consume fermenting fruit in the wild, so it is a reasonable proposition that the primates that preceded humans, and humans when they emerged, also consumed ethanol. A second likely ancient exposure to ethanol is through its production in the human gastrointestinal tract by yeasts (such as candida) and bacteria that colonise the intestine (Lin and Tanaka 2006). In fact, organisms in the intestine may produce sufficient ethanol for it to contribute to liver disease (Zhu et al. 2013).

Human populations differ considerably in their capacity to metabolise ethanol, major differences being found in both ADH (Li et al. 2008) and ALDH (Luo et al. 2009). Deficiencies in ALDH lead to toxic effects of acetaldehyde being manifested, including flushing, headache, and signs of intoxication after ingestion of considerably less alcohol than in people with rapid metabolism via ALDH. An ALDH variant, ALDH2*487Lys, whose carriers are ethanol intolerant, is deduced to have arisen in South China and to have spread in East Asia between 2–3 kya and the present (Luo et al. 2009). This and other variants that reduce ethanol tolerance (Wang et al. 2016) may have had selective advantage by reducing ethanol consumption and thus its damaging health and social effects.

13.6.5 Evolution of the Fatty Acid Desaturase Gene Cluster

The CNS is a fatty structure, 60% lipid by dry weight, contributed to by the large amount of fat-containing membranes relative to cytoplasm in neurons and glial cells. Thus, for its growth the brain requires suitable fatty acids, notably long chain polyunsaturated fatty acids. These are gained from the diet or formed from medium chain fatty acids derived from ingested plant material (Mathias et al. 2012). The conversion requires fatty acid desaturases (FADS). Gene analysis from over 1000 individuals in 14 populations revealed a preferential expression of *FADS* variants favouring rapid conversion of fatty acids of plant origin in people of African origin. The authors conclude that the time of emergence of a common ancestor for genes for a rapid conversion of plant fatty acids was about 85 kya. These alleles for plant fatty acid conversion have been positively selected for in subsequent generations.

13.6.6 Liver Enzymes

Studies of liver enzymes provide genetic evidence of human adaption to cooked diets (Carmody et al. 2016). The authors conducted feeding experiments in mice, and examined differences in liver gene expression profiles between mice fed cooked diets and raw diets, as well as between mice fed meat and fed tubers. There was high correlation between genes for which cooking drove expression changes in mice, and genes known to be differentially expressed between the livers of humans and non-human primates.

These cooking-related genes show evidence of positive selection in the human lineage. By comparing these genes with genes that have potentially undergone positive selection only since humans have split from Neanderthals and Denisovans, the authors conclude that these genes underwent sequence changes *before* this divergence. Since the date of the split is 275 ky at the latest, this indicates that genetic adaptations to a cooked diet may have begun by this time.

13.6.7 Digestive Tract Dimensions and Brain Size

Their dietary history possibly influenced humans to maintain an alimentary tract that is substantially smaller than that of other large primates. In humans, the colon represents only 20% of the total volume of the digestive tract, whereas in apes it is about 50% (Milton 1999; Milton 2003). The sizeable colons of most large-bodied primates permit fermentation of low-quality plant fibres, allowing for extraction of energy in the form of volatile fatty acids (Leonard et al. 2007). Thus, humans are relatively poor in utilising uncooked plant fibre. The human large intestine lies somewhere between that of the pig, an omnivore, and the dog, a carnivore capable of consuming an omnivore diet, which has a reduced caecum and short colon, like humans (Figure 13.1). Evidence for this relationship is that SCFA absorbed in the hindgut provide 2% of maintenance energy for dogs, 6–9% for humans, and 10–31% for pigs (Stevens and Hume 1998). In horses it provides 46% of maintenance energy requirements. By contrast with humans, the New World monkey, *Alouatta palliata*, obtains 31% if its energy requirements from the products of caecal fermentation (Milton and McBee 1983). The hindgut in pigs is 72% of the total volume of the gastrointestinal tract, compared to 20% in humans (Milton and Demment 1988), which relates well to the relative production of SCFA in the two species. The relatively small size of the human colon may limit its capacity for fermentation to the extent that highly fermentable carbohydrates in the modern diet contribute to bowel disease (Marsh et al. 2016; Yao et al. 2016). Highly fermentable carbohydrates that find their way to the colon have been referred to as FODMAPs (fermentable oligosaccharides, disaccharides, monosaccharides, and polyols *or* fermentable, poorly absorbed, short chain carbohydrates). About 15% of the present human population suffer from irritable bowel syndrome, a disorder that is treatable by reducing the intake of foods that lead to fermentable carbohydrates reaching the large intestine (Shepherd et al. 2014).

The smaller gastrointestinal tract of humans requires smaller abdominal and pelvic cavities (Aiello and Wheeler 1995), which may be an advantage to a mammal that stands erect without toppling forward. It has also been speculated that the rather cylindrical rib cage of *Homo*, compared to the inferiorly flared (inverted cone) rib cage of other primates, is an adaptation to a reduced gut volume in humans (Aiello and Wheeler 1995).

The relatively small size of the human stomach and intestines may be related during evolution to the development of a large, energy-hungry brain (Aiello and Wheeler 1995). The idea behind this proposition is that there is not enough energy to spare with a capacious gastrointestinal tract. A hypothetical adaptation would be to consume more energy. However, human populations at subsistence level largely fall within the variation of primate species in the scaling of energy intake to body mass (Simmen et al. 2017); that is, they do not consume significantly more energy than a non-human primate of equivalent mass. A corollary is that nutrient conversion needs to be efficient to feed a large brain from a small gut. Because of their use of prepared foods, modern humans expend an average of 6–7% of meal energy in digestion, compared to the mammalian average of 13–16% (Boback et al. 2007). Thus, a smaller digestive tract, combined with greater efficiency of nutrient extraction, allows energy to be utilised elsewhere. However, fossil evidence indicates that the major change in brain size occurred in the period 1.7–2.3 mya (Schoenemann 2006). Thus, adaptations in size of the digestive system possibly preceded the regular consumption of cooked food. Non-thermal food processing that dates to over 2 my (see Section 13.5) may have contributed to earlier changes. Stone tools were used to access food (shelling nuts, exposing brain tissue and bone marrow), thus increasing the energy available from plant and animal tissues and may have influenced morphological changes that predate cooking. Another human feature that favours smaller gut size may be erect posture and bipedalism, which evolved in hominins some 5–8 mya (Crompton et al. 2010). For humans, mass (M) is related to body length (height, H) squared, leading to body mass index (BMI) being calculated as M/H^2. For other species, M is proportional to H^3. This means that as body length increases, girth increases similarly, in animals other than humans, whereas in humans, girth increases less than proportionately (Do Yi et al. 2017). This is speculated to be a consequence of bipedalism: just imagine a cow or an elephant walking upright, with the centre of mass not being above the feet (hooves) because of the protuberant gut. Thus, it can be speculated that both bipedalism and increased brain size favour a smaller gastrointestinal tract that operates optimally with pre-prepared foods. (For further discussion, see Chapter 17: Brain, Spinal Cord, and Sensory Systems.)

Modern humans combine thermal and non-thermal food processing to maximise energy available for digestion. In the case of the giant panda, a dramatic change in diet 2–2.4 mya has not resulted in a substantial change in morphology relative to the panda's carnivorous forebears. Thus, we can speculate that a reduction in human digestive tract dimensions and increase in brain size occurred between about 6 my, the deduced time of hominin divergence from the chimpanzee lineage, and about 2.6 my. In this period, humans must have had sufficient access to energy-rich food to survive. It is generally believed, primarily from butchery marks on animal bones at sites of human habitation, from dental wear patterns, from plant residues, and by extrapolation from diets of recent hunter-gatherer societies, that humans in the period of approximately 1.8–2.5 my had an adequate diet in terms of energy substrates and nutrient diversity that consisted of meats and fats (perhaps yielding about 50% of energy needs) plus fruits and vegetables (Eaton and Konner 1985; Cordain et al. 2000; Sahnouni et al. 2013).

13.6.8 The Pancreas

Every species must break down organic molecules to feed its needs for energy, growth, and repair. To do this, animals utilise digestive enzymes, which in invertebrates are produced and secreted from cells of the lining of the digestive tract. The most ancient of animals, including corals and hydra, produce enzymes that have homologues in mammals in their digestive tracts (Perillo et al. 2016; Nakayama and Ogasawara 2017). It is only vertebrates that possess a specialised out-pocketing of the digestive tract, the pancreas, which produces the bulk of the digestive enzymes. This produces peptidases, lipases, carbohydrases, and oligonucleotidases. The other major sources of digestive enzymes are saliva (amylases), the stomach (pepsin), and the small intestine (disaccharidases). Over 90% of pancreatic bulk is devoted to the production, storage, and release of digestive enzymes, the other major component being the islets embedded in the pancreas that produce the endocrine hormones, insulin, glucagon, somatostatin, and pancreatic polypeptide. Insulin cells of non-vertebrates are found in the lining of the digestive tract, like other enteroendocrine cells (Falkmer 1979). Gut-associated collections of islet cells (islet organs) appear in agnathia, which do not have an exocrine pancreas, but in other vertebrates, including fish, amphibians, reptiles, and birds, there is a pancreas with both endocrine and exocrine components.

The development of a specialised exocrine pancreas provides a concentrated, regulated, source of digestive enzymes. The pancreas is replete with proteases and other enzymes that degrade tissues. This provides the risk that pancreatic dysfunction can lead to pancreatitis, in which pancreatic proteases digest the pancreas itself, due to inappropriate activation of trypsin within the pancreas that precipitates activation of all pancreatic digestive enzymes (Chen and Férec 2009; Tonsi et al. 2009). To guard against this, enzymes are stored as inactive prohormones (e.g. trypsinogen, procarboxypeptidase, proelastase, prophospholipase) in pancreatic acinar cells. The conversion of procarboxypeptidase and other proenzymes to their active forms relies on proteolytic action of trypsin to remove protective amino acid groups. The conversion of trypsinogen to trypsin, by removal of a protective N-terminal hexapeptide group, is catalysed by enteropeptidase (also known as enterokinase) that is released from cells lining the duodenum. Within the pancreas, autocatalytic or sporadic conversion is inhibited by a locally produced protein, pancreatic secretory trypsin inhibitor (PSTI), which binds and obscures the catalytic site in trypsin. This is not fail-safe, as trypsinogen is not totally devoid of enzymatic activity (Chen and Férec 2009). The precariousness of the balance between insignificant and significant proteolytic activity is exemplified by the occurrence of pancreatitis when there are gain-of-function mutations of the trypsin gene, increased trypsin gene copy numbers, or loss-of-function mutations of the PSTI gene (Chen and Férec 2009). Pancreatitis is more common with increased alcohol use and smoking. There is no evidence that pancreatitis is diet induced.

13.6.9 Muscles of Mastication and Dentition

The properties of the human muscles of mastication and human dentition result in a lesser bite strength than other large primates (Eng et al. 2013), which can be correlated with the habitual consumption of prepared foods. Although these differences were present in Neanderthals, thus predating the time from which there is clear evidence for cooking and the associated ingestion of foods that are softer than in their raw state, it is evident that other preparation techniques preceded cooking (Figure 13.3), some of which softened foods, for example pounding and slicing (Zink and Lieberman 2016). Reduced sizes of masticatory muscles in humans compared to other large primates may include an inheritable mutation, resulting in the loss of transcription of the skeletal muscle gene, *MYH16* (Perry et al. 2015). *MYH16* encodes a muscle myosin, which is part of the contractile machinery. The protein is found only in masticatory muscles of non-human primates, but the protein is not found in humans due to a mutation resulting in the formation of a non-transcribed pseudogene. Analysis of archaic DNA indicates that the pseudogene was present in Neanderthals. This gene loss, plus the reduction in bite surface and muscle mass, may have provided a positive advantage to those humans who cooked or otherwise processed food to make it softer while maintaining its energy content and other nutritional values.

13.7 Human Evolutionary Rate: Was There Time for Digestive System Divergence?

Can we know whether there is likely to have been evolution of human digestive tract form and function in the time since humans were widely exposed to cooked food (approximately < 400 ky) and in the 30 ky that humans have used other preparation techniques that improve the digestibility of food and widen food choices? One line of evidence comes from investigations of genetic divergence between species and between groups of humans with different dietary histories.

Clearly there has been considerable evolutionary change since the chimpanzee–human divergence at approximately 6.5 mya. Using genetic analysis, it has been estimated that a divergence between Neanderthals and modern humans occurred at 410–440 kya (Endicott et al. 2010). This suggests significant evolution of form within the period that humans have been consuming cooked foods. More recent divergences have also been estimated, between the Khoe-San of the Kalahari and Southern Africa and other modern humans at 250–300 kya; between African and non-African populations at 100–120 kya; genetic separation between Europeans and Asians at 100–120 kya; and between Aboriginal Australians and Eurasians at 62–75 kya (Scally and Durbin 2012). Thus, *genetic* differences can be detected within the time period since humans initiated cooking, and there would be time, at least in theory, for evolution of the digestive system during the time that humans have been cucinivores. In fact, analysis of differences between geographically or culturally separated groups of humans indicates that a continuing evolutionary divergence, in which significant differences in the frequencies of occurrence of particular alleles can be detected, occurs over periods of fewer than 10 ky (Laland et al. 2010; Scally and Durbin 2012).

13.7.1 Adaptations Over Short Periods of Time

The vertebrate alimentary system adapts to dietary changes over short periods of time, examples being the induction of carbohydrate transporters within hours of meal ingestion, changes in the size of the gut or its components with altered diet, and post-weaning changes (Karasov et al. 2011; Karasov and Douglas 2013). Some animals reduce the costs of maintaining a digestive system when feeding is interrupted by an atrophy that occurs during periods of low food abundance, such as the winter hibernation of rodents (Carey 2005). This is reversible, and it is characteristic of the vertebrate intestine to experience a 25–50% increase in epithelial mass and potential modest increases (more than two times) in function following feeding after a period of fasting (Dunel-Erb et al. 2001; Secor 2005). More dramatically, intermittently feeding snakes and estivating anurans experience a two- to three-fold increase in small intestinal mass and a two- to ten-fold increase in mass-specific rates of intestinal nutrient transport and hydrolase activities after feeding. These responses are rapidly reversed following the completion of digestion (Secor 2005; Ott and Secor 2007). In a migratory bird, the red knot, a naturally occurring change between a hard and a soft diet caused 40% change in gizzard muscle mass in a week (Dekinga et al. 2001), and Japanese quail fed with alternating diets containing 45% and 1% indigestible fibre reversibly changed gizzard size, with change seen at 1 day and doubling (or halving) in 2 weeks (Starck 1999). Mice and many other mammals digest protein in the stomach through acid hydrolysis and proteases, whereas carbohydrate is largely digested in the small intestine and colon. Consistent with these differences, mice fed a diet with elevated protein exhibit a relative increase in stomach mass, whereas a carbohydrate diet causes a relative increase in intestinal weight (Sørensen et al. 2010).

13.7.2 The Gut Microbiome

The gut microbiome is also influenced by diet. In fact, the microbiomes of different species are more closely related by dietary than phylogenetic similarities (Muegge et al. 2011). For example, a comparison of nutrient metabolising enzymes in the faecal microbiomes of carnivores and herbivores revealed enrichment of enzymes for biosynthesis of amino acids in samples from herbivores, whereas no enzymes for amino acid biosynthesis were enriched in samples from carnivores (Muegge et al. 2011). Moreover, if the diet is changed, the gut microbiota rapidly responds, and within 3–4 days a new bacterial ecology is established, which reverses with reversal of diet (Carmody et al. 2015).

The digestive tracts of invertebrates derived from ancient lineages, including, for example, marine ascidians, sea urchins, and insects, harbour microbiota, including species that are believed to be beneficial to the host (Meziti et al. 2007; Engel and Moran 2013). Social insects (such as ants and bees) appear to pass microbiota between individuals. Thus, two-way interactions between the gut and its flora can be assumed to have shaped the evolution of the digestive system. With change in diet over thousands of years it is certain that the gut microbiome also changed, but what changes this may have made to digestive physiology is unknown.

13.7.3 Diet-Induced Transgenerational Changes

There is increasing evidence that environmental factors can cause epigenetic changes in germ cells that are transferred to the next and in some cases subsequent generations (Bohacek and Mansuy 2015; Cropley et al. 2016), and also non-genetic *in utero* changes that derive from the mother, especially the mother's metabolic state (Archer 2015). The epigenetic changes modify gene transcription, and hence phenotype, without altering DNA nucleotide sequences. Transmissible changes can be mediated by methylation of nucleotides in sperm DNA, post-translational histone modifications, and changes in sperm-derived microRNA (Chen et al. 2016).

Diet-induced paternal obesity at conception causes obese fathers to be more likely to father obese progeny in animal models (Carone et al. 2010; Ng et al. 2010) and in humans (Li et al. 2009). Moreover, reduction in obesity, by diet or exercise, normalises the sperm microRNA profiles of obese fathers and reduces the severity of metabolic syndrome in their female offspring (McPherson et al. 2015). Maternal state during pregnancy and lactation can also influence offspring. Female mice that became obese on a high-fat, high-sugar diet produced offspring with diabetic profiles and fatty liver disease, which was not seen in the offspring of their lean littermates (Oben et al. 2010). Moreover, mouse pups from lean mothers that were cross-fostered to obese mothers also developed the disease phenotype. To ensure that inheritance of obesity-induced changes was due to changes in gametes, *in vitro* fertilisation, using ova and sperm from mice fed high-fat diets, was conducted using healthy, low-fat-diet recipients (Huypens et al. 2016). Female offspring derived from obese parents gained 20% or more body weight compared with mice from lean parents. Male offspring from obese parents tended to be heavier, but the difference was not statistically significant.

These data point to both metabolic and genetic effects on progeny (Archer 2015). Maternal obesity during pregnancy can result in persistent effects, at least to the third generation (Dunn and Bale 2011). Investigations in human subjects indicate the hereditability of physiological measures associated with obesity, and also that obesity influences methylation of genomic DNA in offspring. A very informative study compared DNA of children born to the same mother prior to and after treatment of obesity with gastric bypass surgery (Guénard et al. 2013). The BMI of mothers was changed from 45 before surgery (BS) to 28 after surgery (AS). There were 25 BS and 25 AS offspring. Comparison showed obesity-associated tendencies in BS offspring, including insulin resistance, and over 5000 gene methylation differences. The genes with differential methylation included a preponderance of glucoregulatory, inflammatory, and vascular disease genes. Downstream from the gut is the liver, and there is also evidence for transgenerational transmission of liver disease from mothers provided with a high-fat diet (Wankhade et al. 2017).

The extent that diet-induced transmissible changes manifest across three or more generations in human populations is difficult to assess, because human diet and behaviour is difficult to control and document and generation times are long. Nevertheless, interactions between epigenetic and environmental factors are likely. The children of obese parents will have an epigenetic propensity to obesity, *in utero* metabolic influences, and may also be raised in an environment in which parental example and food availability encourage obesity. It is easy to imagine that these influences could continue over several, even many, generations.

13.7.4 Implications of Recent Dietary Changes for Health

There is little doubt that the incidences of disorders that can be linked to changes in the composition and quantity of human food intake have increased. These include diabetes, alcoholic and non-alcoholic fatty liver disease, cardiovascular disease, and digestive disorders. Obesity, whose incidence continues to increase, can possibly be classified as a disorder of the regulation of food intake. An aspect of this disorder is that obese parents pass on changes that favour obesity and they provide a learning environment, through their example and the food available in the home, that also favours obesity. A clustering of conditions that may lead to diabetes and cardiovascular disease constitute the metabolic syndrome, in which energy storage and utilisation are disturbed, characterised by abdominal obesity, elevated blood pressure, elevated fasting plasma glucose, high serum triglycerides, and low serum high-density lipoprotein levels. These conditions are also influenced by a sedentary lifestyle, transgenerational effects (see Section 13.7.3), genetics, and pollutants.

While unravelling the relative roles of dietary components and environment is difficult for human populations, effects in controlled feeding experiments in animal models can be related to human experience. In humans, reversal of disease when deleterious components are removed provides evidence of their damaging effects. Thus, elevated intakes of fats, particularly polyunsaturated fatty acids, refined and concentrated sugars, advanced glycation end-products, excess sodium chloride, poorly absorbed fermentable carbohydrates, and alcohol all have the potential to lead to disease. Non-nutritive food additives, such as emulsifiers, preservatives, and solvents, may also have deleterious effects (Lerner and Matthias 2015). It is estimated that foods supplying 72% of the dietary calories currently consumed in western diets would not have been found in hunter-gatherer diets (Cordain et al. 2005). (For further discussion, see Chapter 6: Nutrition, Energy Expenditure, Physical Activity, and Body Composition.)

The evolutionary divergence of human populations, including diet-induced divergence (see Section 13.6), takes place over periods of 10 ky or more (Scally and Durbin 2012). These relatively slow rates of adaptation in populations suggest that there may be an inability of humans to adapt adequately to the rapid dietary changes that have occurred in the last hundred years (Lindeberg 2009; Carrigan et al. 2015; Marsh et al. 2016). It can be postulated that humans are inadequately equipped for modern diets high in fats, refined carbohydrate, fructose, and fermentable carbohydrates that are poorly absorbed in the small intestine. Thus, the deleterious effects of modern diets on health, including the rising incidence of liver disease, diabetes, and metabolic syndrome (Williams et al. 2011; Geiss et al. 2014; Leung et al. 2016), may in part reflect limitations of the human digestive system to quickly adapt. On the other hand, the digestive system is sufficiently plastic to adapt to more limited changes in diet.

13.8 Prevention of Digestive and Related Disease

It is apparent from an evolutionary and comparative perspective that the digestive system has evolved, and will continue to evolve, in a species-specific manner, to optimise the assimilation of nutrients and to defend against pathogens and toxins that are inevitably ingested. Food components or products that escape intestinal vigilance enter the wall of the intestine and its venous and lymphatic drainage. Venous drainage flows to the liver, which is the first organ to be challenged after the gut. Also, modern refined foods that are quickly metabolised and cause large increases in plasma glucose challenge the pancreatic islets. The gut has an extensive repertoire of mucosal immune cells, which form its second line of defence after the epithelial lining. These cells must learn their trade, including developing immune tolerance. Early exposure to appropriate microbiota is important in this process of immune system development in the gut (Cerf-Bensussan and Gaboriau-Routhiau 2010; Leung et al. 2016). (For further discussion, see Chapter 10: Immune System.) Our knowledge of evolution and disease processes indicates that to reduce digestive and related diseases, human diets should avoid high fats, foods cooked at excessive temperatures, and highly refined foods that lead to rapid elevation of plasma glucose. In order that there is appropriate maturation of gut immune defence, infants in their first 6 months or so should be exposed to their mother's gut flora and to bacteria in the environment. Antibiotic destruction of the bacteria in the gut in the first 6–12 months of life compromises the gut immune system for the future life of the child.

13.9 Conclusion

Comparative anatomy and physiology of extant animals, and archaeological evidence, points to interactions between diet and the digestive system that have resulted in divergences in form and function. These changes have restricted dietary choices for many species. Humans, through food technology and adaptation, have escaped this narrowing of suitable foods. This has also resulted in a diet-related genetic divergence of human

Table 13.1 Postulated Evolutionary Changes and Divergences in Human Digestive Functions

Digestive feature	Environmental influences	Nature of change	Period of change
Enzymes for milk sugars	Cattle herding, adult consumption of milk	Persistence of lactase gene expression in adults	Approx. 8 ky
Digestion of starches	Increased dietary availability of plant starches (cereal crops, flour production, cooking)	Increased amylase copy numbers	20–30 ky
Gluten enteropathy	Increased incidence of infective challenge	HLA molecules for combating pathogens that coincidently interact with gluten	7–10 ky
Alcohol metabolism	Consumption of fermented fruit	Induction of liver enzymes, population divergence, alcohol tolerance	Precedes controlled fermentation
Fatty acid desaturase	Growth in brain size	Enhanced conversion of fatty acids derived from ingested plant material	85 ky
Liver enzymes	Cooking	Liver enzymes induced by cooked food	Approx. 270 ky
Size of digestive tract	Increased development of brain capacity and associated energy demand, cooking, upright posture.	Reduction in relative size of digestive tract: preceded cooking and advanced food preparation methods	Beginning perhaps 6.5 my, consolidated by later non-thermal food processing and cooking
Dentition and muscles of mastication	Availability of soft foods, cooking	Reduced bite strength compared to apes, changed genes for muscle contractile proteins	Beginning perhaps 6.5 my, consolidated by later non-thermal food processing and cooking

populations. For example, diet-induced divergence in carbohydrate digestion can be seen in human populations whose diets changed 10–30 kya (Table 13.1). In modern societies, there is reliance on cooked foods, and preparation that aids digestibility and storage of foods is pervasive. Humans have evolved to be well adapted to a cooked diet, in terms of the size of the digestive system, dentition, and the energy requirements for digestion. It can be argued that humans are advantaged over other species by the adoption of food preparation and cooking, which allows them to divert time and energy to other activities and to expand the range of foods that can be effectively digested. The observed evolutionary rates of change in digestive processes suggest that the human digestive system is ill-equipped to adapt to substantial changes in diet that have occurred in the last half-century. Humans have evolved to balance the need for the intestine to be accessible and permeable to nutrients and yet defend against pathogens and toxins. Breakdown of this balance can lead to inflammatory bowel disease. Likewise, the intestine exists in equilibrium with microbes in its lumen, an equilibrium whose disturbance can lead to disease.

Acknowledgements

We are indebted to Professor Bernard Wood, Center for the Advanced Study of Human Paleobiology, George Washington University, and Professor Sean Ward, Department of Physiology and Cell Biology, University of Nevada, for valuable comments.

References

Aaron, L. (2011). The last two millennia's echo-catastrophes are the driving forces for the potential genetic advantage mechanisms in celiac disease. *Med Hypotheses* 77, 773–6.

Abadie, V., Sollid, L. M., Barreiro, L. B., et al. (2011). Integration of genetic and immunological insights into a model of celiac disease pathogenesis. *Annu Rev Immunol* 29, 493–525.

Ábrahám, A. (1940). Die Innervation des Darmkanals der Gastropoden. *Cell Tissue Res* 30, 273–96.

Adler, C. J., Dobney, K., Weyrich, L. S., et al. (2013). Sequencing ancient calcified dental plaque shows changes in oral microbiota with dietary shifts of the Neolithic and Industrial revolutions. *Nat Genet* 45, 450–5e.

Aiello, L. C. and Wheeler, P. (1995). The expensive-tissue hypothesis: the brain and the digestive system in human and primate evolution. *Curr Anthropol* 36, 199–221.

Alexandrowicz, J.-S. (1928). Notes sur l'innervation du tube digestif des céphalopodes. *Arch Zool Exp* 67, 69–90.

Archer, E. (2015). The childhood obesity epidemic as a result of nongenetic evolution: the maternal resources hypothesis. *Mayo Clin Proc* 90, 77–92.

Austin, C., Smith, T. M., Bradman, A., et al. (2013). Barium distributions in teeth reveal early-life dietary transitions in primates. *Nature* 498, 216–20.

Barker, C. J., Gillett, A., Polkinghorne, A., et al. (2013). Investigation of the koala (*Phascolarctos cinereus*) hindgut microbiome via 16S pyrosequencing. *Vet Microbiol* 167, 554–64.

Battillo, J. M. and Fisher, A. E. (2015). Reconstructing meat consumption through biomarker analyses of paleofeces. *Ethnobiol Lett* 6, 111–13.

Berna, F., Goldberg, P., Horwitz, L. K., et al. (2012). Microstratigraphic evidence of in situ fire in the acheulean strata of Wonderwerk cave, Northern Cape Province, South Africa. *Proc Natl Acad Sci U S A* 109, E1215–20.

Boback, S. M., Cox, C. L., Ott, B. D., et al. (2007). Cooking and grinding reduces the cost of meat digestion. *Comp Biochem Physiol* 148, 651–6.

Bohacek, J. and Mansuy, I. J. (2015). Molecular insights into transgenerational non-genetic inheritance of acquired behaviours. *Nat Rev Genet* 16, 641–52.

Bradshaw, J. W. S., Goodwin, D., Legrand-Defrétin, V., et al. (1996). Food selection by the domestic cat, an obligate carnivore. *Comp Biochem Physiol* 114A, 205–9.

Braun, D. R., Harris, J. W. K., Levin, N. E., et al. (2010). Early hominin diet included diverse terrestrial and aquatic animals 1.95 Ma in East Turkana, Kenya. *Proc Natl Acad Sci U S A* 107, 10002–7.

Buddington, R. K., Chen, J. W., and Diamond, J. M. (1991). Dietary regulation of intestinal brush-border sugar and amino acid transport in carnivores. *Am J Physiol* 261, R793–801.

Bullock, T. H. and Horridge, G. A. (1965). *Structure and Function in the Nervous Systems of Invertebrates*. San Francisco: W. H. Freeman and Co.

Burger, J., Kirchner, M., Bramanti, B., et al. (2007). Absence of the lactase-persistence-associated allele in Early Neolithic Europeans. *Proc Natl Acad Sci U S A* 104, 3736–41.

Caccia, S., Casartelli, M., Grimaldi, A., et al. (2007). Unexpected similarity of intestinal sugar absorption by SGLT1 and apical GLUT2 in an insect (*Aphidius ervi*, Hymenoptera) and mammals. *Am J Physiol* 292, R2284–91.

Carey, H. V. (2005). Gastrointestinal responses to fasting in mammals: lessons from hibernators. In: Starck, J. and Wang, T. (eds) *Physiological and Ecological Adaptations to Feeding in Vertebrates*. Enfield, NH: Science Publishers.

Carmody, R. N., Gerber, G. K., Luevano, J. M., et al. (2015). Diet dominates host genotype in shaping the murine gut microbiota. *Cell Host Microbe* 17, 72–84.
Carmody, R. N., Dannemann, M., Briggs, A. W., et al. (2016). Genetic evidence of human adaptation to a cooked diet. *Genome Biol Evol* 8, 1091–103.
Carone, B. R., Fauquier, L., Habib, N., et al. (2010). Paternally induced transgenerational environmental reprogramming of metabolic gene expression in mammals. *Cell* 143, 1084–96.
Carrigan, M. A., Uryasev, O., Frye, C. B., et al. (2015). Hominids adapted to metabolize ethanol long before human-directed fermentation. *Proc Natl Acad Sci U S A* 112, 458–63.
Catassi, C., Rätsch, I. M., Gandolfi, L., et al. (1999). Why is coeliac disease endemic in the people of the Sahara? *Lancet* 354, 647–8.
Cerf-Bensussan, N. and Gaboriau-Routhiau, V. (2010). The immune system and the gut microbiota: friends or foes? *Nat Rev Immunol* 10, 735–44.
Chen, J.-M. and Férec, C. (2009). Chronic pancreatitis: genetics and pathogenesis. *Annu Rev Genomics Hum Genet* 10, 63–87.
Chen, Q., Yan, M., Cao, Z., et al. (2016). Sperm tsRNAs contribute to intergenerational inheritance of an acquired metabolic disorder. *Science* 351, 397–400.
Conlon, J. M. (2002). The origin and evolution of peptide YY (PYY) and pancreatic polypeptide (PP). *Peptides* 23, 269–78.
Copeland, L., Blazek, J., Salman, H., et al. (2009). Form and functionality of starch. *Food Hydrocoll* 23, 1527–34.
Copenhaver, P. F. (2007). How to innervate a simple gut: familiar themes and unique aspects in the formation of the insect enteric nervous system. *Dev Dyn* 236, 1841–64.
Copley, M. S., Berstan, R., Dudd, S. N., et al. (2003). Direct chemical evidence for widespread dairying in prehistoric Britain. *Proc Natl Acad Sci U S A* 100, 1524–9.
Cordain, L., Miller, J. B., Eaton, S. B., et al. (2000). Plant–animal subsistence ratios and macronutrient energy estimations in worldwide hunter-gatherer diets. *Am J Clin Nutr* 71, 682–92.
Cordain, L., Eaton, S. B., Sebastian, A., et al. (2005). Origins and evolution of the western diet: health implications for the twenty first century. *Am J Clin Nutr* 81, 341–54.
Cosnes, J., Gower-Rousseau, C., Seksik, P., et al. (2011). Epidemiology and natural history of inflammatory bowel diseases. *Gastroenterology* 140, 1785–94.
Crompton, R. H., Sellers, W. I., and Thorpe, S. K. S. (2010). Arboreality, terrestriality and bipedalism. *Philos Trans R Soc Lond B Biol Sci* 365, 3301–14.
Cropley, J. E., Eaton, S. A., Aiken, A., et al. (2016). Grand paternal inheritance of an acquired metabolic trait induced by ancestral obesity is associated with sperm RNA. *bioRxiv* March 10. doi: https://doi.org/10.1101/042101.
Daly, K., Al-Rammahi, M., Arora, D. K., et al. (2012). Expression of sweet receptor components in equine small intestine: relevance to intestinal glucose transport. *Am J Physiol* 303, R199–208.
De Caro, J., Eydoux, C., Chérif, S., et al. (2008). Occurrence of pancreatic lipase-related protein-2 in various species and its relationship with herbivore diet. *Comp Biochem Physiol* 150, 1–9.
Dekinga, A., Dietz, M. W., Koolhaas, A., et al. (2001). Time course and reversibility of changes in the gizzards of red knots alternately eating hard and soft food. *J Exp Biol* 204, 2167–73.
Diamond, J. (1997). *Guns, Germs and Steel.* New York, London: W. W. Norton and Company.
Dierenfeld, E. S., Hintz, H. F., Robertson, J. B., et al. (1982). Utilization of bamboo by the giant panda. *J Nutr* 112, 636–41.
Do Yi, S., Noh, J. D., Minnhagen, P., et al. (2017). Human bipedalism and body-mass index. *Sci Rep* 7.
Dunel-Erb, S., Chevalier, C., Laurent, P., et al. (2001). Restoration of the jejunal mucosa in rats refed after prolonged fasting. *Comp Biochem Physiol* 129, 933–47.
Dunn, G. A. and Bale, T. L. (2011). Maternal high-fat diet effects on third-generation female body size via the paternal lineage. *J Endocrinol* 152, 2228–36.
Eaton, S. B. and Konner, M. (1985). Paleolithic nutrition: a consideration of its nature and current implications. *N Engl J Med* 312, 283–9.

Endicott, P., Ho, S. Y. W. and Stringer, C. (2010). Using genetic evidence to evaluate four palaeoanthropological hypotheses for the timing of Neanderthal and modern human origins. *J Hum Evol* 59, 87–95.

Eng, C. M., Lieberman, D. E., Zink, K. D., et al. (2013). Bite force and occlusal stress production in hominin evolution. *Am J Phys Anthropol* 151, 544–57.

Engel, P. and Moran, N. A. (2013). The gut microbiota of insects—diversity in structure and function. *FEMS Microbiol Rev* 37, 699–735.

Falkmer, S. (1979). Immunocytochemical studies of the evolution of islet hormones. *J Histochem Cytochem* 27, 1281–2.

Fonseca-Azevedo, K. and Herculano-Houzel, S. (2012). Metabolic constraint imposes tradeoff between body size and number of brain neurons in human evolution. *Proc Natl Acad Sci U S A* 109, 18571–6.

Furness, J. B. and Bravo, D. M. (2015). Humans as cucinivores: comparisons with other species. *J Comp Physiol B* 185, 825–34.

Furness, J. B. and Stebbing, M. J. (2018). The first brain: species comparisons and evolutionary implications for the enteric and central nervous systems. *Neurogast Motil* 30, e13234, 1–6.

Furness, J. B., Rivera, L. R., Cho, H.-J., et al. (2013). The gut as a sensory organ. *Nat Rev Gastroenterol Hepatol* 10, 729–40.

Furness, J. B., Callaghan, B., Rivera, L. R., et al. (2014). The enteric nervous system and gastrointestinal innervation: integrated local and central control. *Adv Exp Med Biol* 817, 39–71.

Furness, J. B., Cottrell, J. J., and Bravo, D. M. (2015). Comparative physiology of digestion. *J Anim Sci* 93, 485–91.

Galliot, B., Quiquand, M., Ghila, L., et al. (2009). Origins of neurogenesis, a cnidarian view. *Dev Biol* 332, 2–24.

García-Arrarás, J. E., Rojas-Soto, M., Jiménez, L. B., et al. (2001). The enteric nervous system of echinoderms: unexpected complexity revealed by neurochemical analysis. *J Exp Biol* 204, 865–73.

Geiss, L., Wang, J., Cheng, Y. J., et al. (2014). Prevalence and incidence trends for diagnosed diabetes among adults aged 20 to 79 years, United States, 1980–2012. *J Am Med Assoc* 312, 1218–26.

Goren-Inbar, N., Alperson, N., Kislev, M. E., et al. (2004). Evidence of hominin control of fire at Gesher Benot Ya'aqov, Israel. *Science* 304, 725–7.

Gowlett, J. A. J. and Wrangham, R. W. (2013). Earliest fire in Africa: towards the convergence of archaeological evidence and the cooking hypothesis. *Azania* 48, 5–30.

Grimmelikhuijzen, C. J. P. and Hauser, F. (2012). Mini-review: the evolution of neuropeptide signaling. *Regul Pept* 177, 56–9.

Guénard, F., Deshaies, Y., Cianflone, K., et al. (2013). Differential methylation in glucoregulatory genes of offspring born before vs. after maternal gastrointestinal bypass surgery. *Proc Natl Acad Sci U S A* 110, 11439–44.

Hansen, G. N., Williamson, M., and Grimmelikhuijzen, C. J. (2002). A new case of neuropeptide coexpression (RGamide and LWamides in hydra), found by whole-mount, two-color double-labeling in situ hybridization. *Cell Tissue Res* 308, 157–65.

Henry, A. G. and Piperno, D. R. (2008). Using plant microfossils from dental calculus to recover human diet: a case study from Tell al-Raqā'i, Syria. *J Arch Sci* 35, 1943–50.

Henry, A. G., Brooks, A. S., and Piperno, D. R. (2014). Plant foods and the dietary ecology of Neanderthals and early modern humans. *J Hum Evol* 69, 44–54.

Heyer, E., Brazier, L., Ségurel, L., et al. (2011). Lactase persistence in Central Asia: phenotype, genotype, and evolution. *Hum Biol* 83, 379–92.

Hu, Y., Wu, Q., Ma, S., et al. (2017). Comparative genomics reveals convergent evolution between the bamboo-eating giant and red pandas. *Proc Natl Acad Sci* 114, 1081–6.

Huypens, P., Sass, S., Wu, M., et al. (2016). Epigenetic germline inheritance of diet-induced obesity and insulin resistance. *Nat Genet* 48, 497–500.

Ierusalimsky, V. N. and Balaban, P. M. (2006). Immunoreactivity to molluskan neuropeptides in the central and stomatogastric nervous systems of the earthworm, *Lumbricus terrestris* L. *Cell Tissue Res* 325, 555–65.

Ingram, C. J. E., Mulcare, C. A., Itan, Y., et al. (2009). Lactose digestion and the evolutionary genetics of lactose persistence. *Hum Genet* 124, 579–91.

Itan, Y., Powell, A., Beaumont, M. A., et al. (2009). The origins of lactase persistence in Europe. *PLoS Comput Biol* 5, e1000491.

Ito, S. and Kurokawa, M. (2007). Coordinated peripheral neuronal activities among the different regions of the digestive tract in *Aplysia*. *Zool Sci* 24, 714–22.

Jin, C., Ciochon, R. L., Dong, W., et al. (2007). The first skull of the earliest giant panda. *Proc Natl Acad Sci U S A* 104, 10932–7.

Johnson, R. N., O'Meally, D., Chen, Z et al. (2018). Adaptation and conservation insights from the koala genome. *Nature Genetics* https://doi.org/10.1038/s41588-018-0153-5

Jönsson, E. (2013). The role of ghrelin in energy balance regulation in fish. *Gen Comp Endocrinol* 187, 79–85.

Kaiya, H., Miyazato, M., Kangawa, K., et al. (2008). Ghrelin: a multifunctional hormone in non-mammalian vertebrates. *Comp Biochem Physiol A* 149, 109–28.

Karasov, W. H. and Douglas, A. E. (2013). Comparative digestive physiology. *Comp Physiol* 3, 741–83.

Karasov, W. H., Martínez Del Rio, C., and Caviedes-Vidal, E. (2011). Ecological physiology of diet and digestive systems. *Annu Rev Physiol* 73, 69–93.

Kienzle, E. (1993). Carbohydrate metabolism of the cat. 1. Activity of amylase in the gastrointestinal tract of the cat. *J Anim Physiol Anim Nutr* 69, 92–101.

Kojima, M. and Kangawa, K. (2005). Ghrelin: structure and function. *Physiological Reviews* 85, 495–522.

Laland, K. N., Odling-Smee, J., and Myles, S. (2010). How culture shaped the human genome: bringing genetics and the human sciences together. *Nat Rev Genet* 11, 137–48.

Landeck, G. and Garriga, J. G. (2016). The oldest hominin butchery in European mid-latitudes at the Jaramillo site of Untermassfeld (Thuringia, Germany). *J Hum Evol* 94, 53–71.

Langille, R. M. and Youson, J. H. (1984). Morphology of the intestine of prefeeding and feeding adult lampreys, *Petromyzon marinus* l: the mucosa of the diverticulum, anterior intestine, and transition zone. *J Morphol* 182, 39–61.

Larhammar, D. (1996). Evolution of neuropeptide Y, peptide YY and pancreatic polypeptide. *Regul Pept* 62, 1–11.

Latorre, R., Mazzoni, M., De Giorgio, R., et al. (2013). Enteroendocrine profile of a-transducin immunoreactive cells in the gastrointestinal tract of the European sea bass (*Dicentrarchus labrax*). *Fish Physiol Biochem* 39, 1555–65.

Leonard, W. R., Snodgrass, J. J., and Robertson, M. L. (2007). Effects of brain evolution on human nutrition and metabolism. *Annu Rev Nutr* 27, 311–27.

Lerner, A. and Matthias, T. (2015). Changes in intestinal tight junction permeability associated with industrial food additives explain the rising incidence of autoimmune disease. *Autoimmun Rev* 14, 479–89.

Leung, C., Rivera, L., Furness, J. B., et al. (2016). The role of the gut microbiota in NAFLD. *Nat Rev Gastroenterol Hepatol* 13, 412–25.

Li, X., Li, W., Wang, H., et al. (2005). Pseudogenization of a sweet-receptor gene accounts for cats' indifference toward sugar. *PLoS Genet* 1, 27–35.

Li, H., Gu, S., Cai, X., et al. (2008). Ethnic related selection for an ADH class I variant within East Asia. *PLoS One* 3, e1881.

Li, L., Law, C., Conte, R. L., et al. (2009). Intergenerational influences on childhood body mass index: the effect of parental body mass index trajectories. *Am J Clin Nutr* 89, 551–7.

Lin, Y. and Tanaka, S. (2006). Ethanol fermentation from biomass resources: current state and prospects. *Appl Microbiol Biotechnol* 69, 627–42.

Lindeberg, S. (2009). Modern human physiology with respect to evolutionary adaptations that relate to diet in the past. In: Hublin, J. J. and Richards, M. P. (eds) *The Evolution of Hominin*

Diets: Integrating Approaches to the Study of Palaeolithic Subsistence. Berlin: Springer Science+Business Media.

Lionetti, E. and Catassi, C. (2014). Co-localization of gluten consumption and HLA-DQ2 and -DQ8 genotypes, a clue to the history of celiac disease. *Dig Liver Dis* 46, 1057–63.

Liu, L., Kealhofer, L., Chen, X., et al. (2014). A broad-spectrum subsistence economy in Neolithic Inner Mongolia, China: evidence from grinding stones. *Holocene* 24, 726–42.

Luo, H. R., Wu, G. S., Pakstis, A. J., et al. (2009). Origin and dispersal of atypical aldehyde dehydrogenase ALDH2*487Lys. *Gene* 435, 96–103.

Malmstrom, H., Linderholm, A., Liden, K., et al. (2010). High frequency of lactose intolerence in a prehistoric hunter-gatherer population in northern Europe. *BMC Evol Biol* 10, 89–95.

Margolskee, R. F. (2002). Molecular mechanisms of bitter and sweet taste transduction. *J Biol Chem* 277, 1–4.

Margolskee, R. F., Dyer, J., Kokrashvili, Z., et al. (2007). T1R3 and gustducin in gut sense sugars to regulate expression of Na^+-glucose cotransporter 1. *Proc Natl Acad Sci U S A* 104, 15075–80.

Marsh, A., Eslick, E. M., and Eslick, G. D. (2016). Does a diet low in FODMAPs reduce symptoms associated with functional gastrointestinal disorders? A comprehensive systematic review and meta-analysis. *Eur J Nutr* 55, 897–906.

Martinez-Pereira, M. A., Franceschi, R. D. C., Antunes, G. D. F., et al. (2013). General morphology and innervation of the midgut and hindgut of *Megalobulimus abbreviatus* (Gastropoda, Pulmonata). *Zool Sci* 30, 319–30.

Mathias, R. A., Fu, W., Akey, J. M., et al. (2012). Adaptive evolution of the *FADS* gene cluster within Africa. *PLoS One* 7, e44926.

McGovern, P. E., Mirzoian, A., and Hall, G. R. (2009). Ancient Egyptian herbal wines. *Proc Natl Acad Sci U S A* 106, 7361–6.

McPherron, S. P., Alemseged, Z., Marean, C. W., et al. (2010). Evidence for stone-tool-assisted consumption of animal tissues before 3.39 million years ago at Dikika, Ethiopia. *Nature* 466, 857–60.

McPherson, N. O., Owens, J. A., Fullston, T., et al. (2015). Preconception diet or exercise intervention in obese fathers normalizes sperm microRNA profile and metabolic syndrome in female offspring. *Am J Physiol Endocrinol Metabol* 308, E805–21.

Meziti, A., Kormas, K. A., Pancucci-Papadopoulou, M.-A., et al. (2007). Bacterial phylotypes associated with the digestive tract of the sea urchin *Paracentrotus lividus* and the ascidian *Microcosmus* sp. *Russ. J Mar Biol* 33, 84–91.

Milton, K. (1999). Nutritional characteristics of wild primate foods: do the diets of our closest living relatives have lessons for us? *Nutrition* 15, 488–98.

Milton, K. (2003). The critical role played by animal source foods in human (*Homo*) evolution. *J Nutr* 133, 3886S–92S.

Milton, K. and Demment, M. W. (1988). Digestion and passage kinetics of chimpanzees fed high and low fiber diets and comparison with human data. *J Nutr* 118, 1082–8.

Milton, K. and McBee, R. H. (1983). Rates of fermentative digestion in the howler monkey, *Alouatta palliata* (primates: ceboidea). *Comp Biochem Physiol* 74, 29–31.

Moran, A. W., Al-Rammahi, M. A., Arora, D. K., et al. (2010). Expression of Na^+/glucose co-transporter 1 (SGLT1) is enhanced by supplementation of the diet of weaning piglets with artificial sweeteners. *Br J Nutr* 32 104, 637–46.

Moran, A. W., Al-Rammahi, M., Zhang, C., et al. (2014). Sweet taste receptor expression in ruminant intestine and its activation by artificial sweeteners to regulate glucose absorption. *J Dairy Sci* 97, 4955–72.

Muegge, B. D., Kuczynski, J., Knights, D., et al. (2011). Diet drives convergence in gut microbiome functions across mammalian phylogeny and within humans. *Science* 332, 970–4.

Nakayama, S. and Ogasawara, M. (2017). Compartmentalized expression patterns of pancreatic- and gastric-related genes in the alimentary canal of the ascidian *Ciona intestinalis*: evolutionary insights into the functional regionality of the gastrointestinal tract in Olfactores. *Cell Tissue Res* 370, 113–28.

Nässel, D. R. and Williams, M. J. (2014). Cholecystokinin-like peptide (DSK) in *Drosophila*, not only for satiety signaling. *Front Endocrinol* 5, 219.

Nässel, D. R., Eckert, M., Muren, J. E., et al. (1998). Species-specific action and distribution of tachykinin-related peptides in the foregut of the cockroaches *Leucophaea maderae* and *Periplaneta americana*. *J Exp Biol* 201, 1615–26.

Nelson, L. E. and Sheridan, M. A. (2006). Gastroenteropancreatic hormones and metabolism in fish. *Gen Comp Endocrinol* 148, 116–24.

Ng, S.-F., Lin, R. C. Y., Laybutt, D. R., et al. (2010). Chronic high-fat diet in fathers programs β-cell dysfunction in female rat offspring. *Nature* 467, 963–6.

Nie, Y., Zhang, Z., Raubenheimer, D., et al. (2015). Obligate herbivory in an ancestrally carnivorous lineage: the giant panda and bamboo from the perspective of nutritional geometry. *Funct Ecol* 29, 26–34.

Nishimura, K., Hiramatsu, K., Monir, M. M., et al. (2013). Ultrastructural study on colocalization of glucagon-like peptide (GLP)-1 with GLP-2 in chicken intestinal L-cells. *J Vet Med Sci* 75, 1335–9.

Oben, J. O., Mouralidarane, A., Samuelsson, A.-M., et al. (2010). Maternal obesity during pregnancy and lactation programs the development of offspring non-alcoholic fatty liver disease in mice. *J Hepatol* 52, 913–20.

Okamoto, T. and Kurokawa, M. (2010). The role of the peripheral enteric nervous system in the control of gut motility in the snail *Lymnaea stagnalis*. *Zool Sci* 27, 602–10.

Olsson, C. and Holmgren, S. (2010). Autonomic control of gut motility: a comparative view. *Auton Neurosci* 165, 80–101.

Osawa, R., Blanshard, W. H., and O'Callaghan, P. G. (1993). Microbiological studies of the intestinal microflora of the koala, *Phascolarctos cinereus*. 2. Pap, a special maternal feces consumed by juvenile koalas. *Aust J Zool* 41, 611–20.

Osawa, R., Rainey, F., Fujisawa, T., et al. (1995). *Lonepinella koalarum* gen. nov., sp. nov., a new tannin-protein complex degrading bacterium. *Syst Appl Microbiol* 18, 368–73.

Ott, B. D. and Secor, S. M. (2007). Adaptive regulation of digestive performance in the genus python. *J Exp Biol* 210, 340–56.

Pante, M. C., Njau, J. K., Hensley-Marschand, B., et al. (2017). The carnivorous feeding behavior of early *Homo* at HWK EE, Bed II, Olduvai Gorge, Tanzania. *J Hum Evol* 120, 215–35.

Payan, F. (2004). Structural basis for the inhibition of mammalian and insect α-amylases by plant protein inhibitors. *Biochim Biophys Acta* 1696, 171–80.

Perillo, M., Wang, Y. J., Leach, S. D., et al. (2016). A pancreatic exocrine-like cell regulatory circuit operating in the upper stomach of the sea urchin *Strongylocentrotus purpuratus* larva. *BMC Evol Biol* 16, 1–15.

Perry, G. H., Dominy, N. J., Claw, K. G., et al. (2007). Diet and the evolution of human amylase gene copy number variation. *Nat Genet* 39, 1256–60.

Perry, G. H., Kistler, L., Kelaita, M. A., et al. (2015). Insights into hominin phenotypic and dietary evolution from ancient DNA sequence data. *J Hum Evol* 79, 55–63.

Raven, H. C. (1936). Notes on the anatomy of the viscera of the giant panda (*Ailuropoda melanoleuca*). *Am Mus Novit* 877, 1–23.

Rehfeld, J. F. (2004). A centenary of gastrointestinal endocrinology. *Horm Metab Res* 36, 735–41.

Ren, G. R., Hauser, F., Rewitz, K. F., et al. (2015). CCHamide-2 is an orexigenic brain–gut peptide in *Drosophila*. *PLoS One* 10, e0133017.

Renard, E., Gazave, E., Fierro-Constain, L., et al. (2013). Porifera (sponges): recent knowledge and new perspectives. *eLS* December.

Revedin, A., Aranguren, B., Becattini, R., et al. (2010). Thirty-thousand-year-old evidence of plant food processing. *Proc Natl Acad Sci U S A* 107, 18815–19.

Roager, H. M., Vogt, J. K., and Kristensen, M. (2017). Whole grain-rich diet reduces body weight and systemic low-grade inflammation without inducing major changes of the gut microbiome: a randomised cross-over trial. *Gut* 1 Nov, pii: gutjnl-2017-314786. doi: 10.1136/gutjnl-2017-314786. [Epub ahead of print].

Rozengurt, E. and Sternini, C. (2007). Taste receptor signaling in the mammalian gut. *Curr Opin Pharmacol* 7, 557–62.

Rubin, D. T., Mody, R., Davis, K. L., et al. (2014). Real-world assessment of therapy changes, suboptimal treatment and associated costs in patients with ulcerative colitis or Crohn's disease. *Aliment Pharmacol Ther* 39, 1143–55.

Sahnouni, M., Rosell, J., Van Der Made, J., et al. (2013). The first evidence of cut marks and usewear traces from the Plio-Pleistocene locality of El-Kherba (Ain Hanech), Algeria: implications for early hominin subsistence activities circa 1.8 Ma. *J Hum Evol* 64, 137–50.

Sakaguchi, M., Mizusina, A., and Kobayakawa, Y. (1996). Structure, development, and maintenance of the nerve net of the body column in hydra. *J Comp Neurol* 373, 41–54.

Scally, A. and Durbin, R. (2012). Revising the human mutation rate: implications for understanding human evolution. *Nat Rev Genet* 13, 745–53.

Schoenemann, P. T. (2006). Evolution of the size and functional areas of the human brain. *Annu Rev Anthropol* 35, 379–406.

Secor, S. M. (2005). Physiological responses to feeding, fasting and estivation for anurans. *J Exp Biol* 208, 2595–609.

Semaw, S., Rogers, M. J., Quade, J., et al. (2003). 2.6-Million-year-old stone tools and associated bones from OGS-6 and OGS-7, Gona, Afar, Ethiopia. *J Hum Evol* 45, 169–77.

Shahack-Gross, R., Berna, F., Karkanas, P., et al. (2014). Evidence for the repeated use of a central hearth at Middle Pleistocene (300 ky ago) Qesem Cave, Israel. *J Arch Sci* 44, 12–21.

Shepherd, S. J., Halmos, E. and Glance, S. (2014). The role of FODMAPs in irritable bowel syndrome. *Curr Opin Clin Nutr Metabol Care* 17, 605–9.

Shimizu, H., Koizumi, O., and Fujisawa, T. (2004). Three digestive movements in *Hydra* regulated by the diffuse nerve net in the body column. *J Comp Physiol* 190, 623–30.

Simmen, B., Pasquet, P., Masi, S., et al. (2017). Primate energy input and the evolutionary transition to energy-dense diets in humans. *Proc R Soc B* 284, 20170577.

Snir, A., Nadel, D., and Weiss, E. (2015). Plant-food preparation on two consecutive floors at Upper Paleolithic Ohalo II, Israel. *J Arch Sci* 53, 61–71.

Sørensen, A., Mayntz, D., Simpson, S. J., et al. (2010). Dietary ratio of protein to carbohydrate induces plastic responses in the gastrointestinal tract of mice. *J Comp Physiol B* 180, 259–66.

Starck, J. M. (1999). Phenotypic flexibility of the avian gizzard: rapid, reversible and repeated changes of organ size in response to changes in dietary fibre content. *J Exp Biol* 202, 3171–9.

Stevens, C. E. and Hume, I. D. (1995). *Comparative Physiology of the Vertebrate Digestive System.* Cambridge: Cambridge University Press.

Stevens, C. E. and Hume, I. D. (1998). Contributions of microbes in vertebrate gastrointestinal tract to production and conservation of nutrients. *Physiol Rev* 78, 303–427.

Swallow, D. M. (2003). Genetics of lactase persistence and lactose intolerance. *Annu Rev Genet* 37, 197–219.

Thomson, J. M., Gaucher, E. A., Burgan, M. F., et al. (2005). Resurrecting ancestral alcohol dehydrogenases from yeast. *Nat Genet* 37, 630–5.

Tonsi, A. F., Bacchion, M., Crippa, S., et al. (2009). Acute pancreatitis at the beginning of the twenty first century: the state of the art. *World J Gastroenterol* 15, 2945–59.

Uyttebroek, L., Shepherd, I. T., Harrisson, F., et al. (2010). Neurochemical coding of enteric neurons in adult and embryonic zebrafish (*Danio rerio*). *J Comp Neurol* 518, 4419–38.

Van Boekel, M., Fogliano, V., Pellegrini, N., et al. (2010). A review on the beneficial aspects of food processing. *Mol Nutr Food Res* 54, 1215–47.

Veenstra, J. A. and Sellami, A. (2008). Regulatory peptides in fruit fly midgut. *Cell Tissue Res* 334, 499–516.

Wang, L.-X., Wen, S., Wang, C.-C., et al. (2016). Molecular adaption of alcohol metabolism to agriculture in East Asia. *Quat Int* 426, 187–94.

Wankhade, U. D., Zhong, Y., Kang, P., et al. (2017). Enhanced offspring predisposition to steatohepatitis with maternal high-fat diet is associated with epigenetic and microbiome alterations. *PLoS One* 12, e0175675.

Warinner, C., Hendy, J., Speller, C., et al. (2014). Direct evidence of milk consumption from ancient human dental calculus. *Sci Rep* 4, 1–6.

Weaver, T. D. (2012). Did a discrete event 200,000–100,000 years ago produce modern humans? *J Hum Evol* 63, 121–6.

Williams, C. D., Stengel, O., Asike, M. I., et al. (2011). Prevalence of nonalcoholic fatty liver disease and nonalcoholic steatohepatitis among a largely middle-aged population utilizing ultrasound and liver biopsy: a prospective study. *Gastroenterology* 140, 124–31.

Wood, B. and Boyle, E. K. (2016). Hominin taxic diversity: fact or fantasy? *Am J Phys Anthropol* 159, S37–78.

Wrangham, R. and Conklin-Brittain, N. (2003). Cooking as a biological trait. *Comp Biochem Physiol A* 136, 35–46.

Wu, S. V., Rozengurt, N., Yang, M., et al. (2002). Expression of bitter taste receptors of the T2R family in the gastrointestinal tract and enteroendocrine STC-1 cells. *Proc Natl Acad Sci U S A* 99, 2392–7.

Yao, C. K., Muir, J. G., and Gibson, P. R. (2016). Review article: insights into colonic protein fermentation, its modulation and potential health implications. *Aliment Pharmacol Ther* 43, 181–96.

Zhao, H., Yang, J.-R., Xu, H., et al. (2010). Pseudogenization of the umami taste receptor gene Tas1r1 in the giant panda coincided with its dietary switch to bamboo. *Mol Biol Evol* 27, 2669–73.

Zhernakova, A., Elbers, C. C., Ferwerda, B., et al. (2010). Evolutionary and functional analysis of celiac risk loci reveals SH2B3 as a protective factor against bacterial infection. *Am J Hum Genet* 86, 970–7.

Zhu, L., Baker, S. S., Gill, C., et al. (2013). Characterization of gut microbiomes in nonalcoholic steatohepatitis (NASH) patients: a connection between endogenous alcohol and NASH. *Hepatology* 57, 601–9.

Zink, K. D. and Lieberman, D. E. (2016). Impact of meat and Lower Palaeolithic food processing techniques on chewing in humans. *Nature* 531, 500–8.

CHAPTER 14

EXCRETORY SYSTEM

PAOLA ROMAGNANI
AND HANS-JOACHIM ANDERS

Abstract

The kidney is the central organ of body fluid and electrolyte homoeostasis and metabolic waste excretion, the conditions of which were settled early in the evolution of multicellular organisms. Life in different environments such as sea water, fresh water, or arid land implies complex challenges for kidneys to maintain body fluid and electrolyte homoeostasis, which is achieved by distinct hormonal pathways regulating water and salt handling via transporter channels across cell membranes. For this to occur there is a tight link between the anatomy and physiology of the nephrons, the independent functional units of the kidney. Adaptation to life on land went along with a loss of *de novo* nephron formation, once nephrons get lost during kidney injury. The limited capacity of mammals to regenerate injured kidneys is one of the reasons for chronic kidney disease being common in humans, especially in ageing populations. This chapter provides an overview on human kidney physiology in view of its evolution across species. A better understanding of the stepwise adaptation processes provides unexpected explanations for numerous human renal diseases and associated health conditions such as hypertension and gout resulting from evolutionary trade-offs. In addition, recent changes in human lifestyle created a mismatch between kidney function and novel environmental challenges, thus representing another source of health conditions affecting the kidney. The evolutionary medicine perspective can offer a better understanding of disease pathogenesis and clues for management and prevention.

Keywords

kidney, nephron, waste excretion, renal diseases, mismatch, evolution, medicine

14.1 Introduction

All organisms invest energy to maintain structure and shape, but energy is also needed to maintain functional metabolic integrity, a process that is named 'homoeostasis'. The driving force for maintaining homoeostasis is a flow of energy, which requires uptake of nutrients and disposal of metabolic waste products. Both nutrients and waste products are solubilised in fluids during uptake and excretion; hence homoeostasis also requires uptake and excretion of fluids—usually water. While the uptake and digestion of nutrients and fluids are discussed in Chapter 13, this chapter focuses on the evolution, physiology, and pathophysiology of the excretory system.

Over evolutionary time, the excretion of waste products and fluids has involved a large number of different body structures or organs, including the skin of starfish and amphibians, the rectal gland of sharks, the salt gland of the albatross, and the sweat glands of certain mammals. However, the central excretory organ has become the kidney, especially in higher vertebrates. Therefore, much of what is discussed here will be about the kidney, kidney development, kidney anatomy, kidney physiology, and kidney diseases. Interestingly, in clinical practice it turned out that using the term 'kidney' has become a hurdle in medical teaching and the understanding of kidney disease. This is why, for the sake of clarity, didactic precision, and conceptual enlightenment, we will avoid using the term 'kidney' as much as possible and rather refer to 'the nephrons of the kidney'. Indeed, nephrons are the working units within kidneys operating independently of each other. Although of similar structure, the nephrons of the kidney are not connected to one another, and each of them independently contributes to what is called in clinical practice 'renal function'. Therefore, excretory function involving the kidneys is based on the number of structurally and functionally intact nephrons, a conceptual conclusion with enormous importance for the integrative understanding of kidney evolution, development, ageing, and disease.

A single nephron in its simplest version is nothing more than a single excretory duct able to excrete fluid and waste products (Vize and Smith 2004). The capacity of such a single nephron to excrete fluid and waste products is limited; hence larger animals need multiple copies of them. Animals that keep growing with age like fish and reptiles have a capacity to produce more nephrons when needed, a process that is also useful once some nephrons get lost due to nephron ('kidney') injury. Higher vertebrates like mammals stop growing beyond a certain age, and they also lose the capacity for *de novo* nephronogenesis, which implies certain limits in responding to accidental nephron losses due to injury, or infectious or inflammatory disease. This simple example may illustrate how the understanding of a frequent medical problem such as 'chronic kidney disease' (CKD) can be improved by replacing the word 'kidney' by 'the nephrons of the kidney'. It immediately implies the need for biologists and physicians to study the evolution and development of 'nephrons' rather than that of 'kidneys', because all kidney functions are conducted by nephrons. Put another way, when kidney function fails, it is because nephrons stop working or get lost. Most importantly, finding innovative cures for 'kidney disease' requires new therapeutic or preventive ways to limit nephron loss and to reinforce nephron regeneration. As the capacity to form new nephrons once existed during evolution and the capacity to regenerate injured nephrons has been maintained to some extent in humans, an important goal should be to recover or pharmacologically enhance nephron regeneration.

Accordingly, we will undertake a journey back in time, that is actually a view across borders, because all previous steps in nephron evolution are still prevalent and can be studied in contemporary species. We aim to demonstrate that enriching the study of kidney anatomy and physiology with knowledge that answers questions related to the origin of nephron anatomy and physiology will help readers to appreciate that in clinical practice the origin and phenotypic presentation of disease can be much better understood when taking an evolutionary medicine approach. Understanding disease requires a sense for evolved physiological processes, which in changing environments may turn out to be less advantageous or even harmful compared to ancestral environments. For example, gills provide great benefits in the aquatic environment but render fish susceptible to dehydration on land. Along similar lines, the loss of the enzyme uricase supports blood pressure (BP) in bipedal primates used to a low-sodium vegetarian diet for millions of years, but may cause hypertension, gout, and kidney stones once primates start to consume sodium (and purine)-rich diets. As such, this chapter intends to introduce the reader to the anatomy and physiology of the nephrons of the kidney in the context of their evolutionary origin to better understand why things work the way they do and why certain diseases of the excretory system are so frequent and difficult to treat.

14.2 Functions and Mechanisms

14.2.1 The Concept of Renal Clearance in Bipedal Primates

Humans are terrestrial primates that, albeit settling close to sources of fresh water, tend to explore arid environments. Three aspects of human biology pose particular demands on their excretory system. First, nitrous waste is not secreted as water-soluble ammonium like in fish, or as water-insoluble uric acid crystals like in birds or reptiles, but as water-soluble urea. This implies the need for excessive filtration of water from the blood and the excretion of a relatively large amount of urine containing this format of nitrous waste. This process takes place in the glomerular compartments of the kidney. Second, as terrestrial life and the unavoidable need for considerable diuresis bear the risk of water depletion, the excretory system must also be able to minimise fluid losses when necessary. This necessitates the capacity to concentrate urine whenever water uptake is not possible (e.g. while sleeping) or to dilute urine whenever water uptake exceeds the need for urine production (e.g. during social drinking). These processes take place in the collecting ducts passing through the hyperosmolar renal medulla. Third, the bipedal posture implies a need for a BP high enough to perfuse a brain that is located 50 cm above the heart. Therefore, humans are adapted to maintain a BP of 110/50 mm Hg or more, for example, by minimising sodium chloride losses, the major determinant of intravascular volume. Sodium chloride reabsorption involves numerous different types of sodium channels at several spots along the nephron.

Maintaining homoeostasis obviously requires restricting losses of important body components, while any uptake exceeding needs, as well as all metabolic waste products must be excreted. Such selective distinctions cannot be sufficiently regulated by simply increasing or decreasing urine volume. Rather it requires a selective enrichment or depletion of elements from the urine. This is achieved via selective transport processes involving specific channels for ions or organic molecules. Indeed, most nutrients, elements, and salts are actively transported across membranes to deplete or enrich them in the urine on demand. This is in contrast to gases or plain water which cross membranes via diffusion or osmosis.

'Clearance' describes the capacity to eliminate an element from a certain volume in a certain time frame. The concept of clearance can be applied to whatever element is passing through the nephrons of the kidney. Depending on the quality of the element, its clearance is determined by the capacity to be reduced in concentration by filtration (for example, insulin that is only filtered via glomeruli but not secreted via transporters into the tubules). Conversely, the clearance of elements that are preferentially secreted, such as p-aminohippurat, is limited by the tubular capacity for directed secretion. In clinical practice, the serum creatinine has been established as a marker of creatinine clearance. For the clinical assessment of renal function it can also be useful to determine the fractional clearance of certain elements calculated on a denominator such as creatinine. For example, a fractional excretion of sodium (Clsod/Clcrea) < 1 implies that sodium is retained as an indicator of active urine concentration as it occurs in shock/hypoperfusion-related oliguric renal failure. In contrast, a fractional excretion of sodium > 1 implies sodium losses as it occurs in oliguric kidney failure due to tubule necrosis, where injured cells can no longer retain sodium as needed.

14.2.2 Nephrons: Number and Structural Organisation

Kidneys are formed by three main structures—the vasculature, the nephrons, and the collecting ducts—which regulate inflow and outflow of blood, generate filtrate in the glomerulus, shape urine composition in tubules, and drain it into the renal pelvis. Each nephron, its associated vasculature, and draining system form an independent functional unit, which implies that each nephron stands alone in terms of its capacity to increase or decrease functional performance and repair after injury. Kidneys are endowed with a certain number of nephrons from birth, but the exact number is difficult to determine. Indeed, a clinically useful marker of nephron number is currently not available. Nephron number is estimated to be around 1 million in each human kidney, but nephron endowment follows a Gaussian distribution across the population, with some people having substantially fewer nephrons from birth, while others are endowed with significantly higher numbers. Such a distribution will underlie the wide distribution of glomerular filtration rate (GFR) that was found in population studies (Benghanem Gharbi et al. 2016). Interestingly, GFR deteriorates with ageing, which is the consequence of age- and injury-related loss of nephrons during one's lifetime (Denic et al. 2016). Monitoring GFR decline with age using percentiles could be a way to distinguish age-related from injury-related nephron loss. That is, when an individual drops below his or her initial percentile, this would indicate that there is additional nephron loss to what would be expected based on that individual's nephron endowment.

The arterial perfusion enters the kidneys at the hilus, rapidly spreading to arcuate arteries that pass along the interface of the renal medulla and cortex. Branches from the arcuate arteries cross the cortex and split into afferent arterioles entering the glomeruli. The fact that also the efferent glomerular vessels are arterioles implies that the glomerular 'capillaries' are branching small arteria, which is illustrated by their perfusion pressure of around 50 mm Hg and a high shear stress enforcing the filtration process. Mesenchymal cells called 'mesangial cells' localise between these glomerular 'capillaries', providing structural and functional support to the vasculature. This first 'capillary network' is followed by a second true capillary network surrounding the renal tubules and facilitating the exchange of solutes by reabsorption and secretion taking place in the tubules. Here, perfusion pressure is much lower than in other tissue capillary networks. Venous outflow parallels the arterial inflow and leaves the kidney in the common conduit of the renal vein (Figure 14.1).

The nephron itself includes the Bowman's capsule around the glomerular capillary network to collect the primary filtrate that drains into the proximal tubule. Within this structure, the glomerular corpuscle spans between the vascular pole and the urinary pole.

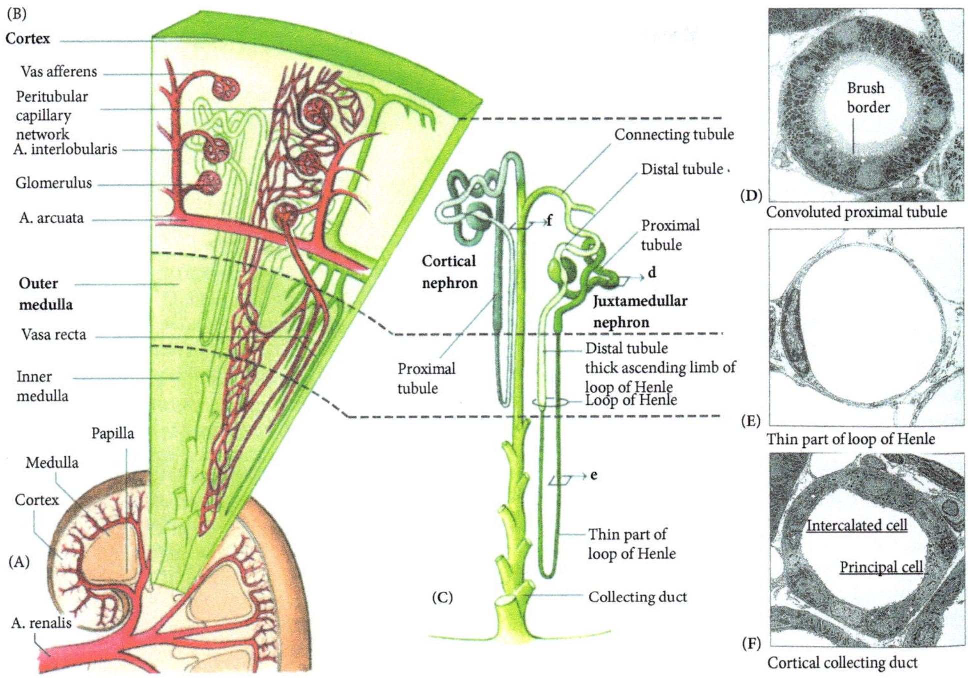

FIGURE 14.1 Anatomical organisation of nephrons. (A and B) Kidney perfusion starts from the cortex where older juxtamedullary nephrons are perfused before younger and smaller cortical nephrons. (C) Loops of cortical nephrons do not reach the inner medulla; hence they do not contribute much to urine concentration. Glomerular 'capillaries' are arterial extensions and only the post-glomerular network of peritubular vessels are true capillaries. Space consumed by convoluted proximal and distal tubules in the cortex defines the pyramidal and bean shape of the kidney. In contrast, the renal medulla contains straight loops of Henle maintaining the osmotic gradient that is needed for urine concentration. Collecting ducts drain urine of numerous nephrons into the renal pelvis.

It is composed of the collagenous matrix of the capsule itself and a monolayer of parietal epithelial cells (PECs). PECs at the vascular pole show transitional characteristics with podocytes, the arborised epithelial cells of the glomerular filtration barrier at the outer aspect of the glomerular capillaries (Shankland et al. 2013). At the vascular pole, PECs show transitional characteristics with proximal tubular cells. The convoluted cortical labyrinth of proximal tubules requires more space than the straight parts of the loop of Henle passing down into outer (cortical nephrons) or inner medulla (juxtamedullary nephrons). The space requirements of the nephron's cortical labyrinth, as well as the cortical nephrons with only short loops determine the bean shape of the kidney (Figure 14.1). The hairpin shape of the loop of Henle implies that the distal nephron returns to the cortex, where first it meets back with its own glomerulus at a touching point named 'macula densa'. This contact point is of great functional importance, and is discussed in Section 14.2.4 in detail. The connecting segment links each nephron with the common draining system, the ureteric bud-derived 'collecting ducts', which descend through the entire medulla and drain the urine into the renal pelvis. As collecting ducts drain many nephrons, any obstruction of these 'ducts of Bellini' at the draining orifice, for example by a calculus, blocks the function of all these nephrons, which cannot be compensated by other nephrons.

14.2.3 Filtrating Blood in the Glomerulus

Although both kidneys represent only 0.4% of body mass, they require 15–25% of the cardiac output, which is around 1.2 l/min, which represents the highest tissue perfusion of all organs. However, this amount of blood is not necessarily needed for oxygen delivery or metabolic demands, but for assuring the filtration process. Accordingly, 90% of renal blood flow (RBF) perfuses the renal cortex where filtration takes place in the glomerular compartment of the nephrons (Figure 14.2). Indeed, filtration itself consumes little oxygen, while the kidney's oxygen demand mostly relates to the reabsorption of sodium in the tubules, which is an active transport process against osmotic gradients. Inside the glomerular compartment, an average renal plasma flow of 600 ml/min produces an average GFR of 120 ml/min (all normalised to 1.73 m^2 of body surface), which implies a filtration fraction of 0.2, 70 times higher compared to capillary leakage in other organs. A GFR of 120 ml/min represents a submaximal GFR that can be transiently increased in moments of higher demands: for example, after a meal (which is nothing else but solute uptake), after fluid intake, or during transient increases of filtration volume like in pregnancy.

Notably, the total GFR is nothing but the result of a single nephron GFR (SNGFR) multiplied by the number of nephrons. In fact, nephron loss with ageing or injury poses additional filtration load to the remaining nephrons that can be compensated up to a maximum SNGFR (Sharma et al. 2014). Hypertrophy of the remnant nephrons is associated with an increased glomerular diameter, which helps to increase SNGFR, while maintaining filtration pressure, the critical determinant of avoiding injury at the delicate glomerular filtration barrier. However, persistent increases of maximal GFR together with increased filtration pressure can cause barotrauma to the filtration barrier, barrier injury, and eventually nephron loss (Kriz and Lemley 2016). This built-in vulnerability is due to the following anatomical and physiological constraints, with the glomerular filtration barrier made of three layers (Figure 14.2) (Haraldsson and Jeansson 2009):

1. The vascular endothelial cells including their glycocalix form a gel-like structure at the inner surface of the glomerular capillaries. Glomerular endothelial cells display unique fenestrations filled with glycocalix, a characteristic that gets lost in diseases like sepsis, diabetes, and autoimmune vascular injury, and this is referred to as 'endothelial dysfunction'. Endothelial dysfunction becomes clinically obvious as microalbuminuria, which reveals that glomerular endothelial cells provide a significant barrier function for serum albumin, probably involving charge-dependent effects of the glycocalix gel layer (Rabelink and De Zeeuw 2015).
2. The glomerular basement membrane (GBM) is the structural backbone of the filtration barrier and produced by both cell types attaching to it from inside and outside. The network of collagen IV, laminin, nidogen, and other matrix proteins is unique in its width and composition. Its main role is to assure vascular integrity. Clinically, breaks in the GBM become evident by bleeding (erythrocyturia). This can occur either due to genetic abnormalities of GBM composition (like in Alport syndrome) or during immune-mediated vascular injuries (e.g. during glomerulonephritis) (Hudson et al. 2003). In turn, erythrocyturia is a diagnostic hallmark of 'nephritic syndrome' caused by inflammatory glomerular injury.

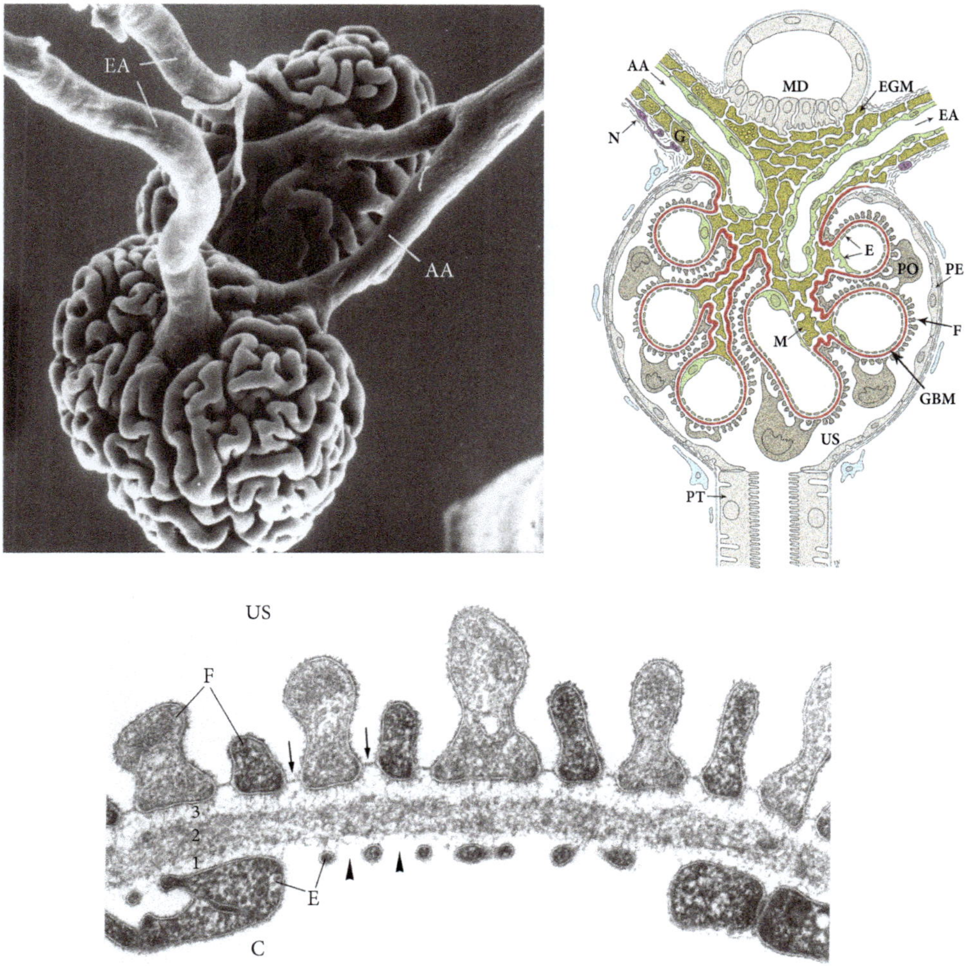

FIGURE 14.2 Glomerular compartment of the nephron. Glomerular 'capillaries' are branching, high-pressure, high shear stress arterioles sealed with fenestrated glycocalix-rich endothelium from inside and arborized visceral epithelial cells (podocytes) from outside. Vascular integrity is maintained by strong glomerular basement membrane (GBM) and interstitial smooth muscle-like mesangial cells. Injury to endothelial cells manifests as microalbuminuria and injury to podocytes as macroproteinuria. Injuries to GBM present as haematuria. Bowman's capsule spans from the vascular to urinary pole of the glomerulus, where ultrafiltrate drains into the proximal tubule. At the vascular pole *macula densa* is a spot sensing sodium chloride concentration of the nephron's own distal tubule. Distinct signalling mechanisms regulate *tubuloglomerular feedback*, directly regulating diameter of the afferent arteriole. Renin release from granulated cells in the afferent arteriole determines glomerular haemodynamics, but also aldosterone-dependent sodium reabsorption is a determinant of extracellular volume (blood pressure) control.

Reproduced from Marlies Elger, and Wilhelm Kriz, 'The renal glomerulus: the structural basis of ultrafiltration', in Neil N. Turner, Norbert Lameire, David J. Goldsmith, Christopher G. Winearls, Jonathan Himmelfarb, Giuseppe Remuzzi, and William G. Bennet, Mark E. de Broe, Jeremy R. Chapman, Adrian Covic, Vivekanad Jha, Neil Sheerin, Robert Unwin, and Adrian Woolf, ed., The Oxford Textbook of Clinical Nephrology (4 ed.), Figures 43.1, 43.2, and 43.5, doi: 10.1093/med/9780199592548.003.0043, © Oxford University Press, 2015.

3. The visceral glomerular epithelial cells at the outer aspect of the glomerular tuft are called 'podocytes' (Figure 14.2). Indeed, the function of podocytes largely depends on their particular structure, a cytoplasm arborised into primary and secondary 'foot processes' that interdigitate to form the slit diaphragm. The slit is another gel-like structure formed by structural proteins like nephron and podocin in a molecular zipper-like cell–cell adhesion between two neighbouring podocytes, together with a glycocalyx made of podocalyxin, podoplanin, and other podocyte-specific proteins (Reiser and Sever 2013). All these components of the filtration barrier determine filter permeability, as the sieving coefficient of blood components depends on molecular size and charge (Table 14.1).

The sophisticated three-dimensional organisation of the visceral glomerular epithelium has been called 'the Achilles' heel' of the kidney, because lost podocytes are difficult to replace, conceptually like a missing piece of a puzzle (Reiser and Sever 2013). Indeed, podocytes are post-mitotic cells that persist for life and are gradually lost with ageing (Hodgin et al. 2015). Parietal epithelial cells of the Bowman's capsule contain progenitors with the potential to differentiate into mature podocytes. However, their capacity for podocyte regeneration is limited and hampered by factors such as proteinuria (Peired et al. 2013). When podocytes are not replaced by local progenitors, adjacent podocytes may undergo hypertrophy, though any insufficient epithelial coverage of the filtration barrier leads to scar formation (focal segmental glomerulosclerosis (FSGS)), a process that often progresses to focal global glomerulosclerosis (FGGS) and loss of the associated nephron (Lasagni et al. 2013). To avoid this damage, evolution has generated mechanisms to maintain filtration pressure within a physiological range as constant as possible.

14.2.4 Autoregulation of Filtration Pressure and the Glomerular Filtration Rate

Arterial pressure declines from around 90 mm Hg in the arcuate arteria to 48 mm Hg in the glomerular capillaries, with a gradient from the glomerular inflow to outflow of as little as

Table 14.1 Sieving Coefficient as a Parameter of Glomerular Filter and Compound Clearance

Substance	Molecular mass (Da)	Radius (nm)	Dimension (nm)	Sieving coefficient
Water	18	0.1		1.0
Urea	60	0.16		1.0
Glucose	180	0.36		1.0
Insulin	5500	1.48		0.98
Myoglobin	17,000	1.95	5.4 × 0.3	0.75
Ovalbumin	43,500	2.85	8.3 × 2.2	0.22
Haemoglobin	68,000	3.25	5.4 × 3.2	0.01
Serum albumin	69,000	3.55	15 × 1.6	< 0.001

1–2 mm Hg. Maintaining intraglomerular perfusion pressure as constant as possible along a wide range of systemic BPs is a function of the resistance of the pre- and postglomerular arterioles, respectively (Carlstrom et al. 2015). If systemic BP is low, the preglomerular arteriole should be open and the postglomerular arteriole should be constricted to maintain glomerular capillary pressure and ultrafiltration. Conversely, at high systemic BP the preglomerular arteriole should be constricted and the postglomerular arteriole dilated to limit perfusion pressure inside the glomerulus. Three mechanisms support this process:

1. The myogenic reaction (Bayliss effect) operates in the Aa. interlobulares and the preglomerular arterioles. Any sudden increase of systemic BP induces a reflectory contraction of the smooth muscle cells of the vascular wall within 1 second, thereby increasing vascular resistance. This way, filtration pressure is kept constant up to a systemic BP of around 170 mm Hg. Higher systemic BP increases RBF and filtration pressure, but does not necessarily increase GFR (Carlstrom et al. 2015).
2. The tubuloglomerular feedback (TGF) mechanism involves cells at the *macula densa* that sense the concentrations of sodium and chloride in the distal tubule and send this information to cells in the vascular wall of the afferent arteriole at the vascular pole of the glomerulus (Singh and Thomson 2010). A decrease in NaCl concentration indicating a reduction in GFR leads to vasodilation of the afferent arteriole, while an increase of NaCl concentration at the *macular densa* has the opposite effect. The TGF assures a vasoregulatory response within 10 seconds to changes in GFR related to an altered sodium balance or volume status, as occurs during eating, drinking, or starving. This mechanism has great pathophysiological importance. For example, hyperglycaemia in diabetes involves massive sodium reabsorption in the proximal tubule, which reduces distal sodium delivery to the *macula densa*, a process that deactivates TGF. Consequently, people with diabetes experience a persistent vasodilation of the afferent arterioles leading to persistent glomerular hyperfiltration, as indicated by persistent consumption of the renal reserve at a maximal GFR. This process is thought to contribute to diabetic kidney disease also because interfering with this process with sodium glucose transporter 2 (SGLT2) inhibitors can reverse this pathomechanism (Anders et al. 2016). Another example is the use of loop diuretics in heart failure, which increases sodium delivery to the *macula densa* and enforces TGF. As a result, vasoconstriction reduces GFR, a process that explains the commonly seen increase in serum creatinine levels upon loop diuretic use in congestive heart failure.
3. A number of vasoactive hormones and substances regulate the renal vasculature and glomerular ultrafiltration. Among them, only angiotensin II will be discussed in detail. TGF activation at the *macula densa* involves release of renin from cells of the wall of the afferent arteriole at the vascular pole of the glomerulus. The enzyme renin processes angiotensinogen from the liver into angiotensin I, which is further processed by the angiotensin converting enzyme into angiotensin II, which has numerous physiological effects including vasoconstriction of afferent and efferent glomerular arterioles as well as inducing aldosterone secretion from adrenal glands eliciting sodium retention and potassium secretion in the distal tubule.

All these mechanisms help to maintain a constant glomerular filtration pressure and GFR.

14.2.5 Sodium Balance Controls Extracellular Volume

Sodium is the most abundant cation in the extracellular space, mirrored by potassium in intracellular compartments. In a state of intact osmoregulation (water balance), changes in sodium directly translate into changes of isotonic fluid, which in clinical practice is often referred to as 'volume' (Bhave and Neilson 2011). Therefore, 'hypervolaemia' or 'hypovolaemia' describe clinical scenarios of gains or losses of sodium in the extracellular space, respectively. It is of note that the extracellular space consists of intravascular and interstitial compartments. Hence, gains of sodium translate into hypervolaemia in the intravascular compartment presenting as arterial hypertension, and as hypervolaemia in the interstitial compartment presenting as pitting oedema. Thus, hypertension and oedema (hypervolaemia) are usually the consequence of a net gain in sodium. In turn, under conditions of intact osmoregulation (water balance), losses of sodium present clinically as arterial hypotension.

Sodium balance is of enigmatic importance for all aquatic and terrestrial animals; hence the nephrons of the kidney have to ensure that the vast majority of the 26,000 mmol/day of sodium in the glomerular ultrafiltrate gets reabsorbed and that urinary sodium losses exactly match sodium uptake ('only what goes in goes out'). Sodium reabsorption starts in the S1 segment of the proximal convoluted tubule where the luminal brush border of the proximal tubular cells spans a surface of 40–80 m^2, allowing the reabsorption of large amounts of solutes and water (Figure 14.3).

The driving force for sodium reabsorption is the basolateral Na^+/K^+ATPase, for which large numbers of mitochondria produce the necessary amounts of adenosine triphosphate (ATP), consuming most of the oxygen delivered to the renal cortex. This mechanism keeps intracellular sodium concentration low and creates an electrical membrane potential, which both foster sodium reabsorption from the ultrafiltrate. Luminal sodium cotransporters (together with sodium or bicarbonate) or antiporters (Na^+/H^+) support this process. A clinically relevant consequence is that in diabetes, large amounts of glucose in the ultrafiltrate affect sodium balance by maximising proximal sodium reabsorption, which reduces distal sodium delivery and deactivates TGF, leading to glomerular hyperfiltration. The transepithelial potential is as low as 2 mV, which is anyway sufficient to drive the paracellular resorption of sodium, chloride, potassium, calcium, and magnesium (Table 14.2).

In the thick ascending limb (TAL) of the loop of Henle, sodium reabsorption is directly linked to active chloride transport and passive paracellular potassium, magnesium, and calcium reabsorption. Thus, the loop diuretic furosemide that blocks this process can cause substantial losses of these ions. While a negative balance for sodium is usually welcome as it reduces arterial hypertension and oedema (extracellular volume), the concomitant losses of the other ions imply metabolic alkalosis, hypolkalaemia, and hypomagnesaemia as common side-effects (Welling 2016). Finally, the distal convoluted tubule (DCT), the connecting tubule (CNT), and the collecting duct (CD) are additional cites of Na^+/K^+ATPase-dependent sodium reabsorption. The luminal Na/Cl symporter of the DCT is a molecular target of thiazide diuretics, while the epithelial Na^+ channel (ENaC) of the CNT and CD is a molecular target of amiloride diuretics. Genetic polymorphisms leading to loss- or gain-of-function to each of the aforementioned sodium channels can cause syndromes of volume depletion or hypervolaemia, respectively.

Aldosterone is an important regulator of the sodium channels in the DCT and CNT; hence the renin–angiotensin–aldosterone system (RAAS) links renal autoregulation with

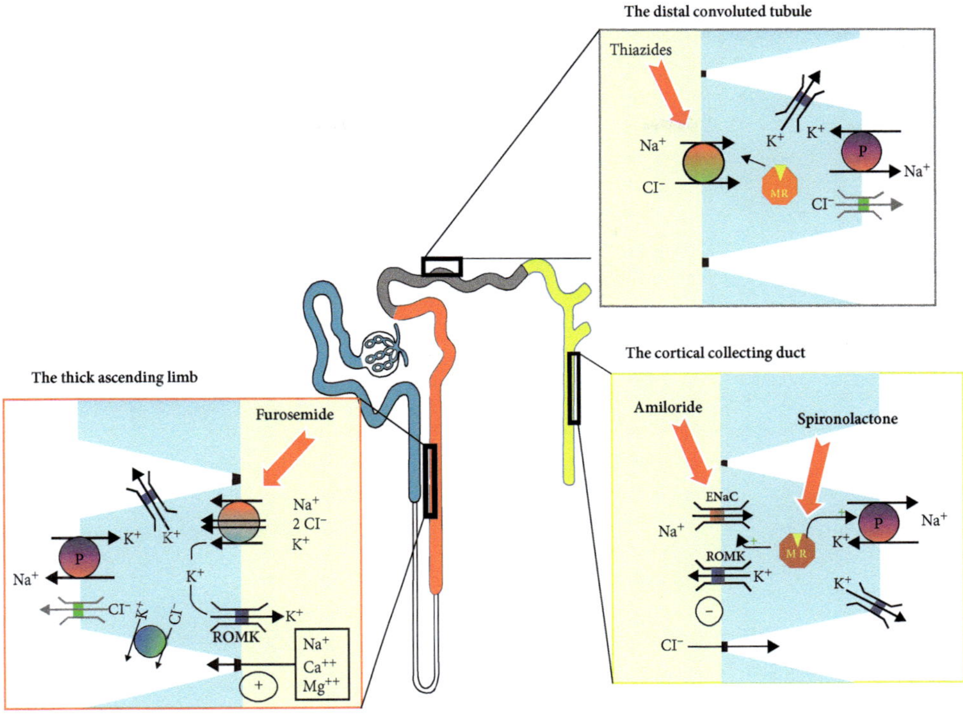

FIGURE 14.3 Sodium reabsorption along the nephron. Sodium channels drive sodium reabsorption from glomerular filtrate at several sites along the nephron. In the connecting segment and collecting duct, sodium reabsorption is under control of aldosterone and natriuretic peptides, which determine fractional excretion of sodium in a range between 0.5 and 5%. In final urine, urea has replaced sodium as the major determinant of osmolarity. ENaC, Epithelial Na^+ channel; MR, mineralocorticoid receptor; P, phosphate; ROMK, renal outer medullary potassium (K) channel.

sodium balance (Rossier et al. 2015). In addition, hyperaldosteronism typically leads to arterial hypertension and hypokalaemia. Natriuretic peptides such as atrial natriuretic peptide (ANP) and brain natriuretic peptide (BNP) act on the medullary CD.

14.2.6 Water Balance Controls Intracellular Volume

In contrast to isotonic fluid (volume), plain water can freely cross cell membranes. Therefore, deficits or excess of water do not become evident as 'hypo- or hypervolaemia' even though gains/losses lead to the equivalent changes in body weight (Sterns 2015). For example, a net gain of 1 l of plain water will distribute by about half each into extra- and intracellular compartments, which has no detectable effect on BP and interstitial volume. The only result of water excess will be a dilution of electrolytes outside and inside cells. Therefore, the clinically apparent manifestations of water excess are hyponatraemia and intracellular

Table 14.2 Transport Processes in the Proximal Tubule

Resorption of	Mechanisms
Na^+	Transcellular: basolateral Na^+/K^+ATPase, Na^+/HCO_3^- symporter Luminal: Na^+ symporter together with glucose (SGLT2), Na^+/H^+ antiporter Paracellular
Cl^-, K^+, Ca_2^+, Mg^+	Paracellular
HCO_3^-	Secreted H^+ plus filtered HCO_3^- converted into CO_2 plus H_2O (carboanhydrase)
Water	Active solute reabsorption drives water reabsorption via osmotic forces
Phosphate	Parathyroid hormone-dependent
Glucose, uric acid, amino acids, and so on	Na^+ symporter
Peptides	Brush border aminopeptidase digestion or H^+ symporter
Proteins	Endocytosis and lysosomal degradation
Urea	Diffusion
Secretion of	**Mechanisms**
H^+	Na^+/H^+ antiporter, H^+-ATPase
NH_3^+, NH_4^-	
Organic acids and bases	

oedema (Sterns 2015). Conversely, net losses of plain water present as hypernatraemia and intracellular volume depletion. Although all body cells increase in size with water excess and shrink with water depletion, the only organ causing clinical symptoms is the brain (Sterns 2015). The solid skull implies that brain swelling rapidly increases intracranial pressure, which gradually impairs cerebral perfusion and causes brain oedema.

Osmoregulation of body fluids is a very ancient homoeostatic mechanism. Only very primitive organisms are osmoconformers, implying that they shrink from water losses in hyperosmolar environments like sea water, and that they expand from water influx in hypoosmolar environments like fresh water. Indeed, inside the human body most cells are osmoconformers, including erythrocytes and brain cells, as they cannot regulate osmolarity on their own. In fact, osmoregulators regulate internal osmolarity at the system level. In humans this has become a complex regulatory system involving the heart, major vessels (volume sensing), the hypothalamus (osmolarity sensing), and the pituitary gland (antidiuretic hormone (ADH) secretion), as well as all parts of the nephron and the collecting ducts (water reabsorption) (Sterns 2015).

Of the 180 l of glomerular ultrafiltrate per day, almost everything is usually recovered and only a small portion is excreted as urine depending on the amount of fluid intake ('all that goes in goes out'). Water reabsorption is usually a passive process linked to the reabsorption of sodium and other ions starting from the proximal tubule (Figure 14.4). The hyperosmolarity of the renal medulla is the driving force of water reabsorption, because it creates an

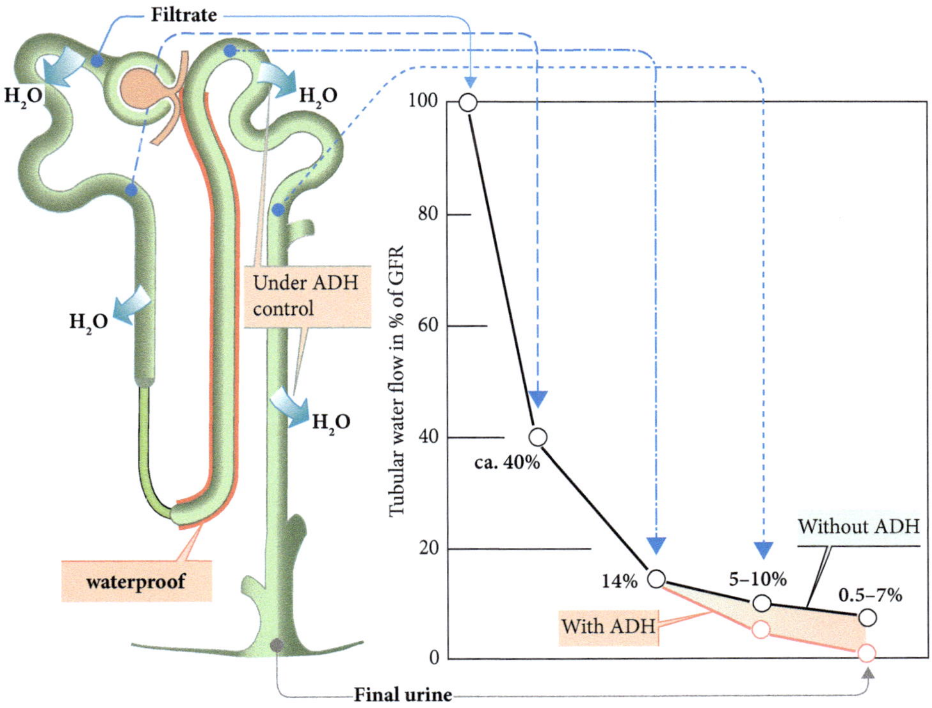

FIGURE 14.4 Water reabsorption along the nephron. Water drains passively together with ion transport in the proximal tubule and descending loop of Henle, while the ascending loop of Henle is impermeable to water, which avoids water reflux into the filtrate on the way out of the hyperosmolar renal medulla back into the iso-osmolar cortex. Water balance is regulated by vasopressin-dependent relocation of aquaporin water channels into the membranes of collecting duct cells by help of osmotic forces of the medulla.

osmotic gradient, dragging water from the tubular lumen into the hyperosmolar interstitium of the renal medulla. This osmotic gradient is maintained by the recirculation of urea in the deep medulla, which increases osmolarity up to 1200 mosmol/kg H_2O. This implies that urine can be maximally concentrated to five times of plasma osmolarity. The hairpin-shape of the loop of Henle causes a countercurrent flow which gradually concentrates the filtrate on its way through the inner medulla, but as the filtrate returns to the kidney cortex the water-impermeable TAL avoids water reflux into the tubular lumen, a process necessary for further sodium reabsorption in the TAL. While finally passing via the CD a second time through the hyperosmolar medulla, water permeability is regulated by ADH-dependent water channels (aquaporins) that allow fine-tuning of water excretion according to the need for the excretion of diluted or concentrated urine (Figure 14.4) (Sands and Layton 2014). As ADH secretion is tightly regulated by plasma osmolarity, serum sodium concentration (osmolarity) and intracellular volume are usually kept constant (Sterns 2015). Indeed, humans are potent osmoregulators, which is required for their life in the arid environment on land, where water and salt loss are a constant threat to homeostasis.

14.2.7 Renal Clearance of Nitrous Waste

Filtered organic molecules are largely reabsorbed in the proximal tubule just like glucose. A number of amino-acid transporters eliminate the filtered free amino acids from the ultrafiltrate. Filtered smaller peptides are digested by aminopeptidases in the brush border of the proximal convoluted tubule. Larger proteins such as albumin or light chains of immunoglobulins are taken up from tubular epithelial cells via various ways of endocytosis involving carrier proteins such as megalin or cubilin in the brush border. Lysosomal proteases then digest these proteins into amino acids. Glomerular diseases associated with defects in the filtration barrier increase the exposure of the proximal tubule to albumin and other plasma proteins. A similar exposure occurs in plasma cell dyscrasia with overproduction of light chains passing the glomerular filter. This can easily exceed the proximal tubule capacity to uptake and break down proteins, causing significant intracellular stress as well as loss of protein in the urine (i.e. proteinuria).

In contrast to the recovery of reusable components of the glomerular ultrafiltrate, renal clearance of nitrous waste refers to the excretion of nitrogen that is taken up with the diet. Nitrogen is an integral component of amino acids and nucleic acids consumed with cellular dietary components of any kind (vegetarian and carnivorous). In contrast to most fish, insects, birds, and reptiles, mammals use the water-soluble molecule urea for nitrogen excretion. Urea is produced mostly in the liver and circulates as blood urea nitrogen. Upon filtration, nitrogen is reabsorbed and secreted back to the lumen, a process that also supports the high osmotic gradient in the renal medulla (Sands and Layton 2014). The distal tubule is largely impermeable for urea; therefore luminal urea concentration increases, together with reabsorption of water and other solutes. In the final CD, some urea is passively reabsorbed in an ADH-dependent manner; hence the fractional excretion depends on water diuresis.

In contrast, birds and reptiles secrete nitrous waste as water-insoluble uric acid crystals, as is obvious from the white colour of bird and reptile droppings. Mammals expressing uricase break down uric acid to the more water-soluble metabolite allantoin. However, humans, together with the other great apes, lost uricase expression and therefore show considerable blood levels of uric acid, which also has potent antioxidative effects (Mandal and Mount 2015). Filtered uric acid is reabsorbed and secreted via the urate transporter 1 (URAT1) and approximately 10% of the filtered uric acid is excreted with the urine.

Another mechanism of nitrous waste clearance is the secretion of ammonium ions built from glutamine in the distal tubule, a process that is tightly linked to acid–base balance, discussed in Section 14.2.8.

14.2.8 How the Nephrons of the Kidney Maintain a Constant pH

The nephrons of the kidney have a major role in maintaining a normal pH because the elements of acid–base balance (bicarbonate and H^+ protons) are readily filtered in the glomerulus. The around 4500 mmol of bicarbonate filtered every day need to be reabsorbed to avoid metabolic acidosis. This happens in the proximal tubule driven by H^+ secretion via H^+ATPase into the tubular lumen, which allows bicarbonate to dissociate into CO_2 and water, driven by carboanhydrase IV in the brush border membrane (Figure 14.5).

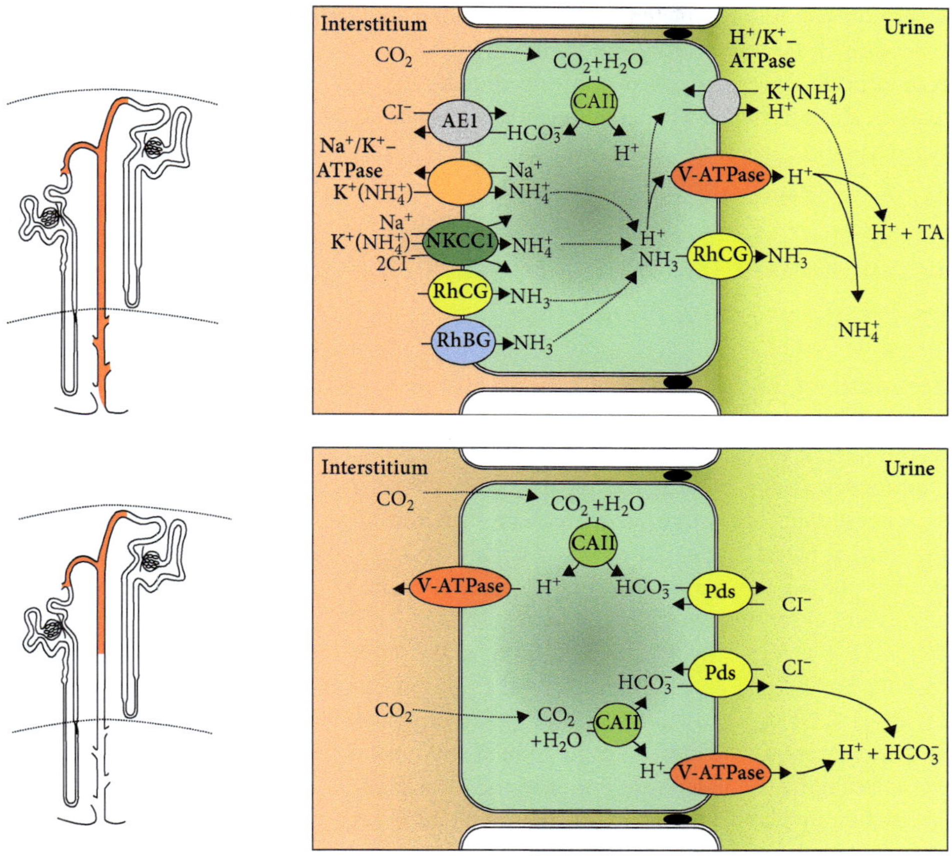

FIGURE 14.5 The kidney in acid–base balance. Filtered bicarbonate is recovered via H^+ secretion in the proximal tubule. Net H^+ secretion for urine acidification is achieved in a stepwise manner along the nephron, mainly in the connecting tubule and collecting duct where H^+ binds to sulfates and phosphates to form titratable acids.

Reproduced from Carsten A. Wagner, and Olivier Devuys, 'Renal acid–base homeostasis', in Neil N. Turner, Norbert Lameire, David J. Goldsmith, Christopher G. Winearls, Jonathan Himmelfarb, Giuseppe Remuzzi, and William G. Bennet, Mark E. de Broe, Jeremy R. Chapman, Adrian Covic, Vivekanad Jha, Neil Sheerin, Robert Unwin, and Adrian Woolf, ed., The Oxford Textbook of Clinical Nephrology (4 ed.), Figures 24.3, doi: 10.1093/med/9780199592548.003.0024, © Oxford University Press, 2015.

This process leads to a gradual drop in urinary pH. Indeed, the dietary uptake of amino acids and other acidic elements requires a net excretion of H^+; therefore urine acidification is necessary. In the distal tubule, a sodium/H^+ antiporter further supports proton excretion, followed by an H^+/K^+ATPase and an H^+ATPase in the CNT and CD that promote H^+ secretion, respectively. These protons bind to phosphate, sulfate, and ammonia, which support elimination as titratable acids for net acid excretion. In case of acidic loads, renal acid excretion can be enhanced within days by increasing transport of glutamin from the liver to the kidney, where mitochondrial glutaminase and glutamate dehydrogenase foster ammonium excretion along the nephron. Loss of nephrons gradually impairs the kidney's capacity to eliminate protons via titratable acids; thus, CKD is associated with renal tubular acidosis (RTA), a metabolic acidosis inducing respiratory compensation (hyperventilation) (Levin and Buck 2015). In contrast to CKD-related RTA that is of the distal type (type I, lack of proton secretion), isolated injury to proximal tubular cells or selective inhibition of carboanhydrase rather induces metabolic acidosis via selective lack of bicarbonate reabsorption. Increased delivery of bicarbonate into the distal tubule induces not only osmotic diuresis but also hyperchloraemic metabolic acidosis, that is, RTA of the proximal type (type II).

14.2.9 Renal Clearance of Bone-Forming Minerals and Kidney Stone Formation

Bone-forming minerals include calcium and phosphate, and their renal clearance is specifically regulated by parathyroid hormone (PTH) (Berndt and Kumar 2007). Most of the filtered phosphate gets reabsorbed in the proximal tubule, but still, its fractional clearance reaches 5–20%, depending on dietary phosphate intake (Figure 14.6). This is possible because in contrast to glucose reabsorption in the PCT which can be up to two-fold of the normal blood levels, phosphate is readily excreted once the filtered amounts exceed normal plasma concentrations of 0.8–1.4 mmol/l, as a kind of spillover mechanism. The expression of sodium-phosphate cotransporters in the PCT is under the control of several factors. For example, hyperparathyroidism, as it occurs during CKD, enhances their proteolytic breakdown and thereby increases phosphate clearance. Also fibroblast growth factor 23 (FGF-23) and Klotho have synergistic effects on phosphate excretion (Berndt and Kumar 2007). Phosphate in the tubular fluid usually presents as HPO_4^{2-}, but proton excretion in the distal nephron produces the titratable acid $H_2PO_4^{2-}$.

Calcium metabolism is based on 99% of calcium being stored in bone and uptake being regulated in the intestinal tract. However, small amounts of ionic calcium are filtered and reabsorbed in various ways. Claudin-16 and various calcium channels contribute to this process which is regulated by PTH, calcitonin, and other hormones.

Calcium often forms complexes with other elements that can form amorphous or crystalline solids in the urine. Such precipitates can attach to the luminal membrane of tubular epithelial cells and grow to the size of crystals, calculi, or even stones. When this happens inside tubules or CD, this process is referred to as nephrocalcinosis (Worcester and Coe 2010). When stone formation occurs inside the renal pelvis this is referred to as urolithiasis. Nephrocalcinosis and urolithiasis can cause significant kidney injury and eventually lead to CKD, for example by obstructing the renal outflow. Once kidney stones detach from the site of growth in the renal pelvis they can mobilise into the ureter and cause acute painful obstructions, a renal colic. Indeed, single episodes of kidney stone disease affect up to 10% of the population (Worcester and Coe 2010). The pathophysiology is complex and involves genetic variants of mineral handling and the absence of crystallisation inhibitors in the nephrons of the kidney, as well as dietary factors. Volume depletion promotes supersaturation and microcrystal formation of stone-forming minerals in concentrated urine (Worcester and Coe 2010). Such microcrystals can attach to luminal membranes of tubular epithelial cells expressing certain binding proteins such as CD44, annexin II, or osteopontin. Crystals have the capacity to directly kill tubular cells or to elicit activation of innate immunity via the NLRP3 inflammasome in intrarenal dendritic cells. However, more commonly crystal plugs obstruct single nephrons eventually leading to nephron atrophy (Figure 14.6).

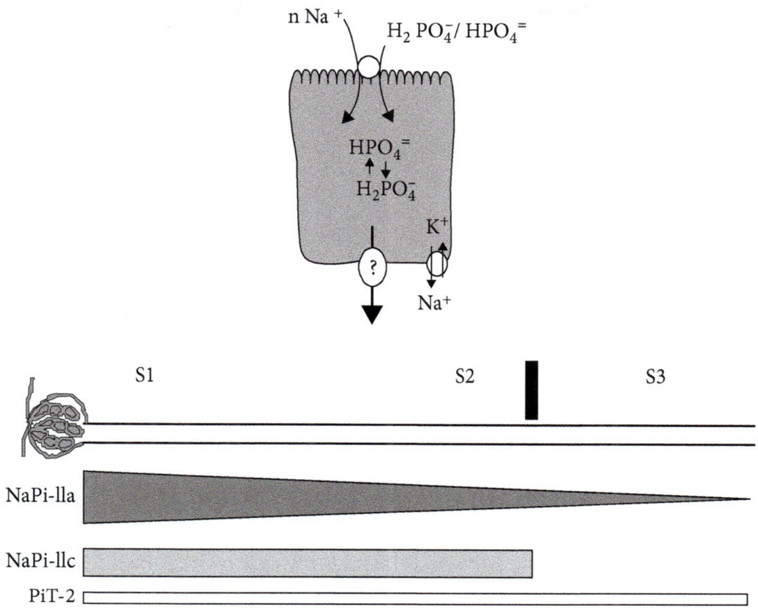

FIGURE 14.6 Renal clearance of bone-forming minerals. Phosphate reabsorption is regulated by parathyroid hormone which drives expression of sodium phosphate cotransporter in the proximal tubule. Nephron loss impairs early phosphate clearance and induces a number of phosphatonins to maintain phosphate excretion. In later stages of CKD, hyperphosphataemia becomes an indicator of hyperparathyroidisms and mineral bone disease.

Reproduced from Heini Murer, Jürg Biber, and Carsten A. Wagner, 'Phosphate homeostasis', in Neil N. Turner, Norbert Lameire, David J. Goldsmith, Christopher G. Winearls, Jonathan Himmelfarb, Giuseppe Remuzzi, and William G. Bennet, Mark E. de Broe, Jeremy R. Chapman, Adrian Covic, Vivekanad Jha, Neil Sheerin, Robert Unwin, and Adrian Woolf, ed., The Oxford Textbook of Clinical Nephrology (4 ed.), Figures 25.1, doi: 10.1093/med/9780199592548.003.0025, © Oxford University Press, 2015.

14.3 Evolutionary Ontogeny of the Excretory System

14.3.1 Kidney Development in Mammals

Variable environmental needs of diverse vertebrate classes have caused the kidney to take very different physical forms and internal organisations. This occurred despite, or perhaps because of, the basic functional unit of the kidney, the nephron, being quite similar across phyletic groups and being generated in a highly conserved manner (Figure 14.7).

Indeed, human kidney development, not only across evolution, but also during ontogeny, is characterised by the development of three successive, bilateral excretory systems: pronephros, mesonephros, and metanephros. All of them develop from the so-called nephrogenic cord, which arises from the intermediate mesoderm. Pronephros and mesonephros are temporary, while metanephros is a permanent excretory organ (Romagnani et al. 2013).

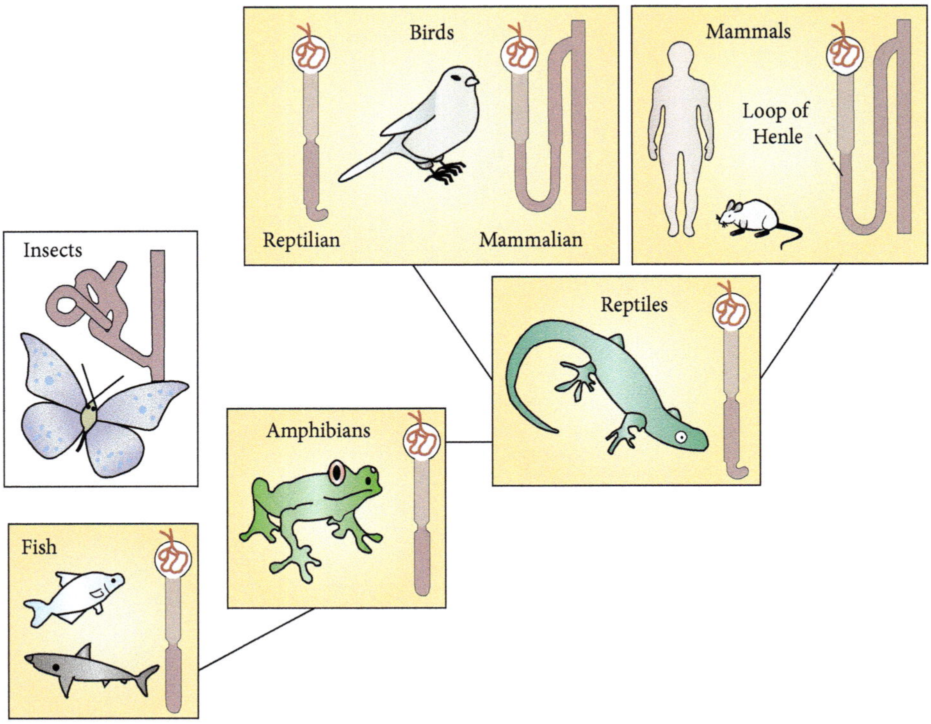

FIGURE 14.7 The nephron is highly conserved across evolutions. In general, it consists of a corpuscle, the filtering unit, which is connected through a neck to a tubule that can be divided into a proximal tubule, intermediate segment, and distal tubule that connects to an excretion unit. Avian and mammalian kidneys show major modification to the basic structure of the nephron—loop of Henle—which enabled urine to be concentrated. Bird kidneys contain nephrons with the loop of Henle ('mammalian type') and nephrons without the loop of Henle ('reptilian type'). By contrast, all nephrons in mammals display the loop of Henle. Even in insects, which are invertebrates, the renal system is recognizable as an analogue of the nephron.

Source: Reprinted by permission from *Nature Reviews Nephrology*, 9 (3), Renal progenitors: an evolutionary conserved strategy for kidney regeneration, Paola Romagnani, Laura Lasagni, Giuseppe Remuzzi, pp. 137–46, Figure 2a, doi.10.1038/nrneph.2012.290. Copyright © 2013, Springer Nature.

14.3.1.1 *Pronephros*

The development of the pronephros begins at the end of the third week of the pre-embryonic period, from the first five cranial segments (nephrotomes) of the nephrogenic cord. Pronephros in a human embryo is first represented by seven to ten solid cell clusters that become vesicular, elongate, and form tubules that connect with the nearest tubule below at their caudal extremities, forming the pronephric duct (Romagnani et al. 2013). At their medial ends these tubules have holes (nephrostomes) used for communication with the coelomic cavity (Romagnani et al. 2013). Pronephric tubules are short—the aorta does not branch into them, so that glomeruli are not developed either and regress very early. Indeed, cranial tubules disappear before the caudal ones appear, while the pronephric duct undergoes regression by the end of the fourth week of embryonic development (Romagnani et al. 2013).

14.3.1.2 *Mesonephros*

The mesonephros appears starting with the fourth week of embryonic life (Romagnani et al. 2013). It stretches from the 6th cranial to the 3rd lumbar segment of the nephrogenic cord, beginning with the successive differentiation of solid cell clusters from the nephrogenic cord, caudally from pronephros. They soon become vesicular. The vesicles then elongate and form tubules (Romagnani et al. 2013). The mesonephric tubule (in the S shape) consists of: the medial expanded and invaginated end (Bowman's capsule), with which it encloses aortic capillaries and forms Malpighi corpuscles; a proximal segment, which has a secretory function; and a distal segment, which through connection creates the mesonephric duct (a continuation of the pronephric duct) (Romagnani et al. 2013). The mesonephric duct elongates caudally and curves ventrally to open into the cloaca. It is believed that about 70–80 glomeruli and tubules develop in the mesonephros, but not all of them develop at the same time (Čukuranović and Vlajković 2005). The maximum number of these glomerular tubular units (nephrons) in each mesonephros amounts to 30 and they are identified in the fifth and sixth weeks of age (Romagnani et al. 2013).

The mesonephros mainly disappears at the end of the eighth week of antenatal life. In female fetuses, the few remaining caudal tubules become non-functional structures, situated in the broad ligaments of the uterus; the mesonephric duct completely disappears (Čukuranović and Vlajković 2005). In male fetuses, the remaining mesonephric tubules build up the efferent ductules of the testes, the rostral and caudal aberrant ductules, and the paradidymis, while the mesonephric duct develops into the canal of the epididymis, ductus deferens, seminal vesicle, and ejaculatory duct (Romagnani et al. 2013).

14.3.1.3 *Metanephros (Definitive Kidney)*

The metanephros represents the final developmental stage of the mammalian kidney. Its development begins in the fifth week of embryonic life. It develops from three sources: an evagination of the mesonephric duct, the *ureteric bud*; a local condensation of mesenchyme termed the *metanephric blastema* from the nephric structure; and *angiogenic mesenchyme*, which migrates into the metanephric blastema slightly later to produce the glomeruli and vasa recta. It is suggested that innervation is necessary for metanephric kidney induction (Romagnani et al. 2013). Indeed, from fish to mammals, induction of the mature kidney occurs starting from a cluster of cells derived from the intermediate mesoderm, the metanephric mesenchyme, through its reciprocal interactions with an epithelial component, the Wolffian duct, which induces the mesenchyme to condense around the tips of the growing bud and convert into an epithelial cell type (Romagnani 2009). These early epithelial cells form a spherical cyst called a renal vesicle (Figure 14.8) (Dressler 2006; Schedl 2007).

The renal vesicle undergoes a series of invaginations and elongations to generate the S-shaped bodies. The proximal end of the S-shaped body becomes invaded by blood vessels and generates the glomerular tuft (Figure 14.8) (Dressler 2006; Schedl 2007). Simultaneously, the middle and distal segments of the S-shaped body that had remained in contact with the ureteric bud epithelium fuse to form a single, continuous epithelial tube and begin to express proteins that are characteristic of tubular epithelia (Figure 14.8) (Dressler 2006; Schedl 2007). At this stage, elongation of the tubule occurs, together with a progressive segmentation that leads to generation of the primordial proximal tubule and distal tubule (Dressler 2006; Schedl 2007). Concurrent with these events, the growth and repeated branching of the ureteric bud form the system of the collecting ducts that join with the

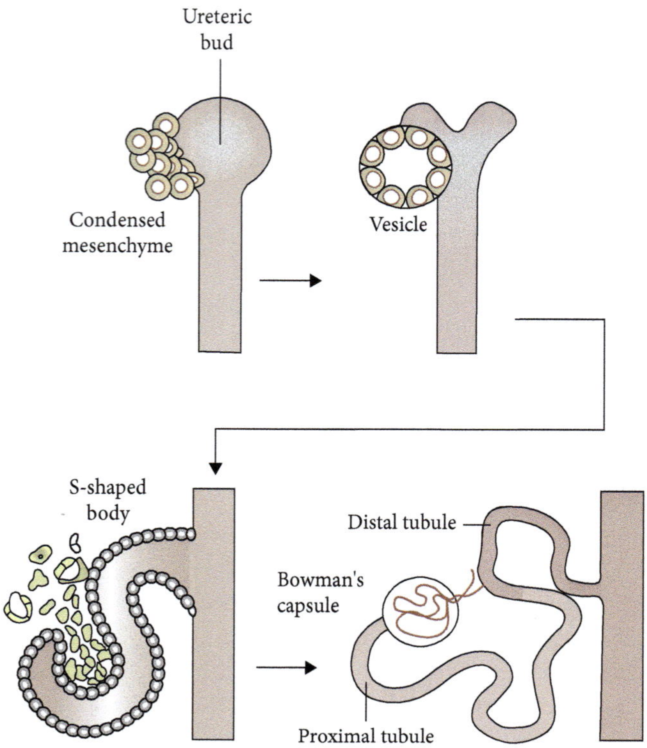

FIGURE 14.8 Nephron development. The metanephric mesenchyme condenses around the ureteric bud and induces metanephric mesenchyme to convert into epithelium, generating in sequence a vesicle and S-shaped body. Then the S-shaped body becomes invaded by blood vessels at one extremity and elongates and segmentates at the other, thus generating the whole nephron. This sequence of events is similar during development across all animal species.

Source: Reprinted by permission from *Nature Reviews Nephrology*, 9 (3), Renal progenitors: an evolutionary conserved strategy for kidney regeneration, Paola Romagnani, Laura Lasagni, Giuseppe Remuzzi, pp. 137–46, Figure 2b, doi.10.1038/nrneph.2012.290. Copyright © 2013, Springer Nature.

nephron and drain into the ureter (Dressler 2006; Schedl 2007). These inductive steps are reiterated throughout the growing kidney, so that older nephrons are located in the centre and newer nephrons are added at the periphery. In fish, amphibians, and reptiles the process of nephron maturation ends at this stage, although new nephrons can be added during adult life (Figure 14.7) (Dantzler and Braun 1997). By contrast, in birds and mammals, nephron maturation stops at this stage at around birth until when, in the perinatal period, generation and progressive maturation of the loop of Henle occurs (Figure 14.7) (Cha et al. 2001; Liu et al. 2001). Indeed, in the developing mammalian kidney, there is no separation of the medulla into an outer and inner zone, and at the time of birth there are no ascending thin limbs (Cha et al. 2001).

However, immediately after birth and concomitant with development of the renal papilla, thin ascending limbs of the loop of Henle are generated as an outgrowth from the S3 segment of the proximal tubule and from the distal tubule anlage of the nephron close to the vascular pole of the glomerulus, and thus confer to the kidney the pyramidal shape and the clear distinction among the cortex and medulla that is characteristic of mammals (Cha et al. 2001). The loop of Henle is thus mostly generated once the fetus is out of water, which suggests that nephron development in individual mammals resembles those steps that allowed life to transition from water onto land (Romagnani et al. 2013).

As a result, the fetus produces an elevated rate of hypotonic urine (1.5 l/day). Urine concentration capacity is blunted in the immature kidney. This physiological delay in development is related to various factors, including a low sensitivity of the collecting duct to arginine vasopressin (AVP), a structural immaturity of the loop of Henle with preferential distribution of blood flow to the inner cortex, a low-gradient concentration in the medulla, due to limited protein intake which generates significant amounts of urea, and a low expression of aquaporin 2 (AQP2). AQP2, a water channel, is located in the apical membrane of collecting duct cells and is involved in water reabsorption (Spitzer and Schwartz 2011). Relative tubular immaturity allows the production of amniotic fluid at a sufficient rate and prepares the fetus for postnatal adaptation (Spitzer and Schwartz 2011). Tubule function depends on structural maturation of the nephron and on various mediators including the renin–angiotensin system (RAS), aldosterone, prostaglandins, atrial natriuretic peptide, and cortisol (Jahnukainen et al. 2001). Cortisol, in particular, has a potent maturational effect and the potential to block nephrogenesis. The sensitivity of renal tubules to such hormonal factors increases during gestation and continues after birth.

14.3.2 Postnatal Development of the Mammalian Kidney

14.3.2.1 *The Kidney in the New-born*

Not only is the function of the kidney different between the fetus and the neonate, but also it continues to 'mature' and adapt as the neonate develops. During the early developmental period, the placenta is primarily responsible for maintaining the fetal fluid–electrolyte homoeostasis, the acid–base balance, and the excretory requirements of the fetus. The fetal kidney during this period is largely involved in maintaining the amniotic fluid level and regulating fetal BP (Jose et al. 1994). Urine production is present when the fetus reaches the 16th week of gestational age. Birth is considered to act as a stimulus for the accelerated postnatal maturation of renal function. However, the neonatal kidneys are still structurally and functionally immature.

Immediately after birth, the new-born infant begins to maintain its homoeostasis on its own. This associates with a rapid structural and functional maturation of its kidneys. Upon completion of nephrogenesis, the mature juxtamedullary nephrons display increased GFR in comparison to the immature superficial cortical nephrons, which, on the other hand, grow intensively and contribute to the greater RBF in this part of the kidney cortex (Čukuranović and Vlajković 2005). Decline of vascular resistance leads to increased kidney perfusion and filtration—that is, 30% of that of adults. In the first 3 months of life, GFR increases fast, then more slowly, until the adult level is reached at the end of the second year of life. Since the fetal kidney receives only 2–3% of the minute heart volume in comparison to 15–18% of the adult heart, this results in a relatively low blood flow in the fetal and neonatal kidney (Čukuranović and Vlajković 2005). The adult values of RBF are reached by the end of the first year of life. One of the significant characteristics of the neonatal kidney is its low urine concentration capacity, which increases the risk of dehydration when intake of liquids is constrained. The human kidney reaches the concentration capacity of the adult level at the age of 18 months (Nigam et al. 1996).

14.3.2.2 *The Kidney during Child Growth*

The kidneys of new-borns are about 4.5 cm long. Kidneys of adults are about 12 cm long (Čukuranović and Vlajković 2005). The growth of the kidney correlates with growth in height rather than child age; hence the kidney length can be estimated based on height. The kidneys grow rapidly in the first year of life, from 4.5 cm to 6.5 cm, and then gradually into adulthood, only about 0.3 cm per year on average (Čukuranović and Vlajković 2005).

Nephrogenesis is considered complete in the term infant. However, there are significant structural and functional changes that continue to take place as the infant matures (Čukuranović and Vlajković 2005). These changes mainly occur between the first and sixth months of life. During the first year of life, the nephrons grow, while tubular structures extend, and the loop of Henle progressively matures and elongates, together with the capacity to concentrate urine to the level observed in adult kidneys (Figure 14.9) (Čukuranović and Vlajković 2005). Postnatal maturation of glomerular structure consists of an increase in glomerular membrane permeability, filtration surface area, corpuscular glomerular diameter, and intrarenal redistribution of blood flow. During the first years of life, glomerular size increases and new podocytes arise from local progenitors along the Bowman's capsule (Lasagni et al. 2015).

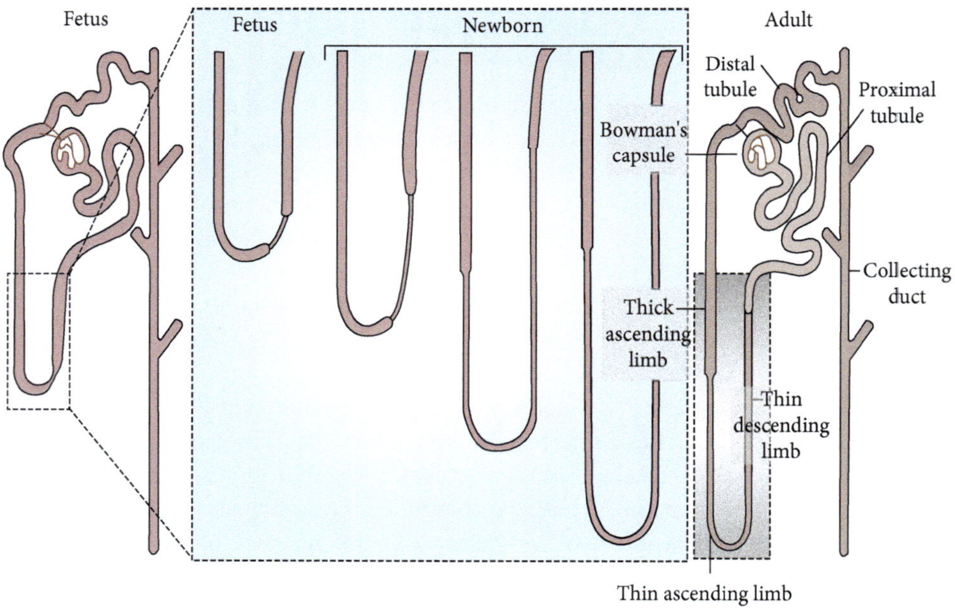

FIGURE 14.9 The loop of Henle is added on the evolutionary conserved nephron structure and observed only in birds and mammals. After nephron development has completed and concomitant with development of renal papilla in the new-born, the thin ascending limb of the loop of Henle is generated as an outgrowth from the S3 segment of the proximal tubule and from the distal tubule anlage of the nephron.

Source: Reprinted by permission from *Nature Reviews Nephrology*, 9 (3), Renal progenitors: an evolutionary conserved strategy for kidney regeneration, Paola Romagnani, Laura Lasagni, Giuseppe Remuzzi, pp. 137–46, Figure 2c, doi.10.1038/nrneph.2012.290. Copyright © 2013, Springer Nature.

14.3.2.3 *The Kidney During Ageing*

During ageing, the weight of both kidneys declines from 250–270 g in early adulthood to 180–200 g in the eighth decade of life. The loss of kidney mass is primarily evident in the cortex, which may reduce up to one-half of its maximum value (Palmer and Levi 1996; Čukuranović et al. 1999; Hill et al. 2003) and is characterised by a reduction in the number of functional nephrons (Palmer and Levi 1996; Tracy et al. 2002; Hill et al. 2003). The glomeruli display a progressive enlargement of the mesangial matrix, thicker basement membranes, and hyalinosis of the arterioles (Palmer and Levi 1996; Tracy et al. 2002; Hill et al. 2003). Glomerular number gradually decreases, while their average surface increases with age. Indeed, from the third to the fifth decade of life, hyalinised or sclerosed glomeruli increase from 1–2% to even 30% in some apparently healthy 80-year-olds (Palmer and Levi 1996). It has been determined that all individuals under normal conditions maintain at least 50% of normal glomeruli to the age of 75 (Denic et al. 2016, 2017). This is a consequence of afferent arteriole obliteration and blood shunting to postglomerular microvessels in juxtamedullar glomeruli (Tracy et al. 2002). In areas with interstitial fibrosis there is an increase of collagen I and III, while the GBM exhibits an increased content of various laminin isoforms (Abrass et al. 1995).

Ageing is also associated with a gradual loss of GFR (Denic et al. 2017). Fortunately, in healthy individuals these age-related changes develop very slowly (Denic et al. 2017). However, an old kidney becomes vulnerable and more sensitive to the toxicity of pharmacologically active substances and their metabolites, which should be taken into account in the treatment of the elderly (Baylis and Schmidt 1996). Notably, women are less affected by the haemodynamic and structural changes in old age. Despite the reduced GFR, the serum creatinine does not change with age, which is explained by a proportional decrease of the muscle tissue (Rodriguez-Puyol 1998). Although one-third of the elderly do not show a decline of renal plasma flow or of GFR (Denic et al. 2017), the renal capacity to concentrate and dilute urine decreases with age (Rowe et al. 1976). Similarly, the ageing kidney can maintain acid–base balance unless exposed to acute acid loads. In that case, acid excretion is not as efficient as in the young (Agarwal and Cabebe 1980). As a consequence of ageing, there is also a lower metabolism of potassium, calcium, phosphorus, and vitamin D (Čukuranović and Vlajković 2005) as well as a reduced synthesis of renin (Rodriguez-Puyol 1998).

14.4 Phylogeny of the Excretory System

The nephron remains the main functional subunit across evolution, even if the kidney topographical complexity increases progressively. The pronephros, at least in frogs and fish, is comprised of two nephrons, while the mesonephros contains tens to hundreds of nephrons (Figure 14.10). The metanephros has a much greater nephron number, although this number is highly divergent, even between members of the same species. In humans, kidneys contain between 200,000 and more than 2.5 million nephrons, with an average of 1 million nephrons (Bertram et al. 2011; Denic et al. 2016).

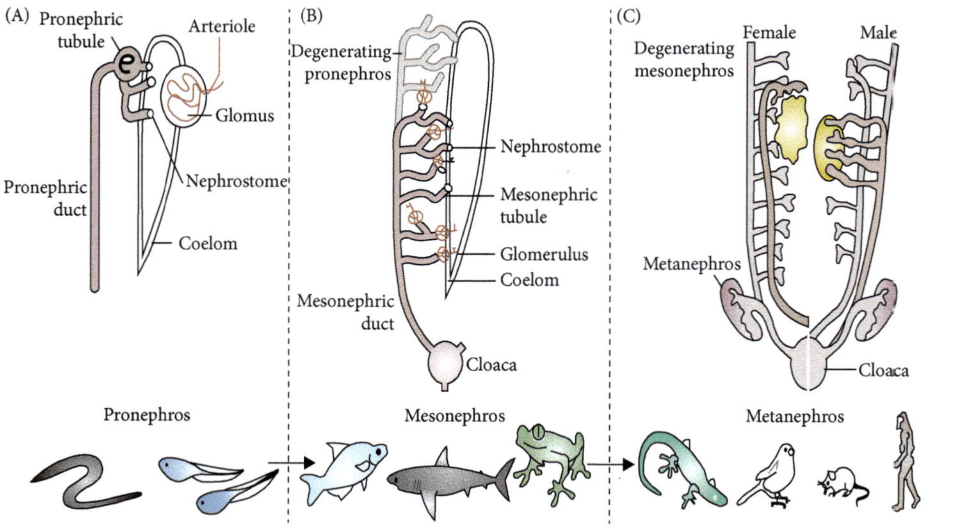

FIGURE 14.10 Kidney development across evolution is characterized by successive phases, each marked by development of a more advanced kidney: the pronephros, mesonephros, and metanephros. (A) The pronephros is the most immature form of kidney and is functional only in ancient fish, such as lampreys or hagfish, or during the larval stage of amphibians. (B) The mesonephros represents the functional mature kidney in most fish and amphibians. It is made up of an increased number of nephrons, usually dozens to hundreds. (C) The metanephros represents the mature kidney in reptiles, birds, and mammals; it consists of a substantially increased number of nephrons, from thousands to millions.

Source: Reprinted by permission from *Nature Reviews Nephrology*, 9 (3), Renal progenitors: an evolutionary conserved strategy for kidney regeneration, Paola Romagnani, Laura Lasagni, Giuseppe Remuzzi, pp. 137–46, Figure 1, doi.10.1038/nrneph.2012.290. Copyright © 2013, Springer Nature.

14.4.1 Life in Water: From Pronephros to Mesonephros

The internal milieu of osmoregulators, which includes almost all vertebrates, is remarkably similar, no matter which environment they live in (Dantzler and Braun 1997). As the concentrations of sodium chloride and magnesium sulfate exceed by far those of the internal milieu of most sea-water animals, mechanisms that maximise the secretion of these minerals and that minimise the loss of water are needed (Dantzler and Braun 1997). Fresh-water fish face the opposite problem as they are exposed to water containing far lower concentrations of minerals compared to their internal milieu. They rely on mechanisms that maximise water excretion and minimise sodium chloride losses (Dantzler and Braun 1997).

In the course of evolution, the number of renal functions has increased. In myxines, the kidney alone is not capable of providing osmoregulation, such as, for example, the excretion of a hypotonic urine; in lampreys and fresh-water fish, the kidneys are capable of excreting a dilute urine (Natochin 1996). The hagfishes, which conform to their marine environment, retain paired apparently non-functional pronephric kidneys as well as paired functional mesonephric kidneys (Dantzler and Braun 1997). The mesonephric kidneys contain 30–35 very large, oval glomeruli arranged segmentally on the medial side of a primitive archinephric duct (ureter) and connected by a short, non-ciliated neck segment. There are no other nephron segments (Dantzler and Braun 1997). Elasmobranchs conform to the osmolality of their environment, but, unlike hagfishes, much of the osmolality of the extracellular fluids is determined by the concentration of urea and trimethylamine oxide (TMAO). Nephrons in these animals contain all the standard vertebrate components noted above, but the arrangement of the nephron is highly complex. Although the elasmobranch kidney is not organised into discrete cortical and medullary regions, the nephrons are arranged to permit countercurrent flow within the dorsolateral region of the kidney and promote retention of urea (Dantzler and Braun 1997). By contrast, the lampreys, which apparently do not conform to their environment, thus maintaining body fluids hypoosmotic to a marine environment and hyperosmotic to a fresh-water environment, have nephrons similar in gross structure to more advanced vertebrates (Dantzler and Braun 1997). Each has a glomerulus, a ciliated neck segment, a proximal tubule, an intermediate segment, a distal tubule, and a collecting duct (Dantzler and Braun 1997). In the sea lamprey, *Petromyzon marinus*, the entire loop consists of a distal tubule, whereas in the river lamprey, *Lampetra fluviatilis*, the descending limb of the loop consists of a proximal tubule and the ascending limb of a distal tubule (Dantzler and Braun 1997).

Grossly, teleost kidneys are divided into an anterior head kidney containing lymphoid, haematopoietic, and glandular tissue, and a posterior trunk kidney containing the renal tissue, but in many species the two kidneys are partially or completely fused (Romagnani et al. 2013). Cartilaginous fishes (sharks, skates, rays, and chimaeras) have body fluids slightly hyperosmotic to surrounding salt water, and thus do not typically suffer dehydration even in the high-osmolality salt-water environment (see also below) (Dantzler and Braun 1997). To maintain high concentration of urea in the body, they produce urea by the ornithine urea cycle in the liver and other extrahepatic organs, such as the muscle and stomach (Hyodo et al. 2014). However, an essential mechanism to maintain the high internal urea level is 'urea retention' by the kidney (Dantzler and Braun 1997). To this end, cartilaginous fishes have developed specialised nephron systems by which they can reabsorb more than 90% of filtered urea from the primary urine (Boylan 1972). Beginning at the glomerulus, the renal tubule has five major segments: neck, proximal, intermediate, distal, and collecting segments. This, together with a series of urea transporters, creates the countercurrent mechanism to promote retention of urea.

By contrast, stenohaline marine teleosts maintain the osmolality of their body fluids well below that of their environment (Dantzler and Braun 1997). The nephrons generally contain, in addition to a glomerulus, a neck segment, two or three proximal segments that constitute the major portion of the nephron, sometimes an intermediate segment between the first and second proximal segments, and a collecting tubule draining into the collecting duct system. As expected in marine animals that do not have to dilute urine to excrete excess water, the distal tubule is absent in almost all species (Dantzler and Braun 1997). By contrast, stenohaline fresh-water teleosts, which maintain the osmolality of their body fluids well above that of their environment, have nephrons that differ and typically contain, in addition to a glomerulus, a ciliated neck segment, an initial proximal segment with a prominent brush border, a second proximal segment with a less prominent brush border, an intermediate segment, and a distal tubule emptying into the collecting duct system (Dantzler and Braun 1997). The kidneys of most stenohaline fresh-water teleosts, as might be expected in animals that need to excrete much water, tend to have more nephrons with larger glomeruli than those of stenohaline marine teleosts (Dantzler and Braun 1997). Interestingly, the glomeruli of some apparently stenohaline fresh-water teleosts that survive adaptation to sea water undergo atrophy and disappear (Dantzler and Braun 1997). The few aglomerular teleosts (apparently species that evolved in sea water and later invaded fresh water) seem to have nephrons entirely like those of marine aglomerular species (Dantzler and Braun 1997). Finally, euryhaline teleosts, which can maintain the osmolality of their body fluids above that of the environment when adapted to fresh water and below that of the environment when adapted to sea water, have nephrons most similar to those of stenohaline fresh-water teleosts (Dantzler and Braun 1997).

14.4.2 From Water to Land: Appearance of the Metanephros and the Growing Problem of Nitrous Waste Excretion

14.4.2.1 *Amphibian Kidney and Metamorphosis*

The kidney in larval amphibians is a pronephros that is normally present and functional throughout larval life, although it regresses and disappears by the end of metamorphosis and is replaced by the mesonephros structure in the adult animal (Figure 14.10) (Dantzler and Braun 1997; Romagnani et al. 2013). Although the kidney shape may appear very different, basic internal renal organisation is rather similar for all amphibians and can be illustrated by that of the frog kidney (Dantzler and Braun 1997).

All nephrons contain a glomerulus, ciliated neck segment, proximal tubule, ciliated intermediate segment, and distal tubule emptying into the collecting duct system (Dantzler and Braun 1997). Of note, there are no discrete cortical and medullary regions and no lengthened intermediate nephron segments arranged parallel to collecting ducts as in avian and mammalian kidneys. Even though the body mass of most amphibians is composed of approximately 80% water, whereas most mammals contain 65% water (Dantzler 1989), amphibians do not drink water to replace fluid loss. Instead, they absorb fluids through the skin by osmosis (Dantzler 1989). In fresh-water habitats, water uptake is passive and under osmotic control (Dantzler 1989). Amphibians excrete large amounts of diluted urine. Like fish, larval and aquatic adult amphibians are exclusively ammonotelic.

It is noteworthy that ammonia is highly water soluble but also highly toxic. Animals that lack access to water environments will decrease their GFR, thereby accumulating ammonia in body tissues which may lead to azotaemia and death (Dantzler 1989). In contrast, most terrestrial amphibians excrete urea as the primary nitrogenous waste product (Dantzler 1989). At times of water deprivation, urea accumulates in the body's fluid compartments, but is excreted rapidly on rehydration (Dantzler 1989; Romagnani et al. 2013). Many terrestrial anurans excrete urea or uric acid to reduce fluid losses. Urea is less toxic than ammonia and may be stored in body tissues until water can be replenished. Some species of tree frogs (e.g. African grey tree frog, *Chiromantis xerampelina*) have adapted to more arid habitats through the excretion of nitrogenous waste as uric acid, as in reptiles and birds (Dantzler 1989; Romagnani et al. 2013). Terrestrial species of amphibians are able to use the urinary bladder as a reservoir and actively reabsorb water to stave off dehydration (Dantzler 1989; Romagnani et al. 2013).

14.4.2.2 *Insects: Finding New Solutions*

The renal system of *Drosophila melanogaster* is composed of two anatomically and functionally discrete organs, nephrocytes and Malpighian tubules. Nephrocytes are specialised groups of cells clustered near the heart and the oesophagus that filter the fly's haemolymph (circulatory fluid) and remove waste products in a manner analogous to the endocytic processes of podocytes in the human glomerulus (Weavers et al. 2009). The Malpighian tubules fulfil similar functions in insects to the human nephron and collecting duct in mammals. They generate urine via active transport of ions, water, and organic solutes from the haemolymph into the Malpighian tubule lumen (Dow et al. 1994; Miller et al. 2013).

D. melanogaster has four Malpighian tubules, one anterior and one posterior pair. Each pair of Malpighian tubules coalesces into a common ureter at the junction of the midgut and the hindgut (Miller et al. 2013). Like the human ureter, longitudinal and circular muscle layers surround the *Drosophila* ureter to facilitate the peristalsis of urine (Wessing and Eichelberg 1978). The Malpighian tubules can be divided into three physiologically distinct domains; the initial, transitional, and main segments (Wessing and Eichelberg 1978). The main segment is primarily responsible for *Drosophila* urine production and is composed of two cell types: principal cells and stellate cells. These cells are comparable in structure and function to the principal cells and the intercalated cells of the human collecting duct tubules (Sozen et al. 1997; Miller et al. 2013).

14.4.2.3 *Reptiles: Appearance of the Metanephros*

Like the mesonephric kidneys of adult amphibians, the metanephric kidneys of adult reptiles, whose habitats also range from aquatic to arid terrestrial, show marked variations in their external morphology (Figure 14.10) (Dantzler 1989). This is due to the extreme variation of their body form. For example, crocodilians have paired, highly lobulated kidneys that lie against the dorsal body wall adjacent to the spinal column, while kidneys of snakes are found in the caudal one-third of the body (Dantzler 1989). Ureters enter the cloaca at the urodeum (Dantzler 1989). Crocodilians do not have a urinary bladder. By contrast, the kidneys of turtles and tortoises are paired structures, generally flattened, symmetrical, and in close association with the gonads (Dantzler 1989). Ureters leave the kidney and enter the bladder. A short urethra connects the bladder to the cloaca, which aids in water conservation. Ureters connect the kidneys to the urodeum. Snakes do not possess a urinary bladder. However, male snakes may have a unique structure to their urinary system known as the sexual segment (Dantzler 1989). Sexual segments are portions of the caudal kidney that enlarge during reproduction. It is believed that secretions from this segment constitute a portion of the seminal fluid (Dantzler 1989). In lizards, kidneys may touch each other or be fused. A ureter leads from each kidney to the urodeum. Most lizard species possess a thin-walled, easily distensible urinary bladder. Species variation in the size and shape of the bladder exists (Dantzler 1989). Although the external shape varies, the internal organisation is reasonably uniform. The nephrons are generally composed of all standard components—glomerulus, ciliated neck segment, proximal tubule, ciliated intermediate segment, and distal tubule—and they empty at right angles in the collecting ducts, although there are no loops of Henle (Dantzler 1989).

14.4.3 The Avian Kidney: Appearance of the Loop of Henle

Avian kidneys are dark red, paired, symmetrical organs located retroperitoneally, but in contrast to the mammalian kidney, the avian kidney is firmly fixed in place and is not movable. The ureters empty into the urodeum, a portion of the dorsal cloaca where the urine and urates are stored before voiding. The avian kidney can be divided into cortical and medullary areas, but the cortex and medulla are intermingled. Moreover, the avian renal cortex, like the snake kidney, contains complete simple nephrons without loops of Henle that empty at right angles into collecting ducts (Figure 14.10) (Dantzler 1989). Because these nephrons resemble reptilian nephrons, they are referred to as reptilian-type nephrons. In their simplest form, they do not even have secondary folds in their proximal and distal tubules like snake nephrons (Dantzler and Braun 1980). These reptilian-type nephrons are arranged in a radiating pattern from central points along the entire surface of the kidney. Underneath these outer units are larger more complex nephrons that do possess loops of Henle and, therefore, have been called mammalian-type nephrons (Dantzler and Braun 1980).

The loops of Henle of the mammalian-type nephrons, the vasa recta, which arise from the efferent arterioles of glomeruli of mammalian-type nephrons, and the collecting ducts, which drain both the reptilian-type and the mammalian-type nephrons from each radial group of cortical lobules, are bound together in parallel by a connective tissue sheath into a tapering structure called a medullary cone (Dantzler and Braun 1980). As in the reptilian kidney there is no renal pelvis. The nephrons, collecting ducts, and ureter are contiguous, and, therefore, each medullary cone terminates as a branch of the ureter (Dantzler 1989). This general pattern of many cortical areas with their associated medullary cones (the lobes of the kidney) resembles that of the extreme renculus type of kidney found in cetaceans and some bovine mammals (Dantzler 1989; Romagnani et al. 2013).

14.4.3.1 *The Problem of Water Reabsorption*

Mammals and birds have developed elaborate mechanisms for renal water reabsorption that is a necessary functional adaptation to survival in terrestrial habitats (Pannabecker 2012). Their kidneys contain juxtamedullary nephrons that are characterised by a countercurrent multiplier system that generates an osmotic gradient from the renal cortex to the inner medulla, enabling urine concentration and minimising water loss. Since the ascending and descending limbs have different permeabilities to salt and water, the loop of Henle generates a hyperosmolar environment in the medulla to concentrate osmolarity in the deep medulla (Pannabecker 2012). Indeed, in the TAL, there is active reabsorption of ions, but this segment is relatively impermeable to water; thus, the interstitial fluid becomes hyperosmotic relative to the filtrate inside the tubule. By contrast, in the descending limb, the cells do not allow ions to leave the filtrate, but water can freely leave, due to the high osmolarity created by the nearby TAL, concentrating the filtrate (Pannabecker 2012). The result is that the medulla maintains a high osmolarity (Pannabecker 2012).

Very long loops of Henle are typical of animals living in extremely dry conditions. In chinchilla, for example, a unique papillary segment beginning 2000 μm or less prior to the bend and terminating some distance after the bend is morphologically distinct from the TAL (Pannabecker 2012). A segment exhibiting some similarity to the chinchilla papillary segment is present in the desert sand rat, *Psammomys* (Pannabecker 2012), a species that can produce a urine having a maximal osmolality similar to that of the mouse and chinchilla (about 6000 mOsm/kg H_2O) (Pannabecker 2012).

14.4.4 The Kidney and Volume/Blood Pressure Control across Evolution

BP control is needed to assure sufficient blood perfusion of all organs. BP is determined by cardiac output and peripheral resistance in a simple equation resembling the law of Ohm: BP = cardiac output × resistance. This implies that to keep BP constant, increases in cardiac output require a drop in vascular resistance and vice versa. According to the Hagen–Poiseuille equation, blood flow is proportional to the radius to the fourth power, meaning that a small decrease in the vascular diameter yields a significant increase in blood flow rate. Therefore, the heart and the vascular system are organised in a dynamic manner to assure perfusion.

The level of BP required for organ perfusion varies considerably between different species. Fish and amphibians have simple circulatory systems, with similar pressures in respiratory organs of gas exchange as well as the musculature and other organs. The delicate vascular structures of gas-exchanging respiratory organs (gills, lungs) set a limit to systemic BP in order to avoid fluid extravasation. For example, increasing BP to more than 40 mmHg in an amphibian will cause pulmonary oedema. This limit in systemic BP constrains the level of physical activity. Evolution has solved this problem by separating the circulatory system into two: a low-pressure system for the respiratory organ (lungs) and a high-pressure system for all other organs. This was already partially achieved in reptiles and completed in birds, crocodilians, sauropods, and mammals mainly via closing the cardiac septum, together with a steady decrease in size of erythrocytes and increases of microvascular density to increase oxygen delivery to peripheral tissues (Glomski et al. 1997). This higher vascular density further increased the range for regulating BP by changing peripheral resistance.

The need for a certain BP also depends on the position of the organs in relation to the heart. For example, brachiosaurus and giraffes both have long necks, but only the giraffe is able to raise the head and feed on plants in the canopies of trees. However, the hydrostatic pressure that is required to pump blood from the giraffe's heart to its brain is enormous and can reach as much as 300 mmHg. This level of BP requires a considerable size of the cardiac muscle generating the necessary cardiac output. For a brachiosaurus with a neck of up to 10 m in length, a cardiac muscle of several tons would have been needed to generate a sufficient BP that can still perfuse a brain up there (Schulte et al. 2015). Therefore, quadripodal amphibia and reptiles usually keep their head at the level of the heart. The same applies to cobra snakes raising to vertical positions, because snake hearts are located right behind the skull. However, humans are bipedal mammals, with their brains being located around 50 cm above the level of the heart. This implies that evolution has endowed us with a BP that assures brain perfusion. However, interestingly, the average BP of other mammals is in a similar range (horse 130/75 mm Hg, cow 140/95 mm Hg, pig 140/80 mm Hg, cat 120/70 mm Hg, rats and mice 110/70 mm Hg, and dog 120/70 mmHg). So, if BP is largely determined by the structure and function of the cardiovascular system, what is the role of the kidney in all this?

Pressure in the cardiovascular system not only requires a pump, vessels, and resistance but also fluid, that is, blood. Blood is made of cellular and fluid components. In anaemia, while reducing the cellular components will reduce the capacity for oxygen delivery (clinically indicated by fatigue), it does not necessarily reduce BP as long as blood 'volume' remains the same and fluids substitute the lack of cells. In contrast, injuries causing blood loss have immediate effects on BP beyond the compensatory mechanisms of the cardiovascular system, such as increasing cardiac output via tachycardia and increasing peripheral resistance by vasoconstriction. It is of note that blood cell numbers can change only within days to weeks, while fluid volume can be regulated within minutes to hours (drinking, isotonic fluid retention in the kidney). Therefore, BP control involves the kidney mainly by controlling sodium and water balance, because both together represent 'volume', that is, isotonic fluid, as the basic component of blood and BP. However, the challenges for volume control are quite different in dehydrating (sea water and terrestrial) versus hydrating (fresh water) habitats (Takei 2015).

14.4.4.1 *Volume Regulation in Aquatic Animals*

Primitive marine animals up to some vertebrates like hagfish are osmoconformers; that is, their internal milieu is identical to that of sea water and hence no energy is needed for osmoregulation. Higher vertebrates including most fish, however, invest large amounts of energy in maintaining an internal osmolarity that is around three times less than that of sea water; thus the potent mechanism of osmoregulation avoids sodium chloride (and magnesium sulfate) overload and plain-water depletion. While for marine fish (teleosts) excess sodium uptake would have the potential to increase BP, the concomitant loss of water has the opposite effect and rather puts marine fish at risk for hypernatraemia and extra- and intracellular volume depletion and hypotension. This happens when shipwrecked humans drink seawater, or in conditions including diseases associated with loss of free water (defects of ADH release or signalling, i.e. diabetes insipidus). Thus, in marine fish, adipsogenic hormones such as atrial natriuretic factor (ANF) promote sodium excretion, while in humans the release of natriuretic factors is accompanied by induction of thirst (Takei 2015). Obviously, uptake of fresh water (but not sea water in teleost fish or shipwrecked humans) supports rehydration (Takei 2015). Marine fish cannot totally avoid the uptake of hypertonic sea water from their stomach during feeding; hence they must secrete equivalent amounts of salt against the osmotic gradient across the gill membranes (Evans 2010). In contrast, their kidneys excrete little sodium, because marine fish have to minimise urine volume to limit water losses. Their little amounts of isotonic urine are rather enriched for magnesium sulfate and other cations present in sea water (Evans 2010).

The internal milieu of elasmobranchs like sharks, rays, and cartilaginous fish is different to that of other fish and marine mammals. They are isotonic to sea water; hence they can drink sea water without problems. However, they do not maintain isotonicity with sea-water electrolytes. Indeed, sea-water electrolytes are rather actively secreted at high concentrations via a special rectal gland (Epstein et al. 1983; Riordan et al. 1994). Elasmobranchs maintain isotonicity via massive retention of urea in the kidney and by production of trimethylamine. Seabirds like cormorants, albatrosses, and penguins, and marine reptiles like sea turtles and some crocodilians have similar salt glands emerging from the nasal cavity or the eye that excrete a highly concentrated sodium chloride solution, which allows them to drink sea water without dehydration. In contrast, marine mammals lack this option; hence they must minimise the uptake of sea water and survive from generating fresh water from metabolism and diet and producing highly concentrated urine, like terrestrial mammals in extremely arid environments. Only sea otters drink sea water without problems as they can produce urine concentrations of sodium chloride similar to those of sea water (Ortiz 2001).

Fresh-water fish (teleosts) need osmoregulation to solve the opposite problem—that is, avoiding drinking fresh water and limiting its osmotic uptake via the gills and skin while minimising salt losses to the environment, a condition that, if not well controlled, can lead to hypotonic hyperhydration (i.e. hyponatraemia, intracellular (brain) oedema, and hypotension). In fresh-water fish, plain-water clearance requires glomerular filtration and optimised sodium chloride reabsorption in the tubules of the nephrons to produce large amounts of hypotonic urine (Vize and Smith 2004). Certain fish like salmon can adapt to both fresh- and salt-water environments in different phases of their lives. This requires spending certain phases of adaptation in mixed environments, as occurs in the mouth of a river or river deltas. During this adaptation period, they change the directional expression of sodium and water channels in their gills.

Terrestrial life combines the two challenges of osmoregulation, the risk of plain-water loss of salt-water fish, and the risk of salt loss of fresh-water fish. For example, water losses via the lungs into the air cannot be entirely avoided during respiration in terrestrial life. Amphibians have largely maintained osmoregulation via the kidney and skin. But after phases of serial adaptions the kidneys of birds and mammals possess optimised organs that can perfectly retain water and salt to maintain a sufficient isotonic volume and BP by producing hypertonic urine with a higher osmotic concentration than their body fluids. Birds, in contrast to mammals, concentrate urine by retaining water via aquaporin water channels in the cloaca (Nishimura 2008), a process that in mammals occurs via similar aquaporin water channels in collecting ducts of the renal medulla (Sterns 2015). Despite different location sites of urine concentration, birds and mammals activate aquaporins in an ADH-dependent manner by activating ADH receptors at the aquaporin-expressing epithelial cells. In addition to urine concentration, their kidneys can respond to sudden increases in water or sodium intake by selectively increasing water or salt excretion, respectively (Sterns 2015). Moreover, thirst is a potent mechanism to increase the uptake of fresh water in terrestrial animals.

Notably, low BP may cause fainting and loss of consciousness; thus, selection operated on the prevention of hypovolaemia and hypotension. In contrast, arterial hypertension is a condition that kills only 10–20 years later via cardiovascular diseases, a consequence never affecting reproduction, and therefore high BP escaped selection. This may be one aspect explaining the high prevalence of arterial hypertension in developing and developed populations consuming salt-rich diets.

14.4.4.2 *The Renin–Angiotensin System*

Volume regulation via sodium channels is regulated by the mineralocorticoid effect of certain steroid hormones such as aldosterone. The activation of aldosterone is downstream of the RAS, an ancient regulator of the cardiovascular system in all vertebrates (Wilson 1984; Nishimura 2016). Indeed, elasmobranch fish express renin and angiotensin, and angiotensin receptors are found in blood vessels, gills, the kidney, and the rectal gland (Hazon et al. 1999).

The RAS is activated in states of salt loss and low BP, a condition that is most important to fresh-water teleost fish (Wilson 1984). Similarly, terrestrial life represents a constant pressure to withstand salt loss and volume depletion. Terrestrial reptiles minimise salt and fluid losses via the skin by an extremely impermeable horny epidermis, scales or scutes. Sweat glands allowed certain mammals and especially humans to revolutionise thermoregulation as the basis for long-distance running (Schulte et al. 2015); however, it adds another source of fluid and salt loss, especially in arid environments. In mammals, volume depletion is sensed at the *macula densa* of the distal tubule and translated into renin secretion into the *vas afferens* of the glomeruli. The downstream production of angiotensin-II elicits vasoconstriction in the gut, skin, and kidney, while aldosterone release from the adrenal gland maximises sodium retention from the glomerular ultrafiltrate at several sites of the nephron. In addition, angiotensin-II regulates cardiac output and BP by stimulating the sympathetic nervous system in agnathans, elasmobranchs, teleosts, amphibians, reptiles, birds, and mammals (Wilson 1984).

14.5 Adaptation and Pervasive Evolutionary Challenges

Recent findings from the Global Burden of Disease Study have highlighted CKD as an important cause of global mortality (Rhee and Kovesdy 2015). The number of reported deaths due to CKD was estimated to be 1.2 million, a 32% increase from 2005. The number of people who will receive renal replacement therapy (RRT) (dialysis or transplantation) worldwide has been projected to more than double from 2.6 to 5.4 million from 2010 to 2030 (Couser et al. 2011). Kidney disease has, therefore, become a global public health priority. End-stage kidney disease (ESKD) is only the tip of the iceberg. In western societies, CKD occurs in approximately 10% of the population (Rhee and Kovesdy 2015). As kidneys are made of nephrons, kidney function can be expressed by the following equation: $GFR_{(Total)} = GFR_{(single\ nephron)} \times$ number of nephrons. This equation has three implications: (1) a low or declining number of nephrons necessitates a compensatory increase of SNGFR to maintain total GFR; (2) total GFR decreases once remnant nephrons are no longer able to (further) increase SNGFR or if they do not at all increase SNGFR; (3) because of compensatory single nephron hyperfiltration, any increase in serum creatinine level has to be dealt with by remnant nephrons.

Nephrons are prepared to cope with transient increases in filtration load such as upon food and fluid intake by transient increases in SNGFR ('renal reserve') (Hostetter et al. 1981; Brenner et al. 1982). However, longer or persistent increases in body mass and respective increases in SNGFR promote nephron hypertrophy. Any injury-related nephron loss may have the same effect. Indeed, either severe kidney injury or combinations of injury-related with ageing-related nephron losses, especially in individuals with poor nephron endowment and/or obesity, accelerate persistent hyperfiltration and loss of remnant nephrons (Benghanem Gharbi et al. 2016).

Persistent elevations of SNGFR and filtration pressure (glomerular hypertension) across the glomerular filtration barrier imply glomerular hyperfiltration, which is potentially harmful, especially to podocytes, the essential epithelial constituent of the filtration barrier. Glomerular hyperfiltration and hypertension trigger tumour growth factor-alpha/epithelial growth factor receptor-mediated remnant nephron hypertrophy (Laouari et al. 2011; Ruggenenti et al. 2012). Further increasing glomerular dimensions causes podocyte shear stress, a process that promotes podocyte detachment, FSGS, global glomerulosclerosis, and subsequent nephron atrophy, a vicious cycle further reducing nephron number and SNGFR of remnant nephrons (Helal et al. 2012; Hodgin et al. 2015; Denic et al. 2016). Nephron hypertrophy is a compensatory response to two problems: (1) it reduces glomerular hypertension by increasing filtration surface (Helal et al. 2012); and (2) by increasing SNGFR it softens the nephron loss-related decline in total GFR. For example, increased SNGFR and hypertrophy in remnant nephrons allows kidney donors to maintain an apparently 'normal' renal function, despite lacking 50% of nephrons. Obviously, kidney donation does not necessarily cause CKD progression when donors are carefully selected for good nephron endowment and the absence of obesity, diabetes, and ongoing nephron injury (Mueller and Luyckx 2012; Grams et al. 2016).

According to Garland and Carter (Chevalier 2017), genetic polymorphisms and environmental stimuli/stressors during development determine the kidney's phenotypic characteristics, including nephron number. Modulated by epigenetics, homoeostatic function of the nephron is determined by its performance capacity and limited by its defences against cell death and capacity for hypertrophy, regeneration, and repair. Modern life has created several environmental conditions that generate mismatches among nephron numbers and filtering requirements. This may be related to the progress of medicine, which has increased survival and lifespan, or to dietary changes, predisposing to obesity and diabetes which enhance filtration needs.

14.5.1 Mismatch of Prematurity/Low Birth Weight

Improvements in emergency measures, intensive care treatments, and neonatology assistance have considerably enhanced survival of infants born prematurely or with low body weight (Luyckx et al. 2017). New-borns with low birth weight when born either preterm or term with signs of intrauterine growth restriction (IUGR) frequently display incomplete kidney development (White et al. 2009; Luyckx and Brenner 2015; Low Birth Weight and Nephron Number Working Group 2017; Luyckx et al. 2017). Moreover, preterm infants are more likely to receive drugs in neonatal intensive care. Many of these drugs, such as vancomycin and aminoglycosides, require renal elimination and undergo glomerular filtration or tubular metabolism, having a high potential of nephrotoxicity, further causing potential nephron loss. Administration of non-steroidal anti-inflammatory drugs (NSAIDs) for the closure of patent ductus arteriosus impairs renal function, induces oliguria, and causes reduced RBF and GFR (Luyckx et al. 2017).

Genetic and fetal environmental factors determine the wide variation in nephron number at birth (range 300,000–1.8 million). Congenital abnormalities of the kidney and the urinary tract (CAKUT) are the most common genetic entity (Nicolaou et al. 2015). CAKUT present a wide spectrum of causes for kidney hypodysplasia, imparting low nephron number and risk of CKD later in life (Nicolaou et al. 2015). The severity of CAKUT ranges from severe forms requiring RRT early in life to milder forms that require RRT in adolescence, when increased filtration is needed to manage body growth, or even later in life, when environmental factors contribute to nephron loss in addition to the congenital low nephron number (Becherucci et al. 2016).

More generally, an ongoing interaction between genes and the environment from prenatal to adult life contribute to an individual's renal potential. Various perinatal factors have been shown to induce reduced nephron number, including low birth weight, IUGR, maternal nutritional deficiency (vitamin A, iron depletion), maternal protein or global nutrition deficit, drugs, maternal and fetal stress, and maternal gestational diabetes (Luyckx et al. 2017). Adverse events occurring before completion of nephrogenesis likely compromise renal growth and produce a long-lasting effect on renal function. In particular, conditions that cause IUGR will cause kidney growth restriction because blood is preferentially shifted from the kidney to the fetus's brain, explaining why CAKUT represent about 30% of all congenital abnormalities (Nicolaou et al. 2015; Becherucci et al. 2016). In contrast, preterm birth with postnatal environmental stressors (malnutrition, stress, and nephrotoxic drugs) or renal/urinary tract malformations are associated with impaired nephrogenesis, which may lead to a reduction in nephron endowment and increase vulnerability for impaired renal function in both the early postnatal period and later in life. Perinatal events are less likely to affect renal function, but are associated with high risk of hypertension and renal disease at adulthood. Depending on the severity of prematurity, poor nephron endowment can cause either early-childhood CKD or CKD later in life (Brenner et al. 1982; De Jong et al. 2012; Luyckx and Brenner 2015; Low Birth Weight and Nephron Number Working Group 2017; Luyckx et al. 2017). CKD onset at puberty is common when rapid body growth exceeds the capacity of nephron number to accommodate the increasing filtration load. In milder cases, poor nephron endowment at birth promotes the development of hypertension, CVD, and CKD later in adults (Becherucci et al. 2016). The global risks of preterm birth and low birth weight are around 10% and 15%, respectively; therefore, millions of children are born at risk of CKD later in life and are found at the lower percentile of age-matched GFR (Charlton et al. 2014; Khalsa et al. 2016).

14.5.2 Mismatch Caused by Obesity and Diabetes

Modern lifestyles characterised by hypercaloric diets rich in fructose and other sugars are associated with increased body mass index (BMI) up to obesity and high risk of diabetes and metabolic syndrome. A larger glomerular size in mildly obese but otherwise healthy individuals suggests an increased SNGFR (Denic et al. 2017). In general, the association between obesity and poorer renal outcomes persists even after adjustments for higher BP and diabetes mellitus, suggesting that obesity-driven hyperfiltration directly contributes to nephron loss (Lu et al. 2015; Kramer et al. 2016). Severe obesity alone or moderate obesity in combination with other factors can lead to development of proteinuria, FSGS, and progressive CKD (Ejerblad et al. 2006; Foster et al. 2008; Vivante et al. 2012; Chang et al. 2013; Lu et al. 2015). Diabetes is also a well-known condition associated with massive hyperfiltration, evident from increased total GFR and renomegaly, as a consequence of increased SNGFR and compensatory increase in nephron size (Tonneijck et al. 2017). Hyperglycaemia drives the SGLT2-driven reabsorption of sodium in the proximal tubule, a process subsequently inactivating TGF and activating the RAS at the *macula densa* (Vallon 2015; Anders et al. 2016). Indeed, hyperglycaemia induces a permanent dilation of the afferent and vasoconstriction of the efferent arteriole, a combination installing a permanent increase in single nephron size and total GFR (Vallon 2015). While diabetes-driven hyperfiltration can be compensated for many years in younger patients with normal nephron number, it serves as a drastic accelerator of CKD progression in those with precedent single nephron hyperfiltration, such as in patients with low nephron endowment, injury- or age-related nephron loss, obesity, or pregnancy (Anguiano Gomez et al. 2017). Unfortunately, this is a highly prevalent combination of risk factors in older patients with type 2 diabetes, for which dual SGLT2/RAS inhibition can elicit potent nephroprotective effects (Wanner et al. 2016).

14.5.3 Kidney Ageing: Beyond Evolutionary Needs

Population studies document an age-related decline in GFR and hence an increasing prevalence of ESKD (Benghanem Gharbi et al. 2016). Studies indicate that the trajectory of decline in GFR with age is quite variable, and often, but not always, independent of BP or cardiac haemodynamics, but usually shows a Gaussian distribution, sometimes with a 'tail' of accelerated loss of GFR (or creatinine clearance, depending on the study). Some longitudinal studies have claimed stable, or even increasing, levels of GFR for long periods in 'healthy' ageing, but these have been criticised for including patients with type 2 diabetes and for limitations of calculating a true slope of GFR over time due to a limited number of observations (Glassock and Winearls 2009). The predominant view is that in otherwise normal and healthy adults, GFR begins to decline after about age 30, with some acceleration in the rate after about age 80 (Denic et al. 2016, 2017). However, the slope of GFR decline varies among individuals depending upon age, genetic factors, BP, diseases causing kidney damage, and body weight.

Histologically, kidney ageing presents as global glomerulosclerosis, the respective atrophy of entire nephrons, and subsequent interstitial fibrosis (Hodgin et al. 2015; Denic et al. 2017). Whether age-related nephron loss is associated with hypertrophy (and hyperfiltration) of remnant nephrons has not consistently been reported in the literature (Hodgin et al. 2015; Denic et al. 2017), but the analytical difficulties of how to precisely assess nephron number and glomerular volume, and how to acknowledge the different functions of juxtamedullary versus cortical nephrons can affect the interpretation of such data (Hodgin et al. 2015; Denic et al. 2017). Age-related nephropathy involves constant as well as incident podocyte losses throughout the person's lifetime, implying hyperfiltration, podocyte stress, and secondary FSGS in remnant nephrons (Glassock and Winearls 2009). This phenomenon presents as a decrease in glomerular podocyte density with age as podocyte number decreases (Hodgin et al. 2015). Massive podocyte detachment occurs upon endomitosis-related mitotic catastrophe, leading to glomerulosclerosis (Lasagni et al. 2013; Liapis et al. 2013; Hodgin et al. 2015).

14.5.4 Kidney Disease as Collateral Damage of Selection Pressures

Several traits providing survival advantages have recently been associated with kidney diseases, suggesting that they may be the result of evolutionary trade-offs.

14.5.4.1 *Trade-off of APOL1 Variants,* Trypanosoma *Infection, and CKD*

African-Americans develop kidney failure at rates four to five times higher than Americans of European descent (Saran et al. 2017). Genetic variants in the apolipoprotein L1 (ApoL1) gene, which is part of a six-membered family of ApoL genes clustered on chromosome 22 that arose by gene duplication, explain a large fraction of this major health disparity (Genovese et al. 2010; Tzur et al. 2010; Friedman and Pollak 2016).

ApoL1 arrived late in mammalian evolution (approximately 30–35 million years ago (mya)) and a functional gene is present only in a few species in the primate lineage (Smith and Malik 2009). The ApoL1 protein circulates in blood at high levels as part of a high-density lipoprotein (HDL) complex and is expressed widely in tissues, including the lung, placenta, pancreas, liver, and kidney (Smith and Malik 2009; Friedman and Pollak 2016). The two ApoL1 coding variants associated with CKD, known as G1 and G2, facilitate parasite-killing activity towards additional *Trypanosoma brucei* subspecies and other pathogens (Friedman and Pollak 2016). This innate immune function is dominant, such that individuals with one G1 or G2 allele (heterozygotes) are resistant to trypanosomiasis. However, individuals with two G1 or G2 alleles (homozygotes or compound heterozygotes, with no reference sequence ApoL1 alleles, known as G0) have an increased risk for developing CKD (Friedman and Pollak 2016). Disease-causing variants tend to be strong or common, but not both (Friedman and Pollak 2016). The ApoL1 high-risk variants are both common and powerful because they carry advantages (enhanced innate immunity) in the heterozygous state, but have a powerful deleterious impact when gene frequency rises sufficiently in a population to generate homozygosity (Friedman and Pollak 2016). In this way, ApoL1 resembles the sickle cell trait, protecting against malaria in heterozygotes but causing sickle cell disease in homozygotes. Unlike sickle cell disease, where there is strong evidence for balancing selection, it is unclear whether and where ApoL1 risk variants have reached equilibrium between increased survival due to superior innate immunity and decreased survival due to kidney disease, or even that the risk variants influence reproductive fitness (Friedman and Pollak 2016).

It is likely that humans in the *Trypanosoma* belt of sub-Saharan Africa who harboured these ApoL1 variants, which emerged around 5000 years ago, had a survival advantage, leading to rapid spread of these alleles, which are now found in as many as 90% of individuals in some African ethnicities (Friedman and Pollak 2016). The risk variants are not found in people without recent (less than 5000 years) African ancestry. Evidence for positive selection is thus very strong, and suggests that ApoL1 is a versatile innate immunity molecule acting not only against *Trypanosoma* but also likely against *Leishmania* and exerting also antiviral activity (Taylor et al. 2014).

Mice with podocyte-specific expression of either the G1 or G2 ApoL1 risk allele, but not the G0 allele, develop functional (albuminuria and azotaemia), structural (foot-process effacement and glomerulosclerosis), and molecular (gene expression) changes that closely resemble human kidney disease (Beckerman et al. 2017). ApoL1 alleles interfere with endosomal trafficking and block autophagic flux, which ultimately leads to inflammatory-mediated podocyte death and glomerular scarring disease, the severity of which correlates with the level of expression of the risk allele (Romagnani et al. 2016; Beckerman et al. 2017). Given the strong associations of diminished estimated GFR (eGFR) and albuminuria with heightened risk of subclinical atherosclerosis, cardiovascular disease, and premature death in older adults, the potential burden of morbidity and mortality related to these *APOL1* variants may be substantial and extend well beyond CKD.

14.5.4.2 *Trade-off of Uromodulin and Urinary Tract Infections*

Uromodulin is a glycoprotein exclusively secreted by the epithelial cells of the TAL of the loop of Henle. It is the most abundant urinary protein in the healthy individual and the main constituent of hyaline urinary casts (Ghirotto et al. 2016). Uromodulin-encoding gene (UMOD) mutations cause hereditary dominant renal diseases, referred to as uromodulin-associated kidney diseases (UAKDs), presenting with tubulointerstitial fibrosis, defective urinary concentration, hyperuricaemia and gout, and progressive renal failure (Scolari et al. 2015).

Multiple genome-wide association studies (GWAS) identified highly prevalent UMOD gene variants that are strongly associated with the risk of CKD, hypertension, and kidney stones (Padmanabhan et al. 2010). These UMOD risk variants directly increase the expression of uromodulin *in vitro* and *in vivo*, while increased uromodulin expression causes salt-sensitive hypertension and kidney lesions in mice and humans (Trudu et al. 2013). The unusually high frequency of UMOD risk variants (70–80% in Africans and Europeans and 92–95% in East Asians) suggests some sort of selective pressure (Manolio et al. 2009).

Uromodulin reduces the risk of urinary tract infection and nephrolithiasis possibly by competing with the binding of *Escherichia coli* to uroplakins and by preventing the aggregation of calcium crystals, respectively (Pak et al. 2001). Besides direct binding to uropathogens, the protective effect of uromodulin may also depend on its ability to regulate innate immunity by activating dendritic cells via Toll-like receptor 4 (TLR4) (Agnese et al. 2002).

The correlations between UMOD ancestral allele frequencies and pathogen diversity or prevalence of urinary tract infections support the hypothesis that this allele was maintained at higher frequencies in areas with stronger selective pressure for its protective effect (Ghirotto et al. 2016). The UMOD situation is therefore different from the paradigm of selection at the ApoL1 locus, where it is the derived allele that confers selective advantage against pathogens (Genovese et al. 2010; Thomson et al. 2014).

In addition to its protective effects for urinary tract infections, UMOD overexpression has been linked with salt-sensitive hypertension. This effect seems to be related to overactivation of the TAL sodium–potassium–chloride co-transporter NKCC2. Indeed, acute treatment of transgenic mice with furosemide, a specific NKCC2-blocker, induces higher natriuretic response and a drop in BP (Trudu et al. 2013). Interestingly, the same response has been observed in human hypertension, as furosemide was effective in reducing BP only in hypertensive patients homozygous for *UMOD* risk variants (Trudu et al. 2013).

14.5.4.3 *Trade-off of Uric Acid and Pressure Control*

Birds and mammals maintain higher BP than other animals but excrete nitrous waste in different ways. Mammals excrete water-soluble urea, which necessitates more urine production and water intake. As a consequence of the switch from uric acid to urea excretion, almost all mammals express the enzyme uricase that facilitates the breakdown of uric acid to allantoin. The presence of uricase sets uric acid levels as low as 1–2 mmol/l in all mammals except the great apes and humans. Interestingly, around 10–15 mya two mutations of the uricase gene, a deletion of codon 18 in exon 2 and a nonsense mutation of codon 33 in exon 2, abolished uricase activity. These mutations occurred independently from each other in two different branches of hominids (Johnson et al. 2013). Around the same time, mutations appeared in the gene encoding for the uric acid transporter 1 (URAT1), which increased its affinity for uric acid and thereby reduced renal uric acid clearance (Tan et al. 2016). It is believed that these mutations provided survival benefits for hominids under selection pressure for climate-related loss of rainforest habitats (Kratzer et al. 2014). Potential benefits of higher uric acid levels in the blood are thought to include a moderate osmotic effect on blood volume (i.e. BP), especially in bipedal great apes on a fruit and vegetarian low-salt diet. Another explanation is the uric acid-related activation of fructokinase that promotes fat-storage and insulin resistance (Johnson et al. 2013). Modern times, however, expose humans to salt- and fructose-rich diets, which can turn these ancient evolutional benefits into hyperuricaemia-related disease, for example arterial hypertension, obesity, gout, kidney stone disease, hyperinsulinaemic diabetes, and even renal failure (Wesseling et al. 2016).

14.5.4.4 *Trade-off of Regeneration: Better Having Many Nephrons or Generating New Ones?*

Maintaining nephron number and the integrity of its structure is critical for kidney function. Although, apart from the appearance of the loop of Henle, the basic structure of the nephron remained constant over evolutionary time, different regenerative strategies emerged in different species (Figure 14.11).

14.5.4.4.1 FISH AND AMPHIBIANS

Fish and amphibians have the capacity to generate new nephrons in response to extensive renal injury (neonephrogenesis) (Elger et al. 2003; Diep et al. 2011), as well as to regenerate partially injured nephrons, which relies on resident renal stem and progenitor cells (Diep et al. 2011; Zhou and Hildebrandt 2012). However, in amphibians, metanephric mesenchyme is not interspersed among haematopoietic tissue, but rather localised in a peripheral area named the 'nephrogenic zone' which persists after metamorphosis and is closely located to the youngest extremities of collecting ducts (Dantzler 1989).

14.5.4.4.2 INSECTS

In insects, the Malpighian tubule undergoes continuous turnover and regeneration through scattered renal stem cells localised within the tubular compartment of the nephron among more differentiated ones that self-regulate their function, without the need to be constrained within a specific environment, usually defined as a stem cell niche (Kiger et al. 2001; Singh et al. 2007).

14.5.4.4.3 REPTILES

In reptiles, the metanephric kidney made its first appearance. Interestingly, a nephrogenic zone generating new nephrons in an identical way to fish and amphibians during adult life has also been reported for reptiles, and nephron numbers increase with size until maximum body mass is attained (Solomon 1985). This demonstrates that a perpetual neonephrogenesis strategy can occur also in the metanephros, suggesting that regenerative strategies analogous to those observed in fish persist in reptiles.

14.5.4.4.4 BIRDS

Similarly to reptiles, birds form a metanephric kidney (Dantzler and Braun 1980). However, in contrast to reptiles and all other non-mammalian vertebrates, the kidneys of birds have

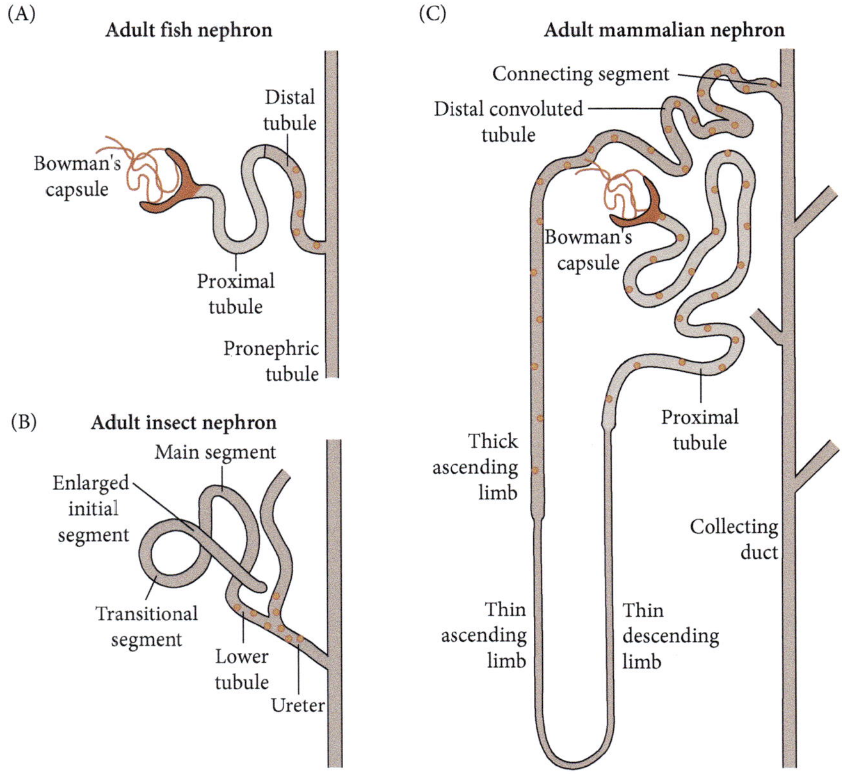

FIGURE 14.11 Renal progenitors are conserved across evolution. They were identified across species as a main strategy for kidney regeneration. Structure and distribution of renal progenitors (marked red) in nephrons of fish (A), insects (B), and mammals (C) are depicted.

Source: Reprinted by permission from *Nature Reviews Nephrology*, 9 (3), Renal progenitors: an evolutionary conserved strategy for kidney regeneration, Paola Romagnani, Laura Lasagni, Giuseppe Remuzzi, pp. 137–46, Figure 3, doi.10.1038/nrneph.2012.290. Copyright © 2013, Springer Nature.

cortical and medullary areas like in mammals (Dantzler and Braun 1980). This is related to the first evolutionary appearance of the loop of Henle, which caused the emergence of a radiating pattern around the collecting ducts and vasa recta and forms the relatively small medullary regions of the avian kidney (Dantzler and Braun 1980). Interestingly, in birds the neonephrogenic strategy of the kidney progressively exhausts. Indeed, new nephrons can be generated for a few weeks of life, but then the nephrogenic zone, which is selectively localised at the cortex periphery, exhausts and new nephrons cannot be generated anymore. That is, only the capacity for cellular regeneration persists as a possible regenerative strategy following renal injury (Dantzler and Braun 1980).

14.5.4.4.5 MAMMALS

The capacity of mammals to generate new nephrons ceases at around birth. However, the mammalian kidney maintains the capacity to provide cellular regeneration of injured portions of the nephrons through renal progenitors. Lineage tracing experiments showed that parietal epithelial cells (PEC) can behave as podocyte progenitors, migrating to the glomerular tuft and becoming fully differentiated and functional podocytes during childhood and adolescence (Appel et al. 2009; Wanner et al. 2014; Lasagni et al. 2015), or in response to injury, determining the outcome of the disease (Lasagni et al. 2015). In addition, adult tubular epithelium undergoes constant cell division through clonal expansion of a subset of tubular cells with segment-specific borders that may function as progenitor cells (Rinkevich et al. 2014). Other studies suggested the presence of progenitors within tubuli also in response to injury (Kang et al. 2016). Whether or not these cells represent a distinct subset of tubular progenitors or whether this represents a functional state that every tubular cell can transiently acquire is still debated. Consistent with observations in mice, human renal progenitors are characterised by coexpression of two markers, CD133 and CD24, and are localised among parietal epithelial cells of the Bowman's capsule, as well as among more differentiated tubular cells in a more scattered pattern. They have also been described and identified as critical players in kidney regeneration (Romagnani et al. 2013).

The observation that renal progenitors represent a critical player in regeneration across different species is apparently surprising, since in mammals the kidney's regenerative capacity does not lead to the generation of new nephrons like in fish (Romagnani et al. 2013). However, a deeper look at renal regeneration from an evolutionary perspective suggests possible explanations for these differences. Indeed, the three main strategies of responses to kidney injury—that is, cellular regeneration, nephron neogenesis, and kidney hypertrophy—have existed since the appearance of our distant fish ancestors (Reimschuessel 2001). Two of these strategies required the capacity to generate new cells and differentiate them into diverse specialised cell types, which happened through selection of a renal stem/progenitor system. By contrast, as in other organs, hypertrophy developed as an efficient support strategy of response to injury, which permitted a quick recovery of kidney function to secure survival, while regenerative processes that required more time took place. Thus, if all these regenerative strategies already existed before the mammalian kidney evolved, why could it be advantageous to forgo the capacity to generate *de novo* nephrons?

From an evolutionary point of view, progressive adaptation of nephrons allowed successful development of mammalian life through one notable anatomical change in its basic structure in comparison to fish: that is, the appearance of the loop of Henle, which was important to solve the problem of water conservation (Romagnani et al. 2013). Interestingly, the thin ascending limb of the loop of Henle does not seem to contain renal progenitors, and is generated only once development of nephrons has already been completed (Romagnani et al. 2013). More importantly, the loop of Henle has imposed on the kidney a structural distinction between a cortex and a medulla and a pyramidal shape that are not observed in 'lower' species.

The progressive acquisition of a pyramidal shape allowed another major evolutionary development: that is, a further increase in the number of nephrons, which were added at the periphery following a radiating pattern. The substantially increased number of nephrons considerably reduced the selective advantage related to the capacity to generate new nephrons (Dantzler and Braun 1980; Bely 2010). The importance of neonephrogenesis thus decreased (Dantzler and Braun 1980; Bely 2010,). Indeed, the regeneration of trivial or redundant structures imparts little or no benefit, whereas the loss of crucial structures might lead to death before the animal can regenerate (Dantzler and Braun 1980; Reichman 1984; Bely 2010). In mammals, nephron number at birth allows the increase of body mass related to childhood growth until adult age (Romagnani et al. 2013). However, the amount of nephrons declines by about 50% after the age of 60, explaining why an inadequate nephron number became a determinant of renal failure in humans, once life expectancy increased beyond evolutionary requirements (Romagnani et al. 2013). Thus, increased kidney complexity induced by progressive adaptations to a changing environment weakened the selection pressure on neonephrogenesis. The renal progenitor system was left and used only to reconstitute injured portions of existing nephrons in the case of injury, and cellular regeneration was retained as a main regenerative strategy (Romagnani et al. 2013). The existence of renal progenitors across different stages of evolution is at least an explanation of how the essential requirement of kidney regeneration was achieved.

14.6 Consequences for Prevention and Treatment of Disease

14.6.1 Drink Sufficient Amounts of Plain Water

The human kidney deals with the challenge of water and salt loss that comes with life in a terrestrial habitat and is further accentuated by the need for excreting nitrous waste as water-soluble urea. Therefore, humans need to drink sufficient amounts of fluid. Although food industries promote isotonic drinks, fructose-rich soft drinks, coffee, and tea as regular habits, the human body requires nothing but plain water to facilitate the excretion of metabolic waste. In fact, sodium-rich western diets absolutely do not need added salt in fluids. Exceptions are scenarios like a marathon run, where losses of water and salt from breathing and sweating might better be replaced by isotonic fluid in case of hyponatraemia-related brain oedema (Noakes 2007). Soft drinks rich in fructose are generally not recommended for rehydration of heat-related fluid losses. Fructose promotes heat stress-related cell injury and hyperuricaemia, the latter supporting uric acid crystal formation and kidney stone disease (Wesseling et al. 2016).

14.6.2 Trust your Thirst on How Much Water to Drink but with a Few Exceptions

Evolution has optimised the sensation of thirst to determine the need for water intake for terrestrial mammals. Thus, we can trust on our subjective desire of thirst to tell us how much fluid our body needs (Noakes 2007). Water intake beyond thirst is only recommended for kidney stone formers. Patients with CKD should avoid polydipsia. There is no need to increase the filtration load for the remaining nephrons already suffering from hyperfiltration. Age-related deterioration of cerebral function is often associated with impaired thirst and insufficient fluid intake, especially in cases of incident fluid losses during episodes of fever, diarrhoea, or vomiting, or simply during heat waves (Sterns 2015). Such individuals may benefit from assisted fluid intake schedules.

14.6.3 Stick to Potassium-Rich and Sodium-/Fructose-Reduced Diets

Modern lifestyle implies a change in diet that differs in several ways to environmental conditions to which our physiology once adapted. Even the relatively recent agricultural 'neolithic' revolution has caused an increased exposure to carbohydrate-rich diets. Nowadays, the global spreading of western diet drastically enriched in salt, purins, fat, and fructose, which at the same time is depleted of potassium-rich fruits and vegetables, represents a challenge to human physiology. Although a healthy kidney is able to excrete large amounts of salt, individuals endowed with fewer nephrons or patients with CKD are susceptible to develop hypertension and hypertension-related (cardiovascular) mortality (Bibbins-Domingo et al. 2010). Fructose-rich soft drinks also contribute to hypertension, gout, and kidney stones (Ha et al. 2013).

14.6.4 Long-term Follow-up of Low Birth Weight and Pre-term Birth

Pre-term children represent a special risk group in which long-term follow-up of renal function is needed. This includes a thorough assessment of BP and markers of renal outcome (creatinine, microalbuminuria) even during early childhood, to prevent adverse cardiovascular and renal diseases. Improvement of maternal nutrition during and before pregnancy, such as preconceptional monitoring, is also suggested to influence renal programming. Early identification of children at high risk of reduced renal reserve allows subsequent monitoring, and also prolonged treatment with renal protective agents such as angiotensin-converting enzyme (ACE) inhibitors or angiotensin receptor blockades. However, more systematic studies should be implemented to investigate the role of these two drugs in managing proteinurias and glomerulosclerosis in children with renal conditions.

Early detection of potential indicators of hyperfiltration, such as impaired renal reserve (low renal volume detected by ultrasound or scintigraphic measurement) and blunted solute clearance, may provide subtle clues to the presence of reduced nephron number, thus providing early objective evidence of hypertension, microalbuminuria, and other renal risks. Unfortunately, to date, no investigation method for early detection is available and validated. Future studies of radiological techniques and biochemical indicators are needed and may provide important advances in long-term follow-up of children and young adults at risk.

14.6.5 Maximise and Protect Nephron Number during Lifetime

Nephron number is the main determinant of kidney health. The increasing lifespan of people living in developing and developed countries makes low nephron endowment or incident nephron losses a major cause of the increasing incidence and prevalence of ESKD in the elderly (Ene-Iordache et al. 2016). There are four ways to avoid this problem: (1) maximise nephron endowment, starting from kidney development in the fetus by assuring optimal nutrition of pregnant women (Luyckx et al. 2013); (2) minimise nephron loss by avoiding exposure to nephrotoxic drugs or environmental toxins, support early diagnosis of kidney disease by sporadic screening including urinary analysis, and ask for rigorous treatment of kidney disease; (3) minimise avoidable glomerular hyperfiltration related to obesity (keep normal BMI), diabetes (rigorously control hyperglycaemia, preferably with SGLT2 inhibitors), avoid anabolic steroids, and use RAS inhibitors rigorously in patients with CKD; and (4) endorse research on kidney regeneration. The kidney has an intrinsic reparative capacity provided by hypertrophy and *de novo* cell production by specialised renal progenitor cells. Understanding how to pharmacologically manipulate these processes may offer innovative ways to avoid nephron loss in kidney disease.

14.6.6 Endorse Research on Identification of a Marker of Nephron Number

Clinical practice in nephrology would completely change if a clinical marker reflecting nephron number would become available. Then, individual risks could be detected early in life, incident nephron losses would become detectable, and renoprotective treatments could be validated more easily.

14.6.7 Cultural Evolution Evolved the Artificial Kidney

Once most nephrons are lost, the kidneys are no longer able to maintain homoeostasis. Retention of uremic toxins including potassium, phosphate, and sodium causes hypertension, fluid overload, cardiac arrhythmias, acidosis, and other potentially life-threatening complications. Haemodialysis machines are products of cultural evolution that have replaced many different functions of the nephrons of the kidneys in a completely unexpected manner. The countercurrent flow of blood and water in the dialyser mimics the countercurrent flow in fish gills or the loop of Henle and maximises the clearance of uremic toxins from the blood at a minimum length of the interface. In contrast, peritoneal dialysis is not based on countercurrent flow, but makes use of the large surface of the peritoneum in the abdominal cavity. Peritoneal dialysis is based on the concept that refilling the abdomen with clean water induces elimination of uremic toxins from the blood across the peritoneal membrane until equilibrium has formed. Repetitive drains of this fluid enriched with uremic toxins and refills with fresh water allow significant clearance of uremic toxins and also of water when the dialysate fluid is kept hyperosmolar to the blood, for example, by being enriched with glucose or other osmolytes. Indeed, haemodialysis or peritoneal dialysis can sufficiently maintain a non-fatal state of uraemia over years and sometimes even decades. However, the associated cardiovascular mortality and procedure-related complications remain challenges in clinical practice.

14.6.8 Nephron Transplantation is the Best Kidney Replacement Therapy

Nephron (kidney) transplantation is an alternative RRT. Indeed, in contrast to dialysis, which can only partially reverse uraemia, a functional transplanted kidney can resolve all clinical symptoms of uraemia. The transplanted organ works 24 hours a day and also generates the hormonal regulations that cannot be provided by either form of dialysis. Transplanted patients enjoy a better quality of life, fitness, independence, and survival (Wolfe et al. 1999). However, kidney transplantation requires kidney donation, which has numerous legal and ethical challenges. Because the number of cadaveric kidney donors is limited, live donation has become popular and is actively promoted in most developed countries. However, live donation implies a 50% nephron loss for the donor, and the associated risk of future health problems including ESKD for the donor remains a concern (Grams et al. 2016). This risk can be minimised by carefully selecting healthy donors with strong nephron endowment.

References

Abrass, C. K., Adcox, M. J., and Raugi, G. J. (1995). Aging-associated changes in renal extracellular matrix. *Am J Pathol* 146, 742–52.

Agarwal, B. N. and Cabebe, F. G. (1980). Renal acidification in elderly subjects. *Nephron* 26, 291–5.

Agnese, D. M., Calvano, J. E., Hahm, S. J., et al. (2002). Human toll-like receptor 4 mutations but not CD14 polymorphisms are associated with an increased risk of gram-negative infections. *J Infect Dis* 186, 1522–5. doi: 10.1086/344893.

Anders, H. J., Davis, J. M., and Thurau, K. (2016). Nephron protection in diabetic kidney disease. *N Engl J Med* 375, 2096–8. doi: 10.1056/NEJMcibr1608564.

Anguiano Gomez, L., Lei, Y., Devarapu, S. K., et al. (2017). The diabetes pandemic suggests unmet needs for 'CKD with diabetes' in addition to 'diabetic nephropathy'. Implications for pre-clinical research and drug testing. *Nephrol Dial Transplant* July 31. doi: 10.1093/ndt/gfx219. [Epub ahead of print].

Appel, D., Kershaw, D. B., Smeets, B., et al. (2009). Recruitment of podocytes from glomerular parietal epithelial cells. *J Am Soc Nephrol* 20, 333–43. doi: 10.1681/ASN.2008070795.

Baylis, C. and Schmidt, R. (1996). The aging glomerulus. *Semin Nephrol* 16, 265–76.

Becherucci, F., Roperto, R. M., Materassi, M., et al. (2016). Chronic kidney disease in children. *Clin Kidney J* 9, 583–91. doi: 10.1093/ckj/sfw047.

Beckerman, P., Bi-Karchin, J., Park, A. S., et al. (2017). Transgenic expression of human APOL1 risk variants in podocytes induces kidney disease in mice. *Nat Med* 23, 429–38. doi: 10.1038/nm.4287.

Bely, A. E. (2010). Evolutionary loss of animal regeneration: pattern and process. *Integr Comp Biol* 50, 515–27. doi: 10.1093/icb/icq118.

Benghanem Gharbi, M., Elseviers, M., Zamd, M., et al. (2016). Chronic kidney disease, hypertension, diabetes, and obesity in the adult population of Morocco: how to avoid 'over'- and 'under'-diagnosis of CKD. *Kidney Int* 89, 1363–71. doi: 10.1016/j.kint.2016.02.019.

Berndt, T. and Kumar, R. (2007). Phosphatonins and the regulation of phosphate homeostasis. *Annu Rev Physiol* 69, 341–59. doi: 10.1146/annurev.physiol.69.040705.141729.

Bertram, J. F., Douglas-Denton, R. N., Diouf, B., et al. (2011). Human nephron number: implications for health and disease. *Pediatr Nephrol* 26, 1529–33. doi: 10.1007/s00467-011-1843-8.

Bhave, G. and Neilson, E. G. (2011). Body fluid dynamics: back to the future. *J Am Soc Nephrol* 22, 2166–81. doi: 10.1681/ASN.2011080865.

Bibbins-Domingo, K., Chertow, G. M., Coxson, P. G., et al. (2010). Projected effect of dietary salt reductions on future cardiovascular disease. *N Engl J Med* 362, 590–9. doi: 10.1056/NEJMoa0907355.

Boylan, J. W. (1972). A model for passive urea reabsorption in the elasmobranch kidney. *Comp Biochem Physiol A Comp Physiol* 42, 27–30.

Brenner, B. M., Meyer, T. W., and Hostetter, T. H. (1982). Dietary protein intake and the progressive nature of kidney disease: the role of hemodynamically mediated glomerular injury in the pathogenesis of progressive glomerular sclerosis in aging, renal ablation, and intrinsic renal disease. *N Engl J Med* 307, 652–9. doi: 10.1056/NEJM198209093071104.

Carlstrom, M., Wilcox, C. S., and Arendshorst, W. J. (2015). Renal autoregulation in health and disease. *Physiol Rev* 95, 405–511. doi: 10.1152/physrev.00042.2012.

Cha, J. H., Kim, Y. H., Jung, J. Y., et al. (2001). Cell proliferation in the loop of Henle in the developing rat kidney. *J Am Soc Nephrol* 12, 1410–21.

Chang, A., Van Horn, L., Jacobs, D. R., Jr., et al. (2013). Lifestyle-related factors, obesity, and incident microalbuminuria: the CARDIA (Coronary Artery Risk Development in Young Adults) study. *Am J Kidney Dis* 62, 267–75. doi: 10.1053/j.ajkd.2013.02.363.

Charlton, J. R., Springsteen, C. H., and Carmody, J. B. (2014). Nephron number and its determinants in early life: a primer. *Pediatr Nephrol* 29, 2299–308. doi: 10.1007/s00467-014-2758-y.

Chevalier, R. L. (2017). Evolutionary nephrology. *Kidney Int Rep* 2(3), 302–17.

Couser, W. G., Remuzzi, G., Mendis, S., et al. (2011). The contribution of chronic kidney disease to the global burden of major noncommunicable diseases. *Kidney Int* 80, 1258–70. doi: 10.1038/ki.2011.368.

Čukuranović, R. and Vlajković, S. (2005). Age-related anatomical and functional characteristics of human kidney. *FU Med Biol* 12, 61–9.

Čukuranović, R., Stefanović, N., and Stojanović, J. (1999). The stereological analysis of age changes of the human renal corpuscle. *Folia Anat* 27, 29–33.

Dantzler, W. H. (1989). *Comparative Physiology of the Vertebrate Kidney*. Berlin: Springer.

Dantzler, W. H. and Braun, E. J. (1980). Comparative nephron function in reptiles, birds, and mammals. *Am J Physiol* 239, R197–213.

Dantzler, W. H. and Braun, J. (1997). Vertebrate renal system. In: Dantzler, W. H. (ed.) *Handbook of Physiology, Section 13: Comparative Physiology*. New York: Oxford University Press.

De Jong, F., Monuteaux, M. C., Van Elburg, R. M., et al. (2012). Systematic review and meta-analysis of preterm birth and later systolic blood pressure. *Hypertension* 59, 226–34. doi: 10.1161/HYPERTENSIONAHA.111.181784.

Denic, A., Lieske, J. C., Chakkera, H. A., et al. (2016). The substantial loss of nephrons in healthy human kidneys with aging. *J Am Soc Nephrol* 28(1), 313–20. doi: 10.1681/ASN.2016020154.

Denic, A., Mathew, J., Lerman, L. O., et al. (2017). Single-nephron glomerular filtration rate in healthy adults. *N Engl J Med* 376, 2349–57. doi: 10.1056/NEJMoa1614329.

Diep, C. Q., Ma, D., Deo, R. C., et al. (2011). Identification of adult nephron progenitors capable of kidney regeneration in zebrafish. *Nature* 470, 95–100. doi: 10.1038/nature09669.

Dow, J. A., Maddrell, S. H., Gortz, A., et al. (1994). The Malpighian tubules of *Drosophila melanogaster*: a novel phenotype for studies of fluid secretion and its control. *J Exp Biol* 197, 421–8.

Dressler, G. R. (2006). The cellular basis of kidney development. *Annu Rev Cell Dev Biol* 22, 509–29. doi: 10.1146/annurev.cellbio.22.010305.104340.

Ejerblad, E., Fored, C. M., Lindblad, P., et al. (2006). Obesity and risk for chronic renal failure. *J Am Soc Nephrol* 17, 1695–702. doi: 10.1681/ASN.2005060638.

Elger, M., Hentschel, H., Litteral, J., et al. (2003). Nephrogenesis is induced by partial nephrectomy in the elasmobranch *Leucoraja erinacea*. *J Am Soc Nephrol* 14, 1506–18.

Ene-Iordache, B., Perico, N., Bikbov, B., et al. (2016). Chronic kidney disease and cardiovascular risk in six regions of the world (ISN-KDDC): a cross-sectional study. *Lancet Glob Health* 4, e307–19. doi: 10.1016/S2214-109X(16)00071-1.

Epstein, F. H., Stoff, J. S., and Silva, P. (1983). Mechanism and control of hyperosmotic NaCl-rich secretion by the rectal gland of *Squalus acanthias*. *J Exp Biol* 106, 25–41.

Evans, D. H. (2010). A brief history of the study of fish osmoregulation: the central role of the Mt. Desert Island Biological Laboratory. *Front Physiol* 1, 13. doi: 10.3389/fphys.2010.00013.

Foster, M. C., Hwang, S. J., Larson, M. G., et al. (2008). Overweight, obesity, and the development of stage 3 CKD: the Framingham Heart Study. *Am J Kidney Dis* 52, 39–48. doi: 10.1053/j.ajkd.2008.03.003.

Friedman, D. J. and Pollak, M. R. (2016). Apolipoprotein L1 and kidney disease in African Americans. *Trends Endocrinol Metab* 27, 204–15. doi: 10.1016/j.tem.2016.02.002.

Genovese, G., Friedman, D. J., Ross, M. D., et al. (2010). Association of trypanolytic ApoL1 variants with kidney disease in African Americans. *Science* 329, 841–5. doi: 10.1126/science.1193032.

Ghirotto, S., Tassi, F., Barbujani, G., et al. (2016). The uromodulin gene locus shows evidence of pathogen adaptation through human evolution. *J Am Soc Nephrol* 27, 2983–96. doi: 10.1681/ASN.2015070830.

Glassock, R. J. and Winearls, C. (2009). Ageing and the glomerular filtration rate: truths and consequences. *Trans Am Clin Climatol Assoc* 120, 419–28.

Glomski, C. A., Tamburlin, J., Hard, R., et al. (1997). The phylogenetic odyssey of the erythrocyte. IV. The amphibians. *Histol Histopathol* 12, 147–70.

Grams, M. E., Sang, Y., Levey, A. S., et al. (2016). Kidney-failure risk projection for the living kidney-donor candidate. *N Engl J Med* 374, 411–21. doi: 10.1056/NEJMoa1510491.

Ha, V., Jayalath, V. H., Cozma, A. I., et al. (2013). Fructose-containing sugars, blood pressure, and cardiometabolic risk: a critical review. *Curr Hypertens Rep* 15, 281–97. doi: 10.1007/s11906-013-0364-1.

Haraldsson, B. and Jeansson, M. (2009). Glomerular filtration barrier. *Curr Opin Nephrol Hypertens* 18, 331–5. doi: 10.1097/MNH.0b013e32832c9dba.

Hazon, N., Tierney, M. L., and Takei, Y. (1999). Renin–angiotensin system in elasmobranch fish: a review. *J Exp Zool* 284, 526–34.

Helal, I., Fick-Brosnahan, G. M., Reed-Gitomer, B., et al. (2012). Glomerular hyperfiltration: definitions, mechanisms and clinical implications. *Nat Rev Nephrol* 8, 293–300. doi: 10.1038/nrneph.2012.19.

Hill, G. S., Heudes, D., and Bariety, J. (2003). Morphometric study of arterioles and glomeruli in the aging kidney suggests focal loss of autoregulation. *Kidney Int* 63, 1027–36. doi: 10.1046/j.1523-1755.2003.00831.x.

Hodgin, J. B., Bitzer, M., Wickman, L., et al. (2015). Glomerular aging and focal global glomerulosclerosis: a podometric perspective. *J Am Soc Nephrol* 26, 3162–78. doi: 10.1681/ASN.2014080752.

Hostetter, T. H., Olson, J. L., Rennke, H. G., et al. (1981). Hyperfiltration in remnant nephrons: a potentially adverse response to renal ablation. *Am J Physiol* 241, F85–93.

Hudson, B. G., Tryggvason, K., Sundaramoorthy, M., et al. (2003). Alport's syndrome, Goodpasture's syndrome, and type IV collagen. *N Engl J Med* 348, 2543–56. doi: 10.1056/NEJMra022296.

Hyodo, S., Kakumura, K., Takagi, W., et al. (2014). Morphological and functional characteristics of the kidney of cartilaginous fishes: with special reference to urea reabsorption. *Am J Physiol Regul Integr Comp Physiol* 307, R1381–95. doi: 10.1152/ajpregu.00033.2014.

Jahnukainen, T., Chen, M., Berg, U., et al. (2001). Antenatal glucocorticoids and renal function after birth. *Semin Neonatol* 6, 351–5. doi: 10.1053/siny.2001.0070.

Johnson, R. J., Stenvinkel, P., Martin, S. L., et al. (2013). Redefining metabolic syndrome as a fat storage condition based on studies of comparative physiology. *Obesity* 21, 659–64. doi: 10.1002/oby.20026.

Jose, P. A., Fildes, R. D., Gomez, R. A., et al. (1994). Neonatal renal function and physiology. *Curr Opin Pediatr* 6, 172–7.

Kang, H. M., Huang, S., Reidy, K., et al. (2016). Sox9-positive progenitor cells play a key role in renal tubule epithelial regeneration in mice. *Cell Rep* 14, 861–71. doi: 10.1016/j.celrep.2015.12.071.

Khalsa, D. D., Beydoun, H. A., and Carmody, J. B. (2016). Prevalence of chronic kidney disease risk factors among low birth weight adolescents. *Pediatr Nephrol* 31, 1509–16. doi: 10.1007/s00467-016-3384-7.

Kiger, A. A., Jones, D. L., Schulz, C., et al. (2001). Stem cell self-renewal specified by JAK-STAT activation in response to a support cell cue. *Science* 294, 2542–5. doi: 10.1126/science.1066707.

Kramer, H., Gutierrez, O. M., Judd, S. E., et al. (2016). Waist circumference, body mass index, and ESRD in the REGARDS (Reasons for Geographic and Racial Differences in Stroke) Study. *Am J Kidney Dis* 67, 62–9. doi: 10.1053/j.ajkd.2015.05.023.

Kratzer, J. T., Lanaspa, M. A., Murphy, M. N., et al. (2014). Evolutionary history and metabolic insights of ancient mammalian uricases. *Proc Natl Acad Sci U S A* 111, 3763–8. doi: 10.1073/pnas.1320393111.

Kriz, W. and Lemley, K. V. (2016). Mechanical challenges to the glomerular filtration barrier: adaptations and pathway to sclerosis. *Pediatr Nephrol* 32(3), 405–17. doi: 10.1007/s00467-016-3358-9.

Laouari, D., Burtin, M., Phelep, A., et al. (2011). TGF-alpha mediates genetic susceptibility to chronic kidney disease. *J Am Soc Nephrol* 22, 327–35. doi: 10.1681/ASN.2010040356.

Lasagni, L., Lazzeri, E., Shankland, S. J., et al. (2013). Podocyte mitosis—a catastrophe. *Curr Mol Med* 13, 13–23.

Lasagni, L., Angelotti, M. L., Ronconi, E., et al. (2015). Podocyte regeneration driven by renal progenitors determines glomerular disease remission and can be pharmacologically enhanced. *Stem Cell Reports* 5, 248–63. doi: 10.1016/j.stemcr.2015.07.003.

Levin, L. R. and Buck, J. (2015). Physiological roles of acid–base sensors. *Annu Rev Physiol* 77, 347–62. doi: 10.1146/annurev-physiol-021014-071821.

Liapis, H., Romagnani, P., and Anders, H. J. (2013). New insights into the pathology of podocyte loss: mitotic catastrophe. *Am J Pathol* 183, 1364–74. doi: 10.1016/j.ajpath.2013.06.033.

Liu, W., Morimoto, T., Kondo, Y., et al. (2001). 'Avian-type' renal medullary tubule organization causes immaturity of urine-concentrating ability in neonates. *Kidney Int* 60, 680–93. doi: 10.1046/j.1523-1755.2001.060002680.x.

Low Birth Weight and Nephron Number Working Group. (2017). The impact of kidney development on the life course: a consensus document for action. *Nephron* 136, 3–49. doi: 10.1159/000457967.

Lu, J. L., Molnar, M. Z., Naseer, A., et al. (2015). Association of age and BMI with kidney function and mortality: a cohort study. *Lancet Diabetes Endocrinol* 3, 704–14. doi: 10.1016/S2213-8587(15)00128-X.

Luyckx, V. A. and Brenner, B. M. (2015). Birth weight, malnutrition and kidney-associated outcomes—a global concern. *Nat Rev Nephrol* 11, 135–49. doi: 10.1038/nrneph.2014.251.

Luyckx, V. A., Bertram, J. F., Brenner, B. M., et al. (2013). Effect of fetal and child health on kidney development and long-term risk of hypertension and kidney disease. *Lancet* 382, 273–83. doi: 10.1016/S0140-6736(13)60311-6.

Luyckx, V. A., Perico, N., Somaschini, M., et al. (2017). A developmental approach to the prevention of hypertension and kidney disease: a report from the Low Birth Weight and Nephron Number Working Group. *Lancet* 390(10092), 424–8. doi: 10.1016/S0140-6736(17)30576-7.

Mandal, A. K. and Mount, D. B. (2015). The molecular physiology of uric acid homeostasis. *Annu Rev Physiol* 77, 323–45. doi: 10.1146/annurev-physiol-021113-170343.

Manolio, T. A., Collins, F. S., Cox, N. J., et al. (2009). Finding the missing heritability of complex diseases. *Nature* 461, 747–53. doi: 10.1038/nature08494.

Miller, J., Chi, T., Kapahi, P., et al. (2013). *Drosophila melanogaster* as an emerging translational model of human nephrolithiasis. *J Urol* 190, 1648–56. doi: 10.1016/j.juro.2013.03.010.

Mueller, T. F. and Luyckx, V. A. (2012). The natural history of residual renal function in transplant donors. *J Am Soc Nephrol* 23, 1462–6. doi: 10.1681/ASN.2011111080.

Natochin, Y. V. (1996). Evolutionary aspects of renal function. *Kidney Int* 49, 1539–42.

Nicolaou, N., Renkema, K. Y., Bongers, E. M., et al. (2015). Genetic, environmental, and epigenetic factors involved in CAKUT. *Nat Rev Nephrol* 11, 720–31. doi: 10.1038/nrneph.2015.140.

Nigam, S. K., Aperia, A. C., and Brenner, B. M. (1996). Development and maturation of the kidney. In: Brenner B. M. (ed.) *The Kidney*. Philadelphia: W. B. Saunders, pp. 72–98.

Nishimura, H. (2008). Urine concentration and avian aquaporin water channels. *Pflugers Arch* 456, 755–68. doi: 10.1007/s00424-008-0469-6.

Nishimura, H. (2016). Renin–angiotensin system in vertebrates: phylogenetic view of structure and function. *Anat Sci Int* 92(2), 215–47. doi: 10.1007/s12565-016-0372-8.

Noakes, T. D. (2007). Hydration in the marathon: using thirst to gauge safe fluid replacement. *Sports Med* 37, 463–6.

Ortiz, R. M. (2001). Osmoregulation in marine mammals. *J Exp Biol* 204, 1831–44.

Padmanabhan, S., Melander, O., Johnson, T., et al. (2010). Genome-wide association study of blood pressure extremes identifies variant near UMOD associated with hypertension. *PLoS Genet* 6, e1001177. doi: 10.1371/journal.pgen.1001177.

Pak, J., Pu, Y., Zhang, Z. T., et al. (2001). Tamm-Horsfall protein binds to type 1 fimbriated *Escherichia coli* and prevents *E. coli* from binding to uroplakin Ia and Ib receptors. *J Biol Chem* 276, 9924–30. doi: 10.1074/jbc.M008610200.

Palmer, B. F. and Levi, M. (1996). Effect of aging on renal function and disease. In: Brenner, B. M. (ed.) *The Kidney*. Philadelphia: W. B. Saunders, pp. 2274–96.

Pannabecker, T. L. (2012). Structure and function of the thin limbs of the loop of Henle. *Compr Physiol* 2, 2063–86. doi: 10.1002/cphy.c110019.

Peired, A., Angelotti, M. L., Ronconi, E., et al. (2013). Proteinuria impairs podocyte regeneration by sequestering retinoic acid. *J Am Soc Nephrol* 24, 1756–68. doi: 10.1681/ASN.2012090950.

Rabelink, T. J. and De Zeeuw, D. (2015). The glycocalyx—linking albuminuria with renal and cardiovascular disease. *Nat Rev Nephrol* 11, 667–76. doi: 10.1038/nrneph.2015.162.

Reichman, J. (1984). Evolution of regeneration capabilities. *Am Nat* 123, 752–63.

Reimschuessel, R. (2001). A fish model of renal regeneration and development. *ILAR J* 42, 285–91.

Reiser, J. and Sever, S. (2013). Podocyte biology and pathogenesis of kidney disease. *Annu Rev Med* 64, 357–66. doi: 10.1146/annurev-med-050311-163340.

Rhee, C. M. and Kovesdy, C. P. (2015). Epidemiology: spotlight on CKD deaths—increasing mortality worldwide. *Nat Rev Nephrol* 11, 199–200. doi: 10.1038/nrneph.2015.25.

Rinkevich, Y., Montoro, D. T., Contreras-Trujillo, H., et al. (2014). In vivo clonal analysis reveals lineage-restricted progenitor characteristics in mammalian kidney development, maintenance, and regeneration. *Cell Rep* 7, 1270–83. doi: 10.1016/j.celrep.2014.04.018.

Riordan, J. R., Forbush, B. 3rd, and Hanrahan, J. W. (1994). The molecular basis of chloride transport in shark rectal gland. *J Exp Biol* 196, 405–18.

Rodriguez-Puyol, D. (1998). The aging kidney. *Kidney Int* 54, 2247–65.

Romagnani, P. (2009). Toward the identification of a 'renopoietic system'? *Stem Cells* 27, 2247–53. doi: 10.1002/stem.140.

Romagnani, P., Lasagni, L., and Remuzzi, G. (2013). Renal progenitors: an evolutionary conserved strategy for kidney regeneration. *Nat Rev Nephrol* 9, 137–46. doi: 10.1038/nrneph.2012.290.

Romagnani, P., Giglio, S., Angelotti, M. L., et al. (2016). Next generation sequencing and functional analysis of patient urine renal progenitor-derived podocytes to unravel the diagnosis underlying refractory lupus nephritis. *Nephrol Dial Transplant* 31, 1541–5. doi: 10.1093/ndt/gfw234.

Rossier, B. C., Baker, M. E., and Studer, R. A. (2015). Epithelial sodium transport and its control by aldosterone: the story of our internal environment revisited. *Physiol Rev* 95, 297–340. doi: 10.1152/physrev.00011.2014.

Rowe, J. W., Shock, N. W., and Defronzo, R. A. (1976). The influence of age on the renal response to water deprivation in man. *Nephron* 17, 270–8.

Ruggenenti, P., Cravedi, P., and Remuzzi, G. (2012). Mechanisms and treatment of CKD. *J Am Soc Nephrol* 23, 1917–28. doi: 10.1681/ASN.2012040390.

Sands, J. M. and Layton, H. E. (2014). Advances in understanding the urine-concentrating mechanism. *Annu Rev Physiol* 76, 387–409. doi: 10.1146/annurev-physiol-021113-170350.

Saran, R., Robinson, B., Abbott, K. C., et al. (2017). US renal data system 2016 annual data report: epidemiology of kidney disease in the United States. *Am J Kidney Dis* 69, A7–8. doi: 10.1053/j.ajkd.2016.12.004.

Schedl, A. (2007). Renal abnormalities and their developmental origin. *Nat Rev Genet* 8, 791–802. doi: 10.1038/nrg2205.

Schulte, K., Kunter, U., and Moeller, M. J. (2015). The evolution of blood pressure and the rise of mankind. *Nephrol Dial Transplant* 30, 713–23. doi: 10.1093/ndt/gfu275.

Scolari, F., Izzi, C., and Ghiggeri, G. M. (2015). Uromodulin: from monogenic to multifactorial diseases. *Nephrol Dial Transplant* 30, 1250–6. doi: 10.1093/ndt/gfu300.

Shankland, S. J., Anders, H. J., and Romagnani, P. (2013). Glomerular parietal epithelial cells in kidney physiology, pathology, and repair. *Curr Opin Nephrol Hypertens* 22, 302–9. doi: 10.1097/MNH.0b013e32835fefd4.

Sharma, A., Mucino, M. J., and Ronco, C. (2014). Renal functional reserve and renal recovery after acute kidney injury. *Nephron Clin Pract* 127, 94–100. doi: 10.1159/000363721.

Singh, P. and Thomson, S. C. (2010). Renal homeostasis and tubuloglomerular feedback. *Curr Opin Nephrol Hypertens* 19, 59–64. doi: 10.1097/MNH.0b013e3283331ffd.

Singh, S. R., Liu, W., and Hou, S. X. (2007). The adult *Drosophila* Malpighian tubules are maintained by multipotent stem cells. *Cell Stem Cell* 1, 191–203. doi: 10.1016/j.stem.2007.07.003.

Smith, E. E. and Malik, H. S. (2009). The apolipoprotein L family of programmed cell death and immunity genes rapidly evolved in primates at discrete sites of host–pathogen interactions. *Genome Res* 19, 850–8. doi: 10.1101/gr.085647.108.

Solomon, S. E. (1985). The morphology of the kidney of the green turtle (*Chelonia mydas* L.). *J Anat* 140 (Part 3), 355–69.

Sozen, M. A., Armstrong, J. D., Yang, M., et al. (1997). Functional domains are specified to single-cell resolution in a *Drosophila* epithelium. *Proc Natl Acad Sci U S A* 94, 5207–12.

Spitzer, A. and Schwartz, G. J. (2011). The kidney during development. *Compr Physiol* 2011, Suppl 25: Handbook of Physiology, Renal Physiology, 475–44.

Sterns, R. H. (2015). Disorders of plasma sodium—causes, consequences, and correction. *N Engl J Med* 372, 55–65. doi: 10.1056/NEJMra1404489.

Takei, Y. (2015). From aquatic to terrestrial life: evolution of the mechanisms for water acquisition. *Zoolog Sci* 32, 1–7. doi: 10.2108/zs140142.

Tan, P. K., Farrar, J. E., Gaucher, E. A., et al. (2016). Coevolution of URAT1 and uricase during primate evolution: implications for serum urate homeostasis and gout. *Mol Biol Evol* 33, 2193–200. doi: 10.1093/molbev/msw116.

Taylor, H. E., Khatua, A. K., and Popik, W. (2014). The innate immune factor apolipoprotein L1 restricts HIV-1 infection. *J Virol* 88, 592–603. doi: 10.1128/JVI.02828-13.

Thomson, R., Genovese, G., Canon, C., et al. (2014). Evolution of the primate trypanolytic factor APOL1. *Proc Natl Acad Sci U S A* 111, E2130-9. doi: 10.1073/pnas.1400699111.

Tonneijck, L., Muskiet, M. H., Smits, M. M., et al. (2017). Glomerular hyperfiltration in diabetes: mechanisms, clinical significance, and treatment. *J Am Soc Nephrol* 28, 1023–39. doi: 10.1681/ASN.2016060666.

Tracy, R. E., Parra, D., Eisaguirre, W., et al. (2002). Influence of arteriolar hyalinization on arterial intimal fibroplasia in the renal cortex of subjects in the United States, Peru, and Bolivia, applicable also to other populations. *Am J Hypertens* 15, 1064–73.

Trudu, M., Janas, S., Lanzani, C., et al. (2013). Common noncoding UMOD gene variants induce salt-sensitive hypertension and kidney damage by increasing uromodulin expression. *Nat Med* 19, 1655–60. doi: 10.1038/nm.3384.

Tzur, S., Rosset, S., Shemer, R., et al. (2010). Missense mutations in the APOL1 gene are highly associated with end stage kidney disease risk previously attributed to the MYH9 gene. *Hum Genet* 128, 345–50. doi: 10.1007/s00439-010-0861-0.

Vallon, V. (2015). The mechanisms and therapeutic potential of SGLT2 inhibitors in diabetes mellitus. *Annu Rev Med* 66, 255–70. doi: 10.1146/annurev-med-051013-110046.

Vivante, A., Golan, E., Tzur, D., et al. (2012). Body mass index in 1.2 million adolescents and risk for end-stage renal disease. *Arch Intern Med* 172, 1644–50. doi: 10.1001/2013.jamainternmed.85.

Vize, P. D. and Smith, H. W. (2004). A Homeric view of kidney evolution: a reprint of H. W. Smith's classic essay with a new introduction. Evolution of the kidney. 1943. *Anat Rec A Discov Mol Cell Evol Biol* 277, 344–54. doi: 10.1002/ar.a.20017.

Wanner, C., Inzucchi, S. E., Lachin, J. M., et al. (2016). Empagliflozin and progression of kidney disease in type 2 diabetes. *N Engl J Med* 375, 323–34. doi: 10.1056/NEJMoa1515920.

Wanner, N., Hartleben, B., Herbach, N., et al. (2014). Unraveling the role of podocyte turnover in glomerular aging and injury. *J Am Soc Nephrol* 25, 707–16. doi: 10.1681/ASN.2013050452.

Weavers, H., Prieto-Sanchez, S., Grawe, F., et al. (2009). The insect nephrocyte is a podocyte-like cell with a filtration slit diaphragm. *Nature* 457, 322–6. doi: 10.1038/nature07526.

Welling, P. A. (2016). Roles and regulation of renal K channels. *Annu Rev Physiol* 78, 415–35. doi: 10.1146/annurev-physiol-021115-105423.

Wesseling, C., Aragon, A., Gonzalez, M., et al. (2016). Kidney function in sugarcane cutters in Nicaragua—a longitudinal study of workers at risk of Mesoamerican nephropathy. *Environ Res* 147, 125–32. doi: 10.1016/j.envres.2016.02.002.

Wessing, A. and Eichelberg, D. (1978). Malpighian tubules, rectal papillae and excretion. In: Ashburner, M. and Wright, T. R. F. (eds) *The Genetics and Biology of* Drosophila. *Vol. 2c*. New York: Academic Press, pp. 1–42.

White, S. L., Perkovic, V., Cass, A., et al. (2009). Is low birth weight an antecedent of CKD in later life? A systematic review of observational studies. *Am J Kidney Dis* 54, 248–61. doi: 10.1053/j.ajkd.2008.12.042.

Wilson, J. X. (1984). The renin–angiotensin system in nonmammalian vertebrates. *Endocr Rev* 5, 45–61. doi: 10.1210/edrv-5-1-45.

Wolfe, R. A., Ashby, V. B., Milford, E. L., et al. (1999). Comparison of mortality in all patients on dialysis, patients on dialysis awaiting transplantation, and recipients of a first cadaveric transplant. *N Engl J Med* 341, 1725–30. doi: 10.1056/NEJM199912023412303.

Worcester, E. M. and Coe, F. L. (2010). Clinical practice. Calcium kidney stones. *N Engl J Med* 363, 954–63. doi: 10.1056/NEJMcp1001011.

Zhou, W. and Hildebrandt, F. (2012). Inducible podocyte injury and proteinuria in transgenic zebrafish. *J Am Soc Nephrol* 23, 1039–47. doi: 10.1681/ASN.2011080776.

CHAPTER 15

ENDOCRINOLOGY

RICHARD G. BRIBIESCAS

Abstract

This chapter on endocrinology aims to shed light on the biology of hormones within the context of human life history evolution. An evolutionary perspective contributes to not only our understanding of human evolution, but also to the contemporary and emerging health challenges across the spectrum of ecologies and environments. Evolutionary endocrinology extends our understanding of human biology and health through the engagement of gene–environment interactions, social dynamics, human variation, and how hormones regulate life history traits such as growth, immune function, metabolism, and ageing. This chapter describes key aspects of endocrinology that are specific to men and women, while also being mindful of the importance of human variation. For example, men and women exhibit reproductive states that deploy specific functions. In women, these are menstruation, gestation, and lactation. These processes are governed largely by the hypothalamic–pituitary–ovarian axis and how it responds to environmental challenges such as nutritional demands, activity, and social stresses. Men also exhibit reproductive states, although they are mostly in the form of investment in sexually dimorphic tissue and behavioural variation. These states are governed by hormones which allocate resources between tissues that are indicative of different forms of reproductive effort. These include sexually dimorphic muscle tissue and adiposity. Spermatogenesis is obviously key but has differential effects on fertility compared to gametogenesis in women. Additional aspects of human evolutionary endocrinology include stress homoeostasis and metabolism, which involve the hypothalamic–pituitary–adrenal axis as well as the thyroid and other metabolic hormones.

Keywords

evolutionary medicine, evolution, hormones, health, adaptation, life history, primates, comparative

15.1 Introduction

Organisms, including humans, are challenged by the need to harvest energy from the environment and convert those resources into fitness. This challenge has resulted in evolutionary processes that aim to do this as well as allow humans to sense and respond to environmental variability. When possible it is also necessary to adjust genetic expression in a manner that optimises lifetime fitness. The primary responsibility of sensing environmental variability and adjusting an organism's physiology in a manner that optimises lifetime fitness falls mainly on the neuroendocrine system, an ancient and overarching collection of biochemical agents and tissue receptors that regulate virtually every life history trait, including growth, development, reproduction, maintenance, and ageing. In this chapter we cover the basic physiology of neuroendocrine function in humans, with a mindful eye towards the comparative biology of hormones, in order to place the reader within an evolutionary context that illuminates the uniqueness of human endocrinology as well as similarities with other organisms. There are many textbooks on human endocrinology. However, few engage this subject matter from an evolutionary perspective. This chapter aims to address this gap. Since evolution by natural selection is largely contingent on variation in reproductive success and the allocation of energy between often competing needs, the endocrinology of reproduction and metabolism will be of particular importance in this chapter.

Reproduction is at the core of what defines life. Other life history traits such as growth and maturation evolved in service of the need to optimise fitness. Reproduction is therefore fundamental to evolutionary biology and is central to the understanding of the evolution of all organisms, including humans. How organisms reproduce, and what constrains reproduction, informs our understanding of the strength and effect of natural selection on the evolution of a species. In this manner, humans are no exception. Reproduction has shaped how humans have evolved and continues to be an important part of our everyday life and continued evolution (Ellison 2001). Reproductive biology is also associated with numerous health challenges. Some are deeply rooted in our evolutionary history, while others are the result of a rapidly changing environment during recent recorded history. This first section aims to review our current state of knowledge of human reproductive endocrinology through the lens of evolutionary and life theory. The goals will be to (1) define our current understanding of basic human reproductive endocrinology; (2) place that understanding within the context of evolutionary and life history theory; and (3) examine sex-specific mechanisms and challenges related to reproductive endocrinology. Subsequent sections will address the evolutionary endocrinology of metabolism.

Reproduction is the duplication of individuals either through direct mitotic division or through the union of gametes from two individuals: asexual or sexual reproduction. The primary purpose of reproduction is to pass genes to subsequent generations. Simply put, individuals who pass genes from one generation to the next will have their genes and traits continue to be favoured and persist. Those that do not pass their genes are more likely to have their traits extinguished from the population. Changes in those genes emerge from mutations, and, in the case of sexual reproduction, the union of unique combinations that often result in phenotypic variability. Clearly, genetics is vital to understanding both the evolution and functional significance of reproductive biology. However, organisms do not evolve or live in a vacuum. They are immersed in an environment of variable stochasticity, and need to deal with variation in food availability, threats, and the need to deploy energy and resources between often competing physiological demands. A more holistic investigative approach is therefore needed to encompass the effects of ecological variation on human reproductive biology. By ecology, we mean environmental factors that influence and are influenced by individual organisms. These factors include the challenges mentioned previously, as well as engagement with other individuals. Indeed, conspecifics are also important since these may include social influences such as kin, offspring, rivals, potential alloparents, and potential mates. All of these may contribute to fitness variation of the individual. To fold these complex variables into our understanding of human reproduction, we turn to the growing field of reproductive ecology.

Reproductive ecology is an area of research within human evolutionary biology in which reproductive physiology is viewed and investigated within the contextual influences of environmental variables, including food availability, diet, activity, and social interactions. Reproductive ecology also involves understanding lifetime variability in reproductive effort, including developmental aspects of reproduction, sexual maturation, and ageing. Not only is this approach pertinent to populations that are especially susceptible to ecological variation, such as foragers, but also it has exhibited utility with non-human primates and contemporary modern human populations (Ellison 2001). Moreover, the importance of reproductive ecology extends to our current global circumstances and is evident in the tremendous population growth of our species. While the total number of other non-human great apes is in the tens of thousands, there are over 7 billion humans. How humans have been able to exhibit such high fertility and evolve the energy-demanding life history traits that we see today is truly remarkable and likely unprecedented in our planet's history.

This perspective allows us to view human reproductive biology as the result of millions of years of evolution that is embedded in the biology of our common ancestors, including hominids, primates, mammals, and vertebrates (Ellison 2001). It allows human evolutionary biologists to gain insights into constraints that emerged more recently over the past 200,000 years with the emergence of modern *Homo sapiens* and our colonisation outside of Africa. Reproduction is contextualised within the need to negotiate trade-offs with other biological demands such as growth and maintenance. Finally, senescence and ageing play a vital role in how we understand human reproductive biology. This is an especially interesting question given the uniquely long period of post-reproductive lifespan in humans. (For further discussion, see Chapter 5: Senescence and Ageing, and Chapter 16: Sexuality, Reproduction, and Birth.)

Human reproduction shares many similarities with other mammals, primates, and great apes. Indeed, human reproductive function is largely governed by the same life history constraints that are faced by virtually every other sexually reproducing organism. Human reproduction is limited by time as well as energy, and it is a reasonable assumption that natural selection has favoured those organisms that most optimally allocated these constraining resources between various life history traits, including reproduction (Stulp et al. 2016). Women in particular are committed to harvesting energy from the environment and converting those calories into offspring and supportive somatic tissue that is deployed to support reproductive effort, both physiologically and through behavioural repertoires that often manifest as offspring care. As iteroparous organisms, women also face the challenge of how much to invest in a particular offspring and when to begin investing in future reproductive efforts.

One particularly unique feature of human female reproduction is the difference between the cessation of reproduction and total decline in fertility and the extensive post-menopausal lifespan. In most other organisms, last reproduction tends to coincide with the species lifespan. However about a third of a woman's life is post-reproductive. Post-reproductive lifespan is also evident in chimpanzees but not to the extent seen in humans, although some cetaceans may display human-like extended post-reproductive lifespan (Emery Thompson et al. 2007; Croft et al. 2017). Understanding this unique aspect of human female life histories requires engagement beyond basic physiology and needs to include evolutionary and life history perspectives on lifespan, kin selection, trade-offs, allocare, and the role of males.

Males also must convert environmentally derived energy into offspring. But the constraints are different compared to females. Differences include the lack of gestation, lactation, and menstruation as well as the risk of paternity uncertainty. Men nonetheless exhibit somatic physical traits that are reflective of reproductive effort in the form of sexually dimorphic muscle, overall larger body mass, and a hormonal milieu that supports this metabolic investment despite the potential costs to survivorship (Bribiescas 2001a). Indeed, the sexually dimorphic somatic lifetime energetic investment that is deployed in male primates is very similar to the caloric investment seen in female lifetime reproductive effort (Key and Ross 1999). Spermatogenesis is on-going in males and does not exhibit the same supply exhaustion that is observed in female gametogenesis. Consequently, male reproductive senescence is not as time limited as females.'

Male reproductive senescence differs from females in that there is no complete or sudden decline in fertility. Men tend to maintain spermatogenesis and the ability to father offspring well into the later decades of life, though older men exhibit a greater degree of sperm defects that can compromise fertility to varying degrees. The ability of older males to procure mates does appear to decline with age, due to changes in attractiveness, the ability to obtain key resources, and impaired competitiveness with rival males. Nonetheless, reproduction by older men is significant in many populations and may have contributed to the evolution of post-menopausal lifespan as well as behavioural strategies such as paternal care.

In Section 15.6 we discuss how an evolutionary perspective on human reproductive biology can inform our understanding of not only reproductive health challenges, but also our entire notion of health itself. In essence, health is largely a function of optimal somatic processes that operate in service of the need to promote reproductive fitness. Future directions and questions for research are suggested.

15.2 Evolutionary Ontogeny of Human Reproduction

The evolution of human reproductive biology is rooted in sexual reproduction, iteroparity, sex-specific constraints, and the challenges faced by mammals compared to other vertebrates. Here, iteroparity is defined as multiple bouts of reproduction over the lifetime of an organism, often resulting in the discounting of offspring investment and the need to allocate resources between present and future reproduction. Human sexual reproduction includes the significant energetic cost of lactation and internal gestation. Despite these costs, the evolution of sexual reproduction presents advantages compared to asexual reproduction. A primary advantage is the ability to generate a much broader range of genetic and phenotypic variation. This is particularly advantageous in the face of environmental variability, pathogenic challenges, and other forms of stochasticity. Another central cost of sexual reproduction is the possibility of breaking up successful and advantageous genetic combinations (Maynard Smith 1978).

Besides the potential genetic costs, other liabilities include the behavioural ecology of the organism, and how organisms negotiate the time and energy needed to navigate and cope with environmental challenges. Reproductive effort, or the time and energy devoted primarily in service to reproduction, often requires resources to be diverted from other needs. Calories that go into gestation, for example, become unavailable for tissue repair or other needs. Similarly, time and energy spent in searching and evaluating potential mates diminishes foraging efficiency and often makes organisms more susceptible to predation (Williams 1966; Bell 1980).

In addition to the costs and benefits of sexual reproduction, internal fertilisation and gestation is a feature of human reproduction that presents unique challenges. For example, internal fertilisation and gestation provides some degree of sanctuary from environmental hazards to the developing offspring. It allows the mother to act as an environmental filter, screening potentially harmful agents and perhaps providing biochemical cues from the environment that can be deployed by the fetus to optimally adjust its physiology after it is born. Internal gestation does, however, introduce tensions between the mother and conceptus. With external gestation, such as in birds, females maintain the option of abandoning the offspring if conditions deteriorate or other factors make it necessary. Internal gestation makes abandonment less feasible, although evidence suggests that the high rate of fetal loss very early during pregnancy may reflect an analogous mode of disinvesting in offspring (Wilcox et al. 1988). Growth, maintenance, and survival are the sole challenges of the growing fetus. Its physiology therefore will be honed to negotiate the allocation of resources between these needs. Mothers however, as iteroparous organisms, must also negotiate the need to sequester and allocate resources towards potential future bouts of reproduction as well as the present gestating offspring. This negotiation can result in challenges.

Investment in present offspring often delays future bouts of reproduction in females. For example, depending on maternal energetic condition, the time delay until the next conception is largely a function of investment in the present offspring. An example of this in humans is in post-lactational amenorrhoea, although maternal condition can attenuate this response (Ellison and Valeggia 2003). Another example is the quantification of the amount of energy devoted to the conceptus, maternal fat deposition, and maternal maintenance. Looking at several different populations living in differing energetic conditions (Figure 15.1), mothers tend to invest a relatively stable amount of energy in each conceptus, with any surplus energy going to maternal fat stores and maintenance, presumably to support future bouts of reproduction (Prentice and Goldberg 2000).

During later stages of gestation, tensions between the mother and fetus resulting from divergent life history agendas can emerge, since mothers must often allocate resources towards future reproduction at the expense of the fetus. This tension can be expressed as various health disorders (Haig 1993). Two examples are gestational diabetes and pre-eclampsia.

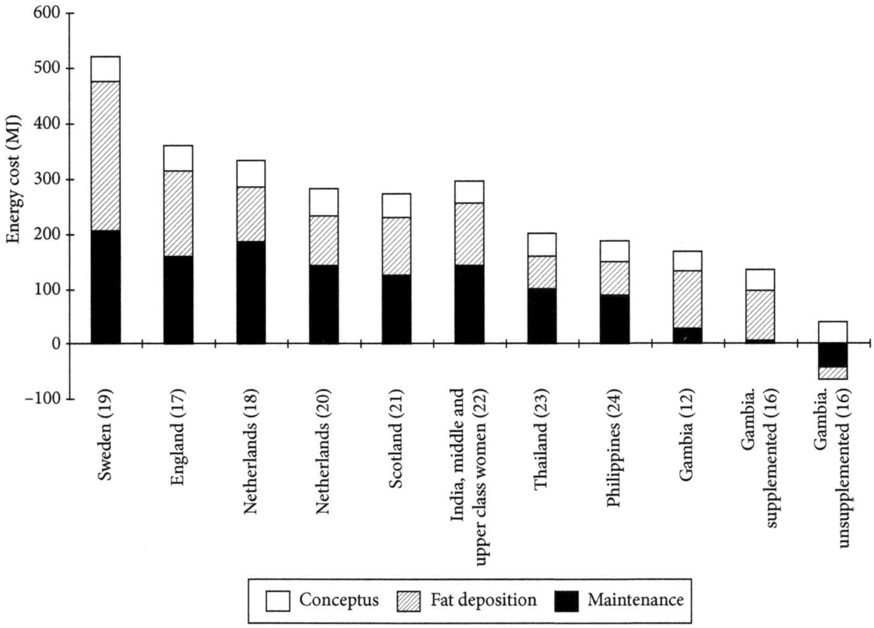

FIGURE 15.1 Energetic resource allotment in pregnancies in several populations living under differing ecological conditions. Although energetic allocation in the conceptus remains relatively constant regardless of ecological context, surplus energy is deployed towards greater maternal fat deposition and maintenance.

Source: Reproduced from *The American Journal of Clinical Nutrition*, 7 (5), Andrew M. Prentice and Gail R. Goldberg, Energy adaptations in human pregnancy limits and long-term consequences, pp. 1226s–1232s, Figure 2. Copyright © 2001, The American Journal of Clinical Nutrition.

Researchers have suggested that the emergence of gestational diabetes may be related to energy-sparing mechanisms that may have evolved in humans in response to unpredictable food availability. For example, three single nucleotide polymorphisms (SNPs) (rs3895874, rs3848460, and rs937301) at the 5′ region of the human glucose-dependent insulinotropic polypeptide (GIP) gene have been implicated in such an adaptive response. Pregnant women with gestational diabetes who are homozygous for the GIP(-1920A/A) genotype display lower levels of GIP than those carrying an ancestral GIP(-1920G) haplotype (Chang et al. 2011). These health issues are particular to women, although men have other challenges.

Male reproduction has been shaped by the evolution of spermatogenesis and internal fertilisation. Males invest in disposable gametes that are produced on a continuous basis. Unlike other seasonally breeding mammals, men produce sperm continually without any significant seasonal changes. This continuous production is the source of significant genetic mutation opportunities that are major contributors to genetic variation which are central to evolution by natural selection. But significant mutation rates also require extensive gene repair mechanisms (Crow 2000; Wang et al. 2012).

Continuous spermatogenesis and the relative buffering of sperm production from energetic stresses provide significant leeway for males to engage in various reproductive strategies that are not constrained by gamete availability or significant direct metabolic investment in offspring (i.e. gestation and lactation). Instead, similar to other mammals, human males are constrained by mating opportunities. This presents challenges that are unique compared to females. Most apparent is that mating opportunities are a much greater constraint on male fitness compared to females. Hypothetically, a male who mates with a hundred different females could potentially father a hundred offspring. A female who mates with a hundred different males will not have a hundred offspring. While increases in the number of different mates can increase female fitness in several ways (Scelza 2013), increased offspring output is not one of them. Nonetheless, it is important to note that male reproduction evolved within the context of the evolution of female reproductive strategies, resulting in selection for a range of behaviours that make human males unique compared to other primates or mammals. (For further discussion, see Chapter 16: Sexuality, Reproduction, and Birth.)

Male reproductive behaviour exhibits the broadest range of variability among all of the great apes. Extrapolating on the common reproductive and social correlates with sexual dimorphism as well as differences in extrinsic mortality, ancestral hominids likely engaged in significant male/male competition that resulted in a broad range of fitness variability between males. Human males can be infanticidal or exhibit diverse investment behaviours that reflect casual interest in offspring to extensive paternal care and investment (Gray and Anderson 2010; Bribiescas et al. 2012). Only a handful of mammals, including some primates, exhibit any form of paternal care. Given the significant evolutionary distance between *Homo* and *Aotus*, a New World primate that commonly exhibits paternal care, it is safe to assume that paternal investment has the capacity to evolve independently on multiple occasions (Fernandez-Duque et al. 2009).

In both males and females, the evolution of reproduction involves selection for genes and physiological mechanisms that regulate investment in reproduction and other life history needs. Among the most important physiological mechanism is endocrinological regulation of reproduction. Understanding the evolution of this system is vital for our present discussion.

15.2.1 Evolutionary Endocrinology

Hormones are evolutionarily ancient. They govern or influence virtually every aspect of reproduction, growth, maintenance, and metabolic regulation in virtually all vertebrates and most invertebrates. The definition of hormones in general is a biochemical agent that is produced and secreted by a gland in a tonic or pulsatile fashion. After release into the bloodstream, cerebral spinal fluid, or intercellular space, hormones bind and activate hormone-specific protein receptors on distant tissues. Some cross-talk or non-specific activation does occur by similar hormones or chemicals that have similar structures, as in the case of many endocrine disruptors such as bisphenol-A. These 'endocrine' functions act in tandem with 'paracrine' function in which hormones bind and activate receptors on tissues that neighbour the site of production. 'Autocrine' function is the binding and stimulation of receptors on/in the same production tissue.

Hormones can be divided into two basic biochemical groups—steroids and proteins—with others such as thyroid hormones being derived from the amino acid tyrosine. Steroids are produced from cholesterol that is either synthesised *de novo* in various glands or occasionally garnered from dietary sources. Steroids are small and lipid soluble with molecular weights under 300 g/mole. They can therefore pass through most cell membranes quite easily. To control the activity of steroids in circulation, they are bound to carrier proteins such as sex hormone binding globulins (SHBG) that inhibit steroid receptor binding and carry the protein/steroid complex throughout the circulatory system often to specific receptors on

Table 15.1 Reproductive Hormones and Carrier Proteins

Name	Effect	Steroid/ protein	Source
Oestradiol	Promotes endometrial growth within uterus; promotes secondary sexual characteristics during puberty	Steroid	Ovary secondarily from adipose tissue; minor production in testes
Dihydrotestosterone (DHT)	Promotes external development of male genitals; necessary for optimal functioning of prostate; converted from testosterone by the enzyme 5 alpha reductase	Steroid	Testes; prostate; various peripheral tissues
Testosterone	Promotes anabolic processes; supports spermatogenesis, secondary sexual characteristics, and libido	Steroid	Testes; minor amounts in ovaries
Progesterone	Supports pregnancy	Steroid	Corpus luteum during luteal phase of menstrual cycle; placenta during gestation; small amounts in adrenal gland
Gonadotropin releasing hormone (GnRH)	Regulates and supports production of gonadotropins (FSH and LH) by pituitary gland	Protein	Hypothalamus
Kisspeptin	Regulates and supports GnRH production	Protein	Hypothalamus; preoptic area (POA); infundibular nucleus
Follicle stimulating hormone (FSH)	Promotes development of ova from follicles as well as spermatogenesis	Protein	Pituitary gland
Luteinising hormone (LH)	Promotes production of sex steroids in both males and females: oestradiol in females, testosterone in males	Protein	Pituitary gland
Sex hormone binding globulin	Serves as carrier binding protein for sex steroids such as testosterone and oestradiol	Protein	Liver

distant tissues. Binding globulins such as SHBG have high specificity and affinity to specific steroids but low capacity. That is, binding globulins create strong molecular bonds to specific steroids but have a limited capacity to carry many steroids. Albumin, on the other hand, can also act as a carrier protein but has much less binding specificity and affinity compared to SHBG, although it can have greater capacity. Once steroids are released from their carrier protein, they are more capable of binding to a specific target receptor (Table 15.1).

Protein hormones such as gonadotropin releasing hormone (GnRH), follicle stimulating hormone (FSH), and luteinising hormone (LH) are much larger, water-soluble molecules, making them less mobile across cell membranes. FSH and LH, two primary reproductive hormones in males and females, consist of one shared alpha dimer and a hormone-specific beta dimer which defines their molecular uniqueness. Interestingly, there is considerable molecular variability in gonadotropins between individuals (Wide 1987; Ulloa-Aguirre et al. 1988). This variation was so significant that until the advent of recombinant DNA synthesis, gonadotropins were measured in international units (IU) which were agreed upon by consensus by early endocrinologists. Some have speculated about the potential evolutionary significance of this variation, but clear evidence is lacking to support any hypothesis of adaptation (Casarini et al. 2015).

Steroid hormones are produced in all eukaryotes in some form. Oestrogens, which include oestradiol, are produced in plants, although this may be an example of convergent evolution. Steroid hormones such as testosterone, oestradiol, and cortisol have identical molecular structures among mammals, primates, and most vertebrates, making comparative studies feasible. Indeed, molecular evidence for oestrogen receptors (OER) include distantly related organisms that include deuterostomes such as sea cucumbers (*Actinopyga echinites*) (Baker 2011). The evolutionary history of GnRH and gonadotropins (LH and FSH) extends back to the earliest vertebrates (Roch et al. 2011). Within vertebrates, there are basically three types of GnRH along with associated receptor variants. Humans and other mammals exhibit type 1 GnRH peptide, while two other variants, type 2 and type 3, are found scattered across various vertebrate taxa, sometimes in addition to type 1. Interestingly, however, this type is also shared with other organisms of differing evolutionary distance, such as the skate and bony fish, including the coelacanth (Roch et al. 2014) (Figure 15.2).

Awareness of these evolutionary relationships in reproductive hormones is useful in that it contributes to our understanding of the evolutionary biology of human reproduction and the origins from which our life history traits evolved. Of particular importance are the regulatory hormones, which are central to our biology, extremely conservative, and exhibit very little variation in biochemical structure across even the most distantly related taxa. This suggests that reproductive endocrinology has been under significant selection and/or that some other suite of physiological constraints disallows the emergence of alternative, perhaps improved systems. One likely constraint is the finely tuned nature of receptors. The difference, for example, between FSH and LH is one dimer. The difference between oestradiol and testosterone, two steroid hormones with contrasting sexual effects, is the modest conversion of the 'A' carbon ring into an aromatic state. The conservative evolutionary biology of reproductive endocrinology is also shared with other hormones such as glucocorticoids and insulin (Baker et al. 2013; Lau and Chalasani 2014). This suggests the importance of these agents extending much further back than the emergence of hominids or primates in general.

Given the molecular conservativism of reproductive hormones, regulatory actions are then reserved for variation in the manner in which receptors are stimulated. This includes variation in levels of a hormone in circulation, the pattern of secretion (i.e. tonic compared to pulsatile), changes in receptor number, density, or sensitivity, as well as changes in carrier proteins. The question then becomes, what environmental cues stimulate such changes in hormone regulatory function? This includes interactions between hormones but also exogenous cues such as changes in circulating glucose due to caloric ingestion, activity,

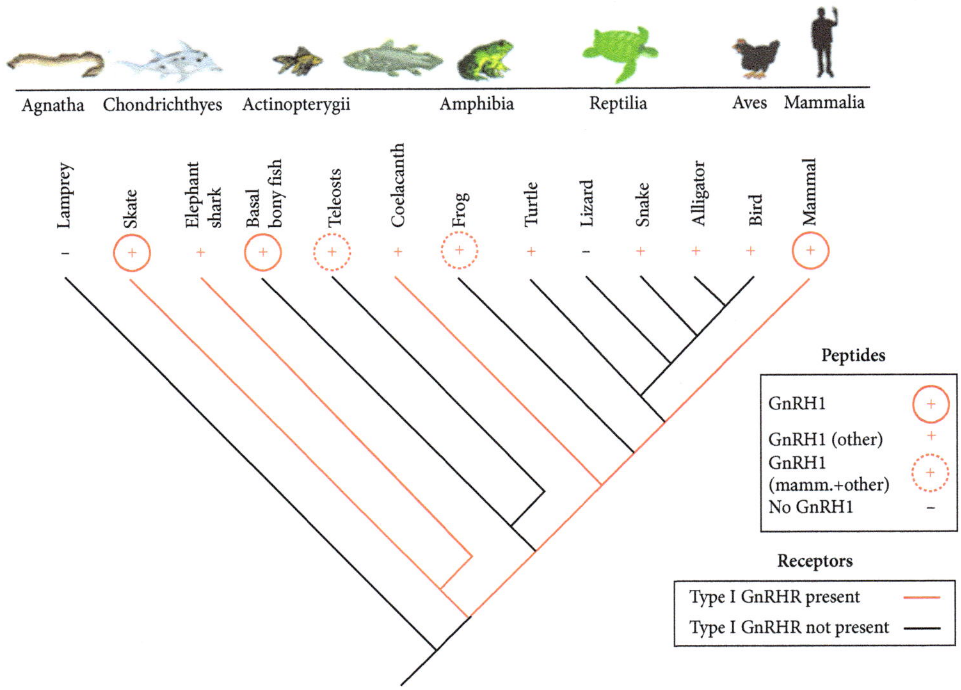

FIGURE 15.2 Evolutionary cladogram for type 1 GnRH receptors and GnRH1 peptides in vertebrates.
Source: Data from *General and Comparative Endocrinology*, 209 (1), Dan Larhammar, Ellen R. Busby, and Nancy M. Sherwood, GnRH receptors and peptides: Skating backward, pp. 118–34, doi.org/10.1016/j.ygcen.2014.07.025, 2009.

sensory cues, or social interactions. To glean a more clear understanding of these mechanisms we move to the basic physiology and mechanisms of hormone function.

15.3 Functions and Mechanisms of the Endocrine System

As with most other aspects of human biology, endocrine function is governed by genes and environment, mostly expressed as gene–environment interactions. While many aspects of endocrine function are coded within the genome, their expression largely depends on signals and cues from the surrounding environment, loosely defined as the physical and social space inhabited by the organism. This is to be expected in any organism that lives in an environment where stochasticity is the norm. Plasticity is therefore important for the timing and investment of various life history needs such as reproduction and growth. While plasticity has clear advantages, it is not without cost. Misread environmental cues and physiological mechanisms that require some degree of instability to be nimble and responsive can result in heavy costs to the organism if those cues are misread (DeWitt et al. 1998).

These miscues and errors in plastic responses are potential costs that can lead to health challenges that are common in many organisms, including humans.

Reproductive endocrinology is of particular importance in describing the physiology of reproduction. The reason is that hormones are in many ways the liaisons between environmental cues and genetic expression. Hormones are also vital agents in monitoring and managing the various physiological processes that govern trade-offs and interactions between the life history demands of growth, maintenance, and reproduction (Finch and Rose 1995).

15.3.1 Reproductive Endocrinology

The human reproductive endocrine system is mostly regulated by a negative feedback loop in which gonadal steroids induce a suppressive effect on the production of GnRH. While many other hormones impact the hypothalamic–pituitary–ovarian/testes (HPO/T) axis, such as those produced in the adrenal and thyroid glands, this discussion will be limited to endocrine factors that are historically and traditionally associated with reproductive function. For a more comprehensive review of human endocrinology the reader is invited to refer to any number of medical textbooks. Given the complex relationships between the hypothalamus, pituitary, and gonads, it is difficult to ascertain where the reproductive endocrine system 'starts'. For the sake of simplicity and convention, we can begin in the brain.

Specific GnRH-producing neurons within the hypothalamus synthesise and release GnRH in a pulsatile fashion. It merits stating that information on GnRH is extremely challenging, given its deep location of production within the brain and the location of target receptors in the pituitary. Much of what is known comes from animal models and natural human 'knockout' conditions (Bianco and Kaiser 2009). While the levels of GnRH are obviously important for basic function, the secretion rate and amplitude of GnRH pulsatility is also important in the HPO/T axis. In essence, the rate and amplitude of GnRH pulsatility appears to be non-random, suggesting the pattern of release may affect target receptors independently of the amount of hormone released into circulation (Pratap et al. 2016). Indeed, GnRH-secreting neurons appear to be inherently primed to release GnRH in a pulsatile fashion even in the absence of neighbouring neuron stimulation (Funabashi et al. 2000). GnRH has a very short half-life of only a few minutes; however, its sole mode of translocation is through the hypophyseal port which connects the hypothalamus to the pituitary gland where it stimulates the production and release of FSH and LH. Since GnRH is released in a pulsatile manner, all downstream hormones that are stimulated by its release are also pulsatile. These include FSH, LH, and sex steroids such as oestradiol and testosterone.

Gonadotropin releasing hormone receptor (GNRHR) is encoded primarily by two genes, GNRHR and GNRHR2. The receptors are located primarily in the pituitary. Defects in the GNRHR are associated with idiopathic hypogonadotropic hypogonadism (IHH), sometimes referred to as Kallman's disease, in which patients fail to undergo puberty and exhibit infertility due to the lack of reproductive hormone production. Variants of this gene appear to be rare, likely due to the acute role it plays in the onset and maintenance of reproductive function (Bianco and Kaiser 2009).

Regulation of GnRH production is left to the recently discovered peptide kisspeptin. Given the lack of clarity on how sex steroids enacted their negative feedback effects on GnRH production, the discovery of kisspeptin and its production neurons within the preoptic area (POA) and infundibular nucleus provided an upstream mechanism for GnRH regulation. Of central importance is the presence of oestradiol and progesterone receptors on the kisspeptin/neurokinin B/dynorphin (KNDy) neuron which plays a role in regulating the production of kisspeptin, its downstream effects on GnRH, and producing neurons that display kisspeptin (Kiss1) receptors. As with GnRH, this appears to be an evolutionarily conservative neuroendocrine trait, since rodents express a very similar regulatory signalling structure, with one of the few contrasts compared to humans being that sex steroids incur a positive feedback effect on kisspeptin neurons within the anteroventral periventricular nucleus (AVPV), the analogous location in rodents being the POA (Skorupskaite et al. 2014) (Figure 15.3).

Stimulated by the production of GnRH and release into the hypophyseal port leading to the pituitary, FSH, along with the gonadal hormone activin, stimulates the development of gametes—ova in females, sperm in males—while LH is responsible for sex steroid production—oestradiol in females and testosterone in males. As with GnRH, gonadotropin disorders are rare, but can result in primary or secondary infertility or subfertility in both men and women. Downstream in the gonads, theca cells are responsible for producing oestradiol within the ovaries, while testosterone is produced in the testes by Leydig cells which sit within the interstitial space between the seminiferous tubules (Figure 15.4). FSH stimulates Sertoli cells to produce sperm, with inhibitory action being induced by inhibin B.

The LH receptor gene (LHR) is located on chromosome 20 and is similar to the FSH receptor gene. These receptors are expressed primarily in the testes and ovaries as part of the negative feedback loop to support the production of sex steroids. Human variation in these genes is rare, since most mutations are deleterious to reproductive function (Desai et al. 2013).

The production of sex steroids involves a series of enzymatic conversions that begin with cholesterol. These enzymes determine if a precursor will ultimately become testosterone, oestradiol, cortisol, or other related steroid hormones. Variation in the genetic coding of the proteins that result in these enzymes can result in changes in steroid hormone production or organisational effects that can alter the sexual phenotype of an individual (Figure 15.5).

The receptor genes for testosterone, oestradiol, and progesterone are androgen receptor (AR) encoded by the NR3C4 gene, OER encoded by the oestrogen receptor 1 (ERS1) gene, and progesterone receptor (PGR) gene. Evidence suggests some population variation in these genes, although the functional significance is unclear. In the case of AR, variants are associated with early-onset alopecia (Hillmer et al. 2005). Ethnic and population variation has been reported in the AR gene; however, the functional or clinical significance remains unclear (Ackerman et al. 2012). Such variation may contribute to population variation in testosterone levels or even differences in prostate cancer risk, but further long-term research is needed.

Variability in the OER gene is also evident, although again the functional and clinical significance continues to be a focus of research (Fjeldheim et al. 2016). OER variants are associated with differences in risk of stroke, Alzheimer's disease, and breast cancer (den Heijer et al. 2004; Shearman et al. 2005). Population variation in the OER gene is also evident. Variants of the CYP17 gene that codes for the enzyme aromatase with converts testosterone

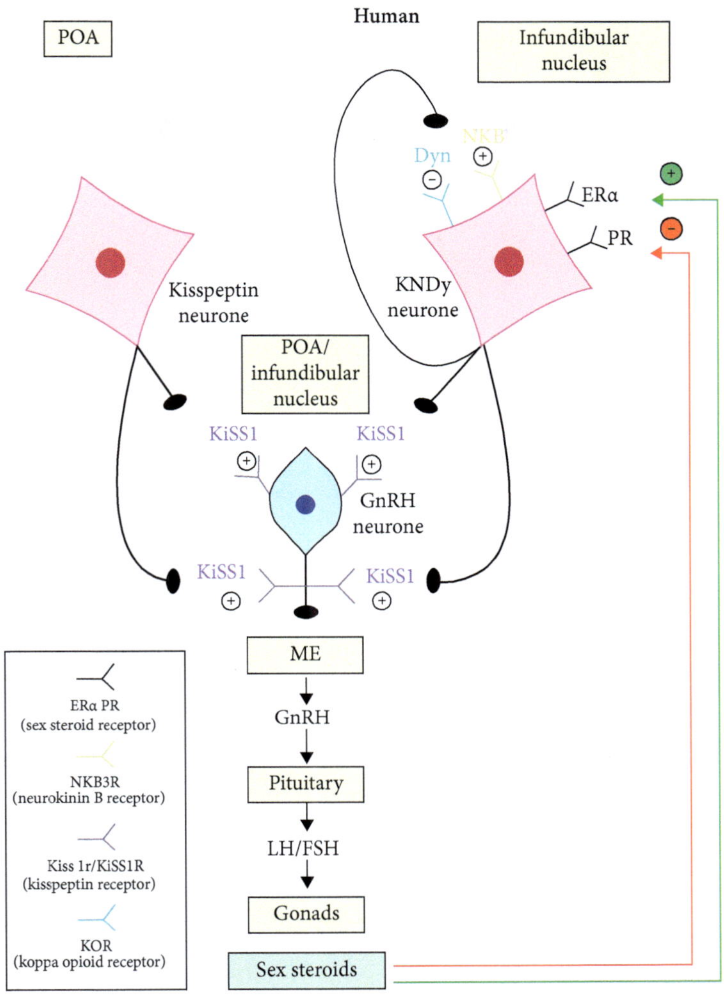

FIGURE 15.3 Neuroanatomic regulation of the HPO/T axis by kisspeptin and associated receptors. POA, Preoptic nucleus; ME, median eminence; KNDy, dynorphin and neurokinin B secreting neuron; Dyn, dynorphin; NKB, neurokinin; ER alpha, oestrogen receptor alpha; PR, progesterone receptor; KiSSI, kisspeptin).

Source: Reproduced from Karolina Skorupskaite, Jyothis T. George, and Richard A. Anderson, The kisspeptin-GnRH pathway in human reproductive health and disease, *Human Reproduction Update*, 20 (4), pp. 485–500, Figure 1, doi.org/10.1093/humupd/dmu009.

to oestradiol have also been reported to be associated with differences in oestradiol levels (Jasienska et al. 2006a). Similarly, variants of the PGR receptor gene for progesterone have been linked to differences in ovarian cancer (Abbasi et al. 2012).

Beyond the biochemical aspects of hormone production and regulation, the question of functional significance is clearly important. Variation in reproductive endocrinology is vital to our understanding of how it affects human fertility. In addition to reproductive

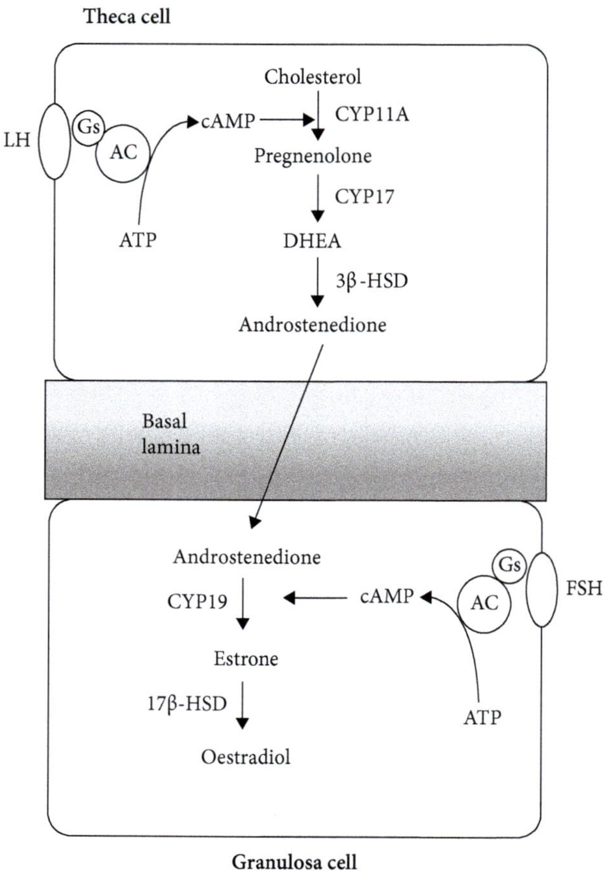

FIGURE 15.4 Metabolic pathway of oestradiol synthesis within the ovary. Cholesterol is converted to androstenedione (an androgen) and passed through basal lamina to adjacent granulosa cells where final conversion is made via aromatase (CYP19) and 17-hydroxysteroid dehydrogenase (17β-HSD) into oestradiol which then supports follicle development into mature ova.

Source: Reprinted from *The International Journal of Biochemistry & Cell Biology*, 37 (7), Denis A. Magoffin, Ovarian theca cell, pp. 1344–49, doi.org/10.1016/j.biocel.2005.01.016. Copyright © 2005 Elsevier Ltd.

endocrinology, other aspects of reproductive biology contribute to lifetime fertility. For this we need to turn to the 'interbirth interval' (IBI) and its sources of variation. The areas of reproductive biology that will be covered include those that are primary contributors to IBI. Since gestation periods vary modestly within humans, the time between births is the primary source of variation in female lifetime fertility. It is therefore important to understand what physiological factors contribute to this variation. Here we will focus on ovarian function, lactation and lactational amenorrhoea, fetal loss, coital frequency, and male factors (Wood 1994; Ellison 2001).

15.3.2 Organisational Aspects of Reproductive Organs

Before turning to factors that contribute to IBI, it is important to understand how human reproductive physiology originates and is organised. Humans follow the basic pattern of

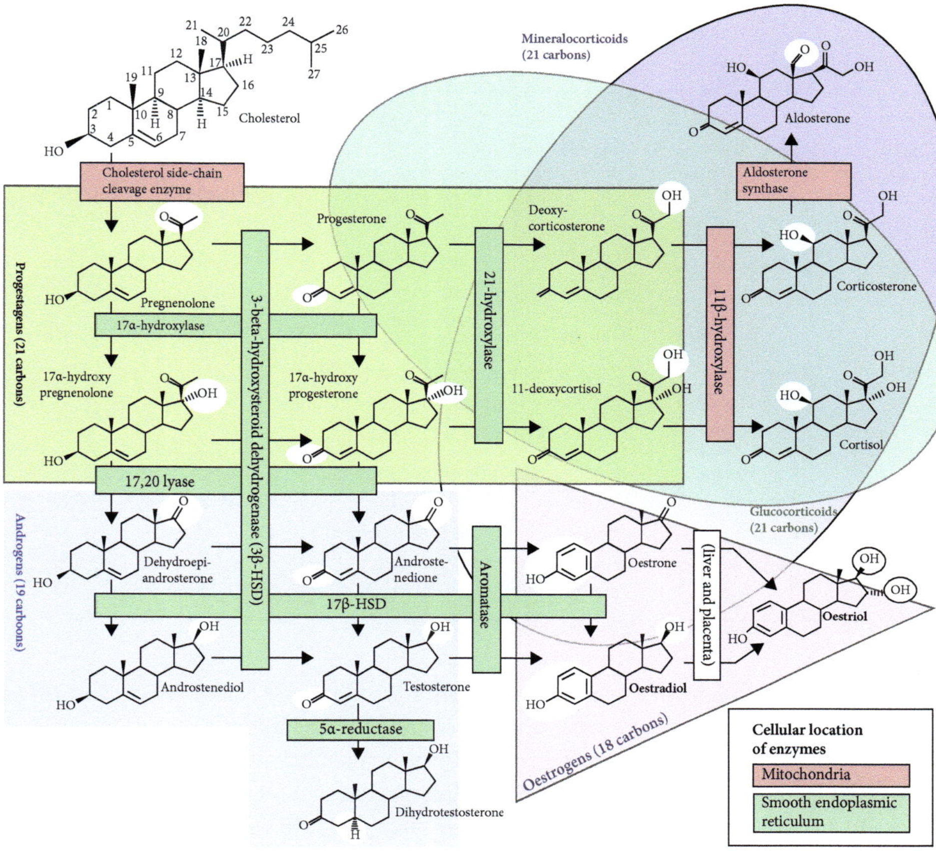

FIGURE 15.5 Synthesis and conversion pathways for steroidogenesis.

reproductive organisation *in utero* as other mammals. Sex differentiation emerges from the union of an ovum with an X- or Y-bearing sperm. If the conceptus is XX the individual is female, and if XY it is male. All humans start with largely undifferentiated internal and external sexual organs. During the first trimester, genetic signals trigger the production of various hormones, such as testes determining factor (TDF), dihydrotestosterone (DHT), and anti-Müllerian hormone (AMH). In males, TDF supports the development of the testes and the subsequent production of AMH, which results in the diminishment of the Müllerian ducts and allows for the growth and development of the Wolffian ducts which go on to form the internal male genital structures. In females, the lack of AMH allows the emergence of the Müllerian ducts which go on to form the internal female reproductive tract. Externally, DHT promotes the development of the penis and fusion of the labial folds. Without DHT, the vaginal opening remains unfused and the clitoris remains small, resulting in a female phenotype. Disturbances in sexual organisation can result in undifferentiated internal and external genital phenotypes. These include the conditions described in Table 15.2.

Organisational aspects of sexual differentiation obviously strongly influence reproductive biology, but they do not completely define an individual's sexual orientation or gender identification. In this manner, humans are among the most malleable and varied organisms

Table 15.2 Major Disturbances in Reproductive Development Arising from Sex Chromosome Disorders

Sexual karyotype	Condition	Symptoms	References
XXY	Klinefelter's syndrome	Underdeveloped genitalia	(Bonomi et al. 2017)
XO	Turner's syndrome	Small stature, cognitive deficits, neck webbing	(Kingery and Wintergerst 2015)
XYY	XYY syndrome	Tall stature	(Bardsley et al. ; 2013)
XY	Androgen insensitivity syndrome	Androgen receptors non-functional. Male karyotype with female somatic and external genitalia phenotype	(Mongan et al. 2015)
XX	Congenital adrenal hyperplasia	Hyperproduction of androgens by adrenal gland often due to non-functional or absent enzymes resulting in masculinisation of external genitalia	(Bulsari and Falhammar 2017)

in regards to sexual identity, preference, and mating strategy (Gray and Garcia 2013). That said, an important aspect of organisational development is the initial maturation of female gametes. During the second trimester, germ cells within the fetal ovary undergo preliminary maturation with meiotic division before undergoing biochemical stasis, until they are subject to cohort recruitment with the onset of menarche. In males, a small amount of spermatogenesis occurs, but there is general quiescence in male reproductive function *in utero* and during infancy.

In both males and females, development and growth of the adrenal gland plays an important role in the onset of puberty. Adrenarche, as it is known, is characterised by the increase in androgen production, in particular dehydroepiandrosterone (DHEA) and dehydroepiandrosterone sulfate (DHEA-S), by the adrenal cortex. This increase in adrenal androgens appears to play a role in the desensitisation of the hypothalamus to circulating sex steroids, as well as initiating the development of secondary sexual characteristics such as pubic and axillary hair (Campbell 2006). Interestingly, adrenarche appears to be present only in humans, other great apes, and perhaps a few other primates, and absent in other mammals; however, very little comparative information on DHEA and DHEA-S levels is available from other mammals (Conley et al. 2012). Adrenarche has been suggested to be evident in bonobos (*Pan paniscus*) and chimpanzees (*Pan troglodytes* sp.) (Behringer et al. 2012; Parker et al. 2014). (For further discussion, see Chapter 4: Growth and Development.)

15.3.3 Reproductive Maturation—Puberty

The transition from non-reproductive child to an adult that is capable of reproducing is among the most important life history transitions in any organism. In essence, puberty involves the reallocation of energy from somatic growth to reproductive effort, at least in organisms with determinant growth like humans. Skeletal growth accelerates with the onset of puberty and declines rapidly upon cessation of the pubertal transition in both males and females; the male pubertal growth spurt tends to occur a year or two later compared to females and is of greater amplitude, which contributes to body size sexual dimorphism in humans (Iuliano-Burns et al. 2009).

Disorders related to sexual maturation include precocious puberty, a condition in which sexual maturation occurs prematurely during childhood. Causes attributed to peripheral precocious puberty include endocrine tumours, brain injury, and genetic disorders (De Sanctis et al. 2015; Luo et al. 2015). Central precocious puberty (CPP) originates within the HPO/T axis with involvement of increased kisspeptin production (Latronico et al. 2016). Although there are numerous disorders associated with sexual maturation, precocious puberty is particularly salient to the evolutionary biology of sexual maturation since it is inherently related to the timing and onset of this life history transition. The hypothalamus becomes prematurely less sensitive to circulating sex steroids and allows for the awakening of the gonads and increased levels of sex steroids in circulation. In severe cases, boys and girls begin to develop secondary sexual characteristics such as axillary and pubic hair, as well as development of breasts in girls and penile growth in boys.

Aside from the obvious ethical, social, and health issues that arise with premature sexual maturation in young boys and girls, precocious puberty provides insights into the importance of skeletal maturation and growth in human reproduction. This is particularly evident in girls, since pregnancy at a young age results in extreme challenges during childbirth, specifically the compromised ability to pass a fetus through the birth canal due to the underdevelopment of the pelvic skeletal structures. In such cases, caesarean births are the only option short of termination of the pregnancy. Interestingly, despite a greater incidence of menstrual irregularities and hyperandrogenism in CPP women, there appears to be no difference in pregnancy outcomes between treated and untreated CPP in adult females, suggesting that the difficulty of pregnancy of CPP females is centred upon the structural challenges of giving birth with an underdeveloped pelvic morphology (Lazar et al. 2014). CPP presents similar social and growth challenges in boys. Because of the increase in testosterone at a young age, skeletal growth is often stunted in untreated boys due to premature closing of the growth plates resulting from high testosterone levels (Carel et al. 2004).

Outside of pathological conditions such as CPP, males and females experience similar challenges regarding reproductive maturation. (For further discussion, see Chapter 4: Growth and Development.) However, because of the different costs of reproduction, the selection forces on each sex will and do vary. What is clear is that while genetics is important to the expression of the timing of pubertal maturation, gene–environment interactions account for much of the variation in the age of reproductive maturation. Since humans and other mammals live in stochastic environments, it is reasonable to assume that physiological mechanisms that are best able to adjust the timing of reproductive maturation in the face of energetic and social challenges would be conferred a selective advantage.

15.3.4 Female Reproductive Maturation

The period between infancy and puberty is marked by slow but steady somatic growth accompanied by HPO quiescence. Depending on the population and environmental conditions, and in the absence of any acute disorders, pubertal development begins between the ages of around 10 and 17 (James-Todd et al. 2010). The onset of reproductive maturation in females is most prominently reflected by first menstruation or menarche. Other secondary sexual characteristics such as the growth of pubic and axillary hair, breasts, pelvic bones, and adipose tissue in the buttocks and thighs are a signal of the onset of reproductive maturation and increases in oestrogens, particularly oestradiol. The trigger for menarche appears to involve the desensitisation of the HPO axis; however, there is a strong genetic component to age at menarche, which may also inform our understanding of downstream health challenges. A recent investigation of over 180,000 European females uncovered evidence for 123 signals at 106 genomic loci associated with age at menarche. Many of the loci were linked to other pubertal traits in both males and females. Some genes were also associated with body mass index and an assortment of diseases (Perry et al. 2014).

Prior to menarche, the hypothalamus, in particular kisspeptin-secreting neurons, are extremely sensitive to circulating sex steroid levels and inhibited by hormones such as oestradiol. For reasons that are somewhat unclear, during the onset of puberty, the hypothalamic sensitivity to sex steroid declines, making the hypothalamus more tolerant of circulating oestradiol levels, which allows the accumulation of oestradiol and the stimulation of target receptors throughout the body. There is also a strong heritability component in which maternal age of menarche accounts for a significant amount of variation in age of menarche in daughters (Towne et al. 2005) (Figure 15.6).

Fertility during adolescence differs from adults in that ovarian function is characterised by lower oestrogen levels and follicles that exhibit a lesser degree of functionality and perhaps smaller size (Anderson et al. 2014). Challenges during adolescent pregnancies, after controlling for prenatal care, smoking, and other adverse behaviours, are also more frequent compared to fully mature adult women (Fraser et al. 1995). In total, this is referred to as adolescent subfecundity (Ellison 1996). From an evolutionary perspective, it would be predicted that adolescent female physiology should negotiate between the emerging needs of reproduction and on-going skeletal growth. While it may be predicted that reproductive investment should not occur until investment in skeletal growth is completed, there is significant selection to commence reproduction sooner, given the limited time window of fertility prior to the onset of menopause. In essence, some degree of fertility is more advantageous than no fertility at all.

There is a tremendous amount of variation in age of menarche between populations. Indeed, age of menarche has decreased significantly over the last century in the United States and Europe in association with greater access to healthcare, the rise of industrialisation, and better nutrition. This secular trend, as it is called, has been the subject of speculation as to the triggers behind the onset of menarche. For some time, the primary driver was believed to be body weight and fat mass, since caloric availability and adiposity are also associated with variation in ovarian function in adults (Frisch and Revelle 1971; Frisch 1974). However, associations between age of menarche, body mass, and adiposity are very poor. Leptin, a hormone that is secreted by adipose tissue and serves as a signal of fat availability, is only marginally correlated with age of menarche (Matkovic et al. 1997).

Given that fetus heads are quite large compared to other great apes, it follows that menarche should not occur until pelvic growth has reached a point where it can accommodate the birth of a large-brained hominid. While fetal cranial size and female pelvic morphology have been suggested to result in an obstetric dilemma (Wittman and Wall 2007), new analysis

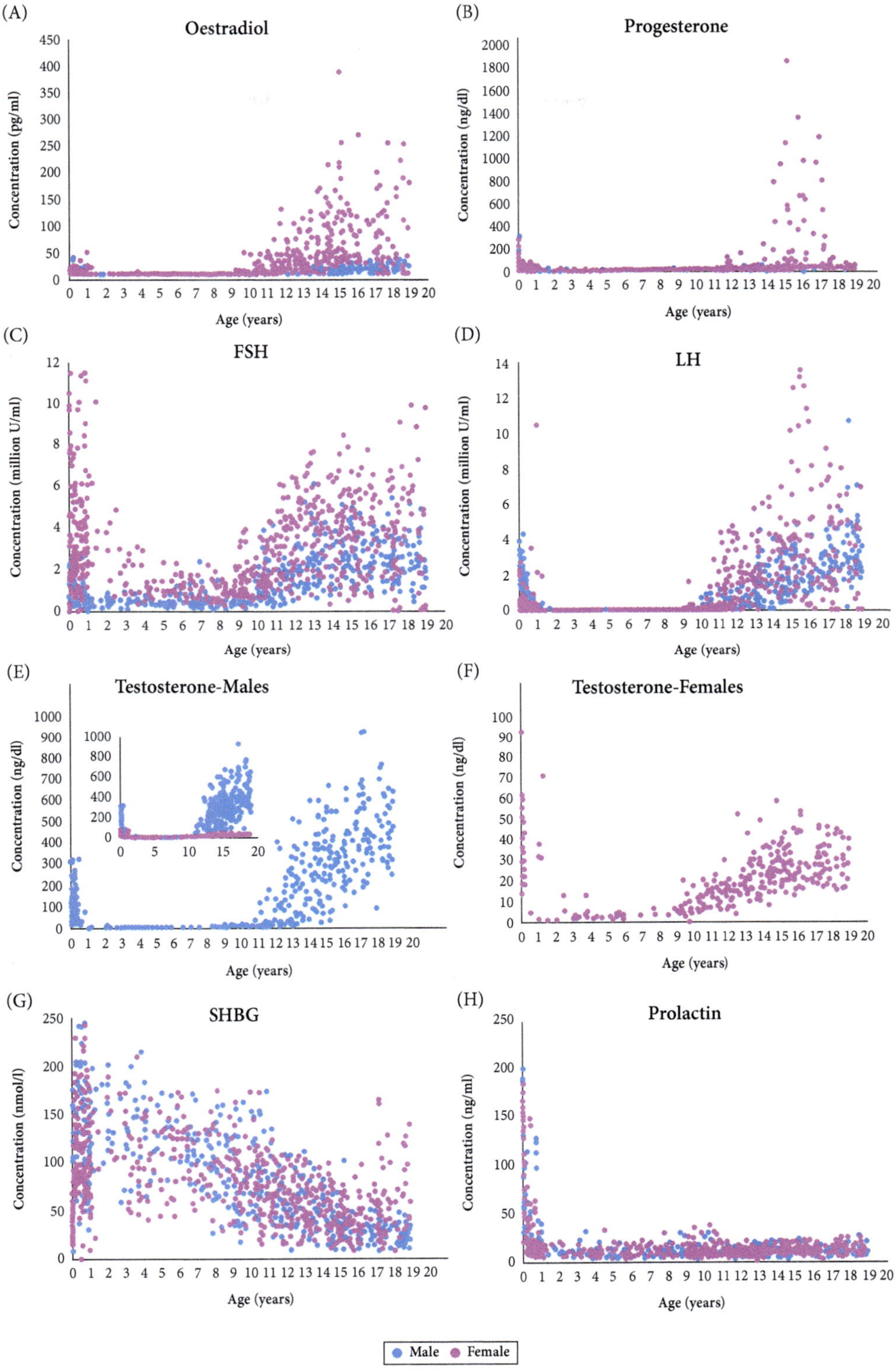

FIGURE 15.6 Reproductive hormone levels during infancy, childhood, adolescence, and early adulthood in healthy males and females.

Source: Reproduced from Danijela Konforte, Jennifer L. Shea, Lianna Kyriakopoulou, David Colantonio, Ashley H. Cohen, Julie Shaw, Dana Bailey, Man Khun Chan, David Armbruster, and Khosrow Adeli, Complex Biological Pattern of Fertility Hormones in Children and Adolescents: A Study of Healthy Children from the CALIPER Cohort and Establishment of Pediatric Reference Intervals, *Clinical Chemistry*, 56 (8), pp. 1215–1227, Figure 2, doi.10.1373/clinchem.2013.204123.

suggests that energetic status plays a more predominant role in parturition (Dunsworth et al. 2012). Indeed, research has shown that age of menarche is more closely associated with bi-iliac width than with adiposity or body mass (Ellison 1982). However, the question of how menarche is triggered by optimal bi-iliac width remains unanswered.

Over the past century there has been a steady decline in the age of menarche in many populations. Age of menarche has declined from about 16 to around 12 in many populations (Morris et al. 2011). In non-western populations undergoing transitions to more sedentary lifestyles, declines in age of menarche have also been observed within a few short years (Prentice et al. 2010). Age of menarche may also be influenced by social factors. That is, environmental stochasticity is a function of not only energetic availability but also social tensions. (For further discussion, see Chapter 4: Growth and Development.) Human reproduction is an incredibly social endeavour, with various potential sources of allocare, as well as conspecific threats. If the social environment is unstable, it has been argued that age of menarche should decrease, since this instability may be indicative of higher extrinsic mortality, for both the mother and the child, as well as for any potential alloparents such as fathers. This hypothesis is difficult to test since it is very challenging to parse out the effects of variation in energetics. Nonetheless, several studies have reported earlier age of menarche, after controlling for energetics, in association with social instability, such as father absence or tensions in the household, as well as predictions of one's own mortality (Amir et al. 2016).

Overall, female reproductive maturation is rooted in hormonal regulation that is common to most other mammals and primates. Energetic availability, social cues, and gene–environment interactions are central to the timing and evolution of this important life history transition. It is important to note that males experience different energetic constraints, but must also face the challenge of timing the onset of reproductive maturation in a manner that is optimal for lifetime reproductive success.

15.3.5 Male Reproductive Maturation

As with females, the period between infancy and puberty is marked by relative hormonal quiescence, with the exception of a neonatal testosterone surge that occurs within the first few days of birth (Corbier et al. 1992). The functional significance of this rise or its cause remain poorly understood, although it is hypothesised that this rise may be related to hypothalamic neurons, specifically in relation to the neuroendocrine agent kisspeptin (Clarkson and Herbison 2016). Similar to females, male reproductive maturation largely involves the desensitisation of the hypothalamus to circulating sex steroids—in this case, testosterone. As testosterone levels rise, secondary sexual characteristics develop, including axillary, pubic, and facial hair growth, as well as enlargement of the penis. Increases in testosterone also cause significant growth of sexually dimorphic skeletal muscle in the arms, chest, back, and legs. In conjunction with skeletal muscle growth, fat catabolism increases, resulting in declines in metabolic hormones such as leptin (Garcia-Mayor et al. 1997).

Very little is known about variation in male reproductive maturation in response to ecological challenges. A primary reason for this question is the lack of a discrete event, such as first menstruation in girls, that marks the beginning of the transition between childhood and reproductive maturity. The initiation and timing of a number of characteristics define the onset of reproductive maturation in males, including the onset of erections, nocturnal emissions, and the emergence of secondary sexual characteristics, all of which provide analogous evidence of the pubertal transition. Differences in gonadotropin secretion (FSH and LH) have been noted between poor and well-off Kenyan boys, indicating the influence of economic and nutritional status (Kulin et al. 1984). A survey of testosterone levels in 441 boys between the ages of 12 and 18 from urban Zimbabwe suggested that abdominal fat was a predictor of maturation (Campbell and Mbizo 2006), while a comparison of urban and rural Zambian boys showed differences in greater somatic growth in the form of weight and height in urban boys, along with greater testicular volume as puberty progressed, but these eventually evened out as adolescence waned (Figure 15.7).

It should be noted that pubertal development in males coincides with a sharp rise in mortality and morbidity as the result of changes in risk assessment. and tolerance Young males, apparently regardless of lifestyle, culture, or environment, exhibit this rise. The physiological underpinning of this behaviour is not well understood. However, given the propensity of similar behavioural changes in other primates and mammals in association with reproductive maturation and increases in testosterone, it is likely that similar processes are in effect in young males (Campbell et al. 2004). There is also evidence of a secular trend in young male mortality during the past century. Causes include the availability of mechanisation, war, and other sources of intense mortality (Mulye et al. 2009) (Figure 15.8). Secular trends in

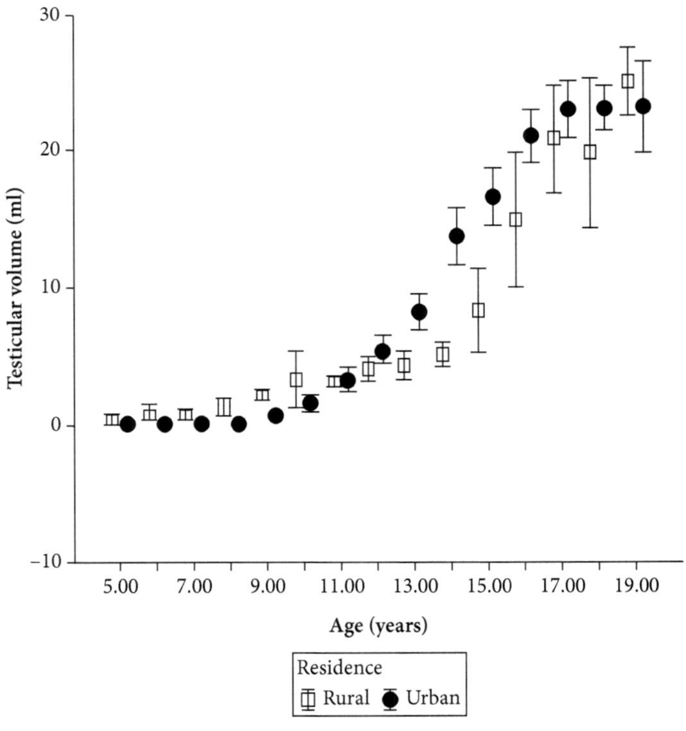

FIGURE 15.7 Differences in testicular volume between urban and rural Zambian boys.

Source: Reproduced from Timing of reproductive maturation in rural versus urban Tonga boys, Zambia, B. C. Campbell, R. Gillett-Netting, and M. Meloy, *Annals of Human Biology*, 31 (2), pp. 213–227, doi.org/10.1080/03014460310001656604. © 2004, Taylor & Francis Ltd (http//www.tandfonline.com).

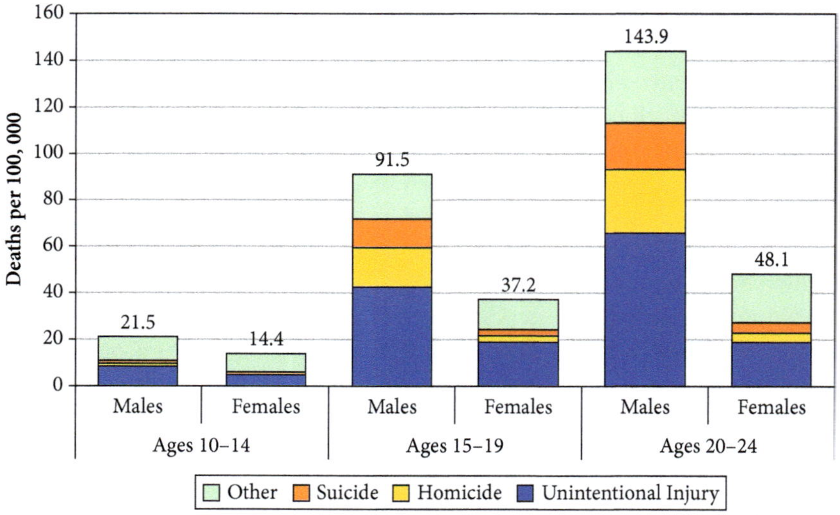

FIGURE 15.8 Sources of mortality in American males.

Source: Reprinted from *Journal of Adolescent Health*, 45 (1), Tina Paul Mulye, M. Jane Park, Chelsea D. Nelson, Sally H. Adams, Charles E. Irwin, and Claire D. Brindis, Trends in Adolescent and Young Adult Health in the United States, pp. 8–24, doi.org/10.1016/j.jadohealth.2009.03.013. Copyright © 2009 Society for Adolescent Medicine. Published by Elsevier Inc.

testosterone in response to better nutrition and environmental conditions in association with industrialisation have also been suggested (Goldstein 2011).

The spike in young adult male mortality is likely linked to changes in reproductive effort. Early in life, males are at their physical prime and are best positioned to leverage their physical fitness into greater competitive ability, attractiveness, and foraging efficiency. It is simplistic to assume that increases in reproductive hormones such as testosterone are the sole cause of this rise. However, it is likely that testosterone does affect the neuroendocrine processes that regulate environmental and social risk assessment (Braams et al. 2015).

15.3.6 Ovarian Function and Fertility

Female fertility is strongly affected by ovarian function and contributes significantly to IBI. This includes ovarian regulation of steroid synthesis, gametogenesis, and functional integration with other endocrine and immunological systems. In conjunction with the hypothalamus and pituitary, the ovaries produce sex hormones and peptide hormones, and regulate oogenesis. After the onset of menarche, the ovaries respond in a classic negative feedback system in coordination with the hypothalamus and pituitary gland. In basic terms, the hypothalamus produces GnRH, which flows in a pulsatile fashion into the hypophyseal port leading to the pituitary. This stimulates the production of gonadotropins by the pituitary; in particular, FSH and LH are synthesised. FSH is responsible for stimulating and supporting follicle development within the ovary and the ultimate maturation of ova. LH is primarily responsible for stimulating the production of oestradiol, which supports ova development and the development and thickening of the endometrium within the uterus.

The menstrual cycle is arguably the most visible aspect of female reproduction. In human females, menstrual cycles are commonly thought to be about 28 days long; however, considerable variation is evident between healthy females, which may contribute to fertility variation since menstrual cycle length has been reported to be inversely associated with probability of conception (Brodin et al. 2008). Menstrual cycles are roughly divided into the follicular and luteal phases, with ovulation occurring around mid-cycle. During the follicular phase, gradual increases in oestradiol promote the growth of the endometrial lining within the uterus. In conjunction with other ovarian hormones such as activin, oestradiol promotes the maturation of a cohort of follicles of which usually only a single dominant follicle matures into an ovum. Other follicles are suppressed by hormones such as inhibin and regress back into the ovary. At mid-cycle, a sudden shift in the negative feedback effect of oestradiol initiates a surge of LH, which triggers the follicle to rupture, thereby releasing ova during the process of ovulation. The remnants of the follicle, known as the corpus luteum, undergo further transformation with extensive lipid formation and the production of the hormone progesterone, which, as the name implies, is responsible for maintaining a conceptus and promoting the process needed to maintain a pregnancy. As the ovum is coaxed down the fallopian tube, it is available for fertilisation. If fertilisation occurs, the ovum will likely embed itself within the endometrial lining and release a hormone known as human chorionic gonadotropin (hCG) which will signal to the corpus luteum to continue producing progesterone in a process known as luteal rescue. If fertilisation does not occur, the corpus luteum eventually exhausts its ability to produce progesterone, the endometrial lining begins to slough off, and the process of bleeding and expulsion of tissue known as menstruation begins (Figure 15.9).

Energetic regulation of ovarian function occurs at both the hypothalamus and ovary. In rodent models, both insulin and leptin receptors are evident in pro-opiomelanocortin (POMC) neurons within the hypothalamus. Female animals lacking these receptors exhibit impaired fertility and reproductive function (Hill et al. 2010). Interestingly, evidence suggests that oestradiol receptors in the hypothalamus may themselves influence energy homeostasis (Xu et al. 2011). Other animal models demonstrate that insulin and circulating glucose levels are vital for optimal ovarian function (Dupont and Scaramuzzi 2016). Indeed, intravenous administration of simple glucose and subsequent increases in insulin resulted in acute increases in the total number of follicles in ewes (Scaramuzzi et al. 2015). Women with hyperinsulinaemia exhibit hyperandrogenism due to excessive stimulation of androgen production in response to insulin. This is usually accompanied by follicular hypertrophy, resulting in polycystic ovarian syndrome (PCOS).

15.3.7 Biodemography and Endocrinology of Female Reproduction

From a biodemographic perspective, ovarian function lies at the heart of fertility variation between populations (Wood 1994). While early anthropological and demographic theories focused on the importance of marital patterns, post-partum sex taboos, and other cultural forces, it is now clear that energetics is central to variation in ovarian function and fertility (Bongaarts and Potter 1983). Such variation has been broadly documented among populations (Henry 1961), with numerous studies providing compelling evidence that

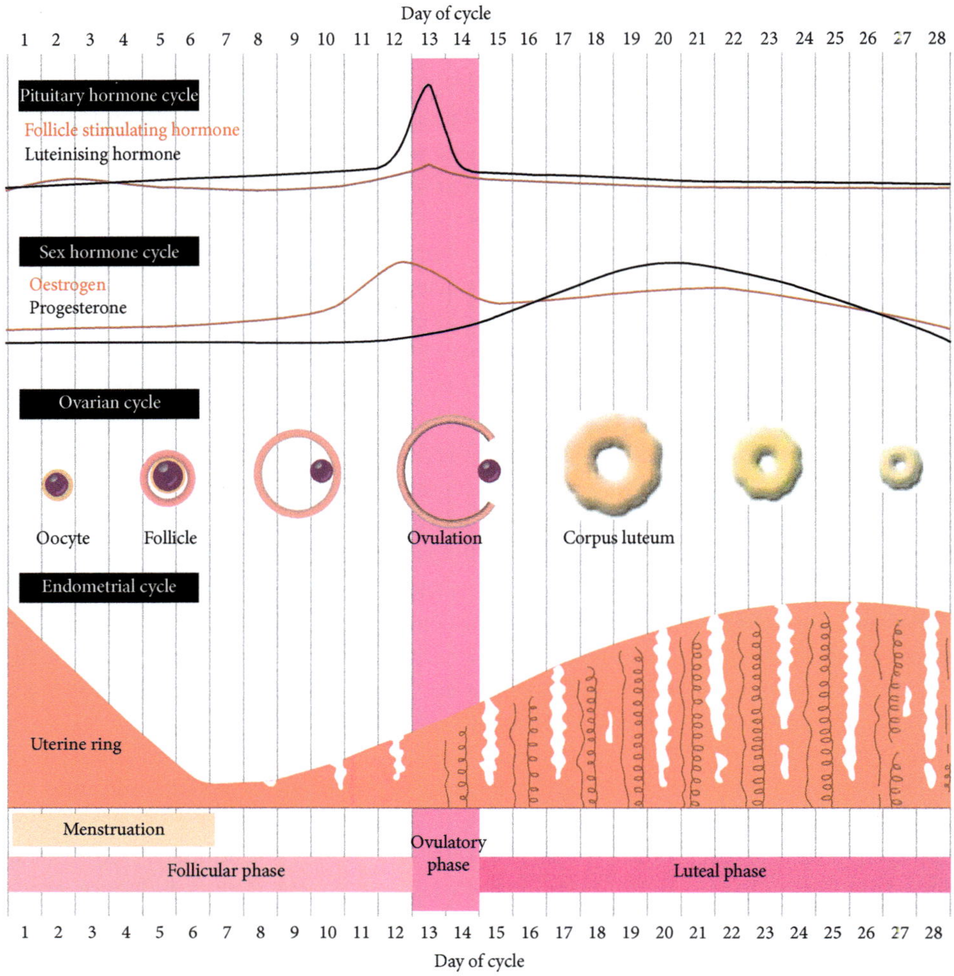

FIGURE 15.9 Summary of the human female menstrual cycle.

Source: Reproduced from R. John Aitken, Mark A. Baker, Gustavo F. Doncel, Martin M. Matzuk, Christine K. Mauck, and Michael J. K. Harper, *Journal of Clinical Investigation*, 118(4), pp. 1330–1343, Figure 1, doi.10.1172/JCI33873.

variation in ovarian function is primarily in response to somatic energy and environmental conditions. (For further discussion, see Chapter 4: Growth and Development.)

In contemporary industrial populations, caloric availability and energetic status is central to ovarian regulation. Other environmental challenges such as high-altitude hypoxia are evident, but energetics remains the most significant contributor to fecundity variability (Vitzthum 2013). Anorectic women consistently exhibit amenorrhoea or oligomenorrhoea and disturbances in leptin and oestrogens (Berner et al. 2017). However, ovarian sensitivity to energetic status is not restricted to extreme weight loss, caloric restriction, or the potential for selection of subjects with other health challenges that may confound results. Non-athletic women on voluntary modest caloric restriction regiments exhibit lower luteal progesterone levels and longer cycles compared to controls (Lager and Ellison 1990).

Women who exercise and burn a significant amount of calories exhibit similar patterns of ovarian down-regulation. Even if the caloric burn is moderate in response to running, progesterone levels tend to still be lower compared to sedentary women (Ellison and Lager 1986). A primary mechanism is likely the availability of circulating glucose and insulin, which stimulates hypothalamic production of GnRH but also ovarian oestradiol. These declines in ovarian steroids are associated with variation in conception (Lipson and Ellison 1996).

While clinical investigations provide valuable information on the basic physiology and responsiveness of ovarian function, studies consistently lack information on the contribution of common human diversity. A growing number of investigations have addressed this gap by incorporating variables related to ethnic or racial diversity which is very useful. However, ethnic definitions are complex and often inconsistent across populations, often relying on self-reported identities, which is useful for designing and promoting greater inclusivity in clinical studies.

Self-reported ethnic identities can have uncertain associations with biological or physiological diversity despite their usefulness in addressing health disparities and the lack of diversity in clinical investigations. To counteract this gap, it has proven useful to extend research beyond populations that are commonly sampled, such as western industrialised societies like the United States. By looking at women who exert significant amounts of energy on a daily basis and are not faced with hyperabundant sources of calories, researchers can not only gain a greater understanding of broader human variation within other ecological contexts but also perhaps glimpse the challenges that were common during human evolution, such as high energetic expenditure and food uncertainty. (For further discussion, see Chapter 16: Sexuality, Reproduction, and Birth.)

Variation in ovarian function is common in populations that do not exercise but who exert significant amounts of energy during everyday activities or who experience significant seasonal swings in caloric availability. These differences in energetic status have been extensively documented in non-western populations. Among Lese foragers of central Africa, salivary progesterone levels are significantly lower compared to western industrial populations (Bentley et al. 1998). Sub-Saharan African populations are, however, subject to non-energetic factors such as high pathogen load that could potentially confound investigations and contribute to female subfecundity. In a population without such burdens, seasonal differences in progesterone have been noted in Nepalese women who engaged in seasonal workloads and weight loss (Figure 15.10). Those that belonged to a social class that did not engage in such activities did not exhibit weight loss or changes in salivary progesterone levels (Panter-Brick and Ellison 1994). Similar effects have been demonstrated in Bolivian women living in the Andes (Vitzthum 2013).

Physical exertion is a significant cause of variability in ovarian function. Rural Polish women who expend significant amounts of energy during the warmer harvest season exhibit concurrent decreases in salivary progesterone levels compared to the winter months when they are more sedentary (Jasienska and Ellison 1998) (Figure 15.11). Indeed, physical exertion can suppress ovarian function independently even without weight loss, suggesting that ovarian function is sensitive to not only decreases of somatic energy reserves (i.e. weight loss) but also the flow of calories even in the absence of body mass loss (Jasienska and Ellison 2004).

Endometrial function is an important aspect of female fertility, since this is vital for embryo implantation. In response to rising oestradiol levels, the endometrial lining within

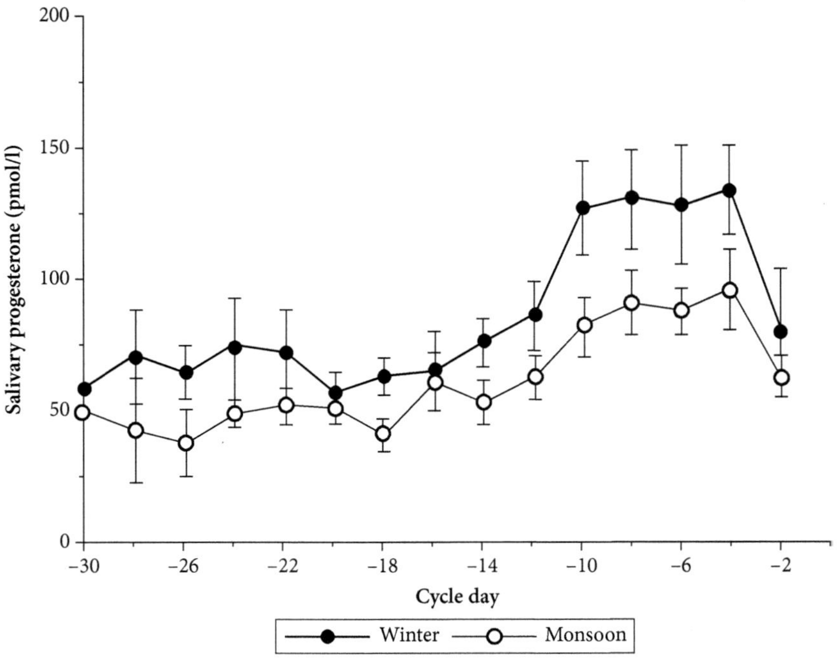

FIGURE 15.10 Salivary progesterone levels in Nepalese women during the winter season with no weight loss and monsoon which is accompanied by significant weight loss.

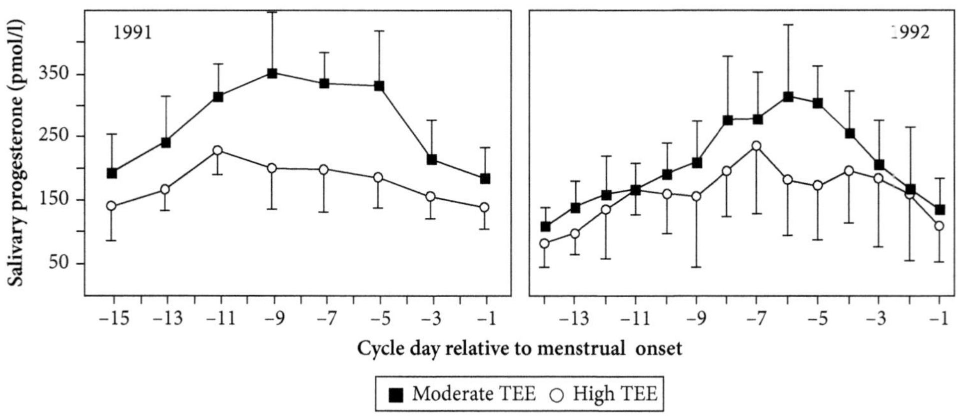

FIGURE 15.11 Salivary progesterone levels in rural Polish women in association with moderate and high total energy expenditure (TEE). High TEE occurs during the harvest season when there is considerable manual labour but no weight loss. Moderate TEE is during winter months when cold and snow inhibit outside activities.

Source: Reproduced from G. Jasieńska and P. T. Ellison, Physical work causes suppression of ovarian function in women, *Proceedings of the Royal Society B Biological Sciences*, 265 (1408), pp. 1847–1851, doi.10.1098/rspb.1998.0511. Copyright © 2017 The Royal Society.

the uterus thickens in preparation for possible conception and embryo implantation. Although endometrial thickness and development is important, very little is known about population variation in endometrial function, although theoretical progress has been made (Clancy 2009). Preliminary evidence suggests that endometrial thickness can vary between populations. Rural Polish women exhibited declines in luteal endometrial thickness (post mid-cycle ovulation) (Clancy 2007). While the significance of this variation remains to be fully elucidated, it is clear that the uniqueness of human menstrual cycles compared to other primates and mammals requires further investigation (Strassmann 1996).

The evolution of menstruation has received a significant amount of attention due to its uniqueness to humans. Most other mammals resorb endometrial tissue if fertilisation does not occur. *Erythrocebus*, *Papio*, *Presbytis*, *Hylobates*, and *Homo* (humans) are the only primates that overtly expel blood and endometrial tissue through the vaginal canal about once a month (Strassmann 1996). Various theories have emerged to explain the adaptive significance, if any, as well as why humans express this unique trait. One theory suggests that menstruation aims to expel pathogens introduced into the reproductive tract by sperm during intercourse (Profet 1993), but evidence for this theory has not been forthcoming. A second theory is that menstruation is not an adaptation but a consequence of uterine evolution (Finn 1998).

The effects of psychosocial stress have also been suggested to contribute to variation in ovarian function. Evidence from women living under extremely stressful conditions such as war has indicated acute disruptions to fertility in response to these stressors (Reese Masterson et al. 2014). More common and everyday stresses remain unclear. An investigation of ovarian steroids and cortisol levels in response to taking the Medical College Admission Test (MCAT) revealed no significant hormonal responses that would be indicative of compromised fertility (Ellison et al. 2007).

15.3.8 Pregnancy

Although gestational length is not believed to be a major source of variation in lifetime fertility, the physiological processes of gestation involve extremely complicated negotiations between the needs of the fetus and the possibility of additional bouts of reproduction by the mother. Pregnancy begins with the fertilisation of an ovum by a sperm. At the time of fertilisation, the ovum blocks out other sperm, allowing only a single sperm to deposit its genetic material to fuse with the genetic components of the ovum. During implantation, the zygote embeds itself into the endometrial lining, drawing blood, nutrients, and oxygen from the mother. A recent investigation reported that the median length of pregnancy was 268 days (Jukic et al. 2013).

The onset of labour and parturition involves the disengagement of progesterone and oestradiol from their role in supporting the developing fetus. The mechanism that initiates the onset of labour is poorly understood, although glucocorticoids have been suggested to play a role. Cortisol, for example, is vital for the maturation of fetal lungs. Premature infants who are suffering from respiratory distress due to pulmonary underdevelopment are sometimes treated successfully with dexamethasone, a synthetic glucocorticoid, as a way to support fetal organ development. (For further discussion, see Chapter 12: Respiratory System.) Maternal and fetal glucocorticoids have also been suggested to play a role in the onset of labour, although numerous questions remain unanswered (Trainer 2002).

Potential health challenges during pregnancy are numerous. Gestational diabetes and pre-eclampsia are among the most common. In some cases, when the fetus attempts to draw too much maternal glucose, the mother deploys significant amounts of insulin in order to maintain her own blood glucose levels, resulting in hyperglycaemia. Gestational diabetes can be treated with diet, exercise, and in many cases the anti-glucose drug metformin. However, gestational diabetes is ultimately resolved with birth and the glucose demands of the fetus being eliminated (Meek 2017). Pre-eclampsia is a potentially life-threatening condition that involves hypertension during gestation (Monte 2011).

The persistence of disorders such as pre-eclampsia during pregnancy appears to fly in the face of an important aspect of human evolution. That is, conditions such as these should be selected out, unless they are maintained by necessity during the course of a life history trade-off. Haig (1993) first suggested that the relationship between a mother and her fetus involves a significant degree of negotiation due to possible conflicts of interest. That is, while selection favours fetuses that demand resources, mothers must sequester some resources for future reproductive effort.

15.3.9 Lactation

As a defining characteristic of mammals, lactation is an important aspect of human reproduction. Besides providing offspring with an important source of calories and essential nutrients, lactation supplies hormones and immunological factors that can be vital to the growing infant. During pregnancy, increased levels of hormones such as progesterone, oestradiol, and prolactin promote the development of breast tissue that involves the production of milk. Prolactin itself is vital for the production of milk, with oxytocin and vasopressin acting in conjunction for the purpose of milk delivery and ejection during nursing. Indeed, the production and delivery of milk is significantly affected by the biomechanical action of suckling on the nipple by the infant. The neurological pathways are from the nipple to the hypothalamus, where the production of prolactin and oxytocin is centred. Lactation and nursing also help to meld the social bond between mother and offspring. Hormones such as oxytocin that are secreted in response to nursing promote and support not only the metabolic aspects of milk production, but also the neuroendocrine processes that result in mother/infant bonding. (For further discussion, see Chapter 16: Sexuality, Reproduction, and Birth.)

Lactation is the greatest source of reproductive energy expenditure in women, even surpassing the caloric demands of gestation. Estimates of caloric expenditure vary depending on the age of the child, frequency of nursing, and other factors, but it is commonly estimated that lactation requires an additional 625 kcal/day on average, which is a greater daily metabolic investment compared to a day of gestation, and adds about 30% to daily energy expenditure on top of basal metabolic rates (BMR), assuming a daily BMR of about 2000 kcal (Butte and King 2005). This significant energetic demand is a major source of variation of IBI.

As iteroparous mammals, the adjustment of ovarian function is vital in optimising the allocation of reproductive effort and energy between reproductive bouts. The neuroendocrine factors that regulate lactation also tend to suppress ovarian function and resumption of menses, commonly known as lactational amenorrhoea. Early research suggested that the frequency and duration of nursing was an important source of variation in female fertility. That is, nursing frequency tended to delay the resumption of menstrual cycling and ovulation.

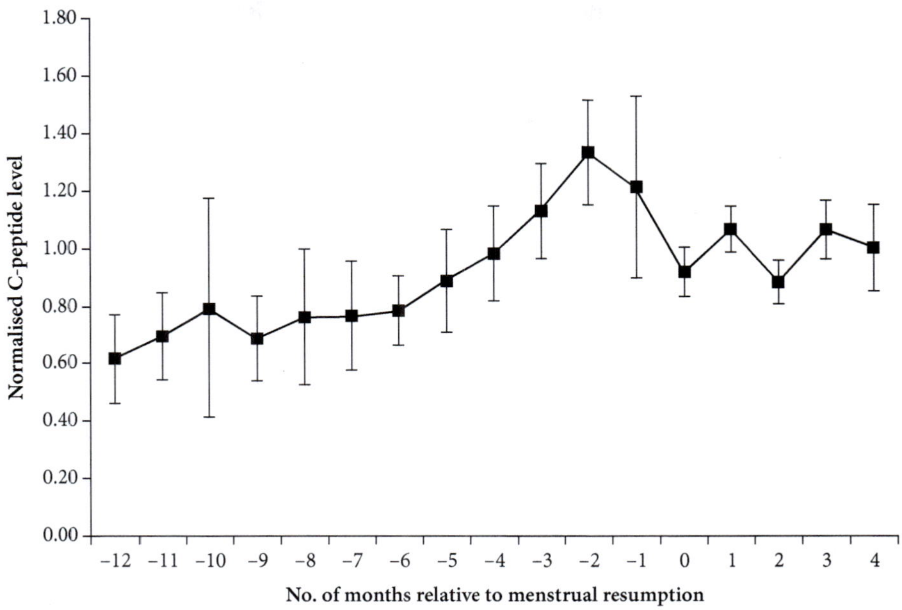

FIGURE 15.12 Normalised C-peptide levels in association with number of months relative to menstrual resumption.

Source: Reprinted from *Fertility and Sterility*, 80 (5), Peter T. Ellison and Claudia R. Valeggia, C-peptide levels and the duration of lactational amenorrhea, pp. 1279–80, http//dx.doi.org/10.1016/S0015-0282(03)02158-7.

Subsequent research demonstrated that suckling by the infant triggered a hormonal cascade that inhibited the production of GnRH and therefore delayed the resumption of ovarian cycling. Studies in Scottish women and among the !Kung support this hypothesis (Konner and Worthman 1980; Howie and McNeilly 1982).

More recently, the mother's energetic status has been recognised as an additional source of variation in IBI and the resumption of menstrual cycling after lactational amenorrhoea. Among the Toba (Qom) of Argentina, women were observed to resume ovarian cycling almost immediately after giving birth despite intensive breastfeeding. It was hypothesised that the length of post-lactational amenorrhoea was influenced by somatic energetic conditions. During the course of lactation, it was revealed that the resumption of ovulation and menstrual cycling was significantly associated with rising levels of C-peptide, the residual protein that is created during the synthesis of insulin after the cleave of proinsulin (Figure 15.12). C-peptide is therefore directly correlated with insulin and is indicative of circulating glucose (Ellison and Valeggia 2003). Other measures of energetic status, such as adiposity as reflected by circulating leptin levels during lactation, are not associated with the resumption of ovarian activity, suggesting that glucose is the more salient signal for the cessation of lactational amenorrhoea (Tennekoon et al. 2005).

15.3.10 Fetal Loss

Fetal loss is an important source of variation for IBI. Each loss results in time and energy expended in a bout of reproductive effort that yields no fitness and disallows investment in additional bouts of reproductive effort including additional conceptions. Indeed, the cost for women is significant due to the constrained lifetime time window of opportunity for reproduction that is available before the onset of menopause. Estimates of fetal loss have varied widely. Prior to the development of hormonal pregnancy tests when clinicians relied on fetal heart beat and an absent menstrual cycle to detect a conception, estimates of fetal loss were about 20%. With increased sensitivity in detecting pregnancies earlier in gestation and with the deployment of hormone assays such as hCG that can detect a pregnancy even before a woman is aware of the pregnancy, estimates of fetal loss have risen to over 30% (Wilcox et al. 1988). Assessing the loss of a conceptus is challenging, but it is thought that many were subject to genetic defects that left them unviable or that the uterine environment was deficient due to hormonal milieu or defects in the endometrium. Indeed, the level of hCG, which is necessary for luteal rescue and the continued production of progesterone by the corpus luteum, immediately after conception is positively correlated with cycle length when fetal loss occurred, suggesting that some aspect of cycle quality may be contributing to fetal loss risk (Figure 15.13).

Stress, broadly defined as dysequilibrium of somatic homeostasis in response to environmental or psychosocial disturbances, may also contribute to fetal loss risk. It is broadly accepted that sudden shocks or traumatic experiences can increase the risk of fetal loss. However, much of this information is anecdotal. Quantification of stress is notoriously challenging. Using neuroendocrine markers of stress derived from changes in the hypothalamic–pituitary–adrenal axis, it is possible to glean some insights from the field as well as the laboratory. Among healthy Mayan women in Guatemala, salivary cortisol levels

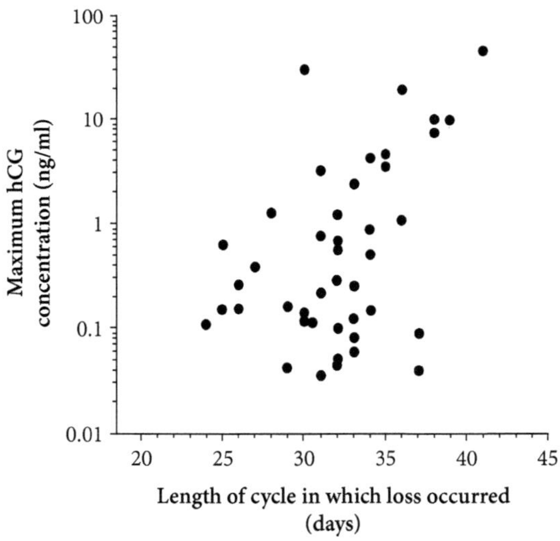

FIGURE 15.13 Length of menstrual cycle in association with hCG levels in cycles where fetal loss occurred.

Source: Reproduced from *The New England Journal of Medicine*, Allen J. Wilcox, Clarice R. Weinberg, John F. O'Connor, Donna D. Baird, John P. Schlatterer, Robert E. Canfield, E. Glenn Armstrong, and Bruce C. Nisula, Incidence of Early Loss of Pregnancy, 319 (4), pp. 189–194, doi.10.1056/NEJM198807283190401. Copyright © 1988, Massachusetts Medical Society. Reprinted with permission from Massachusetts Medical Society.

were significantly higher in failed pregnancies as determined by hCG levels during the first 3 weeks of gestation (Nepomnaschy et al. 2006).

In total, the relatively high rate of fetal loss in humans and other primates may appear to be counterintuitive to an adaptive model of reproduction. That is, it might be assumed that natural selection would have created a suite of physiological mechanisms that would avoid such loss. However, it is important to note that most of the instances of fetal loss are early in the pregnancy before there is significant metabolic investment. Therefore, it may be that natural selection is acting to filter out offspring that might not be reproductively viable or may require a disproportionate amount of care that would detract from future reproductive effort. This may provide one level of understanding, but there is also the issue of allocare and extreme offspring altriciality. All human infants require tremendous amounts of care, not just those that may have physical, cognitive, or genetic deficits. Plus, humans are unique among primates for the significant amount of allocare that is often available from fathers, siblings, extended kin, and even non-kin. Given the similarities, with modest data, in fetal loss rates between humans and non-human primates, it is likely that there are deeply rooted constraints that determine the viability of conceptions.

To assess whether human rates of fetal loss are unique, comparative information would be informative. However, data on fetal loss rates among non-human primates are sparse. Among wild baboons (*Papio cynocephalus*), rates as determined by endocrine assessments of faecal oestrogen, progestin, and glucocorticoids as well as visual evidence were 13.9% (Beehner et al. 2006). A report of fetal loss in captive macaques (*Macaca fuscata*) using similar observational methods as well as endocrine assessments of DHEA-S indicated that out of ten females that became pregnant, three pregnancy losses were determined, or about 30%, although clearly sample sizes were small (Takeshita et al. 2016). In captive marmosets (*Callithrix jacchus*), 35.85% of litters experienced some fetal loss, while 22.15% of pregnancies experienced total loss of litter (Rutherford et al. 2014).

A related issue is the possibility of sex-specific fetal loss or non-random conception rates with X- and Y-bearing sperm. In humans, there is a small but significant and consistent difference in the number of male and female births, with a little over 51% of births resulting in males. This is a robust finding that is evident in virtually every human population surveyed. There is some evidence of a greater number of Y-bearing sperm, especially in older men (Graffelman et al. 1999). It is also possible that there are differences in fetal loss between male and female conceptuses and/or conception success.

15.3.11 Testicular Function

The human testes are responsible for spermatogenesis and the production of sex steroids and protein hormones that support secondary sexual characteristics, libido, and sexually dimorphic muscle mass. As with females, the testes operate in coordination with the hypothalamus and pituitary gland. GnRH from the hypothalamus stimulates the pituitary to produce LH and FSH. LH then stimulates the Leydig cells in the testes to produce testosterone, while FSH supports Sertoli cells with spermatogenesis. Other hormones that contribute to spermatogenesis include inhibin B, which is also secreted by Sertoli cells and acts in an inhibitory manner at the hypothalamus.

Spermatogenesis begins with the spermatogonium undergoing mitosis to form primary spermatocytes, with an initial meiosis to secondary spermatocytes and a secondary meiotic division into four spermatids. Final development in mature sperm occurs within the lumen, where they are sequestered until ejaculation or resorption. The contribution of sperm quality and quantity to human fertility is less clear than for ova. Sperm are obviously necessary for conception, but variation in sperm quantity and quality is less clear-cut than ovarian function. First, assessing sperm count is challenging due to the broad range of variation that is encountered in the literature. For example, between countries, sperm count varies from 102.9 to 52.9 million/ml (Fisch et al. 1996). The World Health Organization defines normal sperm counts to be >20 million/ml, but other studies suggest ranges from 13 to 98 million/ml (Lewis 2007). Nonetheless, various factors can influence sperm count and quality which can contribute to probability of conception (Figure 15.14). These include the number of Sertoli cells in the testes, ejaculatory frequency, and age. Some data suggest that sperm counts have been declining over the past several decades and that this may contribute to changes in fertility in contemporary men. However, research on this potential decline in sperm count and quality requires further investigation due to changes in counting methods, differences in analysis, and sampling regimens (Sharpe 2012).

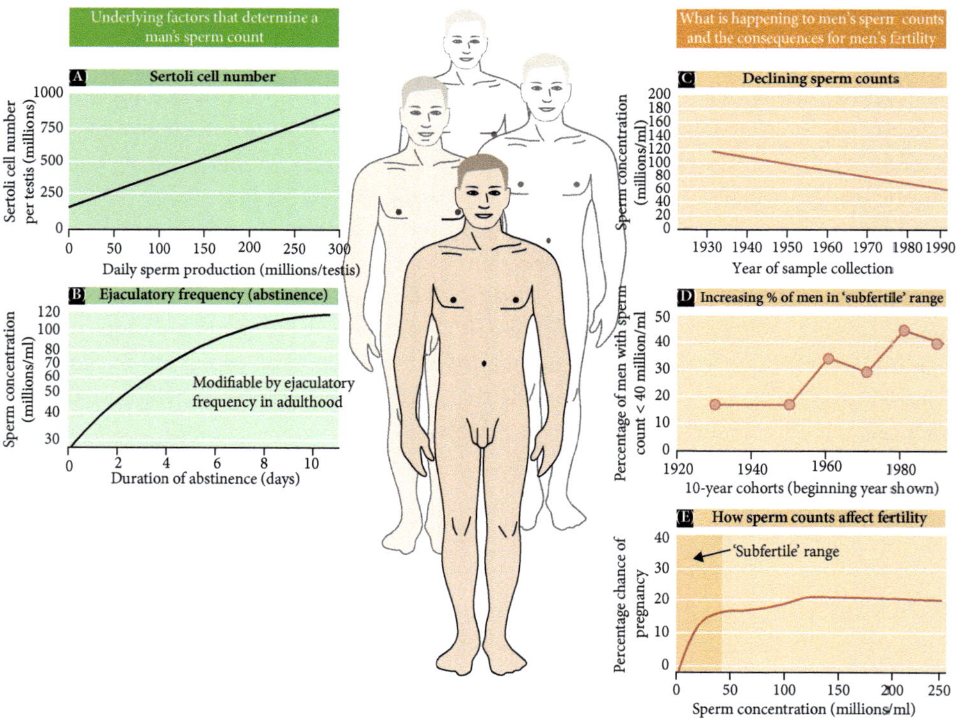

FIGURE 15.14 Factors that contribute to sperm count, quality, and contributions to conception probability.

Source: Reproduced from Richard M. Sharpe, Sperm counts and fertility in men: a rocky road ahead, *EMBO Reports*, 13 (5), pp. 398–403, Figure 1, doi.10.1038/embor.2012.50. Copyright © 2012 European Molecular Biology Organization.

15.3.12 Female Reproductive Senescence

Menopause is among the most vexing questions in human evolutionary biology. The primary issue is that organismal lifespan is often correlated with last reproduction. In humans, however, about one-third of lifespan is post-menopausal. Our common understanding of the evolution of lifespan is that longevity should not be selected for if there is no selective advantage or fitness at those older ages. Otherwise it is challenging to imagine a way in which the genes responsible for longevity would be selected for. (For further discussion, see Chapter 4: Growth and Development, Chapter 16: Sexuality, Reproduction, and Birth, and Chapter 5: Senescence and Ageing.)

Several theories have emerged that attempt to explain the evolution of menopause and post-reproductive lifespan. What is known is that menopause, or the cessation of ovulation and cycling, is the result of the exhaustion of follicles. That is, all women are born with the absolute number of follicles that will ultimately develop into ova. These follicles are formed during gestation and are kept in stasis until menarche, when cohorts of follicles are recruited during ovulation. The pattern of menopause is remarkably constant across populations, and indeed among chimpanzees (Jones et al. 2007; Wallace and Kelsey 2010). Menopause appears to set in around the age of 50. One of the issues is the challenge of defining when menopause occurs, since one is trying to identify the absence of something. In the field, menopause is commonly defined by the lack of a menstrual cycle (bleeding) after a certain amount of time.

Clinical assessments of menopause often deploy the measurement of one or more biomarkers or imaging techniques. With the exhaustion of follicles, gonadotropins such as FSH, LH, and alpha and beta dimers increase dramatically as a result of the decrease in oestradiol and its inhibitory effects on the hypothalamic production of GnRH. A more direct measure of menopause has emerged using the measurement of AMH. AMH is secreted by ovarian follicles, and during the course of menopause begins to decline dramatically (Doroftei et al. 2015) (Figure 15.15).

Arguments for the evolution of post-menopausal lifespan stem from the 'grandmother hypothesis' to suggestions that fertility and fitness at older ages in men might be contributory to human lifespan. In the grandmother hypothesis, Hawkes suggests that older women can contribute to the inclusive fitness of their daughters by investing care and resources in their grandchildren. Investment by grandmothers also tend to alleviate the burdens of care on the mother, allowing her to spend more time foraging and caring for herself, which in turn may bolster her fertility (Hawkes et al. 1998). However, tests of the grandmother hypothesis have been mixed. Hawkes demonstrated how grandmothers contribute significant amounts of care and effort to grandchildren among Hadza foragers of Tanzania and communities in the Gambia (Hawkes et al. 1989; Sear et al. 2000). However, the effects on fertility have not been compelling (Hill and Hurtado 1991; Jamison et al. 2002). The potential effects of grandfathers have also been explored, with no compelling evidence emerging to suggest that they provide any care or act in tandem with grandmothers (Sear et al. 2002; Lahdenpera et al. 2007). The possible effects of older men as fathers may provide new insights.

It has been suggested that fitness at older ages in men may be important for post-menopausal lifespan. Early assumptions by demographers suggested that men ceased reproducing in tandem with their wives. However, this assumption did not take into account the broad

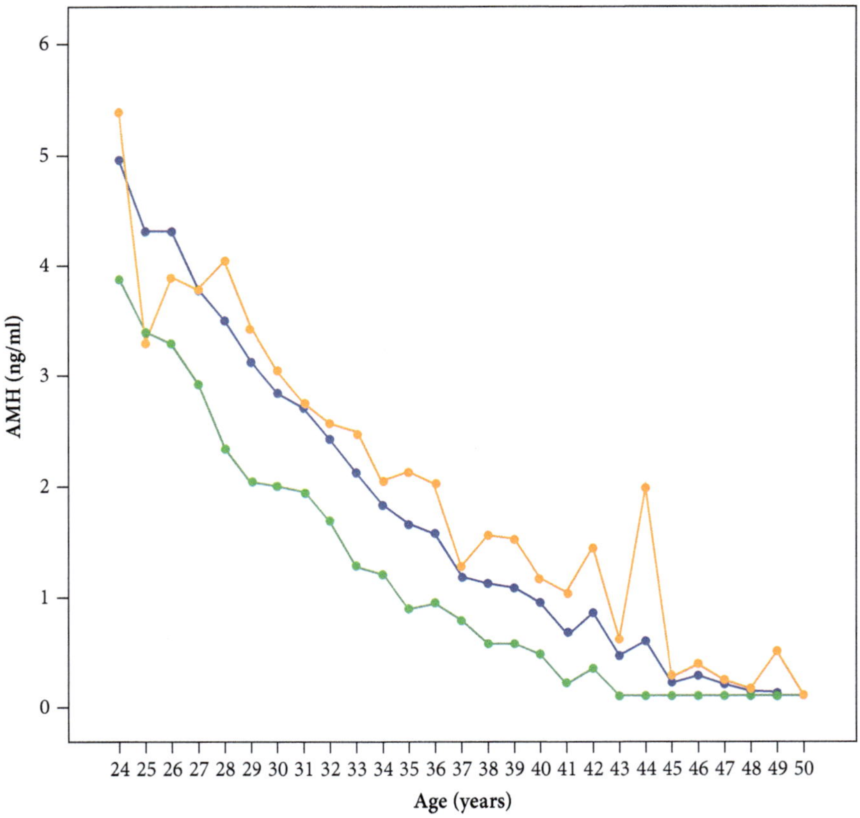

FIGURE 15.15 Serum AMH values in association with age. Mean (blue), median (green), and standard deviation (SD, orange) of AMH values are represented versus age (years).

Source: Reproduced from Bogdan Doroftei, Cristina Mambet, and Mihaela Zlei, It's Never over until It's over: How Can Age and Ovarian Reserve Be Mathematically Bound through the Measurement of Serum AMH—A Study of 5069 Romanian Women, *PLoS ONE*, 10(4), e0125216, Figure 1, https//doi.org/10.1371/journal.pone.0125216. © 2015 Doroftei et al. This is an open access article distributed under the terms of the Creative Commons Attribution License, which permits unrestricted use, distribution, and reproduction in any medium, provided the original author and source are credited.

range of human variation in mating patterns. In fact, there is significant male fertility after the age of 50 in many populations. It has therefore been suggested that the fitness benefits at older ages may select for longevity genes that are passed to not only males but also females (Tuljapurkar et al. 2007).

15.3.13 Male Reproductive Senescence

Male reproductive senescence is distinct from females in that there is no clear demarcation of a decline in reproductive function. In females, menopause is observed through the lack of a menstrual cycle and a complete cessation of fertility, usually around the age of 50.

However, men retain the ability to father children at older ages, although opportunities to procure mates often decline with age. Moreover, spermatogenesis and supportive sex hormone function are often maintained at older ages but not without challenges.

Very little is known about male reproductive senescence in non-western populations. Information on erectile dysfunction, prostate health challenges, and changes in strength and vigour is available from a handful of investigations (Walker and Hill 2003; Campbell 2005; Gray and Campbell 2005). Information on changes in spermatogenesis and the HPT axis is equally sparse. Longitudinal information on changes in sperm count and quality suggests that age has a marginal effect on spermatogenesis and any declines in function are often in conjunction with other health challenges such as cancer, cardiovascular disease, and diabetes.

Among the Ache of Paraguay, a cross-sectional assessment of salivary testosterone and oestradiol revealed no changes with age, although sample sizes were small. In the same study, serum measurements of LH and FSH revealed significant increases with age, which is commonly observed in most other populations. This suggests a decline in the sensitivity of receptors in Leydig and Sertoli cells which may contribute to suboptimal spermatogenesis and the production of testosterone (Bribiescas 2005).

Changes in the hypothalamus are also evident in the loss of circadian patterns of testosterone production. At younger ages, testosterone is often highest in the morning and declines towards the evening. This is a pattern that is widely observed in most populations and is so ubiquitous that investigations involving testosterone need to account for this pattern as do researchers when designing and conducting their research (Bribiescas 2001a). However, at older ages, this circadian pattern tends to diminish. The primary cause of this diminishment appears to be changes in the pattern and pulsatility of GnRH (Bremner et al. 1983). GnRH production is reliant on sensing energetic status. Decreases in circulating glucose and insulin result in decreases in the amplitude and frequency of GnRH (Röjdmark 1987; Röjdmark et al. 1989). The hypothalami in older men appear to lose sensitivity to these energetic cues. While gonadotropin pulse frequency and amplitude decrease in response to fasting and are rekindled by exogenous GnRH in younger men, older men are unresponsive to these same stimuli (Bergendahl et al. 1998). Diminishment of circadian patterns of testosterone and cortisol production has been noted in Ache foragers, suggesting that this pattern is consistent across populations (Bribiescas and Hill 2010; Amir et al. 2015).

Energetic resources are necessary to fuel the needs of reproduction and growth. The management and sequestration of energetic resources such as glucose and fat is therefore vital to the evolution of a species. Pivoting now to the hormones of energy management, endocrine mechanisms that directly affect the management of metabolism are discussed.

15.3.14 Metabolic Endocrinology

A major selection factor in the evolution of all organisms is energy regulation. This involves the breakdown and conversion of food into chemical energy for cellular growth, maintenance, reproduction, and, in the case of endotherms like humans, thermogenesis. Metabolic hormones are vital for regulating the rate at which energy is produced as well as the storage of energetic resources, primarily in the form of fat. Being endothermic—that is, generating body heat, compared to ectotherms (i.e. reptiles) that rely on ambient sources of heat to contribute to metabolic needs—humans consume a considerable amount of energy compared to smaller mammals. BMRs scale at about log 0.75 in relation to body size, meaning that larger animals tend to have slower metabolic rates per unit mass compared to smaller organisms; however, there is significant variation in BMR between great apes independent of body mass (Pontzer et al. 2010). Metabolic hormones are evolutionarily ancient and central to the physiology of all vertebrates and most invertebrates.

Metabolic hormones play an important role in managing other life history traits including reproductive function and growth. For example, the hypothalamus exhibits insulin receptors which complement the influence of kisspeptin, and affect GnRH production and reproductive function. Skeletal muscle maintains thyroid hormone receptors that are vital for growth and energy regulation. Similarly, insulin plays a central role in the resumption of menses after post-partum amenorrhoea. Finally, the ovary contains insulin receptors which affect oestradiol production. Under common circumstances, ovarian sensitivity to insulin results in regulation of fecundity, while overstimulation due to hyperinsulinaemia, often in association with obesity, can contribute to greater risk of PCOS, a growing health concern in developed parts of the world. In this section, we will concentrate on the metabolic aspects of these hormones with a mindful eye towards human variability, gene–environment interactions, evolutionary functional significance, and comparative biology (Table 15.3).

15.3.14.1 *Adrenal hormones*

The adrenal glands are positioned on top of the kidneys, where they produce several vital hormones that are involved in the regulation of metabolism and water balance. Within the outer portion of the adrenals, known as the cortex, a class of hormones known as glucocorticoids are produced that include the most prominent which is cortisol. These steroid hormones are most commonly referred to as 'stress hormones'. However, their primary function is to facilitate the cellular uptake of glucose during periods of energetic or psychological challenges. The production of glucocorticoids is controlled by a classic hormonal negative feedback loop that involves the production of corticotropin-releasing factor (CRF) within the hypothalamus. CRF stimulates the production of adrenocorticotropic hormone (ACTH) within the pituitary, which in turn promotes the production of glucocorticoids. Abnormally low levels of cortisol, often expressed as Addison's disease, create lethargy and hypoglycaemia and can be a serious health challenge. Hypercortisolism is known as Cushing's disease and causes immune disorders, muscle loss, weight gain, and abnormal fat deposition around the face, neck, and midsection. Chronically high glucocorticoids due to chronic psychological stress can result in the desensitisation of receptors and neurological degradation.

Cortisol exhibits a strong diurnal signal—levels being higher at waking in the morning, with declines throughout the day. Interestingly, the diurnal pattern can vary with age and between populations (Amir et al. 2015). Across primates, this diurnal pattern appears to be fairly consistent in captivity and in the wild, although some New World primates may diverge from this pattern (Coe et al. 1992). The literature on cortisol is vast and beyond the limitations of this chapter. From an evolutionary medicine perspective, cortisol has been used as a biomarker of psychosocial stress in humans and a variety of animal models. Cortisol is also valuable in assessing potential trade-offs between life history needs such as between reproduction and maintenance. In human males, glucocorticoids can suppress

Table 15.3 Hormones that Contribute to Metabolic Regulation

Name	Central effect	Steroid/ protein	Primary source of production
Insulin	Regulation of cellular intake and use of glucose. Also has regulatory effects on growth, reproductive function, and immunoregulation	Protein	Pancreas
Leptin	Acts as a lipostat to the hypothalamus	Protein	Adipose tissue, smaller amounts in stomach and gonads
Cortisol	Facilitates cellular glucose uptake and regulates metabolic function during periods of stress	Steroid	Cortical region of adrenal gland
Adiponectin	Regulates glucose and fatty acid metabolism	Protein	Adipose tissue
Ghrelin	Produced when stomach is empty. Stimulates hunger	Protein	Stomach
Orexin	Stimulates hunger	Protein	Hypothalamus
Thyroid-stimulating hormone (TSH)	Serves to stimulate production of thyroid hormone	Protein	Pituitary gland
Triiodothyronine (T3)	Modulates basal metabolic rate and growth	Tyrosine based	Thyroid gland
Thyroxine (T4)	Precursor to T3 and more potent than T3 in affecting BMR	Tyrosine based	Thyroid gland
Thyroxine-binding globulin (TBH)	Serves as a carrier binding protein for T3 and T4	Protein	Liver

reproductive hormone function, leading to the conclusion that stress can contribute to deferring investment in reproductive effort (Bambino and Hsueh 1981), although this association is not consistent and often mediated by other factors (Gettler et al. 2011).

Because of its relatively high circulating levels in blood, it is easily measured in a variety of fluids and biological samples that can be collected non-invasively, such as saliva, urine, and faeces (Whitten et al. 1998; Anestis 2010); it is also a useful hormonal assessment tool for conservation efforts in wild primates (Jaimez et al. 2011). Within the adrenal cortex lies the medulla, which produces hormones that are important for water and mineral balance as well as a small amount of progesterone. Aldosterone is the primary mineralocorticoid hormone, stimulated by levels of ACTH. Little is known about population variation in aldosterone.

15.3.14.2 *Insulin*

Insulin is the primary hormone for regulating glucose uptake by cells in muscle, fat, and various tissues in the body. It is produced by the beta cells of the pancreas and vital for overall health and survival. The production of insulin involves the initial synthesis of preproinsulin which is enzymatically cleaved into proinsulin. This is subject to further enzymatic action by prohormone convertases PC1, PC2, and exoprotease carboxypeptidase E. The final product is insulin, C-peptide, and two residual peptides. The significance of C-peptide, an inactive by-product, is that it is produced in a 1:1 ratio with insulin. In contrast to insulin, C-peptide can be readily measured non-invasively in urine, making it a useful proxy biomarker for insulin assessments in human and non-human primate field studies (Sherry and Ellison 2007). Degradation or failure of the beta cells results in compromised insulin production, hypoinsulinaemia, and type 1 diabetes. Failure or compromised insulin receptors or other forms of cellular insulin resistance results in type 2 diabetes.

Insulin production is stimulated by the introduction of glucose into circulation commonly after the ingestion of a meal. Glucose is introduced to the beta cells in the pancreatic islets by glucose transporters known as GLUT2. Insulin is evolutionarily ancient and found in some form in most eukaryotes. Indeed, human insulin differs from various non-human mammal insulins by only a few amino acids, making mammalian non-human insulin the treatment of choice before the development of modern methods of synthesising human insulin. Most recently, insulin has emerged as a hormone that reflects the dietary and lifestyle changes with sedentism, abundant food, and a westernised lifestyle that is characterised by calorie- and sugar-rich foods. This combination results in overweight, obesity, hypertension, hyperinsulinemia, diabetes, insulin resistance, and overall poor health. (For further discussion, see Chapter 6: Nutrition, Energy Expenditure, Physical Activity, and Body Composition.) In sum, this is commonly known as metabolic syndrome. In many parts of the world, this has surpassed food insufficiency and malnutrition as the primary health challenge in modern times. Metabolic syndrome symptoms, including hyperinsulinaemia and insulin resistance, are also evident in wild primates exploiting human trash dumps and captive animals in zoos (Banks et al. 2003).

15.3.14.3 *Thyroid Hormones*

Thyroid hormones are vital for regulating somatic energy, primarily through their action on adenosine triphosphate (ATP) production, BMR, and thermogenesis. Secreted primarily by the thyroid gland under the regulation of the hypothalamus and pituitary, thyroid hormones are synthesised through the incorporation of the amino acid tyrosine and the basic element iodine. Deficiencies in iodine can result in thyroid dysfunction as well as enlargement of the thyroid gland, commonly known as goitre. Thyrotropin-releasing hormone is produced in the hypothalamus which stimulates production of thyroid-stimulating hormone (TSH) in the pituitary. TSH then stimulates production of thyroxine or T4, a prohormone of the more prevalent and bioactive hormone triiodothyronine, commonly known as T3 within the thyroid gland. About 70% of T3 and T4 is bound to the carrier protein thyroxine-binding globulin (TBG) while in circulation. The rest is either unbound or affiliated with albumin or other proteins.

The primary targets of action for thyroid hormones are white and brown adipose tissue, skeletal muscle, the liver, pancreas, and brain. In addition to regulating BMR and thermogenesis, thyroid hormones are important regulatory agents for cholesterol levels, primarily through hepatic influence. Low levels of thyroid hormone produce hypothyroidism, which is characterised by lowered BMR, weight gain, and lethargy, as well as increases in fat tissue

and cholesterol levels. Abnormally high thyroid hormone levels produce hyperthyroidism, commonly referred to as Grave's disease, characterised by greater BMR, weight loss, decreases in cholesterol levels, and sometimes hyperactivity.

The primary mode of action on BMR, thermogenesis, and weight variability is through the adrenergic nervous system, an autonomic process whose central agents are epinephrine and norepinephrine. Across the human life course, thyroid hormones demonstrate changes with age. Initial increases in TSH, T3, and T4 during childhood are followed by prepubertal surges that decline during early adulthood (Michaud et al. 1991). The physiological significance of this surge is unclear, but it is likely to be associated with the significant increase in the energetic needs of pubertal growth spurts. With ageing, thyroid hormones decline at a steady pace (Mullur et al. 2014).

Thyroid hormone function is sensitive to environmental conditions, specifically seasonal changes in temperature. This has been shown in transient exposure to seasonal cold among individuals living temporarily in frigid conditions, as well as communities who live permanently in environments characterised by significant periods of cold. Increases in TSH and *decreases* in free T4 (fT4) and free T3 (fT3) suggest that the uptake and clearance of fT4 and fT3 tends to outpace production under such conditions. This is often referred to as 'polar T3 syndrome' and is confounded by broad swings in solar periodicity in extreme latitudes (Mullur et al. 2014). Significant human variation is evident in thyroid hormone physiology. Circumpolar populations in particular exhibit increased circulating levels of T3 and T4, most likely as an adaptation to chronic cold stress. However, thyroid hormone variation in circumpolar populations is also subject to lifestyle and seasonal changes (Levy et al. 2013). These increased levels may not be without cost. Populations such as the Yakut in Siberia who have higher levels of T3 and T4 also may be at greater risk of autoimmune disorders. These populations exhibit higher levels of anti-thyroid peroxidase antibody (TPOAb) which attacks key thyroid function proteins, such as thyroid peroxidase which is vital for thyroid hormone synthesis (Cepon et al. 2011).

While the central role of thyroid hormones is well known in apoptosis (programmed cell death) in amphibian metamorphosis, very little is known about the comparative aspects of thyroid hormones, specifically in regard to other great apes and non-human primates. Subtle but significant differences are evident in lifetime thyroid hormone production between common chimpanzees (*Pan troglodytes*) and bonobos (*Pan paniscus*). Total T4 levels have been reported to be higher in humans compared to chimpanzees, whereas fT3 and fT4 as well as total T3 were higher in chimpanzees (Behringer et al. 2014). In contrast, orangutans (*Pongo* sp.) and gorillas (*Gorilla* sp.) exhibit lower T3 and T4 levels (Aliesky et al. 2013). There are also differences between humans and other great apes regarding transthyretin, which acts as a major thyroid hormone regulator in the brain. These differences have been suggested to contribute to the unique brain development trajectories of humans and chimpanzees, although further investigation is needed (Gagneux et al. 2001).

The evolution of thyroid hormones can possibly be traced to the very emergence of life itself. It has been hypothesised that the availability of iodine and its interactions with key amino acids such as tyrosine allowed for the emergence of multicellular organisms that were able to evolve better and more efficient metabolic processes. While invertebrates do not have thyroid glands, virtually all organisms have thyroid hormone-like substances (Crockford 2009).

15.3.14.4 *Leptin*

Leptin is a polypeptide hormone secreted primarily by fat cells and is therefore highly correlated with adiposity in healthy individuals. Leptin has several functions but acts primarily as a lipostat, providing information on the availability of long-term energetic resources to the hypothalamus. Through its interaction with the hypothalamus, leptin has an influence on many aspects of life history physiology including immune function, growth, and reproduction. There is also significant sexual dimorphism in leptin levels, with females exhibiting higher levels after controlling for body fat compared to men. These differences arise during adolescence when sex steroid levels inhibit fat deposition in males and promote adiposity in females (Garcia-Mayor et al. 1997). Age of menarche has been reported to be associated with leptin levels; however, the association was quite modest (Matkovic et al. 1997). While leptin has been reported to be associated with variation in reproductive function in rodent models, similar effects in humans have not been forthcoming. Among the many metabolic hormones, leptin exhibits significant variation between populations, mostly likely due to differences in lifestyle and ecology that affect the production of leptin within adipocytes. In many non-western populations, leptin is significantly lower compared to western industrialised populations even after controlling for adiposity (Bribiescas 2001b; Lindgarde et al. 2004).

Leptin profiles in non-human primates reveal similarities and important differences compared to humans. As with humans, macaques and baboons exhibit positive correlations between adiposity and leptin (Muehlenbein et al. 2003, 2005). In chimpanzees, leptin exhibits typical sexual dimorphism, with leptin being positively associated with body weight in females, but not in males (Bribiescas and Anestis 2010).

15.3.14.5 *Adiponectin*

Adiponectin is also produced by adipose cells with smaller amounts synthesised in the placenta although curiously it is inversely related to adiposity. Initially thought to be primarily engaged in the regulation of adiposity, adiponectin appears to be involved in the regulation of glucose, fatty acid metabolism, and has anti-oxidant properties. Similar to leptin in that it is produced in fat cells although in inverse amounts, adiponectin has many properties that are reciprocally related to leptin. Most prominently, decreases in adiposity promote greater production of adiponectin. The functional significance of adiponectin is still unclear. At low levels, it is associated with type II diabetes, insulin resistance, metabolic syndrome, and hypertension. Little is known about population variation in adiponectin.

15.3.14.6 *Ghrelin*

Ghrelin is a hormone that influences satiety and hunger. It is produced primarily in the stomach and is found in two basic forms in circulation. Total ghrelin is most common but is less bioactive than the free form which has a very short half-life and is therefore more challenging to assess. While levels and actions of other hormones have been demonstrated to vary across human populations living in varied ecologies and lifestyles, very little is known about ghrelin. Only a handful of studies have assessed ghrelin in non-western societies (Shukla et al. 2005; Bribiescas et al. 2008). In these circumstances it appears that dietary habits that involve the timing, frequency, and size of meals have an influence on overall levels and the daily range of variation in ghrelin. The actions of ghrelin are somewhat similar to orexin, a hormone produced in the brain that stimulates hunger but also affects arousal.

15.4 Phylogeny of the Reproductive System

15.4.1 Functional Significance of Comparative Endocrinology

The reproductive biology of our closest evolutionary relatives has only recently begun to be unveiled (Beehner and Lu 2013). Only a handful of studies have provided the most rudimentary information on members of the genera *Pan* (chimpanzees), *Pongo* (orangutans), and *Gorilla* (gorillas). Information on the reproductive biology of lesser apes, the *Hylobates* (gibbons and siamangs), is limited to a single study of a single female (Aramaki et al. 2010). A comparative study of captive great apes (chimpanzees, gorillas, orangutans, and bonobos) revealed reproductive hormone patterns that are similar to other mammals while exhibiting subtle differences in gonadotropins (Shimizu et al. 2003).

Reproductive senescence in chimpanzees appears to be similar to humans, with menopause occurring around the age of 50, although information on key diagnostic hormones such as FSH, LH, and AMH is lacking. Demographic evidence from wild female chimpanzees suggests that this pattern is quite conserved, with declines in oestrogen occurring in a manner that is similar to humans (Emery Thompson 2005; Emery Thompson et al. 2007). Histological assessments of ovaries derived from deceased captive female chimpanzees show that the rates of follicle depletion mirror those in human females (Jones et al. 2007). Age of menopause appears to coincide with lifespan in wild chimpanzees (Hill et al. 2001), although new evidence suggests that chimpanzee lifespan can increase in response to favourable conditions in the wild (Wood et al. 2017).

Female reproductive biology of orangutans, which is known from only a handful of studies, mostly from captivity, indicates physiological functions and hormonal controls that are similar to humans (Lasley et al. 1980; Nadler et al. 1984). In the wild, orangutans respond to extreme environmental stochasticity in the form of wide swings in food availability. The forests of Borneo exhibit masting in which trees fruit in concert, resulting in an abundance of food for various animals including orangutans. During this period of high caloric availability, female orangutans exhibit an increase in urinary oestrogens which is indicative of greater fertility (Knott 1999). Emerging field methods such as near infrared spectroscopy may prove to be useful in providing greater clarity on the reproductive biology of wild great apes (Kinoshita et al. 2016). Female reproductive ecology in gorillas has only recently been addressed (Habumuremyi et al. 2014, 2016). Hormonal mechanisms that are common in other mammals and primates play similar roles in gorillas.

15.4.2 Male Reproductive Ecology

The reproductive ecology of great ape males is similar in many ways to humans. In chimpanzees, testosterone levels vary between captive and wild populations, most likely due to the different energetic and nutritional conditions, with captive populations exhibiting higher testosterone levels compared to wild communities (Muller and Wrangham 2005). However, chimpanzees differ dramatically from human males in that reproductive strategies are more narrow. Human males exhibit a broad range of reproductive strategies ranging from high time and energy investment in mate acquisition to enhanced effort in paternal and mate care (Bribiescas et al. 2012). Variation in testosterone levels in chimpanzee males is acutely sensitive to competitive environments with other males.

Orangutan males exhibit one of the most unique suite of traits within primates with males having a binary set of phenotypes. The larger more robust males have fatty facial flanges, compete heavily with other males, and appear to be preferred by females. However, a second phenotype is smaller and less competitive, and often obtains mating opportunities by forcing themselves onto apparently unwilling females. These two strategies and phenotypes are largely controlled by testosterone. The larger phenotype exhibits higher testosterone, while the smaller phenotype does not (Maggioncalda et al. 1999; Emery Thompson et al. 2012).

Testosterone levels in gorillas are known from captivity and the wild. The selective importance of testosterone in male gorillas lies in the large male size, high degree of sexual dimorphism, and the high male/female ratio in reproductive groups in which a single male often monopolises mating opportunities with multiple females. Group takeovers by rival males often involve violent encounters with the resident male and infanticide if the challenging male succeeds in taking over the group. However, how testosterone plays into these reproductive strategies is unclear (Robbins and Czekala 1997). Interestingly, gorilla males who affiliate with infants in their group, including those that are not their own, exhibit higher reproductive success compared to males who affiliate less. Clearly additional research is necessary to explore the full range of male reproductive strategies in great apes (Rosenbaum et al. 2018).

15.4.3 Hominid Ancestors

The reproductive biology of our hominid ancestors was likely not very different from what is seen today (Bentley 1999). Given the similarities we have with other great apes with whom we share a common ancestor, it is all but certain that other hominids such as australopithecines and other species of our genus *Homo* were influenced by similar environmental challenges and had similar reproductive responses. Clues from the fossil record include the WT15000 which is one of the most complete fossil finds of *Homo erectus* (Brown et al. 1985). This specimen showed that the appearance of many of the life history traits that we see today were already evident, including large body size, growth rates, and brain size. In addition, the Gona pelvis indicates that fetal development resulting in the birth of a relatively large-brained offspring suggested that human life history traits, including those relevant to reproduction, were already well established in early *Homo* (Simpson et al. 2008).

The evolution of pelvic morphology in our hominid ancestors has drawn significant attention due to the impact of negotiating the need to pass a large-brained fetus during birth and positive selection for upright walking, leading some to even argue that it may have contributed to the extinction of australopithecines (Chene et al. 2013). Others suggest that the evolution of female pelvic morphology resulted in a selective negotiation between the demands of childbirth, bipedalism, and thermoregulation (Gruss and Schmitt 2015). The energetic demands and consequences of the evolution of life history traits that define the genus *Homo*, such as large body size, large brains, and, by extension, longer lifespans and offspring altriciality, suggest that female reproductive energy demands were significant. This implies that early *Homo* evolved behavioural foraging and social strategies to meet these large energy needs (Aiello and Key 2002).

Since sexually dimorphic somatic traits are an important part of the reproductive biology of our male hominid ancestors, important information can be gleaned by examining androgen-sensitive areas of skeletal morphology and mating behaviours that are commonly attributed to sexual dimorphism signals which can be detected in the fossil record (Bribiescas et al. 2012). For example, it has been suggested that decreases in testosterone compared to other great apes may have contributed to the evolution of sociality and other life history characteristics (Cieri et al. 2014). However, testing this hypothesis is challenging. Assessing testosterone levels and androgen-sensitive landmarks on primate cranial morphology, it was found that testosterone variation between primate species was correlated with craniofacial morphology. Adding measurements from craniofacial measurements of hominid fossils provides us with an insight into the socioendocrinology of our hominid ancestors (Doman et al. 2013).

15.5 Adaptation and Pervasive Evolutionary Challenges

The resources devoted to reproduction inherently create costs that emerge not only from the conservative nature of evolution by natural selection, but also sometimes by even more basic aspects of biophysics. One such example is oxidative stress. Metabolic investment in reproduction, as in gestation, lactation, or the maintenance of sexually dimorphic muscle mass in males, often requires the incorporation of oxygen in the form of aerobic metabolism. While aerobic metabolism increases the efficiency by which energy becomes available from food compared to anaerobic metabolism, the incorporation of oxygen creates toxic by-products that can compromise genetic, cellular, and tissue integrity and ultimately accelerate ageing and contribute to intrinsic mortality. Since this is present in virtually every organism that relies on aerobic metabolism, including humans, aerobic metabolism-associated increases in reproductive effort are likely contributory to ageing and somatic decline (Alonso-Alvarez et al. 2004).

Evolutionary medicine is vital for understanding these constraints, since it is commonly accepted among evolutionary biologists that organisms are selected for optimal lifetime reproduction and that any aspect of longevity or somatic integrity that can be interpreted as 'health' is a secondary consideration compared to reproductive success. 'Health' and longevity are only favoured if the net result is greater lifetime fitness. To this end, we can turn to how this translates into disease and health challenges. Moreover, consideration of the utility of evolutionary and life history theory for addressing these health challenges certainly merits greater awareness.

15.6 Consequences for Prevention and Treatment of Disease

15.6.1 Evolutionary Endocrinology of Reproductive Health

Reproductive cancers are particularly difficult health challenges, since they inherently involve physiological mechanisms that are essential for reproduction and hence central to natural selection. Reproductive cancers are also examples of trade-offs between reproductive effort and survivorship. That is, they are likely to represent examples of antagonistic pleiotropy in which a trait that supports fitness earlier in life can be a liability to survivorship later in life, usually after reproductive senescence has set in (Jasienska et al. 2015). Hormones such as oestradiol and testosterone are intricately linked to optimal reproductive function but are also central players in breast, ovarian, and prostate cancer. Lifetime exposure to these hormones is likely to contribute to the increasing risk of these cancers later in life (Eaton et al. 1994).

15.6.2 Breast Cancer

Breast cancer is a primary source of mortality in women. Although breast cancer does occur in men, it is rare (Kreiter et al. 2014). The aetiology of breast cancer is complicated, but it is clear that hormones, reproductive history, lifestyle, and genetics all contribute to risk (Desmedt et al. 2016). Some forms of breast cancer are due to specific mutations that can be passed between generations or are the result of exogenous carcinogens (Zhang et al. 2011; Fenton and Birnbaum 2015), although most are heavily influenced by gene–neuroendocrine–environment interactions that are significantly associated with endogenous hormone milieu. Individual hormone levels are of limited value in predicting breast cancer risk; however, population assessments do provide valuable information on the epidemiological levels of risk of breast cancer in specific communities.

Associations between breast cancer risk, diet, and activity have been somewhat uneven, with some studies reporting an effect and others failing to do so (Thomson 2012). Past investigations of dietary assessment have only modestly incorporated human variation outside of western industrialised societies. This is likely to be a contributing factor to these inconsistent findings, since diet and lifestyle incur many of their effects over an entire lifetime, an important consideration that can be addressed by looking beyond western industrialised societies (Ellison 1999). Indeed, the incidence of breast cancer varies between populations. The question then evolves into identifying and unpacking the salient variables that contribute to breast cancer incidence and risk between these populations. Genes that significantly increase breast cancer risk such as BRCA1, BRCA2, p53, PTEN, STK11, and CDH1 have been disproportionately observed in certain populations (Kleibl and Kristensen 2016; Torres et al. 2007), but overall these seem to account for only a small fraction of the variation in breast cancer risk (Chavez-Macgregor et al. 2014). The persistence of these genes is unclear, but it has been suggested that BRCA genes and associated mutations might be associated with greater fertility, which may contribute to their maintenance within the human genome (Kwiatkowski et al. 2015). Nevertheless, the dynamic nature of breast cancer risk when there is migration or changes to the environment suggests that gene–environment interactions are especially important and account for the majority of the proportion of risk.

In general terms, populations who subsist on a more westernised diet of high-caloric intake that includes a substantial amount of sugars, carbohydrates, and saturated fats tend to have higher rates compared to populations that do not exhibit these dietary habits. Similarly, sedentism contributes to hormonal profiles that tend to increase breast cancer risk and has been demonstrated to exhibit a secular trend, with rates rising over the past several decades (DeSantis et al. 2015). Breast cancer is most evident in women from European extraction, while East Asian women exhibit the lowest rates of breast cancer. The reasons behind these differences remain unclear, but it has been suggested that diet and lifestyle are important, since increases in breast cancer risk have been reported in succeeding generations of Japanese and other East Asian immigrants into the United States and urban Brazil (Deapen et al. 2002; Iwasaki et al. 2008).

Evolutionary perspectives on breast cancer have suggested that changes in patterns of childbirth and lifetime exposure to endogenous sex steroids might be contributory to the relatively high degree of breast cancer that is evident today. For example, Eaton et al. (1994) noted that modern women delay reproduction to older ages, have fewer children, breastfeed less, and as a consequence have many more menstrual cycles and exposures to more monthly pulses of ovarian steroids. Moreover, those ovarian steroid pulses will be higher due to greater nutritional status compared to poorer populations or our evolutionary ancestors.

15.6.3 Ovarian and Uterine Cancers

Ovarian and uterine cancers are less common compared to breast cancer but are still important health challenges in women (Figure 15.16). Ovarian cancer is often associated with cysts that can develop into cancerous tissue. Similarly, endometriosis, a thickening of the endometrial lining that is commonly found in older women and those who are nulliparous, sometimes precede cancer diagnoses. The rates of ovarian and uterine cancers are much higher in industrialised populations compared to less-developed regions. Differences in diagnoses and reporting rates probably contribute to some of the variation, but environmental and lifestyle factors are worthy of consideration also (Chornokur et al. 2013).

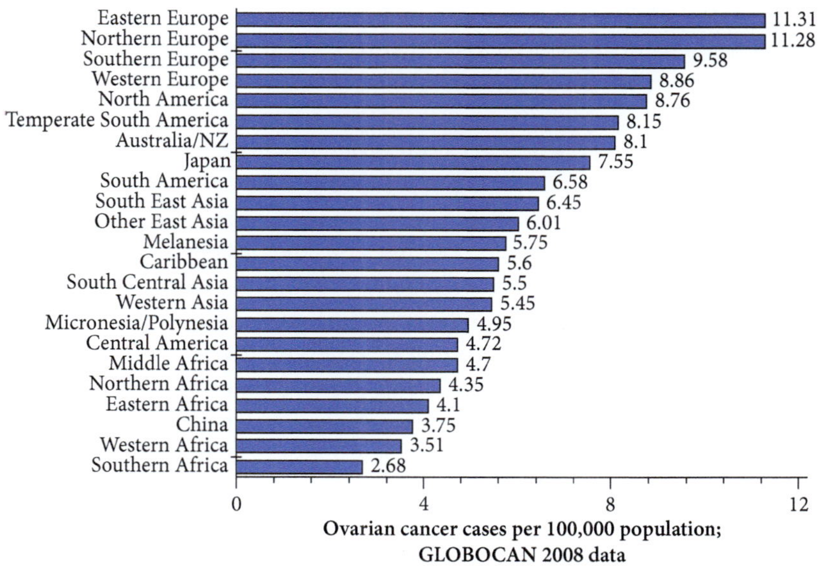

FIGURE 15.16 Ovarian cancer cases from GLOBOCAN data 2008.

Source: Reprinted from *Gynecologic Oncology*, 129 (1), Ganna Chornokur, Ernest K. Amankwah, Joellen M. Schildkraut, and Catherine M. Phelan, Global ovarian cancer health disparities, pp. 258–64, doi.org/10.1016/j.ygyno.2012.12.016.

Ovarian cancer is associated with BRCA1 and BRCA2 genes and is often present as a cancer risk complex along with breast cancer (Couch et al. 2013). Other genes that contribute to ovarian cancer risk are MSH6, MSH2, and MLH1, as well as RAD51D mutations (Girolimetti et al. 2014; Song et al. 2015). Similar to breast cancer, ovarian cancer risk and specific genes associated with specific populations is somewhat rare.

Gene–environment interactions most likely lie at the heart of the variation in ovarian cancer rate between populations. That is, differences in oestrogens due to variations in diet, lifestyle, and activity contribute to ovarian cancer risk. Individual lifestyle differences associated with industrialised populations may also contribute to increases in lifetime oestrogen exposure which may contribute to ovarian cancer risk. This includes the delay in age of first reproduction and fewer pregnancies compared to women in the past. Career and life choices of many professional women result in a greater number of cycles and high levels of oestrogen exposure during those cycles (Eaton et al. 1994). For physicians, a greater awareness of the adaptive nature of ovarian function as well as the prevalence of common human variation may provide a more nuanced perspective when addressing ovarian cancer risk.

15.6.4 Prostate Cancer

The prostate is a small walnut-sized gland that surrounds the male urethra. Most mammals and even some gastropods are equipped with a prostate (Olson 2009), which is very large in some marsupials (Tyndale-Biscoe 1987). The function of the prostate is to produce supporting fluids for male reproductive function, such as to support spermatogenesis and fluids that are deployed during coitus. As men age, usually after the age of 40, the prostate is at greater risk of exhibiting hypertrophy known as benign prostatic hyperplasia (BPH). While producing uncomfortable symptoms such as pain as well as difficulty with and frequent urination, BPH does not always lead to prostate cancer but is a sign of prostate dysfunction that may be indicative of a growth risk of prostate cancer (Boehm et al. 2015). Nonetheless, BPH is indicative of prostate health challenges that emerge with age.

Prostate cancer is one of the primary sources of mortality in men in developed parts of the world (Center et al. 2012). From a comparative anatomical point of view, it is known that male dogs (*Canis familiaris*) suffer from this condition just as humans do. Risk factors in dogs include age, breed, and castration; the latter significantly increases the chance of this cancer, yet it is not the surgical intervention as such but the fact that the hormonal balance of the animal is changed. No data could be found on prostate cancer in wolves (*Canis lupus*); thus, it would be interesting to explore if domestication played a role in producing differential risks for dogs to develop prostate cancer. In regard to humans, the causes of prostate cancer are the same as other reproductive cancers, with important genetic, environmental, and lifestyle factors all contributing to risk and incidence. A few genes have been associated with increased prostate cancer risk; however, their contributions to risk remain uncertain (Punnen and Cooperberg 2013). The aetiology of prostate cancer stems from the growth-promoting properties of androgens, specifically dihydrotestosterone and testosterone, on the prostate. Drugs that block the synthesis of dihydrotestosterone and testosterone, as well as the adrenal androgen androstenedione, are common treatment strategies for prostate cancer (Hoque et al. 2015). Evidence for associations between variation in endogenous testosterone levels and prostate cancer risk has been mixed. The

Endogenous Hormones and Prostate Cancer Collaborative Group conducted a meta-analysis of 18 different studies to assess the evidence for any association between endogenous hormones such as testosterone and prostate cancer. They concluded that there was no compelling evidence for an association between prostate cancer risk and endogenous-free testosterone, dihydrotestosterone, DHEA-S, androstenedione, androstanediol glucuronide, oestradiol, or free oestradiol (Endogenous Hormones and Prostate Cancer Collaborative Group et al. 2008).

These studies, however, were largely limited to western industrialised populations and did not consider variation in androgens or disparities in prostate cancer risk between populations living under different ecologies, especially those with contrasting risks for prostate cancer. When examined through the lens of population-level analyses, evidence emerges that suggests that while individual testosterone and other hormone levels are not indicative of prostate cancer risk, ratios of mean testosterone levels between those with different prostate cancer risk do reveal associations that link testosterone and prostate cancer risk (Calistro Alvarado 2010) (Figure 15.17).

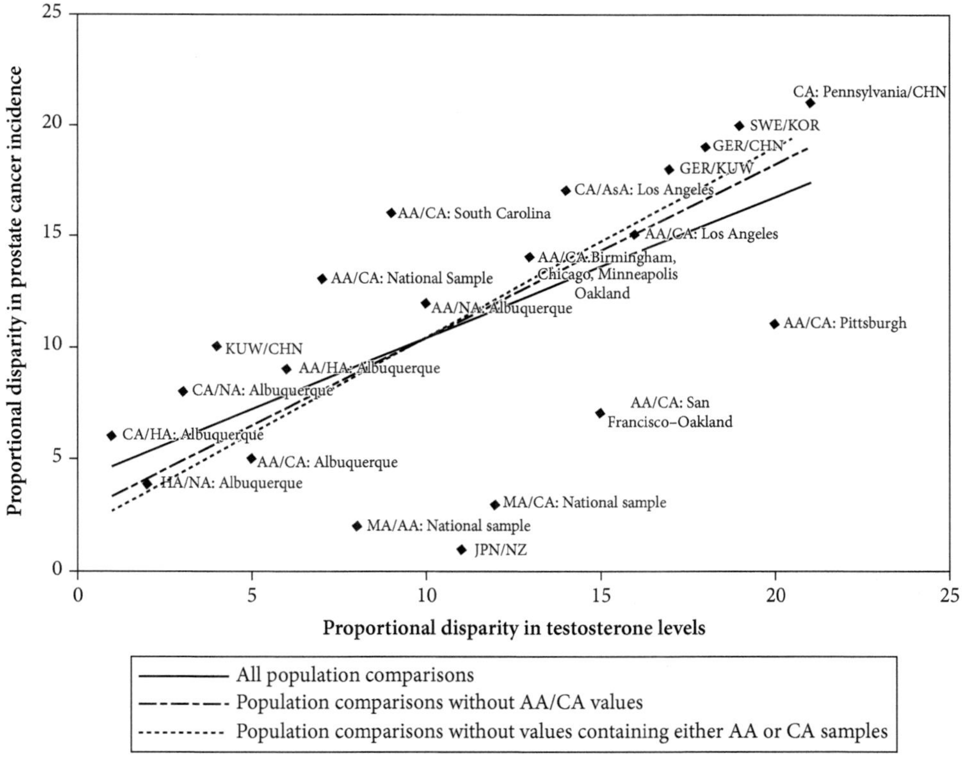

FIGURE 15.17 Proportional disparity in prostate cancer incidence and testosterone. AA, African-American; AsA, Asian-American; CA, Caucasian-American; HA, Hispanic-American; MA, Mexican-American; NA, Native-American; CHN, Chinese; GER, German; JPN, Japanese; KOR, South Korean; NZ, New Zealander; SWE Swedish.

Source: Reproduced from Louis Calistro Alvarado, Population differences in the testosterone levels of young men are associated with prostate cancer disparities in older men, *American Journal of Human Biology*, 22 (4), pp. 449–455, doi.10.1002/ajhb.21016. Copyright © 2010 Wiley-Liss, Inc.

15.6.5 Polycystic Ovarian Syndrome

PCOS is a condition in which ovarian follicles overdevelop and create scarring and lesions that can compromise fertility and lead to a number of other health issues. PCOS is especially common in women who are obese, diabetic, or exhibit other symptoms of metabolic syndrome (Lefebvre et al. 1997). The primary cause of PCOS is overstimulation of cells that support follicle development. In particular, insulin under normal circumstances stimulates theka cells within the ovary to produce androstenedione and testosterone, which are converted by the enzyme aromatase within neighbouring granulosa cells into oestradiol, which then supports follicle development (Agilli et al. 2015). During periods of hyperinsulinaemia, which is common in obese women and those with diabetes and insulin resistance, theka cells become hyperstimulated and produce an overabundance of testosterone which leads to a hyperproduction of oestradiol and hypertrophy of ovarian follicles. This hypertrophy is what leads to the development of cysts and tissue damage. Moreover, sometimes the production of testosterone exceeds the ovary's conversion ability to oestradiol, resulting in excess androgens in general circulation. This leads to symptoms that are often observed in diabetic women and those with insulin resistance, such as hirsutism and alopecia.

It has been suggested that PCOS is evident in pre-modern populations (Azziz et al. 2011). Clinical researchers have expressed interest in the role of natural selection in the emergence of PCOS (Lefebvre et al. 1997). The evolutionary biology of PCOS has also recently been explored (Agilli et al. 2015); however, given the strong association with obesity, diabetes, and metabolic syndrome, it is likely that PCOS is an example of a reproductive health challenge that has emerged due to modern diets and sedentism. It is also important to note that PCOS is especially prevalent in poorly nourished communities that have limited access to high-quality foods and who are more likely to subsist on highly calorically dense, glucose-rich, and high-fat diets (Hillman et al. 2014).

15.6.6 Oestrogen Replacement/Supplementation

Oestrogen replacement in women, especially in those who have undergone or who are undergoing menopause, has been controversial. While the effects of oestrogen supplementation can alleviate some of the less-desirable effects of menopause such as hot flushes, the risks of exacerbating the possibility of reproductive cancers remains worthy of additional research. Of particular concern is lack of awareness of the range of common and normal variability in ovarian hormone physiology between and within populations. Significant non-pathological variation has been shown in numerous populations. The potential sources of this variation lie in differences in production at one or more locales within the HPO axis, differences in clearance rates, or variation in binding proteins that carry steroids while in circulation.

Differences in production are likely to be the major contributor, since changes in metabolic cues such as insulin can upregulate GnRH production or oestradiol output by the ovaries (Navratil et al. 2009). For example, clearance rates of common oral contraceptives and transdermally administered oestradiol varied significantly between populations, suggesting that identical dosages are subject to very different pharmacokinetics and possibly different contraindications such as reproductive cancer risk (Bentley 1994; Huddleston et al. 2011).

15.6.7 Testosterone Replacement/Supplementation

Testosterone supplementation has been deployed as a way to redress the detrimental effects of the loss of endogenous production due to surgery, injury, illness, or treatment for other more serious issues such as testicular cancer. However, since the 2000s, the number of prescriptions and overall use of testosterone and other androgens as supplements has increased dramatically. The driving factor behind this growth is the perception that testosterone can provide relief to common health and well-being challenges in many men. These include weight gain, loss of muscle mass, decline in libido, depressive mood, and erectile dysfunction. Another trend is the deployment of testosterone in men who do not exhibit any of these symptoms or any other evidence of hypogonadism. These men receive testosterone in response to general malaise or as a way of improving their overall outlook on life (Araujo et al. 2007). Other men obtain testosterone without medical guidance for similar purposes or as a way of enhancing athletic performance.

The effects of testosterone on these health challenges are clear. It is well documented that testosterone can alleviate many of these symptoms and there is evidence that low testosterone is associated with greater male mortality, although low testosterone often covaries with other health challenges that compromise health and survivorship (Araujo et al. 2007). However, what has not been clearly delineated is the potential longitudinal detrimental effects of testosterone supplementation. In non-human animal models, testosterone supplementation has been shown to enhance somatic and behavioural reflections of reproductive effort such as greater muscle mass, competitiveness, and libido. Other effects such as catabolism of fat stores often compromise survivorship in the wild. In human males, this issue is especially salient in regard to common variation in testosterone levels. This has been extensively documented by human evolutionary biologists.

As with women, there is also the issue of population variation in the pharmacokinetics of androgens. For example, testosterone clearance rates have been shown to vary significantly between men who self-identify as Caucasian and those of Chinese ancestry (Santner et al. 1998). Genetic variation has also been reported in the androgen receptor in various populations (Butovskaya et al. 2015).

15.6.8 Reproductive Effort and Ageing

Life history theory predicts that investment in reproductive effort could potentially make resources unavailable for somatic tissue repair. Since somatic degradation as the result of ageing is tempered by somatic repair mechanisms, it follows that reproductive effort can accelerate somatic deterioration in association with ageing and potentially shorten lifespan (Kirkwood and Rose 1991). This has been demonstrated in other non-human organisms but is poorly understood in humans (Westendorp and Kirkwood 1998). Recent evidence has emerged that suggests that reproductive effort in females is associated with shorter lifespans, possibly due to the physiological costs of reproduction.

Demographic data garnered from church records in rural Poland show that post-menopausal female lifespan is negatively associated with number of children. This association is observed in both sons and daughters. Daughters actually had a positive effect on paternal lifespan, while sons had no effect on the father's longevity (Jasienska et al. 2006b). Similar results were reported by Helle et al. (2002) in a pre-industrial Finnish population, but the effect was only noted in sons. Other studies have not observed any relationship between reproductive effort and lifespan; however, it has been suggested that this is due to the lack of the incorporation of energetic status which can confound this effect and produce a phenotypic correlation, that is, the existence of sufficient energetic resources to offset the potential cost of the stressor (Jasienska 2009).

One potential physiological cost of reproduction is oxidative stress, which is also a significant contributor to somatic degradation due to ageing (Dowling and Simmons 2009). In essence, reactive oxidative species (ROS), which are toxic by-products resulting from aerobic metabolism, cause cumulative tissue and genetic damage over a lifespan, despite mechanisms to cope with the clearing of ROS and to repair any damage. During pregnancy, oxygen use, aerobic metabolism, and the production of ROS increase significantly due to the demands of the growing fetus and supportive tissues (Hung et al. 2010). Evidence of the physiological cost of reproduction in women is shown in the assessment of biomarkers of oxidative stress in association with measures of reproductive effort in rural Polish women. 8-hydroxy-2'-deoxyguanosine (8-OHdG), a biomarker of genetic repair of ROS-induced damage, and superoxide dismutase, an intracellular protective agent against ROS, were both higher in women who had more than four children compared to those who had fewer (Ziomkiewicz et al. 2016). American women, however, did not exhibit this relationship, suggesting that energetic status may have muted this effect (Ziomkiewicz et al. 2017). Very little is known about oxidative stress in men, although similar trade-offs have been reported in animal models (Alonso-Alvarez et al. 2007).

15.7 Conclusion

Human reproductive biology is more coherently understood through the lens of evolution and life history theory. Moreover, as with many aspects of human biology, reproduction is a function of the interface between genes, environmental challenges, and physiological mechanisms which negotiate these challenges in the face of other somatic needs such as growth and immune function. Human reproduction is also best understood within the context of human variation. While biomedical research has made tremendous strides in our understanding and treatment of reproductive health challenges, much work remains in engaging the full range of human biological variation, as well as environmental diversity in terms of diet, activity, lifestyle, and social stresses.

It is important to note that humans are still evolving. The tempo and pace of our future evolution remains to be seen, but what is certain is that natural selection, that is, the non-random elimination of genes from populations, will almost certainly continue (Bribiescas 2011). This is also true of the pathogens that challenge our health. The primary risk today appears to be environmental stochasticity due to climate change and the accompanying challenges such as migration, food shortages, new pathogenic challenges, and the lack of clean water. Of particular relevance to reproduction is the increase in endocrine disruptors that may affect our reproductive biology in ways that we cannot predict (Grantham and Henneberg 2014; Tapia-Orozco et al. 2017). These are the issues that should bring evolutionary biologists and clinicians together.

References

Abbasi, S., Nouri, M., and Azimi, C. (2012). Estrogen receptor genes variations and breast cancer risk in Iran. *Int J Clin Exp Med* 5, 332–41.

Ackerman, C. M., Lowe, L. P., Lee, H., et al. (2012). Ethnic variation in allele distribution of the androgen receptor (AR) (CAG)n repeat. *J Androl* 33, 210–15.

Agilli, M., Aydin, F. N., Cayci, T., et al. (2015). Insulin sensitivity and leptin in women with PCOS. *Clin Endocrinol* 82, 776.

Aiello, L. C. and Key, C. (2002). Energetic consequences of being a *Homo erectus* female. *Am J Hum Biol* 14, 551–65.

Aliesky, H., Courtney, C. L., Rapoport, B., et al. (2013). Thyroid autoantibodies are rare in nonhuman great apes and hypothyroidism cannot be attributed to thyroid autoimmunity. *Endocrinology* 154, 4896–907.

Alonso-Alvarez, C., Bertrand, S., Devevey, G., et al. (2004). Increased susceptibility to oxidative stress as a proximate cost of reproduction. *Ecol Lett* 7, 363–8.

Alonso-Alvarez, C., Bertrand, S., Faivre, B., et al. (2007). Testosterone and oxidative stress: the oxidation handicap hypothesis. *Proc Biol Sci* 274, 819–25.

Amir, D., Ellison, P. T., Hill, K. R., and et al. (2015). Diurnal variation in salivary cortisol across age classes in Ache Amerindian males of Paraguay. *Am J Hum Biol* 27, 344–8.

Amir, D., Jordan, M. R., and Bribiescas, R. G. (2016). A longitudinal assessment of associations between adolescent environment, adversity perception, and economic status on fertility and age of menarche. *PLoS One* 11, e0155883.

Anderson, R. A., McLaughlin, M., Wallace, W. H., et al. (2014). The immature human ovary shows loss of abnormal follicles and increasing follicle developmental competence through childhood and adolescence. *Hum Reprod* 29, 97–106.

Anestis, S. F. (2010). Hormones and social behavior in primates. *Evol Anthropol* 19, 66–78.

Aramaki, Y., Oae, H., Mouri, Y., et al. (2010). Urinary estrogens, progesterone, and LH changes during normal menstrual cycles of a captive female pileated gibbon (*Hylobates pileatus*). *J Med Primatol* 39, 381–4.

Araujo, A. B., Esche, G. R., Kupelian, V., et al. (2007). Prevalence of symptomatic androgen deficiency in men. *J Clin Endocrinol Metab* 92, 4241–7.

Azziz, R., Dumesic, D. A., and Goodarzi, M. O. (2011). Polycystic ovary syndrome: an ancient disorder? *Fertil Steril* 95, 1544–8.

Baker, M. E. (2011). Origin and diversification of steroids: co-evolution of enzymes and nuclear receptors. *Mol Cell Endocrinol* 334, 14–20.

Baker, M. E., Funder, J. W., and Kattoula, S. R. (2013). Evolution of hormone selectivity in glucocorticoid and mineralocorticoid receptors. *J Steroid Biochem Mol Biol* 137, 57–70.

Bambino, T. H. and Hsueh, A. J. (1981). Direct inhibitory effect of glucocorticoids upon testicular luteinizing hormone receptor and steroidogenesis in vivo and in vitro. *Endocrinology* 108, 2142–8.

Banks, W. A., Altmann, J., Sapolsky, R. M., et al. (2003). Serum leptin levels as a marker for a syndrome X-like condition in wild baboons. *J Clin Endocrinol Metabol* 88, 1234–40.

Bardsley, M. Z., Kowal, K., Levy, C., et al. (2013). 47,XYY syndrome: clinical phenotype and timing of ascertainment. J Pediatr 163(4), 1085–94.

Beehner, J. C. and Lu, A. (2013). Reproductive suppression in female primates: a review. *Evol Anthropol* 22, 226–38.

Beehner, J. C., Nguyen, N., Wango, E. O., et al. (2006). The endocrinology of pregnancy and fetal loss in wild baboons. *Horm Behav* 49, 688–99.

Behringer, V., Hohmann, G., Stevens, J. M., et al. (2012). Adrenarche in bonobos (*Pan paniscus*): evidence from ontogenetic changes in urinary dehydroepiandrosterone-sulfate levels. *J Endocrinol* 214, 55–65.

Behringer, V., Deschner, T., Murtagh, R., et al. (2014). Age-related changes in thyroid hormone levels of bonobos and chimpanzees indicate heterochrony in development. *J Hum Evol* 66, 83–8.

Bell, G. (1980). The costs of reproduction and their consequences. *Am Nat* 116, 45–76.

Bentley, G. R. (1994). Ranging hormones: do hormonal contraceptives ignore human biological variation and evolution? *Ann N Y Acad Sci* 709, 201–3.

Bentley, G. R. (1999). Aping our ancestors: comparative aspects of reproductive ecology. *Evol Anthropol* 7, 175–85.

Bentley, G. R., Harrigan, A. M., and Ellison, P. T. (1998). Dietary composition and ovarian function among Lese horticulturalist women of the Ituri Forest, Democratic Republic of Congo. *Eur J Clin Nutr* 52, 261–70.

Bergendahl, M., Aloi, J. A., Iranmanesh, A., et al. (1998). Fasting suppresses pulsatile luteinizing hormone (LH) secretion and enhances orderliness of LH release in young but not older men. *J Clin Endocrinol Metab* 83, 1967–75.

Berner, L. A., Feig, E. H., Witt, A. A. et al. (2017). Menstrual cycle loss and resumption among patients with anorexia nervosa spectrum eating disorders: is relative or absolute weight more influential? *Int J Eat Disord* 50(4), 442–6.

Bianco, S. D. and Kaiser, U. B. (2009). The genetic and molecular basis of idiopathic hypogonadotropic hypogonadism. *Nat Rev Endocrinol* 5, 569–76.

Boehm, K., Valdivieso, R., Meskawi, M., et al. (2015). BPH: a tell-tale sign of prostate cancer? Results from the Prostate Cancer and Environment Study (PROtEuS). *World J Urol* 33, 2063–9.

Bongaarts, J. and Potter, R. G. (1983). *Natural Fertility and its Proximate Determinants. Fertility, Biology, and Behavior: An Analysis of the Proximate Determinants.* New York: Academic Press.

Bonomi, M., Rochira, V., Pasquali, D., et al. (2017). Klinefelter syndrome (KS): genetics, clinical phenotype and hypogonadism. *J Endocrinol Invest* 40, 123–34.

Braams, B. R., Van Duijvenvoorde, A. C., Peper, J. S., et al. (2015). Longitudinal changes in adolescent risk-taking: a comprehensive study of neural responses to rewards, pubertal development, and risk-taking behavior. *J Neurosci* 35, 7226–38.

Bremner, W. J., Vitiello, M. V., and Prinz, P. N. (1983). Loss of circadian rhythmicity in blood testosterone levels with aging in normal men. *J Clin Endocrinol Metab* 56, 1278–81.

Bribiescas, R. G. (2001a). Reproductive ecology and life history of the human male. *Am J Phys Anthropol* Suppl 33, 148–76.

Bribiescas, R. G. (2001b). Serum leptin levels and anthropometric correlates in Ache Amerindians of eastern Paraguay. *Am J Phys Anthropol* 115, 297–303.

Bribiescas, R. G. (2005). Age-related differences in serum gonadotropin (FSH and LH), salivary testosterone, and 17-beta estradiol levels among Ache Amerindian males of Paraguay. *Am J Phys Anthropol* 127, 114–21.

Bribiescas, R. G. (2011). An evolutionary and life history perspective on the role of males on human futures. *Futures* 43, 729–39.

Bribiescas, R. G. and Anestis, S. F. (2010). Leptin associations with age, weight, and sex among chimpanzees (*Pan troglodytes*). *J Med Primatol* 39, 347–55.

Bribiescas, R. G. and Hill, K. R. (2010). Circadian variation in salivary testosterone across age classes in Ache Amerindian males of Paraguay. *Am J Hum Biol* 22, 216–20.

Bribiescas, R. G., Betancourt, J., Torres, A. M., et al. (2008). Active ghrelin levels across time and associations with leptin and anthropometrics in healthy Ache Amerindian women of Paraguay. *Am J Hum Biol* 20, 352–4.

Bribiescas, R. G., Ellison, P. T., and Gray, P. B. (2012). Male life history, reproductive effort, and the evolution of the genus *Homo*: new directions and perspectives. *Curr Anthropol* 53, S424–35.

Brodin, T., Bergh, T., Berglund, L., et al. (2008). Menstrual cycle length is an age-independent marker of female fertility: results from 6271 treatment cycles of in vitro fertilization. *Fertil Steril* 90, 1656–61.

Brown, F., Harris, J., Leakey, R., et al. (1985). Early *Homo erectus* skeleton from west Lake Turkana, Kenya. *Nature* 316, 788–92.

Bulsari, K. and Falhammar, H. (2017). Clinical perspectives in congenital adrenal hyperplasia due to 11 beta-hydroxylase deficiency. Endocrine 55(1), 19–36.

Butovskaya, M. L., Lazebny, O. E., Vasilyev, V. A., et al. (2015). Androgen receptor gene polymorphism, aggression, and reproduction in Tanzanian foragers and pastoralists. *PLoS One* 10, e0136208.

Butte, N. F. and King, J. C. (2005). Energy requirements during pregnancy and lactation. *Public Health Nutr* 8, 1010–27.

Calistro Alvarado, L. (2010). Population differences in the testosterone levels of young men are associated with prostate cancer disparities in older men. *Am J Hum Biol* 22, 449–55.

Campbell, B. (2005). High rate of prostate symptoms among Ariaal men from northern Kenya. *Prostate* 62, 83–90.

Campbell, B. (2006). Adrenarche and the evolution of human life history. *Am J Hum Biol* 18, 569–89.

Campbell, B. and Mbizo, M. (2006). Reproductive maturation, somatic growth and testosterone among Zimbabwe boys. *Ann Hum Biol* 33, 17–25.

Campbell, B. C., Gillett-Netting, R., and Meloy, M. (2004). Timing of reproductive maturation in rural versus urban Tonga boys, Zambia. *Ann Hum Biol* 31, 213–27.

Carel, J. C., Lahlou, N., Roger, M., et al. (2004). Precocious puberty and statural growth. *Hum Reprod Update* 10, 135–47.

Casarini, L., Santi, D., and Marino, M. (2015). Impact of gene polymorphisms of gonadotropins and their receptors on human reproductive success. *Reproduction* 150, R175–84.

Center, M. M., Jemal, A., Lortet-Tieulent, J., et al. (2012). International variation in prostate cancer incidence and mortality rates. *Eur Urol* 61, 1079–92.

Cepon, T. J., Snodgrass, J. J., Leonard, W. R., et al. (2011). Circumpolar adaptation, social change, and the development of autoimmune thyroid disorders among the Yakut (Sakha) of Siberia. *Am J Hum Biol* 23, 703–9.

Chang, C. L., Cai, J. J., Cheng, P. J., et al. (2011). Identification of metabolic modifiers that underlie phenotypic variations in energy-balance regulation. *Diabetes* 60, 726–34.

Chavez-Macgregor, M., Liu, S., De Melo-Gagliato, D., et al. (2014). Differences in gene and protein expression and the effects of race/ethnicity on breast cancer subtypes. *Cancer Epidemiol Biomarkers Prev* 23, 316–23.

Chene, G., Tardieu, A. S., Trombert, B., et al. (2013). Lucy's parturition, a way towards extinction? *Gynecol Obstet Fertil* 41, 478–84.

Chornokur, G., Amankwah, E. K., Schildkraut, J. M., et al. (2013). Global ovarian cancer health disparities. *Gynecol Oncol* 129, 258–64.

Cieri, R. L., Churchill, S., Franciscus, R. G., et al. (2014). Craniofacial feminization, social tolerance, and the origins of behavioral modernity. *Curr Anthropol* 55, 419–43.

Clancy, K. B. (2007). Unexpected luteal endometrial decline in a healthy rural Polish population. *Eur J Obstet Gynecol Reprod Biol* 134, 133–4.

Clancy, K. B. (2009). Reproductive ecology and the endometrium: physiology, variation, and new directions. *Am J Phys Anthropol* 140(Suppl 49), 137–54.

Clarkson, J. and Herbison, A. E. (2016). Hypothalamic control of the male neonatal testosterone surge. *Philos Trans R Soc Lond B Biol Sci* 371, 20150115.

Coe, C. L., Savage, A., and Bromley, L. J. (1992). Phylogenetic influences on hormone levels across the Primate order. *Am J Primatol* 28, 81–100.

Conley, A. J., Bernstein, R. M., and Nguyen, A. D. (2012). Adrenarche in nonhuman primates: the evidence for it and the need to redefine it. *J Endocrinol* 214, 121–31.

Corbier, P., Edwards, D. A., and Roffi, J. (1992). The neonatal testosterone surge: a comparative study. *Arch Int Physiol Biochim Biophys* 100, 127–31.

Couch, F. J., Wang, X., Mcguffog, L., et al. (2013). Genome-wide association study in BRCA1 mutation carriers identifies novel loci associated with breast and ovarian cancer risk. *PLoS Genet* 9, e1003212.

Crockford, S. J. (2009). Evolutionary roots of iodine and thyroid hormones in cell–cell signaling. *Integr Comp Biol* 49, 155–66.

Croft, D. P., Johnstone, R. A., Ellis, S., et al. (2017). Reproductive Conflict and the Evolution of Menopause in Killer Whales. Curr Biol 27(2), 298–304.

Crow, J. F. (2000). The origins, patterns and implications of human spontaneous mutation. *Nat Rev Genet* 1, 40–7.

De Sanctis, V., Soliman, A. T., Elsedfy, H., et al. (2015). Precocious puberty following traumatic brain injury in early childhood: a review of the literature. *Pediatr Endocrinol Rev* 13, 458–64.

Deapen, D., Liu, L., Perkins, C., et al. (2002). Rapidly rising breast cancer incidence rates among Asian-American women. *Int J Cancer* 99, 747–50.

Den Heijer, T., Schuit, S. C., Pols, H. A., et al. (2004). Variations in estrogen receptor alpha gene and risk of dementia, and brain volumes on MRI. *Mol Psychiatry* 9, 1129–35.

Desai, S. S., Roy, B. S., and Mahale, S. D. (2013). Mutations and polymorphisms in FSH receptor: functional implications in human reproduction. *Reproduction* 146, R235–48.

Desantis, C. E., Bray, F., Ferlay, J., et al. (2015). International variation in female breast cancer incidence and mortality rates. *Cancer Epidemiol Biomarkers Prev* 24, 1495–506.

Desmedt, C., Yates, L., and Kulka, J. (2016). Catalog of genetic progression of human cancers: breast cancer. *Cancer Metastasis Rev* 35, 49–62.

Dewitt, T. J., Sih, A., and Wilson, D. S. (1998). Costs and limits of phenotypic plasticity. *Trends Ecol Evol* 13, 77–81.

Doman, J., Hill, A., and Bribiescas, R. G. (2013). A new osteological approach to inferring hominin social behavior: seeking facial indicators of testosterone level. *Paleoanthropol Soc Meetg Abstr Honolulu* 2013.

Doroftei, B., Mambet, C., and Zlei, M. (2015). It's never over until it's over: how can age and ovarian reserve be mathematically bound through the measurement of serum AMH—a study of 5069 Romanian women. *PLoS One* 10, e0125216.

Dowling, D. K. and Simmons, L. W. (2009). Reactive oxygen species as universal constraints in life-history evolution. *Proc Biol Sci* 276, 1737–45.

Dunsworth, H. M., Warrener, A. G., Deacon, T., et al. (2012). Metabolic hypothesis for human altriciality. *Proc Natl Acad Sci U S A* 109, 15212–16.

Dupont, J. and Scaramuzzi, R. J. (2016). Insulin signalling and glucose transport in the ovary and ovarian function during the ovarian cycle. *Biochem J* 473, 1483–501.

Eaton, S. B., Pike, M. C., Short, R. V., et al. (1994). Women's reproductive cancers in evolutionary context. *Q Rev Biol* 69, 353–67.

Ellison, P. T. (1982). Skeletal growth, fatness and menarcheal age: a comparison of two hypotheses. *Hum Biol* 54, 269–81.

Ellison, P. T. (1996). Developmental influences on adult ovarian function. *Am J Hum Biol* 8, 725–34.

Ellison, P. T. (1999). Reproductive ecology and reproductive cancers. In: Worthman, C. M. (ed.) *Hormones and Human Health*. Cambridge: Cambridge University Press.

Ellison, P. T. (2001). *Reproductive Ecology and Human Evolution*. New York: Aldine de Gruyter.

Ellison, P. T. and Lager, C. (1986). Moderate recreational running is associated with lowered salivary progesterone profiles in women. *Am J Obstet Gynecol* 154, 1000–3.

Ellison, P. T. and Valeggia, C. R. (2003). C-peptide levels and the duration of lactational amenorrhea. *Fertil Steril* 80, 1279–80.

Ellison, P. T., Lipson, S. F., Jasienska, G., et al. (2007). Moderate anxiety, whether acute or chronic, is not associated with ovarian suppression in healthy, well-nourished, western women. *Am J Phys Anthropol* 134, 513–9.

Emery Thompson, M. (2005). Reproductive endocrinology of wild female chimpanzees (*Pan troglodytes schweinfurthii*): methodological considerations and the role of hormones in sex and conception. *Am J Primatol* 67, 137–58.

Emery Thompson, M., Jones, J. H., et al. (2007). Aging and fertility patterns in wild chimpanzees provide insights into the evolution of menopause. *Curr Biol* 17, 2150–6.

Emery Thompson, M., Zhou, A., and Knott, C. D. (2012). Low testosterone correlates with delayed development in male orangutans. *PLoS One* 7, e47282.

Endogenous Hormones and Prostate Cancer Collaborative Group, et al. (2008). Endogenous sex hormones and prostate cancer: a collaborative analysis of 18 prospective studies. *J Natl Cancer Inst* 100, 170–83.

Fenton, S. E. and Birnbaum, L. S. (2015). Timing of environmental exposures as a critical element in breast cancer risk. *J Clin Endocrinol Metab* 100, 3245–50.

Fernandez-Duque, E., Valeggia, C. R., and Mendoza, S. P. (2009). The biology of paternal care in human and nonhuman primates. *Ann Rev Anthropol* 38, 115–30.

Finch, C. E. and Rose, M. R. (1995). Hormones and the physiological architecture of life history evolution. *Q Rev Biol* 70, 1–52.

Finn, C. A. (1998). Menstruation: a nonadaptive consequence of uterine evolution. *Q Rev Biol* 73, 163–73.

Fisch, H., Ikeguchi, E. F. and Goluboff, E. T. (1996). Worldwide variations in sperm counts. *Urology* 48, 909–11.

Fjeldheim, F. N., Frydenberg, H., Flote, V. G., et al. (2016). Polymorphisms in the estrogen receptor alpha gene (ESR1), daily cycling estrogen and mammographic density phenotypes. *BMC Cancer* 16, 776.

Fraser, A. M., Brockert, J. E., and Ward, R. H. (1995). Association of young maternal age with adverse reproductive outcomes [see comments]. *N Engl J Med* 332, 1113–17.

Frisch, R. E. (1974). Menstrual cycles: fatness as a determinant of minimum weight for height necessary for their maintenance or onset. *Science* 185(4155), 949–51.

Frisch, R. E. and Revelle, R. (1971). Height and weight at menarche and a hypothesis of menarche. *Arch Dis Child* 46, 695–701.

Funabashi, T., Daikoku, S., Shinohara, K., et al. (2000). Pulsatile gonadotropin-releasing hormone (GnRH) secretion is an inherent function of GnRH neurons, as revealed by the culture of medial olfactory placode obtained from embryonic rats. *Neuroendocrinology* 71, 138–44.

Gagneux, P., Amess, B., Diaz, S., et al. (2001). Proteomic comparison of human and great ape blood plasma reveals conserved glycosylation and differences in thyroid hormone metabolism. *Am J Phys Anthropol* 115, 99–109.

Garcia-Mayor, R. V., Andrade, M. A., Rios, M., et al. (1997). Serum leptin levels in normal children: relationship to age, gender, body mass index, pituitary-gonadal hormones, and pubertal stage. *J Clin Endocrinol Metab* 82, 2849–55.

Gettler, L. T., Mcdade, T. W., and Kuzawa, C. W. (2011). Cortisol and testosterone in Filipino young adult men: evidence for co-regulation of both hormones by fatherhood and relationship status. *Am J Hum Biol* 23, 609–20.

Girolimetti, G., Perrone, A. M., Santini, D., et al. (2014). BRCA-associated ovarian cancer: from molecular genetics to risk management. *Biomed Res Int* 2014, 787143.

Goldstein, J. R. (2011). A secular trend toward earlier male sexual maturity: evidence from shifting ages of male young adult mortality. *PLoS One* 6, e14826.

Graffelman, J., Fugger, E. F., Keyvanfar, K., et al. (1999). Human live birth and sperm-sex ratios compared. *Hum Reprod* 14, 2917–20.

Grantham, J. P. and Henneberg, M. (2014). The estrogen hypothesis of obesity. *PLoS One* 9, e99776.

Gray, P. B. and Anderson, K. G. (2010). *Fatherhood: Evolution and Human Paternal Behavior*. Cambridge, MA: Harvard University Press.

Gray, P. and Campbell, B. (2005). Erectile dysfunction and its correlates among the Ariaal of northern Kenya. *Int J Impot Res* 17, 445–9.

Gray, P. B. and Garcia, J. R. (2013). *Evolution and Human Sexual Behavior*. Cambridge, MA: Harvard University Press.

Gruss, L. T. and Schmitt, D. (2015). The evolution of the human pelvis: changing adaptations to bipedalism, obstetrics and thermoregulation. *Philos Trans R Soc Lond B Biol Sci* 370, 20140063.

Habumuremyi, S., Robbins, M. M., Fawcett, K. A., et al. (2014). Monitoring ovarian cycle activity via progestagens in urine and feces of female mountain gorillas: a comparison of EIA and LC-MS measurements. *Am J Primatol* 76, 180–91.

Habumuremyi, S., Stephens, C., Fawcett, K. A., et al. (2016). Endocrine assessment of ovarian cycle activity in wild female mountain gorillas (*Gorilla beringei beringei*). *Physiol Behav* 157, 185–95.

Haig, D. (1993). Genetic conflicts in human pregnancy. *Q Rev Biol* 68, 495–532.

Hawkes, K., O'Connell, J. F., and Blurton-Jones, N. G. (1989). Hardworking Hadza grandmothers. In: Foley, R. (ed.) *Comparative Socioecology*. London: Blackwell.

Hawkes, K., O'Connell, J. F., Jones, N. G. B., et al. (1998). Grandmothering, menopause, and the evolution of human life histories. *Proc Natl Acad Sci U S A* 95, 1336–9.

Helle, S., Lummaa, V., and Jokela, J. (2002). Sons reduced maternal longevity in preindustrial humans. *Science* 296, 1085.

Henry, L. (1961). Some data on natural fertility. *Eugen Q* June, 81–91.

Hill, J. W., Elias, C. F., Fukuda, M., et al. (2010). Direct insulin and leptin action on pro-opiomelanocortin neurons is required for normal glucose homeostasis and fertility. *Cell Metab* 11, 286–97.

Hill, K. R. (1993). Life history theory and evolutionary anthropology. *Evol Anthropol* 2, 78–88.

Hill, K. and Hurtado, A. M. (1991). The evolution of reproductive senescence and menopause in human females. *Hum Nat* 2, 315–50.

Hill, K., Boesch, C., Goodall, J., et al. (2001). Mortality rates among wild chimpanzees. *J Hum Evol* 40, 437–50.

Hillman, J. K., Johnson, L. N., Limaye, M., et al. (2014). Black women with polycystic ovary syndrome (PCOS) have increased risk for metabolic syndrome and cardiovascular disease compared with white women with PCOS [corrected]. *Fertil Steril* 101, 530–5.

Hillmer, A. M., Hanneken, S., Ritzmann, S., et al. (2005). Genetic variation in the human androgen receptor gene is the major determinant of common early-onset androgenetic alopecia. *Am J Hum Genet* 77, 140–8.

Hoque, A., Yao, S., Till, C., et al. (2015). Effect of finasteride on serum androstenedione and risk of prostate cancer within the prostate cancer prevention trial: differential effect on high- and low-grade disease. *Urology* 85, 616–20.

Howie, P. W. and Mcneilly, A. S. (1982). Effect of breast-feeding patterns on human birth intervals. *J Reprod Fertil* 65, 545–57.

Hrdy, S. B. (1999). *Mother Nature: A History of Mothers, Infants, and Natural Selection*. New York: Pantheon Books.

Huddleston, H. G., Rosen, M. P., Gibson, M., et al. (2011). Ethnic variation in estradiol metabolism in reproductive age Asian and white women treated with transdermal estradiol. *Fertil Steril* 96, 797–9.

Hung, T. H., Lo, L. M., Chiu, T. H., et al. (2010). A longitudinal study of oxidative stress and antioxidant status in women with uncomplicated pregnancies throughout gestation. *Reprod Sci* 17, 401–9.

Iuliano-Burns, S., Hopper, J., and Seeman, E. (2009). The age of puberty determines sexual dimorphism in bone structure: a male/female co-twin control study. *J Clin Endocrinol Metab* 94, 1638–43.

Iwasaki, M., Mameri, C. P., Hamada, G. S., et al. (2008). Secular trends in cancer mortality among Japanese immigrants in the state of Sao Paulo, Brazil, 1979–2001. *Eur J Cancer Prev* 17, 1–8.

Jaimez, N. A., Bribiescas, R. G., Aronsen, G. P., et al. (2011). Urinary cortisol differences in grey cheeked mangabeys (*Lophocebus albigena*) inhabiting disturbed and undisturbed forest areas of Kibale National Park, Uganda. *Am J Phys Anthropol* 144, 175–5.

James-Todd, T., Tehranifar, P., Rich-Edwards, J., et al. (2010). The impact of socioeconomic status across early life on age at menarche among a racially diverse population of girls. *Ann Epidemiol* 20, 836–42.

Jamison, C. S., Cornell, L. L., Jamison, P. L., et al. (2002). Are all grandmothers equal? A review and a preliminary test of the 'grandmother hypothesis' in Tokugawa Japan. *Am J Phys Anthropol* 119, 67–76.

Jasienska, G. (2009). Reproduction and lifespan: trade-offs, overall energy budgets, intergenerational costs, and costs neglected by research. *Am J Hum Biol* 21, 524–32.

Jasienska, G. and Ellison, P. T. (1998). Physical work causes suppression of ovarian function in women. *Proc R Soc Lond B Biol Sci* 265, 1847–51.

Jasienska, G. and Ellison, P. T. (2004). Energetic factors and seasonal changes in ovarian function in women from rural Poland. *Am J Hum Biol* 16, 563–80.

Jasienska, G., Kapiszewska, M., Ellison, P. T., et al. (2006a). CYP17 genotypes differ in salivary 17-beta estradiol levels: a study based on hormonal profiles from entire menstrual cycles. *Cancer Epidemiol Biomarkers Prev* 15, 2131–5.

Jasienska, G., Nenko, I., and Jasienski, M. (2006b). Daughters increase longevity of fathers, but daughters and sons equally reduce longevity of mothers. *Am J Hum Biol* 18, 422–5.

Jasienska, G., Ellison, P. T., Galbarczyk, A., et al. (2015). Apolipoprotein E (ApoE) polymorphism is related to differences in potential fertility in women: a case of antagonistic pleiotropy? *Proc Biol Sci* 282, 20142395.

Jones, K. P., Walker, L. C., Anderson, D., et al. (2007). Depletion of ovarian follicles with age in chimpanzees: similarities to humans. *Biol Reprod* 77, 247–51.

Jukic, A. M., Baird, D. D., Weinberg, C. R., et al. (2013). Length of human pregnancy and contributors to its natural variation. *Hum Reprod* 28, 2848–55.

Key, C. and Ross, C. (1999). Sex differences in energy expenditure in non-human primates. Proc R Soc Lond 266(1437), 2479–85.

Kingery, S. E. and Wintergerst, K. A. (2015). Turner syndrome and Klinefelter syndrome. *Adolesc Med State Art Rev* 26, 411–27.

Kinoshita, K., Kuze, N., Kobayashi, T., et al. (2016). Detection of urinary estrogen conjugates and creatinine using near infrared spectroscopy in Bornean orangutans (*Pongo pygmaeus*). *Primates* 57, 51–9.

Kirkwood, J. K. and Rose, M. (1991). Evolution of senescence: late survivial sacrificed for reproduction. In: Southwood, L. (ed.) *The Evolution of Reproductive Strategies*. Cambridge: Cambridge University Press.

Kleibl, Z. and Kristensen, V. N. (2016). Women at high risk of breast cancer: molecular characteristics, clinical presentation and management. *Breast* 28, 136–44.

Knott, C. D. (1999). Reproductive, physiological and behavioral responses of orangutans in Borneo to fluctuations in food availability. Ph.D thesis, Harvard University.

Konner, M. and Worthman, C. (1980). Nursing frequency, gonadal function, and birth spacing among !Kung hunter-gatherers. *Science* 207, 788–91.

Kreiter, E., Richardson, A., Potter, J., et al. (2014). Breast cancer: trends in international incidence in men and women. *Br J Cancer* 110, 1891–7.

Kulin, H. E., Bwibo, N., Mutie, D., et al. (1984). Gonadotropin excretion during puberty in malnourished children. *J Pediatr* 105, 325–8.

Kwiatkowski, F., Arbre, M., Bidet, Y., et al. (2015). BRCA mutations increase fertility in families at hereditary breast/ovarian cancer risk. *PLoS One* 10, e0127363.

Lager, C. and Ellison, P. T. (1990). Effect of moderate weight loss on ovarian function assessed by salivary progesterone measurements. *Am J Hum Biol* 2, 303–12.

Lahdenpera, M., Russell, A. F., and Lummaa, V. (2007). Selection for long lifespan in men: benefits of grandfathering? *Proc Biol Sci* 274, 2437–44.

Lasley, B. L., Hodges, J. K., and Czekala, N. M. (1980). Monitoring the female reproductive cycle of great apes and other primate species by determination of oestrogen and LH in small volumes of urine. *J Reprod Fertil* Suppl 28, 121–9.

Latronico, A. C., Brito, V. N., and Carel, J. C. (2016). Causes, diagnosis, and treatment of central precocious puberty. *Lancet Diabetes Endocrinol* 4, 265–74.

Lau, H. E. and Chalasani, S. H. (2014). Divergent and convergent roles for insulin-like peptides in the worm, fly and mammalian nervous systems. *Invert Neurosci* 14, 71–8.

Lazar, L., Meyerovitch, J., de Vries, L., et al. (2014). Treated and untreated women with idiopathic precocious puberty: long-term follow-up and reproductive outcome between the third and fifth decades. *Clin Endocrinol* 80, 570–6.

Lefebvre, P., Bringer, J., Renard, E., et al. (1997). Influences of weight, body fat patterning and nutrition on the management of PCOS. *Hum Reprod* 12(Suppl 1), 72–81.

Levy, S. B., Leonard, W. R., Tarskaia, L. A., et al. (2013). Seasonal and socioeconomic influences on thyroid function among the Yakut (Sakha) of Eastern Siberia. *Am J Hum Biol* 25, 814–20.

Lewis, S. E. (2007). Is sperm evaluation useful in predicting human fertility? *Reproduction* 134, 31–40.

Lindgarde, F., Widen, I., Gebb, M., et al. (2004). Traditional versus agricultural lifestyle among Shuar women of the Ecuadorian Amazon: effects on leptin levels. *Metabolism* 53, 1355–8.

Lipson, S. F. and Ellison, P. T. (1996). Comparison of salivary steroid profiles in naturally occurring conception and non-conception cycles. *Hum Reprod* 11, 2090–6.

Luo, Y., Liu, Q., Lei, X., et al. (2015). Association of estrogen receptor gene polymorphisms with human precocious puberty: a systematic review and meta-analysis. *Gynecol Endocrinol* 31, 516–21.

Maggioncalda, A. N., Sapolsky, R. M., and Czekala, N. M. (1999). Reproductive hormone profiles in captive male orangutans: implications for understanding developmental arrest. *Am J Phys Anthropol* 109, 19–32.

Matkovic, V., Ilich, J. Z., Skugor, M., et al. (1997). Leptin is inversely related to age at menarche in human females. *J Clin Endocrinol Metab* 82, 3239–45.

Maynard Smith, J. (1978). *The Evolution of Sex*. Cambridge: Cambridge University Press.

Meek, C. L. (2017). Natural selection? The evolution of diagnostic criteria for gestational diabetes. *Ann Clin Biochem* 54, 33–42.

Michaud, P., Foradori, A., Rodriguez-Portales, J. A., et al. (1991). A prepubertal surge of thyrotropin precedes an increase in thyroxine and 3,5,3'-triiodothyronine in normal children. *J Clin Endocrinol Metab* 72, 976–81.

Mongan, N. P., Tadokoro-Cuccaro, R., Bunch, T., et al. (2015). Androgen insensitivity syndrome. Best Pract Res Clin Endocrinol Metab 29(4), 569–80.

Monte, S. (2011). Biochemical markers for prediction of preclampsia: review of the literature. *J Prenat Med* 5, 69–77.

Morris, D. H., Jones, M. E., Schoemaker, M. J., et al. (2011). Secular trends in age at menarche in women in the UK born 1908–93: results from the Breakthrough Generations Study. *Paediatr Perinat Epidemiol* 25, 394–400.

Muehlenbein, M. P., Campbell, B. C., Richards, R. J., et al. (2003). Leptin, body composition, adrenal and gonadal hormones among captive male baboons. *J Med Primatol* 32, 320–4.

Muehlenbein, M. P., Campbell, B. C., Richards, R. J., et al. (2005). Leptin, adiposity, and testosterone in captive male macaques. *Am J Phys Anthropol* 127, 335–41.

Muller, M. N. and Wrangham, R. W. (2005). Testosterone and energetics in wild chimpanzees (*Pan troglodytes schweinfurthii*). *Am J Primatol* 66, 119–30.

Mullur, R., Liu, Y. Y., and Brent, G. A. (2014). Thyroid hormone regulation of metabolism. *Physiol Rev* 94, 355–82.

Mulye, T. P., Park, M. J., Nelson, C. D., et al. (2009). Trends in adolescent and young adult health in the United States. *J Adolesc Health* 45, 8–24.

Nadler, R. D., Collins, D. C., and Blank, M. S. (1984). Luteinizing hormone and gonadal steroid levels during the menstrual cycle of orangutans. *J Med Primatol* 13, 305–14.

Navratil, A. M., Song, H., Hernandez, J. B., et al. (2009). Insulin augments gonadotropin-releasing hormone induction of translation in LbetaT2 cells. *Mol Cell Endocrinol* 311, 47–54.

Nepomnaschy, P. A., Welch, K. B., Mcconnell, D. S., et al. (2006). Cortisol levels and very early pregnancy loss in humans. *Proc Natl Acad Sci U S A* 103, 3938–42.

Olson, B. D. (2009). *Understanding Human Anatomy Through Evolution*. Morrisville: Lulu Press.

Panter-Brick, C. and Ellison, P. T. (1994). Seasonality of workloads and ovarian function in Nepali women. In: Wood, J. W. (ed.) *Human Reproductive Ecology: Interactions of Environment, Fertility, and Behavior*. New York: New York Academy of Sciences.

Parker, C. R., Jr., Grizzle, W. E., Blevins, J. K., et al. (2014). Development of adrenal cortical zonation and expression of key elements of adrenal androgen production in the chimpanzee (*Pan troglodytes*) from birth to adulthood. *Mol Cell Endocrinol* 387, 35–43.

Perry, J. R., Day, F., Elks, C. E., et al. (2014). Parent-of-origin-specific allelic associations among 106 genomic loci for age at menarche. *Nature* 514, 92–7.

Pontzer, H., Raichlen, D. A., Shumaker, R. W., et al. (2010). Metabolic adaptation for low energy throughput in orangutans. *Proc Natl Acad Sci U S A* 107, 14048–52.

Pratap, A., Garner, K. L., Voliotis, M., et al. (2016). Mathematical modeling of gonadotropin-releasing hormone signaling. *Mol Cell Endocrinol* 449, 42–55.

Prentice, A. M. and Goldberg, G. R. (2000). Energy adaptations in human pregnancy: limits and long-term consequences. *Am J Clin Nutr* 71, 1226S–32S.

Prentice, S., Fulford, A. J., Jarjou, L. M., et al. (2010). Evidence for a downward secular trend in age of menarche in a rural Gambian population. *Ann Hum Biol* 37, 717–21.

Profet, M. (1993). Menstruation as a defense against pathogens transported by sperm. *Q Rev Biol* 68, 335–86.

Punnen, S. and Cooperberg, M. R. (2013). The epidemiology of high-risk prostate cancer. *Curr Opin Urol* 23, 331–6.

Reese Masterson, A., Usta, J., Gupta, J., et al. (2014). Assessment of reproductive health and violence against women among displaced Syrians in Lebanon. *BMC Womens Health* 14, 25.

Robbins, M. M. and Czekala, N. M. (1997). A preliminary investigation of urinary testosterone and cortisol levels in wild male mountain gorillas. *Am J Primatol* 43, 51–64.

Roch, G. J., Busby, E. R., and Sherwood, N. M. (2011). Evolution of GnRH: diving deeper. *Gen Comp Endocrinol* 171, 1–16.

Roch, G. J., Busby, E. R., and Sherwood, N. M. (2014). GnRH receptors and peptides: skating backward. *Gen Comp Endocrinol* 209, 118–34.

Röjdmark, S. (1987). Influence of short-term fasting on the pituitary–testicular axis in normal men. *Horm Res* 25, 140–6.

Röjdmark, S., Asplund, A., and Rossner, S. (1989). Pituitary–testicular axis in obese men during short-term fasting. *Acta Endocrinol* 121, 727–32.

Rosenbaum, S., Vigilant, L., Kuzawa, C. W., et al. (2018). Caring for infants is associated with increased reproductive success for male mountain gorillas. Scientific Reports 8(1), 15223.

Rutherford, J. N., Demartelly, V. A., Layne Colon, D. G., et al. (2014). Developmental origins of pregnancy loss in the adult female common marmoset monkey (*Callithrix jacchus*). *PLoS One* 9, e96845.

Santner, S. J., Albertson, B., Zhang, G. Y., et al. (1998). Comparative rates of androgen production and metabolism in Caucasian and Chinese subjects. *J Clin Endocrinol Metab* 83, 2104–9.

Scaramuzzi, R. J., Zouaidi, N., Menassol, J. B., et al. (2015). The effects of intravenous, glucose versus saline on ovarian follicles and their levels of some mediators of insulin signalling. *Reprod Biol Endocrinol* 13, 6.

Scelza, B. A. (2013). Choosy but not chaste: multiple mating in human females. *Evol Anthropol* 22, 259–69.

Sear, R., Mace, R., and McGregor, I. A. (2000). Maternal grandmothers improve nutritional status and survival of children in rural Gambia. *Proc Biol Sci* 267, 1641–7.

Sear, R., Steele, F., McGregor, I. A., et al. (2002). The effects of kin on child mortality in rural Gambia. *Demography* 39, 43–63.

Sharpe, R. M. (2012). Sperm counts and fertility in men: a rocky road ahead. Science and Society Series on Sex and Science. *EMBO Rep* 13, 398–403.

Shearman, A. M., Cooper, J. A., Kotwinski, P. J., et al. (2005). Estrogen receptor alpha gene variation and the risk of stroke. *Stroke* 36, 2281–2.

Sherry, D. S. and Ellison, P. T. (2007). Potential applications of urinary C-peptide of insulin for comparative energetics research. *Am J Phys Anthropol* 133, 771–8.

Shimizu, K., Douke, C., Fujita, S., et al. (2003). Urinary steroids, FSH and CG measurements for monitoring the ovarian cycle and pregnancy in the chimpanzee. *J Med Primatol* 32, 15–22.

Shukla, V., Singh, S. N., Vats, P., et al. (2005). Ghrelin and leptin levels of sojourners and acclimatized lowlanders at high altitude. *Nutr Neurosci* 8, 161–5.

Simpson, S. W., Quade, J., Levin, N. E., et al. (2008). A female *Homo erectus* pelvis from Gona, Ethiopia. *Science* 322, 1089–92.

Skorupskaite, K., George, J. T., and Anderson, R. A. (2014). The kisspeptin–GnRH pathway in human reproductive health and disease. *Hum Reprod Update* 20, 485–500.

Song, H., Dicks, E., Ramus, S. J., et al. (2015). Contribution of germline mutations in the RAD51B, RAD51C, and RAD51D genes to ovarian cancer in the population. *J Clin Oncol* 33, 2901–7.

Strassmann, B. I. (1996). The evolution of endometrial cycles and menstruation. *Q Rev Biol* 71, 181–220.

Stulp, G., Sear, R., and Barrett, L. (2016). The reproductive ecology of industrial societies, part I: why measuring fertility matters. *Hum Nat* 27, 422–44.

Takeshita, R. S., Huffman, M. A., Mouri, K., et al. (2016). Dead or alive? Predicting fetal loss in Japanese macaques (*Macaca fuscata*) by fecal metabolites. *Anim Reprod Sci* 175, 33–8.

Tapia-Orozco, N., Santiago-Toledo, G., Barron, V., et al. (2017). Environmental epigenomics: current approaches to assess epigenetic effects of endocrine disrupting compounds (EDCs) on human health. *Environ Toxicol Pharmacol* 51, 94–9.

Tennekoon, K. H., Wasalathanthri, S., Jeevathayaparan, S., et al. (2005). Serum leptin and lactational amenorrhea in well-nourished and undernourished lactating women. *Fertil Steril* 83, 988–94.

Terry, K. L., de Vivo, I., Titus-Ernstoff, L., et al. (2005). Genetic variation in the progesterone receptor gene and ovarian cancer risk. *Am J Epidemiol* 161, 442–51.

Thomson, C. A. (2012). Diet and breast cancer: understanding risks and benefits. *Nutr Clin Pract* 27, 636–50.

Torres, D., Rashid, M. U., Gil, F., et al. (2007). High proportion of BRCA1/2 founder mutations in Hispanic breast/ovarian cancer families from Colombia. *Breast Cancer Res Treat* 103, 225–32.

Towne, B., Czerwinski, S. A., Demerath, E. W., et al. (2005). Heritability of age at menarche in girls from the FELS Longitudinal Study. *Am J Phys Anthropol* 128, 210–19.

Trainer, P. J. (2002). Corticosteroids and pregnancy. *Semin Reprod Med* 20, 375–80.

Tuljapurkar, S. D., Puleston, C. O., and Gurven, M. D. (2007). Why men matter: mating patterns drive evolution of human lifespan. *PLoS One* 2, e785.

Tyndale-Biscoe, H. and Renfree, M. (1987). *Reproductive Physiology in Marsupials*. Cambridge: Cambridge University Press.

Ulloa-Aguirre, A., Espinoza, R., Damian-Matsumura, P., et al. (1988). Immunological and biological potencies of the different molecular species of gonadotrophins. *Hum Reprod* 3, 491–501.

Vitzthum, V. J. (2013). Fifty fertile years: anthropologists' studies of reproduction in high altitude natives. *Am J Hum Biol* 25, 179–89.

Walker, R. and Hill, K. (2003). Modeling growth and senescence in physical performance among the Ache of eastern Paraguay. *Am J Hum Biol* 15, 196–208.

Wallace, W. H. and Kelsey, T. W. (2010). Human ovarian reserve from conception to the menopause. *PLoS One* 5, e8772.

Wang, J., Fan, H. C., Behr, B., et al. (2012). Genome-wide single-cell analysis of recombination activity and de novo mutation rates in human sperm. *Cell* 150, 402–12.

Westendorp, R. G. and Kirkwood, T. B. (1998). Human longevity at the cost of reproductive success. *Nature* 396, 743–6.

Whitten, P. L., Brockman, D. K., and Stavisky, R. C. (1998). Recent advances in noninvasive techniques to monitor hormone–behavior interactions. *Am J Phys Anthropol* Suppl 27, 1–23.

Wide, L. (1987). Evidence for diverse structural variations of the forms of human FSH within and between pituitaries. *Acta Endocrinol* 115, 7–15.

Wilcox, A. J., Weinberg, C. R., O'Connor, J. F., et al. (1988). Incidence of early loss of pregnancy. *N Engl J Med* 319, 189–94.

Williams, G. C. (1966). Natural selection, the cost of reproduction, and a refinement of Lack's principle. *Am Nat* 100, 687–90.

Wittman, A. B. and Wall, L. L. (2007). The evolutionary origins of obstructed labor: bipedalism, encephalization, and the human obstetric dilemma. *Obstet Gynecol Surv* 62, 739–48.

Wood, B. M., Watts, D. P., Mitani, J. C., et al. (2017). Favorable ecological circumstances promote life expectancy in chimpanzees similar to that of human hunter-gatherers. *J Hum Evol* 105, 41–56.

Wood, J. W. (1994). *Dynamics of Human Reproduction: Biology, Biometry, Demography*. New York: Aldine de Gruyter.

Xu, Y., Nedungadi, T. P., Zhu, L., et al. (2011). Distinct hypothalamic neurons mediate estrogenic effects on energy homeostasis and reproduction. *Cell Metab* 14, 453–65.

Zhang, B., Beeghly-Fadiel, A., Long, J., et al. (2011). Genetic variants associated with breast-cancer risk: comprehensive research synopsis, meta-analysis, and epidemiological evidence. *Lancet Oncol* 12, 477–88.

Ziomkiewicz, A., Sancilio, A., Galbarczyk, A., et al. (2016). Evidence for the cost of reproduction in humans: high lifetime reproductive effort is associated with greater oxidative stress in post-menopausal women. *PLoS One* 11, e0145753.

Ziomkiewicz, A., Frumkin, A., Zhang, Y., et al. (2017). The cost of reproduction in women: reproductive effort and oxidative stress in premenopausal and postmenopausal American women. *Am J Hum Biol* 30(1).

CHAPTER 16

SEXUALITY, REPRODUCTION, AND BIRTH

WULF SCHIEFENHÖVEL
AND WENDA TREVATHAN

Abstract

The chapter extends discussion of the human reproductive life course by providing a focus on male and female sexuality and the common result of that sexuality, the production of offspring via childbirth. It begins with an overview of sexuality in non-human primates, with emphasis on variability, considering ways in which sexual behaviours of our closest relatives are both different and similar to sexual behaviours of humans. The chapter addresses sperm competition, male and female orgasmic response, and homosexuality. The approach includes both evolutionary medicine and evolutionary psychology. The chapter considers the hours that surround the production of offspring, beginning with labour and delivery and concluding with the early postpartum treatment of the new-born infant. Topics covered include the role of the hormone oxytocin, the stages and phases of labour, posture in labour and delivery, surgical delivery, postpartum haemorrhage, neonatal hyperbilirubinaemia, treatment of the new-born, mother–infant bonding, postpartum depression, and breastfeeding. Although there are aspects of labour and delivery that benefit from clinical assistance and consideration by evolutionary medicine, most of the time the process progresses with little or no intervention as it has for thousands of generations.

Keywords

sexual behaviour, orgasm, sperm competition, romantic love, homosexuality, labour, parturition, breastfeeding, evolutionary medicine

16.1 Introduction

This chapter covers important behavioural aspects of human sexuality, reproduction, and birth. Specifically, we explore how the human way of mating, pregnancy, and childbirth differs from ways pursued by our closest relatives in the animal kingdom. Thus, we focus on sexuality, parturition, and the postpartum period, all of which are topics of special significance for a number of medical fields like gynaecology, urology, obstetrics, and paediatrics, as well as for fertility medicine and psychology.

16.2 Sexuality

Sex is more important than survival, as can be seen in species in which survival has lasted only long enough for one successful act of reproduction (e.g. praying mantises, the females of which bite the heads off their mates after copulation). There are also species of spiders in which polyandrous females engage in sexual cannibalism, and males at best enjoy only one monogamous copulation (Thornhill 1976; Barry et al. 2008). Most members of our species will probably subjectively rate survival higher than sex, but thoughts and behaviours related to sexuality and related topics are likely among the most powerful in our lives. This is well demonstrated by the role that sexuality plays in the lyrics, legends, stories, literature, music, and visual art of the world. This section is not designed to cover all aspects of complex human sexuality but to highlight some that are of special interest from the viewpoint of evolutionary medicine.

16.2.1 Sexuality in Non-Human Primates

Given the central importance of reproduction, it is not surprising that nature has endowed all living beings, from plants to animals, with powerful mechanisms that ensure that they produce offspring. In many animals, this has led to most extraordinary, and biologically often very costly, behaviours, including spider males and others sacrificing themselves for a reproductive opportunity. For the group of animals phylogenetically most closely related to us, Dixon (1998) has presented a very useful comparative overview of primate sexuality, showing that among eight non-human primate species, females of two have the capacity to experience clitoridal orgasm when stimulated artificially, and also vaginal–cervical–uterine orgasm. Primates exhibit a wide variability of female orgasmic responses—a topic that is discussed below, especially in relation to a new hypothesis concerning the phylogeny of induced very spontaneous ovulation (Pavlicev and Wagner 2016).

Comparing anatomical, physiological, and behavioural features in the animal kingdom, especially in the primates to which we belong, yields important insights into our evolved characteristics. Figure 16.1 sums up, on the basis of comparing body size and sexual anatomy, some of the salient sexual characteristics of the big apes, including the gibbon (biologically and phylogenetically similar to the great apes) and humans. This comparison also allows

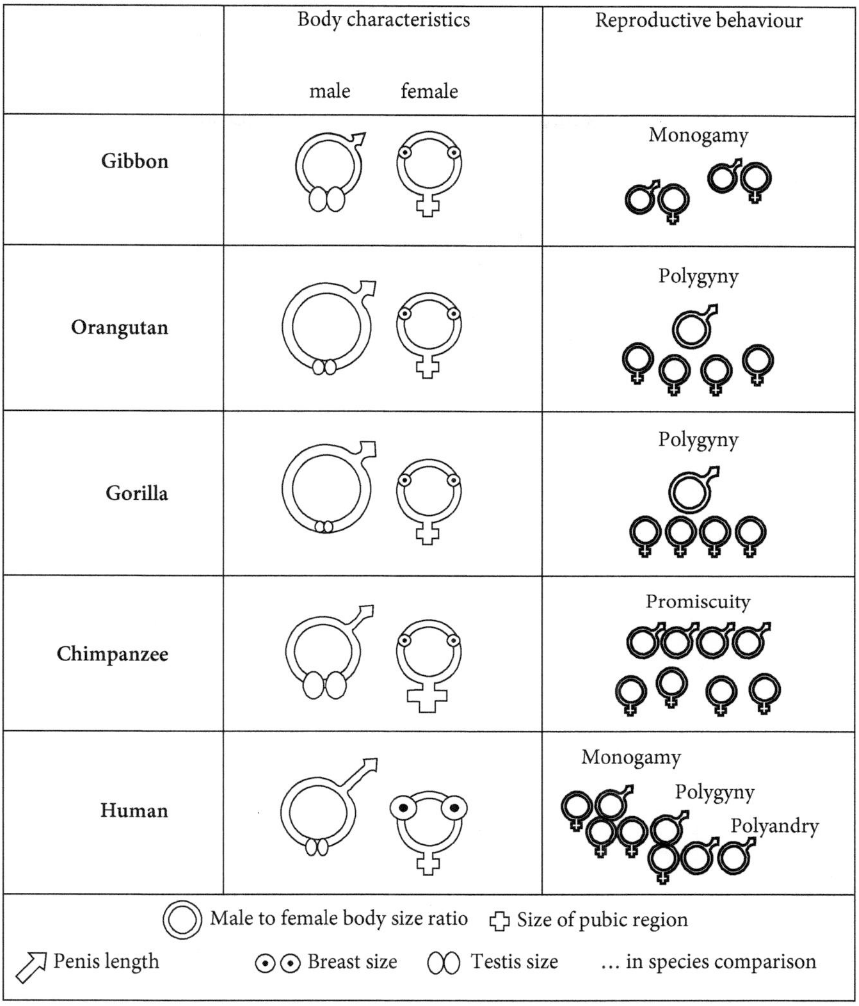

FIGURE 16.1 Schematic comparison of sexual anatomy of four ape species and humans.

some extrapolation backwards into our hominin past: for example, what did our australopithecine ancestors look like and which sexual-reproductive pathway did they likely follow? The figure lists sexual characteristics of the five species shown; the bonobo (*Pan paniscus*) is not listed, even though members of this relative of the chimpanzee (*Pan troglodytes*) exhibit some quite extraordinary sexual behaviours (discussed below), yet their anatomy is very similar to that of the chimpanzee.

In Figure 16.1, the bodies of males and females are symbolised by the signs for Mars and Venus, whereby the arrow in the first and the cross in the second case indicates the size of the genital organ, respectively. In females, the perigenital area including the red swellings on the buttocks of chimpanzee females is also shown. The size of the testes is indicated by the circles at the bottom of the male figures and the size of the female breasts by circles on the chest. The size of the organs is always relative, that is, compared to body mass. The four criteria allow drawing interesting phylogenetic comparisons.

Gibbons (family Hylobatidae) are also named lesser or smaller apes, to indicate their phylogenetic–zoological position between the great apes and monkeys; today, they live in South-East Asia. They have a low degree of sexual dimorphism—that is, females and males are almost equal in size. The penis, the breasts, and the pubic area are small, while the testes, compared to body size, are quite large. The two species of orangutans (*Pongo pygmaeus* from Borneo and *Pongo abelii* from Sumatera) show a very marked sexual dimorphism, with males almost twice the size of females. Adult, dominant males grow impressive cheek flanges and throat pouches, enabling long-distance mating calls. Their appearance is strikingly different from that of females; their penis is a little bigger than that of gibbons, but the testes are very small. A similar constellation is present in the two species of gorillas (*Gorilla gorilla* and *Gorilla beringei*); males, especially dominant 'silver backs', have an impressive body, twice as big as that of females, the penis size is even smaller than that of the orangutan, and their testes are also smaller. Chimpanzees (*Pan troglodytes*) present a very different picture: sexual dimorphism is present but much lower than in orangutan and gorilla; the wider female pubic area (in this case mainly the sexual swellings and reddening on the buttocks) is very large, as are the testes, and the penis is larger (similarly long as that of humans) than in the gorilla and orangutan. Some biologists hypothesise that this last feature is necessary because the penis has to reach the cervix of a female in oestrus through the external swelling. Humans have a similarly small sexual dimorphism (men are about 10–15% taller and heavier than women; in ancestral times the difference may have been 20% as in chimpanzees), the pubic area is comparatively small, the breasts are very big, and the size of the penis is big as well, especially its circumference, while the size of the testes is similar to that of orangutans. Table 16.1 compares the sexual biology of the gorilla, chimpanzee, and orangutan.

This schema allows an understanding of some of the important differences in the sexual and reproductive behaviour of the five species, which are all primates but represent very different and therefore interesting solutions to how to have sex and successfully reproduce. At first sight it seems very odd that the mighty muscular males of gorilla and orangutan should have such small penises and testes—it somehow does not fit our concept of 'wild' sexuality in such big cousins of ours. Yet this is no construction error of evolution but a sign of how highly interlocked adaptive anatomy, sexual behaviour, and reproduction are: those two species are polygynous, and successful adult males have a harem of several females (who play an important role in determining the boss of the harem, who is, in most cases, the father of their offspring). In the case of dominant orangutans, the harem grouping is less evident, as these animals live quite dispersed in the forest, but dominant males still

Table 16.1 Sexual Biology of Three Species of Great Apes

Data	Gorilla (*Gorilla gorilla*)	Chimpanzee (*Pan troglodytes*)	Orangutan (*Pongo pygmaeus*)
Testes weight (g)	29.6	118.8	35.3
Penis length (during erection) (cm)	3	8	4
Volume of ejaculate (ml)	0.4	1.1	1.2
Sperm density (million sperm/ml)	65	548	76
Number of sperm per ejaculate (million)	65	603	91

impregnate the majority of females—in long copulatory acts (14–46 minutes). In order to gain this vital position, males of these two species must fight off competitors, and they do that with their muscle mass, often by inflicting serious injury to the rival, or by just scaring him off due to their sheer impressive sight and aggressive threat behaviour.

Competition for females has resulted in this type of sexual dimorphism: only very big males were and are able to win and defend a harem. The gorilla males enjoy this position on average a mere five years (Fossey 1983). Gorilla females have an ovulatory cycle (not synchronised with the other females in the group) of very similar length as human females (about 28 days) and usually behaviourally indicate their mating interest to the male, occasionally mating with males other than the silverback in the group. As these mating instances are relatively rare, even when the number of adult females is as large as eight, the alpha silverback does not need to ejaculate often, so his testes can be small. Nature is parsimonious. The same principle applies to the orangutan; in this species rape of unwilling females by dominant males has been reported (Mackinnon 1974). Rape occasionally occurs in some other species as well, which has led Thornhill and Palmer (2001) to speculate that rape is an adaptive male strategy. This has been challenged by Drea and Wallen (2003) addressing male–female dominance and control issues. Chimpanzee females apparently cannot be coerced into accepting copulation; the males only succeed when the female cooperates (Fujita and Inoue 2015).

The picture in the polygynandrous (or promiscuous) chimpanzee is quite different: whereas the alpha male monopolises, to a certain degree, copulation with females at the time of mid-oestrus (i.e. the best likelihood of fathering a child), there is still a good chance for males lower in rank to mate and reproduce, especially if they do that out of sight of the alpha male. Sometimes this is achieved by a particular female and particular male forming a temporary bond or consort pair (Fujita and Inoue 2015) during which time they secretly and separately move away from the group and have sex there. This is seen by some evolutionary biologists to be the precursor of typical human mating: it does not take place *coram publico*, as in most primates and mammals (e.g. dogs), but in a private-type environment with no witnesses. Of course, there are examples of public displays of sex, as in 'swinger' clubs. Public orgies involving wife lending and other forms of promiscuity happen at feasts in some cultures and male homosexual intercourse is reported in some (e.g. south coast New Guinea; Knauft 1985), but it seems safe to say that, by far, most humans around the globe prefer intimacy for sex and onlookers are sometimes severely punished. This feature of our biopsychology, the wish to mate in intimacy, needs explanation. The chimpanzee tendency to form consortships for a certain time, which give the female the power of choice, are a good evolutionary model for our species-specific way to mate, even though careful, long-time fieldwork has revealed that the frequency of observed copulations in such consortships was only 1% in the Mahale and 2% in Gombe chimpanzees of all copulations (Fujita and Inoue 2015). This figure could be somewhat higher, as the two individuals could not be followed all the time, but it seems clear that this mating mechanism may not be the most successful one for chimpanzee males.

In general, chimpanzee males have to be prepared to mate opportunistically, perhaps at any moment. In Mahale, males copulated on average two times per daylight hour and females about one time per that time segment; in one case, however, at the height of her oestrus, a female copulated 2.6 times with adult males and 1.25 times with immature males (Fujita and Inoue 2015), suggesting an explanation for the large testes. Their penis is long (up to 14 cm in the erect state), but in contrast to the human penis it tapers off towards the tip, so that its circumference is actually quite small. This feature could serve very fast penetration (hassling by other males is common) and fast copulation (a few seconds) and also has been described as an adaptation to reach the opening of the cervical channel towards the uterus, where the sperm is deposited and soon after coagulates into a rather large gel-like copulatory plug (Tinklepaugh 1930; Dixon and Anderson 2002). This plug fills out the contour of the vagina, thus preventing loss of ejaculate and the sperm of other males moving into the uterus and up the tube. This is a sophisticated form of sperm competition (Parker 1970; see Section 16.2.3) that is also present in some other, phylogenetically distant primate species. Some chimpanzee females have been reported (Dixon and Anderson 2002) to remove the plug, manually, from their vagina, before they copulate with another male—an indication of female choice involving a clever technique (for possible sperm competition in humans, see Section 16.2.3).

Chimpanzee females partly signal their receptivity, like many other species, via chemical signals—that is, sexual hormones, pheromones, or similar substances—which can be extremely effective for the male receptor organs, even in minute concentrations. They also indicate receptivity by their proactive behaviour. But in the chimpanzees the main female signals are the enormously large red swellings on their behinds, a clear visual message that 'I can be impregnated now'. Female chimpanzees, like the other primates, are basically spontaneous ovulators; that is, they have an ovulatory cycle very much like human females.

The most interesting conclusion suggested by the schema is that there is a close relationship between some of the anatomical features and the social as well as cultural aspects of reproduction. Orangutans and gorillas, as mentioned, gain harems and thereby excellent chances for reproduction, by their fighting mass; in other words, sexual dimorphism is a sign of polygyny. The gibbons, in contrast, do not show a difference in female and male body mass; they are monogamous, at least most of the time, although some extra-pair copulations do occur (Reichard 1995). Humans have, as mentioned, a sexual dimorphism of about 10–15%, which would allow the conclusion that we are, by biopsychological inclination, partly monogamous, partly polygynous, and partly polyandrous. Also, our not very pronounced testes size speaks of polygyny, as does the fact that harems of two or more spouses are a very common feature in the cultures of the world—even in traditional Christian groups in the United States (Amish, Hutterer)—and that consecutive polygyny and polyandry (serial monogamy) is an almost normal mode of partnership and reproduction in the industrialised and non-industrialised countries of today (for a theoretical model of forms of marriages see de la Croix and Mariani 2012). Also, the fact that humans fall in love and are powerfully attached, almost addicted, at least for a certain time, to a person of the other (or less commonly same) sex points at monogamy, or better serial monogamy, as a common partnering pattern of our species (see Section 16.2.5).

The bonobo (*Pan paniscus*) was not listed in the schema—their anatomical features are very similar to those of the chimpanzee, yet their sexual behaviour is very different. Observations among captive members of this species (de Waal 1995) have led to the image of *P. paniscus* as a kind of hippie-ape ('make love, not war'), because in conditions of living under human care (zoos), bonobos quite frequently engage in seemingly recreational, non-reproductive sex across the two sexes and all ages, with homosexual behaviour surprisingly often between females. In the wild, the situation is somewhat different and the frequency of these acts is lower. Fruth and Hohmann (2006) showed that males almost never had sexual interactions with infants and juveniles; all such interactions directed at juveniles were initiated by females. They also found that by far the majority of homosexual acts with adult partners followed this female–female pattern: horizontal genito-genital (G-G) rubbing was a common form of female same-sex behaviour, and 86% of all homosexual behaviour was between female individuals.

The differences in sexual behaviour among the closely related chimpanzees of the same genus *Pan* are very striking, and primatologists and evolutionary biologists have been working hard to find explanations for bonobo extraordinary non-reproductive sexual behaviour. The most appealing explanation assumes that having sex with one another is a nice thing and functions as a general social lubricant, defusing tensions which exist, as in other societies of social animals, in *P. paniscus*. Fruth and Hohmann (2006, p. 312) carried out extensive fieldwork on bonobos and showed that 90% of all same-sex female genital interactions are initiated by the lower-ranking individual and that in the majority of these acts the higher-ranking female was physically on top of the other—both signs of a mechanism regulating dominance and hierarchy. They concluded 'that genital contacts among bonobos are not a sign of affiliation, friendship or "sisterhood" per se, instead, female homosexual behaviour is an instrument that allows distantly related individuals to coexist closely and nevertheless peacefully, by functioning as a method of status acknowledgement, tension regulation and, to some extent, reconciliation. From this, bonobo female coalitions might derive ultimate benefits related to resource defence, infanticide avoidance and dominance over males.'

16.2.2 Sexuality in Humans

There does not appear to be jealousy involved in these sexual acts of bonobos, and because these acts of sexual play do not lead to pregnancy and offspring, reproductive success is not at stake. Bonobo sexuality cannot, even though many researchers represent it that way, serve as a functioning model for human sexuality: in our species, genital play usually leads to copulation between the female and male and jealousy is (also in homosexual couples) a very powerful emotion, leading to a large number of aggressive acts (crimes of passion) around the world (Buss 2000). Human societies have the concept of fidelity (which applies also for polygynous couples) and mate guarding is a pronounced trait in our species, exhibited by both males and females. Cultures have experimented with free sex in communes, Indian ashrams, other religious sects, and so on. But few of these experiments have been successful. The only successful official institutions are polygynous or monogamous marriages (polyandrous marriages, as an official institution confined to the Himalayas, are rare and usually not successful), and unofficial or after divorce official serial monogamy. Predominantly female prostitution serves as an often culturally accepted 'valve' for males.

Humans' desire to understand their very nature, the human condition, must have been present very early on in our genus. This drive is the one that keeps science exploring the different traits and states of our species to the present day. One promising avenue for this endeavour has always been comparing ourselves with animals. For evolutionary biology this has been one of the major approaches. The comparison of sexual and reproductive systems undertaken here is in that tradition. Its strength is to show how sexual anatomy, physiology, and behaviour fit the particular physical and social ecologies of the respective species, from promiscuity to harem constellations and monogamy; its weakness is that it is difficult to reconstruct, from these primate models, a set of precise species-typical human sexual behaviours. There is no single sexual reproductive system/pattern of behaviour in extant great apes including the gibbon and ourselves. Are we like chimpanzees, our closest relative, or more like the bonobos, another close relative? And how much gorilla, orangutan, or gibbon behaviour is still stuck in the human male's brain? Cross-cultural comparison, especially of sexual traits, has, despite its power in generating interesting hypotheses, this kind of limitation.

The cross-cultural work of Buss (1989) has shown that all over the globe women prefer male partners who are older than they are, as well as socially and economically powerful (indicating potential for high providing capacity), whereas men prefer female mates who are young and beautiful, that is, sexually attractive—the latter being a shorthand for health, fertility, and vitality. Other authors have stressed the importance of non-biological factors for the differences between women and men (Hrdy 1997; Baumeister and Twenge 2002). While these preferences are likely driven by unconscious or conscious assessments of specific aspects of, in a wide sense, reproductive potential—which ranges from 'good genes' to the capacity and willingness to invest in children—and thus underlines the central importance of reproductive aspects of human sexuality, our species differs dramatically from others by the fact that by far most sexual acts are non-reproductive.

This aspect has been used in an effort to understand the concept of 'hidden' or 'concealed' ovulation and 'lack of oestrus' in human females. Yet this view has been, in various ways, challenged in recent years: some women are able to detect, via the so-called Mittelschmerz, the point in time of actual ovulation, which causes a slight pain ('Schmerz') in the respective ovary. One could argue that a woman who could perceive the signal could directly or indirectly inform her possible partner or partners of this physiologically very important moment in her cycle. Research in traditional societies (Schiefenhövel 2001), however, shows that neither women nor men know precisely about physiological details and timing of the female cycle, that is, fertile versus infertile days. Not surprisingly, there is a rather large window of conception, up to a maximum of ten days as Martin states (2013). It will thus be difficult, even to observant persons, including male partners, to calculate the optimum time of conception this way. And it remains questionable whether our ancestors would and could have done such calculation.

It is more likely that human females, like their primate cousins, send out chemical and behavioural signals to males. It has been argued that in our species the vomero-nasal (Jacobson) organ responsible for male perception of sexual signals, especially ones signalling the best time for conception, is missing or at least does not function any more to detect pheromones. It does so very effectively in many mammals and it would, from an evolutionary point of view, be quite surprising if humans did not have this powerful mechanism. Recent research indicates that this might be indeed the case (D'Anielo et al. 2017). Some men perceive the smell changes connected to mid-oestrus in women (Singh and Bronstad 2001). And as olfactory signals are extremely powerful for sexual arousal also in our species (which is sometimes, but in an unjustified way, said to be microsmic), for men but also for women it is likely that smell (including that of perfume, etc., which often contains the sexual components of other animals (musk)) still plays an important role also for our sexual behaviour and that it provides an additional argument that ovulation in women is not hidden but can be detected by men (Miller and Maner 2011). (For further discussion, see Chapter 17: Brain, Spinal Cord, and Sensory Systems.)

Overt behaviour also counteracts the concept of concealed ovulation: some, perhaps many, women feel a surge of sexual arousal and interest in coitus around ovulation (Morris et al. 1987). Studies (Grammer et al. 2000) have found that exactly at this time women dress up to make themselves sexually attractive and search for (physically) high-quality (high testosterone, 'good genes') males in discotheques and the like. Other studies proposed that the 'hunt for good genes' was directed at men with creative intelligence (Haselton and Miller 2006), rather than broad chins and big muscles. The sociobiological interpretation is that women during the height of their oestrous cycle (which was believed in former research to be missing in human females)—that is, at the moment of highest fertility—intuitively hunt for good genes in one-night stands, even betraying their long-time, but not so testosterone-laden (or intelligent) partner, who would then, so the hypothesis suggests, help, as a good father rather than a good lover, to raise the cuckolded child. Whether this scenario portrays reality around the world, and especially for our evolutionary past, is an open question, but it seems very likely that women behave more sexually at ovulation (e.g. initiate copulation) than at other times, including pregnancy, quasi-pregnancy induced by hormonal contraceptives (Adams et al. 1978), and menopause.

One common misconception regarding human female sexuality is connected to the idea that there is no oestrus—thus the mentioned 'hidden ovulation' in our species and, as a consequence, the statement that women are sexually receptive all the time. As indicated above, women do not spread their erotic interest and activity evenly across the menstrual cycle. There is usually less sex during menstruation (for evolutionarily probably good reasons, see below), during the last phase of pregnancy, and, in some cultures, postpartum and, as mentioned, a peak of sexual interest around the time of ovulation (Morris 1967; Burleson et al. 1995). In short, women are not really detached from the mammal and primate pattern of sexuality.

Humans show, as chimpanzees, a moderate sexual dimorphism, not only in height and weight. Compare the differences in the performances of world-class athletes: in most sports, men are usually about 10–15% ahead of woman. The tendency for women to become relatively taller and heavier in the course of time might be ongoing (skeletons from prehistoric times show that human sexual dimorphism has been declining steadily (Ruff 1987)). Human females have by far the largest breasts of all primates and males have the largest penis, especially concerning girth. To many observers, the breasts of women represent powerful sexual signals. Desmond Morris (1967) hypothesised that breasts are basically buttocks moved to the front because humans, after their fully bipedal stand and the corresponding changes of the pelvis and the birth channel, and thereby the vagina, often have sex face to face. The breasts of the woman could then act as organs of sexual arousal, moved, anatomically, from back to front. This is an interesting idea, but it can hardly be tested other than by the fact that, indeed, the breasts of women are very large and, especially in younger age, very shapely, and that men are indeed very much aroused by them—one only has to look at men's magazines and remember that breast enlargements belong to the most common surgical interventions (Gander 2016). Many well-known icons of our glamour world have had one or more such beauty operations to make their breasts bigger.

Certainly, human female breasts are seen in many cultures as objects of sexual attraction. This is far from universal, however, and there is a great deal of cross-cultural variation in what parts of the body are deemed erotic. In some cultures, female breasts are primarily seen as organs for breastfeeding infants and are not involved in sexual interaction at all. Breasts and the associated ability to feed infants are the defining characteristic of mammals, but for most species they are not very different in size or shape from male nipples, except when the female is pregnant or nursing an infant. An argument that counters the one about sexual attraction is that female breasts are large because they provide an advantage in breastfeeding infants. Although breast size does not appear to be related to milk quantity or quality, the somewhat flat-faced human infant probably finds it easier to nurse from a large and extended breast than from a breast that lies close to the chest. The somewhat small breasts of apes are not problematic for their prognathic infants. One thought is that over the course of human evolution, as snouts have become more orthognathic (flattened), the breasts have protruded further, making it easier for infants to feed.

In some cultures (e.g. the Papuan societies of Highland New Guinea; Schiefenhövel 1988), menstruation is public: women spend the days of bleeding in special huts at the fringe of the village, where they also give birth and spend the puerperal period (see Section 16.4.5). This was a cultural tradition (now widely given up) whereby men could receive information on the reproductive status of women: when they started menarche, whether they repeatedly menstruated (i.e. are not pregnant), and when their cycles would be renewed after birth. This knowledge was easily gained, because the village groups were small, rarely exceeding 200 persons, and it was well known which woman was in the women's house at a given time; a man's decision when to have sex with a particular woman, either as husband or (not infrequently) as lover could be influenced in this way. To make menstruation public provided a mechanism to assess the fertility of individual women in the mountain villages. Managing menstrual discharge of blood was done externally, for example, by sitting on fern leaves, and no tampons or any similar objects were inserted into the vagina, so that the flow of blood remained unimpeded; bacterial infections would have been very serious in the times before disinfectives and antibiotic treatment. The Eipo men were extremely afraid of coming into contact with menstrual blood; it was considered a dangerous fluid, harming male health. Intercourse was therefore strictly avoided during this time, another example of a culturally mediated behaviour pattern preventing possible infections (Tanfer and Aral 1996; Lurie 2010).

16.2.3 Sperm Competition

The human penis is, as mentioned in Section 16.2.2, very differently shaped compared to that of chimpanzees and bonobos, whose sexual organ tapers off at the end to a narrow tip—perhaps a feature adapted for fast penetration: chimpanzees and bonobos have a mean intromission duration of 7 seconds, gorillas 15 seconds, and orangutans 14–46 minutes. Figures for humans (according to Kinsey et al. 1948, 1953) are less than 2 minutes, but there is a high variance concerning copulatory behaviour in our species. Gorillas and orangutans have a small penis (6 cm and 8 cm erect, respectively), which is less developed than in the human species, and their penises are generally of similar shape but smaller in circumference. It has been argued that the pronounced rim of the glans penis in the erect human organ could act as a kind of pump, sucking out sperm of a rival from the vagina (Gallup et al. 2006) and thereby providing an advantage in possible sperm competition (Baker and Bellis 1993).

For the hypothesis that sperm competition was and is at work in humans as well, one would have to count on a certain percentage of ancestral women being so promiscuous that they had intercourse with two or more males shortly one after the other, so that sperm of the first ejaculate would not have already started its long, arduous, but fast race towards the cavum uteri and then up one of the tubes. We cannot know the sexual behaviour of our Plio-Pleistocene Eve, but judging from today's reality one would rather doubt the claim that sperm competition of this kind happened often enough to drive selection in that direction. If males of our species would have to compete with others on, as it were, the battle ground of the vagina, one sure sign would be the genital plug system mentioned in Section 16.2.1, by which chimpanzees and other primates try to secure their sperm against that of rivals (Tinklepaugh 1930; Dixon and Anderson 2002). This is, however, not the case. By way of normal uterine peristalsis, driven by oestradiol, human sperm manages to advance very quickly into the cavum uteri; there are indications that oxytocin can be helpful for this process (Kunz et al. 2007). The sperm of a rival would find it difficult to outspeed the sperm of the first copulation partner, especially in the case where his ejaculate was not placed fairly soon after the first partner. This may not have been the case in our ancestors; promiscuous behaviour (having sex with several men directly one after the other) is not typical for women—except female sex workers. Of course, one could argue that our sexual behaviour underwent substantial change in the course of hominisation and that Eve could have lived in groups with several males and behaved in a polygynandrous fashion, as today's chimpanzees and bonobos do. But from the cross-species perspective, inspired by work which led to the schema shown in Figure 16.1, this is unlikely. Kunz et al. (2007) hypothesise that fast access of sperm to the uterus and towards the Fallopian tube could have been an evolutionary adaptation to human bipedalism, which could, so they say, have led to deposited sperm flowing out of the vagina.

Similar arguments have been put forward in the history of writing about female sexuality; for example, that intercourse, especially oxytocin, makes us sleepy (Morris 1967), so that the woman does not move much after coitus and the sperm can start its race towards the ovum unimpeded. This hypothesis would be, as we would like to argue, contradicted by the fact that women can get pregnant as easily in many other positions as in the supine ventroventral position, among them the woman kneeling over her partner so that the vaginal channel is similarly vertical as in normal bipedal stand. Human sperm, it seems, does not need female immobilisation to ascend towards the ovum. And it may have, in ancestral times, been sufficiently protected against competition from rival sperm because of its demonstrated efficiency in mobility, as well as from a female behaviour pattern, which does not encourage women to run, as it were, from one matrimonial bearskin to the other; that is, copulation with another partner would probably leave, on average, enough time for the first ejaculate to gain a good advantage in the race for impregnation. An interesting question is whether ancestral female sexuality involved elements of romantic love (see Section 16.2.5), which would have counteracted promiscuity to a certain degree. Empathy, typical for our species, and avoiding the risk of acts of jealousy might also have been involved in an ancestral woman's decision not to sleep with a man when the first partner was still around.

Sexual competition among ancestral human males may not have had a mechanism disabling sperm inside the vagina but probably was, as today, primarily a contest of personalities, health, vitality, social competence, intelligence, creativity, and the like. How much the sexual anatomy of males influenced the choice of ancestral women is an interesting question: It is possible (but was not the subject of research until now) that men who until recently lived in cultures of male genital nakedness have larger penises than men in other societies; the Nuba of Africa and the Asmat of south-western New Guinea may serve as intuitive examples. If this hypothesis were correct, then women would have, in ancestral times, preferred men with a more developed genital anatomy over ones who had smaller penises. This may explain the fact that *Homo sapiens* has the conspicuously largest and thickest erect penis of all primates.

In this context, the interesting question arises of when the tradition of covering one's genitalia, so very widespread in the cultures of the world, arose and why it happened. It is obvious that we humans are equipped with the biopsychic potential of feeling genital and sexual shame; this is one of the characteristics that contribute the uniqueness of our species. Why it developed and how it might be connected to social shame (Bischof 2012; Bischof-Köhler 2012) is an interesting question. It may have arisen in the context of sexual intercourse in the intimacy of chimpanzee consortship away from the group, which is rather rare but still regarded as a mechanism ensuring female choice of a particular mate; the typical desire of humans for sexual intimacy (see Section 16.2.1) as well as genital/sexual shame could have come from this root. The development of sexual shame could have also been favoured by the fact that the visual perception of sexual acts is very arousing for humans, particularly for men; hiding one's own sexual behaviour would have avoided arousing other males who could have interfered, as possible rivals, in the copulation (Medicus 2015).

16.2.4 Orgasm

Orgasm in males is a complex yet basically still rather simple biological mechanism. Very few ejaculations occur without orgasm, which has evolved early in phylogeny and is probably present in all mammals (Symons 1979). Males, most probably not only of our species, perceive considerably more enjoyment from orgasm than from sexual stimulation that does not lead to climax: all animal males, including humans, are orgasm-prone beings. In 50 years of fieldwork with free-living chimpanzees, McGrew (personal communication 2017) did not see a single male masturbation that led to ejaculation. It is likely that our ancestors, like probably the majority of men today, preferred regular intercourse versus masturbation, even though it frequently occurs in non-human animals. Patterns of sexual behaviour of chimpanzees have been studied intensively by several groups of researchers (for an overview see Fujita and Inoue 2015); it is not uncommon for females of this species to prolong their sexual swelling and encourage copulations with different males, possibly in an unconscious attempt to confuse paternity. Multiple orgasms/ejaculations are possible for human males, but there is a refractory period of varying duration. In this respect, women, yet not all of them, are remarkably different: they can experience orgasms like pearls on a string (see below). Female sexuality presents a very different spectrum. Kinsey et al. (1953) assured the world that there was female orgasm; their successors (Masters and Johnson 1966), still rather ignorant about female sexual responses, thought that orgasm in women was physiologically the same event in all of them. For some time it was thought that females of non-human species do not have orgasm-like experiences. And, of course, it is not easy to define what exactly that would mean for an animal. Researchers, however, have operationalised definitions and used experimental and ethological approaches to study the sexual responses of female mammals, including primates. There can be no doubt that orgasm exists in females of non-human species; Troisi and Carosi (1998) found that in one out of three of 240 copulations the typical orgasm face was observed in macaque females.

There is a remarkable discrepancy between women who do not at all or very rarely experience orgasm and women who can have several in a row (Hartman and Fithian 1974). Whence this paradox: multiple orgasms versus none? All women belong to the same species; therefore should there not be a specific or at least typical sexual response mechanism like orgasm in males? There isn't, despite the fact that modern functional magnetic resonance imaging (fMRI) techniques have shown that men and women activate very similar brain regions, actually most of this organ, during sexual arousal and that women who had a spinal cord injury, blocking the ordinary transmission of neural activity from the clitoris and vagina to the brain, could climax when being genitally stimulated accordingly; other pathways than the ones depending on the spinal cord are thought to exist as well (Komisaruk et al. 2006).

Women of all times and in all societies most probably were and are very similar in their anatomy. An exception are women of some populations among the Khoisan and other groups in Sub-Saharan Africa who have a marked steatopygia combined with a strong lordosis and often additionally very long labia (Cummings et al. 2014), a trait that is also said to occur among women on the Andaman Islands. It is, however (despite some claims that the famous Venus figurines exhibit symptoms of steatopygia, yet their figures are just like those of generally well-fed women (Figure 16.2)), doubtful that this specific anatomy

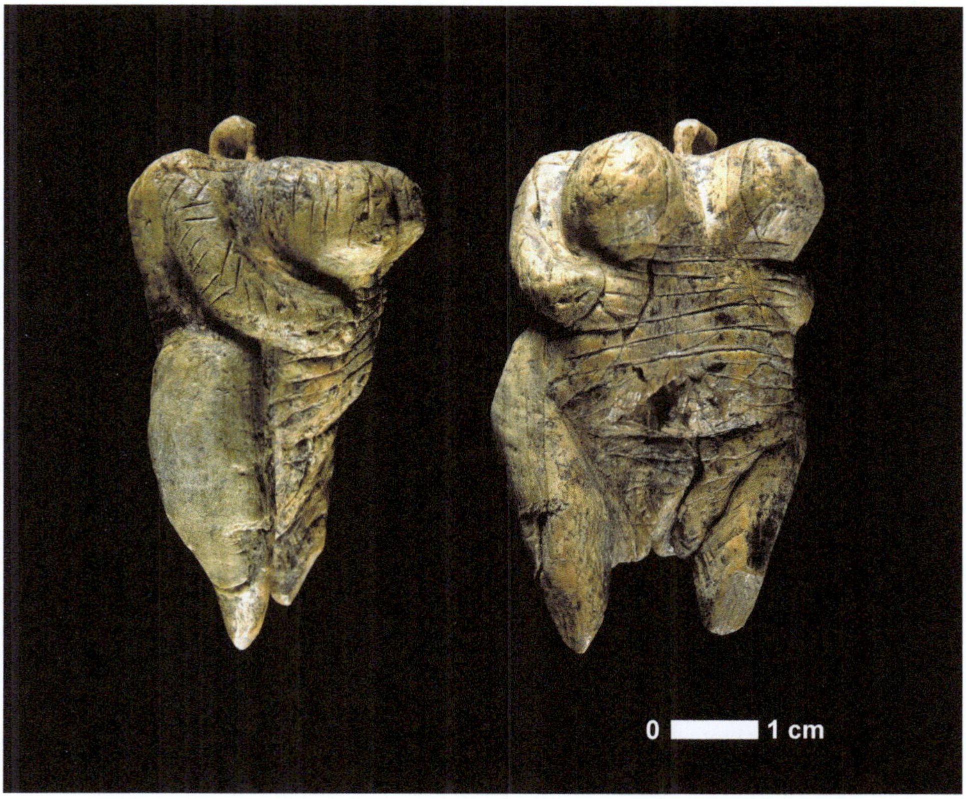

FIGURE 16.2 Venus of Höhle Fels (Germany), 35,000 BP (Aurignacien).

Source: Venus Höhle Fels. Photo © Hilde Jensen/University of Tübingen.

was the one of our early female *Homo* ancestors; skeletal finds should have displayed the strong lordosis in women of this type.

What are the biological, evolutionary foundations of female orgasm? It has been found to exist in many primates and mammals (Symons 1979), so there is a clear phylogenetic base for sexual climax during intercourse, manual or oral stimulation, or masturbation (statistically all more common in human females than orgasm via intercourse). For an evolutionary anthropologist it is problematic that female orgasm not only is no precondition for getting pregnant, but also does not seem to assist getting pregnant. It was thought along adaptationist lines that the strong contractions of the vagina and uterus would dip the cervix into the deposited sperm and help, in the way of a vacuum pump, to suck up sperm into the uterus so that its arduous travel upward to the tube towards the ovum would be assisted (Gould 2002). But this does not really seem to be the case, as is clearly proven by the fact than non-orgasmic women get pregnant—whether as easily as women with orgasm is yet unclear, but possible. So, if the most vital advantage, facilitating pregnancy and producing offspring, is out of the race, why do females orgasm at all? It is not as costly as male orgasm with ejaculated semen, but involves circulatory and hormonal costs which could perhaps be avoided. Some researchers discuss circulatory congestion in the pelvic area as a biological reason for female orgasm, which would release it, but it seems doubtful whether pelvic congestion due to sexual arousal needs, in a physiological and evolutionary sense, orgasm to discharge the congested blood; this condition is mostly caused by pathologies of various kinds (Perry 2001). So, the question arises: why is the clitoris there, and why is it sexually so very sensitive that women who cannot experience vaginal orgasms via intercourse quite often can climax after stimulation of that erectile organ? Why did it evolve, if it can often transmit its powerfully positive signals to the brain only after somewhat artificial stimulation? One hypothesis is that the female genitals and their function as pleasure-providing organs are by-products of the male system (Symons 1979).

Did ancestral men know about the orgasmic potential of their female partners and did they satisfy them by sensitive manual caressing, requiring dedication? Looking at sexual practices around the world, this does not seem necessarily the case. The Trobriand Islanders were made famous by Bronislaw Malinowski's post-Victorian best-seller *The Sexual Life of Savages* (1929), in which he writes about some sophisticated love play, like biting the eyebrow of one's partner, but sexual encounters there hardly involve any manual stimulation, on both sides (Schiefenhövel 2004).

Among the Eipo of Highland West New Guinea, this is different; partners behave passionately and orgasm is credited to both women and men, which is also expressed in surprisingly parallel linguistic terms: '*den ala*', the slimy substance emitted by the penis (i.e. sperm), is compared to '*kwat ala*', the slimy fluid stemming from the vagina—most likely copious vaginal lubrication, possibly fluid ejaculated from the female prostate gland or the bladder. Interestingly, the Trobrianders have the same concept like the Eipo, namely that female and male sexual fluids are the same; at least they are given the same term in Kilivila, their language: *momona*. The Eipo have another telling term: '*kwat futugana/futukana*', the splashing sound made inside the vagina by the penis moving in it; *futugana* is most probably an onomatopoeic term describing very well the sound produced when a fluid is squirted out under pressure: for example, the fluid squirted out between the spread toes when one walks through a wet, muddy path. This term could only have been used in a sexual perspective with a very well-lubricated vagina in mind through which a penis is moved vigorously, or perhaps it is even a term describing the gushing out of fluid from the urethra in what is commonly called 'female ejaculation' (see below). This phenomenon, still hotly discussed and partly denied in western countries, is known to members of other cultures as well (Blackledge 2004).

Back to the enigma of female orgasm. Researchers disagree on a number of important issues. Is the capacity for orgasm in human females a by-product of male anatomy and physiology (Symons 1979), which develops, from a bisexual embryonic stage, into a system with pronounced capacity for orgasm? And would therefore the female analogue, the clitoris, and also the vagina, contain some vestiges of this, but would not be endowed with the same adaptive value because there is no functional ejaculation triggered by orgasm? Is the likelihood of female orgasm reduced because women have, compared with other mammals and also primates, a small clitoris and because this main sexual organ plus the orificium externum urethrae with the (debated) G-spot in its vicinity (Gräfenberg 1950) were, during phylogeny, moved up and away from the movements of the penis during intercourse because in typically difficult birth the large neonatal head would do too much damage otherwise (Wolfe 1991)? These hypotheses are made less likely, however, because some women have the mentioned capacity of quickly repeated orgasms, not only from manual or oral stimulation but also from normal intercourse. They are, compared to males, hyperorgasmic.

Sarah Blaffer Hrdy (1981) has put forward a hypothesis that stresses the adaptive, evolutionary aspect of female orgasm and, as well, the fact that it does not regularly occur. She postulates that ancestral women were similarly difficult to bring to climax as modern ones and would, therefore, try out many partners in search of orgasm. This would have led to obscuring paternity certainty in men, because none of them could be sure of not being the father of the children and thus inhibited to commit infanticide. This behaviour has been shown to exist in some species where it is a mechanism of a new alpha animal killing the infants of his predecessor to ensure a fast set-in of ovulation in the females so that he can impregnate them quickly, that is, before he is overthrown and has to yield reproduction to the next male (in well-defined terms, such adaptive infanticide is not as common as sometimes thought). Hrdy's thesis is appealing as it combines proximate and ultimate aspects of the issue. We cannot know whether ancestral men killed neonates and infants. In some traditional societies of Highland New Guinea it is the mother who commits neonaticide immediately after birth, not the father or other men (Schiefenhövel 1988). Humans have a very acute sense of perceiving physiognomies and detecting family resemblance. Because of this and perhaps due to olfactory signals, ancestral men would probably not have been fooled easily as to who was the father of an infant; in other words, obscuring paternity by promiscuous female behaviour might not have worked; especially as humans very early on would have understood, in contrast to Malinowski's position (1929), the connection between copulation and pregnancy (Schiefenhövel 2004). Also, what would happen if a woman found her first partner to be fully satisfying? Are women who have problems experiencing orgasm more likely to be promiscuous, multibreeding in search for the ideal partner, as Hrdy's hypothesis predicts? Or rather are women who enjoy climax regularly and therefore are more active also looking for extra-pair copulations?

A lack of libido is very common, at least in the western world. Bitzer et al. (2013) state the figure of 20% female sexual arousal disorder or hypoactive sexual syndrome, and Shifren et al. (2008) confirm the magnitude of the problem. One must really wonder whether the sexual response of women to normal (welcomed) sex (including stimulation other than by intromission of the penis) can be seen as evolutionarily non-adaptive because their orgasm is a relatively rare event, or whether one must involve some other lines of thought.

It is possible that non-orgasmic sexual pleasure is the main building block of female sexual experience and that this has been sufficient, in ancestral times, and it is today, to make women get involved in sexual activity, which very often means penile intromission and orgasm/ejaculation by the male partner. It may be that the high degree of empathy, of which women are capable, is an important motor driving female erotic behaviour. This suggests that a woman may not have as a primary goal her own orgasmic climax but a milder form of sexual pleasure while enabling the male partner's orgasm, as a gift of love. Surely this happens. Whether it may have happened sufficiently often in ancestral times to shape the sexual response system of women is a question that can probably not be answered.

This sexual reaction of women to their partners' wishes and expectations would surely be, as mentioned above, a strong element of bonding. Morris (1967), Eibl-Eibesfeldt (1972), and others (see also below) have argued that human sexuality, just because it is, to a large extent, non-reproductive, works as a strong 'glue', binding wife and husband together, two partners of opposite sex, in a very effective way so that they stay together long enough that their children are raised by mother and father in a stable family until they can establish themselves in society. Of course, divorce is common in most cultures of the world and other intervening variables may spoil this image of matrimonial harmony. But it is still true that around the globe women and men stay bonded together for long times. For the first years, the promise of frequent sex may well be the main reason for this staying together, at least for the men. Later, a sense of responsibility for the children is a new bonding element, replacing, to some extent, the bonding function of sex. If this scenario described the evolutionary scenario well, female orgasm would not really have to be an adaptive trait. Yet it could still come into evolutionary play because men feel strongly aroused by the arousal of their female partner, and this could, at least in older men, be important to turn the man's functions of erection and ejaculation on sufficiently. Faking orgasm is a common female strategy; this may not be just a lie (matrimonial beds witness many such lies), but an unconscious, semiconscious mechanism to help the male partner to reach his evolutionary goal and also to increase own arousal.

The listed hypotheses trying to explain the presence (and absence) of female orgasm are, in some aspects, only partly convincing. Recently, a new evolutionary approach to solve the origin of human female orgasm was taken by Pavlicev and Wagner (2016). They contrasted animals in which ovulation is induced by copulation, genital stimulation, or even just visual or olfactory cues, with animals in which ovulation is spontaneous, that is, following a more or less fixed menstrual cycle. The argument is that in order for ovulation and subsequent fertilisation to happen, there must be a powerful mechanism, which starts, in the organism of the female, the cascade of biochemical and physiological events leading to the actual release of an ovum from an ovarian follicle. In female animals with induced ovulation, this mechanism is, according to this view, orgasm or an orgasm-like physiological event, characterised by a strong release of prolactin and oxytocin particularly when the clitoris is stimulated sufficiently—the similarity of orgasm in the human female is obvious.

In humans as well as in other species with spontaneous, chronobiologically fixed ovulation—which according to the authors developed about 75 million years ago during the time the common ancestor of rodents and primates lived—this capacity for orgasm would just be an evolutionary vestige, not needed any more biologically; a similar argument was put forward by Wolfe (1991) who saw female orgasm as an ancient primate characteristic which, in the course of further evolution, had become partially lost in humans, yet other primate females seem to have problems as well. The question is why this characteristic got lost, and for that Pavlicec and Wagner (2016) give a new inspiring answer. These three authors, as well as other authors, agree that orgasm is still lingering there and that the majority of women can sexually climax in some form or other, albeit not usually and regularly by ordinary copulation. The previously mentioned fact that in the human species the clitoris has moved upward (i.e. slightly away from the main thrust of the penis) is seen by the authors as proof that this erectile analogue of the penis does not serve the same absolutely necessary function any more. Clitoris-bound arousal, so their argument goes, could now become co-opted for other functions: selecting a fitting partner and strengthening the pair bond. The human female (and females of other spontaneously ovulating species) would, as suggested by McGrew (personal communication 2017), in this way be better enabled to control impregnation, as ovulation was not a reflex-like event controlled by the male. Until a new hypothesis is advanced, this approach by Pavlicev and Wagner (2016) to explain the orgasm-paradox in women seems a promising one.

One problem with the hypothesis is that not so few women also experience vaginal and cervical–uterine—that is, non-clitoridal— orgasms; thus it would be better (as suggested by Blackledge 2004) to see the whole set of the female sexual organ with its various parts (*corpora venosa*, vulva, clitoris, vagina, *orificium urethrae*, female prostate or Skene's glands, cervix, and uterus) as the receiving end of genital stimulation, which would thus maintain its adaptive function in enabling fertilisation, namely by helping ejaculation and receiving sperm, even in the absence of orgasm. At any rate, this new hypothesis is probably good news for women who do not regularly climax with copulation or do not climax at all: sexual arousal in females of our species does not necessarily have to reach a peak. Another possible weakness of the hypothesis is that the evolutionary history of induced versus spontaneous ovulation is not as phylogenetically clear-cut as often portrayed. Why mice, rats, cats, and horses should have induced ovulation and other mammals of the same families not is probably better explained by their specific physical and social ecology than by position in the phylogenetic tree. It seems, however, possible that, as the authors claim, induced ovulation was the ancestrally earlier mechanism.

But is it really true that women ovulate 'spontaneously' (a somewhat misleading term), that is, according to an internal chronobiological clock, without an external trigger? In the strict sense, this is not the case. Several circumstances can influence the length of the menstrual cycle; among them is the specific age of the woman (towards menopause the cycles often get shorter, a possible mechanism to still enable fertilisation) and sexual intercourse itself. An unexpected effect of wars has kept researchers busy finding explanations: the normal sex ratio of 100 new-born girls to 106 boys (the weaker sex, they die earlier, and the excess at birth is making up for this) rose in the two World Wars to 108 (Spiegel 1950). A number of partly quite complicated models involving protein deficiency, permeability of the zona pellucida of the ovum, and other elements have been hypothesised to be responsible for this demographic outcome. Jöchle (1973), in his work on the question whether the human female ovulates spontaneously in the course of rather fixed cycles or has traits of induced ovulation, put forward another explanation. He found that pilots of fighter planes had a higher number of sons than other soldiers and speculated that this could be an effect of induced ovulation. This could also be true for soldiers in general. They did spend most of their time in difficult situations, separated from their wives, and were only occasionally allowed a few days off where they could visit them. The intensity of long missed and therefore particularly passionate love could have induced ovulation in the women, who would then become pregnant in the optimal window of the ovum being released from the ovary. This is, as is long known, the best time to have a boy, as the male Y spermium has a smaller capacity to reach the egg and needs perfect ecological conditions to do so. This is given at the time of ovulation; X sperm can impregnate the egg under less favourable circumstances, and girls are thus more often conceived outside the optimal window. The Jewish tradition to strictly abstain sexually around menstruation could have been responsible, according to the same principle, for a much higher number of new-born boys in the formerly Jewish society in Russia (Guttentag and Secord 1984). If the mechanism of induced ovulation via unusually intensive female sexual experience were actually existing, it would, as Jöchle states, be a blow against the dogma of cyclical, spontaneous ovulation in the human species and important for gynaecologists and also for specialists of fertilisation, who are often relying on very advanced, quite non-erotic technology and may be likely to disregard evolved biopsychic tendencies still regulating human sexuality.

16.2.5 Romantic Love

In ethnology, cultural anthropology, cultural sociology, and related disciplines, many authors view human sexual behaviour in a particular way: passion (Aries and Bejin 1982; Luhmann 1982; Tyrell 1987; Giddens 1992) may have been present in the human past and can be found in traditional and our own societies, but romantic love is a child of the Renaissance. Only then were public and private life separated, so that, in occidental societies, a particular erotic, intimate bond (the 'romantic love code') could develop between woman and man. Malinowski (1929), the founder of social anthropology and, in many respects, one of the first great fieldworkers, got some of the issues wrong. He stated that in love affairs, all customs, arrangements, and codes of behaviour dictate simple, direct approaches, and that numerous ethnographic studies showed that to be the case also for other traditional societies. The work of Margaret Mead on sexuality in Samoa (Mead 1928), (who did not follow the by then fully established tradition of her discipline to conduct fieldwork in participant observation and using the language of the people studied), as well as that of other authors have contributed to the widely held conviction that romantic love, an erotically charged experience, and the power of accompanying emotions would not be found outside the western world. Couples, according to an often-used argument, would anyway be brought together by parental arrangements and not by love. Marshall, in his book *Island of Passion: Ra'ivavae* (1962), presents a study on the inhabitants of the Polynesian island of Mangaia and claims that there would be no more important issue than coitus, that the local people actually would be sexually obsessed (high coitus frequency as proof), and that the feeling of romantic love was totally unknown. This was convincingly disproved by Harris (1995), who demonstrated that the indigenous people of Mangaia have a love life in which romance plays an important role. She writes that Marshall's informants just boasted about their sexual power and number of conquests.

Numerous other examples could be given. It is quite astounding that western scholars until today would take the position that the 'romantic love code' is culture specific. How could something culturally transmitted have such volcanic power? In the words of political philosopher H. Meier (2001, p. 336), 'the genuine force, from which love draws its power of resisting society, culture and history; this might, which comes from far and roots deeper than anything else which obeys human discretion, is capable of opening realms of happiness in the midst of the most adverse conditions' (translated by W. Schiefenhövel). Is it possible that introspection, an acceptable heuristic tool in the sciences, did not yield the insight that the power which makes one 'fall' in love is an essential element of the human condition, so convincingly expressed in the classic legends of Herakles and Omphale, Jason and Medea, Dido and Aeneas, and other loving pairs, often victims of *l'amour fou*? The latter is an extreme case where the power of evolved reproductive biopsychology is overriding the power of the cortex (Schiefenhövel 2009).

There are a few exceptions in the arts and humanities, for example, the pictures of Paul Gaugin that he painted when he was living (and in love and a father of a child) on Tahiti. He subtitles one of his great works 'How come that you are jealous?' Mead (1928), Malinowski (1929), and other authors had stated that there is free love, not tinged by jealousy, a 'western' emotion—which, however, makes much sense in the perspective of evolutionary psychology: in the whole animal world, mate guarding, that is, guarding one's investment and one's chances for reproduction, is a very common trait.

Bell-Krannhals (1990), who conducted fieldwork on one of the Trobriand Islands (some young men suffered severe wounds during this time, inflicted by jealous partners of the young women they had sex with; Schiefenhövel 2009), found that especially young people take a very circumspect approach in courtship between young men and women (homosexuality was absent in this culture until the advent of white men), involving go-betweens and, today, writing love letters. To be refused by being too direct in one's advances (as suggested by Malinowski) constitutes a high and shameful risk. Courtship in many cultures of the world usually requires careful, often circumlocutary behaviour.

Kohl (2001), an ethnologist with extensive fieldwork experience in South America, states that there is, indeed, romantic love in non-western societies and that the dogma (of missing romantic love) has generally remained unchallenged until now. In a large cross-cultural sample drawn from the Human Area Relations Files, Jankowiak and Fischer (1992) found that of 160 societies 148 showed existence of feelings and behaviours of which one had, for a long time, believed that they are a product of refined western civilisation. Instead, romantic love was found to be a universal—the fact that the databases don't mention it in twelve cases does not mean it does not exist in the respective cultures; the author could just have overlooked the issue or not have talked about it with her/his informants.

The lyrics of *Shir hashirim*, Salomo's (love) song (written 950–750 BCE; King James version, online), is a strong proof that erotic, romantic love has existed long before the times of Enlightenment in Europe. Another famous example is the text of the love song of Walther von der Vogelweide (1170–1230, i.e. written in the 'Dark' Middle Ages). Members of traditional New Guinean societies created, before the white man arrived, lyrics of love songs which are moving to us as well (Schiefenhövel 2014). It is most likely that the capacity to experience romantic love is part of human nature; this does, of course, not mean that all members of our species experience this gift of our evolved biopsychology. Whether our

FIGURE 16.3 Gravesite (Valdaro-S. Giorgi near Mantova) from the Neolithic, *ca.* 4000 BCE, possibly depicting a couple buried face to face.

Source: Pasquale Sorrentino/Science Photo Library.

Homo sapiens ancestors, perhaps even their predecessors, had the same or a similar capacity will probably remain unknown for ever; it seems, however, not unreasonable to assume that they had the capacity for romantic love, and it could represent an honest signal of courtship (Figure 16.3).

Why would evolution have brought about the complex, often bewildering biopsychic mechanism of experiencing romantic love? Why not just have sexual desire, one-night stands, or other non-committing acts and not get involved in time-consuming courtship with its roller-coaster emotional upheavals, love-sickness, as well as bitter–sweet failures and deep, devastating disappointments? The biological functionality of human courtship has been described well by Eibl-Eibesfeldt (1972) and others. Liebowitz (1982), Fisher (2001), Esch and Stefano (2005), Fisher et al. (2005, 2006), and others have convincingly shown that the proximate mechanisms involved in bringing about the sometimes addiction-like falling-in-love experience are deeply rooted in our brain's neurochemistry. These costly mechanisms are there for a good reason: the start of the sexual bond, designed to foster, on the ultimate level, reproduction, is likely to work well. Its power will gradually wane and can give room for the bond to the common children; the breeding success of our species (Morris 1967; Hrdy 1999; Bribiescas, Chapter 15 (this volume)) is well built-in to our biopsychology and it can be argued that ancestral woman and man were probably equipped with these evolved tendencies.

That modern times have brought about problems like loss of libido, virtual sex, and failing reproduction, which in turn sets most sophisticated interventions of the ever-growing fertility medicine in motion, can be well understood (Fleischman 2016) in terms of evolutionary medicine. (For further discussion, see Chapter 15: Endocrinology.)

Romantic love makes sense. The two persons involved are strongly attracted to each other in various ways, they see the other as very special, and speak of her or him in words of praise. The sexual act is often metaphorically talked about, and it is clear that sexuality plays a very important role in the relationship. It is very interesting that erotic feelings, the wish to be together with the other person, and other aspects of such relationships often are not expressed in everyday prose but in poetry. The reasons for this are not really clear (Schiefenhövel 2014), but it is not unreasonable to argue that poetic erotics enhance one's own arousal and emotionality (in contrast to the process of down-regulating them, as suggested by Tooby and Cosmides 2008), facilitates a better and deeper understanding of what is happening in oneself (Johnson-Laird and Oatley 2008), and that speaking, to the partner in question or to the group, of one's desires for a sexual partnership in the symbolically charged way of metaphors increases the attractiveness of individuals who are able to do this (Miller 2000). Given that the principle of female choice, discussed in Section 16.2.1 and addressed also in Chapter 15: Endocrinology, is typical for our own species, the psychological and cognitive investment involved in experiencing and expressing a romantic relationship and signalling one's own honest involvement (Zahavi and Zahavi 1997) to the partner could be seen as male courtship behaviour. In this way, the 'artification' (Dissanayake 1992) of one's erotic and sexual desires would be one of the building blocks of the human propensity to produce art (Sütterlin et al. 2014), in this case poetry. Ancestral men, apart from testosterone-dependent interacting with the environment, with prey, rivals, and enemies, would have had an evolutionary advantage through developing their creativity (Haselton and Miller 2006) and their verbal and social competence.

16.2.6 Sexuality and Modern Times

A recent comprehensive survey (Levine et al. 2017) shows that sperm production in modern societies is decreasing at a high rate; Duty et al. (2003) as well as Swan et al. (2003) had pointed to that somewhat surprising and alarming fact before. Chemicals contained in plastic packing materials and other hormonally active substances, possibly the hormones contained in large amounts of oral contraceptives swallowed and excreted every day, are thought to be the main responsible agents. Other mechanisms, including psychosocial ones, may also have an impact on this unexpected and somewhat alarming result, which seems to be an ongoing process and could lead to sperm production reaching levels below those necessary for fertilisation; this is not yet the case.

A 2012 study found that waning of testosterone levels in men is more likely a result of behavioural and health changes than ageing itself. According to Wittert (2012), declining testosterone levels are not an inevitable part of the ageing process, but other researchers generally agree that increasing age plays a role in slowly decreasing levels of this important male hormone (Shones 2014).

Male libido seems also to be on the decline (Beier and Loewit 2011); excessive viewing of pornographic material has been made responsible for some of these effects in Italian men

(Fleischman 2016). Sexual intercourse was, until recently, a matter of bringing to action a number of physiological, neurochemical mechanisms, which one can class as 'honest signals' (Zahavi and Zahavi 1997). Erection is, to many men's great regret, the result of a complex process not under the power of the will, on the contrary. Ejaculation fits this description as well. This can be seen as evolutionary safeguards signalling the female that she is mating with a suitable male. The sex flush, erection of the mammillae and of the clitoris, lubrication of the vagina walls (whether via Bartholini's glands, Skene's (para-urethral, prostate-like) glands, or the fluid expelled, sometimes in copious gushes, from the opening of the urethra), and vivid contractions of the vagina and uterus are all honest signals which are not or not so easily to be faked—moaning and contraction of perineal muscles do not belong to this category. Our ancestors had honest sex, at least on the male side; otherwise we would not be here. Modern times have changed this dramatically; chemical erection enhancers, implanted plastic or pump-activated enlargement of the penis, electrically driven dildos, genital gels, and so on have changed the game, as have contraceptives, especially hormonally active ones, which have as often the neglected side-effect of dysbalancing the hormonal cocktail regulating libido, arousal, and climax. They work as deceiving devices, signalling to the male partner a different, evolutionarily not so interesting state of the cycle.

There is a general discrepancy of sexual desire versus reality in males, as clearly indicated by the enormous amount of Sildenafil and similar substances sold every day. Sildenafil alone accounted for a global turnover of over 2 billion US$ in the year 2012; it is obvious that men around the world are willing to incur high expenses for the promise that their sexual performance will be improved. In China, various medicines thought to work as aphrodisiacs (sea cucumber or trepang of the genus *Holothuria* as well as pulverised horns of rhinoceros, elephant, and similar animals with suggestive organs or shapes) are sold on all markets. Widespread are love charms to induce a woman to fall in love (and vice-versa), an ancient form of trying to increase one's sexual attractiveness. But particularly men tend to see themselves with regard to their sexual anatomy and performance as biologically defective beings and worry about this. From an evolutionary point of view this is to be expected: one imperative for males is to sire many children; another, contrasting one, is to help a small number of offspring survive and become successful members of society themselves. But for many men, fantasies concerning their sexual power are curtailed and restricted by their relatively small testes and problems connected with erection and ejaculation, and men not only buy sexual enhancers but also consume by far the majority of all pornographic material.

16.2.7 Homosexuality

Although homosexuality is no longer considered a psychiatric disorder, it is sometimes relevant in the medical setting due to the potential for transmission of sexual diseases, and occasional physical and psychiatric morbidity in men and women (Langström et al. 2010). A few aspects of same-sex behaviour will be briefly mentioned here. For evolutionary biology, human homosexuality, especially the more common male form of exclusive same-sex orientation, has always presented a challenge (the 'Darwinian paradox'): how could a behaviour that cannot result in offspring and thereby have no obvious ultimate effects be still existent, and even prominent, in human societies? In many animal species, same-sex sexual acts are common (Sommer and Vasey 2006) and can often be explained on a proximate or ultimate level (Fruth and Hohmann 2006)—for example, female homosexuality in wild bonobos (see Section 16.1.2), which has advantages for the involved females. (Smith et al. (2003) found 0.8% lesbian and 1.4% bisexual women, 1.6% gay, and 0.9% bisexual men for Australia; Gates (2011) 1.1% lesbian and 2.2% bisexual women, 2.2% gay, and 1.4% bisexual men for the United States; figures in large US cities are much higher; another Australian study found 8% homosexual persons (Bailey et al. 2000).)

The human case seems different and is in particular need of an explanation, because in the animal kingdom individuals involved in homosexual acts will, as a rule, behave heterosexually as soon as there is chance, while many homosexual men (and in smaller degree possibly also female homosexuals) have no inclination for heterosexuality at all; their sexual fantasies are, from puberty on, focused on men as attractive sexual partners (Frankowski and Committee on Adolescence 2004). This is different in bisexual men and women. The frequency of practiced homosexuality is very different in different societies. One would expect, on the basis of the fact that humans share (except for genetic differences, which have been shown to exist for some traits) the same biologically mediated physiology and psychology, a small percentage of both male and female homosexuality in every society. There are, however, cultures where there are no reports of adults behaving homosexually, that is, having close, intimate, and/or sexual friendships with a person of the same sex. This is true for the matrilineal Austronesian Trobriand Islanders, where Malinoswki (1929) and Schiefenhövel (2001, 2004) found no signs of male homosexuality for the times before the influx of white tourists; a few men had no female partners as wives or lovers and thus no interactive sexual lives, but they did not seem unhappy. Among the patrilineal Eipo from mountain Papua in the western (Indonesian) half of New Guinea, male homosexual play takes place between boys and adolescents, but there was no case of a male homosexual couple noted in fieldwork carried out between 1974 and the present. Eipo women were said to sometimes engage in same-sex behaviour, rubbing their genitalia on one another—'when they are sexually unsatisfied and hot' was given as explanation. The two cultures had thus little or no place for sexual behaviour other than between woman and man and therefore did not create chances for persons who are genetically or biologically inclined to homosexuality. The small village groups (originally up to 200) had a rather tight network of social control via gossip and other verbal behaviour, so that persons who were different with regard to sexual behaviour would find themselves talked about in a disapproving negative way.

New Guinea is, however, the home of a few societies where male homosexuality was either a kind of normal transgenerational sexual activity (e.g. Knauft 1985, for the Gebusi in the Western Province of Papua New Guinea) or, more noteworthy, a religiously embedded socio-sexual behaviour, which meant that boys and/or adolescents had to go through a phase of passive homosexuality with adult men (described, for example, by Williams (1936) for groups in the Fly River Delta, and by Herdt (1981, 1984, 1987) for the Sambia in Eastern Highland Papua New Guinea; for review see Schiefenhövel 1990). The biologically false, but culturally powerful common emic explanation was that a young male can only grow big and produce sperm by himself if he has received sperm from outside, that is, via the sexual acts of older men. One could hear the argument that the same were true for young women who would flourish and grow into real women because they received sperm in heterosexual intercourse. The homosexual acts, which were usually happening over weeks to months, had to be carried out anally or orally or involved some form of masturbation. Some boys rejected this custom but were usually convinced by the group that it would be absolutely necessary for their becoming adult men. It is interesting and discussed also by Herdt that obviously by far the majority of the boys and adolescents who underwent these initiation ceremonies developed into normal heterosexual fathers, who, at certain times, would become the sexual partner and sperm donor for specific young males. It seems, however, to depend on the cultural embeddedness whether such homosexual acts, even if transient, are perceived as traumatic, as is the case in our culture, where thousands of young boys have been abused by celibate priests. The topic of paedophilia represents one of the big problems for human evolutionary biology: why would such, often irrepressible, tendency evolve, which causes severe trauma in children, including own children? Especially as it is not leading to reproduction. For space constraints, we cannot, however, discuss this issue here.

Various attempts have been made to understand same-sex behaviour, still evoking controversies around the globe (Werner 2006; Janini et al. 2010), and to make homosexual preferences and behaviours compatible with evolutionary biology. This includes the hypothesis that homosexual men are particularly creative and socially influential and could thus support the genetic success of their relatives, who might have the same genetic make-up and would thereby be able to transmit genetically mediated homosexuality to the next generation. These attempts to account for male homosexuality on the ground of kin selection theory advanced especially by sociobiology and evolutionary psychology have not been convincing, except when birth order (Blanchard and Klassen 1997; Janini et al. 2010) and similar effects are measured. Recent genetic studies found, for instance, that Italian mothers who had many children had a higher rate of homosexual sons than women with fewer children; this has been interpreted as genetically (via the female line) transmitted male homosexuality representing the costs for a higher female fertility on balance (Camperio-Ciani et al. 2004; Iemmola and Camperio-Ciani 2009). Other studies have found particular genetic predispositions: for example, a linkage to pericentromeric chromosome 8 and replicated linkage to chromosome Xq28 (Sanders et al. 2014), popularly referred to as 'the gay gene', which could be responsible for producing homosexual phenotypes. Research carried out early on by Dörner and colleagues (1980) showed that a certain degree of feminisation of the male fetus in stress conditions is executed via hormonal pathways. A paper by Blanchard and Klassen (1997) focuses on the effects of H-Y antigen action during pregnancy. Large twin studies (Bailey et al. 2000; Langström et al. 2010) have demonstrated that biological factors indeed explain part of the trait, whereas other factors are stemming from the biological and social environment a person grows up in. Researchers in the developing field of epigenetics are starting to contribute to research on homosexuality as well (Rice et al. 2013). Science has thus helped scholars and the public to understand that male and female homosexuality are variants of sexual orientation. Psychotherapeutic treatment does not reverse the sexual inclination but may temporarily lead to choosing heterosexual partners (Janini et al. 2010). The so-called conversion therapy can lead to severe mental health problems (UK Council for Psychotherapy 2014).

16.3 Pregnancy and Childbirth

No matter how high the quality of the sperm, no matter how lovely the secondary sexual characteristics, no matter the value of orgasm and sexual pleasure, most aspects of sexuality discussed already in this chapter would mean little if they did not lead to the production of offspring. Reproduction is, as noted in Chapter 15, the currency of evolution. A person can have dozens of children, but if none of those children survives to reproduce and transmit their genes into future generations he or she is less successful, in an evolutionary sense, than the person who has only a single child who survives to reproduce. Certainly, this is an over-simplification (one's genes survive in all biological relatives, even if a person never has children), but it reinforces the importance of reproduction to the evolutionary process. And with regard to the other chapters in this book, the experience of pregnancy, birth, and early infancy (including breastfeeding) has a profound impact on health throughout the life course.

For most people, sexual activity eventually leads to conception, pregnancy, and parturition, although occasionally the processes face challenges that require intervention. Chapter 15 about Endocrinology provided both evolutionary and clinical overviews of conception and gestation; the rest of this chapter continues the discussion of human reproduction with a focus on pregnancy, labour, delivery, and the immediate postpartum period. As noted in Chapter 15, reproduction regularly occurs without problems, but there are a number of things that can go wrong with the reproductive process, some of which can be considered pathological, and some of which can be seen to be healthy responses inviting consideration by evolutionary medicine. This chapter focuses on the evolution of the human female reproductive processes related to pregnancy, labour, delivery, the immediate postpartum period, and infancy (the latter defined here as the period of breastfeeding). The moment of birth is probably the single most risky hour in a person's life, which is one reason that it is a focus of concern of parents and clinicians. It is also a point at which natural selection is most powerful as infants and mothers experience physiological, behavioural, and emotional phenomena that result from millions of years of evolutionary history. Most of the time it proceeds uneventfully, but in some cases the evolutionary history of labour, delivery, and early infancy may be at odds with the environmental contexts within which birth and infancy occur today. In other words, childbirth may occasionally be *mismatched* with contemporary environments and practices.

As discussed in Chapter 15, hormones like progesterone and oestrogens coordinate maturation and other reproductive events including ovulation, menstruation, conception, implantation, gestation, labour and delivery, and lactation. These are complex processes that can benefit from medical intervention at several points, although with a current world population greater than seven billion, humans have clearly been successfully reproducing for hundreds of thousands of years with minimal medical assistance. In fact, as we will see, there are many cases of failures in the reproductive process that may be caused or exacerbated by medical intervention.

Our focus will be on the healthy female reproductive system, shaped by millions of years of evolution (*phylogeny* from Tinbergen's Four Questions) and developed over individual lifetimes in various environmental contexts (*ontogeny*) to serve the purpose of conceiving, bearing, and raising offspring (*function*) (Tinbergen 1951). This discussion requires an understanding of what characterises 'normal', healthy reproductive function. Although this characterisation is based on a great deal of research, it is important to understand that most of the research has been conducted in modern industrialised settings based on women whose lives are very different from the lives of their ancestors (Trevathan 2010) or from contemporary people living like our ancestors in fundamental ways (Schiefenhövel 1988).

Furthermore, today's medical notion of 'normal' tends to reflect averages observed in well-nourished (often overnourished) women from resource-rich populations. But the perspective of evolutionary medicine argues that for any trait or behaviour, there is a broad range within which development occurs and within which reproduction is maintained. What may be seen as 'abnormal' in one setting may actually represent healthful responses to local environments. It is important for clinicians and other health practitioners to recognise this variation and to provide medical intervention based on variation and context (including the context of growth) rather than on global averages.

16.3.1 Pregnancy

Despite occasional challenges to becoming pregnant, most women are eventually able to conceive. In a conception cycle when the egg is fertilised, the corpus luteum is 'rescued' and continues to produce oestrogen and progesterone that maintain the pregnancy, until this function is taken over by the placenta, several weeks later. If all goes well, the fertilised egg (zygote) implants into the uterine wall and develops into the embryo and the placenta. Unfortunately, things do not always go well and failure to implant is sometimes the fate of the zygote, which is shed in the next menstruation and may not be detected by the woman. 'Failure to implant' is another common cause of infertility, and it is usually seen as pathological and requiring clinical assistance. In many cases, however, failure to implant is due to abnormalities in the zygote or in the maternal reproductive system, such that conceptions that would develop unsuccessfully are weeded out before too much energy (caloric and emotional) is invested. An evolutionary medicine perspective suggests that many cases of failed conception result from the reproductive system doing what it evolved to do—avoid wasting reproductive effort on offspring that are not viable or are otherwise unable to increase reproductive fitness.

When the zygote is not rejected, it begins the process of burrowing into the endometrium, at which point, human chorionic gonadotropin (HCG) is produced by the placenta and a pregnancy can be detected by a simple urine test. Once implantation occurs, the rate of early pregnancy loss drops and most gestations proceed with few challenges (see Chapter 15). The human placenta is a remarkable organ that directs the course of pregnancy by producing reproductive hormones, transmitting gasses and nutrients to and from the fetus, and shielding the fetus from toxins and other dangerous substances in the mother's environment. Unfortunately, it is not a perfect screen, as can be recognised when drugs, alcohol, and other toxins cross from the maternal to the fetal system, potentially damaging fetal cells in the process.

In general, a woman's physiology becomes more efficient at metabolising food so that she gets more nutrients from the same amount of food than she did before she became pregnant (Cunningham et al. 2014). In this sense, her body is optimising dietary intake so that even if she is not able to increase the volume of the food she consumes, she can still gain the weight needed to support her growing fetus. Clearly, there are advantages to improved metabolism for women who live in marginal environments, but in places where food availability is high, potentially greater problems emerge when women take in more food at the same time they are reducing activity levels, behaviours that may lead to excessive weight gain during pregnancy. Excessive weight gain can lead to gestational diabetes, which, as noted above, may predispose the woman to regular diabetes later in life. Furthermore, weight gain in pregnancy is usually not easy to reverse after delivery, especially in the absence of breastfeeding.

Diet and other factors to which the fetus is exposed are particularly important during the first trimester, the time in which organ systems are developed. Nutrient deficiencies during this period can result in problems in skeletal development, neurological development, and motor development. Unfortunately, this especially vulnerable time occurs when some women are not even aware that they are pregnant. Something that may alert them to the pregnancy is 'morning sickness', also known as nausea and vomiting of pregnancy (NVP). NVP usually results in a woman limiting her food intake and it is so common (see, for example, Kohl et al. 2009) that it may not cause concern unless it becomes extreme in the context of marginal food resources (see Pike 2000 for examples). This 'sickness' is an example of a challenge that may be as beneficial to the health of the mother and her fetus as it is unpleasant. Many of the foods that are especially noxious to women early in pregnancy have been found to have components that are harmful to the developing fetus. Examples are overripe, spicy, fermented, and smoked foods, many of which may be risky for the fetus. Fortunately, low food intake (as long as quality of food is not compromised) is not problematic in the first trimester when fetal weight gain is not of concern. Once the initial period of systems development is over and the second trimester begins, weight gain is the primary goal of a fetus and nausea disappears. At this point, maternal appetite is restored and the woman begins the period of 'eating for two'.

It is during the third trimester that growth really takes off in the fetus. Respiratory and neurological maturation continues throughout pregnancy but increase in body weight and fat deposition is the primary goal of the last trimester of pregnancy as the fetus prepares for life outside the womb. Fat accumulation is particularly important for this fattest of all mammalian babies—fat provides resources for the rapidly growing fetal brain (Kuzawa 1998). (For further discussion, see Chapter 17: Brain, Spinal Cord, and Sensory Systems.) Maternal–fetal competition for nutrients continues and it appears that when resources are limited, priority for allocation of energy goes to the fetus. Photos of near-starving women nursing new-born infants are reminders that once the pregnancy is well established it is usually maintained, even at the expense of maternal health.

Unfortunately, however, all is not well for babies born to mothers who had low nutrient intake during pregnancy. These babies are usually born small, even when they have reached their full gestational age, a phenomenon known as 'intrauterine growth retardation' (IUGR). Some cases of IUGR are of clinical concern and include genetic abnormalities in maternal metabolism and placental problems, but many cases are due to preventable social and socioeconomic factors (Hendrix and Berghella 2008). It was once believed that a small baby born to a mother living in a resource-poor environment was an adaptation to scarce resources because he or she would need fewer resources in the post-uterine environment (see Seckler 1982 for a similar argument about growth in childhood). This was before the realisation that there was much more to being born small than weight and length. When a fetus is deprived of nutrients, the brain is served first, meaning that other organs such as heart, liver, and pancreas are restricted in growth. Babies born with smaller-than-normal organs often face challenges to glucose and cholesterol metabolism and to immune function later in life, which explains the association between low birthweight and adult health revealed in the studies of children born during the 'Dutch Hunger Winter' (Gluckman et al. 2008).

The phrase 'fetal programming' has been used to describe the relationship between IUGR and adult health. If a fetus is 'programmed' during gestation to expect a resource-poor environment after birth, the metabolic system is primed to exploit every available calorie (a 'thrifty metabolism') (Wells 2011). Too often, however, infants today that experience nutrient challenges in pregnancy are born into globalised environments where high-caloric foods with low nutritional quality are readily available. The result is a mismatch of caloric availability and a metabolism primed to get the most out of each calorie, often by storage in adipose tissue. This storage seems to lead to a 'catch-up' in weight, but the organs that were compromised by nutritional stress in pregnancy can never catch up. Thus, metabolism may be challenged by excess carbohydrate intake, leading to diabetes, obesity, hypertension, and cardiovascular challenges in adulthood (Kuzawa and Quinn 2009).

Even women in resource-rich environments who try to limit their food intake in pregnancy to avoid gaining too much weight may find that their infants have been programmed to expect food limitations in the postnatal environment. These infants may then encounter some of the same problems that result from mismatched prenatal and postnatal environments as they try to get everything they can from each calorie with their thrifty metabolisms. (For further discussion, see Chapter 4: Growth and Development.)

The potential for metabolic disorders to develop in mismatched environments argues for public health measures that improve nutrition for all pregnant women in an effort to improve childhood and adult health. But concern cannot stop there. Recall that a woman's diet and health in pregnancy affects not only the fetus but also the eggs that are developing in the ovaries of every female fetus (Thayer and Kuzawa 2011; see Bribiescas, Chapter 15 (this volume)). This means that if her female fetus is affected, it is likely that the health of her grandchildren will be affected as well. Thus the effects of resource availability in pregnancy are transgenerational and all public health measures to improve pregnancy outcomes must take the long view in order to break the cycles of health challenges caused by resource limitations (Nathanielsz 2001; Ellison 2005; Kuzawa 2005).

Advanced pregnancy places unusual stress on the upright, bipedal human. In comparison with quadrupedal mammals in which the weight of the fetus is carried in an abdominal 'sling', evenly distributed along the vertebral column, the human vertebral column supports increasing weight from top to bottom. This is challenging enough for all humans (witness the high frequency of lower back problems that occur with age; see Chapter 7: Musculoskeletal System), but is especially problematic for a pregnant woman who carries a heavy infant during the last month of pregnancy. The weight of the fetus projects outward from her vertebral column, throwing off her centre of gravity and making walking increasingly difficult, not to mention increasing discomfort. Although natural selection has favoured modification of the vertebral column in women to accommodate this challenge to bipedal walking, it puts some women at greater risks for lower back pain and slipped discs (Whitcome et al. 2007).

Although there are numerous physical challenges that women face during pregnancy, psychosocial stresses also play major roles in affecting outcomes. For example, stress hormones (especially glucocorticoids such as cortisol) have been implicated in compromised immune function and in developmental abnormalities of metabolic, cardiovascular, liver, and pancreatic functions (Harris and Seckl 2011). Stress during pregnancy has been linked to behavioural disorders in childhood (Davis et al. 2005), perhaps through epigenetic pathways (Zucchi et al. 2013), suggesting that reduction of stress during pregnancy will likely benefit both mother and infant. (For further discussion, see Chapter 17: Brain, Spinal Cord, and Sensory Systems.) Social support has been shown to be especially important for stress reduction in pregnancy (Mulder et al. 2002). The effects are not surprising, given the dense social networks in which pregnancy was experienced in the past and in many communities today. As we will see in Section 16.3.2, social support that extends to the time of birth may have made the difference between life and death for mothers and infants in the past.

16.3.2 Labour

About nine months after conception, labour begins, marking the end of gestation. The precise trigger is not known, but factors that contribute to the ending of gestation and the onset of labour include placental ageing; hormonal changes; pressure of the presenting part of the fetus (usually the head) on the cervix; stress signals from the fetus as maternal metabolism fails to keep up with fetal needs; and stress on the abdominal cavity as it becomes increasingly stretched to support the growing fetus (Edmonds et al. 2012; Cunningham et al. 2014). All these factors are intertwined and influence each other in the highly complex physiological–biochemical ways that medical science is just beginning to understand. When exactly a pregnancy comes to its natural end cannot be easily predicted. Like other biological processes, the length of pregnancy follows, in principle, a Gaussian/bell curve. Interestingly, it is not perfectly symmetric, mainly because some pregnancies end very early (e.g. in spontaneous abortion), which may be an adaptive event due to genetic or infectious incompatibilities. In these cases, a detection system terminates the pregnancy in a kind of quality control.

The range of 'normality' is actually astoundingly high; birth 'at term' is defined as taking place between the 37th and 42nd weeks of pregnancy. Neonates are able to survive well and usually without any risk of morbidity after 37 weeks of gestation. Usually, physicians are more concerned about the slightly steeper right side of the bell curve; this is expressed in the definition 'term' which is defined as 3 weeks earlier and only 2 weeks later. Obstetricians today tend to induce birth or carry out a caesarean section if pregnancy lasts into the 41st week of pregnancy, even when no signs of danger for the baby or the mother are diagnosed. It is an interesting question whether the human baby's design as physiologically premature (see Section 16.4.4) is such a robust characteristic of the fetus that 3 or even 4 weeks early birth is not harmful, whereas 2 weeks later birth is considered 'post-term' and thereby potentially pathological or at least requiring medical intervention. Galal et al. (2012) suggest that pregnancies that exceed 41 weeks should routinely be terminated. From an evolutionary point of view one would argue that the complex mechanisms regulating the duration of pregnancy and, especially, the onset of birth have been subject to rigorous selection—allowing, of course, for a small percentage of pathology, which happens in all biological systems. Louis and Platt (2011) give a range of 3–12% post-term pregnancy. As with birth in general, modern medicine favours intervention to exclude potential risks; whether iatrogenic risks are incurred at the same time is not known in the case of terminating pregnancies within the 41st week, but it is not unlikely. Here, as in many other aspects of reproduction, more research, inspired by evolutionary medicine, is necessary.

The onset of labour is associated with a drop in progesterone, which has served to keep the uterus from contracting, and an increase in the secretion of oxytocin, which stimulates uterine contractions. Much of the oxytocin derives from the fetus itself, reinforcing the importance of the fetus in initiating its own birth (Liao et al. 2005). In turn, oxytocin finds sufficiently mature receptors in the myometrium of the uterine wall, indicating the importance of uterine and placental maturation in this process. The contractions of labour serve to dilate the cervix and to move the fetus down to begin the process of manoeuvering through the birth canal. The time from onset of regular contractions until birth varies from about 7 hours in multiparous women to as long as 20 hours in those giving birth for the first time. The upper range for primiparas is not fully defined because when labour length becomes excessive (variably defined), clinical intervention usually occurs to terminate it and begin delivery. Contractions are infrequent in the early part of labour and become increasingly stronger and more frequent toward the end. The uterine muscle is one of the strongest muscles in the human body and the force can be doubled when the woman exerts her own strength as she pushes during delivery.

16.3.2.1 *Oxytocin*

A hormone that plays a primary role in almost every aspect of reproduction, including the initiation of labour, is oxytocin. The evolutionary history of oxytocin tells a fascinating story of how a substance, once it appeared, started to fulfill a wide range of roles in the regulation of behaviour over the course of vertebrate evolution. Many of these behaviours are still closely related to reproduction in vertebrates. It is a very old nonaneuropeptide, present in some ancient invertebrates (Mizuno and Takeda 1988), and estimated to be about 600 million years old (Feldman et al. 2016). Oxytocin is associated with a surprisingly wide array of roles in humans: birth, lactation, sex/orgasm, sociality, stress regulation, perhaps autism, and many other physiological phenomena.

As an example of its widespread effects, Stuebe et al. (2013) found that low levels of oxytocin in breastfeeding women, determined 8 weeks after birth, were associated with higher scores of postpartum depression and anxiety. The hormone thus has a protective effect on the health of mothers well after birth. In the complex psychophysiological mechanism in which the baby's sucking and other behaviours towards the mother contribute to the release of oxytocin and prolactin, salutogenic effects are embedded which protect the breastfeeding mother. Taking care of a small infant is usually regarded as a kind of one-way street of heavy maternal investment in the baby. While this is true in many ways (time, energy, producing milk, etc.), this view overlooks the veritable symbiotic nature of the mother–child bond: both partners benefit (see Section 16.4.5). The protective effect of a natural childbirth and a well-developed mother–child bond on the mother is also visible in reduced rates of postpartum mood changes (dysphoria or 'baby blues').

Crockford et al. (2014) shed light on the particular importance of oxytocin and on its role in modulating social behaviour, for example in therapeutic settings. Being able to detect it in non-clinical settings will open new ways to study the effects of behavioural therapies, group therapies, and the like. Bartz et al. (2011) argue that while oxytocin has, on the level of our species, clear specific functions, context and personality characteristics play a role in bringing about and modulating the psychosocial effects of this hormone.

16.3.2.2 *Stages and Phases of Labour*

Clinical writings refer to the labour process as involving power (uterine contractions), the passenger (the fetus, its size and position), and the passage (bony and soft tissues of the birth canal) (Liao et al. 2005). Ideally, the maternal, fetal, and placental systems have matured so that they can simultaneously work to deliver a healthy new-born infant that can survive in the outside world. Labour and delivery are usually described as occurring in three stages: the first stage is from the onset of regular uterine contractions until full dilation of the cervix; the second stage is the period from full dilation to delivery; and the third stage is from birth to delivery of the placenta. Some practitioners also add a fourth stage, which can be described as the time from placental delivery to the initiation of breastfeeding (Sobhy and Mohame 2003), reinforcing the idea that breastfeeding is an important part of the birth process, as discussed in Section 16.4.8.

If labour does not seem to be progressing as expected, following guidelines that often vary from hospital to hospital, a woman in labour may be given an extra 'boost' in the form of synthetic oxytocin. As noted in Section 16.3.2.1, this hormone is produced naturally by the woman in labour and by the fetus itself, who thus plays a very important part in determining the onset of birth, and oxytocin is used to induce or augment the process in more than half of hospital deliveries in the United States (Bell et al. 2014). The effect is often immediate and the resulting contractions are usually more powerful than the 'natural' ones, so many women report it as unpleasant. Because it mimics a natural hormone it is believed to be safe, but its use is associated with higher rates of anxiety, depression, and symptomatology at two months postpartum (Gu et al. 2016). It also seems to interfere with breastfeeding, as shown in lower rates (Bell et al. 2014). There is a suggestion that a dependency can form when women are administered synthetic oxytocin at several deliveries. This suggestion is based on a study of 358 women who had received synthetic oxytocin at previous deliveries. In contrast to expectation, length of the active phase of labour sharply increased for women who had received oxytocin at three or more previous deliveries (Trevathan 1983). Although this study has not been confirmed with additional research, it serves as a reminder that medical intervention can potentially have a long-term effect on labour.

The obstetric use of synthetic oxytocin has increased sharply in the last decades: in 1990 it was used in 10% of birth in the United States, 23% in 2010, and estimates for 2014 were 57% (Bell et al. 2014). For Denmark, a country with generally low rates of obstetric interventions, Stokholm et al. report that in 26% of all births synthetic oxytocin was used (Stokholm et al. 2018). Bell and colleagues write: 'While the judicious use of synthetic oxytocin has many benefits, the biological and behavioural effects of synthetic oxytocin beyond the immediate clinical uses remain largely unknown' (Bell et al. 2014, p. 35). The same surprising research deficit relates to the question whether there are differences between endogenous and exogenous oxytocin and whether these differences may alter the outcome of birth itself, as well as the very complex process of establishing the postnatal symbiotic dyad of mother and child, which is crucial for the start of a new life. In basically all the elements of this process, oxytocin plays a most significant role, including restoring the uterus to normal size and configuration, enabling lactation, and releasing warm, loving emotions facilitating bonding and other evolved patterns of perception and behaviour (Bell et al. 2014). The authors' comprehensive review found negative effects of synthetic oxytocin on basically all major functions associated with naturally released oxytocin: stress reactivity, maternal mood, mothering behaviour, lactation, and consequences for the offspring (e.g. higher rates of attention deficit disorder). They concluded: 'Downstream molecular effects of naturally expressed oxytocin and synthetic oxytocin have not been investigated thoroughly in the context of human birth care. Midwifery and obstetric research should consider the oxytocin system as a whole, not just the immediate clinical result, when investigating the role of physiologic birth as well as birth interventions on biobehavioural outcomes in mothers and infants' (Bell et al. 2014, p. 38). Ekelin et al. (2015), drawing on their experiences in Sweden, also demonstrate the ambiguous nature of using exogenous oxytocin and warn of known and unknown iatrogenic effects.

In an early paper published by Mast et al. (1971), attention was drawn to a possible role of synthetic oxytocin for the management of labour and neonatal jaundice (icterus neonatorum). In a second large-scale study by Chalmers et al. (1975), again a clear correlation between administering synthetic oxytocin and neonatal jaundice was found. These findings have not received much attention; it seems that the medical profession may use oxytocin as a very powerful means to induce or increase uterine contractions during delivery and in the postpartum period, but little attention is given to the possible iatrogenic effects of this hormone. An evolutionary perspective seems needed by which the finely tuned elements of the very complex bio-psycho-endocrino-social system regulating reproduction in general and birth in particular are highlighted, with warnings that our medical interventions may have undesirable side-effects. Interventions are of course justified in well-deliberated cases, but often they have become routinely embedded in the practice of medicine, rather than grounded in well-weighed decisions which have taken the evolutionary nature of the human body into account. From time to time, new standard procedures are defined and formulated by the medical profession. Hopefully evolutionary considerations will be kept in mind when the next round of standards is discussed.

16.3.2.3 *Posture in Labour*

The period from the onset of regular contractions until full dilation of the cervix is referred to as the first stage of labour. When the cervix is fully dilated and the woman feels the urge to push the baby out of the birth canal, she is said to have entered the second stage of labour. This is the most active phase and requires the greatest maternal effort; it lasts about 15 minutes in multiparas and up to 1 hour in primiparas. Maternal exhaustion during this stage may occur, so any behaviour that can reduce its length is welcomed. One of the best ways to shorten the second stage and to ease delivery is for the mother to maintain an upright or close-to-upright position. This position tilts the pelvis forward and enables contractions to work with gravity, increasing their effectiveness. The upright position during delivery helps to relax and stretch the muscles of the pelvic floor, reducing the need for an episiotomy. The bones of the pelvic basin expand as much as 30% in the transverse dimension, which eases the passage of the fetus (Michel et al. 2002). When the baby is in the occiput anterior (OA) position (the position reported for more than 95% of deliveries), the stronger occipital bone bears most of the force of the contractions rather than the more fragile frontal bones, which are subjected to the forces in a lithotomy (on the back) position. If the fetus is in the occiput posterior (OP) position (about 5% of deliveries), the upright posture facilitates rotation to the OA position, making delivery easier. As with walking in labour, women who assume an upright position in delivery report less pain, lower use of analgesics, and fewer perineal tears (Zwelling 2010).

Walking during labour also improves maternal–fetal circulation and avoids compression of the inferior vena cava. Compression of the inferior vena cava has a negative impact on utero-placental circulation and increases the risk of haemorrhage during or after birth (Dunn and Caldeyro-Barcia 2015). Upright movement improves the quality of the contractions, making them stronger with faster cervical dilation. This can, in turn, decrease the length of labour by as much as a third. Frequent changes in posture provide pain relief and improve circulation (Gupta and Nikodem 2000). Maintaining an upright posture helps to get the baby in a better position for delivery by working with gravity to help it move into position. Compared with the force of uterine contractions (measured in Montevideo Units; Caldeyro-Barcia and Poseiro 1960), the force of gravity is not very high, but it is present also during the phases in which the uterus is relaxed and can therefore exercise its physiological effect during all phases of labour. Women who walk in labour report less back pain and less pain in general. All of these benefits yield fewer interventions, including fewer caesarean sections and epidurals. Finally, walking and posture changing is what women *want* to do. This desire probably has deep evolutionary roots to a time when women experienced fewer constraints on their movement during labour. Being able to assume the posture that makes them most comfortable and to change positions frequently also reduces stress on the mother and enhances her self-confidence at a particularly vulnerable time.

Birth is often seen as a more or less mechanical process that can be sped up or facilitated through the interventions typical of modern medicine. An evolutionary understanding of birth will, in contrast, highlight the complexity of anatomical–mechanic, physiological, hormonal, emotional, and cognitive factors, in other words, aim to integrate evolved elements of birth behaviour. This can only be achieved if the parturient has the chance to move and be involved in the decision process. Retaining mobility is a very important part of this. Today, in hospital deliveries, movement is often restricted because of amniotomy, induction, epidurals, fetal heart and contraction monitors, and intravenous (IV) drugs. Data from the cardiotocograph (CTG) can be transmitted telemetrically, but this technology is not very widespread in hospitals. Additionally, it is usually easier for hospital staff to monitor women in labour when they remain in a small space. Allowing a woman freedom of movement during labour is consistent with an evolutionary medicine view of labour and delivery.

16.3.3 Delivery of the Infant—Birth

The fetus moves through the birth canal in a series of rotations with the expulsive forces of the contractions. This movement does not require any assistance and, in fact, interference with the natural movements can be risky for the mother and infant. British paediatrician Peter Dunn (1995) and French obstetrician Michel Odent (2003) argue that most difficulties in labour and delivery today are due to unwarranted medical intervention. Odent proposes that a 'fetus ejection reflex' (1987) allows delivery to proceed just as it does in other mammals that give birth with no assistance (see also Walrath 2006). The fact that women who cannot push during the second stage of labour (i.e. voluntarily activate the muscles of the abdominal wall due to paralysis) can still successfully deliver an infant is support of the concept of a fetal ejection reflex (Liao et al. 2005).

The series of rotations that the human infant must follow to be born is unusual among mammals and is due to another unusual characteristic of humans, bipedalism. In fact, the way in which humans walk, on two legs, is regarded as a hallmark of the family Hominini, humans, and their recent ancestors, and is the primary way in which the species is recognised in the fossil record. There are few aspects of human evolutionary history that have had

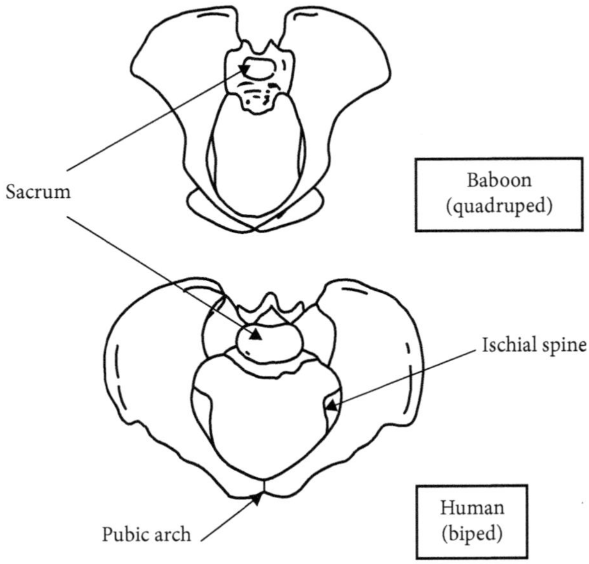

FIGURE 16.4 Comparison of a baboon quadrupedal pelvis with a human bipedal pelvis, emphasising features related to birth.

a greater impact on childbirth than bipedalism. To understand this, compare the pelvis of a quadrupedal primate (baboon) with that of a bipedal human (Figure 16.4). The birth canal of the baboon is an oval passageway that has its broadest dimension front-to-back throughout, at the entrance, midplane, and outlet. The bipedal pelvis, in contrast, is broadest in the side-to-side or transverse dimension at the inlet, but twists in the middle to have the broadest dimension of the outlet in the front-to-back dimension. For both baboons and humans, the infant head (the usual presenting part) is large and has its greatest dimension in the front-to-back direction. The large head makes for a tight squeeze in both baboons and humans, but for humans, the twisted birth canal requires that it undergo a series of rotations as it passes through during delivery. For most large-brained primates, there is a tight squeeze between the neonatal head and the birth canal, making birth challenging (great apes are exceptions because their birth canals are unusually large and their neonates small, so birth is rarely as difficult for them as it appears to be for monkeys and humans). Furthermore, the neonatal shoulders of humans are broadest in a dimension perpendicular to the head, requiring a second set of rotations to pass through the birth canal.

Another neonatal dimension that needs to line up during birth is the large occiput that fits best against the broader anterior portion of the birth canal so that the infant most commonly emerges facing away from the mother (Figure 16.5). This phenomenon of being born facing 'backwards' makes it somewhat challenging for a woman to assist her infant at

FIGURE 16.5 (opposite) Birth of the human infant, from entering the birth canal facing the mother's back (A), to rotation (B–D), emergence (E–G), and manual guiding (H). Each panel is a sagittal section through the body of a mother squatting during labour. Maternal pelvic skeletal elements (pubic bone, sacrum, and vertebrae) are shown in black (other parts of the bony skeleton are not visible in this midline view). In the lower right corner of each panel is a 'midwife's-eye' view of the neonatal head as it rotates within and emerges from the birth canal. This is the most common pathway.

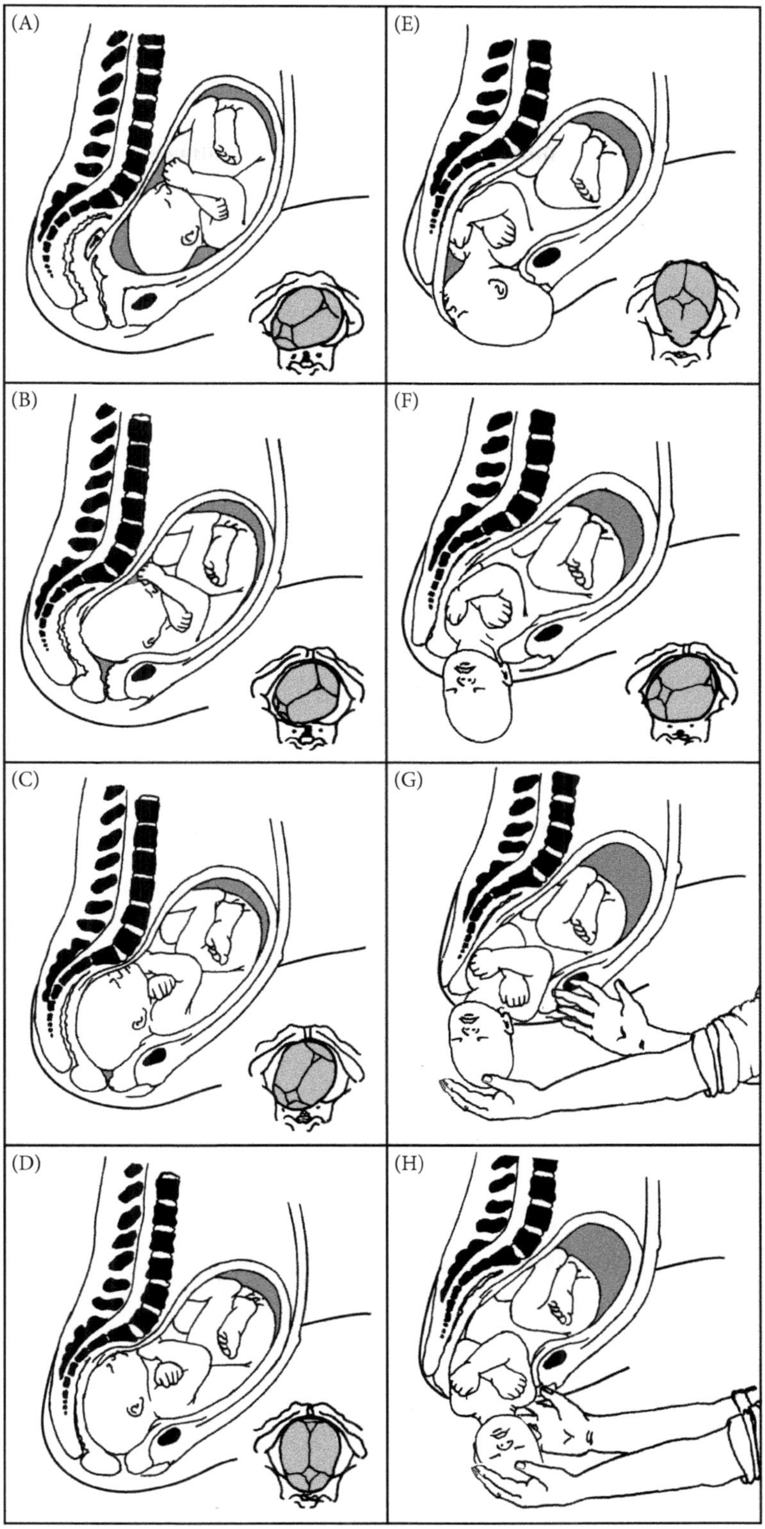
(A)
(B)
(C)
(D)
(E)
(F)
(G)
(H)

birth in the rare occasions when it is necessary, placing a premium on having someone else present. When a monkey is born (usually in a face presentation), it faces the mother and she can reach down and guide it out of the birth canal and up to her chest with her hands (Trevathan 2015). Human mothers can guide their babies out of the birth canal also, but it puts more stress on the perineum, increasing the likelihood of tearing. Pulling the baby against the normal flexion of its body also risks damage to the nerves of the neck region. These risks are minimised when there is another person to 'catch' the baby if necessary and assist the mother in completing the delivery. The other person is usually in a better position to wipe the fluids from the new-born infant's face and nose, check for the umbilical cord, which is often wound around the baby's neck, and help move the baby to the mother's breast. More important than physical assistance, however, the attendant can also provide emotional support to the mother, who usually seeks contact with others when she begins labour, unlike other mammals that more typically seek isolation.

The importance of emotional support during labour and delivery should not be underestimated. Many women report anxiety and even fear as they approach the time when they expect to give birth. Mild anxiety and fear can have positive benefits when it improves the ability to focus cognitively, emotionally, and physically. Similar to a fever, which may be useful for mitigating a viral infection, mild stress can be a physiological 'defence' (Nesse 1991). But when it becomes excessive, it can become a 'defect' that has a negative impact on the birth process, as is true for extreme elevation of body temperature. Probably the best way of keeping anxiety and stress from becoming excessive is to provide emotional support at birth. A companion who was recruited to relieve anxiety would have been coincidentally able to provide assistance if it was needed. In other words, ancestral women likely sought companionship when they began to labour because of anxiety and, perhaps pain, rather than due to an understanding of how the birth would proceed.

The need for this support and occasional minimal physical assistance is probably deeply rooted in our ancestral past when having an assistant at birth made the difference between life and death for some mothers and infants, but in most cases assistance was for emotional support only (Figure 16.6). In observations of Eipo births, if the parturient squats or sits close to the ground there is no need to 'catch' the baby, indicating that the ejection reflex can do its job very well without assistance. Cleaning of the airways, wiping off vernix caseosa, and even severing the umbilical cord is usually done by Eipo mothers themselves (Schiefenhövel 1988). This evidence suggests that, in evolutionary terms, relieving anxiety and pain are the proximate determinants of companionship at birth, but reductions in mortality and morbidity can be seen as ultimate explanations of the value of assistance at birth.

Considering the tight fit between the human neonate and birth canal (and some level of mortality due to cephalopelvic disproportion) begs the question of why natural selection has not favoured slight increases in maternal pelvic dimensions. It was previously thought that this was due to compromised walking that would result from a wider birth canal, but that suggestion has been recently challenged (Warrener et al. 2015). The topic of 'obstructed labour' or 'obstetrical dilemma' is presently attracting evolutionary anthropologists. Huseynov et al. (2016, p. 5227) state: 'The evidence that hormones mediate female pelvic development and morphology supports the view that solutions of the obstetrical dilemma depend not only on selection and adaptation but also on developmental plasticity as a response to ecological/nutritional factors during a female's lifetime.'

FIGURE 16.6 Traditional birth attendants take care of a primiparous parturient among the Eipo, natives of Highland West New Guinea; this social embeddedness of parturition, together with vertical body postures, is typical for birth in many cultures. (Photograph: W. Schiefenhövel).

Another view is that anatomical challenges to delivery and the benefit of practical assistance may have begun to take their toll only 10,000 years ago with the origin of agriculture and associated smaller maternal body size due to poorer nutrition (Wells et al. 2012), meaning that there has not been sufficient time for natural selection to act on maternal pelvic dimensions. Furthermore, even if a larger pelvis does not compromise walking, there are other risks associated with large birth canals that may help to maintain the tight squeeze.

Consider that the muscles and ligaments of the pelvic floor serve to support the internal organs within the pelvic 'bowl' in an upright posture. In quadrupedal mammals, the organs (including the pregnant uterus) are supported by the abdominal wall and are slung beneath the vertebral spine and distributed somewhat evenly, minimising stress at any one point. This difference in the way in which internal organ weight is distributed is especially pronounced in the later stage of pregnancy in a biped when the tissues of the pelvic floor undergo even greater stress. The ischial spines are the primary point of support for most of the ligaments of the pelvic basin, but, unfortunately, they protrude into the birth canal at its narrowest point and restrict passage of the neonate. If selection had favoured a broader birth canal this would have pulled the ischial spines further apart, increasing the risk of pelvic organ prolapse (POP), a disorder that affects as many as 9.3% women today in both the industrialised and non-industrialised nations (Vos et al. 2012). This is an example of an evolutionary trade-off—lowered risk of POP but at the expense of smaller birth canals and more challenging deliveries. As Stewart (1984) describes it, the pelvic basin at the level of the ischial spines is 'adapted for holding things in rather than letting them out'.

The issue of 'obstructed labour' will probably keep researchers in evolutionary medicine and anthropology busy for some time to come. Given the fact that severe perinatal pathology is rare (in the range of 5%) as long as birth is handled in an evolutionary way (Schiefenhövel 1988), the answer to the question why evolution did not provide females with a larger pelvis can probably be answered in this way: the various adaptations typical for human birth (physiological prematurity and the numerous anatomical–physiological mechanisms mentioned above), plus the effectiveness of archaic forms of emotional support by traditional midwives were probably sufficient to make birth in our species reasonably safe. In the medical profession, the opposite view is common: birth is seen as dangerous, thus requiring intervention. As stated above, our modern interventive and invasive forms of obstetrics are creating some of the problems which will then be solved by obstetric technology (Odent 2003; Dunn and Caldeyro-Barcia 2015). Achieving 100% safe solutions is impossible, in modern engineering as well as in biology. Evolution produces, in a long process of optimisation and compromise, solutions that work well enough. Birth is a good example of this. We will most probably never arrive at 0% perinatal mortality and morbidity.

The medical profession has, since obstetrics became a domain of scientific medicine more than 300 years ago, developed many ideas and techniques to make birth safer for child and mother and, in the last decades, emotionally more fulfilling. Ignaz Semmelweis is a good and at the same time very sad example—the Hungarian-born Viennese gynaecologist and obstetrician died in an asylum after having been ferociously attacked by his colleagues who did not want to believe that their hands could bring about death in women who gave birth, in their care, in the big hospitals of the Austrian capital. Yet, his plea for disinfection was finally accepted and immensely successful. The discipline had learned an important lesson.

The question is whether vaginal obstetric examinations and techniques are, as routine manoeuvres, necessary to make birth safe. It seems evident that midwives and doctors in modern hospitals carry out vaginal examinations in much too high frequency (Shepherd and Cheyne 2013). Some obstetricians are warning against this routine. The example of birth in Highland New Guinea demonstrates that no vaginal or other classic obstetric intervention (e.g. assisting the emergence of head and shoulders) at all is necessary for the child to be ejected and to slide on grass or the floor (Schiefenhövel 1988).

Some routines of the past have been given up already. 'Total fetal monitoring', recommended by several authors (Saling and Dudenhausen 1973; Saling 1981) and state of the art some 40 years ago, involving transmission of data generated by the CTGs of individual parturients to monitors in the office of midwives and doctors, plus the possibility to test fetal blood taken from the scalp so that problems, such as a drop in fetal heart rate or neonatal hypoxaemia, could be diagnosed and taken care of immediately, has been given up or at least questioned in a number of countries (Sacco et al. 2015). Perinatal outcome has not worsened with the discontinuation or modification of these practices.

In a similar way, routine episiotomy has been considered obsolete in many western countries (American College of Obstetricians and Gynecologists 2006a) because its consequences, especially genital pain, especially during intercourse, turned out to be an iatrogenic problem. This is again a somewhat unexpected result: would it not be better to surgically widen a too narrow birth channel and then carefully and under sterile conditions stitch the wound *lege artis*? Smaller ruptures of the perineum heal, however, much better and with less after-effects when they are allowed to happen along preformed lines in tissue and cells and heal by themselves. Obviously, evolutionary mechanisms have been selected for the repair of quite common smaller to medium ruptures. Yet this procedure is still one of the most commonly performed interventions in obstetrics (American College of Obstetricians and Gynecologists 2016).

Induced labour by IV injection of synthetic oxytocin was also a common practice in the past, as noted in Section 16.3.2.1. This as well seemed a reasonable method by which the hospital personnel and the wishes of the parturient could be taken into account. Birth would mostly be induced during daytime of a week day. This practice has been given up as well; the outcomes were not as good as when birth happened spontaneously. The 'inner clock' governs the process of birth like many other functions of the human body (Roenneberg and Merrow 2016) and as the nervus vagus, responsible for relaxation, digestion, sexual arousal, and the like, a basically nocturnal nerve, governs also birth, many deliveries naturally take place at night.

It is possible that other obstetrical intervention routines which, today, are believed to be indispensable, will also be given up one day. On the other hand, modern life is characterised by a belief in technical solutions; it will, therefore, always be necessary to weigh pros and cons of surgical and other invasive methods versus evolutionary adaptations for childbirth.

16.3.4 Posture During Delivery

Although there are certainly situations in which the infant's head will not pass through the birth canal without mechanical (e.g. forceps or vacuum extraction) or surgical assistance, many cases of obstructed labour can be relieved if the mother assumes a squatting or other upright position, which helps to widen the birth canal, as discussed in Section 16.3.2.3. This is the most common position assumed by women in societies where medical intervention is not available or not desired. These are usually women who lead very active physical lives in general and spend much of their lives squatting for various activities from elimination to food preparation, however, so they have the strength to maintain that position for a long time, as for the Eipo women. Even these women often have assistance from a support structure (a beam or rope suspended from the ceiling) or from another person. With sufficient assistance women who aren't used to squatting can improve the likelihood of delivery in the position, which is optimal for increasing the diameter of the birth canal and easing delivery.

Women in Highland New Guinea were observed to choose the following body positions during childbirth (Schiefenhövel 1988): walking, standing, kneeling, knee–elbow posture, sitting, squatting, and, rarely, lying on the side or the back for a while. Especially while sitting, standing, and kneeling, asymmetric postures of the body were very common; that is, one leg would be spread out while the other one would be bent, or the whole body would be leaning to one side (Figure 16.7). The women obviously reacted with specific movements and positions of their bodies to the obstetrical events that were occurring during labour. Sometimes the parturient would lean back to a half-reclining posture supported by her own arms and hands or by one of the birth attendants. Generally, positions were changed frequently and almost always upon decision of the parturient herself. In the seven cases that were documented photographically and partly on film, the expulsion of the new-born happened during either sitting (five times) or squatting (two times). In one case a pole was used to support the body while standing; the women also held on to beams or planks sticking out

FIGURE 16.7 Childbirth without obstetric intervention for an Eipo woman from Highland New Guinea. (Photograph: W. Schiefenhövel).

of the roof or the wall of the small women's house. In another case the parturient firmly held onto a horizontal bar, which was fixed at the wall inside of the house especially for this purpose by one of the birth attendants.

Pöschl (1995) conducted ethnomedical fieldwork on Kiriwina, the main island of the Trobriand Archipelago, and was able to document four births. The two main characteristics of natural childbirth—vertical body postures and social embedding of birth—are very well visible. The parturients, like the ones in Highland New Guinea, took a multitude of changing and often asymmetric body positions, surely, as mentioned, an answer to the asymmetric process of the fetus moving down the birth channel. Three of the women had a half-reclining/sitting position and one a standing position at actual birth. Between contractions, the parturients often suspended the weight of their bodies from the beams of the house so that their feet were off the ground; in this position they gently swung their bodies. While the parturient sat on the floor of the house experiencing contractions, the traditional birth attendants (up to four) tightly held the woman in labour and also gave her feet a firm hold so that her body was kept anchored above and below. The abdominal muscles supporting the expulsion of the child can thus be activated most efficiently. Very wide spreading of the legs in half-squatting/half-sitting position is also common; it serves to widen the pelvis. Generally, the obstetrical traditions on the Trobriand Islands are a good example of how natural childbirth can be assisted. The coastal Trobrianders, like the mountain Eipo, do not interfere internally with childbirth: touching the vulva or introducing fingers or a hand into the birth channel does not happen.

From other traditional societies, similar body positions were known. Early on, obstetricians were interested in the ways women gave birth in non-European countries (Ploss 1872;

Engelmann 1881, 1883; Felkin 1884), but the respective reports had little if any effect on obstetrical procedures in Europe or North America. Felkin (1884) reported that he had witnessed a successful caesarean section carried out by a Ugandan healer. Whether this form of surgery has actually been common in this region remains doubtful; mortality rates would have been extremely high. It has been estimated (Sewell 1993) that not a single Parisian woman survived this operation in the 100 years between 1787 and 1876.

Breech birth occurs in 3–5% of all term births, more often preterm when the fetus may not yet have found the normal head-down position. Today, most obstetricians advise surgical delivery, yet studies have found pros and cons for both caesarean section and vaginal delivery in such cases (Hannah et al. 2004). Because the soft tissue of the buttocks is less efficient in dilating the cervix and widening the perineum and other parts of the birth channel, breech birth takes much longer than when the occiput is the presenting part. One can therefore argue that breech birth, while rare, could be considered as a normal birth mechanism. This cannot be said about other positions of the child, for example, when one leg or one arm or shoulder is presented. The latter is often the result of a prior transverse position of the baby. In some societies, midwives (e.g. the *dukun bayi* traditional birth attendants on Bali (Schiefenhövel, unpublished observations)) are able to turn the child by external manoeuvres. Techniques like these are also part of the obstetric repertoire of western midwives.

16.3.5 Surgical Delivery

However, in a certain percentage of births—which does not, according to a recent assessment by leading German obstetricians (Mylonas and Friese 2015), exceed 15% and might actually be not higher than 10%—posture changes and other low-tech methods of assisting delivery of a large baby through a small pelvis fail and surgical delivery is called for. In fact, there are numerous situations in which the mother, infant, or both would die without a caesarean section and this modern medical intervention has saved countless lives. It is, despite contemporary methods, for example, Misgav–Ladach surgery, not without risk, however, and medical as well as cost–benefit analyses must be taken into account when deciding to deliver surgically. For example, the risk of maternal death was seven times greater in a caesarean delivery compared to a vaginal delivery in a Netherlands study (Schuitemaker et al. 1997). Risks include pulmonary embolism, anaesthesia complications, uterine rupture, cardiac arrest, renal failure, haemorrhage, and sepsis (Liu et al. 2007). Mylonas and Fiese (2015) and other obstetricians warn against the seemingly ever-rising tendency towards caesarean section.

For the infant, risks of surgical delivery include respiratory problems, a 20% higher rate of type 1 diabetes (Cardwell et al. 2008; Newhook et al. 2012), asthma (Magnus et al. 2011), cancers, coeliac disease (Decker et al. 2010), inflammatory bowel disease, and obesity (Godfrey et al. 2011; Almgren et al. 2014). Many of these result from a compromised immune function. There is increasing evidence that many of these challenges are due not just to the surgery but, perhaps more importantly, to missed benefits of navigating the birth canal in a vaginal delivery (Montagu 1986; Trevathan 2010). These benefits include stress hormones (e.g. catecholamines) that speed up lung maturation, and exposure to maternal vaginal secretions, which seed the neonatal microbiome with healthier flora and fauna than the hospital-derived ones that populate the microbiome of surgically delivered infants (Dominquez-Bello et al. 2010; Neu and Rushing 2011; Arrieta et al. 2014). Evolutionary medicine recommendations to mitigate the loss of these benefits include allowing labour to begin naturally (and thus exposing the fetus to beneficial stress hormones) (Doherty and Eichenwald 2004) and wiping the new-born's face with vaginal swabs from the mother (Dominguez-Bello et al. 2010; but see Cunnington et al. 2016). In sum, one commentator has stated unequivocally, 'There is no scientific evidence to suggest any benefits to the baby' of elective caesarean delivery (Wagner 2000, p. 1678).

All of these risks are worth taking when the outcome is death or severe damage to the mother and/or the fetus. But, as stated above, by far not all caesarean sections are warranted by medical necessity, as can be seen clearly in Figure 16.8, which shows the rate of caesarean section in selected populations. Many of these deliveries occur because of maternal request, a phenomenon known as caesarean delivery on maternal request (CDMR) or elective caesarean section. Reasons for requesting surgical delivery vary widely and include anxiety about pain and loss of control, concern about urinary or rectal damage, fear of childbirth, concern about trauma and tissue damage that might make the woman less sexually desirable to their partner, and a desire to have a baby on an auspicious day or not on an unlucky day. Perhaps the best reason to consider by evolutionary medicine is anxiety and fear of childbirth. Fear was cited as a reason for elective caesarean section in 7–22% of cases in Finland, Sweden, and the United Kingdom (Saisto and Halmesmäki 2003). Emotional support late in pregnancy and during labour has been shown to have a beneficial effect on lowering fear and anxiety about birth and decreasing the frequency with which women request caesarean section (Waldenström et al. 2006). As noted in Section 16.3.3, the need for

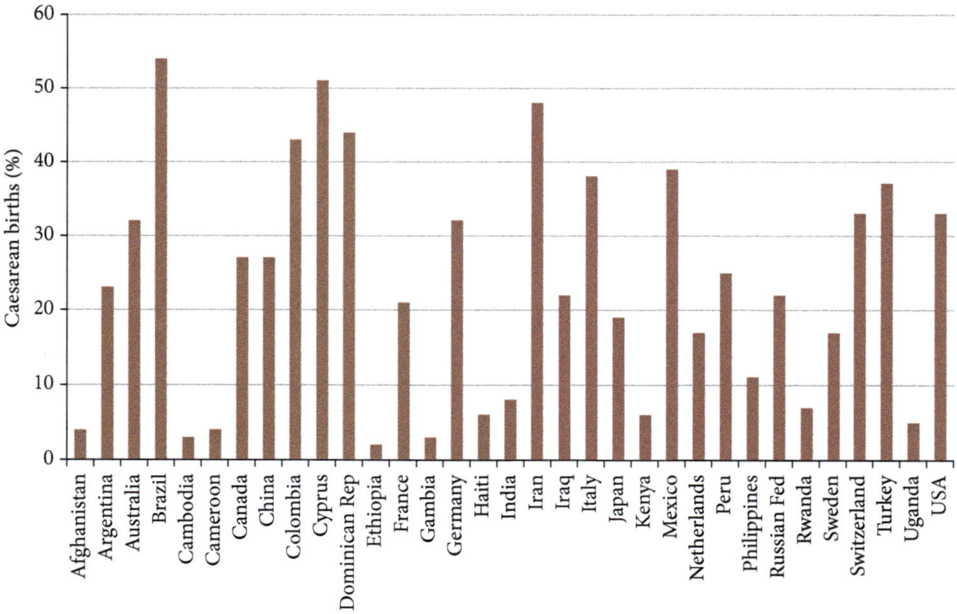

FIGURE 16.8 Histogram showing percentage of births (vertical axis) that are caesarean section for a range of countries.

Source: Data from A. P. Betran, M. R. Torloni, J. J. Zhang, A. M. Gülmezoglu, and the WHO Working Group on Caesarean Section, *WHO Statement on Caesarean Section Rates*, *BJOG: An International Journal of Obstetrics and Gynaecology*, 123 (5), pp. 667–70, doi.10.1111/1471-0528.13526, 2016.

emotional support at the time of birth is probably deeply rooted in the human psyche and it is no wonder that unalleviated fear is characteristic of many hospital-based deliveries. Our view through the lens of evolutionary medicine is that the context of childbirth with a great deal of (often unwarranted) medical intervention is mismatched with deeply rooted emotional needs of women and their families at this especially critical point in the life course.

16.4 The Postpartum Period

Once the infant has been born, its survival is far from assured, and risks to the mother have not ended. The vast majority of births proceed without negative events and result in healthy infants and mothers, but there are occasional events that require medical intervention, some of which have roots in human evolution.

16.4.1 Postpartum Haemorrhage

Postpartum haemorrhage (PPH) is among the most common causes of perinatal death in the world today—about 140,000 women die each year due to PPH (AbouZahr 2003; American College of Obstetricians and Gynecologists 2006b). It is an example of another health problem (and an evolutionary trade-off) that results from the deep invasion of the placenta necessary to support the large and rapidly developing fetal brain (Abrams and Rutherford 2011). When the placenta detaches from the uterine wall in the third stage of delivery (a process that usually takes less than 15 minutes), it leaves thousands of blood vessels exposed and bleeding until they are closed when the uterus clamps down following expulsion of the placenta. Unfortunately, the uterine muscles occasionally fail to contract sufficiently or at all (a phenomenon known as 'uterine atony'), leading to severe bleeding, and, in extreme cases, haemorrhage. Even women who survive PPH often have health complications such as anaemia and, in hospital settings, risks associated with blood transfusions. Drugs such as synthetic oxytocin are usually used to prevent or intervene when PPH occurs, but drugs are not available in many birth settings, nor were they available in the past, although there is reason to believe that PPH has been a risk for humans for millions of years, associated with bipedalism and large brains (Abrams and Rutherford 2011).

Fortunately, a behavioural intervention that has been a part of human evolutionary history is available to help decrease the risk of PPH: the nuzzling and suckling of the breast by the new-born infant. When the infant contacts the mother's breast with hands, mouth, or face, oxytocin is released by the mother, stimulating contractions that help clamp down the uterus and inhibit bleeding (Sobhy and Mohame 2003). This makes good evolutionary sense, but, unfortunately, attempts to assess the effects of nipple stimulation on the incidence of PPH have been inconclusive (Abedi et al. 2016). It is apparent that nipple stimulation is not an adequate substitute for drugs such as Syntometrine® or oxytocin when they are available, but this low-tech intervention may reduce PPH when there are no other options available, as it likely did in the evolutionary past, and provide an additional therapeutic route also in cases where pharmacological treatment is necessary.

Furthermore, there seem to be few if any disadvantages to allowing the new-born access to the mother's breast, and there is evidence that early breast contact facilitates breastfeeding (Moore et al. 2012). Women in Highland New Guinea like the Eipo put the new-born to the breast within minutes after the placenta has appeared. From the evolutionary perspective just discussed this seems a very wise cultural tradition—a win-win situation from which both baby and mother benefit. Surprisingly, women on the Trobriand Islands wait up to one day or even longer before they nurse their new-borns at the breast (Schiefenhövel 1983); they believe that colostrum is 'bad' milk and should not be given to the baby; this conviction is very widespread around the globe and is also present in Europe. It is surprising that this view could ever have persisted: why would a women's body produce a particular fluid if it were not useful for the new-born? Through institutions like La Leche League and lactation advisers the public has become aware of the great physiological importance colostrum has for the baby (https://www.llli.org/). The Trobriand mothers, who create a difficult period for their new-borns, put water or sometimes coconut milk on a finger and thus give small amounts of fluid to the baby. They thus do not utilise the oxytocin-releasing effect of nipple stimulation by the suckling baby and possibly risk dehydration of their child, which would be serious in the hot tropical climate on the islands in the West Pacific. Jüptner (1995), in the 1960s a doctor in the small hospital for the Trobriand Archipelago, found a high incidence of maternal death due to problems connected to the expulsion of the placenta. He explained that as partly being the effect of the very low blood pressure of the Trobriand Islanders (on general hypotony of the Trobriand Islanders see also Lindeberg et al. 1999), but it is probably more likely that the very late onset of nursing was responsible for these perinatal problems. These examples demonstrate that not in all cultures have people found the best ways to interact with biology. From a sociobiological perspective one could argue that traditions like the late giving of breast milk could constitute a test of the new-born's vitality: only those who survive this critical period are accepted. Yet there are no data to support this hypothesis.

16.4.2 Cutting the Umbilical Cord

Before the placenta detaches from the uterine wall, it continues to provide oxygen to the new-born infant and serves as an important backup system, which may be critical under circumstances where oxygen resuscitation is not an option. Even a few minutes without oxygen can cause brain damage, so it may be that this dual system, active for a few minutes after birth, is an adaptation that increased survival and brain health in the past, as well as in many populations today. Eipo women cut the umbilical cord after the placenta has come out, that is, 10–15 minutes after birth of the child. This seems a very wise tradition, as the new-born's circulatory and pulmonary system changes dramatically directly after birth and once the umbilical cord is detached from the blood's pathway so completely, the Eipo women can cut the cord without any tourniquet, either at the baby's umbilicus or at the placenta's side. Only occasionally, a small quantity of blood was seen to ooze out. Allowing the cord to remain attached until it stops pulsating increases infant blood volume by up to 30% (Wiberg et al. 2008), which is especially important if the mother's haemoglobin has been low in pregnancy, characteristic of far more than the 42% of women worldwide who have anaemia (de Benoist et al. 2008). These stores can last as long as four months (van

Rheenen et al. 2007), which is important because little iron is available in breast milk even under optimal nutritional intake. Increased iron stores in infants are associated with reduced incidence of anaemia even in the United States (Chaparro et al. 2006). Despite evidence of the benefits of delayed cord cutting, it still remains controversial, and immediate cutting is common in many hospital settings (Wiberg et al. 2008). The American College of Obstetricians and Gynecologists (2017) now recommends a delay in cutting the cord of at least 30–60 seconds after birth, unless other factors argue against it. The evolutionary medicine perspective concurs with this view that retaining or adopting an ancient pattern provides positive benefits that have probably saved countless infant lives.

16.4.3 Neonatal Hyperbilirubinaemia

Linn et al. (1985) conducted a retrospective study on 12,000 babies born in the Boston area of Massachusetts to determine risk factors for the development of neonatal jaundice (icterus neonatorum). The authors defined this condition as beginning at a cut-off point of 10 mg/decilitre. This is a quite low value, today not seen as connected to much clinical risk, which is thought to commence at 15 mg or above. The following factors were strongly positively correlated to neonatal jaundice (in decreasing strength): birth weight below 2,500 g, being Asian, premature rupture of membranes, breastfeeding, infant infection, male gender, and instrumental delivery. The following factors (in decreasing strength) showed a negative correlation: being black, smoking of the mother, being white, no college education, oxytocin during labour, and parturient's age above 35. That breastfeeding could be a risk factor for the development of neonatal jaundice is stated in almost all papers dealing with this condition. From an evolutionary perspective, this seems odd and needs explanation. Dennery et al. (2001) found similar risk factors as Linn et al. (1985), but in their study 'use of oxytocin in hypotonic solutions during labor' was positively correlated to icterus neonatorum, as was breastfeeding. The authors explain that 'late-onset breast milk jaundice' is the effect of breast milk being a competitive inhibitor of the hepatic enzyme uridine diphosphate glucuronosyltransferase.

Some authors in the past advised interspersing breastfeeding with formula milk; today, mothers are instructed to avoid their baby developing jaundice by starting breastfeeding early and allowing the infant to drink in short intervals. Dennery et al. (2001) stipulate that full-term infants should be subjected to phototherapy when they exceed the following bilirubin levels: 15 mg/decilitre of blood at age 25–48 hours, 18 mg at age 49–72 hours, and 20 mg at more than 72 hours. As hospital stays after birth are shorter than in the past, they urge for clinical supervision of all babies who are discharged before or at 48 hours after birth, thereby establishing a low threshold for medical supervision and intervention (American Academy of Pediatrics 2010). But they also state that 'low concentrations of bilirubin may have antioxidant benefits, suggesting that it should not be completely eliminated'.

What sense can be made out of these and similar studies? Is neonatal hyperbilirubinaemia indeed a common and dangerous condition? Is it caused by events happening during pregnancy (e.g. giving iron tablets because of 'pregnancy anaemia') or by interventions during childbirth? The available literature does not give conclusive answers. A few recent papers (e.g. Xiong et al. 2012) warn of hypermedicalisation and the negative effects of phototherapy (interfering with the bonding process, and possible danger of later developing malign melanoma and other complications). Indeed, to prevent new-borns starting to take in, whenever they feel like it, visual stimuli through their already quite developed eyes, to comprehend the world in this very important way, is a serious impediment. The neonate's brain is a sponge, designed to suck up an enormous amount of information. We must have good reasons to interfere with the visual input of new-borns. Daylight has been used, since centuries in Europe, to 'strengthen' babies. Is exposure to the sun an effective measure against developing hyperbilirubinaemia?

From an evolutionary point of view, jaundice of the new-born is indeed a puzzling topic. There can be no doubt that so-called kernicterus (cell nuclei being stained by a very high level of bilirubin in the blood of the new-born) will have serious consequences for the brain of the infant (Figure 16.9). This is, in our clinics, prevented by common light therapy or, in very pronounced cases, by exchanging the neonate's blood. Yet one wonders, would the new-borns of our ancestors have been exposed to this kind of risk? After all, the physiology of haemoglobin recycling in the baby's body and the functions of the immature liver should have been under powerful natural selection over millions of years. Do we exaggerate the danger of neonatal 'hyper'bilirubinaemia, perhaps because we set cut-off points and are prone (due to various motives) to place them at low thresholds. Does our management of pregnancy cause levels of bilirubin that are higher than they would be in other cultures? Unfortunately, no papers (other than the ones mentioned, stating that members of different populations run different risk of neonatal jaundice) have come to our knowledge which compare the epidemiology. The management of pregnancy in Highland New Guinea is very different from that in Europe. It is not at all impossible that the multiple interventions by the medical system, including caesarean section, create rather than solve a number of problems (Dunn 2003).

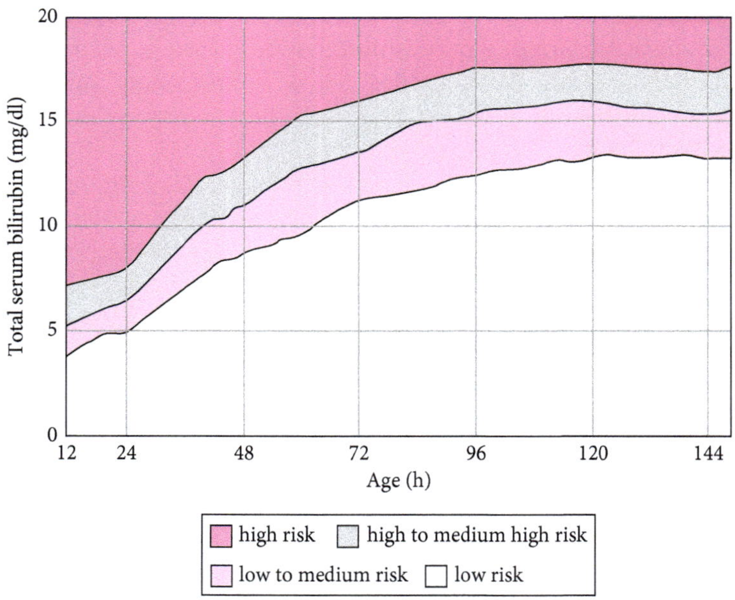

FIGURE 16.9 Total bilirubin serum levels at specific hours after birth confer differential health risks for the neonate.

16.4.4 The New-born Infant

Obvious to everyone who has seen a new-born human infant is how utterly helpless they seem to be. In contrast to new-born monkeys and apes, human new-borns are unable to move about on their own or to cling to their mothers. Their care is entirely up to the mother or other adults and older children. Of course, they are not entirely helpless as they have many ways of attracting mothers and others to them and to entice them to contribute inordinate amounts of time and energy to their care (Trevathan and Rosenberg 2016). Among mammals, new-borns can be considered along a continuum from altricial (entirely helpless and hairless with closed eyes, like new-born puppies and kittens) to precocial (well developed and able to cling to or follow the mother soon after birth, like colts and some monkeys). Human infants show mixed features and are not easily placed at one end of the continuum or the other. It is not correct to call them altricial, because their eyes are wide open and they have hair (albeit sparse), but they are clearly not precocial either, because they cannot move about on their own.

The unusual state of the human infant and its effect on child-rearing and family dynamics have been noted for centuries, even back to the writings of Alexander Pope who, in his *Essay on Man*, wrote, 'A longer care man's helpless kind demands / That longer care contracts more lasting bonds' (cited in Gould 1977). Portmann (1990) referred to the human infant as 'secondarily altricial' in reference to the fact that humans are likely descended from primates who gave birth to precocial young and that the greater helplessness of the human infant is a recently evolved characteristic. He continued his argument by suggesting that the human gestation period is better viewed as lasting 21 months, but the infant is born earlier than that because a longer gestation would result in an infant too large to be born through the birth canal. He referred to the first year of life as the 'extrauterine spring' during which time the human infant develops to the state observed in other primates at birth.

The idea that the first several months of life are better seen as part of gestational development was elaborated upon by Montagu (1961) who proposed that gestation occurred in two phases: uterogestation, the 9 months *in utero*, and exterogestation, the first 9–12 months after birth when the human infant develops in ways more like the processes seen in other primates during gestation. Most especially, the infant brain is increasing from approximately 25% of its adult size to almost 50% by the end of the first year of life.

A contemporary rendering of the idea that the first several months of life are a continuation of gestation is the concept of a fourth trimester (Kitzinger 1975; Jennings and Edmundson 1980; Tully et al. 2017). Until recently, birth was seen as a discrete event that ended with delivery and marked the beginning of the lifecycle phase of infancy. In the view that birth ends with delivery, one health professional, an obstetrician or midwife, takes care of the mother during pregnancy and delivery, and another, perhaps a paediatrician, takes over at birth. Care of the mother was limited to the first few hours of monitoring by those who had attended her during delivery. Once she left the hospital or once the attendants left her home, she was often on her own as she tried to initiate breastfeeding and care for her new-born infant using whatever informal assistance was available to her from family and friends. In the United States, her first postpartum appointment was typically 6 weeks after birth. Noting the challenges that a new mother faces especially with her first infant, several researchers and healthcare providers have suggested that care during pregnancy and delivery should be increased to encompass at least the first three months after birth during a time that should be considered as a continuation of gestation (Tully et al. 2017). This understanding of early motherhood relates to the concept of a fourth stage of labour, discussed above, in which significant physiological and emotional changes are taking place in the mother who has just delivered an infant and in the infant who must quickly adapt to life outside the womb.

16.4.5 Immediate Postpartum Mother–Infant Interaction

Considering that the mother was the primary source of warmth, nutrition, and comfort for most of human evolutionary history, it is surprising that one of the first actions taken after birth in many hospital deliveries was the removal of the baby immediately after the cord was cut. In cases where the infant needed resuscitation or the mother was incapacitated by drugs or anaesthesia, this removal is understandable, but in most normal deliveries, remaining with the mother, in contact with her body, is the safest place for a new-born infant. The abrupt transition from the uterine to the extrauterine environment places a great deal of stress on neonatal regulatory systems. Rather than relying on the placenta for delivery of oxygen and nutrients, the infant lungs must quickly mature and adapt to the external source of oxygen. The maturation process is facilitated by the pressure on the lungs during delivery and by stroking and massaging by the mother as commonly occurs upon first contact (Trevathan 1981). In this way, the mother's behaviour upon first interaction with her infant serves not only as a getting-to-know-you mechanism, but also as a way to enhance infant survival. Because of the importance of touch in human interaction, skin-to-skin contact (Moore et al. 2012) is especially valuable in helping the infant adapt to the extrauterine environment, and it may help stabilise heartbeats (Bergman et al. 2004). This skin-to-skin contact seems especially important for infants born prematurely (Oras et al. 2016; Maitre et al. 2017) and with low birthweight (Luong et al. 2016). It is also associated with improved breastfeeding rates and experience (Moore et al. 2012; Redshaw et al. 2014). Given its benefits, skin-to-skin contact immediately after birth may be especially important in resource-poor populations (Nyqvist 2016). Finally, maternal touch may have epigenetic effects on infant development that we are only just beginning to explore, and those benefits may transcend generations (Champagne et al. 2006; Champagne 2008, 2014).

Associated with the immature state of the human infant is an immature thermoregulatory system, meaning that it is difficult for the neonate to maintain a healthy body temperature. This is one of the reasons that new-born infants are often whisked off to warming blankets in hospital deliveries, but by cradling the infant on her body, the mother provides warmth that has been found to be superior to mechanical sources. When paediatrician Jan Winberg and colleagues compared infants in skin-to-skin contact with their mothers to those in artificially warmed cots, they found that the babies were able to quickly reach stable body temperatures and they were also better able to maintain blood glucose levels (Winberg 2005). Conserving heat and energy contributed to infant survival in the past and mothers have evolved behaviours to facilitate that.

Stroking and massaging the infant immediately after birth also serves to rub the vernix caseosa that covers the infant (and protected it from drying out in the watery environment of the uterus) into the baby's skin, which protects it from dehydration as well as reducing heat loss. Vernix has antimicrobial properties, which provide protection for the infant as well as for the mother, whose nicks and small cuts in the perineal area benefit from anti-infectives even today; their properties were likely far more important for her survival and health in the past before artificial anti-infectives were available. (For further discussion, see Chapter 8: Skin and Integument.) Commonly in hospital births, the infant is immediately bathed to remove the unsightly birth fluids and the vernix, but the moisturising, cleansing, anti-infective, and antioxidant properties are widely recognised now, and the WHO recommends delaying bathing the baby until several hours after birth (WHO 2018; Singh and Archana 2008). Even in contemporary hospital deliveries the vernix may protect infants from hospital-acquired infections, and there is no evidence that leaving it on for an hour or so causes harm.

Infant crying at birth is almost an expected phenomenon, so much so that when the infant takes its first breath, it often begins a crying episode that is welcomed by those present as a sign of life and good health. But crying is energetically expensive, so anything that can soothe the infant and reduce crying is beneficial. In fact, excessive infant crying in the immediate postpartum period has been shown to partially re-establish fetal circulation (Anderson 1989). Furthermore, in the ancient past, a crying infant would have alerted predators of the presence of a vulnerable mother and infant in case she was not protected by the group. By cradling and soothing the infant in the first hour or so after birth, the mother helps it conserve critical nutrient reserves that help to maintain survival in the few days before the mother's milk matures sufficiently to boost resources. The cradling also serves to boost oxytocin, which plays a critical role in affiliation, stress reduction, and infant development, as reviewed by Carter and Porges (2013). As noted in Section 16.3.2.1, when the infant nuzzles the mother's breast, the associated release of oxytocin serves to contract the uterus to stop postpartum bleeding. This behaviour, which infants seem to do almost instinctively (Widström et al. 2011), helps to save the mother's life and may have been critical to maternal health in the past. Uvnäs-Moberg (1996) has referred to the effects of oxytocin as 'biological messages' that have too often been disregarded by modern cultural practices (p. 129).

16.4.6 Mother–Infant Bonding at Birth

There has been a great deal of discussion in past decades about the importance of mother–infant interaction and bonding in the early hours following birth (Trevathan 2010, 2013). Until recently, separation was the norm and mother and infant were treated as two separate individuals by hospital staff as soon as the cord was cut. Babies were bathed, weighed, wrapped, injected with vitamin K, and placed in warming cots; mothers were cleaned, stitched, covered, and moved to clean beds. Finally, sometimes several hours after birth, the infant and the mother were reunited and breastfeeding was attempted. In some cases, the two remained in separate rooms and saw each other only every 4 hours or so. Practices have changed, so that rooming-in is more the routine, but even with that, mothers and babies are not often together in the first hour after birth. This is unfortunate because human infants are unusually alert following unmedicated births (Widström et al. 2011). This period has even been referred to as a 'sensitive period' for mother–infant bonding (Klaus et al. 1982; Trevathan 1987).

Recognition of this special period is seen in other cultures and the beneficial effects of postpartum seclusion: mother and infant can relate to each other in an undisturbed atmosphere. In pre-industrial Europe it was common that a new mother would not leave the house where she had given birth for 40 days (a sacred number in biblical tradition). This was called 'in den 6 Wochen liegen' (to lie for 6 weeks). Puerperal seclusions like this can still be found in some cultures. The Trobriand primipara stays in the house, secluded from the rest of the village except for her mother or other relatives who assist her, for 2 months or more (Schiefenhövel 1983) (see Figure 16.10). Her skin tan disappears somewhat and when she finally comes out into the public sphere, dressed with a beautiful stola of banana leaves with interwoven aromatic leaves, her well-fed baby on her lap, everyone pays respect to her health and beauty and that of her child. She is the focus of attention for some days, surely a very good way to start life as a mother. The Eipo women in Highland New Guinea (Schiefenhövel 1988) also have a puerperal seclusion, albeit much shorter. For primiparae it is about one week, for multiparae often just a few days. In western medicine (Epstein and Fleischer 1927), rest in the puerperal period has been prescribed for medical reasons. The mothers were thought to be exhausted by 'labour' and pain, sick from perineal lacerations

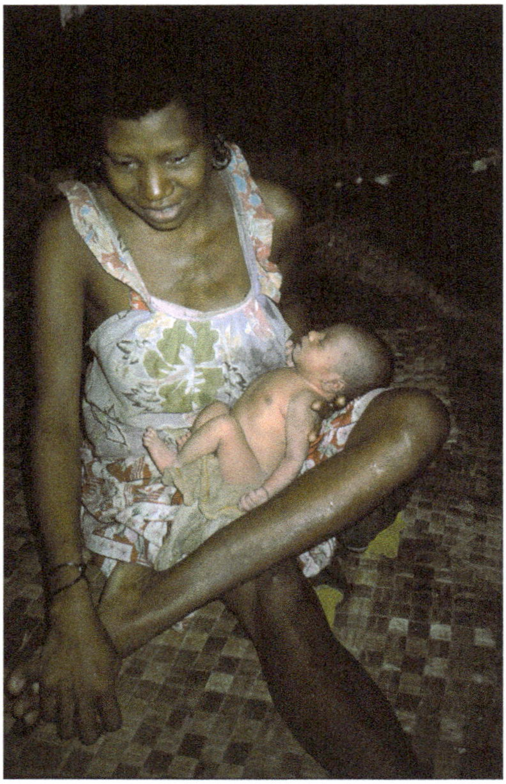

FIGURE 16.10 Primiparous Trobriand woman with her just-born first child. Often childbirth is experienced as an emotional 'high'; the effect of hormones and transmitter substances like oxytocin and endorphine are involved in this biopsychologically positive start of the mother–child bond. (Photograph: W. Schiefenhövel).

and from the lochia, the fluid from the healing uterus. This is, from an evolutionary point of view, a very insufficient perspective. Also, in parts of the world, many mothers feel quite well after birth—except if they suffer from postpartum dysphoria. The widespread tradition of isolating mother and child from ordinary everyday life is probably rooted in the perception that these two human beings have to learn about each other; they haven't really met before. What we call the process of bonding (Bowlby 1969) can best happen in the undisturbed atmosphere of a protected space. Our hospital rooms in the maternity wards are quite different. The door is frequently opened, and personnel go in and out, along with well-wishing visitors. While the latter is a sign of joy and emotional closeness and as thus very positive, these rooms are nothing like the quiet, undisturbed places where mothers of other cultures spend the first days and weeks with their child.

Although motivated mothers can probably overcome most obstacles placed in their way in the immediate postpartum period, it appears that at the very least, this time when the infant is attentive and awake is an especially good time for her to begin to bond (or continue a process that began *in utero*) with her infant. In the ancestral past, being together in the immediate postpartum period probably enhanced infant survival through maternal behaviours of stroking and encompassing the infant, rubbing in the vernix, reducing infant crying, and initiating breastfeeding (which also contributed to maternal recovery following delivery), as discussed in Section 16.4.5. The desire to interact with her infant at this time is probably deeply rooted in human maternal behaviour as a product of selection to enhance survival during what was, and still is, a highly vulnerable time in the life course (Trevathan 2010). Unless there are mitigating circumstances, the evolutionary medicine perspective urges attention to keeping the mother and infant together in the immediate postpartum period as a way of improving maternal and infant health, increasing breastfeeding success, and improving maternal self-esteem and confidence in her abilities to care for her infant. Interaction in this period may even have epigenetic effects in ways we are just beginning to appreciate (Buchen 2010).

16.4.7 Postpartum Depression and 'Baby Blues'

A period of sadness is almost expected following childbirth in many parts of the world, but it is unrecognised in others. Is this an emotional challenge similar to that of post-menstrual syndrome (PMS), which some have argued is a 'culture-bound syndrome' (Johnson 1987; but see Hughes 1998), or is it due to hormonal changes all women experience at this time? Both biological and cultural factors have been implicated in low postpartum mood, although even within a single population there is variation in its expression. Because of the degree of variation, it is important to distinguish between severe depression and relatively milder sadness, often referred to as 'baby blues'. The extreme state is probably best seen as a clinical concern (i.e. a 'defect' in evolutionary medicine terms), whereas there may be selective advantages to a period of low mood following birth (i.e. a 'defence') (Nesse 1991). For example, low mood may elicit assistance and social support from friends and family members (Hagen 1999). Women who report low social support and marital problems experience higher rates of postpartum negative mood, as do those who experienced problems during labour and delivery.

The widespread practice of isolating mother and infant for a period of time following birth (from a few days to 40 days) may be, as has been stated in Section 16.4.6, a cultural adaptation to this period of time when the new mother loses interest in her normal activities. The seclusion provides an opportunity for her to spend time with her new-born infant, learning to care for it, beginning to breastfeed, and bonding. Others are typically available to assist her and to provide food and other necessities, reducing her energy expenditure at a time when much of it goes to the costly physiological process of lactation (Piperata 2008). Another evolutionary view of postpartum blues is that it may have been advantageous in the past for a woman whose new-born was not likely to survive early infancy and may have enabled her to accept what was in the interest of her long-term fitness (Hrdy 1999).

The search for biological factors affecting postpartum mood has proved elusive, although there is little doubt that the combination of hormonal changes, psychosocial challenges, and environmental factors has varying degrees of influence. The precipitous drop in hormones that occurs at delivery may have a more negative effect on mood when the levels during pregnancy are especially high, as they are in women from resource-rich populations (Trevathan 2010). Childbirth also affects immune function and the hormones that regulate the stress response, suggesting that infectious and inflammation factors may have an effect on mood (Kendall-Tackett 2007). Breastfeeding can serve to counter some of these effects, as suggested by the lower incidence of depression in women who breastfeed. In fact, the low rates of postpartum depression (and near absence) in many cultures may be associated with the high rates of breastfeeding. Another factor that may mitigate low mood is getting sufficient sleep, which is easier for breastfeeding women who co-sleep with their infants and breastfeed throughout the night (McKenna and McDade 2005). On the other hand, women who are depressed often stop breastfeeding, which may serve to make the depression worse. Evolutionary medicine recommendations that could decrease the likelihood of experiencing postpartum depression would include breastfeeding, getting a good night's sleep, eating foods with anti-inflammatory properties (e.g. long-chain omega-3 fatty acids), and reducing stress as much as possible.

16.4.8 Lactation, Breastfeeding, and Early Infancy

Perhaps one of the most significant aspects of the period immediately following birth and throughout the fourth trimester is the establishment of successful breastfeeding. In the past and in many parts of the world today, establishing breastfeeding is not seen as a challenge, and whether or not to breastfeed an infant is not viewed as a valid question. Infants who were not breastfed in the past or in much of the world today simply have lower chances of survival. But today, in many parts of the world, breastfeeding does not come easily to new mothers and many abandon their efforts and turn to bottle-feeding when they become frustrated or worried that their infants are not getting sufficient nutrition.

The endocrinology and energetics of lactation are discussed in Chapter 15, but it is also important to reinforce the argument that the act of nursing the infant is a two-way interaction between mother and infant. The first stage of lactation takes place during pregnancy as the mammary glands develop under the influence of hormones such as prolactin; progesterone and oestrogen are also involved and they serve to inhibit secretion of milk. Full milk production does not begin until two to three days after birth, but before that, the breasts secrete colostrum, a fluid that is low in fat and calories but very high in protein and immune factors. As noted in Section 16.4.1 for the Trobriand Islanders, colostrum is withheld from infants in some cultures because of its appearance (more like pus than milk), but it plays a very important role in adaptation of the neonatal immune system during a very vulnerable period (McDade 2005). The neonatal immune system is poorly developed at birth and requires 'seeding' from environmental microbiota (preferably from the mother) and 'priming' from the colostrum.

Successful breastfeeding relies on the hormones prolactin and oxytocin, but, as noted in Chapter 15, more important than either of those is the behaviour of suckling by the infant. Suckling initiates a cascade of hormone secretions and inhibitions that maintains milk production. More suckling means more milk; less suckling means less milk. This is why when infants are supplemented with foods and liquids other than milk, they suckle less frequently and milk production slows down. Mothers often perceive this phenomenon as evidence that they can't produce milk of sufficient quantity and they decide to supplement even more. This initiates a vicious cycle that has been described as 'insufficient milk syndrome' and often leads to termination of breastfeeding (Gussler and Briesemeister 1980). Very few women have real physiological problems with lactation (there would have been strong selective pressures against that in the past), but many experience difficulties for sociocultural and personal reasons.

Human milk is very low in nutrients compared to many other mammalian species, but similar to the milk of most other primates (Figure 16.11). More than 85% of human milk is water. For this reason, human infants need to breastfeed frequently in a pattern referred to as 'on demand' feeding. Leaving an infant behind in a nest and nursing every 4–8 hours like canids and ungulates do was simply not an option for ancestral hominin females and it is not a good option for women today who want to provide optimal feeding for their infants. Equally important to nutrients for infant health are the immune factors found in mother's milk (Hamosh 2001). Synthetic milk substitutes can theoretically be designed to mimic the

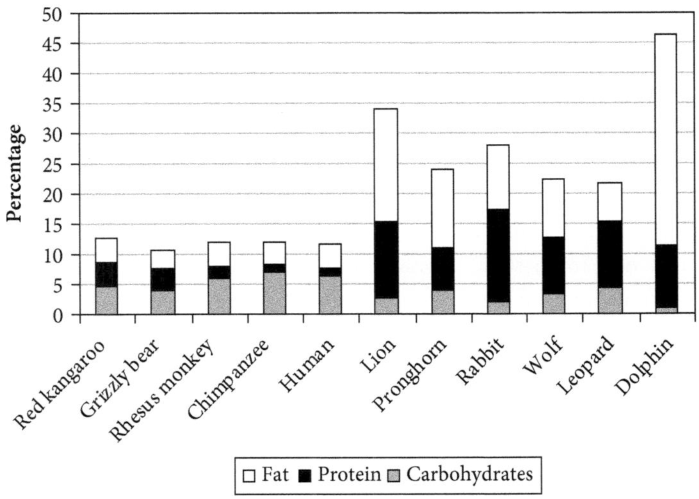

FIGURE 16.11 Nutritional content of milk. Carbohydrate (grey), protein (black), and fat (white) content of milk of selected mammals.

nutrient content of breast milk, but they cannot provide the immune factors. This is probably the reason that formula-fed infants are not as healthy as breastfed infants in infancy (Heinig 2001; Ajetunmobi et al. 2015) and throughout their lives (Martin et al. 2005; Kramer et al. 2007; Victora et al. 2016; but see Colen and Ramey 2014). Furthermore, the health gains are not restricted to infants: women who breastfeed have lower rates of breast and ovarian cancer, as well as lower risk of type 2 diabetes (Labbok 2001; Chowdhury et al. 2015).

The benefits of breastfeeding have been well known for decades, but despite the overwhelming evidence of its positive health impacts, it is surprising how many women terminate breastfeeding early or choose to formula-feed. The benefits are obvious for women in resource-poor populations, but even women and infants from wealthy nations with excellent medical care benefit from the protective measures provided by breastfeeding. The American Academy of Pediatrics (2012) and the American College of Obstetricians and Gynecologists (2016) both recommend at least one year of breastfeeding, with the first 6 months being exclusive breastfeeding. The financial costs alone of not breastfeeding are staggering (Bartick et al. 2017).

Public health agencies and lactation support groups have expended great effort to encourage women to breastfeed, and yet the success rates of these educational campaigns are limited. Reasons for terminating early or not breastfeeding at all include incompatibility with work outside the home, discomfort (both physical and social), lack of support from family members, and lack of education about benefits. Support from clinicians involved with pregnancy and birth and from health educators may be able to make a difference in a woman's decisions about feeding her child. In 2011, the US Surgeon General released a set of recommendations and 'calls to action' to support women who breastfeed and their families and to remove barriers to breastfeeding (US Department of Health and Human Services 2011; http://www.surgeongeneral.gov). Clearly, the evolutionary medicine perspective would argue in support of breastfeeding as the optimal way to improve infant and maternal health.

Successful breastfeeding, especially in the first few weeks of an infant's life, requires that the mother be able to provide for her infant at any time throughout the day or night. Infants who are fed when they are hungry ('on demand') are not expected to sleep through the night without interruption. For the mother who puts her infant in another room to sleep, she must get up from her own bed several times during the night and go to her infant for feeding. One of the biggest challenges that new mothers face in the first few months after birth is lack of sleep. It is therefore no surprise that many women who breastfeed choose to have their infants sleep in the same bed with them in a manner referred to as co-sleeping. Because of its perceived association with sudden infant death syndrome (SIDS), this behaviour sets off alarm bells for social workers and clinicians in some parts of the world, most notably the United States. Well-meaning professionals argue that a mother's body is a 'lethal weapon' (McKenna and McDade 2005, p. 135) and that she should never sleep in the same bed with an infant for fear that she would roll over during the night and smother him or her. Most public health messages about co-sleeping argue that it should never be done under any circumstances. Mothers co-sleep anyway, and messages that emphasise safe ways of co-sleeping would be much more effective at improving infant health and reducing hazards from unsafe bedding materials than outright prohibition.

For most of human history and in much of the world today, there is no question of where a young infant will sleep: in bed with the mother where access to the breast is unrestricted.

The near equivalency of breastfeeding and mother–infant co-sleeping has been recognised in the work of anthropologists Jim McKenna and Lee Gettler (2016), who have coined the phrase 'breast-sleeping' to remind parents and health care practitioners that the two behaviours are so intertwined as to be nearly inseparable, at least for the infant's first few months of life. From an evolutionary viewpoint, the young infant 'expects' to sleep with the mother and does not hide his or her displeasure at being isolated. Infants in the evolutionary past who were not with their mothers would have died from exposure, predators, or hunger (or perhaps at the hands of community members who found the incessant crying intolerable). Co-sleeping is a legacy of human evolutionary history.

16.5 Conclusion

The reproductive life course is a tangled bank of romantic love, orgasmic responses, conception, gestation, parturition, and breastfeeding, followed by almost 20 years of raising a child to the point that he or she begins the cycle again. Culture plays a profound role in influencing who we love, how we make love, when we conceive, how we are treated in pregnancy, how we give birth, and how we feed and care for our new-born infants. But underlying the wide variation in the ways all of these steps are manifested is the long evolutionary process that has shaped behaviour, physiology, and emotions toward the production of offspring. As Darwin famously noted in the last paragraph of *On the Origin of Species* (1859), the immense variability we see around us has been shaped by natural selection via what he termed 'laws': 'These laws, taken in the largest sense, being Growth with Reproduction; Inheritance which is almost implied by reproduction; . . .'.

Nikolaas Tinbergen, a well-known scholar in Darwin's line of thought, formulated the Four Questions evolutionary science has to keep in mind when discussing living beings. *Causation* concerns the 'machinery' and looks at biological, biochemical, biopsychological, and similar *proximate mechanisms*, through which effects are brought about, so that the organism can stay healthy, vital, and reproductive. This approach is well known and often used by medical scientists. As a matter of fact, medicine has contributed a great deal to what we know about the complex inner systems of biological organisms including ourselves. In the field of sexuality, pregnancy, birth, lactation, and infant care a very complex network of organismic, microbiological, and neurobiological events takes place which make sure that everything goes well, in an act of sexual love or in the birth of a child. Given the many intervening parameters, attacks from parasites, and one's own mind and psyche, it is surprising that these meaningful events so often go so well. The task of evolutionary medicine is to address and describe the normal ways of causation and to show where there are problems. Our modern world has created many of them, because the 'machinery' was shaped by evolutionary forces in the Plio-Pleistocene, several million to some hundred thousand years ago. Our modern medicine has created many opportunities and solutions (e.g. modern surgery), but these successes have come at costs. The global obesity epidemic is just one example of this. Wrong, even dangerous developments in medicine itself (e.g. the rising rate of caesarean sections) also need to be understood and taken care of in an intelligent way. *Ontogeny* concerns the process of organismic maturation. The stretch of life from formation of a zygote after fertilisation to birth of a new-born and the ways she/he can be protected and made to grow in all respects has also been dealt with in this chapter. *Adaptive value, the ultimate causes* of evolutionary solutions, is a theme that runs as a red thread through all this volume and has been of particular concern for the authors of this chapter, because it deals with the most essential aspects of all life: to perpetuate itself. The comparative method of including other animals, usually mammals and primates like ourselves, in the discourse—that is, looking at *phylogeny*, the evolutionary history of biological traits—has also been widely used in this chapter. The great apes and the gibbons, the lesser apes, and ourselves have many commonalities, including social life, and hence social behaviour, empathy, so-called theory of mind (i.e. the ability to mind read or emphronesis), and the ability to create culture. In the field of sexuality, the six species are split up into different strategies of mating and reproducing. How can one deduce from this bewildering variance between polygyny, polygynandrous promiscuity, and monogamy what kind of animal we are, sexually and reproductively? Comparison shows the answer is a bit of everything. For the discussion of human birth, the phylogenetic perspective is particularly important: how did the female pelvis change during hominisation? What were the effects of bipedalism and of the very large head of the new-born of our brainy species? We have shown that through evolutionary processes of adaptation a number of effective solutions have been found to make the almost unthinkable possible: the passage of a very large new-born through a very tight channel of bone and tissue.

Medical people can—that is one of the messages of this chapter and the whole volume—trust these evolved principles, and be on careful guard for those cases where something goes wrong. Nature, as any form of technology, cannot produce 100% safe solutions; they are too 'costly'. To perform one of the modern versions of caesarean section saves lives; it should not become, though, the normal way of being born. Some countries are not far from this. Modern medicine has the potential to come to the rescue in cases where pathology happens, but it should not pervade our lives—otherwise, Huxley's *Brave New World* will catch up.

References

Abedi, P., Jahanfar, S., Namvar, F., et al. (2016). Breastfeeding or nipple stimulation for reducing postpartum haemorrhage in the third stage of labour. *Cochrane Database Syst Rev* 1, CD010845.

AbouZahr, C. (2003). Global burden of maternal death and disability. *Br Med Bull* 67(1), 1–11.

Abrams, E. T. and Rutherford, J. N. (2011). Framing postpartum hemorrhage as a consequence of human placental biology: an evolutionary and comparative perspective. *Am Anthropol* 113(3), 417–30.

Adams, D. B., Gold, A. R., and Burt, A. D. (1978). Rise in female initiated sexual activity at ovulation and its suppression by oral contraceptives. *N Eng J Med* 299, 1145–50.

Ajetunmobi, O. M., Whyte, B., Chalmers, J., et al. (2015). Breastfeeding is associated with reduced childhood hospitalization: evidence from a Scottish Birth Cohort (1997–2009). *J Pediatr* 166(3), 620–5.

Almgren, M., Schlinzig, T., Gomez-Cabrero, D., et al. (2014). Cesarean delivery and hematopoietic stem cell epigenetics in the newborn infant: implications for future health? *Am J Obstet Gynecol* 211(5), 502.e1–8.

American Academy of Pediatrics (2010). Policy statement. Hospital stay for healthy term newborns. Committee on Fetus and Newborn. *Pediatrics* 125(2).

American Academy of Pediatrics (2012). Policy statement. Breastfeeding and the use of human milk. Section on Breastfeeding. *Pediatrics* 129, e827–41.

American College of Obstetricians and Gynecologists (2006a). ACOG Practice Bulletin No. 71: Episiotomy. 5. *Obstet Gynecol* 107, 956–62.

American College of Obstetricians and Gynecologists. (2006b). ACOG Practice Bulletin: Clinical Management Guidelines for Obstetrician-Gynecologists Number 76, October 2006: postpartum hemorrhage. *Obstet Gynecol* 108(4), 1039.

American College of Obstetricians and Gynecologists (2016). *Optimizing Support for Breastfeeding as Part of Obstetric Practice*. Committee on Obstetric Practice. Washington, DC: American College of Obstetricians and Gynecologists.

American College of Obstetricians and Gynecologists (2017). Delayed umbilical cord clamping after birth. Committee Opinion No. 684. *Obstet Gynecol* 129, e5–10.

Anderson, G. C. (1989). Risk in mother–infant separation postbirth. *J Nurs Scholarship* 21(4), 196–9.

Ariès, P. and Béjin, A. (1982). *Sexualités Occidentales*. Paris: Point Seul.

Arrieta, M. C., Stiemsma, L. T., Amenyogbe, N., et al. (2014). The intestinal microbiome in early life: health and disease. *Front Immunol* 5, 547. doi: org/10.3389/fimmu.2014.00427.

Bailey, J. M., Dunne, M. P., and Martin, N. G. (2000). Genetic and environmental influences on sexual orientation and its correlates in an Australian twin sample. *J Pers Soc Psychol* 78, 524–36.

Baker, R. R. and Bellis, M. A. (1993). Human sperm competition: ejaculate adjustment by males and the function of masturbation. *Anim Behav* 46, 861–85.

Barry, K. L., Holwell, G. I., and Herberstein, M. E. (2008). Female praying mantids use sexual cannibalism as a foraging strategy to increase fecundity. *Behav Ecol* 19, 710–15.

Bartick, M. C., Schwarz, E. B., Green, B. D., et al. (2017). Suboptimal breastfeeding in the United States: maternal and pediatric health outcomes and costs. *Matern Child Nutr* 13. doi: 10.1111/mcn.12366.

Bartz, J. A., Zaki, J., Bolger, N., et al. (2011). Social effects of oxytocin in humans: context and person matter. *Trends Cogn Sci* 15(7), 301–9.

Baumeister, R. F. and Twenge, J. M. (2002). The cultural suppression of female sexuality. *Rev Gen Psychol* 6(2), 166–203.

Beier, K. M. and Loewit, K. (2011). *Praxisleitfaden Sexualmedizin. Von der Theorie zur Therapie*. Berlin: Springer.

Bell, A. F., Erickson, E. N., and Carter, C. S. (2014). Beyond labor: the role of natural and synthetic oxytocin in the transition to motherhood. *J Midwifery Women's Health* 59(1), 35–42.

Bell-Krannhals, I. N. (1990). *Haben um zu geben. Eigentum und Besitz auf den Trobriand Inseln, Papua New Guinea. Basler Beiträge zur Ethnologie 31*. Basel: Wepf & Co.

Bergman, N. J., Linley, L. L., and Fawcus, S. R. (2004). Randomized controlled trial of skin-to-skin contact from birth versus conventional incubator for physiological stabilization in 1200- to 2199-gram newborns. *Acta Paediatr* 93(6), 779–85.

Bischof, N. (2012). *Moral, Ihre Natur, Ihre Dynamik und Ihr Schatten*. Cologne: Böhlau.

Bischof Köhler, D. (2012). Empathy and self-recognition in phylogenetic and ontogenetic perspective. *Emot Rev* 4, 40–8.

Bitzer, J., Giraldi, A., and Pfaus, J. (2013). Sexual desire and hypoactive sexual desire disorder in women. Introduction and overview. Standard operating procedure (SOP Part 1). *J Sex Med* 10, 36–49.

Blackledge, C. (2004). *The Story of V. A Natural History of Female Sexuality*. New Brunswick: Rutgers University Press.

Blanchard, R. and Klassen, P. (1997). H-Y antigen and homosexuality in men. *J Theor Biol* 185, 373–8.

Bowlby, J. (1969). *Attachment*. New York: Basic Books.

Buchen, L. (2010). In their nurture: can epigenetics underlie the enduring effects of a mother's love? Lizzie Buchen investigates the criticisms of a landmark study and the controversial field to which it gave birth. *Nature* 467(7312), 146.

Burleson, M. H., Gregory, W. L., and Trevathan, W. R. (1995). Heterosexual activity: relationship with ovarian function. *Psychoneuroendocrinology* 20, 405–21.

Buss, D. M. (1989). Sex differences in human mate preferences. Evolutionary hypotheses tested in 37 cultures. *Behav Brain Sci* 12, 1–49.

Buss, D. M. (2000). *Dangerous Passion. Why Jealousy Is as Necessary as Love and Sex*. New York: Free Press.

Caldeyro-Barcia, R. and Poseiro, J. J. (1960). Physiology of the uterine contraction. *Clin Obstet Gynecol* 3(2), 386–410.

Camperio-Ciani, A., Corna, F., and Capiluppi, C. (2004). Evidence for maternally inherited factors favouring male homosexuality and promoting female fecundity. *Proc Biol Sci* 271, 2217–21.

Cardwell, C. R., Stene, L. C., Joner, G., et al. (2008). Caesarean section is associated with an increased risk of childhood-onset type 1 diabetes mellitus: a meta-analysis of observational studies. *Diabetologia* 51, 726–35.

Carter, C. S. and Porges, S. W. (2013). The biochemistry of love: an oxytocin hypothesis. *EMBO Rep* 14(1), 12–16.

Chalmers, I., Campbell, H., and Turnbull, A. (1975). Use of oxytocin and incidence of neonatal jaundice. *BMJ* 2(5963),116–18.

Champagne, F. A. (2008). Epigenetic mechanisms and the transgenerational effects of maternal care. *Front Neuroendocrinol* 29(3), 386–97.

Champagne, F. A. (2014). Epigenetics of mammalian parenting. In: Narvaez, D., Valentino, K., Fuentes, K., et al. (eds) *Ancestral Landscapes in Human Evolution: Culture, Childrearing and Social Wellbeing*. New York: Oxford University Press, pp. 18–37.

Champagne, F. A., Weaver, I. C., Diorio, J., et al. (2006). Maternal care associated with methylation of the estrogen receptor-α1b promoter and estrogen receptor-α expression in the medial preoptic area of female offspring. *Endocrinology* 147(6), 2909–15.

Chaparro, C. M., Neufeld, L. M., Alavez, G. T., et al. (2006). Effect of timing of umbilical cord clamping on iron status in Mexican infants: a randomised controlled trial. *Lancet* 367(9527), 1997–2004.

Chowdhury, R., Sinha, B., Sankar, M. J., et al. (2015). Breastfeeding and maternal health outcomes: a systematic review and meta-analysis. *Acta Paediatr* 104(S467), 96–113.

Colen, C. G. and Ramey, D. M. (2014). Is breast truly best? Estimating the effects of breastfeeding on long-term child health and wellbeing in the United States using sibling comparisons. *Soc Sci Med* 109, 55–65.

Crockford, C., Deschner, T., Ziegler, T. E., et al. (2014). Endogenous peripheral oxytocin measures can give insight into the dynamics of social relationships: a review. *Front Behav Neurosci* 8. doi: 10.3389/fnbeh.2014.00068.

Cummings, V., Jordan, P., and Zvelebil, M. (2014). *The Oxford Handbook of the Archaeology and Anthropology of Hunter-Gatherers*. Oxford: Oxford University Press.

Cunningham, F., Leveno, K., Bloom, S., et al. (2014). *Williams Obstetrics*, 24th ed. New York: McGraw-Hill.

Cunnington, A. J., Sim, K., Deierl, A., et al. (2016). 'Vaginal seeding' of infants born by caesarean section. *BMJ* 352, i227. doi: org/10.1136/bmj.i227.

D'Aniello, B., Semin, G. R., Scandurra, A., et al. (2017). The vomeronasal organ: a neglected organ. *Front Neuroanat* 11, 70. doi: 10.3389/fnana.2017.00070.

Davis, E. P., Hobel, C. J., Sandman, C. A., et al. (2005). Prenatal stress and stress physiology influences human fetal and infant development. In: Power, M. L. and Schulkin, J. (eds) *Birth, Distress, and Disease*. Cambridge: Cambridge University Press, pp. 183–201.

de Benoist, B., McLean, E., Egli, I., et al. (eds) (2008). *Worldwide Prevalence of Anaemia 1993-2005. WHO Global Database on Anaemia*. Geneva: World Health Organization.

Decker, E., Engelmann, G., Findeisen G., et al. (2010). Cesarean delivery is associated with celiac disease but not inflammatory bowel disease in children. *Pediatrics* 125(6), e1433–44.

De la Croix, D. and Mariani, F. (2012). From polygamy to serial monogamy: a unified theory of marriage institutions. *Rev Econ Stud* 82(2).

Dennery, P. A., Seidman, D. S., and Stevenson, D. K. (2001). Neonatal hyperbilirubinemia. *N Eng J Med* 344(8), 581–90.

De Waal, F. (1995). Sex as an alternative to aggression in the bonobo. In: Abramson, P. R. and Pinkerton, S. D. (eds) *Sexual Nature, Sexual Culture*. Chicago: University of Chicago Press, pp. 37–56.

Dissanayake, E. (1992). Homo aestheticus: *Where Art Comes from and Why*. New York: Free Press.
Dixon, A. F. (1998). *Primate Sexuality. Comparative Studies of the Prosimians, Monkeys, Apes and Humans*. Oxford: Oxford University Press.
Dixon, A. F. and Anderson, M. J. (2002). Sexual selection, seminal coagulation and copulatory plug formation in primates. *Folia Primatol* 73, 63–9.
Doherty, E. G. and Eichenwald, E. C. (2004). Cesarean delivery: emphasis on the neonate. *Clin Obstet Gynecol* 47(2), 332–41.
Dominguez-Bello, M. G., Costello, E. K., Contreras, M., et al. (2010). Delivery mode shapes the acquisition and structure of the initial microbiota across multiple body habitats in newborns. *Proc Natl Acad Sci U S A* 107(26), 11971–5.
Dörner, G., Geier, T. Ahrens, L., et al. (1980). Prenatal stress as possible aetiogenetic factor of homosexuality in human males. *Endokrinologie* 75, 365–8.
Drea, C. M. and Wallen, K. (2003). Female sexuality and the myth of male control. In: Travis, C.B. (ed.) *Evolution, Gender and Rape*. Cambridge, MA: MIT Press, pp. 29–60.
Dunn, P. M. (1995). Die Geburt als physiologischer Prozeß—eine pädiatrische Sichtweise der Perinatalzeit. In: Schiefenhövel, W., Sich, D., and Gottschalk-Batschkus, C.E. (eds) Gebären—Ethnomedizinische Perspektiven und neue Wege. Berlin: Verlag für Wissenschaft und Bildung, pp. 225–8.
Dunn, P. M. (2003). The British Association for Perinatal Medicine: the first 25 years (1976–2000). *Archives of Disabled Children*. https://adc.bmj.com/content/archdischild/88/3/181.1.full.pdf.
Dunn, P. M. and Caldeyro-Barcia, R. (2015). Appropriate maternal posture during childbirth: a commentary. *West Engl Med J* 114(1), 1–9.
Duty, S. M., Silva, M. J., Barr, D. B., et al. (2003). Phthalate exposure and human semen parameters. *Epidemiology* 14, 269–77.
Edmonds, K. (ed.) (2012). *Dewhurst's Textbook of Obstetrics and Gynaecology*. New York: Wiley-Blackwell.
Eibl-Eibesfeldt, I. (1972). *Love and Hate. The Natural History of Behaviour Patterns*. New York: Holt, Rinehart and Winston.
Ekelin, M., Svensson, J., Evehammar, S., et al. (2015). Sense and sensibility: Swedish midwives' ambiguity to the use of synthetic oxytocin for labour augmentation. *Midwifery* 31(3), e36–42.
Ellison, P. T. (2005). Evolutionary perspectives on the fetal origins hypothesis. *Am J Hum Biol* 17, 113–18.
Engelmann, G. J. (1881). Pregnancy, parturition, and childbed among primitive people. *Am J Obstet Dis Women Children (1869–1919)* 14(4), 828–43.
Engelmann, G. J. (1883). *Labor Among Primitive Peoples*. Edinburgh: J. H. Chambers & Company.
Epstein, H. J. and Fleischer, A. J. (1927). The disadvantages of the prolonged period of post partum rest in bed. *Am J Obstet Gynecol* 14, 360–3.
Esch, T. and Stefano, G. B. (2005). The neurobiology of love. *Neuroendocrinol Lett* 26, 175–92.
Feldman, R., Monakhov, M., Pratt, M., et al. (2016). Oxytocin pathway genes: evolutionary ancient system impacting on human affiliation, sociality, and psychopathology. *Biol Psychiatry* 79(3), 174–84.
Felkin, R. W. (1884). Notes on labour in Central Africa. *Edinburgh Med J* 29(10), 922–30.
Fisher, H. (2001). Lust, Anziehung und Verbundenheit. Biologie und Evolution der menschlichen Liebe. In: Meier, H. and Neumann, G. (eds) *Über die Liebe*. München: Piper, pp. 81–112.
Fisher, H., Aron, A., and Brown, L. L. (2005). Romantic love: an fMRI study of a neural mechanism for mate choice. *J Comp Neurol* 493, 58–62.
Fisher, H., Aron, A., and Brown, L. L. (2006). Romantic love: a mammalian brain system for mate choice. *Philos Trans R Soc Lond B Biol Sci* 361, 2173–86.
Fleischman, D. S. (2016). An evolutionary behaviorist perspective on orgasm. *Socioaffect Neurosci Psychol* 6. doi: 10.3402/snp.v6.32130.
Fossey, D. (1983). *Gorillas in the Mist*. Boston: Houghton Mifflin Company.
Frankowski, B. L. and Committee on Adolsecence (2004). Sexual orientation and adolescents. *Pediatrics* 113, 1817–32.

Fruth, B. and Hohmann, G. (2006). Social grease for females? Same-sex genital contacts in wild bonobos. In: Sommer, V. and Vasey, P. L. (eds) *Homosexual Behaviour in Animals. An Evolutionary Perspective*. Cambridge: Cambridge University Press, pp. 294–315.

Fujita, S. and Inoue, E. (2015). Sexual behavior and mating strategies. In: Nakamura, M., Hosaka, K., Itoh, N., et al. (eds) *Mahale Chimpanzees*. Cambridge: Cambridge University Press, pp. 485–95.

Galal, M., Symonds, I., Murray, H., et al. (2012). Postterm pregnancy. *Facts Views Vis Obgyn* 4, 175–87.

Gallup, G. G., Burch, R. L., and Mitchell, T. J. (2006). Semen displacement as a sperm competition strategy: multiple mating, self-semen displacement, and timing of in-pair copulations. *Hum Nat* 17, 253–64.

Gander, K. (2016). Cosmetic surgery operations on the rise, with breast enlargements most popular procedure. *Independent* 8 February. http://www.independent.co.uk/life-style/health-and-families/health-news/cosmetic-surgery-up-in-2015-with-breast-enlargement-most-popular-procedure-a6860641.html.

Giddens, A. (1992). *The Transformation of Intimacy. Sexuality, Love and Eroticism in Modern Society*. Stanford: Stanford University Press.

Gluckman, P. D., Hanson, M. A., Cooper, C., et al. (2008). Effect of in utero and early-life conditions on adult health and disease. *N Eng J Med* 359(1), 61–73.

Godfrey, K. M., Sheppard, A., Gluckman, P. D., et al. (2011). Epigenetic gene promoter methylation at birth is associated with child's later adiposity. *Diabetes* 60(5), 1528–34.

Gould, S. J. (1977). *Ontogeny and Phylogeny*. Cambridge, MA: Harvard University Press

Gould, S. J. (2002). *The Structure of Evolutionary Theory*. Cambridge, MA: Harvard University Press.

Gräfenberg, E. (1950). The role of the urethra in female orgasm. *Int J Sexol* 3, 145–8.

Grammer, K., Kruck, J., Juette, A., et al. (2000). Non-verbal behavior as courtship signals: the role of control and choice in selecting partners. *Evol Hum Behav* 21, 371–90.

Gu, V., Feeley, N., Gold, I., et al. (2016). Intrapartum synthetic oxytocin and its effects on maternal well-being at 2 months postpartum. *Birth* 43(1), 28–35.

Gupta, J. K. and Nikodem, C. (2000). Maternal posture in labour. *Eur J Obstet Gynecol Repro Biol* 92(2), 273–7.

Gussler, J. D. and Briesemeister, L. H. (1980). The insufficient milk syndrome: a biocultural explanation. *Med Anthropol* 4(2), 145–74.

Guttentag, M. and Secord, P. F. (1984). The sex ratio question. *Am J Sociol* 90(3), 673–4.

Hagen, E. H. (1999). The functions of postpartum depression. *Evol Hum Behav* 20(5), 325–59.

Hamosh, M. (2001). Bioactive factors in human milk. *Pediatr Clin North Am* 48(1), 69–86.

Hannah, M. E., Whyte, H., Hannah, W. J., et al. (2004). Maternal outcomes at 2 years after planned cesarean section versus planned vaginal birth for breech presentation at term: the international randomized Term Breech Trial. *Am J Obstet Gynecol* 191(3), 917–27.

Harris, A. and Seckl, J. (2011). Glucocorticoids, prenatal stress and the programming of disease. *Horm Behav* 59(3), 279–89.

Harris, H. (1995). Rethinking Polynesian heterosexual relationship: a case study on Mangaia, Cook Islands. In: Jankowiak, W. (ed.) *Romantic Passion. A Universal Experience?* New York: Columbia University Press, pp. 95–112.

Hartman, W. E. and Fithian, M. A. (1974). *Treatment of Sexual Dysfunction*. New York: Jason Aaronson.

Haselton, M. G. and Miller, G. (2006). Women's fertility across the cycle increases the short-term attractiveness of creative intelligence. *Hum Nat* 17, 50–73.

Heinig, M. J. (2001). Host defense benefits of breastfeeding for the infant: effect of breastfeeding duration and exclusivity. *Pediatr Clin N Am* 48(1), 105–23.

Hendrix, N. and Berghella, V. (2008). Non-placental causes of intrauterine growth restriction. *Semin Perinatol* 32, 161–5.

Hrdy, S. B. (1981). *The Woman that Never Evolved*. Cambridge, MA: Harvard University Press.

Hrdy, S. B. (1997). Raising Darwin's consciousness: female sexuality and the prehominid origins of patriarchy. *Hum Nat* 8, 1–49.

Hrdy, S. B. (1999). *Mother Nature: A History of Mothers, Infants, and Natural Selection*. New York: Ballantine.

Hughes, C. C. (1998). The glossary of 'culture-bound syndromes' in DSM-IV: a critique. *Transcult Psychiatry* 35(3), 413–21.

Huseynov, A., Zollikofer, C. P., Coudyzer, W., et al. (2016). Developmental evidence for obstetric adaptation of the human female pelvis. *Proc Natl Acad Sci U S A* 113(19), 5227–32.

Iemmola, F. and Camperio-Ciani, A. (2009). New evidence of genetic factors influencing sexual orientation in men: female fecundity increase in the maternal line. *Arch Sex Behav* 38, 393–9.

Janini, E. A., Blanchard, R., Camperio-Ciani, A., et al. (2010). Male homosexuality: nature or culture. *J Sex Med* 7, 3245–53.

Jankowiak, W. and Fischer, E. F. (1992). A cross-cultural perspective on romantic love. *Ethnology* 31, 149–55.

Jennings, B. and Edmundson, M. (1980). The postpartum period: after confinement: the fourth trimester. *Clin Obstet Gynecol* 23(4), 1093–104.

Jöchle, W. (1973). Coitus-induced ovulation. *Contraception* 7, 523–64.

Johnson, T. M. (1987). Premenstrual syndrome as a western culture-specific disorder. *Cult Med Psychiatry* 11, 337–56.

Johnson-Laird, P. N. and Oatley, K. (2008). Emotions, music, and literature. In: Lewis, M., Havilland-Jones, L., and Feldman, B. (eds) *Handbook of Emotions*, 3rd ed. New York: The Guilford Press, pp. 102–13.

Jüptner, H. (1995). Geburtshilflich-gynäkologische Beobachtungen bei den Trobriandern. In: Schiefenhövel, W., Sich, D., and Gottschalk-Batschkus, C. E. (eds) *Gebären—Ethnomedizinische Perspektiven und neue Wege*. Berlin: Verlag für Wissenschaft und Bildung, pp. 79–82.

Kendall-Tackett, K. (2007). A new paradigm for depression in new mothers: the central role of inflammation and how breastfeeding and anti-inflammatory treatments protect maternal mental health. *Int Breastfeed J* 2(1), 6. doi: org/10.1186/1746-4358-2-6.

Kinsey, A. C., Pomeroy, W. B., and Martin, C. E. (1948). *Sexual Behavior in the Human Male*. Philadelphia: Saunders.

Kinsey, A. C., Pomeroy, W. B., Martin, C. E., et al. (1953). *Sexual Behavior in the Human Female*. Philadelphia: Saunders.

Kitzinger, S. (1975). The fourth trimester? *Midwife Health Visit Community Nurse* 11(4), 118–21.

Klaus, M. H. and Kennell, J. H. (1976). *Mother-Infant Bonding*. St. Louis: Mosby.

Klaus, M. H., Kennell, J. H., and Ballard, R. A. (1982). *Parent-Infant Bonding*. St Louis: Mosby.

Knauft, B. (1985). *Good Company and Violence: Sorcery and Social Action in a Lowland New Guinea Society*. Berkeley: University of California Press.

Kohl, K.-H. (2001). Gelenkte Gefühle. Vorschriftsheirat, romantische Gefühle und Determinanten der Partnerwahl. In: Meier, H. and Neumann, G. (eds) *Über die Liebe*. München: Piper, pp. 113–37.

Kohl, S., Kainer, F., and Schiefenhövel, W. (2009). Nausea and vomiting as evolutionary mechanisms of the complex adaptation reaction to pregnancy. *Z Geburtshilfe Neonatol* 213(5), 186–93.

Komisaruk, B. R., Beyer-Flores, C., and Whipple, B. (2006). *The Science of Orgasm*. Baltimore: Johns Hopkins University Press.

Kramer, M. S., Matush, L., Vanilovich, I., et al. (2007). Effects of prolonged and exclusive breastfeeding on child height, weight, adiposity, and blood pressure at age 6.5 y: evidence from a large randomized trial. *Am J Clin Nutr* 86(6), 1717–21.

Kunz, G., Beil, D., Huppert, P., et al. (2007). Oxytocin—a stimulator of directed sperm transport in humans. *Reprod Biomed Online* 14(1), 32–9.

Kuzawa, C. W. (1998). Adipose tissue in human infancy and childhood: an evolutionary perspective. *Am J Phys Anthropol* 107, 177–209.

Kuzawa, C. W. (2005). Fetal origins of developmental plasticity: are fetal cues reliable predictors of future nutritional environments? *Am J Hum Biol* 17, 5–21.

Kuzawa, C. W. and Quinn, E. A. (2009). Developmental origins of adult function and health: evolutionary hypotheses. *Annu Rev Anthropol* 38, 131–47.

Labbok, M. H. (2001). Effects of breastfeeding on the mother. *Pediatr Clin N Am* 48(1), 143–58.

Langström, N., Rahman, Q. Carlström, E., et al. (2010). Genetic and environmental effects of same-sex behavior: a population study of twins in Sweden. *Arch Sex Behav* 39, 75–80.

Levine, H., Jørgensen, N., Martino-Andrade, A., et al. (2017). Temporal trends in sperm count: a systematic review and meta-regression analysis. *Hum Reprod Update* 23(6), 646–59.

Liao, J. B., Buhimschi, C. S., and Norwitz, E. R. (2005). Normal labor: mechanism and duration. *Obstetr Gynecol Clin N Am* 32(2), 145–64.

Liebowitz, M. R. (1982). *The Chemistry of Love*. Boston: Little, Brown and Co.

Lindeberg, S., Eliasson, M., Lindahl, B., et al. (1999). Low serum insulin in traditional Pacific Islanders—the Kitava Study. *Metabolism* 48, 1216–19.

Linn, S., Schoenbaum, S. C., Monson, R. R., et al. (1985). Epidemiology of neonatal hyperbilirubinemia. *Pediatrics* 75(4), 770–4.

Liu, T. C., Chen, C. S., Tsai, Y. W., et al. (2007). Taiwan's high rate of cesarean births: impacts of national health insurance and fetal gender preference. *Birth* 34(2), 115–22.

Louis, G. M. B. and Platt, R. W. (2011). *Reproductive and Perinatal Epidemiology*. New York: Oxford University Press.

Luong, K. C., Long Nguyen, T., Huynh Thi, D. H., et al. (2016). Newly born low birthweight infants stabilise better in skin-to-skin contact than when separated from their mothers: a randomised controlled trial. *Acta Paediatrica* 105(4), 381–90.

Luhmann, N. (1982). *Liebe und Passion. Zur Codierung von Intimität*. Frankfurt: Suhrkamp.

Lurie, S. (2010). Does intercourse during menses increase the risk for sexually transmitted disease? *Arch Gynecol Obstet* 282, 627–30.

Mackinnon, J. (1974). *In Search of the Red Ape*. New York: Holt, Rinehart and Winston.

Magnus, M. C., Håberg, S. E., Stigum, H., et al. (2011). Delivery by cesarean section and early childhood respiratory symptoms and disorders: the Norwegian Mother and Child Cohort Study. *Am J Epidemiol* 174(11), 1275–85.

Maitre, N. L., Key, A. P., Chorna, O. D., et al. (2017). The dual nature of early-life experience on somatosensory processing in the human infant brain. *Curr Biol* 27(7), 1048–54.

Malinowski, B. (1929). *The Sexual Life of Savages in North-Western Melanesia. An Ethnographic Account of Courtship, Marriage and Family Life Among the Natives of the Trobriand Islands, British New Guinea*. London: Routledge & Kegan Paul.

Marshall, D. (1962). *Island of Passion: Ra'ivavae*. London: George Allen & Unwin.

Martin, R. (2013). *How We Do It. The Evolution and Future of Human Reproduction*. New York: Basic Books.

Martin, R. M., Gunnell, D., and Davey Smith, G. (2005). Breastfeeding in infancy and blood pressure in later life: systematic review and meta-analysis. *Am J Epidemiol* 161(1), 15–26.

Mast, H., Quakernack, K., Lenfers, M., et al. (1971). Influence of the course of labor on icterus neonatorum. *Geburtshilfe Frauenheilkd* 31(5), 443–53.

Masters, W. H. and Johnson, V. E. (1966). *Human Sexual Response*. Boston: Little Brown.

McDade, T. W. (2005). The ecologies of human immune function. *Annu Rev Anthropol* 34, 495–521.

McKenna, J. J. and Gettler, L. T. (2016). There is no such thing as infant sleep, there is no such thing as breastfeeding, there is only breastsleeping. *Acta Paediatr* 105, 17–21.

McKenna, J. J. and McDade, T. (2005). Why babies should never sleep alone: a review of the co-sleeping controversy in relation to SIDS, bedsharing and breast feeding. *Paediatr Respir Rev* 6(2), 134–52.

Mead, M. (1928). *Coming of Age in Samoa. A Psychological Study of Primitive Youth for Western Civilization*. New York: Morrow & Co.

Medicus, G. (2015). *Being Human. Bridging the Gap between the Sciences of Body and Mind*. Berlin: VWB.

Meier, H. (2001). Epilog. Über Liebe und Glück. In: Meier, H. and Neumann, G. (eds) *Über die Liebe*. München: Piper, pp. 333–43.

Michel, S. C. A., Rake, A., Treiber, K., et al. (2002). MR obstetric pelvimetry: effect of birthing position on pelvic bony dimensions. *Am J Roentgenol* 179, 1063–7.

Miller, G. F. (2000). *The Mating Mind: How Sexual Choice Shaped the Evolution of Human Nature*. New York: Doubleday.

Miller, S. L. and Maner, J. K. (2011). Ovulation as a male mating prime: subtle signs of women's fertility influence men's mating cognition and behavior. *J Personality Social Psychol* 100(2), 295–308.

Mizuno, J. and Takeda, N. (1988). Phylogenetic study of the oxytocin-like immunoreactive system in invertebrates. *Comp Biochem Physiol A Physiol* 91(4), 733–8.

Montagu, A. (1961). Neonatal and infant immaturity in man. *JAMA* 178, 56–7.

Montagu, A. (1986). *Touching: The Human Significance of the Skin*, 3rd ed. New York: Harper and Row.

Moore, E. R., Anderson, G. C., Bergman, N., et al. (2012). Early skin-to-skin contact for mothers and their healthy newborn infants. *Cochrane Database Syst Rev* 5(5).

Morris, D. (1967). *The Naked Ape: A Zoologist's Study of the Human Animal*. London: Jonathan Cape.

Morris, N. M., Udry, J. R., Khan-Dawood, F., et al. (1987). Marital sex frequency and midcycle female testosterone. *Arch Sex Behav* 16, 27–37.

Mulder, E. J., De Medina, P. R., Huizink, A. C., et al. (2002). Prenatal maternal stress: effects on pregnancy and the (unborn) child. *Early Hum Dev* 70(1), 3–14.

Mylonas, I. and Friese, K. (2015). Indications for and risks of elective cesarean section. *Deutsches Ärzteblatt Int* 112(29–30), 489.

Nathanielsz, P. W. (2001). *The Prenatal Prescription*. New York: HarperCollins

Nesse, R. M. (1991). What good is feeling bad? *The Sciences* November/December, 30–7.

Neu, J. and Rushing, J. (2011). Cesarean versus vaginal delivery: long-term infant outcomes and the hygiene hypothesis. *Clin Perinatol* 38(2), 321–31.

Newhook, L. A., Penney, S., Fiander, J., et al. (2012). Recent incidence of type 1 diabetes mellitus in children 0–14 years in Newfoundland and Labrador, Canada climbs to over 45/100,000: a retrospective time trend study. *BMC Res Notes* 5(1), 628.

Nyqvist, K. H. (2016). Given the benefits of Kangaroo mother care, why has its routine uptake been so slow? *Acta Pædiatr* 105(4), 341–2.

Odent, M. (2003). *Birth and Breastfeeding*. West Hoathly, UK: Clairview.

Oras, P., Thernström Blomqvist, Y., Hedberg Nyqvist, K., et al. (2016). Skin-to-skin contact is associated with earlier breastfeeding attainment in preterm infants. *Acta Paediatr* 105(7), 783–9.

Parker, G. (1970). Sperm competition and its evolutionary consequences in the insects. *Biol Rev* 45, 525–67.

Pavlicev, M. and Wagner, G. (2016). The evolutionary origin of female orgasm. *J Exp Zool B Mol Dev Evol* 326, 326–37.

Perry, C. P. (2001). Current concepts of pelvic congestion and chronic pelvic pain. *JSLS* 5, 105–10.

Pike, I. L. (2000). The nutritional consequences of pregnancy sickness: a critique of a hypothesis. *Hum Nat* 11, 207–32.

Piperata, B. A. (2008). Forty days and forty nights: a biocultural perspective on postpartum practices in the Amazon. *Soc Sci Med* 67, 1094–103.

Ploss, H. (1872). *Über die Lage und Stellung der Frau Während der Geburt bei Verschiedenen Völkern. Eine Anthropologische Studie*. Leipzig: Veit and Comp.

Portmann, A. (1990). *A Zoologist Looks at Humankind*. Translated by Schaefer, J. Original edition 1944. New York: Columbia University Press.

Pöschl, U. (1995). Geburten bei den Trobriandern. In: Schiefenhövel, W., Sich, D., and Gottschalk-Batschkus, C. E. (eds) *Gebären—Ethnomedizinische Perspektiven und neue Wege*. Berlin: Verlag für Wissenschaft und Bildung, pp. 67–78.

Redshaw, M., Hennegan, J., and Kruske, S. (2014). Holding the baby: early mother–infant contact after childbirth and outcomes. *Midwifery* 30(5), e177–87.

Reichard, U. (1995). Extra-pair copulations in a monogamous gibbon (*Hylobates lar*). *Ethology* 100, 99–112.

Rice, W. R., Friberg, U., and Gavrilets, S. (2013). Homosexuality via canalized sexual development: a testing protocol for a new epigenetic model. *Bioessays* 35, 764–70.

Roenneberg, T. and Merrow, M. (2016). The circadian clock and human health. *Current Biology* 26(10), R432–43.

Ruff, C. (1987). Sexual dimorphism in human lower limb bone structure: relationship to subsistence strategy and sexual division of labour. *J Hum Evol* 16, 391–416.

Sacco, A., Muglu, J., Navaratnarajah, R., et al. (2015). ST analysis for intrapartum fetal monitoring. *Obstet Gynaecol* 17(1), 5–12.

Saisto, T. and Halmesmäki, E. (2003). Fear of childbirth: a neglected dilemma. *Acta Obstet Gynecol Scand* 82(3), 201–8.

Saling, E. (1981). Fetal scalp blood analysis. *J Perinat Med* 9, 165–17.

Saling, E. Z. and Dudenhausen, J. W. (1973). The present situation of clinical monitoring during labor. *J Perinat Med* 1, 75–103.

Sanders, A. R., Martin, E. R., Beecham, G. W., et al. (2014). Genome-wide scan demonstrating significant linkage for male sexual orientation. *Psychol Med* 45, 1379–88.

Schiefenhövel, W. (1983). Weitere Informationen zur Geburt auf den Trobriand-Inseln. In: Schiefenhövel, W. and Sich, D. (eds) *Die Geburt aus ethnomedizinischer Sicht*. Wiesbaden: Vieweg, Braunschweig, pp. 143–50.

Schiefenhövel, W. (1988). *Geburtsverhalten und reproduktive Strategien der Eipo—Ergebnisse humanethologischer und ethnomedizinischer Untersuchungen im zentralen Bergland von Irian Jaya (West-Neuguinea), Indonesien*. Berlin: Reimer.

Schiefenhövel, W. (1990). Ritualized adult-male/adolescent-male sexual behavior in Melanesia: an anthropological and ethological perspective. In: Feierman, J. R. (ed.) *Pedophilia. Biosocial Dimensions*. New York: Springer, pp. 394–421.

Schiefenhövel, W. (2001). Sexualverhalten in Melanesien. Ethnologische und humanethologische Aspekte. In: Sütterlin, Ch. and Salter, F. (eds) *Irenäus Eibl-Eibesfeldt. Zu Person und Werk. Bibliotheca Aurea*. Frankfurt: Peter Lang, pp. 274–88.

Schiefenhövel, W. (2004). Trobriands. In: Ember, Carol, R., and Ember, M. (eds) *Encyclopedia of Sex and Gender. Men and Women in the World's Cultures*, 2 Volumes. New York: Kluwer Academic/Plenum Publishers, pp. 912–21.

Schiefenhövel, W. (2009). Romantic love. A human universal and possible honest signal. *Human Ontogenet* 3(2) (July), 39– 50.

Schiefenhövel, W. (2014). The quest for beauty and powerful expression. Examples from New Guinean poetry. In: Sütterlin, C., Schiefenhövel, W., Lehmann, C., et al. (eds) *Art as Behaviour. An Ethological Approach to Visual and Verbal Art, Music and Architecture. Hanse Studies Vol. 10*. Oldenburg: Bis-Verlag, University Oldenburg, pp. 407–35.

Schuitemaker, N., Van Roosmalen, J., Dekker, G., et al. (1997). Maternal mortality after cesarean section in the Netherlands. *Acta Obstet Gynecol Scand* 76, 332–4.

Seckler, D. (1982). Small but healthy? A basic hypothesis in the theory, measurement and policy of malnutrition. In: Sukhatme, P. V. (ed.) *Newer Concepts in Nutrition and their Implications for Policy*. Pune, India: Maharashtra Association for the Cultivation of Science Research Institute, pp. 127–37.

Sewell, J. E. (1993). *Cesarean Section—A Brief History*. (Brochure to accompany an exhibition on the history of cesarean section.) Washington, DC: National Library of Medicine.

Shepherd, A. and Cheyne, H. (2013). The frequency and reasons for vaginal examinations in labour. *Women Birth* 26, 49–53.

Shifren, J. L., Monz, B. U., Russo, P. A., et al. (2008). Sexual problems and distress in United States women: prevalence and correlates. *J Obstet Gynecol* 112(5), 970–8.

Shones, M. M. (2014). The implication of low testosterone on mortality in men. *Curr Sex Health Rep* 6, 235–43.

Singh, D. and Bronstad, P. M. (2001). Female body odour is a potential cue to ovulation. *Proc Royal Soc B* 268, 797–801.

Singh, G. and Archana, G. (2008). Unraveling the mystery of vernix caseosa. *Indian J Dermatol* 53, 54–60.

Smith, A. M., Rissel, C. E., Richters, J., et al. (2003). Sex in Australia: sexual identity, sexual attraction and sexual experience among a representative sample of adults. *Aust N Z J Public Health* 27(2), 138–45.

Sobhy, S. I. and Mohame, N. A. (2003). The effect of early initiation of breast feeding on the amount of vaginal blood loss during the fourth stage of labor. *J Egypt Public Health Assoc* 79(1–2), 1–12.

Sommer, V. and Vasey, P. L. (eds) (2006). *Homosexual Behavior in Animals*. Cambridge: Cambridge University Press.

Stewart, D. B. (1984). The pelvis as a passageway. I. Evolution and adaptations. *Br J Obstet Gynaecol* 91, 611–17.

Stokholm, L., Juhl, M., Lonfeld, N., et al. (2018). Obstetric synthetic oxytocin use and subsequent hyperactivity/inattention problems in Danish children. *Acta Obstet Gynecol Scand* 97(7), 880–9.

Stuebe, A. M., Grewen, K., and Meltzer-Brody, S. (2013). Association between maternal mood and oxytocin response to breastfeeding. *J Women's Health* 22(4), 352–61.

Sütterlin, C., Schiefenhövel, W., Lehmann, C., et al. (eds) (2014). *Art as Behaviour. An Ethological Approach to Visual and Verbal Art, Music and Architecture. Hanse Studies Vol. 10*. Oldenburg: Bis-Verlag, University Oldenburg.

Swan, S. H., Kruse, R. L., Liu, F. et al. (2003). Semen quality in relation to biomarkers of pesticide exposure. *Environ Health Perspect* 111(12), 1478–84.

Symons, D. (1979). *The Evolution of Human Sexuality*. New York: Oxford University Press.

Tanfer, K. and Aral, S. O. (1996). Sexual intercourse during menstruation and self-reported sexually transmitted disease history among women. *J Sex Transm Dis* 23, 395–401.

Thayer, Z. M. and Kuzawa, C. W. (2011). Biological memories of past environments: epigenetic pathways to health disparities. *Epigenetics* 6(7), 798–803.

Thornhill, R. (1976). Sexual selection and paternal investment in insects. *Am Nat* 110, 153–63.

Thornhill, R. and Palmer, C. T. (2001). *A Natural History of Rape. Biological Bases of Sexual Coercion*. Cambridge, MA: MIT Press.

Tinbergen, N. (1951). *The Study of Instinct*. Oxford: Clarendon Press.

Tinklepaugh, O. L. (1930). Occurrence of a vaginal plug in a chimpanzee. *Anat Rec* 46, 329–32.

Tooby, J. and Cosmides, L. (2008). The evolutionary psychology of the emotions and their relationship to internal regulatory variables. In: Lewis, M., Havilland-Jones, L., and Feldman, B. (eds) *Handbook of Emotions*, 3rd ed. New York: The Guilford Press, pp. 114–37.

Trevathan, W. R. (1981). Maternal touch at first contact with the newborn infant. *Dev Psychobiol* 14, 549–58.

Trevathan, W. R. (1983). Analysis of out-of-hospital births—suggested dependency on artificial oxytocin. *Am J Phys Anthropol* 60, 262.

Trevathan, W. R. (1987). *Human Birth: An Evolutionary Perspective*. Hawthorne, NY: Aldine de Gruyter. Reissued in 2011 by Transaction, New Brunswick, NJ.

Trevathan, W. R. (2010). *Ancient Bodies, Modern Lives: How Evolution Has Shaped Women's Health*. New York: Oxford University Press.

Trevathan, W. R. (2015). Primate pelvic anatomy and implications for birth. *Philos Trans R Soc Lond B Biol Sci* 370, 20140065.

Trevathan, W. R. and Rosenberg, K. R. (eds) (2016). *Costly and Cute: Helpless Infants and Human Evolution*. Albuquerque: University of New Mexico Press.

Troisi, A. and Carosi, M. (1998). Female orgasm rate increases with male dominance. *Anim Behav* 56, 1261–6.

Tully, K. P., Stuebe, A. M., and Verbiest, S. B. (2017). The fourth trimester: a critical transition period with unmet maternal health needs. *Am J Obstet Gynecol* 217, 37–41.

Tyrell, H. (1987). Romantische Liebe. Überlegungen zu ihrer 'quantitativen Bestimmtheit'. In: Bäcker, D. (ed.) *Theorie als Passion. Niklas Luhmann zum 60*. Frankfurt: Geburtstag, Suhrkamp, pp. 570–99.

Uvnäs-Moberg, K. (1996). Neuroendocrinology of the mother–child interaction. *Trends Endocrinol Metabol* 7(4), 126–31.

Van Rheenen, P., De Moor, L., Eschbach, S., et al. (2007). Delayed cord clamping and haemoglobin levels in infancy: a randomised controlled trial in term babies. *Trop Med Int Health* 12(5), 603–16.

Victora, C. G., Bahl, R., Barros, A. J., et al. (2016). Breastfeeding in the twenty first century: epidemiology, mechanisms, and lifelong effect. *Lancet* 387(10017), 475–90.

Vos, T., Flaxman, A. D., Naghavi, M., et al. (2012). Years lived with disability (YLDs) for 1160 sequelae of 289 diseases and injuries 1990–2010: a systematic analysis for the Global Burden of Disease Study 2010. *Lancet* 380(9859), 2163–96.

Wagner, M. (2000). Choosing caesarean section. *Lancet* 356(9242), 1677–80.

Waldenström, U, Hildingsson, I., and Ryding, E. L. (2006). Antenatal fear of childbirth and its association with subsequent caesarean section and experience of childbirth. *BJOG* 113, 638–46.

Walrath, D. (2006). Gender, genes, and the evolution of human birth. In: Geller, P. L. and Stockett, M. K. (eds) *Feminist Anthropology: Past, Present, and Future*. Philadelphia: University of Pennsylvania Press, pp. 55–69.

Warrener, A. G., Lewton, K. L., Pontzer, H., et al. (2015). A wider pelvis does not increase locomotor cost in humans, with implications for the evolution of childbirth. *PLoS One* 10(3), e0118903.

Wells, J. C. (2011). The thrifty phenotype: an adaptation in growth or metabolism? *Am J Hum Biol* 23(1), 65–75.

Wells, J. C., DeSilva, J. M., and Stock, J. T. (2012). The obstetric dilemma: an ancient game of Russian roulette, or a variable dilemma sensitive to ecology? *Am J Phys Anthropol* 149(S55), 40–71.

Werner, D. (2006). The evolution of male homosexuality and its implications for human psychological and cultural variations. In: Sommer, V. and Vasey, P. L. (eds) *Homosexual Behaviour in Animals*. Cambridge: Cambridge University Press, pp. 316–46.

Whitcome, K. K., Shapiro, L. J., and Leiberman, D. E. (2007). Fetal load and the evolution of lumbar lordosis in bipedal hominins. *Nat Genet* 450, 1075–8.

WHO (2018). *WHO Recommendations: Intrapartum Care for a Positive Childbirth Experience*. Licence: CC BY-NC-SA 3.0 IGO. Geneva: World Health Organization.

Wiberg, N., Källén, K., and Olofsson, P. (2008). Delayed umbilical cord clamping at birth has effects on arterial and venous blood gases and lactate concentrations. *BJOG* 115(6), 697–703.

Widström, A. M., Lilja, G., Aaltomaa-Michalias, P., et al. (2011). Newborn behaviour to locate the breast when skin-to-skin: a possible method for enabling early self-regulation. *Acta Paediatr* 100(1), 79–85.

Williams, F. E. (1936). *Papuans of the Trans-Fly*. Oxford: Clarendon Press.

Winberg, J. (2005). Mother and newborn baby: mutual regulation of physiology and behavior—a selective review. *Dev Psychobiol* 47(3), 217–29.

Wittert, G. (2012). Declining testosterone levels in men not part of normal aging. *ScienceDaily* 23 June. https://www.sciencedaily.com/releases/2012/06/120623144944.htm.

Wolfe, L. D. (1991). Human evolution and the sexual behavior of female primates. In: Loy, J. and Peters, C.B. (eds) *Understanding Behavior: What Primate Studies Tell Us about Human Behavior*. New York: Oxford University Press, pp. 121–51.

Xiong, T., Tang, J., and Mu, D. Z. (2012). Side effects of phototherapy for neonatal hyperbilirubinemia (in Chinese with English abstract). *Chinese J Contemporary Pediatr* 14, 396–400.

Zahavi, A. and Zahavi, A. (1997). *The Handicap Principle—A Missing Piece of Darwin's Puzzle*. Oxford: Oxford University Press.

Zucchi, F. C., Yao, Y., Ward, I. D., et al. (2013). Maternal stress induces epigenetic signatures of psychiatric and neurological diseases in the offspring. *PLoS One* 8(2), e56967.

Zwelling, E. (2010). Overcoming the challenges: maternal movement and positioning to facilitate labor progress. *MCN Am J Matern Child Nurs* 35(2), 72–8.

CHAPTER 17

BRAIN, SPINAL CORD, AND SENSORY SYSTEMS

MARTIN BRÜNE

Abstract

The central nervous system (CNS) comprising the brain and the spinal cord is the control system of bodily functions and processes incoming information from the external environment via diverse sensory systems. Incoming information from the body (interoception) and from the environment (exteroception) is integrated and utilised in complex patterns of action and re-action (allostasis) to maintain homoeostasis and ensure survival and reproduction. The brains of primates are larger than expected for their body weight, whereby the human brain is not exceptional in terms of the number of neurons. However, connectivity and cross-talk between different brain areas has increased during human evolution. The adult human brain consumes about 20% of total energy intake, which is needed to maintain complex excitatory and inhibitory mechanisms with the aid of glia cells and microglia. Abnormal amounts of stress during different stages of development may expose the CNS to an excess of toxic and inflammatory metabolic products, which may cause neuropsychiatric disease and disorders. Such stressors may include, among other factors, adverse events such as abuse or neglect during childhood, immunological mismatches between ancestral and current environments, and a plethora of social challenges related to 'modern' cultural peculiarities. Gene–environment interactions involved in CNS disease processes are far from being fully understood. Preventive measures to protect the CNS from premature functional deterioration may include safe-guarding child development, a reduction of toxic waste products, and mental and physical exercise.

Keywords

brain, spinal cord, interoception, exteroception, allostasis, excitatory and inhibitory processes, toxic and inflammatory metabolic products, microglia, neuropsychiatric disease and disorders, gene–environment interaction, evolution, medicine

17.1 Introduction

The human central nervous system (CNS), comprising the spinal cord and the brain, is a tremendously complex organ. It virtually controls all body functions via multiple and interacting feedback loops, and regulates how the human organism navigates through time and the environment. The manifold functions that the human CNS entertains are the result of a long and intricate evolutionary history. This chapter focuses on the evolutionary trajectories of the human CNS, as far as they are relevant for the understanding of human physiology and medicine. Specifically, the aim of this chapter is to highlight important interactions of the CNS with other organ systems, and to emphasize that the human CNS—like all other organic systems—evolved in response to specific socioecological exigencies in relation to life history patterns that are typical for our own species (Stearns 1976). Following this scope, relative emphasis will be put on brain anatomy and function, including the evolution of the sensory systems, with somewhat less attention paid to the organisation and function of the spinal cord.

Nervous systems appeared in animals as organised collectives of specialised, excitable cells, with a molecular machinery that allowed for synaptic communication that probably evolved from pre-existing vesicular systems and membrane ionic channels some 1 billion years ago (Ryan and Grant 2009). Remarkably, it now appears that complex nervous systems evolved several times independently from those excitable cells (Moroz 2009). The mammalian nervous system is not simply built 'on top' of pre-existing reptilian structures (Herculano-Houzel 2016a), but rather is the result of a complex reorganisation involving all parts of the brain, not just the evolutionarily most recent neocortical areas.

The CNS allows animals to move and maintain body functioning as an integrated whole—something that is feasible by simple diffusion in unicellular or small multicellular beings, but impractical without fast communication networks once past a few millimetres or so of body size. While there is a loose direct relationship between the size of the body and the size of the brain, to this day it remains to be determined how come larger bodies are usually accompanied by larger centralised brains: is it because more neurons are required to operate larger bodies (Jerison 1973), or because larger bodies allow the maintenance of more neurons (Watson et al. 2012), or simply because of joint genetic mechanisms that regulate the size of body parts? Whatever the underlying cause, brains made of larger numbers of components—neurons, glial cells, fibres, synapses, extracellular matrix—must benefit from greater complexity, even if solely because of the rapidly increasing number of combinatorial possibilities of information processing pathways.

Even though the human brain does not contain the largest number of neurons (the larger African elephant brain, for instance, contains 257 billion neurons compared with the average 86 billion neurons in the human brain; Herculano-Houzel et al. 2014a), it does hold the largest number of neurons in the cerebral cortex, the structure responsible for complex associative processing, reasoning, and planning (Azevedo et al. 2009). The prefrontal region of the human cerebral cortex, in particular, is thought to contain the largest number of associative neurons of any brain, even though that number seems to represent the same 8% of all cortical neurons as in other primate cortices (Gabi et al. 2016). In fact, determining between-species differences in prefrontal grey and white matter volume critically depends on the applied methodology. Accordingly, a recent magneto-resonance imaging study reported significantly larger grey and white matter volumes in humans compared to chimpanzees and macaques (Donahue et al. 2018).

In any event, larger brains come at a cost, however. In the first place, CNSs probably evolved as devices that allow avoidance of hazards, and hence of threats, to the organism's survival, as well as evaluation of the environment for extractable resources (food, shelter, mates). At some point during evolution, however, the brain expanded its capabilities from simply allowing an organism to 'react' to environmental fluctuations in space and time to actively exploring the environment (Allman 1999). In a sense, the evolution of brains is the history of diversification of an organism's behavioural capacity in ways that often favour exploitation and manipulation of the environment (including the social environment, which is, as we will see, particularly important for social animals like ourselves), improving the organism's reproductive fitness.

However, the evolution of brains has always been constrained by the brain's energy consumption, because neurons are highly expensive in energetic terms. It is at the cost of energy that they maintain the ionic balance between themselves and their environment, and synthesise transmitters used in the communication between nerve cells (Fehm et al. 2006). Dendrites, for example, contain large numbers of mitochondria that maintain energy supply. Neurons communicate via neurotransmitters and electric currents called action potentials. Action potentials can be propagated along great distances, and the evolutionary novelty of myelination greatly accelerated the information transfer from one neuron to the other (Hofman 2001). Even though the large cost of excitatory synaptic transmission can be kept down by lower firing frequencies, the cost of maintaining membrane polarisation is enormous, and grows together with the surface area of the membrane (Attwell and Laughlin 2001). All these expensive mechanisms must have had an evolutionary advantage that led to their selection, otherwise brains, especially large ones, would not exist, simply because energy supplies are not limited by feeding, and evolutionary processes are essentially 'thrifty' (Northcutt 2001).

Large brains pose several biological problems, among other reasons (such as fitting in a head still small enough to pass through the birth canal; see Chapter 16: Sexuality, Reproduction, and Birth) because they require large energetic input. Also, a big brain develops slowly, because the wiring process requires a lot of environmental input over extended periods of time, and thus constrains the reproductive potential of its bearer. In order to reach adulthood, an immature organism has to be protected from environmental hazards such as predation, starvation, and so forth by adult individuals in whose genetic interest it must be that the offspring organism reproduces. In other words, the parental effort in raising offspring increases proportionally with the immaturity of descendants (Trivers 1974; Allman 1999). All these biological problems have come to a head in humans. So why does the human CNS exist at all? How and why is it organised the way it is? What makes it sometimes vulnerable to dysfunction? Answers to these questions shall be sought in the following sections.

17.2 Macroanatomical Features of the Human Central Nervous System

17.2.1 Anatomy of the Spinal Cord

The spinal cord is the dorsal extension of the hindbrain and runs dorsally along the longitudinal axis in all bilaterians (bilaterally symmetric organisms). It consists of white and grey matter. The grey matter is composed of sensory and motor neurons, preganglionic neurons, and interneurons, whereas the white matter comprises several tracts that are somatotopically organised. These tracts contain bundles of axons which send sensory (ascending from the periphery to the brain) or effector (descending from the brain to the periphery) signals. Tracts starting with the syllable 'spino-' are afferent to the brain (sensory), whereas the opposite direction ends on '-spinalis'.

The main ascending tracts are the dorsal or posterior column tract (sending information about proprioception, vibration, pressure, and fine touch), the spinothalamic tract (the lateral portion transmitting information about pain and temperature, the anterior portion sending information about crude touch and pressure), both of which project to the thalamus, and the spinocerebellar tracts (sending information about proprioception to the cerebellum). As a rule of thumb, more medially localised neurons carry information from lower body parts, whereas more laterally located axons send information from the upper part of the body. The neuronal chain of information starts with the axon of the first-order neuron in the peripheral tissue sending to the cell body in the dorsal root ganglion, which then connects to the second-order interneuron in the spinal cord, the centripetal information of which is directed to various brainstem nuclei or further relayed by a third-order neuron in the thalamus to specific cortical areas (Guyton and Hall 2006).

Motor commands are sent from the brain to the periphery via several routes. The corticospinal tracts transfer signals for voluntary movement of the striated muscles via the corticobulbar tract (eye movements and facial muscles), the lateral corticospinal tract, and the anterior corticospinal tract (skeletal muscles). Involuntary ('extrapyramidal') movements are controlled by the vestibulospinal (body posture and balance), the tectospinal (startle reflex), the reticulospinal (automatic eye movements), and the rubrospinal tracts (balancing the innervation of flexor and extensor muscles). Some primitive movements are under direct control of the spinal cord, without involving the brain. For example, monosynaptic reflexes such as the knee-jerk reflex operate at the level of the spinal cord. Other reflexes in response to cutaneous stimulation involve the activation of a spinal interneuron, but not conscious awareness of the reflexive response.

In addition to the pyramidal and 'extrapyramidal' motor systems, the autonomic (visceral) nervous system (ANS) controls the involuntary movement of the intestines and the blood vessels. It also innervates the other organs, notably the heart, and is thus critically involved in the regulation of homoeostasis of the organism. Preganglionic sympathetic neurons located in different points of the spinal cord innervate visceral effector neurons in the sympathetic chain of ganglia. The connection to the effector sympathetic neuron is variable and can occur in the paravertebral ganglia, in one of the other ganglia, or synapse in the periphery, which is the case in those sympathetic neurons whose axons travel to the adrenal medulla located on top of the kidneys (Guyton and Hall 2006). The bilaterally symmetrical adrenal medullae are composed of modified postganglionic neuronal cells which secrete catecholamines (epinephrine and norepinephrine) and are critically involved in stress regulation. The parasympathetic system originates in the brainstem. Parasympathetic preganglionic axons travel with the oculomotor, the facial, the glossopharyngeal, and the vagus nerve. The latter is the most important parasympathetic nerve and innervates ganglionic neurons in the thoracic and abdominal organs. The parasympathetic system is cholinergic (like all preganglionic visceral neurons), and has a dampening effect on heart rate and anabolic effects on metabolism (Guyton and Hall 2006). Thus, the ANS is the most important neuronal network for stress regulation and cross-talk between the CNS and other organs.

An interesting feature of the vertebrate spinal cord is that many ascending and descending tracts cross to the other side of the CNS at some level. For example, information about pain and temperature cross at the level of the spinal segment to the other side and travel upwards via the lateral spinothalamic tract. In contrast, the axons of the posterior column tracts cross at the level of the brainstem. Other tracts do not cross at all, such as the reticulospinal and the vestibulospinal tract. Accordingly, the localisation of a spinal cord injury or malperfusion can be clinically determined by a characteristic (dissociative) pattern of sensory and motor deficits. Similarly, the corticospinal ('pyramidal') tract decussates in the brainstem, such that lesions of the tract are either contra- or ipsilateral to the lesion, depending on its localisation (above or below the decussation).

From an evolutionary point of view, it has been debated why decussations exist at all, as they do not seem to have an adaptive value (Kinsbourne 2013). In fact, decussations are unique to vertebrates, whereas invertebrates completely lack crossings of nerve tract at any level (that is, all information runs ipsilaterally). Put another way, chordates are unique with regard to their *bauplan*, which, among other features, is characterised by a dorsally located vertebral column and spinal cord, whereas all invertebrates possess a ventral nerve cord or two parallel nerve cords (as in flatworms). Anatomically, the nerve cord of invertebrates originates dorsally from a ganglionic nerve cell mass from where it continues along the ventral aspect of the body's longitudinal axis. In proto-chordates, the earliest ancestors of vertebrates which evolved from invertebrate phyla, a dorsal notochord is formed that organises the vertebral column and spinal cord. Vertebrates thus have a different *bauplan* that apparently evolved in parallel to the invertebrate *bauplan*, and the first stages of endoskeletons, as marked by the evolution of the cartilaginous notochord, may have developed, because the notochord helped protect the inner organs from damage and improved movement of the trunk (Kinsbourne 2013).

It has been suggested that, at some point in vertebrate evolution, a 'somatic twist' occurred dorsally from the oropharynx, by 180 degrees (Kinsbourne 2013). This model provides an explanation for the existence of the optic chiasm and other decussations further down the hindbrain and spinal column. It also accounts for the observation that the olfactory nerve is the only cranial nerve that runs ipsilaterally (Kinsbourne 2013). There are, however, neural tracts that do not decussate, possibly because they evolved later than the major anatomical twist. Interestingly, and compatible with the model, evolutionarily highly conserved regulatory genes expressed in the ectodermal midlines (ventral and dorsal, respectively) that are involved in programming what is ventral and what is dorsal swapped their function (Lowe et al. 2006). That is, 'ventral' in invertebrates became 'dorsal' in vertebrates, and vice versa. Other indications for a somatic twist come from observations of blood flow, which also changed direction and, in vertebrates, runs now from ventral to dorsal (Kinsbourne 2013).

A modification of the twist hypothesis proposes two 90-degree twists in opposite directions (de Lussanet and Osse 2015), yet both models rely on the conclusion that decussation is a by-product of major evolutionary adaptations in chordates, rather than an adaptation in its own right. This view has recently been challenged, based on three-dimensional computer models of brain and body structures, suggesting that decussations are associated with a greater robustness against wiring errors, and mandatory for maintaining somatotopic organisation and motor coordination (Shinbrot and Young 2008). This view of decussation as a mandatory adaptation may have important ramifications for the treatment of spinal cord injury, because sprouting observed in undamaged cerebrospinal axons crosses the midline, and, hence, may be detrimental to sensorimotor coordination (Shinbrot and Young 2008). Figure 17.1 illustrates the main ascending and descending tracts of the spinal cord and their way from and towards the brain, including decussations.

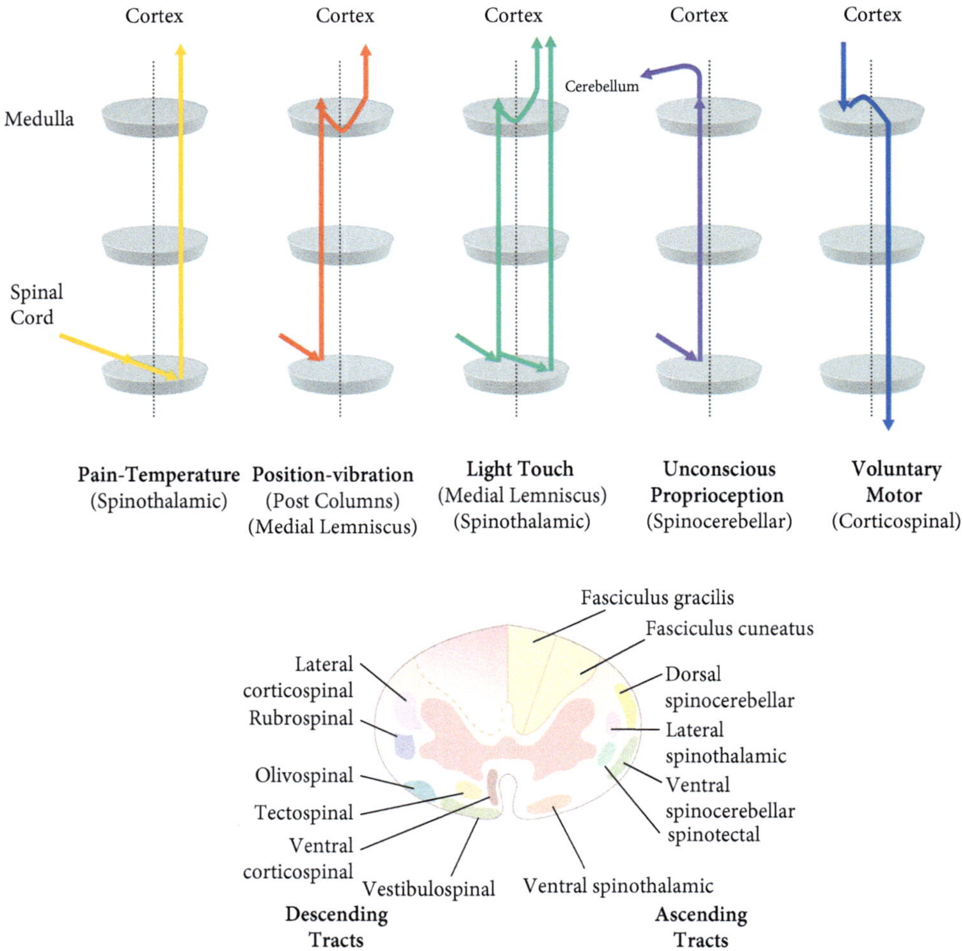

FIGURE 17.1 Ascending and descending tracts and their decussations at different levels of the spinal cord or brainstem. Decussation of the optic nerve (chiasma opticum) is not shown here.

The importance of superior sensorimotor coordination in human evolution is also visible at the macroanatomical level. Comparative anatomical studies of the CNS have revealed that in humans the spinal cord has increased in both absolute and relative size compared to other apes. This evolutionary change is further substantiated by comparisons of the diameter of the spinal canal. The width of the bony spinal canal is proportional to the size of the spinal cord, so estimates of spinal cord size can be deduced from fossil vertebral columns (Meyer and Haeusler 2015). Thus, this work suggests that the spinal canal in humans has increased in size particularly in cervical and thoracic regions, possibly reflecting an improvement in motor coordination of hand and arm movements. Importantly, in humans the size of the spinal cord is unrelated to brain size, and the fossil evidence suggests that spinal cord enlargement clearly preceded brain enlargement, presumably related to the evolution of bipedalism, which freed the hands from their role in locomotion (Meyer 2016).

17.2.2 Anatomical Subdivisions of the Brain

The human brain, like other mammalian brains, can be divided into several parts that are intensely connected with one another, although arranged sequentially along the neuraxis, given that the brain is derived from the rostral part of the neural tube. The nomenclature and relationships used here follow the gene expression-based ontology of the brain of Watson et al. (2016).

The most caudal part of the brain, continuous with the spinal cord, is called hindbrain, and it is sufficient to maintain basic physiological functions such as respiration, blood pressure, sleep–wakefulness rhythm, and simple, reflex behavioural responses. The next part is called the mesencephalon or midbrain, and while responsible for organising movements of the eyes and limbs, it adds a level of signal integration to behaviour. Next, the diencephalon comprises the thalamus, a necessary gateway between information originating from hindbrain or midbrain structures towards the cerebral cortex, and a structure that serves modulatory loops across cortical sites.

Anterior to the diencephalon lie the telencephalon and hypothalamus. Although the adult telencephalon appears to be the anterior-most derivative of the neural tube, it is not; the most rostral (or anterior) division of the vertebrate brain is the hypothalamus, a structure that exerts direct or indirect control over body physiology by controlling the visceral peripheral nervous system (sympathetic and parasympathetic branches) and the entire endocrine system via modulation of the hypophysis. The telencephalon originates from a part of the alar plate of the second prosencephalic neuromere, which, together with the first neuromere, gives rise to the hypothalamus. The telencephalon has expanded tremendously in vertebrates, including humans, both anteriorly and posteriorly, thus appearing to be rostral to the hypothalamus. The telencephalon in subdivided into pallial (cortical) and subpallial (subcortical) structures such as the striatum and pallidum. The word pallium means 'cloak' or 'coat'. This outer part of the brain can be subdivided into allocortex and isocortex, which can be distinguished on the basis of their cytoarchitecture. The isocortex comprises a six-layered structure containing large pyramidal cells that can be subdivided into primary sensory cortices (which receive massive driving projections from sensory nuclei of the thalamus), secondary and tertiary sensory areas (which receive sensory thalamic input as well as input from primary cortical areas), motor cortex (which receives modulatory input from thalamic nuclei and massive connections from sensory cortical areas), and associative cortex (which also receives modulatory input from thalamic nuclei and forms reciprocal connections with multiple cortical areas). The allocortex comprises the piriform (olfactory) cortex, the hippocampal region, and entorhinal cortex, as well as the cingulate cortex, and parts of the amygdalae. The isocortex makes up about 96% of the human cortical surface: some 32% belong to the frontal lobes, 30% to the temporal lobes, 23% to the parietal cortex, and 15% to the occipital lobes (Zilles 1987).

Overall, the total number of neurons of the prefrontal cortex of primates has increased over evolutionary time. Humans possess approximately four times more neurons compared with other apes, comprising 14 cm^3 in humans compared with 2.8 cm^3 in chimpanzees. By contrast, the *density* of neurons reveals an inverse relationship: humans have the lowest density per cubic centimetre due to an increase in connecting tissue; thus, an increasing number of neurons constrains the density of white matter. In humans, grey matter is about 50% of total brain volume, and the human neocortex contains 40% white matter. Thus, although brain size matters, patterns of brain reorganisation in terms of function are at least equally important (Hofman 2001; Jerison 2001). For example, further growth of cortical surface would require such an increase in size of the skull that the passage through the birth canal would become impossible (or require timing of birth to be even more displaced preterm). Moreover, if the cortex expanded along the horizontal axis only, the length of connecting fibres would inevitably increase, such that the flow of information from one cell or cell assembly to another would be slowed. These physical constraints may have been one reason why cortical folding (gyrification) occurred in brain evolution—to facilitate neural transmission by reducing the number and length of connecting axons. The number of connecting fibres was reduced by compartmentalisation of neurons into modular circuits, whereas the length of connecting tissue was diminished by cortical folding (Hofman 2001). In other words, gyrification represents a solution for large brains to pack a maximum of surface into a minimal volume. Thus, in the adult human brain, about two-thirds of cortical surface is buried in the sulci, and only one-third is superficially exposed. All gyri and sulci are already visible at birth, but it takes two decades of growth to conceal two-thirds of the cortex surface in the sulci (Allman 1999; Roth and Wullimann 2001).

A derivative of the structures that form the hindbrain, the cerebellum, which lies dorsally to the neuraxis, comprises an external cerebellar cortex, subdivided into lobes like the cerebral cortex, and subcortical nuclei. The human cerebellum contains about 80% of all neurons in the human brain (Azevedo et al. 2009), the vast majority of them in the granule cell layer, which receives input from virtually all parts of the cerebral cortex. Although known best for its role in eye movement, vestibular control, and motor coordination, the cerebellum contributes to a variety of cognitive functions including attention, action planning, visual-spatial cognition, and memory, as well as emotion regulation (Katz and Steinmetz 2002).

17.2.3 Allometric Growth

Brain size varies greatly between species, even within mammals. The brain of the tiniest mammals weighs less than 1 g, whereas the brain of whales can weigh up to 10 kg. As a rule of thumb, larger animals have larger brains. Brain mass is, however, not a universal function of body mass; brain mass scales as a power function of body mass with exponents that vary across clades, from 0.548 in artiodactyls and 0.602 in carnivores to a nearly linear 0.903 in primates (Herculano-Houzel 2016b). Compared to non-primate mammals, primates as a whole—and not just humans—have much larger brains for their body mass; the observation that the human species is the most encephalised, with a brain 7.5 times larger than expected for its body mass (Jerison 1973), is a simple consequence of the difference in brain allometry between primates and other mammals—and the fact that, within primates, humans are the largest-brained species.

Another common comparison, among humans and great apes (orangutans and gorillas), has been the basis for the claim that the human brain is exceptionally large: if larger animal species have larger brains, then it should be the gorilla, up to three times as large in body mass as a human of average body weight, that has the largest brain among primates, yet the gorilla brain is only about one-third the mass of the human brain. However, given the large energetic cost of adding neurons to the brain (Herculano-Houzel 2011), particularly in face of the metabolic limitation imposed by the raw diet consumed by non-human primates, gorillas and orangutans could not afford the larger brain that one might expect to accompany their large bodies; primate evolution seems to be under energetic constraint such that when bodies become too large, there is a trade-off between body mass and the number of brain neurons that can be afforded (Fonseca-Azevedo and Herculano-Houzel 2012). Figure 17.2 shows coronal sections of different primate brains (including the human brain) in comparison to a dolphin brain. Note that the organisation of the dolphin brain is entirely different from primate brain organisation.

The anatomically modern human brain represents about 2% of body mass, as in smaller primates. In the large great apes, in contrast, the brain may represent only as little as 0.2% of body mass; thus, the small relative brain mass of large great apes is a consequence of the metabolic constraint imposed by their raw diet. These animals feed on average 8 hours/day, and the finding that they lose mass when food becomes scarce strongly indicates that feeding for longer than 8 hours/day is not feasible for these primates—particularly given that they sleep 6–7 hours/day (Fonseca-Azevedo and Herculano-Houzel 2012). With our average number of brain neurons, humans should spend over 9 hours/day feeding if we had the same diet as other primates—which gorillas and orangutans show not to be feasible. What allowed the evolution of the larger human brain, the largest among primates, and thus associated with a larger energetic cost proportional to its number of neurons, was a radical change in diet brought about by technologies that can be linked to the process of 'cooking': first, stone tools that allowed cutting, chopping, grinding, and mashing foods outside the mouth; the controlled use of fire that further softened meat and tendons, facilitated mastication, and thus in one strike both increased caloric yield (now that meat and other foodstuffs could be completely digested) and decreased the amount of time that had to be dedicated to obtaining calories per day (Herculano-Houzel 2016a). Liberated from the metabolic constraint that still applied to other primates, humans gradually increased the scaling of their brain due to selection pressures that partly resided in the increasing complexity of their social environments (Dunbar and Shultz 2007), and the growing body of culturally transmitted technologies (Henrich 2015). Thus, the average 86 billion neurons that are found to compose the modern human brain are expected both for the mass of the human brain and the mass of the human body when compared to other, non-great-ape primate species (Azevedo et al. 2009; Herculano-Houzel 2009). (For further discussion, see Chapter 13: Digestive System.)

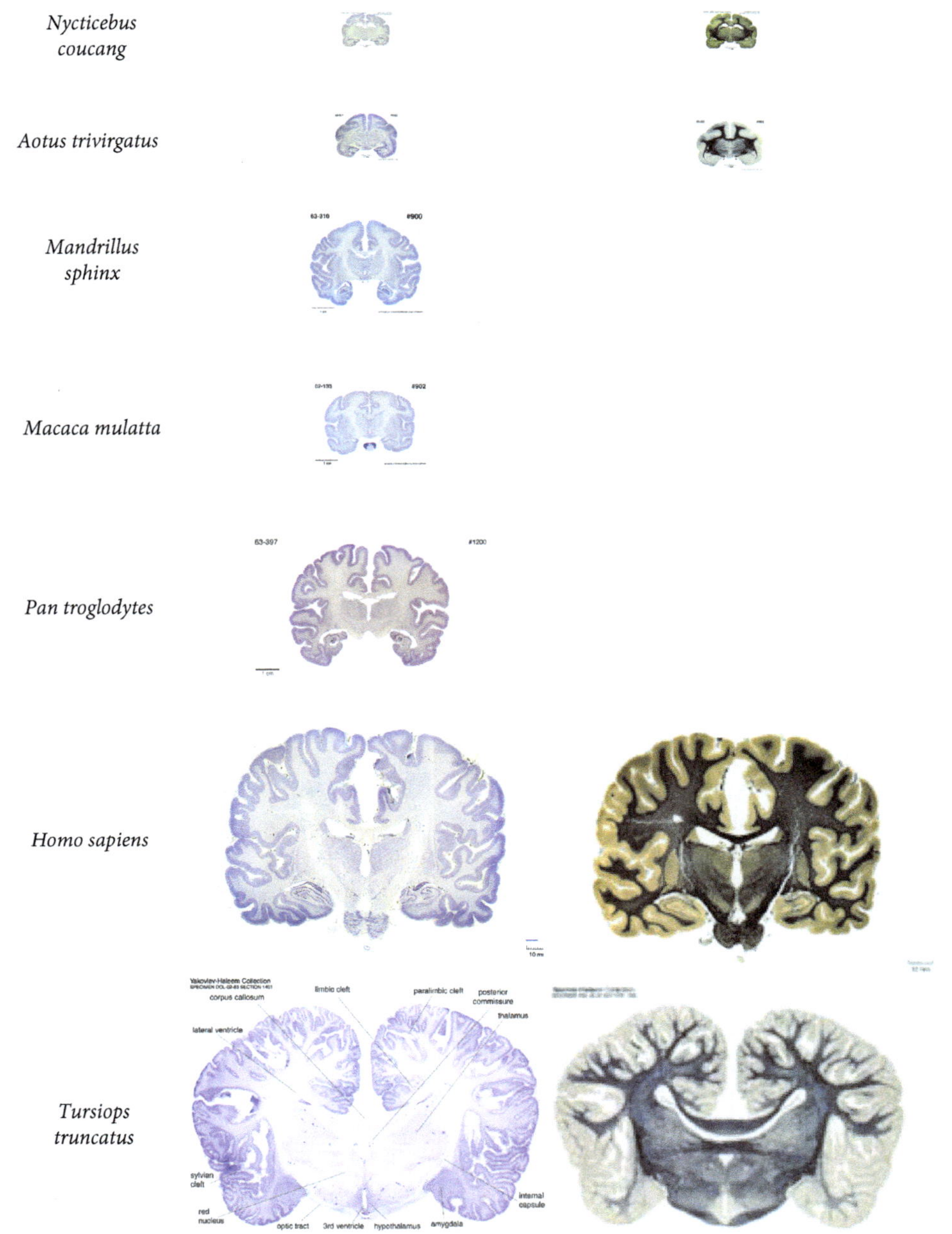

FIGURE 17.2 Coronal sections of primate brains (including human brain) in comparison to a non-primate brain (dolphin). From top to bottom: prosimian, slow loris (*Nycticebus coucang*), a New World monkey; three-striped (owl) monkey (*Aotus trivirgatus*); mandrill (*Mandrillus sphinx*); rhesus monkey (*Macaca mulatta*); chimpanzee (*Pan troglodytes*); human (*Homo sapiens*); and bottlenose dolphin (*Tursiops truncatus*). Brains are shown in proportional size. Note increase in gyrification and connective tissue (shown in brown-coloured images on right-hand side), buried insula in chimpanzee and human brains, and entirely different cortical organisation of dolphin brain compared to primate brains. Scale bar 10 mm.

Source: Adapted with permission from http//www.brains.rad.msu.edu, supported by the US National Science Foundation.

Along the same lines, it has been argued that the evolutionary expansion of the human brain involved a relative decrease in the size of another energy-expensive organ, the gut (Aiello and Wheeler 1995). However, a more recent analysis of a larger dataset of primate and non-primate mammalian species did not find the expected negative correlation between residual brain mass and residual intestinal mass after controlling for body size (Navarrete et al. 2011). Instead, it is more likely that the relatively reduced intestinal mass of humans was not a condition for the evolution of a larger brain, but rather a consequence of the adoption of processed, more easily digestible diet (Herculano-Houzel 2016a). (For further discussion, see Chapter 13: Digestive System.)

Still, a major portion of the current literature compares the human brain and its structures to great apes and other primate and non-primate species after normalising for body mass—a procedure that is inaccurate for two reasons: first, because brain mass does not scale universally with body mass across all mammals; and second, because the allometric rules for primates do not apply to great apes for reasons outlined above. Another contentious issue concerns the calculation of 'progression indices' (PI), obtained by dividing the observed brain mass in a species by the brain mass expected for an 'ancestral species' resembling the last common ancestor, assuming a common allometric scaling rule between brain and body. Here, the problem is that much of the existing literature from the 1960s–1990s suggested that the earliest primates descended from insectivore-like ancestors (e.g. Stephan 1979). The prevailing logic was that when comparing the size of particular parts of the brain relative to absolute brain size between closely and more remotely related species, controlling for body mass, one could get an impression of which brain areas enlarged over time during human evolution and which ones became relatively smaller. For example, based on PIs it was concluded that the importance of the olfactory system declined in the primate lineage over evolutionary time. That is, in 'insectivores' the olfactory bulb makes 18% of the total brain volume, in tupaias it figures around 7%, in the galago 4%, in monkeys around 0.2%, and in humans 0.01% of total brain volume (Rapoport 1990). Expressed in PI, the size of the human olfactory bulb is roughly merely 1/40 to 1/50 as expected for an 'insectivore' of comparable body weight. By contrast, the PI of the human neocortex is 156, and in our closest extant relative, the chimpanzee, the neocortex PI is 58 (Rapoport 1990).

However, 'insectivores' are neither a monophyletic group nor the closest relatives of primates. The group previously known as 'insectivores' is now known to have lumped together evolutionarily distant afrotherians and eulipotyphlans, many of which are indiscriminately referred to as 'shrews' or 'moles'. Moreover, relative size should not be taken as an indication of function. For example, a recent study reported that the 'microosmatic' human species (for its 'regressed' olfactory bulbs, according to PI) has as many neurons in the olfactory bulb as the highly olfactory-dependent eulipotyphlan shrews and moles (Ribeiro et al. 2014). The ratio between numbers of neurons in the cerebral cortex and the olfactory bulb is large in primates, but also in artiodactyls (Herculano-Houzel et al. 2014b). Similarly, the ratio between numbers of neurons in the cerebral cortex and the 'rest of brain' (the ensemble of brainstem, diencephalon, and striatum) is larger in primates than in other mammalian species, but not the largest in humans. The most distinguishing characteristic of the human brain in comparison to any other brain is the sheer, absolute number of neurons in its cerebral cortex, larger in our species than in any other. It is similarly likely that other human advantages in terms of information processing involving the hippocampus, entorhinal cortex, amygdala, and striatum, rather than reflecting large PIs (Rapoport 1990), are due instead to the large number of neurons in these structures of the human brain.

Along similar lines, it seemed perplexing at first sight when the human frontal lobes—relative to neocortex size—were found not bigger than expected for an ape of our body size (Semendeferi et al. 1997, 2002). However, the human prefrontal cortex is significantly more convoluted or gyrified than expected for a primate of the same brain size, and the same is probably true for the posterior temporal and parietal lobes, whereas the motor, premotor, and primary visual areas are the least convoluted (Rilling and Insel 1999; Rilling and Seligman 2002). Interestingly, the prefrontal cortex and the temporo-parietal junction are the last to myelinate during ontogeny. Due to complex reorganisations, some areas located in the orbitofrontal cortex such as Brodmann area (BA) 13 are even smaller in humans, relative to other ape species (Semendeferi et al. 1998). BA 10, by contrast, has undergone an evolutionary shift in location by moving from the orbital frontal cortex to cover the entire frontal pole in apes (Semendeferi et al. 2001). In terms of function, the frontal poles are engaged in future action planning, initiation of action, and 'prospective memory' (keeping something in mind for future action), whereas the orbitofrontal cortex is involved in evaluating the emotional significance of social stimuli (Bechara et al. 2000). Damage to the medial part of the orbitofrontal cortex produces a profound impairment of empathy, and dysfunction of the dorsolateral part of the orbitofrontal cortex leads to lack of drive and motivation (Bechara et al. 2000). Moreover, the medial part of the prefrontal cortex is specifically involved in social cognitive processes including the representation of one's own and others' thoughts and intentions (Frith and Frith 1999, 2001). Interestingly, the paracingulate sulcus, which can be found in only 50% of individuals, is thought to be under ongoing adaptive modification (Paus et al. 1996; Walter et al. 2004).

The temporal lobes have disproportionately increased in the human lineage relative to other apes, particularly with respect to white matter volume, possibly reflecting its functional significance in speech comprehension and production, face processing, and recognition of intentional movements (Rilling and Seligman 2002). The temporal poles contribute to the storage of autobiographical memory. The fusiform gyrus at the lower surface of the temporal lobe is critical for recognising invariant aspects of faces, such that individuals can be identified irrespective of the angle from which their faces are visible (Kanwisher and Yovel 2006; see also Section 17.4.3.1).

For some reason, the evolutionary history of the parietal lobes has remained, to some extent, obscure. However, the infraparietal lobule, which comprises the supramarginal gyrus and the adjacent angular gyrus, has probably undergone recent evolutionary reorganisation. It has rich connections to the prefrontal and temporal cortices and plays an important role in multimodal sensory integration, self-perception, and differentiation of self from others (Torrey 2007). Thus, damage to this brain area may lead to bodily neglect (e.g. anosognosia after damage to the right parietal lobe), lack of insight and self-reflection, and misattribution of thoughts and behaviours as alien-made (Torrey 2007). In contrast to the more anterior parts of the neocortex, the primary visual cortex has been displaced posteriorly and its relative size is reduced in humans relative to other primates, yet with large interindividual differences (Aboitiz et al. 2003).

17.2.4 Heterochrony

The study of evolutionary ontogeny, or heterochrony, refers to the observation that changes in timing or rate of developmental events, relative to the homologous patterns in the ancestor, are the key mechanisms by which evolutionary modifications are due to ontogenetic changes—that is, to changes in development (McKinney and McNamara 1991). Not only can developmental events themselves change in nature, but also a given developmental process can be accelerated and terminated earlier, or delayed and set off later during ontogeny. (For further discussion, see Chapter 4: Growth and Development.)

There has been a persistent trend to consider that several features of the human body and brain represent delayed development or even the retention of juvenile characteristics into adulthood, a concept captured by the words 'neoteny' and 'paedomorphosis'. Indeed, several human body features including hairlessness and the anterior position of the vagina have been interpreted as the result of retention of juvenile characteristics of the ancestral species into adulthood of the extant human species (paedomorphosis) (McKinney and McNamara 1991). Moreover, persistent playfulness and curiosity into adulthood have been likened to behaviours found in juvenile apes (and other animals), hence giving rise to the hypothesis of behavioural paedomorphosis (Montagu 1989). Put slightly differently, the hypothesis suggests that paedomorphic organisms reach sexual maturity while retaining features of a juvenile body. Paedomorphosis is one of several developmental timing processes, which are embraced under the term 'heterochrony'. Heterochrony explains why during ontogeny the growth of different parts of the body (including the brain) diverges. Some tissues grow faster (and longer) than others (McKinney and McNamara 1991). Whether or not some aspects of brain enlargement and organisation are associated with paedomorphic timing of maturation is a matter of controversy. Indeed, the recognition that the temporal relationship between brain and body development (growth) can vary across primate species is a strong argument against paedomorphosis and for a much looser relationship in different species (Riska and Atchley 1985).

Importantly, many of the current stands in the literature regarding the exceptionality of human development are due to expectations based on the scaling of brain mass with body mass. As seen above, whereas the human brain was long considered an outlier in its mass, given the expected brain mass for a generic mammal of our body mass (Jerison 1973), it is now likely that it is the non-human great apes that are outliers, with brains too small for their bodies (Herculano-Houzel and Kaas 2011; Herculano-Houzel 2016a). For example, considering the age of sexual maturity in primates to be a function of brain weight relative to body weight, the inclusion of great apes makes it so that in a primate with a human-sized brain, sexual maturity should be reached by age 44. The actual time point at which humans become sexually mature is, however, much earlier (with differences between men and women). Similarly, the eruption of wisdom teeth correlates with brain weight. The projection of the regression line would predict the eruption of the third molars to occur at about 38 years of age, whereas the wisdom teeth actually erupt in humans on average around age 20 (Allman 1999).

Comparisons of growth processes in teeth between fossil human species and anatomical modern humans suggest that the typical human-like growth pattern emerged relatively late in human evolution. Taken together, these characteristics indicate that humans become sexually mature earlier than predicted from their relative brain weight (again, depending on the relationship with body mass), whereas development in general is delayed (Allman 1999). This has important ramifications for growth and development across the lifespan, which is dealt with extensively in Chapter 4.

Similarly, it has long been considered that cortical development, including synaptic pruning and myelination, is overly extended in humans compared to other primates

(Geschwind and Rakic 2013), though it seems more intuitive to expect the largest brain, irrespective of body mass, to take longer to develop.

17.2.5 Laterality

The human brain comprises many functions that are carried out by neural networks, most of which are differentially represented in the left or right cerebral hemisphere. This division of labour by the two hemispheres is referred to as lateralisation (Saugstad 1998). An important caveat to keep in mind is that 'lateralisation' is rarely complete, all-or-none: typically, reports of lateralised function refer to small advantages of one hemisphere over the other in a given task.

Language and handedness have been considered prime examples of functional lateralisation. Whereas 'linear' functions such as speech production and grammar as well as auditory processing of spoken language are lateralised to the left (in right-handed individuals), the 'holistic' comprehension of content of speech such as intonation, metaphor, and emotional prosody, that is, the ability to recognise the affective overtone of spoken language, is lateralised to the right (Mitchell and Crow 2005). Handedness is also indicative of a functional brain asymmetry. Although it has been suggested that language and handedness are tightly related, because in most right-handed individuals language is lateralised to the left ('dominant') hemisphere, recent genetic research has revealed that the two domains are genetically distinct (Schmitz et al. 2017). Interestingly, genes involved in language lateralisation are more strongly associated with genes that play a role in CNS dysfunction, including schizophrenia and autism spectrum disorders, whereas genes involved in handedness are more strongly linked to anatomical structural abnormalities (Schmitz et al. 2017).

In other domains, lateralisation refers only to a slight advantage of one hemisphere over the other, as in the case of spatial orientation, which is usually lateralised to the right hemisphere. Recent research has revealed that even the capacity to distinguish self from others is to some degree lateralised. The first-person perspective is represented preferentially in the left inferior parietal cortex, whereas the third-person perspective is represented preferentially in the corresponding region on the right side of the human brain. For example, the imitation of another person's action activates the left inferior parietal cortex; conversely, the opposite side is involved when subjects view their actions being imitated (Decety and Chaminade 2005).

Anatomically, there is a leftward occipital and rightward frontal lobe asymmetry, known as the cerebral torque, and perhaps a mild leftward planum temporale asymmetry (Crow 1997). Asymmetry of the planum temporale, a portion of Wernicke's speech area, however, is not human-specific (Gannon et al. 1998; Hopkins et al. 1998). Rather it has been shown to be already present in great apes, a finding suggesting that apes may have some functional specialisation located in the temporal lobe indicative of gestural proto-language. However, handedness is much less pronounced in apes compared to humans, such that there is still uncertainty about how to interpret these results.

In any event, increasing functional and anatomical specialisation of the cerebral hemispheres of a large brain causes some problems regarding the connectivity of anatomically distributed neural networks. Cortical connectivity, defined as the percentage of cortical neurons connected through the white matter, decreases with increasing numbers of cortical neurons in primates (Herculano-Houzel et al. 2010), with the consequence that in large primate brains, such as the human brain, the interhemispheric connectivity is reduced relative to other primates, whereas intrahemispheric connectivity is augmented. This may have been a constraint that led to the evolution of brain lateralisation as an inevitable side-effect of increasing brain size with decreasing connectivity. Indeed, the ratio of the corpus callosum volume to neocortex surface size decreases in primate species as hemispheric asymmetry and handedness (cerebral dominance) increase, suggesting a relationship between directional asymmetry and interhemispheric connectivity (Hopkins and Rilling 2000). However, exactly because asymmetry may increase specialisation and thus decrease redundancy, it may also make human brains more vulnerable to brain damage, due to the lack of contralateral 'back-up systems' as in more symmetrical brains (Striedter 2006).

17.2.6 Sex Differences

Males have slightly larger brains in terms of cortical volume by about 11–18%, and overall brain weight differs between men and women by about 110 g, even when body weight is covaried out. The total number of cortical neurons seems to be slightly lower in women compared with men, with a difference of about 15.5% (Pakkenberg and Gundersen 1997). The density of cortical neurons is reportedly the same in both sexes. However, these numbers were obtained using a stereological design with very low sampling and very small optical disectors, which makes it difficult to estimate variations in neuronal density accurately, at the same time making overall results dependent on the volume of the cortex.

The overall ratio of cortex to the whole brain volume is about 46% in both men and women, and the ratio of cortical to subcortical brain mass is identical in both sexes. The right hemisphere is usually slightly larger than the left hemisphere, with differences of about 3.5 g for both sexes. The thickness of the corpus callosum is similar for men and women. However, in light of the smaller brain size of women, the commissural connectivity is about 10% larger, which may support the assumption of a lesser hemispheric lateralisation in women (Gur et al. 2002).

The functional significance of these differences is obscure. It is well known that women have superior verbal fluency compared with men, whereas men, on average, are better at visuospatial orientation. However, women are more skilled than men in object location. Again, it must be kept in mind that these differences, if significant, amount to only a few percentage points—that is, they are small advantages of one sex over the other, which additionally are now known to be highly dependent on cultural expectations. Still, many authors have speculated that these functions have been associated with divergent selection pressures for men and women in ancestral times, where early humans lived as hunter-gatherers, and women with infants and juveniles formed the core of the community (as 'cooperative breeders'; Hrdy 2016), whereas men were engaged in travelling large distances in order to hunt large game. In line with this, research has revealed that women tend to have greater orbitofrontal cortices (involved in emotion regulation and empathy) compared with men, whereas no sex differences have been found regarding the size of the amygdalae, hippocampi, and dorsolateral prefrontal cortices. However, the ratio of orbitofrontal cortex volume and amygdala volume has been found greater in women, which may explain sex differences in emotion processing and affect-driven behaviour (Gur et al. 2002).

Considering how much is now known about how use can lead to experience-dependent modifications, it is crucial to keep in mind that what one strives to make of one's own brain is likely to be much more consequential for one's abilities than the impact of one's gender on the brain.

17.2.7 Blood and Energy Supply

The human brain is a metabolically expensive organ, consuming (at the adult stage) about 20% of the total resting metabolic rate (while making only 2% of the body weight). Blood, as the carrier of oxygen and nutrients such as glucose, is transported to the brain via the two carotid arteries and, to a lesser degree, via a posterior route which is taken by the vertebral arteries. The anterior and posterior routes communicate with one another via the arterial circle of Willis. The two internal carotid arteries provide some 85% of blood supply to the brain, while the vertebral arteries forming the basilar artery provide about 15% (Guyton and Hall 2006). The internal carotid artery on each side splits up to form the middle cerebral artery supplying blood to the lateral part of the frontal lobe, the parietal and the temporal lobe, and the anterior cerebral artery servicing the medial parts of the frontal and parietal lobes. The vertebral arteries merge to form the basilar artery, which supply structures beneath the tentorium cerebelli and the occipital lobes with blood (Seymour et al. 2016).

Research in hominoids has shown an allometric relationship of the width of the carotid canal with brain size (Braga and Hublin 1998). The lumen radius of the internal carotid artery is a reliable estimate of cerebral blood flow and energy consumption, because arterial size is closely related to blood flow and metabolic rate. Moreover, perfusion and metabolic rate remain relatively constant in the human brain, irrespective of cognitive or physical activity. Also, the internal carotid artery is not accompanied by veins or nerves when entering the skull via the carotid foramen (Seymour et al. 2016). Figure 17.3 depicts the diameter of the carotid canal in three different hominid species (*Australopithecus africanus, Homo sapiens Neanderthalensis*, and archaic *Homo sapiens*).

Comparative research of hominid species has revealed that the perfusion rate of the brain via the carotid arteries has increased disproportionately over evolutionary time, suggesting that brain metabolism has also increased in relation to brain size (Figure 17.3). Specifically, it has been estimated that cerebral perfusion increased by factor 1.7, such that brain perfusion increased six-fold, if one assumes an increase in brain size by 3.5 in the last 2 million years or so. This rise in metabolic rate may be related to the increase of glia cell activity and neuroplasticity (Campbell 2010; Seymour et al. 2016; see also Section 17.3.2), because the human brain is linearly scaled-up in terms of number of neurons (Azevedo et al. 2009; Herculano-Houzel 2009), so neuronal activity alone is unlikely to account for the disproportional rise in metabolism. In fact, other comparative research among the great apes (including humans) has shown that humans have the largest total energy expenditure as a function of the highest basal metabolic rate among the hominoid species studied. This could explain why humans are able to not only maintain larger brains, but also reproduce at a higher rate compared to other (extant) apes, despite their slower life history patterns and greater longevity (Pontzer et al. 2016). (For further discussion, see Chapter 4: Growth and Development.) Among the great ape species, humans have the largest amount of body fat (Pontzer et al. 2016), which may serve as energy storage for the brain. In fact, in times of

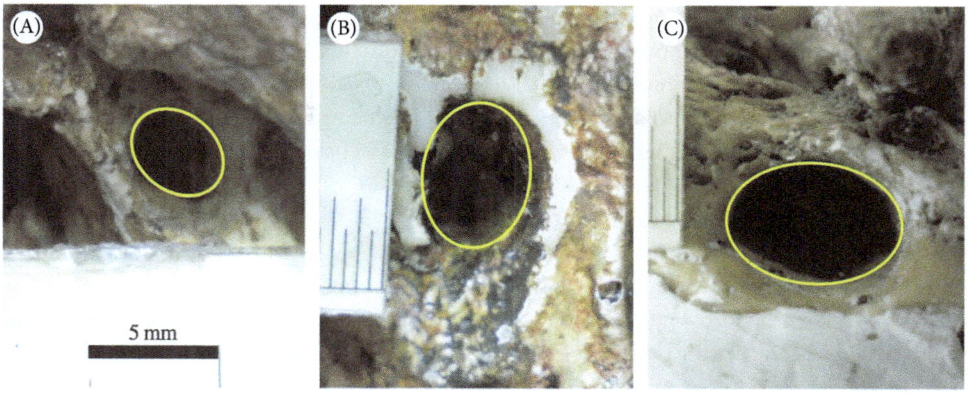

FIGURE 17.3 Size of the internal carotid canal in three different hominid species—(A) *Australopithecus africanus*, (B) *Homo sapiens Neanderthalensis*, and (C) archaic *Homo sapiens*—depicted on the same scale. Increments in (B) and (C) equal 5 mm.

Source: Reproduced from Roger S. Seymour, Vanya Bosiocic, and Edward P. Snelling, Fossil skulls reveal that blood flow rate to the brain increased faster than brain volume during human evolution, *Royal Society Open Science*, 3, p. 160305, Figure 1, doi.10.1098/rsos.160305.

starvation, all organs lose about 40% of their weight, except the brain, where weight loss is just 1% (Peters 2011).

There is evidence from comparative studies of endocasts of extinct hominid species and modern humans suggesting adaptive alterations in meningeal vessel anatomy (Bruner et al. 2005, 2011). For example, the middle meningeal artery, which derives from the maxillary artery off the external carotid artery, supplies the middle cranial fossa with blood and has undergone some significant anatomical changes when compared with the homologous arteries of apes and premodern humans. Overall, the number of anastomoses within the meningeal arterial network seems to have increased over evolutionary time, although its functional and physiological significance is not entirely clear (Kunz and Iliadis 2007; Bruner et al. 2011).

Aside from blood supply, it is important to note that the brain is constrained in its ability to store energy. In fact, 50% of the entire glucose uptake goes to the brain, and brain metabolism heavily relies on the utilisation of glucose (reviewed in Campbell 2010). An exception to this occurs during physical exercise, during which the brain is able to substitute glucose with lactate, at least in the short term. In fact, recent research suggests that astrocytes actively produce lactate and transfer it to aid neuronal metabolism, because lactate seems to be relevant for brain homoeostasis and can excitotoxic effects of glutamate (Magistretti and Allaman 2018). (For further discussion, see Chapter 6: Nutrition, Energy Expenditure, Physical Activity, and Body Composition.) Similarly, during times of starvation, the brain can rely on metabolising ketones synthesised from (abdominal) fat (Campbell 2010). Developmentally, it is important to note that the infant brain relies more on the consumption of ketone bodies, which are utilised to synthesise myelin by oligodendrocytes. In contrast, glucose metabolism in the CNS is relatively low at birth, but increases dramatically around age 4 to 11, when brain growth and synaptic pruning peaks. At the same time, body growth is attenuated during this critical period, presumably to allocate more resources to brain development. The opposite occurs during puberty, where resource allocation to somatic growth is balanced against reduced glucose metabolism in the CNS (Campbell 2010). (For further discussion, see Chapter 4: Growth and Development.)

In adult humans, functional activation of the brain has little influence on brain metabolism and may account for just 1% of energy consumption, suggesting that 99% of energy is consumed when the brain is apparently at rest. Campbell (2010) suggests that a large proportion of brain glucose is metabolised by astrocytes to convert glutamate to glutamine. Glutamine is then further metabolised by neurons, where it is reconverted, by the aid of lactate, into glutamate for synaptic transmission, whereby a smaller amount of glutamine is metabolised to gamma-butyric acid (GABA). This occurs across the brain, as 90% of neurons express glutamate receptors, though probably it is more pronounced in the hippocampus. The hippocampus is a metabolically active region of the brain and possesses glycogen stores, even though the entire glycogen stored in the brain may not exceed 0.1% of total brain weight, such that maintenance of brain function may last for only a few minutes (Campbell 2010). Sleep seems to be important, as animal models demonstrate that glycogen synthesis is increased during slow-wave sleep, but suppressed in conditions of sleep deprivation (reviewed in Campbell 2010).

Other work has suggested that the brain is able to manipulate the intake of energy by regulating insulin secretion. Put another way, the brain seems to compete with other organs for energy, specifically when stressed, by suppressing the action of insulin to keep blood glucose at high levels (thus facilitating the availability of glucose for the brain). Depending on individual differences in stress responsivity, some people lose body weight when chronically stressed, whereas others compensate for their elevated stress levels by overeating (Peters 2011). Paradoxically, the 'leaner' individuals may be at greater risk of developing cardiovascular disease, because they do not habituate to chronic stress and keep their adrenergic tone high, whereas the 'habituators' show an attenuated tone of the sympathetic nervous system (Peters 2011). These insights may be relevant for the understanding of stress-dependent excessive calorie intake in binge-eating disorder and the impact of stress on people with type 2 diabetes.

17.2.8 Drainage and Waste Clearance

High metabolic rate and blood perfusion of the human brain necessitates adaptive solutions of drainage and clearance of metabolic products. In addition, cerebral blood flow is critically involved in the thermal regulation of the brain (Wang et al. 2016). This is of eminent importance, because the brain is extremely heat sensitive. For example, a temperature rise by only 4 or 5°C may cause seizures and convulsions, or states of confusion. As humans evolved to become endurance runners in arid environments (see Chapter 6: Nutrition, Energy Expenditure, Physical Activity, and Body Composition), the risk of overheating of the brain increased (Hublin 2005). However, humans are capable of maintaining normal brain temperature even if body temperature rises to over 40°C, for example during exercise (Hublin 2005). This is due to an effective cooling system of the venous system of the brain (Falk 1990).

Accordingly, compared to other apes, the human dural system of venous sinuses has undergone significant adaptive changes. Intracranial venous blood is largely drained by the internal jugular vein (and by the vertebral venous plexus), into which the superior sagittal sinus, the transverse/sigmoid sinus, and the occipital/marginal sinus flow (Kunz and Iliadis 2007). Blood temperature in the internal jugular vein is higher than in the carotid artery, suggesting that removal of excess heat is one critical aspect of the brain's drainage system. Upright posture (bipedalism) has shifted blood drainage to the vertebral venous plexus (Falk 1990). Importantly, there is evidence to suggest that brain cooling preceded brain enlargement. Bipedalism itself reduces the risk of hyperthermia by exposure of less body surface to the heat of the day, a scenario that appears plausible for early hominids who foraged as diurnal creatures in African savannas (Wheeler 1988). Additional support for the hypothesis that brain cooling was present before brain size started to increase in size during hominid evolution comes from the emergence and increasing frequency of emissary veins in early hominids. These emissary veins connect intracranial veins and sinuses with extracranial blood vessels, thus helping to cool the cranium, effectively akin to a brain 'radiator' (Falk 1990; Kunz and Iliadis 2007). In addition, small cortical capillaries traverse the subarachnoidal space and allow thermal exchange with the cerebrospinal fluid (CSF) (Bruner et al. 2011; Wang et al. 2016).

Other adaptive changes occurred in the orbital-temporal sinuses, which are well developed in apes, but decreased in size or even disappeared in the human lineage, as the large temporal chewing muscles became smaller (Diamond 1992). Furthermore, the nasal cavity contributes to thermoregulation (compare Section 17.4.3.4 about the role of the nasal cavity in brain cooling), and forehead sweating is particularly effective in cooling the cerebrum. (For further discussion, see Chapter 8: Skin and Integument.) The cerebral venous system is clinically important, as blood clotting may cause cerebral venous thrombosis, a life-threatening disease, occasionally occurring during pregnancy or in association with oral contraceptive medication (Silvis et al. 2017).

Another important and highly effective way of clearing the CNS of potentially metabolic products is maintained by the brain's system of perivascular tunnels, which is functionally equivalent to the peripheral lymphatic system. These perivascular tunnels are formed by astrocytic endfeet surrounding the entire CNS vasculature, including arterioles, capillaries, and venoles. The perivascular space (called Virchow–Robin space) between the wall of the vasculature and the astrocytic endfeet is filled with CSF. The flow of the CSF into the perivascular space is maintained by arterial pulsatility and respiration. Convective movement of CSF into the brain parenchyma is responsible for the clearance of metabolic products such as beta-amyloid, which are enriched in the interstitial fluid and potentially toxic to the CNS in higher concentrations. In other words, there is intensive exchange of substances between the interstitial fluid and the CSF. The latter is mainly produced in the choroid plexus of the lateral, third, and fourth ventricles, but perhaps also by filtration of fluid through the capillaries by osmotic and hydrostatic gradients. The production of CSF and metabolic clearance is aided by aquaporins, which are 'water channel' proteins located on perivascular and ependymal cells (Jessen et al. 2015). Importantly, the total volume of the CSF of about 150 ml is renewed every 6 hours, such that under normal conditions, the clearance of metabolic products is highly effective, whereby the total volume of the CSF is kept constant by efflux via the spinal and cranial nerves, as well as the olfactory bulb (Jessen et al. 2015).

In a variety of pathological conditions, including Alzheimer's disease and stroke, the elimination of metabolic products via the 'glymphatic system' (so-called due to the specific role of astrocytes) is compromised. In fact, it is intriguing that apolipoprotein E (ApoE), of which the isoform ApoE4 is a risk factor for Alzheimer's disease, appears in high concentrations in astrocytic processes around blood vessels, and seems to be produced in the choroid plexus. ApoE is necessary for the transport of cholesterol and beta-amyloid. The close anatomical relationship of ApoE synthesis and the (g)lymphatic system therefore suggests that the (g)lymphatic system is involved in the clearance of excess cholesterol and delivery of lipids to neuronal structures (Jessen et al. 2015). (For further discussion, see Chapter 3: Genetics and Epigenetics.)

Sleep seems to play a decisive role in clearance of metabolic products. In fact, among mammals, sleep is balanced against foraging time, whereby more feeding time is necessary for supplying large bodies and brains with energy at the expense of sleep (Herculano-Houzel 2015). Conversely, larger brains produce more metabolic waste, which may create selection pressures towards longer sleep duration. However, sleep requirements seem to be related to neural density such that lower neuronal density in cortical areas is associated with less sleep, possibly, because lower neural density could be associated with less metabolic waste (Herculano-Houzel 2015). Interestingly, the brain's metabolic rate decreases only slightly during sleep, while the convective flux of interstitial and cerebrovascular fluid is greatly enhanced during sleep (Herculano-Houzel 2013). Conversely, during wakefulness the action of the (g)lymphatic system seems to be suppressed, possibly mediated by the action of norepinephrine (Jessen et al. 2015). Accordingly, while sleep puts organisms at risk of predation and reduces feeding time, it seems to have an important role in the regulation of brain metabolism (Jesse et al. 2015). In addition, sleep is involved in inflammatory activity of interleukins and other immunologically active substances (Irwin and Opp 2017). Sleep deprivation, on the other hand, may cause both immunological responses such as an increase of tumour necrosis factor (TNF) in the periphery (i.e. blood) and, via adrenergic signalling, alterations is the function of the (g)lymphatic system.

Thus, the efficient elimination of potentially toxic metabolic products by the (g)lymphatic system of the CNS may be compromised by several pathological conditions, including degenerative brain disease, stroke, intracerebral haemorrhage, and traumatic brain injury, which may disrupt the convective flow of the fluid in the interstitial compartment and the CSF. Ageing is another important factor, because the production of CSF declines with age, as does the quality and duration of sleep, which together may impact on the brain's capacity to eliminate toxic metabolic products such as beta-amyloid, tau protein, and excess glutamate (Jessen et al. 2015; Irwin and Opp 2017). Figure 17.4 schematically illustrates the functioning of the (g)lymphatic system.

17.3 Microanatomical Features of the Human Brain

17.3.1 Comparative Cytoarchitecture

Aside from differences in cortical folding (gyrification) and macroscopic changes in shape of the human brain relative to other primate brains, there are a number of distinctive

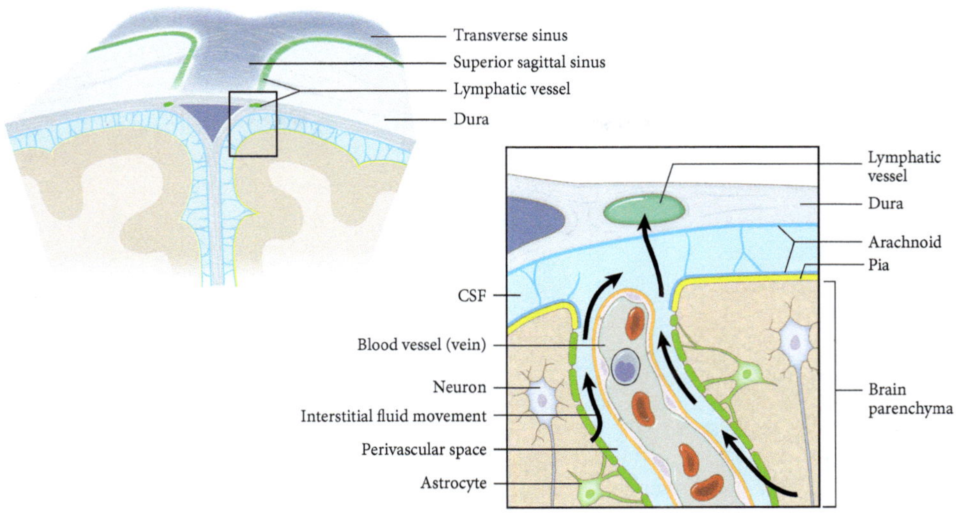

FIGURE 17.4 Schematic representation of a connection between the (g)lymphatic system, responsible for collecting of the interstitial fluids from within the central nervous system parenchyma to cerebrospinal fluid, and meningeal lymphatic vessels (mouse model).

Source: Reprinted by permission from *Nature*, 523, Antoine Louveau, Igor Smirnov, Timothy J. Keyes, Jacob D. Eccles, Sherin J. Rouhani, J. David Peske, Noel C. Derecki, David Castle, James W. Mandell, Kevin S. Lee, Tajie H. Harris, and Jonathan Kipnis, Structural and functional features of central nervous system lymphatic vessels, pp. 337–341, Figure 10, doi:10.1038/nature14432, Copyright © 2015, Springer Nature.

cytoarchitectonic features of the human neocortex that emerged in response to selective pressures unique to our species. The motor and premotor cortex of all primates is characterised by a loss of an inner granular layer IV that disappears before and after birth. Layer V of the primary motor cortex contains Betz cells or giant pyramidal cells, which possess large apical and basal dendrites, as well as numerous additional dendrites attached to the cell body (Zilles 2005). Similarly, the primary sensory cortices of primates are characterised by a six-layered 'granular' structure, named so for its conspicuous layer IV, which contains densely packed neurons receiving input from the thalamus and the lateral geniculate nucleus (LGN; vision) and medial geniculate nucleus (MGN; audition).

The cytoarchitectonic features of the hippocampus seemed to be even more conserved among primates. The Ammon's horn is subdivided into four subdivisions, of which the CA1 region is most extensively connected with the neocortex. The CA2 region comprises only a small portion interposed between the larger CA1 and CA3 regions, but it seems to be particularly relevant for social memory (Hitti and Siegelbaum 2014). The dentate gyrus is believed to be involved in adult neurogenesis. Compromised adult neurogenesis may play a role in several neuropsychiatric disorders including depression, schizophrenia, and various forms of dementia (Kempermann 2012).

Novel approaches investigating transmitter receptor densities have revealed some functionally important differences in cytoarchitecture between humans and nonhuman primates. Specifically, receptor autoradiography could demonstrate that receptors for acetylcholine, norepinephrine, dopamine, serotonin, glutamate, and GABA are heterogeneously distributed across the human neocortex and archicortex (Toga et al. 2006). The

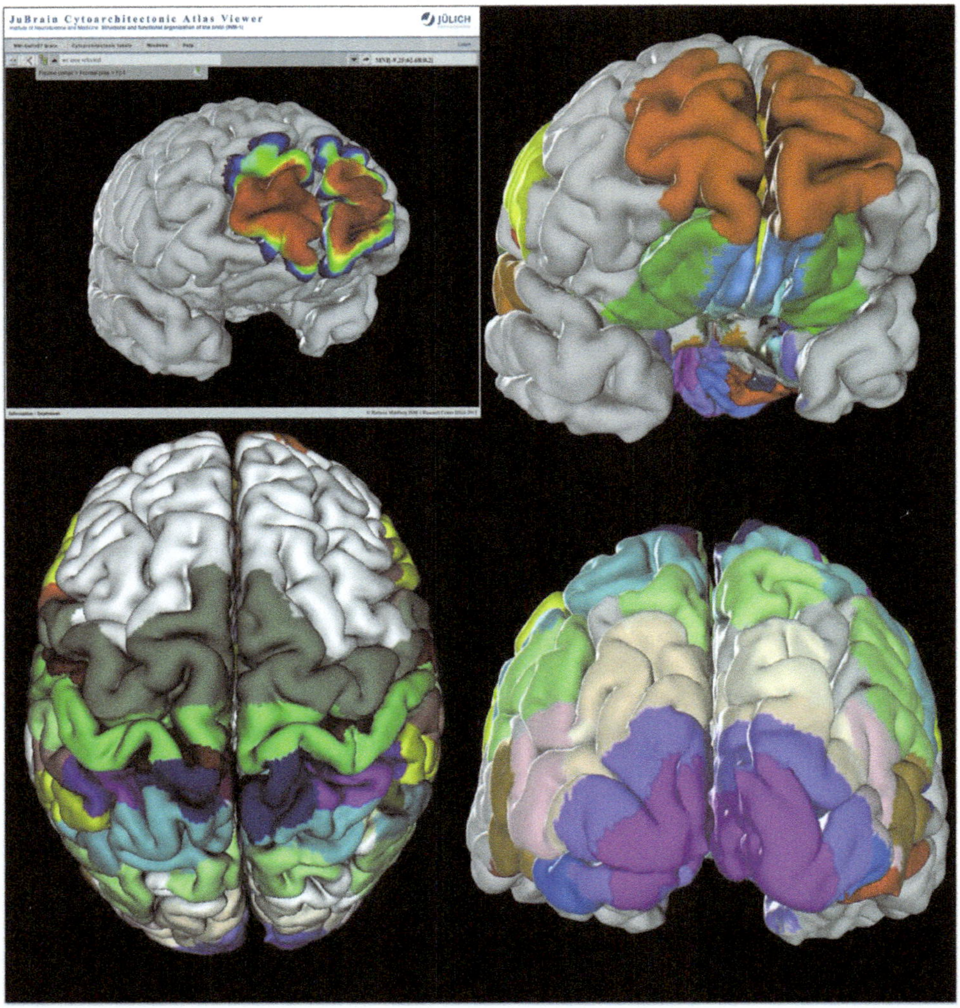

FIGURE 17.5A Probabilistic JuBrain Cytoarchitectonic Atlas. Maps are based on observer-independent definitions of areal borders and quantitative cytoarchitectonics in serial histological sections of ten human post-mortem brains. Upper left is a screenshot from the website showing probabilistic map of frontal pole area Fp1. Upper right and lower row present the maximum probability map (MPM), where each position of the reference space is associated to the area showing the highest probability. Different areas are shown by different colours.

Source: Published maps are available for download at https//www.jubrain.fz-juelich.de/apps/cytoviewer/cytoviewer-main.php. Reprinted from *Neuron*, 88 (6), Katrin Amunts and Karl Zilles, Architectonic Mapping of the Human Brain beyond Brodmann, pp. 1086–1107, Figure 1, doi.10.1016/j.neuron.2015.12.001. Copyright © 2015 Elsevier Inc.

patterns of distribution of different receptors in anatomically defined areas of the cortex produce 'receptor fingerprints' specific to individual regions of the brain. Differences between macaque and human brains concern, for example, the expression of glutamatergic kainate receptors in the dentate gyrus, muscarinic receptors in CA3, and alpha adrenergic receptors and inhibitory GABA receptors in CA1 (Zilles 2005). In addition, differences in

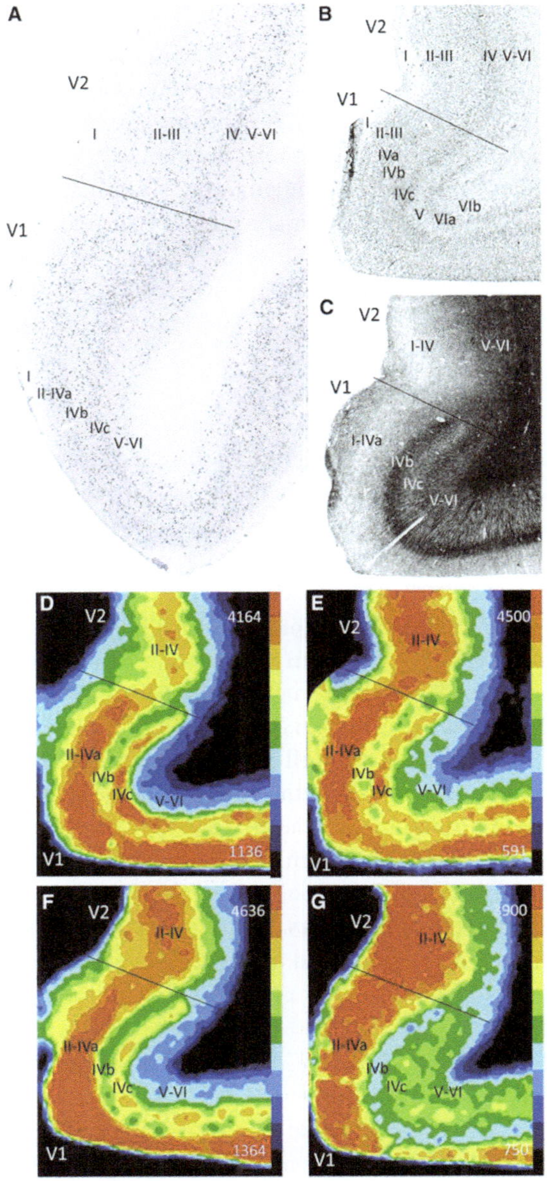

FIGURE 17.5B Corresponding areal borders across different modalities. (A) Section from the Allen Brain Atlas showing parvalbumin gene expression in neurons of human primary visual area V1 and secondary visual area V2 (https//www.alleninstitute.org, specimen RP_070313_01_C07), with delineation between V1 and V2 and laminar labelling by the authors of this volume. (B–G) Cyto-, myelo-, and receptor-architecture of human areas V1 and V2 from a different brain (Institute of Neuroscience and Medicine, INM-1, Research Centre Jülich). (B) Cell-body-stained section. (C) Myelin-stained section. (D) Agonistic binding sites of GABAA receptor labelled with $[^3H]$ muscimol. (E) Antagonistic binding sites of GABAA receptor labelled with $[^3H]$ SR95531. (F) Benzodiazepine binding sites of GABAA receptor labelled with $[^3H]$ flumazenil. (G) Binding sites of GABAB receptor labelled with $[^3H]$ CGP 54626. Parvalbumin-positive cortical neurons have their termination field in surroundings of their cell bodies, where they release inhibitory transmitter GABA. Notably, laminar density distribution of parvalbumin-positive neurons and GABA receptor binding sites are similar in both V1 and V2, with the exception of GABAB receptors, which are also present at intermediate densities in layers V and VI of V1. Roman numerals indicate cortical layers. Scale bars code receptor densities in femtomoles per milligram of protein.

Source: Reprinted from *Neuron*, 88 (6), Katrin Amunts and Karl Zilles, Architectonic Mapping of the Human Brain beyond Brodmann, pp. 1086–1107, Figure 5, doi.10.1016/j.neuron.2015.12.001. Copyright © 2015 Elsevier Inc.

receptor fingerprints pertain to the primary sensory areas and parietal association cortices such as the intraparietal lobule (Zilles 2005; Amunts and Zilles 2015). Figures 17.5A and 17.5B show modern cartographic approaches to map the human neocortex.

This novel way of brain mapping provides important information in addition to the conventional brain atlases according to Brodmann and others, and may help define more precisely the functional representation of cognitive and emotional processes in healthy subjects and patients with neurological and psychiatric disorders (Toga et al. 2006).

17.3.2 Synaptic Transmission

Neurons communicate with one another via specific junctions called synapses. Action potentials that reach the terminal end of one nerve cell, the presynaptic neuron, induce the secretion of a particular substance, a neurotransmitter, into the synaptic cleft, from where it travels to the postsynaptic neuron. Most axosomatic synapses are inhibitory, whereas the majority of axodendritic synapses are excitatory, and axoaxonal synapses are double inhibitory, and thus disinhibitory (Zilles 1987).

Neurotransmitters are stored in vesicles in the presynaptic ending and the quantity released into the synaptic cleft is related to the strength of the electric signal that opens presynaptic calcium channels. Most neurotransmitters are evolutionarily ancient substances that have been 'co-opted' for the purpose of cell–cell communication in the CNS (see Section 17.3.3 for further details). Receptor proteins on the postsynaptic membrane to which the neurotransmitters bind can excite, inhibit, or otherwise modify the excitability of the postsynaptic neuron (Guyton and Hall 2006; Raven et al. 2017a). This happens either directly via cation channels that excite the postsynaptic neuron, or via anion channels that are opened by inhibitory neurotransmitters. Another method of synaptic transmission utilises second messengers for prolonged postsynaptic neuronal changes via G-proteins that can activate metabolic process by enzymatic activity or gene transcription (Guyton and Hall 2006).

Note that a minority of synapses in the CNS are electrical, that is, the transmission of an action potential from one neuron to another via gap junctions utilises electric impulses that travel via open fluid channels (Guyton and Hall 2006). The action of neurotransmitters is terminated by reuptake mechanisms, enzymatic degradation, or diffusion. Reuptake is the most important mechanism for many neurotransmitters, because their synthesis is metabolically expensive. Thus, it is in the organism's interest not to waste such precious molecules. Some neurotransmitters produce rather short, sharply peaking action potentials at ionotropic receptors, whereas others produce much slower alterations at metabotropic receptors that last much longer (Panksepp 1998).

Comparative proteomic composition of synapses suggests that proto-synapses evolved even prior to the advent of the earliest nervous systems (Ryan and Grant 2009). Some of the protein families found in synapses such as calcium transporters and protein kinases are also found in unicellular organisms. For example, in yeast, calmodulin and calcineurin are involved in the response to environmental changes. In multicellular organisms which possess a CNS, these substances contribute to postsynaptic signalling pathways (Ryan and

Grant 2009). Several genome duplications in the chordate lineage (in comparison to invertebrates) have produced a significant expansion of synaptic gene families, including ones pertaining to glutamate and GABA receptors and postsynaptic membrane proteins (Ryan and Grant 2009; Emes and Grant 2012). In primates and humans, several genes coding for neurotransmitters have undergone positive selection, which is described in more detail in Section 17.5.

Comparative neuroanatomy has revealed that the mean number of synapses per neuron has increased from 2000 to 5600 in non-human primates and from 6800 to 10,000 synapses in humans (Changeux 2005). The number of dendritic spines of prefrontal pyramidal cells has also increased relative to the number of dendritic spines found in the primary visual cortex, probably reflecting the expansion of corticocortical connections (Changeux 2005). Roughly 80% of presynaptic terminals are located on dendrites, and 20% on the soma of the postsynaptic cell (Guyton and Hall 2006).

In addition to synaptic communication from neuron to neuron, recent discoveries have highlighted the significance of glia cells and extracellular molecules for the maintenance of structure and function of the CNS. Neuroglia comprise oligodendrocytes and astrocytes, which embryologically derive from the ectodermal plate. In the CNS, oligodendrocytes are responsible for the production of myelin sheaths around the axons (this function is maintained in the peripheral nervous system by Schwann cells). Small gaps in the myelin sheath (Ranvier nodes) help accelerate the transmission of electric impulses by saltatory conduction (Guyton and Hall 2006). A single oligodendrocyte can myelinate up to 60 axons, or repair the myelin sheath following damage, for example by stroke (Edgar and Sibille 2012). In addition to controlling saltatory conduction, oligodendrocytes help promote neuronal plasticity. In contrast to astrocytes, they are particularly vulnerable to oxidative stress due to their content of iron, and to glutamatergic toxicity (Haroon et al. 2017).

Astrocytes, too, have manifold functions in the CNS. They maintain local ion concentrations and pH homoeostasis, provide metabolites such as glucose and cholesterol, synthesise the elements of the extracellular matrix, mainly proteoglycans and other important substances (Maeda 2015), and clear metabolic 'waste' by reuptake mechanisms, including the recycling of glutamate (Nedergaard et al. 2003; Faissner et al. 2010; Haroon et al. 2017). In addition, astroglia seem to actively release neurotransmitters, foremost glutamate, steroids, lipids, neuropeptides, and growth factors. Neurotrophic and growth factors synthesised by astroglia are also believed to contribute to the maintenance of the blood–brain barrier (Nedergaard et al. 2003). These newly revealed functions assign to astroglia a dynamic role in neuronal plasticity, as well as synapse formation and regeneration. In fact, Nedergaard et al. (2003) have referred to astroglia as 'multifunctional housekeeping cells', whereas Faissner et al. (2010) have suggested redefining the classic structure of synapses as a 'tripartite' unit due to the close contact of astroglia processes with pre- and postsynaptic neuronal elements. Haroon et al. (2017) distinguish between at least three types of astrocytes. Protoplasmic astrocytes are primarily involved in glutamate clearance and protection of the extrasynaptic space from a 'spillover' of excess glutamate by converting it into glutamine. Note that the mechanism of glutamate transport into astrocytes is immature in young infants, which may explain the elevated rate of seizures during this developmental period (Campbell 2010).

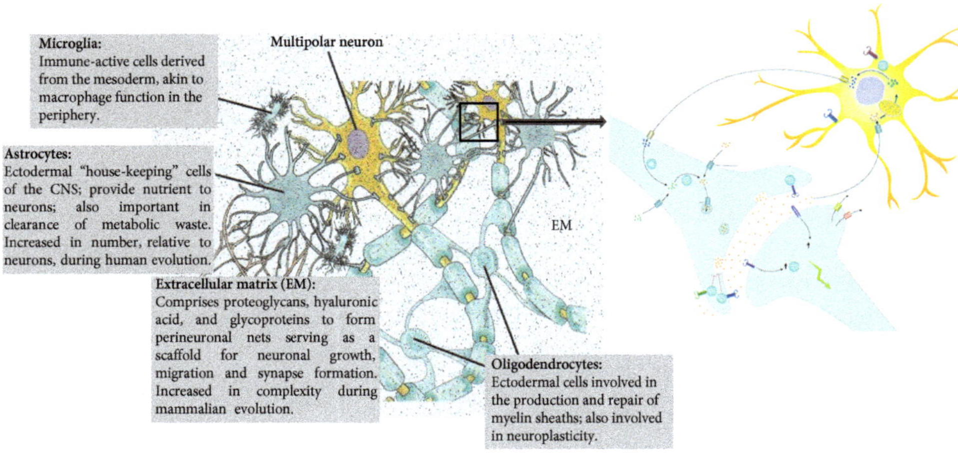

FIGURE 17.6 Interplay of glia cells with neuronal structures showing the 'tripartite synapse' (enlargement on right).

Fibrous astrocytes are engaged in the protection of the nodes of Ranvier, while radial astrocytes support neuronal migration (Haroon et al. 2017).

In pathological conditions such as stroke or inflammation, astrocytes can become phagocytotic, whereas chronic activation can lead to both regenerative and apoptotic activity, whereby senescence increases their reactivity to inflammatory processes (Haroon et al. 2017; Niraula et al. 2017). Figure 17.6 illustrates the different roles of glia cells and the extracellular matrix, and their impact on neuron function.

From an evolutionary perspective, the importance of astroglia for the maintenance of complex neuronal networks is reflected in an increasing number of astrocytes per neuron over evolutionary time. For example, the neuron to astroglia ratio is about 6:1 in nematodes and about 3:1 in small rodents, whereas in humans the ratio is 1:1.4. That is, there are more astroglia than neurons in the human CNS (Nedergaard et al. 2003). This shift in neuron to astrocyte ratio cannot solely be explained by metabolic demands (which is relatively similar among higher vertebrates), but is presumably related to the regulatory function of astroglia with regard to the strength of synaptic transmission, synaptogenesis, and neurogenesis (Nedergaard et al. 2003).

Specifically, synaptic plasticity seems to critically depend on the extracellular matrix. The extracellular matrix contains proteoglycans, hyaluronic acid, glycoproteins, and other large proteins. Extracellular matrix is present in all organs, but its composition differs between tissues. Comparative genetics has revealed that the extracellular matrix has increased in complexity over evolutionary time. Mammalian extracellular matrix contains more than 300 different proteins, over 30 proteoglycans, and more than 200 glycoproteins, some of which are very ancient, while others appeared more recently in chordates, such as reelin, fibronectin, and the tenascins (Hynes and Naba 2012). In the brain, chondroitin sulfate proteoglycans comprise the most abundant component of the extracellular matrix. As already mentioned, molecules that compose the extracellular matrix are synthesised by glia cells, and the synthesis of these large proteins is metabolically expensive. Proteoglycans and glycoproteins including the tenascins form so-called perineuronal nets that serve as a scaffold for neuronal growth and synapse formation. Proteoglycans are also pivotal in the regulation of neuronal migration in the developing brain (Maeda 2015). Perineuronal nets are believed to buffer against oxidative stress and to be involved in the synthesis of myelin sheaths. Animal models suggest that the extracellular matrix contributes to the establishment and maintenance of long-term memories such as conditioned fear (reviewed in Faissner et al. 2010).

Abnormalities in perineuronal nets and mutation of genes coding for various components of the extracellular matrix are involved in several neuropsychiatric disorders. For example, it has been shown that the density of perineuronal nets in the amygdala, the entorhinal cortex, and the prefrontal cortex is diminished in schizophrenia (Mauney et al. 2013). Along similar lines, alterations of genes coding for different proteoglycans have been associated with an increased risk for schizophrenia, bipolar disorder, autism, and intellectual disability (reviewed in Maeda 2015).

17.3.3 Neurotransmitters

All neurotransmitters are ancient molecules that have existed long before nervous systems evolved. Many of these substances evolved to promote bacterial growth, to which plants evolved counter-strategies utilised by animals through pharmacophagy. A broader and perhaps more appropriate term for neurotransmitters may, therefore, be 'biomodulators' (Roshchina 2010; St John-Smith et al. 2013). The long evolutionary history of biomodulators is underscored by the fact that acetylcholine, the biogenic amines (epinephrine, norepinephrine, dopamine, and serotonin), amino acid transmitters (e.g. glycine and adenosine), enzymatically modified amino acids such as glutamate and GABA, and various neuropeptides have already been found in protozoa (Roshchina 2010). (For further discussion, see Chapter 2: Cellular Signalling Systems.)

The synthesis of neurotransmitters in the vertebrate brain mainly takes place in nuclei located in phylogenetically old parts of the brain. Developmentally, neurotransmitter receptors emerge before birth. Postnatally, they are overexpressed with a peak around 2–4 months of age, and receptor density subsequently declines to adult levels similar across all cortical areas at age 3 years. Sections 17.3.3.1–17.3.3.6 provide an overview of the major (known) neurotransmitters. There are, however, many more substances that may act as neurotransmitters, many of which are probably yet to be discovered. Likewise, the number of different receptor types to which neurotransmitters bind is probably incomplete, and new discoveries will modify the present picture substantially.

17.3.3.1 *Acetylcholine*

The largest acetylcholine-producing nuclei are found in the basal forebrain (i.e. the nucleus basalis of Meynert). They mainly control hippocampal and cortical functions. Other nuclei in the midbrain are involved in the regulation of thalamic and hypothalamic functions. Acetylcholine is synthesised from the nutrient choline and acetyl coenzyme A in the presence of choline acetyltransferase, and degraded enzymatically by acetylcholine esterase. Acetylcholine binds to two types of receptors, called nicotinic and muscarinic, the functions of which are influenced by cholesterol, which helps stabilise the cell membrane. Nicotinic receptors are found in skeletal muscle cells, whereas muscarinic receptors modify visceral parasympathetic activity (Panksepp 1998). Developmentally, cholinergic projections from the basal forebrain reach the cortex in the early prenatal period.

Cognitive dysfunction in the context of Alzheimer's disease is related to a deficit in cortical acetylcholine due to a progressive degeneration of the nucleus basalis of Meynert. Therapeutically, one strategy to ameliorate the symptoms of memory loss and poor attention is to prevent the degradation of acetylcholine by acetylcholine esterase inhibitors (which may have significant side-effects on the intestines and cardiac function).

17.3.3.2 *Catecholamines*

Catecholamines are synthesised from the amino acid tyrosine by hydroxylation. This enzymatic step produces L-DOPA, which can cross the blood–brain barrier. Within the CNS, L-DOPA is transformed into dopamine by decarboxylation, and another hydroxylation produces norepinephrine. Epinephrine results from methylation of norepinephrine, though this is more important outside the brain. Ontogenetically, monoamine-synthesising neurons or clusters of neurons in the brainstem are generated in the first trimester of embryogenesis.

Catecholamines have widespread functions and bind to several receptors that are differentially distributed across the brain. Norepinephrine binds to alpha 1 and 2 receptors, as well as to beta 1 to 3 receptors. Dopamine binds to five different dopamine receptor types that are functionally distinct. They segregate into two classes, D1-like and D2-like receptors. The former stimulate second-messenger systems, while the latter act in inhibitory ways on second-messenger systems (Stahl 2017).

The action of catecholamines is mainly terminated by reuptake into the presynaptic neuron. In addition, catecholamines are enzymatically degraded by catechol-O-methyl transferase (COMT), as well as by monoamino-oxidase (MAO-A and MAO-B), both of which are encoded by polymorphic genes. These allelic variations determine the speed at which catecholamines are degraded. Either allelic variant is associated with functional advantages and disadvantages, such that they are balanced in prevalence. The 'fast' version of the gene coding for COMT is putatively associated with increased risk for psychosis, at least in some populations (González-Castro et al. 2016).

Norepinephrine-synthesising cells are abundant in the locus coeruleus from which pathways connect with the neocortex, hypothalamus, cerebellum, brainstem, and the spinal cord. Norepinephrine increases the signal-to-noise ratio such that sensory input is filtered according to its biological relevance to the organism. It is thus involved in regulating arousal and attention. In human psychiatric disorders, deficiency of norepinephrine is, for example, thought to be involved in attention deficit hyperactivity disorder (ADHD), somatisation, lowered pain threshold, and impulsivity.

Dopamine synthesis takes place in the pars compacta of the substantia nigra from which dopaminergic projections are sent to the caudate nuclei. Dopamine deficiency in this circuit causes Parkinson's disease. Dopaminergic neurons are abundant in the ventral tegmental area connecting to the nucleus accumbens in the ventral striatum and the cortex via mesolimbic and mesocortical pathways. The former circuit is crucial for motivated behaviour including reward anticipation. That is, dopamine turnover increases in the ventral striatum when a reward is expected, whereas it decreases upon receiving the reward. Projections to the cortex are involved in motor arousal. A reduced dopaminergic activity in the frontal cortex may produce avolition and apathy. Excessive dopamine in the frontal cortex, by contrast, may lead to stereotyped behaviours that are carried out repetitively and purposeless, which may be part of drug-induced states (Panksepp 1998). Mesolimbic overactivity of dopamine is thought to be involved in the causation of positive symptoms associated with schizophrenia, whereas a prefrontal deficit in dopamine causes negative symptoms.

17.3.3.3 *Serotonin*

Serotonin is synthesised in a two-step process from the amino acid tryptophan via 5-hydroxytryptophan, which can cross the blood–brain barrier and is then transformed into serotonin by decarboxylation (another important derivative of tryptophan is melatonin, which in plants acts as an antioxidant, and in animals is involved in sleep regulation; Azmitia 2010). Serotonin is produced in the raphe nuclei of the brainstem, which have maintained the same location in the vertebrate brain perhaps for 500 million years. In primates, serotonergic neurons occur in clusters, from where ascending axons project to the lower brainstem and spinal cord, and to the striatum, hypothalamus, amygdala, hippocampus, and cortex. In contrast to plants, the ability to synthesize tryptophan was lost in animal evolution. Accordingly, tryptophan is an essential amino acid that animals need to take up orally (Azmitia 2010).

Serotonin controls many body functions, including food intake and digestion, sexual behaviour, aggression, and explorative behaviour. About 95% of the body's serotonin is in the intestines. In the brain, serotonin modulates the response of neurons to other neurotransmitters rather than exerting an excitatory function itself, except in pyramidal neurons in the cerebral cortex. There are at least 16 types of serotonin receptors that evolved by a series of gene duplications. The different serotonin receptors are distributed across different brain regions in different densities. The reuptake of serotonin from the synaptic cleft into the terminal axon is under control of a serotonin transporter gene. The transporter gene is controlled by a DNA promoter sequence, which evolved in primates some 40 million years ago (mya). Variations of this promoter gene in humans are associated with differences in personality traits such as anxiousness, impulsivity, hostility, and depression-proneness.

Serotonin deficiency is discussed with regard to a broad spectrum of emotional disorders including pathological anxiety, obsessive-compulsive disorder, depression, and eating disorders. Interestingly, serotonergic activity in primate brains is associated with social status as well as explorative and aggressive behaviour (Suomi 2003). Experimental evidence suggests that individuals with low social status have lower serotonin concentrations, as measured using its main metabolite, 5-hydroxyindoleacetic acetate (5-HIAA). Moreover, individuals with low serotonin turnover show a greater sensitivity towards reward and risk-taking, and it is well established that humans with low levels of serotonin are more likely to behave aggressively towards self and others. Conversely, aggression, impulsivity, and antisocial behaviour have been linked to low-activity variants of both MAO-A and COMT, which contribute to the enzymatic degradation of serotonin. Thus, high levels of dopamine, norepinephrine, and serotonin put individuals at risk of behaving aggressively, particularly if the genetic disposition is associated with adverse experiences during early childhood such as abuse or neglect. Importantly, the indole structure of tryptophan has light-absorbing properties. It is therefore straightforward to speculate about the interaction between serotonergic (and melatonergic) action and exposure to sunlight with regard to mood regulation and sleep (Azmitia 2010).

17.3.3.4 *Glutamate and GABA*

Glutamate is synthesised from the amino acid glutamine. Glutamate is the major excitatory transmitter and globally distributed across the brain. It binds to three major receptors:

N-methyl-D-aspartic acid (NMDA), kainate, and alpha-amino-3-hydroxy-5-methylisoxazole-4-proprionic acid (AMPA). Glutamate is involved in the control of thalamic input to the cortex and probably contributes to learning processes and consciousness. It has been experimentally shown that the administration of glutamate antagonists such as ketamine may produce psychotic symptoms, and that agonists at the NMDA receptor such as glycine may reduce psychotic symptoms (Panksepp 1998).

GABA is the most abundant inhibitory neurotransmitter in the brain. It is synthesised from glutamate in a single step through decarboxylation. GABA binds to $GABA_{A-C}$ receptors that are preferentially localised at the soma and proximate axon of GABAergic interneurons, which makes inhibitory control most efficient. In the human brain, GABA receptors are particularly abundant in the primary sensory cortex, Heschl's gyrus (temporal cortex), and the anterior cingulate cortex (ACC). Both glutamatergic projection neurons and GABAergic interneurons are largely generated before birth. GABAergic substances effectively control cortical and subcortical excitation. Benzodiazepines, which bind to $GABA_A$ receptors, reduce anxiety and aggression and suppress pathological excitation of the brain. These substances are widely used for the treatment of panic attacks, withdrawal symptoms, potentially life-threatening catatonic syndromes, and epileptic seizures (Panksepp 1998).

17.3.3.5 *Neuropeptides*

Neuropeptides are synthesised in the hypothalamus, the pituitary gland, and other areas of the brain. Neuropeptides, though often simple in structure, are metabolically expensive products. Neuropeptides act as modulators of neurotransmitter activities rather than exerting direct excitatory or inhibitory effects. Beta-endorphin, for example, has a calming effect and induces a feeling of pleasure. It binds to opioid receptors in the brain and reduces separation distress in young animals. It is therefore conceivable that the artificial stimulation of endogenous opioid receptors by illicit drugs helps overcome negative feelings resembling separation distress (Panksepp 1998).

Oxytocin, which is produced under the control of oestrogen in the nucleus paraventricularis of the thalamus and nucleus supraopticus and stored in the pituitary gland, is involved in parturition and lactation. (For further discussion, see Chapter 16: Sexuality, Reproduction, and Birth.) Oxytocin has profound effects on social behaviour. It reduces aggression (except in females with young offspring) and promotes parental bonding and partner bonding (Donaldson and Young 2008). Moreover, oxytocin improves 'mindreading', empathy, and trust, although such prosocial effects very likely depend on the quality of early attachment experiences and experience of childhood adversity (Domes et al. 2007; Bartz et al. 2011; Brüne 2016). Oxytocin is released during orgasm in large quantities, and has been found to facilitate (at physiological doses) social memories and empathy. Oxytocin is more abundant in female brains, and is released upon infant suckling, a process that strengthens the mother–child dyad (Feldman 2012). Oxytocin receptors are highly expressed in the nucleus basalis of Meynert, pons, amygdala, and nucleus accumbens (Hammock and Young 2006; Ross et al. 2009), and it interacts with dopamine, serotonin, and the opioidergic system in manifold ways (Walker and Glone 2013), whereby the interaction of oxytocin and serotonin seems to be crucial for the experience of social reward (Dölen et al. 2013).

Vasopressin, which differs from oxytocin in only one out of nine amino acids, is more prevalent in the male brain and its synthesis is facilitated by testosterone. Vasopressin promotes male sexuality and aggression. In contrast to oxytocin, vasopressin peaks during sexual arousal (Panksepp 1998). The role of these peptides in psychopathology is only emerging, but it is most likely that disruptions of early infant attachment have profound effects on the expression of these social affiliation-promoting neuropeptides (Meyer-Lindenberg et al. 2011; Hammock 2015).

Another class of neuropeptides called neurotrophins controls the proliferation, migration, and replacement of damaged neurons by regulating neural stem cell proliferation. Neurotrophins bind to tyrosine kinase receptors. Nerve growth factor (NGF), for example, is involved in the differentiation and protection of nerve cells. Similarly, the brain-derived neurotrophic factor (BDNF) is a peptide that is active in the hippocampus, basal forebrain, and cortex. It exerts protective effects and is perhaps the most important neurotrophin for neurogenesis from neural stem cells. BDNF is involved in cell migration during ontogeny. Of note, stress has been shown to reduce BDNF levels in the brain, such that BDNF deficiency has been suggested to play a role in several psychiatric disorders including depression, obsessive-compulsive disorder, autism, and Alzheimer's disease.

Neuregulins are proteins that contribute to the development of Schwann cells involved in myelination of neurons, and to the survival of oligodendrozytes (Panksepp 1998). Neuregulin1 is perhaps involved in the pathogenesis of schizophrenia by down-regulating glutamatergic NMDA receptor activity in the prefrontal cortex (Sei et al. 2007).

17.3.3.6 *Endocannabinoids*

It has long been known that ingestion or inhalation of the plant *Cannabis sativa* exerts psychotropic effects. Endogenous G-protein-coupled cannabinoid receptors, CB1 and CB2, are stimulated by anandamide and 2-arachidonoylglycerol (2-AG), which acts as a retrograde messenger and down-regulates the activity of other neurotransmitters. Endocannabinoid receptors are widely distributed in the nervous system of vertebrates. CB1 receptors are mainly located on GABA-ergic neurons in the basal ganglia, hippocampus, neocortex, spinal cord, and periphery (including the heart), which may account for the dampening effect of cannabinoids on movement (catalepsy), memory (impairment), nociception (analgesia), and the cardiovascular system (bradycardia and hypotension via blockade of noradrenaline release; Elphick and Egertová 2001). A smaller proportion of CB1 receptors are located on glutamatergic terminals. Activation of these receptors seems to be associated with neuroprotective effects (Chiarlone et al. 2014). Animal studies suggest that experimental up-regulation of CB1 receptors in the medial prefrontal cortex of the rat induces marked changes in social behaviour and impairs cognitive flexibility (Klugmann et al. 2011). In addition, anandamide has been shown to be involved in the reward processing of social play in rats (Trezza et al. 2012), which is consistent with findings from imaging studies in humans showing that genetic variation of the CB1 receptor modulates physiological responses to happy faces (Chakrabarti and Baron-Cohen 2011).

From a clinical perspective, it is interesting that delta-9-tetrahydrocannabinol (THC) is the main psychoactive component of *C. sativa*, which binds to CB1 receptors. The widespread abuse of THC may be linked to its calming effect, and perhaps its influence on the ANS, though chronic THC abuse may increase the risk for psychosis in (genetically) vulnerable individuals (McLoughlin et al. 2014). Moreover, chronic stress seems to cause down-regulation of CB1 receptors, whereby anandamide activates the HPA axis, while 2-AG dampens the stress response (Morena et al. 2016).

17.3.4 Microglia and Innate Brain Immunity

The term microglia is, at first sight, misleading because it suggests a common origin with oligodendrocytes and astroglia. However, microglia derive from the embryonic (myeloid) mesoderm and migrate to the neural plate to form the resident immune cells of the CNS (Haroon et al. 2017). Functionally, microglia have similarities with macrophages in the periphery, though microglia reside in the brain parenchyma and act independent of circulation (Haroon et al. 2017). Microglia are highly sensitive to inflammatory signals and thus survey the entire CNS. Once microglia detect microbial antigens, they transform into an activated amoeboid state, whereby pathogenic molecules bind to toll-like receptors on the surface of microglia. This elicits an inflammatory cascade associated with the secretion of interleukins, interferon, TNF, and chemokines, attracting monocytes from the bloodstream. Aside from this pro-inflammatory pathway, there seems to be an alternative activation pattern, which is associated with the synthesis of anti-inflammatory cytokines, phagocytic activity restricted to damaged tissue and misfolded proteins, and neuroprotection (reviewed in Haroon et al. 2017).

Microglial priming occurs if a chronic state of increased inflammation is combined with an acute inflammatory stimulus, which can have multiple causes, including traumatic brain injury, systemic inflammation, ageing and neurodegeneration, and a number of neuropsychiatric disorders such as depression and schizophrenia (Hodes et al. 2015; Fleshner et al. 2017; Haroon et al. 2017; Miller and Goldsmith 2017; Niraula et al. 2017).

Microglia are a key player in synaptic pruning, whereby animal models suggest that early life stress and intrauterine infection can cause aberrant pruning. In physiological conditions, astrocytes have the potential to limit the pro-inflammatory response of microglia by secreting anti-inflammatory cytokines, though senescence can inhibit the counteraction of astrocytes (Haroon et al. 2017). Caloric restriction and physical exercise in particular can attenuate the age-related microglia proliferation and priming, hence dampening the pro-inflammatory disposition associated with senescence (Niraula et al. 2017). (For further discussion, see Chapter 6: Nutrition, Energy Expenditure, Physical Activity, and Body Composition.) Of note, in a wide range of species including rodents and primates, chronic psychosocial stress such as repeated social defeat produces a (sterile) pro-inflammatory microglia response (Weber et al. 2017). Similarly, patients with depression exhibit increased peripheral levels of cytokines including interleukin 1-beta, interleukin-6, and TNFα, which can pass the blood–brain barrier and directly act on neurons and glia (Hodes et al. 2015), possibly because stress is assumed to cause leakage of the blood–brain barrier. Another pathway for immunologically active molecules to enter the brain is via the area postrema, where no blood–brain barrier exists (note that nutritional toxins causing vomiting and nausea use this rapid route). Similar to depression, obesity has been found to be associated with a pro-inflammatory state (Capuron et al. 2017). Finally, early life stress including childhood trauma has been identified as a vulnerability factor for excessive inflammatory responses to social stress later in life and may impact on microglia priming (Danese and Lewis 2017).

Moreover, since ageing is associated with a pro-inflammatory disposition, stress in aged individuals can potentiate pro-inflammatory effects, which, however, can be mitigated by caloric restriction and physical exercise (Niraula et al. 2017). It is unclear whether the counteracting effect of exercise on inflammation and microglia activation is partly related to reward; however, animal studies have shown that reward in general has a booster effect on innate and adaptive immunity (Ben-Shaanan et al. 2016).

Thus, experience of adversity can have profound effects on immunity, including the innate brain microglial immune system, and can activate ancient evolved defensive responses to environmental threats that may run out of control in pathological conditions. Importantly, the cross-talk between the gut microbiota and the brain in relation to immune function has only begun to be better understood (Dinan and Cryan 2017). Accumulating evidence from studies of germ-free laboratory animals (mainly mice) has demonstrated that the absence of a functional gut microbiome is associated with an increased number of immature microglia cells and aberrant myelination of prefrontal neurons, though the exact mechanisms by which the microbiota influence microglia maturation are unknown (Hoban et al. 2016; Fung et al. 2017). Moreover, the gut microbiota influences astrocyte activity, possibly by metabolising dietary tryptophan to indole derivatives. Whether findings from experimental studies in rodents can be transferred to the human condition needs to be explored in future studies. In any event, the gut microbiota seems to impact on a wide range of condition, including stroke, autoimmune processes, and neuropsychiatric disorders such as anxiety disorders, depression, and dementia (reviewed in Fung et al. 2017). (For further discussion, see Chapter 13: Digestive System.)

Importantly, research in humans supports the notion that inflammation can have a profound impact on one's social behaviour. Specifically, a pro-inflammatory state seems to foster certain threat-related behaviours including increased sensitivity to rejection and social exclusion, but also help-seeking behaviour of close others, while minimising contact with strangers (Eisenberger et al. 2017). In fact, it makes perfect sense from an evolutionary perspective to seek proximity of significant others in times of increased vulnerability. Conversely, social stressors including parental separation, bereavement, and intense social competition are associated with a pro-inflammatory state, whereby individual differences in rejection sensitivity seem to influence the magnitude of inflammatory responses to social stress (Eisenberger et al. 2017). Interestingly, deliberate prosocial behaviour can attenuate pro-inflammatory gene expression, suggesting that social belongingness may have beneficial effects on immunity and health (Nelson-Coffey et al. 2017). These insights have profound implications for the understanding of psychological well-being as well as psychiatric conditions (e.g. depression, social anxiety disorder, borderline personality disorder) in which heightened social sensitivity is a key factor of pathophysiology (Eisenberger et al. 2017). Figure 17.7 schematically shows the interaction between stress at different life stages, brain immunity, and social behaviour.

17.3.5 Neuronal Specialisation?

The recent finding that the human brain is a scaled-up version of other primate brains in its neuronal composition (Azevedo et al. 2009; Herculano-Houzel 2009) fits the current scenario of discoveries that place the human brain as an unexceptional version of other large primate brains.

Single cell recordings in macaque monkeys have revealed that neurons in the middle portion of the temporal lobe, particularly in the superior temporal sulcus (STS), selectively

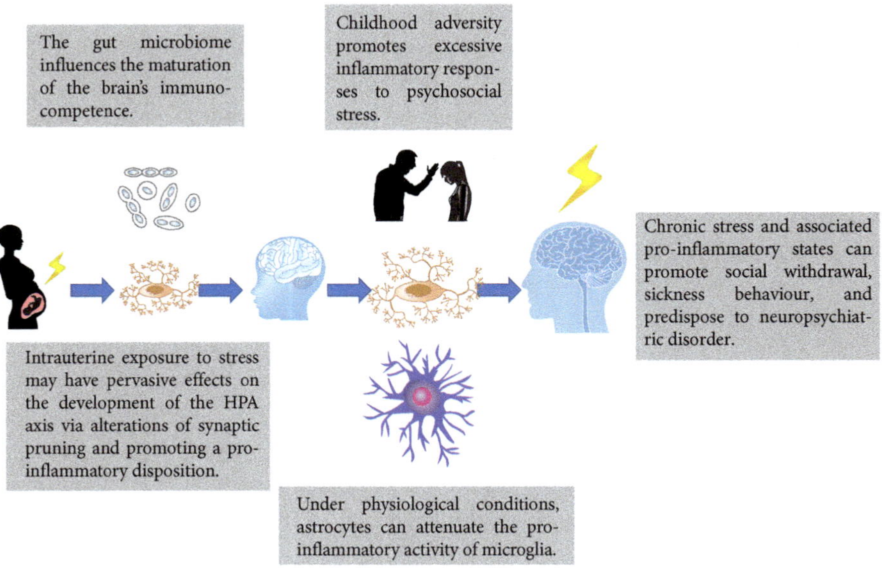

FIGURE 17.7 Schematic and simplified illustration of the impact of stress on brain immunity and function at different developmental stages. HPA, Hypothalamic–pituitary–adrenal.

fire when monkeys observe the gaze direction of other monkeys. These neurons are also active when animals observe goal-directed behaviour. In humans, a homologous area of the temporal lobe is activated by observation of seemingly purposeful movements of inanimate objects (as opposed to random movements), even when still photographs depict 'implied' motion (Kourtzi and Kanwisher 2000). Activity in parts of the STS, therefore, is linked in humans and other primates alike to the observation of intentional movements. Although this does not imply conscious awareness, the representation of 'intentions' is certainly a critical aspect of complex social interactions.

The temporal and the frontal lobes of primates also contain a specific type of cell called 'mirror neurons' due to their unique quality to discharge both during the execution of a certain hand or mouth action and by the mere observation of the same behaviour carried out by another individual, even if the terminal part of the movement is hidden from observation (Umiltà et al. 2001; Rizzolatti and Craighero 2004). This suggests that mirror neurons provide a basic mechanism for predicting behaviour and imitating the actions of others (Gallese and Goldman 1998). Mirror neurons have been found in great density in the ventral premotor cortex (area F5) of monkeys, an area that is probably homologous to Broca's speech area in humans (Rizzolatti et al. 2002). Human language might therefore have evolved from the ability to use gestures for communication (Corballis 2003). Interestingly, the mirror neuron system is elegantly accessible by measuring the suppression of alpha and beta bands of the electroencephalogram, referred to as 'mu rhythm' suppression (due to similarities of the alpha waves with the Greek letter 'mu') (Pineda 2005), which may open new avenues for the study of such basic social cognitive processes in neuropsychiatric conditions, including the role of reward mechanisms (Brown et al. 2013; Brown and Brüne 2014).

Another brain area that was previously hailed as having undergone human-specific modification is the ACC. The ACC receives input from the motor cortex and the spinal cord, from the ipsilateral prefrontal cortex, and from the thalamus and brainstem nuclei. It is highly heterogeneous in terms of its cytoarchitecture and functional organisation. The ACC is believed to serve as an important mediator of motor control, cognition, and arousal, as well as an inhibitory control device to suppress impulsive reactions in favour of 'rational' decisions (Devinsky et al. 1995; Allman et al. 2001). Bilateral damage to the ACC may produce akinetic mutism and other complex abnormalities including disinhibition of primitive behaviours. The human ACC inconsistently forms a paracingulate sulcus that is present in only 30–50% of individuals and is perhaps still under selection pressure (Paus et al. 1996). The ACC contains a spindle-shaped cell type (thus somewhat misleadingly termed 'spindle cells', also known as 'von Economo neurons' (VENs)) that was initially thought to be unique to apes and humans; however, investigation of a wider range of species showed that VENs exist in other mammalian species, including whales (Butti and Hof 2010). VENs are also densely located in the anterior part of the insular cortex (AI) (Nieuwenhuys 2012) and in a more scattered distribution in the dorsolateral prefrontal cortex (Fajardo et al. 2008). They mature fully after birth and seem to be more abundant in the right hemisphere (Allman et al. 2011).

Although the exact function of VENs is not yet known, there is some evidence to suggest that they play an important role in self-awareness (interoception), empathy, and emotion processing, including disgust and more complex emotions such as guilt and shame (Craig 2003; Allman et al. 2011). Both ACC and AI are sensitive to social exclusion (Eisenberger 2012; Powers and Heatherton 2012). Interestingly, the density of VEN in the ACC has been found to be reduced in several neuropsychiatric conditions (Allman et al. 2010), including schizophrenia (Brüne et al. 2010), frontotemporal and Alzheimer's dementia, and possibly autism (reviewed in Butti et al. 2013). Conversely, the density is apparently increased in the ACC of psychiatric patients who committed suicide, which may justify speculations about the functional role of VENs in regard to complex emotions (Brüne et al. 2011). There is also preliminary evidence to suggest that VENs are more vulnerable to oxidative stress than adjacent pyramidal cells, as the former accumulate more lysosomal aggregates, particularly in neuropsychiatric disorders such as schizophrenia (Krause et al. 2017a) (Figure 17.8).

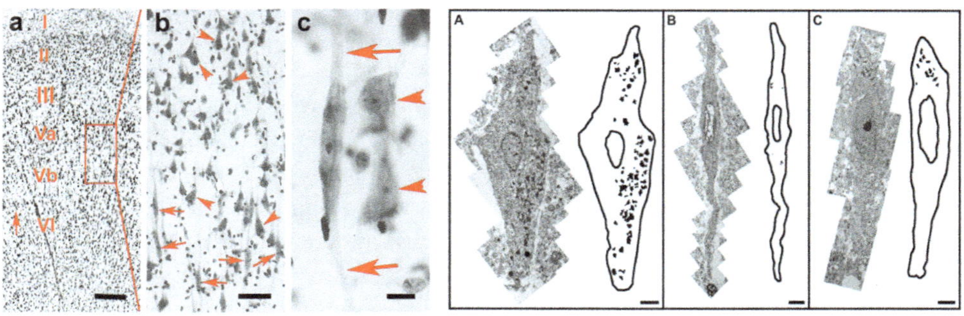

FIGURE 17.8 Von Economo neurons (VENs) in layer V of the anterior cingulate cortex (left panel, a–c). (b) and (c) are enlargements of (a) and (b), respectively. Right panel shows electron microscopic images depicting lysosomal aggregations in a VEN from post-mortem tissue of a patient with schizophrenia (A), a patient with bipolar disorder (B), and a psychologically unaffected individual (C).

17.4 Evolution of Sensory Systems

17.4.1 General Remarks

The evolution of CNS is intricately linked to the need of organisms to process information from the internal and external environment, which helped the organism to avoid hazards (toxicity, predation), find food, or identify potential mates.

Multicellular organisms possess cells that are exposed to the external environment, while other cells lie within the body. Thus, sensory cells must have evolved properties for exteroception or interoception. Sensory organs, by definition, form macroscopic structures composed of neuronal as well as non-neuronal cells. Interestingly, sensory organs and other structures seem to have evolved from common ancestral structures. For example, sensory organs may share attributes with endocrine formations, as evidenced by the common expression of ancient regulatory genes. The adenohypophyseal part of the pituitary gland has probably evolved from an external chemosensory organ. Gustatory receptors likely evolved from cutaneous receptors (Hodos and Butler 2001), and so forth. Another source of evidence for common ancestry of sensory organs and non-neural structures comes from observations of diseases that may, for instance, affect the ear and the kidney (reviewed in Jacobs et al. 2007). Even the gut has sensory properties, including classic taste receptors, nutrient receptors, chemical and mechanoreceptors, and other extrinsic and intrinsic sensory neurons which form an integral part of the gut defence system (Furness et al. 2013). (For further discussion, see Chapter 13: Digestive System.)

Traditionally, exteroceptors (stimulated by cues from the external environment) are distinguished from interoceptors (stimulated by internal including visceral processes). Thus, exteroception includes vision, audition, taste, smell, motion, and gravity (as well as other sensory qualities that humans do not possess, such as electricity, magnetism, and heat), whereas interoception comprises touch, pain, vibration, temperature, muscle tension, and posture, as well as blood chemistry and pressure (Raven et al. 2017b). Receptors involved in vision (as well as electricity and magnetism) process electromagnetic stimuli, while taste and smell typically depend on the processing of chemical information. Audition, touch, vibration, and gravity are sensory qualities that make use of mechanical information. Mechanoreceptors and chemoreceptors were probably the earliest receptor types (Hodos and Butler 2001). In any event, the traditional conceptualisation suggesting that humans possess five senses ignores the fact that important information is processed by other senses, largely outside conscious awareness, however. Likewise, in contrast to traditional medical textbooks, comparative anatomists distinguish at least 26, rather than 12, cranial nerves involved in the transmission of stimuli from sensory organs to the brain (Hodos and Butler 2001).

It is of utmost importance to emphasise that different sensory modalities evolved in close connection to one another, though for didactical reasons they are described in separate sections. In humans, for instance, vision, audition, touch, and smell are all involved in the processing of social information and may process a huge amount of information about an individual encounter at a time, the integration of which into a coherent impression of the other individual requires fine-tuned and coordinated processing in a cross-modal fashion.

Sensory specialisation, in contrast, may incur costs to the acuity of other senses, which in our own species may apply to trichromatic vision and smell (Hodos and Butler 2001; Gilad et al. 2004). By and large, the evolution of new sensory modalities is regulated by the expression of structural and regulatory genes, often involving gene duplication (Hodos and Butler 2001).

All sensory systems seem to follow very similar organisational principles. Receptor cells project to bipolar neurons which in turn connect to first-order multipolar neurons. The first-order multipolar neurons project either to the thalamus via a direct route (called the lemniscal pathway) or to the optic tectum via an indirect route (the collothalamic pathway). The only exception to this is the olfactory system which projects directly to the telencephalon (see Section 17.4.2.4). The formation of sensory pathways and sensory nuclei is under the control of cell adhesion molecules, among which the group of cadherins seems to be involved in aggregation of neuroblasts in the developing brain (Hodos and Butler 2001).

Another common feature of the senses is their somatotopic representation in the optic tectum and in specialised cortical areas of the brain, called isomorphic representation (Hodos and Butler 2001). However, there is remarkable plasticity of the cortical representation of the senses, as has been demonstrated in cases of sensory deprivation. For example, mole rats live underground in complete darkness, and although born with eyes, they lose vision as their eyes degenerate. Interestingly, their visual cortex becomes increasingly responsive to auditory stimuli and the auditory cortical area 'invades' the visual cortex area as the animal develops (Hodos and Butler 2001). Similar occupation of visual sensory cortex areas has been observed in humans deprived of vision (Guerreiro et al. 2016).

17.4.2 Interoception

The skin contains several different types of receptors that are classified as interoceptors. Mechanoreceptors provide information about the localisation and duration of cutaneous stimuli. Phasic receptors include the Meissner corpuscles and hair follicle cells. The Meissner corpuscles are necessary for touch and object discrimination. They are located beneath the ridges of glabrous hair-free skin such as the fingertips, palm, and nipples. The number of Meissner corpuscles has increased during primate evolution, possibly in relation to frugivorous foraging, presumably because they provide information about the softness of the texture of ripe fruit through palpation (Hoffmann et al. 2004). Meissner corpuscles may, however, be important in arboreal locomotion by assisting the individual to identify slippery surfaces of branches (Martin 1990). Ruffini endings and Merkel cells are tonic receptors which sense the duration of touch. Vater-Pacini corpuscles are mainly located subcutaneously palmar and plantar, and provide information about pressure and vibration (Raven et al. 2017b). Another type of receptor, called proprioceptor, located in muscle spindles, tendons, and joints is responsible for the determination of relative position and movement of limbs or other body parts. (For further discussion, see Chapter 8: Skin and Integument.)

The information from the different interoceptors is sent to the primary somatosensory cortex in the postcentral gyrus via the dorsal root and spinal ganglion up the spinal cord to the thalamus. The somatosensory cortex is organised in a somatotopic fashion; that is, different body parts (of the contralateral side of the body) are represented on the surface of the postcentral gyrus in a specific order. According to differences in size of the areas in which

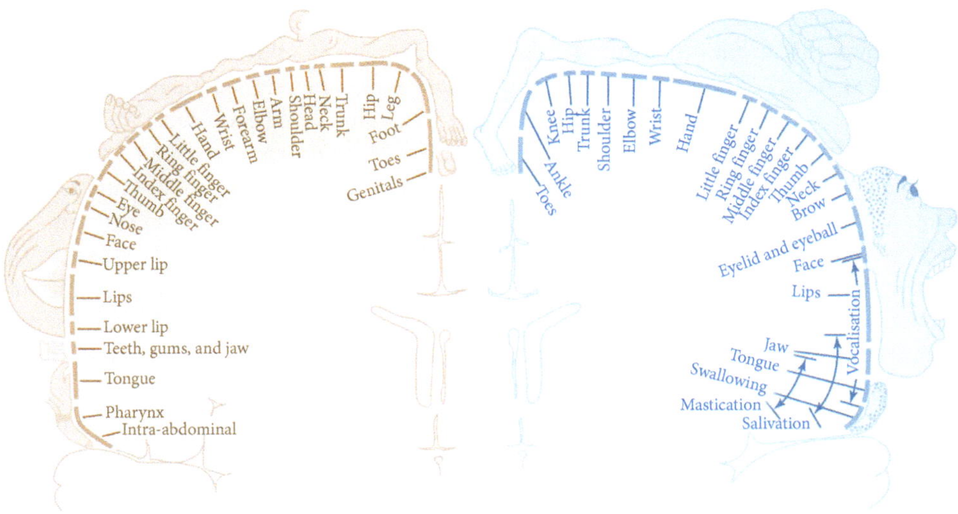

FIGURE 17.9 Human sensory (left) and motor (right) homunculus depicting cortical representations of different body parts. Areas representing facial muscles and vocal tract have substantially enlarged during human evolution.

different body parts are represented, anatomists have drawn maps of a 'homunculus', where for the somatosensory cortex, the facial area and the area for individual finger movements are greatly enlarged (a different 'homunculus' emerges for motor representation) (Figure 17.9).

The somatosensory cortex possesses a six-layered structure (referred to as granular cortex because of its prominent layer IV which is lacking in the motor cortex). Each layer contains a different type of neuron. Layer IV receives most of the input, from where the incoming excitation is transmitted to other layers or to the contralateral hemisphere. Neurons in layer V connect to the basal ganglia, brainstem, and spinal cord, whereas neurons from layer VI project to the thalamus. Along the horizontal axis, the somatosensory cortex is organised in columns, which contain up to 10,000 neurons each. These columns are functionally independent from one another and receive input from different modalities. Accordingly, there are columns for touch and proprioception (more anterior position within the somatosensory cortex and connecting to the motor cortex); others receive input from pressure receptors (more dorsally located) (Guyton and Hall 2006). The somatosensory association cortex in the parietal lobe is dorsally adjacent to the primary sensory cortex and integrates information from different sensory modalities.

Comparative research in mammals, particularly primates, has shown that a primary somatosensory cortex is present across all major clades. Early mammals also seem to have possessed a secondary somatosensory cortex and there is some evidence for a ventroparietal region involved in information processing from cutaneous receptors. Research in prosimians suggests that the representation of glabrous skin of hands and feet in the primary somatosensory cortex enlarged over time, while differentiation of the parietal association cortex occurred in anthropoids (Kaas 2004).

Other interoceptors are sensitive to changes in temperature (thermoreceptors) or pain (nociceptors). Pain receptors are free nerve endings in the skin, with two systems being responsible for the transmission of 'fast' pain via A-delta fibres (mainly mediated by glutamate) and 'slow' pain via C-fibres (mediated by substance P). The 'slow' pathway is phylogenetically much older than the 'fast' pathway and slow chronic pain is much more difficult to identify than fast sharp pain. This could be the reason why many patients with chronic pain conditions, including individuals with psychiatric or psychosomatic disorders, often experience difficulties in localising the origin of pain. However, a given stimulus can produce a sensation of both fast and slow pain (Guyton and Hall 2006). Importantly, C-fibres are sensitive to hypoxia, hypoglycaemia, hypoosmolarity, and products from muscle metabolism (reviewed in Craig 2003).

Nociception is probably phylogenetically very old, as responses to painful stimuli are among the most crucial to avoid or at least minimise damage to the body. In fact, acute or chronic pain can exert huge effects on homoeostasis, including pervasive arousal due to activation of the ascending reticular system, thus causing sleeplessness or clinical depression.

In mammals, the somatosensory attributes of pain pass the ventral posteromedial and posterolateral nuclei of the thalamus and then relay the information to the primary and secondary somatosensory cortices (Craig 2003; Singer et al. 2004). 'Affective' attributes—that is, identifying painful stimuli as unpleasant—are processed by the mediodorsal and parafascicular nuclei of the thalamus, which send projections to the ACC and the amygdala. In primates, the projections extend to the posterior part of the insular cortex. Craig (2003) has suggested that in humans (and perhaps other hominoids) this primary interoceptive representation in the posterior insula embodies a wide range of sensations including pain, temperature, itch, touch, and muscular and visceral sensations. Humans, in addition, have 'a metarepresentation of the primary interoceptive activity' which involves the AI (possibly lateralised to the right hemisphere) (Craig 2003). In addition, there is considerable top-down modulation of pain perception, influenced by learnt experiences (prior exposure to nociceptive stimuli), expectations, current mood, and attention (Bingel and Tracey 2008). The neuronal network involved in the processing of subjective pain, and specifically the AI, seems also critical for the representation of another's painful experience, that is, empathic responses that may trigger prosocial behaviour (reviewed in Gonzalez-Liencres et al. 2013).

Baroreceptors in the carotid sinus and the aortic arch sense changes in blood pressure. They are connected to the ANS which responds via sympathetic or parasympathetic (vagal) pathways to an increase or decrease in blood pressure and heart rate (Raven et al. 2017b).

An interesting speculation about the evolution of the ANS concerns the role of the vagus nerve in the regulation of homeostasis. Porges (1995, 2009) has suggested that over evolutionary time the ANS underwent some significant adaptations in response to threat. Specifically, most vertebrates 'freeze' or 'play dead' when severely threatened and escape is impossible. Freezing and tonic immobility are mediated by the action of an unmyelinated part of the vagus nerve. Fight or flight responses, in contrast, depend on the action of the sympathetic (adrenergic) nervous system. According to Porges (2009), mammals evolved a myelinated vagus which is not only involved in the variability of heart rate. He suggests that in humans this evolutionarily more recent vagal system is concerned with the 'neuroception' of bodily sensations and emotions that may signal the absence or presence of a threat and thus cause social approach or avoidance behaviour. A dysfunction of the myelinated vagus may lead to greater activity of the more primitive systems, which is accompanied by a reduction of the heart rate variability. Interestingly, it seems that the ontogenetic development of the myelinated vagus is dependent on environmental input. For example, intrauterine distress (e.g. during the second trimester of gestation) seems to compromise the maturation of the 'polyvagal' system, which may have profound ramifications for stress regulation, immunology, and mental health (Porges 2009). Indeed, the parasympathetic transmission from the periphery to the CNS entails information about the immunological state of the organism. For example, cytokines and other inflammatory substances stimulate the vagus nerve, which in turn inhibits excessive inflammation via activation of a 'cholinergic brake'. This could be of clinical importance, as electric vagus nerve stimulation seems to exert beneficial effects on autoimmune diseases such as Crohn's disease and rheumatoid arthritis (reviewed in Pavlov and Tracey 2017).

17.4.3 Exteroception

17.4.3.1 *Vision*

Vision evolved hundreds of million years ago (mya), possibly several times independently in different phyla. For primates and humans, vision is perhaps the most important source of environmental information. Light travels extremely fast, and vertebrate eyes are specialised to determine the direction and distance of an object (Allman 1999; Raven et al. 2017b).

It is believed that during the Cambrian explosion some 500 mya the gene coding for the photoreceptor protein duplicated, which produced two types of receptors: one called a rod, for vision in dim light conditions, and the other called a cone, for processing visual information under bright light conditions, including colour vision and sharpness (visual acuity).

The photopigment utilised by rods to absorb light is called rhodopsin. Rhodopsin binds to a molecule that is evolutionarily derived from carotene, a pigment found in carrots and other plants involved in photosynthesis (vitamin A deficiency is widespread among malnourished children in developing countries and produces night blindness, among other symptoms, due to an inhibition of rhodopsin synthesis). Cones utilise three different pigments called photopsins, which differ slightly in their amino acid structure, thus preferably absorbing blue, green, or red light (hence the distinction between blue, green, and red cones). In primates, trichromatic colour vision evolved about 30–45 mya by gene duplication and epistatic effects among the visual pigments (Yokoyama et al. 2014). It is estimated that in humans each retina contains about 100 million rods and 3 million cones (Sharpe et al. 1999). Colour blindness can be genetic in origin (with substantial differences in prevalence between ethnicities) or acquired (e.g. through trauma, stroke, or inflammatory brain diseases). The most frequent anomalies are protanopia and deuteranopia, where people have difficulties in telling red from green. Males are more frequently affected than women, because the trait is located on the X-chromosome. However, anomalous trichromacy, dichromacy, and monochromacy may also occur, and the molecular genetics of the opsins is highly complex (Sharpe et al. 1999).

Rods and cones are localised in the outermost layer of the retina of the vertebrate eye (which is the opposite in the octopus). Once stimulated, they send electrical signals to bipolar cells, which in turn are connected to the innermost layer of ganglion cells, which have dramatically increased in number during human evolution (Allman 1977). Cones have the highest density in an area of the retina called fovea, where there is a one-to-one connection of each cone to a single bipolar cell and to a single ganglion cell. Accordingly, the fovea has the highest visual acuity. Rods are almost absent from the fovea (Raven et al. 2017b).

The activation of the visual system is under inhibitory control. For example, under conditions of darkness, cyclic guanosine monophosphate (cGMP) controls the openness of Na^+ channels which depolarise the membrane. Light, in contrast, leads to a chemical reaction whereby opsins change their shape, causing Na^+ channels to close in the process, involving the dissociation of G-proteins (Raven et al. 2017b).

The axons of the ganglion cells, after forming the optic nerve, send projections to the LGN of the thalamus bilaterally from where projections connect to the visual cortex, whereby the medial fibres of the optic nerve cross at the optic chiasm to the contralateral side of the brain (a brain tumour in the sella turcica region can lead to bitemporal hemianopsia). Other projections from the optic nerve travel to the optic tectum, which has undergone adaptive reorganisation in relation to binocularity. That is, in primates the optic tectum strictly receives input from the contralateral visual hemifield, which has not been found in any other vertebrate order, not even other species with frontally directed eyes (Allman 1977). The LGN comprises distinct layers of smaller neurons (parvocellular) involved in the recognition of fine details and colour, and magnocellular layers (larger neurons) which are smaller in number and process information about movement and contrast. Parvo- and magnocellular neurons project to different sublayers of layer IV of the primary visual cortex (V1) and V1 is organised in a retinotopic fashion. Accordingly, lesions to V1 result in segmental deficit of the visual field. Also, Anton's syndrome, a type of anosognosia, can occur following damage of the primary visual cortex. This condition is characterised by the denial of visual problems in spite of actual blindness (Greene 2005).

Magnocellular projections connect to the middle temporal visual area (MT, also termed V5), which then projects to the posterior parietal lobe involved in goal-directed eye and hand movement. Neurons in area MT are also sensitive to the direction of movement and background movement (Allman 1999). Additional connections of V1 exist to secondary visual areas V2 and V4. These regions are involved in the perception of size and shape. Area V4 also contains neurons that seem to process cues related to the distance of an object. Such 'nearness' and 'farness' neurons are possibly involved in the construction of a three-dimensional image from the two-dimensional images produced in each visual cortex of the left and right hemisphere (Allman 1999). V4 is connected to the inferotemporal visual cortex which is involved in visual memory of objects. Adjacent to this lies an area called the fusiform face area (FFA), which is suggested to be specialised for face recognition (Kanwisher and Yovel 2006), though this is controversial. Damage to the inferotemporal cortex is associated with impairment in object recognition (agnosia) and prosopagnosia (inability to recognise faces), and occasionally with achromatopsia (Greene 2005). As a rule of thumb, a more dorsal stream of information processing seems to be involved in the detection of 'where' (spatial information), whereas a more ventral stream is involved in the recognition of 'what' (object recognition) (Mishkin and Ungerleider 1982). The inferotemporal cortex connects to the amygdala, which is involved in the processing of emotional content (of facial expressions and other stimuli). Bilateral damage to the amygdala leads to impaired emotion recognition, especially those emotions that may signal danger (i.e. fear and anger) (Adolphs et al. 1994).

Evolutionarily, there is evidence to suggest that the primary visual cortex (V1) enlarged in anthropoids (compared to strepsirrhines, i.e. lemurs) in relation to the size of the LGN (Bush and Allman 2004). This reflects the increased projection of the foveal area (10 degrees of the visual field) to 60% of V1. By comparison, in lemurs this portion of the visual field is represented in only 20% of V1 (Rosa et al. 1997).

Image-forming eyes evolved several times, and the genetic control of eye formation has been remarkably conservative throughout evolutionary times. For example, Pax-6 is an ancient regulatory gene sequence that controls eye development in fruit flies and vertebrates, suggesting a common evolutionary origin of vertebrate and invertebrate eyes (Erclik et al. 2009).

Aside from capturing a relatively narrow spectrum of visible light in the retina, the vertebrate eye is endowed with physical properties which produce an inverted sharp image of objects on the retina. Light travels through the pupil, refracted by the cornea and the lens. The pupil size adapts to the light intensity by contraction of the muscles attached to the iris. Ciliary muscles round or flatten the lens to produce sharp images on the retina for close and far vision, respectively.

Eye movement control is executed by two regions in the frontal cortex, one called the frontal eye field (FEF) and the other the supplementary eye field. The FEF is involved in generating coordinated (saccadic) eye movements, whereas the other integrates eye and limb movements (Tehovnik et al. 2000). These regions project to the brainstem, where the oculomotor nuclei are connected to the vestibular system, which is important to stabilise gaze direction while moving. Lesions to the FEF can produce conjugate eye deviation (or 'deviation conjugée'), and damage to the medial longitudinal fasciculus in the pons may produce internuclear ophthalmoplegia (often due to multiple sclerosis).

Eyesight in primates (including humans) has undergone some remarkable adaptations. The first mammals were largely nocturnal animals, so vision was poorly developed. Most early mammals were prey animals with eyes being located to the sides of the head. Thus, they had an almost panoramic visual field which allowed swift recognition of predators, but very little depth perception (many birds' eyes are also placed laterally; they have two foveas, one for frontal and the other for lateral visual acuity). The ancestors of primates and early primates (emerging some 55 mya) were solitary-living predators, however, feeding on insects and other small invertebrates. They developed large forward-facing eyes adapted to stereoscopy. Binocularity is necessary for the perception of three-dimensional space and depth. This is brought about by the largely overlapping visual fields of both retinas, though the two images are slightly displaced, producing an effect called parallax (Raven et al. 2017b), whereby the visual cortex translates these small differences into depth (see above). This is advantageous in terms of predation, as well as locomotion through the canopy of trees, that is, precision grip involving superior visuomotor coordination. This comes, however, at the cost of losing a panoramic view. Allman (1999) has suggested that this may have created selective pressure on the formation of social groups and the emergence of novel communication systems (both vocal and via facial expression). In fact, alarm calls in primates typically occur upon sighting a predator (avian or terrestrial), even at the cost to the individual making the call to draw the predator's attention toward itself (Seyfarth et al. 1980).

Thus, adaptation of the visual system in 'crown anthropoids' (which includes all living and fossil species, which descended from the last common ancestor of extant anthropoids; Williams et al. 2010) comprises evolution of a retinal fovea almost exclusively containing cones for trichromatic colour vision (Dominy et al. 2004); increasing number of retinal ganglion cells and size of the optic nerve (Allman 1977; Kirk and Kay 2004); elongation of the eye bulb (Ross 2000); decreasing diameter of the cornea (Kirk 2004); increased orbital convergence; and evolution of a postorbital septum separating the orbita from the temporal chewing muscles—altogether improving visual acuity (summarised in Dominy et al. 2004). These changes are probably linked to a switch from nocturnalism to diurnal activity. Trichromatic vision, it is suggested, evolved due to dietary changes with increasing amounts of leaves and fruit, because colour vision allows the discerning of ripe fruit (in addition to superior palpation abilities; see Section 17.4.2).

Modern environment can cause disproportionate growth of the eye bulb during childhood, for example, by excessive exposure to near objects such as computers, TV screens, and so on. Elongation of the eye bulb causes nearsightedness, which, in part, can be prevented in children by increasing the amount of outdoor activity (Rose et al. 2008).

Stereoscopy may also be vulnerable to dysfunction, for example, by strabismus, anisometropia, high refractive error, or cataract (Birch 2013). In young children, amblyopia or 'lazy eye' syndrome may occur and go undetected. This is tragic, as the critical window for getting the lazy eye to learn to see is limited, and chances of restitution decrease after age 7 (Birch 2013).

17.4.3.2 *Audition*

Hearing is an ancient sensory capacity, which processes mechanical waves that travel through a transmission medium. In natural environments, this is either water or air. Sound travels better in water than air. Accordingly, the hearing apparatus of terrestrial animals underwent several significant adaptations, including ones that involve the transformation from air to water conduction of sound waves.

Audition allows the perception of cues that are out of sight. Moreover, acoustic information can travel farther than chemical information (such as odour), though audition is limited in precision in terms of distance estimation (Raven et al. 2017b). Many natural sounds are evoked by either movement or vocalisation of animals. As the first primates (and their ancestors) dwelled on insects which often make high-frequency humming sounds when flying, and were also threatened to fall prey to larger predators, superior audition has therefore certainly been adaptive for nocturnal animals living in dense rainforest environments (Martin 1990; Heffner 2004). In addition, sound is an important source of information exchanged in intraspecies communication, which—last, but not least—includes human speech (Ghazanfar and Santos 2004; Rauschecker and Scott 2009).

A small spectrum of mechanical waves, usually in the range between 20 Hz and 40 kHz at an amplitude of 60 dB, can be detected as audible sounds by most terrestrial vertebrates. These waves are captured by outer ears (which in many species can be oriented towards the source of the sound) and channelled through an ear canal where the acoustic stimuli hit a tympanic membrane.

Among primates, the shape of the outer ear has undergone significant changes over evolutionary time. In all anthropoids, the ratio of height to width of the pinna is smaller than in non-anthropoid primates, that is, they became wider over time relative to height. These changes in shape may be associated with increased sensitivity to low-frequency sounds, while anthropoids gradually lost to some degree high-frequency perception (see below; Coleman and Ross 2004). Moreover, the outer ear became less flexible in primates, and has no function in intraspecies communication (in contrast to most carnivores and hooved mammals).

The tympanic membrane demarcates the border between the outer and the middle ear, and transfers the vibrations to the ossicles of the middle ear, the malleus, incus, and stapes, so called for the resemblance of their form to a hammer, anvil, and stirrup, respectively. In mammals, two of these ossicles, the malleus and incus, are derived from jaw bones of cynodonts, evolutionary predecessors of early mammals, while the stapes has remained the only ossicle in extant amphibians, reptiles, and birds (Allman 1999). The adaptive advantage of the evolved chain of ossicles lies in the superior perception of high-frequency sounds (above 10 kHz) evoked by flying insects. It is also plausible to assume that mammals evolved the ability to send high-pitched sounds signalling separation distress in immature offspring. Distress vocalisations of small rodents lie above 25 kHz, which reptilian predators cannot perceive (Allman 1999).

The middle ear forms a bony cavity filled with air. It is connected to the pharynx via the Eustachian tube, which equalises the air pressure in the middle ear with the environment when swallowing or yawning. The basis of the stapes is connected to the oval window, which borders the inner ear, filled with fluid. This suggests that the middle ear needs to solve the problem of an impedance mismatch between air and water, which impacts the mechanical relationship of the manubrium of the malleus and the long process of the incus. In fact, the lever arm ratio of the two ossicles has significantly changed during primate evolution, mainly driven by the size of the manubrium, though the functional significance of these changes is not entirely clear (Coleman and Ross 2004). Two small muscles, the M. stapedius and the M. tensor tympani, regulate the stiffness of the connection among the ossicles. A contraction of the M. stapedius pulls the stapes outward, whereas the M. tensor tympani moves the manubrium of the malleus inward. This reflexive muscle response occurs if loud sound hits the tympanon (eardrum). The muscle contraction reduces the transmission of low-frequency sounds significantly to protect the inner ear from damage. It also helps to mask the effect of loud background noise and reduces the sensitivity to one's own voice (Guyton and Hall 2006).

Clinically, the intactness of the conduction of sound by air (middle ear) and fluid or bone (inner ear) can elegantly be screened using a vibrating tuning fork placed on the vertex or mastoid (after nineteenth-century physicians Weber and Rinne). For example, if the middle ear is filled with fluid due to infectious disease, the vibration is perceived as lateralised to the affected side (Weber test) and a negative Rinne test (air conduction compromised).

The inner ear is formed by the labyrinth comprising the cochlea, three semicircular canals, and the utricle and saccule. The cochlea comprises three canals: the cochlear duct (or scala media), the vestibular canal (scala vestibularis), and the tympanic canal (scala tympani). The scala media is filled with endolymph, which resembles intracellular fluid, while the other two canals are filled with perilymph, which is more similar in composition to extracellular fluid. This difference is important with regard to the stimulation of the organ of Corti. Importantly, the scala vestibularis and the scala tympani are in direct connection with the subarachnoid space and therefore nearly identical with the CSF.

The cochlear duct vibrates in a frequency-dependent manner as the pressure waves travel from the oval window through the cochlear to the round window, which again connects to the middle ear. The organ of Corti is formed by flexible membranes, the basilar membrane and the tectorial membrane. In between lie the outer and inner hair cells. The outer hair cells are uniquely mammalian, and may also have evolved to improve high-frequency perception. Their impulses are sent to the inner hair cells, which depolarise and send electric signals to the auditory nerve, which are then transmitted further, via the brainstem, to the primary auditory cortex (Brownell et al. 1985; Guyton and Hall 2006; Raven et al. 2017b).

Efferent connections impact the sensitivity of the outer hair cells, which helps distinguish a particular sound from background noise. The flexibility of the basilar membrane increases from its base to its apex, such that high pitch is transmitted to the brain near the base and low-frequency sounds closer to the apex (Raven et al. 2017b).

Audition in primates has undergone some significant changes in comparison to non-primate mammals, which, in part, depend on head size, shape of the outer ear (pinna), and reorganisation of the middle and inner ear. All primates are relatively good at hearing low frequencies, at the expense of hearing high-frequency sounds, and there seems to be an evolutionary trend towards improved hearing of low-frequency sound and reduced ability to perceive high-frequency sounds, possibly in relation to a change in diet towards greater herbivory. Comparisons of audiograms have revealed, for example, that chimpanzees are better at detecting high-frequency sounds than humans, while the opposite is true for low frequencies (Heffner 2004). However, the limited high-frequency perception in humans is not likely linked to the evolution of articulated speech (Heffner 2004). In any event, primate vocalisations carry information about body size, dominance, reproductive status, group membership, and individuality (Ghazanfar and Santos 2004).

Interestingly, auditory acuity, the ability to localise sound, is associated with the width of visual acuity. Put another way, a relatively narrow field of best (foveal) vision (as is the case in humans) requires superior sound localisation abilities to direct the visual orienting reflex to the source of sound as quickly as possible (Heffner 2004), suggesting significant coevolution of sensory modalities. In addition, the system has adapted to capturing a conspecific's gaze direction (Ghazanfar and Santos 2004), which in humans further evolved to form joint attention, which is an important developmental step towards visual perspective taking and mentalising (compare Section 17.6.1). Figure 17.10 illustrates sound localisation thresholds for different mammalian species and the 'trade-off' between high-frequency and low-frequency hearing.

The primary auditory cortex (A1) of primates, located on the superior temporal plane involving Heschl's gyrus and the planum temporale (roughly equivalent to Wernicke's sensory speech area), receives input from the superior colliculus in the brainstem via the medial geniculate nucleus (MGN) of the thalamus. Akin to the visual cortex, A1 is tonotopically organised. Another similarity to the visual system is the existence of a dorsal pathway concerned with the processing of spatial information of auditory stimuli ('where'), and a ventral stream involved in pattern or auditory object recognition ('what') (Rauschecker and Scott 2009). In the evolution of human articulated speech, the hierarchical organisation of secondary auditory cortex areas became increasingly complex, with speech-specific regions of the anterior temporal cortex encoding specific sound classes such as vowels (Rauschecker and Scott 2009). Perception of the invariance of particular phonemes in spite of variation of its acoustic properties is maintained by the ventral pathway which involves the inferior frontal cortex and its reciprocal connections to the anterior part of the superior temporal cortex.

As regards the dorsal stream, the planum temporale seems to be involved in the processing of spatial information which entails the perception of music. Moreover, the inferior parietal lobule and the angular gyrus contribute to speech comprehension, but may also play an important role in the integration of speech perception and articulation. Rauschecker and Scott (2009) have suggested that projections from Broca's area and the premotor cortex to the inferior parietal lobule provide an efference copy, whereby the

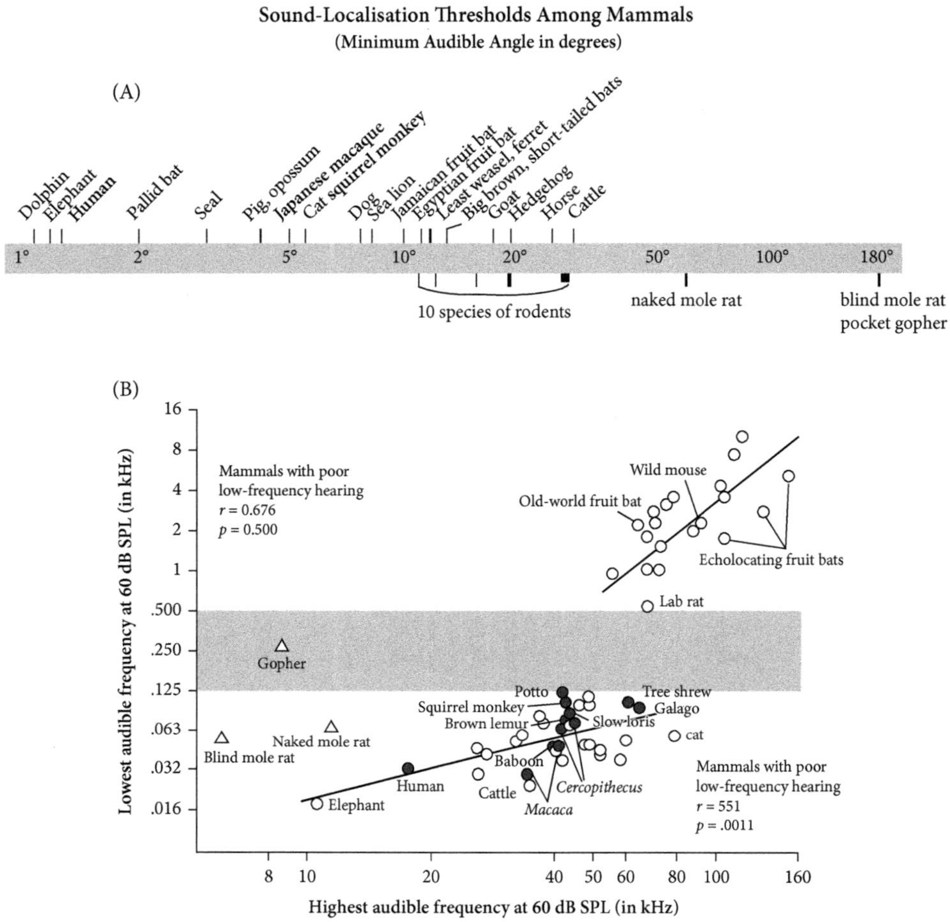

FIGURE 17.10 (A) Thresholds for sound localisation depicted for 35 mammalian species, with minimum audible angle in degrees. Note that primates are particularly good at localising sound. However, with regard to humans, this development is probably independent of the evolution of articulated speech. (B) Detection threshold at 60 dB for highest and lowest audible frequencies. Note the 'trade-off' between good high-frequency hearing versus low-frequency hearing. Primates including humans are better at low-frequency hearing. Grey tinted rectangle indicates a gap in distribution of low-frequency hearing limits. SPL, Sound pressure level. Subterranean species (indicated by open triangles) are not included in the analysis; closed circles indicate primate species; open circles refer to other mammalian species.

Source: Reproduced from Rickye S. Heffner, Primate hearing from a mammalian perspective, *The Anatomical Record: Advances in Integrative Anatomy and Evolutionary Biology*, 281a (1), pp. 1111–1122, Figures 8 and 9, doi.10.1002/ar.a.20117. Copyright © 2004 Wiley-Liss, Inc.

inferior parietal lobule and the planum temporale contribute to the anticipation of a self-generated sensorial event (speech). In turn, these areas may send back feedback in the form of an afference copy at short latencies and with high temporal precision that is compared with the incoming predictive signal (Rauschecker and Santos 2009). As is known from clinical studies, lesions to one or more of these regions can produce not only 'classic' sensory or motor aphasia, but also more complex impairments. For example, while Wernicke's aphasia is an auditory agnosia for words, auditory agnosia for environmental sounds may occur after bilateral damage to the auditory cortex (Greene 2005). Conduction aphasias occur upon damage to the arcuate fasciculus connecting frontal, temporal, and parietal areas (Damasio 1992). Primary progressive aphasias (which can be classified into a semantic variant, a non-fluent/agrammatic variant, and a logopenic variant) are associated with degenerative brain diseases, which involve both speech production and comprehension (Vandenberghe 2016).

17.4.3.3 *Vestibular System*

Maintaining postural balance and equilibrium is essential to organisms exposed to gravity. Terrestrial and aerial animals are more concerned with it than aquatic creatures. Most primates are arboreal, so precise determination of one's body position while moving through the canopy of trees is critical for survival. Though this task, strictly speaking, involves exteroception, there is some overlap with the perception of vibration, which is considered interoceptive. This is because the small organs that evolved to detect acceleration of the head or body relative to the environment are functionally similar to the lateral line system that evolved in fishes to process distant vibration produced by conspecifics swimming in shoals or, as in sharks, by prey (Raven et al. 2017b).

The vestibular apparatus is part of the inner ear (and connected to the hearing system), but functionally quite different. It utilises information from gravity and acceleration to stabilise body posture and equilibrium. Acceleration is measured by an otolith organ or macula located in the utricle and saccule. The orientation of the otolith organs differs between utricle and saccule, such that the utricle is more sensitive to horizontal acceleration, whereas the saccule processes information from vertical acceleration (Raven et al. 2017b). The maculae consist of hair cells which are covered by a gelatinous layer. This layer contains the statoconia or otoliths. Upon horizontal or vertical movement of the head, the gravity of the statoconia embedded in the gelatinous layer bends the cilia of the hair cells, thereby causing them to depolarise or hyperpolarise, depending on the direction of the movement (Guyton and Hall 2006). The hair cells are oriented in different directions such that forward and backward movements depolarise and hyperpolarise different hair cells.

Angular acceleration is detected by the three semicircular canals, each of which ends in an ampulla. The canals are filled with endolymph. The ampulla contains the cupula which protrudes into the cavity of the ampulla. The ampulla contains hair cells, too, and a gelatinous coat. The inertia of the endolymph causes the hair cells of the cupula to depolarise when a rotation movement is initiated or terminated. This is important for a terrestrial (or arboreal) animal to maintain balance; thus the semicircular canals can be assigned a kind of 'anticipatory' function (Guyton and Hall 2006). In any event, the six canals, three on each side of the head, provide information about angular acceleration in a three-dimensional space (Graf 2007).

The information from the vestibular apparatus is sent to the vestibular nuclei located in the brainstem. From here, connections exist to the oculomotor system via the medial longitudinal fascicle, which is essential for the stability of eye gaze. Moreover, other neurons connect to the cerebellum, whereby the flocconodular lobes of the cerebellum receive input from the semicircular canals, whereas the uvula of the cerebellum is involved in maintaining static equilibrium. Additional connections to the spinal cord are involved in automatic responses by means of activation of antigravity musculature (Guyton and Hall 2006). Importantly, the two vestibular systems on each side cooperate perfectly in that excitation of hair cells in one canal leads to an inhibition of the corresponding hair cells on the other side (Graf 2007).

Comparative evolution of the vestibular apparatus is difficult as the fossil record is sparse. There is evidence to suggest that early (jawless) vertebrates (evolving around 350–400 mya) possessed vertical, but not horizontal canals (Graf 2007). It is unclear when exactly and why the horizontal canals evolved in jawed vertebrates, though one can imagine that orientation in a three-dimensional space was associated with increased fitness. The embryological development of the labyrinth and ear is under control of multiple ancient regulatory genes that are also involved in the development of the kidney, lungs, and extremities (Graf 2007). The extraocular muscle apparatus that moves the eye balls is a perfect match to the orientation of the semicircular canals of the vestibular apparatus; that is, the pulling directions of the eye muscles correspond exactly with the direction of the three canals (Graf 2007). The vestibulo-oculomotor connections conjugate the movement of the eye muscles such that innervation of one muscle on one side, for example, the M. rectus lateralis, is associated with innervation of the M. rectus medius on the other side (Graf 2007). The vestibulo-oculomotor coordination is shown in Figure 17.11 (from Graf 2007).

Interestingly, bipedalism, as opposed to quadrupedalism, involved only minor changes in the orientation of the vestibular apparatus, because in quadrupedal mammals the cervical vertebral column has a vertical orientation, and bipedalism requires a postural change of the thoracic part of the vertebral column into an upright position (Graf 2007). However, Spoor et al. (1994) discovered an increase in size of the anterior and posterior semicircular canal, while the lateral canal decreased in size in humans compared to other hominids. They argued that increased sensitivity of the vertically oriented canals could have been beneficial for upright body posture and endurance running. Such morphological changes of the bony labyrinth are absent in australopithecines, which possessed a more ape-like shape of the labyrinth. Also, Neanderthals seem to have retained some of the more 'primitive' features concerning the shape and size of the bony labyrinth. A speculative interpretation of this finding is that Neanderthals were more endurance walkers, rather than runners (Spoor et al. 2003), though this explanation has been disputed by some authors (Graf and Vidal 1996).

The vestibular system usually operates smoothly and outside conscious awareness. Conversely, dysfunction or lesions of the vestibular system—peripheral or central—causes massive subjective discomfort and distress due to nausea, vertigo, and/or vomiting. Clinically, peripheral vestibular disorders include, among other things, paroxysmal positional vertigo, vestibular migraine, and Menière disease, all of which can be mimicked by central vestibular syndromes caused by brainstem or cerebellar lesions. The distinction between peripheral and central vestibular syndromes is not always easy, and warrants careful clinical examination of the vestibulo-ocular reflex, spontaneous and gaze-evoked nystagmus, and smooth pursuit of eye movements in horizontal and vertical directions (Brandt and Dieterich 2017).

17.4.3.4 *Olfaction*

Olfaction concerns the perception of volatile particles or odorants, thus requiring the evolution of chemoreceptors that are able to detect airborne stimuli, sometimes at extremely low concentrations. In contrast to the visual and auditory systems, olfaction is more about the presence or absence of a particular odour, and this may be an explanation for the much

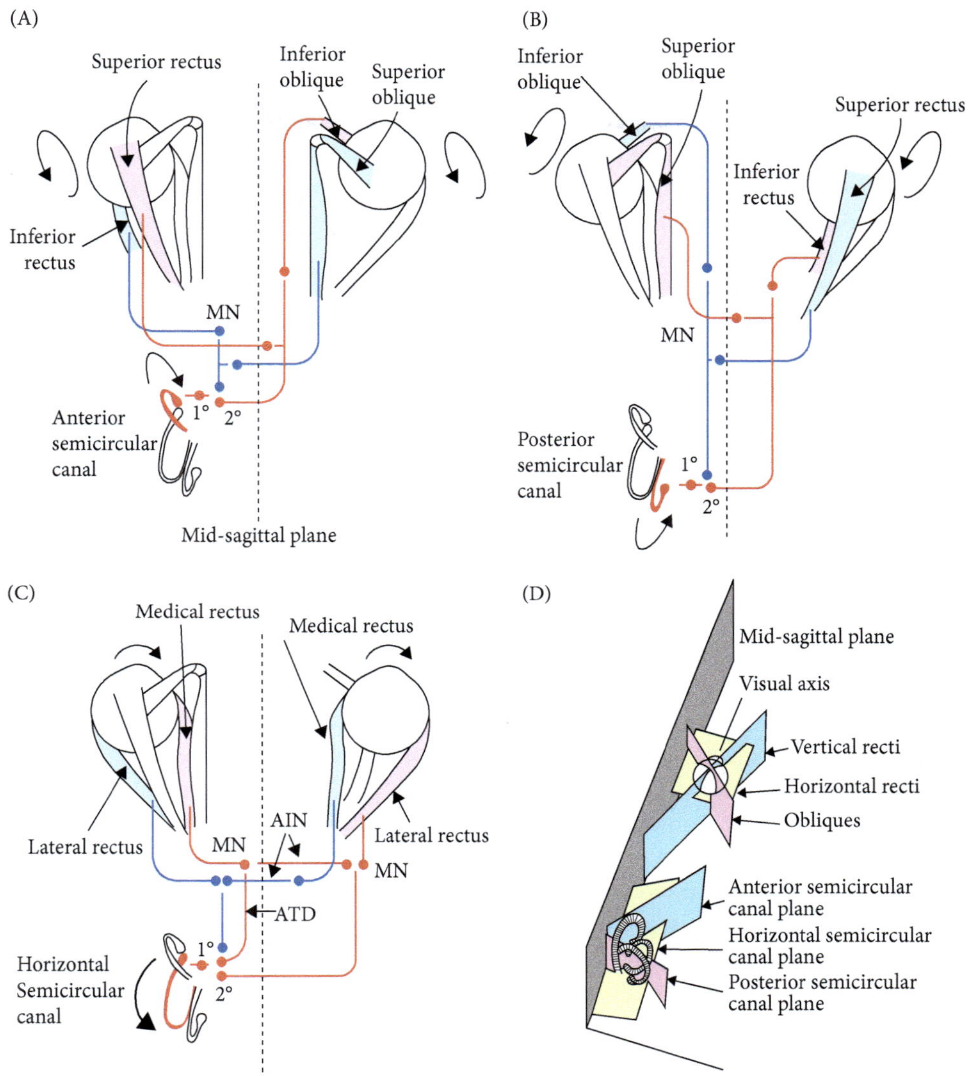

FIGURE 17.11 Vestibulo-oculomotor connections, the neural and geometrical basis for spatial coordination of compensatory eye movements. The reflex arc between individual semicircular canals and four extraocular muscles typically consists of three neurons: primary neuron (1°, vestibular nerve), second-order vestibular neuron (2°, vestibular nucleus neurons), and oculomotor neuron (MN, in oculomotor, trochlear, and abducens nuclei). Excitatory connections are shown in red, inhibitory connections in blue. Contralaterally projecting vestibular neurons are in general excitatory, and ipsilaterally projecting ones inhibitory. Respective semicircular canals ((A) anterior canal; (B) posterior canal; (C) horizontal canal) and their primary nerve pathways are marked in red. On-directions of semicircular canals are illustrated by thick black arrows. (D) Semicircular canals and extraocular muscles form a three-dimensional intrinsic reference frame system for production of vestibulo-ocular reflexes. This connectivity constitutes the basis for compensatory eye movement coordination in physical space. AIN, Abducens internuclear neuron pathway; ATD, ascending tract of Deiters.

Source: Reproduced from Graf, Werner M., 'Vestibulo-Oculomotor Connections', in Marc D. Binder, Nobutaka Hirokawa, Uwe Windhorst eds., *Encyclopedia of Neuroscience*, pp. 4228–4235, Figure 1, doi.10.1007/978-3-540-29678-2_6311.

lower threshold at which maximum intensity of smell is reached (Guyton and Hall 2006). Another distinctive feature of the sense of smell compared to the other senses is that the number of different chemicals eliciting a sensorial response is much higher, perhaps comprising around 100 to 1000 different qualities (Guyton and Hall 2006; Raven et al. 2017b).

Olfactory cells are located in the superior part of the nasal cavity. The olfactory cilia are covered by a layer of mucus, where odorants are solved in fluid. Accordingly, volatile particles are ideally water soluble and slightly lipid soluble to pass the two barriers between air and mucus and between mucus and ciliae (Guyton and Hall 2006). There the odorant binds to a receptor protein, which itself is coupled to a G-protein complex in similar ways as described in Section 17.4.3.1. When the olfactory cells depolarise they project to the olfactory bulb via the olfactory nerve. The border of the nasal cavity to the olfactory bulb is formed by the lamina cribrosa which the small nerves perforate upwards. In continuity with the olfactory bulb lies the olfactory tract, which divides to project to phylogenetically older parts of the brain such as the hypothalamus, and via the piriform cortex and amygdala to the hippocampus. The former is involved in immediate behavioural responses to smell such as licking and salivating, whereas the latter concerns learnt responses including food aversion associated with the experience of nausea and vomiting (Guyton and Hall 2006). Moreover, direct connection from the piriform cortex to the orbitofrontal cortex is involved in the recognition of the pleasantness or unpleasantness of odours, as well as reward processing and associated learning associated with visual and olfactory input and taste (Rolls 2004).

In early mammals and their ancestral cynodonts, the olfactory bulb was a large structure of the basal forebrain, suggesting that olfaction, for a very long period of time, was perhaps the most important sensory system for foraging and mating. It is important to stress that smell is intricately connected to the limbic system, including its emotion processing properties. Smell can easily be distinguished according to its perception as pleasant or unpleasant, an extremely important feature for the discrimination of palatable from inedible food, as well as mate quality, including the fit of the mate's human leucocyte antigen (HLA) system with one's own (see below).

Many mammals possess two olfactory systems: a main and an accessory (or vomeronasal) olfactory system. The two systems seem to preferentially process odorants that differ in molecular weight, whereby the accessory system detects less volatile substances, but does not seem to be specialised in the detection of pheromones (Dominy et al. 2004). The presence of a vomeronasal organ in humans is inconsistent; it seems to develop during embryogenesis, but loses its sensory function in adults, though an endocrine function has been hypothesised (discussed in D'Aniello et al. 2017). In any event, following the change from nocturnalism to diurnalism, it seems that primates lost some of their olfactory acuity, which may have come as a cost of evolving trichromatic vision (Gilad et al. 2004). In fact, research has shown that about 60% of human olfactory receptor genes are functionally impaired by coding region disruption, while the number of such 'pseudogenes' for chimpanzees is about half of that. Still, the number of 30% pseudogenes is higher than in dogs and mice (reviewed in Gilad et al. 2004). The loss of fully functional olfactory receptor genes is paralleled by the evolution of trichromatic vision in two distinct primate lineages, suggesting some trade-off between the two sensory modalities (Gilad et al. 2004).

Macroanatomically, the external nose of humans protrudes much more outward of the facial plane than the external noses of other apes and many other primate species. It is unclear whether this relates to the humidification of air inhaled through the nostrils. An alternative explanation for the evolution of the human nasal cavity pertains to its function as a brain cooling device. In fact, it has been argued that hunting during the day was advantageous for early hominids, and that this created the need to effectively regulate brain temperature independently of body temperature. In fact, the intracranial sinus cavernosus communicates with the facial venous system including the veins draining the mucous nasal cavity, suggesting that the independence of the brain's thermoregulation from bodily thermoregulation evolved in response to increased physical activity (reviewed in Dean 1988; compare Section 17.2.8).

In any event, as already noted in Section 17.3.1, the decrease in size of the olfactory bulb and the reduction of the number of olfactory receptor genes does not implicitly suggest that olfaction is irrelevant in humans. In fact, the close connection of olfaction to the limbic system, which is critical for caregiving, attachment formation, and pair-bonding, suggests that olfaction plays an important role in human parenting and in detecting fertility (Miller and Maner 2010; Feldman 2015). In addition, people can discriminate the smell of sweat produced by anxiety from sweat produced by physical exercise (Prehn-Kristensen et al. 2009), and even immunological sickness responses are detectable by olfactory cues. This makes perfect sense in evolutionary terms, because this mechanism may help avoid body contact of healthy people with sick ones (Olsson et al. 2014). Hypersensitivity to certain odours during pregnancy may also be protective against the ingestion of contaminated food (Kohl et al. 2009). Pathologies of smell may occur in patients with orbitofrontal tumours, as well as patients with degenerative brain diseases including Huntington's disease, Parkinson's disease, and Alzheimer's disease (Kohl et al. 2017). Thus, olfaction is far from unimportant for humans, and future research will certainly expand on the understanding of the role of odour for interpersonal communication (McGann 2017).

17.4.3.5 *Gustatory System*

Similar to olfaction, the sense of taste utilises chemoreceptors that serve to detect palatable or poisonous food. In terrestrial vertebrates these chemoreceptors are located in the oral cavity and on the tongue. Eighty to a hundred chemosensitive cells form a taste bud, which are connected to nerve fibres. The taste buds are placed on three types of papillae that differ in location on the tongue. A small portion of taste buds are located on the palate, the epiglottis, and the proximal oesophagus (Guyton and Hall 2006). Humans can distinguish four taste modalities—sweet (elicited by sugars, localised at the tip of the tongue), sour (produced by hydrogen, localised at the sides of the tongue), bitter (produced by nitrogen-containing organic substances or alkaloids, localised at the back of the tongue), and salty (natrium chloride, the least localised taste). A fifth quality, 'umami' (or savory) seems to sense the presence of glutamate (Rolls 2004; Breslin 2013). The number of receptor genes encoding the sense of 'bitter' is larger than for the other taste qualities, because the substances evoking a sense of bitter taste are chemically the most diverse (Breslin 2013). Signals from the nerve fibres are sent to the brainstem via three cranial nerves: the anterior two-thirds of the tongue via the facial nerve; the posterior part of the tongue, mouth, and pharynx via the glossopharyngeal nerve; and a small portion via the vagus nerve. From the brainstem, neuronal projections connect to the thalamus, and from there are relayed to cortical and limbic areas (Rolls 2004; Guyton and Hall 2006). The gustatory system is closely linked to olfaction, and a sense of flavour is produced by both gustatory and olfactory cues.

Akin to olfaction, the rewarding properties of taste (and flavour) are neuronally represented in the orbitofrontal cortex, and this region seems to be implicated in conscious taste perception (Rolls 2004). As Breslin (2013) pointed out, taste perception is particularly important for omnivorous species such as humans, because the risk of accidentally ingesting poisonous substances is much greater than in species that are highly specialised on specific diets. Accordingly, leaf eaters like koalas, but also carnivores, especially aquatic carnivores, have lost many functional genes coding for taste receptors (summarised in Breslin 2013). Of note, most natural foods comprise a mixture of different taste (and flavour) qualities, suggesting continuous taste–taste interaction. Moreover, taste perception is influenced by other perceptual modalities such as temperature, auditory cues from chewing, and proprioception from the teeth. The integration of these multimodal sensations takes place in the orbitofrontal cortex and the insular cortex (Breslin 2013). Aside from conscious evaluation of food qualities, taste buds are present in the intestines and regulate hormonal secretion, but also anticipatory responses to potential toxins including nausea, sickness, and vomiting (for further discussion, see Chapter 13: Digestive System). What's more, toxins are readily detected in the area postrema of the brainstem, which is involved in nausea and vomiting responses (note that the area postrema lacks a blood–brain barrier). Figure 17.12 shows the connections of different sensory modalities with brain centres involved in the emotional appraisal of the sensory input (after Rolls 2004).

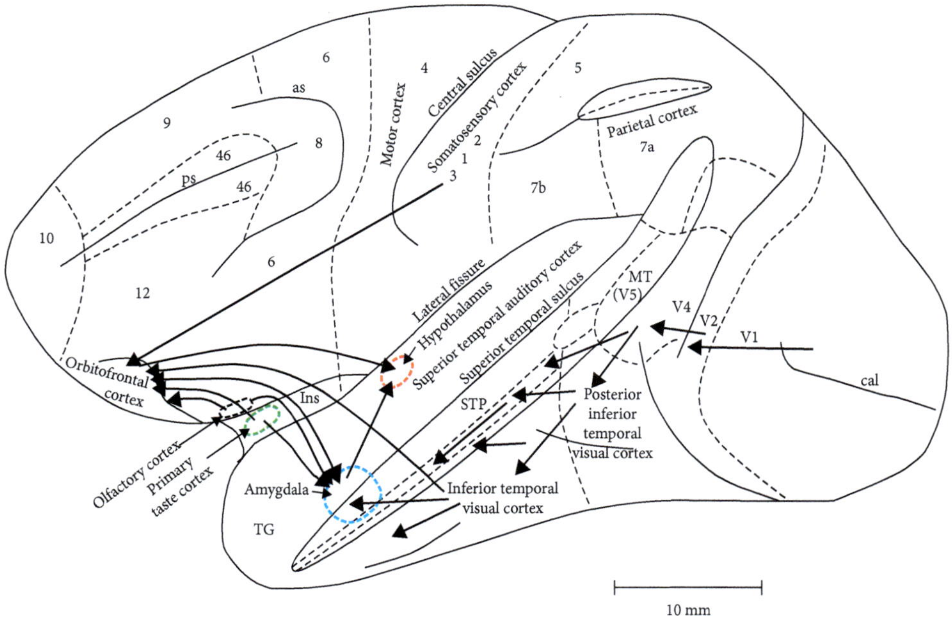

FIGURE 17.12 Pathways of sensory systems to brain regions involved in emotional evaluation of stimuli. The secondary taste cortex and secondary olfactory cortex are within the orbitofrontal cortex. Numbers indicate Brodmann areas. V1, Primary visual cortex; V4, secondary visual cortex; MT, middle temporal visual area (also referred to as V5); TG, temporal gyrus; ps, principal sulcus; Ins, insula; cal, calcarine sulcus; as, arcuate sulcus.

Source: Reproduced from Edmund T. Rolls, *Emotion Explained*, p. 64, Figure 4.1 © Edmund T. Rolls, 2007. Published by Oxford University Press https//global.oup.com/academic/product/emotion-explained-9780198570042?lang=en&cc=gb. For permission to reuse this material, please visit http//global.oup.com/academic/rights.

The human diet has substantially changed since the human lineage split from the last common ancestor with chimpanzees. However, humans, like other apes, have retained a preference for sweet–sour taste found in ripe fruit, possibly due to the fact that monkeys and apes lost the ability to synthesise vitamin C. Moreover, in apes (including humans) alpha amylase can be produced in the salivary glands, possibly by a retroviral insertion of the pancreatic amylase gene into the salivary glands, though the high number of copies of the amylase gene in the human lineage may have evolved due to the increased ingestion of starchy fibres (Breslin 2013). In addition, like other omnivores, humans possess a preference for mildly salty foods, which may, in part, be related to the high amount of salt in human sweat, because living in hot environments and being physically active may bear the risk of salt depletion. Unlike chimpanzees, humans have evolved a special sense for 'umami', which can be found in processed (e.g. fermented) or aged, but not fresh, meat. Of note, human milk contains high amounts of glutamate, which could arguably be linked to the brain's need of glutamate for maintaining synaptic plasticity (Campbell 2010).

Pregnancy, particularly the first trimester, is associated with heightened sensitivity and aversion of bitter taste, similar to the aversion of certain olfactory stimuli (see Section 17.4.3.4). This is adaptive because it protects both mother and fetus from potentially toxic substances (Breslin 2013). The consequences of human preferences for sugar, salt, and glutamate in modern environments are dealt with in Chapter 6: Nutrition, Energy Expenditure, Physical Activity, and Body Composition.

17.5 Gene Expression in the CNS

Research suggests that about 55% of all coding genes of the human genome are expressed in CNS tissue. This insight is of outstanding importance, because it puts the CNS in a central position for mutation events to occur and selective forces to act upon (Caceres et al. 2003; Myers et al. 2007). When comparing the human genome with that of the chimpanzee, our closest extant relative, it seems that the differences are relatively small. Both species share about 98% of DNA sequence, suggesting that the human genome differs from the chimpanzee's genome by only about 35 million bases, of which about 5 million bases are inserted or deleted (together referred to as 'indels') in humans or chimpanzees, and another 3 million bases differ in protein-coding genes or other functional areas of the genome (Khaitovich et al. 2005). However, more rapid genome changes seem to have occurred in the human lineage compared to the chimpanzee lineage since the two split from a common ancestor (Pollard et al. 2006). (For further discussion, see Chapter 3: Genetics and Epigenetics.) Examples of protein-coding mutations in the human lineage include the FOXP2 gene, known for its involvement in articulated language. Of the three mutations that have occurred since human and mouse split from a common ancestor, thus illustrating the highly conserved nature of this gene, two occurred after the human lineage diverted from the chimpanzee lineage.

In addition, duplications occurred in genes involved in synapse formation and maturation (e.g. SRGAP2), neurodevelopment (e.g. ARHGAP11B), and neurogenesis (e.g. DUF1220), which may play a role in some neurodevelopmental disorders. Complex up- and down-regulations of gene expressions, including methylation of promoter regions, are also relevant in explaining differences in brain function between humans and non-human primates (Geschwind and Rakic 2013; Sousa et al. 2017a). These up- and down-regulations are also region-specific. For example, evidence suggests that genes involved in dopamine synthesis are up-regulated in the human striatum, hippocampus, and amygdala, associated with a selectively higher number of tyrosine hydroxylase-positive interneurons in the human striatum, but not olfactory bulb neurons, compared to non-human primates. Conversely, comparative research in dopamine receptor expression has demonstrated a down-regulation of DRD1, DRD2, and DRD3 coding genes in the human striatum compared to non-human primates (Sousa et al. 2017b).

As regards genetics of stress-related disorders, specifically psychiatric conditions, it is important to note that many do not follow the logic of the diathesis stress model, suggesting that some genetic variants predispose to developing a disorder (Monroe and Simons 1991), if associated with adverse life events, but are otherwise phenotypically silent. From an evolutionary point of view, the diathesis stress model seems to be incomplete or even implausible, given the evidence that several 'risk' alleles have undergone recent positive selection in human evolution. For example, cross-cultural genetic studies suggest that the 7-repeat variant of the DRD4 gene (a 'risk' variant for ADHD) emerged some 50,000 years ago, with the 4-repeat variant being the ancestral form (Ding et al. 2002). Notably, the 7-repeat allele has been found to be more prevalent in migratory, as opposed to sedentary, populations (Chen et al. 1999), which is consistent with the variant's association with the personality trait 'novelty seeking'. This personality trait may have conferred a reproductive advantage in human history, particularly for migrating populations (Reist et al. 2007; Matthews and Butler 2011). In modern environments, however, this trait may be a risk factor for ADHD.

Another observation germane to the understanding of stress-related conditions suggests that the same allelic variation can confer both risk for and protection from psychopathology, depending on the quality of the early social environment. Put another way, the theory of 'differential genetic susceptibility' to environmental conditions posits that a certain variant can have detrimental effects on health if associated with adverse experiences, while the same variant can be associated with a reduced risk for psychiatric conditions if interacting with more emotionally supportive early environments (Boyce et al. 1995; Belsky 1997). For example, children carrying the 7-repeat variant of the DRD4 gene develop ADHD and externalising problems *less often than average* if their mothers act in emotionally responsive ways to their children's needs (Bakermans-Kranenburg and van Ijzendoorn 2006). This and other examples suggest that allelic variation involved in dopamine and serotonin turnover convey differential susceptibility to environmental conditions, possibly mediated by one's responsivity to reward and punishment, as well as stress regulation (Bakermans-Kranenburg and van Ijzendoorn 2007; Ellis et al. 2011). Whether analogous scenarios exist for neurological disorders is currently unknown.

17.6 An Integrative View of the Role of the CNS in Health and Disease

17.6.1 'Social Brains' and Expensive Tissues

It used to be thought that the fact that the human brain comprises about 2% of adult body weight but consumes between 15 and 20% of total energy required an exceptional evolutionary explanation, because such 'expensive tissue' would not have been selected had its benefits not outweighed its costs (Aiello and Wheeler 1995). What makes the human brain so costly compared to the rest of the body is the combination of a very large number of neurons (the human brain being the largest primate brain) with the fact that humans are primates, and, as such, possess a large number of neurons for their body mass (Herculano-Houzel 2011). Compared to non-great ape primates, humans have a brain with as many neurons as expected for both brain and body mass, and that, at an average of 6000 calories per billion neurons per day, costs as much energy as expected for its 86 billion neurons. Because primates have more neurons for their body mass than other mammals (Herculano-Houzel et al. 2014b), the relative cost of the primate brain, expressed as a percentage of body metabolic cost, is larger than the relative cost of non-primate brains. The human brain, it turns out, is not special in its metabolic cost.

That is not to say that the energetic cost of the human brain is trivial to support. When calculating the energy intake of primates based on their body mass and number of hours spent feeding and foraging per day, and the energetic cost depending on body mass and number of brain neurons, human ancestors would have had to feed for about 9.5 hours/day to afford our modern number of brain neurons (Fonseca-Azevedo and Herculano-Houzel 2012). Such long daily hours of feeding are unlikely to be viable, given that 8 hours seems to be the maximum that orangutans can spend feeding, even when food is scarce (Fonseca-Azevedo and Herculano-Houzel 2012). Instead, affording our modern number of neurons must have required a radical change in how calories were consumed—which can be accomplished by cooking in the widest sense, with or without fire (that is, through tool use) (Herculano-Houzel 2016a). Modern humans indeed spend one order of magnitude less time feeding than predicted from other primate species (Organ et al. 2011).

Given that large numbers of neurons still come at a sizable cost, investing in them must confer some advantage. A hypothesis first put forth in the 1960s suggests that it was foremost the social environment that led to an increase in brain size during primate evolution, eventually culminating in the evolution of human psychological mechanisms that are specialised in the processing of social information (Jolly 1966; Humphrey 1976). The 'social brain' hypothesis (Brothers 1990; Dunbar 2003) is probably complementary rather than incompatible with the assumption that the evolution of technical intelligence and cultural transmission of knowledge has been a driving force in human brain evolution (Tomasello and Call 1997). In fact, a recent study suggests that ecological factors were more important in human brain size evolution than social drivers (Gonzáles-Forero and Gardner 2018). However, even if not the only driving force, the idea of a 'social brain' nicely integrates an evolutionary scenario of human brain development, comparative research into non-human primates, and recent findings from neuroscience including functional brain imaging and genetics of social behaviour (Ebstein et al. 2010).

However, an increased capacity to deal with social information was probably just one of many factors leading to positive selection of larger brains, once they became affordable. Evolutionary psychology suggests that the human brain's enormous capacity to store information, to flexibly interact with the environment, and to quickly affiliate with others is rooted in the fact that selection favoured the formation of larger social groups in which social interactions became increasingly complex (Chance 1988). The increase in size of ancestral hominin groups was adaptive, probably because—as the climate in East Africa became cooler—trees were more sparsely dispersed in the open savannah such that larger social groups yielded better protection from large predators (Jones et al. 1992; Cerling et al. 2011).

On the other hand, this situation posed greater pressure on the individual to successfully compete for food and sexual partners. The dilemma between the need for growing gregariousness and social competition may have led to a runaway selection of higher cognitive functions, increasing social competence, and emotional systems of attachment and bonding, all of which ultimately increased the chance of survival and reproduction of those who had such capacities over others who were less well endowed with these abilities. Accordingly, selection pressures may have increased the demands of greater computational resources and thus conferred an advantage to bigger brains with more neurons.

Two independent predictors strongly support the social brain hypotheses. In primates, average group size correlates with the ratio of the neocortex volume (excluding the visual cortex) to the rest of the brain (Dunbar 1995). In other words, the larger the neocortex ratio, the larger the average group size in different primate species. Similarly, the maximum lifespan of different primate species is highly correlated with the relative size of the neocortex, but not with group size, suggesting that group size and longevity are independent factors predicting the relative size of the neocortex in primates (Allman 1999). In addition, neocortex size seems to correlate with the length of childhood and adolescence (Joffe 1997). Interestingly, however, the neocortex ratio does not predict group size in apes, because apes usually live in much smaller groups than do many other primate species, but have larger brains relative to body size. However, when looking at strategic social interactions, including what has been called 'tactical deception'—that is, the ability to intentionally manipulate the behaviour of other individuals to one's own advantage—relative neocortex size correlates with the number of deceptive acts in apes (Byrne and Whiten 1992). As predicted from these approaches, humans are expected to live in groups of 150 people on average, a number that has been found to strikingly match contemporary hunter-gatherer groups (Dunbar 1998).

A recent re-formulation of the 'social brain' hypothesis posits that the driving force behind increased brain size might be social complexity, rather than group size alone, whereby social complexity is reflected in the ability to form coalitions and maintain cooperative relationships (Dunbar and Shultz 2007). In fact, human societies, compared with chimpanzees for example, are characterised by much more cooperative behaviour (Tomasello and Vaish 2013), which is essentially linked to a sophisticated emotional repertoire, including trust, sympathy, liking, and love, but also shame, guilt, envy, and schadenfreude (Trivers 1971; Dvash et al. 2010; Damasio and Carvalho 2013). In fact, fairness behaviour is associated with the activation of the reward system, which may serve as a proxy to maintain altruistic behaviour in social groups (Tabibnia et al. 2008). However, in order to maintain cooperation, individuals must be able to detect non-cooperative behavioural strategies of others. Due to the need for group cohesion and cooperation between genetically unrelated or distantly related individuals, it probably paid off not only to cooperate to a high degree, but also to develop rules of collective punishment of individuals who disobeyed the rules of cooperation (Trivers 1971). Collective punishment and the evolution of morality must have been important landmarks of human cognitive and emotional evolution, because, most remarkably, people tend to punish others for their misdemeanour even if the punishment incurs costs to the punisher (Camerer 2003). On the other hand, selection has also operated on mechanisms involved in reparative altruism, for example forgiveness, because restoration of mutual cooperation in social groups can be advantageous, depending on the estimated risk of further exploitation by a defector (McCullough et al. 2013). Accordingly, many rules of moral punishment do not follow any mathematical logic; on the contrary, they often appear highly irrational at first sight, which underscores their value for early and contemporary human societies—otherwise such rules would not have evolved.

The brain comprises, therefore, neural representations of cognitive and emotional capacities that make humans experts in reflecting own and others' mental states and feelings, which thus contribute to the ability to maintain complex social relationships. There are two domains that can theoretically be separated from one another, although they broadly overlap, behaviourally and neurally: empathy and 'theory of mind' or 'mentalising'.

Empathy concerns the ability to intuitively feel what others are feeling and to understand cause and effect. Phylo- and ontogenetically, empathy is predated by emotional contagion, where the understanding of a causal relationship between the contextual factors that elicited a specific emotion is not required (Gonzalez-Liencres et al. 2013). Empathy-related processes probably evolved early during mammalian evolution as a consequence of intensive maternal care and nurturance (Preston 2013). Empathy is therefore not specific to primates, let alone humans. Generally, empathy can be elicited by every emotion, which includes not only the 'classic' basic emotions such as happiness, sadness, fear, surprise, anger, and disgust (as well as contempt), but also complex emotions such as shame, guilt, envy, and schadenfreude (see above). As a prerequisite, it is essential to be able to decipher observable social signals such as facial expressions of emotions, body posture, and gestures. Emotion processing networks develop slowly during human infancy, childhood, and adolescence, whereby amygdala and orbitofrontal function seem to emerge as early as around 5–7 months of age (Leppänen and Nelson 2009).

A large body of research has examined empathic concern elicited by the observation of others who are exposed to physical or emotional pain. This research has shown that the neural representations of physical and emotional pain broadly overlap (Eisenberger 2012), and similar brain activation can be measured when one is observing another person in a painful situation (Singer et al. 2004). Key structures involved in empathy for pain include the ACC, the AI, and the somatosensory cortex. Empathic responses to another person's sickness behaviour and suffering vary widely in intensity, however, and such responses are also greatly influenced by the way the sufferer actually expresses his or her way to cope with illness (Preston et al. 2013). Empathy for another's pain is compromised in several neuropsychiatric conditions, including antisocial personality disorder, frontotemporal dementia, and other conditions affecting frontal lobe and amygdala functioning (reviewed in Gonzalez-Liencres et al. 2013). Conversely, some individuals may overempathise with another's psychological pain caused by rejection or social exclusion, sometimes before a background of own adverse childhood experiences, which has been found to be relevant in borderline personality disorder (Flasbeck et al. 2017).

The other domain, 'theory of mind', concerns the ability to reflect upon one's own and other persons' mental states in terms of beliefs, knowledge, intentions, feelings, and so forth (Premack and Woodruff 1978) (the terms 'mentalising', 'mental state attribution', and 'reflexive functioning' are used more or less interchangeably). Theory of mind seems to be quite human-specific (as compared to empathy). Comparative research suggests that this is achieved in similar ways in every known culture (Avis and Harris 1991), though culture modulates the speed and timing of social competence development (Greenfield et al. 2003). Neurally, theory of mind is associated with activation of the medial prefrontal cortex, the precuneus, and regions containing mirror neurons (Frith and Frith 2001).

The representation of own and others' minds is a prerequisite for mutual cooperation, but also for cheating and the detection of deceptive intentions (Trivers 1971). Moreover, the ability to deceive others may even require the capacity for deceiving the self. Self-deception may increase individual fitness, because it may improve one's success in deceiving others; in other words, an individual that is unaware of his or her (perhaps morally unacceptable) wishes may appear more sincere and trustworthy to others (Trivers 2000; von Hippel and Trivers 2011). From a medical point of view, it is highly relevant to recognise the widespread negative consequences for health and well-being that may arise from impairments in empathy and/or mentalising (Brüne and Brüne-Cohrs 2006; Gonzalez-Liencres et al. 2013). In fact, many psychiatric disorders and neurological disease affecting the neural circuits involved in empathy and mentalising profoundly impact on patients' ability to effectively interact with their social environments. Likewise, trust and cooperation, as well as understanding social rules involving fairness and reciprocity may be severely compromised in neuropsychiatric conditions in manifold ways (Brüne 2016).

17.6.2 Stress Regulation and the Brain

As shown in the previous sections, the CNS is critically concerned with maintaining homoeostasis of the organism. The more complex the environment to which an organism is exposed, the greater the demands imposed on the CNS in terms of responding to environmental challenges. Accordingly, the human brain is equipped with sophisticated means not only to detect threats (both social and non-social), to find suitable partners and allies, and to protect and nurture dependent offspring, but also to respond and cope with dysregulation of homoeostasis, whereby dysregulation is commonly referred to as 'stress'.

The term 'stress' was first used in a scientific context by Selye (1936) who defined stress as 'the non-specific response of the body to any demand for change'. Stress therefore is not necessarily pathological, but can lead to pathology if the organism is excessively exposed to a stressor. That is, stress intensity, timing, and/or duration of exposure matters, whereby both under- and overstimulation may be detrimental to the organism's homoeostasis, following an inverted U-shaped curve (Sapolsky 2015).

Adaptive regulation of physiological processes in response to stress exposure has been termed 'allostasis' (a recent review is given in McEwen 2017). Put another way, the organism seeks to maintain stability through active adjustment of bodily functions to both predictable and unpredictable events (McEwen 2017). The systems involved in stress regulation entail the HPA axis, the ANS, and the immune system. Thus, adrenal hormones, neurotransmitters, and cytokines are among the main molecules that generate short-term adaptive responses. However, the same molecules can exert detrimental effects, termed 'allostatic load'. Accordingly, if allostatic load is chronically high, pathologies develop, including trauma and stressor-related psychiatric disorders, as well as cardiovascular disease, insulin resistance, and accumulation of amyloid (summarised in McEwen 2017).

The neuroendocrine cascade elicited by the activation of the HPA axis is evolutionarily ancient (at least some 250 million years) and thus highly conserved. (For further discussion, see Chapter 15: Endocrinology.) The molecular cascade in vertebrates comprises a 'first wave' characterised by the secretion of catecholamines from the sympathetic nervous system (SNS) within seconds following the recognition of a threat. A few seconds later, corticotrophin-releasing hormone (CRH) is released from the hypothalamus, followed by the pituitary secretion of adrenocorticotropic hormone (ACTH). This reaction is accompanied by a decreased hypothalamic release of gonadotropin-releasing hormone and decreased secretion of pituitary gonadotropins. Conversely, a pituitary secretion of prolactin and (in primates) growth hormone occurs, as well as secretion of glucagon to raise the level of glucose in the bloodstream. The brain also actively suppresses insulin secretion to maintain its increased demand for energy (Peters 2011). CRH and cortisol-like molecules as well as ACTH have been found in the haemocytes of some invertebrates (e.g. mollusks) (Ottaviani 2011), and it is assumed that these systems evolved in marine animals as an excretory mechanism to eliminate excess sodium absorbed from salt-water environments, whereas the renal system evolved later, enabling sodium retention in the body. (For further discussion, see Chapter 14: Excretory System.) Over evolutionary time, the thymus became the main organ in which this conserved stress response was concentrated, while only later it moved to the HPA system. (For further discussion, see Chapter 10: Immune System.)

Glucocorticoids, when released in a 'second wave', suppress the stress response and help prevent the organism's stress axis from being pathologically overactivated. Cortisol, for instance, has important regulatory effects on the prefrontal cortex, amygdala, and hippocampus. Functionally, it contributes to increased arousal, vigilance, focused attention, and memory formation; glucocorticoids also inhibit growth and reproduction, and exert immune-activating effects. Conversely, chronic exposure to stress causes damage to these systems; that is, chronic stress impairs prefrontal and hippocampal function, and causes immune suppression (McEwen 2017). Chronic stress, therefore, is causal in the accumulation of deleterious effects on multiple organ systems, as indicated by an increased allostatic load.

Consistent with the view that CNS function is intricately linked to the microbiome and immunity, there is mounting evidence suggesting that dysfunction of the interaction between these systems is involved in neurological and psychiatric disorders in manifold ways. Examples include, but are probably not limited to depression (Miller and Raison 2016), anxiety disorders and post-traumatic stress disorder (Michopoulos et al. 2016), autism (Meltzer and van de Water 2017), schizophrenia (Miller and Goldsmith 2017), and multiple sclerosis (Embry 2004; Conradi et al. 2011), whereby timing and exposure to stress and dysregulation of immune responses may vary greatly. (For further discussion, see Chapter 3: Genetics and Epigenetics.)

An intriguing idea based on evolutionary considerations suggests that the vulnerability to many of these disorders and diseases comes from a reduced exposure to pathogens early in life (referred to as the 'old friends' hypothesis), which explains why such pathologies are—apparently counter-intuitively—more prevalent in developed countries and urban societies. Accordingly, exaggerated inflammatory responses in conjunction with stress constitute risk factors for neuropsychiatric disorders, due to reduced contact with the symbiotic (coevolved) microbiota on the skin, in the respiratory tract, and in the intestines, or reduced exposure to pathogens like helminths, which were abundant in hunter-gatherer populations, or reduced exposure to those microbiota to which our ancestors were habitually exposed, but which were tolerated by the immune system (reviewed in Lowry et al. 2016). (For further discussion, see Chapter 10: Immune System.) As the interaction between brain function, immunity, and stress is unfolding, these insights may have profound implications for future treatment strategies (Miller et al. 2017).

Stress can deregulate the innate 'zeitgebers' located in the suprachiasmatic nucleus (SCN) of the hypothalamus. Together with the pineal gland (a homologous structure to the parietal eye in early vertebrates) (Allman 1999) and the retina, the SCN forms a part of the photoendocrine system, whereby clock genes are expressed in the SCN and putatively regulate melatonin synthesis and the activity of the HPA axis (Stehle et al. 2011; Kassi and Chrousos 2013). In addition, it has been shown that circadian rhythms are also present in all types of glia cells, whereby dysfunctional expression of clock genes may cause glutamatergic imbalance (Chi-Castañeda and Ortega 2016).

Circadian rhythms have been under strong selection pressures to help adapt the organism to oscillating environmental changes, including changing exposure to daylight, temperature, and season (see overview in Frank et al. 2013). Sleep seems to act on the expression of clock genes by feedback mechanisms, whereby variation of clock genes may impact on individual vulnerability to sleep disturbances (Frank et al. 2013). In addition, rapid eye movement (REM) sleep, which correlates with brain size in mammals, has been found to be important for memory consolidation and emotion regulation, and is also linked in the developing brain to maturation of neuronal circuits (which is why neonates and infants have larger sleep requirements; Joiner 2016). Conversely, chronic sleep deprivation can contribute to an increase in allostatic load and impact on several organ systems, including the cardiovascular system, endocrinology, and immune function (compare Sections 17.2.7 and 17.2.8; reviewed in Joiner 2016). Furthermore, sleep deprivation has far-reaching effects on cognitive functioning. It causes attentional problems and working memory performance, and increases impulsivity and antisocial behavioural tendencies, including aggression, and may also alter reward sensitivity and risk-taking, which thus has important ramifications for the treatment of psychiatric and neurological conditions associated with chronic sleep disturbances (Krause et al. 2017b).

Interestingly, however, recent research in a hunter-gatherer population (the Hadza of northern Tanzania) has revealed that asynchrony of sleep patterns in human social groups is the norm, rather than an exception. The adaptive value of sleep asynchrony among members of a social group is that vigilance during sleep is shared. In other words, sentinel-like behaviour and chronotype variation may have been favoured by selection. This could have implications for the understanding of common psychiatric disorders and age-related sleep disturbances, which are often accompanied by disorders of the circadian rhythm and distinct genetic chronotype variations (Samson et al. 2017).

Coming back to the relevance of stress, it is important to note that stress can have transgenerational consequences for health and disease vulnerability. For example, the experience of early-life stress such as abuse can change the HPA axis in pervasive ways. As adults, these individuals may be less emotionally responsive to their infants' needs. Moreover, exposure to famine *in utero* (i.e. during gestation) has been found to increase the risk for somatic (cardiovascular disease, etc.) and psychological malfunction (depression, schizophrenia). Furthermore, children of people with post-traumatic stress disorder are more likely to develop as adults psychiatric disorders than children from unexposed parents (reviewed in Cowan et al. 2016). The transgenerational transmission of risk for somatic and mental disorder is possibly linked to epigenetic alterations such as methylation of genes that are relevant to brain development and stress regulation, though our understanding is still in its infancy with regard to the stability of methylation patterns in the germline over successive generations (Szyf 2015).

17.6.3 Cross-Talk between the CNS and Other Organs

The CNS, though superficially appearing as an anatomically isolated organ, is in intensive exchange with virtually all other organ systems. In fact, the CNS is the great orchestrator of the organism's homoeostasis, whereby feedback comes from within the body, which needs to be integrated with incoming information from the environment. Much of the information from the periphery to the brain travels via the bloodstream (humoral pathway) or via the ANS. Conversely, brain development critically depends on input from the periphery, including signals from the microbiota, which colonises all tissues that are in direct contact with the external environment, that is, the intestines, respiratory tract, and skin (Dinan and Cryan 2017), either as commensals, mutualists, or parasites. (For further discussion, see Chapter 13: Digestive System, Chapter 12: Respiratory System, and Chapter 8: Skin and Integument.) Succinctly put, even though we perceive ourselves as individuals, our bodies operate more like an ecosystem of incredible complexity.

This chapter has put emphasis on how the CNS interacts with the external environment, foremost our social environments, but also with other organ systems such as the intestines (i.e. via the gut–brain axis), the immune system (both innate and adaptive), and the cardiovascular system (i.e. via the ANS). It is remarkable how flexible and malleable the interaction of the CNS with other organs is, and how quickly adaptive responses emerge due to environmental challenges including social stressors, infectious agents, and so forth. However, as natural systems are never optimal by design, dysfunction and diseases are common. For example, psychiatric disorders such as depression and anxiety disorders are frequently associated with autoimmunological disease including asthma, rheumatoid arthritis, and thyroiditis, to name just a few (Michopoulos et al. 2016). Likewise, stress-related disorders such as post-traumatic stress disorder are often linked to symptoms that manifest in the guts and the skin (Burges Watson et al. 2016).

Understanding human physiology and pathophysiology, including the complex interactions between different organ systems, necessitates a profound knowledge of evolutionary mechanisms, that is, how and why the systems evolved the way they did. Medicine, in addition, constantly seeks for improvement of existing or development of new treatment strategies. This can only be a successful endeavour if constraints, trade-offs, and evolved functions of organic systems are scrutinised in all dimensions, as suggested by Tinbergen. (For further discussion, see Chapter 1: Core Principles for Evolutionary Medicine.)

17.6.4 Prevention of CNS Disease

The human CNS is an extraordinarily complex organ that is influenced in many ways by the internal and external environment. As discussed in this chapter, the CNS is exposed to the action of gut microbiota, immunological processes, and nutritional, potentially toxic substances. Due to the manifold interactions with other organ systems, it is also exposed to metabolic waste products, though to some degree protected by the blood–brain barrier. Even more importantly, the CNS is exposed to a vast array of stressors, most of which interfere with important biosocial goals such as finding a supportive peer group, finding a mate, successfully raising children, and keeping a rewarding job. In fact, research has shown that maintaining reciprocal social relationships is the most relevant factor in decreasing one's mortality risk (Holt-Lunstad et al. 2010). Vulnerability to CNS dysfunction is, of course, also genetically and epigenetically moderated, such that there is no silver bullet for preventive measures that helps keep CNS dysfunction at bay throughout the lifespan.

Beyond individual factors, however, there are several means by which allostasis can be maintained, not all of which are under personal control, but fate. Intrauterine exposure to maternal stress, malnutrition, and infection, for example, are important risk factors in the development of disease postnatally. Along similar lines, growing up in harsh conditions both emotionally and economically can have detrimental effects on mental health, as well as on somatic health. Abuse and neglect are probably early-life conditions to which no adaptive coping mechanisms evolved, because anthropological data suggest that they were absent in ancestral societies (Hrdy 2000). These issues are outside personal control, though nonetheless important for health policy.

Aside from these thoughts about primary prevention, there is abundant evidence for the effectiveness of secondary and tertiary prevention of CNS dysfunction and disease. Both animal and human studies suggest that exercise, including brain exercise (i.e. cognitive training), sleep regulation, stress coping strategies, social support, avoidance of toxins such as nicotine, and possibly also the consumption of micronutrients and probiotics may help keep the brain healthy (Cowan et al. 2016; McEwen 2017).

Acknowledgements

I am grateful to Professor Suzana Herculano-Houzel for her thoughtful comments on an earlier draft of this chapter. In addition, I am indebted to Professor (Emeritus) Jay Feierman, who, as the webmaster of the World Psychiatric Association (WPA) Group of Evolutionary Psychiatrists, has shared heaps of interesting links and published work on brain function and evolution.

References

Aboitiz, F., Morales, D., and Montiel, J. (2003). The evolutionary origin of the mammalian isocortex: towards an integrated developmental and functional approach. *Behav Brain Sci* 26, 535–52; discussion 552–85.

Adolphs, R., Tranel, D., Damasio, H., et al. (1994). Impaired recognition of emotion in facial expressions following bilateral damage to the human amygdala. *Nature* 372, 669–72. doi: 10.1038/372669a0.

Aiello, L. C. and Wheeler, P. (1995). The expensive-tissue hypothesis—the brain and the digestive-system in human and primate evolution. *Curr Anthropol* 36, 199–221. doi: 10.1086/204350.

Allman, J. M. (1999). *Evolving Brains*. New York: Scientific American Library.

Allman, J. M., Hakeem, A., Erwin, J. M., et al. (2001). The anterior cingulate cortex. The evolution of an interface between emotion and cognition. *Ann N Y Acad Sci* 935, 107–17.

Allman, J. M., Tetreault, N. A., Hakeem, A. Y., et al. (2010). The von Economo neurons in frontoinsular and anterior cingulate cortex in great apes and humans. *Brain Struct Funct* 214, 495–517. doi: 10.1007/s00429-010-0254-0.

Allman, J. M., Tetreault, N. A., Hakeem, A. Y., et al. (2011). The von Economo neurons in the frontoinsular and anterior cingulate cortex. *Ann N Y Acad Sci* 1225, 59–71. doi: 10.1111/j.1749-6632.2011.06011.x.

Amunts, K., and Zilles, K. (2015). Architectonic mapping of the human brain beyond Brodmann. *Neuron* 88, 1086–107. doi: 10.1016/j.neuron.2015.12.001.

Attwell, D. and Laughlin, S. B. (2001). An energy budget for signaling in the grey matter of the brain. *J Cereb Blood Flow Metab* 21, 1133–45. doi: 10.1097/00004647-200110000-00001.

Avis, J. and Harris, P. L. (1991). Belief-desire reasoning among Baka children—evidence for a universal conception of mind. *Child Dev* 62, 460–7. doi: 10.1111/j.1467-8624.1991.tb01544.x.

Azmitia, E. C. (2010). Evolution of serotonin: sunlight to suicide. In: Müller, C. P. and Jacobs, B. L. (eds) *Handbook of the Behavioral Neurobiology of Serotonin*. London: Academic Press, Elsevier.

Bakermans-Kranenburg, M. J. and Van Ijzendoorn, M. H. (2006). Gene–environment interaction of the dopamine D4 receptor (DRD4) and observed maternal insensitivity predicting externalizing behavior in preschoolers. *Dev Psychobiol* 48, 406–9. doi: 10.1002/dev.20152.

Bakermans-Kranenburg, M. J. and Van Ijzendoorn, M. H. (2007). Research review: genetic vulnerability or differential susceptibility in child development: the case of attachment. *J Child Psychol Psychiatry* 48, 1160–73. doi: 10.1111/j.1469-7610.2007.01801.x.

Bartz, J., Simeon, D., Hamilton, H., et al. (2011). Oxytocin can hinder trust and cooperation in borderline personality disorder. *Soc Cogn Affect Neurosci* 6, 556–63. doi: 10.1093/scan/nsq085.

Bechara, A., Damasio, H., and Damasio, A. R. (2000). Emotion, decision making and the orbitofrontal cortex. *Cereb Cortex* 10, 295–307.

Belsky, J. (1997). Theory testing, effect-size evaluation, and differential susceptibility to rearing influence: the case of mothering and attachment. *Child Dev* 68, 598–600.

Ben-Shaanan, T. L., Azulay-Debby, H., Dubovik, T., et al. (2016). Activation of the reward system boosts innate and adaptive immunity. *Nat Med* 22, 940–4. doi: 10.1038/nm.4133.

Bingel, U. and Tracey, I. (2008). Imaging CNS modulation of pain in humans. *Physiology (Bethesda)* 23, 371–80. doi: 10.1152/physiol.00024.2008.

Birch, E. E. (2013). Amblyopia and binocular vision. *Prog Retin Eye Res* 33, 67–84. doi: 10.1016/j.preteyeres.2012.11.001.

Boyce, W. T., Chesney, M., Alkon, A., et al. (1995). Psychobiologic reactivity to stress and childhood respiratory illnesses: results of two prospective studies. *Psychosom Med* 57, 411–22.

Braga, J. and Hublin, J. J. (1998). What do carotid canals tell us about human brain evolution? *Am J Phys Anthropol* Suppl 26, 112.

Brandt, T. and Dieterich, M. (2017). The dizzy patient: don't forget disorders of the central vestibular system. *Nat Rev Neurol* 13, 352–62. doi: 10.1038/nrneurol.2017.58.

Breslin, P. A. (2013). An evolutionary perspective on food and human taste. *Curr Biol* 23, R409–18. doi: 10.1016/j.cub.2013.04.010.

Brothers, L. (1990). The social brain: a project for integrating primate behavior and neurophysiology in a new domain. *Concepts Neurosci* 1, 27–51.

Brown, E. C. and Brüne, M. (2014). Reward in the mirror neuron system, social context, and the implications on psychopathology. *Behav Brain Sci* 37, 196–7. doi: S0140525X13002240.

Brown, E. C., Wiersema, J. R., Pourtois, G., et al. (2013). Modulation of motor cortex activity when observing rewarding and punishing actions. *Neuropsychologia* 51, 52–8. doi: 10.1016/j.neuropsychologia.2012.11.005.

Brownell, W. E., Bader, C. R., Bertrand, D., et al. (1985). Evoked mechanical responses of isolated cochlear outer hair cells. *Science* 227, 194–6.

Brüne, M. (2016). On the role of oxytocin in borderline personality disorder. *Br J Clin Psychol* 55, 287–304. doi: 10.1111/bjc.12100.

Brüne, M., Schöbel, A., Karau, R., et al. (2010). Von Economo neuron density in the anterior cingulate cortex is reduced in early onset schizophrenia. *Acta Neuropathol* 119, 771–8. doi: 10.1007/s00401-010-0673-2.

Brüne, M., Schöbel, A., Karau, R., et al. (2011). Neuroanatomical correlates of suicide in psychosis: the possible role of von Economo neurons. *PLoS One* 6, e20936. doi: 10.1371/journal.pone.0020936.

Bruner, E., Mantini, S., Perna, A., et al. (2005). Fractal dimension of the middle meningeal vessels: variation and evolution in *Homo erectus*, Neanderthals, and modern humans. *Eur J Morphol* 42, 217–24. doi: 10.1080/09243860600746833.

Bruner, E., Mantini, S., Musso, F., et al. (2011). The evolution of the meningeal vascular system in the human genus: from brain shape to thermoregulation. *Am J Hum Biol* 23, 35–43. doi: 10.1002/ajhb.21123.

Burges Watson, I. P., Brüne, M., and Bradley, A. J. (2016). The evolution of the molecular response to stress and its relevance to trauma and stressor-related disorders. *Neurosci Biobehav Rev* 68, 134–47. doi: 10.1016/j.neubiorev.2016.05.010.

Bush, E. C. and Allman, J. M. (2004). Three-dimensional structure and evolution of primate primary visual cortex. *Anat Rec A Discov Mol Cell Evol Biol* 281, 1088–94. doi: 10.1002/ar.a.20114.

Butti, C. and Hof, P. R. (2010). The insular cortex: a comparative perspective. *Brain Struct Funct* 214, 477–93. doi: 10.1007/s00429-010-0264-y.

Butti, C., Santos, M., Uppal, N., et al. (2013). Von Economo neurons: clinical and evolutionary perspectives. *Cortex* 49, 312–26. doi: 10.1016/j.cortex.2011.10.004.

Byrne, R. W. and Whiten, A. (1992). Cognitive evolution in primates—evidence from tactical deception. *Man* 27, 609–27. doi: 10.2307/2803931.

Caceres, M., Lachuer, J., Zapala, M. A., et al. (2003). Elevated gene expression levels distinguish human from non-human primate brains. *Proc Natl Acad Sci U S A* 100, 13030–5. doi: 10.1073/pnas.2135499100.

Camerer, C. F. (2003). Psychology and economics. Strategizing in the brain. *Science* 300, 1673–5. doi: 10.1126/science.1086215.

Campbell, B. C. (2010). Human biology, energetics, and the human brain. In: Muehlenbein, M. P. (ed.) *Human Evolutionary Biology*. Cambridge: Cambridge University Press.

Capuron, L., Lasselin, J., and Castanon, N. (2017). Role of adiposity-driven inflammation in depressive morbidity. *Neuropsychopharmacology* 42, 115–28. doi: 10.1038/npp.2016.123.

Cerling, T. E., Wynn, J. G., Andanje, S. A., et al. (2011). Woody cover and hominin environments in the past 6 million years. *Nature* 476, 51–6. doi: 10.1038/nature10306.

Chakrabarti, B. and Baron-Cohen, S. (2011). Variation in the human cannabinoid receptor CNR1 gene modulates gaze duration for happy faces. *Mol Autism* 2, 10. doi: 10.1186/2040-2392-2-10.

Chance, M. R. A. (ed.) (1988). *Social Fabrics of the Mind*. Hove: Lawrence Erlbaum.

Changeux, J.-P. (2005). Genes, brains, and culture: from monkey to human. In: Dehaene, S., Duhamel, J.-R., Hauser, M., et al. (eds) *From Monkey Brain to Human Brain*. Cambridge, MA: MIT Press.

Chen, C. S., Burton, M., Greenberger, E., et al. (1999). Population migration and the variation of dopamine D4 receptor (DRD4) allele frequencies around the globe. *Evol Hum Behav* 20, 309–24. doi: 10.1016/S1090-5138(99)00015-X.

Chiarlone, A., Bellocchio, L., Blazquez, C., et al. (2014). A restricted population of CB1 cannabinoid receptors with neuroprotective activity. *Proc Natl Acad Sci U S A* 111, 8257–62. doi: 10.1073/pnas.1400988111.

Coleman, M. N. and Ross, C. F. (2004). Primate auditory diversity and its influence on hearing performance. *Anat Rec A Discov Mol Cell Evol Biol* 281, 1123–37. doi: 10.1002/ar.a.20118.

Conradi, S., Malzahn, U., Schroter, F., et al. (2011). Environmental factors in early childhood are associated with multiple sclerosis: a case-control study. *BMC Neurol* 11, 123. doi: 10.1186/1471-2377-11-123.

Corballis, M. C. (2003). From mouth to hand: gesture, speech, and the evolution of right-handedness. *Behav Brain Sci* 26, 199–208; discussion 208–60.

Cowan, C. S., Callaghan, B. L., Kan, J. M., et al. (2016). The lasting impact of early-life adversity on individuals and their descendants: potential mechanisms and hope for intervention. *Genes Brain Behav* 15, 155–68. doi: 10.1111/gbb.12263.

Craig, A. D. (2003). Interoception: the sense of the physiological condition of the body. *Curr Opin Neurobiol* 13, 500–5.

Crow, T. J. (1997). Schizophrenia as failure of hemispheric dominance for language. *Trends Neurosci* 20, 339–43.

Damasio, A. R. (1992). Aphasia. *N Engl J Med* 326, 531–9. doi: 10.1056/NEJM199202203260806.

Damasio, A., and Carvalho, G. B. (2013). The nature of feelings: evolutionary and neurobiological origins. *Nat Rev Neurosci* 14, 143–52. doi: 10.1038/nrn3403nrn3403.

Danese, A. and Lewis, S, J. (2017). Psychoneuroimmunology of early-life stress: the hidden wounds of childhood trauma? *Neuropsychopharmacology* 42, 99–114. doi: 10.1038/npp.2016.198.

D'Aniello, B., Semin, G. R., Scandurra, A., et al. (2017). The vomeronasal organ: a neglected organ. *Front Neuroanat* 11, 70. doi: 10.3389/fnana.2017.00070.

Dean, M. C. (1988). Another look at the nose and the functional significance of the face and nasal mucous membrane for cooling the brain in fossil hominids. *J Hum Evol* 17, 715–18.

Decety, J. and Chaminade, T. (2005). The neurophysiology of imitation and intersubjectivity. In: Hurley, S. and Chater, N. (eds) *Perspectives on Imitation. Vol. 1. Mechanisms of Imitation and Imitation in Animals*. Cambridge, MA: MIT Press.

De Lussanet, M. H. and Osse, J. W. (2015). Decussation as an axial twist: a comment on Kinsbourne (2013). *Neuropsychology* 29, 713–14. doi: 10.1037/neu0000163.

Devinsky, O., Morrell, M. J., and Vogt, B. A. (1995). Contributions of anterior cingulate cortex to behaviour. *Brain* 118(Part 1), 279–306.

Diamond, M. K. (1992). Homology and evolution of the orbitotemporal venous sinuses of humans. *Am J Phys Anthropol* 88, 211–44. doi: 10.1002/ajpa.1330880209.

Dinan, T. G. and Cryan, J. F. (2017). Microbes, immunity, and behavior: psychoneuroimmunology meets the microbiome. *Neuropsychopharmacology* 42, 178–92. doi: 10.1038/npp.2016.103.

Ding, Y. C., Chi, H. C., Grady, D. L., et al. (2002). Evidence of positive selection acting at the human dopamine receptor D4 gene locus. *Proc Natl Acad Sci U S A* 99, 309–14. doi: 10.1073/pnas.012464099.

Dölen, G., Darvishzadeh, A., Huang, K. W., et al. (2013). Social reward requires coordinated activity of nucleus accumbens oxytocin and serotonin. *Nature* 501, 179–84. doi: 10.1038/nature12518 nature12518.

Domes, G., Heinrichs, M., Michel, A., et al. (2007). Oxytocin improves 'mind-reading' in humans. *Biol Psychiatry* 61, 731–3.

Dominy, N. J., Ross, C. F., and Smith, T. D. (2004). Evolution of the special senses in primates: past, present, and future. *Anat Rec A Discov Mol Cell Evol Biol* 281, 1078–82. doi: 10.1002/ar.a.20112.

Donaldson, Z. R. and Young, L. J. (2008). Oxytocin, vasopressin, and the neurogenetics of sociality. *Science* 322, 900–4. doi: 10.1126/science.1158668322/5903/900.

Dunbar, R. I. M. (1995). Neocortex size and group-size in primates—a test of the hypothesis. *J Hum Evol* 28, 287–96. doi: 10.1006/jhev.1995.1021.

Dunbar, R. I. M. (1998). The social brain hypothesis. *Evol Anthropol* 6, 178–90. doi: 10.1080/03014460902960289.

Dvash, J., Gilam, G., Ben-Ze'ev, A., et al. (2010). The envious brain: the neural basis of social comparison. *Hum Brain Mapp* 31, 1741–50. doi: 10.1002/hbm.20972.

Dunbar, R. I. and Shultz, S. (2007). Evolution in the social brain. *Science* 317, 1344–7. doi: 10.1126/science.1145463.

Ebstein, R. P., Israel, S., Chew, S. H., et al. (2010). Genetics of human social behavior. *Neuron* 65, 831–44. doi: 10.1016/j.neuron.2010.02.020.

Edgar, N. and Sibille, E. (2012). A putative functional role for oligodendrocytes in mood regulation. *Transl Psychiatry* 2, e109. doi: 10.1038/tp.2012.34.

Eisenberger, N. I. (2012). The pain of social disconnection: examining the shared neural underpinnings of physical and social pain. *Nat Rev Neurosci* 13, 421–34. doi: 10.1038/nrn3231.

Eisenberger, N. I., Moieni, M., Inagaki, T. K., et al. (2017). In sickness and in health: the co-regulation of inflammation and social behavior. *Neuropsychopharmacology* 42, 242–53. doi: 10.1038/npp.2016.141.

Ellis, B. J., Boyce, W. T., Belsky, J., et al. (2011). Differential susceptibility to the environment: an evolutionary—neurodevelopmental theory. *Dev Psychopathol* 23, 7–28. doi: 10.1017/S0954579410000611.

Elphick, M. R. and Egertova, M. (2001). The neurobiology and evolution of cannabinoid signalling. *Philos Trans R Soc Lond B Biol Sci* 356, 381–408.

Embry, A. F. (2004). The multiple factors of multiple sclerosis: a Darwinian perspective. *J Nutr Environ Med* 14, 1–11.

Emes, R. D. and Grant, S. G. (2012). Evolution of synapse complexity and diversity. *Annu Rev Neurosci* 35, 111–31. doi: 10.1146/annurev-neuro-062111-150433.

Erclik, T., Hartenstein, V., Mcinnes, R. R., et al. (2009). Eye evolution at high resolution: the neuron as a unit of homology. *Dev Biol* 332, 70–9. doi: 10.1016/j.ydbio.2009.05.565.

Faissner, A., Pyka, M., Geissler, M., et al. (2010). Contributions of astrocytes to synapse formation and maturation—potential functions of the perisynaptic extracellular matrix. *Brain Res Rev* 63, 26–38. doi: 10.1016/j.brainresrev.2010.01.001.

Fajardo, C., Escobar, M. I., Buritica, E., et al. (2008). Von Economo neurons are present in the dorsolateral (dysgranular) prefrontal cortex of humans. *Neurosci Lett* 435, 215–18. doi: 10.1016/j.neulet.2008.02.048.

Falk, D. (1990). Brain evolution in *Homo*: the 'radiator' theory. *Behav Brain Sci* 13, 333–81.

Fehm, H. L., Kern, W., and Peters, A. (2006). The selfish brain: competition for energy resources. *Prog Brain Res* 153, 129–40. doi: 10.1016/S0079-6123(06)53007-9.

Feldman, R. (2012). Oxytocin and social affiliation in humans. *Horm Behav* 61, 380–91. doi: 10.1016/j.yhbeh.2012.01.008.

Feldman, R. (2015). The adaptive human parental brain: implications for children's social development. *Trends Neurosci* 38, 387–99. doi: 10.1016/j.tins.2015.04.004.

Flasbeck, V., Enzi, B., and Brüne, M. (2017). Altered empathy for psychological and physical pain in borderline personality disorder. *J Pers Disord* 31, 689–708. doi: 10.1521/pedi_2017_31_276.

Fleshner, M., Frank, M., and Maier, S. F. (2017). Danger signals and inflammasomes: stress-evoked sterile inflammation in mood disorders. *Neuropsychopharmacology* 42, 36–45. doi: 10.1038/npp.2016.125.

Fonseca-Azevedo, K. and Herculano-Houzel, S. (2012). Metabolic constraint imposes tradeoff between body size and number of brain neurons in human evolution. *Proc Natl Acad Sci U S A* 109, 18571–6. doi: 10.1073/pnas.1206390109.

Frank, E., Sidor, M. M., Gamble, K. L., et al. (2013). Circadian clocks, brain function, and development. *Ann N Y Acad Sci* 1306, 43–67. doi: 10.1111/nyas.12335.

Frith, C. D. and Frith, U. (1999). Interacting minds—a biological basis. *Science* 286, 1692–5.

Frith, U. and Frith, C. (2001). The biological basis of social interaction. *Curr Dir Psychol Sci* 10, 151–5. doi: 10.1111/1467-8721.00137.

Fung, T. C., Olson, C. A., and Hsiao, E. Y. (2017). Interactions between the microbiota, immune and nervous systems in health and disease. *Nat Neurosci* 20, 145–55. doi: 10.1038/nn.4476.

Furness, J. B., Rivera, L. R., Cho, H. J., et al. (2013). The gut as a sensory organ. *Nat Rev Gastroenterol Hepatol* 10, 729–40. doi: 10.1038/nrgastro.2013.180.

Gabi, M., Neves, K., Masseron, C., et al. (2016). No relative expansion of the number of prefrontal neurons in primate and human evolution. *Proc Natl Acad Sci U S A* 113, 9617–22. doi: 10.1073/pnas.1610178113.

Gallese, V. and Goldman, A. (1998). Mirror neurons and the simulation theory of mind-reading. *Trends Cogn Sci* 2, 493–501.

Gannon, P. J., Holloway, R. L., Broadfield, D. C., et al. (1998). Asymmetry of chimpanzee planum temporale: humanlike pattern of Wernicke's brain language area homolog. *Science* 279, 220–2.

Ghazanfar, A. A. and Santos, L. R. (2004). Primate brains in the wild: the sensory bases for social interactions. *Nat Rev Neurosci* 5, 603–16. doi: 10.1038/nrn1473.

Gilad, Y., Przeworski, M., and Lancet, D. (2004). Loss of olfactory receptor genes coincides with the acquisition of full trichromatic vision in primates. *PLoS Biol* 2, E5. doi: 10.1371/journal.pbio.0020005.

Gonzalez-Castro, T. B., Hernandez-Diaz, Y., Juarez-Rojop, I. E., et al. (2016). The role of a catechol-O-methyltransferase (COMT) Val158Met genetic polymorphism in schizophrenia: a systematic review and updated meta-analysis on 32,816 subjects. *Neuromolecular Med* 18, 216–31. doi: 10.1007/s12017-016-8392-z.

Gonzalez-Liencres, C., Shamay-Tsoory, S. G., and Brüne, M. (2013). Towards a neuroscience of empathy: ontogeny, phylogeny, brain mechanisms, context and psychopathology. *Neurosci Biobehav Rev* 37, 1537–48. doi: 10.1016/j.neubiorev.2013.05.001.

Graf, W. M. (2007). Vestibular system. In: Kaas, J. H. and Krubitzer, L. A. (eds) *Evolution of Nervous Systems. A Comprehensive Reference*. Amsterdam: Elsevier.

Graf, W. and Vidal, P.-P. (1996). Semicircular canal size and upright stance are not interrelated. *J Hum Evol* 30, 175–81.

Greene, J. D. (2005). Apraxia, agnosias, and higher visual function abnormalities. *J Neurol Neurosurg Psychiatry* 76(Suppl 5), v25–34. doi: 10.1136/jnnp.2005.081885.

Greenfield, P. M., Keller, H., Fuligni, A., et al. (2003). Cultural pathways through universal development. *Annu Rev Psychol* 54, 461–90. doi: 10.1146/annurev.psych.54.101601.145221.

Guerreiro, M. J., Putzar, L., and Roder, B. (2016). The effect of early visual deprivation on the neural bases of auditory processing. *J Neurosci* 36, 1620–30. doi: 10.1523/JNEUROSCI.2559-15.2016.

Gur, R. C., Gunning-Dixon, F., Bilker, W. B., et al. (2002). Sex differences in temporo-limbic and frontal brain volumes of healthy adults. *Cereb Cortex* 12, 998–1003.

Guyton, A. C. and Hall, J. E. (2006). *Textbook of Medical Physiology*. Philadelphia: Elsevier.

Hammock, E. A. (2015). Developmental perspectives on oxytocin and vasopressin. *Neuropsychopharmacology* 40, 24–42. doi: 10.1038/npp.2014.120.

Hammock, E. A. and Young, L. J. (2006). Oxytocin, vasopressin and pair bonding: implications for autism. *Philos Trans R Soc Lond B Biol Sci* 361, 2187–98. doi: 10.1098/rstb.2006.1939.

Haroon, E., Miller, A. H., and Sanacora, G. (2017). Inflammation, glutamate, and glia: a trio of trouble in mood disorders. *Neuropsychopharmacology* 42, 193–215. doi: 10.1038/npp.2016.199.

Heffner, R. S. (2004). Primate hearing from a mammalian perspective. *Anat Rec A Discov Mol Cell Evol Biol* 281, 1111–22. doi: 10.1002/ar.a.20117.

Henrich, J. (2015). *The Secret of Our Success. How Culture Is Driving Human Evolution, Domesticating Our Species, and Making Us Smarter*. Princeton, NJ: Princeton University Press.

Herculano-Houzel, S. (2009). The human brain in numbers: a linearly scaled-up primate brain. *Front Hum Neurosci* 3, 31. doi: 10.3389/neuro.09.031.2009.

Herculano-Houzel, S. (2011). Scaling of brain metabolism with a fixed energy budget per neuron: implications for neuronal activity, plasticity and evolution. *PLoS One* 6, e17514. doi: 10.1371/journal.pone.0017514.

Herculano-Houzel, S. (2013). Neuroscience. Sleep it out. *Science* 342, 316–7. doi: 10.1126/science.1245798.

Herculano-Houzel, S. (2015). Decreasing sleep requirement with increasing numbers of neurons as a driver for bigger brains and bodies in mammalian evolution. *Proc Biol Sci* 282, 20151853. doi: 10.1098/rspb.2015.1853.

Herculano-Houzel, S. (2016a). *The Human Advantage: A New Understanding of How Our Brain Became Remarkable*. Cambridge, MA: MIT Press.

Herculano-Houzel, S. (2016b). What modern mammals teach about the cellular composition of early brains and mechanisms of brain evolution. In: Kaas, J. H. (ed.) *Evolution of Nervous Systems*, 2nd ed. London: Academic Press.

Herculano-Houzel, S. and Kaas, J. H. (2011). Gorilla and orangutan brains conform to the primate cellular scaling rules: implications for human evolution. *Brain Behav Evol* 77, 33–44. doi: 10.1159/000322729.

Herculano-Houzel, S., Mota, B., Wong, P., et al. (2010). Connectivity-driven white matter scaling and folding in primate cerebral cortex. *Proc Natl Acad Sci U S A* 107, 19008–13. doi: 10.1073/pnas.1012590107.

Herculano-Houzel, S., Avelino-De-Souza, K., Neves, K., et al. (2014a). The elephant brain in numbers. *Front Neuroanat* 8, 46. doi: 10.3389/fnana.2014.00046.

Herculano-Houzel, S., Manger, P. R., and Kaas, J. H. (2014b). Brain scaling in mammalian evolution as a consequence of concerted and mosaic changes in numbers of neurons and average neuronal cell size. *Front Neuroanat* 8, 77. doi: 10.3389/fnana.2014.00077.

Hoban, A. E., Stilling, R. M., Ryan, F. J., et al. (2016). Regulation of prefrontal cortex myelination by the microbiota. *Transl Psychiatry* 6, e774. doi: 10.1038/tp.2016.42.

Hodes, G. E., Kana, V., Menard, C., et al. (2015). Neuroimmune mechanisms of depression. *Nat Neurosci* 18, 1386–93. doi: 10.1038/nn.4113.

Hodos, W. and Butler, A. B. (2001). Sensory system evolution in vertebrates. In: Roth, G., Wullimann, M. F. (eds) *Brain Evolution and Cognition*. New York: Wiley.

Hoffmann, J. N., Montag, A. G., and Dominy, N. J. (2004). Meissner corpuscles and somatosensory acuity: the prehensile appendages of primates and elephants. *Anat Rec A Discov Mol Cell Evol Biol* 281, 1138–47. doi: 10.1002/ar.a.20119.

Hofman, M. A. (2001). Evolution and complexity of the human brain: some organizing principles. In: Roth, G. and Wullimann, M. F. (eds) *Brain Evolution and Cognition*. New York: Wiley.

Holt-Lunstad, J., Smith, T. B., and Layton, J. B. (2010). Social relationships and mortality risk: a meta-analytic review. *PLoS Med* 7, e1000316. doi: 10.1371/journal.pmed.1000316.

Hopkins, W. D. and Rilling, J. K. (2000). A comparative MRI study of the relationship between neuroanatomical asymmetry and interhemispheric connectivity in primates: implication for the evolution of functional asymmetries. *Behav Neurosci* 114, 739–48.

Hopkins, W. D., Marino, L., Rilling, J. K., et al. (1998). Planum temporale asymmetries in great apes as revealed by magnetic resonance imaging (MRI). *Neuroreport* 9, 2913–18.

Hrdy, S. B. (2000). *Mother Nature*. London: Vintage.

Hrdy, S. B. (2016). Variable postpartum responsiveness among humans and other primates with 'cooperative breeding': a comparative and evolutionary perspective. *Horm Behav* 77, 272–83. doi: 10.1016/j.yhbeh.2015.10.016.

Hublin, J.-J. (2005). Evolution of the human brain and comparative paleoanthropology. In: Dehaene, S., Duhamel, J.-R., Hauser, M., et al. (eds) *From Monkey Brain to Human Brain*. Cambridge, MA: MIT Press.

Humphrey, N. K. (1976). The social function of intellect. In: Bateson, P. P. G. and Hinde, R. A. (eds) *Growing Points in Ethology*. Cambridge: Cambridge University Press.

Hynes, R. O. and Naba, A. (2012). Overview of the matrisome—an inventory of extracellular matrix constituents and functions. *Cold Spring Harb Perspect Biol* 4, a004903. doi: 10.1101/cshperspect.a004903.

Irwin, M. R. and Opp, M. R. (2017). Sleep health: reciprocal regulation of sleep and innate immunity. *Neuropsychopharmacology* 42, 129–55. doi: 10.1038/npp.2016.148.

Jacobs, D. K., Nakanishi, N., Yuan, D., et al. (2007). Evolution of sensory structures in basal metazoa. *Integr Comp Biol* 47, 712–23. doi: 10.1093/icb/icm094.

Jerison, H. J. (1973). *Evolution of the Brain and Intelligence*. New York: Academic Press.

Jerison, H. J. (2001). The evolution of neural and behavioral complexity. In: Roth, G. and Wullimann, M. F. (eds) *Brain Evolution and Cognition*. New York: Wiley.

Jessen, N. A., Munk, A. S., Lundgaard, I., et al. (2015). The glymphatic system: a beginner's guide. *Neurochem Res* 40, 2583–99. doi: 10.1007/s11064-015-1581-6.

Joffe, T. H. (1997). Social pressures have selected for an extended juvenile period in primates. *J Hum Evol* 32, 593–605. doi: S0047-2484(97)90140-8.

Joiner, W. J. (2016). Unraveling the evolutionary determinants of sleep. *Curr Biol* 26, R1073–87. doi: 10.1016/j.cub.2016.08.068.

Jolly, A. (1966). Lemur social behavior and primate intelligence. *Science* 153, 501–6.

Jones, S., Martin, R. D., and Pilbeam, D. R. (1992). *The Cambridge Encyclopedia of Human Evolution*. Cambridge: Cambridge University Press.

Kaas, J. H. (2004). Evolution of somatosensory and motor cortex in primates. *Anat Rec A Discov Mol Cell Evol Biol* 281, 1148–56. doi: 10.1002/ar.a.20120.

Kanwisher, N. and Yovel, G. (2006). The fusiform face area: a cortical region specialized for the perception of faces. *Philos Trans R Soc Lond B Biol Sci* 361, 2109–28. doi: 10.1098/rstb.2006.1934.

Kassi, E. N. and Chrousos, G. P. (2013). The central CLOCK system and the stress axis in health and disease. *Hormones (Athens)* 12, 172–91.

Katz, D. B. and Steinmetz, J. E. (2002). Psychological functions of the cerebellum. *Behav Cogn Neurosci Rev* 1, 229–41.

Khaitovich, P., Hellmann, I., Enard, W., et al. (2005). Parallel patterns of evolution in the genomes and transcriptomes of humans and chimpanzees. *Science* 309, 1850–4. doi: 10.1126/science.1108296.

Kinsbourne, M. (2013). Somatic twist: a model for the evolution of decussation. *Neuropsychol* 27, 511–15. doi: 10.1037/a0033662.

Kirk, E. C. (2004). Comparative morphology of the eye in primates. *Anat Rec A Discov Mol Cell Evol Biol* 281, 1095–103. doi: 10.1002/ar.a.20115.

Klugmann, M., Goepfrich, A., Friemel, C. M., et al. (2011). AAV-mediated overexpression of the CB1 receptor in the mPFC of adult rats alters cognitive flexibility, social behavior, and emotional reactivity. *Front Behav Neurosci* 5, 37. doi: 10.3389/fnbeh.2011.00037.

Kohl, S., Kainer, F., and Schiefenhövel, W. (2009). Nausea and vomiting as evolutionary mechanisms of the complex adaptation reaction to pregnancy. *Z Geburtshilfe Neonatol* 213, 186–93. doi: 10.1055/s-0029-1224188.

Kohl, Z., Schlachetzki, J. C., Feldewerth, J., et al. (2017). Distinct pattern of microgliosis in the olfactory bulb of neurodegenerative proteinopathies. *Neural Plast* 2017, 3851262. doi: 10.1155/2017/3851262.

Kourtzi, Z. and Kanwisher, N. (2000). Activation in human MT/MST by static images with implied motion. *J Cogn Neurosci* 12, 48–55.

Krause, M., Theiss, C., and Brüne, M. (2017a). Ultrastructural alterations of Von Economo neurons in the anterior cingulate cortex in schizophrenia. *Anat Rec (Hoboken)* 300, 2017–24. doi: 10.1002/ar.23635.

Krause, A. J., Simon, E. B., Mander, B. A., et al. (2017b). The sleep-deprived human brain. *Nat Rev Neurosci* 18, 404–18. doi: 10.1038/nrn.2017.55.

Kunz, A. R. and Iliadis, C. (2007). Hominid evolution of the arteriovenous system through the cranial base and its relevance for craniosynostosis. *Childs Nerv Syst* 23, 1367–77. doi: 10.1007/s00381-007-0468-5.

Leppanen, J. M. and Nelson, C. A. (2009). Tuning the developing brain to social signals of emotions. *Nat Rev Neurosci* 10, 37–47. doi: 10.1038/nrn2554.

Lowe, C. J., Terasaki, M., Wu, M., et al. (2006). Dorsoventral patterning in hemichordates: insights into early chordate evolution. *PLoS Biol* 4, e291. doi: 10.1371/journal.pbio.0040291.

Lowry, C. A., Smith, D. G., Siebler, P. H., et al. (2016). The microbiota, immunoregulation, and mental health: implications for public health. *Curr Environ Health Rep* 3, 270–86. doi: 10.1007/s40572-016-0100-5.

Maeda, N. (2015). Proteoglycans and neuronal migration in the cerebral cortex during development and disease. *Front Neurosci* 9, 98. doi: 10.3389/fnins.2015.00098.

Matthews, L. J. and Butler, P. M. (2011). Novelty-seeking DRD4 polymorphisms are associated with human migration distance out-of-Africa after controlling for neutral population gene structure. *Am J Phys Anthropol* 145, 382–9. doi: 10.1002/ajpa.21507.

Mauney, S. A., Athanas, K. M., Pantazopoulos, H., et al. (2013). Developmental pattern of perineuronal nets in the human prefrontal cortex and their deficit in schizophrenia. *Biol Psychiatry* 74, 427–35. doi: 10.1016/j.biopsych.2013.05.007.

McCullough, M. E., Kurzban, R., and Tabak, B. A. (2013). Cognitive systems for revenge and forgiveness. *Behav Brain Sci* 36, 1–15. doi: 10.1017/S0140525X11002160.

McEwen, B. S. (2017). Neurobiological and systemic effects of chronic stress. *Chronic Stress* 1, 1–11.

McGann, J. P. (2017). Poor human olfaction is a 19th-century myth. *Science* 356(6338). doi: 10.1126/science.aam7263.

McKinney, M. L. and McNamara, K. (1991). *Heterochrony: The Evolution of Ontogeny*. New York: Plenum Press.

McLoughlin, B. C., Pushpa-Rajah, J. A., Gillies, D., et al. (2014). Cannabis and schizophrenia. *Cochrane Database Syst Rev* CD004837. doi: 10.1002/14651858.CD004837.pub3.

Meltzer, A. and Van De Water, J. (2017). The role of the immune system in autism spectrum disorder. *Neuropsychopharmacology* 42, 284–98. doi: 10.1038/npp.2016.158.

Meyer, M. R. (2016). The spinal cord in hominin evolution. In: eLS *Evolution and Diversity of Life*. Chichester: John Wiley and Sons. doi: 10.1002/9780470015902.a0027058.

Meyer, M. R. and Haeusler, M. (2015). Spinal cord evolution in early *Homo*. *J Hum Evol* 88, 43–53. doi: 10.1016/j.jhevol.2015.09.001.

Meyer-Lindenberg, A., Domes, G., Kirsch, P., et al. (2011). Oxytocin and vasopressin in the human brain: social neuropeptides for translational medicine. *Nat Rev Neurosci* 12, 524–38. doi: 10.1038/nrn3044.

Michopoulos, V., Vester, A., and Neigh, G. (2016). Posttraumatic stress disorder: a metabolic disorder in disguise? *Exp Neurol* 284, 220–9. doi: 10.1016/j.expneurol.2016.05.038.
Miller, A. H. and Raison, C. L. (2016). The role of inflammation in depression: from evolutionary imperative to modern treatment target. *Nat Rev Immunol* 16, 22–34. doi: 10.1038/nri.2015.5.
Miller, A. H., Haroon, E., and Felger, J. C. (2017). Therapeutic Implications of brain-immune interactions: treatment in translation. *Neuropsychopharmacology* 42, 334–59. doi: 10.1038/npp.2016.167.
Miller, B. J. and Goldsmith, D. R. (2017). Towards an immunophenotype of schizophrenia: progress, potential mechanisms, and future directions. *Neuropsychopharmacology* 42, 299–317. doi: 10.1038/npp.2016.211.
Miller, S. L. and Maner, J. K. (2010). Scent of a woman: men's testosterone responses to olfactory ovulation cues. *Psychol Sci* 21, 276–83. doi: 10.1177/0956797609357733.
Mishkin, M. and Ungerleider, L. G. (1982). Contribution of striate inputs to the visuospatial functions of parieto-preoccipital cortex in monkeys. *Behav Brain Res* 6, 57–77.
Mitchell, R. L. and Crow, T. J. (2005). Right hemisphere language functions and schizophrenia: the forgotten hemisphere? *Brain* 128, 963–78. doi: 10.1093/brain/awh466.
Monroe, S. M. and Simons, A. D. (1991). Diathesis-stress theories in the context of life stress research: implications for the depressive disorders. *Psychol Bull* 110, 406–25.
Montagu, A. (1989). *Growing Young*. Granby, MA: Bergin & Garvey.
Morena, M., Patel, S., Bains, J. S., et al. (2016). Neurobiological interactions between stress and the endocannabinoid system. *Neuropsychopharmacology* 41, 80–102. doi: 10.1038/npp.2015.166.
Moroz, L. L. (2009). On the independent origins of complex brains and neurons. *Brain Behav Evol* 74, 177–90. doi: 10.1159/000258665.
Myers, A. J., Gibbs, J. R., Webster, J. A., et al. (2007). A survey of genetic human cortical gene expression. *Nat Genet* 39, 1494–9. doi: 10.1038/ng.2007.16.
Navarrete, A., Van Schaik, C. P., and Isler, K. (2011). Energetics and the evolution of human brain size. *Nature* 480, 91–252. doi: 10.1038/Nature10629.
Nedergaard, M., Ransom, B., and Goldman, S. A. (2003). New roles for astrocytes: redefining the functional architecture of the brain. *Trends Neurosci* 26, 523–30. doi: 10.1016/j.tins.2003.08.008.
Nelson-Coffey, S. K., Fritz, M. M., Lyubomirsky, S., et al. (2017). Kindness in the blood: a randomized controlled trial of the gene regulatory impact of prosocial behavior. *Psychoneuroendocrinology* 81, 8–13. doi: 10.1016/j.psyneuen.2017.03.025.
Nieuwenhuys, R. (2012). The insular cortex: a review. *Prog Brain Res* 195, 123–63. doi: 10.1016/B978-0-444-53860-4.00007-6.
Niraula, A., Sheridan, J. F., and Godbout, J. P. (2017). Microglia priming with aging and stress. *Neuropsychopharmacology* 42, 318–33. doi: 10.1038/npp.2016.185.
Northcutt, R. G. (2001). Changing views of brain evolution. *Brain Res Bull* 55, 663–74.
Olsson, M. J., Lundstrom, J. N., Kimball, B. A., et al. (2014). The scent of disease: human body odor contains an early chemosensory cue of sickness. *Psychol Sci* 25, 817–23. doi: 10.1177/0956797613515681.
Pakkenberg, B. and Gundersen, H. J. (1997). Neocortical neuron number in humans: effect of sex and age. *J Comp Neurol* 384, 312–20.
Panksepp, J. (1998). *Affective Neuroscience: The Foundations of Human and Animal Emotions*. New York: Oxford University Press.
Paus, T., Tomaiuolo, F., Otaky, N., et al. (1996). Human cingulate and paracingulate sulci: pattern, variability, asymmetry, and probabilistic map. *Cereb Cortex* 6, 207–14.
Pavlov, V. A. and Tracey, K. J. (2017). Neural regulation of immunity: molecular mechanisms and clinical translation. *Nat Neurosci* 20, 156–66. doi: 10.1038/nn.4477.
Peters, A. (2011). The selfish brain: competition for energy resources. *Am J Hum Biol* 23, 29–34. doi: 10.1002/ajhb.21106.
Pineda, J. A. (2005). The functional significance of mu rhythms: translating 'seeing' and 'hearing' into 'doing'. *Brain Res Brain Res Rev* 50, 57–68. doi: 10.1016/j.brainresrev.2005.04.005.
Pollard, K. S., Salama, S. R., King, B., et al. (2006). Forces shaping the fastest evolving regions in the human genome. *PLoS Genet* 2, e168. doi: 10.1371/journal.pgen.0020168.

Pontzer, H., Brown, M. H., Raichlen, D. A., et al. (2016). Metabolic acceleration and the evolution of human brain size and life history. *Nature* 533, 390–2. doi: 10.1038/nature17654.

Porges, S. W. (1995). Orienting in a defensive world: mammalian modifications of our evolutionary heritage. A polyvagal theory. *Psychophysiology* 32, 301–18.

Porges, S. W. (2009). The polyvagal theory: new insights into adaptive reactions of the autonomic nervous system. *Cleve Clin J Med* 76(Suppl 2), S86–90. doi: 10.3949/ccjm.76.s2.17.

Powers, K. E. and Heatherton, T. F. (2012). Characterizing socially avoidant and affiliative responses to social exclusion. *Front Integr Neurosci* 6, 46. doi: 10.3389/fnint.2012.00046.

Prehn-Kristensen, A., Wiesner, C., Bergmann, T. O., et al. (2009). Induction of empathy by the smell of anxiety. *PLoS One* 4, e5987. doi: 10.1371/journal.pone.0005987.

Premack, D. and Woodruff, G. (1978). Does the chimpanzee have a Theory of Mind? *Behav Brain Sci* 1, 515–26.

Preston, S. D. (2013). The origins of altruism in offspring care. *Psychol Bull* 139, 1305–41. doi: 10.1037/a0031755.

Preston, S. D., Hofelich, A. J., and Stansfield, R. B. (2013). The ethology of empathy: a taxonomy of real-world targets of need and their effect on observers. *Front Hum Neurosci* 7, 1–13. doi: 10.3389/Fnhum.2013.00488.

Rapoport, S. I. (1990). Integrated phylogeny of the primate brain, with special reference to humans and their diseases. *Brain Res Rev* 15, 267–94. doi: 10.1016/0165-0173(90)90004-8.

Rauschecker, J. P. and Scott, S. K. (2009). Maps and streams in the auditory cortex: nonhuman primates illuminate human speech processing. *Nat Neurosci* 12, 718–24. doi: 10.1038/nn.2331.

Raven, P., Johnson, G., Mason, K., et al. (2017a). *Biology. Chapter 43: The Nervous System.* New York: McGrath-Hill Education.

Raven, P., Johnson, G., Mason, K., et al. (2017b). *Biology. Chapter 44: Sensory Systems.* New York: McGrath-Hill Education.

Reist, C., Ozdemir, V., Wang, E., et al. (2007). Novelty seeking and the dopamine D4 receptor gene (DRD4) revisited in Asians: haplotype characterization and relevance of the 2-repeat allele. *Am J Med Gen B Neuropsychiatr Genet* 144B, 453–7. doi: 10.1002/ajmg.b.30473.

Ribeiro, P. F., Manger, P. R., Catania, K. C., et al. (2014). Greater addition of neurons to the olfactory bulb than to the cerebral cortex of eulipotyphlans but not rodents, afrotherians or primates. *Front Neuroanat* 8, 23. doi: 10.3389/fnana.2014.00023.

Rilling, J. K. and Insel, T. R. (1999). The primate neocortex in comparative perspective using magnetic resonance imaging. *J Hum Evol* 37, 191–223. doi: 10.1006/jhev.1999.0313.

Rilling, J. K. and Seligman, R. A. (2002). A quantitative morphometric comparative analysis of the primate temporal lobe. *J Hum Evol* 42, 505–33. doi: 10.1006/jhev.2001.0537.

Riska, B. and Atchley, W. R. (1985). Genetics of growth predict patterns of brain-size evolution. *Science* 229, 668–71. doi: 10.1126/science.229.4714.668.

Rizzolatti, G. and Craighero, L. (2004). The mirror-neuron system. *Ann Rev Neurosci* 27, 169–92. doi: 10.1146/annurev.neuro.27.070203.144230.

Rizzolatti, G., Fogassi, L., and Gallese, V. (2002). Motor and cognitive functions of the ventral premotor cortex. *Curr Opin Neurobiol* 12, 149–54.

Rolls, E. T. (2004). Convergence of sensory systems in the orbitofrontal cortex in primates and brain design for emotion. *Anat Rec A Discov Mol Cell Evol Biol* 281, 1212–25. doi: 10.1002/ar.a.20126.

Rose, K. A., Morgan, I. G., Ip, J., et al. (2008). Outdoor activity reduces the prevalence of myopia in children. *Ophthalmologyl* 115, 1279–85. doi: 10.1016/j.ophtha.2007.12.019.

Roshchina, V. V. (2010). Evolutionary considerations of neurotransmitters in microbial, plant, and animal cells. In: Lyte, M. and Freestone, P. P. E. (eds) *Microbial Endocrinology: Interkingdom Signaling in Infectious Disease and Health.* Berlin: Springer, pp. 17–52.

Ross, H. E., Freeman, S. M., Spiegel, L. L., et al. (2009). Variation in oxytocin receptor density in the nucleus accumbens has differential effects on affiliative behaviors in monogamous and polygamous voles. *J Neurosci* 29, 1312–18. doi: 10.1523/JNEUROSCI.5039-08.2009.

Roth, G. and Wullimann, M. F. (eds) (2001). *Brain Evolution and Cognition.* New York: Wiley.

Ryan, T. J. and Grant, S. G. (2009). The origin and evolution of synapses. *Nat Rev Neurosci* 10, 701–12. doi: 10.1038/nrn2717.

Samson, D. R., Crittenden, A. N., Mabulla, I. A., et al. (2017). Chronotype variation drives night-time sentinel-like behaviour in hunter-gatherers. *Proc Biol Sci* 284(1858), pii: 20170967. doi: 10.1098/rspb.2017.0967.

Sapolsky, R. M. (2015). Stress and the brain: individual variability and the inverted-U. *Nat Neurosci* 18, 1344–6. doi: 10.1038/nn.4109.

Saugstad, L. F. (1998). Cerebral lateralisation and rate of maturation. *Int J Psychophysiol* 28, 37–62. http://dx.doi.org/10.1016/S0167-8760(97)00063-9.

Schmitz, J., Lor, S., Klose, R., et al. (2017). The functional genetics of handedness and language lateralization: insights from gene ontology, pathway and disease association analyses. *Front Psychol* 8, 1144. doi: 10.3389/fpsyg.2017.01144.

Sei, Y., Ren-Patterson, R., Li, Z., et al. (2007). Neuregulin1-induced cell migration is impaired in schizophrenia: association with neuregulin1 and catechol-O-methyltransferase gene polymorphisms. *Mol Psychiatry* 12, 946–57. doi: 10.1038/sj.mp.4001994.

Semendeferi, K., Damasio, H., Frank, R., et al. (1997). The evolution of the frontal lobes: a volumetric analysis based on three-dimensional reconstructions of magnetic resonance scans of human and ape brains. *J Hum Evol* 32, 375–88. doi: 10.1006/jhev.1996.0099.

Semendeferi, K., Armstrong, E., Schleicher, A., et al. (1998). Limbic frontal cortex in hominoids: a comparative study of area 13. *Am J Phys Anthropol* 106, 129–55. doi: 10.1002/(SICI)1096-8644(199806)106:2<129::AID-AJPA3>3.0.CO;2-L

Semendeferi, K., Armstrong, E., Schleicher, A., et al. (2001). Prefrontal cortex in humans and apes: a comparative study of area 10. *Am J Phys Anthropol* 114, 224–41. doi: 10.1002/1096-8644(200103)114:3<224::AID-AJPA1022>3.0.CO;2-I.

Semendeferi, K., Lu, A., Schenker, N., et al. (2002). Humans and great apes share a large frontal cortex. *Nat Neurosci* 5, 272–6. doi: 10.1038/nn814nn814.

Seyfarth, R. M., Cheney, D. L., and Marler, P. (1980). Monkey responses to three different alarm calls: evidence of predator classification and semantic communication. *Science* 210, 801–3.

Seymour, R. S., Bosiocic, V., and Snelling, E. P. (2016). Fossil skulls reveal that blood flow rate to the brain increased faster than brain volume during human evolution. *R Soc Open Sci* 3, 160305. doi: 10.1098/rsos.160305.

Sharpe, L. T., Stockman, A., Jägle, H., et al. (1999). Opsin genes, cone photopigments, color vision, and color blindness. In: Gegenfurter, K. R. and Sharpe, L. T. (eds) *Color Vision. From Genes to Perception*. Cambridge: Cambridge University Press.

Shinbrot, T. and Young, W. (2008). Why decussate? Topological constraints on 3D wiring. *Anat Rec (Hoboken)* 291, 1278–92. doi: 10.1002/ar.20731.

Silvis, S. M., De Sousa, D. A., Ferro, J. M., et al. (2017). Cerebral venous thrombosis. *Nat Rev Neurol* 13, 555–65. doi: 10.1038/nrneurol.2017.104.

Singer, T., Seymour, B., O'Doherty, J., et al. (2004). Empathy for pain involves the affective but not sensory components of pain. *Science* 303, 1157–62. doi: 10.1126/science.1093535303/5661/1157.

Sousa, A. M. M., Meyer, K. A., Santpere, G., et al. (2017a). Evolution of the human nervous system: function, structure, and development. *Cell* 170, 226–47. doi: 10.1016/j.cell.2017.06.036.

Sousa, A. M. M., Zhu, Y., Raghanti, M. A., et al. (2017b). Molecular and cellular reorganization of neural circuits in the human lineage. *Science* 358, 1027–32. doi: 10.1126/science.aan3456.

Spoor, F., Wood, B., and Zonneveld, F. (1994). Implications of early hominid labyrinthine morphology for evolution of human bipedal locomotion. *Nature* 369, 645–8. doi: 10.1038/369645a0.

Spoor, F., Hublin, J.-J., Braun, M., et al. (2003). The bony labyrinth of Neanderthals. *J Hum Evol* 44, 141–65.

Stahl, S. M. (2017). Dazzled by the dominions of dopamine: clinical roles of D3, D2, and D1 receptors. *CNS Spectr* 22, 305–11. doi: 10.1017/S1092852917000426.

Stearns, S. C. (1976). Life-history tactics: a review of the ideas. *Q Rev Biol* 51, 3–47.

Stehle, J. H., Saade, A., Rawashdeh, O., et al. (2011). A survey of molecular details in the human pineal gland in the light of phylogeny, structure, function and chronobiological diseases. *J Pineal Res* 51, 17–43. doi: 10.1111/j.1600-079X.2011.00856.x.

Stephan, H. (1979). Comparative volumetric studies on striatum in insectivores and primates. Evolutionary aspects. *Appl Neurophysiol* 42, 78–80.

St John-Smith, P., McQueen, D., Edwards, L., et al. (2013). Classical and novel psychoactive substances: rethinking drug misuse from an evolutionary psychiatric perspective. *Hum Psychopharmacol* 28, 394–401. doi: 10.1002/hup.2303.

Striedter, G. F. (2006). Precis of principles of brain evolution. *Behav Brain Sci* 29, 1–12; discussion 12–36. doi: 10.1017/S0140525X06009010.

Suomi, S. J. (2003). Gene–environment interactions and the neurobiology of social conflict. *Ann N Y Acad Sci* 1008, 132–9.

Szyf, M. (2015). Nongenetic inheritance and transgenerational epigenetics. *Trends Mol Med* 21, 134–44. doi: 10.1016/j.molmed.2014.12.004.

Tabibnia, G., Satpute, A. B., and Lieberman, M. D. (2008). The sunny side of fairness: preference for fairness activates reward circuitry (and disregarding unfairness activates self-control circuitry). *Psychol Sci* 19, 339–47. doi: 10.1111/j.1467-9280.2008.02091.x.

Tehovnik, E. J., Sommer, M. A., Chou, I. H., et al. (2000). Eye fields in the frontal lobes of primates. *Brain Res Brain Res Rev* 32, 413–48.

Toga, A. W., Thompson, P. M., Mori, S., et al. (2006). Towards multimodal atlases of the human brain. *Nat Rev Neurosci* 7, 952–66. doi: 10.1038/Nrn2012.

Tomasello, M. and Call, J. (1997). *Primate Cognition*. New York: Oxford University Press.

Tomasello, M. and Vaish, A. (2013). Origins of human cooperation and morality. *Ann Rev Psychol* 64, 231–55. doi: 10.1146/annurev-psych-113011-143812.

Torrey, E. F. (2007). Schizophrenia and the inferior parietal lobule. *Schizophr Res* 97, 215–25. doi: 10.1016/j.schres.2007.08.023.

Trezza, V., Damsteegt, R., Manduca, A., et al. (2012). Endocannabinoids in amygdala and nucleus accumbens mediate social play reward in adolescent rats. *J Neurosci* 32, 14899–908. doi: 10.1523/JNEUROSCI.0114-12.2012.

Trivers, R. L. (1971). Evolution of reciprocal altruism. *Q Rev Biol* 46(1), 35–57. doi: 10.1086/406755.

Trivers, R. L. (1974). Parent–offspring conflict. *Am Zool* 14, 249–64.

Trivers, R. (2000). The elements of a scientific theory of self-deception. *Ann N Y Acad Sci* 907, 114–31.

Umilta, M. A., Kohler, E., Gallese, V., et al. (2001). I know what you are doing. A neurophysiological study. *Neuron* 31, 155–65.

Vandenberghe, R. (2016). Classification of the primary progressive aphasias: principles and review of progress since 2011. *Alzheimers Res Ther* 8, 16. doi: 10.1186/s13195-016-0185-y.

Von Hippel, W. and Trivers, R. (2011). The evolution and psychology of self-deception. *Behav Brain Sci* 34, 1–16; discussion 16–56. doi: 10.1017/S0140525X10001354.

Walker, S. C. and McGlone, F. P. (2013). The social brain: neurobiological basis of affiliative behaviours and psychological well-being. *Neuropeptides* 47, 379–93. doi: 10.1016/j.npep.2013.10.008.

Walter, H., Adenzato, M., Ciaramidaro, A., et al. (2004). Understanding intentions in social interaction: the role of the anterior paracingulate cortex. *J Cogn Neurosci* 16, 1854–63. doi: 10.1162/0898929042947838.

Wang, H., Kim, M., Normoyle, K. P., et al. (2016). Thermal regulation of the brain—an anatomical and physiological review for clinical neuroscientists. *Front Neurosci* 9, 528. doi: 10.3389/fnins.2015.00528.

Watson, C., Provis, J., and Herculano-Houzel, S. (2012). What determines motor neuron number? Slow scaling of facial motor neuron numbers with body mass in marsupials and primates. *Anat Rec (Hoboken)* 295, 1683–91. doi: 10.1002/ar.22547.

Watson, C., Mitchelle, A., and Puelles, L. (2016). A new mammalian brain ontology based on developmental gene expression. In: Kaas, J. H. (ed.) *Evolution of Nervous Systems*, 2nd ed. London: Elsevier.

Weber, M. D., Godbout, J. P., and Sheridan, J. F. (2017). Repeated social defeat, neuroinflammation, and behavior: monocytes carry the signal. *Neuropsychopharmacology* 42, 46–61. doi: 10.1038/npp.2016.102.

Wheeler, P. (1988). Stand tall and stay cool. *New Scientist* 12 May, 62–5.

Williams, B. A., Kay, R. F., and Kirk, E. C. (2010). New perspectives on anthropoid origins. *Proc Natl Acad Sci U S A* 107, 4797–804. doi: 10.1073/pnas.0908320107.

Yokoyama, S., Xing, J., Liu, Y., et al. (2014). Epistatic adaptive evolution of human color vision. *PLoS Genet* 10, e1004884. doi: 10.1371/journal.pgen.1004884.

Zilles, K. (1987). Graue und weiße Substanz des Hirnmantels. In: Leonhardt, H., Tillmann, B., Töndury, G., et al. (eds) *Anatomie des Menschen. Band III. Nervensystem und Sinnesorgane.* Stuttgart: Thieme.

Zilles, K. (2005). Evolution of the human brain and comparative cyto- and receptor architecture. In: Dehaene, S., Duhamel, J.-R., Hauser, M., et al. (eds) *From Monkey Brain to Human Brain.* Cambridge, MA: MIT Press.

PART III

FUTURE DIRECTIONS

CHAPTER 18

THE FUTURE OF MEDICINE

MARTIN BRÜNE AND WULF SCHIEFENHÖVEL

Abstract

Ever since the dawn of humankind, people have been preoccupied with their physical and mental health. Even though many causes of disease remained obscure for millennia, humans have been quite inventive to find remedies for many ailments. Only quite recently, discoveries of pathogenic agents such as bacteria and viruses have paved the way for the development of modern antimicrobial therapies and preventive measures including hygiene. We now know, however, that the indiscriminate use of antibiotics has produced new problems, and excessive hygiene may be causally involved in the emergence of autoimmune diseases. Chapter 18 proposes that evolutionary medicine may be informative in understanding seemingly unpredictable consequences of advances in some areas of medical treatment. Moreover, the authors believe that evolutionary medicine could help avoid many of the pitfalls associated with the negligence of important evolutionary concepts by contemporary medicine.

Keywords

physical and mental health, pathogenic agents, antimicrobial treatment, treatment resistance, hygiene, autoimmune diseases, medical advancements, evolutionary medicine

18.1 Introduction

Ever since the dawn of self-consciousness, and the awareness of one's own death and vulnerability, early humans have been preoccupied with their physical and mental health. Healers or shamans and midwives were thus perhaps the first professions in ancestral human societies. Biologically, this makes perfect sense, as all organisms have evolved species-typical defence mechanisms against injury, predation, and infectious attacks to ensure survival and procreation, which, at times, may break down, in spite of their exquisite design. So, clearly, in social and cooperative species like ourselves, mutual help in times of dearth or illness has had selective advantages for both the helping individual and the recipient of support, such that 'reciprocal altruism' could prevail (Trivers 1971). Human concerns with health and disease surely predated knowledge about how germs travel from one individual to another; and the use of medicinal plants, as well as support of physically disabled conspecifics have already been observed in nonhuman primates (van Lawick Goodall 1971; Huffman 1997). However, the evolved ability to mentally travel in time, that is, to use knowledge about past events to predict future contingencies, is most likely more human-specific (Suddendorf 2013), and definitively key to the prevention, recognition, and treatment of disease.

In a way, this is the route we aim to take here. Specifically, we intend to discuss what evolutionary approaches can contribute to future medicine, what the limitations are, and what we cannot predict from our present-day stance. We dare to take a hypothetical look into the future, in a hundred years' time from now, to speculate about how practicing medicine may have changed, what novel challenges may have occurred, and what new therapeutic developments may have emerged. In doing so, we focus on aspects relevant for prevention, diagnosis, and treatment of diseases, as well as public health issues and medical policy, always seen through the lens of evolutionary medicine and acknowledging our evolved mental and physical heritage. Specifically, we are interested in questions such as future cancer therapy, treatment developments for currently intractable conditions like Alzheimer's disease, the emergence of new epidemics and other novel threats to health, availability of new antibiotics, gene therapy, longevity, and artificial reproduction, as well as questions pertaining to the yawning gulf between rich and poor countries and what this will mean for access to healthcare.

Before delving into these areas of medical practice in greater detail, however, we believe that a look back into the past—say, around a hundred to fifty years from now—may be informative with regard to the prospects of medical achievements, because, as we assert, the 'fate' of current and future therapeutic approaches will share some important characteristics with past developments.

18.2 Historical Considerations

In the early decades of the twentieth century, and as a matter of fact for all prehistoric and historic times until then, infectious diseases were the most common causes of deaths worldwide (this is still the case in many countries around the globe), and were only replaced around 1950 by coronary heart disease as the number one cause of death, at least in so-called developed countries. Moreover, cancers surged in number, partly due to changes in lifestyle such as smoking, but also due to environmental pollution and workplace-associated exposure to toxic substances like asbestos and anilin. Rickets became 'endemic' among children working in coal mines and living in industrial cities due to vitamin D deficiency caused by the lack of exposure to sunlight, and silicosis was almost inevitable for mine workers in Britain and the European continent.

Conversely, when looking at medical advancements of the time, one could compile a formidable list of important discoveries. Here are a few exemplars, with no claim of completeness whatsoever. The pain killer acetylsalicylic acid could be produced on a large scale by the second half of the nineteenth century, though the bark and leaves of willow trees rich in acetylsalicylic acid had been used for more than 2000 years. Similarly, general anaesthesia and analgesia constantly improved and became safer. Around the same time, research on pathogenic bacteria became possible with technical improvements in microscopy. Robert Koch discovered *Bacillus anthracis* and *Mycobacterium tuberculosis*, to name just a few of his merits. Louis Pasteur promoted vaccination (the first smallpox vaccination being credited to Edward Jenner much earlier) and disproved the doctrine of 'spontaneous generation' of infectious agents. Another important discovery of the outgoing nineteenth century was made by Wilhelm Conrad Röntgen who managed to produce the first X-ray images, on which Pierre and Marie Curie elaborated to discover the origins of radioactivity and how to make use of it medically in diagnosing bone fractures and in treating tumour cells by radium. In 1928, Alexander Fleming discovered (by chance) penicillin, which became clinically available in the early 1940s. Sulfonamides, however, were the first antibiotics to be commercially produced and available. Other researchers succeeded in insulin extraction in the 1920s to treat diabetes mellitus. The (independent) rediscoveries of the Mendelian rules of inheritance by Carl Erich Correns, Erich Tschermak von Seysenegg, Hugo Marie de Vries, and William Jasper Spillman were another breakthrough in biology with relevance to medicine, yet the structure of the double-helix was only deciphered in the early 1950s and published by James D. Watson and Francis H. C. Crick in 1953. A further true revolution that changed the sexual lives of women (and men) was associated with the development of hormonal contraceptives in 1960, though their use for everyone was not legalised before 1972. We leave out even more recent milestones like the first genetic cloning, *in vitro* fertilisation, and non-invasive intervention technologies, which future textbook authors may recapitulate when, in the year 2100, looking back on medical discoveries a century ago.

All these late nineteenth/early twentieth century discoveries were associated with hopes and expectations to eliminate many life-threatening diseases, foremost the ones that were infectious or malignant in nature. Alas, this turned out to be too optimistic. In fact, antibiotic resistance due to spontaneous or random mutations of bacteria was already demonstrated by Salvador Luria and Max Delbrück in the 1940s. This was clearly an indication of ongoing and observable evolution in the Petri dish or glass tube, given the high rate of reproductive cycles of bacteria. Notably, Fleming, in his Nobel laureate speech delivered in 1945, already warned that the development of antibiotic resistance could be a severe drawback for the treatment of infectious diseases.

18.2.1 Contemporary Practical Implications of Historical Discoveries

There is no doubt that the discovery of antimicrobial substances and other milestones of medical advancement such as radiotherapy and general anaesthesia, as well as treatments for metabolic diseases such as diabetes, saved millions of lives worldwide. Unfortunately, with regard to antimicrobial treatment, early warnings were largely disregarded or ignored in medical practice, such that generations of physicians overprescribed antibiotics and thus contributed to the development of resistant strains, and continue to do so (Lee et al. 2014). In addition, the massive use of antibiotics in industrialised agriculture and livestock increased the rate of resistance dramatically. The reasons for the overuse of antimicrobial substances in human and veterinary medicine are manifold, but one possibility is, as we may speculate, that evolutionary principles in medical schools have not been taught in depth as they should have. Recently, a new debate has been instilled about limiting the duration of antibiotic treatment, instead of giving a full 7- to 10-day course, to fight the development of antibiotic resistance (Llewelyn et al. 2017). Moreover, the World Health Organization (WHO 2014) has devoted a 219-page volume to the global threats posed by increasing and alarming rates of antimicrobial resistance worldwide, suggesting that 'greater emphasis should be placed on prevention, including strengthening hygiene and infection prevention and control measures, improving sanitation and access to clean water, and exploring a more widespread use of vaccines. Although preventive vaccines have become available for several bacterial infections, their application is still limited' (p. 2).

Similar stories could be told about the widespread prescription of powerful analgesic substances, specifically opioids, for unwarranted periods of time, causing kidney and liver disease, drug dependency, and other side-effects including dysphoria (Zellner et al. 2011). According to the Centers for Disease Control and Prevention (CDC), drug overdose is now the leading cause of deaths in the United States in people aged 50 or under (www.cdc.gov). This clearly is an iatrogenic epidemic incompatible with the Hippocratic dictum of 'nihil nocere' (do not cause harm).

Likewise, the prescription of hormonal contraceptives and hormone replacement therapy, including testosterone for anti-ageing purposes, can be associated with side-effects (Sansone et al. 2017), including increased risk for cancer depending on the hormone or hormonal combination and the tissue (breast, endometrium) (Chlebowski and Anderson 2014), though hormone replacement therapy has the potential to reduce the risk of coronary heart disease and osteoporotic fractures in women (Lobo 2017).

It is not our intention to denigrate or disprove the huge health benefits these medical innovations have brought about. The point we wish to make here is that unwarranted and unindicated prescription of pharmacological treatment can be associated with severe side-effects, sometimes adverse consequences that are poorly known at the time a new drug is thrown on the market, and sometimes covered up by the pharmaceutical industry. More importantly, we want to emphasise that evolutionary approaches to medicine can help reduce the risk of unwanted effects of any medical treatment. A simple example in this regard may illustrate this claim: the human appendix has long been regarded as a useless evolutionary vestige, the surgical removal of which has no adverse effects. There is no doubt that appendectomy is a safe and indicated treatment of acute appendicitis (the first reported and successful operation was carried out in the eighteenth century); however, recent evidence suggests that appendectomy may increase the risk for *Clostridium difficile*-associated colitis (e.g. Sanders et al. 2013), and that the appendix may exert important immunological functions, thus dismissing its vestigial status (Kooij et al. 2016).

Together, these historical examples illustrate, as we contend, that many medical discoveries and innovations may have downsides or side-effects that may better be understood when taking into account evolutionary explanations of their impact on biological tissue, organisms, or ecosystems. We predict that this will happen in the future to present-day advancements for the same reasons, with potentially huge impact on prevention, diagnosis, and treatment of disease, and medical policy-making. These predictions are also based on and compatible with evidence suggesting that humans continue to evolve (Stearns et al. 2010), whereby selection for pathogen resistance has played an important role (Karlsson et al. 2014), a fact that is often overlooked or insufficiently acknowledged in contemporary medicine.

18.3 Prevention

The WHO promotes two different, though related strategies to improve health. One is called promotion, which is the attempt to increase well-being, strengthen resilience, and improve living and environmental conditions. This approach explicitly assumes that well-being is not just the absence of disease. The other strategy concerns prevention *sensu stricto*, which focuses on the reduction of symptoms, frequently combined with measures of health promotion. Preventive measures differ according to timing of the intervention and the target population. Primary prevention aims at reducing the risk of health problems at the population level, whereby the population is not specifically at an increased risk for disease. From an evolutionary point of view, and in light of what has been outlined in the chapters of this volume, promotion of well-being and primary prevention of illness could include ways of controlling stress exposure during pregnancy, supply of clean water, careful choice of diet, reduction of environmental pollution, and improved hygiene. We are, however, sceptical that this will happen in the next hundred years, as environmental destruction, reduction of biodiversity close to mass extinction, and climate change continue to threaten whole ecosystems in a dangerous, accelerating spiral. Thus, we expect that a broad spectrum of old and new stressors will lead to a deterioration of conditions for many people in the world, perhaps sparing just a few.

Aside from these worldwide consequences of what we would like to call 'habitat destruction' with disastrous impact on large proportions of the world population, particularly children, it is also realistic to predict that in rich countries, embryos will be genetically manipulated (the recent controversy about genetic manipulation in utero carried out by the Chinese researcher He Jiankui shows that this has obviously already become reality) (Cyranoski and Ledford 2018) to promote or prevent the expression of a certain trait. We believe that this is going to happen in the not so distant future, in spite of serious ethical concerns. It is equally likely that this will have terrible consequences for some, because of the lack of control of epistatic (gene–gene interaction) and pleiotropic (genes impacting on more than one trait) effects, as well as insufficient knowledge about gene–environment interactions, and epigenetic control of gene expression.

Another reason for our scepticism of the success of promotion and primary prevention concerns the failure of health education programmes in western societies. In fact, as we have learnt in Chapter 6, only 5% of adult Americans exercise 150 minutes per week, as recommended by the American College of Sports Medicine (Troiano et al. 2008). (For further discussion, see Chapter 6: Nutrition, Energy Expenditure, Physical Activity, and Body Composition.) In a more general vein, treatment adherence with regard to dietary recommendations or medication is relatively poor, independent of diagnosis, hovering around 20–50% of patients with chronic diseases (WHO 2003). We assert that the disregard of medical advice resides, in part, in the human capacity to repress undesired feelings. According to psychoanalytic theory, individuals possess the ability to conceal unacceptable feelings or desires from the self by keeping them unconscious. Put another way, there are several immature and more mature repression mechanisms that help to actively distort and keep out of conscious awareness cognitive appraisals of one's real motives (Northoff et al. 2007). Repression, therefore, shares several features with self-deception (Trivers 2000; von Hippel and Trivers 2011), and is by no means necessarily pathological. The cognitive ability to deceive oneself may have evolved to increase the ability to deceive others (Trivers 2000). The logic behind this idea is that an individual who is unaware of his or her selfish motives can send more convincing signals to others and thus conceal his or her real intention. As deception and cheater detection may have played a significant role in human cognitive evolution (Trivers 1971), self-deception, too, increases individual fitness. Intriguingly, it seems that natural selection has not favoured cognitive abilities to produce accurate images of the world, but has favoured cognitive abilities to systematically distort conscious awareness and to block inadvertent access to non-conscious information processing (Nesse and Lloyd 1992). Accordingly, the failure of health education programmes may exactly be due to distortions of people's self-images and false appraisals of one's own health status and vulnerability.

Another causal factor in the failure of health education programmes relates to the human capacity to restrict energy expenditure. Physical activity of hunter-gatherers may have been substantially greater than in other hominids because (as we have learnt from Chapter 6 about Nutrition, Energy Expenditure, Physical Activity, and Body Composition, and Chapter 17 about Brain, Spinal Cord, and Sensory Systems) the human basal metabolic rate is exceptionally high, whereby a substantial amount of energy goes to the large human brain. So, humans have evolved physical traits for endurance running to hunt large game, for instance, but also to save energy in times of food scarcity. Physical inactivity, though detrimental to health in modern environments due to a dysbalance between food (i.e. calorie) intake and energy expenditure, was adaptive in ancestral conditions in which future resource availability was unpredictable (see Chapter 6 about Nutrition, Energy Expenditure, Physical Activity, and Body Composition for further details). Consistent with life history models of energy allocation (Stearns 1976, 1992), one would predict that primary prevention programmes are particularly difficult to sell to those who are in need most. These are individuals who lead unhealthy lives, consume unhealthy ('fast food') nutrients like saturated fats and large amounts of sugar, smoke, are overweight, and so on, based on their unconscious predictions that future resource availability, including financial and workplace security, and social integration in the community, is uncertain. So, why routinely exercise if it doesn't pay off anyway? If policy makers want this to change, they need to address the whole bundle of socioeconomic and biobehavioural problems to make primary prevention programmes work.

In contrast to primary prevention, selective prevention seeks to counteract the development of disease in populations at risk. This could become a target of new technologies in cancer therapy and prevention—for example, in individuals with elevated risks for certain malignant tumours that run in their families.

Similar scenarios may apply to indicated prevention, which targets individuals at high risk of developing a certain disease in the sense of personalised medicine, and who already express minimal signs or symptoms. Precancerous tissue or other biological markers which are likely associated with the emergence of serious illness may be an example. A similar strategy may also work for Alzheimer's disease.

Likewise, secondary prevention, which aims to reduce prevalence rates of diseases by early detection and treatment, could benefit from new technological developments such as genetic engineering. Thus, it could be, optimistically seen, that the burden of cancer will diminish in the future, for example, by controlling the microenvironment of cancers. That is, one may not be able to avoid cancer development by preventing oncogenic mutations altogether, but it may be possible to control these mutations by altering the parameters of a tissue to make the microenvironment more unfavourable for somatic cancer evolution.

Finally, tertiary prevention seeks to reduce disability, relapse, and recurrence of illness and to enhance rehabilitation, which, ideally, would become less necessary if health promotion and primary and secondary prevention became more successful.

It is clear that any means of health promotion and prevention of disease critically depends on the identification of risk factors and protective factors. Risk factors concern the probability of manifestation, and the severity and duration of illness. Protective factors, in contrast, increase resilience.

Socioeconomic risk factors, such as poverty, poor nutrition and hygiene, lack of access to clean water, exposure to between-group aggression, including war, flight, and displacement, racial discrimination, and access to healthcare, have great impact on well-being and health. With regard to *mental* health, a growing concern worldwide, individual and family-related risk factors have the greatest impact on psychological well-being. Childhood adversity and parental illness are non-specific risk factors for psychiatric disorders. There is also growing evidence that the fetal environment and maternal stress during pregnancy leave pervasive biological marks on the immunology and stress axis of the next generation. Preventive factors, therefore, include promotion of secure attachment, parental emotional responsivity, and social support. From an evolutionary perspective, vulnerability to disease depends on complex interactions between genetic factors, developmental exposures to pathogens, psychological distress, immunology, and stress responsivity, with substantial differences between populations and geographic regions.

As the economic divide between developed and developing countries is widening, there is little hope that access to healthcare and more sophisticated treatment strategies will be attainable for the majority of the world population. Programmes developed by the WHO to utilise elements and engage practitioners of traditional medicine were largely given up, even though the involvement of traditional birth attendants (TBAs), for example, seemed to be a promising avenue.

18.4 Diagnosis and Treatment

Diagnosis and treatment of many diseases will benefit from personalised approaches. Technical improvements and lower costs for genome-wide sequencing will allow us to not only detect individual risk genes, but also sequence cancer genomes to tailor individualised immune interventions.

Digitalisation will change access to diagnostic tools, and in some areas of medicine it may not even be necessary to see a doctor, even though it is hardly imaginable that face-to-face contact, therapeutic compassion, and consolation will be abandoned or regarded superfluous, given the desire for closeness and trust in times of sickness or disease, which is so deeply rooted in our evolutionary primate heritage (the German term *Behandlung*, i.e. treatment, literally means being treated by the hand of the doctor). Alas, even today there is little time for this kind of humaneness in modern clinics and practices, which are trimmed for speed and efficiency. Technical innovations will perhaps replace human surgeons by robots in some domains. It is at present unclear if norm values for blood pressure, body mass index, blood sugar, and so on will change in the future. From an evolutionary vantage point, it is interesting to note that population averages for many metabolic products (with the notable exception of cholesterol), blood pressure, and body mass index are below the corresponding figures for people of Caucasian descent. (For further discussion, see Chapter 6: Nutrition, Energy Expenditure, Physical Activity, and Body Composition.) Evolutionary medicine would accordingly argue that our 'western' standards are inadequate, and that it might be more appropriate, biologically, to adjust current normative values to those found in pre-industrial hunter-gatherer or horticultural societies.

We believe that these examples dispute the widespread prescription of medication for lifestyle purposes, as pointed out in Section 18.2.1. Consistent with these concerns, 'disease mongering' (Moynihan and Henry 2006)—that is, the widening of illness boundaries, defining new thresholds for 'normalcy', or the declaration of normal coping or defence mechanisms as diseases, as has happened for the premenstrual syndrome, bereavement, and other useful responses to stress—has become fashionable. The risk that such developments prevail is substantial, as monetary interests of the pharmaceutical and bioengineering industry, as well as medical institutions continue to push the pathologisation of adaptive defences.

Medicine has a long tradition in the support of healing with minimal intervention. This is useful from an evolutionary vantage point, as selection has endowed organisms with powerful self-repair mechanisms, including blood clotting, inflammation, and other immune responses to pathogenic intruders. (For further discussion, see Chapter 10: Immune System, and Chapter 9: Haematopoetic System.)

Some evolutionary medical principles are present in early classical texts, such as the corpus hippocraticum (https://www.loebclassics.com), the basis of scientific medicine. For example, the Hippocratic corpus emphasised the self-healing properties of the human body—a very modern and evolutionary approach (Schiefenhövel 1995). Rest, advised by an uninterrupted row of doctors from ancient to modern times, is another very popular therapeutic measure. While this is reasonable in many cases, for example, during acute stages of infectious and non-infectious disease, it is ill-advised for a number of conditions. Halhuber, an Austrian internist, was first heavily criticised by his fellow doctors when he suggested not to immobilise patients with angina pectoris and after ischaemic heart attacks, but to get them back on their feet as early as possible to exercise. 'Cardio training' is now standard in tertiary prevention of cardiovascular diseases.

Along similar lines, rethinking has taken place in other fields of medicine. Certain forms of pelvic, clavicular, and other fractures are not necessarily treated by splint or osteosynthetic surgery, but by some moderate form of immobilisation. The commonly practiced 'RICE' (rest, ice, compression, elevation above heart level) treatment of sprains and similar conditions has not been found to be as effective as previously thought (Bekerom et al. 2012). These examples assign a role to modern medicine that focuses on evidence-based treatments that are effective and, almost equally important, do not have unwanted side-effects, for example, in the case of skeletal damage, muscular atrophy, and joint problems after longer immobilisation.

Research in wild populations of non-human primates suggests behavioural programmes of self-protection and intuitive usage of medical plants (Huffman 1997; Fruth et al. 2014). Chimpanzees swallow whole leaves which have been shown to function as controlling intestinal parasites; the leaves are defecated *in toto*, that is, they do not serve a nutritional purpose. Chimpanzees also ingest bitter-tasting pith of *Vernonia amydalina*, which is effective in reducing nematode infestation of the gut; the local human people in the area use the same plant, that is, they have the same ethnomedicinal knowledge (Huffman 1997). This was the only avenue to restore health for our early primate ancestors. The question how our ancestors got phytotherapeutic, chiropractic, and other medically relevant know-how is fascinating but poorly researched. Only so much can be said here: like other animals, early humans will have used acute perception of olfactory stimuli coming from plants, an experimental mindset, and a good sense of causality to find medicinally active principles, and passed this knowledge on to the next generation. A number of very efficient pharmacological substances in modern medicine have been borrowed from the pharmacopoeia of people in traditional societies around the globe (Schiefenhövel and Prinz 1984). At some point in our phylogeny, perhaps with the evolution of *Homo erectus* some two million years ago, mental concepts of care for injured and sick members of family and the group developed further, motivated by the high degree of empathic concern (for further discussion, see Chapter 17: Brain, Spinal Cord, and Sensory Systems). Most probably, this greatly improved our chances for survival—at the same time it bore a risk: caring for someone with a dangerous infectious disease can easily kill the healer. Still, the new concepts and techniques of keeping and restoring health, developed in those ancient times and still present in all cultures of the world, paid off: we became able to fight disease with new methods.

The present opioid crisis in the United States, a veritable tragedy stemming from a combination of social problems and severe iatrogenic damage by inducing addiction, is already mentioned in Section 18.2.1. Findings that nociception is 'fixed' in the synapses, which represent part of our memory function (Zieglgänsberger and Tölle 1993), led to the conclusion that even minor pains need to be counteracted by analgesic means to prevent 'fixation' of pain memories and the development of chronic pain (a big problem in modern societies). This concept runs diametrically against the ones of evolutionary medicine. All animals are subjected to pain and it would be most surprising if such episodes would cripple an individual for life. Also, our human bodies are equipped with highly effective mechanisms to fight pain—the well-known beta-endorphin is one of the elements in this system. Research in traditional societies suggests that it is wrong to take refuge in pharmacological analgesia: when children are exposed to pain and live in an environment without readily available pain killers, they develop an astounding capacity to tolerate even severe pain (Schiefenhövel 1995). This 'pain training' empowers them to manage life with its many risks of getting physically injured. In a similar way, by experiencing and mastering injury, damage, loss, and grief (supported by family and group; Schiefenhövel 2002), as well as dangerous, frightening situations, members of these kinds of societies are psychologically 'immunised' against anxiety disorders, which are so common in our societies and are likely to become even more of a problem in the future—despite the fact that at present we live, in western countries, a more sheltered and secure life than ever before in our history.

18.5 Global Healthcare Issues

Global healthcare and international medical policy ought to consider insights from evolution in medicine. We have learnt in the past what can happen if population differences in genetic make-up and vulnerability to disease are insufficiently reflected. For example, the United Nations (UN) well-intended famine-relief programmes in the 1950s and 1960s inadvertently caused sickness in sub-Saharan populations due to widespread lactose intolerance. This failure occurred because policy makers were ignorant of the fact that lactase persistence differs between ethnic groups. It is very common in northern Europe and in people of Scandinavian descent, but the figures are very low in sub-Saharan Africa. The first description of lactose intolerance in a European infant dates back to the late 1950s (Holzel et al. 1959), while reliable population estimates of lactase persistence became only recently available (Itan et al. 2010).

We are now witnessing that acculturation, that is, the adoption of a 'western' lifestyle and diet, is associated with mushrooming incidence rates of 'diseases of civilisation' of all kinds. These changes have partly occurred within a single generation in hitherto highly athletic and healthy peoples. One of us (WS) has carried out research in the highlands of Indonesian New Guinea (province of Papua) in an ethnic group, the Eipo, who until recently lived as horticulturalists without access to western culture. They present an interesting example of how adaptable our mind and how conservative our physiological mechanisms are. While they lived Neolithic lives in the isolated Star Mountains of western New Guinea, almost 90% of their diseases were caused by infections and infestations, while everyone had the body and endurance of athletes. There was no obesity and no anorexia, no high blood pressure, and thus no cardiovascular consequences like coronary heart disease or stroke (Schiefenhövel 1982, 1994). This has changed dramatically, within only one generation (so it cannot be genetic). Those of the Eipo who now live a sedentary life on the coast often suffer from overweight and problems secondary to this—our body and mind are made for a life of high physical activity and meagre supply of healthy fresh food. These ethnic groups may possess 'thrifty genes', which are responsible for members of Melanesian and Polynesian societies developing massive bodies once they are exposed to western food and living conditions. The situation in populations like the Eipo demonstrates, in a kind of unplanned biocultural experiment, the central principles of evolutionary medicine: for example, the age of menarche has dropped dramatically from about 17 to about 13–14; the onset of female reproductive functions in our species is, as to be expected from a life history perspective, ecologically sensitive. (For further discussion, see Chapter 15: Endocrinology.) Most impressive is the cognitive ability of the Eipo to cope with the completely new life they are immersed in now: almost everyone speaks Bahasa Indonesia, the *lingua franca*, and can handle money; an impressive number manage computers, university education, and the complex political structure they live in very well; they have begun, quite self-confidently, to play a role in the local parliament and in decisions for their future. Their mothers and fathers were Stone Age people who had no knowledge of the world outside their mountains.

In Papua New Guinea, the eastern half of the island of New Guinea, which was an Australian protectorate until 1975, the first report of an ischaemic stroke was published in the mid-1970s; atherosclerotic and other cardiovascular diseases were previously unknown also there (except malaria-associated incidents) (Mathews 1974).

Another pressing problem that warrants recognition of evolutionary insights pertains to sexually transmitted diseases and birth control programmes. Although the number of human immunodeficiency virus (HIV) infections in the developed countries has declined over the last years, it has not, to the same degree, in developing countries. According to WHO data, there were about 37 million people infected with HIV worldwide and about one million died from the disease. New infections mainly occur in key populations such as homosexual men, sex workers, intravenous drug abusers, and prisoner inmates, but also in heterosexual people, whereas the highest rates of new infections are still observed in sub-Saharan Africa (WHO 2017).

Likewise, birth control programmes have failed to keep the world population constant. Instead, there are now 7 billion people inhabiting the globe, most likely too many, in the long run, to feed in ecologically sustainable ways. The biological impetus intrinsic in all organisms to 'produce offspring' is detached from the living conditions of our ancestors, where procuring enough food for survival and reproduction was often a very difficult task and infant mortality high, so that the number of children usually did not exceed the carrying capacity of the respective environment. Today, even in parts of the world where people are suffering from the disastrous effects of civil or other war, fertility is often high. The question is whether all massive governmental and non-governmental efforts to establish birth control, for example, by condom use (which would at the same time be an effective protection against the transmission of sexual diseases), will work well enough. China has been very successful in cutting birth rates, at high psychological and social costs. In our own societies, we are experiencing the opposite trend: birth rates are so low that deaths are by far more common than births. A large role in this development is the decision of couples to postpone, for various reasons, first pregnancy; this results in pronounced loss of fertility. (For further discussion, see Chapter 15: Endocrinology, and Chapter 16: Sexuality, Reproduction, and Birth.) Modern life (where one factor probably relates to the high survival rates of infants and children, which could have led to having fewer children in whom one would invest more) has obviously changed people's psychology and minds, so that they became detached from evolved reproductive mechanisms, which have worked in the past and are still at work in many parts of the world.

From life history theory (LHT) it is known that people adjust their resource allocation to somatic growth or reproduction depending on the safety and predictability of future environments. Put another way, if prospects are such that future resources are likely to be scarce, individuals tend to choose (unconsciously) a 'faster' life history strategy (LHS), that is, they tend to mature earlier, reproduce earlier, are more indiscriminate in regard to mate choice, invest less in own children, and are psychologically more impulsive and exploitative in terms of trustful and cooperative relationships (Stearns 1976; Ellis et al. 2011). Slow LHS are characterised by the opposite pattern—slower somatic maturation, later reproduction, higher investment in mate choice, own children, and so on. There are certainly within-population differences in regard to LHS; for example, in psychopathological conditions like borderline personality disorder, many clinical features are compatible with a 'fast' LHS, which may be promoted by childhood adversity and attachment insecurity (Brüne 2016). The point we wish to make here, however, is that there are most likely between-population differences, based on differential availability of resources. Put differently, poverty, homelessness, environmental destruction, high pathogen exposure, and so on contribute to the adoption of 'fast' LHS. The political consequences and public health measures that are needed to approach these problems are clear: less exploitation of developing countries by developed countries, financial investment in improvement of living condition in economically unstable societies, improvement of access to healthcare, investment in education, and so on, which together help people slow down their LHS.

As regards healthcare, there is certainly need for improvement with regard to not only malnourishment, clean water, and other hygiene measures, but also vaccination. It is a shame that malaria vaccination is still in its infancy and not available for those who are in need of support most. The same applies to cholera vaccination and other potentially lethal infections. Ironically, as climate change progresses, some of the deadliest diseases may migrate from the tropics north (and south) to the wealthier countries in moderate climate zones, such that almost certainly more serious endeavours will be undertaken to develop vaccines or other more powerful treatments.

Poignantly, in some western societies, there is some kind of vaccination 'fatigue', sometimes based on esoteric or religious concerns, which has repeatedly led to endemic outbreaks of measles and other highly contagious diseases. It is not clear why politicians are reluctant to introduce compulsory vaccination programmes in the developed world. One of the greatest success stories in this respect was certainly the worldwide eradication of smallpox in 1980 by means of consequent vaccination (and low mutation rate of the *Variola* virus), though there are certainly stocks of the virus for biological warfare purposes.

18.6 Unpredictable Issues

As noted in Section 18.2.1, humans continue to evolve (Stearns et al. 2010), but we have little clue where this will go. Modern medicine contributes to alterations in selection pressures by changing microbial environments, improving metabolic constraints on reproduction, and by superior treatments of cancer. There is also evidence for increasing variation, as shown for the M. palmaris longus. (For further discussion, see Chapter 7: Musculoskeletal System.) In contrast, environmental contingencies seem to increase vulnerability for depression worldwide. This may partly be linked to changes in exposure to pathogens (for further discussion, see Chapter 10: Immune System) but certainly also to increases in exposure to psychosocial stressors. As Gilbert (1998) once put it, never in human history have we been forced to express ourselves as 'socially attractive packages'. As the world population continues to move from rural areas to megacities, an evolutionary reason for increasing psychiatric problems resides in greater pressure on intrasexual competition. Furthermore, in countries in which the ratio of male to female newborns is artificially shifted to the male sex (like in India and China), the intrasexual competition among young men increases dramatically for access to a reduced number of females. (For further discussion, see Chapter 1: Core Principles for Evolutionary Medicine.) How this impacts human evolution is unclear.

Other uncertainties concern the ongoing environmental destruction. This not only reduces biodiversity, but also may prevent the discovery of new antibiotic or anticancer drugs. Moreover, human-made disasters based on the use or misuse of nuclear power, bioterrorism, and other horrific scenarios are also entirely unpredictable.

A final issue that is hard to envision is the extension of the human lifespan. It may be a biologically attainable goal for a few individuals in wealthy countries, particularly if the pathways of tissue replacement and maintenance are better understood and the potential costs of continuing cell division (i.e. cancer) can be controlled. If it can also be a desirable aim will likely be hotly debated in philosophical circles.

18.7 Conclusion

Evolutionary medicine has the potential to help prevent, diagnose, and treat a broad array of diseases. There is, however, anti-evolutionism in the midst of the medical profession. It almost looks as if authors avoid using the term 'evolution' in publications, and one has to concede the possibility that the desire for 'meaning' (an evolved human trait) is so powerful that religious and esoteric beliefs (which cost very little, as we have learnt in Chapter 1: Core Principles for Evolutionary Medicine) may prevent scientific progress from happening.

Phylogenetic and functional perspectives typical for evolutionary biology and anthropology as well as for evolutionary medicine can shed light on many facets of human life. Why are we the way we are, what is our inner essence, our *conditio humana*? These are not just philosophical questions. They concern our daily lives and therefore also medicine and the other professions on which we depend for being well. Looking back into our animal past, understanding the Darwinian principles of natural and sexual selection as well as modern ecological perspectives like LHT, utilising cross-cultural epidemiology and similar approaches, integrating the results of empirical modern science and medicine, and trying to sketch strategies for the future is the path we have taken in this book. We hope it will contribute to understanding and solving some of the big tasks ahead of us.

Acknowledgement

We are grateful to our fellow authors of this volume for sharing their valuable insights into how medicine may look in a hundred years' time.

References

Bekerom, M. P. J. van den, Struijs, P. A. A., Blankevoort, L., et al. (2012). What is the evidence for rest, ice, compression, and elevation therapy in the treatment of ankle sprains in adults? *J Athletic Train* 47(4), 435–43.

Brüne, M. (2016). Borderline personality disorder: why 'fast and furious'? *Evol Med Public Health* 2016, 52–66. doi: 10.1093/emph/eow002.

Chlebowski, R. T. and Anderson, G. L. (2014). Menopausal hormone therapy and cancer: changing clinical observations of target site specificity. *Steroids* 90, 53–9. doi: 10.1016/j.steroids.2014.06.001.

Cyranoski, D. and Ledford, H. (2018). International outcry over genome-edited baby claim. The revelation from a Chinese scientist represents a controversial leap in genome editing. *Nature* 563, 607–8.

Ellis, B. J., Shirtcliff, E. A., Boyce, W. T., et al. (2011). Quality of early family relationships and the timing and tempo of puberty: effects depend on biological sensitivity to context. *Dev Psychopathol* 23, 85–99. doi: 10.1017/S0954579410000660 S0954579410000660 [pii].

Fruth, B., Ikombe, N. B., Matshimba, G. K., et al. (2014). New evidence for self-medication in bonobos: *Manniophyton fulvum* leaf- and stemstrip-swallowing from LuiKotale, Salonga National Park, DR Congo. *Am J Primatol* 76(2), 146–58. doi: 10.1002/ajp.22217. Epub 2013 Sep 30.

Gilbert, P. (1998). Evolutionary psychopathology: why isn't the mind designed better than it is? *Br J Med Psychol* 71(Part 4), 353–73.

Holzel, A., Schwarz, V., and Sutcliffe, K. W. (1959). Defective lactose absorption causing malnutrition in infancy. *Lancet* 1, 1126–8.

Huffman, M. A. (1997). Current evidence for self-medication in primates: a multidisciplinary perspective. *Yearb Phys Anthropol* 40, 171–200.

Itan, Y., Jones, B. L., Ingram, C. J., et al. (2010). A worldwide correlation of lactase persistence phenotype and genotypes. *BMC Evol Biol* 10, 36. doi: 10.1186/1471-2148-10-36.

Karlsson, E. K., Kwiatkowski, D. P., and Sabeti, P. C. (2014). Natural selection and infectious disease in human populations. *Nat Rev Genet* 15, 379–93. doi: 10.1038/nrg3734.

Kooij, I. A., Sahami, S., Meijer, S. L., et al. (2016). The immunology of the vermiform appendix: a review of the literature. *Clin Exp Immunol* 186, 1–9. doi: 10.1111/cei.12821.

Lee, G. C., Reveles, K. R., Attridge, R. T., et al. (2014). Outpatient antibiotic prescribing in the United States: 2000 to 2010. *BMC Med* 12, 96. doi: 10.1186/1741-7015-12-96.

Llewelyn, M. J., Fitzpatrick, J. M., Darwin, E., et al. (2017). The antibiotic course has had its day. *BMJ* 358, j3418. doi: 10.1136/bmj.j3418.

Lobo, R. A. (2017). Hormone-replacement therapy: current thinking. *Nat Rev Endocrinol* 13, 220–31. doi: 10.1038/nrendo.2016.164.

Mathews, C. L. (1974). Cardiovascular disease in Lae—a five year review. *Papua New Guinea Med J* 17, 251–62.

Moynihan, R. and Henry, D. (2006). The fight against disease mongering: generating knowledge for action. *PLoS Med* 3, e191.

Nesse, R. M. and Lloyd, A. T. (1992). The evolution of psychodynamic mechanisms. In: Barkow, J. H., Cosmides, L., and Tooby, J. (eds) *The Adapted Mind. Evolutionary Psychology and the Generation of Culture*. Oxford: Oxford University Press.

Northoff, G., Bermpohl, F., Schoeneich, F., et al. (2007). How does our brain constitute defense mechanisms? First-person neuroscience and psychoanalysis. *Psychother Psychosom* 76, 141–53. doi: 10.1159/000099841.

Sanders, N. L., Bollinger, R. R., Lee, R., et al. (2013). Appendectomy and *Clostridium difficile* colitis: relationships revealed by clinical observations and immunology. *World J Gastroenterol* 19, 5607–14. doi: 10.3748/wjg.v19.i34.5607.

Sansone, A., Sansone, M., Lenzi, A., et al. (2017). Testosterone replacement therapy: the emperor's new clothes. *Rejuvenation Res* 20, 9–14. doi: 10.1089/rej.2016.1818.

Schiefenhövel, W. (1982). Results of ethnomedical fieldwork among the Eipo, Daerah Jayawijaya, Irian Jaya, with special reference to traditional birthgiving. *Medika* 11(8), 829–43.

Schiefenhövel, W. (1994). Krankheit, Altern, Tod. In: Schiefenhövel, W., Vogel, Ch., Vollmer, G., et al. (eds) *Zwischen Natur und Kultur. Der Mensch und seine Beziehungen*. Stuttgart: Trias, pp. 217–44.

Schiefenhövel, W. (1995). Perception, expression, and social function of pain: a human ethological view. *Sci Context* 8(1), 31–46.

Schiefenhövel, W (2002). Evolutionäre und transkulturelle Perspektiven in der Psychiatrie. Trauer und Depression. *Nervenheilkunde* 3, 119–26.

Schiefenhövel, W. and Prinz, A. (1984). Ethnomedizin und Ethnopharmakologie—Quellen wichtiger Arzneimittel. In: Czygan, F. C. (ed.) *Biogene Arzneistoffe*. Wiesbaden: Vieweg, Braunschweig, pp. 223–38.

Stearns, S. C. (1976). Life-history tactics: a review of the ideas. *Q Rev Biol* 51, 3–47.

Stearns, S. C. (1992). *The Evolution of Life Histories*. Oxford: Oxford University Press.

Stearns, S. C., Byars, S. G., Govindaraju, D. R., et al. (2010). Measuring selection in contemporary human populations. *Nat Rev Genet* 11, 611–22. doi: 10.1038/nrg2831 nrg2831 [pii].

Suddendorf, T. (2013). *The Gap: The Science of what Separates us from Other Animals*. New York: Basic Books.

Trivers, R. L. (1971). Evolution of reciprocal altruism. *Q Rev Biol* 46(1), 35–57. doi: 10.1086/406755.

Trivers, R. (2000). The elements of a scientific theory of self-deception. *Evol Persp Hum Reprod Behav* 907, 114–31.

Troiano, R. P., Berrigan, D., Dodd, K. W., et al. (2008). Physical activity in the United States measured by accelerometer. *Med Sci Sports Exerc* 40(1), 181–8.

Van Lawick-Goodall, J. (1971). *In the Shadow of Man*. Boston: Houghton Mifflin.

Von Hippel, W. and Trivers, R. (2011). The evolution and psychology of self-deception. *Behav Brain Sci* 34, 1–16; discussion 16–56. doi: S0140525X10001354 [pii] 10.1017/S0140525X10001354.

WHO (2003). Adherence to long-term therapies: evidence for action. Geneva: World Health Organization.

WHO (2014). Antimicrobial resistance. Global report on surveillance. Geneva: World Health Organization.

WHO (2017). Policy brief: consolidated guidelines on HIV prevention, diagnosis, treatment and care for key populations, 2016 update (WHO/HIV/2017.05). Licence: CC BY-NC-SA 3.0 IGO. Geneva: World Health Organization.

Zellner, M. R., Watt, D. F., Solms, M., et al. (2011). Affective neuroscientific and neuropsychoanalytic approaches to two intractable psychiatric problems: why depression feels so bad and what addicts really want. *Neurosci Biobehav Rev* 35, 2000–8. doi: 10.1016/j.neubiorev.2011.01.003.

Zieglgänsberger, W. and Tölle, R. (1993). The pharmacology of pain signalling. *Curr Opin Neurobiol* 3(4), 611–18.

Glossary

ABO blood groups: Blood groups in humans that refer to the A and B antigens expressed on red blood cells (and other cells). Serum antibodies naturally raised against antigens not present on an individual's red blood cells influence whether an infusion of blood is rejected or accepted.

Acetylation: In epigenetics, the attachment of an acetyl molecule into a histone to activate the expression of a gene.

Achromatopsia: A condition characterised by absence of colour vision.

Adaptation: A dynamic evolutionary process that fits a population of organisms to its environmental niche, enhancing its evolutionary fitness.

Adaptive immunity: Cellular and humoral immunity that is specifically targeted to antigens, such as through gene rearrangements of antigen receptors, and which results in memory of antigen exposure.

Adaptive landscape: A description of the possible fitness values for individuals, whether organisms or cells, of different genotypes in a particular environment.

Adiponectin: A protein hormone produced in fat cells (adipocytes) that is involved with energy regulation. Production is usually inversely related to leptin production.

Adolescence: A stage in the human life cycle covering the years after the onset of puberty until the onset of adulthood (approximately ages 9–18 years for girls and 11–21 years for boys). The adolescent phase is characterised by a growth spurt in height and weight, the development of secondary sexual characteristics, sociosexual maturation, and intensification of interest and practice of adult social, economic, and sexual activities.

Adrenal gland: A gland that is situated on top of the kidneys that secretes hormones that contribute to the regulation of metabolism, and water and electrolyte balance, with indirect effects on growth, reproduction, and immunological function. Metabolic hormones that are produced in the adrenal cortex include glucocorticoids, such as cortisol. Water and electrolyte balance hormones produced in the adrenal cortex include aldosterone. Other hormones such as dehydroepiandrosterone (DHEA) and DHEA-sulfate (DHEA-S) are produced and secreted from the zona reticularis of the adrenals and are vital for adrenarche and sexual maturation.

Adrenarche: Postnatal onset of secretion of the androgen hormones dehydroepiandrosterone (DHEA) and DHEA-sulfate (DHEA-S) from the zona reticularis of the adrenal gland.

In humans and chimpanzees adrenarche occurs between the ages of 6 and 10 years. In some other primates, such as the rhesus monkey, the up-regulation of DHEA and DHEA-S begins perinatally.

Adrenocorticotropic hormone (ACTH): A protein hormone produced in the pituitary gland in response to stimulation by corticotropic-releasing factor (CRF) from the hypothalamus. ACTH promotes the production and secretion of glucocorticoids such as cortisol in the adrenal gland.

Aerobic metabolism: A process to obtain energy through the combustion of carbohydrates, amino acids, and fats in the presence of oxygen.

Afrotherians: A clade of mammals comprising elephants, sea cows, hyraxes, aardvarks, sengis, tenrecs, and golden moles—that is, some insectivores and ungulates that share anatomical features such as non-descending testes, morphology of the placenta, and variable number of vertebrae.

Ageing/senescence: The intrinsic, progressive, and irreversible deterioration of most physiological processes as organisms get chronologically older.

Agnathans: A class of jawless fish, like lampreys; possibly the earliest vertebrates.

Air capillaries: Terminal minuscule respiratory units of the bird lung.

Aldosterone: A steroid hormone produced in the adrenal cortex that contributes to the regulation of water and electrolyte balance.

Allele: One of two alternative sequences of a gene.

Allicin: A compound with antimicrobial properties obtained from garlic.

Allometry: The study of the relationship between body size and shape, anatomy, and physiology: for example, between brain size relative to body size (or weight).

Alloparental care: The feeding, protection, transport, and general care of immature members of the social group by those who are not their own biological parents.

Allostasis: Adaptive regulation of physiological processes in response to stress exposure.

Allostatic load: Accumulation of physiological consequences ('wear and tear') of exposure to chronic stress exceeding the body's ability to maintain allostasis.

α-thalassaemia: A hereditary haemolytic disease caused by mutations in the HBA1/HBA2 genes, leading to faulty haemoglobin synthesis (and milder malaria disease).

Altricial offspring: Young that are hatched or born in a very immature and helpless condition, requiring care and feeding from parents or others for some time after birth.

Alveoli: Small air sacs located at the end of the air passageways of the lung that are the respiratory unit of ventilation.

Alzheimer's disease: The most common form of dementia, characterised by the presence of amyloid plaques and tau tangles in the brain, with insidious onset of memory loss, disorientation, neurological symptoms such as dyspraxia, visuo-constructive deficits, and behavioural signs.

Amniotomy: Mechanical rupture of the membranes of the amniotic sac to induce or hasten labour.

Amyloid: Aggregates of proteins that become folded into a shape that allows many copies of that protein to stick together forming fibrils. Insoluble accumulations of these proteins give rise to amyloidosis, which impairs the function of the affected organ.

Amyloid plaques: Proteinaceous deposits in the brain that are associated, perhaps causally, with Alzheimer's disease.

Anaerobic respiration: A process to obtain energy by the incomplete intracellular breakdown of sugar or other organic compounds in the absence of oxygen.

Anoxaemia: Absence of oxygen in the arterial blood.

Anoxia: Total lack or absence of oxygen.

Antagonist pleiotropy: An allele that causes disadvantages can be maintained in the gene pool if it also gives advantages. Alleles that are strongly selected for benefits in youth despite costs at advanced ages when selection is weaker offer the usual example.

Antagonistic pleiotropy theory: An evolutionary explanation of ageing which argues that ageing has evolved as a result of selection favouring the spread of mutations with beneficial effects at early ages but deleterious effects at late ages.

Antibiotic: An antimicrobial drug used in the treatment of bacterial infections. Some of these drugs can generate an allergic reaction and increased use has also led to the development of 'antibiotic-resistant' strains.

Aortic stenosis: Progressive narrowing across the aortic valve, commonly caused by dystrophic calcification in elderly patients.

Apnoea: A sleep disorder characterised by pauses in breathing or periods of shallow breathing during sleep.

ApoE: Abbreviation for an apolipoprotein of the epsilon class.

Apolipoprotein: A protein that bind lipids, such as triglycerides and cholesterol, into amalgams that transport the lipids through the lymphatic and circulatory systems.

Apoptosis: A programmed mechanism of cellular suicide.

Ashkenazi: A person belonging to a Jewish population of central or eastern European descent that coalesced around a millennium ago.

Astrocyte: (aka astroglia) A special type of glia cell characterised by its star-shaped appearance.

Atelectasis: Complete or partial collapse of a lung or lobe of a lung that develops when the alveoli become deflated.

Atherosclerosis: The progressive thickening and hardening of the inner layer of arteries with deposition of fat and cholesterol.

Atopy: A predisposition, which can be both genetic and environmental, towards development of allergic reactions.

ATP (adenosine triphosphate): A molecule that stores the energy produced by cellular respiration or photosynthesis, which is utilised in biological processes. It consists of adenosine, a nucleoside, associated with three phosphate radicals attached in a chain.

Atrial fibrillation: A common irregular heart beat affecting the heart's atria, which can predispose to blood clots, stroke, heart failure, and other heart-related complications.

Atrium: Either of the two upper chambers on each side of the heart that receive blood from the veins and in turn force it into the ventricles.

Autocrine: Autocrine signalling involves the secretion of a chemical messenger, whereby the autocrine receptor is located on the same cell.

Autoreactive: Elements of the immune system's attack mechanisms that recognise and target components of self: that is to say, cells or antibodies that attack the host's own tissues.

Autosomal recessive inheritance: Genetics where two unaffected parents each carry one copy of a gene mutation for a non-sex chromosome recessive disorder that can generate either an affected, or a carrier, or a non-carrier offspring.

Auxologist: A student or practitioner of auxology, the study of all aspects of human physical growth. Auxology is a multidisciplinary science involving paediatrics, neuroendocrinology, epidemiology, nutrition, genetics, anthropology, economic history, economics, sociology, public health, psychology, and other fields.

Balancing selection: Multiple different alleles can be maintained at a locus if they give increasing advantages when rare, or if different alleles give fitness advantages in different environments. Heterozygote advantage is the usual example.

Basal cell carcinoma: A non-melanoma skin cancer forming from the epidermis basal cells and contributing about 70% of non-melanoma skin cancers, with a higher frequency among Caucasians.

Bauplan: A German term in biology referring to morphological features shared by all members of a phylum (between the class level and the kingdom level according to the biological classification of organisms).

B-cell receptors (BCR): Receptors expressed on B cells that directly recognise antigens and alert other immune cells to the presence of a pathogen. BCR activation leads to maturation of the antigen-specific B cell clone and subsequent antibody production.

BCG: Acronym for Bacille Calmette Guerin, a bacterium used in a vaccine that stimulates marginal protection against tuberculosis.

Berylliosis: An inflammatory disease caused by exposure to the element beryllium and characterised by non-caseinating granulomas and fibrosis of the lungs (aka chronic beryllium disease).

Bichorial: Refers to two twins with each enclosed in its own chorionic sacs.

Bilaterians: Animals with a symmetrical *bauplan*, often synonymously used with triploblasts.

Bimodal breathing: Exchanging respiratory gases with water and air using different organs.

Biocultural perspective: The scientific exploration of the relationships between human biology and culture, based on the principle that culture is the sum of human technology, sociology, and ideology and that culture is a major human adaptation, permitting individuals and populations to adapt to widely varying local ecologies.

Biocultural reproduction: The human behavioural system for feeding and care of pregnant women and their offspring. Many members of the social group pool resources, such as food and labour, to increase the reproductive success of most adults.

Biological age: A measure of how well or poorly the body is functioning relative to the organism's chronological age.

Biomarkers of ageing: One or more molecular or physiological parameters that would describe physical vitality, and would predict the future onset of age-related diseases and risk of mortality more accurately than one's chronological age.

Biomodulator: Any substance that can change biological processes. An alternative (and broader) term for neurotransmitter.

Biosphere: Consists of all the regions of the surface and atmosphere of the Earth or another planet occupied by living organisms.

Blaschko lines (lines of Blaschko): Within skin development, this may represent pathways of epidermal cell migration and proliferation. A specific type of lupus erythematosus shows this distinctive pattern. Named after Alfred Blaschko, a German dermatologist who first described the feature in 1901.

Blood–gas barrier: A thin tissue barrier through which gases are exchanged between the alveolar air and the blood in the pulmonary capillaries.

Breech birth: Delivery of a fetus with the feet and legs first; occurs in 3–4% of deliveries.

Brodmann area: Cartographical mapping of the cerebral cortex according to microscopic cytoarchitectonic features developed by Korbinian Brodmann.

Bronchopulmonary dysplasia (BPD): Respiratory distress syndrome that is common in preterm infants.

Caecum: From Latin *caecum* meaning blind, this is a blind tube of the large bowel located at the junction of the small bowel and large bowel. Its size varies markedly in different animal species.

Cancer: Abnormal growth of cells characterised by invasive spread.

Candida: A family of skin-residing yeasts (fungi) causing skin and genital-tract infections.

Carpal tunnel syndrome: Occurs when the median nerve is compressed in its distal part close to the wrist. Typical symptoms are pain, numbness, and tingling in the thumb, index finger, and middle finger, as well as pain in the forearm and hand, often during the night (meralgia paraesthetica).

CD14: A receptor for the LPS that acts in concert with TLR4 to generate an inflammatory response to bacteria.

Cellular senescence: Permanent withdrawal of cells from the cell cycle, while remaining viable and metabolically active.

Cephalopelvic disproportion: A condition in which the fetal head is larger than the maternal birth canal, presenting challenges to delivery that may require caesarean section.

CFTR: *See* cystic fibrosis transmembrane conductance regulator.

Chemokine: Small signalling proteins secreted by cells.

Childhood: A stage in the human life cycle that occurs between the end of infancy and the start of semi-independent feeding. Children require intense care by older members of the social group due to the child's motor, neurological, and cognitive immaturity. Children must be provided specially prepared foods, because they cannot process the adult-type diet due to immaturity of their dentition and digestive systems.

Chitin: A polymer of the sugar N-acetylglucosamine found in the exoskeletons of insects, the cell walls of fungi, and certain hard structures in invertebrates and fish.

***Chlamydia* (aka *Chlamydophila*) *pneumoniae*:** An intracellular bacterium that infects the respiratory tract and has been associated with atherosclerosis and Alzheimer's disease.

Cholera: A diarrhoeal disease caused by *Vibrio cholerae* and associated with dehydration.

Cholera toxin: A protein produced by *Vibrio cholerae* that causes cholera by stimulating the cells that line the small intestine to secrete fluids.

Cholestyramine: A drug that is used to lower serum cholesterol levels by sequestering bile acids.

Chordates: Animals possessing a vertebral column or a notochord (dorsally located nervous system).

Chromatin: Chromosomal material, which is composed of DNA, RNA, and protein.

Chronic obstructive pulmonary disease (COPD): Progressive lung disease characterised by emphysema, chronic bronchitis, asthma, and some forms of bronchiectasis.

Chronological age: The age of an individual, measured in days, months, or years from the time the individual was born.

Clearance: The process of 'clearing' certain solutes or water from the blood, for example by excretion into the urine. Clearance is an important (but not the only) function of the kidney. For each substance, a certain clearance can be defined, which depends upon parameters such as serum concentration, size, charge, protein binding, filtration coefficient, and secretion via specific transporters.

Cliff-edged fitness landscapes: If fitness for a trait peaks close to a cliff edge, then natural selection will stabilise the population mean for the trait at a value that, because of stochastic variation, inevitably leaves some individuals with low fitness.

Clonal haematopoiesis: Expansions of cells bearing a particular genotype, such as occurs with oncogenically initiated haematopoietic cells in old age.

Coelom: A fluid-filled cavity in simple animals that can contribute to circulation.

Collaborative pretence: Creation of shared fictional realities between members of a social group based on common intentions.

Collagen: The main protein of the connective tissues, and present in tendons, ligaments, skin, cartilage, and bone, depending on the degree of mineralisation.

Commensal: A replicating agent that lives in or on a host organism, but neither harms nor benefits that host, which from an evolutionary perspective is indicated by no net effect on host fitness.

Commensalism: An association of a commensal with its host.

Compensating benefit of an allele: A positive effect of an allele on host fitness that offsets negative effects of the allele.

Complementarity-determining regions: The parts of the variable regions of antibodies and of the receptors expressed by lymphocytes that bind to antigens.

Concealed ('hidden') ovulation: In contrast to other primates, human females do not have obvious signs of ovulation. There are, however, indications that they behave differently at the peak of fertility around ovulation, dress differently, smell differently for males, and actively seek sexual intercourse. Ovulation in humans would thus be only partly concealed.

Conduction system: The group of specialised cardiac muscle cells in the heart that conduct signals in a synchronised manner to cause organised contraction of the heart.

Consortship: A bond between a female and a male animal (e.g. chimpanzees), who secretly leave the group and thus avoid control of the alpha male (and possibly alpha female); they have sex together in seclusion.

Cooperative breeding: A social system that is centred around alloparental care, a reproductive strategy by which social group members other than biological parents help to protect, care for, and provision young. Such alloparental care and provisioning allows mothers to breed at a faster pace without sacrificing offspring survival.

Corneocytes: Skin terminally differentiated keratinocytes forming the outer barrier of the stratum corneum.

Corticotropic-releasing factor (CRF): A protein hormone produced in the hypothalamus that stimulates production of ACTH in the pituitary gland.

Cortisol: A steroid hormone belonging to a class of glucocorticoids produced by the adrenal glands. Commonly associated with stress responses, cortisol facilitates the uptake of cellular glucose and can affect various physiological functions including reproduction, metabolism, growth, and immune function.

Counter-current exchange: A system in which there is movement (crossover) of some property or material when a concentration gradient is created between two masses, normally fluids, flowing in opposite directions.

C-reactive protein: A plasma protein that increases in response to inflammation.

Critical period: A time during the life cycle when one or more properties of the organism is constrained to grow or develop, or when this property develops most rapidly. Failure for proper growth or development cannot be fully recovered after this time.

Cross-current exchange: A system in which there is a crossover of some property or material when a concentration gradient is created between two masses flowing in perpendicular directions.

Cucinivore: A species that relies primarily on non-thermally and/or thermally prepared (cooked) foods for sustenance.

Cutis: An alternative skin term for the combined layers of the epidermis and the dermis layers.

Cystic fibrosis (CF): A multisystem disease affecting the airways, pancreas, intestines, liver, and reproductive system. CF is caused by a mutation in the transmembrane regulator gene (CFTR).

Cystic fibrosis transmembrane conductance regulator (CFTR): A protein that regulates the flow of ions from cells and thereby influences the viscosity of mucus.

Cytoarchitecture: Definition and categorisation of the cellular composition of body tissue, with particular relevance for the central nervous system.

Decussation: Crossings of spinocerebral or cerebrospinal tracts at different levels that putatively evolved with the emergence of the first notochord; also includes the chiasma opticum.

Defences: Useful but costly and aversive defences are shaped by natural selection in conjunction with regulation mechanisms that express responses in situations where benefits are greater than their costs. Pain, fever, and vomiting are examples.

Depression: A psychiatric condition characterised by low mood, lack of initiative or drive, psychomotor retardation, and rumination, often accompanied by somatic symptoms such as sleep disturbances and weight loss. In severe cases, suicidal ideation or behaviour may occur, as well as psychotic symptoms.

Dermatoglyphic patterns: The unique skin patterns found on the fingers, palms of the hand, toes, and soles of the feet.

Dermatosis: Refers to diseases of the integumentary system.

Dermis: The connective tissue middle layer of the skin, consisting of two sub-layers, papillary and reticular, that form the physical strength of the skin. Many different cell types, blood vessels, sensory structures, and nerves can be found in this layer.

Development: A progression of changes, either quantitative or qualitative, that lead from an undifferentiated or immature state to a highly organised, specialised, and mature state.

Developmental Origins of Health and Disease (DOHaD): A field of research into the fetal, neonatal, infant, and childhood origins of disease, especially chronic metabolic disease, such as diabetes, obesity, cardiovascular, and dementia, of later adulthood.

Developmental plasticity: *See* differential susceptibility.

Dialysis: A technical procedure enabling people with end-stage kidney disease to survive by means of clearing uremic toxins via a membrane. This can be either by the peritoneal membrane (peritoneal dialysis) with regular refills of the peritoneal cavity with clean dialysate (fresh hypertonic fluids) or by an extracorporeal circulation of blood crossing the filter membrane of a dialysator with fresh dialysate passing on the other side. In both techniques osmotic forces support the clearance of uremic solutes into the dialysate.

Diathesis–stress hypothesis: Refers to the liability to develop a psychopathological condition through the interaction between a genetic disposition and a precipitating causal event, most frequently acute or chronic stress.

Dietary restriction: An environmental intervention of ageing which involves reduction in nutrient availability without malnutrition.

Differential susceptibility: (aka developmental plasticity) This suggests that individuals are genetically susceptible to environmental conditions (mainly in the context of parental emotional availability), which, under adverse conditions, may increase the risk for psychological problems, or *decrease* the risk (relative to genetically unresponsive or 'resilient' individuals), when combined with supportive environments.

Diffusion: Movement of molecules or atoms from a region of high concentration to one of low concentration.

Disposable soma theory: This theory of ageing is based on the notion that there is a trade-off between investment in somatic maintenance and repair, on the one hand, and reproduction on the other (*see also* Life history theory). The optimal allocation that maximises fitness is one that leads to suboptimal investment in maintenance and repair with respect to survival rate.

Dizygotic: Twins derived from two different fertilised ova.

Duchenne's muscular dystrophy: A disease characterised by progressive muscle degeneration and caused by a mutation in the dystrophin gene.

Duffy antigen receptor: A membrane protein that serves as the receptor for chemical communication of the immune system but is also used by *Plasmodium vivax* to enter red blood cells.

Dysbiosis: Disturbance of the equilibrium between the microbiome and its host, which compromises health.

Dysplasia: Presence of cells of an abnormal type within a tissue, which may signify a stage preceding the development of cancer or abnormal growth or development of a tissue or organ (congenital dysplasia).

Dyspnoea: Breathlessness or shortness of breath occurring in a variety of respiratory disorders as a primary indicator of inadequate ventilation or insufficient amounts of oxygen in the circulating blood.

Dystrophin: A protein that connects the interior architecture of a muscle cell to the matrix outside the cell.

Eccrine sweat glands: Skin sweat glands that only in humans are required in large numbers for body thermoregulation.

Ectoparasites: Parasites that live on or outside the body, typically ticks, fleas, mites, flies, mosquitoes, and lice.

Ectothermy: A physiological condition in which body temperature is regulated by exposure to environmental energy, for example sunlight. Ectothermal animals therefore have varying body temperatures.

Embryonic period: Refers in humans to the first 8 weeks after fertilisation (10 gestational weeks), when most organs within the embryo have begun to form. The remainder and the majority of time of prenatal development is called the fetal period.

Empathy: Refers to the ability to appreciate another's emotions, while knowing the causes of the other's affective state, but without necessarily sharing the emotion. Empathy, narrowly defined, is therefore distinct from emotional contagion.

Emphysema: A condition in which the air sacs of the lungs are damaged and enlarged, causing breathlessness.

Endocrine: Secretion of hormones from glands directly into the circulatory system, to be carried towards distant target organs.

Endosome: A membrane-bound compartment inside a cell.

Endosymbiotic event: An endosymbiont is an organism living symbiotically (i.e. with mutual benefit) *inside* another organism. This state becomes irreversible if genes encoding functions that are duplicated in the two genomes are lost from one or both partners or transferred to the partner's genome. It is thought that mitochondria and chloroplasts (in plants) evolved from free-living bacteria in this way.

Endothermy: A physiological condition in which body temperature is regulated by metabolic processes and kept relatively stable (as in birds and mammals).

Endotoxin tolerance: Endotoxins are lipopolysaccharides (LPS) from the outer layers of Gram-negative bacteria such as *Escherichia coli.* LPS can drive powerful inflammatory responses, but repeated low dose exposure leads to a degree of tolerance, and also, particularly in early life, sets up immunoregulatory mechanisms that modulate subsequent immune responses.

Energy balance: A state in which energy intake and expenditure are equal and weight is stable.

Enteric nervous system: Intrinsic nervous system of the digestive tract that contains full reflex circuits that can operate independently of the central nervous system.

Enteroendocrine: Belonging to the hormonal control system for digestion and ingestive behaviour: hence enteroendocrine cell and enteroendocrine hormone.

Environments of Evolutionary Adaptiveness (EEAs): The adaptation-relevant properties of ancestral environments that provided selection pressures and influenced the evolution of a given trait.

Epidermal differentiation complex (EDC): Human chromosome (1q2) containing linked 63 genes within four gene families that are molecular markers for stratified epidermis terminal differentiation.

Epidermis: Skin external cellular epithelial layer, formed by keratinocytes, covering the entire body that forms the waterproof barrier.

Epigenetic: Literally, above the level of the gene. Epigenetics is the study of cellular and physiological phenotypic variations caused by external or environmental factors that switch genes on and off and affect the phenotype without any changes in the DNA sequence.

Epigenetic drift: Divergence of the epigenome as a function of age due to stochastic changes in modifiers of chromatin structure.

Epigenetics: The study of changes in gene function that result from changes in the proteins associated with DNA rather than the DNA itself.

Epiphysis: The end of long bones. If two bones form a joint, the epiphysis is covered with cartilage. The epiphysis might be filled with bone marrow, which produces erythrocytes.

Episiotomy: Surgical incision of the perineal area to expand the vaginal opening during childbirth and to reduce the incidence of tearing.

Epsilon 4: A protein encoded by an allele of the ApoE gene that is associated with vulnerability to atherosclerosis and Alzheimer's disease.

Eukaryote: Any organism that, unlike the prokaryotes, has a cell nucleus (which contains the genetic material) and other organelles enclosed within membranes.

Eulipotyphla: Order of mammals which includes hedgehogs, hairy hedgehogs, moles, shrew-like moles, and true shrews, based on molecular phylogenetic studies.

Eumelanin: The most common of the types of skin pigment formed from melanin, found in hair, skin, and dark areas around the nipples, and highly abundant in black populations.

Evolution: In biology, the changes in an organism over time that are associated with genetic changes.

Evolutionary fitness: The net success in passing on of alleles across generations.

Evolutionary selection: A category subsuming natural selection, sexual selection, kin selection, and artificial selection; a synonym of genetic selection.

Exon: An exon is any part of a gene that encodes part of the final mature RNA produced by that gene after 'introns' have been removed by RNA splicing.

Exosome: A membrane-bound extracellular vesicle released by cells and found in most, perhaps all eukaryotic fluids such as blood and urine. Exosomes are involved in communication between cells, but their roles are not fully elucidated.

Exteroception: Sensitivity to and perception of stimuli that originate outside the body (e.g. vision, audition).

Extrinsic mortality rate: Mortality rates attributable to external factors, such as disease or predation.

Falciparum malaria: Malaria caused by *Plasmodium falciparum*.

Familial hypercholesterolaemia: A heritable condition characterised by high serum levels of cholesterol.

Fatty acid: A lipid component that is composed of a hydrocarbon chain with a terminal carboxyl group.

Fecundity: The biological capacity to reproduce.

Female choice: In many animal species, and in humans as well, males advertise their qualities and eagerness to find a female partner (courtship); females choose whom they will mate with.

Fetal haemoglobin: A variant of haemoglobin that is present in fetal blood and has a greater affinity of oxygen than haemoglobin variants normally found in adults and children.

Fetal period: The prenatal period of development after the first 8 weeks, 10 gestational weeks, also called the second and third trimester.

Fitness (evolutionary): A measure of the evolutionary success of an organism or genotype relative to others in the population.

Follicle stimulating hormone (FSH): Encoded by the alpha subunit gene CGA and chorionic beta subunit gene FSHB, FSH is a heterodimer gonadotropin protein hormone produced in the pituitary in both males and females. FSH is responsible for supporting the production of gametes, ova in females and sperm in males.

Founder effect: Deviation of the genetic characteristic of a population from the overall genetic make-up in a species that is attributed to the genetic composition of a small population from which the larger population was derived.

FOXP3 (forkhead box P3): A protein involved in the development and function of regulatory T lymphocytes. Therefore, defects in FOXP3 are associated with severe inflammatory disorders, including enteropathy and autoimmunity.

Fractal: A complex structure exhibiting self-similarity across all scales of magnification.

Functional residual capacity (FRC): The volume of air present in the lungs at the end of passive expiration.

Gas exchange: Biological process by which gases move passively by diffusion over a tissue surface and across a barrier.

Gas exchange organs: Organs involved in the biological process in which gases move passively by diffusion across a surface. Typically, this surface is—or contains—a biological membrane that forms the boundary between an organism and its extracellular environment.

Gaucher's disease: An autosomal recessive disorder in which the breakdown of a particular type of lipid is compromised, resulting in pathological accumulation of the lipid; it results from a mutation in the gene encoding glucocerebrosidase betahexoseaminidase.

Genetic drift: The genetic characteristics of a small population that are attributable to the loss of alleles due to chance events rather than the action of evolutionary selection.

Genetic selection: The process by which certain traits become more prevalent in a species than other traits. Natural selection favours the preferential survival and reproduction of individuals with certain genotypes (genetic compositions) that allow the best adaptation of the organism to the environment.

Germline: The genetic material (DNA) that is transferred via the gametes (sperm and ova) to the next generation, while the DNA from the two parents undergoes recombination during formation of the genome of the offspring.

Gestational age: A clinical term usually given in weeks and days to describe human development timed from the first day of the last menstrual period (LMP).

G–G (genital–genital) rubbing: A form of homosexual genital stimulation common in female bonobos.

Ghrelin: A protein hormone produced by the stomach that stimulates hunger.

Glabrous skin: The skin without hair, it refers to the palms of the hands and soles of the feet that is also structurally different from other skin.

Glomerular filtration rate (GFR): The volume filtered by the glomeruli of the kidney is usually applied to describe renal excretory function in humans. We distinguish total GFR by all the nephrons of the kidneys from single nephron GFR. For example, after uninephrectomy (e.g. for kidney donation), total GFR declines but single nephron GFR of the remnant nephrons increases to meet the body's filtration load. GFR would be less appropriate to define renal functions in animals in which urine production depends more on tubular secretion than glomerular filtration (e.g. in worms or insects that do not have glomeruli or in salt-water fish that minimise glomerular filtration to avoid the loss of fresh water).

Glucose-6-phosphate dehydrogenase (G6PD): An enzyme that is essential for flux of glucose into the pentose phosphate pathway; deficiency can lead to haemolytic anaemia, resistance to lethal effects of malaria, but potentially lethal vulnerability to toxins and drugs.

Glycosylation: The enzymatic process that attaches glycans (sugars, carbohydrates) to proteins, lipids, or other molecules.

Gnathostomes: Jawed vertebrates.

Gonadotropic-releasing hormone (GnRH): A protein hormone secreted in the hypothalamus under regulation of the hormone kisspeptin that is vital to reproductive function in both males and females. GnRH is responsible for stimulating the production of gonadotropins, follicle stimulating hormone (FSH), and luteinising hormone (LH) in the pituitary.

G protein-coupled receptor: The largest and most diverse group of membrane receptors in eukaryotes. They detect molecules outside the cell and activate internal signal transduction pathways by coupling intracellular with G proteins and, ultimately, cellular responses. They are also called seven-transmembrane receptors because they pass through the cell membrane seven times.

Graft versus host disease: Pathology due to the immunological response to the difference between the antigens in cells of the host tissue relative to the transplanted tissue, resulting in damage to the host caused by the transplanted cells.

Granuloma: A mass that may arise in response to infection, containing immune cells such as macrophages.

Growth: A quantitative increase in size or mass. Measurements of height in centimetres or weight in kilogrammes, for example, indicate how much growth has taken place in a child.

Haematopoietic: Relating to the process of generating the cellular components of blood.

Haematopoietic system: The hierarchical system that generates all blood cell types, starting with self-renewing haematopoietic stem cells (HSCs) in the bone marrow.

Haemocytes: Thrombocyte- or macrophage-like cells of invertebrates that function in clotting and immunity.

Haemoglobin: The oxygen-transporting protein found in red blood cells.

Haemoglobinopathies: Diseases attributable to genetic alterations of haemoglobin genes.

Hair: Skin specialisations that in humans consist of either vellus or terminal hairs.

Heart failure: A clinical syndrome in which the heart is unable to efficiently pump enough blood to keep up with the body's metabolic demands.

Herpes simplex virus type 1: A virus that initially infects the skin and spreads to the nervous system as a latent infection, reactivating with stresses, such as sun exposure.

Heterochrony: A developmental change in the timing or rate of events, leading to changes in size and shape. There are two main components, namely (1) the onset and offset of a growth process, and (2) the rate at which the process operates.

Heterodimers: A dimer is a protein composed of two non-covalently linked polypeptide chains. In a heterodimer these two polypeptide chains are different.

Heterozygous advantage: When a genotype is most advantageous in the heterozygous state.

High-density lipoprotein (HDL): A lipid-transporting amalgam in which the proportion of proteins relative to lipid is high; aka good cholesterol.

Histones: Proteins that interact with and arrange DNA into structural units.

HIV: The human immunodeficiency virus that weakens the human immune system and can lead to acquired immune deficiency syndrome (AIDS). The virus is most commonly spread by sexual intercourse or by blood.

HLA (human leukocyte antigen): Proteins that display processed peptide antigens to T-cell receptors; synonym of major histocompatibility molecules.

Holobiont: Assemblages of different species as an ecological unit, often in the form of symbiosis.

Hominids: A family of mammals comprising the great apes (including humans).

Hominins: The primate group consisting of modern humans, extinct human species, and all immediate ancestors who were capable of some degree of bipedal locomotion (after the split from the last common ancestor with chimpanzees). The hominin group includes the genera *Homo*, *Australopithecus*, *Paranthropus*, *Ardipithecus*, *Neanderthals*, and *Denisovans*.

Hominoids: A superfamily of mammals comprising the lesser apes (gibbons) and the great apes (including humans), which have in common a greater freedom of motion at the shoulder joint (brachiation).

Homology: Similar biological trait acquired by species due to their shared ancestry, such as the skeletal elements of the human arm and hand and the whale flipper.

Homoplasy: Similar biological trait acquired by species from distantly related lineages and not through a common ancestor, such as the bird wing and the bat wing (synonym: convergence).

Honest signal: An anatomical, physiological, or behavioural feature, usually in the service of reproduction, which is difficult or impossible to fake. In human sexuality, honest signals were/are important, such as erection, vaginal lubrication, sex flush, orgasm-related ejaculation, and vaginal and uterine contractions, and could not be influenced deliberately. This has changed dramatically with availability of erection-enhancing drugs like Sildenafil.

Horizontal gene transfer (HGT): Movement of genetic material between organisms other than via vertical transmission of DNA from parent to offspring. This phenomenon is frequent between the bacteria that form the gut microbiota, even between species that are only distantly related. Many mechanisms are involved, some mediated by viruses.

Hormones: Biochemical agents that are common throughout most organisms, contributing to the regulation of many life history traits such as growth, development, reproduction, and metabolism. Secreted by glands and occasionally by other tissues with local or distant activation effects on specific tissue receptors.

Human chorionic gonadotropin (hCG): Encoded by the alpha subunit gene CGA and chorionic beta subunit gene CGB2, hCG is a heterodimer protein hormone similar in structure to LH that is produced by the conceptus prior to implantation. Its role is to stimulate the continued production of progesterone by the corpus luteum in order to support implantation and pregnancy.

Hypocapnia: The state of reduced carbon dioxide in the blood.

Hypodermis: Also called subcutis or subcutaneous adipose lower layer of the skin that binds it to underlying structures and acts as an adipose (fat) repository for cushioning and energy storage.

Hypothalamus: A small collection of neurons at the base of the brain that contributes to the regulation of numerous functions including growth, reproduction, and metabolism through the pulsatile release of various protein hormones.

Hypoxaemia: An abnormally low level of oxygen in the arterial blood, often caused by respiratory disorders.

Hypoxia: A condition in which the body or a region of the body is deprived of adequate oxygen supply.

IgG: An immunoglobulin of class G, generally present after the immunological response to an infection has become established.

IgM: An immunoglobulin of class M, generally present early during the establishment of an immunological response to an infection.

Induced ovulation: Eggs are released from the ovaries only when the female is in contact (visual, olfactory, through genital manipulation or copulation) with a male; this type of ovulation, brought about by, for example, oxytocin leading to an orgasm-like state, is thought by some researchers to be phylogenetically older than spontaneous ovulation.

Infancy: Period of life from birth to the end of feeding by lactation (or bottle feeding); for humans this is at about age 3 years in pre-industrial, traditional societies.

Infanticide: The act of killing an infant. This behaviour has been shown to be adaptive in some species (e.g. langurs, lions), where males take over a harem and kill the infants of their predecessor. In humans, infanticide is usually not carried out by men, but by the mothers, often directly at birth (neonaticide). They thus control the health and vitality of their newborns as well as their sex.

Infection: A state in which a parasite is living in or on a host.

Inflammation: An immunological response pathogens, cellular damage, or molecules that can protect or damage tissues.

Innate immunity: The immune system that orchestrates responses to pathogens using germline-encoded recognition systems that do not need to be specifically tailored for each new infection or invasion; this system also mediates responses to endogenous damage-associated signals and activates tissue repair.

Integumentary: A term used for the tissue barrier of the skin and its appendages between the body and the surrounding environment.

Interferon: A cytokine that is critical for orchestrating responses to intracellular bacteria and viruses by inducing an 'antiviral state' that limits a cell's ability to be used as a host for pathogen replication.

Interleukin: Cytokines that contribute to adaptive and innate immune responses.

Interoception: Sensitivity to and perception of stimuli that originate inside the body (e.g. touch, pain, vibration, temperature, muscle tension, and posture, as well as blood chemistry and pressure).

Intragenomic conflict: Genes that distort segregation or otherwise get an advantage at the expense of the individual are generally well suppressed by mechanisms shaped by natural selection, but exceptions are notable.

Involucrin: A skin protein required for the development of the cell envelope protecting corneocytes in the skin.

Isoforms: Protein isoforms are members of a set of similar proteins formed from a single gene by alternative splicing or other post-translational modifications.

Iteroparity: Organismal life history trait where an organism commonly has multiple bouts of reproductive effort (i.e. gestation) during its lifetime.

Juvenile: For non-human mammals this is the period of life from the end of lactation until puberty and the onset of sexual maturation; for humans this is the period of life from the end of the childhood stage to the onset of the adolescence stage, from about age 7–9 years for girls and 11 years for boys.

Keratinocyte: The cell type forming the epidermal outer layer of the skin.

Kin selection: An allele that imposes costs on an individual can nonetheless be selected for if it gives sufficient advantages to children or other relatives who have that allele in common by descent.

Kisspeptin: Encoded by the KISS1 gene, it is protein hormone that is secreted by neurons in the hypothalamus and regulates the production and release of GnRH.

Lactase persistence: Continuation of the production of enzymatically active lactase beyond weaning, permitting the consumption of milk and milk products by adults.

Langerhans cell: The skin immune dendritic cell or antigen presenting cell, Langerhans cells can, depending on the immunological setting, elicit immunity or tolerance. Named after Paul Langerhans.

Langer's lines: (aka skin cleavage lines, cleavage lines) A clinical term for the orientation of reticular dermis collagen bundles causing tensions on skin and subcutaneous tissues. Lines tend to be horizontal in the trunk and neck, and longitudinal in the skin and limbs.

Lateralisation: Representation of brain functions in one dominant hemisphere relative to the other. The concept is well established for language functions and handedness, but pertains also to other functions such as emotion recognition and self–other distinction.

LDL (low-density lipoprotein): A lipid-transporting amalgam in which the proportion of protein relative to lipid is low; aka bad cholesterol.

Leptin: Encoded by the LEP gene, it is a protein hormone produced primarily by fat cells. It serves many functions, but most prominently serves as a lipstat or fat reporter to the hypothalamus and is involved in energy regulation.

Lewis secretor status: The Lewis antigens constitute one of the blood group systems. The FUT2 or secretor gene determines whether the Lewis (b) antigen is formed. If present, this is passively adsorbed onto red blood cells. Secretor status also determines whether soluble forms of the A, B, and H blood group molecules are secreted.

Life course approach: In epidemiology and health sciences this emphasises the progression of disease aetiology, risk factors, and disease outbreak during the lifespan of an individual.

It encompasses the prenatal period (fetal programming), early childhood, adolescence, adulthood, and senescence.

Life history theory: A branch of biology that studies the selective forces that have guided the evolution of the schedule and duration of key events in an organism's lifetime related to investments in growth, reproduction, and survivorship.

Lifetime reproductive effort: A measure of total amount of metabolised energy diverted to reproduction during the lifespan.

Lipid: Substances, such as fats and steroids, that contain carbon–carbon bonds and are soluble in organic solvents but insoluble in water.

Loop of Henle: Henle first described the thin connection between the proximal and distal tubule which implies a hairpin-like shape of the nephron. The loop of Henle is an essential component of urine concentration in the mammalian kidney. It elongates mainly after birth.

Lungs: The two saclike respiratory organs in the thorax of humans and higher vertebrates.

Luteinising hormone (LH): Encoded by the alpha subunit gene CGA and chorionic beta subunit gene LHB, LH is a heterodimer gonadotropin protein hormone produced in the pituitary. It supports the production of sex steroids in both males and females, oestradiol in females, testosterone in males.

Lysosome: A membrane-bound organelle containing hydrolytic enzymes that can break down almost all biomolecules. Lysosomes are present in nearly all animal cells.

Macrobe: Since 'microbe' refers to microscopic organisms, the term 'macrobe' is useful in contexts where some of the organisms involved are visible to the naked eye. Helminths are obvious examples.

Major histocompatibility complexes (MHC): Proteins that display processed peptide antigens to T-cell receptors, thus alerting T cells to the presence of particular proteins that may represent non-self origin, such as pathogens, or altered self, such as tumour cells.

Major histocompatibility (MHC) molecule: Binds antigens or fragments of antigens and displays them on the cell surface for recognition by the appropriate lymphocytes.

Malaria: Disease caused by a *Plasmodium* parasite carried by mosquitoes, and transferred to humans by their bite.

Malpighian tubules: Excretory ducts of insects that drain into the gut.

Mammary gland: Functional structure of the female breast that develops after puberty initially as a skin specialisation. Breast growth and appearance in male and female children are virtually identical prior to puberty.

Maturation: A measure of functional capacity; for example, maturation of bipedal walking results from changes with age in skeletal, muscular, and motor capacities of the infant and child.

Mediterranean diet: A traditional diet of Mediterranean countries characterised by abundant vegetables, olive oil, garlic, and onions, and moderate amounts of protein, and associated with protection from chronic diseases such as atherosclerosis.

Meissner corpuscle: A skin somatosensory structure acting as a rapidly adapting mechanoreceptor located mainly in the glabrous skin dermal papillae of digital skin. Meissner corpuscles are receptors for sensitivity to light touch.

Melanin: From Greek *melanos* meaning black, a pigment produced by melanocytes of the skin that provide photo-protection, preventing cellular DNA damage and colouring of basal epithelial cells that absorb the pigment. The two types are the brownish-black eumelanin and the reddish-yellow pheomelanin.

Melanocyte: From Greek *melanos* meaning black, a pigmented skin cell, neural crest in origin, differentiating from melanoblasts located in the skin and other tissues that produce melanin. These cells transfer melanin to keratinocytes to give skin colour and to the hair follicle to give hair colour. This is the cell type that proliferates in the cancer melanoma.

Melanoma: A serious skin cancer derived from melanocytes in the skin, associated with sun exposure.

Melatonin: The pineal gland is the main source of melatonin, a hormone that is the body's diurnal clock. The skin can also produce melatonin. Levels increase at night before sleep and are low during the day.

Menarche: The first menstrual period.

Menopause: The sudden or gradual cessation of the menstrual cycle subsequent to the loss of ovarian function.

Merkel cell: A skin epidermal-derived cell located in touch-sensitive areas that mediate sensory mechano-transduction in the skin.

Mesonephros: An ancient version of the kidney still present in certain species (fish, amphibians) that is still formed in the mammalian embryo but gets ultimately replaced during development by the nephrons of the kidney.

Meta-analysis: A statistical analysis of several studies that evaluate the same hypothesis to evaluate the statistical significance of the overall trend.

Metabolic endotoxaemia: Raised levels of endotoxin (also known as lipopolysaccharide or LPS) are commonly found in obese individuals, and can be elicited experimentally by a high-fat diet. The phenomenon involves changed gut barrier permeability. LPS drives chronic low-grade inflammation and pro-oxidative stress.

Metabolic syndrome: A cluster of disorders associated with increased risk of diabetes and cardiovascular disease. It is generally defined as central obesity that coexists with at least two of the following: elevated blood pressure, elevated fasting glucose, raised triglycerides (dyslipidaemia), or insulin resistance.

Metabolite: A small molecule product of metabolic processes.

Metanephros: Developmental origin of the mammalian kidney from mesenchymal cells around the aorta.

Methylation: A process by which a methyl group is added to a larger molecule; it can alter gene expression when added to DNA or histones.

Microbiome: The collective genome of all the bacteria, archaea, fungi, protozoa, and viruses that live within hollow organs and on body surfaces or within tissues.

Microbiota: The total sum of bacteria, archaea, fungi, protozoa and viruses that live within hollow organs and on body surfaces and in some cases, within the tissues.

Microglia: A type of neuroglia with immunological functions, much like macrophages in the periphery. Embryologically of mesodermal origin, in contrast to oligodendroglia and astroglia.

MicroRNA (miRNA): A small non-coding RNA molecule (about 22 nucleotides) found in plants, animals, and some viruses. miRNA plays an important role in post-transcriptional regulation of gene expression.

Microvilli: Tiny membrane-covered projections from cell surfaces that greatly increase the surface area of the cell. Notably, they increase the capacity for absorption of intestinal epithelial cells.

Miocene: A geological epoch from approximately 23 to 5 million years ago.

Mirror neurons: A type of neuron putatively involved in action–observation matching.

Mismatch/discordance: Differences between western, industrialised environments and those faced by humans during the vast majority of evolutionary history that result in formerly adaptive traits or characteristics being ill-suited for current environments.

Monochorial twins: Twins that are enclosed in a single chorionic membrane.

Monogamy: Common version of reproduction and (in humans) a form of institutionalised marriage, where a female and a male reproduce and stay together for an extended time.

Monozygotic twins: Twins derived from the same fertilised ovum, which then divides to form two genetically identical embryos.

Morphological microevolution: Describes secular changes in anatomical features in modern humans, over a timespan of a few to several generations. These alterations are understood as a combination of changes in gene variant frequencies, phenotypic plasticity, and environmental influences.

Multicellularity organism: An organism that consists of more than one cell.

Multiple sclerosis: A chronic inflammatory condition of unknown origin affecting the central nervous system. The neuropsychiatric signs and symptoms are disseminated, depending on the localisation of the demyelinating process.

Mutation accumulation theory: The first modern evolutionary theory of ageing which hypothesises that ageing is a result of the inefficiency of natural selection to get rid of mutations that are deleterious only at late ages.

Mutualism: An association between two individuals of two different species through which both individuals accrue increased fitness.

Mutualist: Either of the two interacting individuals in a mutualism.

***Mycobacterium tuberculosis*:** The bacterium that causes tuberculosis.

Myeloid: Cells derived from the bone marrow cell lineage.

Myocardial infarction: Heart muscle death that occurs typically when blood flow to the heart is abruptly cut off as a result of rupture of a plaque within a coronary artery.

Myosin: An ATP-dependent protein in muscle cells that is responsible for muscle contraction, together with the protein actin.

Nail: Skin specialisations, fingernails and toenails, that form on the dorsum of the digits. In other species these develop as claws and hooves. Each nail is convex on its outer surface and concave within, and is implanted by a portion, called the root, into a groove in the skin; the exposed portion is called the body, and the distal extremity the free edge.

Natural selection: Broadly defined as the process of differential survival and reproduction that results in changes in gene frequencies and in the characteristics that the genes encode; a narrow definition restricts the selection to the effect of the natural environment.

Neonatal hyperbilirubinaemia: Jaundice in the new-born associated with elevation of bilirubin level in the blood, which results from the breakdown of red blood cells that cannot be effectively metabolised by the immature liver. When it appears 2–5 days after birth, it is viewed as a normal process, but when it appears at birth it is viewed as a medical condition requiring treatment.

Neoplasia: New abnormal growth of tissue, often developing into a tumour or cancer.

Nephron: Functional unit of the kidney including the filter (glomerulus) and the excretory tube (tubule) that ultimately connects with the drainage system (collecting duct–renal pelvis–ureter–urinary bladder–urethra). A single healthy kidney contains around 800,000–1,200,000 nephrons. As nephrons get lost during human lifetime, renal function declines.

Neuropeptide: A class of relatively simple, evolutionarily highly conserved peptides utilised for the modulation of neurotransmission.

Neurotransmitter: A class of messenger substances (often synthesised in the body) that are active at chemical synapses and bind to postsynaptic receptors.

Niemann–Pick disease: An autosomal recessive disorder in which the breakdown of a particular type of lipid is compromised, resulting in pathological accumulation of the lipid; this results from a mutation in the gene encoding acid sphingomyelinase.

Nitrous waste: Nitrogen-containing metabolites, for example from ingested protein (amino acids) as well as from chromatin (purins), that need special water-soluble formulations to endorse urinary excretion (urea and uric acid). Fish excrete nitrogen as ammonium, which is toxic but gets immediately diluted in the surrounding water environment.

Nociceptors: Skin receptors for encoding and processing noxious or painful stimuli.

NOD-like receptors: Abbreviation of nucleotide-binding oligomerisation domain-like receptors. These are intracellular sensors of molecules derived from organisms or from cellular stress and damage. They are involved in the activation and regulation of the innate immune response.

Notochord: A cartilagenous structure typical of chordates derived from the mesoderm that functions as a scaffold for muscle attachment and precursor of the vertebral column. It is present at a certain embryological stage of development, and in vertebrates becomes the nucleus pulposus.

Obstetrical dilemma: Describes the trade-off in human anatomy resulting from adaptations in the female pelvis to bipedal gait, parturition, and large brains in neonates. This dilemma was partially answered by a shift towards premature births in human neonates.

Oestradiol: A sex steroid produced by the ovaries, although it is also produced in small amounts in the testes as well as by adipocytes (fat cells) through the conversion of testosterone by the enzyme aromatase. It is involved in the regulation of reproduction, primarily through the promotion and support of endometrial tissue in the uterus.

Oligodendrocyte: A type of neuroglia in the central nervous system producing myelin sheaths for neurons.

Oligomer: A molecule consisting of a small number of monomer (i.e. single) subunits, in contrast to a polymer that has many such subunits, or a dimer which has two.

Omega 3 fatty acid: A class of essential fatty acids that may protect against heart attacks in people who have a history of cardiovascular disease.

Omega 6 fatty acid: A class of essential fatty acids that can be pro-inflammatory.

Ontogeny: Development of an organism from the time of fertilisation of the egg, through to maturity.

Organic complexity: The complexity of evolved organisms is different in fundamental ways from the complexity of designed machines.

Osmiophilic lamellated bodies: Intracellular organelles in which pulmonary surfactant is stored before it is secreted.

Osteoarthritis: The most common chronic condition of the joints in humans. It occurs when the joint cartilage is consumed, leading to pain, stiffness, and swelling. Apart from genetic predisposition, overweight, injuries, or overuse are the main causes.

Osteoporosis: In osteoporosis bones are demineralised with an increased susceptibility to fracture. Bone density decreases with age, but bone loss occurs most rapidly in postmenopausal women. Risk factors for osteoporosis are genetic predisposition, lack of exercise, lack of calcium and vitamin D, smoking, alcohol consumption, and low body weight.

Ovary: A central component of the reproductive system, it is responsible for the production, maturation, and release of ova as well as supporting hormones such as oestradiol.

Oxytocin: Encoded by the OXT gene, it is a protein hormone that is involved in smooth muscle contraction and is central to lactation, parturition, and orgasm. It is also involved with behaviours that are central to intimate social bonding (e.g. between mother and infant), empathy, and trust, but also with maternal aggression.

Pacinian corpuscle: A skin somatosensory vibration receptor that produces rapidly adapting responses.

Palaeolithic: The time from the earliest use of stone tools by human ancestors, 2.6 million years to 10,000 years BP.

Palaeopathography: A research field combining palaeopathology, the analysis of historical data such as written or figurative sources, and archaeological sources. The aim is to reveal medical conditions from past individuals from all sources at disposition, especially where the human remains themselves are not available.

Palaeopathology: The study of ancient diseases, investigating human remains such as bones or mummies, using an interdisciplinary approach, involving medical experts and anthropologists.

Paracrine: The secretion of hormones or other signalling molecules from a cell into its extracellular environment to induce changes in nearby cells, altering the function of those cells.

Parallel evolution: The independent evolution of a trait that occurs in two or more species that have previously diverged from a common ancestor.

Parasite: A replicating agent that lives in or on a host organism to its own benefit at the expense of its host, thus reducing the evolutionary fitness of the host.

Parasitism: An association between a parasite and a host.

Path dependence: Initial states or pathways, some resulting from chance, can lock in a system in ways that constrain future changes. The typewriter keyboard and the path of the recurrent laryngeal nerve are examples.

Pathogen: A parasite with structural organisation that is at or below the unicellular level.

Pathogen-associated molecular patterns (PAMPs): Molecules on pathogens that are recognised by receptors (often on innate immune cells) to alert these cells to the presence of the pathogens.

Penetrance: The extent to which an allele is expressed, measured by the individuals who exhibit the characteristic phenotype relative to the total number of individuals harbouring the allele.

Periconceptual: An event that occurs close to the time of conception.

Periodontal disease: An inflammatory disease that erodes the tissues that support teeth.

Periodontitis: Inflammation of the tissues that support the teeth.

Phenotype: A set of observable traits that influence appearance, development, or behaviour that were shaped by genetic expression and gene–environment interactions.

Phenotypic plasticity: The ability of one genotype to result in a range of phenotypes depending on interactions with the environment throughout development.

Pheomelanin: A melanin skin pigment reddish-yellow in colour and also responsible for red hair.

Phylogenetic tree: A diagram that shows the inferred evolutionary relationships among different species based on genetic and physical characteristics.

Phylogeny: History of organism lineages as they change through evolutionary time. It implies that different species arise from previous forms via common descent, and that all organisms, from the smallest microbe to the largest plants and vertebrates, are connected by the passage of genes along the branches of the phylogenetic tree that links all life.

PI3K/Akt/mTORC1 signalling system: An intracellular signalling pathway that is involved in a vast range of cellular functions, including regulation of the cell cycle.

Pituitary gland: A small collection of tissue found at the base of the brain. Under the regulation of various hormones released by the hypothalamus, it is responsible for producing and releasing several protein hormones that are vital for growth, metabolism, and reproduction.

***Plasmodium falciparum*:** A protozoan parasite that causes malaria, which spends part of its life in mosquitoes and other phases in humans as an intracellular parasite in red blood cells.

***Plasmodium vivax*:** The parasite that causes a milder form of falciparum malaria.

Podocytes: Octopus-shaped cells covering the outer aspect of the glomerular capillaries. By forming primary and secondary foot processes they interdigitate with neighbouring podocytes and thereby form the epithelial component of the glomerular filtration barrier. The ultrafiltrate passes through pores between the secondary foot processes. Podocyte loss during ageing or injury is difficult to replace and often leads to scar formation (glomerulosclerosis).

Polyandry: A version of reproduction and (in humans) rare form of institutionalised marriage, where one female has two or more male partners (usually brothers). An official marriage type confined to the Himalayas.

Polygynandry (promiscuity): A version of reproduction where females and males in the group copulate with each other and have offspring together (multibreeding); paternity can thus be obscured.

Polygyny: A common version of reproduction and (in humans) form of institutionalised marriage, where one male has two or more female partners with whom he reproduces and stays together for an extended time.

***Porphyromonas gingivalis*:** A bacterium that causes periodontitis.

Power grip: One of the two primary human hand prehensile grips; the other is the precision grip. The fingers and the thumb tightly clamp an object, allowing a strong hold.

Precision grip: One of the two primary human hand prehensile grips; the other is the power grip. The fingertips and the thumb press against each other, allowing fine manipulation.

Preterm birth or premature birth: Birth prior to 37 weeks gestational age.

Primipara(ae)/multipara(ae): A woman giving birth for the first time; a multipara has given birth one or more times previously.

Progenitor: Stem cell-like cell type residing in adult tissues with the potential to divide and differentiate into parenchymal cells in high turnover tissues (intestinal epithelia) or upon injury-related losses of parenchymal cells in low turnover tissues (kidney).

Progesterone: A steroid hormone produced by the corpus luteum of the ovary during the luteal (second half) phase of the menstrual cycle and by the placenta that is involved in ovarian function, implantation, maintenance of pregnancy, and preparation of the mammary glands for breastfeeding.

Prokaryote: Unicellular organisms that contain no membrane-bound organelles, notably no nucleus or mitochondria. Fossil prokaryotes have been dated to at least 3.5 billion years ago and were among the earliest life forms.

Prolactin: Encoded by the PRL gene, it is a protein hormone produced in the pituitary gland that stimulates and maintains lactation. It has also been associated with behaviours that are supportive of offspring care.

Pronephros: An ancient version of an excretory duct still used by certain species (worms, snails, insects) that is still formed in the mammalian embryo but gets replaced during development by later versions of the excretory system including ultimately the nephrons of the kidney.

Protease: An enzyme that can digest proteins by hydrolysing the bonds between individual constituent amino acids.

Protein hormone: A hormone that consists of numerous amino acid and peptide chains. Protein hormones are genetically encoded, larger compared to steroid hormones, and water soluble.

Protists: A term of convenience for a heterogeneous collection of eukaryotic organisms that are not plants, animals, or fungi. They are usually unicellular and even if they form aggregates, they do not differentiate into tissues.

Proximate explanations: Mechanistic explanations; aka 'how' explanations.

Pseudogene: A genomic DNA sequence similar to a normal gene but non-functional. A pseudogene is different from a normal gene due to either a lack of encoding a functional protein resulting from a variety of disabling mutations (e.g. premature stop codons or frameshifts), a lack of transcription, or inability to encode RNA (such as with ribosomal RNA pseudogene).

Psychoses: Mental disorders characterised by thought and emotional impairments that are so severe that the individual loses contact with external reality.

Quorum sensing: The regulation of gene expression in response to fluctuations in cell-population density. Quorum sensing bacteria produce and release chemical signal molecules called autoinducers that increase in concentration as a function of cell density.

Receptor: A protein complex that can be found within and on the surface of cells that selectively binds to specific hormones. Activation of receptors by hormones triggers the specific metabolic or transcription activities.

Recombinase: Enzymes that can modify the structure of genomes by excision, insertion, inversion, or translocation of segments of DNA.

Relaxed natural selection: Relaxed natural selection occurs when former positive or negative selective pressures do not act anymore (or act less) on an organism, normally due to changes in the environment. In humans, relaxed natural selection is assumed in response to man-made changes in the modern world, increased life conditions, and better hygiene and medical care.

Respiratory system: The system of organs involved in the exchange of carbon dioxide and oxygen between an organism and its environment.

Rete ridge: The skin extensions of the epidermis into the dermis. These epidermal surface thickenings extend downward between underlying connective tissue dermal papillae and are also the site of initial sweat gland differentiation.

Reward system: A phylogenetically old system regulating motivated behaviour by evaluating the incentive salience of stimuli ('wanting') to achieve a state of pleasure ('liking'). Reward is mainly modulated by dopaminergic neurons in the ventral tegmental area. Reward pathways can be 'usurped' by dopaminergic chemicals, such as alcohol and several other psychoactive drugs.

Risk factor: A variable that is associated with an altered probability of developing a disease.

Ritualised homosexuality: In a minority of societies in New Guinea, boys and adolescents had to pass through a period of being passive (anal, oral, or masturbatory) partners for adult men's sexual activity. Thus religiously embedded ceremonies were part of male initiation and the process of becoming adult.

Romantic love: A particular emotional state present in two partners (usually a woman and man), involving the (sometimes addiction-like) tendency to see this person as a special, uniquely wonderful individual, and to express and experience erotic attraction, as well as the strong desire to be with the other person and engage in exclusive sexual intercourse with her or him. Endocrinological and physiological parameters connected to experiencing romantic love stress the biopsychic, universal nature of this emotional state.

***Salmonella typhi*:** The bacterium that causes typhoid fever.

Scavenger receptors: A diverse assortment of cell membrane receptors that recognise and scavenge a wide variety of ligands, including microbial components, polyionic ligands including lipoproteins, phospholipids, proteoglycans, ferritin, apoptotic cells, cholesterol ester, and carbohydrates.

Schizophrenia: The term was first introduced by Swiss psychiatrist Eugen Bleuler to describe a group of heterogeneous disorders characterised by thought disorder, affective disturbances, loss of 'willed' action, and psychomotor signs including catatonia.

Sebaceous gland: A gland associated with both the hair follicle and hairless parts of the skin (lips, cheek oral surface, and external genitalia), embedded in the dermis, which can be sites of infections.

Secondary sex characteristics: The male and female developmental surface features forming after puberty: pubic hair and breasts in females, and facial hair in males.

Second messenger: An intracellular substance that mediates cellular activity by transmitting and amplifying a signal from an extracellular molecule—first messenger (e.g. hormone)—bound to a receptor on the cell's surface. Many second messenger molecules are small and therefore diffuse rapidly through the cytoplasm, enabling information to move quickly throughout the cell.

Self-renewal: For stem cells, this refers to the ability of stem cells to generate at least one daughter cell that remains an undifferentiated stem cell following a cell division.

Semelparity: An organismal life history trait where an organism commonly has one or very few bouts of reproductive effort (i.e. gestation) during its lifetime.

Sepsis: A life-threatening response to an infection.

Sex hormone binding globulin (SHBG): Encoded by the Shbg gene, SHBG is a protein that acts as a carrier for steroid hormones in circulation. It selectively binds to sex steroids such as testosterone, oestradiol, and progesterone, rendering it inactive. Facilitation of steroid hormone activity is enabled when the steroid is released from SHBG at or near the site of receptor activation.

Sexual dimorphism: Biological difference (e.g. in body mass, height, size of teeth) between females and males of a species.

Shared intentionality: The ability to participate with others in collaborative activities with shared goals and intentions.

Sickle cell disease: A genetic disease of humans caused by mutations in the HBB haemoglobin gene, which results in crescent-shaped red blood cells at low oxygen levels (and resistance to the lethal effects of malaria).

Signal transduction: The cascade of processes by which an extracellular chemical or physical signal is transmitted through a cell by causing a change in the level of a second messenger and, eventually, triggering a change in the cell's function.

Smoke detector principle: According to signal detection theory the smoke detector principle explains why mechanisms that regulate responses give rise to normal false alarms when the costs of a false alarms are low compared to the cost of failing to respond when a response is needed.

Social brain hypothesis: First put forth by Leslie Brothers to suggest that primate brains are relatively specialised with regard to processing social information, which evolved due to the complexity of social interactions in group-living species.

Social selection: Selection resulting from social choices. Sexual selection is one example, selection of cooperation partners is another; both can shape extreme traits.

Somatic cell fitness: The relative ability of a cell of a particular genotype (or epigenetic profile) to pass this type on to future cell generations.

Somatic evolution: Changes in the frequency of cells with particular genotypes, depending on their somatic fitness, mutation rates, and drift.

Somatic selection: Selection that takes place among cells within an individual's body.

Somatosensory: Refers to the sense of touch, heat, and cold associated with the skin. This sensory system consists of many different specialised receptors throughout the body, with many located within the hands and lips.

Somite: A division of the embryonic body of an animal. Somites are of mesodermal origin and placed bilaterally along the body axis. During embryological development, they differentiate into sclerotomes, myotomes, and dermatomes that further develop into bones, muscles, cartilage, tendons, and skin.

Sperm competition: In most animal species, males compete for access to fertile females. This competition can take many forms. One of them is the fight of sperm from two or more males in the vagina of a female. In humans, sperm competition does not seem to play an important role; the spermia gain comparatively quick access to the cervix and the tube so that the ejaculate of a second or third partner is at a disadvantage.

Spirochaete: A flexible spirally shaped bacterium in the order Spirochaetales.

Spontaneous ovulation: Eggs are released from the ovaries in a chronobiological cycle, in women approximately every 28 days.

Squamous: A term used to describe cells of the skin epithelium where the upper cell layer shape is flattened. A squama is also this cell's residue that is shed from the skin.

Squamous cell carcinoma: A non-melanoma skin cancer forming from the epidermis upper cells which accounts for about 30% of non-melanoma skin cancers.

Statin: A class of lipid-lowering drugs that lower cholesterol by targeting HMG-CoA reductase.

Stem cell niche: The cellular, structural, and soluble components of a tissue that support a particular type of stem cell.

Steroid: A class of lipid-soluble hormones derived from cholesterol. Steroids are evolutionarily ancient and conservative, contributing to the regulation of numerous life history functions including growth, reproduction, and energy regulation.

Stress: Non-specific physiological response to a 'stressor' to maintain homoeostasis involving the autonomic nervous system and the hypothalamic–pituitary axis. Stress has become an umbrella term for a broad spectrum of mainly adverse experiences and stimuli.

Stroke: The sudden death of brain cells caused by blockage of blood flow or rupture of an artery.

Surfactant: A surface-active lipoprotein complex (phospholipoprotein) formed by type II cells of the lung.

Symbiont: An organism that lives intimately with a host.

Symbiosis: The intimate living together of two organisms of two different species.

Synovia: A viscous fluid in joint spaces. The main function is to reduce friction between articular cartilages during joint movement.

Takotsubo cardiomyopathy: A type of heart muscle weakness often triggered by emotional stress which commonly results in dilatation of the left ventricular apex with normal functioning basal segments in the setting of no coronary obstruction. This syndrome is also known as 'stress-induced cardiomyopathy' and the word *takotsubo* is Japanese for an octopus trap, wherein the ventricle resembles the fishing pot.

Tangles: Pathological sign of Alzheimer's disease, arising from the coalescence of tau proteins.

Tanner stages (Tanner scale): An anatomical staging system for measuring male/female sexual development at puberty.

Tanning: The facultative darkening of the skin colour by increased melanin production in response to sun exposure.

Tay-Sachs disease: An autosomal recessive disorder in which the breakdown of a particular type of lipid is compromised, resulting in pathological accumulation of the lipid; it results from a mutation in the gene encoding beta-hexoseaminidase.

T-cell receptors (TCR): Receptors expressed on T cells that recognise antigens as small peptides presented on MHC proteins by antigen-presenting cells. TCR ligation leads to subsequent T-cell activation.

Telomeres: A region of repetitive DNA sequence at the ends of chromosomes that protects functional sequences from being truncated during cell division. Age-dependent shortening of telomeres has been suggested as one proximate cause of ageing.

Terminal hairs: The hair seen in obviously hairy skin regions of the body: the scalp, armpits, and genital region. Humans have lost these hairs from the remainder of the body.

Testes: A central component of the male reproductive system, testes are responsible for the production, maturation, and release of sperm as well as the production of supporting hormones such as testosterone.

Testosterone: A sex steroid protein that is produced by the testes and in the ovaries in smaller amounts. Testosterone supports the regulation of reproductive maturation, metabolism, and anabolism of sexually dimorphic muscle tissue, as well as libido.

Tetrapod: Means 'four feet' and includes all species alive today that have four feet but also many animals that do not have four feet.

Theory of Mind: A misleading term to describe the cognitive ability to infer the mental state of another individual and to consciously reflect one's own mental states such as beliefs, desires, intentions, and feelings.

Thermoreceptors: Skin sensory receptors for heat (warmth) and cold (chill) detection.

Thick skin: Refers to the skin found on the palms of the hand and soles of the feet that does not contain hair. Note that this is used as a histological description not a measurement of overall skin thickness.

Thin skin: Refers to the skin structure on all body regions other than palms of the hands and soles of the feet which are thick skin.

Thrombin: An enzyme in the coagulation cascade that is activated by proteolysis, and that subsequently cleaves other proteins in the cascade to effect coagulation.

Thrombocyte: A cell type in vertebrates that mediates clot formation; in mammals, these are platelets.

Thyrotropin-releasing hormone (TRH): A protein hormone that is produced in the hypothalamus that stimulates the production of thyroid-stimulating hormone (TSH) in the pituitary gland.

Thyroxine (T4): A hormone based on iodine and the amino acid tyrosine. It is produced in the thyroid gland in response to TRH. It is a precursor (prohormone) of T3. It is involved in the regulation and management of basal metabolic rate and thermogenesis. Although found in greater quantities compared to T3, it is less potent than T3.

Tidal volume (TV): The lung volume of air displaced between normal inhalation and exhalation when extra effort is not applied.

Tinbergen's Four Questions: In a 1963 paper published by ethologist Nikolaas Tinbergen, he defined four major categories for explanations of animal behaviour: mechanism, adaptive value, ontogeny, and phylogeny. Current scholars typically separate these into 'proximate' and 'ultimate' (evolutionary) causes. Tinbergen's Four Questions provide a comprehensive, logical approach to studying behaviour that is particularly useful for in-depth analysis.

Tissue factor: A protease in the coagulation cascade, which activates Factor VII in the cascade.

TLR4: A receptor for bacterial lipopolysaccharide.

Toll-like receptors: A series of receptors that resemble a gene studied in fruit flies, named 'toll'. This fruit fly (*Drosophila*) gene has a role in immunity to infection. The mammalian Toll-like receptors are germline-encoded receptors that alert the innate immune system to the presence of microbial components.

Transcriptional profiles: The profile of the genes that are actually being transcribed. Transcription is the first stage in gene expression, when the DNA encoding the gene product is copied into RNA.

Transgenerational: Effects that are transmitted across generations. These include epigenetic changes, *in utero* influences that are not epigenetic, and behavioural influences.

Transposon: A piece of chromosomal DNA that can translocate from one position to another, typically requiring a transpose encoded by the same fragment; an example of 'selfish DNA'.

Trichromacy: Perception of different wavelengths of light by three types of cone cells in the retina.

Triiodothyronine (T3): A hormone based on iodine and the amino acid tyrosine. It is produced through the conversion of the hormone thyroxine (T4) in response to TRH and contributes to energy regulation, primarily through the management of basal metabolic rate and thermogenesis.

Triploblasty: Three germ layers, comprising ectoderm, mesoderm, and endoderm.

Tuberculosis: An infectious disease caused by *Mycobacterium tuberculosis* characterised by the growth of nodules in lungs and other tissues.

Tubulo-glomerular feedback: An important component of renal autoregulation in which the kidney is able to maintain a constant GFR along a wider range of systemic blood pressures by translating sodium chloride concentrations in the distal tubule into either vasoconstriction or vasodilation of the arterioles entering and leaving the glomerulus. In this manner, filtration pressure and GFR are kept constant in phases of high or low fluid intake.

Typhoid fever: An infectious disease caused by *Salmonella typhi*, and characterised by fever, diarrhoea, prostration, headache, intestinal inflammation, and red spots on the chest and abdomen.

Ultimate explanations: Evolutionary explanations for the origin of characteristics, including allele frequencies within a population; aka 'why' explanations.

Ultraviolet (UV): Non-visible light electromagnetic radiation classified by increasing wavelength as UVA (400–320 nm), UVB (320–290 nm), or UVC (100–280 nm). Only UVA and UVB pass through the ozone layer and then skin to damage cell DNA and cause skin cancer.

Unicellular organism: An organism that consists of only one cell.

Vascular system: The circulatory system of vertebrates that uses endothelium-lined blood vessels to transport blood and lymph (and thus cells, oxygen, and nutrients) throughout the body.

Vellus hair: The fine short hairs of the skin present in 'hairless' regions of the body; they are lightly pigmented.

Ventricle/ventricular: A chamber of the heart that receives blood from one or more atria and pumps it by muscular contraction into the arteries.

Vernix caseosa: A protective coating covering the fetal human skin consisting of sebum, skin cells, and lanugo hair. Forming late in only human fetal development, in a head to tail sequence associated with epithelium differentiation. The waxy substance may have protective properties during the postpartum period.

***Vibrio cholerae*:** The bacterium that causes cholera.

Virchow–Robin space: Perivascular space between an artery, a vein, and the pia mater forming a subarachnoidal tunnel-like structure containing interstitial fluid.

Vitamin D: A hormone formed within the skin by sun exposure, required for many physiological functions requiring calcium including bone formation, and muscle, neural, and overall health. Low vitamin D has adverse health effects.

Volume control: Control of isotonic extracellular fluid. Do not mix 'volume' with 'water'. Water is salt-free and can enter cells easily. 'Volume' is isotonic fluid (= sodium chloride-enriched water). As sodium chloride cannot enter cells it remains in the extracellular space and therefore determines the fluid content in intravascular and interstitial compartments. Several mechanisms contribute to volume control such as hormones regulating sodium channels (aldosterone), water balance (antidiuretic hormone), osmoregulation, and thirst.

Von Economo neuron: A type of spindle-shaped projection neuron in layer V of the anterior cingulate cortex and the anterior insula of the human brain, with some scattered populations in the dorsolateral prefrontal cortex. These neurons are supposed to play a role in the processing of complex emotions and interoception.

WEIRD societies: Western, Educated, Industrialised, Rich, and Democratic nations of the world. These societies provide most human participants in research, but represent only a minority of the world's population. Results of research with WEIRD societies are likely not representative of human nature as lived for most of our evolutionary history.

X chromosome: A sex-determining chromosome of mammals; two are present in females and one in males.

Zinc finger transcription factor: Zinc finger proteins contain folds that are stabilised by zinc ions. Transcription factors, some which contain zinc finger domains, bind to DNA and regulate transcription of the gene.

Zoonosis: An animal disease, or zoonotic infection, that can be transmitted to humans. This can be through contact with animals (pets, farm animals, wildlife) or their products (milk, meat, waste).

Author Index

B

D

E

G

H

I

J

K

L

M

N

O

P

Q

S

T

U

V

X

Y

Subject Index

B

C

D

E

G

H

M

O

P

Q

R

T

U

V

W

X

Y

Z